W9-CJS-756

ENCYCLOPEDIA OF STATISTICAL SCIENCES

VOLUME 7

Plackett Family of Distributions to Regression, Wrong

ENCYCLOPEDIA OF STATISTICAL SCIENCES

VOLUME 7

PLACKETT FAMILY OF DISTRIBUTIONS to REGRESSION, WRONG

A WILEY-INTERSCIENCE PUBLICATION

John Wiley & Sons

NEW YORK · CHICHESTER · BRISBANE · TORONTO · SINGAPORE

Library of Congress Cataloging in Publication Data:
Main entry under title:

Encyclopedia of statistical sciences.

"A Wiley-Interscience publication."
Includes bibliographies.
Contents: v. 1. A to Circular probable error—v. 3. Faà di Bruno's formula to Hypothesis testing—[etc.]—v. 7. Plackett family of distributions to Regression, wrong.
1. Mathematical statistics—Dictionaries.
2. Statistics—Dictionaries. I. Kotz, Samuel.
II. Johnson, Norman Lloyd. III. Read, Campbell B.

QA276.14.E5 1982 519.5′03′21 81-10353
ISBN 0-471-05555-7 (v.7)

Printed in the United States of America

10 9 8 7 6 5 4 3 2 1

CONTRIBUTORS

J. R. Abernathy, *University of North Carolina, Chapel Hill, North Carolina.* Randomized Response

R. J. Adler, *The Technion, Haifa, Israel.* Random Fields

K. Alam, *Clemson University, Clemson, South Carolina.* Polarization Test

E. B. Andersen, *University of Copenhagen, Denmark.* Rasch, Georg.

R. L. Anderson, *University of Kentucky, Lexington, Kentucky.* Plateau Models, Linear

J. R. Ashford, *University of Exeter, Exeter, England.* Quantal Response Analysis

A. C. Atkinson, *Imperial College, London, England.* Regression Diagnostics

S. S. Atkinson, *University of North Carolina, Chapel Hill, North Carolina.* Poisson Regression

R. A. Bailey, *Rothamsted Experimental Station, Harpenden, England.* Randomization, Constrained

T. A. Bancroft, *Iowa State University, Ames, Iowa.* Pooling Data

M. L. Berenson, *Baruch College, New York, New York.* Puri's Expected Normal Scores Test

S. Berg, *University of Lund, Lund, Sweden.* Reed–Frost Chain Binomial Model

S. Blumenthal, *Old Dominion University, Norfolk, Virginia.* Population or Sample Size Estimation

M. T. Boswell, *Pennsylvania State University, University Park, Pennsylvania.* Poisson–Markov Process

K. R. W. Brewer, *Commonwealth Schools Commission, Woden, A.C.T., Australia.* Proportional Sampling; Proportional Allocation; Primary Sampling Unit; Quota Sampling

G. W. Brown, *CSIRO, Lindfield, New South Wales, Australia.* Regression, Inverse

J. Burbea, *University of Pittsburgh, Pittsburgh, Pennsylvania.* Probability Spaces, Metrics, and Distances on

D. P. Byar, *National Cancer Institution, Bethesda, Maryland.* Play-the-Winner Rules

S. Cambanis, *University of North Carolina, Chapel Hill, North Carolina.* Radon–Nikodym Theorem

V. Chew, *University of Florida, Gainesville, Florida.* Radial Error

R. S. Chhikara, *National Aeronautics and Space Center, Houston, Texas.* Proportion Estimation in Surveys Using Remote Sensing

L. R. Chow, *Tamkang University, Tamkang, Taiwan, Republic of China.* Policy and Information, International Journal on

E. Çinlar, *Princeton University, Princeton, New Jersey.* Regenerative Processes

N. Cliff, *University of Southern California, Los Angeles, California.* Psychometrika

C. C. Clogg, *Pennsylvania State University, University Park, Pennsylvania.* Quasi-Independence

M.-D. Cohen, *SAS Institute, Cary, North Carolina.* Pseudo-Random Number Generators

D. A. Conway, *University of Chicago, Chicago, Illinois.* Plackett Family of Distributions

P. Coughlin, *University of Maryland, College Park, Maryland.* Probabilistic Voting Models

P. R. Cox, *Mayfield, Sussex, England.* Population Pyramid; Population Projections

K. L. Coxe, *Washington, D.C.* Principal Components Regression Analysis

M. Csörgő, *Carleton University, Ottawa, Canada.* Quantile Processes

R. C. Dahiya, *Old Dominion University, Norfolk, Virginia.* Population or Sample Size Estimation

J. N. Darroch, *The Flinders University, Bedford Park, South Australia.* Quasi-Symmetry

A. W. Davis, *CSIRO, Glen Osmond, South Australia.* Polynomials of Matrix Arguments

C. E. Davis, *University of North Carolina, Chapel Hill, North Carolina.* Regression to the Mean; Quantile Estimation

A. P. Dawid, *University College, London, England.* Probability Forecasting

J. P. M. de Kroon, *Nederlandse Philips Bedrijven B.V., Eindhoven, The Netherlands.* Rank Interaction

M. H. DeGroot, *Carnegie-Mellon University, Pittsburgh, Pennsylvania.* Record Linkage/Matching Systems

W. E. Deming, *Washington, D.C.* Principles of Professional Statistical Practice

B. Dennis, *Pennsylvania State University, University Park, Pennsylvania.* Profiles of Diversity

W. S. DeSarbo, *University of Pennsylvania, Philadelphia, Pennsylvania.* Redundancy Analysis

D. Disch, *Rose-Hulman Institute of Technology, Terre Haute, Indiana.* Ramsey's Prior

J. B. Douglas, *University of New South Wales, Kensington, NSW, Australia.* Pólya–Aeppli Distribution

F. Drasgow, *University of Illinois, Champaign, Illinois.* Polychoric and Polyserial Correlations

A. J. Duncan, *Johns Hopkins University, Baltimore, Maryland.* Quality Control, Statistical

B. S. Duran, *Texas Tech University, Lubbock, Texas.* Regression Polynomials

J. F. Early, *U.S. Bureau of Labor Statistics, Washington, D.C.* Producer Price Index

E. S. Edgington, *University of Calgary, Alberta, Canada.* Randomization Tests

A. W. F. Edwards, *University of Cambridge, England.* Problem of Points

A. S. C. Ehrenberg, *London Business School, London, England.* Reduction of Data

R. C. Elandt-Johnson, *University of North Carolina, Chapel Hill, North Carolina.* Rates, Standardized

R. L. Eubank, *Southern Methodist University, Dallas, Texas.* Quantiles

J.-C. Falmagne, *New York University, New York, New York.* Psychophysics, Statistical Methods in

S. E. Fienberg, *Carnegie-Mellon University, Pittsburgh, Pennsylvania.* Rasch Model

P. D. Finch, *Monash University, Clayton, Victoria, Australia.* Randomization—I

D. A. S. Fraser, *University of Toronto, Toronto, Ontario, Canada.* Reduced Model

L. S. Freedman, *Medical Research Council Centre, Cambridge, England.* Pocock and Simon Method

G. H. Freeman, *University of Warwick, Coventry, England.* Plaid and Half-Plaid Squares; Quasi-Factorial Designs

P. Gács, *Boston University, Boston, Massachusetts.* Randomness and Probability—Complexity of Description

J. Galambos, *Temple University, Philadelphia, Pennsylvania.* Probabilistic Number Theory; Raikov's Theorem

J. L. Gastwirth, *George Washington University, Washington, D.C.* Pyramid Scheme Models

S. Geisser, *University of Minnesota, Minneapolis, Minnesota.* Predictive Analysis

J. D. Gibbons, *University of Alabama, University, Alabama.* *P* Values; Randomized Tests; Randomness, Tests for; Ranking Procedures; Rank Tests (Excluding Group Rank Tests)

M. Goldstein, *University of Hull, Hull, England.* Prevision

B. G. Greenberg*, *University of North Carolina, Chapel Hill, North Carolina.* Randomized Response

R. S. Greenberg, *Emory University, Decatur, Georgia.* Prospective Studies

K. H. Gross, *California Survey Research, Sherman Oaks, California.* Random-Digit Dialing, Sampling Methods

E. Grosswald, *Temple University, Philadelphia, Pennsylvania.* Rademacher Functions

L. Guttman, *Israel Institute of Applied Social Research, Jerusalem, Israel.* Polytonicity and Monotonicity, Coefficients of

R. Haag, *University of Hamburg, Hamburg, West Germany.* Quantum Mechanics: Statistical Interpretation

C.-P. Han, *University of Texas, Arlington, Texas.* Pooling Data

C. C. Heyde, *University of Melbourne, Parkville, Victoria, Australia.* Probability Theory (Outline); Quantile Transformation Methods; Random Matrices; Random Sum Distributions

K. Hirano, *Institute of Statistical Mathematics, Tokyo, Japan.* Randomized Gamma Distribution; Rayleigh Distribution

A. B. Hoadley, *Bell Communications Research, Holmdel, New Jersey.* Quality Measurement Plan (QMP)

W. Hoeffding, *University of North Carolina, Chapel Hill, North Carolina.* Probabilistic Inequalities for Sums of Bounded Random Variables; Range-Preserving Estimators

D. G. Horvitz, *Research Triangle Institute, Research Triangle, North Carolina.* Randomized Response

J. R. M. Hosking, *Institute of Hydrology, Wallingford, England.* Portmanteau Tests; Quenouille Test

A. G. Houston, *University of Houston, Clear Lake, Houston, Texas.* Proportion Estimation in Surveys Using Remote Sensing

*Deceased

R. Hultquist, *Pennsylvania State University, University Park, Pennsylvania.* Regression Coefficients

N. T. Jazairi, *York University, Downsview, Ontario, Canada.* Productivity Measurement; Purchasing Power Parity

K. Jedidi, *University of Pennsylvania, Philadelphia, Pennsylvania.* Redundancy Analysis

L. V. Jones, *Center for Advanced Study in the Behavioral Sciences, Stanford, California.* Psychological Scaling

E. W. Jordan, *University of Texas, Austin, Texas.* Problem Solving in Statistics

J. D. Kalbfleisch, *University of Western Ontario, London, Ontario, Canada.* Pseudo-Likelihood

G. Kallianpur, *University of North Carolina, Chapel Hill, North Carolina.* Prediction and Filtering, Linear

G. Kalton, *The University of Michigan, Ann Arbor, Michigan.* Question Wording Effects in Surveys

D. Kannan, *University of Georgia, Athens, Georgia.* Processes, Discrete

M. Karoński, *Adam Mickiewicz University, Poznań, Poland.* Random Graphs

A. F. Karr, *Johns Hopkins University, Baltimore, Maryland.* Poisson Processes; Point Process, Stationary

O. Kempthorne, *Iowa State University, Ames, Iowa.* Randomization—II

D. G. Kendall, *University of Cambridge, Cambridge, England.* Quantum Hunting; Random Sets of Points

N. Keyfitz, *International Institute for Applied Systems Analysis, Laxenburg, Austria.* Population Growth Models

G. G. Koch, *University of North Carolina, Chapel Hill, North Carolina.* Poisson Regression

J. M. Landwehr, *AT & T Bell Laboratories, Murray Hill, New Jersey.* Projection Plots

J. Ledolter, *University of Iowa, Iowa City, Iowa.* Prediction and Forecasting

L. Lefeber, *York University, Downsview, Ontario, Canada.* Productivity Measurement

A. M. Liebetrau, *Batelle Pacific Northwest*

Laboratories, Richland, Washington. Proportional Reduction in Error (PRE) Measures of Association

F. M. Lord, *Educational Testing Service, Princeton, New Jersey.* Psychological Testing Theory

T. Matsunawa, *Institute of Statistical Mathematics, Tokyo, Japan.* Poisson Distribution

D. W. Matula, *Southern Methodist University, Dallas, Texas.* Random Graphs

P. McCullagh, *Imperial College, London, England.* Quasi-Likelihood Functions

P. W. Mielke, *Colorado State University, Fort Collins, Colorado.* Quantit Analysis

R. E. Miles, *The Australian National University, Canberra, Australia.* Random Tesselations

S. G. Nash, *Johns Hopkins University, Baltimore, Maryland.* Quasi-Random Sampling; Quasi-Random Sequences

B. Natvig, *University of Oslo, Oslo, Norway.* Priority Queue

L. S. Nelson, *Nashua Corporation, Nashua, New Hampshire.* Quality Technology, Journal of; Precedence Life Test

J. V. Noble, *University of Virginia, Charlottesville, Virginia.* Quantum Mechanics and Probability: an Overview

W. R. S. North, *Northwick Park Hospital, Harrow, England.* Quangle

D. J. Pack, *Union Carbide Corporation, Oak Ridge, Tennessee.* Posterior Distributions; Posterior Probabilities; Prior Distributions; Prior Probabilities

A. G. Pakes, *University of Western Australia, Nedlands, Western Australia.* Pollaczek–Khinchin Formula; Preemptive Discipline; Queueing Theory

G. P. Patil, *Pennsylvania State University, University Park, Pennsylvania.* Pólya Distribution, Multivariate; Power Series Distributions; Profiles of Diversity

B. F. Pendleton, *University of Akron, Akron, Ohio.* Ratio Correlation

A. N. Pettitt, *University of Technology, Loughborough, England.* Rank Likelihood

D. Pfeifer, *Institute of Statistics and Insurance, Aachen, Federal Republic of Germany.* Pólya-Lundberg Process

E. C. Pielou, *University of Lethbridge, Lethbridge, Alberta, Canada.* Quadrat Sampling

H. O. Posten, *University of Connecticut, Storrs, Connecticut.* Quincunx

R. F. Potthoff, *Burlington Industries, Greensboro, North Carolina.* Potthoff–Whittinghill Tests

M. B. Priestley, *University of Manchester, Manchester, England.* Priestley's Test for Harmonic Components

S. Pruzansky, *AT & T Bell Laboratories, Murray Hill, New Jersey.* Proximity Data

R. E. Quandt, *Princeton University, Princeton, New Jersey.* Regressions, Switching

C. E. Quesenberry, *North Carolina State University, Raleigh, North Carolina.* Probability Integral Transformations

D. Raghavarao, *Temple University, Philadelphia, Pennsylvania.* Random Balance Designs

B. R. Rao, *University of Pittsburgh, Pittsburgh, Pennsylvania.* Pólya Type 2 Frequency (PF_2) Distributions

C. R. Rao, *University of Pittsburgh, Pittsburgh, Pennsylvania.* Rao's Axiomatization of Diversity Measures

J. N. K. Rao, *Carleton University, Ottawa, Canada.* Ratio Estimators

A. Rapoport, *University of Haifa, Haifa, Israel.* Psychological Decision Making

M. V. Ratnaparkhi, *Wright State University, Dayton, Ohio.* Pólya Distribution, Multivariate

R. F. Raubertas, *National Institutes of Health, Bethesda, Maryland.* Pool-Adjacent Violators Algorithm

C. B. Read, *Southern Methodist University, Dallas, Texas.* Ranges; Rao-Blackwell Theorem

H. T. Reynolds, *University of Delaware, Newark, Delaware.* Political Science, Statistics in

J. L. Rosenberger, *Pennsylvania State University, University Park, Pennsylvania.* Recovery of Interblock Information

W. D. Rowe, *The American University, Washington, D.C.* Rare Event Risk

Analysis

A. K. M. E. Saleh, *Carleton University, Ottawa, Canada.* Rank Tests, Grouped Data

P. Schmidt, *Michigan State University, East Lansing, Michigan.* Quandt-Ramsey (MGF) Estimator

H. T. Schreuder, *U.S. Forest and Range Experiment Station, Fort Collins, Colorado.* Quenouille's Estimator

C. Scott, *World Fertility Survey, London, England.* Population Sampling in Less Developed Countries

P. K. Sen, *University of North Carolina, Chapel Hill, North Carolina.* Progressive Censoring Schemes; Progressively Censored Data Analysis; Rank Order Statistics

E. Seneta, *University of Sydney, Sydney, Australia.* Poisson, Siméon-Denis; Probability, History of (Outline)

L. R. Shenton, *University of Georgia, Athens, Georgia.* Quasibinomial Distributions

R. Shibata, *Tokyo Institute of Technology, Tokyo, Japan.* Regression Variables, Selection of

A. F. Siegel, *University of Washington, Seattle, Washington.* Rarefaction Curves

G. L. Sievers, *Western Michigan University, Kalamazoo, Michigan.* Probability Plotting

B. Simon, *California Institute of Technology, Pasadena, California.* Quantum Physics and Functional Integration

R. Singh, *Indian Agricultural Statistics Research Institute, New Delhi, India.* Predecessor-Successor Method

C. J. Skinner, *University of Southampton, Southampton, England.* Probability Proportional to Size (PPS) Sampling

H. Smith, *Mt. Sinai School of Medicine, New York, New York.* Regression Models, Types of

E. Spjøtvoll, *Norwegian Institute of Technology, Trondheim, Norway.* Preference Functions

S. M. Steinberg, *University of North Carolina, Chapel Hill, North Carolina.* Quantile Estimation

S. M. Stigler, *University of Chicago, Chicago, Illinois.* Quetelet, Adolphe

M. E. Stokes, *University of North Carolina, Chapel Hill, North Carolina.* Poisson Regression

S. L. Stokes, *University of Texas, Austin, Texas.* Ranked Set Sampling

D. J. Strauss, *University of California, Riverside, California.* Random Utility Models

H. Strecker, *University of Tübingen, Tübingen, Federal Republic of Germany.* Quotient Method

D. F. Stroup, *University of Texas, Austin, Texas.* Problem Solving in Statistics

S. Sudman, *University of Illinois, Urbana, Illinois.* Public Opinion Polls

L. Takács, *Case Western Reserve University, Cleveland, Ohio.* Reflection Principle

W. Y. Tan, *Memphis State University, Memphis, Tennessee.* Quadratic Forms

R. Thakkar, *York University, Downsview, Ontario, Canada.* Purchasing Power Parity

G. Tintner*, *University of Tubingen, Tubingen, Federal Republic of Germany.* Quotient Method

D. S. Tracy, *University of Windsor, Windsor, Ontario, Canada.* Polykays

R. L. Trader, *University of Maryland, College Park, Maryland.* Quasi-Bayesian Inference; Regression, Bayesian

R. L. Tweedie, *Siromath Pty., Sydney, New South Wales, Australia.* Recurrence Criterion

D. S. Watkins, *Washington State University, Pullman, Washington.* QR Algorithm

L. J. Wei, *George Washington University, Washington, D.C.* Play-the-Winner Rules

G. H. Weiss, *National Institutes of Health, Bethesda, Maryland.* Random Walks

J. A. Wellner, *University of Washington, Seattle, Washington.* Processes, Empirical

H. Wold, *University of Uppsala, Uppsala, Sweden.* Regression: Confluence Analysis

*Deceased

ENCYCLOPEDIA OF STATISTICAL SCIENCES

VOLUME 7

Plackett Family of Distributions to Regression, Wrong

P *continued*

PLACKETT FAMILY OF DISTRIBUTIONS

The Plackett family of bivariate distributions includes all cumulative distribution functions $F_{X_1,X_2}(x_1,x_2) = F(x_1,x_2)$ that satisfy the quadratic equation

$$\psi = \frac{F(x_1,x_2)[1 - F_1(x_1) - F_2(x_2) + F(x_1,x_2)]}{[F_1(x_1) - F(x_1,x_2)][F_2(x_2) - F(x_1,x_2)]}, \tag{1}$$

where $F_1(x_1)$ and $F_2(x_2)$ are the marginal distribution functions of X_1 and X_2, respectively, and $\psi \in (0,\infty)$ is an association parameter. Plackett [10] constructed the family by extending the cross-product measure of association* in a 2×2 contingency table* to general marginal distributions, $F_1(x_1)$ and $F_2(x_2)$. Mardia [4] solved (1) for the root, $F_{X_1,X_2}(x_1,x_2)$, satisfying the Fréchet inequalities*,

$$\begin{aligned}\max[0, F_1(x_1) + F_2(x_2) - 1] &\leqslant F_{X_1,X_2}(x_1,x_2)\\ &\leqslant \min[F_1(x_1), F_2(x_2)].\end{aligned} \tag{2}$$

He showed that when $\psi = 1$, $F_{X_1,X_2}(x_1,x_2) = F_1(x_1)F_2(x_2)$, and for $\psi \neq 1$, the distribution function has the form

$$F_{X_1,X_2}(x_1,x_2) = \frac{S - [S^2 - 4\psi(\psi - 1)F_1(x_1)F_2(x_2)]^{1/2}}{2(\psi - 1)}, \tag{3}$$

where $S = 1 + (\psi - 1)[F_1(x_1) + F_2(x_2)]$. When $F_1(x_1)$ and $F_2(x_2)$ are absolutely continuous, the bivariate density function is given by

$$f_{X_1,X_2}(x_1,x_2) = \frac{\psi f_1 f_2[S - 2(\psi - 1)F_1F_2]}{[S^2 - 4\psi(\psi - 1)F_1F_2]^{3/2}}. \tag{4}$$

PROPERTIES OF BIVARIATE PLACKETT DISTRIBUTIONS

Similar to the Farlie–Gumbel–Morgenstern (FGM) distributions*, bivariate distributions in the Plackett family are expressed in terms of the univariate marginals. The two families are related, with FGM distributions approximating those in the Plackett family. By letting $\alpha = \psi - 1$ and expanding the square root in (3), we obtain the approximation

$$F_{X_1,X_2}(x_1,x_2) = F_1(x_1)F_2(x_2)\left[1+\alpha(1-F_1(x_1)) \times(1-F_2(x_2))\right]+o(\alpha).$$

Unlike the FGM distributions, the association parameter ψ characterizes the full range of dependence between X_1 and X_2. When $F_1(x_1)$ and $F_2(x_2)$ are fixed, it follows from (1) that ψ is a monotonic increasing function of $F_{X_1,X_2}(x_1,x_2)$. The random variables X_1 and X_2 are independent whenever $\psi = 1$, positively associated when $\psi > 1$, and negatively associated when $\psi < 1$. Fréchet's lower and upper boundary distributions in (2) are obtained as ψ approaches the limits 0 and ∞, respectively.

The regression curves for bivariate Plackett distributions generally do not have simple, analytic forms. One exception is the case when $F_1(x_1)$ and $F_2(x_2)$ correspond to uniform distributions on the interval (0, 1). The regressions are then linear and have the form

$$E(X_1 \mid X_2 = x_2) = \tfrac{1}{2} + \rho_U(\psi)(x_2 - \tfrac{1}{2}),$$

where $\rho_U(\psi)$ is the correlation coefficient for two uniform random variables and equals

$$\rho_U(\psi) = (\psi^2 - 1 - 2\psi\log\psi)/(\psi-1)^2. \quad (5)$$

The median regression curves are often used, since they have a simple, analytic form. The conditional distribution function of X_1 given $X_2 = x_2$ is

$$F_{X_1\mid X_2}(x_1 \mid x_2) = \frac{\psi F_1(x_1) + (1-\psi)F_{X_1,X_2}(x_1,x_2)}{S + 2(1-\psi)F_{X_1,X_2}(x_1,x_2)}. \quad (6)$$

By equating (6) to $\frac{1}{2}$, the median regression curves satisfy

$$(\psi+1)F_1(x_1) = 1 + (\psi-1)F_2(x_2). \quad (7)$$

Equation (7) specifies that the median regressions have the form of a linear regression of $F_1(x_1)$ on $F_2(x_2)$. As $F_2(x_2)$ varies from 0 to 1, $F_1(x_1)$ varies from $1/(\psi+1)$ to $\psi/(\psi+1)$.

The maximum likelihood* estimate $\hat{\psi}$ of ψ depends on the marginal distributions and is computed using numerical methods from the density function in (4). If ψ is close to 1, the analytic approximation for $\hat{\alpha} = \hat{\psi} - 1$ for the FGM distributions (Volume 3, p. 29) can be used. Plackett [10] considered the consistent estimator,

$$\psi^\dagger = (ad)/(bc), \quad (8)$$

where a, b, c, and d represent the observed frequencies of the pairs (x_{i1}, x_{i2}) in the quadrants $(X_1 \leqslant p,\ X_2 \leqslant q)$, $(X_1 \leqslant p,\ X_2 > q)$, $(X_1 > p,\ X_2 \leqslant q)$, and $(X_1 > p,\ X_2 > q)$, respectively. Mardia [4] showed that the optimum choice of the point (p,q) minimizes the asymptotic variance of $\psi^\dagger$ and is given by the population median vector [i.e., when $F_1(p) = F_2(q) = \frac{1}{2}$]. If (x_{m1}, x_{m2}) denotes the sample median vector and $F_{X_1,X_2}(x_{m1},x_{m2}) = M$, then (8) simplifies to

$$\psi^\dagger = 4M^2(1-2M)^2,$$

and the asymptotic variance of $\psi^\dagger$ is given by

$$\operatorname{var}(\psi^\dagger) \simeq (4/n)\psi^{3/2}(1+\psi)^2.$$

Mardia [4] proposed an efficient estimator of ψ that is asymptotically equivalent to the maximum likelihood estimate. Since the marginal distributions $F_1(x_1)$ and $F_2(x_2)$ are known, we can compute the sample value of $\operatorname{corr}[F_1(x_1), F_2(x_2)]$. Equating the sample correlation, denoted by $\bar{r}$, to the expression for $\rho_U(\psi)$ in (5), we obtain the estimator $\tilde{\psi}$ as the solution of the equation

$$\bar{r} = (\tilde{\psi}^2 - 1 - 2\tilde{\psi}\log\tilde{\psi})/(\tilde{\psi}-1)^2.$$

Mardia [4] showed that in the region of greatest practical interest, when $|\rho_U(\psi)| < 0.99$, the asymptotic efficiency $e(\psi^\dagger, \tilde{\psi})$ lies between 0.46 and 0.56.

RELATIONSHIP TO CROSS-PRODUCT RATIO AND MULTIVARIATE EXTENSIONS

A bivariate Plackett distribution has the property that when it is cut anywhere by two lines parallel to the X_1 and X_2 axes, the cross-product ratio of the probabilities in the

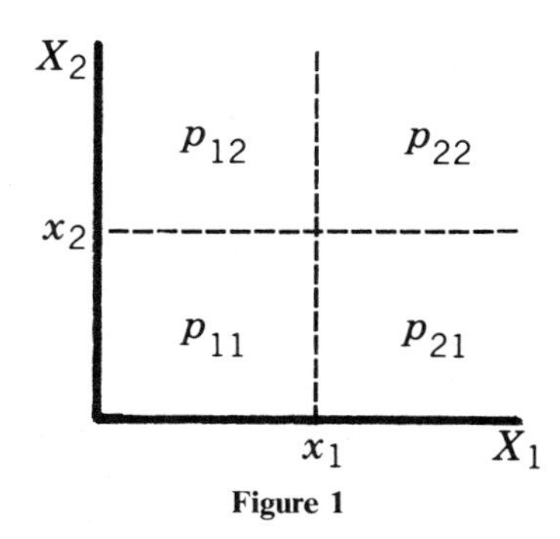

Figure 1

four quadrants is constant. Specifically, for any vertex (x_1, x_2), let p_{11}, p_{12}, p_{21}, and p_{22} denote the probabilities of the four regions in Fig. 1. The defining property of the Plackett family given by (1) is equivalent to

$$\psi = (p_{11}p_{22})/(p_{12}p_{21}). \quad (9)$$

Because (9) must hold for any point (x_1, x_2), the Plackett family describes bivariate distributions with constant quadrant or Yulean association. This property does not hold, for example, for the bivariate normal distribution*.

Pearson [9] briefly considered bivariate distributions with constant quadrant association and felt that they were unlikely to arise in practice. Plackett [10] proposed the general family as a one-parameter class of bivariate distributions with given marginals. Mardia [4, 6] and Steck [12] examined additional properties of the family. Mosteller [8] showed that the Plackett family arises naturally in the process of standardizing the margins of a 2×2 contingency table. In view of the correspondence with the cross-product measure of association for a 2×2 table, bivariate distributions in the Plackett family are also referred to as *contingency-type* or *C-type* distributions.

One drawback of the Plackett family is that it does not extend easily to the multivariate case. Anscombe [1] suggested a trivariate extension, based on the cross-product ratio for a $2 \times 2 \times 2$ contingency table. Analogous to (9), the ratio of the eight quadrant probabilities is constant and the triviariate distribution is defined by

$$\psi = (p_{111}p_{122}p_{212}p_{221})/(p_{112}p_{121}p_{211}p_{222}).$$

Plackett and Paul [11] suggest a different trivariate extension in which the trivariate cross-product ratio is 0, and three parameters describe the constant bivariate cross-product ratios. The properties of the two trivariate distributions have not been studied.

APPROXIMATION OF BIVARIATE NORMAL* PROBABILITIES

A distinct advantage of the bivariate Plackett distributions is that quadrant probabilities are easily computed from (3) and the univariate marginal distributions. For the case when X_1 and X_2 are standard normal random variables, various authors have investigated whether a C-type bivariate normal distribution can be used to approximate the probabilities of the standard bivariate normal. Mardia [4] showed that the correlation coefficient for the C-type bivariate normal distribution is given by

$$\rho_N(\psi) = \frac{1}{\psi - 1}\int_0^\infty \int_0^\infty \{\psi + 1 - A[\Phi(x_1), \Phi(x_2)] - A[1 - \Phi(x_1), \Phi(x_2)]\}\, dx_1\, dx_2,$$

where $\Phi(\cdot)$ is the standard normal distribution function and

$$A(u, v) = \left\{\left[1 + (\psi - 1)(u + v)\right]^2 - 4\psi(\psi - 1)uv\right\}^{1/2}.$$

The expression cannot be further simplified and a number of approximations have been developed.

Plackett [10] equated the C-type and bivariate normal distribution functions at the median value of $(0, 0)$ to obtain the approximation

$$\rho_1 = \cos\left[\pi/\left(1 + \sqrt{\psi}\right)\right].$$

Using this approximation, he compared the quantiles of a standard normal bivariate distribution when $\rho = 0.5$ with the C-type quantiles for $\psi = 4$ and found close agreement in the center of the distribution with minor departures at the tails. Mardia [4] related the values of $\operatorname{corr}(F_1(x_1), F_2(x_2))$ for the two bivariate normal distributions and obtained the approximation

$$\rho_2 = 2\sin\left[\pi(\psi^2 - 1 - 2\psi\log\psi)/\left\{6(\psi-1)^2\right\}\right].$$

Comparing the probabilities of the C-type normal for $\psi = 0.25$ with those of the bivariate normal when $\rho_1 = -0.50$ and $\rho_2 = -0.45$, he found that the value of ρ_2 gave more accurate approximations. Plackett's approximation ρ_1 gave somewhat better approximations near the center of the distribution. In the same paper, Mardia showed that the Spearman correlation coefficient ρ_U in (5) provided a more accurate approximation of ρ_N than either ρ_1 or ρ_2 across a wide range of different values for ψ. In general, the approximations had the following relationship:

$$|\rho_N - \rho_U| < |\rho_N - \rho_2| < |\rho_N - \rho_1|.$$

Anscombe [1] compared the contours of the density functions for C-type and bivariate normal distributions when ρ equals 0.6 and 0.875, and ψ equals 4 and 15, respectively. The contours of the C-type normal density are approximately elliptical in the center and become more circular as the distance from the center increases. This is due to the slight skewness of the conditional distributions and the nonlinearity of the regressions. However, he notes that the bivariate probabilities for the two distributions are very close and that a large sample size would be required to distinguish them.

APPLICATIONS OF BIVARIATE PLACKETT DISTRIBUTIONS

The Plackett family of distributions has been used as an alternative to the bivariate normal distribution to examine the power and robust properties of various statistical tests. Mardia [5] used bivariate Plackett distributions having uniform*, normal, Laplace*, exponential*, and Pareto* marginals to compare the power of Kendall's τ and the Spearman rank correlation coefficient* for tests of independence. Ferguson [2] used bivariate Plackett distributions with gamma and lognormal marginals to compare three nonparametric tests of bivariate exchangeability. Mardia [7] used a bivariate uniform Plackett distribution to examine the effects of nonnormality on multivariate regression tests.

Bivariate Plackett distributions have also been used as the underlying continuous models for the discrete entries in contingency tables. Mardia [6] used a bivariate logistic C-type distribution to fit the observed frequencies in a two-way table giving the breadth and length of 9440 beans. He found that the expected frequencies from the Plackett distribution were reasonable but showed some sharp departures. The median regression curves provided a close fit to the data but the scedastic curves indicated a poor fit. Anscombe [1] found a good fit of a C-type bivariate normal to Gilby's data that cross-classifies 1725 schoolboys by clothing and intelligence rating. However, he found poor fits using the C-type normal distribution to three other contingency tables of data. Goodman [3] also noted the poor fit of a bivariate C-type distribution to British and Danish data on social mobility that cross-classifies the occupational status of fathers and sons. Significant departures from the expected and observed frequencies occurred in the cells where the status was the same. Finally, Wahrendorf [13] found that a bivariate C-type normal distribution approximated the observed frequencies closely in a table cross-classifying 898 English schoolboys by the number of newspapers skimmed versus the number read carefully.

References

[1] Anscombe, F. (1981). *Computing in Statistical Science through APL*. Springer-Verlag, New York, Chap. 12.

[2] Ferguson, N. L. (1973). *Austral. J. Statist.*, **15**, 191–208.

[3] Goodman, L. A. (1981). *Biometrika*, **68**, 347–355.

[4] Mardia, K. V. (1967). *Biometrika*, **54**, 235–249.

[5] Mardia, K. V. (1969). *Biometrika*, **56**, 449–451.

[6] Mardia, K. V. (1970). *J. R. Statist. Soc. B*, **32**, 254–264.

[7] Mardia, K. V. (1971). *Biometrika*, **58**, 105–121.
[8] Mosteller, F. (1968). *J. Amer. Statist. Ass.*, **63**, 1–28.
[9] Pearson, K. (1913). *Biometrika*, **9**, 534–537.
[10] Plackett, R. L. (1965). *J. Amer. Statist. Ass.*, **60**, 516–522.
[11] Plackett, R. L. and Paul, S. R. (1978). *Commun. Statist. A*, **7**, 939–952.
[12] Steck, G. P. (1968). *Biometrika*, **55**, 262–264.
[13] Wahrendorf, J. (1980). *Biometrika*, **67**, 15–21.

(ASSOCIATION, MEASURES OF
BIVARIATE NORMAL DISTRIBUTION
FARLIE–GUMBEL–MORGENSTERN DISTRIBUTIONS
FREQUENCY SURFACES, SYSTEMS OF)

DELORES CONWAY

PLACKETT'S IDENTITY

For the multivariate normal density

$$\phi_m(\mathbf{x}, \mathbf{R}) = (2\pi)^{-m/2}|\mathbf{R}|^{-1/2}\exp(-\tfrac{1}{2}\mathbf{x}'\mathbf{R}^{-1}\mathbf{x})$$

with correlation matrix $\mathbf{R} = (\rho_{ij})$,

$$\frac{\partial}{\partial \rho_{ij}} \phi_m(\mathbf{x}, \mathbf{R}) = \frac{\partial^2}{\partial x_i \, \partial x_j} \phi_m(\mathbf{x}, \mathbf{R})$$

for all $i \neq j$.

This identity is useful for establishing inequalities involving multivariate normal distributions and computation of multivariate integrals. *See* MULTINORMAL DISTRIBUTION.

Bibliography

Johnson, N. L. and Kotz, S. (1972). *Distributions in Statistics: Continuous Multivariate Distributions*. Wiley, New York.

Plackett, R. L. (1954). *Biometrika*, **41**, 351–360.

PLAID AND HALF-PLAID SQUARES

Among the class of *row and column designs** the *Latin square** is perhaps the most widely used. The commonest use of Latin squares as experimental designs occurs when there is no particular treatment structure, but there is no reason why they should not be used for *factorial experiments** and indeed they often are. If some of the *interactions** between the treatment factors are *confounded* with rows or columns, the design is then said to be a *quasi-Latin square*, such designs being introduced by Yates [3]. However, it is quite possible to apply some factors to complete rows or columns of a Latin square, and then the corresponding *main effects* rather than interactions are confounded with rows or columns. If treatment factors are applied either to complete rows or to complete columns of a Latin square the design is said to be a *half-plaid square*: if the factors are applied to both complete rows and complete columns, the design is a *plaid square*. The name "plaid" comes from the supposed resemblance to a Scottish tartan, with a half-plaid bearing half as much resemblance. Since, as with any other design, it is essential to randomize the rows or columns receiving particular treatment factors, the similarity with the tartan might be apparent only after consuming another well-known Scotch product!

Plaid and half-plaid squares are due to Yates [2], who noted that if a particular treatment factor is applied to whole rows, then certain interactions between that factor and those within the Latin square are automatically confounded with columns. Most practical plaid and half-plaid designs have six, eight, or nine rows and columns, and it is usually possible to arrange that this confounding is confined to high-order interactions. The simplest possible half-plaid square is a 4×4 Latin square having three factors A, B, and C with two levels each, A being confounded with rows, the design being as follows before randomization:

$$\begin{array}{cccc} b & c & (1) & bc \\ c & b & bc & (1) \\ a & abc & ab & ac \\ abc & a & ac & ab. \end{array}$$

Here the first two rows receive the lower level of A and the last two rows the higher level; ABC is confounded with columns. Such a design using only one square is too small for practical use, so several replicates would be needed for real experiments. It is also possible to have partial confounding of effects; examples include a 6×6 Latin square having a 2×3 factorial arrangement within the square and a further factor with either two or three levels applied to complete rows. Plans for plaid and half-plaid squares are available in Yates [3] and Cochran and Cox [1].

In practice, the designs might be of use where some factors cannot be readily applied to the individual plots of a Latin square. For example, this could well happen in an agricultural field trial where one of the factors was presence or absence of irrigation, a treatment difficult to apply on small areas but easier on a long strip.

The analysis of plaid and half-plaid squares is similar to that of a *criss-cross* or *strip-plot** design. In the analysis of variance* there will be one error term for the treatments within the square, one for rows in the half-plaid square, and one each for rows and columns in the plaid square. Account must be taken in the analysis of those terms which have been confounded as a consequence of the design, such as ABC in the example shown above. Since inevitably there will be very few degrees of freedom in the analysis for the error term in rows or columns, the corresponding main effects will not be well estimated. Indeed, in practice they are usually included mainly to determine their interactions with the other treatments.

References

[1] Cochran, W. G. and Cox, G. M. (1957). *Experimental Designs*, 2nd ed. Wiley, New York. (The most readily available reference to plans for these designs.)

[2] Yates, F. (1933). *J. Agric. Sci.*, **23**, 108–145. (The paper that introduced these designs.)

[3] Yates, F. (1937). *Imp. Bur. Soil Sci. Tech. Commun. 35*.

(CONFOUNDING
DESIGN OF EXPERIMENTS
FACTORIAL EXPERIMENTS
ROW AND COLUMN DESIGNS)

G. H. Freeman

PLANCK DISTRIBUTION

A distribution with probability density function

$$f_X(x) = \frac{Kx^3}{e^{\alpha x} - 1} \qquad (x > 0;\ \alpha > 0)$$

is known as *Planck's radiation formula*. The family of distributions

$$f_X(x) = \alpha^{\gamma+1} C_\gamma X^\gamma (e^{\alpha x} - 1)^{-1}$$

$$(x > 0;\ \alpha > 0;\ \gamma < 0)$$

with

$$C_\gamma = \{\Gamma(\gamma + 1)\zeta(\gamma + 1)\}^{-1},$$

where $\Gamma(\cdot)$ and $\zeta(\cdot)$ denote gamma* and Riemann zeta functions, respectively, may be called *Planck distributions*.

The variable $Y = \alpha X$ has the *standard* Planck distribution

$$f_Y(y) = C_\gamma y^\gamma (e^y - 1)^{-1}.$$

This distribution can be regarded as a mixture of $(2J)^{-1} X^2_{2(\gamma+1)}$ distributions, with J having the zeta distribution*

$$\Pr[J = j] = \{\zeta(f + 1)\}^{-1} - j^{-(f+1)}.$$

The mode is approximately equal to

$$\gamma\{1 - (\gamma + 1)e^{-\gamma}\}/(1 - \gamma e^{-\gamma}) \qquad \text{for } \gamma > 1.496,$$

$$2(\gamma - 1) \qquad \text{for } 1 < \gamma \leqslant 1.496.$$

The mode is zero for $\gamma \leqslant 1$.

For further details, see ref. 1.

Reference

[1] Johnson, N. L. and Kotz, S. (1970). *Distributions in Statistics*, Vol. 2: *Continuous Univariate Distributions*. Wiley, New York.

PLATEAU MODELS, LINEAR

For many experimental situations, especially crop experiments in developing countries, quadratic surfaces do not fit the responses to applied treatments. As a result, there may be costly biases in the estimates of the optimal treatment levels. There are two basic causes of these results:

1. There is usually a large increase in yield due to the initial application of treatment followed by a much smaller increase for subsequent applications.
2. Eventually, the response reaches a plateau.

A family of linear-plateau (LP) models, consisting of intersecting straight lines, the last of which is usually a plateau, is prepared for fitting such response data. The general model is

$$\hat{Y}_h = b_0X_{0h} + b_1X_{1h} + b_2X_{2h} + b_3X_{3h},$$

where $\hat{Y}$ is the column vector of predicted values of the mean responses, b_0 is the intercept, the b_k are slopes of intersecting straight lines, and the X_k are column vectors ($X_{0h} = 1$), which are specially constructed for the particular model being used. (For certain models, specific b_k are assumed to be zero. It is assumed that $b_1 > b_2 > b_3 \geqslant 0$ and that an experiment has been conducted with p replications of each of the n treatment levels.) Procedures [1] were prepared initially for equally spaced nutrient levels ($N_0, N_1, \ldots, N_{n-1}$) which have been coded to $h = 0, 1, \ldots, n-1$.

It is often necessary to estimate h_e, the coded input level for which there is the greatest economic return. It is possible to make economic interpretations by comparing the regression coefficients obtained in the linear-plateau analysis with r, the ratio of the cost of a unit of input to the price of a unit of response.

The models and the estimated values $\hat{h}_e$ of h_e are (number of b's in parentheses):

I(1): Horizontal line; $\hat{h}_e = 0$.

II(2): Single sloping line; $\hat{h}_e \geqslant n-1$ if $0 \leqslant r < b_1$.

III_j(2): Single sloping line intersecting a plateau at $h = j$; $\hat{h}_e = j$ if $0 \leqslant r < b_1$.

IV_j(3): Single sloping line intersecting a plateau between $h = j$ and $h = j + 1$; $\hat{h}_e = j + (b_2/b_1)$ if $0 \leqslant r < b_1$.

V_i(3): Two sloping lines intersecting at $h = i$; $\hat{h}_e \geqslant n-1$ if $0 \leqslant r < b_2$; $\hat{h}_e = i$ if $b_2 \leqslant r < b_1$.

VI_{ij}(3): Two sloping lines intersecting at $h = i$ and second sloping line intersecting a plateau at $h = j$; $\hat{h}_e = j$ if $0 \leqslant r < b_2$; $\hat{h}_e = i$ if $b_2 \leqslant r < b_1$.

VII_{ij}(4): As for VI_{ij} except second sloping line intersecting a plateau between $h = j$ and $j + 1$; $\hat{h}_e = j + (b_3/b_2)$ if $0 \leqslant r < b_2$; $\hat{h}_e = i$ if $b_2 \leqslant r < b_1$.

VIII_{ij}(4): Three sloping lines with intersections at $h = i$ and j; $\hat{h}_e \geqslant n-1$ if $0 \leqslant r < b_3$; $\hat{h}_e = j$ if $b_3 \leqslant r < b_2$; $\hat{h}_e = i$ if $b_2 \leqslant r < b_1$.

In all cases if $r \geqslant b_1$, $\hat{h}_e = 0$.

As an example, consider the following Tennessee corn-yield data presented in the 1975 article, for which a model IV_2 was used (see also Fig. 1):

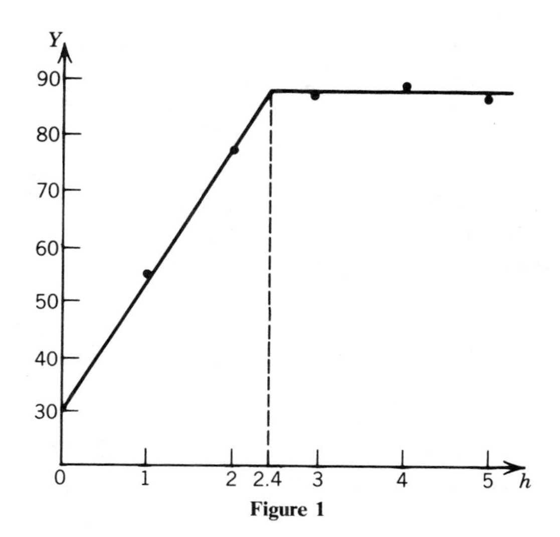

Figure 1

h	0	1	2	3	4	5
Y_h	29.3	55.2	77.3	88.0	89.4	87.0
X_{1h}	0	1	2	2	2	2
X_{2h}	0	0	0	1	1	1

$b_0 = 29.933$; $b_1 = 24.0$, $b_2 = 10.2$.

If $r < 24.0$, $\hat{h}_e = 2 + b_2/b_1 = 2.425$; if $r \geqslant 24.0$, $\hat{h}_e = 0$. The residual mean square is MSE = 1.771.

The strategy used to choose the appropriate models is as follows:

1. Screen the data for outliers*.
2. Obtain an estimate of the error variance, s^2, from the original np observations.
3. Calculate the treatment mean levels, $Y_0, Y_1, \ldots, Y_{n-1}$.
4. Calculate the vector products $g_k = \sum X_{kh} Y_h$ for all possible models and submodels. This process is shortened by considering only those models for which there are m observations on the plateau (see below).
5. For each model, calculate the regression coefficients (b_k) by least squares and $\text{MSE} = \sum_h Y_h^2 - \sum_k b_k g_k$.
6. Choose the model with minimal MSE for which $b_1 > b_2 > b_3 \geqslant 0$.

The determination of m utilizes a modification of the isotonic regression* procedure: Compute successive moving averages* of the Y_h, starting with Y_{n-1} and moving toward Y_0; designate these as MA_h where $\text{MA}_h = (Y_h + \cdots + Y_{n-1})/(n-h)$. The plateau stops at the last point before MA_h begins to decrease monotonically. For the Tennessee data above, successive values of MA_h are 87.0, 88.2, 88.133, 85.425; $m = 3$.

For some biological responses, especially for drugs, the plateau may occur within a range of the lowest treatment levels as well as for the highest levels; the point at which the first plateau intersects a sloping line is often designated as a *threshold* input. The usual LP models can be used by making a slight modification of notation. Suppose that the threshold is at $h = t$, followed by $n - 1$ observations. A model IV_2 would now be designated as $\text{IV}_2(t, n)$; the X-vectors would be as follows:

h	0	$\cdots$	t	$t+1$	$t+2$	$t+3$	$\cdots$	$n+t-1$
X_1	0	...	0	1	2	2	...	2
X_2	0	...	0	0	0	1	...	1

For the threshold data, b_2 may be larger than b_1. The LP procedure can be extended to unequally spaced observations and to models for which $b_k < 0$. Cady et al. [3] have compared for 27 experimental sites in Hawaii, Indonesia, and the Philippines the amounts of soil phosphate (T) in ppm as measured by the Truog method and the linear response (P_b) of maize yields in kilograms per hectare to coded amounts of applied phosphates. Since there is no real response to additional applied phosphate when the amount of soil phosphate is adequate, one expects a plateau for all values of $T \geqslant$ some T_0. Using the moving-average procedure for Cady's data, demonstrated below, $T_0 = 30$.

T	126	113	112	109	107	79
P_b	224	41	127	122	114	181
MA	224	132	131	128	126	135

T	75	70	42	40
P_b	(−1; 37)	142	(99; 227; 409)	299
MA	106	110	148	159

T	36	31	30	24	22
P_b	(78; 462)	73	117	399	554
MA	174	168	165	178	198

The data shown in Table 1 were used to determine the relationship between P_b and T; p = number of sites. It appears that the initial rapid decrease in P_b per unit increase

Table 1

			VI		VII		
T	P_b	p	X_1	X_2	X_1	X_2	X_3
3.7	2432	1	3.7	0	3.7	0	0
4.7	2228	1	4.7	0	4.7	0	0
4.9	1430	1	4.9	0	4.9	0	0
5.3	1647	1	5.3	0	5.3	0	0
5.8	1737	1	5.8	0	5.8	0	0
6.9	1142	1	6.9	0	6.9	0	0
9.9	1107	1	6.9	3.0	6.9	3.0	0
18	746	1	6.9	11.1	6.9	11.1	0
22	554	1	6.9	15.1	6.9	15.1	0
24	399	1	6.9	17.1	6.9	17.1	0
30	164.765	17	6.9	23.1	6.9	17.1	6.0

in T stops at $T = 6.9$; thereafter, using Model VI the slower decrease continues to $T = 30$, where the plateau starts. The solution is: $b_0 = 3{,}642.45$; $b_1 = -356.507$; $b_2 = -43.978$; MSE = 25,605.2. If we use a Model VII, the solution is $b_0 = 3{,}661.07$; $b_1 = -360.523$; $b_2 = -42.200$; $b_3 = -47.845$; MSE = 26,671.3. This model is rejected because $|b_2| < |b_3|$ and MSE is increased.

Some recent developments in LP procedures are presented by Anderson and Nelson [2].

References

[1] Anderson, R. L. and Nelson, L. A. (1975). *Biometrics*, **31**, 303–318.

[2] Anderson, R. L. and Nelson, L. A. (1982). *Proc. 10th Int. Biom. Conf.*, Guaruja, Brazil, Aug. 6–10, 1979.

[3] Cady, F. B., Chan, C. P. Y., Garver, C. L., Silva, J. A., and Wood, C. L. (1982). Quantitative Evaluation of Agrotechnology Transfer: A Methodology Using Maize Response to Phosphorus on Hydric Dystrandepts in the Benchmark Soils Project. *HITAHR Res. Ser. 015*, College of Tropical Agriculture, University of Hawaii, Honolulu.

(AGRICULTURE, STATISTICS IN
MULTIPLE LINEAR REGRESSION
REGRESSION (various entries)
RESPONSE SURFACE DESIGNS)

R. L. ANDERSON

PLATYKURTIC CURVE

A frequency curve with a negative coefficient of kurtosis,

$$\gamma_2 = \frac{\mu_4}{\mu_2^2} - 3 < 0.$$

(KURTOSIS
LEPTOKURTIC CURVE
MESOKURTIC CURVE)

PLAUSIBILITY FUNCTION

A term introduced by Barndorff-Nielsen [1] to denote the ratio

$$l(T \mid \boldsymbol{\theta}) \Big/ \sup_T l(T \mid \boldsymbol{\theta})$$

where $l(\cdot)$ is a likelihood* function, $\boldsymbol{\theta}$ a parameter vector, and T is a statistic. (Note that the supremum is taken with respect to T and not $\boldsymbol{\theta}$, as in a likelihood ratio*.) The relation between this function and surprise indexes* has been discussed by Good [2].

References

[1] Barndorff-Nielsen, O. (1976). *J. R. Statist. Soc. B*, **38**, 103–131.

[2] Good, I. J. (1983). *J. Statist. Comp. Simul.*, **18**, 215–218.

PLAY-THE-LOSER SAMPLING

Analogously to play-the-winner (PW) sampling*, the first treatment (see PW for terminology) in play-the-loser sampling is determined by randomization and the trials carried out one at a time. For the case of two treatments, a failure with a given treatment calls for a subsequent trial on the *same* treatment, while success results in a switch to an alternative treatment. In the case when success probabilities are small, the PL-rule is superior to the PW-rule. For a detailed discussion of this rule, see, for example, Büringer et al. [1].

Reference

[1] Büringer, H., Martin, H., and Schriever, K.-H. (1980). *Nonparametric Sequential Selection Procedures*. Birkhäuser Boston, Cambridge, Mass.

(PLAY-THE-WINNER-RULES)

PLAY-THE-WINNER RULES

Consider a clinical trial* to compare two treatments A and B where response is dichotomous but not necessarily instantaneous. Patients enter the trial sequentially and must be treated when they arrive. The major goal of this trial is to gather sound data from current patients to derive information about the effectiveness of these two treatments for the benefit of future patients. Another goal of this trial is to treat each current patient in the best way that we can. This is due to the ethical problem of studies on human beings. These two goals are contradictory to some extent. Our purpose is to provide a treatment assignment scheme which tends to put more current patients on better treatments but is also able to give us reliable information about treatment effectiveness after the trial is over.

To meet the ethical requirement, Zelen [14] introduced the play-the-winner (PW) rule, which prescribes that a success with a given treatment generates a future trial with the same treatment, whereas a failure generates a trial with the alternative treatment. The PW rule can be implemented by placing in an urn a ball marked with an A whenever a success is obtained with treatment A or a failure with treatment B. Similarly, a ball marked with a B is placed in the urn whenever a success is obtained with treatment B or a failure with treatment A. When a new patient enters the trial, the treatment assignment is determined by drawing a ball randomly from the urn *without* replacement; if the urn is empty, the assignment is determined by tossing a fair coin. In practice, a table of random numbers can be used to simulate the urn process. This procedure does not require that all previous responses be known before the next treatment assignment. In actual trials the time required to observe the response of a patient to treatment may be much longer than the times between patient entries. Therefore, most of the time the urn contains no balls and the PW rule is of little value because it will assign approximately equal numbers of patients to each treatment.

If responses from all previous patients are known before the arrival of the next patient, the PW rule specifies that after each success we continue to use the same treatment and after each failure we switch to the other treatment. Zelen [14] called this the "modified play-the-winner" (MPW) rule, although it is really just a special case of the PW rule. Many authors ignore the distinction between the PW and MPW rules and most theoretical investigation deals with the MPW rule, that is, with the case where responses for all previous patients are known before the next treatment assignment, and they have referred to this as the PW rule. This rule has been studied extensively in selection and ranking theory. An excellent review of this area is given by Hoel et al. [6]. The MPW rule tends to put more patients on the better treatment. For example, suppose that the probability of a single trial success for treatment i is p_i, where $0 < p_i < 1$, $i = A, B$, and $q_i = 1 - p_i$. Then the asymptotic proportion of patients treated by A for the MPW rule is $q_B/(q_B + q_A)$. This quantity equals 0.5 when $p_A = p_B$ and is an increasing function of p_A for fixed $\Delta = p_A - p_B > 0.0$, and of Δ for fixed p_A regardless of the sign of Δ. However, the proportions assigned to the two treatments are not very markedly different unless p_A, Δ, or both are large. In clinical trials, it is unusual to study two treatments whose success rates differ by more than 0.3. For values of $\Delta \leqslant 0.3$ and for $p_A \leqslant 0.7$, the proportion of patients assigned to the better treatment never exceeds 0.667.

Since the MPW rule is deterministic, it may bias the trial in various ways. For example, if the experimenter prefers treatment A and knows or guesses which treatment will be the next assignment, he or she may introduce bias into the trial through the selection of patients. This type of bias is called *selection bias* by Blackwell and Hodges [2]. With the MPW rule the possibility of selection bias is maximal because the experimenter knows the next assignment with certainty. Simon et al. [9] proposed a nondeterministic plan based on the likelihood* function, but it is rather complicated for practical use.

Wei and Durham [13] proposed the "randomized play-the-winner rule," which can be described as follows: An urn contains balls that are marked either A or B. We start with μ balls of each type. When a patient is available for an assignment, a ball is drawn at random and *replaced*. If it is type i, then treatment i is assigned to the patient, where $i = A, B$. When the response of a previous patient to treatment i is available, we change the contents of the urn according to the following rule: If this response is a success, then additional α balls of type i and β additional balls of type j are put in the urn; if this response is a failure, then additional β balls of type i and additional α balls of type j are put in the urn, where $\alpha > \beta \geqslant 0$; $i, j = A, B$; and $j \neq i$. This rule is denoted by RPW (μ, α, β). Obviously, this nondeterministic rule is applicable whether responses of patients are immediate or delayed. If we choose μ sufficiently larger than α, the RPW rule tends toward simple random assignment. To increase the proportion of patients on the better treatment, we would choose a small value for μ and α much larger than β.

It is interesting to note that the difference between the PW rule and the RPW (0, 1, 0) is that the balls are drawn without replacement in the former case and with replacement in the latter case. If responses are instantaneous, asymptotically the RPW $(\mu, \alpha, 0)$ places the same proportion of patients on the better treatment as does the MPW rule (see Wei and Durham [13]).

The MPW rule has been extended by Hoel and Sobel [5] to the case where more than two treatments are compared. At the outset of the trial, we order the K treatments at random and use this ordering in a cyclic manner. After each success, we treat the next patient with the same treatment; after each failure, we switch to the next treatment in the ordering scheme; after completing the cycle, we go back to the first treatment. This rule is called the cyclic play-the-winner rule (PWC). Wei [12] proved that the PWC rule tends to put more patients on better treatments and also generalized the RPW rule to the case of K treatment comparisons. Again, this nondeterministic assignment rule tends to place the same proportions of patients on better treatments as does the PWC rule when the trial size is sufficiently large.

Until now we have only described the PW rule and its variants for patient allocation. Various stopping procedures in the setting of selection and ranking approach (as opposed to hypothesis testing) have been reviewed by Hoel et al. [6]. One such stopping rule is based on inverse sampling* [4, 10, 11] where one continues assigning patients to treatment until a prespecified number r of successes is obtained on treatment A or B. For illustration let $p_A = 0.9$, $p_B = 0.7$, and $r = 12$. A possible realization might be as follows:

A:		SSSF	SSSSF		SSSSS
B:	SSSF		F	SF	

where S and F denote success and failure, respectively. We stopped because we have observed 12 successes on treatment A and we are at least 90% certain of having selected the better treatment [11].

Appropriate methods for hypothesis testing* have not been studied for the PW, RPW rules. But for the MPW rule with a predetermined and equal number of failures f on each treatment group, Zelen [14] proposed a method to test the hypothesis that $p_A = p_B$. If N_i is the number of patients on treatment $i = A, B$ at the end of the trial, then for this particular stopping rule, N_i is the sum of f independent and identically distributed geometric* random variables. Zelen derives the conditional null distribution of N_A given $N_A + N_B$, which is a hypergeometric waiting-time distribution.

These PW rules, although of great interest to statisticians, have found little use in practice. Byar et al. [3], Bailar [1], Pocock [7], and Simon [8] have discussed the reasons. In most real trials patients are heterogeneous with respect to important prognostic factors and PW rules have not been adapted to stratification. Also these methods (especially the MPW and PWC rules) do not protect against bias introduced by changes over time in the types of patients entering into the trial. In chronic diseases like cancer, responses are usually so long delayed that the advantages of the PW procedures are effectively lost. In most real trials multiple end points are of interest and it may be inappropriate to base the entire allocation process on a single response.

PW rules might be useful in certain specialized medical situations where ethical problems are paramount and one is reasonably certain that time trends and patient heterogeneity are unimportant. The greatest benefit from using PW rules in such situations would occur when response times are short compared to times between patient entries.

References

[1] Bailar, J. C., III. (1976). *Proc. 9th Int. Biom. Conf.*, Vol. I, pp. 189–206.

[2] Blackwell, D. and Hodges, J. L., Jr. (1957). *Ann. Math. Statist.*, **28**, 449–460.

[3] Byar, D. et al. (1976). *N. Eng. J. Med.*, **295**, 74–80.

[4] Hoel, D. G. (1972). *J. Amer. Statist. Ass.*, **67**, 148–151.

[5] Hoel, D. G. and Sobel, M. (1972). *Proc. 6th Berkeley Symp. Math. Statist. Prob.*, Vol. 4. University of California Press, Berkeley, Calif., pp. 53–69.

[6] Hoel, D. G., Sobel, M., and Weiss, G. H. (1975). In *Perspectives in Biometrics*, Vol. I, R. M. Elashoff, ed. Academic Press, New York, pp. 29–61.

[7] Pocock, S. J. (1977). *Biometrics*, **33**, 183–197.

[8] Simon, R. (1977). *Biometrics*, **33**, 743–749.

[9] Simon, R., Weiss, G. H., and Hoel, D. G. (1975). *Biometrika*, **62**, 195–200.

[10] Sobel, M. and Weiss, G. H. (1971). *J. Amer. Statist. Ass.*, **66**, 545–551.

[11] Thionet, P. D. A. (1975). *Proc. 40th Sess. Indian Statist. Inst.*, pp. 822–825.

[12] Wei, L. J. (1979). *Ann. Statist.*, **7**, 291–296.

[13] Wei, L. J. and Durham, S. (1978). *J. Amer. Statist. Ass.*, **73**, 840–843.

[14] Zelen, M. (1969). *J. Amer. Statist. Ass.*, **64**, 131–146.

(ADAPTIVE METHODS
CLINICAL TRIALS)

L. J. Wei
David Byar

PLUG-IN RULES *See* NONPARAMETRIC DISCRIMINATION

POCHHAMMER'S SYMBOL

Pochhammer's symbol for the product of r successively increasing factors starting with n is given by

$$(n)_r = n(n+1)(n+2)\cdots(n+r-1),\qquad r = 1, 2, \dots .$$

In terms of gamma functions it is expressed as

$$(n)_r = \frac{\Gamma(n+r)}{\Gamma(n)}.$$

This quantity is also known as the *rth ascending factorial of n*, commonly written $n^{[r]}$.

(DIFFERENCE EQUATIONS)

POCOCK AND SIMON METHOD

Pocock and Simon [4] propose a method of assigning one of two or more treatments to patients who enter a clinical trial* sequentially. The method is designed to achieve a balance of the effects of several concomitant factors which are thought to influence the outcome of treatment.

Suppose that there are K treatments and M concomitant factors and the number of levels of the mth factor is L_m ($m = 1, \dots, M$). Consider an arbitrary time during the trial when the number of patients with level l of factor m assigned treatment k is n_{klm} ($k = 1, \ldots, K$; $l = 1, \ldots, L_m$; $m = 1, \ldots, M$).

Let $D(n_{1lm}, \ldots, n_{klm})$ be a function mapping $\mathbb{R}^K \to \mathbb{R}$ which measures the "imbalance" between the number of patients in each treatment group having level l of factor m. For example, D may be the range*, variance*, or standard deviation*. Suppose that a new patient enters with levels λ_m of factor m ($m = 1, \ldots, M$). If treatment t is allocated, the numbers n_{klm} change to $n_{klm}^{(t)}$, where

$$n_{klm}^{(t)} = n_{klm} + 1 \qquad \text{if } l = \lambda_m \text{ and } k = t$$

$$n_{klm}^{(t)} = n_{klm} \qquad \text{otherwise}$$

$$(k = 1, \ldots, K;\ l = 1, \ldots, L_m;$$

$$m = 1, \ldots, M)$$

Let $d_{mt} = D(n_{1\lambda_m m}^{(t)}, n_{2\lambda_m m}^{(t)}, \ldots, n_{K\lambda_m m}^{(t)})$. This represents the imbalance for factor m if treatment t is allocated. Let $G_t = G(d_{1t}, \ldots, d_{Mt})$ be a function mapping $\mathbb{R}^M \to \mathbb{R}$ which combines the imbalance d_{mt} for each factor m into a "total amount of imbalance."

Then the Pocock and Simon rule is to allocate treatment t with probability p_t, where $\sum_{t=1}^{K} p_t = 1$ and

$$G_s \leqslant G_t \Rightarrow p_s \geqslant p_t$$

$$(s = 1, \ldots, K;\ t = 1, \ldots, K).$$

The choice of G is arbitrary. Pocock and Simon suggest

$$G(d_{1t}, \ldots, d_{Mt}) = \sum_{m=1}^{M} w_m d_{mt},$$

where w_m are weights chosen to reflect the relative importance of factor m. The choice of $p_t (t = 1, \ldots, K)$ is also arbitrary, subject to the constraints given above. Pocock and Simon suggest $p_t = p$ for treatment t with minimum G_t and $p_t = (1 - p)/(K - 1)$ for all other t, where p ($1/K \leqslant p \leqslant 1$) is at the discretion of the investigator. The choice of $p = 1$ yields a scheme that is deterministic, that is, allocate treatment t with minimum G_t, except when two or more treatments have equal minima G_t, in which case a random choice between these treatments may be introduced.

Pocock and Simon investigated the performance of their method with D equal to the range, G an unweighted sum ($w_m = 1$; $m = 1, \ldots, M$), and $p = 1$ or 0.75. They simulated by computer a simplified clinical trial with 50 patients, $K = 2$, $M = 1, 2, \ldots, 8$, and $L_m = 2$ for all m. They found that their method, with $p = 1$ or 0.75, considerably reduced the chance of treatment imbalance compared with simple randomization* or random permuted blocks. The method of random permuted blocks within strata achieved results comparable to Pocock and Simon's for $m \leqslant 4$ but was inferior for larger M.

A numerical example based on the simulation above, with $M = 3$, follows. At a certain point in the trial a new patient presents with $\lambda_1 = 1$, $\lambda_2 = 2$, and $\lambda_3 = 1$. Suppose that for factor 1, $n_{111} = 15$, $n_{211} = 16$; for factor 2, $n_{122} = 24$, $n_{222} = 27$; and for factor 3, $n_{113} = 17$, $n_{213} = 13$. If treatment 1 were allocated, these six numbers would change to 16, 16, 25, 27, 18, and 13, respectively. Since D is the range, we have $d_{11} = 16 - 16 = 0$, $d_{21} = 27 - 25 = 2$, and $d_{31} = 18 - 13 = 5$. Then G_1 is the unweighted sum $d_{11} + d_{21} + d_{31} = 7$. Similarly, G_2 is calculated to be 9. Thus with $p = 1$ we allocate treatment 1, since G_1 is smaller. If $p = 0.75$, we allocate treatment 1 with probability 0.75 and treatment 2 with probability 0.25.

Freedman and White [3] show that when the "imbalance" function D is the variance and G is a weighted sum,

$$G_t = 2 \sum_{m=1}^{M} w_m n_{t\lambda_m m}/(N - 1) + \text{const.},$$

where the constant is independent of treatment t. If $p = 1$, the rule is to allocate the treatment t which has a minimum value of $\sum_{m=1}^{M} w_m n_{t\lambda_m m}$. This simplified rule had been proposed earlier by Taves [6] in the special case where $w_m = 1$ for all m, and is now, even for general w_m, termed "minimization." White and Freedman [7] give an example where $K = 2$, $M = 4$, $L_1 = 2$, $L_2 = 2$, $L_3 = 4$, and $L_4 = 3$. Suppose that at some time during the trial a new patient presents with $\lambda_1 = 1$, $\lambda_2 = 1$, $\lambda_3 = 3$, and $\lambda_4 = 3$. Suppose also that for factor 1, $n_{111} = 12$, $n_{211} = 8$; for factor 2, $n_{112} = 11$, $n_{212} = 12$; for factor 3, $n_{133} = 4$, $n_{233} = 3$; and for factor 4, $n_{134} = 4$, $n_{234} = 6$. Then with $w_m = 1$ $(m = 1, \ldots, 4)$, $\sum_{m=1}^{4} w_m n_{t\lambda_m m}$ equals $12 + 11 + 4 + 4 = 31$ for $t = 1$ and equals $8 + 12 + 3 + 6 = 29$ for $t = 2$. Thus treatment 2 would be allocated.

Minimization may be operated easily by trained clerks without recourse to a digital computer, an important practical advantage in many clinical trials. Pocock and Lagakos [5] found in a survey of 15 major cancer centers that 4 had used minimization in at least one clinical trial.*

Begg and Iglewicz [2] propose an alternative method of allocation which aims to minimize the variance of the estimated treatment effect. They found, using a criterion of performance based on the standardized post-trial variance of the treatment effect, that their method was consistently more efficient than three versions of the Pocock and Simon method, namely with G an unweighted sum and D equal to (1) the range, (2) the variance, and (3) the range standardized by the number of patients in the corresponding factor level. Although the improvements were too small to be of real importance, their method is nevertheless a practical alternative to Pocock and Simon's.

Atkinson [1] describes another approach based on the theory of D_A-optimum design of experiments*. Unlike Pocock and Simon's, this method takes account of correlations between the concomitant factors. However, no comparisons between the methods are yet published.

References

[1] Atkinson, A. C. (1982). *Biometrika*, **69**, 61–67.

[2] Begg, C. B. and Iglewicz, B. (1980). *Biometrics*, **36**, 81–90.

[3] Freedman, L. S. and White, S. J. (1976). *Biometrics*, **32**, 691–694.

[4] Pocock, S. J. and Simon, R. (1975). *Biometrics*, **31**, 103–115. (The source paper.)

[5] Pocock, S. J. and Lagakos, S. W. (1982). *Brit. J. Cancer*, **46**, 368–375. (An international survey of currently used methods of treatment allocation.)

[6] Taves, D. R. (1974). *Clin. Pharmacol. Ther.*, **15**, 443.

[7] White, S. J. and Freedman, L. S. (1978). *Brit. J. Cancer*, **37**, 849–857. (A nonmathematical account of methods of allocating treatments.)

(CLINICAL TRIALS
PLAY-THE-WINNER RULES)

L. S. FREEDMAN

POINCARÉ RECURRENCE THEOREM

See ERGODIC THEOREMS

POINT BISERIAL CORRELATION

See BISERIAL CORRELATION

POINT ESTIMATION

See ESTIMATION, POINT

POINT MULTISERIAL (PMS) CORRELATION

Let X be an ungrouped quantitative variable and Y be an r-level qualitative variable with given levels $y_1, \ldots, y_r$ and corresponding probabilities $p_1, \ldots, p_r$. If m_i and σ_i^2 are the conditional mean and variance of X for the jth individual in the ith level of Y, the PMS correlation coefficient is defined as

$$\rho_{\text{PMS}} = \frac{\sum_{i=1}^{r}\sum_{j=1}^{r} p_i p_j (m_i - m_j)(y_i - y_j)}{\left\{\left[\sum_{i=1}^{r} p_i \sigma_i^2 + \sum_{i=1}^{r}\sum_{j=1}^{r} p_i p_j (m_i - m_j)^2\right]^{1/2} \times \left[\sum_{i=1}^{r}\sum_{j=1}^{r} p_i p_j (y_i - y_j)^2\right]^{1/2}\right\}}$$

(Das Gupta [1]). The coefficient ρ_{PMS} is maximized if y_i are linear functions of m_i. An alternative, simpler measure was proposed by Hamdan and Schulman [2]. This measure is useful when the data are available in the form of an $r \times c$ contingency table with the quantitative variables grouped at given levels $x_1, \ldots, x_c$ and the qualitative variable Y is given in terms of r descriptive categories (e.g., no and did not, yes but did not, yes and did).

References

[1] Das Gupta, S. (1960). *Psychometrika*, **25**, 393–408.

[2] Hamdan, M. A. and Schulman, R. S. (1975). *Austral. J. Statist.*, **17**, 84–96.

(BISERIAL CORRELATION)

POINT OF AN UMBRELLA *See* UMBRELLA ALTERNATIVES

POINT PROCESSES *See* STOCHASTIC PROCESSES, POINT

See also POINT PROCESS, STATIONARY; PROCESSES, DISCRETE

POINT PROCESS, STATIONARY

Stationarity for point processes on Euclidean space means distributional invariance under simultaneous translation of all points. Its consequences resemble well-known properties of stationary stochastic processes* and time series*, and theories of inference are similarly parallel, but the fundamental role of the Palm measure is without analog for other stationary processes. Heuristically, the Palm measure corresponds to conditioning the point process on the presence of a point at the origin, but the conditioning cannot be effected in an elementary manner. For a point process on $\mathbb{R}^1$ the Palm measure represents observation initiated at the time of an event; failure to distinguish this synchronous observation from asynchronous observation (begun at an arbitrary point in time) has engendered confusion and errors in the past. Stationarity is a natural assumption for many physical situations; stationary point processes have been used to model earthquakes, precipitation, queueing systems, and structure of granular objects. The latter requires marked point processes that are stationary in only one component.

FORMULATION

Some measure-theoretic machinery is essential to describe stationary point processes properly; we have minimized it here. Let E denote d-dimensional Euclidean space ($d \geqslant 1$); Lebesgue measure (length, area, or volume) on E is written as $\lambda(dx)$. A (simple) point process* on E is a stochastic system of distinct, indistinguishable points in E with random locations X_i; we write it as a sum $N = \sum \epsilon_{X_i}$ of point masses [Dirac measures ϵ_x, where $\epsilon_x(A)$ is 1 or 0 according as $x \in A$ or not], so that $N(A) = \sum \epsilon_{X_i}(A)$ is the random number of points located in A. When $d = 1$ we take the points as times T_i of events ("arrival" times), with the convention that $\cdots < T_{-1} < T_0 \leqslant 0 < T_1 < \cdots$; the differences $U_i = T_i - T_{i-1}$ are "interarrival" times.

For each $x \in E$ we define the point process $N_x = \sum \epsilon_{X_i - x}$, in which all points of N are translated by the vector x. Then, N is *stationary* if for every x, N_x has the same distribution as N. Stationarity is distributional invariance under bodily translation of all points of the process, with neither relative displacement nor rotation.

Two classes of stationary point processes are particularly well understood: stationary Poisson processes and stationary renewal processes. A Poisson process* N on $\mathbb{R}^d$ with mean measure a scalar multiple of λ is stationary; its characteristic feature is independent increments: numbers of points in disjoint sets are independent random variables. Given a distribution function F on $(0, \infty)$ with finite mean m, one constructs a stationary renewal process $N = \sum \epsilon_{T_i}$ as follows: (a) the random vector $(-T_0, U_1)$ and the interarrival times U_i, $i \neq 1$, are mutually independent; (b) the U_i, $i \neq 1$, have distribution F; (c) $(-T_0, U_1)$ has distribution $m^{-1}1(u < v)\,du\,dF(v)$. In particular, U_1 has distribution $m^{-1}v\,dF(v)$; the location of the origin within it is conditionally uniform. Because it contains the origin this interval is longer than others in several probabilistic senses, which some have thought paradoxical.

MOMENTS AND SPECTRA

Similarity of stationarity for point processes to other forms suggests that key descriptors of the law of a stationary point process are appropriate first and second moments, which is true. However, since stationarity for point processes is analogous to strong stationarity rather than L^2 (weak) stationarity, this description is incomplete except in special cases. The *first moment measure* $A \to E[N(A)]$ is translation invariant, hence a scalar multiple of Lebesgue measure; the multiplier ν, if finite, is the *intensity* of N and admits in addition local and asymptotic interpretations. Similarly, the *second moment measure* $(A \times B) \to E[N(A)N(B)]$ on E^2 admits a representation

$$E[N(A)N(B)] = \int\int 1_A(z+x)1_B(x)\lambda(dx)\mu_*^2(dz),$$

with μ_*^2, the *reduced second moment measure*, a measure on E. Equally important is the *reduced covariance measure* $\rho_*(dz) = \mu_*^2(dz) - \nu^2\lambda(dz)$. More generally one can define moment measures, reduced moment measures, cumulant measures (the covariance measure is one), and reduced cumulant measures of all orders (see Krickeberg [4]).

Use of reduced measures is not simply a matter of economy; finiteness of them has the same implications concerning near independence of remote parts of the process as in other contexts, and estimation of reduced measures is natural, so on both grounds their role in inference is important.

A stationary point process has stationary increments and consequently (Itô [2]) admits a *spectral representation*

$$\int \psi(x)N(dx) = \sum \psi(X_i) = \int \tilde{\psi}(v)Z(dv), \qquad (1)$$

where $\tilde{\psi}$ is the Fourier transform of ψ and Z is a complex-valued random measure on $\mathbb{R}^d$ with orthogonal increments. The measure $F(dv) = E[|Z(dv)|^2]$, the *spectral measure* of N, is related to the reduced second moment measure via the Parseval identity $F(\psi) = \mu_*^2(\hat{\psi})$, where the caret denotes the inverse Fourier transform. If the covariance spectral measure, associated analogously to the centered process $M(A) = N(A) - \nu\lambda(A)$, is absolutely continuous, its derivative f is the *spectral density function* of N.

DISTRIBUTIONAL ASPECTS

One consequence of the "clean" moment theory of stationary point processes is a difficult distribution theory. Even in one dimension objects such as interarrival-time distributions usually cannot be calculated, at least with respect to the underlying probability; however, proper use of the Palm measure often simplifies calculations substantially. An interesting and general "zero-infinity" law holds for stationary point processes: every finitely definable configuration of points occurs (except with probability zero) either infinitely often or not at all. Thus for a stationary point process on $\mathbb{R}$, almost surely there are either no points or infinitely many on each half-line.

PALM MEASURES

Indispensable to description of distributional and asymptotic aspects of stationary point processes, to applications (in which a point of the process is typically taken as origin of the coordinates), and to inference is the *Palm measure*. Given a stationary point process N there exists ([6]; see also Neveu [7]) a unique σ-finite measure P^* on Ω satisfying

$$E\left[\int H(N_x, x)N(dx)\right] = E^*\left[\int H(N, x)\lambda(dx)\right] \qquad (2)$$

for each functional H. (Although P^* need not be a probability it is usual to employ expectation notation.) Despite the complexity of this characterization the heuristic interpretation is simple: The *Palm distribution* $P^*(N \in (\cdot))/P^*(\Omega)$ is the law of N given the (probability zero) event that $N(\{0\}) = 1$.

More precisely, if H is a bounded functional such that $x \to H(N_x)$ is continuous, then for open sets $V_n \downarrow \{0\}$,

$$E^*[H(N)] = \lim E[H(N)1(N(V_n) \neq 0)]/\lambda(V_n) \tag{3a}$$

and

$$E^*[H(N)]/P^*(\Omega) = \lim E[H(N) \mid N(V_n) = 1]. \tag{3b}$$

Important classical results can be deduced immediately. The Khintchine–Koroljuk theorem: If ν is finite, then

$$\lim \Pr\{N(V_n) > 0\}/\lambda(V_n) = \lim \Pr\{N(V_n) = 1\}/\lambda(V_n) = \nu,$$

which provides the local interpretation of the intensity, follows by taking $H = 1$ in (3a). The Palm–Khintchine equations, which for $d = 1$ express *Palm functions** $p_k^*(t) = P^*\{N(0,t) = k\}$ as derivatives of tail probabilities: for each k, $p_k^*(t) = (d/dt)\Pr\{N(0,t) > k\}$, result from putting $H(N,s) = 1(N(0,t-s) = k)1(0 < s < t)$ in (2). Choosing $H \equiv 1$ in (2) shows that $\nu = P^*(\Omega)$, yet another interpretation of the intensity.

Although the Palm measure determines the underlying probability uniquely there is no useful inversion formula, which is unfortunate because there are natural, effective estimators of P^* [see (5)].

ASYMPTOTICS

Limit theorems for a stationary point process concern behavior of $N(A)$ for large sets A. Stationarity replaces the "identically distributed" part of classical "i.i.d." hypotheses; independence is replaced by asymptotic independence of $N(A)$ and $N(B)$ for distantly separated sets A and B, deduced in turn from (assumed) finiteness of reduced cumulant measures (akin to summability or integrability of covariance functions). Ergodicity [7] suffices for a strong law of large numbers* [8]: for suitable functionals H, almost surely

$$\lim_{r\to\infty} \lambda(B_r)^{-1} \int_{B_r} H(N_x) N(dx) = E^*[H(N)], \tag{4}$$

where B_r is the ball of radius r centered at the origin. For example, with $H \equiv 1$, (4) gives $\lim N(B_r)/\lambda(B_r) = \nu$, the asymptotic interpretation of the intensity. A central limit theorem holds under rather severe assumptions, namely restrictions on the form of H and finite total variation of reduced cumulant measures of *all* orders (see Jolivet [3]).

STATISTICAL INFERENCE

For statistical models of stationary point processes satisfying the limit theorems just mentioned, inference can be based on single realizations rather than multiple copies; asymptotics arise from observation over increasingly large sets. Estimation of the intensity is straightforward: the estimators $\hat{\nu} = N(B_r)/\lambda(B_r)$ are unbiased and (given suitable assumptions) strongly consistent and asymptotically normal. Choosing $H(N_x, x) = 1(x \in B_r)\int f(y)N_x(dy)$ in (2) shows that the estimators $\hat{\mu}_*^2 = \lambda(B_r)^{-1}\int_{B_r} N(dx) \int N(dy) f(y-x)$ of the reduced second moment measure (i.e., of $\int f d\mu_*^2$) are unbiased; again the asymptotic theory applies. These and analogous estimators of higher-order reduced moment measures are special cases of estimators of the Palm measure, which is the natural object to estimate in nonparametric cases. By (2) the estimators

$$\hat{E}^*[H(N)] = \lambda(B_r)^{-1} \int_{B_r} H(N_x) N(dx) \tag{5}$$

are unbiased. Moreover, there is a natural interpretation: only points of N in B_r contribute to the integral in (5): For x a point of N, N_x has a point at the origin, so the estimator is simply an average of evaluations of H at translations of N that place each point at the origin.

The spectral density function is estimated with periodogram* estimators $\hat{f}(v) = |[2\pi\lambda(B_r)]^{-1}\int_{B_r} e^{-i\langle v,x\rangle}N(dx)|^2$ (see Brillinger [1]).

Because of difficult distributional computations, rather fewer testing problems have been examined; among them the most important and best understood is that of testing whether a stationary point process on $\mathbb{R}$ is Poisson (see Lewis [5]).

Linear prediction is well developed for stationary point processes on $\mathbb{R}$, with most techniques based on spectral or backward moving-average representations; nonlinear state estimation is inchoate.

References

[1] Brillinger, D. R. (1975). In *Stochastic Processes and Related Topics*, M. L. Puri, ed. Academic Press, New York.

[2] Itô, K. (1955). *Proc. Mem. Sci. Univ. Kyoto A*, **28**, 209–223.

[3] Jolivet, E. (1981). In *Point Processes and Queueing Problems*, P. Bartfai and J. Tomko, eds. North-Holland, Amsterdam.

[4] Krickeberg, K. (1982). *Lect. Notes Math.*, **929**, 205–313. Springer-Verlag, New York.

[5] Lewis, P. A. W. (1972). In *Stochastic Point Processes: Statistical Analysis, Theory and Applications*, P. A. W. Lewis, ed. Wiley, New York.

[6] Matthes, K. (1963). *Jb. Dtsch. Math.-Ver.*, **66**, 66–79.

[7] Neveu, J. (1977). *Lect. Notes Math.*, **598**, 249–447. Springer-Verlag, New York.

[8] Nguyen, X. X. and Zessin, H. (1979). *Z. Wahrscheinl. verw. Geb.*, **48**, 133–158.

Bibliography

See the following works, as well as the references just given, for more information on the topic of stationary point processes.

Brillinger, D. R. (1972). *Proc. 6th Berkeley Symp. Math. Statist. Prob.*, Vol. 1. University of California Press, Berkeley, Calif., pp. 483–513. (Spectral analysis.)

Brillinger, D. R. (1978). In *Developments in Statistics*, P. R. Krishnaiah, ed. Academic Press, New York. (Extended analogy between stationary point processes and time series.)

Cox, D. R. and Isham, V. (1980). *Point Processes*. Chapman & Hall, London. (Modern but elementary presentation of the basic theory.)

Cox, D. R. and Lewis, P. A. W. (1966). *The Statistical Analysis of Series of Events*. Chapman & Hall, London. (Inference for stationary point processes on the line; many specific models and procedures.)

Daley, D. J. (1971). *J. R. Statist. Soc. B*, **33**, 406–428. (Spectra of stationary point processes and random measures.)

Daley, D. J. (1974). In *Stochastic Geometry*, E. F. Harding and D. G. Kendall, eds. Wiley, New York. [Forms of orderliness (e.g., $\Pr\{N(0,t) \geqslant 2\} = o(t)$, $t \to 0$); the Khintchine–Koroljuk theorem.]

Franken, P., König, D., Arndt, U., and Schmidt, V. (1980). *Point Processes and Queues*. Akademie-Verlag, Berlin. (Application of marked stationary point processes to queueing problems.)

Jolivet, E. (1981). In *Point Processes and Queueing Problems*, P. Bartfai and J. Tomko, eds. North-Holland, Amsterdam. (Ordinary and functional central limit theorems.)

Jowett, J. H. and Vere-Jones, D. (1972). In *Stochastic Point Processes: Statistical Analysis, Theory and Applications*, P. A. W. Lewis, ed. Wiley, New York. (Linear prediction.)

Karr, A. F. (1986). *Point Processes and Their Statistical Inference*. Marcel Dekker, New York. (Theory and inference in a measure-theoretic setting.)

Krickeberg, K. (1982). *Lect. Notes Math.*, **929**, 205–313. [Very well written analysis of the estimators (5), their special cases, and related inference problems.]

Leadbetter, M. R. (1972). *Proc. 6th Berkeley Symp. Math. Statist. Prob.*, Vol. 3. University of California Press, Berkeley, Calif., pp. 449–462. (Orderliness and Khintchine–Koroljuk theorems in higher dimensions.)

Lewis, P. A. W. (1972). In *Stochastic Point Processes: Statistical Analysis, Theory and Applications*, P. A. W. Lewis, ed. Wiley, New York. (Specific inference problems; elementary level.)

Matthes, K., Kerstan, J., and Mecke, J. (1978). *Infinitely Divisible Point Processes*. Wiley, New York. (Stationarity in the presence of other assumptions; complete but difficult.)

Mecke, J. and Stoyan, D. (1983). *Stochastische Geometrie*. Akademie-Verlag, Berlin. (Applications, especially to stereology.)

Neveu, J. (1977). *Lect. Notes Math.*, **598**, 249–447. (Rigorous, elegant, and marvelously readable exposition of the theory.)

Ogata, Y. (1978). *Ann. Inst. Statist. Math.*, **30**, 243–251. (Parametric inference in a stochastic intensity setting.)

Ripley, B. (1981). *Spatial Statistics*. Wiley, New York. ("Distance" and other methods for real data; many references.)

Ryll-Nardzewski, C. (1961). *Proc. 4th Berkeley Symp. Math. Statist. Prob.*, Vol. 2. University of California Press, Berkeley, Calif., pp. 455–465. (Key early work on Palm distributions.)

Vere-Jones, D. (1970). *J. R. Statist. Soc. B*, **32**, 1–62. (Stationary point process model of earthquake occurrences.)

(MEASURE THEORY IN PROBABILITY
PALM FUNCTIONS, PALM–KHINCHIN EQUATIONS
QUEUEING THEORY
RENEWAL THEORY)

ALAN F. KARR

POISSON-BETA DISTRIBUTION

Two types of compound Poisson* distribution, obtained by ascribing to the parameter θ of the Poisson distribution either (1) a beta distribution, with probability density function (PDF)

$$f(t) = \left\{(b-a)^{\alpha+\beta-1}B(\alpha,\beta)\right\}^{-1} \times (t-a)^{\alpha-1}(b-t)^{\beta-1} \quad (a < t < b)$$

or (2) a beta distribution of the second kind, with PDF

$$f(t) = \frac{1}{B(\alpha,\beta)} \frac{t^{\alpha-1}}{(1+t)^{\alpha+\beta}} \qquad (t > 0).$$

(They should, strictly speaking, be called beta-Poisson distributions.) Properties of these distributions are presented in ref. 2. A special case of (1), when the distribution of θ is rectangular*, is discussed in ref. 1.

References

[1] Bhattacharyya, S. K. and Holla, M. S. (1965). *J. Amer. Statist. Ass.*, **60**, 1060–1066.
[2] Holla, M. S. and Bhattacharyya, S. K. (1965). *Ann. Inst. Statist. Math., Tokyo*, **17**, 377–384.

POISSON BINOMIAL DISTRIBUTION

This is the distribution of the number of successes in n independent trials, when the probability of success varies from trial to trial. It is to be distinguished from the Poisson-binomial* (note the hyphen!) distribution, which is a compound binomial distribution with a Poisson compounding (mixing) distribution.

The probability of exactly n successes in the n trials is the coefficient of t^n in the expansion of

$$\prod_{i=1}^{n} (q_i + p_i t),$$

where $q_i = 1 - p$.

If $p_1 = p_2 = \cdots = p_n = p$, the distribution is binomial with parameters n and p.

The Poisson binomial distribution is a Lexian distribution* with $m = 1$. Equivalently, one can say that a Lexian distribution is a Poisson binomial distribution with $p_1 = \cdots = p_m = p_{(1)}$, $p_{m+1} = \cdots = p_{2m} = p_{(2)}$, and so on.

(BINOMIAL DISTRIBUTION
LEXIAN DISTRIBUTION)

POISSON-BINOMIAL DISTRIBUTION

This is a compound binomial distribution obtained by replacing the parameter N of a binomial distribution* (with parameters N and p) by n times a Poisson variable with parameter θ. Symbolically, the distribution is

$$\text{binomial}(N, p) \bigwedge_{N/n} \text{Poisson}(\theta).$$

Its probability mass function is

$$\Pr[X = x] = e^{-\theta}(p/q)^x (x!)^{-1} \times \sum_{j \geqslant x/n} (nj)^{(x)} [(\lambda q)^n / j!] \qquad (x = 0, 1, \ldots),$$

where $q = 1 - p$ and $(nj)^{(x)} = nj(nj-1)\ldots(nj-x+1)$. Numerical values of the probabilities may be calculated from the recurrence relation

$$\Pr[X = x] = \frac{n\theta p}{x+1} \sum_{j=0}^{x} \binom{n-1}{j} p^j q^{n-j-1} \times \Pr[X = x - j].$$

Further details can be found in refs. 1 and 2.

This distribution is to be distinguished from the Poisson binomial* (no hyphen!) distribution discussed in the immediately preceding entry. The name "Poisson-binomial" is sometimes also applied to the distribution

$$\text{Poisson}(\theta) \bigwedge_{\theta/\varphi} \text{binomial}(n, p),$$

which should be more aptly called "binomial-Poisson."

References

[1] Johnson, N. L. and Kotz, S. (1969). *Distributions in Statistics: Discrete Distributions*. Wiley, New York.

[2] Shumway, R. and Gurland, J. (1960). *Skand. Aktuarietidskr.*, **43**, 87–108. (See also *Biometrics*, **16**, 522–533.)

POISSON DISTRIBUTION

A random variable X is said to have a Poisson distribution with parameter λ ($\lambda > 0$) if

$$P(X = x) = p_x(\lambda) \qquad (x = 0, 1, \ldots),$$

where $p_x(\lambda)$ is the probability function of the distribution defined by

$$p_x(\lambda) = e^{-\lambda}\lambda^x/x! \qquad (x = 0, 1, \ldots),$$

which may be also called the Poisson discrete density function with respect to the counting measure over the set of all nonnegative integers.

The Poisson random variables are assumed in describing many phenomena. For example, (1) the number of α particles emitted from radioactive substance in a fixed time interval with the emission rate λ, (2) the number of yeast cells per unit volume of a suspension placed in hemacytometer, or (3) the number of telephone calls arriving at a telephone switchboard per unit time. The typical examples of the Poisson distribution above were considered by Rutherford et al. [16], Student [18], and Erlang [4], respectively.

In the subsequent sections topics on the distribution are inevitably limited. For further information the reader is referred to the books by Haight [7] and Johnson and Kotz [8].

POISSON APPROXIMATION TO THE BINOMIAL DISTRIBUTION

The Poisson distribution is derived as the limit distribution of the binomial distribution* under certain limiting conditions. Denote the probability function of the binomial distribution with parameters n and p by

$$b_x(n, p) = \binom{n}{x} p^x (1-p)^{n-x} \qquad (x = 0, 1, \ldots, n).$$

Under the limiting process

$$n \to \infty, p \to 0, np \to \lambda \ (\lambda\text{: positive constant}),$$

it is well known that $b_x(n, p) \to p_x(\lambda)$.

S. D. Poisson* (1837) gave the limiting distribution in art. 81 in his book *Recherches sur la Probabilité des Jugements en Matière Criminelle et en Matière Civile, Précédées des Règles Générales du Calcul des Probabilitiés*. In the derivation of the result Poisson proved that the lower tail probability of the binomial distribution, $\sum_{i=0}^{x} b_i(n, p)$, equals the upper tail probability of the corresponding negative binomial distrubution*, although the property had already been found by P. R. de Montmort in his *Essai d'analyse sur les jeux de hazards* (1708). Poisson then proved that the latter probability converges to the quantity $\sum_{i=0}^{x} p_i(\lambda)$ under the limiting process before mentioned. So, after his name, the limit distribution is called the Poisson distribution. In his *Exposition de la Théorie des Chances des Probabilitiés* A. A. Cournot* (1843) evaluated the Poisson approximation to a binomial distribution numerically when binomial parameters $n = 200$, $p = 0.01$, and supported its accuracy of approximation. The limit approximation given by Poisson, however, had already been discovered by A. De Moivre* (1718). In his book *The Doctrine of Chances*, he considered the problem to be essentially the same problem as that solved by the equation in x, $\sum_{i=0}^{x-1} b_i(n, p) = \frac{1}{2}$, under the conditions $n \to \infty$, $p \to 0$, and $np \to \lambda$ (finite). He solved it approximately by giving the approximate equation $\lambda = \ln 2 + \ln\{1 + \sum_{i=1}^{x-1} \lambda^i / i!\}$, instead of the equation $e^{-\lambda}(1 + \sum_{i=1}^{x-1} \lambda^i / i!) = \frac{1}{2}$, because the exponential symbol e had never been used until the days of Euler (1707–1783).

Concerning the Poisson-binomial approximation, the following holds:

$$\sum_{x=0}^{\infty} |b_x(n, p) - p_x(np)| \leqslant 2p$$

$$(0 \leqslant p \leqslant 1, n = 0, 1, 2, \dots),$$

and for the Poisson-negative binomial approximation

$$\sum_{x=0}^{\infty} |c_x(a, q) - p_x(aq/p)| < \sqrt{2}\, q/p,$$

where

$$c_x(a, q) = \binom{a - 1 + x}{x} p^a q^x,$$

$$a > 0, \quad 0 < p < 1, \quad q = 1 - p.$$

From these inequalities Poisson's limit results automatically follow. More accurate bounds to the above are given by Matsunawa [11].

POISSON DISTRIBUTION IN POISSON PROCESS

The Poisson distribution is also derived from knowledge of a random point process*. In such an aspect the Poisson distribution is the counting distribution for the corresponding Poisson process*. The process continues over a certain index set of time or space where specified rare events occur at random in the index with some fixed mean rate. Let $X(t)$ be the number of occurrence of the specified events in the time interval $[0, t)$ with $t \geqslant 0$ and $X(0) = 0$. For example, the number of arrivals of customers, the number of breakdowns of a machine, and so on, may be considered. Usually, the following postulates are made to define the Poisson process:

1. The probability $q_x(t)$ of exactly x events occurring in the time interval $[\tau, \tau + t)$ depends only on the number x and on the length t of the interval, but not on the time τ. Thus the random variables $X(\tau + t) - X(\tau)$ and $X(t)$ are equidistributed. (Stationarity)
2. The numbers of events occurring on nonoverlapping time intervals are mutually independent. Namely, for any choice of indices $0 \leqslant s_1 < s_2 \leqslant t_1 < t_2$, the random variables $X(s_2) - X(s_1)$ and $X(t_2) - X(t_1)$ are independently distributed. (Independency)
3. In a small time interval of length h the probability of one event occurring is $\lambda h + o(h)$ and that of multiple events occurring is $o(h)$, where λ is some fixed positive number and $o(h)$ means a positive quantity such that $o(h)/h \to 0$ as $h \to 0$. Thus $P\{X(t + h) - X(t) = 1\} = \lambda h + o(h)$ and $P\{X(t + h) - X(t) \geqslant 2\} = o(h)$, for any index t. (Rareness)

Under the postulates above the form of $q_x(t)$ is obtained by solving the following differential-difference equations:

$$\frac{d}{dt} q_0(t) = -\lambda q_0(t),$$

$$\frac{d}{dt} q_x(t) = -\lambda q_x(t) + \lambda q_{x-1}(t) \qquad (x \geqslant 1),$$

with the initial conditions $q_0(0) = 1$ and $q_x(0) = 0$. This leads to

$$q_x(t) = (\lambda t)^x e^{-\lambda t} / x! \qquad (x = 0, 1, 2, \dots),$$

which is nothing but the Poisson distribution with parameter λt, and hence $X(t)$ is called the Poisson process with intensity (or mean rate) λ.

To get the form of $q_x(t)$ heuristically, the following intuitive approach is available. Dividing the time interval $[0, t)$ into n disjoint intervals of equal length $h = t/n$, then, under the postulates above, we have a limiting relation for a sequence of n Bernoulli trials such that

$$q_x(t) = \lim_{n\to\infty} \binom{n}{x}\left(\lambda \frac{t}{n}\right)^x \left(1 - \lambda \frac{t}{n}\right)^{n-x}$$
$$= \frac{(\lambda t)^x e^{-\lambda t}}{x!}, \qquad (x = 0, 1, 2, \dots).$$

MOMENTS, GENERATING FUNCTIONS, AND SOME PROPERTIES

For the Poisson random variable X with parameter λ, the rth factorial moment* is given by

$$E(X^{(r)}) = E[X(X-1)\dots(X-r+1)] = \lambda^r,$$

from which moments* of X about zero can be calculated as

$$E(X) = \lambda, \qquad E(X^2) = \lambda + \lambda^2,$$
$$E(X^3) = \lambda + 3\lambda^2 + \lambda^3,$$
$$E(X^4) = \lambda + 7\lambda^2 + 6\lambda^3 + \lambda^4,$$
$$E(X^5) = \lambda + 15\lambda^2 + 25\lambda^3 + 10\lambda^4 + \lambda^5, \quad \dots .$$

The moment generating function* of X is given by

$$E(e^{tX}) = \exp[\lambda(e^t - 1)],$$

from which moments about the mean can be obtained as

$$\operatorname{var}(X) = \lambda, \quad \mu_3(X) = \lambda, \quad \mu_4(X) = \lambda + 3\lambda^2,$$
$$\mu_5(X) = \lambda + 10\lambda^2, \quad \dots .$$

Further, the cumulant generating function* is $\lambda(e^t - 1)$, and cumulants of X are $\kappa_r(X) = \lambda$ for all $r \geqslant 2$. Since $E(X) = \lambda$ we sometimes call that X is distributed as the Poisson distribution with mean λ. We see also $E(X) = \operatorname{var}(X) = \lambda$, which is a remarkable property of the Poisson distribution. The shape of every Poisson density $p_x(\lambda)$ increases as x increases from zero, and has a maximum value near the mean λ, and then decreases. This is immediately obtained through

$$\frac{p_{x+1}(\lambda)}{p_x(\lambda)} = \frac{\lambda}{x+1}, \qquad (x = 0, 1, \dots),$$

from which it follows precisely that if λ is not an integer then $p_x(\lambda)$ has a unique maximum at $x = [\lambda]$, the largest integer less than λ, and if λ is an integer, then $p_x(\lambda)$ takes maximum values both at $x = \lambda$ and $x = \lambda - 1$. The identities above are useful to calculate Poisson individual probabilities successively, too. Concerning the shape of the distribution Samuels [17] gave the following bounds on the cumulative distribution function (CDF) of the Poisson distribution, $P(X \leqslant x) = P_x(\lambda)$:

$$P_{x-1}(\lambda) > 1 - P_x(\lambda) \qquad \text{if } \lambda \leqslant x,$$
$$P_x(\lambda) \geqslant e^{-\lambda/(x+1)} \qquad \text{if } \lambda \leqslant x + 1,$$

and in particular

$$P_\lambda(\lambda) > \tfrac{1}{2} \qquad (\lambda \text{ an integer})$$
$$P_x(\lambda) > e^{-1} \qquad \text{if } \lambda < x + 1.$$

The Poisson distribution has a reproductive property: Let X_1 and X_2 be independent Poisson random variables with parameters λ_1 and λ_2, respectively. Then the sum of these variables $X_1 + X_2$ is distributed according to the Poisson distribution with parameter $\lambda_1 + \lambda_2$. The following property is also known: if X_1 and X_2 are two independent Poisson random variables with respective parameters λ_1 and λ_2, then the conditional distribution of X_1 given $X_1 + X_2 = x$ is a binomial distribution with parameters x and $\lambda_1/(\lambda_1 + \lambda_2)$. The last two properties can be extended to the case of $n \geqslant 3$ Poisson random variables.

SOME CHARACTERIZATIONS OF THE POISSON DISTRIBUTION

The following theorems are well known:

1. If X_1 and X_2 are independent random variables and $X_1 + X_2$ is distributed as a Poisson distribution, then X_1 and X_2 must each have Poisson distributions [14].
2. If X_1 and X_2 are independent nonnegative integer-valued random variables

and if

$$P(X_1 = x_1 \mid X_1 + X_2 = x) = \binom{x}{x_1} p_x^{x_1}(1-p_x)^{x-x_1}, \quad (x_1 = 0, 1, \ldots, x),$$

then (a) $p_x \equiv p$ for all x, and (b) both X_1 and X_2 have Poisson distributions with parameters in the ratio $p/(1-p)$ [3].

3. Let X be a nonnegative integer-valued random variable and Y be another random variable such that for all x,

$$P(Y = y \mid X = x) = \binom{x}{y} p^y (1-p)^{x-y} \quad (y = 0, 1, \ldots, x),$$

where p is some constant satisfying $0 < p < 1$ and not being dependent on x. Then the condition

$$P(Y = y) = P(Y = y \mid X = Y) = P(Y = y \mid X > Y)$$

is necessary and sufficient for X to have a Poisson distribution [15].

4. Let X and Y be random variables satisfying the same setup as in theorem 3. Further, let $E(X) < \infty$. Then the condition

$$E(X \mid Y = y) = y + b$$

is necessary and sufficient for X to have the Poisson distribution with parameter $b/(1-p)$ [10].

APPROXIMATIONS TO POISSON DISTRIBUTION BY CONTINUOUS DISTRIBUTIONS

1. Let X be a Poisson-distributed random variable with parameter λ. Then the distribution function of X can be represented as

$$P_x(\lambda) = P(X \leqslant x) = P\left(\chi^2_{2(x+1)} > 2\lambda\right),$$

where $\chi^2_{2(x+1)}$ is the random variable distributed according to a chi-square distribution* with $2(x+1)$ degrees of freedom. Approximating the right-hand-side probability with the aid of the Wilson–Hilferty approximation*, we have

$$P_x(\lambda) \doteqdot 1 - \Phi(z),$$

where $\Phi(\cdot)$ is the distribution function of the standard normal distribution* and

$$z = 3(x+1)^{-1/2}\left[\left(\frac{\lambda}{x+1}\right)^{3/2} - 1 + \frac{1}{9(x+1)}\right].$$

2. The following approximation by Peizer and Pratt [13] is a little complex but very accurate.

$$P_x(\lambda) \doteqdot \Phi(z),$$

$$z = \left(x - \lambda + \frac{2}{3} + \frac{\epsilon}{x+1}\right) \times \left[1 + T\left(\frac{x + 1/2}{\lambda}\right)\right]^{1/2} \lambda^{-1/2},$$

where $T(y) = (1 - y^2 + 2y \ln y)(1 - y)^{-2}$, $T(1) = 0$, and where $\epsilon = 0$ for simplicity or $\epsilon = 0.02$ for more accurate approximation. Molennar [12] proposes to take $\epsilon = 0.022$. The error of this approximation is expressed as

$$\phi(\xi)\lambda^{-3/2}(-\xi^2 + 1620\epsilon - 32)/1620 + O(\lambda^{-2}),$$

where $\phi(\cdot)$ is the standard normal probability density function and ξ is the exact normal deviate defined by $P_\lambda(x) = \Phi(\xi)$.

3. By the formal Edgeworth expansion we have

$$\begin{aligned} P_\lambda(x) = \Phi(z) - \phi(z)[&\lambda^{-1/2}(z^2 - 1)/6 \\ &+ \lambda^{-1}(z^5 - 7z^3 + 3z)/72 \\ &+ \lambda^{-3/2}(5z^8 - 95z^6 + 384z^4 - 129z^2 \\ &- 123)/6480] + O(\lambda^{-2}), \end{aligned}$$

where $z = (x - \lambda + 1/2)\lambda^{-1/2}$.

4. The following bounds on $P_x(\lambda)$ were

given by Bohman [1]:

$$P_x(\lambda) \leqslant \Phi(z) \quad \text{with } z = (x - \lambda + 1)\lambda^{-1/2},$$

$$P_x(\lambda) \geqslant \int_0^x t^\lambda e^{-t}\, dt / \Gamma(\lambda + 1).$$

POISSON'S LAW OF SMALL NUMBERS

Bortkiewicz* [2] published in 1898 his famous monograph *Das Gesetz der kleinen Zahlen*, where he called attention to Poisson's exponential limit, which had been forgotten for a long time. In the monograph he gave many important characteristics and properties of the Poisson distribution. His work, however, did not make clear what the so-called "law of small numbers" meant. So there were many disputes on the matter in those days.

Nowadays, there is a belief that the law of small numbers does not mean the Poisson distribution itself but it should be understood as follows. Consider n independent trials with probabilities p_i of success in the ith trial, $i = 1, \ldots, n$ (Poisson trials). If n is sufficiently large and all p_i's look nearly equal, then the trials can be regarded as the sampling of size n from the Poisson distribution with parameter $\sum_{i=1}^n p_i$.

This interpretation on the law of small numbers is closely related to von Mises' work [19]. He considered an approximation to probabilities of rare events that occurred in the Poisson trials mentioned above. Let ω_x be the probability of x successes in n independent Poisson trials, and suppose that there exists a constant p which may depend on n such that $p_i \leqslant p$ $(i = 1, \ldots, n)$. Further, let ψ_x be the usual Poisson individual probability with the parameter $\lambda = \sum_{i=1}^n p_i$. Then von Mises* proved that

$$\left(1 - \frac{\lambda p}{1-p}\right)\left\{1 - \left(\frac{np}{\lambda(1-p)}\right)^x \times\left[1 - \left(1 - \frac{1}{n}\right) \cdots \left(1 - \frac{x-1}{n}\right)\right]\right\} \leqslant \frac{\omega_x}{\psi_x} \leqslant (1-p)^{-x}.$$

Hence, if, as $n \to \infty$, λ, x, np remain finite and $p \to 0$, then the approximation by the Poisson distribution is valid for the Poisson trials.

Fuchs and Roby [5] gave more general result. Let $X_{k,n}$ $(k = 1, \ldots, n;\ n = 1, 2, \ldots)$ be an infinite triangular array of random variables with success probabilities $p_{k,n}$ $(0 < p_{k,n} < 1)$ of the kth trial in the nth set of trials. Put $S_n = \sum_{k=1}^n X_{k,n}$, $\lambda_n = \sum_{k=1}^n p_{k,n} = E(S_n)$, $\alpha_n = \max_{1 \leqslant k \leqslant n} p_{k,n}$ $(0 < \alpha_n < 1)$; then the simultaneous conditions $\lim_{n\to\infty} \lambda_n = \lambda$ $(0 < \lambda < \infty)$ and $\lim_{n\to\infty} \alpha_n = 0$ are necessary and sufficient for the S_n to be asymptotically Poisson-distributed random variables with parameter λ.

POISSON'S LIMIT THEOREM FOR A SEQUENCE OF DEPENDENT RARE EVENTS

The Poisson limit theorem to binomial distributions is a fairly robust approximation. It can be extended to the case where underlying trials are not necessarily independent. Let $\{A_i\}$ $(i = 1, \ldots, n)$ be an arbitrary sequence of events on a given probability space* $(\Omega, \mathscr{A}, P)$ and put

$$S_k(n) = \sum P(A_{i_1} \cap A_{i_2} \cap \cdots \cap A_{i_k}), \qquad k \geqslant 1,$$

where the summation is over all k-tuples $(i_1, i_2, \ldots, i_k)$ with $1 \leqslant i_1 < i_2 < \cdots < i_k \leqslant n$. Define

$$p_k(n) = S_k(n) \Big/ \binom{n}{k},$$

and designate by $X_n(A)$ the number of events that occur among $\{A_i\}$ $(i = 1, \ldots, n)$. A sequence $C_1, C_2, \ldots, C_n$ of elements of $\mathscr{A}$ is called *exchangeable* when for any choice of indices $1 \leqslant i_1 < i_2 < \cdots < i_k \leqslant n$ the probabilities

$$\alpha_k = P(C_{i_1} \cap C_{i_2} \cap \cdots \cap C_{i_k})$$

depend on k alone. The following result modified by Galambos [6] is due to D. G. Kendall [9]. Let $(\Omega_t, \mathscr{A}_t, P_t)$ (for each $t = 1,$

2, ...) be a probability space carrying a finite (or infinite) number M_t of exchangeable events C_{rt}, $r = 1, 2, \ldots, M_t$, with $t \leqslant M_t$. Let X_t be the number of the first events $C_{1t}, C_{2t}, \ldots, C_{tt}$ which occur, and let α_{kt} be defined as α_k mentioned above, $P(\cdot)$ being replaced by $P_t(\cdot)$. Suppose that $\lim_{t\to\infty} t\alpha_{1t} = \lambda$ and $\lim_{t\to\infty} t^2\alpha_{2t} = \lambda^2$ $(0 < \lambda < \infty)$, and if M_t is infinite, suppose that $\lim_{t\to\infty} M_t/t = \infty$. Then

$$\lim_{t\to\infty} P_t(X_t = x) = \lambda^x e^{-\lambda}/x! \qquad (x = 0, 1, 2, \ldots).$$

Galambos [6] extended the result above to a more general case. Let $\{A_i\}$ $(i = 1, \ldots, n)$ be an arbitrary sequence of events on the probability space $(\Omega, \mathscr{A}, P)$, and assume that the set $\{p_i(n)\}$ $(i = 1, \ldots, n)$ can be enlarged to the set $\{p_j(n)\}$ $(j = 1, 2, \ldots, M_n)$, so that for $n > n_0$ this enlarged set can be associated with a sequence of M_n exchangeable events with $p_j(n) = \alpha_j$. Suppose that $\lim_{n\to\infty} S_1(n) = \lambda$ and $\lim_{n\to\infty} S_2(n) = \lambda^2/2$ $(0 < \lambda < \infty)$, and if M_n is infinite, suppose further that $\lim_{n\to\infty} M_n/n \to \infty$; then $\lim_{n\to\infty} P(X_n(A) = x) = \lambda^x e^{-\lambda}/x!$ $(x = 0, 1, 2, \ldots)$.

References

[1] Bohman, H. (1963). *Scand. Aktuarietidskr.*, **46**, 47–52.

[2] Bortkiewicz, L. von. (1898). *Das Gesetz der kleinen Zahlen*. Teubner, Leipzig, Germany.

[3] Chatterji, S. D. (1963). *Amer. Math. Monthly*, **70**, 958–964.

[4] Erlang, A. K. (1909). *Nyt Tidsskr. Mat. B*, **20**, 33–39 (in Danish). (Probability calculus and telephone conversations.)

[5] Fuchs, A. and Roby, N. (1960). *Publ. Inst. Statist. Univ. Paris*, **9**, 391–394.

[6] Galambos, J. (1973). *Duke Math. J.*, **40**, 581–586.

[7] Haight, F. A. (1967). *Handbook of the Poisson Distribution*. Wiley, New York. (The most substantial handbook on the Poisson and related distributions, with an almost complete list of references up to 1966.)

[8] Johnson, N. L. and Kotz, S. (1969). *Distributions in Statistics: Discrete Distributions*. Wiley, New York, Chap. 4. (This chapter is written compactly and is a useful source for basic information about the Poisson distribution.)

[9] Kendall, D. G. (1967). *Studia Sci. Math. Hung.*, **2**, 319–327.

[10] Korwar, R. M. (1975). *Commun. Statist.*, **4**, 1133–1147.

[11] Matsunawa, T. (1982). *Ann. Inst. Statist. Math.*, **34**, 209–224.

[12] Molennar, W. (1970). *Approximations to the Poisson, Binomial and Hypergeometric Distribution Functions*, Math. Centre Tracts No. 31, Mathematische Centrum, Amsterdam.

[13] Peizer, D. B. and Pratt, J. W. (1968). *J. Amer. Statist. Ass.*, **63**, 1417–1456.

[14] Raikov, D. (1938). *Izv. Akad. Nauk USSR*, **2**, 91–124.

[15] Rao, C. R. and Rubin, H. (1964). *Sankhyā A*, **26**, 295–298.

[16] Rutherford, E., Geiger, H., and Bateman, H. (1910). *Philos. Mag., 6th Series*, **20**, 698–707.

[17] Samuels, S. M. (1965). *Ann. Math. Statist.*, **36**, 1272–1278.

[18] Student (1907). *Biometrika*, **5**, 351–360.

[19] von Mises, R. (1921). *Zeit. angew. Math. Mech.*, **1**, 121–124.

(BINOMIAL DISTRIBUTION
NEGATIVE BINOMIAL DISTRIBUTION
NORMAL DISTRIBUTION
POISSON PROCESS)

T. MATSUNAWA

POISSON INDEX OF DISPERSION

A statistic used in testing whether a distribution is Poisson*, based on n observed values $X_1, \ldots, X_n$ from a random sample. The index of dispersion is

$$D = \sum_{i=1}^{n} \left(X_i - \bar{X}\right)^2 / \bar{X}$$

where $\bar{X} = n^{-1} \sum_{i=1}^{n} X_i$.

If the population distribution is Poisson, the distribution of D is approximately chi-square* with $(n - 1)$ degrees of freedom.

Usually, large values $(D > \chi^2_{n-1,\alpha})$ are regarded as evidence of departure from a Poisson form of distribution, but low values ($D < \chi^2_{n-1,1-\alpha}$) can also be indicative of such departure.

(INDEX OF DISPERSION)

POISSON-INVERSE GAUSSIAN DISTRIBUTION

This is a compound Poisson distribution*, generated by ascribing an inverse Gaussian distribution* with probability density function

$$f_\theta(t) = \sqrt{\frac{\lambda}{2\pi t^3}} \exp\left[-\frac{\lambda(t-\xi)^2}{2\xi^2 t} \right] \qquad (t > 0, \lambda > 0) \quad (1)$$

to the expected value, θ, of a Poisson distribution [2]. If X has such a distribution, then

$$\begin{aligned} \Pr[X = x] &= (x!)^{-1} \int_0^\infty e^{-t} t^x f_\theta(t)\, dt \\ &= (2\lambda/\pi)^{1/2} (x!)^{-1} (\lambda/\mu)^{(x-1/2)/2} \\ &\quad \times K_{x-1/2}\left(\sqrt{(\lambda\mu)}\right) \\ &\qquad (x = 0, 1, 2, \ldots), \end{aligned}$$

where $\xi = 2 + \lambda\xi^{-2}$ and $K_\nu(y)$ is the modified Bessel function* of the third kind [see, e.g., ref. 1 (p. 374 et seq.)]. As for all compound Poisson distributions, the factorial cumulants* of X are equal to the cumulants* (of corresponding order) of θ. The moment generating function* is

$$\exp\left[\lambda\xi^{-1} \left\{ 1 - \sqrt{[1 + 2\lambda^{-1}\xi^2(1 - e^t)]} \right\} \right].$$

In particular,

$$E[X] = \xi; \ \mathrm{var}(X) = \lambda^{-1}\xi(\xi^2 + \lambda).$$

A multivariate Poisson-inverse Gaussian distribution is generated by taking variables $X_1, X_2, \ldots, X_k$ which, for given θ, are conditionally independent with a common Poisson distribution with expected value θ and ascribing distribution (1) to θ.

For this distribution

$$\begin{aligned} \Pr\left[\bigcap_{j=1}^k (X_j = x_j) \right] &= E\left[e^{-k\theta} \prod_{j=1}^k (\theta^{x_j}/x_j!) \right] \\ &= (2\lambda/\pi)^{1/2} \left(\prod_{j=1}^k x_j! \right)^{-1} \\ &\quad \times e^{\lambda/\xi} (\lambda/\mu_k)^{(\Sigma x_j - 1/2)/2} \\ &\quad \times K_{\Sigma x_j - 1/2}\left(\sqrt{\lambda\mu_k} \right) \\ &\qquad (x_j = 0, 1, \ldots), \end{aligned}$$

where $\mu_k = 2k + \lambda\xi^{-2} = \mu + 2(k-1)$.

References

[1] Abramowitz, M. and Stegun, I. A., eds. (1964). Handbook of Mathematical Functions. *Natl. Bur. Stand. (U.S.) Appl. Math. Ser. 55* (Washington, D.C.).

[2] Holla, M. S. (1966). *Metrika*, **11**, 115–121.

POISSON LIMIT

This states that the limit of the probability

$$\begin{aligned} \Pr[X = x] &= b(x; n, p) \\ &= \binom{n}{x} p^x (1-p)^{n-x} \\ &\qquad (0 \leqslant x \leqslant n; 0 < p < 1) \end{aligned}$$

for a binomial distribution* with parameters n, p, as $n \to \infty$ and $p \to 0$ with $np = \theta$ is the probability

$$\Pr[X = x] = \mathscr{P}(x; \theta) = e^{-\theta}\theta^x/x! \qquad (0 \leqslant x; \theta > 0)$$

for a Poisson distribution* with parameter θ. (See Feller [1] for a detailed proof.)

Sheu [5] gives a simple proof of the result of Khinchin [3] and Prohorov [4] that, with $\theta = np$,

$$\begin{aligned} S &= \sum_{x=0}^\infty |b(x; n, p) - \mathscr{P}(x; \theta)| \\ &\leqslant \min(2np^2, 3p) \end{aligned}$$

[taking $b(x; n, p) = 0$ for $x > n$].

Further results of this kind include [7]

$$S \leqslant p\sqrt{2/q}, \qquad \text{where } q = 1 - p,$$

and [2]

$$S \leqslant 6p/5 \qquad \text{for } n \geqslant 4.$$

Simons and Johnson [6] have shown that for θ fixed, with $np = \theta$,

$$\sum_{x=0}^{\infty} h(x)|b(x;n,p) - \mathscr{P}(x;\theta)| = 0$$

if and only if the sum $\sum_{x=0}^{\infty} h(x)\mathscr{P}(x;\theta)$ converges.

References

[1] Feller, W. (1968). *An Introduction to Probability Theory and Its Applications*, 3rd ed. Wiley, New York.

[2] Kerstan, J. (1964). *Z. Wahrscheinl. verw. Geb.*, **2**, 173–179.

[3] Khinchin, A. Ya. (1933). *Asymptotische Gesetze der Wahrscheinlichkeitsrechnung*. Springer-Verlag, Berlin.

[4] Prohorov, Yu. V. (1953). *Russ. Math. Surv.*, **8**, 135–142.

[5] Sheu, S. S. (1964). *Amer. Statist.*, **38**, 206–207.

[6] Simons, G. and Johnson, N. L. (1971). *Ann. Math. Statist.*, **42**, 1735–1736.

[7] Vervaat, W. (1970). *Statist. Neerlandica*, **23**, 79–86.

POISSON-LOGNORMAL DISTRIBUTION

This is a compound Poisson distribution, generated by ascribing a lognormal distribution* with probability density function

$$f_\theta(t) = \left(\sigma t\sqrt{2\pi}\right)^{-1}\exp\left[-\frac{1}{2}\left(\frac{\log t - \xi}{\sigma}\right)^2\right] \qquad (t > 0, \sigma > 0) \quad (1)$$

to the expected value, θ, of a Poisson distribution* [3]. It has also been called a "discrete lognormal" [1, pp. 362–363; 2], but this is a rather misleading name. If X has such a distribution, then

$$\Pr[X = x] = (x!)^{-1}\int_0^\infty e^{-t}t^x f_\theta(t)\,dt.$$

Unfortunately, this cannot be expressed in a simple form. As for all compound Poisson distributions, the factorial cumulants* of X are equal to the cumulants* (of corresponding order) of θ. In particular,

$$E[X] = \exp\left(\xi + \tfrac{1}{2}\sigma^2\right)$$

$$\operatorname{var}(X) = \exp(2\xi + 2\sigma^2) + \exp\left(\xi + \tfrac{1}{2}\sigma^2\right) - \exp(2\xi + \sigma^2).$$

A multivariate Poisson-lognormal distribution can be generated by taking variables $X_1, X_2, \ldots, X_k$, which, for given θ, are conditionally independent with a common Poisson distribution with expected value θ, and ascribing distribution (1) to θ.

Reference 4 contains a useful summary of properties of the Poisson-lognormal distribution, and methods of fitting it to data.

References

[1] Anscombe, F. J. (1950). *Biometrika*, **37**, 358–382.

[2] Grundy, P. M. (1951). *Biometrika*, **38**, 427–434.

[3] Preston, F. W. (1948). *Ecology*, **29**, 254–283.

[4] Reid, D. D. (1980). In *Statistical Distributions in Scientific Work*, Vol. 6, G. P. Patil and B. Baldessari, eds. D. Reidel, Dordrecht, The Netherlands, pp. 303–316.

POISSON–MARKOV PROCESS

The Poisson–Markov process, a multivariate process in continuous time, was introduced by Bartlett [1] as a generalization of a process in discrete time. This process was just one of many used for modeling evolutionary stochastic processes. The name *Poisson–Markov process* was used by Patil [3], who studied it to see how well it could explain density fluctions in, for example, spermatozoa counts. Both these authors give derivations showing the *equilibrium* distribution to be that of independent Poisson processes*.

Boswell and Patil [2] give the solution for finite time as well as the equilibrium distribution for the univariate model.

THE MODEL

Let $N_i(t)$ be the number of particles in region R_i at time t, $i = 1, 2, \ldots, k$. Particles immigrate from outside these regions, move from region to region, and emigrate from the system (or die). The instantaneous rates of change are:

1. Immigration into region R_i from outside the system with a constant rate λ_i
2. Movement from region R_i into region R_j at a rate $n_i\lambda_{ij}$ proportional to the number n_i of particles in region R_i at that instant
3. Emigration from region R_i at a rate $n_i\mu_i$ proportional to the number n_i in region R_i at that instant

There are no other instantaneous changes possible. The process is assumed to be a Markov process*.

THE EQUILIBRIUM SOLUTION TO THE GENERAL MODEL

Let the matrix $\mathbf{\Lambda}$ be

$$\begin{bmatrix} \mu_1 + S_1 & -\lambda_{12} & \cdots & -\lambda_{1k} \\ -\lambda_{21} & \mu_2 + S_2 & \cdots & -\lambda_{2k} \\ \vdots & \vdots & \ddots & \vdots \\ -\lambda_{k1} & -\lambda_{k2} & \cdots & \mu_k + S_k \end{bmatrix},$$

where $S_i = \sum_{j \neq i} \lambda_{ij}$; $i = 1, \ldots, k$.

Let $\boldsymbol{\lambda} = (\lambda_1, \lambda_2, \ldots, \lambda_k)$, and let $\mathbf{m} = \boldsymbol{\lambda}\mathbf{\Lambda}^{-1}$. Then as time goes to infinity the joint distribution of $(N_1(t), N_2(t), \ldots, N_k(t))$ becomes that of independent Poisson random variables with means $(m_1, m_2, \ldots, m_k)$ (see Bartlett [1] or Patil [3]).

THE ONE-DIMENSIONAL MODEL

Let $N(t)$ be the number of particles in a region (population size) at time t. In this case $N(t)$ is a birth-and-death process* with constant birth rate λ and death (emigration) rate $n\mu$ proportional to the number of particles n in the region at that instant. Assuming that $N(t) = 0$, then the number in the region at time t has the Poisson distribution with mean $m(t) = (1 - e^{-\mu t})\lambda/\mu$ (see Boswell and Patil [2]). If $N(0) = n_0$, since individuals do not give birth to new individuals, the number left at time t follows a pure death process and has a binomial distribution with parameters n_0 and $p = e^{-\mu t}$ (see Patil and Boswell [5]). Therefore, if $N(0) = n_0$, then the number of particles in the region at time t has the convolution of binomial and Poisson distributions given above.

THE EQUILIBRIUM DISTRIBUTION OF THE ONE-DIMENSIONAL PROCESS

The equilibrium distribution is found by taking the limit as time goes to infinity. The initial population size is no longer important, as all individuals eventually emigrate from the region. The resulting distribution of $N(t)$ as t goes to infinity is the Poisson distribution with mean λ/μ.

CONCLUDING REMARKS

The Poisson–Markov process is a special case of a multivariate birth-and-death process with linear birth and death rates. Properties of the process can be obtained from birth-and-death process and from the Poisson distribution. Many of these and interrelations with other processes can be found in Patil et al. [6].

The univariate Poisson–Markov process has been generalized to a nonhomogeneous process with:

1. Immigration rate $\lambda(t)$ depending on time
2. Emigration rate $n\mu(t)$ depending on time and proportional to the population size n at time t

Let $\Lambda(t) = \int_0^t \lambda(s)\,ds$ and let $m(t) = e^{-\Lambda(t)} \int_0^t \mu(s) e^{\Lambda(s)}\,ds$. Assuming that the population size at time $t = 0$ is $N(0) = 0$, the population size at time t has a Poisson distribution with mean $m(t)$ (see Patil and Boswell [4]).

References

[1] Bartlett, M. S. (1949). *J. R. Statist. Soc. B*, **11**, 211–229.

[2] Boswell, M. T. and Patil, G. P. (1972). In *Stochastic Point Processes*. Wiley, New York, pp. 285–298.
[3] Patil, V. T. (1957). *Biometrika*, **44**, 43–56.
[4] Patil, G. P. and Boswell, M. T. (1972). *Sankhyā A*, **34**, 293–296.
[5] Patil, G. P. and Boswell, M. T. (1975). In *Statistical Distributions in Scientific Work*, Vol. 2. G. P. Patil and B. Baldessari, eds. D. Reidel, Dordrecht, The Netherlands, pp. 11–24.
[6] Patil, G. P., Boswell, M. T., Joshi, S. W., and Ratnaparkhi, M. V. (1984). *A Modern Dictionary and Classified Bibliography of Statistical Distributions*, Vol. 3: *Discrete Models*. International Cooperative Publishing House, Fairland, Md.

Bibliography

Adke, S. R. (1969). *J. Appl. Prob.*, **6**, 689–699.
Brown, M. (1970). *Ann. Math. Statist.*, **41**, 1935–1941.
Karlin, S. and McGregor, J. (1958). *J. Math. Mech.*, **7**, 643–661.
Milch, P. R. (1968). *Ann. Math. Statist.*, **39**, 727–754.
Morgan, R. W. and Welsh, D. J. A. (1965). *J. R. Statist. Soc. B*, **27**, 497–504.
Renshaw, E. (1972). *Biometrika*, **59**, 49–60.

(BIRTH-AND-DEATH PROCESSES
MARKOV PROCESSES)

M. T. BOSWELL

POISSON MATRIX

An elementary stochastic matrix* with at most one positive off-diagonal element. These matrices are used in the theory of embeddable matrices, that is, matrices that can occur as transition matrices in nonhomogeneous Markov chains. See, for example, refs. 1 and 2 for details.

References

[1] Frydman, H. (1980). *Math. Proc. Camb. Philos. Soc.*, **87**, 285–294.
[2] Frydman, H. (1983). *J. Multivariate Anal.*, **13**, 464–472.

(EMBEDDED PROCESSES
MARKOV PROCESSES
STOCHASTIC PROCESSES)

POISSON PROCESSES

Among point processes the most important theoretically and in applications are the Poisson processes. Let $N = (N_t)$ be a point process on $\mathbb{R}_+$ with arrival times $T_1 < T_2 < \cdots$ and interarrival times $U_i = T_i - T_{i-1}$, so that N_t is the number of arrivals in $[0, t]$ and more generally $N_{t+s} - N_t$ is the number of arrivals in the interval $(s, t]$. Then N is a homogeneous Poisson process with rate λ if N has *independent increments* in the sense that for $0 \leqslant t_0 < t_1 < \cdots < t_k$ the random variables $N_{t_1}, N_{t_2} - N_{t_1}, \ldots, N_{t_k} - N_{t_{k-1}}$ are independent and if for each t and s, $N_{t+s} - N_t$ has a Poisson distribution with mean λt. Since $E[N_t] = \lambda t$ for each t and $\lim N_t/t = \lambda$ almost surely, λ is known as the rate of N. In particular, N also has *stationary increments*; the distribution of the number of arrivals in an interval $(t, t+s]$ depends only on its length s. Typical applications are models of arrivals at a queueing system and decay of radioactive material; the Poisson limit theorem mentioned below explains why Poisson process models fit so many situations.

The class of homogeneous Poisson processes on $\mathbb{R}_+$ admits three significant characterizations. A point process N is a (homogeneous) Poisson process with rate λ if and only if any of the following conditions is fulfilled:

1. The interarrival times are independent and identically exponentially distributed with $E[U_i] = \lambda^{-1}$ for each i (i.e., N is a renewal process).
2. For each t

$$\lim_{h \to 0} \Pr[N_{t+h} - N_t = 1 \mid N_u : u \leqslant t]/h = \lambda, \quad (1)$$

while

$$\lim_{h \to 0} \Pr[N_{t+h} - N_t \geqslant 2 \mid N_u : u \leqslant t]/h = 0. \quad (2)$$

3. The process $M_t = N_t - \lambda t$ is a martingale*.

A fundamental property of homogeneous Poisson processes is conditional uniformity: for each t, given that $N_t = k$ and regardless of the rate λ the conditional distribution of the arrival times $T_1, \ldots, T_k$ is that of order statistics $X_{1,k}, \ldots, X_{k,k}$ engendered by independent random variables $X_1, \ldots, X_k$, each uniformly distributed on $[0, t]$. From this property many computational relationships can be deduced. Conversely, application of Poisson processes has been made to strong approximation of uniform empirical processes* (see Brillinger [1]).

For a *nonhomogeneous Poisson process* on $\mathbb{R}_+$ the independent increments property is retained, but the arrival rate is a function of time, so that $N_t - N_s$ has a Poisson distribution with mean $\int_s^t \lambda(u)\,du$, where λ is known as the arrival-rate function, since in this case (1) becomes

$$\lim_{h \to 0} \Pr\left[N_{t+h} - N_t = 1 \mid N_u : u \leqslant t\right]/h = \lambda(t), \tag{1'}$$

with (2) remaining as is; thus $\lambda(t)$ is the instantaneous arrival rate at time t. The nonhomogeneous Poisson model is appropriate for situations in which the independent increments property holds but that of stationary increments fails; these include, for example, output processes of certain queueing systems and arrival processes with periodicity structure.

The first definition above generalizes to other spaces, whereas the alternative characterizations do not. Let $\{N(A) : A \subset E\}$ be a point process* on a general space E (e.g., $\mathbb{R}^d$ for some $d \geqslant 1$); that is, N is a distribution of indistinguishable points at random locations X_i in E such that only finitely many fall in any bounded set, with $N(A) = \sum 1\,(X_i \in A)$, the number of points in the set A. Then N is a *Poisson process* with mean measure ν, where ν is a measure on E (finite on bounded sets), provided that:

1. N has independent increments: whenever $A_1, \ldots, A_k$ are disjoint sets, the numbers $N(A_1), \ldots, N(A_k)$ of points in them are independent random variables.
2. For each set A, $N(A)$ has a Poisson distribution with mean $\nu(A)$.

Applications include distribution of stars, locations of trees, times and magnitudes of flood peaks, and geometrical probability.

For f a function on E, let $N(f) = \sum f(X_i)$ be the integral of f with respect to N viewed as a purely atomic random measure on E. Important computational properties are as follows.

1. For each f, $E[N(f)] = \int f\,d\nu$.
2. For each f and g, $\operatorname{cov}(N(f), N(g)) = \int fg\,d\nu$.
3. The Laplace functional of N is given by

$$\begin{aligned} L_N(f) &= E\left[\exp(-N(f))\right] \\ &= \exp\left[-\int (1 - e^{-f})\,d\nu\right]. \end{aligned}$$

Conditional uniformity generalizes, and retains its power as a computational and theoretical tool: if $0 < \nu(A) < \infty$, then conditional on $N(A) = k$, the restriction of N to A has the same distribution as the empirical process engendered by k independent, identically distributed random elements of A, each with distribution given by $F(B) = \nu(B \cap A)/\nu(A)$.

The class of Poisson processes is invariant under many transformations, including mapping, superposition (the sum of independent Poisson processes is Poisson), thinning (independent random deletion of points), random translation of points, and marking; such properties are not only important theoretically but are also useful in modeling real-world phenomena such as traffic flow on highways and networks of queues*.

Theoretical significance and ubiquity of Poisson processes in physical situations stem mainly from the Poisson limit theorem for point processes, due originally to Franken [6]; it generalizes the classical Poisson limit theorem for binomial distributions. Let N_{nk} be point processes such that for each n, $N_{n1}, \ldots, N_{nk_n}$ are independent and uniformly sparse in the sense that

$$\lim_{n\to\infty} \max_{k \leqslant k_n} \Pr\left[N_{nk}(B) > 1 \right] = 0$$

for each bounded set B. Then the row sums $N_n = N_{n1} + \cdots + N_{nk_n}$ converge in distribution (see Kallenberg [7, Chap. 7]) to a Poisson process with mean measure ν if and only if

$$\lim_{n\to\infty} \sum_{k=1}^{k_n} \Pr\left[N_{nk}(B) = 1 \right] = \nu(B)$$

for each bounded B. Generalizations, variations, and inverse theorems are discussed in Çinlar [3]. The variety of phenomena, for example, arrival of telephone calls at a central exchange or optical transmission of signals, satisfying these rather mild conditions explains the success with which Poisson models have been applied.

An important theoretical role of Poisson processes concerns structure of other stochastic processes; for example, every random measure with independent increments, except for atoms at nonrandom locations, has a Poisson cluster representation as the sum of points with random masses U_i at random locations X_i, where $N = \{(X_i, U_i)\}$ is a Poisson process on $E \times \mathbb{R}_+$. More generally, every infinitely divisible random measure has a Poisson cluster representation in terms of a Poisson process on a space of measures on E (see Matthes et al. [8, Chap. 4]). Related representations of Markov processes* as stochastic integrals* with respect to Poisson and Wiener processes* have been developed; see Çinlar and Jacod [4].

Concerning statistical inference*, if N is a homogeneous Poisson process on $\mathbb{R}_+$ with unknown rate λ, observed over $[0, t]$, then the log-likelihood function is $L(\lambda) = t - \lambda t + N_t(\log \lambda)$, which is evidently maximized for $\hat{\lambda} = N_t/t$; these estimators are strongly consistent and asymptotically normal. Note that N_t is a sufficient statistic* for λ by conditional uniformity. Observation over intervals $[0, t]$ is asynchronous observation; a Poisson process can also be observed synchronously (i.e., for the random length T_k required in order that k arrivals be recorded); the data are the interarrival times $U_1, \ldots, U_k$, which in the homogeneous case are independent and identically distributed and can be analyzed by classical methods. A central hypothesis-testing* problem is to determine whether a point process known, for example, to be a renewal process or a nonhomogeneous Poisson process is a homogeneous Poisson process (see Cox and Lewis [5, Chap. 9]).

More generally, inference for Poisson processes is based on likelihood ratios* and equivalence/singularity properties of probability laws of Poisson processes (see Brown [2]). For every bounded set B the probabilities P_0, P_1 with respect to which a point process N is Poisson with equivalent mean measures ν_0, ν_1, respectively, are equivalent on the σ-algebra corresponding to observation of N over B, with likelihood ratio

$$dP_1/dP_0 = \exp\left[\int_B (1 - d\nu_1/d\nu_0)\, d\nu_0 + \int_B \log(d\nu_1/d\nu_0)\, dN \right]. \quad (3)$$

To illustrate, for a nonhomogeneous Poisson process N with arrival times T_i, observed over $[0, t]$, and arrival rate functions λ_0, λ_1, (3) becomes

$$dP_1/dP_0 = \exp\left[\int_0^t (\lambda_0(u) - \lambda_1(u))\, du + \sum_{T_i \leqslant t} \log(\lambda_1(T_i)/\lambda_0(T_i)) \right].$$

Under standard dominated family hypotheses maximum likelihood estimation* of the mean measure can be effected in parametric and nonparametric settings, as can construction of likelihood ratio tests*.

Nearly all important classes of point processes contain the Poisson processes; these include Cox processes (doubly stochastic Poisson processes), infinitely divisible point processes (Poisson cluster processes), and in the case of Poisson processes on Euclidean spaces whose mean measure is a multiple of Lebesgue measure, stationary point processes*.

References

[1] Brillinger, D. R. (1969). *Bull. Amer. Math. Soc.*, **75**, 545–547.

[2] Brown, M. (1972). In *Stochastic Point Processes*, P. A. W. Lewis, ed. Wiley, New York.

[3] Çinlar, E. (1972). In *Stochastic Point Processes*, P. A. W. Lewis, ed. Wiley, New York.

[4] Çinlar, E. and Jacod, J. (1981). In *Seminar on Stochastic Processes, 1981*, E. Çinlar, K. L. Chung, and R. K. Getoor, eds. Birkhäuser Boston, Cambridge, Mass.

[5] Cox, D. R. and Lewis, P. A. W. (1966). *The Statistical Analysis of Series of Events*, Chapman & Hall, London.

[6] Franken, P. (1963). *Math. Nachr.*, **26**, 101–114.

[7] Kallenberg, O. (1983). *Random Measures, 3rd ed.*, Akademie-Verlag, Berlin.

[8] Matthes, K., Kerstan, J., and Mecke, J. (1978). *Infinitely Divisible Point Processes*. Wiley, New York.

Bibliography

See the following works, as well as the references just given, for more information on the topic of Poisson processes.

Brémaud, P. (1980). *Point Processes and Queues: Martingale Dynamics*. Springer-Verlag, New York. (Martingale approach to Poisson processes; role of Poisson processes in queueing networks; advanced but very readable.)

Cox, D. R. and Isham, V. (1980). *Point Processes*. Chapman & Hall, London. (Accessible, elementary treatment.)

Cox, D. R. and Lewis, P. A. W. (1966). *The Statistical Analysis of Series of Events*. Chapman & Hall, London. (Key work on pre-martingale inference for Poisson processes.)

Doob, J. L. (1953). *Stochastic Processes*. Wiley, New York. (Early treatment of Poisson processes on spaces other than the line.)

Kallenberg, O. (1983). *Random Measures, 3rd ed.* Akademie-Verlag, Berlin. (Poisson processes as random counting measures; Poisson limit theorems; advanced level.)

Karr, A. F. (1986). *Point Processes and Their Statistical Inference*. Marcel Dekker, New York. (Theory; statistical inference; state estimation; measure theoretic.)

Khintchine, A. Y. (1960). *Mathematical Methods in the Theory of Queueing*. Charles Griffin, London. (First rigorous treatment of many theoretical questions.)

Kingman, J. F. C. (1963). *Ann. Math. Statist.*, **34**, 1217–1232. (Poisson sampling of other stochastic processes.)

Krickeberg, K. (1982). *Lect. Notes Math.*, **929**, 205–313. [Modern treatment of inference (in French).]

Kutoyants, Yu. A. (1979). *Prob. Control Inf. Theory*, **8**, 137–149. (Detailed analysis of parametric estimation problems.)

Matthes, K., Kerstan, J., and Mecke, J. (1978). *Infinitely Divisible Point Processes*. Wiley, New York. (Encyclopedic coverage of Poisson processes and their generalizations; difficult for all but experts.)

Snyder, D. L. (1975). *Random Point Processes*. Wiley-Interscience, New York. (Inference and engineering applications, especially to communication theory.)

(POINT PROCESSES
POINT PROCESSES, STATIONARY
PROCESSES, DISCRETE
STOCHASTIC PROCESSES)

ALAN F. KARR

POISSON REGRESSION

Poisson regression encompasses statistical methods for the analysis of the relationship between an observed count with a Poisson distribution and a set of explanatory variables. Examples of application include colony counts for bacteria or viruses for a set of varying dilutions and/or experimental conditions [21, 25, 49, 51]; numbers of failures (or accidents) for equipment during varying conditions of operation [29, 33, 38, 52]; vital statistics pertaining to infant morbidity or mortality [42, 46] or to cancer incidence [2, 6, 20, 24, 26, 27, 34, 54] for a cross-classified sample according to demographic and other characteristics. For such situations, a Poisson regression model has the general form

$$\mu(\mathbf{x}) = \{N(\mathbf{x})\}\{g(\boldsymbol{\beta}|\mathbf{x})\}; \qquad (1)$$

here $\mu(\mathbf{x})$ is the expected value of the number of events $n(\mathbf{x})$ from the subpopulation corresponding to the known vector $\mathbf{x} = (x_1, x_2, \ldots, x_t)'$ of t explanatory variables; $N(\mathbf{x})$ is the known total (or relative) exposure to risk of this subpopulation in the time, subject, and/or space units of the environment in which the events occur (e.g., the volume or dilution per unit volume in which bacteria are counted, subject-days of operation in a work or usage setting, the number of eligible subjects for vital events); and $g(\cdot)$ is the known functional form which specifies the relationship of the rates $\lambda(\mathbf{x}) = \{\mu(\mathbf{x})/N(\mathbf{x})\}$ to $\mathbf{x}$ and the unknown $(u_t \times 1)$ vector of nonredundant parameters $\boldsymbol{\beta} = (\beta_1, \beta_2, \ldots, \beta_{u_t})'$. Frome et al. [23] have reviewed methodology for estimating the parameters $\boldsymbol{\beta}$ for models like (1) and have provided several illustrations from the biological and physical sciences. When the counts $n(\mathbf{x})$ have independent Poisson distributions for the respective $\mathbf{x}$, they discuss the extent to which maximum likelihood*, minimum chi-square*, and weighted least squares* procedures yield equivalent results. More generally, Charnes et al. [8] give conditions under which iterative weighted least-squares computations yield maximum likelihood estimates for data from the regular exponential family.

An historically relevant application of Poisson regression was presented by Cochran [10], who considered the analysis of Poisson distributed data from designed experiments. Attention was given to the $u_t = t$ parameter linear model $g(\boldsymbol{\beta} \mid \mathbf{x}) = \mathbf{x}'\boldsymbol{\beta}$ and the "square-root" model $g(\boldsymbol{\beta} \mid \mathbf{x}) = (\mathbf{x}'\boldsymbol{\beta})^2$, with the latter being indicated as involving more straightforward computations. Also, model appropriateness was emphasized as an important issue; and in this spirit, the product (or log-linear) model was suggested as being of potential interest.

POISSON LOG-LINEAR REGRESSION MODEL

At the present time, the log-linear model (*see* CONTINGENCY TABLES) is the best known type of Poisson regression. Its specification with $u_t = t$ parameters is

$$\mu(\mathbf{x}) = \{N(\mathbf{x})\}\{\exp(\mathbf{x}'\boldsymbol{\beta})\} \tag{2}$$

for counts $n(\mathbf{x})$ with independent Poisson distributions. More specifically, let $i = 1, 2, \ldots, s$ index a set of samples for which $\mathbf{x}_i = (x_{i1}, x_{i2}, \ldots, x_{it})'$ denotes the vector of t linearly independent, explanatory variables where $t \leqslant s$; let $n_i = n(\mathbf{x}_i)$ denote the number of events for the ith sample; and let $N_i = N(\mathbf{x}_i)$ denote the corresponding exposure. Under the assumption that the $\{n_i\}$ have independent Poisson distributions with expected value parameters $\{\mu_i = \mu(\mathbf{x}_i)\}$, the likelihood function for the data is

$$\phi(\mathbf{n} \mid \boldsymbol{\mu}) = \prod_{i=1}^{s} \mu_i^{n_i}\{\exp(-\mu_i)\}/n_i! \tag{3}$$

where $\mathbf{n} = (n_1, n_2, \ldots, n_s)'$ and $\boldsymbol{\mu} = (\mu_1, \mu_2, \ldots, \mu_s)'$. The maximum likelihood (ML) estimates $\hat{\boldsymbol{\beta}}$ for the parameters of the log-linear model (2) can be expressed as the solution of the nonlinear equations obtained from substituting the model counterparts $\{N_i[\exp(\mathbf{x}_i'\boldsymbol{\beta})]\}$ for the $\{\mu_i\}$ into the likelihood (3), differentiating $\log_e \phi$ with respect to $\boldsymbol{\beta}$, and equating the result to 0. The equations have the form

$$\mathbf{X}'\mathbf{n} = \mathbf{X}'\hat{\boldsymbol{\mu}} = \mathbf{X}'\{\mathbf{D_N}[\mathbf{exp}(\mathbf{X}\hat{\boldsymbol{\beta}})]\} \tag{4}$$

where $\mathbf{X} = [\mathbf{x}_1, \mathbf{x}_2, \ldots, \mathbf{x}_s]'$ is the $(s \times t)$ explanatory variable matrix, $\mathbf{D_N}$ is the diagonal matrix with the exposures $\mathbf{N} = (N_1, N_2, \ldots, N_s)'$ on the main diagonal, and **exp** is the operation which exponentiates the elements of a vector. The equations (4) usually do not have an explicit solution, so iterative procedures are necessary for the computation of $\hat{\boldsymbol{\beta}}$. One useful approach for obtaining $\hat{\boldsymbol{\beta}}$ is iterative weighted least squares* as described in Nelder and Wedderburn [44] and Frome et al. [23]. Its use involves adjusting an lth step estimate $\hat{\boldsymbol{\beta}}_{*,l}$ to an $(l+1)$th step $\hat{\boldsymbol{\beta}}_{*,(l+1)}$ via

$$\hat{\boldsymbol{\beta}}_{*,(l+1)} = \hat{\boldsymbol{\beta}}_{*,l} + [\mathbf{V}(\hat{\boldsymbol{\beta}}_{*,l})]\mathbf{X}'(\mathbf{n} - \hat{\boldsymbol{\mu}}_{*,l}), \tag{5}$$

where $\hat{\boldsymbol{\mu}}_{*,l} = \mathbf{D_N}[\mathbf{exp}(\mathbf{X}\hat{\boldsymbol{\beta}}_{*,l})]$ is the lth step predicted value vector and $\mathbf{V}(\hat{\boldsymbol{\beta}}_{*,l}) = \{\mathbf{X}'\mathbf{D}_{\hat{\boldsymbol{\mu}}_{*,l}}\mathbf{X}\}^{-1}$ is the lth step estimate for the asymptotic covariance matrix for $\hat{\boldsymbol{\beta}}$. Such adjustments are initiated with a preliminary estimate $\hat{\boldsymbol{\beta}}_0$; and they are terminated after a convergence criterion is reached (e.g., maximum distance between two successive sets of values $\leqslant 0.0001$) or a specified maximum number of iterations (e.g., $l \leqslant 10$). A useful preliminary estimate is

$$\hat{\boldsymbol{\beta}}_0 = [\mathbf{X}'\mathbf{D}_{\hat{\boldsymbol{\mu}}_{*,0}}\mathbf{X}]^{-1}\mathbf{X}'\mathbf{D}_{\hat{\boldsymbol{\mu}}_{*,0}}[\mathbf{log}_e(\mathbf{D}_{\mathbf{N}}^{-1}\hat{\boldsymbol{\mu}}_{*,0})] \tag{6}$$

with $\hat{\boldsymbol{\mu}}_{*,0} = \mathbf{n}$ if all $n_i > 0$ and $\boldsymbol{\mu}_{*,0} = (\mathbf{n} + \mathbf{1}_s)$ if otherwise, where $\mathbf{1}_s$ is the $(s \times 1)$ vector of 1's; also, $\mathbf{log}_e(\cdot)$ is the operation which forms natural logarithms for the elements of a vector. The iterative computation of $\hat{\boldsymbol{\beta}}$ via (5) usually converges quickly when the following conditions hold:

1. The model $\mathbf{X}$ is nonredundant in the sense that the submatrix $\tilde{\mathbf{X}}$ corresponding to samples with $n_i > 0$ has full rank t.
2. The model provides a good fit to the data in the sense that the residuals $(\mathbf{n} - \hat{\boldsymbol{\mu}})$ are small and uncorrelated with other potential explanatory variables.
3. The counts $\mathbf{n}$ are sufficiently large for the linear functions $\mathbf{X}'\mathbf{n}$ approximately to have a multivariate normal distribution by virtue of central limit* theory.

Moreover, conditions 1 to 3 can be viewed as necessary for the usage of $\hat{\boldsymbol{\beta}}$ and the model (2) to be reasonable. In this regard, conditions (2) and (3) provide the basis for $\hat{\boldsymbol{\beta}}$ to have approximately a multivariate normal distribution for which the covariance matrix can be considered known through its consistent estimator $\mathbf{V}(\hat{\boldsymbol{\beta}}) = (\mathbf{X}'\mathbf{D}_{\hat{\boldsymbol{\mu}}}\mathbf{X})^{-1}$. When all counts are large (e.g., $n_i \geqslant 10$), related considerations imply the asymptotic equivalence of $\hat{\boldsymbol{\beta}}_0$ and $\hat{\boldsymbol{\beta}}$. In such situations, $\hat{\boldsymbol{\beta}}_0$ is the minimum modified chi-square estimate in the sense of Neyman [45] and Grizzle et al. [28]; also, a convenient estimate for its covariance matrix is $\mathbf{V}(\hat{\boldsymbol{\beta}}_0) = (\mathbf{X}'\mathbf{D}_{\mathbf{n}}\mathbf{X})^{-1}$.

The goodness of fit* of a model $\mathbf{X}$ can be assessed by investigating whether the residuals $\{(n_i - \hat{\mu}_i)\}$ are uncorrelated with other explanatory variables. More specifically, let $\mathbf{X}_{\mathbf{W}} = [\mathbf{X}, \mathbf{W}]$ denote the expansion of $\mathbf{X}$ to include w additional explanatory variables $\mathbf{W}$ such that $\mathbf{X}_{\mathbf{W}}$ satisfies conditions 1 to 3 and rank$(\mathbf{X}_{\mathbf{W}}) = (w + t)$. The expansion $\mathbf{W}$ can be effectively tested by using the *score statistic*

$$Q_S = (\mathbf{n} - \hat{\boldsymbol{\mu}})'\mathbf{W}\{\mathbf{W}'\mathbf{V}_{(\mathbf{n}-\hat{\boldsymbol{\mu}})}\mathbf{W}\}^{-1}\mathbf{W}'(\mathbf{n} - \hat{\boldsymbol{\mu}}), \tag{7}$$

where $\mathbf{V}_{(\mathbf{n}-\hat{\boldsymbol{\mu}})} = \mathbf{D}_{\hat{\boldsymbol{\mu}}} - \mathbf{D}_{\hat{\boldsymbol{\mu}}}\mathbf{X}(\mathbf{X}'\mathbf{D}_{\hat{\boldsymbol{\mu}}}\mathbf{X})^{-1}\mathbf{X}'\mathbf{D}_{\hat{\boldsymbol{\mu}}}$ is the estimated covariance matrix for the residuals $(\mathbf{n} - \hat{\boldsymbol{\mu}})$. Another appropriate criterion is the *likelihood ratio statistic*

$$Q_L = \sum_{i=1}^{s} 2n_i[\log_e(\hat{\mu}_{i,\mathbf{W}}/\hat{\mu}_i)], \tag{8}$$

where the $\{\hat{\mu}_{i,\mathbf{W}}\} = N_i[\mathbf{exp}(\hat{\mathbf{x}}'_{\mathbf{W},i}\hat{\boldsymbol{\beta}}_{\mathbf{W}})]$ are the expanded model predicted values from the maximum likelihood estimates $\hat{\boldsymbol{\beta}}_{\mathbf{W}}$ for the model $\mathbf{X}_{\mathbf{W}}$. If the model $\mathbf{X}$ provides an adequate description of the variation among the $\{n_i\}$, then each of these test statistics has an approximate chi-square distribution with degrees of freedom (d.f.) $= w$. Thus significantly large values of Q_S or Q_L contradict the model $\mathbf{X}$. However, smaller values can only be viewed as supporting it rather than substantiating it because the scope of potential $\mathbf{W}$ for which large sample chi-square approximations are applicable is limited by the available data through conditions 1 to 3. Since the criteria Q_S and Q_L are asymptotically equivalent, the choice between them is mostly a matter of personal preference; a computational advantage of Q_S is that only the estimates $\hat{\boldsymbol{\mu}}$ from the model $\mathbf{X}$ are involved, whereas Q_L requires results from the fitting of the expanded model $\mathbf{X}_{\mathbf{W}}$. Additional information concerning Q_S for regression models is given in Chen [9] and SCORE STATISTICS; and concerning Q_L in CHI-SQUARE TESTS and LIKELIHOOD RATIO TESTS. Other aspects of analysis which are potentially pertinent to goodness of fit* issues include regression diagnostic* procedures as discussed in Frome [20] and explanatory variable selection as discussed in Lawless and Singhal [41].

Example 1: Application of Variance Test to Salmonella Data. Some aspects of the application of Poisson regression with log-linear models are illustrated via three exam-

Table 1 ***Salmonella*** **Counts at Two Laboratories**

Laboratory	Salmonella Counts
A	63, 64, 65, 68, 69, 70, 72, 73, 75, 80, 82, 83, 83, 84, 84, 85, 90, 91
B	168, 171, 174, 175, 185, 189, 190, 191, 195, 197, 198, 198, 203, 205, 205, 207, 210, 214, 216, 218

ples. The data in Table 1 are replicated counts pertaining to *Salmonella* at two laboratories which participated in a study reported by Margolin et al. [43]. For each laboratory $h = \text{A}, \text{B}$, the compatibility of the data with a common Poisson distribution with corresponding expected value $\exp(\beta_h)$ can be investigated by fitting the log-linear model (2) with

$$\mathbf{N}_h = \mathbf{1}_{s_h} \quad \text{and} \quad \mathbf{X}_h = \mathbf{1}_{s_h}; \qquad (9)$$

here $s_h = 18, 20$ denotes the number of replicates for laboratories A, B, respectively. In this case, the ML equations (4) have the explicit solution

$$\hat{\beta}_h = \log_e \left\{ \sum_{i=1}^{s_h} (n_{hi}/s_h) \right\}, \qquad (10)$$

where n_{hi} denotes the count for the ith replicate at the hth laboratory; the corresponding estimated variance is then $V(\hat{\beta}_h) = \{s_h[\exp(\hat{\beta}_h)]\}^{-1}$. Thus the resulting $\hat{\beta}_h$ and their estimated standard errors for the data in Table 1 are

$$\begin{aligned} \hat{\beta}_\text{A} &= 4.340, \text{ s.e. } (\hat{\beta}_\text{A}) = 0.027 \\ \hat{\beta}_\text{B} &= 5.275, \text{ s.e. } (\hat{\beta}_\text{B}) = 0.016 \end{aligned} \qquad (11)$$

Since all the n_{hi} are large, the goodness of fit of the model (9) can be assessed with respect to any maximal set of explanatory variables $\mathbf{W}_h$ such that $\text{rank}(\mathbf{X}_{\mathbf{W}_h}) = s_h$; for example, $\mathbf{W}_h = [\mathbf{I}_{w_h}, -\mathbf{1}_{w_h}]'$ where $w_h = (s_h - 1)$. In general, Q_S in (7) simplifies for such $\mathbf{W}$ to the well-known Pearson criterion

$$Q_S = Q_P = (\mathbf{n} - \hat{\boldsymbol{\mu}})'\mathbf{D}_{\hat{\boldsymbol{\mu}}}^{-1}(\mathbf{n} - \hat{\boldsymbol{\mu}}); \qquad (12)$$

and for the model (9), it becomes the *Poisson variance test*

$$Q_{V,h} = \sum_{i=1}^{s_h} (n_{hi} - \bar{n}_{h*})^2 / \bar{n}_{h*} \qquad (13)$$

where $\bar{n}_{h*} = \exp(\hat{\beta}_h)$. Usage of the *variance test* as an effective goodness-of-fit method for the Poisson distribution was recommended by Cochran [11], who noted that it had a long history dating as far back as Fisher [15]. Since $Q_{V,\text{A}} = 17.98$ and $Q_{V,\text{B}} = 22.03$ are nonsignificant ($\alpha = 0.10$) relative to chi-square approximations with $(\text{d.f.})_\text{A} = 17$ and $(\text{d.f.})_\text{B} = 19$, the data for each laboratory in Table 1 are interpreted as being compatible with a common Poisson distribution. However, aside from this illustration, an important point emphasized by Margolin et al. [43] was that the variance test contradicted the Poisson distribution for three other laboratory studies of a similar nature.

An exact method for judging the significance of the variance test in small samples was presented in Fisher [16]; and Frome [19] has provided an algorithm for implementing it. This procedure and some extensions were discussed by Rao and Chakravarti [48]; and from this work it was suggested that chi-square approximations were reasonable when $\bar{n} = \sum_{i=1}^{s} n_i/s \geqslant 3$. Additional discussion of methods for testing goodness of fit for the Poisson distribution are reviewed in Gart [26].

POISSON TREND TEST

A somewhat more general situation for which explicit expressions can be given for estimators and test statistics involves independent counts n_i having Poisson distributions with expected values $\mu_i = N_i\{\exp(\beta)\}$; here the N_i could correspond to a measure of exposure, or to a background explanatory variable like dose, or both. Since $\mathbf{X} = \mathbf{1}_s$, it follows from (4) that

$$\hat{\beta} = \log_e\left\{ \sum_{i=1}^{s} n_i \Big/ \sum_{i=1}^{s} N_i \right\}; \qquad (14)$$

the corresponding estimated variance is $V(\hat{\beta}) = \{[\sum_{i=1}^{s} N_i][\exp(\hat{\beta})]\}^{-1}$. The extension of the variance test to this setting was noted by Gart [25] to have the form

$$Q_V = \sum_{i=1}^{s} N_i(\hat{\lambda}_i - \hat{\lambda})^2/\hat{\lambda} \qquad (15)$$

where $\hat{\lambda}_i = (n_i/N_i)$ and $\hat{\lambda} = \exp(\hat{\beta})$. However, often the *trend* test* for the association of the residuals* $\{N_i(\hat{\lambda}_i - \hat{\lambda})\}$ with some other explanatory variable $\{w_i\}$ is of more interest. In this case, the score statistic (7) has the form

$$Q_T = \frac{\left\{ \sum_{i=1}^{s} N_i w_i (\hat{\lambda}_i - \hat{\lambda}) \right\}^2}{\hat{\lambda}\left\{ \sum_{i=1}^{s} N_i (w_i - \bar{w})^2 \right\}} \qquad (16)$$

where $\bar{w} = \{\sum_{i=1}^{s} w_i N_i / \sum_{i=1}^{s} N_i\}$; it has an approximate chi-square distribution with d.f. = 1 when the model $\mathbf{X} = \mathbf{1}_s$ applies and the counts $\{n_i\}$ are sufficiently large for $\{\sum_{i=1}^{s} n_i, \sum_{i=1}^{s} n_i w_i\}$ approximately to have a bivariate normal distribution. The trend test based on (16) is essentially the same as that in Armitage [3] and simplifies when all $N_i = 1$ to that in Cochran [11]. Gart [25] discusses its usage for a test concerning a quadratic parameter and a test of zero intercept. Tarone [51] has shown that it asymptotically has local optimality properties with respect to a general class of monotone alternatives. Some other methods for assessing goodness of fit and other hypotheses with respect to the model $\mu_i = N_i\{\exp(\beta)\}$ are given in Pyne [47].

Example 2: Application of Log-Linear Model to Vital Rates. The data in Table 2 are based on an example used by Gail [24] to illustrate the application of Poisson regression with log-linear models to vital statistics. The counts $\{n_{hi}\}$ are the numbers of new melanoma cases reported during 1969–1971 among white males for the hth age group and ith area where $h = 1,2,3,4,5,6$ and $i = 1,2$; and the exposures $\{N_{hi}\}$ are corresponding estimated populations at risk. The underlying information source for these data is the Third National Cancer Survey, for which documentation is given in Cutler and Young [12]. For this type of example, it is of interest to investigate whether the ratio of rates (n_{hi}/N_{hi}) across areas (or age groups) tends to be homogeneous across age groups (or areas). Such a structure, which is usually called a multiplicative (or product) model, can be expressed in the log-linear form

$$E\{n_{hi}\} = \mu_{hi} = N_{hi}\left\{ \exp\left[\lambda x_{hi1} + \sum_{k=1}^{5} \tau_k x_{hi,(k+1)} + \xi x_{hi7} \right] \right\}; \qquad (17)$$

here λ is a reference value for the < 35 age group in the northern area; the τ_k are age-

Table 2 Age × Region Cross-Classification of New Melanoma Cases among White Males during 1969–1971 and Estimated Populations at Risk

	Melanoma Cases, n_{hi}		Estimated Populations at Risk, N_{hi}	
Age Group	Northern	Southern	Northern	Southern
< 35	61	64	2,880,262	1,074,246
35–44	76	75	564,535	220,407
45–54	98	68	592,983	198,119
55–64	104	63	450,740	134,084
65–74	63	45	270,908	70,708
⩾ 75	80	27	161,850	34,233

group-effect parameters relative to indicator variables $x_{hi,(k+1)}$, which are 1 for the $(k+1)$th age group and 0 otherwise; and ξ is a southern-area-effect parameter relative to the indicator variable x_{hi7}, which is 1 for that area and 0 otherwise.

Since the ML equations (4) for the model (17) were nonlinear, the iterative method (5) was used to obtain $\hat{\boldsymbol{\beta}}$. These estimates and their estimated standard errors were

Parameter	λ	τ_1	τ_2	τ_3
ML estimate	-10.66	1.80	1.91	2.24
Estimated s.e.	0.10	0.12	0.12	0.12

Parameter	τ_4	τ_5	ξ	
ML estimate	2.37	2.94	0.82	(18)
Estimated s.e.	0.13	0.13	0.07	

The goodness of fit of this model was supported by the nonsignificance ($\alpha = 0.10$) of the Pearson criterion

$$Q_P = \sum_{h=1}^{6} \sum_{i=1}^{2} (n_{hi} - \hat{\mu}_{hi})^2 / \hat{\mu}_{hi} = 6.12 \quad (19)$$

and the log-likelihood ratio criterion

$$Q_L = \sum_{h=1}^{6} \sum_{i=1}^{2} 2n_{hi}\left[\log_e(n_{hi}/\hat{\mu}_{hi})\right] = 6.21 \quad (20)$$

with respect to chi-square approximations with d.f. = 5. Such statistics are identical to the criteria (7) and (8), respectively, when $\mathbf{W}$ is a maximal set of $(s - t)$ additional explanatory variables. The estimates $\exp(\hat{\tau}_k)$ express the ratio of the incidence of melanoma for the $(k+1)$th age group relative to the < 35 age group and $\exp(\hat{\xi})$ expresses the ratio of the incidence of melanoma for the southern area relative to the northern area. Thus usage of the model (17) can provide results similar in spirit to standardized rates*; see Breslow and Day [6], Gart [26], Osborn [46], and Gail [24] for further discussion.

For many applications dealing with vital rates, the analysis may not be straightforward because of one or more of the following issues:

1. A large array of counts from the cross-classification of several dimensions is to be analyzed. When attention is restricted to the class of hierarchical models of main effects and their interactions, then the iterative proportional fitting (or raking) algorithm described in Bishop et al. [5] and Imrey et al. [32] can be a substantially less costly method for ML estimation than (5).
2. The exposure measures are unknown but can be assumed to be compatible with a log-linear model in their own right (e.g., a model which expresses their homogeneity across one or more dimensions). Since the information in the variation among the counts is being split between the exposures and the count /exposure ratios, the choice of potential relationships for each can be restricted in ways which prevent the analysis of some quantities of interest. Nevertheless, the investigation of such models and their properties can still be worthwhile. See Breslow and Day [6], Koch et al. [39], and Imrey et al. [33] for additional discussion.
3. The counts are small, so exact methods for confidence intervals* or tests of significance* concerning $\boldsymbol{\beta}$ are needed. Some discussion of available strategies for this purpose is given in Gart [26, 27].
4. The counts and exposures are from a sample survey with a complex probability selection process, so the Poisson likelihood function (3) does not apply. Nevertheless, when such quantities have approximately a multivariate normal distribution for which a consistent estimate of the covariance matrix is available, weighted least-squares methods can be effectively used to fit the log-linear model (2). Such methodology is described in Freeman and Holford [17]; its nature is analogous to that illustrated for Example 1 in CHI-SQUARE TESTS, NUMERICAL EXAMPLES.

5. A log-linear model does not provide a satisfactory framework for addressing the questions of interest for a study. Some situations where the use of certain nonlinear models was emphasized include Whittemore and Altshuler [54], James and Segal [34], Frome [20], and Frome and DuFrain [22].

Example 3: Application of Piecewise Exponential Model to Survival Data. Another important application of Poisson regression which is similar in spirit to that for vital rates is the fitting of piecewise exponential models to survival data. The rationale for such analysis has been discussed in Holford [31], Whitehead [53], Laird and Olivier [40], and Aitkin et al. [1]. Its nature can be seen by considering the data in Table 3, which summarizes the follow-up experiences of duodenal ulcer patients with one of four randomly assigned operations. These patients were evaluated at 6 months, 24 months, and 60 months for ulcer recurrence; other patient outcomes of interest were death, reoperation, and lost to follow-up. For purposes of discussion here, death and recurrence are treatment failure events; and reoperation and lost to follow-up are withdrawal from risk (i.e., censoring) events. Other details concerning this clinical trial are given in Johnson et al. [37] and Johnson and Koch [36].

Usage of the piecewise exponential model involves several assumptions concerning the experiences of the subjects in each treatment group $i = 1, 2, 3, 4$ for (V + D), (V + A), (V + H), GR, respectively, during each time interval $j = 1, 2, 3$ for 0–6 months, 7–24 months, 25–60 months, respectively. These include:

1. The withdrawal events are unrelated to treatment failure events and occur uniformly.
2. The treatment failure events have independent exponential distributions and their within interval probabilities are small.

Given this background, the piecewise exponential likelihood function for the data in Table 3 is

$$\phi_{\text{PE}} = \prod_{i=1}^{4} \prod_{j=1}^{3} \lambda_{ij1}^{n_{ij1}} \{\exp[-\lambda_{ij1} N_{ij}]\}; \quad (21)$$

here n_{ij1} is the number of deaths or recurrences during the jth interval for the ith

Table 3 Follow-Up Data for Comparison of Four Operations for Patients with Duodenal Ulcer

		Observed Frequencies				ML Model Predicted Survival Rates	
Operation[a]	Time (months)	Death or Recur.	Reop. or Lost	Satisfactory	Exposure (months)	Estimate	S.E.
V + D	0–6	10	10	317	1962	0.9683	0.0052
	7–24	13	16	288	5445	0.9274	0.0065
	25–60	26	36	226	9252	0.8506	0.0117
V + A	0–6	9	9	313	1932	0.9683	0.0052
	7–24	16	7	290	5427	0.9274	0.0065
	25–60	18	36	236	9468	0.8506	0.0117
V + H	0–6	9	5	329	2016	0.9846	0.0038
	7–24	5	17	307	5724	0.9642	0.0073
	25–60	10	24	273	10440	0.9247	0.0148
GR	0–6	9	8	329	2025	0.9683	0.0052
	7–24	15	11	303	5688	0.9274	0.0065
	25–60	24	37	242	9810	0.8506	0.0117

[a] V + D, vagotomy and drainage; V + A, vagotomy and antrectomy; V + H, vagotomy and hemigastrectomy; GR, gastric resection.

group, N_{ij} is the total person months of exposure, and λ_{ij1} is the hazard parameter. Also, the exposures N_{ij} are determined as

$$N_{ij} = a_j(n_{ij0} + 0.5n_{ij1} + 0.5n_{ij2}), \quad (22)$$

where $a_j = 6, 18, 36$ denotes the length of the jth interval, n_{ij0} denotes the number of subjects with satisfactory status, and n_{ij2} denotes the number withdrawn due to reoperation or loss to follow-up*. The Poisson counterpart to (21) is

$$\phi_{\mathrm{PO}} = \prod_{i=1}^{4} \prod_{j=1}^{3} (N_{ij}\lambda_{ij1})^{n_{ij1}} \{\exp(-N_{ij}\lambda_{ij1})\} / n_{ij1}!$$

$$= \phi_{\mathrm{PE}} \left\{ \prod_{i=1}^{4} \prod_{j=1}^{3} N_{ij}^{n_{ij1}} / n_{ij1}! \right\} \quad (23)$$

Its nature can be motivated by viewing the numbers of deaths $\{n_{ij1}\}$ conditional on their exposures $\{N_{ij}\}$ as having independent Poisson distributions with mean parameters $\mu_{ij} = N_{ij}\lambda_{ij1}$. However, this structure is not being specifically assumed. The role of (23) is to clarify the applicability of Poisson regression computing procedures to estimate parameters $\boldsymbol{\beta}$ for log-linear models concerning the $\{\mu_{ij}\}$ and hence the $\{\lambda_{ij1}\}$ in (21); that is, the $\hat{\boldsymbol{\beta}}$ which maximizes (23) under the model (2) also maximizes (21). For the data in Table 3, a model of interest has the specification matrix

$$\mathbf{X} = \begin{bmatrix} 1 & 1 & 1 & 1 & 1 & 1 & 1 & 1 & 1 & 1 & 1 & 1 \\ 0 & 1 & 1 & 0 & 1 & 1 & 0 & 1 & 1 & 0 & 1 & 1 \\ 0 & 0 & 0 & 0 & 0 & 0 & 1 & 1 & 1 & 0 & 0 & 0 \end{bmatrix}', \quad (24)$$

for which β_1 is a reference value for (V + D) during the 0–6 months interval, β_2 is a common effect for the 7–24 months, and 25–60 months intervals, and β_3 is an effect for the (V + H) treatment. The ML estimates $\hat{\boldsymbol{\beta}}$ from (5) and their estimated covariance matrix are

$$\hat{\boldsymbol{\beta}} = \begin{bmatrix} -5.23 \\ -0.80 \\ -0.73 \end{bmatrix},$$

$$\mathbf{V}(\hat{\boldsymbol{\beta}}) = \begin{bmatrix} 0.0280 & -0.0270 & -0.0069 \\ & 0.0349 & -0.0003 \\ & & 0.0488 \end{bmatrix}. \quad (25)$$

The goodness of fit of the model (24) is supported by the nonsignificance ($\alpha = 0.10$) of the Pearson criterion $Q_P = 6.47$ and the likelihood ratio criterion $Q_L = 6.07$ with respect to chi-square approximations with d.f. $= 9$; such statistics are analogous to (19) and (20), respectively. Thus the follow-up experience of the four treatment groups can be usefully summarized in terms of the predicted survival rates

$$S_{ij} = \prod_{k=1}^{j} \exp\{-a_k \exp(\mathbf{x}'_{ik}\hat{\boldsymbol{\beta}})\}, \quad (26)$$

where $a_j = 6, 18, 36$ for the intervals $j = 1, 2, 3$. These quantities are shown on the right side of Table 3 together with their estimated standard errors from linear Taylor series methods like those described in Imrey et al. [32].

The computations for the examples in this entry were undertaken with the SAS MACRO CATMAX documented in Stokes and Koch [50]. Other available procedures which are potentially applicable include GLIM* [4], PREG [18], and BMD P4F [7].

Finally, it should be noted that the concepts concerning Poisson regression which have been included in this entry can be extended in several directions. Some of these can be summarized briefly as follows:

1. Methods for models other than the log-linear model; see Jorgenson [38], Weber [52], and Gustavsson and Svensson [29] for a discussion of linear models and Frome et al. [23] for a discussion of nonlinear models*.
2. Methods for counts which have compound (or mixed Poisson) distributions [e.g., the negative binomial distribution or the discrete (Poisson) lognormal distribution]. Some models of interest along these lines are discussed in Engen [14], Margolin et al. [43], and Tarone [51]. Bayesian analysis as considered in El-Sayyad [13] is a related topic.

General references for background information concerning the Poisson distribution are

Haight [30], Johnson and Kotz [35], and the other Poisson entries of this encyclopedia.

References

[1] Aitkin, M., Laird, N., and Francis, B. (1983). *J. Amer. Statist. Ass.*, **78**, 264–274.

[2] Andersen, E. B. (1977). *Scand. J. Statist.*, **4**, 153–158.

[3] Armitage, P. (1955). *Biometrics*, **11**, 375–386.

[4] Baker, R. J. and Nelder, J. A. (1978). *The GLIM System Manual (Release 3)*. The Numerical Algorithms Group/Royal Statistical Society, Oxford, England.

[5] Bishop, Y. M. M., Fienberg, S. E., and Holland, P. W. (1975). *Discrete Multivariate Analysis: Theory and Practice*. MIT Press, Cambridge, Mass.

[6] Breslow, N. E. and Day, N. E. (1975). *J. Chronic Dis.*, **28**, 289–303.

[7] Brown, M. B. (1981). In *BMDP Statistical Software*, W. J. Dixon et al., eds. University of California Press, Los Angeles, CA, Chap. 11.

[8] Charnes, A., Frome, E. L., and Yu, P. L. (1976). *J. Amer. Statist. Ass.*, **71**, 169–172.

[9] Chen, C. (1983). *J. Amer. Statist. Ass.*, **78**, 158–161.

[10] Cochran, W. G. (1940). *Ann. Math. Statist.*, **11**, 335–347.

[11] Cochran, W. G. (1954). *Biometrics*, **10**, 417–451.

[12] Cutler, S. and Young, J., eds. (1975). *Third National Cancer Survey: Incidence Data*. NCI Monograph 41, DHEW No. NIH-75-787. National Cancer Institute, Bethesda, Md.

[13] El-Sayyad, G. M. (1973). *J. R. Statist. Soc. B*, **35**, 445–451.

[14] Engen, S. (1978). *Stochastic Abundance Models*. Chapman & Hall, London.

[15] Fisher, R. A. (1925). *Statistical Methods for Research Workers*. Hafner, New York.

[16] Fisher, R. A. (1950). *Biometrics*, **6**, 17–24.

[17] Freeman, D. H., Jr. and Holford, T. R. (1980). *Biometrics*, **36**, 195–205.

[18] Frome, E. L. (1981). *Amer. Statist.*, **35**, 262–263.

[19] Frome, E. L. (1982). *Appl. Statist.*, **31**, 67–71.

[20] Frome, E. L. (1983). *Biometrics*, **39**, 665–674.

[21] Frome, E. L. and Beauchamp, J. J. (1968). *Biometrics*, **24**, 595–605.

[22] Frome, E. L. and DuFrain, R. J. (1983). Maximum Likelihood Estimation for Cytogenetic Dose-Response Curves. *Res. Rep. ORNL/CSD-123*, Oak Ridge National Laboratory, Oak Ridge, Tenn.

[23] Frome, E. L., Kutner, M. H., and Beauchamp, J. J. (1973). *J. Amer. Statist. Ass.*, **68**, 935–940.

[24] Gail, M. (1978). *J. R. Statist. Soc. A*, **141**, 224–234.

[25] Gart, J. J. (1964). *Biometrika*, **51**, 517–521.

[26] Gart, J. J. (1975). In *Statistical Distributions in Scientific Work*, Vol. 2, G. P. Patil, S. Kotz, and J. K. Ord., eds. D. Reidel, Boston, pp. 125–140.

[27] Gart, J. J. (1978). *Commun. Statist. A*, **7**, 917–937.

[28] Grizzle, J. E., Starmer, C. F., and Koch, G. G. (1969). *Biometrics*, **25**, 489–504.

[29] Gustavsson, J. and Svensson, A. (1976). *Scand. J. Statist.*, **3**, 49–60.

[30] Haight, F. A. (1967). *Handbook of the Poisson Distribution*. Wiley, New York.

[31] Holford, T. R. (1980). *Biometrics*, **36**, 299–306.

[32] Imrey, P. B., Koch, G. G., and Stokes, M. E., et al. (1981). *Int. Statist. Rev.*, **49**, 265–283; ibid., **50**, 35–64 (1982).

[33] Imrey, P. B., Koch, G. G., and Davis, G. W. (1984). In *Topics in Applied Statistics*, Y. G. Chaubey and T. D. Dwivedi, eds. Concordia University Press, Montreal, Canada, pp. 123–128.

[34] James, I. R. and Segal, M. R. (1982). *Biometrics*, **38**, 433–443.

[35] Johnson, N. L. and Kotz, S. (1969). *Discrete Distributions*. Houghton Mifflin, Boston.

[36] Johnson, W. D. and Koch, G. G. (1978). *Int. Statist. Rev.*, **46**, 21–51.

[37] Johnson, W. D., Grizzle, J. E., and Postlethwait, R. W. (1970). *Arch. Surg.*, **101**, 391–395.

[38] Jorgenson, D. W. (1961). *J. Amer. Statist. Ass.*, **56**, 235–245.

[39] Koch, G. G., Gillings, D. B., and Stokes, M. E. (1980). *Ann. Rev. Public Health*, **1**, 163–225.

[40] Laird, N. and Olivier, D. (1981). *J. Amer. Statist. Ass.*, **76**, 231–240.

[41] Lawless, J. F. and Singhal, K. (1978). *Biometrics*, **34**, 318–327.

[42] Mantel, N. and Stark, C. R. (1968). *Biometrics*, **24**, 997–1005.

[43] Margolin, B. H., Kaplan, N., and Zeiger, E. (1981). *Proc. Nat. Acad. Sci. USA*, **78**, 3779–3783.

[44] Nelder, J. A. and Wedderburn, R. W. (1972). *J. R. Statist. Soc. A*, **135**, 370–384.

[45] Neyman, J. (1949). *Proc. First Berkeley Symp. Math. Statist. Prob.*, J. Neyman, ed. University of California Press, Berkeley, Calif., pp. 230–273.

[46] Osborn, J. (1975). *Appl. Statist.*, **24**, 75–84.

[47] Pyne, D. A. (1979). *J. Amer. Statist. Ass.*, **74**, 489–493.

[48] Rao, C. R. and Chakravarti, I. M. (1956). *Biometrics*, **12**, 264–282.

[49] Roberts, E. A. and Coote, G. G. (1965). *Biometrics*, **21**, 600–615.

[50] Stokes, M. E. and Koch, G. G. (1983). *Proc. 8th Annu. SAS Users Group Int. Conf.*, pp. 795–800.

[51] Tarone, R. E. (1982). *Biometrics*, **38**, 457–462.

[52] Weber, D. C. (1971). *J. Amer. Statist. Ass.*, **66**, 285–288.

[53] Whitehead, J. (1980). *Appl. Statist.*, **29**, 268–275.

[54] Whittemore, A. and Altshuler, B. (1976). *Biometrics*, **32**, 805–816.

Acknowledgements

The authors would like to thank Edward L. Frome for reviewing an earlier version of this entry and providing several helpful comments with respect to its revision, Barry H. Margolin for permission to use the data in Example 1, and Mitchell H. Gail for permission to use the data in Example 2. They also would like to express appreciation to Ann Thomas and Lori Turnbull for editorial assistance. This research was partially supported by the U.S. Bureau of the Census through Joint Statistical Agreements JSA 83-1 and 84-1.

(CHI-SQUARE TESTS
CONTINGENCY TABLES
GENERALIZED LINEAR MODELS
ITERATIVELY REWEIGHTED LEAST SQUARES
POISSON DISTRIBUTIONS
POISSON PROCESSES
REGRESSION (various entries)
SCORE STATISTICS
STANDARDIZED RATES
SURVIVAL ANALYSIS FOR GROUPED DATA)

GARY G. KOCH
SUSAN S. ATKINSON
MAURA E. STOKES

POISSON, SIMÉON-DENIS

Born: June 21, 1781, in Pithiviers, Loiret, France.

Died: April 25, 1840, in Paris, France.

Contributed to: mathematical physics, mechanics, probability theory.

Poisson's formative years were spent in the provinces, where he came of modest family. Encouraged by a dedicated teacher at the École Centrale at Fontainebleau, he was admitted to the École Polytechnique in Paris in 1798, where through the backing of Laplace* he was appointed to the academic staff in 1800. He stayed at the École Polytechnique, replacing Fourier as professor in 1806, while gaining other posts, notably a professorship of mechanics at the Faculty of Sciences at the Sorbonne in 1816. He had been elected to the Paris Academy of Sciences at the Sorbonne in 1816. Through his life, characterized by hard work and dedication to scientific research and the teaching of science, Poisson accommodated himself to the various changes of political regime. In his research he treated a very broad range of subjects in a great number of publications, of which those on probability and its applications are relatively few, the most important being the book, published close to the end of his life, *Recherches sur la Probabilité des Jugements en Matière Criminelle et en Matière Civile* (1837). In this area, he may have influenced A. A. Cournot*, in the development of whose career he played a large part.

In the *Recherches* the Poisson distribution* is derived as the limit of the distribution function of the (Pascal) distribution with mass function

$$\binom{m+t-1}{m-1} p^m q^t,$$

$$t = 0, 1, 2, \ldots \text{ as } m \to \infty, \quad q \to 0$$

in such a way that $qm \to \omega = \text{const.} > 0$ [8, 9]. Poisson arrived at this distribution starting from the binomial, from the equivalence {number of failures in μ trials is $\leqslant n$} = {number of trials until mth success is $\leqslant \mu$}, where success probability is $p = 1 - q$, and $\mu = m + n$ [8], which gives

$$\sum_{r=0}^{n} \binom{m+n}{r} q^r p^{m+n-r} = \sum_{t=0}^{n} \binom{m+t-1}{m-1} p^m q^t.$$

Hence the connection with what is termed the Poisson approximation* to the binomial for a large number of trials and a small success probability, which was Poisson's intent, although it seems that this should be ascribed to De Moivre* [2, 8]. Indeed, the Poisson mass function $e^{-\omega}\omega^t/t!$, $t = 0, 1, 2, \ldots$ occurs significantly earlier than in Poisson's work [5], even though he himself used it earlier, in a demographic context in

1830 [3]. On the other hand, he considered the "Cauchy" distribution* with density $f(x) = 1/[\pi(1 + x^2)]$, $-\infty < x < \infty$, some 20 years before Cauchy* [3, 7], and showed using characteristic functions* that the arithmetic mean* of a sample from this distribution has the same distribution, a result which Cauchy generalized.

The use of the notion of a random variable, of the cumulative distribution function, and the definition of the density as its derivative, may be original with Poisson [6].

One of Poisson's lasting contributions is his law of large numbers*. It refers to a sequence of independent trials where the success probability in the ith trial is p_i, $i \geqslant 1$. If X is the number of successes in n trials, then the law states that $\Pr(|X/n - \bar{p}(n)| < \epsilon) \to 1$ as $n \to \infty$ for arbitrary $\epsilon > 0$, where $\bar{p}(n) = \sum_{i=1}^{n} p_i/n$, and thus, since $\bar{p}(n)$ may not even approach a limit as $n \to \infty$, expresses loss of variability of X/n about the "floating" mean $p(n)$ rather than stability of X/n. Although the law was proved rigorously by Chebyshev* in 1846, this passed unnoticed and it remained an object of controversy in France. The context of binomial trials with unequal success probabilities was later taken up to develop into dispersion theory*.

Poisson worked on the central limit* problem in the preceding context, and this work to some extent influenced Liapunov*. He also developed an asymptotic theory, of central limit type with remainder, for hypergeometric sampling, and applied his results to the study of the French electoral system.

His probabilistic work was severely criticized, essentially unjustly, by Bienaymé* [4]. Through his *Recherches*, Poisson popularized probability theory, and through Chebyshev and Ostrogradsky had some influence on the St. Petersburg School.

Apart from his many scientific articles and memoirs, Poisson was heavily involved in administration and pedagogical activity. He is quoted as saying: "Life is good for only two things: to study mathematics and to teach it." His contributions to science overall have yet to be properly evaluated [1].

References

[1] Costabel, P. (1978). Poisson, Siméon-Denis. In *Dictionary of Scientific Biography*, C. C. Gillespie, ed., Vol. XV, Suppl. 1. Scribners, New York, pp. 480–490. (Good biography and general appraisal. Emphasis on mathematical physics.)

[2] David, F. N. (1962). *Games, Gods and Gambling: The Origins and History of Probability and Statistical Ideas from the Earliest Times to the Newtonian Era*. Charles Griffin, London.

[3] Gnedenko, B. V. and Sheynin, O. B. (1978). In *Matematika XIX Veka (Mathematics of the 19th Century)*. Nauka, Moscow, pp. 184–240. (Pages 199–205 contain, with a portrait, a good Russian-language summary of Poisson's probabilistic contributions.)

[4] Heyde, C. C. and Seneta, E. (1977). *I. J. Bienaymé: Statistical Theory Anticipated*. Springer-Verlag, New York. (Contains accounts of Bienaymé's attitude to Poisson's work, especially Poisson's law of large numbers, within the general context of nineteenth-century probability.)

[5] Kendall, M. G. (1968). Thomas Young on coincidences. *Biometrika*, **55**, 249–250. (The Poisson distribution in the game of rencontre in 1819.)

[6] Sheynin, O. B. (1977–78). S. D. Poisson's work in probability. *Arch. History Exact. Sci.*, **18**, 245–300. (The most extensive account of Poisson's work in probability and statistics.)

[7] Stigler, S. M. (1974). Cauchy and the Witch of Agnesi: An historical note on the Cauchy distribution. *Biometrika*, **61**, 375–380.

[8] Stigler, S. M. (1982). Poisson on the Poisson distribution. *Statist. Prob. Lett.*, **1**, 33–35.

[9] Ulbricht, K. (1980). *Wahrscheinlichkeitsfunktionen im neunzehnten Jahrhundert*. Minerva, Munich. (Probability functions in the nineteenth century. Includes Poisson's derivation of the Poisson distribution. Good bibliography.)

Bibliography

See the following works, as well as the references just given, for more information on the topic of Siméon-Denis Poisson.

Arago, F. (1854). *Oeuvres complètes*, II. (Pages 591–698 are on Poisson and his work, including Poisson's bibliography.)

Métivier, M., Costabel, P., and Dugac, P., eds. (1981). *Siméon-Denis Poisson et la science de son temps*. École Polytechnique, Palaiseau. [Includes papers by B. Bru (Poisson, le calcul des probabilités et l'instruction publique) and O. B. Sheynin (Poisson and statistics).]

Seneta, E. (1983). Modern probabilistic concepts in the work of E. Abbe and A. De Moivre. *Math. Sci.*, **8**,

75–80. (Includes a discussion of the origins of the Poisson approximation to the binomial.)

(CAUCHY DISTRIBUTION
CHEBYSHEV, PAFNUTY LVOVICH
DISPERSION THEORY
POISSON DISTRIBUTION
POISSON'S LAW OF LARGE NUMBERS
PROBABILITY, HISTORY OF (OUTLINE)
STABLE LAWS)

E. SENETA

POISSON'S LAW OF LARGE NUMBERS

In a sequence of independent trials, denote by p_j the probability that the event E occurs at the jth trial $(j = 1, 2, \ldots)$ and let X_n denote the number of times E occurs in the first n trials. Poisson's law of large numbers states that for any $E > 0$,

$$\lim_{n\to\infty} \Pr\left[\left|n^{-1}X_n - n^{-1}\sum_{j=1}^{n} p_j\right| > \epsilon\right] = 0.$$

Verbally: the difference between the relative frequency of occurrence of E and the arithmetic mean of the probabilities p_j ("average probability of E") tends to zero as n tends to infinity.

It is noteworthy that this result is valid even when $\lim_{n\to\infty} n^{-1}\sum_{j=1}^{n} p_j$ does not exist. The result can be proved by noting that $X_n = \sum_{j=1}^{n} Y_j$, where $\Pr[Y_j = 1] = p_j = 1 - \Pr[Y_j = 0]$ and the Y_j's are mutually independent, and applying Markov's inequality*. [Y_j represents the number of occurrences (0 or 1) of E at the jth trial.]

(BERNSTEIN'S INEQUALITY
CHEBYSHEV'S INEQUALITY
LAWS OF LARGE NUMBERS
MARKOV INEQUALITY
PROBABILITY INEQUALITIES FOR SUMS OF BOUNDED RANDOM VARIABLES)

POISSON TREND TEST *See* POISSON REGRESSION

POISSON VARIANCE TEST *See* POISSON REGRESSION

POLAR CHART

A means of graphical representation for seasonal or directional data.* It represents the values of a variable by distance from the origin at the time or direction represented by the angle made with a horizontal axis.

(SNOWFLAKES
TIME SERIES)

POLARIZATION TEST

Some common tests of statistical hypotheses, such as Pearson's chi-square tests* for goodness of fit* and tests for independence of classifications in a contingency table*, concern testing of hypotheses relating to the parameters (cell probabilities) of a multinomial distribution*. In a goodness-of-fit test, for example, the parameters of the underlying multinomial distribution from which the sample is drawn are specified under the null hypothesis. A multinomial distribution is said to be *even* if the cell probabilities are nearly equal. It is said to be *polarized* if the total probability mass of the distribution is essentially concentrated in a few cells. The term *polarization* connotes the opposite of diversity (*see* DIVERSITY, INDICES OF) and a polarization test reduces to a test of the diversity of a population.

APPLICATION

Suppose that k political parties are contesting an election. Let p_i denote the proportion of voters in favor of the ith party $(i = 1, \ldots, k)$ at a certain time before the election. It might be interesting to know at a latter time before the election whether the voting preference had polarized, in the sense that a single party or only a few of the k parties would share almost all the votes. The change in the voting preference may be due to the emergence of new issues or the occurrence of certain events.

POLARIZATION RELATION

An intrinsic measure of polarization is provided by the partial ordering of multinomial populations with respect to the majorization* relation. Let π_1 and π_2 be two multinomial populations with the associated probability vectors $p = (p_1, \ldots, p_k)$ and $q = (q_1, \ldots, q_k)$, respectively. Let $p_{[1]} \geqslant \cdots \geqslant p_{[k]}$ denote the ordered values of $p_1, \ldots, p_k$. Similarly, let $q_{[1]} \geqslant \cdots \geqslant q_{[k]}$ denote the ordered values of $q_1, \ldots, q_k$. We say that q *majorizes* p (written $p \prec q$) if $\sum_{i=1}^{j} p_{[i]} \leqslant \sum_{i=1}^{j} q_{[i]}$, $\forall\ j = 1, \ldots, k$ with equality for $j = k$. We say that π_1 is *less polarized* than π_2 if $p \prec q$. As the majorization relation is a partial ordering on the space of all multinomial populations, so is the polarization relation.

MEASURES OF POLARIZATION

Let $\phi = \phi(p)$ be a symmetric function of p which preserves the majorization ordering. That is, $\phi(p) \leqslant \phi(q)$ if $p \prec q$. Such a function is called Schur convex; see Schur [3]. If ϕ is symmetric and convex, then it is also Schur convex. However, a Schur convex function need not be a convex function. It is natural to consider a Schur convex function of p as a suitable index of the polarization of the multinomial distribution associated with the probability vector p. For example, consider two functions, (i) $\phi(p) = \sum_{i=1}^{k} p_i^2$ and (ii) $\psi(p) = \sum_{i=1}^{k} p_i \log p_i$. Note that both ϕ and ψ are maximized when the multinomial population is degenerate and minimized when the population is even. The transforms $1 - \phi(p)$ and $-\psi(p)$ are known as the Simpson index of diversity* and Shannon's entropy*, respectively (see Simpson [5] and Shannon [4]). *See* also MAJORIZATION AND SCHUR CONVEXITY.

SAMPLING THEORY

The sampling theory of diversity measures has hardly been touched in the literature. Let $x = (x_1, \ldots, x_k)$ be a sample from a multinomial distribution with the associated probability vector $p = (p_1, \ldots, p_k)$, where

$$\sum_{i=1}^{k} p_i = 1, \qquad \sum_{i=1}^{n} x_i = n.$$

Let

$$T = \sum_{i=1}^{k} x_i^2,$$

$$S = \sum_{i=1}^{k} (x_i/n)\log(x_i/n).$$

The statistic $(T - n)/\{n(n-1)\}$ is an unbiased estimator of the Simpson index $\phi(p)$ and the statistic S is a consistent estimator of the Shannon index $\psi(p)$. Alam and Mitra [1] have derived the sampling distribution of T and considered its application in tests of hypotheses concerning the diversity of a multinomial population. Alam and Taneja [2] have tabulated the sampling distributions of T and S.

As T and S are symmetric and convex functions of x, they are also Schur convex functions of x. The multinomial distribution preserves Schur convexity in the sense that if $f(x)$ is a Schur convex function of x, then its expected value $E_p f(x)$ is Schur convex in p. It follows that $f(x)$ is stochastically ordered with respect to the majorization relation. Therefore, both T and S are stochastically ordered with respect to the majorization relation. The choice of T and S for testing hypotheses relating to the polarization of a multinomial distribution is largely based on this property.

References

[1] Alam, K. and Mitra, A. (1981). *J. Amer. Statist. Ass.*, **76**, 107–109.

[2] Alam, K. and Taneja, V. (1982). Sampling Distributions of Two Measures of Diversity of a Multinomial Population. *Tech. Rep. No. 405*, Dept. of Mathematical Sciences, Clemson University, Clemson, S.C.

[3] Schur, I. (1911). *J. Reine angew. Math.*, **140**, 1–28.

[4] Shannon, C. E. (1948). *Bell Syst. Tech. J.*, **27**, 379–423.

[5] Simpson, E. H. (1949). *Nature*, **163**, 688–691.

Bibliography

See the following works, as well as the references just given, for more information on the topic of polarization tests.

Patil, G. P. and Taillie, C. (1982). *J. Amer. Statist. Ass.*, **77**, 548–567.

Rao, C. R. (1982). *Theor. Popul. Biol.*, **21**, 24–43.

(DIVERSITY, INDICES OF
MAJORIZATION AND SCHUR CONVEXITY
MULTINOMIAL DISTRIBUTION)

K. ALAM

POLICY AND INFORMATION, INTERNATIONAL JOURNAL ON

HISTORY

The *International Journal on Policy and Information* (*PAI*) branched out from the *International Journal on Policy Analysis and Information Systems* (*PAIS*). *PAIS* was jointly founded in July 1977 by the Knowledge Systems Laboratory, University of Illinois at Chicago Circle, and the School of Engineering, Tamkang University, Taiwan, Republic of China. The founding and chief editor was S. K. Chang. *PAIS* appeared biannually in July and January. To broaden the areas of *PAIS* and promote its international service, *PAI* branched out from *PAIS* as an independent publication in 1980. Since then the publishers of *PAI* have been changed to the Graduate School of Information Engineering, Tamkang University, Taiwan, Republic of China, and Knowledge Systems Institute, Glencoe, Illinois. Beginning in 1980, *PAI* has been issued biannually in June and December starting with Vol. 4, No. 1, two issues constituting one volume.

POLICY AND AREA

The editorial policy of *PAI* focuses on policymaking processes, computer-processable data bases, and knowledge-based models for socioeconomic systems. It lays special emphases on applications, especially on policy analysis, information management, and case studies which benefit the developing countries. It presently covers the following areas: (1) economic systems, (2) social and urban systems, (3) industrial engineering, (4) industrial management, (5) stochastic systems, (6) information systems, (7) regional science, (8) management science*, and (9) strategic planning. It is intended to serve those in academic institutions, business, industry, and government.

STATUS QUO

The majority of contributions since 1980 come from outside Taiwan. So far, the countries of authors of published papers include the United States, West Germany, Canada, Japan, Korea, Thailand, India, Australia, Saudi Arabia, Singapore, Denmark, the Philippines, and the Republic of China. All papers are reviewed, normally by one specialist. Reviewers receive a modest payment for their work. The editor tries to inform authors of the acceptance of papers within two to three months. Authors are requested to check the first proof of their papers and to fill out an order slip for reprints. After publication, authors may have a 50 percent discount for two years. The circulation for each issue is 400 copies.

Beginning in 1980 the chief editor has been Louis R. Chow. He is responsible for all publication affairs and goes over each manuscript to select a reviewer. All correspondence with authors is done in the editorial office. Contributions should be sent to the current editorial address: Dr. Louis R. Chow, Editor-in-Chief, *International Journal on Policy and Information*, Office of the Vice-President, Tamkang University, Taipei, Taiwan 106, Republic of China.

SIZE AND CONTENTS

The journal uses 19.2 cm $\times$ 26 cm paper, and the size of each issue is about 120 pages.

The contents of a recent issue (Vol. 7, No. 2) included: (1) "Behavioral and Organizational Models for Human Decision-Making" by Andrew P. Sage; (2) "The Use of Scientific Knowledge for Public Decisions: The Problem of Epistemological Divergence" by David A. Gulley; (3) "A New Ranking Technique of Fuzzy Alternatives and Its Applications to Decision-Making" by Louis R. Chow and Wen-kui Chang; (4) "Optimal Information Gathering and Planning Policies of the Profit-Maximizing Firm" by Christophe Deissenberg and Siegmar Stöppler; (5) "An Analysis of Factors Affecting the Components of Human Resource Information Systems" by Helen LaVan and Nicholas J. Mathys; (6) "Bridging the Gap between Modeling and Data Handling in a Decision Support System Generator" by Michael Szu-Yuan Wang; (7) "The Design of a Computerized Document Processing System for Government Organizations" by I-Ming Shen; (8) "Simulation Experiments with a Canadian Macroeconometric Model" by John Banasik and Carl-Louis Sandblom; (9) "(T_i, S_i) Inventory Policy for Decaying Items with Time Proportional Demand" by Kyung S. Park and Dae H. Kim; (10) "Application of Multiobjective Linear Programming in Deriving Preferential Tariff Adjustments" by M. T. Tabucanon, P. Adulbhan, and R. A. Alivio; (11) "Bar Code Scanning Information Entry Technology and Application" by I. F. Chang, S. J. Chu, and S. T. Liu.

LOUIS R. CHOW

POLICY SPACE METHOD *See* MARKOV DECISION PROCESSES

POLITICAL SCIENCE, STATISTICS IN

Political science is now universally recognized as a quantitative discipline that uses statistical and mathematical techniques extensively [1]. The widespread application of quantitative techniques, however, is a relatively recent phenomenon. Indeed, apart from a few notable exceptions, statistical analyses do not appear in the literature much before 1950 and did not become common until the following decade. Ironically, this statement remains true even in view of the fact that the science of statistics itself originated in the study of politics [23]. As might be expected, the initial works were rather rudimentary, relying for the most part on descriptive methods rather than classical inference.

The situation had changed dramatically by the 1960s, when most subfields of political science were widely using statistical procedures with considerable sophistication. Although political scientists made few original contributions to formal statistics and borrowed heavily from related fields such as sociology and economics, they nevertheless produced innumerable novel and fruitful applications of a vast array of quantitative methods. Indeed, the topic is so integral to the profession that its study has become virtually a subfield in its own right. Statistics, it is safe to say, is the main analytic tool employed today in political science.

Yet its coming was not altogether peaceful. The rapid rise of quantification sparked considerable controversy. For years, particularly between 1955 and 1970, the discipline found itself divided into two camps, behavioral and traditional. Among other things, behavioralists emphasized systematic observation, rigorous and precise concept formation, empirical verification of proposition, and theory building. Statistics seemed to be ideal for accomplishing these ends, especially since voluminous census* and survey data were becoming available. Statistics also seemed appropriate for dealing with the indeterminacies that seem to characterize much of human behavior.

On the other hand, critics charged that quantification often leads to gross oversimplifications and that the subtle but essential vagaries of politics get lost too easily in mathematical models. Although the behavioralists' findings may be statistically correct, they are usually trivial or irrelevant or both. Furthermore, the traditionalists despaired that so much time was spent on "technique" that the equally important task of judgment was ignored.

This debate has now become quieter, as each side concedes points to the other without entirely abandoning its original position. Political scientists generally see the advantages but also the very real limitations of statistical analysis. Despite the considerable progress in understanding made possible by statistics, the mysteries of political phenomena seem as deep as ever.

STATISTICAL ANALYSIS IN POLITICAL RESEARCH

The role that statistics currently plays is perhaps nowhere better illustrated than in the area of public opinion* and elections*. To be sure, the methods are found in virtually all subfields: public administration, policy analysis, international relations, legislative and judicial behavior, and comparative politics. But in the study of attitudes and voting, one of the first topics to be approached in this fashion, one sees the diversity and sophistication with which statistics is applied to political science. One can also appreciate some of the achievements and failures, the hopes and disappointments, and the future prospects of a humanistic discipline trying to become scientific.

The earliest attempts to study voting and public opinion, like so many other parts of political science, relied heavily on intuitition, subjective observation, and untested generalizations. There were, of course, path breakers in the 1930s and 1940s whose work encompassed multivariate methods. Gosnell and Page [11], for example, studied voting patterns in Chicago with the help of simple and partial correlation* coefficients and a new "method called multiple factor analysis, which is now being perfected by Professor L. L. Thurstone" (p. 979). Similarly, V. O. Key's seminal work, *Southern Politics* [16], gave political scientists fresh new ideas about how to approach public opinion data.

What really encouraged the development of statistical applications, however, was the advent of the sample survey*, a technique that soon produced an abundance of quantifiable data. The first uses of this information were rather elementary. Most emphasis fell on descriptive statistics and frequency distributions. The early voting studies, for example, relied primarily on complex bar* graphs rather than, say, regression coefficients* to present multivariate results. (See among others Lazarsfeld et al. [17], Berelson et al. [2], and Campbell et al. [4].) Then, and even to a large extent now, there was little classical hypothesis testing*. Except for the commonly calculated goodness-of-fit* chi-square* statistic, used to test for an association between two variables, political analysts treated hypothesis testing as secondary to parameter estimation. Indeed, they were taught early on that tests such as the chi-square statistic could be misleading since large samples, which were increasingly prevalent, inflated the numeric values of the results. Consequently, political scientists have not spent much effort on test theory. One seldom sees simultaneous inference or Bayesian* techniques in political analysis. Furthermore, there is relatively little experimentation and hence only modest use of experimental designs* and analysis of variance* methods.

No, instead of stressing test or decision theory* political scientists in the early period turned to measures of central tendency, variation, frequency distributions, cross-classifications, and especially measures of association between two variables. When truly quantitative (interval-level) data were available, they calculated the Pearsonian (product-moment) correlation coefficient; when they had only categorical data*, they calculated nominal and ordinal measures of association* developed by among others Goodman and Kruskal [10], Kendall [15], and Stuart [26]. These indices showed how strongly two variables—party affiliation and occupation, for example—were related. Knowing the strength of an association was considered more important than knowing only that the variables had a statistically significant relationship.

Soon regression analysis became popular because it allowed one to study a single dependent variable (e.g., whether people voted for a Democratic or Republican candidate) as a function of a set of independent variables (e.g., partisanship, socioeconomic status, and attitudes). But even simple one-equation regression was not entirely satisfactory because there was also an interest in explaining the relationships among the predictors. Partly for these reasons, political scientists in the early 1960s began to turn to causal and path analysis.

Based on the work of Wright [28], Simon [24], and Blalock [3], causal analysis requires that the investigator hypothesize causal dependencies among a set of variables. If certain assumptions and conditions hold, one can derive predictions and test them against observed data. The predictions usually involve partial correlation coefficients which are readily calculated from a basic set of two variable correlation. Causal models which are graphical representations of the interrelationships among the variables have both analytic and heuristic value since they require one to make explicit assumptions that are all too often left implicit. The procedure provides a method for translating a verbal theory into a mathematical one. *See* also CAUSATION.

An important example of causal analysis is Goldberg's [8] study of American voting behavior. Although Goldberg believed that a person's vote depended causally on various socioeconomic and political characteristics, he recognized that the antecedent factors also had a causal ordering among themselves.

Figure 1 (an example of this work) shows a model that seemed to fit best his data. Arrows, representing direct causal linkages, point in one direction because this model, as in most of the original models developed by political scientists, allowed only for one-way, as opposed to reciprocal, causation. The model makes very stringent assumptions about error or disturbance terms (here represented by e's) such as that they are not directly interrelated among themselves. If these assumptions and other conditions hold, one can derive predictions about the magnitude of certain partial correlation or path coefficients. By comparing the observed and predicted values, one can decide whether or not the model is tenable. If it is not, arrows representing direct causal linkages are added or deleted and new predictions derived.

Causal and path analyses of these sorts have appeared frequently in the social sciences since their original development in the early 1960's. Besides being helpful analytic devices, they have heuristic purposes as well. Causal models have been employed in the investigation of panel data* (i.e., surveys of respondents at different times), measurement error*, and unmeasured or unobserved variables. As useful as the method is, however, it has several drawbacks.

Concentrating on partial correlation coefficients can obscure the importance of estimating the underlying parameters. After all, since a causal model represents a system of equations, one wants to know more than how well the data fit a particular model; one also needs good estimates of its structural coefficients. Others complained that the necessary assumptions were unrealistic. For instance, is a person's partisan attitude only an effect of his party identification? Could there not exist a reciprocal relationship in which the variables are both causes and effects of each other? Finally, these models were analyzed by ordinary least-squares* (OLS) regression which give inefficient or inexact results when the assumptions about the error terms are not satisfied.

For these and other reasons, political scientists along with other social scientists turned to general structural equation models [6]. This more general approach contains causal and path analysis as special cases.

Once again the motivation and limitations of structural equation modeling are best illustrated by an example. Page and Jones [22] in what may become another methodological milestone in political research

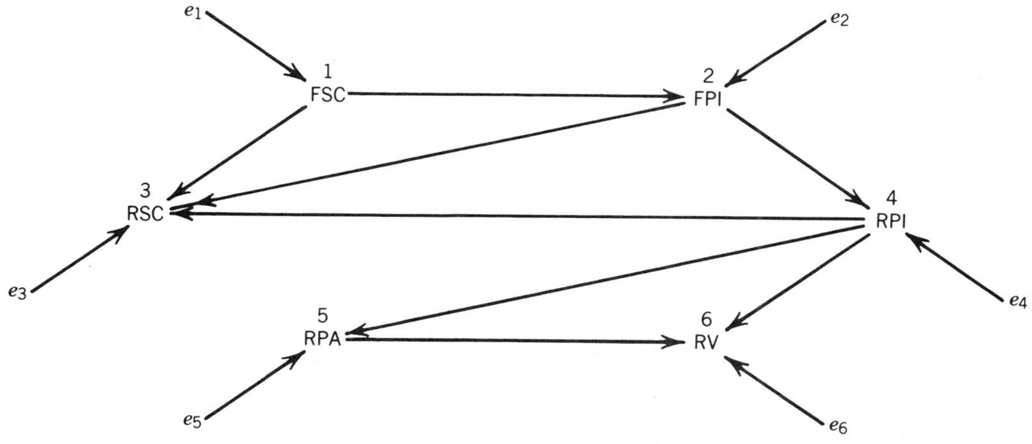

Figure 1 Goldberg's causal model of voting behavior.
Key:
1 FSC, father's sociological characteristics
2 FPI, father's party identification
3 RSC, respondent's sociological characteristics
4 RPI, respondent's party identification
5 RPA, respondent's partisan attitude
6 RV, respondent's vote for president in 1956

Predicted Equations	Observed Values
$r_{41.23} = 0$	− 0.017
$r_{51.234} = 0$	0.083
$r_{61.2345} = 0$	− 0.019
$r_{52.134} = 0$	0.032
$r_{62.1345} = 0$	0.053
$r_{53.124} = 0$	− 0.073
$r_{63.1245} = 0$	− 0.022

wanted to explain voting behavior using explanatory variables similar to Goldberg's. But troubled by what they considered an "incorrect" assumption of one-way causation, they developed a model that allowed reciprocal causal effects among the main variables. (See Fig. 2 in which each arrow stands for a causal path and represents a structural coefficient. The sets of exogenous variables are assumed to be prior to the endogenous variables and orthogonal to one another.) Since the structural equations contained too many endogenous unknowns to be estimated from the observed data, the authors had to add exogenous variables—that is, variables which were presumably causally unaffected by the endogenous factors—and use two-stage* and three-stage least squares to obtain the estimates.

What is noteworthy about this work is not its substantive conclusions, which will surely be debated, but its demonstration of how far the study of politics has progressed. Starting with simple cross tabulation, political scientists then advanced to the analysis of single-question models, then to multiequation recursive (one-way causal) systems, and now to dynamic simultaneous equation models. Along the way, they recognized the limitations of OLS regression and are now using advanced regression estimating techniques: multistage least squares, full-information*

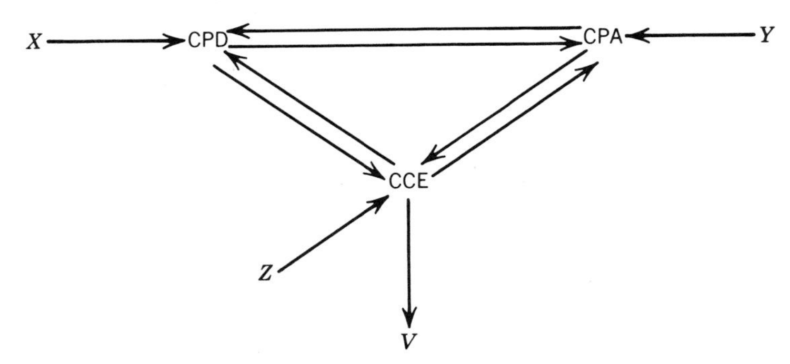

Figure 2 Page and Jones' nonrecursive voting model.
Key:
CPD, comparative policy distances
CPA, current party attachment
CCE, comparative candidate evaluation
V, vote
X, Y, Z, exogenous variables

maximum likelihood, instrumental variables*, and the like.

They are also increasingly utilizing time-series* analysis because many variables of interest have an order based on time. Time-series data appear regularly in the study of defense expenditures, the amount of violence in the international system, economic conditions and election outcomes, and public policy analysis.

Although the growing sophistication in the use of regression analysis has brought many results, it has also raised troublesome questions. To estimate the causal processes, for example, Page and Jones [22] introduced a number of presumably exogenous variables. But in doing so they introduce additional assumptions about how these variables are related to those already in the system and to each other. This is a perennial problem in political research: the phenomena of most interest are quite complex and any effort to describe them mathematically inevitably leads to restrictive and sometimes unrealistic assumptions. One difficulty may be overcome at the cost of introducing new ones. The authors criticized previous investigators for assuming one-way causation while making a host of other assumptions. Not surprisingly, then, their results have not found universal acceptance and the definitive explanation of voting seems as elusive as ever.

Paralleling the application of regression techniques has been the widespread use of factor analysis*. Its acceptance has been motivated by two related considerations. First, there is a belief that although political behavior manifests itself in numerous ways —it can still be explained by reference to a much smaller number of factors and the researcher's objective is to use factor analysis to identify and name the underlying factors. A second, perhaps more practical motivation is the utility of factor analysis as a data reduction technique. A typical public opinion survey contains scores of questions. If they are all intercorrelated, the correlation matrix may be reduced to a smaller number of factors thereby simplifying the interpretation of the data.

Typical is Finifter's [7] study of political alienation. Starting with 26 items that appear to measure the concept, she used factor analysis to isolate two factors, labeled "powerlessness" and "perceived normalessness."

Factor analysis and related techniques—discriminant*, cluster, and canonical analysis*—are becoming quite popular for a variety of purposes. They are used in legislative and judicial studies to identify voting factions or clusters of issues that define policy dimensions, in public opinion research to find ideological structures among individuals or groups, and in comparative politics to compare the attributes of nations on various political and socioeconomic variables. Factor analysis is also used as a tool for improving political measurement and exploring the consequences of measurement error.*

Another relatively recent development is the emergence of multivariate procedures for the analysis of categorical (as opposed to quantitative) data. A common problem in political research takes this form: Suppose that one wants to analyze candidate preference (Democrat or Republican), partisanship (Democratic, Republican, independent), income (high, medium, low), region (South and non-South), and race (white, nonwhite). If a sufficiently large number of cases are available, one can make a $2 \times 3 \times 3 \times 2 \times 2$ cross-classification or contingency table. But what is the best way to analyze it?

Two general approaches have been proposed. The first, based on log-linear models (i.e., model for the logarithm of cell frequencies or some function of them), uses maximum likelihood estimation [9]. It permits one to estimate and test the significance of "main" effects and various types of interaction. One might want to know, for example, if the relationship between preference and partisanship is the same for different combinations of the demographic factors. The second approach also answers this type of question and others as well, but uses weighted least squares as the estimating procedure [12].

At first sight, the multivariate analysis of categorical data appears to be a godsend to social scientists who have an abundance of ordinal and nominal scales to contend with. They are also useful in analyzing two-way cross classifications such as mobility tables and panel studies.

Neither approach has displaced regression and factor analysis, however. Perhaps they have arrived on the scene too recently. Furthermore, both require large sample sizes and lead to the estimation of innumerable parameters. Political scientists may prefer the relative simplicity of least squares. Nevertheless, it seems certain that weighted least squares and maximum likelihood analysis of categorical data analysis will increase substantially in the future.

STATISTICS AND MEASUREMENT IN POLITICAL SCIENCE

In attempting to achieve the rigor and precision of natural sciences, political science faces a formidable obstacle. The problems of greatest interest usually involve considerable complexity and subtlety. Indeed, there is often very little agreement about what certain concepts mean, much less about how to measure them. How, for instance, does one conceptualize and measure "power," "equality," or "democracy"? Nearly everyone agrees, then, that empirical measures of political concepts are fallible indicators subject to at least three types of errors.

Errors may arise in the first place from an inappropriate level of measurement. Most statistical tools assume that the variables are measured quantitatively on interval or ratio scales. But political phenomena are not so easily quantified, and the best one may achieve is a classification of the subjects on a nominal or ordinal scale. There has been considerable to-do about whether such scales should be treated as interval level or analyzed by methods designed explicitly for categorical data. Many investigators simply assign numbers to categories, especially if they have dichotomous data, and use the usual statistical formulas. Others prefer to rely strictly on categorical statistical procedures. *See* also NOMINAL DATA and ORDINAL DATA.

A particularly troublesome question remains, however. Suppose that an aptitude or behavior is actually continuous, rather than discrete, but is measured as though it consisted of only two categories. Such a case might arise in a sample survey in which people's preferences are classified as "pro" or "con," when, in fact, innumerable shades of opinion may exist. Is it legitimate to make inferences about substantive phenomena on the basis of such data—no matter what statistic is used? This is still an unresolved issue.

A second source of error is random and nonrandom measurement error. Sample surveys and census materials, the source of much data, frequently reflect selective retention, biases, disinterest, incomplete record keeping, or other mistakes. One must assume that empirical data at best imperfectly represent the underlying properties and that error variance will be a sizable portion of a variable's total variance. Thus it is not surprising that reliability and validity have become important issues in the scientific study of politics (*see* PSYCHOLOGICAL TESTING THEORY).

They have been dealt with in a variety of ways. There is perhaps less concern in the literature with "classical measurement theory" than one would imagine for a discipline that relies so heavily on questionnaires. In fact, many early investigators took the reliability of their measures for granted as they seldom bothered to compute or report standard reliability checks such as test–retest, split halves, alternative form, or reliability coefficients (e.g., Cronbach's alpha) [18, 20, 21]. Fortunately, the last decade has witnessed a reaction against this laissez-faire attitude toward measurement, and findings that had for years been accepted as dogma are now being challenged on the grounds that the data were faulty.

Another source of measurement error is more troublesome. Empirical measures often serve as indirect or substitute indicators of the true concepts. Political scientists find themselves in the place of a Boy Scout who must measure the height of a tree from its shadow. Only they frequently have only a vague notion of how the shadow's dimension relates to the tree's size. Not surprisingly, then, elaborate statistical manipulations cover but do not hide the profoundities of politics.

Confronted with these sources of error, political scientists have followed several paths. Many have tried to overcome them by using multiple indicators of a single theoretical concept. The earliest efforts in this direction were the construction of measurement scales from Likert (agree–disagree) items and scalogram analysis [25]. In the latter procedure, questions are related to one another in such a way that, ideally, an individual who replies favorably to item 1 also responds favorably to item 2; an individual who favors item 3 would also prefer items 1 and 2. A Guttman scale is both cumulative and presumably unidimensional in that it purports to measure a single underlying continuum. Guttman scales were used to measure individual behavior (e.g., attitudes, policy dimensions in Congress and the judiciary, the attributes of nations, and a host of other concepts). Although scalogram analysis was popular in the formative years of quantitative political research, it is seen less widely now because it offers little guidance for selecting items that are likely to form a scale and because criteria for assessing the adequacy of the scale sometimes give very misleading results. More important reasons for its decline are the assumption of unidimensionality and the emergence of multidimensional scaling techniques.

Multidimensional scaling*, a product of psychometric research, has found a natural audience in political science. It permits one to locate individuals on several attitudinal dimensions instead of just one. Weisberg and Rusk [27], as an example, wanted to know which factors (partisanship, ideology, issues, personality) affected people's evaluation of 12 candidates. They also wanted to know how many dimensions people use to evaluate candidates. Employing a nonmetric multidimensional scaling routine, the investigators found that candidate evlauations as well as certain attitude preferences required a two-dimensional space with one axis representing traditional left–right politics, the other a newer left–right division.

Although multidimensional scaling, along with factor analysis, are currently the most widely employed techniques for the construction of measurement scales, causal and path analysis have become a main means to assess the reliability of measures and to explore the causes, effects, and possible remedies of various kinds of measurement errors. Using path analysis, Costner [5] provided criteria for identifying biases due to certain types of nonrandom measurement error through the use of multiple indicators. Similarly, Jöreskog's [13, 14] analysis of covariance structures permits one to identify errors and assess their statistical significance. Variations of both approaches are found throughout political science.

THE FUTURE OF STATISTICS IN POLITICAL SCIENCE

Statistics in political science, it seems safe to say, will flourish in the coming decades. Quantitative methods have shown their value and researchers steadily expand the areas to which they are applied. Obviously, this survey barely touches the breadth of applications. At the same time, political scientists have become more knowledgeable and careful users.

A factor that encouraged the proliferation of statistical applications in political science was the availability of computers* and especially prewritten program packages. These systems, which will surely increase in power and sophistication, relieved social scientists of the burden of learning the computational and theoretical underpinnings of many procedures. By relying on these preprogrammed instructions, one could produce reams of computer output without having an extensive training in statistics. Textbooks that supplied computing formulas but few proofs and little theory also fueled the growth. But the discipline discovered that although "results" come early and profusely, it is often hard to make theoretical sense out of them.

Today political scientists, having a solid background in mathematics, statistics, and computer science, tend to be more self-conscious and responsible about their methods and methodology. They have learned to be very thoughtful about the assumptions, theory, and interpretation of the techniques they use, and when problems arise they are better able to consult with specialists in the statistical sciences. Being more confident, they have also become more eclectic and adaptable, borrowing from whatever field promises the best solutions to particular problems. The result is more rigorous and precise research.

At the same time, quantitative research is becoming less dogmatic. In its infancy, empirical political scientists tended to dismiss nonquantitative studies as too impressionistic, unsystematic, and parochial. Even worse, technique at times dominated content; findings were taken as proven not because evidence necessarily supported them but because they flowed from extremely esoteric multivariate procedures involving seemingly endless equations and matrix operations. Statistical significance passed for substantive significance, while intuition and common sense fell by the wayside.

Fortunately, by now everyone seems to realize the very real limits to which the statistical sciences can be pushed in human affairs. The heart of the matter is of course the multifarious nature of the topic. It is not simply that politics is difficult to conceptualize and measure, although that is certainly true because measurement continues to be the Achilles heel of the discipline; and without improvement in this area progress will be slow indeed. But just as important, statistical generalizations often obscure the nuances, the idiosyncrasies, the special cases—in short, the very things that make politics so interesting and significant. Many investigators have found that however elaborate the research design, their results are seldom convincing unless they are embedded in a contextual grasp of the problem, and that one simply cannot analyze disembodied numbers without grossly misreading the phenomena they measure.

In the future, therefore, the discipline will continue to borrow heavily from the statistical sciences, but it will apply this wherewithal with greater care, imagination, and most important, appreciation of the limits of its valid application.

References

[1] Alker, H. R., Jr. (1975). In *Handbook of Political Science*, Vol. 7: *Strategies of Inquiry*, F. I. Greenstein and N. W. Polsby, eds. Addison-Wesley, Reading, Mass., pp. 139–210. (A good history and survey of statistical application in political science.)

[2] Berelson, B., Lazarsfeld, P. F., and McPhee, W. N. (1954). *Voting*. University of Chicago Press, Chicago.

[3] Blalock, H. M., Jr. (1964). *Causal Inferences in Nonexperimental Research*. University of North Carolina Press, Chapel Hill, N.C. (A landmark in the social sciences, this work introduced political scientists to causal analysis.)

[4] Campbell, A., Gurin, G., and Miller, W. E. (1954). *The Voter Decides*. Row and Peterson, Evanston, Ill.

[5] Costner, H. L. (1969). *Amer. Polit. Sci. Rev.*, **75**, 245–263.

[6] Duncan, O. D. (1975). *Introduction to Structural Equation Models*. Academic Press, New York.

[7] Finifter, A. W. (1970). *Amer. Polit. Sci. Rev.*, **64**, 389–410.

[8] Goldberg, A. S. (1966). *Amer. Polit. Sci. Rev.*, **60**, 913–922.

[9] Goodman, L. A. (1970). *J. Amer. Statist. Ass.*, **65**, 226–256.

[10] Goodman, L. A. and Kruskal, W. H. (1954). *J. Amer. Statist. Ass.*, **49**, 732–764.

[11] Gosnell, H. F. and Gill, N. N. (1935). *Amer. Polit. Sci. Rev.*, **29**, 967–984.

[12] Grizzle, J. E., Starmer, C. F., and Koch, G. G. (1969). *Biometrics*, **25**, 489–504.

[13] Jöreskog, K. G. (1969). *Psychometrika*, **34**, 183–202.

[14] Jöreskog, K. G. (1970). *Biometrika*, **57**, 239–251.

[15] Kendall, M. G. (1955). *Rank Correlation Methods*, 2nd ed. Charles Griffin, London.

[16] Key, V. O., Jr. (1949). *Southern Politics*. Alfred A. Knopf, New York.

[17] Lazarsfeld, P., Berelson, B., and Gaudet, H. (1944). *The People's Choice*. Columbia University Press, New York. (One of the first works to apply quantitative analysis in a systematic way to the study of electoral politics.)

[18] Lord, F. M. and Novick, M. R. (1968). *Statistical Theories of Mental Test Scores*. Addison-Wesley, Reading, Mass.

[19] Miller, R. G., Jr. (1966). *Simultaneous Statistical Inference*. McGraw-Hill, New York.

[20] Nunnally, J. C. (1964). *Educational Measurement and Evaluation*. McGraw-Hill, New York.

[21] Nunnally, N. C. (1978). *Psychometric Theory*. McGraw-Hill, New York.

[22] Page, B. I. and Jones, C. C. (1979). *Amer. Polit. Sci. Rev.*, **73**, 1071–1089.

[23] Pearson, K. (1978). *The History of Statistics in the 17th and 18th Centuries*, E. S. Pearson, ed. Macmillan, New York.

[24] Simon, H. (1957). *Models of Man: Social and Rational*. Wiley, New York.

[25] Stouffer, S. A. et al. (1949). *Measurement and Prediction*. Studies in Social Psychology during World War II, Vol. 4. Princeton University Press, Princeton, N.J.

[26] Stuart, A. (1953). *Biometrika*, **40**, 106–108.

[27] Weisberg, H. F. and Rusk, J. G. (1970). *Amer. Polit. Sci. Rev.*, **64**, 1167–1185.

[28] Wright, S. (1934). *Ann. Math. Statist.*, **5**, 161–215.

Journal Sources

American Political Science Review. (Many articles rely heavily on statistical methods.)

American Journal of Political Science. (Most issues contain a "workshop" that often explains a statistical application in political science.)

Political Methodology. (Contains mostly articles on statistical applications in political science.)

(CAUSATION
DEMOGRAPHY
ECONOMETRICS
ELECTION FORECASTING
PATH ANALYSIS
PUBLIC OPINION POLLS
SOCIOLOGY, STATISTICS IN
SURVEY SAMPLING)

H. T. REYNOLDS

POLLACZEK–KHINCHIN FORMULA

This term refers to a group of formulae associated with equilibrium distributions of queue length and waiting time of an $M/G/1$ queueing* system. Here individual customers arrive at event times of a Poisson* process having rate λ, queue in order of arrival and wait for service from the single server. Service times are mutually independent, are independent of the arrival process, and have the same distribution with Laplace–Stieltjes transform $\beta(\cdot)$ (*see* INTEGRAL TRANSFORMS) and finite mean μ.

Suppose that the traffic intensity $\rho = \lambda\mu < 1$; that is, on average, customers are served at a faster rate than they arrive. Let Q and W be random variables representing the equilibrium queue length and waiting time, respectively, and S represent service times. Each of the following expressions has been called the Pollaczek–Khinchin (PK) formula:

$$E(Q) = \rho + \left[\rho^2 + \lambda^2 \mathrm{var}(S)\right]/\left[2(1-\rho)\right], \tag{1}$$

$$E(W)/E(S) = \left[1 + \mathrm{var}(S/\mu)\right]/\left[2(1-\rho)\right], \tag{2}$$

$$E(e^{-\theta W}) = (1-\rho)\left[1 - \rho\left(\frac{1-\beta(\theta)}{\mu\theta}\right)\right]^{-1}, \tag{3}$$

$$E(z^Q) = (1-\rho)\frac{(1-z)\beta(\lambda(1-z))}{\beta(\lambda(1-z)) - z}. \tag{4}$$

Formulae (2) and (3) were derived by Pollaczek [9] in 1930 and independently, and more directly, by Khinchin [6] in 1932. Seal [12, p. 120] observed that a result tantamount to (3) can be found in early works on risk theory* [2, 8]. Many general queueing theory texts (e.g., Kleinrock [7]) call (1) the PK-*formula* and (4) is so named by Gnedenko and Kovalenko [4]. Kleinrock [7, p. 177] suggests distinguishing (1) and (3) by naming them the PK-*mean value formula* and PK-*transform equation*, respectively.

Useful insights into the steady state $M/G/1$ system follow immediately from (1) to (3). If ρ and λ are constant, then the mean queue length $E(Q)$ is least when $\mathrm{var}(S)$ is zero. Similarly, Kendall [5, p. 155] observed that the ratio on the left of (2) is a "figure of demerit," which is least when $\mathrm{var}(S)$ is zero.

Both (1) and (2) show that $E(Q)$ and $E(W)$ grow rapidly as ρ approaches unity. It is rarely possible to invert (3) or (4) to get closed expressions for the distributions of W or Q, although (3) has been used to obtain approximate expressions for the distribution of W; see refs. 3 and 13 (p. 224). However, it is possible to obtain from (3), for example, an expression for the distribution function of W in the form of an infinite series of geometrically weighted convolution powers of a certain distribution function; see [10, p. 38] for discussion. Related to this is Runnenberg's observation [11, p. 164] that the form of (3) shows immediately that W has an infinitely divisible* distribution. Starting with (3), Cohen [1] obtained results showing that $P(W > t)$ decays at an algebraic rate, as $t \to \infty$, if $P(S > t)$ does so. He used this result to obtain limit theorems for the maximum waiting time over many busy periods.

References

[1] Cohen, J. W. (1972). *Ann. Inst. Henri Poincaré*, **8**, 255–263.

[2] Cramér, H. (1930). On the mathematical theory of risk. *Festskrift Skandia 1855–1930*, Stockholm.

[3] Delbrouck, L. E. N. (1978). *J. Appl. Prob.*, **15**, 202–208.

[4] Gnedenko, B. V. and Kovalenko, I. N. (1968). *Introduction to Queueing Theory*. Israel program for Scientific Translations, Jerusalem.

[5] Kendall, D. G. (1951). *J. R. Statist. Soc. B*, **13**, 151–185.

[6] Khinchin, A. Y. (1932). *Mat. Sb.*, **39**, 73–83.

[7] Kleinrock, L. (1975). *Queueing Systems*, Vol. 1: *Theory*. Wiley, New York.

[8] Lundberg, F. (1926). *Försäkringsteknisk Riskutjämning: I. Theori*. Englund, Stockholm.

[9] Pollaczek, F. (1930). *Math. Zeit.*, **32**, 64–100, 729–750.

[10] Prabhu, N. U. (1980). *Stochastic Storage Processes—Queues, Insurance Risk, and Dams*. Springer-Verlag, Berlin.

[11] Runnenburg, J. Th. (1965). In *Proc. Symp. Congestion Theory*, W. L. Smith and W. E. Wilkinson, eds. University of North Carolina Press, Chapel Hill, N.C., p. 164.

[12] Seal, H. (1969). *Stochastic Theory of a Risk Business*. Wiley, New York.

[13] Srivastava, H. M. and Kashyap, B. R. K. (1982). *Special Functions in Queueing Theory: And Related Stochastic Processes*. Academic Press, New York.

(QUEUEING THEORY)

ANTHONY G. PAKES

PÓLYA-AEPPLI DISTRIBUTION

In published form, the Pólya–Aeppli distribution first appeared in Pólya [10], where urn models* were used to derive various distributions. When the outcome x_n at stage n depended only on $x_1 + x_2 + \cdots + x_{n-1}$, the process was called one of "contagion"; when x_n depended only on x_{n-1}, "heredity." (This nomenclature has not been generally continued with.) The Pólya–Aeppli distribution, with a reference to a thesis of Aeppli [1], was obtained as a case of the latter, being described as one of rare events with weak dependence. It is a two-parameter distribution with probability generating function* (PGF) in one form as

$$\exp\left(-\mu + \frac{\mu}{1+\rho-\rho z}\right) = \sum_{x=0}^{\infty} p_x z^x,$$

in which μ and ρ are positive, with an alternative, sometimes more convenient, parameterization as

$$\exp\left(-\frac{\theta}{1-\omega} + \frac{\theta}{1-\omega z}\right),$$

where θ is positive and $0 < \omega < 1$, with $\theta = \mu/(1+\rho)$, $\omega = \rho/(1+\rho)$. It may be regarded as a special case of the Poisson–Pascal distribution, sometimes referred to as the generalized Pólya–Aeppli distribution (rather unfortunately, in view of the many other meanings of "generalized"), with PGF

$$\exp\{-\mu + \mu(1+\rho-pz)^{-\kappa}\},$$

in which $\kappa = 1$, and in some applications it would nowadays be so treated. Prior to the availability of electronic computing, the numerical labor associated with more than two parameters was often regarded as excessive; even now, many applied workers appear more comfortable with the interpretation of two-parameter models.

The distribution can be obtained by combining "simpler" distributions, two ways being set out below.

1. As a *stopped* or *generalized* distribution (*see* CONTAGIOUS DISTRIBUTIONS) it is a

$$\text{Poisson}(\mu) - \text{stopped geometric}\left(\frac{\rho}{1+\rho}\right)$$

distribution

[or in Gurland's notation

$$\text{Poisson}(\mu) \vee \text{geometric}\left(\frac{\rho}{1+\rho}\right)\Bigg].$$

2. As a *mixture**, it is

Pascal(u,ρ) mixed on u with Poisson(μ)

or

$$\text{Pascal}(u,\rho) \wedge \text{Poisson}(\mu),$$

where Pascal is used synonymously with negative binomial*. [In these, the PGF of the Pascal(κ,ρ) distribution is taken as

$$(1+\rho-\rho z)^{-\kappa},$$

with $\quad \text{PF} \dbinom{\kappa+x-1}{x}\rho^x(1+\rho)^{-\kappa-x};$

that of the geometric(ω) as $(1-\omega)/(1-\omega z)$, and that of the Poisson(μ) as $\exp(-\mu+\mu z)$.] Its expectation is $\mu\rho$, and variance $\mu\rho(1+2\rho)$; higher-order cumulants are given later.

It is also a limiting case of the two-parameter Neyman type A, B, C* considered as a three-parameter family with PGF:

$$\exp\left[-\mu + \mu\frac{e^{\Delta} - 1 - \Delta - \cdots - \dfrac{\Delta^{\beta}}{(\beta-1)!}}{\Delta^{\beta}/\beta!}\right]$$

with $\Delta = \nu(z-1)$, where $\beta = 0,1,2$ correspond to types A, B, and C. If $\beta \to \infty$ with $\nu/(\beta+1) \to \rho$, so that $\nu \to \infty$ also, this leads to the Pólya–Aeppli distribution [3]. (This limiting procedure is equivalent to holding the mean and variance finite as $\beta \to \infty$.)

In the Pólya–Aeppli distribution itself, if $\rho \to 0$ as $\mu \to \infty$ in such a way that $\mu\rho = \lambda$, say, is constant, then a Poisson(λ) distribution is obtained.

An explicit formula for the probability function p_x is

$$p_x = p_0 \omega^x \sum_{r=1}^{x} \binom{x-1}{r-1} \frac{\theta^r}{r!}, \qquad x = 1, 2, \ldots$$

with

$$p_0 = \exp\left[-\left(\frac{\theta\omega}{1-\omega} \right) \right];$$

it can also be expressed in terms of a confluent hypergeometric function* [9]. But recurrence relations are frequently more convenient:

$$p_{x+1} = \frac{\theta\omega}{x+1} \sum_{r=0}^{x} (r+1)\omega^r p_{x-r}, \qquad x = 0, 1, \ldots$$

or

$$p_{x+1} = \frac{\omega}{x+1} \{(\theta + 2x)p_x - \omega(x-1)p_{x-1}\}, \qquad x = 1, 2, \ldots$$

with $p_1 = \theta\omega p_0$. (It may be noted that the second of these can quickly run into numerical difficulties because of the differencing of small terms.) There are also rather cumbersome asymptotic expressions and inequalities available. The recurrence relations are easily implemented on a computer: an implementation of the former in APL is

```
∇  R←X  PAI  P;A;B;C;D;E;F;G;H
[1]  B←⌈/X
[2]  R←R,(A←C×(D←E÷F)÷F)
        ×R←*(C←P[1])×(÷F←
        1+E←P[2])−G←,1+H←0
[3]  L_:→(B≥ρR←R,(A÷H+1)
        ×+/(G←G,G[H]×(1+H)
        ×D÷H←H+1)×⌽R)/L_
[4]  R←R[1+X]  ∇
```

Once this has been entered in a computer, typing

0 1 2 3 8 *PAI* 2.4 2.5

calculates the probabilities for a Pólya–Aeppli distribution with $\mu = 2.4$ and $\rho = 2.5$ for $x = 0$, 1, 2, 3, and 8, for example.

The rth factorial cumulant* is simple:

$$\kappa_{(r)} = \mu \cdot r!\rho^r,$$

from which the power cumulants follow:

$$\kappa_1 = \mu\rho, \qquad \kappa_2 = \mu\rho(1 + 2\rho),$$

$$\kappa_3 = \mu\rho(1 + 6\rho + 6\rho^2),$$

$$\kappa_4 = \mu\rho(1 + 14\rho + 36\rho^2 + 24\rho^3),$$

with a recurrence relation

$$\kappa_{r+1} = \rho\kappa_r + \rho(\rho+1)\frac{\partial\kappa_r}{\partial\rho}.$$

For data with sample mean $\overline{X}$ and second moment about the mean M_2, the moment estimators, denoted by $^\circ$ over the parameter, are

$$\mathring{\mu} = \frac{2\overline{X}^2}{M_2 - \overline{X}}, \qquad \mathring{\rho} = \frac{M_2 - \overline{X}}{2\overline{X}}$$

$$\left[\text{or } \mathring{\theta} = \frac{\overline{X}^3}{M_2^2 - \overline{X}^2}, \qquad \mathring{\omega} = \frac{M_2 - \overline{X}}{M_2 + \overline{X}} \right],$$

where, for example, $M_2 > \overline{X}$ for these to exist. The large-sample covariance matrix is obtainable by the usual asymptotic expansions*.

Maximum likelihood* estimators, denoted by ^, follow from the likelihood equations:

$$\hat{\mu}\hat{\rho} = \overline{X},$$

$$\sum F_x \Pi_x(\hat{\mu}, \hat{\rho}) = n\overline{X},$$

where the sample of size n is of frequencies $F_0, F_1, \ldots$ on $x = 0, 1, \ldots$ and

$$\Pi_x(\mu, \rho) = (x+1) \cdot \frac{p_{x+1}(\mu, \rho)}{p_x(\mu, \rho)},$$

the solution requiring iterative numerical procedures. Thus, if $\hat{\mu}$ is eliminated with $\hat{\mu} = \overline{X}/\hat{\rho}$, the single equation is

$$H(\hat{\rho}) = 0$$

where $H(\rho) \equiv \sum F_x \Pi_x - n\overline{X}$ with

$$H'(\rho) = \frac{2}{\rho} \sum F_x \Pi_x - \frac{1 + 2\rho}{\rho^2} \sum F_x \Pi_x (\Pi_{x+1} - \Pi_x).$$

For a starter value ρ_1, the second iterate is

$$\rho_2 = \rho_1 - H(\rho_1)/H'(\rho_1),$$

whence $\mu_2 = \overline{X}/\rho_2$, with further iteration. The large-sample covariance matrix may be found numerically at the same time.

It is also possible to use, for example, the proportion in the zero class (if this is large) and the sample mean; or various functions of this zero class and the moments, along the lines of Katti and Gurland's [7] treatment of the Poisson–Pascal distribution.

As with all the contagious distributions, the use of large-sample approximations for the standard errors and correlation of the estimators needs caution; see Shenton and Bowman [11] for more detail.

The Pólya–Aeppli, Pascal (or negative binomial), and Neyman (type A) distributions, all with two parameters, are competitors for data fitting. The Pascal can have at most one mode or one half-mode at the origin; the Pólya–Aeppli can have a half-mode, a mode, or a half-mode and a mode; while the Neyman can have any number of modes with or without a half-mode. (For $\mu = 4$, $\rho = 1$, the Pólya–Aeppli, unusually, has probability 0.1353 ... at $x = 0$, 1 and 2.) Fixing the mean and variance, as a general statement for these three distributions, the Pascal has the heavier right-hand tail (has most probability in it), with the Pólya–Aeppli and Neyman following in that order, although the differences are well out in the tail. For example, the choice of a mean and standard deviation both of 6 leads to the Pascal ($\kappa = 1.2$, $\rho = 5$), Pólya–Aeppli ($\mu = 2.4$, $\rho = 2.5$), and Neyman ($\mu = 1.2$, $\nu = 5$), where the Neyman (μ, ν) distribution has PGF $\exp\{-\mu + \mu\exp(-\nu + \nu z)\}$, with the means and variances of these three $\kappa\rho$ and $\kappa\rho(1 + \rho)$, $\mu\rho$ and $\mu\rho(1 + 2\rho)$, and $\mu\nu$ and $\mu\nu(1 + \nu)$, respectively. Then a cumulative sum of 0.9 is just obtained with the first 16 probabilities of all three distributions. For 0.999, the first 42, 38, and 35 terms are needed; for 0.999999, the first 81, 68, and 57, all respectively. Reducing the standard deviation reduces the differences: for a cumulative sum of 0.999999 with a mean of 6 and standard deviation of 3, the numbers of terms are 30, 29, and 28; with a mean of 3 and standard deviation of 6, they are 126, 95, and 68. These numerical results are illustrated in Fig. 1, in which the initial spike of the Neyman distribution is very apparent; the substantial tails beyond the fifteenth term are also clear.

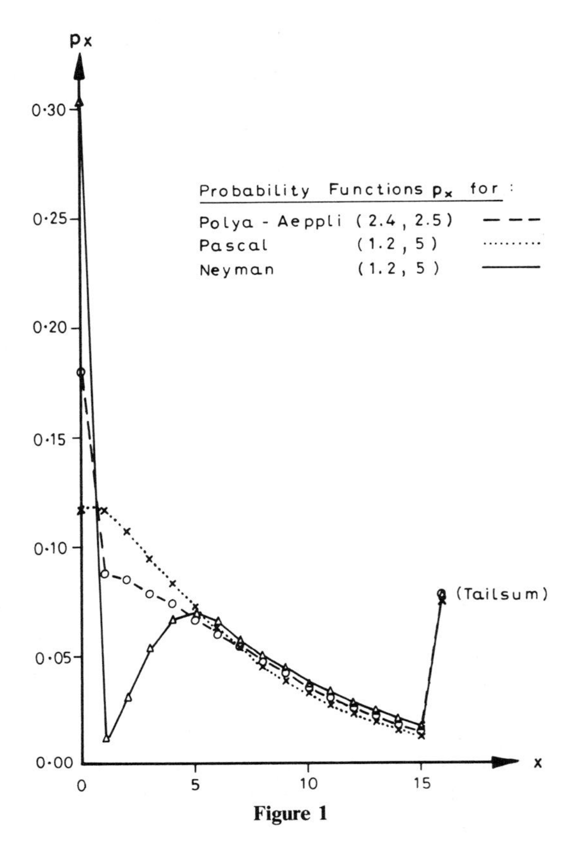

Figure 1

Many examples of data fitting are given in Beall [2], Beall and Rescia [3], Evans [5], and Katti and Rao [8]. (Katti and Rao's report includes no Pólya–Aeppli fits, but covers in detail the fitting of 35 observed distributions by the Neyman, Pascal, and other models.) Note that the Pólya–Aeppli parameters were confused in Beall [2] and that the fits are in fact good. For *Loxostege sticticalis L.* in his Table IX, treatment 1, the results are typical: see Table 1.

The high estimated correlations of the estimators reflect the instability of the estimates, in that compensating changes in these can

Table 1

Moment Estimates	M.L. Estimates
$\mathring{\mu} = 4.261$, s.e.$(\mathring{\mu}) = 0.968$	$\hat{\mu} = 3.940$, s.e.$(\hat{\mu}) = 0.790$
corr.$(\mathring{\mu}, \mathring{\rho}) = -0.966$	corr.$(\hat{\mu}, \hat{\rho}) = -0.955$
$\mathring{\rho} = 0.329$, s.e.$(\mathring{\rho}) = 0.076$	$\hat{\rho} = 0.355$, s.e.$(\hat{\rho}) = 0.075$

lead to fitted frequencies which differ very little in spite of substantial changes.

More details concerning many of the results above and additional material are to be found in Johnson and Kotz [6] and Douglas [4].

References

[1] Aeppli, A. (1924). Thesis, University of Zurich.

[2] Beall, G. (1940). *Ecology*, **21**, 460–474.

[3] Beall, G. and Rescia, R. R. (1953). *Biometrics*, **9**, 344–386.

[4] Douglas, J. B. (1980). *Analysis with Standard Contagious Distributions*. International Co-operative Publishing House, Fairland, Md.

[5] Evans, D. A. (1953). *Biometrika*, **40**, 186–211.

[6] Johnson, N. L. and Kotz, S. (1969). *Distributions in Statistics. Discrete Distributions*. Wiley, New York.

[7] Katti, S. K. and Gurland, J. (1961). *Biometrics*, **17**, 527–538.

[8] Katti, S. K. and Rao, A. V. (1965). The Log-Zero-Poisson Distribution. *Florida State Univ. Statist. Rep. M106*, Tallahassee, Fla.

[9] Phillipson, C. (1960). *Skand. Aktuarietidskr.*, **43**, 136–162.

[10] Pólya, G. (1930–31). *Ann. Inst. Henri Poincaré*, **1**, 117–161.

[11] Shenton, L. R. and Bowman, K. O. (1977). *Maximum Likelihood Estimation in Small Samples*. Charles Griffin, London.

(CONTAGIOUS DISTRIBUTIONS
NEGATIVE BINOMIAL DISTRIBUTIONS
NEYMAN'S TYPE A, B, C DISTRIBUTIONS)

J. B. DOUGLAS

PÓLYA DISTRIBUTION, MULTIVARIATE

The Pólya distribution [12] is well known as a model for contagion in statistics literature. Its multivariate generalization, known as the multivariate Pólya distribution (MPD), is discussed in Steyn [13]. Further modifications of MPD and related distributions, such as the multivariate inverse Pólya and multivariate quasi-Pólya distributions, are discussed in Janardan [1] and Janardan and Patil [2, 3]. The MPD, as noted by Mosimann [8], also arises in the analysis of pollen data.

DEFINITION AND STRUCTURE

The random vector $\mathbf{X}' = (X_1, X_2, \ldots, X_s)$ has the s-variate MPD with parameters $\mathbf{a}' = (a_1, a_2, \ldots, a_s)$, N, n, and c if its joint probability density function (PDF) is

$$f_{X_1, X_2, \ldots, X_s}(x_1, x_2, \ldots, x_s) = \frac{n! \prod_{j=0}^{s} a_j^{(x_j, c)}}{\left(\prod_{j=0}^{s} x_j!\right) N^{(n,c)}}, \qquad (1)$$

$x_j = 0, 1, 2, \ldots, n$ for $j = 0, 1, 2, \ldots, n$, $\sum_{j=0}^{s} x_j = n$, $n = 1, 2, \ldots$; $a_j = 0, 1, 2, \ldots, N$, for $j = 0, 1, 2, \ldots, s$, where $\sum_{j=0}^{s} a_j = N$, $a_0 = N - \sum_{i=1}^{s} a_i$, and

$$a^{(x,c)} = a(a+c)(a+2c) \cdots [a + (x-1)c].$$

For $c \neq 0$, writing $p_j = a_j/N$, $j = 0, 1, 2 \ldots, s$, and $r = c/N$, the PDF given by (1) reduces to

$$f_{X_1, X_2, \ldots, X_s}(x_1, x_2, \ldots, x_s) = \prod_{j=0}^{s} \binom{-p_j/r}{x_j} \bigg/ \binom{-1/r}{n},$$

$$0 < p_j < 1, \quad 0 < r < \infty, \quad \sum_{j=0}^{s} p_j = 1. \qquad (2)$$

The PDF (2) can be expressed in an alternative form, given by

$$f_{X_1,X_2,\ldots,X_s}(x_1,x_2,\ldots,x_s) = \frac{n!}{(\prod_{j=0}^{s} x_j!)} \frac{\Gamma(m)}{\Gamma(m+n)} \prod_{j=0}^{s} \frac{\Gamma(m_j+x_j)}{\Gamma(m_j)}, \qquad (3)$$

where $m_j = a_j/c$, $m = \sum_{j=0}^{s} m_j = N/c$, and $\Gamma(\cdot)$ is the gamma function.

GENESIS

1. **(True Contagion).** Suppose that an urn contains N balls of $(s+1)$ different colors, a_i being of the ith color, $i = 1, 2, \ldots, s$ and $a_0 = N - \sum_{i=1}^{s} a_i$ of the $(s+1)$th color. Suppose that n balls are drawn at random one by one with replacement, such that at each replacement c new balls of the same color are added to the urn. If X_i equals the number of balls of the ith color in the sample, $i = 1, 2, \ldots, s$, then the random vector $(X_1, X_2, \ldots, X_s)$ has the s-variate MPD with parameters $a_1, a_2, \ldots, a_s$, N, n, and c.
2. **(Apparent Contagion).** Consider a family of multinomial distributions with parameters n and $\mathbf{p}$, where $\mathbf{p}$ itself is a random vector having the Dirichlet distribution*. Then the resulting mixture distribution* is MPD (Mosimann [8]).
3. **(Random Walk Model).** Consider an s-dimensional random walk* with the following characteristics.
   (a) The s-dimensional space consists of points with nonnegative intervalued coordinates.
   (b) The origin is the starting point of the random walk.
   (c) Each step is of unit distance.
   (d) The probability of the first jump along the ith axis $(i = 1, 2, \ldots, s)$ is equal to p_i, $0 < p_i < 1$, $\sum_1^s p_i = 1$.
   (e) The transition probability from an arbitrary point A to an adjacent point B equals $(p_i + cx_i)/(1 + c\sum_{j=1}^{s} x_l)$, where x_i $(i = 1, 2, \ldots, s)$ is the ith coordinate of A, and c is a scale parameter.
   (f) $\sum_{i=1}^{s} x_i = n$ (fixed) is a stopping boundary.

   Then the transition probabilities to a boundary point are the multivariate Pólya probabilities (see Lumel'skii [7]).

The probability generating function* and the moments of the MPD are discussed in Janardan and Patil [2] and Patil et al. [10, 11].

Property 1. If $(X_1, X_2, \ldots, X_s)$ has the MPD with parameters $a_1, a_2, \ldots, a_s$, N, n, and c, then:

(a) The subset $(X_1, X_2, \ldots, X_k)$, $1 \leqslant k \leqslant s$ has the MPD with parameters $a_1, a_2, \ldots, a_k$, N, n, and c.

(b) $(X_1 + X_2 + \cdots + X_s)$ has the Pólya distribution with parameters $\sum_{i=1}^{s} a_i$, N, n, and c.

(c) $(X_1, X_2, \ldots, X_k \mid X_{k+1}, \ldots, X_s)$ has the MPD with parameters $a_1, a_2, \ldots, a_k$, M, m, and c, where $M = N - \sum_{i=k+1}^{s} a_i$ and $m = n - \sum_{i=k+1}^{s} x_i$.

For proofs of the foregoing properties and further details, see Janardan and Patil [2].

Property 2. If $(X_1, X_2, \ldots, X_s)$ has the multinomial distribution* with parameters n and $\mathbf{p}' = (p_1, p_2, \ldots, p_s)$, where $\mathbf{p}$ is a random vector having the Dirichlet distribution with parameters m_j, $j = 0, 1, 2, \ldots, s$, then the resulting mixture distribution is the MPD with PDF given by (3). The derivation of MPD in this form and its application in the analysis of pollen data are discussed in Mosimann [8].

Property 3. The Bose–Einstein distribution is a special case of the MPD having the PDF given by (2) when $p_i = 1/(1+s)$, $i = 1, 2, \ldots, s$, and $r = 1/(1+s)$.

Property 4. The multinomial distribution with parameters n and $\mathbf{p}' = (p_1, p_2, \ldots, p_s)$ is a special case of MPD given by (2) when $c = 0$.

Property 5. The multivariate hypergeometric distribution* with parameters $a_1, a_2,$

$\ldots, a_s$, N, and n is a special case of the MPD given by (1) when $c = -1$.

Property 6. The multivariate negative hypergeometric distribution with parameters $a_1, a_2, \ldots, a_s$, N, and n is a special case of the MPD given by (1) when $c = 1$.

Property 7. The MPD with parameters $p_1, p_2, \ldots, p_s$, and n, given by (2), tends to the multiple Poisson distribution with parameters λ_i, $i = 1, 2, \ldots, s$, as $n \to \infty$, $p_i \to 0$ such that $np_i \to \lambda_i$.

For the details regarding the multinomial, multivariate hypergeometric, multivariate negative hypergeometric and multiple Poisson distributions, see Patil et al. [10, 11].

ESTIMATION OF PARAMETERS

Let $\mathbf{X}' = (X_1, X_2, \ldots, X_s)$ have the s-variate MPD with parameters $a_1, a_2, \ldots, a_s$, N, n, and c. If $c > 0$ and n is known, then:

1. For known N the estimators of a_i are given by $\hat{a}_i = N\overline{X}_i/n$, $i = 1, 2, \ldots, s$, where $\overline{X}_i$ is the sample mean of x_i, $i = 1, 2, \ldots, s$.
2. If N is unknown, then the estimators of a_i are given by

$$\hat{a}_i = \overline{X}_i \left[\frac{c(n - \hat{R})}{n(\hat{R} - 1)} \right], \qquad i = 1, 2, \ldots, s,$$

where $\hat{R}$ is an estimator of

$$R = (N + nc)/(N + c).$$

The methods for estimation of R are discussed in Mosimann [8] and Janardan and Patil [2].

OTHER RELATED DISTRIBUTIONS

Considering a variation in the urn model described above as a genesis of MPD, researchers have obtained either the generalization or the modification of MPD. These models are described here for completeness.

MULTIVARIATE INVERSE PÓLYA DISTRIBUTION (MIPD)

A random vector $\mathbf{X}' = (X_1, X_2, \ldots, X_s)$ is said to have the MIPD with parameters $a_1, a_2, \ldots, a_s$, N, k, and c if its joint PDF is

$$f_{X_1, X_2, \ldots, X_s}(x_1, x_2, \ldots, x_s) = \frac{(k + x - 1)!\, a_0^{(k,c)} \prod_{i=1}^{s} a_i^{(x_i, c)}}{(k-1)! \left(\prod_{i=1}^{s} x_i!\right) N^{(k+x,c)}}, \qquad (4)$$

$x_i = 0, 1, 2, \ldots$, $i = 1, 2, \ldots, s$, $x = \sum_1^s x_i$, $a_i = 1, 2, \ldots, N$, for $i = 1, 2, \ldots, s$, $a_0 = n - \sum_{i=1}^{s} a_i$, $k = 1, 2, \ldots$, and $a^{(k,c)} = a(a + c) \cdots a + (k-1)c$.

GENESIS. Consider a finite population consisting of N individuals of which a_i have characteristic C_i, $i = 1, 2, \ldots, s$, and $a_0 = N - \sum_{i=1}^{s} a_i$ have none of the characteristics C_i. The individuals are selected randomly one by one with replacement, until k individuals having none of the characteristics C_i [i.e., not-$(C_1, C_2, \ldots, C_s)$] are observed. Further, we note the characteristics of the individual selected and c additional individuals of the same characteristics before the selection. If X_i equals the number of individuals of character C_i $(i = 1, 2, \ldots, n)$, then the distribution of $(X_1, X_2, \ldots, X_s)$ is MIPD with PDF (4).

The PDF of MIPD could be expressed in different forms as in the case of MPD. See Janardan and Patil [3] for these different forms and the properties of MIPD, which are similar to those of MPD. The estimation of parameters is discussed in Janardan and Patil [3] and Mosimann [9].

MULTIVARIATE MODIFIED PÓLYA DISTRIBUTION (MMPD)

Suppose that an urn contains M balls, a_i of the ith color, $i = 1, 2, \ldots, s$ $(\sum_{i=1}^{s} a_i = m)$ and a_0 white balls. A ball is drawn and if it is colored, it is replaced with additional $c_1, c_2, \ldots, c_s$ balls of s colors, and if it is white, it is replaced with $d = \sum_{i=1}^{s} c_i$ white balls. This is repeated n times. Let X_i denote the number of balls of the ith color and let X_0

denote the number of white balls such that $\sum_{i=1}^{s} x_i = n$. Then the random vector $\mathbf{X}' = (X_1, \ldots, X_s)$ has the MMPD with parameters $a_1, \ldots, a_s$, d, n, and N. Its joint PDF is given by

$$f_{X_1,X_2,X_s}(x_1, x_2, \ldots, x_s)$$

$$= \frac{n!}{\prod_{j=0}^{s} x_j!} \frac{a_0^{(x_0,d)} M^{(x,d)}}{N^{(n,d)}} \prod_{i=1}^{s} \left(\frac{a_i}{M} \right)^{x_i},$$

$x_i = 0, 1, 2, \ldots, n$ for $i = 1, 2, \ldots, s$, $x = \sum_{i=1}^{s} x_i$, and $N = a_0 + M$. The properties of MMPD are discussed in Janardan and Patil [4].

MULTIVARIATE QUASI-PÓLYA DISTRIBUTION (MQPD)

The MQPD arises as a multivariate generalization of the quasi-Pólya distribution (see Janardan [1]).

Suppose that there are two boxes, say B_1 and B_2, with the following characteristics. Box B_1 has s ($\geqslant 1$) compartments. The ith compartment ($i = 1, 2, \ldots, s$) has a_i white balls in it. Box B_2 has only one compartment containing N balls of which a_0 are white and a_i are of color C_i, $i = 1, 2, \ldots, s$. Let m be a positive integer and c be an integer. Then n balls are drawn from B_2 as described below in steps S_1 to S_4.

S_1. Select s integers r_i, $i = 1, 2, \ldots, s$, at random such that $0 \leqslant r_i \leqslant n$ and $0 \leqslant r = \sum_1^s r_i \leqslant n$.

S_2. Add rm colored balls to B_1 such that the ith compartment receives $r_i m$ balls of color C_i, $i = 1, 2, \ldots, s$. Also, add $r_0 m$ white balls and $r_i m$ balls of color C_i, $i = 1, 2, \ldots, s$, to B_2.

S_3. Draw a set of s balls consisting of one ball from each compartment of B_1. If this set has all white balls, proceed to step S_4; otherwise, discontinue this experiment.

S_4. Draw n balls from B_2 one by one and with replacement. For each draw, note the color of the ball and add c additional balls of the same color before the next draw. Let X_i equal the number of balls of color C_i, $i = 1, 2, \ldots, s$. If $X_i = ri$, $i = 1, 2, \ldots, s$, the event is a "success"; otherwise, it is a "failure." Then the distribution of $(X_1, X_2, \ldots, X_s)$ is called the MQPD having the joint PDF

$$f_{X_1,X_2,\ldots,X_s}(x_1, x_2, \ldots, x_s)$$

$$= \frac{n! \prod_{i=1}^{s} \{ a_i (a_i + x_i m)^{-1} \}}{\prod_{i=1}^{s} x_i!} \times \prod_{j=0}^{s} \frac{J_{x_i}(a_i, m, c)}{J_n(n, m, c)},$$

where $J_{x_i}(a_i, m, c) = (a_i + x_i m)^{(x_i, c)}$ and $b^{(0,c)} = 1$.

Special Cases of MQPD

1. If $m = 0$, the PDF of MQPD reduces to that of MPD.
2. If $c = 0$, the MQPD is called the quasi-multinomial distribution.
3. If $c = -1$, the MQPD is called the multivariate quasi-hypergeometric distribution.

For further details regarding the MQPD and the related multivariate quasi-inverse Pólya distribution, see Janardan [1].

POLYTOMOUS MULTIVARIATE PÓLYA DISTRIBUTION (PMPD)

A generalization of MPD that arises in the analysis of contingency tables was obtained by Kriz [6] by considering the simultaneous drawings from m urns.

Consider m urns, each containing N balls, with the following characteristics.

1. The ith urn ($i = 1, 2, \ldots, m$) has a_{ij} balls of jth color $\{C_{ij}, j = 1, 2, \ldots, k_i\}$ such that $\sum_{j=1}^{k_i} a_{ij} = N$.
2. Let the m sets of colors, $\{C_{ij}, j = 1, 2, \ldots, k_i\}$ representing the set for the ith urn ($i = 1, 2, \ldots, m$), be disjoint. Thus the total number of different colors is $\sum_{i=1}^{m} k_i$.
3. Suppose that n balls are drawn one by one from each urn with replacement such that c balls of the same color as the

ball drawn are added before the next draw. Then the joint distribution of $\prod_{i=1}^{m} k_i$ random variables $X_{l_1, l_2, \ldots, l_m}$ ($l_i = 1, 2, \ldots, k_i$, $i = 1, 2, \ldots, m$) which represent the number of sets of m balls (from the total of n sets), one from each urn, has the PMPD. We refer the reader to Johnson and Kotz [5] for the joint PF of the PMPD and related discussion.

References

[1] Janardan, K. G. (1975). *Gujarat Statist. Rev.*, **3**, 17–32.

[2] Janardan, K. G. and Patil, G. P. (1970). In *Random Counts in Scientific Work*, Vol. 3, G. P. Patil, ed. Pennsylvania State University Press, University Park, PA, pp. 143–161.

[3] Janardan, K. G. and Patil, G. P. (1972). *Studi di Probabilità Statistica e Ricerca Operative in Omore de G. Pompilj*. Oderisi, Gubbie, pp. 1–15.

[4] Janardan, K. G. and Patil, G. P. (1974). *Ann. Inst. Statist. Math.*, **26**, 271–276.

[5] Johnson, N. L. and Kotz, S. (1977). *Urn Models and Their Applications*. Wiley, New York, pp. 197–198.

[6] Kriz, J. (1972). *Statist. Hefte*, **13**, 211–224.

[7] Lumel'skii, Ja. P. (1973). *Sov. Math. Dokl.*, **14**, 628–631.

[8] Mosimann, J. E. (1962). *Biometrika*, **49**, 65–82.

[9] Mosimann, J. E. (1963). *Biometrika*, **50**, 47–53.

[10] Patil, G. P., Boswell, M. T., Joshi, S. W., and Ratnaparkhi, M. V. (1984). *Dictionary and Classified Bibliography of Statistical Distributions in Scientific Work*, Vol. 1: *Discrete Models*. International Co-operative Publishing House, Fairland, MD, p. 458.

[11] Patil, G. P., Boswell, M. T., Ratnaparkhi, M. V., and Roux, J. J. J. (1984). *Dictionary and Bibliography of Statistical Distributions in Scientific Work*, Vol. 3: *Multivariate Models*. International Co-operative Publishing House, Fairland, MD, p. 431.

[12] Pólya, G. (1931). *Ann. Inst. Henri Poincaré*. **1**, 117–161.

[13] Steyn, H. S. (1951). *Indag. Math.*, **17**, 588–595.

(CONTAGIOUS DISTRIBUTIONS
LAGRANGE DISTRIBUTIONS
OCCUPANCY PROBLEMS
URN MODELS)

G. P. PATIL
M. V. RATNAPARKHI

PÓLYA–EGGENBERGER DISTRIBUTION *See* NEGATIVE BINOMIAL DISTRIBUTIONS; OCCUPANCY PROBLEMS

PÓLYA–LUNDBERG PROCESS

The Pólya–Lundberg process, frequently abbreviated as *Pólya process*, is a stochastic process* with various applications in nuclear physics and insurance mathematics. Depending on the context in which it is used, the Pólya–Lundberg process can be characterized as a pure birth Markov process*, a mixed Poisson process*, or a limit of contagion-type urn models*. Due to its Markovian character, the first approach is especially appropriate for the modeling and description of physical processes such as electron–photon cascades (see Arley [2]; see also Bharucha-Reid [4] and the references given therein). As a weighted Poisson process, the Pólya–Lundberg process plays an important role in non-life insurance, as was shown by Lundberg [8], who fitted the Pólya process to sickness and accident statistics. Here the urn model approach provides a simple interpretation of the contagion property of the Pólya–Lundberg process (see also Beard et al. [3]; for a more advanced exposition of the corresponding urn process, see also Hill et al. [6]). Recently, some connections between the Pólya–Lundberg process and records* have been pointed out; for instance, the Pólya process can be considered as a counting process of record values coming from independent Pareto*-distributed random variables (see Pfeifer [10]). Conversely, the study of record values paced by a Pólya–Lundberg process gives an interesting insight into the probabilistic behavior of this process from a very different point of view (see Orsingher [9]). Lately, the Pólya–Lundberg process has been employed to illuminate structural properties of infinitely divisible* stochastic point processes with respect to the representation of the probability generating functional (see Waymire and Gupta [13]).

In the Markovian setting, a Pólya–Lundberg process $\{N(t), t \geqslant 0\}$ is a nonhomogeneous birth process (*see* BIRTH-AND-DEATH

PROCESSES) with birth rates

$$\lambda_n(t) = \lambda \frac{1 + \alpha n}{1 + \alpha \lambda t},$$
$$t \geqslant 0, \quad n = 0, 1, 2, \ldots, \quad (1)$$

where $\lambda, \alpha > 0$ are scale and shape parameters, respectively; that is, the probability of a new birth in the time interval $(t, t + h)$ is given by $\lambda_n(t)h + o(h)$, while the probability of two or more births in this interval is $o(h)$ for $h \to 0$. Here $o(h)$ is a remainder term with $o(h)/h \to 0$ for $h \to 0$. As solutions of Kolmogorov's backward differential equations, the marginal distributions of the process are obtained, given by the Pólya distributions

$$\Pr(N(t) = n) = \frac{(\lambda t)^n}{n!}(1 + \alpha\lambda t)^{-(n+1/\alpha)} \times \prod_{k=1}^{n-1}(1 + \alpha k) \quad (2)$$

for $n = 0, 1, 2, \ldots$ and $t \geqslant 0$ with mean and variance

$$E(N(t)) = \lambda t, \qquad \text{var}(N(t)) = \lambda t(1 + \alpha\lambda t), \quad (3)$$

which also illustrates the meaning of the parameters λ and α. As suggested by the formulas above, the Pólya–Lundberg process approaches a Poisson process* if α approaches zero.

As a birth process, the Pólya–Lundberg process can also equivalently be described by the sequence T_n, $n = 1, 2, \ldots,$ of birth occurrence times which form a Markov chain with transition probabilities

$$\Pr(T_{n+1} > s \mid T_n = t) = \left(\frac{1 + \alpha\lambda t}{1 + \alpha\lambda s}\right)^{n+1/\alpha},$$
$$0 \leqslant t \leqslant s \quad (4)$$

(see Albrecht [1] and Pfeifer [11]). As a special property of the Pólya–Lundberg process, the sequence $S_n = n/(1 + \alpha\lambda T_n)$ forms a mean-bounded submartingale (*see* MARTINGALES). This provides a simple proof of the fact that for the time averages

$$\frac{N(t)}{t} \to \Lambda, \qquad \frac{T_n}{n} \to \frac{1}{\Lambda}$$
$$\text{almost certainly} \quad (t, n \to \infty) \quad (5)$$

(*see* CONVERGENCE OF SEQUENCES OF RANDOM VARIABLES and Pfeifer [12]). Here Λ is a random variable following a gamma distribution* with mean λ and variance $\alpha\lambda^2$.

In the setting of mixed Poisson processes, (5) gives a limit representation of the mixing random variable; that is, a Pólya–Lundberg process can be considered as a weighted Poisson process whose parameter is chosen at random according to the distribution of Λ. In risk theory*, the distribution function of Λ is also called structure function or unconditioned risk distribution. Characteristically, the Pólya–Lundberg process is the only mixed Poisson process whose birth rates for fixed time t are a linear function of n, as was proven by Lundberg [8].

Finally, the Pólya–Lundberg process can be represented as a limit of urn processes of contagion type introduced by Eggenberger and Pólya [5] (*see also* URN MODELS and Johnson and Kotz [7]). Here a certain number of white and black balls is collected in an urn where p denotes the proportion of white balls and $q = 1 - p$ the proportion of black balls. When a ball is drawn at random, it is replaced along with a fixed proportion β (of the initial total number of balls) of the same color, which causes the contagious effect. If N_m denotes the number of white balls drawn in m trials, the probability distribution of N_m is given by the Pólya–Eggenberger distribution

$$\Pr(N_m = k) = \binom{m}{k}\frac{p^{(k,\beta)}q^{(m-k,\beta)}}{1^{(m,\beta)}} \quad (6)$$

for $k = 0, 1, \ldots, m$, where $p^{(k,\beta)}$ denotes $p(p + \beta) \cdots (p + (k - 1)\beta)$, etc., with

$$E(N_m) = mp, \qquad \text{var}(N_m) = mpq\frac{1 + m\beta}{1 + \beta}. \quad (7)$$

Now if for $m \to \infty$, the portions $p = p_m$ and $\beta = \beta_m$ are chosen such that

$$mp_m \to \lambda t, \qquad m\beta_m \to \alpha\lambda t, \quad (8)$$

then N_m tends to $N(t)$ in distribution [i.e., the probabilities (6) approach the Pólya probabilities (2)]; similarly for the moments

(7). A vivid interpretation of this could also be given as follows. Suppose that for a fixed time $t > 0$ a series of m drawings at times $h, 2h, \ldots, mh$ is made, where $h = t/m$ and $p_m = \lambda h$, $\beta_m = \alpha p_m$. Further, let $N^*(s)$, $s \geqslant 0$, denote the number of white balls drawn up to time s. Then $N^*(t)$ approximately behaves like a Pólya–Lundberg process with parameters λ and α at time t in that

$$\Pr(N^*(t+h) = n+1 \mid N^*(t) = n) = \frac{p_m + n\beta_m}{1 + m\beta_m} = \lambda_n(t)h \tag{9}$$

with birth rates $\lambda_n(t)$ given by (1).

References

[1] Albrecht, P. (1982). Zur statistischen Analyse des gemischten Poissonprozesses, gestützt auf Schadeneintrittszeitpunkte. *Blä. Dtsch. Ges. Versich.-Math.*, **15**, 249–257.

[2] Arley, N. (1949). *On the Theory of Stochastic Processes and Their Applications to the Theory of Cosmic Radiation*. Wiley, New York.

[3] Beard, R. E., Pentikäinen, T., and Pesonen, E. (1969). *Risk Theory*. Methuen, London.

[4] Bharucha-Reid, A. T. (1960). *Elements of the Theory of Markov Processes and Their Applications*. McGraw-Hill, New York.

[5] Eggenberger, F. and Pólya, G. (1923). Über die Statistik verketteter Vorgänge. *Zeit. angew. Math. Mech.*, **1**, 279–289.

[6] Hill, B. M., Lane, D., and Sudderth, W. (1980). A strong law for some generalized urn processes. *Ann. Prob.*, **8**, 214–226.

[7] Johnson, N. L. and Kotz, S. (1977). *Urn Models and Their Applications*. Wiley, New York.

[8] Lundberg, O. (1940). *On Random Processes and Their Applications to Sickness and Accident Statistics*. Almqvist & Wiksell, Uppsala, Sweden (2nd ed., 1964).

[9] Orsingher, E. (1980). Extreme values of a sequence of random variables associated with the linear birth and Pólya process. *Biom. Praxim.*, **20**, 47–58.

[10] Pfeifer, D. (1982). The structure of elementary pure birth processes. *J. Appl. Prob.*, **19**, 664–667.

[11] Pfeifer, D. (1982). An alternative proof of a limit theorem for the Pólya–Lundberg process. *Scand. Actuarial J.*, **15**, 176–178.

[12] Pfeifer, D. (1983). A note on the occurrence times of a Pólya–Lundberg process. *Adv. Appl. Prob.*, **15**, 886.

[13] Waymire, E. and Gupta, V. K. (1983). An analysis of the Pólya point process. *Adv. Appl. Prob.*, **15**, 39–53.

(BIRTH-AND-DEATH PROCESSES
CONTAGIOUS DISTRIBUTIONS
MIXED POISSON PROCESSES
POISSON PROCESS
RISK THEORY
STOCHASTIC PROCESSES
URN MODELS)

DIETMAR PFEIFER

PÓLYA'S THEOREM

If a continuous, even and concave function $\phi(t)$, $t \in R^1$, is such that $\phi(t) \geqslant 0$, $\phi(0) = 1$, and $\phi(t) \to 0$ as $t \to \infty$, then $\phi(t)$ is the characteristic function of some distribution.

This theorem provides a convenient procedure for constructing characteristic functions*. For example, using this theorem, one can assert that

$$\phi_1(t) = e^{-|t|} \quad \text{and} \quad \phi_2(t) = \begin{cases} 1 - |t|, & |t| \leqslant 1 \\ 0, & |t| > 1 \end{cases}$$

are both characteristic functions. For more details, see, for example, Shiryaev [1].

Reference

[1] Shiryaev, A. N. (1980). *Probability*. Nauka, Moscow (in Russian). [English translation (1984), Springer-Verlag, New York.]

(CHARACTERISTIC FUNCTIONS)

PÓLYA TYPE 2 FREQUENCY (PF_2) DISTRIBUTIONS

The wide class of Pólya-type distributions was first studied by Schoenberg [17]. These have been applied extensively in several domains of mathematics, statistics, economics, and mechanics. These functions arise naturally in developing procedures for inverting, by differential polynomial operators, integral transformations defined in terms of convolution kernels. Pólya-type distributions are fundamental in characterizing best statistical procedures for decision problems [7, 8, 13]. These are encountered in clarifying the structure of stochastic processes* with continuous path functions and in the stability of certain models in mathematical economics.

A nonnegative measurable function $p(x)$ defined for all real x is a PF_2 function if the determinant

$$\begin{vmatrix} p(x_1 - y_1) & p(x_1 - y_2) \\ p(x_2 - y_1) & p(x_2 - y_2) \end{vmatrix} \geqslant 0$$

$$\text{for all } -\infty < x_1 < x_2 < \infty$$

and $-\infty < y_1 < y_2 < \infty$ and $p(x) \neq 0$ for at least two distinct values of x (see refs. 2, 3, 5, 6, 10, 14, 15, and 17). Two alternative definitions are (1) $p(x)$ is PF_2 if $\log p(x)$ is concave on $-\infty < x < \infty$, and (2) $p(x)$ is PF_2 if, for fixed $\Delta > 0$, $p(x + \Delta)/p(x)$ is a decreasing function of x in the interval (a, b), where

$$a = \inf_{p(y)>0} \{y\} \qquad \text{and} \qquad b = \sup_{p(y)>0} \{y\}$$

A PF_2 is not necessarily a probability density function (PDF), in the sense that $\int_{-\infty}^{\infty} p(x)\,dx$ need not be 1 or even finite [2, 3].

PF_2 arises in statistics as follows: Let the continuous random variable $X > 0$ denote the life length of a unit, which is subject to failure. Suppose that X has the PDF $f(x)$ and CDF $F(x)$. Then the reliability or survival function [3] of the unit at time x is defined as $S(x) = 1 - F(x) = P(X \geqslant x)$. The failure rate (FR) of the PDF $f(x)$ is defined by the ratio $\lambda(x) = f(x)/S(x)$. This is also called the hazard rate function (HRF) (*see* HAZARD RATE AND OTHER CLASSIFICATIONS OF DISTRIBUTIONS).

A distribution is IFR (DFR) if, and only if (see refs. 2, 10, and 14), $\{F(t + x) - F(x)\}/S(x)$ is increasing (decreasing) in x for $t > 0$, $x > 0$ such that $S(x) > 0$. Equivalently, a CDF $F(x)$ is IFR (DFR) if, and only if, the ratio $S(x + t)/S(x)$ is decreasing in $-\infty < x < \infty$ for each $t > 0$. A probabilistic interpretation is that if a unit has an IFR distribution, it is more likely to survive a given amount of time t at an earlier age than at an older age.

A density $f(x)$ for which the ratio $f(x)/[F(x + t) - F(x)]$ is nondecreasing in x, whenever the denominator is nonzero for all real values of t, is a Pólya frequency density of order 2. This and other properties of PF_2 are summarized below.

1. A density $f(x)$ is PF_2 if, and only if, for all t the ratio $[F(x + t) - F(x)]/f(x)$ is decreasing in x.
2. If $f(x)$ is PF_2, then $f(x)$ is unimodal.
3. If a CDF $F(x)$ is IFR, then its survival function $S(x)$ is PF_2, and conversely.
4. If $f(x)$ is a PF_2 density of a positive random variable, then its CDF $F(x)$ is IFR. The converse is not true; see refs. 10 and 13.
5. If $f(x)$ is a PF_2 density of a positive random variable X, as in (4), for which $\log f(x)$ is convex on $(0, \infty)$, then the corresponding CDF $F(x)$ is DFR.

As an example, consider the truncated normal distribution, with only positive values for the underlying random variable, with the PDF $f(x) = (a\sigma\sqrt{2\pi})^{-1}\exp\{-(x - \mu)^2/(2\sigma^2)\}$, $0 < x < \infty$, where a is the normalizing constant. It is easily verified that $\log f(x)$ is concave on $(0, \infty)$, showing that $f(x)$ is a PF_2 density and the corresponding CDF $F(x)$ is IFR [3].

The following densities are PF_2:

(a) exponential*: $f(x) = \theta e^{-\theta x}$, $x \geqslant 0$;

(b) gamma*: $f(x) = \{\theta^\alpha/\Gamma(\alpha)\}e^{-\theta x}x^{\alpha-1}$, $x \geqslant 0$, $\alpha > 1$;

(c) Weibull*: $f(x) = \alpha\theta(\theta x)^{\alpha-1}e^{-(\theta x)^\alpha}$, $x \geqslant 0$, $\alpha > 1$;

(d) normal*:

$$f(x) = \left(\sigma\sqrt{2\pi}\right)^{-1}\exp\{-(x - \mu)^2/(2\sigma^2)\},$$
$$-\infty < x < \infty;$$

(e) Laplace* (or double exponential):

$$f(x) = \tfrac{1}{2}e^{-|x|}, \; -\infty < x < \infty;$$

and

(f) the truncated normal distribution above.

On the other hand, the Weibull distribution with $0 < \alpha < 1$ and the Cauchy distribution* are not PF_2.

The class of PF_2 functions enjoys many useful properties, such as variation diminishing, closure, convolution, unimodality* and some moment properties [7, 8, 14]. When the underlying density is PF_2 (more generally, a monotone likelihood ratio* density), statistical decision procedures are particularly simple; see refs. 4 and 12. The concept of PF_2 functions where the argument is the difference of two variables is extended by Karlin and Rubin [13] to functions of two separate variables. A function $f(x, y)$ of two real variables ranging over linearly ordered one-dimensional sets X and Y is said to be totally positive of order k (TP_k) if for all $x_1 < x_2 < \cdots < x_m$, $y_1 < y_2 < \cdots < y_m$ ($x_i \in X$, $y_i \in Y$) and all $1 < m < k$, the determinant

$$f\begin{pmatrix} x_1, x_2, \dots, x_m \\ y_1, y_2, \dots, y_m \end{pmatrix} = |f(x_i, y_j)| \geqslant 0$$

(*see* TOTAL POSITIVITY).

Moment inequalities are derived by Karlin et al. [14] for Pólya frequency functions of various orders. These functions on the positive axis are characterized in terms of inequalities on the moments. Several results are also developed describing the rate of decrease of Pólya frequency functions. For example, it is proved that if $f = 0$ for $x < 0$ and if f is a PF_2 and $\not\equiv (1/\mu_1)e^{-t/\mu_1}$, there exists t_0 such that for $t < t_0$, then $0 < f(t) < (1/\mu_1)e^{-t/\mu_1}$. The key device in these studies is to compare the PF_2 density with an approximately selected exponential density. This device exploits the general variation diminishing property (VDP) of Pólya frequency functions.

The VDP is an important feature of PF_k and TP_k functions. This is defined below for the sake of completeness. Suppose that $V(g)$ denotes the number of variations of sign of a function $g(x)$ as x traverses the real line from left to right. Let $h(x)$ be given by the absolutely convergent integral

$$h(x) = \int f(x, w) g(w)\, d\sigma(w),$$

where $f(x, w)$ is TP_k, $\sigma(w)$ is a nonnegative σ-finite measure, and $V(g) \leqslant k - 1$. Then $V(h) \leqslant V(g)$. Moreover, if $V(h) = V(g)$, then h and g have the same arrangement of signs.

Many of the structural properties of PF_k and TP_k functions are based on the following identity; see Pólya and Szegö [15, p. 48, Prob. 68]:

If $r(x, w) = \int p(x, t) q(t, w)\, d\sigma(t)$, then

$$r\begin{pmatrix} x_1, x_2, \dots, x_m \\ y_1, y_2, \dots, y_m \end{pmatrix}$$

$$= \int\int_{t_1 < t_2 < \cdots < t_m} \cdots \int p\begin{pmatrix} x_1, x_2, \dots, x_m \\ t_1, t_2, \dots, t_m \end{pmatrix}$$

$$\times q\begin{pmatrix} t_1, t_2, \dots, t_m \\ w_1, w_2, \dots, w_m \end{pmatrix} d\sigma(t_1) \cdots d\sigma(t_m).$$

More recently, Alam [1] and Efron [5] have obtained some interesting results concerning PF_2 functions.

References

[1] Alam, K. (1968). *Ann. Math. Statist.*, **39**, 1759–1761.

[2] Barlow, R. E. and Proschan, F. (1965). *Mathematical Theory of Reliability*. Wiley, New York.

[3] Barlow, R. E. and Proschan, F. (1975). *Statistical Analysis of Reliability and Life Testing*. Holt, Rinehart and Winston, New York.

[4] Barlow, R. E., Marshall, A. W., and Proschan, F. (1963). *Ann. Math. Statist.*, **34**, 375–389.

[5] Efron, B. (1965). *Ann. Math. Statist.*, **36**, 272–279.

[6] Johnson, N. L. and Kotz, S. (1970). *Distributions in Statistics. Continuous Univariate Distributions—(1)*. Wiley, New York.

[7] Karlin, S. (1956). *Proc. 3rd Berkeley Symp. Math. Statist. Prob.*, Vol. 1. University of California Press, Berkeley, Calif., pp. 115–128.

[8] Karlin, S. (1957). *Ann. Math. Statist.* **28**, 281–308.

[9] Karlin, S. (1964). *Trans. Amer. Math. Soc.*, **111**, 33–107.

[10] Karlin, S. (1968). *Total Positivity*, Vol. I. Stanford University Press, Stanford, Calif.

[11] Karlin, S. and McGregor, J. (1957). *Trans. Amer. Math. Soc.*, **86**, 366–400.

[12] Karlin, S. and Proschan, R. (1960). *Ann. Math. Statist.*, **31**, 721–736.

[13] Karlin, S. and Rubin, H. (1956). *Ann. Math. Statist.*, **27**, 272–299.

[14] Karlin, S., Proschan, F., and Barlow, R. E. (1961). *Pacific J. Math.*, **11**, 1023–1033.

[15] Pólya, G. and Szegö, G. (1925). *Aufgaben und Lehrsatze aus der Analysis*, Vol. I. Julius Springer, Berlin.

[16] Proschan, F. (1960). *Pólya Type Distributions in Renewal Theory, with an Application to an Inventory Problem*. Prentice-Hall, Englewood Cliffs, N.J.

[17] Schoenberg, I. J. (1951). *J. Anal. Math. Jerusalem*, **1**, 331–374.

B. Raja Rao

POLYCHORIC AND POLYSERIAL CORRELATIONS

The *polyserial* and *polychoric* correlations are measures of bivariate association (*see* ASSOCIATION, MEASURES OF) arising when one or both observed variates are ordered, categorical variables that result from polychotomizing one or two underlying continuous variables. The origin of these coefficients can be traced to Pearson [11, 12], whose work was motivated by difficulties in obtaining exact, quantitative measurements of some variables. He noted that in some cases it is very laborious to measure precisely the value of a variable on a quantitative scale, but very easy to place observations into ordered categories. For other variables such as eye color, it is easy to place observations into ordered categories, but difficult to assign numerical values to observations that preserve their relative relations. Pearson assumed that these types of categorical variables were functions of underlying (approximately) normally distributed variables (*see* NORMAL DISTRIBUTION) and then estimated the product moment correlation* between the normal variables. The sample tetrachoric correlation* [11] was developed as an estimate of the correlation parameter of the bivariate normal distribution* from data in the form of a 2×2 table*. In Tate's [19] development of the biserial correlation*, the correlation parameter of the bivariate normal distribution was estimated from data in which one of the normal variables had been dichotomized and the other variate remained in its continous form. (Weaker properties for the observed quantitative variable were assumed by Pearson [14] in his original work on the biserial; see ref. 7, pp. 337–338.) *See* also ORDINAL DATA.

Estimation of the tetrachoric correlation from a 2×2 table can be generalized to estimation of the polychoric correlation from an $r \times s$ table obtained from polychotomizing both variables of a bivariate normal distribution. Early work in this area was conducted by the authors of refs. 13, 16, and 17. More recently developed estimation methods can be found in refs. 6, 8, 9, and 18.

The polyserial correlation is a generalization of the biserial correlation*. It is estimated from data in which one of the two normal variables has been polychotomized into r ordered categories and the other normal variable remains untransformed. Pearson [15] described a method for estimation under restrictive conditions and more general approaches to estimation are presented in refs. 1 and 10.

In the remainder of this article the statistical models for the polyserial and polychoric correlations will be presented. Then estimation methods for each type of correlation will be summarized and examples will be provided.

STATISTICAL MODELS

The categorical variables C and D are assumed to be related to underlying continuous variables X and Y by

$$C = c_i \quad \text{if } \gamma_{i-1} \leqslant X < \gamma_i, \; i = 1, 2, \ldots, r$$

and

$$D = d_j \quad \text{if } \tau_{j-1} \leqslant Y < \tau_j, \; j = 1, 2, \ldots, s,$$

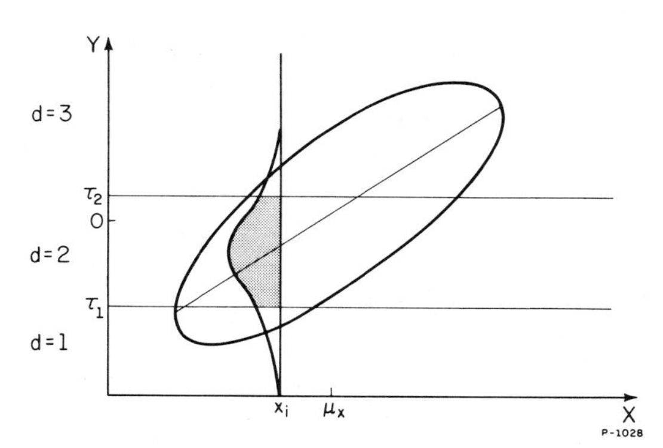

Figure 1. Example of two ordered, categorical variables formed by imposing thresholds on underlying continuous variables.

where the γ_i and τ_j are referred to as *thresholds*. For convenience, define $\gamma_0 = \tau_0 = -\infty$ and $\gamma_r = \tau_s = +\infty$. The thresholds and values of the categorical variables are taken to be strictly increasing; thus $\gamma_1 < \gamma_2 < \cdots < \gamma_{r-1}$ and $c_1 < c_2 < \cdots < c_r$. Analogous expressions are assumed to hold for the τ_j and d_j. Figure 1 illustrates the relations between the categorical variables C and D, the continuous variables X and Y, and the thresholds γ_i and τ_j.

The joint distribution of X and Y has been assumed to be bivariate normal in the development of the polyserial and polychoric correlations. Denote the parameters of this distribution by $E[X] = \mu$, $\text{var}(X) = \sigma^2$, $E[Y] = 0$, $\text{var}(Y) = 1$, and $\text{cov}(X, Y)/\sigma = \rho$. Note that ρ is the product-moment correlation between X and Y.

Assume that D and X are observed. Then the polyserial correlation ρ_{ps} between the categorial variable D and the normal variable X is defined as the product-moment correlation ρ between X and the latent normal variable Y underlying D. The product-moment correlation ρ_{pp} between D and X is termed the *point polyserial correlation*, which is a generalization of the point biserial correlation (*see* BISERIAL CORRELATION). Now assume that C and D are observed. Then the polychoric correlation ρ_{pc} between C and D is defined as the product-moment correlation between the underlying normal variables X and Y. Then [10]

$$\rho_{\text{pp}} = \frac{\rho_{\text{ps}}}{\sigma_d} \sum_{j=1}^{s-1} \phi(\tau_j)(d_{j+1} - d_j), \tag{1}$$

where σ_d^2 is the variance of D and $\phi(\tau) = (2\pi)^{-1/2}\exp(-\tau^2/2)$ is the standard normal density. The relation between the point biserial and biserial correlations presented by Tate [19] and by Lord and Novick [7, p. 340] is a special case of (1). No closed-form expression for the relation between ρ_{pc} and the product-moment correlation between C and D is available.

ESTIMATION OF THE POLYSERIAL CORRELATION

The likelihood of a sample of n observations (x_i, d_i) is

$$L = \prod_{k=1}^{n} f(x_k, d_k)$$

$$= \prod_{k=1}^{n} f(x_k)\Pr(D = d_k \mid x_k), \tag{2}$$

where

$$f(x_k) = \frac{1}{\sqrt{2\pi}\,\sigma} \exp\left[-\frac{1}{2}\left(\frac{x_k - \mu}{\sigma}\right)^2\right]$$

and d_k is used in this section only to indicate the observed value of D for the kth observation. The conditional distribution of Y given $X = x_k$ is normal with mean ρz_k and variance $(1 - \rho^2)$, where $z_k = (x_k - \mu)/\sigma$. Therefore

$$\Pr(D = d_k \mid x_k) = \Phi(\tau_j^*) - \Phi(\tau_{j-1}^*), \qquad j = 1, \ldots, s,$$

where

$$\tau_j^* = \frac{\tau_j - \rho z_k}{(1 - \rho^2)^{1/2}}, \quad \Phi(\tau) = \int_{-\infty}^{\tau} \phi(t)\, dt. \tag{3}$$

The maximum likelihood* estimate of the polyserial correlation is obtained by maximizing $l = \log L$ with respect to μ, σ^2, ρ, and the τ_j: differentiate l with respect to each parameter, set each derivative equal to zero, and solve the resulting set of equations. For example, the partial derivative of l with respect to ρ is

$$\frac{\partial l}{\partial \rho} = \sum_{k=1}^{n} \left\{ \frac{1}{\Pr(d_k \mid x_k)} \frac{1}{(1-\rho^2)^{3/2}} \times \left[\phi(\tau_j^*)(\tau_j\rho - z_k) - \phi(\tau_{j-1}^*)(\tau_{j-1}\rho - z_k) \right] \right\},$$

where τ_{j-1}^* and τ_j^* are the thresholds surrounding d_k. The remaining partial derivatives are given in refs. 1 and 10. The maximum likelihood estimates of μ and σ^2 are the usual expressions for a sample drawn from a normal population; estimates of the other parameters must be obtained by iterative methods. For example, l may be maximized by using the first derivatives and the method of Fletcher and Powell [2]. Alternatively, the estimates can be obtained by a maximization technique that uses the expectations of the second derivatives of l. These expressions, given in ref. 1, are most easily evaluated by Gauss–Hermite quadrature. Both approaches to computing the maximum likelihood estimates can be used to obtain large-sample estimates of the parameter estimates' variance-covariance matrix. The maximum likelihood estimators of ρ_{ps} and the τ_j are consistent*, asymptotically efficient*, and asymptotically multivariate normal [1].

Two polyserial estimators that require fewer calculations than the maximum likelihood estimator have also been examined. The first of these, which has been termed the *two-step estimator*, estimates μ and σ^2 as does the maximum likelihood estimator. However, the estimates of the thresholds τ_j are taken to be the values of the inverse standard normal distribution function evaluated at the cumulative marginal proportions of D. If, for example, d_1 were observed 20 times in a sample of 100 observations, then $\hat{\tau}_1 = \Phi^{-1}(0.20) = -0.8416$. The two-step estimate of ρ_{ps} is then obtained by equating to zero the derivative of l with respect to ρ_{ps} and solving for ρ_{ps}. Thus it is necessary only to solve one equation in one unknown. The other estimator of the polyserial requires even fewer calculations. The thresholds here are also estimated by applying Φ^{-1} to the cumulative marginal proportions. In addition, the sample point polyserial correlation between D and X and the sample variance of the categorical variable are calculated. These statistics are then substituted into (1) to yield an explicit expression for this estimate of ρ_{ps}. One simulation study [10] found the two estimators described in this paragraph to be nearly unbiased and to have sampling variances that were only slightly larger than the maximum likelihood estimator.

Example 1: The Polyserial Correlation. Table 1 contains fertility data presented by Cox [1] for a flock of four-year Romney ewes. The continuous observed variable X corresponds to weight at mating and the categorical variable D corresponds to the number of lambs born to a ewe. The mean weight of the ewes is 90.680 and the standard deviation is 11.808. The mean and standard deviation of the number of lambs born are 1.080 and 0.560, and the sample point polyserial correlation is 0.185. The marginal proportions of the categorical variable are 0.120 and 0.800, so the threshold estimates for the two-stage method are $\hat{\tau}_1 = \Phi^{-1}(0.120) = -1.175$ and $\hat{\tau}_2 = 0.842$. Substitution into (1) yields $\hat{\rho}_{ps} = (0.185) \times (0.560)[0.200 + 0.280]^{-1} = 0.216$. The two-stage estimate of ρ_{ps} can be shown to be

Table 1 Weights (in Pounds) and Lambs Born for 25 Ewes

X	*d*	*X*	*d*	*X*	*d*	*X*	*d*
72	0	69	1	80	1	104	1
88	0	101	1	99	1	103	2
112	0	108	1	92	1	85	2
93	1	104	1	72	1	96	2
86	1	92	1	81	1	96	2
87	1	77	1	88	1	104	2
78	1						

Source. N. R. Cox, *Biometrics*, **30**, 177 (1974). With permission from the Biometric Society.

0.211. Finally, Cox [1] shows that the maximum likelihood estimate of ρ_{ps} is 0.211 with an estimated standard error of 0.224. Cox also shows that maximum likelihood estimates of the thresholds are $\hat{\tau}_1 = -1.169$ and $\hat{\tau}_2 = 0.843$, and the corresponding estimated standard errors are 0.324 and 0.286.

ESTIMATION OF THE POLYCHORIC CORRELATION

Let $\phi(x, y; \rho)$ denote the standard bivariate normal density with correlation ρ:

$$\phi(x, y; \rho) = \frac{1}{2\pi(1-\rho^2)^{1/2}} \exp\left\{ \frac{-1}{2(1-\rho^2)} (x^2 - 2\rho xy + y^2) \right\}, \tag{4}$$

where without loss of generality the mean and variance of X and Y are taken to be $\mu = 0$ and $\sigma^2 = 1$. Then the probability of an observation with $C = c_i$ and $D = d_j$ is

$$P_{ij} = \int_{\gamma_{i-1}}^{\gamma_i} \int_{\tau_{j-1}}^{\tau_j} \phi(x, y; \rho)\, dy\, dx.$$

Let n_{ij} denote the number of observations with $C = c_i$ and $D = d_j$ in a sample of size $n = \sum_i^r \sum_j^s n_{ij}$; then the likelihood of the sample is

$$L = K \prod_{i=1}^{r} \prod_{j=1}^{s} P_{ij}^{n_{ij}},$$

where K is a constant.

The maximum likelihood estimate of ρ_{pc} is obtained by first differentiating the logarithm l of the likelihood function L with respect to all model parameters $(\rho, \gamma_1, \ldots, \gamma_r, \tau_1, \ldots, \tau_s)$, equating to zero the resulting partial derivatives, and then solving this equation system. For example, the partial derivative of l with respect to ρ is

$$\frac{\partial l}{\partial \rho} = \sum_{i=1}^{r} \sum_{j=1}^{s} \frac{n_{ij}}{P_{ij}} \left[\phi(\gamma_i, \tau_j; \rho) - \phi(\gamma_{i-1}, \tau_j; \rho) - \phi(\gamma_i, \tau_{j-1}; \rho) + \phi(\gamma_{i-1}, \tau_{j-1}; \rho) \right].$$

The partial derivatives of l with respect to the γ_i and τ_j are given in ref. 9. The terms in the information matrix* are also provided in ref. 9 and can be used to solve the likelihood equations by iterative methods. The inverse of the information matrix evaluated at the maximizing value of the parameter vector yields a large-sample estimate of the parameter estimate variance–covariance matrix.

To reduce the amount of calculation, Martinson and Hamdan [8] suggested fitting univariate standard normal distributions to the marginal distributions of C and D and then maximizing l with respect to ρ only. If, for example, c_1 were observed 20 times and d_1 were observed 30 times in a sample of 100 observations, then $\hat{\gamma}_1 = \Phi^{-1}(0.20) = -0.842$ and $\hat{\tau}_1 = \Phi^{-1}(0.30) = -0.524$. The value of ρ that maximizes l given these fixed values for the thresholds would be taken as the estimate of ρ_{pc}. This estimate, termed the two-step estimate, requires the solution of one equation—the partial derivative of l with respect to ρ set equal to zero—in one unknown. Martinson and Hamdan gave an expression for the asymptotic variance of the two-step estimator. In the derivation of this equation, however, the thresholds were treated as known rather than as sample estimates. Olsson [9] presented a more complicated expression for the variance of the two-step estimator based on the assumption that the thresholds are estimated from sample data.

A third estimate of ρ_{pc}, termed the polychoric series estimate, was developed by Lancaster and Hamdan [6]. They used the theory of orthonormal functions to generalize Pearson's [11] tetrachoric series to a polychoric series. The tetrachoric series is obtained by expanding the right side of

$$\frac{n_{22}}{n} = \int_{\hat{\gamma}_1}^{\infty} \int_{\hat{\tau}_1}^{\infty} \phi(x, y; \rho)\, dy\, dx$$

into a series in ρ. Here

$$\hat{\gamma}_1 = \Phi^{-1}\left[(n_{11} + n_{12})/n\right],$$

$$\hat{\tau}_1 = \Phi^{-1}\left[(n_{11} + n_{21})/n\right].$$

Then [5, (pp. 325–326)]

$$\frac{n_{22}}{n} = \sum_{m=0}^{\infty} \rho^m g_m(\hat{\gamma}_1) g_m(\hat{\tau}_1), \tag{5}$$

where

$$g_m(k) = \frac{H_{m-1}(k)\phi(k)}{(m!)^{1/2}}, \qquad m = 1, 2, \ldots,$$

$H_{m-1}(k)$ is the Chebyshev–Hermite polynomial* of degree $m - 1$ given in ref. 4 (p. 167), and $g_0(k) = 1 - \Phi(k)$. To generalize (5), let $n_{i\cdot}$ and $n_{\cdot j}$ denote the marginal row and column totals of the $r \times s$ table of observed frequencies, and

$$p_{i\cdot} = n_{i\cdot}/n \qquad q_{i\cdot} = p_{1\cdot} + \cdots + p_{i\cdot}$$

$$p_{\cdot j} = n_{\cdot j}/n \qquad q_{\cdot j} = p_{\cdot 1} + \cdots + p_{\cdot j}$$

Lancaster and Hamdan defined step functions $\zeta_i(x)$, $i = 1, \ldots, r - 1$, by

$$\zeta_i(x) = \begin{cases} \zeta_i^+ = (p_{i+1\cdot}/(q_{i\cdot}q_{i+1\cdot}))^{1/2}, & \gamma_0 < x \leqslant \gamma_i \\ \zeta_i^- = -(q_{i\cdot}/(p_{i+1\cdot}q_{i+1\cdot}))^{1/2}, & \gamma_i < x \leqslant \gamma_{i+1} \\ 0, & \text{otherwise} \end{cases}$$

and

$$\eta_j(y) = \begin{cases} \eta_j^+ = (p_{\cdot j+1}/(q_{\cdot j}q_{\cdot j+1}))^{1/2}, & \tau_0 < y \leqslant \tau_j \\ \eta_j^- = -(q_{\cdot j}/(p_{\cdot j+1}q_{\cdot j+1}))^{1/2}, & \tau_j < y \leqslant \tau_{j+1} \\ 0, & \text{otherwise} \end{cases}$$

These step functions are then expanded as Fourier series using Chebyshev–Hermite polynomials with coefficients

$$\begin{aligned} a_{im} &= \int_{-\infty}^{\infty} \zeta_i(x)H_m(x)\phi(x)\,dx \\ &= -m^{-1/2}\left[(\zeta_i^+ - \zeta_i^-)H_{m-1}(\hat{\gamma}_i)\phi(\hat{\gamma}_i) + \zeta_i^- H_{m-1}(\hat{\gamma}_{i+1})\phi(\hat{\gamma}_{i+1})\right] \end{aligned}$$

and

$$\begin{aligned} b_{jm} &= \int_{-\infty}^{\infty} \eta_j(y)H_m(y)\phi(y)\,dy \\ &= -m^{-1/2}\left[(\eta_j^+ - \eta_j^-)H_{m-1}(\hat{\tau}_j)\phi(\hat{\tau}_j) + \eta_j^- H_{m-1}(\hat{\tau}_{j+1})\phi(\hat{\tau}_{j+1})\right]. \end{aligned}$$

Lancaster and Hamdan showed that Pearson's corrected ϕ is related to the a_{im}, b_{jm}, and ρ by

$$\begin{aligned} \phi^2 &= \left[\chi^2 - (r-1)(s-1)\right]/n \\ &\doteq \sum_{i=1}^{r-1}\sum_{j=1}^{s-1}\left[\sum_{m=1}^{\infty} a_{im}b_{jm}\rho^m\right]^2, \end{aligned} \qquad (6)$$

where χ^2 is the ordinary chi-square for the $r \times s$ contingency table. The right-hand term in (6) is called the *polychoric series*. Equating ϕ^2 with the polychoric series truncated after the first few terms and solving for ρ yields the polychoric series estimate of the polychoric correlation. Unfortunately, no expression for the asymptotic variance of this estimate is available. Thresholds are estimated here as in the two-step estimator.

Hamdan [3] showed that the polychoric series estimate of ρ_{pc} is identical to the maximum likelihood estimate when $r = s = 2$. These two estimates may differ, however, when $r \neq 2$ or $s \neq 2$.

Example 2. Table 2, originally given by Tallis [18], presents the first and second lambing records of 227 Merino ewes. Here C represents the number of lambs born during the ewes' lambings in 1952 and D corresponds to the number born in 1953.

The maximum likelihood estimate of ρ_{pc} is 0.418 with an estimated standard error of 0.076. The estimated thresholds (with estimated standard errors in parentheses) are

$$\hat{\gamma}_1 = -0.242\ (0.084) \qquad \hat{\tau}_1 = -0.030\ (0.083)$$

$$\hat{\gamma}_2 = 1.594\ (0.134) \qquad \hat{\tau}_2 = 1.133\ (0.106).$$

The intermediate calculations that lead to these estimates are given in ref. 18.

The two-step estimate of ρ_{pc} is 0.420 and the standard error is estimated to be 0.074 by Martinson and Hamdan's equation. The threshold estimates are

$$\hat{\gamma}_1 = \Phi^{-1}(92/227) = -0.240 \qquad \hat{\tau}_1 = -0.027$$

$$\hat{\gamma}_2 = 1.578 \qquad \hat{\tau}_2 = 1.137.$$

Finally, the polychoric series of ρ_{pc} is 0.556. The threshold estimates for the polychoric series method are identical to the two-step threshold estimates.

Table 2 First and Second Lambing Records

		1952			
		No Lambs	1 Lamb	2 Lambs	Total
1953	No lambs	58	52	1	111
	1 Lamb	26	58	3	87
	2 Lambs	8	12	9	29
	Total	92	122	13	227

Source. G. M. Tallis, *Biometrics*, **18**, 349 (1962). With permission from the Biometric Society.

It is apparent that the two-step estimates provide close approximations to the maximum likelihood estimates. A Monte Carlo* study [9] obtained similar results across a variety of conditions.

References

[1] Cox, N. R. (1974). *Biometrics*, **30**, 171–178.

[2] Fletcher, R. and Powell, M. J. D. (1963). *Computer J.*, **2**, 163–168.

[3] Hamdan, M. A. (1970). *Biometrika*, **57**, 212–215.

[4] Kendall, M. G. and Stuart, A. (1977). *The Advanced Theory of Statistics*, Vol. 1: *Distribution Theory*, 4th ed. Macmillan, New York.

[5] Kendall, M. G. and Stuart, A. (1979). *The Advanced Theory of Statistics*, Vol. 2: *Inference and Relationship*, 4th ed. Macmillan, New York.

[6] Lancaster, H. O. and Hamdan, M. A. (1964). *Psychometrika*, **29**, 383–391.

[7] Lord, F. M. and Novick, M. R. (1968). *Statistical Theories of Mental Test Scores*. Addison-Wesley, Reading, Mass.

[8] Martinson, E. O. and Hamdan, M. A. (1971). *J. Statist. Comp. Simul.*, **1**, 45–54.

[9] Olsson, U. (1979). *Psychometrika*, **44**, 443–460.

[10] Olsson, U., Drasgow, F., and Dorans, N. J. (1982). *Psychometrika*, **47**, 337–347.

[11] Pearson, K. (1900). *Philos. Trans. R. Soc. Lond. A*, **195**, 1–47. (A technical paper that includes the original derivation of the tetrachoric series.)

[12] Pearson, K. (1900). *Philos. Trans. R. Soc. Lond. A*, **195**, 79–150.

[13] Pearson, K. (1904). *Drapers' Company Research Memoirs. Biometrics Series, 1*. University College, London. (On the theory of contingency and its relation to association and normal correlation.)

[14] Pearson, K. (1909). *Biometrika*, **7**, 96–105. (Presents Pearson's original work on the biserial correlation.)

[15] Pearson, K. (1913). *Biometrika*, **9**, 116–139.

[16] Pearson, K. and Pearson, E. S. (1922). *Biometrika*, **14**, 127–156.

[17] Ritchie-Scott, A. (1918). *Biometrika*, **12**, 93–133.

[18] Tallis, G. M. (1962). *Biometrics*, **18**, 342–353. (This paper presents the equations for maximum likelihood estimates for 2-by-2 and 3-by-3 tables. It also has a worked example.)

[19] Tate, R. F. (1955). *Biometrika*, **42**, 205–216. (Derives the maximum likelihood estimator of the biserial correlation and its asymptotic properties.)

Bibliography

See the following works, as well as the references just given, for more information on the topic of polychoric and polyserial correlations.

Goodman, L. A. (1981). *Biometrika*, **68**, 347–355. (Presents association models that achieve some reconciliation between the models of Pearson and Yule.)

Hamdan, M. A. (1971). *Psychometrika*, **36**, 253–259. (This paper shows how to compute the polychoric series estimate of ρ_{pc}. It includes a worked example.)

IMSL Library 1, 5th ed. (1975). International Mathematical and Statistical Libraries, Houston, Tex. (The subroutine CBNRHO computes the two-step estimate of ρ_{pc}.)

Martinson, E. O. and Hamdan, M. A. (1975). *Appl. Statist.*, **24**, 272–278. (A Fortran subroutine that computes the polychoric series estimate of ρ_{pc} is described and the source code is listed.)

Pearson, K. and Heron, D. (1913). *Biometrika*, **9**, 159–315. (Pearson's vitriolic response to Yule's approach to the analysis of categorical data.)

Yule, G. U. (1912). *J. R. Statist. Soc.*, **75**, 579–642. (This article contains Yule's critique of Pearson's approach to categorical data and presents Yule's models for categorical data.)

(BISERIAL CORRELATION
CONTINGENCY TABLES

TETRACHORIC CORRELATION COEFFICIENT)

FRITZ DRASGOW

POLYGAMMA FUNCTIONS

These are derivatives of $\log \Gamma(x)$.

Generally, the s-gamma function is

$$d^{s-1}\log \Gamma(x)/dx^{s-1} = \Psi^{(s-2)}(x),$$

where $\Psi(x) = d\log\Gamma(x)/dx$ is the digamma* or psi function.

Moments* and cumulants* of some distributions are conveniently expressed in terms of polygamma functions. For example, if X has a standard beta distribution* with parameters α, β, then the rth cumulant of $\ln X$ is $\Psi^{(r-1)}(\alpha) - \Psi^{(r-1)}(\alpha+\beta)$.

POLYKAYS

Polykays are symmetric functions* of sample observations, having the basic property of being unbiased estimators of products of population cumulants*, that is, $E(k_{p_1 p_2 \cdots p_s}) = \kappa_{p_1}\kappa_{p_2}\cdots\kappa_{p_s}$, for infinite populations. Unbiased estimation of population moments* involves unbiased estimation of products of population cumulants, and it was in this connection that Dressel [11] introduced them first, following the work of Dwyer [12], calling them L seminvariants. They were reintroduced by Tukey [41], who called them polykays, and obtained them [42] by a symbolic multiplication of Fisher's k-statistics*, written as linear combinations of angle brackets* (symmetric means). Thus [19, p. 104]

$$\begin{aligned}
k_{11} &= k_1 \circ k_1 = \langle 1\rangle \circ \langle 1\rangle = \langle 11\rangle \\
k_{21} &= k_2 \circ k_1 \\
&= \{\langle 2\rangle - \langle 11\rangle\} \circ \langle 1\rangle = \langle 21\rangle - \langle 111\rangle \\
k_{22} &= k_2 \circ k_2 \\
&= \{\langle 2\rangle - \langle 11\rangle\} \circ \{\langle 2\rangle - \langle 11\rangle\} \\
&= \langle 22\rangle - 2\langle 211\rangle + \langle 1111\rangle.
\end{aligned}$$

A general expression for polykay $k_{p_1 p_2 \cdots p_s} = k_{p_1} \circ k_{p_2} \circ \cdots \circ k_{p_s}$ is obtained from $k_p = \sum_P (-1)^{\pi-1}(\pi-1)!\, C(P)\langle P\rangle$, where the summation is over all partitions P of p, with π as the number of parts in P, and $C(P)$ is the associated combinatorial coefficient. Then

$$\begin{aligned}
k_{p_1 p_2 \cdots p_s} &= \sum (-1)^{\Sigma(\pi_i - 1)} \prod (\pi_i - 1)! \\
&\quad \times \prod C(P_i)[P_I]/n^{(\Sigma \pi_i)},
\end{aligned}$$

where the summation extends over all partitions P_I obtained by partitioning one or more of the p_i [37] and $[P_I]$ denotes the power product sum* (or augmented monomial symmetric function, AMSF). The number p is the *weight* of the polykay.

Wishart [46] gives tables for expressing polykays through weight 6 in terms of AMSFs. Kendall [20], calling them generalized k-statistics, remarks that by using the Irwin–Kendall principle [18], for a finite population of size N, $E_N(k_{p_1 p_2 \cdots p_s}) = K_{p_1 p_2 \cdots p_s}$, where the K-parameter $K_{p_1 p_2 \cdots p_s}$ is the same function of the N members of the population as $k_{p_1 p_2 \cdots p_s}$ is of the n sample observations, and E_N denotes the expected value for the finite population. Thus polykays are "inherited on the average," and when N becomes infinite, $K_{p_1 p_2 \cdots p_s}$ tends to the cumulant product $\kappa_{p_1}\kappa_{p_2}\cdots\kappa_{p_s}$. As an example,

$$\kappa_3\kappa_1^2 = \mu_3'\mu_1'^2 - 3\mu_2'\mu_1'^3 + 2\mu_1'^5$$

yields, as in Wishart [47],

$$k_{31^2} = \langle 31^2\rangle - 3\langle 21^3\rangle + 2\langle 1^5\rangle$$

and if $\langle\cdot\rangle_N$ denotes angle brackets for the finite population,

$$K_{31^2} = \langle 31^2\rangle_N - 3\langle 21^3\rangle_N + 2\langle 1^5\rangle_N.$$

Writing angle brackets in terms of AMSFs, and expressing these as products of k-statistics, one obtains

$$\begin{aligned}
k_{31^2} &= k_3 k_1^2 - \frac{2}{n} k_4 k_1 \\
&\quad - \frac{n-1}{n(n+5)} k_3 k_2 + \frac{2(n+2)}{n^2(n+5)} k_5.
\end{aligned}$$

Wishart's [46] tables are extended by Abdel-

Aty [1] through weight 12, and reproduced by David et al. [8]. Kendall and Stuart [21] give more examples, suggesting "l-statistics" for "polykays," which they consider "linguistic miscegenation."

To obtain the moments and cumulants of k-statistics, one needs formulae for their powers and products expressed as linear combinations of polykays. Thus, when sampling from a finite population, the variance of the sample variance is

$$\begin{aligned}
\operatorname{var}(k_2) &= E_N(k_2^2) - E_N^2(k_2) \\
&= E_N\left(\frac{1}{n}k_4 + \frac{n+1}{n-1}k_{22}\right) - K_2^2 \\
&= \left(\frac{1}{n}K_4 + \frac{n+1}{n-1}K_{22}\right) \\
&\quad - \left(\frac{1}{N}K_4 + \frac{N+1}{N-1}K_{22}\right) \\
&= \left(\frac{1}{n} - \frac{1}{N}\right)K_4 \\
&\quad + 2\left(\frac{1}{n-1} - \frac{1}{N-1}\right)K_{22}
\end{aligned}$$

as in Tukey [41], and its unbiased estimate is [35]

$$\begin{aligned}
\widehat{\operatorname{var}}(k_2) &= \left(\frac{1}{n} - \frac{1}{N}\right)k_4 \\
&\quad + 2\left(\frac{1}{n-1} - \frac{1}{N-1}\right)k_{22}.
\end{aligned}$$

Wishart [46] gives several results for products of polykays through weight 8, and these are extended by Schaeffer and Dwyer [35], who also give formulae for moments and estimates of moments of k-statistics through weight 8. Higher-weight formulae are given by Tracy [36]. For infinite populations, one obtains Fisher's [15] results.

Tukey [42] gives a method for finding expressions for products of two polykays by using a table for multiplication of angle brackets and a rule involving the number of unit parts. Dwyer and Tracy [14] modified Tukey's method to a direct combinatorial method, and provided rules to obtain products of two polykays as a linear combination of polykays. Tracy [37] extended these rules for products of several polykays, and Tracy [38] and Nagambal and Tracy [31–33] give general formulas for multiplying any polykay by polykay products through weight 5.

Carney [4] uses ordered partitions for the multiplication of two polykays. This method was extended to multiple products by Tracy and Gupta [39]. Carney [5] discusses some computing problems encountered when dealing with polykays and ordered partitions.

Methods for computing polkays are given by Schaeffer and Dwyer [35], who call them multiple k-statistics. Kinney [23, 24] provides another method for expressing polykays in terms of k-statistics. Dwyer [13] studies polykays of deviates from the mean, and applies them to obtain finite moment formulae involving the sample mean. Koop [25] uses polykays to provide an expression for the bias of the estimate of the correlation coefficient of a finite population.

A finite population of particular interest, which occurs when dealing with ranks, is that of natural numbers $1, 2, \ldots, N$. Barton et al. [2] obtain the polykays of the first N natural numbers by using expressions for generalized Bernoulli polynomials*. They illustrate the use of their formulae, which are also listed in David and Barton [7], in obtaining the moments of Mood's statistic [30].

Hooke [16] introduced bipolykays as extensions of polykays. These are symmetric functions of a two-way array or matrix, in the sense that they are invariant under permutation of rows and/or columns of the matrix. Hooke shows that these are also inherited on the average and develops formulae for use in random pairing. He also obtains formulae for multiplication of bipolykays and for variances of bipolykays. This helps in finding expressions for sampling moments of (a) functions of the elements of a matrix, and (b) functions associated with the analysis of variance* of a two-way table with systematic interactions, such as estimates of variance components*, as also in finding unbiased estimators for the variances and covariances of estimated variance components in a two-way table without interactions [17].

Tukey [43] uses polykays in studying the

variances of variance components in balanced single and double classification, Latin squares*, and balanced incomplete block designs. He [44] also applies polykays to a family of weighted estimates to obtain the variances and covariances of the estimates of between and within variance component* in an unbalanced single classification. Tukey [45] extends the results to third moments of the estimates of the between variance component in a balanced single classification, using polykays.

Lamoš [26] studies generalized symmetric means of a three-way array, and introduces [27] tripolykays as symmetric functions of a three-way array. He also develops formulae for use in random pairing. He [28] gives conversion formulae for generalized symmetric means and tripolykays through degree 3, and also some multiplication formulae for these tripolykays, which enable one to find their variances and covariances.

Dayhoff [9] extended bipolykays of degree 2 to n-way polykays, calling them generalized polykays, to encompass all balanced population structures defined by Zyskind [47], and established the equivalence of generalized polykays and $\sum$ functions. Dayhoff [10] presented a more general formalization of generalized symmetric means and polykays of arbitrary degree and some of their sampling properties. Carney [3] studied the relationship of generalized polykays to unrestricted sums for balanced complete finite populations.

Kinney [22] gives bivariate polykays through weight 8 in terms of bivariate k-statistics. Mikhail [29] gives trivariate polykays in terms of AMSFs, and products of bivariate polykays as linear combinations of bivariate polykays, through weight 4.

David [6] introduces factorial polykays $k_{[\cdot]}$ to estimate products of factorial cumulants. Thus $E(k_{[31^2]}) = \kappa_{[3]}\kappa_{[1]}^2$, and one obtains $k_{[31^2]} = k_{31^2} - 3k_{21^2} + 2k_{1^3}$. These are useful for discrete populations.

Generalized h-statistics are defined as unbiased estimates of products of moments. Their relationship with polykays is studied by Tracy and Gupta [41], who also provide tables through weight 12 for these relationships. Robson [35] defines multivariate polykays and uses them in the theory of unbiased ratio-type estimation.

References

[1] Abdel-Aty, S. H. (1954). *Biometrika*, **41**, 253–260.

[2] Barton, D. E., David, F. N., and Fix, E. (1960). *Biometrika*, **47**, 53–59.

[3] Carney, E. J. (1968). *Ann. Math. Statist.*, **39**, 643–656.

[4] Carney, E. J. (1970). *Ann. Math. Statist.*, **41**, 1749–1752.

[5] Carney, E. J. (1972). In *Symmetric Functions in Statistics*, D. S. Tracy, ed. University of Windsor, Windsor, Ontario, pp. 87–99.

[6] David, F. N. (1972). In *Symmetric Functions in Statistics*, D. S. Tracy, ed. University of Windsor, Windsor, Ontario, pp. 53–69.

[7] David, F. N. and Barton, D. E. (1962). *Combinatorial Chance*. Hafner, New York, Chap. 17.

[8] David, F. N., Kendall, M. G., and Barton, D. E. (1966). *Symmetric Function and Allied Tables*. Cambridge University Press, Cambridge, England.

[9] Dayhoff, E. (1964). *Ann. Math. Statist.*, **35**, 1663–1672.

[10] Dayhoff, E. (1966). *Ann. Math. Statist.*, **37**, 226–241.

[11] Dressel, P. L. (1940). *Ann. Math. Statist.*, **11**, 33–57.

[12] Dwyer, P. S. (1938). *Ann. Math. Statist.*, **9**, 1–47, 97–132.

[13] Dwyer, P. S. (1964). *Ann. Math. Statist.*, **35**, 1167–1173.

[14] Dwyer, P. S. and Tracy, D. S. (1964). *Ann. Math. Statist.*, **35**, 1174–1185.

[15] Fisher, R. A. (1929). *Proc. Lond. Math. Soc.*, **30**, 199–238.

[16] Hooke, R. (1956). *Ann. Math. Statist.*, **27**, 55–79.

[17] Hooke, R. (1956). *Ann. Math. Statist.*, **27**, 80–98.

[18] Irwin, J. O. and Kendall, M. G. (1944). *Ann. Eugen. (Lond.)*, **12**, 138–142.

[19] Keeping, E. S. (1962). *Introduction to Statistical Inference*. D. Van Nostrand, Princeton, N.J., Chap. 5.

[20] Kendall, M. G. (1952). *Biometrika*, **39**, 14–16.

[21] Kendall, M. G. and Stuart, A. (1969). *The Advanced Theory of Statistics*, Vol. 1: *Distribution Theory*, 3rd ed. Charles Griffin, London, Chap. 12.

[22] Kinney, J. J. (1971). *Multivariate and Generalized Polykays in Statistical Structures*. Doctoral dissertation, Iowa State University, Ames, IA.

[23] Kinney, J. (1972). In *Symmetric Functions in Statistics*, D. S. Tracy, ed. University of Windsor, Windsor, Ontario, pp. 101–119.

[24] Kinney, J. (1976). *Sankyha A*, **38**, 271–286.

[25] Koop, J. C. (1972). In *Symmetric Functions in Statistics*, D. S. Tracy, ed. University of Windsor, Windsor, Ontario, pp. 121–129.

[26] Lamoš, F. (1972). *Acta Fac. Rerum Natur. Univ. Comenian. Math.*, **27**.

[27] Lamoš, F. (1979). *Acta Fac. Rerum Natur. Univ. Comenian. Math.*, **32**, 17–27.

[28] Lamoš, F. (1979). *Acta Fac. Rerum Natur. Univ. Comenian. Math.*, **35**, 29–41.

[29] Mikhail, N. N. (1970). *Egypt. Statist. J.*, **14**, 15–36.

[30] Mood, A. M. (1954). *Ann. Math. Statist.*, **25**, 514–522.

[31] Nagambal, P. N. and Tracy, D. S. (1970). *Ann. Math. Statist.*, **41**, 1114–1121.

[32] Nagambal, P. N. and Tracy, D. S. (1974). *Canad. J. Statist.*, **2**, 107–112.

[33] Nagambal, P. N. and Tracy, D. S. (1974). *Canad. J. Statist.*, **2**, 169–180.

[34] Robson, D. S. (1957). *J. Amer. Statist. Ass.*, **52**, 511–522.

[35] Schaeffer, E. and Dwyer, P. S. (1963). *J. Amer. Statist. Ass.*, **58**, 120–151.

[36] Tracy, D. S. (1963). *Finite Moment Formulae and Products of Generalized k-Statistics with a Generalization of Fisher's Combinatorial Method*. Doctoral dissertation, University of Michigan, Ann Arbor, MI.

[37] Tracy, D. S. (1968). *Ann. Math. Statist.*, **39**, 983–998.

[38] Tracy, D. S. (1969). *Ann. Math. Statist.*, **40**, 1297–1299.

[39] Tracy, D. S. and Gupta, B. C. (1973). *Ann. Statist.*, **1**, 913–923.

[40] Tracy, D. S. and Gupta, B. C. (1974). *Ann. Statist.*, **2**, 837–844.

[41] Tukey, J. W. (1950). *J. Amer. Statist. Ass.*, **27**, 501–519.

[42] Tukey, J. W. (1956). *Ann. Math. Statist.*, **27**, 37–54.

[43] Tukey, J. W. (1956). *Ann. Math. Statist.*, **27**, 722–736.

[44] Tukey, J. W. (1957). *Ann. Math. Statist.*, **28**, 43–56.

[45] Tukey, J. W. (1957). *Ann. Math. Statist.*, **28**, 378–384.

[46] Wishart, J. (1952). *Biometrika*, **39**, 1–13.

[47] Zyskind, G. (1962). *Sankhya A*, **24**, 115–148.

(ANGLE BRACKETS
CUMULANTS
FISHER'S *k*-STATISTICS
MOMENTS
SYMMETRIC FUNCTIONS)

D. S. TRACY

POLYNOMIAL REGRESSION *See* REGRESSION POLYNOMIALS

POLYNOMIALS OF MATRIX ARGUMENTS

An extension of the zonal polynomials* to invariant polynomials in r matrix arguments was defined by Davis [6] for $r = 2$ and generalized to $r > 2$ by Chikuse [1]. The class $P_k(X)$ of homogeneous polynomials of degree k in the elements of the $m \times m$ symmetric matrix X decomposes into the direct sum

$$P_k(X) = \bigoplus_{\kappa} \mathscr{V}_{\kappa}(X)$$

of irreducible invariant subspaces, under the representation of the real linear group $Gl(m, R)$ of nonsingular $m \times m$ matrices L induced by $X \to L'XL$. $\mathscr{V}_{\kappa}$ carries the irreducible representation $[2\kappa]$ of $Gl(m, R)$, κ being an ordered partition of k into $\leqslant m$ parts. It contains a one-dimensional subspace which is invariant under the orthogonal group $O(m)$, and generated by the zonal polynomial $C_{\kappa}(X)$. More generally, $P_{k[r]}(X_{[r]})$, the class of polynomials of degree $k(1), \dots, k(r)$ in the elements of $X_1, \dots, X_r$, respectively, is given by decomposition of the Kronecker product*

$$P_{k[r]}(X_{[r]}) = \bigotimes_{i=1}^{r} P_{k(i)}(X_i)$$

$$= \bigoplus_{\kappa[r]} \bigoplus_{\Phi} \mathscr{V}_{\Phi}^{\kappa[r]}(X_{[r]}),$$

where $\kappa[r]$ runs through all ordered partitions $\kappa(1), \dots, \kappa(r)$ of $k(1), \dots, k(r)$, respectively, into $\leqslant m$ parts, and Φ through the irreducible representations in the decomposition of $\bigotimes_{i=1}^{r}[2\kappa(i)]$. If ϕ is an ordered partition of $f = \sum_{i=1}^{r} k(i)$ into $\leqslant m$ parts, then $\mathscr{V}_{2\phi}^{\kappa[r]}$ contains a polynomial $C_{\phi}^{\kappa[r]}(X_{[r]})$, invariant under the simultaneous transformations $X_i \to H'X_iH$ $(H \in O(m))$ $(i = 1, \ldots, r)$. A basis for the polynomials is provided by all distinct products of traces of products of the X_i, total degree $k(i)$ in X_i $(i = 1, \ldots, r)$. They have been constructed for $r = 2, f \leqslant 6$; $r = 3, f \leqslant 5$ [5, 7], using an extension of the original group-theoretic method of James [13] for zonal polynomials.

Example. The polynomials of degree 2 in X and 1 in Y are as follows $[(A) = \operatorname{trace} A]$:

$$C_3^{2,1}(X,Y) = \tfrac{1}{15}\{(X)^2(Y) + 4(XY)(X) + 2(X^2)(Y) + 8(X^2Y)\}$$

$$C_{21}^{2,1}(X,Y) = \tfrac{2}{5}\{(X)^2(Y) - (XY)(X) + 2(X^2)(Y) - 2(X^2Y)\}$$

$$C_{21}^{1^2,1}(X,Y) = 5^{-1/2}\{(X)^2(Y) + 2(XY)(X) - (X^2)(Y) - 2(X^2Y)\}$$

$$C_{1^3}^{1^2,1}(X,Y) = \tfrac{1}{3}\{(X)^2(Y) - 2(XY)(X) - (X^2)(Y) + 2(X^2Y)\}.$$

For example,

$$(X)^2(Y) = C_3^{2,1} + \frac{2}{3}C_{21}^{2,1} + \frac{5^{\frac{1}{2}}}{3}C_{21}^{1^2,1} + C_{1^3}^{1^2,1}.$$

The fundamental property of the polynomials is

$$\int_{O(m)} \prod_{i=1}^{r} C_{\kappa(i)}(A_iH'X_iH)\,dH \tag{1}$$
$$= \sum_{\phi \in \kappa(1)\ldots\kappa(r)} C_\phi^{\kappa[r]}(A_{[r]})C_\phi^{\kappa[r]}(X_{[r]})/C_\phi(I),$$

where dH is the invariant measure on $O(m)$, I is the $m \times m$ unit matrix, and the sum extends over all ϕ described above.

From James [14], the zonal polynomials are eigenfunctions of the Laplace–Beltrami operator Δ_X on the symmetric space of real positive-definite symmetric X. This has the general form

$$\Delta_X = \operatorname{tr}\left(X\frac{\partial}{\partial X}\right)^2,$$

where $\left(\frac{\partial}{\partial X}\right)_{ij} = \frac{1}{2}(1 + \delta_{ij})\frac{\partial}{\partial x_{ij}}, X = (x_{ij})$.

From (1), $C_\phi^{\kappa[r]}(X_{[r]})$ is an eigenfunction of the Laplace–Beltrami operator Δ_{X_i} in each argument, with the same eigenvalue as $C_{\kappa(i)}$. Further, since

$$C_\phi^{\kappa[r]}(X_{[r]} \cdot A) = \sum_{\sigma \in \phi \cdot \phi} \zeta_\sigma^{\kappa[r];\phi} C_\sigma^{\kappa[r],\phi}(X_{[r]}, A)$$

for suitable coefficients ζ, the left-hand side is an eigenfunction of Δ_A with the same eigenvalue as C_ϕ. The operation of Δ_X on the basis is specified by

$$(XA)(XB) \to (XAXB) + (XAXB'),$$
$$(XAXB) \to (XA)(XB) + (XAXB'),$$

where $(X) = \operatorname{tr} X$. However, this approach to constructing the polynomials has not yet been implemented.

Unlike zonal polynomials, the $C_\phi^{\kappa[r]}$ are not uniquely defined in general, since $\mathscr{V}_{2\phi}^{\kappa[r]}$ is not unique when $[2\phi]$ has multiplicity > 1. However, the direct sum of equivalent subspaces $\bigoplus_{\phi' \equiv \phi} \mathscr{V}_{[2\phi']}^{\kappa[r]}$ is unique, and (1) holds provided that the $C_{\phi'}^{\kappa[r]}$ are constructed to be "orthogonal" in this space in a certain sense.

Further properties of zonal polynomials extending to the invariant polynomials include [1, 5, 11]

$$\int_{O(m)} C_\phi^{\kappa[r]}(A'HXHA, A_2, \ldots, A_r)\,dH$$
$$= C_\phi^{\kappa[r]}(A'A, A_2, \ldots, A_r)C_{\kappa(1)}(X)/C_{\kappa(1)}(I).$$

$$\int_{R>0} \exp(-\operatorname{tr} WR)|R|^{a-(m+1)/2} \times C_\phi^{\kappa[r]}(X_{[r]} \cdot R)\,dR$$
$$= \Gamma_m(a,\phi)|W|^{-a}C_\phi^{\kappa[r]}(X_{[r]} \cdot W^{-1}),$$

where $\Gamma_m(a,\phi)$ was defined by Constantine [2]. Various properties are specific to the invariant polynomials:

$$C_\phi^{\kappa[r]}(X, \ldots, X) = \theta_\phi^{\kappa[r]}C_\phi(X),$$

where $\theta_\phi^{\kappa[r]} = C_\phi^{\kappa[r]}(I, \ldots, I)/C_\phi(I)$.

$$C_\phi\left(\sum_{i=1}^{r} X_i\right) = \sum_{\kappa[r]}\sum_{\phi' \equiv \phi}\binom{f}{k_{(1)}, \ldots, k_{(r)}} \times \theta_{\phi'}^{\kappa[r]}C_{\phi'}^{\kappa[r]}(X_{[r]}).$$

$$\prod_{i=1}^{r} C_{\kappa(i)}(X_i) = \sum_{\phi \in \kappa(1)\ldots\kappa(r)} \theta_\phi^{\kappa[r]}C_\phi^{\kappa[r]}(X_{[r]}).$$

Extensions are possible for the invariant polynomials, but care is required because of nonuniqueness. Constantine's [3] Laguerre polynomials have various generalizations [1, 5]. Hayakawa's [10] polynomials may be expanded

$$P_{\phi}(T,A) = (\tfrac{1}{2}m)_{\phi} \sum_{\kappa,\lambda} \sum_{\phi' \equiv \phi} \binom{f}{k} \theta_{\phi'}^{\kappa,\lambda} \times C_{\phi'}^{\kappa,\lambda}(-A, T'TA)/(\tfrac{1}{2}m)_{\lambda}, \quad (2)$$

where $(a)_{\kappa} = \Gamma_m(a;\kappa)/\Gamma_m(a)$. Khatri's [15] generalized Laguerre polynomial

$$L_{\lambda}^{\tau}(X,A) = \exp(\operatorname{tr} X) \int_{R>0} \exp(-\operatorname{tr} R)|R|^{\tau} \times C_{\kappa}(AR)A_{\tau}(XR)\,dR,$$

where A_{τ} is the Bessel function* of matrix argument, is given by (2) with $\frac{1}{2}m, T'T$ replaced by $\tau + \frac{1}{2}(m+1), X$, respectively. Formal expansions in terms of the invariant polynomials have been given for a number of multivariate distributions, including the latent roots* [6] and trace [16] of the noncentral Wishart matrix*, the multivariate noncentral quadratic form [5], doubly noncentral multivariate F [1, 5], and MANOVA F under model II assumptions [8]. The latter provides a direct derivation of conditions [18] under which F has the usual multivariate F distribution. Phillips [17] gives the density of the two-stage least-squares estimator b of the coefficient vector β in the structural equation in econometrics*, with $n+1$ endogenous variables, as

$$\text{const.}|I + bb'|^{-(L+n+1)/2} \times \sum_{j,k=0}^{\infty} \sum_{\phi \in j \cdot \kappa} \left(\frac{1}{2}L\right)_j \left(\frac{1}{2}(L+n)\right)_{\kappa} \theta_{\phi}^{j,\kappa} \times \frac{C_{\phi}^{j,\kappa}\left(A, (I+\beta b')(I+bb')^{-1}(I+b\beta')G\right)}{j!\,k!\,\left(\frac{1}{2}(L+n)\right)_{\phi}}$$

where A, G are constant matrices, and L is the degree of overidentification*. The mean and covariance matrix of b have been expressed in terms of the polynomials [12]. Problems of construction and convergence are, of course, far more serious for invariant polynomials than they already are for zonal polynomials. The $C_{\phi}^{\kappa[r]}$, however, originated in an investigation of the effects of moderate nonnormality on the MANOVA statistics. A formal procedure [4] requires the normal-theory distribution of the doubly noncentral beta distribution with n_1, n_2 degrees of freedom,

$$\text{const.}|B|^{(n_1-m-1)/2}|I-B|^{(n_2-m-1)/2} \times \prod_{i<j}^{m}(b_i - b_j) \times \sum_{k,l=0}^{\infty} \sum_{\phi \in \kappa \cdot \lambda} (-1)^{k+l} C_{\phi}^{\kappa,\lambda}(\Omega_1, \Omega_2) \times p_{\phi}^{\kappa,\lambda}(n_1, n_2, B)/[k!\,l!\,(\tfrac{1}{2}n_1)_{\kappa}(\tfrac{1}{2}n_2)_{\lambda}], \quad (3)$$

where $0 < b_m < \cdots < b_1 < 1$ are the roots of B; Ω_1, Ω_2 are noncentrality matrices, and $p_{\phi}^{\kappa,\lambda}$ are multivariate generalizations of Tiku's [19] polynomials. The distribution of B under moderate nonnormality may be expanded to first order using the early terms of (3), and results have been derived for various MANOVA statistics (see ref. 9 and references therein).

References

[1] Chikuse, Y. (1980). In *Multivariate Analysis*, R. P. Gupta, ed. North-Holland, Amsterdam, pp. 53–68.

[2] Constantine, A. G. (1963). *Ann. Math. Statist.*, **34**, 1270–1285.

[3] Constantine, A. G. (1966). *Ann. Math. Statist.*, **37**, 215–225.

[4] Davis, A. W. (1976). *Biometrika*, **63**, 661–670.

[5] Davis, A. W. (1979). *Ann. Inst. Statist. Math.*, **31**, **A**, 465–485.

[6] Davis, A. W. (1980). In *Multivariate Analysis V*, P. R. Krishnaiah, ed. North-Holland, Amsterdam, pp. 287–299.

[7] Davis, A. W. (1981). *Ann. Inst. Statist. Math.*, **33**, **A**, 297–313.

[8] Davis, A. W. (1982). *Ann. Inst. Statist. Math.*, **34**, **A**, 517–521.

[9] Davis, A. W. (1982). *J. Amer. Statist. Ass.*, **77**, 896–900.

[10] Hayakawa, T. (1969). *Ann. Inst. Statist. Math.*, **21**, **A**, 1–21.

[11] Hayakawa, T. (1982). *J. Statist. Plann. Infer.*, **6**, 105–114.

[12] Hillier, G. H., Kinal, T. W., and Srivastava, V. K. (1984). *Econometrica*, **52**, 185–202.

[13] James, A. T. (1961). *Ann. Math. Statist.*, **32**, 874–882.

[14] James, A. T. (1968). *Ann. Math. Statist.*, **39**, 1711–1718.

[15] Khatri, C. G. (1977). *S. Afr. Statist. J.*, **11**, 167–179.

[16] Mathai, A. M. and Pillai, K. C. S. (1982). *Commun. Statist. Theor. Meth.*, **11**, 1077–1086.

[17] Phillips, P. C. B. (1980). *Econometrica*, **48**, 861–878.

[18] Roy, S. N. and Gnanadesikan, R. (1959). *Ann. Math. Statist.*, **30**, 318–339.

[19] Tiku, M. L. (1964). *Biometrika*, **51**, 83–95.

(MATRIX-VALUED DISTRIBUTIONS
ZONAL POLYNOMIALS)

A. W. DAVIS

POLYSERIAL CORRELATION *See* POLYCHORIC AND POLYSERIAL CORRELATION

POLY-*t* DISTRIBUTIONS

These are continuous distributions with density functions which are proportional to products of probability density functions (PDFs) for Pearson type VII distributions. The distributions may be univariate or multivariate. A general form of PDF for k dimensional poly-t distributions is

$$f_{\mathbf{T}}(\mathbf{t}) \propto \prod_{j=1}^{g} \left\{ 1 + (\mathbf{t} - \boldsymbol{\mu}_j)' \boldsymbol{\Sigma}_j^{-1} (\mathbf{t} - \boldsymbol{\mu}_j) \right\}^{-(1/2)\nu_j}, \tag{1}$$

where $\mathbf{T}$, $\mathbf{t}$, and $\boldsymbol{\mu}_j$ are $k \times 1$ vectors and $\boldsymbol{\Sigma}_j$ is a $k \times k$ positive definite matrix. If $\boldsymbol{\mu}_j = \mathbf{0}$, then (1) is proportional to a product of multivariate t PDFs*. If $k = 1$, we have a univariate poly-t distribution.

The distributions arise in connection with Bayesian* analysis of (multivariate and univariate) linear models*. Practical use is handicapped by considerable computational difficulties. Broemeling and Abdullah [1] describe a method of approximation which gives promising results, although it still requires considerable calculation.

The name "poly-t" appears to have been coined by Dreze [3], in a paper containing a summary of properties of these distributions (see also Dickey [2]). An interesting result is that the conditional distribution of any subset of the T's, given any disjoint subset, is also a poly-t distribution.

An extension to the class of poly-t distributions is formed by taking the PDF proportional to the ratio of two poly-t PDFs. These are called *ratio-form* poly-t distributions. To have a proper distribution, it is necessary for the sum of ν_j''s for the numerator PDFs to exceed the sum of ν_j's for the denominator PDFs by more than 1.

References

[1] Broemeling, L. and Abdullah, M. Y. (1984). *Commun. Statist. Theor. Meth.*, **13**, 1407–1422.

[2] Dickey, J. (1968). *Ann. Math. Statist.*, **39**, 1615–1627.

[3] Dreze, J. H. (1977). *J. Econometrics*, **6**, 329–354.

(MULTIVARIATE *t* DISTRIBUTION)

POLYTOMOUS VARIABLES

Variables which can take more than two possible values. Although this would include, formally, all random variables except Bernoulli-type variables, the term is conventionally used for variables taking only a finite number (greater than two) of possible values, and almost always these are nonnumerical (geographical region, temperament, etc.).

(CONTINGENCY TABLES)

POLYTONICITY AND MONOTONICITY, COEFFICIENTS OF

The concepts of monotonicity and polytonicity are among the most widespread in science and mathematics. In all disciplines there appear—in one form or another—many problems involving two ordered variables, say x and y—numerical or nonnumerical. Interest is in ascertaining what systematic change—if any—occurs in y as x increases. Sometimes there is also interest in any trend* of change in x as y increases. Data analysis involving the concepts occurs in diverse forms and contexts such as elementary descriptive statistics, multiple regression and discriminant analysis*, time series*, models for representing data structures, and algorithm construction for multivariate computer programs.

In empirical problems, perfect trend rarely occurs. Extent of scatter, or efficacy of fit of the trend to the empirical data, can be assessed in many ways. Popular types of coefficients for this purpose are correlation coefficients, which vary in absolute value between 0 and 1, where 1 expresses perfect fit. Coefficients of monotonicity and polytonicity are examples of such efficacy coefficients. Despite—or because of—their generality, their structure is simpler than that of traditional specialized correlation coefficients which assume linear or polynomial trend. The more general coefficients treat the concepts of monotonicity and polytonicity *intrinsically*, in terms of inequalities. The relevant inequalities will be presented next for the case of monotone trend; extension to polytone trend is rather immediate. Efficacy coefficients based on the inequalities are presented thereafter.

PERFECT TREND

By a *monotone trend* for y on x is meant that, as x increases, any change in y is one-

Table 1 Four Levels of Strictness of Positive Monotone Trend, for y on x

Level	Verbal Definition	Algebraic Definition ($i, j = 1, 2, \ldots, n$)	Graphic Example	Monotonicity Coefficient[a]
1. Weak	As x increases, y does not decrease; and if x stands still, y may increase, decrease, or also stand still	If $x_i > x_j$, then $y_i \geqslant y_j$ and if $x_i = x_j$, then $y_i < y_j$	y, x	$\mu^{(1)}$
2. Semiweak	As x increases, y increases; x is a single-valued function of y	If $x_i > x_j$, then $y_i > y_j$; if $y_i = y_j$, then $x_i = x_j$	y, x	$\mu^{(2)}$
3. Semistrong	As x increases, y does not decrease; y is a single-valued function of x	If $x_i > x_j$, then $y_i \geqslant y_j$; if $x_i = x_j$, then $y_i = y_j$	y, x	$\mu^{(3)}$
4. Strong	As x increases, y increases; y is a single-valued function of x (and x is a single-valued function of y)	If $x_i > x_j$, then $y_i > y_j$; if $x_i = x_j$, then $y_i = y_j$	y, x	$\mu^{(4)}$

[a] For efficacy of fit

directional: always upward (positive trend) or always downward (negative trend), with the possibility that y can occasionally stand still. If both upward and downward movements occur for y as x continues to increase, then the trend of y on x is called *polytone*. These definitions allow either or both of x and y to be ordered only qualitatively. They apply equally well to items with discrete attitudinal response categories (such as "strongly agree," "agree," "disagree," "strongly disagree") and to numerically valued continuous physical variables such as temperature and speed. Each of x and y need possess only an internal ordering for its values, so that if i and j are two individuals with pairs of values (x_i, y_i) and (x_j, y_j), respectively, from x and y, it is specified in advance whether or not $x_i > x_j$ and whether or not $y_i > y_j$.

In many cases, tied values may occur *in principle*: It is possible that for some observational pairs $x_i = x_j$ and/or $y_k = y_l$. Such occurrences give rise to four levels of *strictness* of a perfect monotone trend: weak, semiweak, semistrong, and strong. Definitions and graphic examples of these levels are given in Table 1, which presents only the cases for *positive* trend. Reversing the inequalities (and diagrams) yields the corresponding strictness levels for the cases of perfect *negative* trend.

Linear trend is a classical example of monotonicity. Greatly exaggerated use has been made of linear equations as a first approximation to a monotone trend, sometimes with quadratic or higher-degree corrections. Such treatment is *extrinsic*, unnecessarily introducing error of approximation of its own. It may be much simpler—and always correct—to treat monotonicity *intrinsically* in terms of inequalities, without need for an explicit algebraic formula for the trend.

IMPERFECT TREND: EFFICACY COEFFICIENTS FOR MONOTONICITY

Monotonicity coefficients that have been widely used include Spearman's* coefficient of rank correlation ("rho"), Yule's Q* for dichotomies, Kendall's "tau"*, Goodman–Kruskal's* γ, and Guttman's μ_2. Each of these is actually a special case of an infinitely large family of monotonicity coefficients—defined by Guttman's general formula for μ presented below—as is spelled out in Table 2.

The general coefficient μ allows specifying the desired strictness levels, and selection of any desired weights for the loss function for violation of the specified inequalities. To distinguish among the four strictness levels, define the *signature* functions α and β to have the following values for each i and j:

$$\alpha_{ij} = \begin{cases} 1 & \text{if } x_i > x_j, \\ 0 & \text{if } x_i = x_j, \\ -1 & \text{if } x_i < x_j; \end{cases}$$

$$\beta_{ij} = \begin{cases} 1 & \text{if } y_i > y_j, \\ 0 & \text{if } y_i = y_j, \\ -1 & \text{if } y_i < y_j. \end{cases}$$

Let $\theta_{ij}^{(s)}$ be defined in terms of the absolute values of α and β, the superscript s designating the strictness level to be assessed:

$$\theta_{ij}^{(s)} = \begin{cases} |\alpha_{ij}\beta_{ij}| & (s = 1), \\ |\alpha_{ij}| & (s = 2), \\ |\beta_{ij}| & (s = 3), \\ |\alpha_{ij}| + |\beta_{ij}| - |\alpha_{ij}\beta_{ij}| & (s = 4). \end{cases}$$

Let w_{ij} be any nonnegative numbers (weights) such that $w_{ij} > 0$ whenever $\theta_{ij}^{(s)} > 0$. Then the general μ formula is given by

$$\mu^{(s)} = \frac{\sum_{i=1}^{n}\sum_{j=1}^{n} w_{ij}\alpha_{ij}\beta_{ij}}{\sum_{i=1}^{n}\sum_{j=1}^{n} w_{ij}\theta_{ij}^{(s)}},$$

$$\begin{cases} s = 1\text{: weak}, \\ s = 2\text{: semiweak for } y \text{ on } x, \\ s = 3\text{: semistrong for } y \text{ on } x, \\ s = 4\text{: strong}. \end{cases}$$

It is easily seen that, for all choices of w_{ij},

$$-1 \leqslant \mu^{(s)} \leqslant 1 \qquad (s = 1, 2, 3, 4).$$

Table 2 Examples of Monotonicity Coefficients of the μ-Family, by Strictness and Weight

	Strictness of Monotone Trend of y on x			
Weight[a]	$\mu^{(1)}$: Weak	$\mu^{(2)}$: Semiweak	$\mu^{(3)}$: Semistrong	$\mu^{(4)}$: Strong
$\mu_0 : w_{ij} \equiv 1$	$\mu_0^{(1)} = \frac{\sum\sum_{ij} \alpha_{ij}\beta_{ij}}{\sum\sum_{ij} \lvert\alpha_{ij}\beta_{ij}\rvert}$ (Goodman–Kruskal's γ)	$\mu_0^{(2)} = \frac{\sum\sum_{ij} \alpha_{ij}\beta_{Ij}}{\sum\sum_{ij} \lvert\alpha_{ij}\rvert}$	$\mu_0^{(3)} = \frac{\sum\sum_{ij} \alpha_{ij}\beta_{ij}}{\sum\sum_{ij} \lvert\beta_{ij}\rvert}$	$\mu_0^{(4)} = \frac{\sum\sum_{ij} \alpha_{ij}\beta_{ij}}{n(n-1)}$ (Kendall's tau)
$\mu_1 : w_{ij} = \lvert y_i - y_j\rvert$	$\mu_1^{(1)} = \frac{\sum\sum_{ij} \alpha_{ij}(y_i - y_j)}{\sum\sum_{ij} \lvert\alpha_{ij}(y_i - y_j)\rvert}$	$\mu_1^{(2)} = \frac{\sum\sum_{ij} \alpha_{ij}(i-j)}{\sum\sum_{ij} \lvert\alpha_{ij}(i-j)\rvert}$	$\mu_1^{(3)} = \frac{\sum\sum_{ij} \alpha_{ij}(y_i - y_j)}{\sum\sum_{ij} \lvert y_i - y_j\rvert}$	$\mu_1^{(4)} = \frac{\sum\sum_{ij} \alpha_{ij}(i-j)}{\sum\sum_{ij} \lvert i-j\rvert}$ (Spearman's rho)
$\mu_2 : w_{ij} = \lvert x_i - x_j\rvert \cdot \lvert y_i - y_j\rvert$	$\mu_2^{(1)} = \frac{\sum\sum_{ij}(x_i - x_j)(y_i - y_j)}{\sum\sum_{ij} \lvert x_i - x_j\rvert \lvert y_i - y_j\rvert}$ (Guttman's weak μ_2)	$\mu_2^{(2)} = \frac{\sum\sum_{ij}(x_i - x_j)(i-j)}{\sum\sum_{ij} \lvert x_i - x_j\rvert \lvert i-j\rvert}$	$\mu_2^{(3)} = \frac{\sum\sum_{ij}(i-j)(y_i - y_j)}{\sum\sum_{ij} \lvert i-j\rvert \cdot \lvert y_i - y_j\rvert}$	$(\mu_2^{(4)})$

[a]See the discussion in the text.

EXAMPLES OF WEIGHTS IN TABLE 2

For the μ_1 series, it is assumed that y is numerical, so that $\operatorname{sgn}(y_i - y_j) = \beta_{ij}$, whence $|y_i - y_j| \beta_{ij} = y_i - y_j$. For $\mu_1^{(2)}$ and $\mu_1^{(4)}$—where no ties are supposed to occur for y—we have for simplicity assumed no ties actually occur and that the y_i are ranked from low to high by their index i, so that $|i - j|$ can be used for w_{ij}. A surprising result is that such a $\mu_1^{(4)}$ turns out to equal Spearman's rho when ties are absent also for x ($\alpha_{ij} = 0$ only if $i = j$). By definition, $\mu_1^{(4)}$ is asymmetric in x and y, but turns out to be symmetric when neither of x or y has ties. The numerators of each of the four preceding strictness examples of μ_1 are actually equal to $2n$ times the covariance of the y_i (or rank of y_i) with the $\alpha_{i\cdot}$, where $\alpha_{i\cdot} = \sum_{j=1}^{n} \alpha_{ij}$. The $\alpha_{i\cdot}$ are translations of the ranks of the x_i, whether or not ties occur in x.

For the μ_2 series, both x and y are assumed numerical, where $\operatorname{sgn}(x_i - x_j) = \alpha_{ij}$ and again $\operatorname{sgn}(y_i - y_j) = \beta_{ij}$. The case of $\mu_2^{(2)}$ is illustrated by assuming no ties actually occur in y and that each index i is the rank of y_i; and the case of $\mu_2^{(3)}$ is illustrated by assuming no ties in x and that the indices i are assigned so as to indicate the ranks of x_i. For space reasons an explicit example for $\mu_2^{(4)}$ does not appear in Table 2. The numerators of each of the three given examples of μ_2 are equal to $2n^2$ times the covariance of y (or rank of y) with x (or rank of x). For a fixed choice of the w_{ij}, the levels of strictness are partly ordered among themselves according to the following inequalities for the monotonicity coefficient:

$$\mu^{(4)} \leqslant \left\{ \begin{array}{c} \mu^{(2)} \\ \mu^{(3)} \end{array} \right\} \leqslant \mu^{(1)}.$$

Thus, if $\mu^{(4)}$—the strongest coefficient—is large, so must be all the weaker ones; but if $\mu^{(4)}$ is small, $\mu^{(1)}$ can nevertheless be close to 1.

Yule's Q is a special case for all weak coefficients $\mu^{(1)}$. Different weights do not affect the result when x and y are dichotomies: they cancel out to behave as if all weights equal 1.

Since the numerator of each preceding example of μ_2 is proportional to the covariance of x and y, μ_2 and Pearson's linearity correlation coefficient r must always have the same sign. For Guttman's $\mu_2^{(1)}$, the denominator is relatively smaller than Pearson's, so $|\mu_2^{(1)}| \geqslant |r|$ always (the equality holding only when $r = 0$ or $r = \pm 1$). The other monotonicity coefficients have no such systematic inequality with r.

Considerable attention has been given in the past to the strong coefficients, especially Spearman's and Kendall's. Much of the discussion has been on "what to do in the case of ties," since the "rho" and "tau" formulas assume no ties. Corrections have been attempted by using sampling considerations (see Kendall [5]). The traditional discussions and corrections are beside the point if ties can occur in principle; nonstrong coefficients may be more appropriate.

CATEGORICAL DATA AND STEP FUNCTIONS

Categorical data* and grouped data* are examples where ties generally occur in principle: more than one case generally falls into each category or interval. Typical of this are contingency tables* of a rows and c columns, or where x has a categories and y has c categories. The regression of y on x (or of x on y) is typically not linear for such data. No less important, there is the theorem that the Pearson coefficient r_{xy} can never equal $+1$ unless the marginal distributions of x and y are identical. In particular, if the frequency table is not square ($a \neq c$), and each category has at least one case, then r_{xy} can never equal 1. Attempts to correct for such "artifacts" due to marginal distributions are guided by a usually unnecessary insistence on strong monotonicity, and in particular on linearity. Weak coefficients can equal ± 1 no matter how much a differs from c. The (perfect) underlying trend is usually that of a step function.

RANK-IMAGE PRINCIPLE FOR MONOTONICITY COEFFICIENTS

Spearman's rho is actually a special case of Pearson's product-moment correlation coefficient r for linear regression*, where the x and y variables are specialized to have only the integer values from 1 to n. Strong monotonicity in such a case implies linearity. Thus Spearman's coefficient belongs both to the μ family generated by the absolute principle, and to the Pearson linearity family generated by the least-squares principle. It also belongs to a family of monotonicity coefficients generated by a third principle, namely the *rank-image* principle.

By the rank image y_i^* of y_i is meant a value of x that has the same rank among the n x-values that y_i has among the n y-values:

$$\text{rank}(y_i^* \mid x) = \text{rank}(y_i \mid y).$$

Thus the y_i^* are essentially a permutation of the n x-values, as in Spearman's special case. A perfect monotone relation means there is no need for a permutation: $y_i^* = x_i$ $(i = 1, 2, \ldots, n)$ yielding $r_{y^*x} = r_{xx} = 1$. Hence the Pearson coefficient r_{y^*x} is a monotonicity coefficient: $r_{y^*x} = 1$ if and only if y has a perfect (positive) monotone trend on x. This discussion implies that proper definition of the rank images is made in the case of ties: r_{y^*x} is appropriate for weak as well as for strong monotonicity. The discussion here has also been only for positive trend; reversal of rank order is needed for negative trend.

The rank-image principle was formulated first (in Guttman [2]) for the computing algorithms of smallest space analysis (SSA, better called similarity structural analysis). The discrete process of permutation helps avoid the trap of local minima which plagues other algorithms for "multidimensional scaling"* which rely only on smooth gradient methods. All such algorithms are concerned with determining a best monotone trend between input (similarity coefficients) and output (Euclidean distances). The rank-image principle is, of course, appropriate more widely, being closely related to the absolute value principle.

DISCRIMINANT ANALYSIS* AND MULTIPLE REGRESSION

One of the most elementary problems of data analysis is that of comparing differences among two or more populations in terms of the overlaps of their respective frequency distributions on some numerical variable y. So-called analysis of variance* is widely used for this problem, focusing on the differences among the means of the several populations. The classical correlation coefficient for comparing means is Pearson's correlation ratio* *eta*, and traditional tests of statistical significance are based on Fisher's statistics t and F. Pearson's *eta* is an *efficacy coefficient*, concerned with predicting the y-values from knowledge of the populations to which the individuals belong. The predicted y-values are the respective means of the population, so *eta* can never equal $+1$ unless the variance *within* populations is zero. However, the population distributions can be completely nonoverlapping, whether or not the within-variance vanishes: perfect discrimination between populations can hold, no matter what the size of *eta*. What is needed, then, is a *discrimination coefficient* which will equal 1 for the case of perfect discrimination (complete nonoverlap) among population distributions. Such is provided by special cases of μ_2 [4]. For example, let x_i for each individual be defined to be the mean $\bar{x}$ of the population to which the individual belongs. Then μ_2 becomes a discrimination coefficient, varying between 0 and 1. In this case, $\mu_2 = 0$ if and only if there is no difference in means among the populations (when also *eta* is zero) and equals 1 if and only if there is no overlap among the population distributions (no matter what *eta* is). Intermediate values of μ_2 indicate intermediate extent of overlap.

The statistics t and F are not efficacy coefficients, and indeed estimate no population parameter. They are only part of an abstract scaffolding for statistical inference concerning a null hypothesis of no difference among means. They are irrelevant to discussing efficacy—or completeness—of nonoverlap among population distributions. It is recommended that along with every t and F statistic, there be published the two correlation coefficients *eta* and μ, to give consistent estimates of the respective efficacies of predicting y from the population values $\bar{x}$, and discriminating among the populations from knowledge of y.

Extension is immediate to discriminant functions based on several numerical variables, say $y^{(1)}, y^{(2)}, \ldots, y^{(m)}$. For the linear case, let $y = b_0 + b_1 y^{(1)} + b_2 y^{(2)} + \cdots + b_m y^{(m)}$, and determine the b's to maximize $\mu_{\bar{x}y}$, where the $\bar{x}$ are conditional means of the populations for y. Other variations on this theme are possible.

If there is *a priori* order among the populations according to some variable x, then it suffices to determine the b's in the discriminant function to maximize μ_{xy} as is, without need for conditional means. A special case of this is when there is no *a priori* division into subpopulations, but each individual i has a numerical value x_i. Then the problem is analogous to that of least-squares multiple regression: determine the b's to maximize μ_{xy}. Using μ instead of r here has the same property of avoiding unnecessary assumptions about linearity as discussed above for the simpler cases.

COEFFICIENTS OF POLYTONICITY

Returning to the case of two simply ordered variables x and y, by the number of *tones* in the trend of y on x is meant the minimum number of intervals into which the x-axis can be divided such that monotonicity holds within each interval. The simplest case is of monotone trend itself, which requires only one interval (the entire x-axis). Duotone trend has two intervals, with opposing signs of direction. Tritone trend requires three intervals, with alternating sign of direction, and so on.

It is easy to modify μ to assess the efficacy of polytone trend for any specified number of intervals, provided that division points between intervals are also specified in advance [8]. To do this, define the function γ to take on the values

$$\gamma_{ij} = \begin{cases} 1 & \text{if } i \text{ and } j \text{ are both in the same interval, which is odd numbered} \\ -1 & \text{if } i \text{ and } j \text{ are both in the same interval, which is even numbered} \\ 0 & \text{if } i \text{ and } j \text{ are not in the same interval.} \end{cases}$$

Then $\Pi^{(s)}$ is a polytonicity coefficient for the specified intervals, with strictness s, where

$$\Pi^{(s)} = \frac{\sum_{i=1}^{n}\sum_{j=1}^{n} w_{ij}\alpha_{ij}\beta_{ij}\gamma_{ij}}{\sum_{i=1}^{n}\sum_{j=1}^{n} w_{ij}\theta_{ij}^{(s)}|\gamma_{ij}|} \qquad (s = 1,2,3,4).$$

If perfect polytonicity holds for the given partition into intervals, then $\Pi^{(s)} = \pm 1$, the sign being that of the odd-numbered intervals.

Polytonicity coefficient $\Pi^{(s)}$ weights error only *within* intervals, and makes no comparisons of signs *across* intervals. A polytonicity coefficient for the same data that does make cross-interval comparisons is $\Pi^{*(s)}$, where

$$\Pi^{*(s)} = \frac{\sum_{i=1}^{n}\sum_{j=1}^{n}\sum_{k=1}^{n}\sum_{l=1}^{n} w_{ijkl}\alpha_{ij}\beta_{ij}\alpha_{kl}\beta_{kl}\gamma_{ij}\gamma_{kl}}{\sum_{i=1}^{n}\sum_{j=1}^{n}\sum_{k=1}^{n}\sum_{l=1}^{n} w_{ijkl}\theta_{ij}^{(s)}\theta_{kl}^{(s)}|\gamma_{ij}\gamma_{kl}|}.$$

The weighting possible here is more flexible than previously: it is *not* necessary to have $w_{ijkl} = w_{ij}w_{kl}$. If perfect polytonicity holds, $\Pi^{*(s)} = 1$.

An important special case of polytonicity is that of *cyclicity*. Adi Raveh has developed an ingenious conversion of μ into a family of cyclicity coefficients, giving a powerful means of studying time series* without unnecessary extrinsic assumptions [7].

If an optimal a priori specification of number of intervals—and dividing points therefore—is not available, the data analysis problem can be reversed to seek the optimum by successive trials, each yielding its Π or Π^*. The least polytonicity (smallest number of intervals) can be sought which will yield a satisfactorily large Π or Π^*.

References

[1] Goodman, L. A., and Kruskal, W. H. (1954). *J. Amer. Statist. Ass.*, **49**, 732–764.

[2] Guttman, L. (1968). *Psychometrika*, **33**, 469–506.

[3] Guttman, L. (1979). In *Geometric Representations of Relational Data*, 2nd ed., J. C. Lingoes et al., eds. Mathesis Press, Ann Arbor, Mich., pp. 707–712.

[4] Guttman, L. (1981). In *Multidimensional Data Representations: When and Why*, I. Borg, ed. Mathesis Press, Ann Arbor, Mich., pp. 1–10.

[5] Kendall, M. G. (1963). *Rank Correlation Methods*, 3rd ed. Hafner, New York; Charles Griffin, London.

[6] Kruskal, W. H. (1958). *J. Amer. Statist. Ass.*, **53**, 814–861.

[7] Raveh, A. (1978). In *Theory Construction and Data Analysis in the Behavioral Sciences*, S. Shye, ed. Jossey-Bass, San Francisco, pp. 371–387.

[8] Raveh, A. (1978). In *Theory Construction and Data Analysis in the Behavioral Sciences*, S. Shye, ed. Jossey-Bass, San Francisco, pp. 387–389.

(ASSOCIATION, MEASURES OF
CORRELATION
CORRELATION RATIO
GOODMAN–KRUSKAL TAU AND GAMMA
KENDALL'S TAU
RANK CORRELATIONS
SPEARMAN'S RANK CORRELATION COEFFICIENT
TREND
YULE'S *Q*)

LOUIS GUTTMAN

POOL-ADJACENT-VIOLATORS ALGORITHM

The pool-adjacent-violators algorithm is a procedure that converts a finite sequence of numbers $x_1, x_2, \ldots, x_n$ into a nondecreasing sequence $x_1^* \leqslant x_2^* \leqslant \cdots \leqslant x_n^*$. The new sequence is the isotonic regression of $x_1, x_2, \ldots, x_n$ with respect to simple order (*see* ISOTONIC INFERENCE). It is characterized by the following property. Let w_i be a positive weight associated with x_i. Then the sequence $x_1^*, \ldots, x_n^*$ obtained from the pool-adjacent-violators algorithm with weights $w_1, \ldots, w_n$ minimizes $\sum_{i=1}^n w_i(x_i - x_i^*)^2$ over all nondecreasing sequences.

The pool-adjacent-violators algorithm was first described by Ayer et al. [1] as a means of obtaining maximum likelihood estimates* of Bernoulli parameters $p_1, p_2, \ldots, p_n$ under the constraint $p_1 \leqslant p_2 \leqslant \cdots \leqslant p_n$. With proper choice of weights the algorithm produces maximum likelihood estimates of similarly constrained parameters for many other distributions [2, Ex. 2.1, Sec. 2.4].

To use the algorithm, find a pair of adjacent values x_i and x_{i+1} which violate the restriction $x_i \leqslant x_{i+1}$. (If no such pair exists, then $x_i^* = x_i$ for $i = 1, \ldots, n$, and the algorithm stops.) Replace this pair by the weighted average $(w_i x_i + w_{i+1} x_{i+1})/(w_i + w_{i+1})$, and assign weight $w_i + w_{i+1}$ to the new value. Again search the sequence for adjacent values that are not nondecreasing, replace them by their weighted average, and give the new value a weight equal to the sum of the weights of the pooled values. When no more violating pairs exist, restore the sequence to its original length by setting x_i^* equal to whichever average contains x_i.

Example. Consider the sequence 1, 4, 1, 0, 5, with weights 1, 2, 1, 3, 1. The pair 1, 0 is not nondecreasing, so replace it by the weighted average $(1 \times 1 + 3 \times 0)/(1 + 3) = 0.25$, with weight 4. The new sequence is 1, 4, 0.25, 5, with weights 1, 2, 4, 1. Move on to the pair 4, 0.25, which is replaced by $(2 \times 4 + 4(0.25))/(2 + 4) = 1.5$, with weight 6. The sequence 1, 1.5, 5 (weights 1, 6, 1) has no violating pairs, so the algorithm ends by producing the sequence 1, 1.5, 1.5, 1.5, 5. The end result does not depend on the order in which violating pairs are pooled.

J. B. Kruskal has developed a version of the algorithm that can be easily programmed for a computer; a FORTRAN implementation is given in ref. 3.

References

[1] Ayer, M., Brunk, H. D., Ewing, G. M., Reid, W. T., and Silverman, E. (1955). *Ann. Math. Statist.*, **26**, 641–647.

[2] Barlow, R. E., Bartholomew, D. J., Bremner, J. M., and Brunk, H. D. (1972). *Statistical Inference under Order Restrictions*. Wiley, New York.

[3] Cran, G. W. (1980). *Appl. Statist.*, **29**, 209–211.

(ISOTONIC INFERENCE)

RICHARD F. RAUBERTAS

POOLING DATA

It is generally true that more sample data usually provide more information about the population under study unless under unusual circumstances, for example, when data are "too noisy" (see Chernoff [12]). Using this general rule, statisticians and research workers tend to pool data from various sources when it is believed that the different samples are taken from the same population. In practice, these samples may be taken at different places or at different times. For example, an algebra test is given to eighth-grade students in two schools. Suppose that a random sample of 12 students taken from school A has test scores 86, 95, 78, 81, 64, 73, 92, 46, 59, 77, 75, 68. An independent random sample of 16 students taken from school B yields test scores 48, 82, 68, 91, 68, 77, 74, 69, 86, 78, 59, 90, 79, 64, 59, 80. The sample mean for school A is $\overline{X}_1 = 74.5$ and the sample mean for school B is $\overline{X}_2 = 73.25$. Suppose that the researcher is interested in estimating the population mean score for school A; the usual estimator is $\overline{X}_1 = 74.5$. However, if the researcher believes that the two schools have the same population mean and is correct, it would be advantageous to pool the two samples and use $(12\overline{X}_1 + 16\overline{X}_2)/28 = 73.79$ as the estimate of the population mean. In each case, the precision of the estimate is increased with more data.

In certain situations, the researcher may not be certain whether the different samples are taken from the same population. In the example above, school A may be located in a city and school B located in a rural area. In such a case, he or she may test H_0: $\mu_1 = \mu_2$. If H_0 is accepted, the samples indicate that the population means of the two schools are equal, and the pooled estimate is used. If H_0 is rejected, the researcher uses only the sample mean $\bar{x}_1$. In the example a test statistic for H_0 is the t-test. The calculated t value is 0.25 with 26 degrees of freedom, which is not significant at the 0.05 level. So the final estimate is 73.79.

The test for H_0 is referred to as the *preliminary test*. It is evident that the inference procedure is conditional on the result of the preliminary test. For this reason, Bancroft and Han [3] have designated such an inference procedure as "inference based on conditional specification." They also provided a bibliography in this area of study.

In the next section we consider the pooling of means in detail. The pooling of variances and regression equations are discussed later.

POOLING MEANS

Suppose that a random sample $X_{11}, X_{12}, \ldots, X_{1n_1}$ is taken from a normal distribution $N(\mu_1, \sigma^2)$ and a second independent random sample $X_{21}, X_{22}, \ldots, X_{2n_2}$ is taken from $N(\mu_2, \sigma^2)$. We are interested in estimating μ_1. The sample means are $\overline{X}_1 = \sum X_{1j}/n_1$ and $\overline{X}_2 = \sum X_{2j}/n_2$. The usual estimator of μ_1 is $\overline{X}_1$. However, if $\mu_1 = \mu_2$, a better estimator is the pooled sample mean $\overline{X}_{12} = (n_1\overline{X}_1 + n_2\overline{X}_2)/(n_1 + n_2)$. When the experimenter suspects equality but is uncertain whether $\mu_1 = \mu_2$, he may use a preliminary test to resolve the uncertainty. The choice of the preliminary test depends on whether σ^2 is known or unknown. If σ^2 is known, a normal test is used, and the test statistic for $H_0: \mu_1 = \mu_2$ is

$$Z = (\overline{X}_1 - \overline{X}_2)\Big/\sqrt{\sigma^2(1/n_1 + 1/n_2)}\ .$$

The *sometimes-pool estimator* (also known as the preliminary test estimator) is defined as

$$\overline{X} = \begin{cases} \overline{X}_1 & \text{if } |Z| \geqslant Z_\alpha \\ \overline{X}_{12} & \text{if } |Z| < Z_\alpha, \end{cases} \tag{1}$$

where Z_α is the $100(1 - \alpha/2)\%$ point of the standard normal distribution. The rationale of using $\bar{X}$ in (1) is that when the preliminary test accepts H_0, a pooled sample mean is used; so that the precision of the estimator is increased when the difference between μ_1 and μ_2 is zero or negligible.

The bias* and the mean square error* (MSE) of the sometimes-pool estimator has been studied by Bennett [9] and Mosteller [25]. Kale and Bancroft [20] have extended this work to discrete data by using a proper transformation*. They also studied the relative efficiency of the sometimes-pool estimator to the *never-pool estimator* $\bar{X}_1$. For fixed n_1, n_2, and α, the bias and relative efficiency* are functions of $\delta = (\mu_2 - \mu_1)/\sigma$. The general behavior of the bias is that it equals 0 when $\delta = 0$; it increases to a maximum and then decreases to 0 as $|\delta|$ increases to infinity. The general behavior of the relative efficiency is that it has the maximum value at $\delta = 0$, decreases to a minimum (below unity), and then increases to unity as $|\delta|$ increases from 0 to infinity.

When σ^2 is unknown, the preliminary test is a t-test. The test statistic is

$$t = (\bar{X}_1 - \bar{X}_2)\Big/\sqrt{S_p^2(1/n_1 + 1/n_2)}\ ,$$

where S_p^2 is the pooled sample variance

$$S_p^2 = \frac{\sum(X_{1j} - \bar{X}_1)^2 + \sum(X_{2j} - \bar{X}_2)^2}{n_1 + n_2 - 2}. \tag{2}$$

The sometimes-pool estimator of μ_1 is defined as

$$\bar{X}^* = \begin{cases} \bar{X}_1 & \text{if } |t| \geqslant t_\alpha \\ \bar{X}_{12} & \text{if } |t| < t_\alpha, \end{cases} \tag{3}$$

where t_α is the $100(1 - \alpha/2)\%$ point of the t distribution with $n_1 + n_2 - 2$ degrees of freedom.

Kitagawa [21] has found the distribution of the sometimes-pool estimator. He also gave the bias and mean square error (MSE) in infinite sums. Han and Bancroft [14] derived the bias and MSE in closed forms. They studied the bias and relative efficiency and gave the following criterion for selecting the level of significance of the preliminary test.

Denote the relative efficiency of the sometimes-pool estimator to the never-pool estimator by $e(\alpha, \delta)$, which is a function of α and δ. If the experimenter does not know the size of δ and is willing to accept an estimator which has a relative efficiency of no less than e_0, then among the set of estimators with $\alpha \in A$, where $A = \{\alpha : e(\alpha, \delta) \geqslant e_0 \text{ for all } \delta\}$, the estimator is chosen to maximize $e(\alpha, \delta)$ over all α and δ. Since $\max_\delta e(\alpha, \delta) = e(\alpha, 0)$, he selects the $\alpha \in A$ (say α^*), which maximizes $e(\alpha, 0)$ (say e^*). This criterion will guarantee that the relative efficiency of the chosen estimator is at least e_0 and it may become as large as e^*. The values of α^* and e^* are tabulated in Han and Bancroft [14], which are also given in Bancroft and Han [5].

In another study, Hirano [17] used Akaike's* information criterion to determine the level of significance of the preliminary test. Bancroft and Han [7] studied the robustness of the sometimes-pool estimator. They found that when the population is either uniform* or exponential*, the sometimes-pool estimator is quite robust. Ohta [27] proposed a graphical procedure for pooling data by the analysis of means. Bancroft and Han [2] studied the sometimes-pool estimator in multivariate normal distributions.

Mosteller [25] and Han and Bancroft [14] considered a Bayesian pooling estimator. Let the prior distribution of δ be normal with mean 0 and variance a^2; then the maximum likelihood estimator is

$$\hat{\mu}_1 = \frac{n_1(n_2a^2 + 1)\bar{X}_1 + n_2\bar{X}_2}{n_1(n_2a^2 + 1) + n_2}. \tag{4}$$

The MSE of $\hat{\mu}_1$ after averaging over the distribution of δ is smaller than the MSE of the sometimes-pool estimator. This is expected because more information is assumed to be available in the Bayesian* analysis.

POOLING VARIANCES

The first study in conditional specification inference is the paper by Bancroft [1], who considered the pooling of two sample variances. Let S_1^2 and S_2^2 be two independent estimators of σ_1^2 and σ_2^2, respectively, with $\nu_i S_i^2/\sigma_i^2$ distributed as χ^2 with ν_i degrees of freedom, $i = 1, 2$. We are interested in estimating σ_1^2. When $\sigma_1^2 = \sigma_2^2$, the pooled estimator $S_p^2 = (\nu_1 S_1^2 + \nu_2 S_2^2)/(\nu_1 + \nu_2)$ should be used. In practice the experimenter may be uncertain whether $\sigma_1^2 = \sigma_2^2$. In such a case, he may use an F test to test $H_0: \sigma_1^2 = \sigma_2^2$. Let the alternative hypothesis be $H_1: \sigma_1^2 > \sigma_2^2$. The sometimes-pool estimator is defined as

$$\hat{\sigma}_1^2 = \begin{cases} S_p^2 & \text{if } (S_1^2/S_2^2) < F(\alpha; \nu_1, \nu_2) \\ S_1^2 & \text{if } (S_1^2/S_2^2) \geqslant F(\alpha; \nu_1, \nu_2) \end{cases} \tag{5}$$

where $F(\alpha; \nu_1, \nu_2)$ is the $100(1-\alpha)\%$ point of the F distribution* with ν_1 and ν_2 degrees of freedom.

The bias and MSE of the sometimes-pool estimator are given in Bancroft [1]. The relative efficiency of $\hat{\sigma}_1^2$ to the never-pool estimator S_1^2 was studied by Bancroft and Han [6]. They found that if the significance level of the preliminary F test is carefully selected, the sometimes-pool estimator is always better than S_1^2. An approximate rule for the optimal significance level is that one selects $\alpha = 0.40$ when $\nu_1 \doteq \nu_2$, $\alpha = 0.35$ when $\nu_2 \doteq 2\nu_1$ and $\alpha = 0.45$ when $\nu_1 \doteq 2\nu_2$.

The methodology of pooling variances also occurs in analysis of variance* (ANOVA). When two or more mean squares in an analysis-of-variance table have the same expected value, these mean squares are pooled to obtain a better estimate of the error variance. To illustrate this, let us consider a two-way classification (which may either be a fixed model or a random model):

$$y_{ijk} = \mu + \alpha_i + \beta_j + (\alpha\beta)_{ij} + \epsilon_{ijk}, \tag{6}$$

$i = 1, \ldots, I, \quad j = 1, \ldots, J, \quad k = 1, \ldots, K,$

where $\epsilon_{ijk} \sim \text{NID}(0, \sigma^2)$. The experimenter may be uncertain whether the interaction term $(\alpha\beta)_{ij}$ is zero. To resolve the uncertainty, he performs an F-test for testing the interaction term. If the test accepts the null hypothesis, that the interaction* is zero, the interaction mean square is pooled with the error mean square; otherwise, it is not pooled. Hence the interaction mean square is viewed as a doubtful error term. A description of the ANOVA is given in Table 1.

The usual test statistic for $H_0: \sigma_3^2 = \sigma_2^2$ is V_3/V_2, which has an $F(n_3, n_2)$ distribution. However, if $\sigma_1^2 = \sigma_2^2$, the experimenter, from sound theoretical considerations, may pool the two mean squares V_2 and V_1 and use $V = (n_1 V_1 + n_2 V_2)/(n_1 + n_2)$ as the error term in the test procedure. The test statistic would be V_3/V. When the experimenter is uncertain whether $\sigma_2^2 = \sigma_1^2$, he may use a preliminary F test. The test procedure under the conditionally specified model is to reject the main hypothesis $H_0: \sigma_3^2 = \sigma_2^2$ if either

$$\{V_2/V_1 \geqslant F(\alpha_1; n_2, n_1) \quad \text{and} \quad V_3/V_2 \geqslant F(\alpha_2; n_3, n_2)\}$$

or

$$\{V_2/V_1 < F(\alpha_1; n_2, n_1) \quad \text{and} \quad V_3/V \geqslant F(\alpha_3; n_3, n_1 + n_2)\}.$$

This test procedure is referred to as the *sometimes-pool test*.

The effect of the preliminary test on the size and power depends on whether the model in (6) is random or fixed. For the random model, Bozivich et al. [10] and Paull [28] have studied the size and power of the sometimes-pool test. The fixed model was studied by Bechhofer [8] and Mead et al. [24]. See also Bancroft and Han [4] for other references; *see also* FIXED-, RANDOM-, AND MIXED-EFFECTS MODELS.

Table 1 Analysis of Variance

Source of Variation	Degrees of Freedom	Mean Square	Expected Mean Square
Treatment	n_3	V_3	σ_3^2
Error	n_2	V_2	σ_2^2
Doubtful error	n_1	V_1	σ_1^2

POOLING REGRESSION EQUATIONS

When two or more regression equations are the same, it is advantageous to pool the data for making inferences about the population regression model of interest. If the investigator is uncertain whether the regression equations are the same, he may use a preliminary test to decide whether to pool the data. As an example, the Vital Statistics of the United States gave the death rate for males (per 1000 population) by age for 1969 and 1970. With the age coded from 0 to 9, the regression line of $\log_{10}$(death rate) $= y$ on age $= x$ are given as

$$1970: \quad \hat{y}_1 = 1.5017 + 0.1732(x - 4.5),$$

with residual mean square

$$s_1^2 = 0.000646$$

$$1969: \quad \hat{y}_2 = 1.5087 + 0.1745(x - 4.5),$$

with residual mean square

$$s_2^2 = 0.000793.$$

The two lines are almost identical. The investigator may use a preliminary test to decide whether to pool the 1969 line with the 1970 line in prediction. Such a procedure is referred to as the *sometimes-pool predictor*. Symbolically, we are given two regression models:

$$y_{ij} = \beta_{i0} + \beta_{i1}(x_{ij} - \bar{x}_i) + \epsilon_{ij}, \quad i = 1, 2; \quad j = 1, \ldots, n_i, \quad (7)$$

where x_{ij}'s are fixed known constants and $\epsilon_{ij} \sim \text{NID}(0, \sigma^2)$. We are interested in estimating the line

$$y_{1j} = \beta_{10} + \beta_{11}(x_{1j} - \bar{x}_1),$$

when it is suspected that $\beta_{10} = \beta_{20}$ and $\beta_{11} = \beta_{21}$. A preliminary test for testing the equality of the two lines is an F-test with test statistic

$$F = \left[(b_{10} - b_{20})^2 n_1 n_2/(n_1 + n_2) + (b_{11} - b_{21})^2 c_1 c_2/(c_1 + c_2)\right]/(2s^2),$$

where the b's are the least-squares estimate of the β's, $c_i = \sum(x_{ij} - \bar{x}_i)^2$, $i = 1, 2$, and s^2 the pooled estimate of σ^2 from the two models. The sometimes-pool predictor is defined as

$$y_{1j}^* = \begin{cases} \bar{b}_0 + \bar{b}_1(x_{1j} - \bar{x}_1) & \text{if } F \leqslant F(\alpha; 2, n_1 + n_2 - 4) \\ b_{10} + b_{11}(x_{1j} - \bar{x}_1) & \text{if } F > F(\alpha; 2, n_1 + n_2 - 4), \end{cases} \quad (8)$$

where $\bar{b}_0 = (n_1 b_{10} + n_2 b_{20})/(n_1 + n_2)$ and $\bar{b}_1 = (c_1 b_{11} + c_2 b_{21})/(c_1 + c_2)$. The relative efficiency of the sometimes-pool predictor to the never-pool predictor $\hat{y}_1$ was given by Johnson et al. [19], who also studied the pooling of multiple regression* equations.

In the vital statistics* example above, the calculated F value from the data is 0.20 with 2 and 16 degrees of freedom and this is not significant at the 0.40 level. In such cases, the rule of procedure tells us to pool the 1969 line with the 1970 line to provide a "better" prediction. The resulting prediction line is $y^* = \bar{b}_0 + \bar{b}_1(x - \bar{x}) = 1.5052 + 0.1739(x - 4.5)$.

If the investigator is interested in estimating the regression coefficient β_{11}, the sometimes-pool estimator is given as

$$\hat{\beta}_{11} = \begin{cases} \bar{b}_1 & \text{if } \sum c_i(b_{i1} - \bar{b}_1)^2/s^2 \leqslant F(\alpha; 1, n_1 + n_2 - 4) \\ b_{11} & \text{if } \sum c_i(b_{i1} - \bar{b}_1)^2/s^2 > F(\alpha; 1, n_1 + n_2 - 4). \end{cases} \quad (9)$$

The properties of $\hat{\beta}_{11}$ were studied by Han and Bancroft [15].

In determining the sometimes-pool predictor in (8) or the sometimes-pool estimator in (9), the investigator must decide an appropriate significance level of the preliminary test. As stated in the criterion given in the section "Pooling Means," the level should be selected so that the relative efficiency is high. This usually indicates that the value of α is selected at a moderate value other than the traditional 0.05 or 0.01 level.

In econometrical models, cross-sectional and time-series* data may be pooled. Suppose that data are collected on N economic or cross-sectional units and over T time periods. The model can be written as

$$y_{jt} = \sum_{i=1}^{k} \beta_i x_{ijt} + \delta_j + \gamma_t + \epsilon_{jt},$$

$$j = 1, \ldots, N, \quad t = 1, \ldots, T, \quad (10)$$

where δ_j is the cross-sectional unit effect, γ_t is the time period effect, and ϵ_{jt} is the error term. The effects may be treated either as fixed effects or random effects. The choice would depend on the particular practical situation. Once the model assumptions are determined, estimators of β_i can then be obtained. The various models were studied by Maddala [22], Mundlak [26], and Wallace and Hussain [29]. Applications were given in Chang [11], and Johnson and Oksanen [18]. A computer program for analyzing the data was given by Havenner and Herman [16]. Dielman [13] gave a comprehensive review of pooling cross-sectional and time-series* data.

The model in (10) assumes that the regression coefficients β_i are the same for all cross-sectional units. In practice the investigator may not be certain whether the β_i are equal for all the units. In such a case Maddala [23] advised checking the consistency of the information given about parameters by the data before pooling. This can be done by using preliminary tests of significance. When the regression coefficients are different in cross-sectional units, β_i in (10) is replaced by β_{ij}. The null hypothesis to be tested is $\beta_{i1} = \beta_{i2} = \cdots = \beta_{iN}$ for all i. The level of significance of the preliminary test should be selected appropriately at a moderate value.

References

[1] Bancroft, T. A. (1944). *Ann. Math. Statist.*, **15**, 190–204. (A pioneering paper on pooling data.)

[2] Bancroft, T. A. and Han, C. P. (1976). In *Essays in Probability and Statistics*, S. Ikeda, ed. Shinko Tsusho, Tokyo, pp. 353–363. (Extension of pooling means to multivariate normal case.)

[3] Bancroft, T. A. and Han, C. P. (1977) *Int. Statist. Rev.*, **45**, 117–127. (A chronological bibliography and a subject index are included.)

[4] Bancroft, T. A. and Han, C. P. (1980). In *Handbook of Statistics, Vol. I: Analysis of Variance*, P. R. Krishnaiah, ed. Elsevier North Holland, New York, pp. 407–441. (A review of pooling data in analysis of variance and regression.)

[5] Bancroft, T. A. and Han, C. P. (1981). *Statistical Theory and Inference in Research*. Marcel Dekker, New York.

[6] Bancroft, T. A. and Han, C. P. (1983). *J. Amer. Statist. Ass.*, **78**, 981–983.

[7] Bancroft, T. A. and Han, C. P. (1984). In *Impact of P. V. Sukhatme on Agricultural Statistics and Nutrition*, P. Narain, ed. Indian Soc. Agricult. Statist., pp. 103–114.

[8] Bechhofer, R. E. (1951). Unpublished Ph.D. thesis, Columbia University, New York.

[9] Bennett, B. M. (1952). *Ann. Inst. Statist. Math.*, **4**, 31–43.

[10] Bozivich, H., Bancroft, T. A., and Hartley, H. O. (1956). *Ann. Math. Statist.*, **27**, 1017–1043.

[11] Chang, H. S. (1975). *Amer. Statist. Ass. Proc. Bus. Econ. Sec.*, pp. 269–274. (A study of location of the cotton textile industry in the United States in pooling cross-sectional and time-series data.)

[12] Chernoff, H. (1983). In *A Festshrift for Erich L. Lehmann*. Wadsworth, Belmont, Calif., pp. 115–130. (Uses astronomical data to demonstrate when to ignore data.)

[13] Dielman, T. E. (1983). *Amer. Statist.*, **37**, 111–122. (A survey of pooled cross-sectional and time-series data with a large number of references.)

[14] Han, C. P. and Bancroft, T. A. (1968). *J. Amer. Statist. Ass.*, **63**, 1333–1342.

[15] Han, C. P. and Bancroft, T. A. (1978). *Commun. Statist. A*, **7**, 47–56.

[16] Havenner, A. and Herman, R. (1977). *Econometrics*, **45**, 1535–1536.

[17] Hirano, K. (1978). *Ann. Inst. Statist. Math.*, **30**, 1–8.

[18] Johnson, J. A. and Oksanen, E. H. (1977). *Rev. Econ. Statist.*, **59**, 113–118. (Example of demand for alcoholic beverages in Canada from cross-sectional and time-series data.)

[19] Johnson, J. P., Bancroft, T. A., and Han, C. P. (1977). *Biometrics*, **33**, 57–67.

[20] Kale, B. K. and Bancroft, T. A. (1967). *Biometrics*, **23**, 335–348.

[21] Kitagawa, T. (1963). *Univ. Calif. Publ. Statist.*, **3**, 147–186. (A comprehensive and broad study of estimation after preliminary tests of significance.)

[22] Maddala, G. S. (1971). *Econometrica*, **39**, 341–358.

[23] Maddala, G. S. (1971). *Econometrica*, **39**, 939–953.

[24] Mead, R., Bancroft, T. A., and Han, C. P. (1975). *Ann. Statist.*, **3**, 797–808.

[25] Mosteller, F. (1948). *J. Amer. Statist. Ass.*, **43**, 231–242.

[26] Mundlak, Y. (1978). *Econometrica*, **46**, 69–85.

[27] Ohta, H. (1981). *J. Quality Tech.*, **13**, 115–119. (Uses graphical method in pooling means.)

[28] Paull, A. E. (1950). *Ann. Math. Statist.*, **21**, 539–556.

[29] Wallace, T. D. and Hussain, A. (1969). *Econometrica*, **37**, 55–72.

(ANALYSIS OF VARIANCE
VARIANCE COMPONENTS)

T. A. Bancroft
Chien-Pai Han

POOLS, OPINION

A method introduced by Stone [3] to combine the experts' distributions using weighted averages. Suppose that there are n individual assessors who have assessed distributions $\Pi_1, \ldots, \Pi_n$ over a space Ω. The decision makes them choose weights $\alpha_1, \ldots, \alpha_n$, nonnegative and summing to 1, and the *consensus* distribution Π_c over Ω. The method of opinion pools is defined by putting for any event A in Ω:

$$\Pi_c(A) = \sum_{i=1}^{n} \alpha_i \Pi_i(A).$$

Bacharach [1] calls this the *linear* opinion pool to distinguish it from the logarithmic opinion pool, which is essentially a weighted geometric mean*. A detailed discussion of this method and its consequences is presented in McConway [2].

References

[1] Bacharach, M. (1974). Bayesian Dialogues. Unpublished manuscript, Christ Church, Oxford, England.

[2] McConway, K. J. (1981). *J. Amer. Statist. Ass.*, **76**, 410–414.

[3] Stone, M. (1961). *Ann. Math. Statist.*, **32**, 1339–1342.

(DECISION THEORY)

POPULATION GROWTH MODELS

Population growth models start with the simple exponential* and proceed to more elaborate forms that propose to represent the trajectory of populations.

The exponential applies to a population in which all rates are fixed. That births, deaths, and net migrants, and hence net increase in absolute numbers, grow by the same fraction in each period of given duration defines the exponential population model. Recognizing overall growth at fraction $r = b - d$, where b is the crude birth rate, d the crude death rate, both applying over the short period dt, we have for the population $P(t)$ at time t the equation

$$dP(t) = rP(t)\,dt, \tag{1}$$

whose integral is

$$P(t) = P_0 e^{rt}. \tag{2}$$

Conversely, if we know the population number, the rate of growth is expressible as

$$r = \ln(P(t)/P_0)/t,$$

where ln stands for the natural logarithm.

For some purposes the logarithm to base 2 is useful; it is equal to the number of doublings contained in the growth from P_0 to $P(t)$. The time for doubling is very nearly $t = 70/100r$, where $100r$ is the percent of annual increase. Study of such data as exist on world population over a long period of time shows that until recently doubling times were centuries, even millennia, and our epoch is unique in the lowering of doubling time to as little as 20 years for some countries. World growth reached a peak of just under 2% per year in the 1970s and is now probably 1.8% or lower, with a doubling time of about 40 years. The decades after World War II have been called a time of population explosion in popular writings. Tapering off is so far uneven: Western countries are now below replacement; China and other East Asian countries show a precipitate drop in birth rates, as do several countries of Latin America; Africa remains almost as high as ever.

The exponential progression through time is associated with the name of Malthus*, who called it "geometric," and contrasted it with the arithmetic (constant absolute difference) progression that he considered the most rapid possible expansion of food supplies [26]. Malthus did not believe that a population could continue in an exponential growth trajectory for long—indeed, the impossibility of continued geometric growth is the essence of Malthusian theory—and most of what he wrote is dedicated to showing that stationarity, or at least fluctuation about a mean, is the most common condition.

Whether growth is limited by the natural environment has been a public scholarly issue since the time of Malthus. Such a limit is suggested by current press accounts of drought and starvation in the Sahel, Nigeria, and other parts of Africa. The strong expression of the Malthusian view is that either births will fall or there will be such drastic shortages that mortality will rise sharply. Either way population will approach a ceiling. Research showing this, conducted under the aegis of the Club of Rome [28] and other writing [1] as well as writings opposing the view that world population is now pressing against a ceiling [44] have enlivened the publishing scene over the past decade.

A model in which growth is an intermediate stage between two stationary conditions was developed by Verhulst [49], and later rediscovered by Pearl [30] and others. Verhulst, in effect, modified the simple exponential model $dP(t) = rP(t)\,dt$ by multiplying on the right by a factor $1 - P(t)/a$, where a is the population ultimately attained:

$$dP(t) = rP(t)\left(1 - \frac{P(t)}{a}\right)dt. \tag{3}$$

Evidently, the growth will diminish to zero when $P(t)$ has reached a, which is therefore an asymptote. Solving (3) by partial fractions gives

$$P(t) = \frac{a}{1 + be^{-rt}}. \tag{4}$$

This form has been found useful by ecologists, as portraying the growth of vertebrate and, in general, nonhuman populations, as well as in applications far from the field of population.

Pearl [30] argued that the simple logistic* is not adequate—that population history moves through a series of logistics. The North American Indians had reached an asymptote that accorded with their technology, and in a sense "filled" the continent. Europeans came with more advanced technology, and thus initiated a period of growth that provided Malthus with his example of a population doubling every 25 years, but that then reaches its own asymptote. Further phases of expansion may appear in the future. Such a schematization attracted scholars in its day, with the implication that technology is the engine of population growth. In recent years, Boserup [2] has shown cases where the causation is in the opposite direction: population pressure induces technical advance.

At one time the logistic was thought ideal for forecasting*, and a logistic based on the U.S. censuses from 1790 [31] turned out to fit the 1930 U.S. census* population almost exactly—probably with less departure than the census from the true figure. Following a period of euphoria on the applicability of the logistic, disillusion set in among demographers, especially after the 1940 census came out far below the logistic that had fitted 1930 so well.

One would like to test out such growth models, to see which accord with the course of past experience. Can we not allow the statistical record to discriminate among them? The answer to this question appears to be no [55]. If we fit a logistic, a cumulative normal, and an inverse tangent to a population growth series, we find that they all fit indifferently well. More than that: if we fit a hyperbola,

$$P(t) = \frac{a}{t_e - t},$$

and ascertain the constants t_e and a, it also fits, and t_e, the date of the population explosion, turns out to be in the first half of the twenty-first century [5, 52]. That a perfectly impossible curve fits past data reasonably well is an example of the inability of historical data to discriminate among models.

Recognizing age complicates the trajectory in the short run even if all the age-specific rates* are fixed. With an arbitrary initial age distribution the trajectory over a future period of a century or more will in general show waves, of which the principal one has length equal to the generation, about a quarter century for human beings. Starting with an arbitrary age distribution, say for females, and applying rates unchanging over time generates a trajectory [20, 22] in which the rate of increase is ultimately stable and hence the population numbers exponential.

The model is expressed most readily in matrix form, with the survivorship matrix defined as

$$S = \begin{bmatrix} 0 & 0 & \cdots & 0 & 0 \\ s_0 & 0 & \cdots & 0 & 0 \\ 0 & s_1 & \cdots & 0 & 0 \\ \cdots & \cdots & \cdots & \cdots & \cdots \\ \cdots & \cdots & \cdots & \cdots & \cdots \\ 0 & 0 & \cdots & s_{w-1} & 0 \end{bmatrix}, \tag{5}$$

where s_x is the probability that an individual aged x at last birthday survives to age $x + 1$. The sole nonzero elements are in the subdiagonal, as required by the fact that the only possible transition from age x is to age $x + 1$ if the person survives. The birth matrix has its nonzero elements in the first row, since the newborn start at age zero:

$$B = \tag{6}$$

$$\begin{bmatrix} 0 & \cdots & b_{15} & b_{16} & \cdots & 0 \\ 0 & \cdots & 0 & 0 & \cdots & 0 \\ 0 & \cdots & 0 & 0 & \cdots & 0 \\ \cdots & \cdots & \cdots & \cdots & \cdots & \cdots \\ \cdots & \cdots & \cdots & \cdots & \cdots & \cdots \\ 0 & \cdots & 0 & 0 & \cdots & 0 \end{bmatrix},$$

where $b_{15}, b_{16}, \ldots,$ are age-specific birth rates (i.e., births during a year divided by the midyear population if the data come from actual observation). The initial population is expressed as a (vertical) vector P_0 in which the number of persons in the several ages are listed in ascending order of age.

The growth model then consists in finding the survivors at the end of the first period, SP_0, averaging that with the initial population to find the exposure during the period to the risk of childbearing, $(P_0 + SP_0)/2$, premultiplying this by the birth matrix and finally multiplying by the chance L_0 that each birth survives to the end of the first time unit. Adding this element for the first age interval to the other ages given by SP_0, we have for the complete distribution at the end of the first time interval,

$$SP_0 + \{B(P_0 + SP_0)/2\}L_0 = \{S + B(I + S)L_0/2\}P_0, \tag{7}$$

that can be written as simply MP_0 if we define M as $M = S + B(I + S)L_0/2$.

This representation of population growth by age can be modified to apply to age and sex; age, sex, and marital status; region of residence; and so on [45].

If the parameters of birth, death, and migration* change from time period to time period, those for the ith time period making up the matrix M_i, then the growth process is

$$P_t = M_t \cdots M_2 M_1 P_0. \tag{8}$$

With growth under the operation of a fixed regime of birth and death the age distribution that is ultimately attained can be expressed in terms of the birth and death parameters. If the probability of surviving from birth to exact age x is $l(x)$, then the number of individuals that will be found in the ultimate, stable, age distribution between age x and $x + dx$ is

$$c(x)\,dx = e^{-rx} l(x)\,dx \tag{9}$$

per person just born (i.e., where the unit or radix is one individual at exact age 0). This ergodic property, that the age distribution under the action of the fixed projection matrix forgets its past, is due to Lotka [22] in its modern form, although the idea is clearly present in Euler [8]. Its generalization to the changing projection matrix of (8) is expressed somewhat differently: that two arbitrary and different age distributions acted on by the same sequence of projection matrices $M_1, M_2, \ldots$, will ultimately move through the same sequence of age distributions, a proposition suggested by Coale [4] and demonstrated by Lopez [21].

The age-dependent population model can be expressed in other equivalent ways. Lotka's continuous formulation involved two stages; in the first [22] the expression (9) contains r as an arbitrary parameter; in the second [41] r emerges in the course of solving the equation

$$B(t) = \int_{\alpha}^{\beta} B(t-a)l(a)m(a)\,da + G(t),$$

where $B(t)$ is the births at time t, $m(a)\,da$ the chance of a woman aged a having a child before she attains age $a + da$, and $G(t)$ the number of births at time t to women alive at the outset. The solution by trying $B(t) = e^{rt}$, and Feller's [9] more sophisticated solution using the Laplace transform, both produce the equation for r:

$$\int_{\alpha}^{\beta} e^{-rx} l(x)m(x)\,dx = 1.$$

Cole [6] presented a difference equation*, and A. G. McKendrick [27] a partial differential equation, rediscovered by von Foerster [51]. It may be shown [12] that these are mathematically equivalent.

Sex is more difficult to model than age. The usual practice in population forecasts is to make the population female-dominant, which is to calculate the male births as a simple multiple (about 1.05) of the female. Proper realistic treatment of the sexes leads to non-linear equations for which no closed-form solution is to be found [16, p. 11; 38]. Attempts [13, 33, 40] to show the evolution of a two-sex population continue, inevitably based on simplifying assumptions whose realism is not easily tested.

Notice that rate of growth is ambiguous in a population in which age is recognized. The United States now has about 3.6 million births per year and 1.9 million deaths; it is growing by natural increase at 1.7 million or 0.7% per year. But yet it is said not to be replacing itself; if the present age-specific rates of birth and death continue, it will start to decline once the baby boomers pass out of the reproductive ages. In different words, the net reproduction rate, the expected number of girl children that will be born to a girl child, is less than unity. Replacement requires an average of 2.2 or so children per couple, as against the present 1.8. (These statements abstract from migration.) One aspect of the problem of determining "real" growth is resolved by treating the cohort—the collection of individuals born at the same time and going through life together—as the unit, rather than the cross section of those alive at all ages at a particular moment [36, 54].

The population history of the twentieth century has been dominated by the demographic transition, in advanced countries now past, in others currently being undergone, or anticipated for the near future. In the transition model death rates decline first, followed after a longer or shorter interval by birth rates. If the time interval by which the fall of deaths precedes that of deaths is long, the population greatly increases. Death rates have now fallen almost universally, and the principal population question of this century is when birth rates will fall in those countries where they are still high.

The Lotka–Leslie–Cole–McKendrick model that has been described above generates waves whenever the initial population is different in age distribution from the final stable condition, and the most prominent wave is of length one generation. This growth path has been modified by Easterlin [7], who suggested that small cohorts have better job opportunities and incomes in relation to their parents, and hence tend to have higher birth rates, with the opposite applying to large cohorts. The relation can easily be shown to generate waves of two generations in length [19, p. 271]. The effect of cohort size, in abstraction from all else, seems here to be correctly set forth, and the waves described in a sense exist, but in most historical epochs they do not appear because they are swamped by other influences on fertility*. In the baby boom of the 1950s the Easterlin effect emerged from the complex of factors in the system that determines births. Subsequent to that a different effect —the disposition of women to seek jobs and careers outside the home, along with divorce associated with that disposition—has caused births to fall and to stay low, and so could well inhibit the emergence of an Easterlin rise of birth rates about the end of the century.

A frequent question concerns the growth of a population in which not only age and sex but labor force status, marital status, geographical subareas, and other characteristics are recognized. A method for this is due to Rogers [35] and Schoen [39]. In effect, they apply a theory due to Kolmogorov and are able to derive matrix analogs of the common one-dimensional formulas of demography*.

Lotka [23] and Volterra [50] took up growth in which subpopulations were interacting in various nonlinear ways. The Volterra model is visualized as foxes and their prey, say rabbits, on an island; the foxes live off the rabbits and have no other sustenance. Volterra showed that the population evolves in a series of waves; the foxes overeat on the rabbits, whose numbers diminish; this is followed by diminution of the fox population, which allows the regeneration of the rabbits, and so forth. If the rabbits have no sanctuary, the foxes will eat the last of them and then themselves die out. Such a model has implications for the human population and the resources it requires. Kendall [18] and Goodman [11] applied the ideas of Volterra to the interaction of the sexes, as well as to marriage*, in a human population. They have been extended by developing submodels of population processes such as reproduction, competition, and resource acquisition, which themselves have been tested and generalized in a variety of natural and laboratory studies. When such submodels are combined to form models of population dynamics, the full range of population behaviours occurring in nature are duplicated —stable, oscillatory, epidemic, chaotic, and discontinuous (Holling [15], Clark and Holling [3]).

The growth models here discussed are applicable beyond human populations, to populations of vertebrates and of insects. Lotka [24] applied his formulation to a population of items of industrial equipment.

The models above are in essence deterministic, although the word probability occurs, for instance in the definition of $l(x)$ as the probability of survival from birth to age x, yet they allow for no random variation among individuals. In large aggregates such as nations the random variation is trifling in comparison with uncertainty regarding the parameters. Various attempts have been made to translate the uncertainty of the parameters into a probability distribution of the growth trajectory, but none has come into wide use.

One place where random variation has been recognized is the problem of extinction —what is the probability that a population evolving under the operation of fixed rates will become extinct [10, 25]? Suppose that a man (here the literature turns to the male side, since Galton* defined the problem as extinction of surnames) has a probability p_0 of having no sons, p_1 of having one son, p_2 of having two sons, and so on, and that all lines develop independently. Suppose that the required probability of extinction is x; then that probability is equal to the chance that the man has no sons, p_0; that he has one son and that one ultimately becomes extinct, p_1x; that he has two sons and the lines of both become extinct, p_2x^2; and so on. Having no sons, one son, and so on, being mutually exclusive, the probabilities involved are additive, and their sum is equal to the probability that the line of the original ancestor becomes extinct:

$$x = p_0 + p_1x + p_2x^2 + \cdots, \qquad (10)$$

an equation readily solved by iteration. Waugh [53] extended the mathematics to the age-dependent case, and Pollard [32] created a branching process* model that incorporates variances and covariances in an elegant extension of the Leslie matrix; his method gives not only mean values but variances and higher moments age by age as the population evolves.

Other stochastic models were provided in papers of Kendall [18] and Goodman [13] and in Sheps and Menken [43]. As one progresses toward realism in stochastic models, the mathematics becomes intractable, and resort to simulation* is inevitable. Simulation programs have been developed by Hammel et al. [14], Horvitz et al. [17], Mode [29], Ridley et al. [34], and others.

Most growth models suppose that all individuals of a given category have identical probabilities. That simplifies computation at the cost of realism. In projecting a population forward in time, recognition of subgroups that vary in rate of increase will produce a higher result than projection of the whole population at the average rate of increase.

A series of recent papers [42, 47, 48] show how recognition of heterogeneity can turn up unexpected effects. For instance, a medical advance can diminish the probability of dying for any individual to which it is applied, and yet have the effect of raising the overall death rate.

Growth models serve to estimate future population, but that may not be their most important use. They are used for inferring the rates of birth and death in a population lacking birth and death registration, in which two censuses have been taken. Where only one census has been taken and death rates can be surmised, the birth rate can likewise be inferred. Sophisticated methods have been developed by a National Academy Committee chaired by Ansley Coale [46]. Advanced countries are preoccupied with the difficulties that their social security systems will face over the next 50 years; population growth models show the relation between growth rates and the ratio of older persons to workers. Again, with the approach to stationarity promotion in offices and factories will slow down; the amount by which promotion will be delayed for average individuals is found through the stable model referred to above, incorporated in a comparative statistics argument [19, p. 107].

References

[1] Barney, G. O. (1980). *The Global 2000 Report to the President: Entering The Twenty-first Century*, Vol. 3: *The Government's Global Model*. Council on Environmental Quality. U.S. Government Printing Office, Washington, D.C.

[2] Boserup, E. (1981). *Population and Technological Change: A Study of Long-Term Trends*. Chicago.

[3] Clark, W. C. and Holling, C. S. (1979). In *Population Ecology. Fortschritte der Zoologie*, U. Halbach and J. Jacobs, eds. Fischer Verlag, Stuttgart, Federal Republic of Germany, **25**, pp. 29–52.

[4] Coale, A. (1972). *The Growth and Structure of Human Populations: A Mathematical Investigation*. Princeton University Press, Princeton, N.J.

[5] Cohen, J. E. (1984). Demographic doomsday deferred. *Harvard Mag.*, **86**(3), 50–51.

[6] Cole, L. (1954). The population consequences of life history phenomena. *Quart. Rev. Biol.*, **19**, 103–137.

[7] Easterlin, R. A. (1980). *Birth and Fortune: The Impact of Numbers on Personal Welfare*. Basic Books, New York.

[8] Euler, L. (1760). Recherches générales sur la mortalité et la multiplication (General researches on mortality and multiplication). *Mém. Acad. R. Sci. Belles Lett.*, **16**, 144–164. [English transl.: N. Keyfitz and B. Keyfitz, *Theor. Popul. Biol.*, **1**, 307–314 (1970).]

[9] Feller, W. (1941). On the integral equation of renewal theory. *Ann. Math. Statist.*, **12**, 243–267.

[10] Galton, F. and Watson, H. W. (1874). On the probability of extinction of families. *J. Anthropol. Inst.*, **6**, 138–144.

[11] Goodman, L. A. (1953). Population growth of the sexes. *Biometrics*, **9**, 212–225.

[12] Goodman, L. A. (1967). Reconciliation of mathematical theories of population growth. *J. R. Statist. Soc. A*, **130**, 541–553.

[13] Goodman, L. A. (1968). Stochastic Models for the Population Growth of the Sexes. Unpublished manuscript.

[14] Hammel, E. A. et al. (1976). *The SOCSIM Demographic–Sociological Microsimulation Program: Operating Manual*. Research Ser. No. 27, University of California.

[15] Holling, C. S. (1973). *Annu. Rev. Ecol. Systems*, **4**, 1–23.

[16] Hoppensteadt, F. (1975). *Mathematical Theories of Populations: Demographics, Genetics, and Epidemics*. SIAM, Philadelphia.

[17] Horvitz, D. G., Giesbrecht, F., and Lachenbruch, P. A. (1967). Microsimulation of vital events in a large population. (Paper presented at meeting of

Population Association of America, Cincinnati, 1967).

[18] Kendall, D. G. (1949). Stochastic processes and population growth. *J. R. Statist. Soc. B*, **11**, 230–264.

[19] Keyfitz, N. (1977). *Applied Mathematical Demography*. Wiley, New York.

[20] Leslie, P. H. (1945). On the use of matrices in certain population mathematics. *Biometrika*, **35**, 213–245.

[21] Lopez, A. (1961). *Problems in Stable Population Theory*. Office of Population Research, Princeton University, Princeton, N.J.

[22] Lotka, A. J. (1907). Relation between birth rates and death rates. *Science*, N.S., **26**, 21–22.

[23] Lotka, A. J. (1925). *Elements of Physical Biology*. Williams & Wilkins, Baltimore, MD. (Republished by Dover, New York, 1956.)

[24] Lotka, A. J. (1933). Industrial replacement. *Skand. Aktuarietidskr.*, 51–63.

[25] Lotka, A. J. (1939). *Théorie Analytique des Associations Biologiques:* Part II. *Analyse Démographique avec Application Particulière à l'Espèce Humaine*. Actualités Scientifiques et Industrielles No. 780. Hermann, Paris.

[26] Malthus, T. R. (1798). *Essay on the Principle of Population As It Affects the Further Improvement of Society*. Royal Economic Society, Facsimile Edition. (Also *Population: The First Essay*, Foreword by K. Boulding. University of Michigan Press, Ann Arbor, MI, 1959.)

[27] McKendrick, A. G. (1926). Applications of mathematics to medical problems. *Proc. Edinb. Math. Soc.*, **44**(1), 98–130.

[28] Meadows, D. H. et al. (1972). *The Limits to Growth*. Universe Books, New York.

[29] Mode, C. J. (1983). *Stochastic Processes in Demography and Their Computer Implementation*. Dept. of Mathematical Sciences and Institute for Population Studies, Drexel University, Philadelphia.

[30] Pearl, R. (1939). *The Natural History of Population*. Oxford University Press, New York.

[31] Pearl, R. and Reed, L. J. (1920). On the rate of growth of the population of the United States since 1790 and its mathematical representation. *Proc. Natl. Acad. Sci. (USA)*, **6**, 275–288.

[32] Pollard, J. H. (1966). On the use of the direct matrix product in analysing certain stochastic population models. *Biometrika*, **53**, 397–415.

[33] Pollard, J. H. (1973). *Mathematical Models for the Growth of Human Populations*. Cambridge University Press, New York.

[34] Ridley, J. C. et al. (1975). *Technical Manual—Reproductive Simulation Model: REPSIM-B*. Kennedy Institute Center for Population Research, Georgetown University, Washington, D.C.

[35] Rogers, A. (1975). *Introduction to Multiregional Mathematical Demography*. Wiley, New York.

[36] Ryder, N. B. (1964). The process of demographic translation. *Demography*, **1**, 74–82.

[37] Samuelson, P. A. (1958). An exact consumption-loan model of interest with or without the social contrivance of money. *J. Polit. Econ.*, **66**(6).

[38] Samuelson, P. A. (1977). Generalizing Fisher's reproduction value—non-linear, homogeneous, bi-parental systems. *Proc. Natl. Acad. Sci. (USA)*, **74**, 5772–5775.

[39] Schoen, R. (1975). Constructing increment–decrement life tables. *Demography*, **12**, 313–324.

[40] Schoen, R. (1978). Standardized two-sex stable populations. *Theor. Popul. Biol.*, **14**, 357–370.

[41] Sharpe, F. R. and Lotka, A. J. (1911). A problem in age-distribution. *Philos. Mag., 6th Ser.*, **21**, 435–438.

[42] Shepard, D. S. and Zeckhauser, R. J. (1980). Long-term effects of interventions to improve survival in mixed populations. *J. Chronic Dis.*, **33**, 413–433.

[43] Sheps, M. C. and Menken, J. A. (1973). *Mathematical Models of Conception and Birth*. University of Chicago Press, Chicago.

[44] Simon, J. L. (1981). *The Ultimate Source*. Martin Robertson, Oxford.

[45] Tabah, L. (1968). Représentations matricielles de perspectives de population active (Matrix representations of projections of active population). *Population (Paris)*, **23**, 437–476.

[46] United Nations (1983). *Manual X: Indirect Techniques for Demographic Estimation*. U.N., New York.

[47] Vaupel, J. W. and Yashin, A. I. (1985). *The Deviant Dynamics of Death in Heterogeneous Populations. Sociological Methodology 1985*. Jossey–Bass, San Francisco, pp. 179–211.

[48] Vaupel, J. W., Manton, K. G., and Stallard, E. (1979). The impact of heterogeneity in individual frailty on the dynamics of mortality. *Demography*, **16**, 439–454.

[49] Verhulst, P. F. (1838). Notice sur la loi que la population suit dans son accroissement. *Corresp. Math. Phys. Publ. par A. Quételet* (Brussels), **10**, 113–121.

[50] Volterra, V. (1926). Variazioni e fluttuazioni del numero d'individui in specie animali conviventi. *Mem. R. Accad. Naz. Lincei*, anno CCCXXIII, **2**, 1–110.

[51] von Foerster, H. (1959). In *The Kinetics of Cellular Proliferation*, F. Stohlman, ed. Grune & Stratton, New York.

[52] von Foerster, H., Mora, P. M., and Amiot, L. W. (1960). Doomsday: Friday, 13 November, A.D. 2026. *Science*, **132**, 1291–1295.

[53] Waugh, W. O'N. (1955). An age-dependent birth and death process. *Biometrika*, **42**, 291–306.

[54] Whelpton, P. K. (1954). *Cohort Fertility: Native White Women in the United States*. Princeton University Press, Princeton, N.J.

[55] Winsor, C. P. (1932). A comparison of certain symmetrical growth curves. *J. Washington Acad. Sci.*, **22**, 73–84.

(BIRTH-AND-DEATH PROCESSES
FERTILITY
LIFE TABLES
MARRIAGE
MATHEMATICAL THEORY OF POPULATION
POPULATION PROJECTION)

N. KEYFITZ

POPULATION OR SAMPLE SIZE ESTIMATION

Suppose that $X_1, \ldots, X_N$ are independent random variables with a common probability density function (PDF) $f(x \mid \theta)$, where θ is a scalar or vector parameter. Let X_i be observable only if it lies outside a given region R. Thus the number M of observed X's is a binomial (N, p) variate, $p = 1 - P(X \in R)$. Here we consider a survey of the recent work where N itself is of considerable interest and is estimated, along with θ, from observed values of M and X's. We give several examples below where estimation of the sample size, N, is of primary interest. In some of these situations, N represents the population size, but the problem of estimation is similar in both the situations.

In some life-testing situations the total number of items being tested may not be known and if the test is carried out for a fixed amount of time, then this situation gives rise to a truncated sample. This would happen when, among the items put on a life test, there is a certain unknown number of items with a specific defect identifiable only after the item fails. Blumenthal and Marcus [8, 9] consider this case when the interest is in estimating the number of remaining defectives of a particular type after an initial burn-in period. Jelinski and Moranda [30] consider the problem of estimating the initial number of errors N in a program after running the program for a fixed period of time and thus obtaining time of detection for M distinct errors. This is exactly the same problem as estimating the number of remaining defectives in the above life test situation. Another application of the life-testing* situation is estimating the number of responses to an advertising campaign as considered by Anscombe [2]. Here N would represent the number of responsive people among the K contacted in the campaign.

Wittes [50] and Wittes et al. [51] consider estimating the size of a subpopulation of persons who have a trait that occurs rarely in the population at large. Ascertainment of the study group is often done by combining incomplete lists of members of the target population*. Some of the cases, however, may not be listed by any source and the problem is to estimate the number of such cases. Similarly, in visual scanning experiments in particle physics, film containing photographs of particles is scanned independently by two or more scanners and particles with poor visibility may be missed by all scanners. The situations cited give rise to the problem of estimating N in the multinomial $(N, p_1, \ldots, p_k)$ distribution* when some of the class frequencies are missing. This is investigated in detail by Sanathanan [40, 41].

Another interesting application is in the Poisson case when the zero class is unobservable. One such example is given in Dahiya and Gross [16] and Blumenthal et al. [11] and pertains to an epidemic of cholera where the number of households which are infected but did not have any active case of cholera was unobservable. Another application of the Poisson distribution* with zero class missing is in estimating the population of factory workers from the hospital records of the number of accidents per worker.

The truncated negative binomial distribution* has been fitted to field failure data for several types of systems by Goel and Joglekar [23]. For each system type, N units were in use for a known period of time T, and a count of the number (n_x) of units failing x times, $x = 1, 2, \ldots,$ was available but n_0 and N both were unknown. Sampford [39] also considers the truncated negative binomial for the distribution of chromosome breaks per cell, when cells not susceptible to breakage are indistinguishable from susceptible cells in which no breaks occur. Another application of the truncated negative binomial is given by Rao et al. [36], where counts of the number of children per family are considered but families with no children cannot be distinguished from the sterile group of familes. Sanathanan [43] has investigated the problem of estimating N when the zero class of the negative binomial is truncated.

The case of the binomial distribution* with unknown sample size arises when only the record of successes is available in r independent experiments consisting of N (unknown) Bernoulli trials. This problem has been investigated by Binet [3], Blumenthal and Dahiya [7], Draper and Guttman [19], Feldman and Fox [20], Ghosh and Meeden [22], Haldane [25], and Olkin et al. [34]. Alden et al. [1] and Blumenthal and Dahiya [7] utilize this model for estimating the size of a zooplankton population from subsamples.

When sampling from an infinite population where only the X's with values outside R are observable, one could sample until M X's are observed. The total number N of X's generated will now be a geometric* random variable. If N is not known, the problem of estimating N from "inverse binomial" sampling has been examined by Blumenthal and Sanathanan [10] and will not be reported herein.

The examples given here exemplify what we shall call truncated samples with a binomial sampling rule, and it is this case that we describe in detail later. We also discuss briefly the situation when θ is known, or when M is observed but not the X's, and comment on estimation of integer parameters in general.

THE ESTIMATORS AND SOME ASYMPTOTIC RESULTS

Let $X_1, \ldots, X_N$ be independent, identically distributed real-valued random variables with values in Ω having common distribution $F_X(x \mid \theta)$ which has a density $f_X(x \mid \theta)$ with respect to (w.r.t.) a σ-finite measure μ (written in the sequel as Lebesgue measure for notational simplicity) with the real-valued parameter $\theta \in \Theta$. Let $R \in \Omega$ be the measurable region such that X is not observable if $X \in R$. Thus a random number M of the N random variables is observed and denoted by $X_1, \ldots, X_M$. We wish to estimate N and θ on the basis of M and $X_1, \ldots, X_M$. We have assumed θ to be scalar, which, in fact, is the case for most of the specific cases that we consider here. The likelihood* function of the observations m, $x_1, \ldots, x_m$, is given by

$$L(\theta, N \mid m, x) = \binom{N}{m} q^{N-m}(\theta) \prod_{i=1}^{m} f_X(x_i \mid \theta)$$
$$= L_1(\theta, N) L_2(\theta \mid m), \qquad (1)$$

where

$$L_1(\theta, N) = \binom{N}{m} p^m(\theta) q^{N-m}(\theta),$$
$$L_2(\theta \mid m) = \prod_{i=1}^{m} f_t(x_i \mid \theta),$$
$$p(\theta) = 1 - P(X \in R), \qquad (2)$$
$$f_t(x \mid \theta) = \begin{cases} f_X(x \mid \theta)/p(\theta), & x \in \overline{R} \\ 0 & x \in R \end{cases}$$

and $q(\theta) = 1 - p(\theta)$. Here $L_2(\theta \mid m)$ denotes the conditional likelihood function of $x_1, \ldots, x_m$ for given m.

The traditional approach for estimating θ has been to obtain the conditional maximum likelihood estimator (CMLE) by maximizing $L_2(\theta \mid m)$ and thus eliminating the problem of unknown sample size from consideration. Here we consider different estimators when both N and θ are parameters of interest.

Since N is an integer parameter, we first discuss the method given in Dahiya [15] for estimating an integer parameter by maximum likelihood. Let $L(N,\theta)$ be the likelihood function, where N is an unknown integer parameter and θ is a parameter vector. Also, let $\hat{\theta}(N)$ denote the value of θ which maximizes $L(N,\theta)$ for given N and denote

$$L^*(N) = L(N, \hat{\theta}(N)). \tag{3}$$

If $L^*(N)$ is unimodal and if V, real valued, is the solution of

$$L^*(V) = L^*(V-1), \tag{4}$$

then the integer MLE $\hat{N}$ of N is given by $\hat{N} = [V]$, where $[a]$ denotes the greatest integer $\leqslant a$ (the integer part of a). If V, the solution of (4) happens to be an integer, then both V and $V-1$ are the integer MLE's of N.

To estimate N and θ for the problem defined earlier, the CMLE's $\tilde{N}$ and $\tilde{\theta}$ are found by maximizing L_2 w.r.t. θ and then maximizing $L_1(\tilde{\theta}, N)$ w.r.t. N. On using (4), this gives

$$\tilde{N} = \left[m/p(\tilde{\theta})\right] \tag{5}$$

and

$$\sum_1^m s(x_i, \tilde{\theta}) = 0, \tag{6}$$

where $s(x,\theta) = \partial \log f_t(x \mid \theta)/\partial\theta$. This conditional approach has been used by Hartley [26] for the missing zero class of the negative binomial, by Dahiya and Gross [16] for the Poisson and by Sanathanan [40] for the multinomial. The unconditional maximum likelihood estimators (UMLE), considered by Sanathanan [40] and Blumenthal and Marcus [9], are obtained by maximizing the likelihood function $L(\theta, N \mid m, x)$ w.r.t. θ and N simultaneously. Let, for fixed N, $\hat{\theta}(N)$ be the solution of

$$\begin{aligned} 0 &= \partial \log L/\partial\theta & (7)\\ &= T(\theta)\left[N - m/p(\theta)\right] + \sum_1^m s(x_i, \theta), \end{aligned}$$

where $T(\theta) = \partial \log q(\theta)/\partial\theta$. Also, let

$$L^*(N) = L(\hat{\theta}(N), N \mid m, x). \tag{8}$$

Then $\hat{N}$, the integer UMLE of N, is given by $[V]$, where V is the solution of $L^*(V) = L^*(V-1)$.

As we shall see later, for some distributions, $L^*(N)$ turns out to be an increasing function of N if the sample lies in a specific set with positive measure. This gives rise to an infinite solution for the MLE of N. A class of modified MLE's which includes the UMLE as a special case and has better properties for estimating N than the UMLE has been investigated by Blumenthal and Marcus [9], Blumenthal [5], Blumenthal et al. [11], and Watson and Blumenthal [47]. Formally, these are derived as Bayes modal estimates w.r.t. a generalized prior density $h(\theta)$ for θ and a uniform prior for N. Now the modified or weighted likelihood function is given by

$$L(\theta, N, h \mid m, x) = h(\theta) L(\theta, N \mid m, x)$$

and the modified MLEs are denoted by $\hat{N}(h)$ and $\hat{\theta}(h)$.

Asymptotic distributions and asymptotic expansions for the second-order terms are derived in Blumenthal [5, 6] for the general case and we summarize these here. Let $\xi(\theta) = h'(\theta)/h(\theta)$, where prime stands for derivative w.r.t. θ, and assume the following conditions:

(a) $\xi'(\theta)$ bounded for all $\theta \in \Theta$.

(b) $S''(x,\theta)$ is continuous in θ for all $\theta \in \Theta$, $x \in \overline{R}$.

(c) $E[S''(X,\theta)]$ is continuous in θ for all $\theta \in \Theta$ [expectation being w.r.t. $f_t(x \mid \theta)$].

(d) $q(\theta)$ and all related functions, $q'(\theta)$, up to and including $T''(\theta)$ are continuous in θ, all $\theta \in \Theta$ and $T(\theta) < \infty$.

(e)

$$(\partial^i/\partial\theta^i) \int_{\overline{R}} f_X(x \mid \theta)\, dx = \int_{\overline{R}} (\partial^i/\partial\theta^i) f_X(x \mid \theta)\, dx, \quad i = 1,2,3,$$

$$(\partial/\partial\theta) \int_{\overline{R}} \left[f_X'(x \mid \theta)\right]^2 / f_X(x \mid \theta)\, dx = \int_{\overline{R}} (\partial/\partial\theta)\left[f_X'(x \mid \theta)\right]^2 / f_X(x \mid \theta)\, dx.$$

We now state the following theorem from Blumenthal [5].

Theorem 1. Under the assumptions stated above, the following expansions hold:

$$\tilde{\theta} - \theta - \tau Z/\sqrt{N} = A/N + o_p(1/N), \quad (9a)$$

$$\hat{\theta}(h) - \theta - \tau Z/\sqrt{N} = \left[A - q'/(2pq\mu_2) + \xi/(p\mu_2)\right]/N + o_p(1/N), \quad (9b)$$

$$\tilde{N} - N - (\rho Y + \sigma Z)\sqrt{N} = B + o_p(1), \quad (10a)$$

$$\hat{N}(h) - N - (\rho Y + \sigma Z)\sqrt{N} = B - q'^2/(2p^2 q\mu_2) + q'\xi/(p^2\mu_2) + o_p(1), \quad (10b)$$

where

$$Z = \sum_{i=1}^{M} S(X_i, \theta)/\sqrt{M\mu_2},$$

$$\mu_2 = E\left[S^2(X,\theta)\right], \qquad \tau = 1/\sqrt{\mu_2 p},$$

$$Y = (M - Np)/\sqrt{Npq},$$

$$\sigma = \tau q'/p, \qquad \rho = \sqrt{q/p},$$

$$A = (1/(2p\mu_2))\left[2ZW/\sqrt{\mu_2} - ZY\sqrt{\mu_2 q} + \left(Z^2/(2\mu_2)\right)(2\mu_3 - 3\mu_{11})\right],$$

$$B = (q'/p^2)\left[pA + ZY(q/\mu_2)^{1/2}\right] + \left(Z^2/(2\mu_2 p^3)\right)(2q'^2 + pq'') - \tfrac{1}{2},$$

$$\mu_{11} = E\left[f_t'(X\mid\theta)f_t''(X\mid\theta)/f_t^2(X\mid\theta)\right],$$

$$\mu_3 = E\left[S^3(X,\theta)\right],$$

$$W = \left(\sum_{1}^{M} S'(X_i,\theta) + M\mu_2\right)\Big/\sqrt{M}.$$

All expectations are w.r.t. $f_t(x\mid\theta)$. The expansions for the UMLEs $\hat{N}$ and $\hat{\theta}$ are given by (9a) and (10b) as a special case corresponding to $h = 1$.

On making use of the expansions above and the fact that the asymptotic distribution of (Z, Y, W) is multivariate normal, we have the following result.

Corollary. Under the conditions of Theorem 1, $\sqrt{N}(\hat{\theta} - \theta)$ and $(\hat{N} - N)/\sqrt{N}$ are asymptotically bivariate normal with means zero and covariances given by $\text{var}[\sqrt{N}(\hat{\theta} - \theta)] = \tau^2$, $\text{var}[(\hat{N} - N)/\sqrt{N}] = \sigma^2 + \rho^2$, $\text{cov}(\hat{\theta}, \hat{N}) = \tau\sigma$. The result remains the same with $\tilde{\theta}$, $\tilde{N}$ or $\hat{\theta}(h), \hat{N}(h)$ replacing $\hat{\theta}, \hat{N}$.

Even though all the normalized estimators of N have the same asymptotic distribution, it can be seen from Theorem 1 that

$$\lim_{N\to\infty} P\left\{\tilde{N} - \hat{N} = q'^2/(2p^2 q\mu_2)\right\} = 1. \quad (11)$$

Also, some information regarding the asymptotic bias of the estimators can be obtained from

$$E(A) = -\mu_{11}/(2p\mu_2^2) \quad (12)$$

and

$$E(B) = p'\mu_{11}/(2p^2\mu_2^2) + \left[(2q'^2 + pq'')/(2p^3\mu_2)\right] - \tfrac{1}{2}. \quad (13)$$

Since $\tilde{N} - N - \sqrt{N}(\rho Y + \sigma Z)$ converges in law to the same random variable as does B and $E(Y) = E(Z) = 0$, $E(B)$ can be considered as the "asymptotic bias" of $\tilde{N}$. Similar expressions for the "asymptotic bias" of $\hat{N}$ and $\hat{N}(h)$ can be obtained from the expansions above.

TRUNCATED SAMPLES FROM DISCRETE DISTRIBUTIONS

We now consider different discrete distributions where estimation of N has been investigated.

Multinomial Distribution*

Examples of censored and truncated multinomial data appear in Blumenthal [4], Chen and Fienberg [12, 13], Fienberg [21], Hocking and Oxspring [27, 28], and Sanathanan [40–42]. These examples arise when some observations in contingency tables* are partially classified or from incomplete contingency tables. Visual scanning models and multiple recapture census type data also give rise to truncated multinomial data. Estimation of the missing number of observations, however, has been investigated only recently by Fienberg [21] and Sanathanan [40–42]. Here we summarize the results given by Sanathanan.

Investigation by Sanathanan has been motivated by experiments in particle physics which often involve scanning of film containing photographs of particles. Scanning is carried out to count the number N of particles of a specified type. Due to poor visibility, some of the particles are missed during the scanning process. The interest lies in estimating N from the incomplete data. This problem, in general, can be formulated in the multinomial model now to be described.

Let $(n_1, \dots, n_s)$ be distributed as multinomial $(N, p_1, \dots, p_s)$. Suppose that only a subset $(n_1, \dots, n_{k-1})$, $2 \leqslant k \leqslant s$, is observable and N itself is unknown. Let $m = \sum_1^{k-1} n_i$ and $n_k = N - m$. Now $(n_1, \dots, n_k)$ is multinomial $(N, p_1, \dots, p_k)$, where p_k denotes $1 - \sum_1^{k-1} p_i$. Let p_i be a known function of $r < k - 1$ parameters given by

$$p_i = p_i(\theta), \qquad i = 1, \dots, k, \tag{14}$$

where $\theta' = (\theta_1, \dots, \theta_r)$ is a vector of r independent parameters. The likelihood function is given by (1), where $L_1(\theta, N)$ is given in (2) and $L_2(\theta \mid m)$ specializes to

$$L_2(\theta \mid m) = \left[m!/(n_1! \cdots n_{k-1}!) \right] \prod_{i=1}^{k-1} (\pi_i(\theta))^{n_i}, \tag{15}$$

with $\pi_i(\theta) = p_i(\theta)/p(\theta)$, and $p(\theta) = 1 - p_k(\theta)$.

To obtain the MLE $\tilde{\theta}$, we need, in general, numerical solution of r equations

$$\frac{\partial \log L_2(\theta)}{\partial \theta_j} = 0, \qquad j = 1, \dots, r. \tag{16}$$

Finally, $\tilde{N}$ is given by (5). Obtaining the UMLEs $\hat{\theta}$ and $\hat{N}$ is much more involved and requires numerical solution of $r + 1$ equations, unless $p_i(\theta)$ are simple functions of θ. The asymptotic properties of these estimators are given in the following modification of a theorem obtained by Sanathanan [40].

Theorem 2. As $N \to \infty$, we have

(a) $\hat{\theta} \xrightarrow{\text{a.s.}} \theta$.

(b) $(\hat{N} - m/p(\hat{\theta}))/\sqrt{n} \xrightarrow{\text{a.s.}} 0$.

(c) $[\sqrt{N}(\hat{\theta} - \theta)', (\hat{N} - N)/\sqrt{N}]$ is asymptotically $N(0, \Sigma)$, where the $(r + 1) \times (r + 1)$ covariance matrix Σ is given as follows:

$$p(\theta)\Sigma = \begin{pmatrix} A & -A\rho \\ -\rho' A & q(\theta) + \rho' A \rho \end{pmatrix}$$

and

$$p(\theta)\rho' = (a_1, \dots, a_r), \qquad a_i = -\frac{\partial}{\partial \theta_i} p_k(\theta),$$

$$A^{-1} = \left(A_{ij}^{-1}\right),$$

$$A_{ij}^{-1} = \sum_{s=1}^{k-1} (1/\pi_s(\theta)) \frac{\partial \pi_s(\theta)}{\partial \theta_i} \frac{\partial \pi_s(\theta)}{\partial \theta_j}.$$

Note that A is the usual covariance matrix for the estimate of θ based on the multinomial $(m, \pi_1, \dots, \pi_{k-1})$, i.e., the CMLE $\tilde{\theta}$ of θ. In fact, the asymptotic results for $\tilde{\theta}$ and $\tilde{N}$ are identical to those stated above for $\hat{\theta}$ and $\hat{N}$.

Sanathanan [41] has obtained the specific form of the foregoing results for several different models utilized in visual scanning experiments. These models are compared, in Sanathanan [42], for some specific data sets from the scanning experiments.

The Binomial Distribution

Binomial samples with unknown N arise when in a sequence of Bernoulli trials only the successes are observable. Let X be the number of successes in N (unknown) Bernoulli trials with probability of success p. If p itself is unknown, then we need to observe X in more than one sequence of N Bernoulli trials in order to estimate both N and p. The problem of estimating N and p has been investigated by Whittaker [49], Haldane [25], and Binet [3], where moment method estimators (MMEs) and MLEs are investigated. Although the cited papers investigate the asymptotic properties of the MME and MLE, the results contain a variety of errors. These properties have recently been reexamined by Blumenthal and Dahiya [7], where an exhaustive investigation of the asymptotic properties for different estimators is undertaken.

When p is known, Feldman and Fox [20] examine asymptotic properties of the MLE and MME of N and of two modifications of these estimators. Olkin et al. [34] give a summary of early literature and propose some stabilized estimators. Additional papers of interest are Draper and Guttman [19], who consider a Bayes estimator, and Robinson [38].

We now summarize the results given by Blumenthal and Dahiya [7] (denoted B-D), who also discuss the results of the other authors referenced above. Only the situation where zero values of x are observable are described here. For details on the case in which zero values of x are truncated, see B-D.

KNOWN p CASE. We first consider the case when $X_1, \dots, X_r$ is a random sample from the binomial (N, p), where p is known and the interest lies in the estimation of N. The likelihood is

$$L(N \mid p, x) = \prod_{i=1}^{r} \binom{N}{x_i} p^{x_i} q^{N-x_i}, \quad (17)$$

where x represents the vector $(x_1, \dots, x_r)$. Following the procedure in Dahiya [15], the integer MLE $\hat{N}$ of N is given by $[V]$, where V is the solution of

$$r \log(Vq) - \sum_{1}^{r} \log(V - x_i) = 0. \quad (18)$$

The MME of N is given by $\tilde{N}^*$, where $\tilde{N}^*$ is the nearest integer to $\tilde{N} = \overline{X}/p$, and

$$\overline{X} = \sum_{1}^{r} X_i / r. \quad (19)$$

The following results (from B-D) give the asymptotic properties of $\hat{N}$.

Theorem 3.

(a) For any sample, the estimator $\hat{N}$ is finite.

(b) For n fixed as $r \to \infty$, $P(\hat{N} = N) \to 1$ (the probability tends to 1 exponentially fast).

(c) Let $N \to \infty$ and let r be either fixed or increasing.

(1) $(\hat{N} - N)/N^{\alpha} \xrightarrow{P} 0$ for $\alpha > \frac{1}{2}$ and $\log(r/N) \to 0$.

(2) $(\sqrt{r/N})(\hat{N} - N) \xrightarrow{L} N(0, q/p)$ if $r/N \to 0$.

All the results in Theorem 3 also hold true for the MME $\tilde{N}^*$. Also, Feldman and Fox [20] obtained (c)(1) for r fixed and (c)(2) for r increasing.

The asymptotic properties of the minimum chi-square estimator and another estimator suggested by Feldman and Fox [20] have also been given by B-D. Furthermore, the result in the following lemma is also given by B-D for testing the goodness of fit* for the binomial model.

Lemma. Let $N \to \infty$, r fixed. Then the asymptotic distribution of

$$\hat{X}^2 = \sum_{1}^{r} (x_i - \overline{N}p)^2 / (\overline{N}pq)$$

is χ^2_{r-1} where $\overline{N}$ is $\hat{N}$, $\tilde{N}$, or the minimum chi-square* estimator.

The lemma is used in B-D to determine the fit of the binomial model to counts for five repeated subsamples of plankton, each representing a known proportion of the original sample. The binomial model is invalidated if clumping occurs in the original sample (see Alden et al. [1]).

Simulation results given by B-D suggest that the asymptotic results hold at least qualitatively for moderate r and n and that generally the MLE is to be preferred over its competitors.

UNKNOWN p CASE. For p unknown, the MMEs $\tilde{N}^*$ and $\tilde{p}$ are obtained from

$$\tilde{N} = \overline{X}/\tilde{p}, \qquad \tilde{p} = 1 - S^2/\overline{X}, \quad (20)$$

where $S^2 = \sum_1^r (X_i - \bar{X})^2/r$. Clearly, this leads to negative $\tilde{p}$ if $S^2 > \bar{X}$. For this same situation, the MLE is infinite, and the modified MLE is suggested by B-D to avoid this difficulty. Using the conjugate prior density for p, proportional to $p^a q^b$, we define the modified likelihood

$$L_m(N, p \mid x, r) = cp^a q^b L(N, p \mid x, r), \quad a, b \geqslant 0, \quad (21)$$

where $L(N, p \mid x, r)$ is given by (17) and c is a constant depending on a and b. The modified MLE of p for any given N, denoted by $\hat{p}_m(N)$, is now given by

$$\hat{p}_m(N) = (r\bar{X} + a)/(rN + a + b) \quad (22)$$

and the MLE of N is given by (5), where V is the solution of

$$L(V, \hat{p}_m(V) \mid x, r) = L(V-1, \hat{p}_m(V-1) \mid x, r). \quad (23)$$

The MLE is a special case when $a = b = 0$. Olkin et al. [34] demonstrate the existence of a finite $\hat{N}$ iff $S^2 \leqslant \bar{X}$ and discuss uniqueness, which remains unresolved. We now state the following results regarding $\hat{N}_m$ given by B-D.

Theorem 4.

(a) $\hat{N}_m < \infty$ for all samples if $a > 0$.
(b) For N fixed, $\lim_{r\to\infty} P(\hat{N}_m = N) = 1$.
(c) For r fixed, $N \to \infty$, $\hat{N}_m$ is not consistent and, in fact,

$$\hat{N}_m / N \xrightarrow{L} (pV/(4a)) \times \left[-1 + \sqrt{1 - 8a(2a + 2b - r)/V^2} \right],$$

where $V = rp - q\sqrt{2r}\, W - 4a - 2b$ and where $W\sqrt{2/r} + 1$ is distributed as χ^2_{r-1}. For $a = b = 0$, corresponding to $\hat{N}_m = \hat{N}$, the limiting random variable simplifies to $p[1 - (q/r) - \sqrt{2}\, qW/r^{3/2}]^{-1}$.

(d) If $N \to \infty$, $r \to \infty$, with $(\sqrt{r}/N) \to 0$, then

$$(\sqrt{r}/N)(\hat{N}_m - N) \xrightarrow{L} N(0, 2q^2/p^2).$$

The normalization by $1/N$ instead of $1/\sqrt{N}$ is somewhat surprising and atypical. Also, the results (b)–(d) in Theorem 4 hold for the MLE $\hat{N}$ and MME $\tilde{N}^*$.

Results from B-D regarding the estimation of p are given below.

Theorem 5. Let $(\bar{N}, \bar{p})$ represent $(\tilde{N}^*, \tilde{p})$ or $(\hat{N}_m, \hat{p}_m)$.

(a) Let $r \to \infty$ for fixed N. Then $\sqrt{rN}\,(\bar{p} - p) \xrightarrow{L} N(0, pq)$.
(b) Let $N \to \infty$, r fixed. Then $\bar{p}$ is not a consistent estimator of p.
(c) Let $N \to \infty$, $r \to \infty$ so that $(\sqrt{r}/N) \to 0$. Then $\sqrt{r}\,(\bar{p} - p) \xrightarrow{L} N(0, 2q^2)$.

As pointed out earlier, the MLE of N is infinite if $S^2 > \bar{X}$ and the MME of N does not exist or becomes infinite if $\tilde{p}$ is taken to be 0 when $S^2 > \bar{X}$. This behavior causes instability in the estimators when $\bar{X}$ is close to S^2 even if $\bar{X} > S^2$. Olkin et al. [34] have investigated a jackknife estimator and stabilized moment estimator of N, and give simulation results to compare the two modified estimators.

Poisson Distribution

The truncated Poisson sample typically arises when the zero class is unobservable, which is the case in the several examples given in the introduction. Most of the papers on truncated Poisson (see David and Johnson [17], Hartley [26], Moore [32, 33], Plackett [35], Rider [37], Tate and Goen [46], Irwin [29], Cohen [14], Subrahmaniam [45], and Selvin [44]) deal with the estimation of the intensity parameter. The problem of estimating the sample size N and the frequency of the missing class has only recently been investigated by Dahiya and Gross [16] and Blumenthal et al. [11].

Let $X_1, \ldots, X_N$ be a random sample from Poisson (λ). The truncated Poisson sample $X_1, \ldots, X_M$ arises when all the zeros of the original sample are missing and N is unknown. Let n_x denote the number of sample values for which $X = x$ and let $M = \sum_{x \geqslant 1} n_x$. The likelihood function for the truncated sample is given by

$$L(\lambda, N) = L_1(\lambda, N) L_2(\lambda), \qquad (24)$$

where L_1 is given by (2),

$$L_2(\lambda) = (q/p)^m \prod_{x=1}^{l} (\lambda^x / x!)^{n_x}, \qquad (25)$$

and where $p = 1 - e^{-\lambda}$, $q = 1 - p$, and l is the maximum observed value of X.

The CMLEs considered by Dahiya and Gross [16], are given by (5) and

$$\tilde{\lambda}/(1 - e^{-\tilde{\lambda}}) = \bar{x}, \qquad (26)$$

where $\bar{x} = \sum_1^m x_i / m$ and $\tilde{p} = 1 - e^{-\tilde{\lambda}}$. The UMLEs, obtained by Blumenthal et al. [11], are given by

$$\hat{\lambda} = m\bar{x}/\hat{N}, \qquad (27)$$

where $\hat{N} = [V]$, and V is the solution of

$$-V \ln(1 - m/V) = m\bar{x}. \qquad (28)$$

For $\bar{x} = 1$, both the CMLEs and UMLEs are infinite and these estimators are unstable for values of $\bar{x}$ near 1. The modified estimators which circumvent this difficulty are obtained in ref. 16 by considering the gamma prior density for λ and uniform prior for N. The modified estimators are given by

$$\hat{\lambda}^* = (m\bar{x} + \rho)/(\hat{N}^* + r); \qquad \hat{N}^* = [V], \qquad (29)$$

where V is the solution of

$$\ln(1 - m/V) = (m\bar{x} + \rho)\ln[1 - 1/(V + r)], \qquad (30)$$

and where ρ and r are the parameters of the prior density of λ. The choice of $\rho = 1$ and $r = \frac{1}{3}$ is shown to minimize the maximum asymptotic bias of $\hat{N}^*$ by Blumenthal et al. [11], where the asymptotic expansions* for different estimators are considered. The asymptotic distribution of the different estimators of N is given in the following theorem.

Theorem 6. Let $\bar{N}$ denote any of the estimators $\tilde{N}$, $\hat{N}$, or $\hat{N}^*$. Then

$$(\bar{N} - N)/\sqrt{N} \xrightarrow{L} N(0, \sigma^2),$$

where $\sigma^2 = q/(p - \lambda q)$.

The asymptotic distribution of the estimators of λ can be obtained by using the result of Theorem 1.

Simulation* results show that the modified estimator $\hat{N}^*$ is most attractive from its overall performance of simulated bias and mean square error. An example referring to an epidemic of cholera is also given where estimation of N is of importance.

TRUNCATED SAMPLES FROM THE EXPONENTIAL DISTRIBUTION

Examples of truncated exponential sampling where an estimate of N is needed are given by Blumenthal and Marcus [9] (to be abbreviated B-M), Jelinski and Moranda [30], and Anscombe [2]. All of these examples are characterized by having a fixed time T available for observation, an unknown number of items N which might respond in that period and a set of response times $X_1, \ldots, X_m$ for the m responses actually obtained.

In this section, the likelihood function for the truncated sample is given by (1) and (2) with

$$f_X(x \mid \theta) = \begin{cases} (1/\theta)e^{-(x/\theta)}, & x \geqslant 0 \\ 0, & x < 0 \end{cases} \qquad (31)$$

and

$$p(\theta) = 1 - e^{-T/\theta}. \qquad (32)$$

The CMLE, given in B-M and by Deemer and Votaw [18], is found from (5) and

$$\begin{aligned} \bar{x} &= \tilde{\theta} - T(e^{T/\tilde{\theta}} - 1)^{-1} && \text{if } \bar{x} < T/2 \\ \tilde{\theta} &= +\infty && \text{if } \bar{x} \geqslant T/2. \end{aligned} \qquad (33)$$

Deemer and Votaw give a table to facilitate the solution of (33).

The UMLE obtained in B-M is given by

$$\hat{\theta} = \bar{x} + ((\hat{N}/m) - 1)T \qquad (34)$$

and $[V]$, where

$$(V/m) - \{m(1 - (1 - (m/V))^{1/m})\}^{-1} = 1 - (\bar{x}/T). \qquad (35)$$

This gives an infinite value for $\hat{N}$ whenever $2\bar{x} \geqslant T + (1/m)$.

Modified estimators which avoid the instability in both $\hat{N}$ and $\tilde{N}$ are based on a conjugate type prior $h(\theta)$. For the CMLE $\tilde{N}$, we take $h(\theta)$ proportional to $(1/\theta)^b(1 - e^{-(T/\theta)})^a$. The modified CMLE is now found by solving

$$\bar{x} = \tilde{\theta}(h)(1 + (b/m)) - T(1 - (a/m)) \times (e^{T/\tilde{\theta}(h)} - 1)^{-1}, \quad (36)$$

where (36) will have a solution whenever $b > 0$.

To obtain the modified UMLE, we use $h(\theta)$ proportional to $(1/\theta)^b e^{-(a/\theta)}$, giving

$$\hat{\theta}(h) = (1 + (b/m))^{-1} \times \left[\bar{x} + (a/m) + \left((\hat{N}(h)/m) - 1\right)T\right], \quad (37)$$

and for $\hat{N}$ we use $[V]$ with

$$(V/m) - \left\{m\left(1 - (1 - (m/V))^{1/(b+m)}\right)\right\}^{-1} = 1 - (1/T)(\bar{x} + a/m). \quad (38)$$

Equation (38) has a finite solution for any $b > 0$.

The maximum asymptotic bias of $\tilde{N}(h)$ is minimized by using $a = -2/(3T^2)$ and $b = 3/(4T^2)$ while the same effect for $\hat{N}(h)$ is achieved with $a = (-T/2)$, $b = 1$. There is little to choose between these two modified estimators. The asymptotic distribution of the estimators of N is given as follows.

Theorem 7. Let $\bar{N}$ denote any of the estimators $\tilde{N}$, $\tilde{N}(h)$, $\hat{N}$, or $\hat{N}(h)$. Then as N increases, $(\bar{N} - N)/\sqrt{N} \xrightarrow{L} N(0, \sigma^2)$, where $\sigma^2 = pq/(p^2 - q(T/\theta)^2)$, $p = p(\theta)$, and $q = 1 - p$.

Recently, Goudie and Goldie [24] have shown that for finite sample sizes, no exactly unbiased nonnegative estimators of N exist.

Instead of expanding $E(\hat{N})$ to get asymptotic bias, we could expand $E(\hat{N} - N)^2$ to get a correction to σ^2, the leading term in the mean squared error (MSE). Then the constants a and b of the prior could be chosen to minimize the MSE correction. This was done by Watson and Blumenthal [47] by keeping an additional term in the stochastic expansion (10b). Blumenthal [6] showed how to obtain the same MSE correction without the extra term in (10b) when a certain identity among the terms in the stochastic expansion of $\hat{N}$ is valid. Monte Carlo results in Watson and Blumenthal suggest that the minimax* bias values of a and b are close to optimal for MSE also over a wide range of θ and N.

From Theorem 7, it is easy to see that as $\theta \to \infty$, $\sigma^2 \propto (\theta/T)^3$ which becomes infinite regardless of the choice of a and b and the sampling period T. Thus there is no modified MLE which can control the maximum σ^2 over the entire θ scale. To achieve the goal of bounding the limiting variance σ^2 for all θ, Watson and Blumenthal [48] propose a two-stage procedure in which observations are taken in $(0, T_0)$, estimates $\hat{\theta}$ and $\hat{N}$ are computed, and a sampling time $\hat{T} = c/\hat{\theta}$ is determined where c is a constant. If $T_0 < \hat{T}$, additional observations are taken for a period of time $(\hat{T} - T_0)$, and an estimate N^* is computed based on all the data available up to time $\hat{T}$. The estimator of N, say $\tilde{N}$, is then given as

$$\tilde{N} = \begin{cases} 0 & \text{if no failures occur in } (0, T_0), \\ \hat{N} & \text{if } \hat{T} < T_0, \\ N^* & \text{if } \hat{T} > T_0. \end{cases}$$

The main result is the following.

Theorem 8. As $N \to \infty$, $(\tilde{N} - N)/\sqrt{N} \xrightarrow{L} N(0, \tilde{\sigma}^2)$ where

$$\tilde{\sigma}^2 = \begin{cases} \sigma^2, & \theta \leqslant (T_0/c) = \theta_c \\ \sigma_c^2, & \theta > \theta_c, \end{cases}$$

σ^2 is given in Theorem 7, and σ_c^2 is obtained by using θ_c in the σ^2 formula. Further, $\sigma^2 \leqslant \sigma_c^2$, $\theta \leqslant \theta_c$, so that $\tilde{\sigma}^2 \leqslant \sigma_c^2$ for all θ.

Monte Carlo* results show that the procedure works reasonably well for small N values also. Note that for small θ values, the sampling time T_0 is larger than what we would use to achieve variance σ_c^2 if θ were known, but the resulting variance is correspondingly smaller than σ_c^2. For large θ, we do as well as if θ were known. Results on choosing T_0 and the distribution of the total test time are given by Watson and Blumenthal.

CONCLUDING REMARKS

If θ is known, then M is a sufficient statistic* for N with a binomial distribution and known p. Hence the results of the section "The Binomial Distribution" apply. A simple sequential estimation* procedure is also possible in this case. Let $F(x)$ be $F(x \mid \theta)$ and $Y_i = -\log(1 - F(X_i))$, so that the Y's are standard exponential random variables. Assume that observations are obtained ordered by size as in life testing, so that $Y_{(1)} \leqslant Y_{(2)} \leqslant \cdots$ are observed in that order. The rule is to continue sampling until no observation falls in the interval $(Y_{(m)}, Y_{(m)} + t^*)$, where $Y_{(m)}$ was the time of the last observation obtained. We estimate N by m. With a proper choice of t^*, we can guarantee $P(N - k \leqslant \hat{N} \leqslant N) \geqslant 1 - c$ for any preassigned pair k and c. Details are given by Marcus and Blumenthal [31].

If only the value of M is recorded when θ is unknown, we have a binomial random variable with unknown p. If only one observation is available, only a trivial estimator of N and p can be obtained, unless a Bayesian approach is taken and a prior distribution is used for p or θ. Some aspects of this are examined in Blumenthal and Marcus [8].

References

[1] Alden, R. W., Dahiya, R. C., and Young, R. J., Jr. (1982). *J. Exper. Mar. Biol. Ecol.*, **59**, 185–206.

[2] Anscombe, F. J. (1961). *J. Amer. Statist. Ass.*, **56**, 493–502.

[3] Binet, F. E. (1953). *Ann. Eugen. (Lond.)*, **18**, 117–119.

[4] Blumenthal, S. (1968). *J. Amer. Statist. Ass.*, **63**, 542–551.

[5] Blumenthal, S. (1977). *Commun. Statist. A*, **6**, 297–308.

[6] Blumenthal, S. (1982). *Sankhyā A*, **44**, 436–451.

[7] Blumenthal, S. and Dahiya, R. C. (1981). *J. Amer. Statist. Ass.*, **76**, 903–909.

[8] Blumenthal, S. and Marcus, R. (1975). *IEEE Trans. Rel.*, **R-24**, 271–273.

[9] Blumenthal, S. and Marcus, R. (1975). *J. Amer. Statist. Ass.*, **28**, 913–922.

[10] Blumenthal, S. and Sanathanan, L. P. (1980). *Commun. Statist. A*, **9**, 997–1017.

[11] Blumenthal, S., Dahiya, R. C., and Gross, A. J. (1978). *J. Amer. Statist. Ass.*, **73**, 182–187.

[12] Chen, T. and Fienberg, S. E. (1974). *Biometrics*, **30**, 629–642.

[13] Chen, T. and Fienberg, S. E. (1976). *Biometrics*, **32**, 133–144.

[14] Cohen, A. C. (1960). *J. Amer. Statist. Ass.*, **55**, 139–143.

[15] Dahiya, R. C. (1981). *Amer. Statist.*, **35**, 34–37.

[16] Dahiya, R. C. and Gross, A. J. (1973). *J. Amer. Statist. Ass.*, **68**, 731–733.

[17] David, F. N. and Johnson, N. L. (1952). *Biometrics*, **8**, 275–285.

[18] Deemer, W. L. and Votaw, D. F. (1955). *Ann. Math. Statist.*, **26**, 498–504.

[19] Draper, N. and Guttman, I. (1971). *Technometrics*, **13**, 667–673.

[20] Feldman, D. and Fox, M. (1968). *J. Amer. Statist. Ass.*, **63**, 150–158.

[21] Fienberg, S. E. (1972). *Biometrika*, **59**, 591–603.

[22] Ghosh, M. and Meeden, G. (1975). *Sankhyā A*, **37**, 523–529.

[23] Goel, A. L. and Joglekar, A. M. (1976). Reliability Acceptance Sampling Plans Based upon Prior Distribution: III. Implications and Determination of the Prior Distribution. *Tech. Rep. No. 76-3*, Dept. of Industrial Engineering and Operations Research, Syracuse University, Syracuse, N.Y.

[24] Goudie, I. B. J. and Goldie, C. M. (1981). *Biometrika*, **68**, 543–50.

[25] Haldane, J. B. S. (1941). *Ann. Eugen. (Lond.)*, **11**, 179–181.

[26] Hartley, H. O. (1958). *Biometrics*, **14**, 174–194.

[27] Hocking, R. R. and Oxspring, H. H. (1971). *J. Amer. Statist. Ass.*, **66**, 65–70.

[28] Hocking, R. R. and Oxspring, H. H. (1974). *Biometrics*, **30**, 469–483.

[29] Irwin, J. O. (1959). *Biometrics*, **15**, 324–326.

[30] Jelinski, Z. and Moranda, P. (1972). In *Statistical Computer Performance Evaluation*, W. Frieberger, ed. Academic Press, New York, pp. 465–484.

[31] Marcus, R. and Blumenthal, S. (1974). *Technometrics*, **16**, 229–234.

[32] Moore, P. G. (1952). *Biometrika*, **39**, 247–251.

[33] Moore, P. G. (1954). *Biometrics*, **10**, 402–406.

[34] Olkin, I, Petkau, A. J., and Zidek, J. V. (1981). *J. Amer. Statist. Ass.*, **76**, 637–642.

[35] Plackett, R. L. (1953). *Biometrics*, **9**, 485–488.

[36] Rao, B. R., Mazumdar, S., Waller, J. H., and Li, C. C. (1973). *Biometrics*, **29**, 271–279.
[37] Rider, P. R. (1953). *J. Amer. Statist. Ass.*, **48**, 826–830.
[38] Robinson, J. (1976). *Biometrics*, **32**, 61–68.
[39] Sampford, M. R. (1955). *Biometrika*, **42**, 58–69.
[40] Sanathanan, L. P. (1972). *Ann. Math. Statist.*, **43**, 142–152.
[41] Sanathanan, L. P. (1972). *Technometrics*, **14**, 813–830.
[42] Sanathanan, L. P. (1973). *Technometrics*, **15**, 67–78.
[43] Sanathanan, L. P. (1977). *J. Amer. Statist. Ass.*, **72**, 669–672.
[44] Selvin, S. (1974). *J. Amer. Statist. Ass.*, **69**, 234–237.
[45] Subrahmaniam, K. (1965). *Biometrika*, **52**, 279–282.
[46] Tate, R. F. and Goen, R. L. (1958). *Ann. Math. Statist.*, **29**, 755–765.
[47] Watson, D. and Blumenthal, S. (1980). *Commun. Statist. A*, **9**, 1535–1550.
[48] Watson, D. and Blumenthal, S. (1980). *Austral. J. Statist.*, **22**, 317–327.
[49] Whittaker, L. (1914). *Biometrika*, **10**, 36–71.
[50] Wittes, J. T. (1970). Estimation of Population Size: The Bernoulli Census. Unpublished Ph.D. thesis, Harvard University, Cambridge, Mass.
[51] Wittes, J. T., Cotton, T., and Sidel, V. W. (1974). *J. Chronic Dis.*, **27**, 25–36.

(ESTIMATION, POINT
TRUNCATION)

RAM C. DAHIYA
SAUL BLUMENTHAL

POPULATION PROJECTION

Ever since the time when modern censuses and vital registration began, nearly 200 years ago, it has been possible to make some sort of reasoned attempt to forecast the population of the future. In recent years, there has been much demand for information of this kind. Plans for the production of food and the provision of schools and hospitals, for example, must be related to the numbers of people likely to need them. In the short term, simple linear extrapolation from recent census results may be able to lead to a good enough idea of the requirements, classified by sex, age, region, and other categories; and it may be possible to confirm or refine such assessments on the basis of the trends in births, deaths, marriages, and migration. For the longer term, however—say over five years ahead—more complex calculations are normally desirable.

The earliest predictions were based on the fear that population growth*, then rapid in the most advanced countries, would outstrip the development of resources and lead to poverty and starvation. These ideas were given literary expression, for example, by Thomas Malthus* [2], who thought of the growth of numbers as a geometrical progression but of food supply only as an arithmetic progression. Verhulst [3] later suggested, however, that human populations must reach an upper limit of size, and therefore that the growth rate must fall; on the basis that the initial pace of increase should diminish to zero in a manner proportionate to the size of the population, that is, that

$$\frac{1}{{}^tp}\frac{d\,{}^tp}{dt} = \rho - k\,{}^tp,$$

where ${}^tp$ is the population size at time t, and ρ and k are constants, he arrived at the logistic curve ${}^tp = (Ce^{-\rho t} + k/\rho)^{-1}$, where C is constant. On this basis, ${}^tp$ rises to a maximum of ρ/k. Logistic* curves have been fitted to population sizes over the nineteenth and twentieth centuries, obtained from censuses, but projections derived from them have needed material revision as new census* data have become available from time to time. The use of formulae such as the above has the disadvantage of not providing any analysis by age, which is normally required. It also implies a definite link between the trends of mortality, fertility, and migration which is not at all closely in accord with actual experience—although it is not impossible that in the long run, population growth could be self-correcting, demographic behavior being influenced by warnings based on projected trends.

TECHNIQUE

A different method of forecasting future population was developed by Arthur Bowley [1] in the 1920s. He first set out the numbers of people classified by age and then, by the application of constant mortality rates, assessed the numbers dying and surviving at various future periods. On the assumption that the numbers of births would remain constant from year to year, he was able with the aid of survival rates to calculate how many of these infants would still be alive after different lengths of time. Bringing the results of the two sets of calculations together, he arrived at an estimate of future population, classified by age, at various future times over a 50-year span. This process is illustrated in Table 1; usually known as the "component" method, it has been employed, with various refinements, for the great majority of population forecasts published up to the present day. Some of the technical refinements are:

1. The application of fertility* rates (not necessarily constant ones) to the fertile part of the population (i.e., excluding children and the aged), at any time, to estimate the numbers of future births
2. Allowance for changing mortality and for migration*; also, immigrants may be assumed to experience mortality and fertility rates different from those of the resident population
3. Analysis by marital status and the use of marriage-dependent fertility rates, possibly classified by duration of marriage* rather than by age

It is a simple matter to compare in detail such forecasts with the actual subsequent populations, and the usual result of such comparisons is to find substantial accuracy in the part of the forecast population which depends only on the initial numbers and the rates of mortality, but to find material inaccuracy, growing with the lapse of time, in the portion dependent on the numbers of future births.

In modern societies, the control over family planning which is available to each married couple for their personal use, coupled with the influence on human behavior of economic variations, social changes, government policy, and the vagaries of fashion, has made fertility forecasting unreliable. For this reason, calculations of future population are now usually termed "projections," illustrating the working out of the demographers' assumptions—sometimes adopted as a pointer or a warning rather than as a considered belief. Such calculations are often presented in sets with alternative combinations

Table 1

Age	Current Population	Survival Factor for One Year	Population One Year Hence	Population Two Years Hence
0			$B_0 p_0'^{a}$	$B_1 p_1'$
1			$P_0 p_0$	$B_0 p_0' p_1$
⋮				$P_0 p_0 p_1$
$x-1$	P_{x-1}	p_{x-1}		
x	P_x	p_x	$P_{x-1} p_{x-1}$	
$x+1$	P_{x+1}	p_{x+1}	$P_x p_x$	$P_{x-1} p_{x-1} p_x$
$x+2$	P_{x+2}	p_{x+2}	$P_{x+1} p_{x+1}$	$P_x p_x p_{x+1}$
⋮				

[a] More precisely, the numbers of births not counted in P_0 but resulting in survivors aged 0 last birthday in year 1.

of bases. The work is easy to program for a computer, which can quickly provide all the figures required. The starting population, classified by age, can be described in matrix algebra as a "vector" (**v**). All that need happen, basically, is that this should be multiplied by a matrix (**m**) consisting of:

1. In the top row, zeros for infertile ages and fertility factors for other ages
2. In the second row, the first-year survival factor p_0, then all zeros
3. In the third row, zero, p_1, then all zeros (p_1 is the chance of survival from age 1 to age 2)

and so on.

ACCURACY

Population projections made for 14 European countries in the year 1944 showed *inter alia* the total numbers of people expected for the year 1965. Comparison with the populations actually present in 1965 shows that (to the nearest 100,000) only one was accurate. In the other 13 instances the expected populations were all too low, with a shortfall averaging 21% and varying from 2 to 48%. These "errors" are particularly big because (1) World War II greatly hampered the collection of statistics and caused large population (and boundary) changes, and (2) a standard method was used which did not pay close enough attention to circumstances in individual countries.

Projections on 16 alternative bases made in 1947 forward to 1977 in respect to Great Britain produced the following results in comparison with the actual 1977 numbers:

Age Group	Projections (millions)	Actual (millions)
0–14	$8\frac{1}{4}$–$11\frac{1}{2}$	12
15–64	30–$35\frac{1}{2}$	$34\frac{1}{2}$
65 and over	$7\frac{1}{4}$–$8\frac{1}{4}$	8
Total	$46\frac{1}{2}$–$54\frac{1}{2}$	$54\frac{1}{2}$

and such disparities are fairly typical.

PROJECTIONS OF WORLD POPULATION

Since its inception after World War II, the United Nations Organization has published many volumes on the subject of world population, and these have included a number of projections made by its Population Division. Reliable data adequate for the use of the component method are not available for many countries, and in consequence simpler approaches have been adopted; the finer details of these approaches have not been disclosed. The total numbers of people expected for the year 1980 have been:

Projection Published in:	Number (millions)
1951	3600
1954	3300–4000
1958	3900–4300
1967	4100–4600

whereas it can now be estimated that the actual 1980 total must have been within the range 4300 to 4500 millions. A more exact count is rendered impossible because of uncertainty, particularly about the size of China's population (not far short of 1000 million). The Chinese Population Research Institute in Peking has itself produced some forecasts recently showing a large further rise in the national numbers from now on to be likely unless the birth rate can be reduced below its present level. These types of calculation confirm that, even if inaccurate in detail and based on only partial evidence, projections are used as an aid to the formulation of general policy.

The forecasts described above treat population as an independent variable, that is, without attempting to measure the effects on human numbers of general economic and social change. This is because, in the main, even more uncertainty attaches to projections of economic and social developments than to demographic assessments. Since 1970, however, a number of computerized models of world dynamics have been produced which examine the interactions with population of such elements as capital equipment, technical knowledge, land area, stocks of raw materials, systems of beliefs, and official policies. Such models combine established knowledge with a good deal of speculation or assumption, but once set up they can show how the future would unfold on the stated bases. The predictions of different operators of these models have ranged from warnings of ruin, unless certain policies are adopted or practices changed, to hopes of a golden future—much or even all depending on how inflexible mankind is assumed to be. Studies are proceeding to improve the quality of the bases and assumptions of such calculations, which clearly will be of great importance in years to come.

BACK PROJECTION

In a recent assessment by Wrigley and Schofield [4] of population history in England, it was possible to estimate (from parish registers) the number of births and deaths for the country as a whole for a period of some 250 years before the first censuses were taken. On the basis of those censuses, and the reckoned deaths and migratory movement for earlier decades, a "population projection" was carried out in reverse in order to arrive at the supposed numbers of people at each age alive in those earlier decades. "Entrants" were those aged over 90, and their predecessors at earlier ages were calculated right back to "exit" at birth. This method, similar in principle to that introduced by Bowley (see above), was tested by other means and seemed to give reasonable results; thus it became possible to estimate rates of fertility, marriage, and mortality throughout the whole of the historical period studied.

References

[1] Bowley, A. L. (1924). *Econ. J.*, **34**, 188–192.

[2] Malthus, T. R. (1803). *An Essay on the Principle of Population*. London.

[3] Verhulst, P. F. (1838). In *Correspondence mathématique et physique* (*Quetelet, A.*). Brussels, pp. 113–121.

[4] Wrigley, E. A. and Schofield, R. S. (1981). *The Population History of England, 1541–1871—A Reconstruction*. Harvard University Press, Cambridge, Mass.

Bibliography

Brass, W. (1974). *J. R. Statist. Soc. A*, **137**, 532–570. (A very clear assessment of the doubts and difficulties associated with projections, with some ideas for technical improvements.)

Leach, D. (1981). *J. R. Statist. Soc. A*, **144**, 94–103. (A defense of the use of mathematical curves for projection purposes.)

Meade, J. E. (1973). In *Resources and Population*, B. Benjamin, P. R. Cox, and J. Peel, (eds.) Academic Press, London. p. 119. (An authoritative discussion of computerized models of demographic–economic interrelationships.)

Tapinos, G. and Piotrow, P. T. (1978). *Six Billion People*. McGraw-Hill, New York. (Contains an account of recent perspectives and a critical evaluation of projections.)

United Nations (1979). *Propects of Population: Methodology and Assumptions*. U.N., New York. (Papers by experts designed to improve the quality and utility of the U.N. national, regional, and global projections.)

United Nations Secretariat (1974). *Recent Population Trends and Future Prospects*. World Population Conference, Bucharest. (A description of the facts and prospects with some account of the assumptions made.)

(DEMOGRAPHY
FERTILITY
LIFE TABLES
MARRIAGE
MATHEMATICAL THEORY OF POPULATION
MIGRATION
POPULATION GROWTH MODELS
SURVIVAL ANALYSIS
VITAL STATISTICS)

P. R. Cox

POPULATION PYRAMID

A population pyramid is not a three-dimensional solid but a diagram designed to show the composition of a human population by sex and age at a given time. It consists of a pair of histograms*, one for each sex, laid on their sides with a common base. This base, now vertical, shows the age in years, rising from zero at the bottom to 100 at the top. To the right, the width of the diagram at any age is proportionate to the numbers of one sex at that age; to the left, the width relates similarly to the other sex. The whole area comprised within the diagram thus represents the total population size at the given time. (If the sizes of the age groups are not uniform, the areas of the rectangles must be proportionate to the population.) An example is given in Fig. 1 in bold lines, from which it may be seen that the biggest numbers for either sex are in the age group 10–20, and at higher ages there is a reduction in size as the age advances; but also the numbers of children below age 10 are lower than those at 10–20.

One advantage of such pyramids is their usefulness in comparing two populations. In Fig. 1 the fainter outlines give the data for a second population, which is, compared with the first, (1) a little smaller in size over all, (2) larger at ages under 10 but smaller at most other ages, and (3) more disparate between the sexes in middle life (possibly because of predominantly male emigration). The difference between the slender and bold outlines could indeed relate to an earlier and a later count of the same population, in which a growth of total numbers occurred, coupled with a cessation of emigration and a fall in fertility.

It is possible to superimpose, on a pyramid, curves to represent a population of equal size, the numbers of which decline with advancing age in accordance with a life table*. Such a curve is illustrated in Fig. 2, and it provides an idea of how the varying history of the actual population in regard to fertility, mortality, and migration has shaped it away from the theoretical model. Figure 3 shows how pyramids compare for a developing population in (say) Africa and an economically advantaged population of roughly equal size in (say) Western Europe. The contrast emphasizes the high proportion of youth in the poorer country and the older ages of the Europeans, probably in association with higher fertility and mortality in the developing area. Pyramids for India (1971) and Australia (1966) showing percentages of total population appear in Fig. 4: it will be

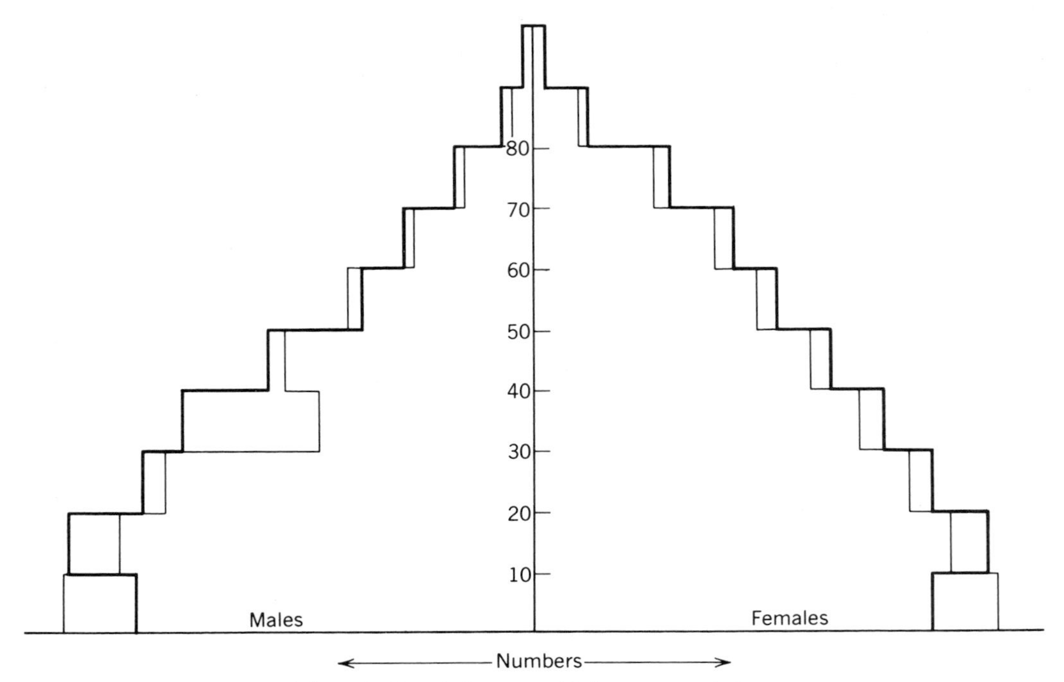

Figure 1 Population pyramids for two populations.

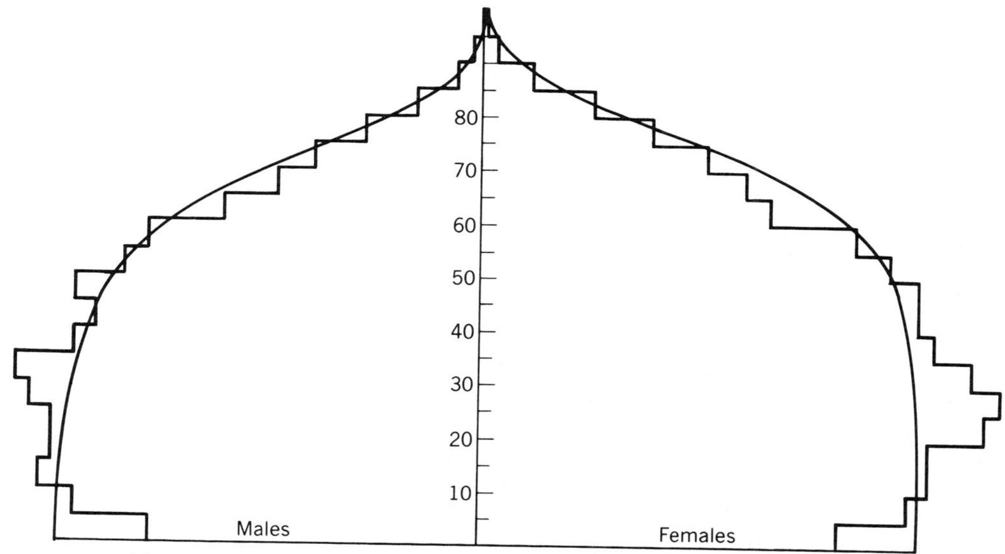

Figure 2 Population pyramid compared with stationary (life table) population.

seen that the Indian data show signs of a transition from the totally undeveloped state, attributable mainly to a decline in mortality but giving evidence also of a recent fall in fertility.

Population pyramids came into use mainly in the 1930s, when the important effects of a long decline in fertility in Western countries on their age distributions began to be generally realized.

The use of pyramids such as these need not be confined to human populations but could illustrate animal or even inanimate numbers. In the human field, they could show the composition of a part population such as that of a particular occupation or

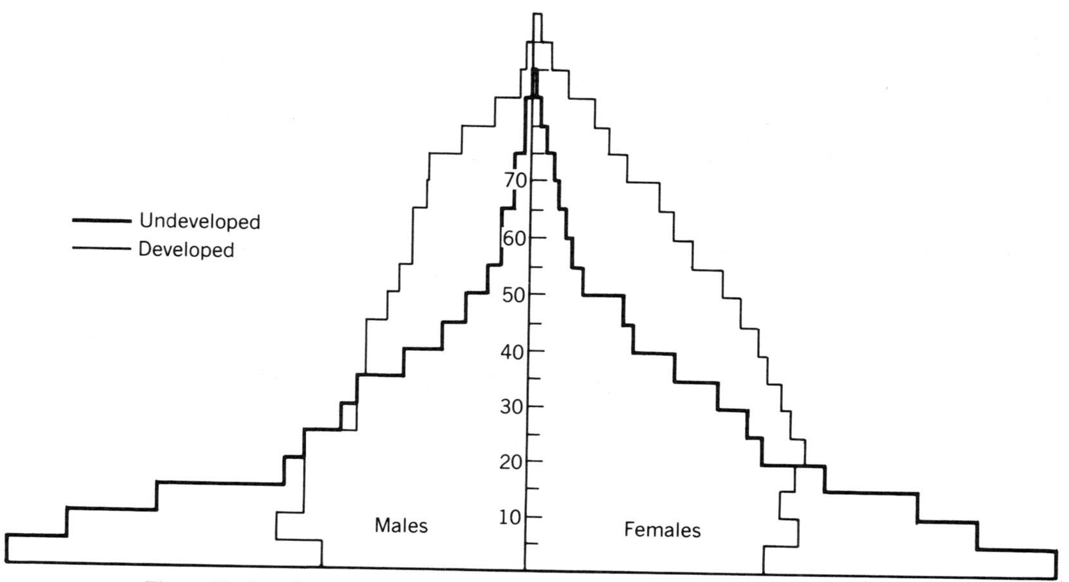

Figure 3 Population pyramids typical of developed and undeveloped countries.

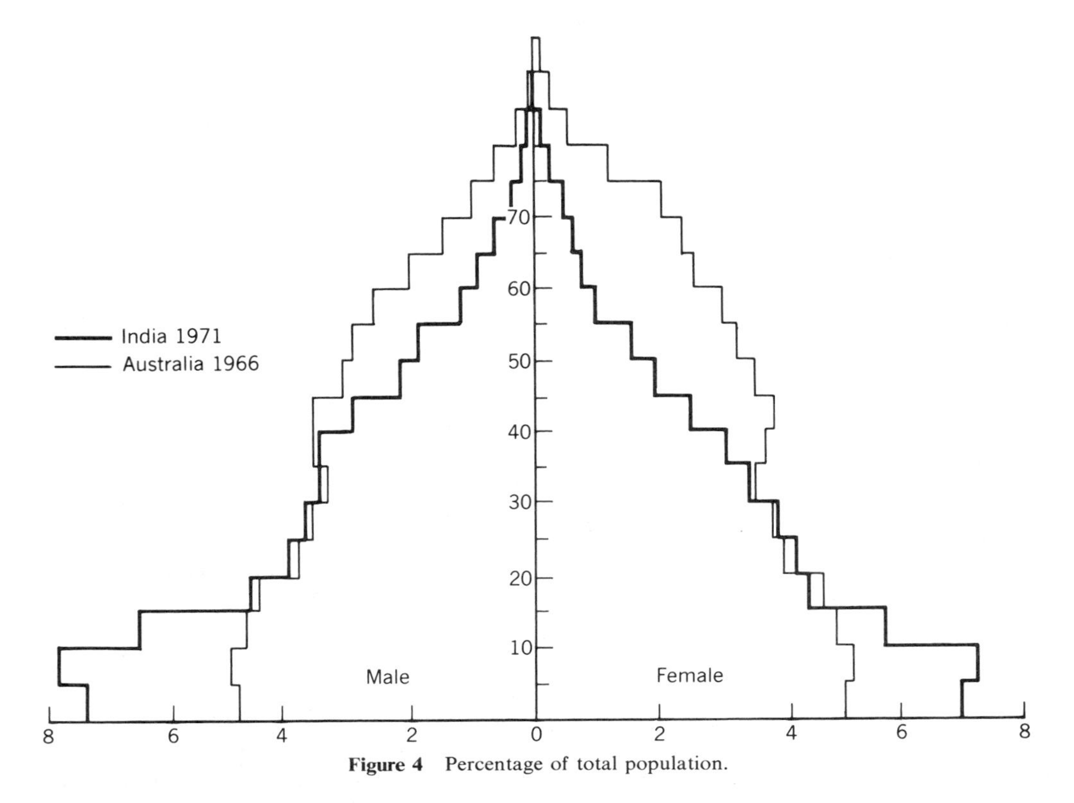

Figure 4 Percentage of total population.

district, the numbers born outside the country, or those suffering from a disease. Pyramids represent only one of many possible methods of graphic illustration in demography* but they are especially valuable in issuing a reminder of the importance of age distribution in influencing crude rates of fertility*, mortality*, marriage* and migration*. Yet, as the articles on these subjects in this encyclopedia show, more specific methods of analysis are necessary for refined comparisons of experience in regard to these elements in population change.

Bibliography

Pressat, R. (1974). *A Workbook in Demography*. Methuen, London. [Although nearly all demographic textbooks illustrate the use of population pyramids, this book gives several examples of how they may be constructed (pp. 11–13, 37–39, and 224–226).]

(DEMOGRAPHY
FERTILITY
MARRIAGE
MIGRATION
POPULATION PROJECTION)

P. R. Cox

POPULATION SAMPLING IN LESS DEVELOPED COUNTRIES

Almost all less developed countries (LDCs) have some experience of sampling their population for statistical surveys. In the typical LDC the national statistical office carries out surveys from time to time on such topics as demographic characteristics, fertility*, mortality, migration*, health, household income and expenditure, manpower, peasant agriculture, and so on. These surveys, and the sampling operations on which they are based, do not differ in any fundamental way from those conducted in developed countries; only the parameters and some of the operational details differ.

Most LDCs possess an adequate sampling frame* of small areas, consisting of a list of census* enumeration districts (EDs) each with its census population (typically around 1000) and usually with rough maps showing boundaries and main features. The list is generally grouped in terms of the country's administrative divisions, but apart from this geographical information and the population size of the locality, information useful for stratifying the EDs before selection is in most cases not readily available, at least for rural areas. Thus, typically, the main effective stratification in the rural sample is provided by systematic sampling* of EDs from the list. In the urban sector, socioeconomic differences between different districts of the same city are often very marked and these may provide a useful basis for stratification.

In many surveys the census EDs are too large to serve as efficient units for the final stage of area sampling. Arrangements are therefore often made to subdivide the selected EDs (or, in some cases, only those that exceed a certain threshold population size) into segments, which then provide a further sampling stage. In urban areas such subdivision may be done in the office, working on maps or aerial photographs, but in rural areas segmentation will generally require field visits.

In most countries demarcation of EDs is not based on strict cadastral surveying and serious errors or ambiguities are not uncommon. The statistician needs to consider the probable impact of such errors in the light of the planned estimation* procedure. In general, if totals are being estimated and no use is to be made of supplementary information, errors in area demarcation will have a greater impact and it may be necessary to mount an expensive checking operation in the field. More often the survey is planned to provide only estimates of ratios and means from within the sample, and where this is so demarcation errors can be more readily tolerated. Obviously, if demarcation is to be checked, economy argues strongly for limiting the checking procedure to the selected EDs, although such a procedure can never be fully rigorous. If, however, this is judged inadequate, there is an overwhelming case for introducing a sampling stage prior to that of the ED, using as a unit the smallest type of area whose boundaries are known accurately; checking of ED boundaries can then be limited to the selected sample of these higher-stage units.

Apart from the latter case, the main motive for introducing a sampling stage prior to that of the ED would be to cluster the ED sample to reduce travel between EDs—whether in the interests of economy or to facilitate supervision. Statisticians more familiar with developed countries, or with large countries, may be tempted to overestimate the gain from multistage area sampling* in the typical small LDC. In the latter the saving in travel cost or time through clustering of the ED sample is often very small, especially since in LDCs, long-distance travel is generally much cheaper and faster than local travel, while the loss in sampling precision* due to clustering in the rural sector is likely to be greater than in a developed country. Thus in many LDC surveys the ED may serve as an efficient primary sampling unit.*

In general, LDCs lack any systematic address system for dwellings and there is no available list of households, much less of persons. Thus any sampling within the ultimate area unit (whether the ED or the segment) requires a specially mounted *listing operation* to provide a sampling frame, whether of dwellings, households, or individuals. The cost of this operation is an important consideration in optimizing sample design. In many LDCs mobility is high, particularly in urban areas, so that any listing of households or individuals can get seriously out of date within a few months. Thus a new list is likely to be needed for every survey. Considerably more durable, besides cheaper to produce, is a listing of structural units of housing, which may loosely be termed "dwellings." Such units can generally be identified and listed without an interview; in some surveys the lister affixes a sticker to the wall or door for later identification. In the best case (which is not uncommon) the dwelling corresponds closely with the household. However, in some areas the only identifiable structural units are very large, or very variable in size, and where this is so listing of dwellings may have to be abandoned in favor of listing of households or individuals. In most (but not all) cultures a "household" concept can be realistically identified, defined as a group who live and eat together (exact definitions vary).

A common, and recommended, sample design is the following. First, a sample of area units, typically EDs, is selected by systematic sampling with probability proportional to size* (PPS), the measure of size being the number of households or persons reported in the census. If necessary, the selected EDs, or at least the larger ones, may be segmented and the segment sizes estimated; one segment is then selected with PPS in each selected ED. A listing operation (of dwellings, households, or persons) is then carried out in each selected area unit and a sample of the listing units is selected with the reciprocal of the ED (or segment) selection probability. This yields a self-weighting sample with an approximately fixed "take" in each area unit. (A common error is to select an exactly fixed take in each area unit and to assume the resulting sample to be self-weighting. In the real world, measures of size are unlikely to be accurate enough anywhere, and particularly in LDCs, to justify this assumption.)

Perhaps because transport is difficult for the general population and information travels slowly, intracluster correlation* tends to be higher in LDC rural populations than in developed countries for variables related to knowledge, education, and attitudes. That is, the variance between small area units is relatively high in proportion to the variance within them. Other things being equal, this should imply that optimal rural cluster size (i.e., the sample taken per area unit) is smaller in LDCs. However, the relatively high cost of travel for field workers between clusters in LDCs operates in the opposite sense, as does the relatively high cost of listing and map checking. The latter factors are probably dominant in most cases, so that rural cluster sizes are in practice generally larger in LDCs than in developed countries.

The high level of between-area variance and the paucity of information for explicit stratification also imply that simple geographical stratification (e.g., by systematic sampling from the census list of EDs) can bring very substantial gains in LDCs. Urban–rural differentials are also more marked than in developed countries on most variables, as are the differentials between types of urban habitat; these differences offer further scope for effective stratification.

In most LDCs interviewers require close supervision in the field, and this can affect sample design. An effective arrangement is to organize field workers in teams under a supervisor with an accompanying field editor. This implies a sufficiently large sample take in each cluster to justify the team's visit. In some surveys such a large cluster size may be uneconomical in terms of sampling efficiency, and in this case the team method may be inappropriate.

Turning to sample implementation, the problems are rather different from those encountered in developed countries. Refusal rates in LDCs are usually negligible. On the other hand, defects of mapping may be frequent and are not easily dealt with—or even, perhaps, recognized. A further problem common in LDC sampling is the necessity to exclude (often at the last minute) a substantial area due to inaccessibility, whether because of floods, other transport breakdowns, or civil disturbance.

Mention should be made of sampling error computation. Most LDCs have published data on sampling errors and design effects for at least one survey—and indeed the record of LDCs in this respect may be marginally less lamentable than that of developed countries. Such computations present no essential difficulty, and several convenient packages are available, notably the CLUSTERS program of the World Fertility Survey.

Finally, a problem characteristic of LDCs is the high rate of turnover among senior staff. One consequence of this is the difficulty of retaining in the "memory" of the survey organization crucial details of the sampling procedures, or at least of communicating these effectively to the data analysts. This argues for simple designs and self-weighting samples.

Effective sampling in LDCs depends on careful attention to numerous practical constraints such as those mentioned above. Concern solely with minimizing sampling error per unit of cost is unlikely to prove an adequate basis for a successful design.

Bibliography

Kish, L. (1965). *Survey Sampling*. Wiley, New York.

U.S. Bureau of the Census (1968). *Sampling Lectures: Supplemental Courses for Case Studies in Surveys and Censuses*. U.S. Government Printing Office, Washington, D.C.

Verma, V. and Pearce, M. (1978). *Users' Manual for CLUSTERS*. World Fertility Survey, London.

Verma, V., Scott, C., and O'Muircheartaigh, C. (1980). *J. R. Statist. Soc. A*, **143**, 431–463.

World Fertility Survey. (1975). *Manual on Sample Design*. International Statistical Institute, The Hague.

(CENSUS
SURVEY SAMPLING)

CHRISTOPHER SCOTT

PORTMANTEAU TEST

The aim of time-series* modeling may be thought of as, given an observed series, finding a transformation which when applied to the series yields a residual series whose members are independent and identically distributed. For if this be so, then no further information can be extracted from the series for purposes of prediction or control. To test the goodness of fit of the model, one may therefore test the residual series $\{\hat{a}_t : t = 1, \ldots, n\}$ for independence. In a time series model relationships between observations separated by one or several time periods are of particular interest, so natural test statistics for independence of the residuals are the residual correlations

$$\hat{r}_k = \sum_{t=k+1}^{n} \hat{a}_t \hat{a}_{t-k} \Big/ \sum_{t=1}^{n} \hat{a}_t^2, \quad k = 1, 2, \ldots,$$

which one would expect to be close to zero for an independent series. For the simplest case, in which the observed time series is white noise* (i.e., consists of independent identically distributed observations), the residual correlations $\hat{r}_k$ are independent with mean zero and variance $(n-k)/\{n(n+2)\}$ [1].

A particularly important and widely used class of time-series models is the autoregressive integrated moving-average (ARIMA) processes* introduced by Box and Jenkins [2] and defined by

$$\nabla^d y_t = w_t,$$

$$w_t - \phi_1 w_{t-1} - \cdots - \phi_p w_{t-p}$$
$$= a_t - \theta_1 a_{t-1} - \cdots - \theta_q a_{t-q},$$
$$t = 1, \ldots, n,$$

where $\{y_t\}$ is the observed series and $\{a_t\}$ is a white noise* series. Here ∇^d is the dth power of the backward-differencing* operator ∇ defined by $\nabla y_t = y_t - y_{t-1}$, and $\phi_1, \ldots, \phi_p$ and $\theta_1, \ldots, \theta_q$ are constants (the autoregressive and moving-average* coefficients of the process). For these processes the distribution of the residual correlations depends on the model parameters and the correlations are themselves correlated. Box and Jenkins, drawing on the work of Box and Pierce [3], were nonetheless able to show how inferences for goodness of fit* testing could be based on the residual correlations, either individually or by combining a large number of correlations into one statistic to give what Box and Jenkins called, because of its all-inclusive nature, a "portmanteau lack of fit test." The test statistic for the portmanteau test is

$$Q = n \sum_{k=1}^{m} \hat{r}_k^2$$

and the test is to compare Q with the significance levels of a χ^2 distribution with $(m - p - q)$ degrees of freedom. A significantly large value of Q indicates inadequacy in the fit of the ARIMA(p, d, q) model.

The choice of m, the number of residual correlations used to define the portmanteau statistic, requires some care. Too few, and the χ^2 approximation to the distribution of Q may fail; too many, and "end effects" in the definition of correlations (such as there being n terms in the sum in the denominator in the definition of $\hat{r}_k$, but only $n - k$ in the numerator) may become important, causing too few significant values of Q to be observed. There are many examples of surprisingly low values of Q being reported (e.g., of the 31 values of Q calculated by Prothero and Wallis [11], none exceeds the 50% point of its theoretical χ^2 distribution), and many of these are likely to be due to choosing m too large. A reasonable compromise is to choose m approximately equal to $2n^{1/2}$ provided that the impulse response function of the fitted model decays fairly rapidly.

The chi-squared distribution of the portmanteau statistic is only a large-sample asymptotic approximation, and much work has been done [5, 6] in investigating the adequacy of this approximation when the test is applied to the fairly short series (50 to 100 observations) frequently encountered in industrial and economic applications. Again a tendency for significant values to occur too rarely has been observed, and modifications to the portmanteau test have been suggested in consequence. A simple but effective modification, recommended by Ljung and Box [8], is to replace the multiplier n in the definition of Q by $n(n+2)/(n-k)$, the reciprocal of the variance of the lag-k correlation of a white noise process, giving a new statistic

$$Q' = n(n+2) \sum_{k=1}^{m} \hat{r}_k^2/(n-k).$$

Significance levels of Q' agree much more closely with a χ^2_{m-p-q} distribution than do those of Q.

As an example of the use of the portmanteau test, consider Series C of Box and Jenkins [2]. This consists of 226 observations, made at intervals of 1 minute, of the temperature of a chemical reaction. Study of the correlations of the series yields two possible models:

1. ARIMA(1, 1, 0): $\nabla y_t - 0.82\nabla y_{t-1} = a_t$
2. ARIMA(0, 2, 2): $\nabla^2 y_t = a_t - 0.13a_{t-1} - 0.12a_{t-2}$

Portmanteau statistics were calculated for these models using $m = 30$ residual correlations. For model 2 the value of Q' is 45.2, which, compared with a χ^2 distribution with 28 degrees of freedom, has a significance level of 2.1%, strongly indicating an inadequate model. For model 1 the value of Q' is 36.9; this corresponds to a significance level of 15%, which one would normally accept as indicating a reasonable fit of the model to the data.

Because the portmanteau statistic is constructed from a large number of correlation* statistics each of which may be regarded as a basis for a test against a particular deviation from the fitted model, the portmanteau test is not very sensitive to specific deviations from the model. Davies and Newbold [4] have shown that the power of the portmanteau test can be disturbingly low, particularly when the sample size is less than 100.

The portmanteau test was originally derived as a pure significance test*; its distribution was derived on the assumption that the fitted model gave a true description of the observed process and no alternative was taken into account. However, the portmanteau test may also be thought of as a Lagrange-multiplier test* of the null hypothesis that the observed series is an ARIMA(p, d, q) process against the alternative hypothesis that it is an ARIMA$(p + m, d, q)$ process [9]. When m is large, this alternative model is very general, illustrating the nonspecific nature of the portmanteau test.

The term "portmanteau test" originated in

the testing of ARIMA models but can equally well be applied to other tests of significance which combine information from several statistics. As examples, there are the extensions of the portmanteau test to seasonal ARIMA models and transfer-function* models [2], regression models with dynamic disturbances [10], and multivariate ARIMA models [7]. All these tests are based on quadratic forms in residual correlations.

In view of the popularity of the portmanteau test and its extensions some warning about their use seems appropriate. They test for a wide range of deviations from the fitted model without being particularly sensitive to any of them. If a particular form of deviation (e.g., an extra autoregressive parameter, or a seasonal term to explain cyclic variation) is of interest then a test specifically for this should be used, and the portmanteau test treated more as a final check which may reveal unexpected inadequacies in the model. In short, a portmanteau test is a necessary stage in diagnostic checking of a model, but is not in itself sufficient.

References

[1] Anderson, R. L. (1942). *Ann. Math. Statist.*, **13**, 1–13.

[2] Box, G. E. P. and Jenkins, G. M. (1970). *Time Series Analysis Forecasting and Control*. Holden-Day, San Francisco. (Section 8.3 contains a discussion of the portmanteau test and diagnostic checking of time-series models.)

[3] Box, G. E. P. and Pierce, D. A. (1970). *J. Amer. Statist. Ass.*, **65**, 1509–1526.

[4] Davies, N. and Newbold, P. (1979). *Biometrics*, **66**, 153–155.

[5] Davies, N., Triggs, C. M., and Newbold, P. (1977). *Biometrics*, **64**, 517–522.

[6] Hosking, J. R. M. (1978). *J. R. Statist. Soc. B*, **40**, 340–349.

[7] Hosking, J. R. M. (1980). *J. Amer. Statist. Ass.*, **75**, 602–608.

[8] Ljung, G. M. and Box, G. E. P. (1978). *Biometrika*, **65**, 297–303.

[9] Newbold, P. (1980). *Biometrika*, **67**, 463–465.

[10] Pierce, D. A. (1972). *J. Amer. Statist. Ass.*, **67**, 636–640.

[11] Prothero, D. L. and Wallis, K. F. (1976). *J. R. Statist. Soc. A*, **139**, 468–500.

(AUTOREGRESSIVE-INTEGRATED MOVING AVERAGE (ARIMA) MODELS
LAGRANGE MULTIPLIER TEST
SIGNIFICANCE TESTS
TIME SERIES)

J. R. M. HOSKING

POSITIVITY, TOTAL *See* TOTAL POSITIVITY

POSTERIOR DISTRIBUTIONS

Posterior distributions are probability distributions or probability density functions (PDFs) that summarize information about a random variable or parameter after or *posterior* to having obtained new information from empirical data, and so on. The use of the designation "posterior distribution" occurs almost entirely within the context of statistical methodology generally labeled "Bayesian inference"*. Bayes theorem* and a 1763 paper by Thomas Bayes [1] are generally cited as the beginnings of this methodology. However, Stigler [10] cites evidence that the little-known English mathematician Saunderson may have been the true builder of the foundations of what is now termed Bayesian inference. Posterior distributions relate to corresponding prior distributions, and one should logically understand the prior distribution concept before they consider the posterior distribution concept (*see* PRIOR DISTRIBUTIONS).

Suppose that $\mathbf{Y}' = (Y_1, Y_2, \ldots, Y_n)$ is a vector of n independent Y_i's representing sample observations, each with PDF $f_{Y_i \mid \Theta}(y_i \mid \boldsymbol{\theta})$ dependent on the value of m parameters in the vector $\boldsymbol{\Theta}$, and joint probability distribution $f_{\mathbf{Y} \mid \Theta}(\mathbf{y} \mid \boldsymbol{\theta}) = \prod_i f_{Y_i \mid \Theta}(y_i \mid \boldsymbol{\theta})$. If a prior PDF $f_{\Theta}(\boldsymbol{\theta})$ summarizes knowledge and assumptions about the unknown $\boldsymbol{\Theta}$ before taking the sample, Bayes theorem says that the posterior distribution $f_{\Theta \mid \mathbf{Y}}(\boldsymbol{\theta} \mid \mathbf{y})$ summarizing the knowledge and assumptions about $\boldsymbol{\Theta}$ after taking the sample is

$$f_{\Theta \mid \mathbf{Y}}(\boldsymbol{\theta} \mid \mathbf{y}) = \frac{f_{\mathbf{Y} \mid \Theta}(\mathbf{y} \mid \boldsymbol{\theta}) f_{\Theta}(\boldsymbol{\theta})}{f_{\mathbf{Y}}(\mathbf{y})},$$

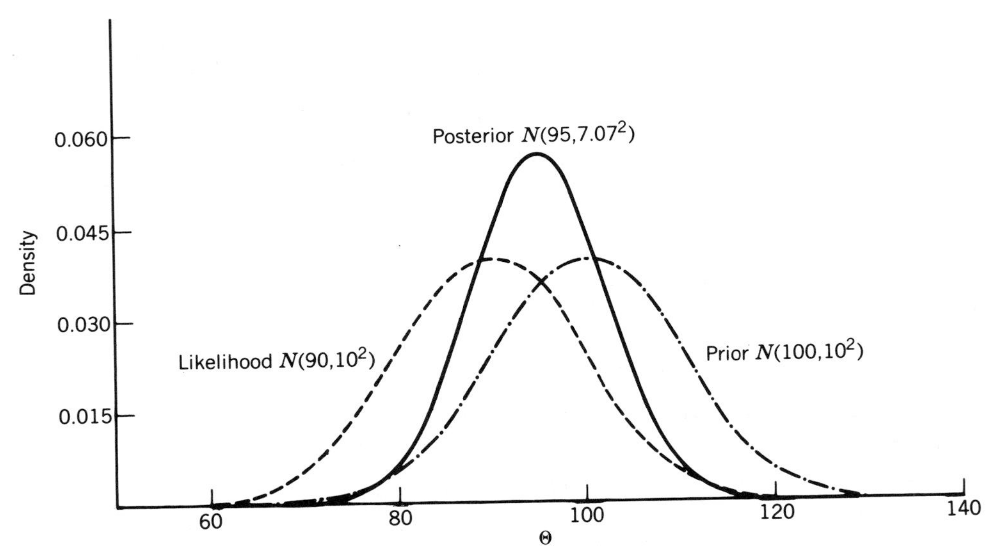

Figure 1 Example of posterior distribution.

the conditional distribution of $\boldsymbol{\Theta}$ given $\mathbf{Y}$, expressed as the joint distribution of $\boldsymbol{\Theta}$ and $\mathbf{Y}$ divided by the marginal distribution of $\mathbf{Y}$, following the definition of a conditional distribution. Now, given an observed $\mathbf{y}$, $f_{\mathbf{Y}|\boldsymbol{\Theta}}(\mathbf{y}|\boldsymbol{\theta})$ is the likelihood* function for $\boldsymbol{\Theta}$, $l(\boldsymbol{\theta}|\mathbf{y})$, and $f_{\mathbf{Y}}(\mathbf{y})$ is a constant. Thus one can see that

$$\text{posterior distribution} \propto \text{likelihood} \times \text{prior distribution}.$$

The data $\mathbf{y}$ modify the prior distribution* through the likelihood function, producing the resultant posterior distribution. Imagining a sequence of n observation samples, it is clear that the posterior distribution from one sample becomes the prior distribution for the next sample, where, modified by another likelihood, another posterior distribution is produced, and so on.

Consider the example illustrated in Fig. 1. Suppose that a real estate analyst summarizes current knowledge and assumptions about the average house price in a neighborhood, Θ, with a normal distribution* centered at 100 (thousands of dollars) and a standard deviation of 10, denoted $N(100, 10^2)$ in Fig. 1. Thus

$$f_{\Theta}(\theta) = \frac{1}{\sqrt{2\pi}\,(10)} \exp\left[-\frac{1}{2}\left(\frac{\theta - 100}{10}\right)^2\right].$$

A sample of $n = 4$ houses is taken randomly, each with assumed probability distribution of house price Y_i of the normal form with mean Θ and known standard deviation 20. Thus

$$f_{Y_i|\Theta}(y_i|\theta) = \frac{1}{\sqrt{2\pi}\,(20)} \times \exp\left[-\frac{1}{2}\left(\frac{y_i - \theta}{20}\right)^2\right].$$

Now given $\mathbf{y}' = (y_1, y_2, y_3, y_4)$, the likelihood function for Θ based on the sample readily follows, and is of the normal form centered about $\bar{y}$ with standard deviation $20/\sqrt{4} = 10$. Assuming that $\bar{y} = 90$, the likelihood is the $N(90, 10^2)$ form in Fig. 1. The combination of the $N(100, 10^2)$ prior and the $N(90, 10^2)$ likelihood through Bayes' theorem results in a $N(95, 7.07^2)$ posterior distribution for Θ, as shown by Box and Tiao [3, p. 74] and illustrated in Fig. 1. Because the variances associated with the prior and the likelihood are equal, the posterior mean of 95 is simply the average of the prior mean of 100 and the sample $\bar{y}$ of 90. The posterior variance of 50 (7.07^2) is smaller than that associated with the prior and the likelihood, effectively reflecting the averaging of two independent observations each having variance 10^2 ($50 = 10^2/2$).

Given the posterior distribution in Fig. 1, one could, for example, state that the probability that Θ is between 81.14 and 108.86 (within 1.96 standard deviations) is 0.95. This interval is a *probability interval* and stands in contrast to the confidence interval* of sampling theory inference, which would *not* be a probability statement on Θ. This interval would also be termed a 95% highest posterior density (HPD) region, although in this one-parameter example the region is a simple interval. The interval is a 95% HPD interval because there is no *shorter* interval which contains 95% of the total probability mass.

Box and Tiao [3] provide complete coverage of the posterior distributions relevant to many practical statistics problems. Bernardo [2] and Dickey [4] discuss posterior distribution choice in a broad context. Laird and Lewis [5], Martz and Waller [6], and Raiffa and Schlaifer [8] present posterior distributions in the context of particular application problems. Methods for numerical computation of posterior distributions are considered by Naylor and Smith [7] and Reilly [9].

References

[1] Bayes, T. (1958). *Biometrika*, **45**, 298–315. (Reprint of the famous 1763 paper generally credited with giving birth to Bayesian statistical inference.)

[2] Bernardo, J. M. (1979). *J. R. Statist. Soc. B*, **41**, 113–147. (Reference for posterior distributions.)

[3] Box, G. E. P. and Tiao, G. C. (1973). *Bayesian Inference in Statistical Analysis*. Addison-Wesley, Reading, Mass. (An excellent general statistical reference on Bayesian inference.)

[4] Dickey, J. M. (1976). *J. Amer. Statist. Ass.*, **71**, 680–689. (Proposes approximate posterior distributions resulting from operational priors chosen with regard to the realized likelihood function.)

[5] Laird, N. M. and Louis, T. A. (1982). *J. R. Statist. Soc. B*, **44**, 190–200. (Approximate posterior distributions for censored, truncated, grouped, or otherwise incomplete data.)

[6] Martz, H. F. and Waller, R. A. (1982). *Bayesian Reliability Analysis*. Wiley, New York. (Chapters 5 and 6 contain good general knowledge about Bayesian inference.)

[7] Naylor, J. C. and Smith, A. F. M. (1982). *Appl. Statist.*, **31**, 214–225. (Efficient computation of posterior distributions by numerical integration. Claim advance over Reilly [9].)

[8] Raiffa, H. and Schlaifer, R. (1961). *Applied Statistical Decision Theory*. Harvard University Press, Cambridge, Mass.

[9] Reilly, P. M. (1976). *Appl. Statist.*, **25**, 201–209.

[10] Stigler, S. (1983). *Amer. Statist.*, **37**, 290–296. (A witty search into who really discovered Bayes theorem.)

Bibliography

See the following works, as well as the references just given, for more information on the topic of posterior distributions.

Lindley, D. V. (1965). *Introduction to Probability and Statistics from a Bayesian Viewpoint, Part 1: Probability; Part 2: Inference*. Cambridge University Press, Cambridge, England.

Savage, L. J. (1954). *The Foundation of Statistics*. Wiley, New York. (General reference with philosophical basis for Bayesian statistical inference.)

(BAYESIAN INFERENCE
BAYES' THEOREM
LIKELIHOOD
POSTERIOR PROBABILITIES
PRIOR DISTRIBUTIONS
PRIOR PROBABILITIES)

DAVID J. PACK

POSTERIOR MOMENTS *See* CONJUGATE FAMILIES OF DISTRIBUTIONS

POSTERIOR PROBABILITIES

Posterior probabilities are probabilities that are applicable at a point in time which is *posterior* to one obtaining additional information provided by empirical studies, experiments, surveys, and so on. In the context of a particular random variable or unknown parameter, posterior probabilities follow directly from the determination of a *posterior distribution*, as the provision of a probability distribution or probability density function in any statistical endeavor automatically leads to the ability to determine probabilities of various events of interest. (*See* POSTERIOR DISTRIBUTIONS for a full discussion.) Posterior probabilities relate to corresponding prior probabilities and thus prior distributions, and one should logically first understand those concepts (*see* PRIOR DISTRIBUTIONS and PRIOR PROBABILITIES).

Suppose for the sake of illustration that you are concerned with the probability of the event E that a particular person is a college graduate. When you knew only the person's name, you judged the probability of E to be $\Pr[E] = 0.05$, based on your assumption that 1 in 20 people in the general population was a college graduate. This was your *prior* probability of E. Now you are given additional information indicating that the person reads *Newsweek* magazine. Your probability of E given this new information is your *posterior* probability of E.

Denoting the event "reads *Newsweek*" by N, the posterior probability sought is the conditional probability* $\Pr[E \mid N]$. By definition, this probability is

$$\Pr[E \mid N] = \frac{\Pr[E \cap N]}{\Pr[N]},$$

which can also be written

$$\Pr[E \mid N] = \frac{\Pr[N \mid E]\Pr[E]}{\Pr[N \mid E]\Pr[E] + \Pr[N \mid \bar{E}]\Pr[\bar{E}]},$$

where $\bar{E}$ represents the complement of E (i.e., noncollege graduate). This is a representation of Bayes theorem* expressed in terms of events. The posterior probability $\Pr[E \mid N]$ is produced through the prior probability $\Pr[E]$ and the "likelihood" of a college graduate reading *Newsweek* $\Pr[N \mid E]$. Assuming that *Newsweek's* own surveys have shown that 30% of all college graduates and 2% of all noncollege graduates read the magazine, this posterior probability is

$$\Pr[E \mid N] = \frac{0.30 \times 0.05}{0.30 \times 0.05 + 0.02 \times 0.95} = 0.44,$$

a considerable increase from the prior probability $\Pr[E] = 0.05$.

The illustration above is limited because both the object of concern and the additional information are of a discrete form. Many practical posterior probability problems focus on the posterior probability of a parameter Θ given a vector $\mathbf{Y}' = (Y_1, Y_2, \ldots, Y_n)$ representing n independent sample observations. Kleyle [1] speaks of the case where Θ is continuous and the Y_i are discrete (e.g., observations from a Poisson distribution* with parameter Θ) and representing the degree of uncertainty in a particular situation with a system of upper and lower posterior probabilities, rather than a unique posterior probability. If Θ is continuous and the Y_i are either discrete or continuous, a unique posterior probability follows from integration over the conditional probability density function

$$f_{\Theta \mid \mathbf{Y}}(\theta \mid \mathbf{y}) = \frac{f_{\mathbf{Y} \mid \Theta}(\mathbf{y} \mid \theta) f_\Theta(\theta)}{f_{\mathbf{Y}}(\mathbf{y})},$$

that is, the posterior distribution of Θ given $\mathbf{Y}$.

Reference

[1] Kleyle, R. (1975). *Ann. Statist.*, **3**, 504–511.

(BAYES' THEOREM
CONDITIONAL PROBABILITY AND EXPECTATIONS
POSTERIOR DISTRIBUTIONS
PRIOR DISTRIBUTIONS
PRIOR PROBABILITIES)

DAVID J. PACK

POTTHOFF–WHITTINGHILL TESTS

For the binomial*, multinomial*, and Poisson* distributions, Potthoff and Whittinghill [5, 6] proposed certain tests of homogeneity* that were aimed at optimizing power* in given ways. The standard homogeneity tests apparently were not specifically constructed with such an aim, although they seem to be reasonably powerful. The most definitive results in refs. 5 and 6 are for the cases with known parameters and for a Poisson case with unknown parameter.

In the binomial model, the ith of k samples has n_i elements, of which x_i are observed to belong to one class and $y_i = n_i - x_i$ are in the complementary class. Under the null hypothesis of homogeneity, each x_i is binomially distributed with the same parameter p. Let $q = 1 - p$. If p is known (as it is in some genetics applications, among others), the standard homogeneity test is based on the statistic

$$\sum_{i=1}^{k} \frac{(x_i - n_i p)^2}{n_i pq}, \tag{1}$$

whose null distribution is approximately that of chi-square* with k degrees of freedom. However, ref. 5 shows that against an alternative under which the binomial parameter for each i is drawn from a beta distribution* with mean p and variance close to zero, the critical region* of the most powerful test of homogeneity is composed of large values of

$$V = \frac{1}{p} \sum_{i=1}^{k} x_i(x_i - 1) + \frac{1}{q} \sum_{i=1}^{k} y_i(y_i - 1). \tag{2}$$

In the more general multinomial model, there are $m \geqslant 2$ classes. The ith of k samples has n_i elements, of which x_{ij} are in the jth class ($j = 1, 2, \ldots, m$). Under the null hypothesis of homogeneity, each set $x_{i1}, x_{i2}, \ldots, x_{im}$ is multinomially distributed with the same parameter vector $\mathbf{p} = (p_1, p_2, \ldots, p_m)'$, where $\sum_{j=1}^{m} p_j = 1$. Against an alternative under which the set of multinomial parameters for each i is drawn from a Dirichlet distribution* with mean vector $\mathbf{p}$ and variances and covariances close to zero in a certain sense (see ref. 5), the most powerful homogeneity test when $\mathbf{p}$ is known has critical region composed of large values of

$$V = \sum_{j=1}^{m} X_j / p_j, \tag{3}$$

where $X_j = \sum_{i=1}^{k} x_{ij}(x_{ij} - 1)$.

Define $N_1 = \sum_{i=1}^{k} n_i(n_i - 1)$ and $N_2 = \sum_{i=1}^{k} n_i(n_i - 1)(n_i - 2)$. Under the null hypothesis, V of (3) has expectation and variance given by

$$E(V) = N_1, \qquad \sigma^2(V) = 2(m-1)N_1. \tag{4}$$

An approximate significance test for V can be obtained by dividing $V - E(V)$ by $\sigma(V)$ and referring the quotient to a table of the normal distribution*. Alternatively, one can use a more refined approximation and refer $eV + f$ to tables of the chi-square distribution* with degrees of freedom equal to ν (generally not an integer), where the values

$$e = \frac{N_1}{\frac{1}{2(m-1)}\left[\left(\sum_{j=1}^{m} \frac{1}{p_j}\right) - 3m + 2\right] N_1 + N_2}, \tag{5}$$

$$f = e[(m-1)e - 1]N_1, \tag{6}$$

and

$$\nu = e^2(m-1)N_1 \tag{7}$$

are chosen so that the null distribution of $eV + f$ and the distribution of χ_ν^2 have the same first three moments. Of course, all the results of this paragraph can be applied to V of (2) (binomial case) by setting $m = 2$.

In the Poisson model, x_i, the ith of k observations, is drawn from a Poisson distribution with mean $b_i\lambda$ under the null hypothesis. The b_i's are given constants. If λ is known, one would base the standard homogeneity test on the statistic

$$\sum_{i=1}^{k} \frac{(x_i - b_i\lambda)^2}{b_i\lambda}, \tag{8}$$

whose null distribution is approximately that of chi-square with k degrees of freedom. However, ref. 6 showed that against an alternative under which the Poisson parameter for each i is drawn from a gamma distribution* with mean λ and variance close to zero, the most powerful homogeneity test has critical region composed of large values of

$$U = \sum_{i=1}^{k} x_i(x_i - 1) - 2\lambda \sum_{i=1}^{k} b_i x_i. \tag{9}$$

Under the null hypothesis, $E(U) = -\lambda^2 \sum_{i=1}^{k} b_i^2$ and $\sigma^2(U) = 2\lambda^2 \sum_{i=1}^{k} b_i^2$. Thus

$$z = \left(U + \lambda^2 \sum_{i=1}^{k} b_i^2\right) \Big/ \left(2\lambda^2 \sum_{i=1}^{k} b_i^2\right)^{1/2} \tag{10}$$

can be referred to a table of the normal curve to obtain an approximate significance test for U. Alternatively, one can use a more refined approximation and refer $gU + h$ to tables of the chi-square distribution with ν degrees of freedom, where

$$g = \sum_{i=1}^{k} b_i^2 \Big/ \left(\tfrac{1}{2} \sum_{i=1}^{k} b_i^2 + \lambda \sum_{i=1}^{k} b_i^3 \right), \quad (11)$$

$$h = g(g+1)\lambda^2 \sum_{i=1}^{k} b_i^2, \quad (12)$$

$$\nu = g^2 \lambda^2 \sum_{i=1}^{k} b_i^2. \quad (13)$$

For the important special case where all b_i's are equal to 1, formulas (11) to (13) simplify to $g = 1/(\lambda + \frac{1}{2})$, $h = (\lambda + \frac{3}{2})\nu$, $\nu = k\lambda^2/(\lambda + \frac{1}{2})^2$.

For the Poisson case with unknown λ and all b_i's equal to 1, the standard homogeneity test is based on

$$\sum_{i=1}^{k} (x_i - \bar{x})^2 / \bar{x}, \quad (14)$$

where $\bar{x} = x_{.}/k$ and $x_{.} = \sum_{i=1}^{k} x_i$. The null distribution of (14) is approximately that of chi-square with $k - 1$ degrees of freedom. Reference 6 established an optimal power property for this test: Against alternatives under which the Poisson parameter for each i is drawn from a gamma distribution with an unknown mean λ and variance close to zero, the test based on (14) is, locally (in the sense of the variance of the gamma distribution being close to zero), a most powerful test among all unbiased tests (*see* NEYMAN-PEARSON LEMMA and UNBIASEDNESS).

The tests for the cases with known parameters can be adapted for use when the parameters are not known, but only imperfectly. Thus, for example, if the p_j's are unknown in the multinomial case and V of (3) is considered as a function of the p_j's, then the values of the p_j's that minimize V of (3) are

$$p_j = X_j^{1/2} \Big/ \sum_{J=1}^{m} X_J^{1/2}, \quad (15)$$

for which

$$V = V_{\min} = \left(\sum_{j=1}^{m} X_j^{1/2} \right)^2. \quad (16)$$

Hence, since V of (3) can be no smaller than $V_{\min}$ of (16) no matter what the values of the unknown p_j's, this suggests that a conservative test based on V can be obtained by using (15) for the p_j's and (16) for V. The disadvantage is that the test loses power by being conservative, so that it no longer has optimal power.

Similarly, if z of (10) is minimized with respect to λ, then a conservative test based on U of (9) can be obtained for the Poisson case when λ is unknown. Two other tests for the Poisson case when the b_i's are not all equal and λ is unknown (neither of them conservative) may be mentioned. Define $b_{.} = \sum_{i=1}^{k} b_i$ and $b_i^* = b_i / b_{.}$. The standard homogeneity test is based on the statistic

$$\sum_{i=1}^{k} \frac{(x_i - b_i^* x_{.})^2}{b_i^* x_{.}}, \quad (17)$$

whose null distribution is approximately that of chi-square with $k - 1$ degrees of freedom; of course, (14) is a special case of (17). The remaining test is obtained by noting that since the conditional distribution of $x_1, x_2, \ldots, x_k$ given $x_{.}$ is multinomial with known parameters $b_1^*, b_2^*, \ldots, b_k^*$ under the null hypothesis of homogeneity, the test of (3) and (5) to (7), or possibly of (3) and (4), can be applied, using 1 for k, k for m, i for j, b_i^* for p_j, $x_i(x_i - 1)$ for X_j, $x_{.}(x_{.} - 1)$ for N_1, and $x_{.}(x_{.} - 1)(x_{.} - 2)$ for N_2.

Other authors have obtained results that are related to refs. 5 and 6 in different ways. Some of these results have indicated that a test described above has locally optimal power against a class of alternatives considerably more general than what is given by the beta mixing distribution (for the binomial case) or the gamma mixing distribution (for the Poisson case); some authors have aimed to obtain improved homogeneity tests for cases with unknown parameters. The papers of Wisniewski [10] and Tarone [8] relate to the binomial distribution. Smith [7] dealt with both the binomial and multinomial cases. For work on Poisson homogeneity testing, see Bühler et al. [1] and Moran [4], and references given by the latter. Methods like those that have been used for the binomial and Poisson distributions have also been applied to construct homogeneity tests for the negative binomial distribution* and for its special case, the geometric distribution*; see Meelis [3], Vit [9], and Chi [2]. Among the 10 papers shown in the References, there are a number of partial overlaps.

Numerical examples will illustrate the use of the tests based on (2), (3), and (9).

Example 1. In a genetic experiment in which the probability is $p = \frac{1}{4}$ for a recessive offspring and $q = \frac{3}{4}$ for a dominant offspring, suppose that data on $k = 104$ families are available. Suppose that the results show 41 families with one offspring, of which 9 have a recessive offspring and 32 have a dominant one; 33 families with two offspring, of which 3 have two recessive offspring, 7 have one recessive, and 23 have no recessives; and 30 families with three offspring, of which 2 have all recessives, 5 have two recessives, 8 have one recessive, and 15 have no recessives. To apply the binomial test based on (2), first calculate $\sum_i x_i(x_i - 1) = 3(2) + 2(6) + 5(2) = 28$ and $\sum_i y_i(y_i - 1) = 23(2) + 8(2) + 15(6) = 152$. Then (2) gives $V = (28/\frac{1}{4}) + (152/\frac{3}{4}) = 944/3$. Also, $N_1 = 33(2) + 30(6) = 246$ and $N_2 = 30(6) = 180$. The denominator of (5) is $\frac{1}{2}(4 + \frac{4}{3} - 6 + 2)(246) + 180 = 344$. Thus $e = 0.71512$, $f = -50.12$, and $\nu = 125.80$, from (5)–(7). For the normal test, $(944/3 - 246)/492^{1/2} = +3.10$. For the more refined approximation, $eV + f = 174.91$ is referred to the chi-square distribution with 125.80 degrees of freedom. Either result is highly significant. Note that the 41 families with only one offspring were effectively not used at all in the calculations.

Example 2. A 14-year-old tossed a pair of dice 360 times and recorded the results that appear in the middle two rows of Table 1. The multinomial test based on (3) can be used as a goodness-of-fit test. Note first that $k = 1$, $m = 11$, $X_j = x_{1j}(x_{1j} - 1)$, $n_1 = 360$, $N_1 = n_1(n_1 - 1) = 129{,}240$, and $N_2 = 358N_1$. Then $V = (20/\frac{1}{36}) + (72/\frac{2}{36}) + \cdots + (56/\frac{1}{36}) = 131{,}226$ from (3). The sum of the reciprocals of the 11 p_j's is 170.4. Hence $e = 1/364.97$, $f = -344.41$, and $\nu = 9.70$, from (5)–(7). The normal test gives

$$\frac{131{,}226 - 129{,}240}{(20 \times 129{,}240)^{1/2}} = +1.24.$$

The more refined approximation refers $eV + f = 15.14$ to the chi-square distribution with 9.70 degrees of freedom. Neither result is great enough to be significant at the 0.10 level.

Example 3. From a large container holding a substance that is known to have 0.24 particle of a certain type per unit volume, suppose that $k = 5$ samples have been taken at different places. Suppose that three of the samples are 10 units in volume and the other two are 25 units, with 1, 5, and 0 particles counted in the first three samples and 3 and 11 in the last two. Are the particles homogeneously distributed? To apply the Poisson test based on (9), note first that $\lambda = 0.24$, $b_1 = b_2 = b_3 = 10$, $b_4 = b_5 = 25$, and $x_1 = 1$, $x_2 = 5$, $x_3 = 0$, $x_4 = 3$, $x_5 = 11$. Then $\sum_i x_i(x_i - 1) = 136$, $\sum_i b_i x_i = 410$, $\sum_i b_i^2 = 1550$, and $\sum_i b_i^3 = 34{,}250$. Thus $U = -60.8$ from (9), and $g = 0.17232$, $h = 18.04$, $\nu = 2.65$ from (11)–(13). The normal approximation yields $z = +2.13$ from (10), and the more refined approximation refers $gU + h$

Table 1 Data from Tosses of Two Dice

												Total
Class, j	1	2	3	4	5	6	7	8	9	10	11	
Sum of pips	2	3	4	5	6	7	8	9	10	11	12	
Number of tosses, x_{1j}	5	9	34	42	41	63	60	45	29	24	8	360
Probability, p_j	$\frac{1}{36}$	$\frac{2}{36}$	$\frac{3}{36}$	$\frac{4}{36}$	$\frac{5}{36}$	$\frac{6}{36}$	$\frac{5}{36}$	$\frac{4}{36}$	$\frac{3}{36}$	$\frac{2}{36}$	$\frac{1}{36}$	1

= 7.56 to the chi-square distribution with 2.65 degrees of freedom. The null hypothesis of homogeneity can be rejected at the 0.05 level.

References

[1] Bühler, W., Fein, H., Goldsmith, D., Neyman, J., and Puri, P. S. (1965). *Proc. Natl. Acad. Sci. (USA)*, **54**, 673–680.

[2] Chi, P. Y. (1980). *Biometrika*, **67**, 252–254.

[3] Meelis, E. (1974). *J. Amer. Statist. Ass.*, **69**, 181–186.

[4] Moran, P. A. P. (1973). *Biometrika*, **60**, 79–85.

[5] Potthoff, R. F. and Whittinghill, M. (1966). *Biometrika*, **53**, 167–182.

[6] Potthoff, R. F. and Whittinghill, M. (1966). *Biometrika*, **53**, 183–190.

[7] Smith, C. A. B. (1951). *Ann. Eugen. (Lond.)*, **16**, 16–25.

[8] Tarone, R. E. (1979). *Biometrika*, **66**, 585–590.

[9] Vit, P. (1974). *Biometrika*, **61**, 565–568.

[10] Wisniewski, T. K. M. (1968). *Biometrika*, **55**, 426–428.

(CATEGORICAL DATA
CHI-SQUARE TESTS
CONTINGENCY TABLES
HOMOGENEITY AND TESTS OF HOMOGENEITY
HYPOTHESIS TESTING)

RICHARD F. POTTHOFF

POWER

In statistical theory, *power* is the probability that a test results in rejection of a hypothesis*, H_0 say, when some other hypothesis, H say, is valid. This is termed the power of the test "with respect to the (alternative) hypothesis H."

If there is a set of possible alternative hypotheses, the power, regarded as a function of H, is termed the *power function* of the test. When the alternatives are indexed by a single parameter θ, simple graphical presentation is possible. If the parameter is a vector $\boldsymbol{\theta}$, one can visualize a *power surface*.

If the power function is denoted by $\beta(\theta)$ and H_0 specifies $\theta = \theta_0$, then the value of $\beta(\theta)$—the probability of rejecting H_0 when it is in fact valid—is the significance level*.

(CRITICAL REGION
HYPOTHESIS TESTING)

POWER FUNCTION DISTRIBUTION

This is a family of distributions with cumulative distribution functions

$$F_X(x) = \begin{cases} 0 & \text{for } x \leqslant a \\ \left(\dfrac{x-a}{b-a}\right)^c & \text{for } a < x \leqslant b \\ 1 & \text{for } x > b, \\ & (a < b; c > 0). \end{cases}$$

The corresponding probability density function is

$$f_X(x) = \begin{cases} \dfrac{c}{b-a}\left(\dfrac{x-a}{b-a}\right)^{c-1}, & a < x \leqslant b \\ 0, & \text{elsewhere.} \end{cases}$$

This family is a special subclass of beta distributions*.

POWER FUNCTIONS *See* POWER

POWER INDEX OF A GAME

An alternative name for the *value* of the game in the case of simple n-person games (namely, a procedure which assigns for each such game a number to *each* player, representing the player's "power" in the game). Various power indices have been constructed by Shapley and Shatnik [4], Banzhaf [1], and Deagan and Packel [2]. These have been used in law* courts to settle disputes involving fair representation on voting bodies. See Packel [3] for additional details.

References

[1] Banzhaf, J. S. (1965). *Rutgers Law Rev.*, **19**.

[2] Deagan, J. and Packel, E. (1978). *Int. J. Game Theory*, **7**.

[3] Packel, E. (1981). *The Mathematics of Games and Gambling*. Mathematical Association of America, Washington, D.C.

[4] Shapley, L. S. and Shatnik, M. *Amer. Polit. Sci. Rev.*, **48**.

(SIMPLE n-PERSON GAMES)

POWEROIDS

The class of finite difference* operators $\{Q\}$ such that

$$QEf(x) \equiv EQf(x)$$

(symbolically $QE \equiv EQ$), where the function $f(x)$, which is operated upon, is a polynomial and E is the displacement operator $(E^y f(x) \equiv f(x+y))$ is called *shift invariant*. Such operators can be expressed symbolically

$$Q \equiv \sum_{j=0}^{\infty} c_j D^j$$

when $D(\equiv d/dx)$ is the differential operator. The members of the subclass for which $c_0 = 0$ and $c_1 = 1$ are called *delta operators*. Each delta operator Q has a unique sequence of polynomials $\{p_n(x)\}$ corresponding to it, with the properties

$$p_0(0) = 1, \qquad p_n(0) \quad \text{for } n \geqslant 1$$

and

$$Qp_n(x) = np_{n-1}(x) \qquad (n = 1, 2, \ldots)$$

The *poweroids* corresponding to the operator D are the simple powers x^n. The relation

$$Q^n p_n(x) = n!$$

holds for all delta operators and their corresponding poweroids. The Taylor expansion, symbolically

$$E^x \equiv \sum_{j=0}^{\infty} (x^j/j!)D^j,$$

generalizes to

$$E^x \equiv \sum_{j=0}^{\infty} \{p_j(x)/j!\}\, Q^j.$$

Many finite difference formulas, such as interpolation* formulas, can be derived as special cases.

To find the poweroid corresponding to a given Q, the formulas

$$p_n(x) = \left(\frac{dQ}{dD}\right)\left(\frac{D}{Q}\right)^{n+1} x^n, \tag{1}$$

$$p_n(x) = x\left(\frac{D}{Q}\right)^{n} x^{n-1}, \tag{2}$$

and the recurrence formula

$$p_{n+1}(x) = x\frac{dD}{dQ}\, p_n(x) \tag{3}$$

are often useful. Table 1 summarizes some common poweroids and their corresponding delta operators.

Bibliography

Shiu, E. S. W. (1982). *Scand. Actuarial J.*, 123–238.

Steffensen, J. F. (1941). *Acta. Math.*, **73**, 333–366.

(BACKWARD DIFFERENCE
FINITE DIFFERENCES, CALCULUS OF
FORWARD DIFFERENCE
INTERPOLATION)

POWER PRODUCT SUMS

Given n sets of values $(x_{1i}, x_{2i}, \ldots, x_{mi})$ $(i = 1, \ldots, n)$, the product sums are sums of products of powers of the x's, that is,

$$\sum_{i=1}^{n}\left(\prod_{j=1}^{m} x_{ji}^{\alpha_i}\right).$$

Table 1

	Delta Operator, Q	Poweroid, $p_n(x)$
Differential	D	x^n
Forward difference[a]	Δ	$x^{(n)}$
Backward difference[a]	∇	$x^{[n]}$
Central difference[a]	δ	$x^{\{n\}}$
Apel	$E^\alpha D$	$x(x - n\alpha)^{n-1}$
Hermite	$E^\alpha d^{\beta D^2} D$	$x(2n\beta)^{1/2(n-1)} H_{n-1}\left(\frac{n\alpha - x}{(2n\beta)^{1/2}}\right)$

[a] $H(\cdot)$ is a Hermite polynomial.

Sample moments and product moments are multiples [(sample size)$^{-1}$] of product sums of sample values.

Power product sums are special symmetric functions*.

POWER SERIES DISTRIBUTIONS

In certain theoretical work on cosmic rays, it was found directly from unsolved differential equations that a needed discrete distribution should have equal mean and variance. It is known that the Poisson distribution* satisfies this; the question then arose: Does the equality of the mean and the variance suffice to exclude any other one-parameter distribution? It was in this context that Kosambi [8] introduced the power series distribution and studied some of its properties. Independently, Noack [10] defined the power series probability function purely as a mathematical function, and investigated recurrence relations of its moments and cumulants. Earlier, Tweedie [27, 28] had considered discrete and continuous linear exponential families*. In a series of papers, Patil and his co-workers studied generalized power series distributions from a variety of configurations. For extensive details and references, see Douglas [1], Johnson and Kotz [3], Ord [11], and Patil et al. [21–23].

The family of power series distributions provides a very elegant and perceptive formulation of several classical discrete distributions that are used in statistical research and teaching.

DEFINITIONS AND PROPERTIES OF STRUCTURE

Let T be a subset of the set I of nonnegative integers. Define $f(\theta) = \sum a(x)\theta^x$ where the summation extends over T and $a(x) > 0$, $\theta \geqslant \Theta$, the parameter space, such that $f(\theta)$ is finite and differentiable. One has $\Theta = \{\theta : 0 \leqslant \theta < \rho\}$, where ρ is the radius of convergence of the power series of $f(\theta)$. Then a random variable X with probability function (PF)

$$\Pr\{X = x\} = p(x) = p(x;\theta) = a(x)\theta^x / f(\theta), \quad x \in T \tag{1}$$

is said to have the *(generalized) power series distribution* (PSD) with range T and the series function $f(\theta)$. The parameter θ is called the series parameter, whereas $a(x)$ is known as the coefficient function [14].

Property 1. (Kosambi [8], Noack [10]). The binomial*, Poisson*, negative binomial*, and the logarithmic* distributions are special cases of PSD as follows:

$f(\theta) = (1+\theta)^n$, n a positive integer, for binomial,

$f(\theta) = e^\theta$, for Poisson,

$f(\theta) = (1-\theta)^{-k}$, k positive, for negative binomial,

$f(\theta) = -\log(1-\theta)$, for logarithmic series.

Property 2. (Patil [17]). A truncated PSD is itself a PSD in its own right and hence the properties that hold for a PSD continue to hold for its truncated form.

Property 3. (Kosambi [8]).

$$E[X] = \mu = \theta \frac{d}{d\theta}[\log f(\theta)] \tag{2}$$

$$V(X) = \mu_2 = \theta \frac{d\mu}{d\theta} = \mu + \theta^2 \frac{d^2}{d\theta^2}[\log f(\theta)] \tag{3}$$

Property 4. (Noack [10]).

$$E[X^{r+1}] = m_{r+1} = \theta \frac{d}{d\theta}[m_r] + \mu m_r \tag{4}$$

$$E[(X-\mu)^{r+1}] = \mu_{r+1} = \theta \frac{d}{d\theta}[\mu_r] + r\mu_2\mu_{r-1} \tag{5}$$

Property 5. (Kosambi [8]). Characterization of Poisson distribution. The equality of mean and variance characterizes the Poisson distribution among the power series distributions.

Property 6. (Patil [14]). The rth factorial moment is

$$E[X(X-1)\cdots(X-r+1)]$$

$$= \mu_{(r)} = \frac{\theta^r}{f(\theta)} \frac{d^r}{d\theta^r}[f(\theta)] \quad (6)$$

$$\mu_{(r+1)} = (\mu - r)\mu_{(r)} + \theta \frac{d}{d\theta}[\mu_{(r)}] \quad (7)$$

Let X represent a random count of individuals on a unit. The index $\psi_r = \mu_{(r+1)}/\mu_{(r)}$ measures the expected number of individuals in excess of a randomly selected group of size r, and is therefore called rth-order crowding.

For a PSD

$$\psi_r = \theta f^{(r+1)}(\theta)/f^{(r)}(\theta),$$

which turns out to be linear in r for the four classical discrete distributions as follows:

$$\psi_r = \begin{cases} \mu & \text{for Poisson} \\ \mu - pr & \text{for binomial} \\ \mu + \dfrac{q}{p} r & \text{for negative binomial} \\ \dfrac{\mu}{\alpha} r & \text{for logarithmic series.} \end{cases}$$

A plot of $\{(r, \psi_r)\}$ provides a graphical method to discriminate between discrete distributions and also to estimate parameters (see Ottestad [12] and Patil and Stiteler [20]).

Property 7. Let κ_r denote the rth cumulant, and $\kappa_{(r)}$ the rth factorial cumulant*

$$\text{Patil [14]:} \quad \kappa_{r+1} = \theta \frac{d}{d\theta}[\kappa_r] \quad (8)$$

$$\text{Khatri [7]:} \quad \kappa_{(r+1)} = \theta \frac{d}{d\theta}[\kappa_r] - r\kappa_{(r)} \quad (9)$$

Khatri [7] has shown that the first two moments (cumulants/factorial cumulants) as functions of an arbitrary parameter determine the PSD uniquely.

Property 8. (Joshi [5, 6]). Integral expressions for tail probabilities. Let $T = I_n$ or I, where $I_n = \{x : 0, 1, 2, \ldots, n\}$. Given the PSD defined by (1), there exists a family of densities of absolutely continuous distributions

$$g(\theta; x) = \begin{cases} c(x, \theta) p(x; \theta), & 0 < \theta < \rho \\ 0, & \text{otherwise,} \end{cases} \quad (10)$$

where $x \in T$, such that

$$P(x; \theta) = \sum_{r=0}^{x} p(r; \theta) = \int_{\Theta}^{\rho} g(\theta; x)\, d\theta \quad (11)$$

if and only if $f(\rho) = \infty$, in which case $c(x, \theta)$ is defined by

$$c(x, \theta) p(x; \theta) + \sum_{r=0}^{x} \frac{r - \mu}{\theta} p(r; \theta) = 0. \quad (12)$$

The well-known examples of the above are embodied in the dualities between binomial and beta of the second kind, Poisson and gamma, and negative binomial and beta of the first kind. To be specific, consider, for example, the usual Poisson process* with intensity parameter λ. Let $X(t)$ denote the number of events until time t, and let $T(k)$ denote the waiting time for k events to occur. Let $\theta = \lambda t$. Then $X(t) \sim \text{Poisson}(\theta) \equiv \text{PSD}(\theta, e^{\theta})$, $T(k) = \text{gamma}(\lambda, k)$, and $\Pr(X(t) \leqslant x) = \Pr(T(x+1) \geqslant t)$. It is clear that the duality argument provides distribution theoretic proofs of the corresponding mathematical proofs.

MINIMUM VARIANCE UNBIASED ESTIMATION*

It is fascinating to note that the statistical theory of minimum variance unbiased estimation and the additive number theory in mathematics combine to throw light on the interdependence of the MVU estimability for a PSD with the structure of its range T. The MVU estimability of θ depends only on the structure of the range T; it is curious that it has nothing to do with the specific form of a PSD as determined by its coefficient function $a(x)$. The following results are of interest in this connection. For details, see Patil [18] and Patil and Joshi [19].

Property 1. Estimability* of θ. Let $A \subseteq I$ and $B \subseteq I$. A is called a *basis* for B if the n-fold sum of A defined by $n[A] = \{\sum_{i=1}^{n} a_i, a_i \in A\}$ is equal to B for some n, called an *order* of the *basis* A for B.

Let $T \subseteq I$. The displaced set $D(T)$ is defined as $D(T) = T - \{\min(T)\}$. Thus $D(T)$ is the difference between T and the singleton $\{\min(T)\}$.

A necessary and sufficient condition for the parameter θ of a PSD with range T to be MVU estimable is that the displaced set $D(T)$ be a basis of I (i.e., $n[D(T)] = I$ for some n). When it exists, the MVU estimator of θ is given by

$$h(z,n) = \begin{cases} b(z-1,n)/b(z,n), & \\ \qquad z \in n[T] + \{1\} & \\ 0, \quad \text{otherwise,} & \end{cases}$$

where $b(z,n)$ is the coefficient of θ^z in the series expansion of $f_n(\theta) = [f(\theta)]^n$.

Property 2. Nonestimability of θ. θ is not MVU estimable if T is finite. This follows from the finiteness of $n[T]$ inherited from T. To cite a few applications, one may note that θ is MVU estimable for every sample size n for complete or left-truncated Poisson and logarithmic distributions. One may note further that the right-truncated Poisson and logarithmic distributions do not have unbiased estimators for θ (see Roy and Mitra [24], Tate and Goen [26], and Patil [13, 15, 16].

Property 3. Estimability of $g(\theta)$. Let $T \subset I$. T is said to be the index set of the function $f(\theta) = \sum a(x)\theta^x$ with $a(x) \neq 0$, $x \in T$, and is denoted by $T = W[f(\theta)]$. Clearly, the range of a PSD is also the index set of its series function.

Let $g(\theta)$ be a given function of θ which is such that $g(\theta)f_n(\theta)$ admits a power series expansion in θ, where $f_n(\theta) = [f(\theta)]^n$. A necessary and sufficient condition for $g(\theta)$ to be MVU estimable on the basis of a random sample of size n from the PSD given by (1) is that $W[g(\theta)f_n(\theta)] \subseteq W[f_n(\theta)]$. Also, whenever it exists, the MVU estimator for $g(\theta)$ is given by $h(z,n) = c(z,n)/b(z,n)$ for $z \in W[g(\theta)f_n(\theta)]$ and $= 0$ otherwise, where $c(z,n)$ is the coefficient of θ^z in the expansion of $g(\theta)f_n(\theta)$, and $b(z,n)$ is the coefficient of θ^z in the expansion of $f_n(\theta)$.

We observe that the variance of $h(z,n)$ is given by

$$V\{h(z,n)\} = E(\{h(z,n)\}^2) - \{g(\theta)\}^2.$$

Writing $g_2(\theta) = \{g(\theta)\}^2$, the MVU estimator of $g_2(\theta)$ can be written down on the usual lines whenever it exists.

We may note that on the basis of an observation from the binomial distribution with parameters $p = \theta/(1+\theta)$ and n, a polynomial in p is MVU estimable only if the degree of the polynomial does not exceed n. For the negative binomial distribution, however, $p = 1 - \theta$ and a polynomial of any degree in p is MVU estimable.

Property 4. The MVU estimation of the probability function of a PSD. Let $g(\theta) = p(k;\theta)$, where $k \in T$ is known. One has $W[g(\theta)f_n(\theta)] = (n-1)[T] + \{k\}$, whereas $W[f_n(\theta)] = n[T]$. Clearly, as a result of Property 3, the probability function of a PSD is estimable for every sample size n. Its MVU estimate is given by

$$h(z,n;k) = a(k)b(z-k,n-1)/b(z,n)$$
$$\text{for } z \in (n-1)[T] + \{k\},$$
$$\text{and } = 0, \text{ otherwise.}$$

The applications of this result are of interest for distributions such as Poisson, zero-truncated Poisson, binomial, negative binomial, logarithmic, and others (see Patil [18]).

MAXIMUM LIKELIHOOD* ESTIMATION

Let x_i, $i = 1, 2, \ldots, n$, be a random sample of size n from the PSD defined by (1). Then the logarithm of the likelihood function L is $\log L = \text{constant} + \sum_{i=1}^{n} x_i \log\theta - n\log\theta$, so that the "efficient score" for θ is

$$\psi(\theta) = \frac{\partial}{\partial\theta}\log L = n(\bar{x} - \mu)/\theta,$$

where $\bar{x} = \sum x_i/n$ is the sample mean and μ

is the mean of the PSD. The likelihood equation $\psi(\hat{\theta}) = 0$ for estimating θ thus reduces to

$$\bar{x} = \mu(\hat{\theta}) = (= \hat{\mu}, \text{ say}),$$

which means equating the sample mean to the population mean. The method of maximum likelihood and the method of moments* thus lead to the same estimate in the case of a PSD.

Denoting this estimate by $\hat{\theta}$, its asymptotic variance then simplifies to $\text{var}(\hat{\theta}) = \theta/[n(d\mu/d\theta)] = \theta^2/[n\mu_2(\theta)]$, where $\mu_2(\theta)$ is the variance of the PSD.

If the likelihood equation does not readily give an algebraic solution, one may use an iterative process of solution by starting with an approximation θ_0. An improved approximation θ_1 is then obtained from

$$\theta_1 = \theta_0 + (\bar{x} - \mu(\theta_0))/(\theta_0 d\mu/d\theta)$$
$$= \theta_0\left[1 + \frac{\bar{x} - \mu(\theta_0)}{\mu_2(\theta_0)}\right].$$

Alternatively, if it is possible to invert the likelihood equation, its solution is quickly available as $\hat{\theta} = \mu^{-1}(\bar{x}) = \hat{\theta}(\bar{x})$, say, and tables may be made available for the estimating function $\hat{\theta}$. Next, the asymptotic variance $\text{var}(\hat{\theta})$ of the estimator $\hat{\theta}$ can be estimated from such tables by $\hat{\theta}/[n(\Delta\hat{\theta}/\Delta\bar{x})]$. The maximum likelihood estimate may not be necessarily unbiased. The bias to order n^{-1} can be obtained as

$$b(\hat{\theta}) = -\frac{1}{n}\frac{\theta}{2}\frac{\mu_3 - \mu_2}{\mu_2^2}.$$

Next, to estimate a differentiable function of θ, such as $g(\theta)$, the method of maximum likelihood leads to the estimate

$$\hat{g} = g(\hat{\theta})$$

with asymptotic variance

$$\text{var}(\hat{g}) = \frac{\theta}{n}\frac{(dg/d\theta)^2}{(d\mu/d\theta)^2} = \theta^2 \Big/ \left[n\mu_2\left(\frac{dg}{d\theta}\right)^2\right].$$

The amount of bias in $\hat{g}$ to order n^{-1} is given by

$$b(\hat{g}) = \frac{1}{n}\frac{\theta}{2}\frac{dg}{d\theta}\frac{\mu_3 - \mu_2\left(1 + \theta\frac{d^2g}{d\theta^2}\Big/\frac{dg}{d\theta}\right)}{\mu_2^2}.$$

For further details and applications, see Patil [16, 17].

CONCLUDING REMARKS

Several publications have appeared on families of distributions that are related to power series distributions in one way or the other. These include modified power series distributions* and also the truncated power series distributions with or without an additional parameter present. The analogous version called factorial series distribution* may also be of interest. Regarding multivariate versions, *see* SUM-SYMMETRIC POWER SERIES DISTRIBUTIONS *and* MULTIVARIATE POWER SERIES DISTRIBUTIONS. For other details, see Gokhale [2], Johnson and Kotz [3, 4], Lin [9], Patil et al. [23], Snider [25], and Wani [29].

References

[1] Douglas, J. B. (1979). *Analysis with Standard Contagious Distributions*. International Co-operative Publishing House, Fairland, Md.

[2] Gokhale, B. V. (1980). On power-series distributions with mean-variance relationships, *J. Indian Statist. Ass.*, **18**, 81–84.

[3] Johnson, N. L. and Kotz, S. (1969). *Discrete Distributions*. Wiley, New York.

[4] Johnson, N. L. and Kotz, S. (1982). Developments in discrete distributions, 1969–1980. *Intern. Statist. Rev.*, **50**, 43–73.

[5] Joshi, S. W. (1974). Some recent advances with power series distributions. In *Statistical Distributions in Scientific Work*, Vol. 1, G. P. Patil, S. Kotz, and J. K. Ord, eds. D. Reidel, Boston, pp. 9–18.

[6] Joshi, S. W. (1974). Integral expressions for the tail probabilities of the power series distributions. *Sankhyā B*, **36**, 462–465.

[7] Khatri, C. G. (1959). On certain properties of power-series distributions. *Biometrika*, **46**, 486–490.

[8] Kosambi, D. D. (1949). Characteristic properties of series distributions. *Proc. Natl. Inst. Sci. India*, **15**, 109–113.

[9] Lin, C. C. (1981). Conjugate Priors and Characterizations of Conjugate Priors for Power Series Distributions, Ph.D. thesis, Texas Tech University, Lubbock, TX.

[10] Noack, A. (1950). A class of random variables with discrete distribution. *Ann. Math. Statist.*, **21**, 127–132.

[11] Ord, J. K. (1972). *Families of Frequency Distributions*. Charles Griffin, London.

[12] Ottestad, P. (1939). On the use of the factorial moments in the study of discontinuous frequency distributions. *Skand. Aktuarietidskr.*, **22**, 22–31.

[13] Patil, G. P. (1957). *Problems of Estimation in a Class of Discrete Distributions*. AISI thesis, Indian Statistical Institute, Calcutta, India.

[14] Patil, G. P. (1959). *Contributions to Estimation in a Class of Discrete Distributions*. Ph.D. thesis, University of Michigan, Ann Arbor, MI.

[15] Patil, G. P. (1961). Asymptotic bias and variance of ratio estimates in generalized power series distributions and certain applications. *Sankhyā A*, **23**, 269–280.

[16] Patil, G. P. (1962). Some methods of estimation for the logarithmic series distribution. *Biometrics*, **18**, 68–75.

[17] Patil, G. P. (1962). Maximum likelihood estimation for generalized power series distributions and its application to a truncated binomial distribution. *Biometrika*, **49**, 227–237.

[18] Patil, G. P. (1963). Minimum variance unbiased estimation and certain problems of additive number theory. *Ann. Math. Statist.*, **34**, 1050–1056.

[19] Patil, G. P. and Joshi, S. W. (1970). Further results on minimum variance unbiased estimation and additive number theory. *Ann. Math. Statist.*, **41**, 567–575.

[20] Patil, G. P. and Stiteler, W. M. (1974). Concepts of aggregation and their quantification: a critical review with some new results and applications. *Res. Popul. Ecol.*, **15**, 238–254.

[21] Patil, G. P., Boswell, M. T., Joshi, S. W., and Ratnaparkhi, M. V. (1985). *Dictionary and Classified Bibliography of Statistical Distributions in Scientific Work*, Vol. 1: *Discrete Models*. International Co-operative Publishing House, Fairland, Md.

[22] Patil, G. P., Boswell, M. T., and Ratnaparkhi, M. V. (1985). *Dictionary and Classified Bibliography of Statistical Distributions in Scientific Work*, Vol. 2: *Univariate Continuous Models*. International Cooperative Publishing House, Fairland, Md.

[23] Patil, G. P., Boswell, M. T., Ratnaparkhi, M. V., and Roux, J. J. J. (1985). *Dictionary and Classified Bibliography of Statistical Distributions in Scientific Work*, Vol. 3: *Multivariate Models*. International Co-operative Publishing House, Fairland, Md.

[24] Roy, J. and Mitra, S. (1958). Unbiased minimum variance estimation in a class of discrete distributions. *Sankhyā*, **18**, 371–378.

[25] Snider, L. D. (1981). *Properties of Certain Generalized Power Series Distributions*. Ph.D. thesis, Texas Tech University, Lubbock, TX.

[26] Tate, R. F. and Goen, R. L. (1958). Minimum variance unbiased estimation for the truncated Poisson distribution. *Ann. Math. Statist.*, **29**, 755–765.

[27] Tweedie, M. C. K. (1947). Functions of a statistical variate with given means, with special reference to Laplacian distributions. *Proc. Camb. Philos. Soc.*, **43**, 41–49.

[28] Tweedie, M. C. K. (1965). Further results concerning expectation-inversion technique. In *Classical and Contagious Discrete Distributions*, G. P. Patil, ed. Statistical Publication Society, Calcutta Pergamon Press, Oxford, 195–218.

[29] Wani, J. K. and Lo, H. P. (1979). Characterizing constants for biological populations following certain power series abundance distributions. *J. Indian Statist. Ass.*, **47**, 189–195.

(MODIFIED POWER SERIES DISTRIBUTIONS
MULTIVARIATE POWER SERIES DISTRIBUTIONS
SUM-SYMMETRIC POWER SERIES DISTRIBUTIONS)

G. P. PATIL

POWER TRANSFORMATIONS *See* BOX AND COX TRANSFORMATION

P-P PLOTS *See* GRAPHICAL REPRESENTATIONS OF DATA

PRECEDENCE LIFE TEST

A well-known, distribution-free* procedure for detecting a difference in the locations of two populations is based on the number of exceedances, which is the number of observations in one sample that exceed the observation of rank r in the other sample. Tables for this procedure were given by Epstein [2], who also reported the results of some simulations relating to the power with respect to normal populations [3]. Closely related work on exceedances was published by Sarkadi [6] with extended results reported in ref. 7.

A mathematically equivalent test involves the number of observations in one sample that precede the observation of rank r in the other sample. Nelson [4] called this a *precedence test*. Such a test can be highly economical in life testing*.

Example. A manufacturer of light bulbs wishes to compare two filament designs A and B with respect to life. Specifically, he wishes to abandon design B if it gives an indication at the 0.05 level that it is shorter lived. He has available $n_1 = 8$ samples of design A and $n_2 = 15$ samples of design B to be placed on simultaneous life test. What is the smallest number, x, of design B failures that can occur before the first failure of design A items for this test specification to be met?

Let $Z(n_1, n_2)$ be the random variable associated with the number of values in one sample that precede the smallest value in the other sample. Then

$$P_i[Z(n_1, n_2) \geqslant x] = \binom{n_i}{x} \Big/ \binom{n_1 + n_2}{x},$$

$$i = 1, 2 \qquad x = 0, 1, \ldots, n_i,$$

where $i = 1$ when one considers the number of failures in the smaller sample (n_1) that precede the first failure in the larger sample (n_2), and $i = 2$ when the reverse is of interest. It is found (by trial) that the probability of having $x = 6$ or more failures of design B items before the first failure of a design A item, given that the two life distributions are identical, is 0.04958.

In the general case the null hypothesis of equal life distributions is rejected if x failures among the n_1 (or n_2) items occurs before the rth failures among the n_2 (or n_1) items. Choosing $r = 1$ as in the example just given results in the shortest but not necessarily the most powerful test. The precedence test procedure involves choice of the values of n_1, n_2, r, and x so as to achieve a significance level α determined by

$$\alpha \leqslant \left[1 \Big/ \binom{n_1 + n_2}{n_1}\right] \sum_{j=x}^{n_i} \binom{x + r - 1}{j} \times \binom{n_1 + n_2 - r + 1 - x}{n_i - j},$$

where $i = 1$ if x is the number of failures in the smaller sample n_1 which precede the rth failure in the larger sample n_2, and $i = 2$ otherwise. This calculation is the cumulative probability of the following 2×2 table and all tables with the same marginal values but larger frequencies in the lower left-hand cell.

$r - 1$	$n_1 - r + 1$	n_1
x	$n_2 - x$	n_2
$x + r - 1$	$n_1 + n_2 - x - r + 1$	

Since these cumulative probabilities are equivalent to those of a fourfold table with fixed marginal frequencies, tables for Fisher's exact test* can be used. A number of special tables have been prepared [2, 4, 5, 9].

Nelson [4] also gives an approximation not involving factorials that yields critical values for all combinations of all sample sizes greater than five for any significance levels from 0.005 to 0.10 (one-sided).

It can be shown that as the sample sizes approach infinity, the value of $P[Z(n_1, n_2) \geqslant x]$ approaches $[n_1/(n_1 + n_2)]^x$. For large n_1 (say at least 20) and $n_1 = n_2$ this reduces to finding the smallest value of x for which $(\frac{1}{2})^x \leqslant \alpha$. Thus for significance levels of 0.05 and 0.01, the corresponding values of x are 5 and 7.

Some work has been done on the power of this test. Simulation results assuming normality were reported by Epstein [3]. Analytical studies for the exponential distribution* were reported by Eilbott and Nadler [1]. Shorack [8] showed that the results of Eilbott and Nadler can be extended to a large class of distributions that includes the exponential as well as other parametric families important in life testing.

References

[1] Eilbott, J. and Nadler, J. (1965). *Technometrics*, **7**, 359–377.

[2] Epstein, B. (1954). *Ann. Math. Statist.*, **25**, 762–768.

[3] Epstein, B. (1955). *J. Amer. Statist. Ass.*, **50**, 894–900.

[4] Nelson, L. S. (1963). *Technometrics*, **5**, 491–499.

[5] Rosenbaum, S. (1954). *Ann. Math. Statist.*, **25**, 146–150.

[6] Sarkadi, K. (1957). *Ann. Math. Statist.*, **28**, 1021–1023.

[7] Sarkadi, K. (1957). *Magy. Tudo. Akad. Mat. Kutató Intéz. Közl.*, **2**, 59–68 (in English).

[8] Shorack, R. A. (1967). *Technometrics*, **9**, 154–158.

[9] Westlake, W. J. (1971). *Technometrics*, **13**, 901–903.

(LIFE TESTING)

LLOYD S. NELSON

PRECISE MEASUREMENT, PRINCIPLE OF

It has been noted by many authors (Barnett [1], de Finetti [2], Geertsema [3], *inter alia*) that conclusions reached by classical Bayesian analysis are not very sensitive to the prior distribution* employed if a sufficiently large amount of sample data is available. This is called the *principle of precise measurement*. It reflects the fact that the data tend to "swamp" the prior information.

References

[1] Barnett, V. D. (1982). *Comparative Statistical Inference*, 2nd ed. Wiley, New York.

[2] de Finetti, B. (1974). *Int. Statist. Rev.*, **42**, 117–130.

[3] Geertsema, J. C. (1973). *S. Afr. Statist. J.*, **17**, 121–146.

(BAYESIAN INFERENCE)

PRECISION, GAUSS'S MEASURE OF

This is $\{(\text{standard deviation}) \times \sqrt{2}\}^{-1}$. [The quantity (standard deviation) $\times \sqrt{2}$ is called the *modulus*.]

PRECISION, INDEX OF

Suppose that for a particular bioassay* system the mean response y is linearly related to log dose (x_S or x_T, where S denotes standard and T denotes test preparation). For a log dose, x_S of S, the expected response is

$$E(y) = \alpha + \beta x_s .$$

The same expected response would be obtained by a log dose, x_T of T, where $x_S - x_T = \log \rho$ (the *log potency ratio* of T in terms of S).

The regression* is estimated by fitting two parallel lines giving the equations

$$Y_S = \bar{y}_S + b(x_S - \bar{x}_S)$$

$$Y_T = \bar{y}_T + b(x_T - \bar{x}_T)$$

and the estimate of $\log \rho$ is given by the difference $x_S - x_T$ when $Y_S = Y_T$: that is,

$$M = \bar{x}_S - \bar{x}_T - \frac{\bar{y}_S - \bar{y}_T}{b} .$$

The variance of M is approximated by

$$\text{var}(M) \simeq \frac{S^2}{b^2}\left(\frac{1}{n_S} + \frac{1}{n_T}\right),$$

where n_S and n_T are the number of observations on S and T and the residual mean square about parallel lines is S^2. Thus the "precision" of the estimate of potency depends on the number of observations and the ratio $\lambda = S/b$, which is called *index of "precision,"* representing the inherent imprecision of the bioassay method. More details can be found in, for example, Armitage [1].

Reference

[1] Armitage, P. (1971). *Statistical Methods in Medical Research*. Blackwell, Oxford.

(BIOASSAY, STATISTICAL METHODS IN)

PRECISION MATRIX

In a regression model*

$$\mathbf{Y} = \mathbf{X}\boldsymbol{\beta} + \boldsymbol{\epsilon},$$

let $\mathbf{b}$ be the least-squares estimator of $\boldsymbol{\beta}$.

The covariance* matrix

$$E\left[(\mathbf{b} - \boldsymbol{\beta})(\mathbf{b} - \boldsymbol{\beta})'\right],$$

under the assumptions $E(\boldsymbol{\epsilon}) = \mathbf{0}$ and $E(\boldsymbol{\epsilon\epsilon}') = \sigma^2\mathbf{I}$, where $\mathbf{I}$ is the unit matrix and σ^2 is a scalar equal to $\sigma^2(\mathbf{X}'\mathbf{X})^{-1}$, is sometimes referred to as the *precision matrix.*

(GENERAL LINEAR MODELS
LEAST SQUARES
LINEAR REGRESSION)

PREDECESSOR–SUCCESSOR METHOD

The predecessor–successor method described by Hansen et al. [1] offers a procedure to obtain information on omissions in a frame. The incompleteness is common to almost all frames, owing mainly to the dynamic nature of most populations. If an incomplete frame is used for selecting a random sample, the units that are not listed in the frame will have zero probability of being included in the sample. Therefore, some alternative procedures have to be used for sampling from an incomplete frame. Seal [4] proposed the use of successive frames to allow for the changes in the population when sampling is done from outdated frames, by considering the changes in the population as a continuous stochastic process*. Hartley [2, 3] proposed the use of two or more frames to overcome the problem of incomplete frames such that the entire population is covered by the use of these frames.

The predecessor–successor method provides all such units of the population which are not listed in the frame a probability of inclusion equal to those which are listed in the frame. It is assumed here that a geographical ordering of the units can be established in principle and further that the rules of ordering are such that given any one unit in the population, we can uniquely determine its successor by following the defined path of travel. Consider a finite population divided into two classes, those that are included in the available frame and those that are not available on this frame. We can select a random sample from this population using the predecessor–successor method as follows:

1. Select a random sample from the units in the frame.
2. For each unit in the sample from the available frame, determine its successor unit and see if it is on the frame. If the successor is on the frame, discard it. If the successor is not on the frame, include it in the sample; then identify its successor and proceed in the same way until a successor is found to be on the frame.

Thus the sample will consist of units in the sample selected from the frame plus all sequences of units not on the frame which immediately follow these units in the path of travel. The probability of selection of any unit not listed in the frame is therefore the same as that of the first listed unit immediately preceding it in the path of travel.

Singh [5] gave a mathematical formulation of the problem for estimating (a) the total number of units of the target population missing from the frame, and (b) the total for the character under study (y) for the target population under the following two situations.

1. When the units missing from the frame are random, that is, these do not differ significantly from the units available in the frame with respect to the character under study
2. When missing units differ from the units in the frame

ESTIMATION PROCEDURE

Case 1: When Missing Units Are Random

Consider a finite target population of size N'. Let N units be available in the frame and let M units be missing from the frame in the form of N gaps. Let m_i denote the number of units missing between the ith unit and $(i + 1)$th unit in the frame. Let y_i denote the value of the character under study for the

ith unit in the frame. Select a random sample of size n from the existing frame. From these selected units observe the y values and also note the number of missing units in between the selected units and the next unit in the frame.

Now an unbiased estimator of M is given by

$$\hat{M} = \frac{N}{n}\sum^{n} m_i = N\bar{m},$$

$$V(\hat{M}) = \frac{N(N-n)S_m^2}{n},$$

$$S_m^2 = \frac{1}{N-1}\sum^{N}(m_i - \bar{m})^2.$$

Further, as the units are assumed to be missing at random from the frame, an unbiased estimator of the population total for y is given by

$$\hat{Y} = N(1+\bar{m})\bar{y}_n, \qquad \bar{y}_n = \sum y_i/n,$$

$$V(\hat{Y}) = \frac{N(N-n)}{n}\left(\frac{N'^2}{N^2}S_y^2 + \bar{y}_n^2 S_m^2\right).$$

Case II: When Units Missing from the Frame Are Not Random

Select a random sample of size n from the available frame and also obtain all the additional units included in the sample from the units not listed in the frame by the predecessor–successor method. Let m_i denote the number of such units after the ith unit and let $m = \sum^n m_i$. Let z_i denote the sum of the values for y and for the ith unit and all the m_i units following it.

In this case an unbiased estimator of Y is given by

$$\hat{Y} = \frac{N}{n}\sum^{n} z_i,$$

$$V(\hat{Y}) = \frac{N(N-n)}{n}S_z^2$$

$$\text{where } S_z^2 = \frac{1}{N-1}\sum^{N}(z_i - \bar{Z})^2$$

The estimation procedure can easily be modified when instead of observing all the m units from the missing part, only a random sample is observed.

EMPIRICAL INVESTIGATION

There are four industrial units employing more than 100 employees registered with the registrar of industries in a district. But two of these units have started three additional industries (two by unit 1 and one by unit 3) which have not been registered, as their registration is not compulsory under the Industries Act. To study the socioeconomic conditions of the employees, a sample of two industries is to be selected from each district using the frame available with the registrar of industries. Thus we have

Actual industries in the target population (district)	1 $\bar{2}$ $\bar{3}$ 4 5 $\bar{6}$ 7
Industries listed in the frame	1 4 5 7

(An overbar indicates industries not registered.) The possible samples as obtained from the frame, and obtained by using the predecessor–successor method are given as follows:

Sample Number	Sample from the Frame	Sample Obtained Using the Predecessor–Successor Method
1	1 4	1 $\bar{2}$ $\bar{3}$ 4
2	1 5	1 $\bar{2}$ $\bar{3}$ 5 $\bar{6}$
3	1 7	1 $\bar{2}$ $\bar{3}$ 7
4	4 5	4 5 $\bar{6}$
5	4 7	4 7
6	5 7	5 $\bar{6}$ 7

When additional units do not differ from those units available in the frame, it is not necessary to observe them and it is sufficient only to know the number of additional units of each registered unit selected in the sample. But if the additional units differ from the registered units, we have to observe all those additional units that appear in the sample.

References

[1] Hansen, M. H., Hurwitz, W. N., and Jabine, T. B. (1963). *Bull. Int. Statist. Inst.*, **40**, 497–517.

[2] Hartley, H. O. (1962). *Proc. Social Statist. Sect. Amer. Statist. Ass.*

[3] Hartley, H. O. (1974). *Sankhyā C*, **36**, 113.

[4] Seal, K. C. (1962). *Calcutta Statist. Ass. Bull.*, **11**, 68–84.

[5] Singh, R. (1983). *Biom. J.*, **25**(6), 545–549.

(RATIO ESTIMATORS
SURVEY SAMPLING)

RANDHIR SINGH

PREDICTION AND FILTERING, LINEAR

It is possible to give no more than a few glimpses of this vast subject in a general article. In view of the close conceptual relationship that exists between the Kolmogorov–Wiener (KW) theory of linear prediction and filtering, the theory of canonical representations due to Hida and Cramér and the Kalman–Bucy theory, we shall begin by giving a very brief overview of the three theories. It is convenient to do this for continuous parameter processes.

The second part of the article will be devoted to a description of the main results of the KW theory of prediction of multidimensional stationary sequences, followed by references to some recent developments. References to individual papers are kept to a minimum and the reader is referred to survey articles and standard books which contain an extensive bibliography. Equations are independently numbered within each section or subsection.

LINEAR PREDICTION AND FILTERING: FROM KOLMOGOROV–WIENER TO THE KALMAN–BUCY THEORY

Statement of the Problem

Before formulating the linear estimation problems for second-order stochastic processes, it is convenient to introduce the following terminology and notation. A real- or complex-valued stochastic process $\xi = (\xi_t)$, $a < t < b$ (the values $-\infty$ for a and $+\infty$ for b being permitted) is of second order if $E\xi_t = 0$ and $E|\xi_t|^2 < \infty$ for every t. Each ξ_t can be viewed as an element of the (real or complex) Hilbert space $L^2 = L^2(\Omega, P)$ of square-integrable random variables. For u, v in L^2 the inner product $(u,v) = E(uv)$ in the real case and $E(u\bar{v})$ if u and v are complex valued. The norm $\|u\|$ is then $\sqrt{E|u|^2}$ (E denotes expected value). Using the symbol $\overline{\text{sp}}$ to mean "closed linear span of" or "closed linear subspace generated by," we define the following Hilbert spaces connected with ξ:

$$H(\xi) = \overline{\text{sp}}\ \{\xi_t, a < t < b\},$$

$$H(\xi; t) = \overline{\text{sp}}\ \{\xi_s, a < s \leqslant t\},$$

$$H(\xi; a) = \bigcap_{a<t<b} H(\xi; t).$$

For any closed linear subspace $\mathscr{M}$ of $H(\xi)$, the symbol $P_{\mathscr{M}}$ will denote orthogonal projection with domain $H(\xi)$ and range $\mathscr{M}$.

Let $X = (X_t)$ be the estimation or signal process, $Y = (Y_t)$ be a process on which observations are made, and assume that

$$Y_s = X_s + N_s \tag{1}$$

for each s. Here $N = (N_t)$ is a "noise"* process, all three processes X, Y, and N are taken to be second order, and assumed real-valued for simplicity. The problem is to obtain the optimal (i.e., least squares*) linear estimate of the state X_{t_1} of the signal process given the observations $\{Y_s,\ a < s \leqslant t\}$ or, what is the same thing, given the Hilbert space $H(Y; t)$. The optimal estimate $\hat{X}_{t_1|t}$ is to satisfy

$$\|X_{t_1} - \hat{X}_{t_1|t}\| = \inf_{u \in H(Y;t)} \|X_{t_1} - u\|. \tag{2}$$

The solution is well known, and $\hat{X}_{t_1|t}$ is characterized as the projection

$$\hat{X}_{t_1|t} = P_{H(Y;t)} X_{t_1}. \tag{3}$$

Equation (3) is equivalent to the equation

$$(\hat{X}_{x_1|t}, Y_s) = (X_{t_1}, Y_s), \qquad a < s \leqslant t. \tag{3'}$$

All the algorithms for finding $X_{t_1|t}$ (under additional assumptions) take (3′) as the starting point. What has been termed the linear estimation problem encompasses the following three problems:

(i) (a) Prediction if $t_1 > t$; (b) pure prediction if $t_1 > t$ and $N = 0$
(ii) Filtering if $t_1 = t$
(iii) Smoothing if $t_1 < t$

The theory that solves problems (i)–(iii) for second-order stationary processes [i.e., such that $R_X(t,s)$, $R_Y(t,s)$, and $R_{XY}(t,s)$ are functions of $t-s$] is called the *Kolmogorov–Wiener theory*. We shall concentrate on (ii) and occasionally discuss (i, b) when it is convenient to do so.

Equation (3′) can be used to derive a prototype of what is known as a *Wiener–Hopf equation*. For this, taking $t_1 = t$ and writing $\hat{X}_t$ for $\hat{X}_{t|t}$, assume that $\hat{X}_t = \int_a^t h(t,u)Y_u du$, where h is the impulse transfer function, which is unknown. Here R_Y and R_X are the covariance functions of Y and X, respectively, and R_{XY} is the cross-covariance of X and Y [i.e., $R_{XY}(t,s) = E(X_tY_s)$]. From (3′) we have

$$\int_a^t h(t,u)R_Y(s,u)\,du = R_{XY}(t,s). \tag{4}$$

The problem is to find h from (4). In the case when X, N are weakly stationary and stationarily correlated and $a = -\infty$, (4) leads to the equation more commonly identified as the Wiener–Hopf equation.

Let us now consider the problem for weakly stationary and stationarily correlated processes X and N and $a = -\infty$. Then setting

$$\hat{X}_t = \int_{-\infty}^t h(t-u)Y_u\,du$$

$$\text{where } \int_0^\infty |h(u)|^2\,du < \infty,$$

(4) takes the form

$$\int_{-\infty}^t h(t-u)R_Y(s-u)\,du = R_{XY}(t-s),$$
$$-\infty < s \leqslant t. \tag{4′}$$

Assume that the spectral density f_Y and the cross-spectral density f_{XY} exist, that is,

$$R_Y(\tau) = \int_{-\infty}^\infty e^{i\tau\omega} f_Y(\omega)\,d\omega,$$

$$R_{XY}(\tau) = \int_{-\infty}^\infty e^{i\tau\omega} f_{XY}(\omega)\,d\omega,$$

and that

$$\int_{-\infty}^\infty \frac{\log f_Y(\omega)}{1+\omega^2}\,d\omega > -\infty. \tag{5}$$

Then under further conditions which will not be stated here, the function $H(i\omega) = \int_0^\infty h(t)e^{-i\omega t}\,dt$ is determined and from it the impulse transfer function is calculated. The point is that condition (5) is a necessary and sufficient condition for the density to be factorable in the form

$$f_y(\omega) = |H(i\omega)|^2,$$

$$\text{where } H(i\omega) = \int_0^\infty h(\tau)e^{-i\omega\tau}\,d\tau,$$

$$\int_0^\infty |h(\tau)|^2\,d\tau < \infty. \tag{6}$$

The spectral factorization (6) is the analog of the covariance factorization discussed in the section "The Prediction Problem for Discrete-Time Multivariate Time Series" and forms a link between the Wiener–Kolmogorov theory and the later Kalman theory of prediction and filtering.

The solution of (4′) was undertaken first by Wiener and later by several writers in the engineering literature [8]. The problem is also discussed in ref. 5 (Chap. 5). In general, the optimal filter is not of the form assumed above. Instead, it can be shown that we always have $\hat{X}_t = \int_\infty^\infty e^{it\omega}G(\omega)\,dZ_y(\omega)$, where Z_y is the process with orthogonal increments appearing in the spectral representation $Y_t = \int_{-\infty}^\infty e^{it\omega}\,dZ_y(\omega)$. In ref. 14, it is shown that the spectral characteristic of $\hat{X}_t$ [i.e., $G(\omega)$] can be obtained for the case useful in applications, that is, when X and Y have rational spectral densities and $f_y(\omega)$ is strictly positive for all ω.

Time-Domain Analysis and Prediction

In principle, only the knowledge of the covariances is needed for solving equation (4). To put it another way, (4) does not depend on the choice of a particular version of the processes themselves. By a version of a stochastic process $\xi = (\xi_t)$ is meant a process $\tilde{\xi}$ such that for each t, we have $\tilde{\xi}_t = \xi_t$ a.s. For the prediction or filtering problem it is not unreasonable to expect that the equation which determines the impulse response function of $\hat{X}_{t_1|t}$ can be considerably simplified by choosing an appropriate version of Y. As an important illustration, let us consider an extreme example from the problem of pure prediction.

Let us suppose that $\tilde{Y}$ is a version of Y given by

$$\tilde{Y}_s = \int_{t_0}^{s} f(s,u)\, d\zeta_u \qquad \text{for all } s \geqslant t_0, \quad (7)$$

where ζ is a process of orthogonal increments with $E|d\zeta_u|^2 = d\rho(u)$, ρ being a finite measure. Assume that the process ζ has the further important property

$$H(\zeta; s) = H(\tilde{Y}; s) \qquad \text{for each } s. \quad (8)$$

Furthermore, the nonrandom function f satisfies

$$\int_{t_0}^{s} |f(s,u)|^2\, d\rho(u) < \infty \qquad \text{for every } s. \quad (9)$$

The process given by (7) is called a *representation* of Y. If (7) is our observation model for the problem, then the optimal linear predictor of Y_{t_1} is given by

$$\begin{aligned}\hat{Y}_{t_1|t} &= P_{H(\tilde{Y};t)} \int_{t_0}^{t_1} f(t_1,u)\, d\zeta_u \\ &= P_{H(\zeta;t)} \int_{t_0}^{t_1} f(t_1,u)\, d\zeta_u \\ &= \int_{t_0}^{t} f(t_1,u)\, d\zeta_u. \qquad (10)\end{aligned}$$

The prediction error

$$\begin{aligned}\| Y_{t_1} - \hat{Y}_{t_1|t} \|^2 &= \left\| \int_{t}^{t_1} f(t_1,u)\, d\zeta_u \right\|^2 \\ &= \int_{t}^{t_1} |f(t_1,u)|^2\, d\rho(u). \qquad (11)\end{aligned}$$

Thus we obtain the optimal predictor together with the error of prediction directly from (9) and (10) without having to solve (3′). However, we have a new problem on our hands: (a) to derive conditions under which Y does have a representation of the form (7), and (b) to derive algorithms for finding f. The theory of representing a second-order process Y in terms of one or more "simpler" processes (called *innovation processes* in the discrete parameter case and *differential innovation processes* in the continuous parameter case) is due to Cramér and to Hida (see ref. 9) and may be considered as the starting point of the general "time-domain" methods in prediction theory. It must be pointed out, however, that before the work of Cramér and Hida, Karhunen (and, independently, Hanner) had shown that (7) holds with $t_0 = -\infty$ and Y a continuous in quadratic mean, weakly stationary process which is purely nondeterministic [i.e., such that $L(Y; -\infty) = \{0\}$]. The latter holds if and only if the spectral density exists and satisfies condition (5). Thus, to derive the representation, we are led once again to the factorization of the spectral density.

In posing the general problem leading to (7), it has not been assumed that either Y is stationary or that the interval of observation is infinite. The derivation of the result which we state below relies on Hilbert space techniques and ideas from multiplicity theory.

Let $Y = (Y_t)$, $t_0 \leqslant t \leqslant T$, be a vector-valued zero mean, second-order process assumed to be continuous in quadratic mean and purely nondeterministic. Then $Y_t = (Y_t^1, \ldots, Y_t^q)$ has the representation (called "canonical" representation)

$$Y_t^i = \sum_{j=1}^{N} \int_{t_0}^{t} f_{ij}(t,u)\, d\zeta_j(u) \ \text{(a.s.)} \qquad (i = 1, \ldots, q). \quad (12)$$

(a) The processes ζ_j are mutually orthogonal and have orthogonal increments, $E|d\zeta_j(u)|^2 = d\rho_j(u)$, where ρ_j's are finite Borel measures such that $\rho_1 \gg \rho_2 \gg \cdots \gg \rho_N$. (The symbol $\rho_1 \gg \rho_2$ means that ρ_2 is absolutely continuous with respect to ρ_1.) Furthermore,

$$\sum_{i=1}^{q} \sum_{j=1}^{N} \int_{t_0}^{t} |f_{ij}(t,u)|^2\, d\rho_j(u) < \infty.$$

(b) $L(Y; t) = \sum_{j=1}^{N} \oplus L(\zeta_j; t)$ for each t.

(c) In (12) neither the kernels f_j nor the processes ζ_j are uniquely determined. The spectral types of ρ_j and N, the multiplicity of the representation, are uniquely determined. If Y is weakly stationary, the multiplicity N equals the rank of the spectral density matrix. The principal difficulty lies in determining N and the kernels f_{ij} given the covariance of Y. In the nonstationary case Cramér has shown that a real- or complex-valued process can have a representation with infinite multiplicity. Property (b) of (12) is crucial since it enables us to write down the optimal predictor (taking the scalar case for illustration),

$$\hat{Y}_{t_1 | t} = \sum_{j=1}^{N} \int_{t_0}^{t} f_j(t_1, u)\, d\zeta_j(u).$$

The prediction error is easily calculated.

For reasons of space we shall not explore the connection between the Cramér–Hida canonical representation theory and the Kalman–Bucy approach except to remark that the latter can be regarded as a further development of time-domain methods.

Kalman–Bucy Theory

In the Kolmogorov–Wiener theory, the problem of finding the optimal linear predictor or filter is formulated essentially in terms of the covariances (alternatively, the spectral functions) of the signal and noise processes. The approach of the Kalman–Bucy theory is to replace it by a state-space description where the signal process is the output of a linear dynamical system.

$$\begin{aligned} z_t &= H(t)x_t && (13a)\\ x_t &= F(t)x_t + G(t)u_t, \qquad t \geqslant t_0 \\ x_{t_0} &= x_0 \end{aligned}$$

and the observation process is given by

$$y_t = z_t + v_t. \tag{13b}$$

The state vector x_t is n-dimensional, y_t and z_t are m-vectors, and u_t and v_t are Gaussian "white noise" processes. There is no stationarity assumption and t varies over the interval (t_0, T), which is usually taken to be finite.

A rigorous formulation of the state-space model above is obtained by the use of stochastic differential equations*. Here we consider a somewhat more general setup than (13): The signal process $X = (X_t)$ and the observation process $(Y = Y_t)$ are both stochastic processes which are unique solutions of the stochastic differential system

$$\begin{aligned} dX_t &= \left[A_0(t) + A_1(t)X_t + A_2(t)Y_t\right]dt \\ &\quad + B(t)dW_t && (14a)\\ dY_t &= \left[C_0(t) + C_1(t)X_t + C_2(t)Y_t\right]dt \\ &\quad + D(t)dW_t. \end{aligned}$$

X and Y are m- and n-dimensional and $W = (W_t)$ is a q-dimensional standard Wiener process* $(q = m + n)$. It is further assumed that

$$\begin{aligned} &X_0 \text{ is a Gaussian random variable} \\ &\text{independent of } (W_t); && (14b)\\ &Y_0 = 0 \text{ a.s.} \end{aligned}$$

We will not discuss the conditions under which the system (14) has a unique solution except to draw attention to the fact that the coefficients in (14a) are nonrandom and that (14) is a *linear* system.

For each t, let F_t^Y be the sigma field (with null sets added) generated by $\{Y_s, 0 \leqslant s \leqslant t\}$. The family (F_t^Y) represents the observations or data on the basis of which the state-space vector is to be estimated. The following facts about the solution of (14) are well known:

(1) (X_t, Y_t) is jointly Gaussian.

(2) The conditional distribution $L(X_t | F_t^Y)$ of X_t given F_t^Y is Gaussian with mean vector $\hat{X}_t = E(X_t | F_t^Y)$ and covariance $P(t) = E[X_t - \hat{X}_t)(X_t - \hat{X}_t)^* | F_t^Y]$. The least-squares estimate of the state X_t is given by $\hat{X}_t$ and the posterior distribution of X_t given the observations up to t is uniquely determined by the knowledge of $\hat{X}_t$ and $P(t)$.

These facts are not surprising and are not a

special feature of the Kalman theory. In fact, the conclusions above do not exploit the dynamics of the state vector and are valid under the following more general conditions: We assume only the second equation in (14a) connecting the signal and observation processes. In addition, to preserve the essential Gaussian properties of the filtering problem, we suppose that:

(i) (X, W) is a Gaussian process.
(ii) For each s, $\{X_u, W_u, 0 \leqslant u \leqslant s\}$ and the "future" noise $\{W_v - W_u, s \leqslant u < v \leqslant T\}$ are independent.

It is well known that the solution of the filtering problem in the Kalman theory does not involve factorization but is achieved by solving certain differential equations. This point is worth further comment. The development thus far, i.e., the filtering problem under only the assumptions (14b, i, ii) is essentially equivalent to solving a generalized Wiener–Hopf equation. There is, of course, no spectral factorization since, in general, the processes under consideration are nonstationary and observation is over a finite time interval. However, solving the Wiener–Hopf equation is, in turn, equivalent to a special (covariance) factorization of the Gohberg–Krein type. We shall have more to say about this in the next section.

The most important contribution of Kalman filtering and one that distinguishes it from the Kolmogorov–Wiener theory is that it permits *recursive* calculation of the estimates. In other words, the estimates $\hat{X}_{t+\Delta t}$ and $P(t+\Delta t)$ for $\Delta t > 0$ can be computed using the previously calculated values of $\hat{X}_t$ and $P(t)$, an enormous advantage from the computational point of view. What makes recursive filtering possible is the dynamical description of the signal process, which is the first equation in (14a) of the model given above.

Under the conditions imposed on the model,

$$\nu_t = Y_t - \int_0^t \alpha(s)\left[C_0(s) + C_1(s)\hat{X}_s + C_2(s)\right] ds,$$

where $\alpha(t) = [D(t)D^*(t)]^{-1/2}$ is a F_t^Y-adapted Wiener process called the *innovation process*. The equations for the optimum filter (Kalman filter) are given by the following (* denotes transpose of a matrix):

$$d\hat{X}_t = \left[A_0(t) + A_1(t)\hat{X}_t + A_2(t)Y_t\right] dt + \left[P(t)C_1^*(t) + B(t)D^*(t)\right] \times \left[D(t)D^*(t)\right]^{-1/2} d\nu_t \qquad (15a)$$

with $\hat{X}_0 = E(X_0)$;

$$\frac{dP(t)}{dt} = \left[A_1(t)P(t) + P(t)A_1^*(t) + B(t)B^*(t)\right] - \left[P(t)C_1^*(t) + B(t)D^*(t)\right] \times \left[D(t)D^*(t)\right]^{-1} \times \left[C_1(t)P(t) + D(t)B^*(t)\right]; \qquad (15b)$$

$$P(0) = E\left[(X_0 - EX_0)(X_0 - EX_0)^*\right]. \qquad (15c)$$

Equation (15b) for the conditional covariance matrix $P(t)$ is a matrix Riccati equation whose solution is unique subject to the given initial condition (15c). Equation (15a) has a unique solution $\hat{X}_t$ which can easily be expressed explicitly in terms of $\{Y_s, s \leqslant t\}$ and $\{\nu_s, s \leqslant t\}$ [see ref. 9 (Chap. 10)].

Consider the following special case when X and Y are one-dimensional and satisfy the equations

$$dx_t = a_1(t)x_t dt + b(t)dw_t^1,$$
$$dy_t = c_1(t)x_t dt + dw_t^2$$

(lowercase letters denote scalars). x_0 is independent of w^1 and w^2 and the latter are independent Wiener processes. The Kalman filter equations then take the form:

$$d\hat{x}_t = a_1(t)\hat{x}_t dt + c_1(t)p(t)d\nu_t$$
$$\hat{x}_0 = E(x_0); \qquad (16a)$$

$$\frac{dp}{dt} = b^2(t) + 2a_1(t)p(t) - c_1^2(t)p^2(t), \qquad (16b)$$

where $p(0)$ is given.

STEADY-STATE BEHAVIOR. Suppose that the system (14) is time invariant, that is, $A_1(t)$, $B(t)$, $C_1(t)$, and $D(t)$ are constant matrices independent of t. In addition, assume that $A_0 = A_2 = C_0 = C_2 = 0$ and $DD^* = I$. The following result gives information about the asymptotic behavior $P(t)$ as $t \to \infty$.

Suppose that all eigenvalues of A_1 have strictly negative real parts (i.e., A_1 is stable). If matrix

$$\int_0^\infty e^{A_1^* t} C_1^* C_1 e^{A_1 t}\, dt \text{ is nonsingular, then}$$

$$P(\infty) = \lim_{t \to \infty} P(t) \text{ exists}$$

and satisfies the equation

$$A_1 P(\infty) + P(\infty) A_1^* + BB^* - P(\infty) C_1^* C_1 P(\infty) = 0.$$

It is instructive to look at the scalar case. Noting that b, a_1, and c_1 are now constants independent of t and differentiating (16b), we get

$$\ddot{p}(t) = 2\left[a_1 - c_1^2 p(t)\right] \dot{p}(t).$$

Hence

$$\dot{p}(t) = \phi(p(0)) c^{\int_0^t 2[a_1 - c_1^2 p(s)]\, ds}, \qquad (17)$$

where $\phi(p) = b^2 + 2a_1 p - c_1^2 p^2$. It follows that $p(t)$ is monotone increasing if $\phi(p(0)) > 0$ and monotone decreasing if $\phi(p(0)) < 0$. From (16b) the steady-state equation is seen to be $\phi(p) = 0$. Integrating (17),

$$p(t) = \phi(p(0)) \int_0^t e^{2a_1 s - \int_0^s c_1^2 p(u)\, du}\, ds + p(0)$$

and recalling that $p(\cdot)$ is nonnegative [we are only interested in nonnegative solutions of (16b)], we have

$$p(t) \leqslant \phi(p(0)) \frac{e^{2a_1 t} - 1}{2a_1} + p(0) \qquad \text{for all } t \geqslant 0.$$

Let us now assume that $a_1 < 0$ and $c_1^2 > 0$. From the inequality for $p(t)$ it follows that $p(t)$ is bounded for all $t \geqslant 0$. Since $p(t)$ is always a monotonic function of t, $p(\infty) = \lim_{t \to \infty} p(t)$ exists and is finite. The steady-state equation has then the nonnegative solution

$$p(\infty) = \left(a_1 + \sqrt{a_1^2 + b^2 c_1^2}\right) / c_1^2.$$

Finally, if the initial value $p(0)$ is less than $p(\infty)$, then $p(t)$ increases monotonically to $p(\infty)$; if $p(0)$ is greater than $p(\infty)$, then $p(t)$ decreases monotonically to $p(\infty)$ (see ref. 2).

It is worth commenting briefly on the methods used in establishing the Kalman filter equations. Perhaps the most important and fruitful approach is the one based on the theory of Ito's stochastic differential equations and the techniques of martingale* theory. A rigorous treatment of this aspect of the subject can be found in several books (e.g., [1, 9, 11]). The martingale approach has been eminently successful also in the development of nonlinear filtering theory. The optimal filter, then, is governed by a nonlinear stochastic differential equation.

Some remarks on the scope of the Kalman theory are in order. The choice between the "covariance formulation" and the "state-space formulation" (which is at the heart of the Kalman theory) is not clear cut and often not easy to make. The most successful applications of the theory appear to have been in problems of satellite tracking and in aerospace engineering where the dynamical equations of motion of the system in terms of physical variables can be more easily converted to the state-space description [see, e.g., ref. 2 (Chap. 2)]. In general, however, it is not easy to discover state-space models for nonstationary processes or for stationary processes given over a finite time interval. Problems of this kind have led to recent research by engineers in stochastic "realization theory." For more information on this and related questions, see ref. 8.

Equations of Wiener–Hopf Type, Gohberg–Krein Factorization, and Linear Filtering

We have seen in earlier sections that the algorithm for the prediction problem for purely nondeterministic stationary processes rests on a suitable factorization of the spectral density. In the filtering theory of nonstationary second-order processes (with finite or infinite time interval of observation) the optimal filter (more precisely, its impulse transfer function) is obtained by solving what we may call a generalized Wiener–Hopf equation. We shall now give an indication of the proof and simultaneously show the equivalence of this technique with that of covariance factorization. The latter term will be used in this section synonymously with the special factorization of Gohberg and Krein [6].

Let $H = L^2[0, T]$ and let H_t be the increasing family of subspaces defined by

$$H_t = \{ f \in H : f(s) = 0 \text{ a.e.}, t < s \leqslant T \}.$$

We consider integral operators A on H given by

$$(Af)(t) = \int_0^T A(t,s) f(s)\, ds \qquad (f \in H),$$

where $A(t,s)$ is a square-integrable kernel. For any Hilbert–Schmidt A we can write

$$A = A_+ + A_-,$$

where A_+ has (H_t) $(0 < t < T)$ as invariant subspaces (i.e., $A_+ H_t \subset H_t$) and A_- has the orthogonal complement family $(H_t^{\perp})$ as invariant subspaces. Suppose that R is a covariance operator of the form $I + K$, where K is a Hilbert–Schmidt integral operator. A result of Gohberg and Krein (proved in far greater generality than needed for our purpose here) states that if R is invertible, then there is a unique factorization of the form

$$R = (I + K) = (I + L)(I + L^*), \qquad (18)$$

where $L = L_+$, $L^* = L_-$ ($L^* =$ adjoint of L), and L is Hilbert–Schmidt.

Moreover, the factorization (18) holds if and only if L is the unique solution of the Wiener–Hopf equation

$$L + (HL)_+ = H_+. \qquad (19)$$

In (19) H is the Fredholm resolvent of K [i.e., $(I + H) = (I + K)^{-1}$]. For the general model defined by (14b, i, ii, scalar case), the optimal estimate

$$\hat{x}_t = \int_0^t G(t,u)\, d\nu_u, \qquad (20)$$

ν being the innovation process. (For the derivation, see ref. 9, where the notation is slightly different.) One of the main concerns of filtering (or prediction) theory is to devise an algorithm for finding the impulse transfer function G. Despite the mathematical convenience of (20), expressing $\hat{x}_t$ as a stochastic integral* with respect to a Wiener process, it is more natural to express it in terms directly of the observations. We shall do this for the Kalman model (14), using the white noise formulation for convenience of exposition. Writing

$$\hat{x}_t = \int_0^t G(t,u) y_u\, du,$$

it is seen that $G(t,s)$ must satisfy the equation

$$G(t,s) + \int_0^s G(t,u) K(u,s)\, du, \qquad 0 \leqslant s \leqslant t. \qquad (21)$$

Since only the values $0 \leqslant s \leqslant t$ enter into the problem, we may set $G(t,s) = 0$ if $s > t$. In terms of the notation introduced earlier, (21) becomes the operator (Wiener–Hopf) equation

$$G + (GK)_+ + K_+. \qquad (22)$$

Comparing (22) with (21) it follows that G, the impulse transfer of the optimal filter, is the unique solution of the Wiener–Hopf equation (21) and is equivalently obtained by the Gohberg–Krein special factorization (18) of the covariance R.

The foregoing solution of the filtering problem does not assume a state-space model for the system. If we do have the state-space description of the Kalman model, the Wiener–Hopf equation leads to the Riccati differential equation. For a derivation of the latter directly from the Wiener–Hopf equation, see ref. 9 or 11.

PREDICTION PROBLEM FOR DISCRETE-TIME MULTIVARIATE TIME SERIES

The foundations of linear prediction theory for one-dimensional stationary sequences were laid and the basic problems solved in the famous and beautifully written 1941 paper by Kolmogorov. Wiener's independent work followed a little later. The extension to multidimensional stationary sequences began in the 1950s (however, a reference is made in Wiener and Masani's work to an early paper by Zasuhin, who died during World War II). The important papers in this area are by Helson and Lowdenslager, Wiener and Masani, Masani, Rosenblatt, and Rozanov. These papers contain references to the work of Kolmogorov and Wiener. The reader will find in ref. 12 precise references to the papers cited above and a more complete bibliography. It also contains a series of illuminating articles on the subject, notably the ones by Masani and Salehi. The notation and terminology of this section are borrowed from the latter. We shall content ourselves with describing only a few of the most basic results.

Let $X_n = (X_n^1, \ldots, X_n^q)$ $n \in \mathbb{Z}$ be a zero-mean, second-order stationary process (or sequence), that is, $EX_n = 0$, $E|X_n|^2 < \infty$ for all n, and $E(X_n^j \overline{X}_m^k) = \Gamma_{n-m}^{ik}$. $\Gamma_n = (\Gamma_n^{ik})$ is the covariance matrix of (X_n). Then

$$\Gamma_n = (2\pi)^{-1} \int_0^{2\pi} e^{-ik\lambda}\, dF(\lambda).$$

The covariance matrix Γ_n is positive definite and the matrix-valued (spectral) function $F \geqslant 0$. If F is absolutely continuous, $f(\lambda) = F'(\lambda)$ is called the *spectral density matrix*.

For purposes of linear prediction, we may define the Hilbert spaces $H(X)$, $H(X, n)$, and $H(X; -\infty) = \cap_n H(X; n)$ as in the one-dimensional case [e.g., $H(X; n) = \overline{\text{sp}}\{X_k^j,\ k \leqslant n,\ j = 1, \ldots, q\}$]. The latter is the closed linear subspace taken with scalar (or complex) coefficients. Equivalently, we may take $H(x; n) = H_n$ to be the closed linear space where the finite linear combinations of the X_k's are taken with $q \times q$ matrix coefficients.

(X_n) is called *regular* or *purely nondeterministic* (PND) if $H_{-\infty} = 0$; it is called *singular* if $H_{-\infty} = H_\infty$ $[= H(X)]$. Two elements $u, v \in H$ are *orthogonal* $(u \perp v)$ if the Gramian $[u, v] = 0$. Let $\hat{X}_{p,n}$ denote the orthogonal projection of X_p onto H_n $(p > n)$. Define the "innovation" process $\xi_n = X_n - \hat{X}_{n,n-1}$ $(n \in \mathbb{Z})$. Then we have $[\xi_m, \xi_n] = \delta_{mn} G$, where $G = [\xi_0, \xi_0]$. G is called the *prediction-error matrix* with lag 1. The rank ρ of G is defined to be the rank of (X_n). We say that (X_n) has full rank if $\rho = q$.

WOLD DECOMPOSITION. Let (X_n) be nondeterministic (i.e., $H_{-\infty} \neq H_\infty$). Then there is a unique decomposition

(a) $X_n = X_n' + X_n''$ $(n \in \mathbb{Z})$, where

(b) (X_n') is PND and (X_n'') is deterministic.

(c) The Hilbert spaces $H(X')$ and $H(X'')$ of the processes (X_n') and (X_n'') are mutually orthogonal.

One of the main results in the spectral theory is the following: (X_n) is regular and of full rank if and only if (i) F is absolutely continuous, and (ii)

$$\int_0^{2\pi} \log \det f(\lambda)\, d\lambda > -\infty.$$

Moreover,

$$\log \det G = \frac{1}{2\pi} \int_0^{2\pi} \log \det f(\lambda)\, d\lambda.$$

We also have the following factorization of the spectral density in the full-rank case: $f(\lambda) = \psi(\lambda)\psi^*(\lambda)$, where

$$\psi(\lambda) = \sum_{k=0}^{\infty} C_k e^{ik\lambda} \in L_2[0, 2\pi] \quad \text{with } C_0 \geqslant 0$$

and

$$\log \det C_0 = \frac{1}{4\pi} \int_0^{2\pi} \log \det f(\lambda)\, d\lambda.$$

In the general case when (X_n) is not assumed to be of full rank, it is still true that (X_n) is regular if and only if the spectral density f exists and has a factorization $f = \psi\psi^*$, where $\psi(\lambda) = \sum_{k=0}^{\infty} C_k e^{ik\lambda} \in L_2[0, 2\pi]$. However, the necessary and sufficient analytical conditions (replacing the condition $\log \det f \in L_1[0, 2\pi]$) for factorability are more complicated and are to be found in the article referred to above.

We conclude this discussion with a brief reference to an important problem in linear prediction, still awaiting a complete solution: Find an algorithm to compute the best linear predictor in terms of the data. In other words, suppose that $\hat{X}_n$ is the orthogonal projection of X_n onto H_0. Under what conditions can we obtain a formula of the type

$$\hat{X}_n = \sum_{k=0}^{\infty} E_{n,k} X_{-k}, \tag{23}$$

where the $E_{n,k}$'s are computable from a knowledge of ψ? It has been shown that (23) is valid under the condition that $f \in L_\infty[0, 2\pi]$ and $f^{-1} \in L_1[0, 2\pi]$. The coefficients $E_{n,k}$ are finite sums involving the Fourier coefficients of ψ and ψ^{-1} (see Salehi's article in ref. 12).

SOME RECENT DEVELOPMENTS

Prediction on the Finite Past

The problem of prediction of $X(t)$ $(t > 0)$ based on the *finite* past of X: $\{X(s),\ -2T \leqslant s \leqslant 0\}$ is much more difficult. It was first solved by M. G. Krein in 1954. A full discussion of this problem, together with the corresponding interpolation problem of predicting $X(t)$ for $|t| < T$ given $\{X(s) : |s| \geqslant T\}$, is given in the recent book of Dym and McKean [4], which contains references to other work on these questions. The main difficulty here is that the theory of Hardy class functions is not the appropriate tool. Instead, one uses the theory of "strings" developed by Krein, which regards the spectral distribution function of the process as the principal spectral function Δ of a string with which is associated a suitable differential operator G. The required projections that yield the solutions to the prediction and the interpolation problems are obtained in terms of eigen differential expansions of G. See ref. 4 for an exhaustive treatment of the subject.

K-W Theory and Discrete Kalman Filters

The linear least-squares predictor of the Kolmogorov–Wiener theory is not a recursive estimator, that is, if $\hat{X}_{t+m}$ or $\hat{X}_t(m)$ is the best linear predictor of X_{t+m} based on the past up to t, then the predictor $\hat{X}_{t+m+1}$ cannot, in general, be computed by "updating" $\hat{X}_{t+m}$ and has to be recomputed. An algorithm, due to Box and Jenkins, does indeed yield the Kolmogorov–Wiener predictor in a recursive form. The procedure applies to certain linear, finite parametric models, for example, when X_t is given by a stationary invertible model

$$X_t + a_1 X_{t-1} + \cdots + a_k X_{t-k} = \epsilon_t + b_1 \epsilon_{t-1} + \cdots + b_l \epsilon_{t-l}, \quad (24)$$

where the ϵ's are independent random variables with zero mean and finite variance. In view of the invertibility assumption, (24) implies that X_t has a "proper canonical representation" in the sense described in the section "The Kalman–Bucy Theory." (Normality assumptions for the ϵ_t's are not required.) In fact, $\epsilon_t = X_t - \hat{X}_{t-1}(1)$ and represents the "innovation" process. It is an easy exercise to show that (assuming that $k \geqslant l$, $m \geqslant 1$) $\hat{X}_t(m)$ can be computed recursively from $\hat{X}_t(1), \ldots, \hat{X}_t(m-1)$ and the observations $X_t, X_{t-1}, \ldots$. Finally, the prediction error

$$\sigma_m^2 = E|X_{t+m} - \hat{X}_{t+m}|^2 = E(\epsilon_1^2)\left(\sum_{j=0}^{m-1} g_j^2\right),$$

where the g_j's are calculated by equating coefficients in the identity

$$\left(\sum_{j=0}^{k} a_j z^j\right)\left(\sum_{j \geqslant 0} g_j z^j\right) = \sum_{j=0}^{l} b_j z^j \qquad (a_0 = b_0 = 1).$$

The Box–Jenkins approach is similar in spirit to the discrete Kalman filter. The latter, however, seems to be more powerful and flexible, in that it can be applied easily to a variety of multivariate problems which can be cast in the form of a "state-space" representation. As an illustration, consider the univariate AR(2) model

$$X_t - a_1 X_{t-1} - a_2 X_{t-2} = \epsilon_t. \quad (25)$$

Setting $X_t^{(2)} = X_t$, $X_t^{(1)} = -a_2 X_{t-1}$, (24) may be recast in the form

$$\begin{pmatrix} X_t^{(1)} \\ X_t^{(2)} \end{pmatrix} = \begin{pmatrix} 0 & -a_2 \\ 1 & -a_1 \end{pmatrix}\begin{pmatrix} X_{t-1}^{(1)} \\ X_{t-1}^{(2)} \end{pmatrix} + \begin{pmatrix} 0 \\ 1 \end{pmatrix}\epsilon_t. \quad (26)$$

The advantage is that (26) defines a vector-valued *Markov* process*, whereas the original model (25) is non-Markovian. Thus, in the case of a filtering problem where X_t is a stationary ARMA process, the state-space formulation leads to

$$x_{t+1} = Fx_t + G\epsilon_{t+1}$$
$$y_t = Hx_t + n_t,$$

which is in the standard Kalman form. It is to be noted that the state-space formulation given above is not unique and it is desirable to obtain a representation with the smallest possible dimension. A discussion of these matters together with a general comparison of the Box–Jenkins and the discrete Kalman models is given in the two-volume work by M. B. Priestley [13], on which the material here is based.

Stationary Random Fields

A sequence of random variables $X = (X_{m,n})$, $m, n \in \mathbb{Z}$, is called a (weakly) stationary second-order random field (SORF) if $E|X_{mn}|^2 < \infty$, $EX_{mn} = 0$, and $E(X_{mn}\overline{X}_{m'n'}) = R(m - m', n - n')$. Time-domain analysis of a SORF has recently led to results extending the well-known Wold or Halmos decomposition of a second-order stationary sequence of a single discrete parameter. Define the Hilbert spaces $L(X; m, n) = \overline{\text{sp}}\{X_{uv},\ u \leqslant m,\ v \leqslant n\}$, $L^1(X; m) = \overline{\text{sp}}\{X_{uv},\ u \leqslant m,\ v \in \mathbb{Z}\}$, and $L^2(X; n) = \overline{\text{sp}}\{X_{u,v},\ u \in \mathbb{Z},\ v \leqslant n\}$. Almost all the analysis of random fields* so far carried out makes the following commutativity assumption: For all $m, n \in \mathbb{Z}$,

$$P_{L^1(X;m)}P_{L^2(X;n)} = P_{L(X;m,n)}, \qquad (27)$$

where the notation P_M denotes orthogonal projection with range M.

Three distinct kinds of innovation subspaces occur: the two-dimensional innovations and the one-dimensional innovations coming from each of the two directions—horizontal and vertical. Under assumption (27), a "fourfold" Wold decomposition of X is obtained, the components being (respectively), PND in the two-dimensional sense, horizontally PND, vertically PND, and deterministic. For definitions and full details, including results on multiplicity, we refer the reader to ref. 10. An early paper devoted to the topic is the one by Chiang Tse-Pei [3]. A different approach to discrete parameter random fields (which is not discussed here due to lack of space) with different definitions of the nondeterministic property, is to be found in Helson and Lowdenslager [7].

References

[1] Balakrishnan, A. V. (1973). Stochastic Differential Systems I. *Lect. Notes Econ. Math. Syst.*, **84**. Springer-Verlag, New York.

[2] Balakrishnan, A. V. (1981). *Stochastic Filtering and Control*, Vol. 2. Lecture Notes in System Science. Optimization Software, Los Angeles.

[3] Chiang, T. (1957). *Theory Prob. Appl.*, **2**, 58–89.

[4] Dym, H. and McKean, H. P. (1976). *Gaussian Processes, Function Theory and the Inverse Spectral Problem*. Academic Press, New York.

[5] Gikhman, I. I. and Skorokhod, A. V. (1969). *Introduction to the Theory of Random Processes* (English transl.). W. B. Saunders, Philadelphia.

[6] Gohberg, I. C. and Krein, M. G. (1970). Theory and Applications of Volterra Operators in Hilbert Space. *Amer. Math. Soc. Transl.*, **24**. Amer. Math. Soc., Providence, RI.

[7] Helson, H. and Lowdenslager, D. (1959). *Acta Math.*, **99**, 165–202.

[8] Kailath, T. (1974). *IEEE Trans. Inf. Theory*, **IT-20**, 145–181.

[9] Kallianpur, G. (1980). *Stochastic Filtering Theory*. Springer-Verlag, New York.

[10] Kallianpur, G. and Mandrekar, V. (1983). In *Prediction Theory and Harmonic Analysis*, V. Mandrekar and H. Salehi, eds. North-Holland, Amsterdam, pp. 165–190.

[11] Liptser, R. and Shiryaev, A. N. (1978). *Statistics of Random Processes*, Vol. I: *General Theory*. Springer-Verlag, New York.

[12] Masani, P., ed. (1981). *Collected Works of Norbert Wiener*, Vol. III. MIT Press, Cambridge, Mass.

[13] Priestley, M. B. (1981). *Spectral Analysis and Time Series*, Vols. 1 and 2. Academic Press, New York.

[14] Yaglom, A. M. (1962). *An Introduction to the Theory of Stationary Random Functions*. Prentice-Hall, Englewood Cliffs, N.J.

(KALMAN FILTERING
NOISE
STOCHASTIC PROCESSES
TIME SERIES)

G. KALLIANPUR

PREDICTION AND FORECASTING

A prediction is a statement about an unknown and uncertain event—most often, but not necessarily, a future event. If time is involved, a prediction (or forecast) is an assertion about a future outcome that is based on observed regularities among consecutive events in the past.

Prediction is usually not an end in itself. The ability to form predictions is important in many disciplines, since decisions and control strategies are frequently based on predictions of uncertain events. For example, the economist forecasts future economic variables in order to better direct the economy, and the control engineer needs predictions in order to adjust input variables to maintain future values of output variables close to specified targets.

Predictions are by-products of the quantitative understanding of a situation. This understanding is usually expressed in a model that relates the variables of interest. In some cases the relationships are derived from physical laws and represent an *explanation* of a regularity. In other cases they represent the *recognition* of a regularity and express statistical relationships (correlations) among the variables. There is yet another category where predictions are based on *extrapolations*. There the forecaster has past observations on only the variable that has to be predicted and extrapolates historic trends.

REGRESSION MODELS

Regression* models are commonly used to describe the relationships among a dependent response variable Y, which we try to predict, and a set of explanatory (predictor) variables $X_1, X_2, \dots, X_p$. Linear regression models are written as

$$\begin{aligned} Y_t &= \beta_0 + X_{t1}\beta_1 + \cdots + X_{tp}\beta_p + \epsilon_t \\ &= \mathbf{X}_t'\boldsymbol{\beta} + \epsilon_t, \end{aligned} \tag{1}$$

where t is an index that represents the experimental or survey unit, or time if the observations are collected sequentially over time; $\mathbf{X}_t = (1, X_{t1}, \dots, X_{tp})'$ is a vector that contains the predictor variables, $\boldsymbol{\beta} = (\beta_0, \beta_1, \dots, \beta_p)'$ is a vector of unknown coefficients, and the ϵ_t are uncorrelated identically distributed mean zero random variables.

Least-squares* estimates $\hat{\boldsymbol{\beta}}$ of the unknown coefficients can be calculated from past data $(Y_1, \mathbf{X}_1), \dots, (Y_n, \mathbf{X}_n)$. These estimates minimize the sum of squares $\sum_{t=1}^{n}(Y_t - \mathbf{X}_t'\boldsymbol{\beta})^2$. The least-squares estimates are linear functions of the observations $Y_1, \dots, Y_n$.

For given values of the explanatory variables $X_1, \dots, X_p$ one predicts the corresponding response variable from

$$\hat{Y}^{\text{pred}} = \hat{\beta}_0 + X_1\hat{\beta}_1 + \cdots + X_p\hat{\beta}_p. \tag{2}$$

This prediction is unbiased and has the smallest prediction error variance among all linear unbiased predictions. We call it a *minimum mean square error prediction*. The variance of the corresponding prediction error and prediction intervals for the future value Y are easily calculated. The prediction error variance depends on the variance of the future unpredictable error and on the variability of the coefficient estimates.

If the observations are collected over time and if one wishes to predict the future Y_{n+l}, then the necessary predictor variables at time $n + l$ are usually unknown and have to be replaced by forecasts. Extrapolation methods are frequently used to obtain forecasts of future explanatory variables.

Equation (1) represents a relationship that is linear in the coefficients. Some models, especially those arising from theoretical relationships, are nonlinear. In such cases closed form expressions for the least squares estimates can usually not be found; iterative nonlinear minimization procedures have to be used in their calculation (see Draper and Smith [4]).

REGRESSION AND SMOOTHING METHODS FOR EXTRAPOLATING A SINGLE TIME SERIES

Here we assume that observations Z_t on a single series are made at consecutive time periods, and that there are no other series that can help in their prediction.

Regression on Functions of Time

The historical extrapolation approach is to regress Z_{t+l} on functions of the forecast lead time l. The coefficients in the model

$$Z_{t+l} = f_1(l)\beta_1 + \cdots + f_m(l)\beta_m + \epsilon_{t+l}$$
$$= \mathbf{f}'(l)\boldsymbol{\beta} + \epsilon_{t+l} \quad (3)$$

are estimated by ordinary least squares ($\hat{\boldsymbol{\beta}}$), and the l-step-ahead forecast of Z_{n+l} from time origin n is given by

$$\hat{Z}_n(l) = \mathbf{f}'(l)\hat{\boldsymbol{\beta}}. \quad (4)$$

The fitting (forecast) functions $f_i(l)$ are usually polynomials, and seasonal indicators or trigonometric functions if the series is seasonal. The linear trend model $Z_{t+l} = \beta_0 + \beta_1 l + \epsilon_{t+l}$ is a special case.

General Exponential Smoothing

The regression approach assumes that the trend model stays constant over time; least squares treats all observations as equally relevant to parameter estimation. For many economic and business series this is an unrealistic assumption. There one expects the trend* to shift with time and be adaptive. Thus the estimation method should put more weight on recent observations and less on the ones in the past. In *discounted least squares* (or *general exponential smoothing*) the estimates $\hat{\boldsymbol{\beta}}_n$ minimize the discounted sum of squares $\sum_{j\geqslant 0}\omega^j[Z_{n-j} - \mathbf{f}'(-j)\boldsymbol{\beta}]^2$; $0 < \omega < 1$ is a known discount coefficient and $\alpha = 1 - \omega$ is called the smoothing constant. The estimates can be updated recursively (see Brown [3]). With an additional observation Z_{n+1} the estimate $\hat{\boldsymbol{\beta}}_{n+1}$ is a linear combination of the previous estimate $\hat{\boldsymbol{\beta}}_n$ and a fraction of the most recent forecast error $Z_{n+1} - \hat{Z}_n(1)$. The weights that are given to this forecast error depend on the chosen smoothing constant.

Special cases of general exponential smoothing lead to simple and double exponential smoothing.

In *simple exponential smoothing* the forecasts are calculated from $\hat{Z}_n(l) = S_n$. The smoothed statistic S_n is an exponentially weighted moving average* of past observations. With a new observation Z_{n+1} it is updated according to $S_{n+1} = \alpha Z_{n+1} + (1 - \alpha)S_n$. The smoothing constant $\alpha = 1 - \omega$ determines how much weight is given to the most recent observation. Simple exponential smoothing is appropriate if the level of the series is subject to random shifts.

In *double exponential smoothing* the forecasts lie on the linear trend line

$$\hat{Z}_n(l) = \left[2S_n - S_n^{[2]}\right] + \frac{\alpha}{1-\alpha}\left[S_n - S_n^{[2]}\right]l. \quad (5)$$

S_n is the smoothed statistic discussed above, and $S_n^{[2]} = \alpha S_n + (1-\alpha)S_{n-1}^{[2]}$ is obtained by smoothing the observations twice.

Holt's Two-Parameter Double Exponential Smoothing

Double exponential smoothing uses only one smoothing constant. A more general approach is to consider forecasts from the linear trend model

$$\hat{Z}_n(l) = \hat{\beta}_0(n) + \hat{\beta}_1(n)l, \quad (6)$$

where the coefficients are updated according to

$$\hat{\beta}_0(n+1) = \alpha_1 Z_{n+1} + (1 - \alpha_1) \times\left[\hat{\beta}_0(n) + \hat{\beta}_1(n)\right]$$
$$\hat{\beta}_1(n+1) = \alpha_2\left[\hat{\beta}_0(n+1) - \hat{\beta}_0(n)\right] + (1 - \alpha_2)\hat{\beta}_1(n) \quad (7)$$

There are two smoothing constants; one updates the level, the other updates the slope. The new estimate of the level is a weighted average of the most recent observation Z_{n+1} and of $[\hat{\beta}_0(n) + \hat{\beta}_1(n)]$ which is the projected level at time $n + 1$, using data up to time period n. The slope estimate is a weighted average of the difference of the last two level estimates and of the previous slope estimate.

Winters' Additive and Multiplicative Seasonal Smoothing Methods

These are seasonal extensions of Holt's method. In the multiplicative version the trend coefficients and the seasonal factors (corresponding to s seasons) in the forecast function

$$\hat{Z}_n(l) = [\hat{\beta}_0(n) + \hat{\beta}_1(n)l]\hat{S}_{n+l-s} \quad (8)$$

are updated according to

$$\begin{aligned}\hat{\beta}_0(n+1) &= \alpha_1[Z_{n+1}/\hat{S}_{n+1-s}] \\ &\quad + (1-\alpha_1)[\hat{\beta}_0(n) + \hat{\beta}_1(n)], \\ \hat{\beta}_1(n+1) &= \alpha_2[\hat{\beta}_0(n+1) - \hat{\beta}_0(n)] \\ &\quad + (1-\alpha_2)\hat{\beta}_1(n), \\ \hat{S}_{n+1} &= \alpha_3[Z_{n+1}/\hat{\beta}_0(n+1)] \\ &\quad + (1-\alpha_3)\hat{S}_{n+1-s}. \quad (9)\end{aligned}$$

In the additive version the forecast is the sum of trend and seasonal components; the ratios $Z_{n+1}/\hat{S}_{n+1-s}$ and $Z_{n+1}/\hat{\beta}_0(n+1)$ in the updating equations have to be replaced by the respective differences.

Comment

Exponential smoothing procedures have received considerable attention in the business forecasting literature. Two main reasons for the popularity of these techniques among business forecasters are (a) easy updating relationships that allow the forecaster to update the forecasts without storing all past observations, and (b) the fact that these procedures are said to be automatic and easy to use. These procedures, however, are automatic only if one has already chosen a particular smoothing method and knows the smoothing constant and the initial values for the recursive smoothing equations. Several recommendations on how to choose those are given in the literature. The smoothing constants are frequently picked by minimizing the sum of the squared historic one-step-ahead forecast errors $\sum_{t=1}^{n}[Z_t - \hat{Z}_{t-1}(1)]^2$ (*see also* GRADUATION *and* FORECASTING).

FORECASTS FROM UNIVARIATE TIME-SERIES* MODELS

Smoothing procedures are heuristic *methods* that are applied to past data. They differ from the stochastic *model approach* to forecasting, which first constructs appropriate statistical models and then derives the resulting forecasts.

The subject of forecasting has occupied a central part in the theory of stochastic processes and time-series analysis. Two of the earliest treatments of this subject are by Kolmogorov [17] for discrete-time stationary stochastic processes and by Wiener [30] for continuous-time processes. Kolmogorov uses a decomposition suggested by Wold [32], who showed that any stationary nondeterministic stochastic process can be written as a one-sided moving average of an uncorrelated sequence. Wiener considers a frequency-domain representation and reduces the prediction problem to the solution of a Wiener–Hopf equation. A comprehensive discussion of these approaches is given in Whittle [29] and in the review paper by Kailath [13].

Autoregressive integrated moving-average* (ARIMA) models approximate the coefficients in the one-sided moving-average representation as functions of a smaller number of parameters. These models have proved very useful in representing many stationary, as well as certain nonstationary time series. ARIMA(p, d, q) models are usually written as

$$\begin{aligned}(1 - \phi_1 B - \cdots - \phi_p B^p)(1-B)^d Z_t \\ = (1 - \theta_1 B - \cdots - \theta_q B^q)\epsilon_t, \quad (10)\end{aligned}$$

where B is the backshift operator ($B^m Z_t = Z_{t-m}$), $1 - \phi_1 B - \cdots - \phi_p B^p$ is a stationary autoregressive operator, $(1-B)^d$ is the differencing operator, and $1 - \theta_1 B - \cdots - \theta_q B^q$ is a moving-average operator. The ϵ_t are uncorrelated, identically distributed mean-zero random variables with constant variance. Many papers and books have been written on how to specify, estimate, and validate such models (see Box and Jenkins [2]).

The minimum mean square error forecast of the future Z_{n+l} is given by the conditional expectation

$$Z_n(l) = E[Z_{n+l} \mid Z_n, Z_{n-1}, \ldots]. \quad (11)$$

It minimizes the expected value of the squared future forecast error. Forecasts are conveniently calculated from the difference equation (10), and prediction intervals are readily derived. To obtain more insight into the nature of the forecasts, one can express the forecasts as

$$Z_n(l) = \sum_{j \geqslant 1} \pi_j^{(l)} Z_{n+1-j}. \qquad (12)$$

The weights $\pi_j^{(l)}$ in this linear combination depend on the forecast lead time l and on the particular model and its parameters. Yet another representation expresses the eventual behavior of the forecasts as

$$Z_n(l) = f_1(l)\beta_1^{(n)} + \cdots + f_{p+d}(l)\beta_{p+d}^{(n)}. \qquad (13)$$

The forecast functions $f_i(l)$ depend on the autoregressive and differencing operators and are usually exponentials, polynomials, and trigonometric functions of the forecast lead time l. The coefficients in this equation, which is called the *eventual forecast function* since it is valid only for $l > p + d - q$, depend on the data up to time n; they are updated with each new observation.

Relationships between Forecasts from ARIMA Time-Series Models and Exponential Smoothing

ARIMA models imply the various smoothing forecast methods as special cases. A detailed discussion is given in Abraham and Ledolter [1]. For example, the ARIMA(0, 1, 1) model

$$(1 - B)Z_t = (1 - (1 - \alpha)B)\epsilon_t$$

leads to the same forecasts as simple exponential smoothing with smoothing constant α. The ARIMA(0, 2, 2) model

$$(1 - B)^2 Z_t = (1 - (1 - \alpha)B)^2\epsilon_t$$

leads to the same forecasts as double exponential smoothing. The ARIMA(0, 2, 2) model

$$(1 - B)^2 Z_t = (1 - \theta_1 B - \theta_2 B^2)\epsilon_t$$

with $\theta_1 = 2 - \alpha_1(1 + \alpha_2)$ and $\theta_2 = -(1 - \alpha_1)$ leads to the same forecasts as Holt's two-parameter double exponential smoothing with smoothing constants α_1 and α_2. The seasonal ARIMA model $(1 - B)(1 - B^s)Z_t = (1 - \theta_1 B - \cdots - \theta_{s+1}B^{s+1})\epsilon_t$ leads to the same forecasts as the additive Winters' method; the moving-average coefficients are functions of the smoothing constants. However, since forecasts from ARIMA models are linear in the observations, it is not possible to find an ARIMA model that leads to the same forecasts as the multiplicative Winters' method.

ARIMA models are more flexible than the traditional smoothing methods. Furthermore, the forecast functions in ARIMA models are the result of a careful model building approach, not just the result of a cursory visual inspection of the data, as is often the case in the smoothing approach.

Adaptive Forecast Methods

The usual smoothing methods assume fixed smoothing constants. Concern about parameter instability has led to a number of methods that relax this assumption and adjust the smoothing constants according to the size of the most recent forecast errors. For example, in adaptive response rate simple exponential smoothing, Trigg and Leach [28] increase the smoothing constant whenever there is an indication of a rapid change in the level.

Also, the usual time-series models assume constant coefficients. Methods that relax this assumption have been considered, such as adaptive filtering and time-series models with stochastically changing coefficients.

Forecasts from Nonlinear Time-Series Models

The ARIMA models lead to forecasts which are linear functions of previous observations. There are many possible nonlinear extensions. Priestley [24], for example, has developed a general class of nonlinear time-series models, called the state-dependent models. They include bilinear models [5], threshold autoregressive models [27], exponential autoregressive models [6], and ARIMA models with time-varying coefficients as special cases. Several empirical studies suggest that in some cases a careful nonlinear time-series modeling approach can lead to additional forecast improvements.

Seasonal Adjustment

If a series exhibits seasonality, one can use forecast methods that handle seasonality directly, such as seasonal ARIMA models or Winters' forecast procedures. Alternatively, one can use a seasonal adjustment method, such as Census X-11 or Census X-11 ARIMA, to obtain a deseasonalized series; predict the seasonally adjusted series using nonseasonal forecast procedures; and then reseasonalize the forecasts by applying the seasonal factors. In this approach the forecaster assumes that the seasonal aspects of the data are better captured by the seasonal adjustment procedure than by an explicit model representation. Comparisons of these two approaches are given in Plosser [23] and Makridakis et al. [19]. From these studies one may conclude that it is rather difficult to assess the effect of seasonal adjustment on the forecast accuracy.

FORECASTS FROM MULTIVARIATE TIME-SERIES MODELS

In univariate time-series modeling one predicts future values of a particular series from its own past history. If there are additional related series, one can use this information and incorporate it into the forecasts.

Transfer Function Models

The simplest situation occurs when past values of a series X are useful in predicting Y, but past values of Y are irrelevant to the prediction of X. In such situations one can use transfer function models of the form

$$Y_t = \frac{\omega_0 - \omega_1 B - \cdots - \omega_s B^s}{1 - \delta_1 B - \cdots - \delta_r B^r} X_{t-b} + \frac{1 - \theta_1 B - \cdots - \theta_q B^q}{(1 - \phi_1 B - \cdots - \phi_p B^p)(1 - B)^d} \epsilon_t. \quad (14)$$

These models are extensions of regression models; they represent dynamic relationships and allow for correlation among the errors. Box and Jenkins [2] discuss a model-building procedure that specifies the error model and the lag structure among the two series from past data.

Minimum mean square error forecasts of the future Y_{n+l} are given by the conditional expectation

$$Y_n(l) = E[Y_{n+l} \mid Y_n, Y_{n-1}, \ldots ; X_n, X_{n-1}, \ldots]. \quad (15)$$

These forecasts can be calculated from the difference equation of the model in (14). These models are especially useful if there is a delay in the system ($b > 0$), since then the X variable is a leading indicator. Here we have considered the case of a single input series; however, it is not difficult to extend this analysis to more than one input series.

Multiple Time-Series Models

If feedback is present (which means that past values of Y are useful in predicting X, and past values of X are useful in predicting Y), then one can use a multiple (vector) time-series model of the form

$$\mathbf{Z}_t = \mathbf{\Phi}_1 \mathbf{Z}_{t-1} + \cdots + \mathbf{\Phi}_p \mathbf{Z}_{t-p} + \boldsymbol{\epsilon}_t - \boldsymbol{\theta}_1 \boldsymbol{\epsilon}_{t-1} - \cdots - \boldsymbol{\theta}_q \boldsymbol{\epsilon}_{t-q}. \quad (16)$$

Here $\mathbf{Z}_t = (Y_t, X_t)'$ is a vector of two time series, and $\boldsymbol{\epsilon}_t = (\epsilon_{t1}, \epsilon_{t2})'$ consists of two white noise* sequences that are only contemporaneously correlated; $\mathbf{\Phi}_i$ ($i = 1, \ldots, p$) and $\boldsymbol{\theta}_j$ ($j = 1, \ldots, q$) are (2×2) coefficient matrices. If these matrices are lower triangular, then the model simplifies to the transfer function model given above.

The difficulty with multiple time-series* models is not the prediction of future values. In fact, the minimum mean square error forecast of $\mathbf{Z}_{n+l}$,

$$\mathbf{Z}_n(l) = E[\mathbf{Z}_{n+l} \mid \mathbf{Z}_n, \mathbf{Z}_{n-1}, \ldots], \quad (17)$$

is easily obtained from the difference equation. The difficulty is in the selection of a particular model from this general class, based on past data alone. Despite the fact that a number of specification procedures have been recently developed (see, e.g., Tiao and Box [26]), empirical model building is still a difficult task. Especially if there are several series involved, the parameters in and the difficulties with these models increase rapidly. In such cases it appears likely that a pure time-series approach (i.e., an approach that uses only past data to infer the form of the relationships among the series) may face difficulties. In many situations, however, theory may lead the model builder to certain structural relationships. These relationships should impose restrictions on the class of models that need to be considered and should simplify the model specification.

STATE-SPACE MODELS, KALMAN FILTERING*, AND BAYESIAN* FORECASTING

A state-space model consists of two equations. An observation (measurement) equation which relates a vector of dependent variables $\mathbf{Y}_t$ to a state vector $\mathbf{S}_t$,

$$\mathbf{Y}_t = \mathbf{H}_t\mathbf{S}_t + \mathbf{v}_t, \tag{18}$$

and a system equation,

$$\mathbf{S}_t = \mathbf{G}\mathbf{S}_{t-1} + \mathbf{w}_t, \tag{19}$$

which describes the evolution of the state vector. The unobserved state vector $\mathbf{S}_t$ characterizes the position of the process at time t. The disturbances $\mathbf{v}_t$ and $\mathbf{w}_t$ are two independent uncorrelated sequences with mean vectors zero and covariance matrices $\boldsymbol{\Sigma}_v$ and $\boldsymbol{\Sigma}_w$. Kalman [14] and Kalman and Bucy [15] have shown that, provided the transition matrix $\mathbf{G}$, the covariance matrices $\boldsymbol{\Sigma}_v$ and $\boldsymbol{\Sigma}_w$, and the matrix $\mathbf{H}_t$ are known, the estimate $\hat{\mathbf{S}}_t$ of the current state vector $\mathbf{S}_t$ can be calculated recursively from the previous estimate at time $t-1$, $\hat{\mathbf{S}}_{t-1}$, and the most recent observation $\mathbf{Y}_t$. These updating equations are known as the Kalman filter. A detailed intermediate-level discussion is given by Jazwinski [10].

A forecast for the future state vector $\mathbf{S}_{n+l}$ is needed to predict the future $\mathbf{Y}_{n+l} = \mathbf{H}_{n+l}\mathbf{S}_{n+l} + \mathbf{v}_{n+l}$, for given $\mathbf{H}_{n+l}$. From the Markovian model in (19) this forecast is given by $\mathbf{G}^l\hat{\mathbf{S}}_n$, and the forecast of $\mathbf{Y}_{n+l}$ is $\hat{\mathbf{Y}}_n(l) = \mathbf{H}_{n+l}\mathbf{G}^l\hat{\mathbf{S}}_n$.

The generality of the Kalman filter and the ease of its implementation on computers have been main factors in its popularity in control and forecasting applications.

State-space models are very general and include, among others, ARIMA time-series models and time-varying regression models as special cases. Let us consider the time-varying regression model, which Harrison and Stevens [7] have called the dynamic linear regression model. There the time-varying coefficients $\boldsymbol{\beta}_t$ take the place of the state vector $\mathbf{S}_t$, and the single dependent variable Y_t that is to be predicted is modeled as

$$\begin{aligned} Y_t &= \mathbf{X}_t'\boldsymbol{\beta}_t + v_t, \\ \boldsymbol{\beta}_t &= \mathbf{G}\boldsymbol{\beta}_{t-1} + \mathbf{w}_t. \end{aligned} \tag{20}$$

For known transition matrix $\mathbf{G}$, variance σ_v^2, and covariance matrix $\boldsymbol{\Sigma}_w$, one can use the Kalman filter equations to obtain the coefficient estimates at time t, $\hat{\boldsymbol{\beta}}_t$. For given independent variables $\mathbf{X}_{n+l}$, we can predict the future Y_{n+l} from $\hat{Y}_n(l) = \mathbf{X}_{n+l}'\mathbf{G}^l\hat{\boldsymbol{\beta}}_n$.

Within the extrapolation context one can use these methods to describe observations that follow polynomial trend models whose coefficients change with time. For example, a time-varying linear trend model can be written as

$$Z_t = \begin{bmatrix} 1 & 0 \end{bmatrix} \begin{bmatrix} \beta_0(t) \\ \beta_1(t) \end{bmatrix} + v_t, \tag{21}$$

where

$$\begin{bmatrix} \beta_0(t) \\ \beta_1(t) \end{bmatrix} = \begin{bmatrix} 1 & 1 \\ 0 & 1 \end{bmatrix} \begin{bmatrix} \beta_0(t-1) \\ \beta_1(t-1) \end{bmatrix} + \begin{bmatrix} w_{t1} \\ w_{t2} \end{bmatrix},$$

and where w_{t1} and w_{t2} are two independent sequences of random shocks that affect the level and the slope of the linear trend line. To carry out the Kalman filter equations one needs to know their variances, relative to σ_v^2. These variances enter the recursive Kalman filter equations that update the level and slope estimates with each new observation Z_t. Also the ARIMA(0, 2, 2) model leads to a linear trend forecast function in which the coefficients adapt with each new observation; there the moving-average coefficients determine the weights in these updating recursions. In fact, the Kalman filter recursions for the model in (21) and for the ARIMA(0, 2, 2) model are the same, provided that the variances are certain functions of the moving-average coefficients.

Harrison and Stevens consider multistate models. There the process in (21) may at any time be in any one of several different submodels. The submodels are characterized by different assumptions about the variances of v_t, w_{t1}, and w_{t2}, and reflect stable situations, changes in level, changes in slope, and transient changes. For given variances and prior probabilities for the various states, Harrison and Stevens derive the posterior probabilities for each submodel at each point in time. These probabilities determine the weights that are given to the forecasts from the various submodels. Harrison and Stevens call this the *Bayesian forecast approach*.

ECONOMETRIC* MODELS

Instead of restricting the analysis to the past history of a particular series or to some closely related series, the econometrician constructs a set of relationships based on economic theory. Econometric models use a system of algebraic equations to represent the basic quantitative relationships among, and the behavior over time of, major economic variables. Some of the equations describe the behavior of certain economic agents, and are called *behavioral equations*. Others are definitional equations representing relationships that follow from the definition of the variables. There are two different types of variables: endogenous and exogenous variables. An exogenous variable is a datum that is predetermined in the sense that its value must be specified before the model is solved. The variables that are explained by the model are called endogenous. Econometric models are usually large-scale, simultaneous, dynamic systems that can involve hundreds of relationships. For example, the Wharton model of the U.S. economy combines over 1000 equations.

Much effort has been devoted to the statistically efficient estimation of parameters in simultaneous equation systems. This has led to various estimation methods, such as two-stage least squares*, three-stage least squares, and full-information maximum likelihood (see Theil [25], Johnson [11], and Judge et al. [12]). In practice, however, large-scale econometric models are commonly estimated by single-equation ordinary least squares, despite the fact that such estimators are usually inconsistent.

The computation of the predictions from structural models amounts to the evaluation of the conditional expectations of future endogenous variables in the equation system. This requires as inputs the historical values of endogenous and exogenous variables, as well as projected values of exogenous variables.

A good description of the practice of econometric forecasting is given by Klein and Young [16]. Their discussion shows that the finally published forecasts from econometric models are the results of numerous judgmental adjustments of the initial computer solutions. Constant terms in behavioral equations, especially, are subject to frequent adjustments.

INPUT–OUTPUT TABLES

Input–output tables describe the relationships and interdependencies among various sectors of the economy (see Leontief [18]). In its basic form the underlying theory is as follows. Assume that the economy is divided into k sectors. The output of sector i, Y_i, may be either bought by firms in any of the k sectors (interindustry demand; Y_{ij} is the output of sector i sold to sector j), or by economic agents outside the business system, such as consumers, government, or foreign firms (final demand f_i). The output of sector i is the sum of interindustry demand

and final demand

$$Y_i = \sum_{j=1}^{k} Y_{ij} + f_i$$
$$= \sum_{j=1}^{k} a_{ij} Y_j + f_i, \qquad (22)$$

where $a_{ij} = Y_{ij}/Y_j$ is the (i, j) input coefficient which measures the flow from sector i to sector j as a fraction of the total output of the buying sector j. Writing the equation above for $i = 1, \ldots, k$ leads to the system

$$\mathbf{Y} = \mathbf{AY} + \mathbf{f}$$

and its solution

$$\mathbf{Y} = (\mathbf{I} - \mathbf{A})^{-1}\mathbf{f}, \qquad (23)$$

where $\mathbf{Y} = (Y_1, \ldots, Y_k)'$ and $\mathbf{f} = (f_1, \ldots, f_k)'$ are vectors of total output and final demand, and $\mathbf{A}$ is the matrix of input coefficients. Equation (23) can be used to make a conditional prediction of the output $\mathbf{Y}$ that is necessary to sustain a given final demand $\mathbf{f}$.

Input–output models reflect the relatedness of the various sectors. Limitations, however, stem from their extensive data requirements and the fact that current information is usually difficult to obtain. Furthermore, the sector interactions are usually assumed constant, which is a questionable assumption given today's rapidly changing environment.

TURNING POINTS AND BUSINESS CYCLE INDICATORS

Turning points are points in time where a series that had been increasing (decreasing) reverses and decreases (increases) for some time. One way of predicting turning points in the economy is to identify leading indicators whose behavior anticipates the movements in the series that has to be predicted. The National Bureau of Economic Research has identified various business indicators and has studied their relationships (leading, lagging, coincident) to the general business cycle. The business indicators, and various composite indices, are selected from a large collection of economic series for the relative consistency of their timing at cyclical recessions and revivals. They are intended to reduce the lag in the recognition of cyclical turning points.

SURVEYS OF ANTICIPATIONS AND INTENTIONS

There is yet another approach to forecasting —through the sample survey method. Individuals or firms may be approached directly and asked about their future plans, hopes, and aspirations. These surveys are frequently designed to ascertain voting intentions, household buying plans, and hiring, inventory, capital formation, and production plans of businesses. An example of such a survey is the economic outlook survey conducted by the National Bureau of Economic Research and the American Statistical Association*.

COMBINATION OF FORECASTS

When developing a forecast model, most analysts examine several different forecasting techniques, finally choosing the one that generates the most satisfactory measure of forecast accuracy. Many papers conclude that the forecasting accuracy of the best single forecasting model can usually be improved by combining the forecasts that are generated by competing forecast models. This is based on the premise that a discarded forecast contains some useful information that is independent of that supplied by the chosen model. Various approaches for combining the forecasts, some simple and some complicated, are proposed in the literature. For further information, see Winkler and Makridakis [31].

PREDICTION OF QUALITATIVE CHARACTERISTICS

The prediction of quantitative variables can be expressed in a variety of functional forms and squared or absolute differences of observed and predicted observations are appropriate. In the social sciences one is often interested in predicting qualitative variables. For example, in social mobility studies one predicts children's career choices as a function of their parents' occupations. In such a context a squared error framework is not feasible and one needs to adopt a different approach. The book by Hildebrand et al. [8] shows how propositions that predict states of a qualitative variable can be evaluated.

FORECAST QUALITY AND THE EVALUATION OF FORECASTS

Forecasting is certainly a risky undertaking, since eventually any prediction $\hat{Y}_t$ can be compared with the actual realization Y_t and the associated forecast error $e_t = Y_t - \hat{Y}_t$ can be calculated. Forecast accuracy must be concerned with postsample periods. If sufficient observations are available, one should divide the series into two parts, derive estimates of the parameters that are necessary to construct the forecasts from the first part, and evaluate the accuracy of the *ex ante* forecasts from the observations in the holdout period. This is important, since models that fit the data well (i.e., have small *ex post* forecast errors) sometimes perform badly in forecasting, and vice versa.

Various summary measures can be calculated from forecast errors, such as mean error, mean percent error, mean square error, mean absolute error, and mean absolute percent error. These measures are symmetric and treat overpredictions and underpredictions alike. If one has information on the cost associated with the forecast errors, one can construct more general nonsymmetric measures. In most cases, however, such information is not available.

There are many empirical studies that compare: (a) forecasts from econometric models with extrapolation forecasts, in particular those from ARIMA models; (b) forecasts from quantitative methods with human judgment forecasts; and (c) forecasts from various extrapolation methods.

Several studies conclude that in terms of forecast accuracy univariate ARIMA models provide results that are comparable to, if not better than, econometric models [20, 21]. However, the scope of econometric modeling usually goes beyond deriving unconditional forecasts. Econometric models improve our understanding of economic relationships and can be used to perform sensitivity analyses, evaluate the effects of alternative policy decisions, and derive conditional forecasts.

Several studies show that judgmental forecasts are less accurate than forecasts from quantitative methods. A review of this literature is given by Hogarth and Makridakis [9]. They make a strong case for the superior predictive performance of simple quantitative models over the judgmental ability of experts.

A frequent complaint among users of extrapolation methods has not been that there are not enough methods, but that there are too many, and that it becomes sometimes difficult to choose among the various options. Empirical comparisons of many extrapolation methods (see Newbold and Granger [22] and Makridakis et al [19]) have led to somewhat conflicting results. However, these studies agree that no single method is best in all cases and that sophisticated methods are not always more accurate than simple ones.

In general, researchers have found that the prediction of economic and business data was much easier in the 1960s, since the relationships were fairly stable. In the 1970s and 1980s, however, accurate prediction became more difficult. Many changes contributed to make the models increasingly obsolete and nonrepresentative under future conditions.

References

[1] Abraham, B. and Ledolter, J. (1983). *Statistical Methods for Forecasting*. Wiley, New York.

[2] Box, G. E. P. and Jenkins, G. M. (1970). *Time Series Analysis, Forecasting and Control*. Holden-Day, San Francisco (2nd ed., 1976).

[3] Brown, R. G. (1962). *Smoothing, Forecasting and Prediction*. Prentice-Hall, Englewood Cliffs, N.J.

[4] Draper, N. R. and Smith, H. (1966). *Applied Regression Analysis*. Wiley, New York (2nd ed., 1981).

[5] Granger, C. W. J. and Andersen, A. P. (1976). *An Introduction to Bilinear Time Series Models*. Vandenhoeck and Ruprecht, Göttingen, Federal Republic of Germany.

[6] Haggan, V. and Ozaki, T. (1981). *Biometrika*, **68**, 189–196.

[7] Harrison, P. J. and Stevens, C. F. (1976). *J. R. Statist. Soc. B*, **38**, 205–247.

[8] Hildebrand, D. K., Laing, J. D., and Rosenthal, H. (1977). *Prediction Analysis of Cross Classifications*. Wiley, New York.

[9] Hogarth, R. M. and Makridakis, S. (1981). *Management Sci.*, **27**, 115–138.

[10] Jazwinski, A. H. (1970). *Stochastic Processes and Filtering Theory*. Academic Press, New York.

[11] Johnston, J. (1972). *Econometric Methods*, 2nd ed. McGraw-Hill, New York.

[12] Judge, G. G., Hill, R. C., Griffiths, W., Lütkepohl, H. and Lee, T.-C. (1982). *An Introduction to the Theory and Practice of Econometrics*. Wiley, New York.

[13] Kailath, T. (1974). *IEEE Trans. Inf. Theory*, **IT-20**, 146–181.

[14] Kalman, R. E. (1960). *Trans. ASME, Ser. D, J. Basic Eng.*, **82**, 35–45.

[15] Kalman, R. E. and Bucy, R. S. (1961). *Trans. ASME, Ser. D, J. Basic Eng.*, **83**, 95–107.

[16] Klein, L. R. and Young, R. M. (1980). *An Introduction to Econometric Forecasting and Forecasting Models*. D. C. Heath, Lexington, Mass.

[17] Kolmogorov, A. (1941). *Bull. Acad. Sci. (Nauk) URSS, Ser. Math.*, **5**, 3–14.

[18] Leontief, W. (1966). *Input–Output Economics*. Oxford University Press, New York.

[19] Makridakis, S., Andersen, A., Carbone, R., Fildes, R., Hibon, M., Lewandowski, R., Newton, J., Parzen, E., and Winkler, R. (1982). *J. Forecasting*, **1**, 111–153.

[20] Naylor, T. H. and Seaks, T. G. (1972). *Int. Statist. Rev.*, **40**, 113–137.

[21] Nelson, C. R. (1972). *Amer. Econ. Rev.*, **62**, 902–917.

[22] Newbold, P. and Granger, C. W. J. (1974). *J. R. Statist. Soc. A*, **137**, 131–165.

[23] Plosser, C. I. (1979). *J. Amer. Statist. Ass.*, **74**, 15–24.

[24] Priestley, M. B. (1980). *J. Time Series Anal.*, **1**, 47–71.

[25] Theil, H. (1971). *Principles of Econometrics*. Wiley, New York.

[26] Tiao, G. C. and Box, G. E. P. (1981). *J. Amer. Statist. Ass.*, **76**, 802–816.

[27] Tong, H. (1983). Threshold Models in Non-linear Time Series Analysis. *Lect. Notes Statist.*, **21**. Springer-Verlag, New York.

[28] Trigg, D. W. and Leach, A. G. (1967). *Operat. Res. Quart.*, **18**, 53–59.

[29] Whittle, P. (1963). *Prediction and Regulation by Linear Least-Squares Methods*. University of Minnesota Press, Minneapolis (2nd ed., 1983).

[30] Wiener, N. (1949). *Extrapolation, Interpolation and Smoothing of Stationary Time Series*. Wiley, New York.

[31] Winkler, R. G. and Makridakis, S. (1983). *J. R. Statist. Soc. A*, **146**, 150–157.

[32] Wold, H. (1938). *The Analysis of Stationary Time Series*. Almqvist and Wiksell, Uppsala, Sweden (2nd. ed., 1954).

(AUTOREGRESSIVE-INTEGRATED
MOVING AVERAGE (ARIMA) MODELS
ECONOMETRICS
FORECASTING
GRADUATION
KALMAN FILTERING
PROBABILITY FORECASTING
REGRESSION (VARIOUS ENTRIES)
SEASONALITY
TIME SERIES)

JOHANNES LEDOLTER

PREDICTION SPECIFICATION *See* SPECIFICATION, PREDICTOR

PREDICTION, STRUCTURAL *See* STRUCTURAL PREDICTION

PREDICTIVE ANALYSIS

Statistical prediction is the process by which values for unknown observables (potential observations yet to be made or past ones which are no longer available) are inferred based on current observations and other information at hand. An analysis of this type appears first in Bayes'* 1763 posthumous essay, wherein the following problem was addressed and solved. A ball was rolled on a unit square table and the horizontal coordinate of the final resting place was then assumed to be uniformly distributed in the unit interval. A second, or the same ball, is then rolled N times and one is informed as to the number of times the second ball came to rest to the left of the first ball without the actual horizontal coordinate of the first being disclosed. The problem is to infer a plausible range of values for the horizontal coordinate of the first ball.

In the appendix by Price, who communicated the essay, a calculation is made of the chance that the $(N+1)$st toss of the second ball will exceed the first, based on the previous knowledge. Laplace* and Condorcet in the last quarter of the eighteenth century began to make more general calculations of this kind (e.g., the chance that r out of the next M observations will be successes). Such efforts continued into the first quarter of the twentieth century, so that Karl Pearson* [29] could state in 1907 that the fundamental problem of statistics was prediction.

Jeffreys [21] and Fisher [6] calculate predictive distributions but by different modes of inference. Although de Finetti [5] did not directly contribute to the methodology of statistical prediction, he clearly provided the major philosophical underpinning for the observabilistic or predictivistic view. In fact, next to Bayes' theorem, de Finetti's exchangeability* theorem is second to none in its importance for the predictive view. It is also useful in demonstrating that a good deal of parametric inference can be viewed as a special or limiting case of observabilistic inference. The emphasis on observables also has the effect of shifting the focus in statistics from testing and estimation to model selection and prediction.

Predictive analyses are currently applied to a variety of problems and are executed in several different modes, which we now elucidate.

FREQUENCY APPROACH

Let the set of random variables $(X_1, \ldots, X_N; X_{N+1}, \ldots, X_{N+M})$ or in a more compact notation $(X^{(N)}; X_{(M)})$ reflect a partition of past (or to be observed) and future (or to be predicted) variables. Then the classical frequency approach to prediction takes the form of a tolerance region*. Here we assume that $(X^{(N)}; X_{(M)})$ has sampling distribution $F(x^{(N)}; x_{(M)} \mid \alpha)$ with sufficient structure such that $\Pr[X_{(M)} \in A(X^{(N)})] = p$ independent of α, which represents the chance that the random set $X_{(M)}$ is included in the random region $A(X^{(N)})$. Hence p is the long-run relative frequency of the event for random $(X^{(N)}; X_{(M)})$ and is interpreted as a measure of the confidence induced in the statement that $X_{(M)}$ will be included in the observed tolerance region $A(x_{(M)})$. As an example we illustrate these ideas for the case of $(X^{(N)}; X_{(M)})$ being a set of independently and identically distributed random variables with density function

$$f\left(x^{(N)}; x_{(M)}\right) = \prod_{i=1}^{N+M} \alpha e^{-\alpha x_i}.$$

The sampling distribution of $2\alpha S$, where $S = \sum_{i=1}^{N} X_i$, is χ^2_{2N}, while $2\alpha X_{N+i}$ is χ^2_2 and all $M+1$ variables are mutually independent. Transforming to $Z_i = X_{N+i}/S$ $i = 1, \ldots, M$ yields the joint density

$$f(z_1, \ldots, z_M) = \frac{\Gamma(M+N)}{\Gamma(N)} \times \left(1 + \sum_{i=1}^{M} z_i\right)^{-(M+N)}.$$

If $M = 1$, then a tolerance interval for the next observation can be obtained through the relationship

$$\Pr\left[X_{N+1} \leqslant N^{-1} S F_p(2, 2N)\right] = p,$$

where $\Pr[F \leqslant F_p] = p$ and F represents an F variate.

Often it is of interest to predict the number R of M future X's that lie in some interval (e.g., if X represents survival time we might be interested in the fraction that survive until a given time t). For the more general problem the calculation is more difficult.

Letting $I = [u, \infty)$, $Z = (Z^{(r)}; Z_{(M-r)})$, $z = (z^{(r)}; z_{(M-r)})$,

$$\int_{I(z^{(r)})} = \int_I, \qquad \int_{I^c(z_{(M-r)})} = \int_{I^c},$$

the r and $M-r$ fold integrals, where

$z_1, \ldots, z_r$ and $z_{r+1}, \ldots, z_M$ are integrated over I and I^c, respectively, then

$$P_r = \Pr[R = r \text{ of } M\ Z\text{'s} \in I]$$
$$= \binom{M}{r} \int_I \int_{I^c} f(z^{(r)}; z_{(M-r)})\, dz^{(r)}\, dz_{(M-r)}.$$

The right-hand side is a function of r, M, N, and u. For a tolerance interval on R or, say, $R \geqslant r_0$, one computes $P_{r_0} + \cdots + P_M = p$ at the value $u = t/s$, where s is the observed value of $S = X_1 + \cdots + X_N$ and

$$P_r = \binom{M}{r} \sum_{j=0}^{M-r} \binom{M-r}{j} (-1)^j \times [1 + u(n+j)]^{-n}.$$

One can fix r_0 and calculate p or fix p and calculate r_0 to the nearest integer.

In the frequency mode a highly distribution-robust procedure is also available. We need only assume that the underlying distribution of a set of exchangeable variables $X_1, \ldots, X_N$ is absolutely continuous (note that independence is not even necessary).

Let the ordered values of $X_1, \ldots, X_N$ be $X_1' \leqslant X_2' \leqslant \cdots \leqslant X_N'$; then for the interval

$$I_{jk} = (X_j', X_{j+k}'),$$

defining $I_{j,N+1-j} = (X_j', \infty)$ and $I_{0,k} = (-\infty, X_k')$, it can be shown by combinatorial methods [30] that

$$\Pr[\text{exactly } R = r \text{ out of } M\ X_{N+i}\text{'s lie in } I_{jk}]$$
$$= \frac{\binom{k+r-1}{r}\binom{N+M-k-r}{M-r}}{\binom{N+M}{M}}$$
$$= P_{r,k}$$

and

$$\Pr[\text{exactly } R = r \text{ out of } M\ X_{N+i}\text{'s exceed } X_j']$$
$$= \frac{\binom{N+r-j}{r}\binom{M-r+j-1}{M-r}}{\binom{N+M}{M}}$$
$$= P_{r,N+1-j}$$

with distribution function

$$\Pr\left[\frac{R}{M} \leqslant \frac{r}{M}\right] = \sum_{x=0}^{r} P_{x,N+1-j}.$$

In this case, only probabilities for fractions exceeding order statistics* can be exactly computed. For the special case of a single future observation X_{N+1}, the result is simply

$$\Pr[X_j' < X_{N+1} < X_{j+k}'] = \frac{k}{N+1}. \quad (1)$$

PREDICTIVE SAMPLE REUSE APPROACH

A predictive mode which makes no distributional assumptions, termed *predictive sample reuse** PSR [13, 14], requires the following ingredients:

1. An arbitrarily chosen predictive function of a future observable
$$x = x(x^{(N)}, \alpha) \qquad \alpha \in \mathscr{A},$$
where α is a set of values to be determined.
2. A schema $\mathscr{S} = \mathscr{S}(N, n, \Gamma)$ of partitions where
$$P_i = (x_{ir}^{(N-n)}, x_{io}^{(n)})$$
is the ith partition of $x^{(N)}$ into $x_{ir}^{(N-n)}$ the set of $N-n$ retained values and $x_{io}^{(n)}$ the set of n omitted values of $x^{(N)}$. The defined set of such partitions for a given n is Γ, say, and the number of such partitions $P_i \in \Gamma$ is P. The predictive function is applied to the retained observation set $x_{ir}^{(N-n)}$ and used to predict $x_{io}^{(n)}$, the deleted set for each P_i yielding $\hat{x}_{io}^{(n)}(\alpha)$, which is a function of α.
3. A discrepancy measure
$$D_n(\alpha) \propto \sum_{P_i \in \Gamma} d(x_{io}^{(n)}, \hat{x}_{io}^{(n)}(\alpha)),$$
where $d(a, b)$ is a defined measure of the discrepancy between two n-dimensional vectors.
4. To obtain the final predictor $\hat{x} = x(x^{(N)}, \hat{\alpha})$ of the future value, $D_n(\alpha)$ is minimized with respect to α, which yields $\hat{\alpha}$.

As a simple example consider as a predictive function a linear combination $x = \alpha h + (1-\alpha)m$, $\alpha \in [0,1]$, of the median m and the average of symmetric order statistics

$$h = \tfrac{1}{2}\left(x'_{[pN+1]} + x'_{N-[pN]}\right),$$

where $[A]$ represents the largest integer in A and $0 < p < 0.5$, for N observations. Use of a squared error discrepancy based on a one-at-a-time omission schema requires minimization of

$$D_1(\alpha) \propto \sum_{j=1}^{N} (\alpha h_j + (1-\alpha)m_j - x_j)^2,$$

where h_j and m_j are h and m, respectively, with x_j deleted. The solution yields for the predictor

$$x = \begin{cases} h & \text{if } \hat{\alpha} \geqslant 1 \\ \hat{\alpha}h + (1-\hat{\alpha})m & \text{if } 0 < \hat{\alpha} < 1 \\ m & \text{if } \hat{\alpha} \leqslant 0, \end{cases}$$

where

$$\hat{\alpha} = \sum_{j=1}^{N} (h_j - m_j)(x_j - m_j) \Big/ \sum_{j=1}^{N} (h_j - m_j)^2.$$

Sample reuse intervals may also be obtained using similar ingredients [3]. They are:

1. A predictive interval function

$$\text{P.I.}(x^{(N)}; \alpha).$$

2. A criterion function (assuming a simple one-at-a-time omission schema)

$$D_1(\alpha) \propto \sum_{j=1}^{N} \mathscr{L}\left\{\text{P.I.}\left(x_j^{(N-1)}; \alpha\right)\right\},$$

where $\mathscr{L}\{\cdot\}$ is defined as the length of the jth interval based on all the observations but the jth, denoted by $x_j^{(N-1)}$.

3. A relative frequency of coverage of $1-\beta$ in a predictive simulation is then obtained by minimizing $D_1(\alpha)$ with respect to α subject to

$$\frac{1}{N} \sum_{j=1}^{N} I\left[x_j \notin \text{P.I.}\left(x_j^{(N-1)}; \alpha\right)\right] \leqslant \beta,$$

where I is the indicator of the event in brackets. The resulting solution for $\hat{\alpha}$ is then substituted in the predictive interval function to obtain P.I.$(x^{(N)}; \hat{\alpha})$.

As a very simple illustration, consider the predictive interval function which uses the symmetric order statistics

$$\text{P.I.}(x^{(N)}; \hat{\alpha}) = (x'_{\alpha}, x'_{N+\alpha+1}).$$

Minimizing the criterion function subject to the constraint and setting $\beta = 2p$, we obtain as solution

$$\text{P.I.}(x^{(N)}; \hat{\alpha}) = \left(x'_{[Np]}, x'_{N+1-[Np]}\right),$$

with coverage $1-2p$. For $p = 1/N$ the simulated coverage that the $(N+1)$st observation lies within the range of the previous N observations is $(N-2)/N$. If we compare this with the result (1) of the more structured situation for which the tolerance coefficient is $(N-1)/(N+1)$, it is clear that it is as if the loosening of the structure manifests itself in the loss of a single observation.

BAYESIAN MODE

Assume that the joint probability function of $(X^{(N)}, X_{(M)})$ is

$$f\left(x^{(N)}; x_{(M)} \mid \alpha\right) = f\left(x_{(M)} \mid x^{(N)}, \alpha\right) f\left(x^{(N)} \mid \alpha\right),$$

indexed by a set of parameters α. A prior density for α, $p(\alpha \mid \beta)$, indexed by β is also part of the structure assumed. For known β the posterior probability* function of α, for observed $X^{(N)} = x^{(N)}$, is

$$p(\alpha \mid x^{(N)}, \beta) = \frac{f(x^{(N)} \mid \alpha) p(\alpha \mid \beta)}{f(x^{(N)} \mid \beta)},$$

where

$$f(x^{(N)} \mid \beta) = \int f(x^{(N)} \mid \alpha) p(\alpha \mid \beta)\, d\alpha.$$

The predictive probability function of $X_{(M)}$ is then obtained as

$$f\left(x_{(M)} \mid x^{(N)}, \beta\right) = \int f\left(x_{(M)} \mid x^{(N)}, \alpha\right) p(\alpha \mid x^{(N)}, \beta)\, d\alpha.$$

Hence any probability statements about the future values $X_{(M)}$ or functions thereof depend on the given probability function and any utilities, losses, costs, and so on, that are brought to bear on a specific prediction problem. To illustrate this, consider again the case where $(X^{(N)}; X_{(M)})$ is a set of independently and identically distributed exponential variables with density function $f(x \mid \alpha) = \alpha e^{-\alpha x}$. Further assume a prior gamma density for α,

$$p(\alpha \mid \delta, \gamma) \propto \alpha^{\delta - 1} e^{-\gamma\alpha}.$$

Now, if among the observed $x^{(N)} = (x^{(d)}, x^{(N-d)})$, the second set of $N - d$ observations have been censored, it is not difficult to calculate the predictive density of $X_{N+1}, \dots, X_{N+M}$ given $X^{(N)} = (x^{(d)}, x^{(N-d)})$:

$$f(x_{N+1}, \dots, x_{N+M} \mid x^{(N)}, \gamma, \delta) = \frac{\Gamma(x + M + \delta)}{\Gamma(d + \delta)} \times \frac{(s + \gamma)^{d+\delta}}{[s + \gamma + x_{N+1} + \cdots + x_{N+M}]^{d+M+\delta}},$$

where $s = (x_1 + \cdots + x_N)$. Of course, if interest is to be focused on only the next value X_{N+1} we merely set $M = 1$ above to obtain its probability function. We note also that this predictive density is exchangeable; that is, in our assessment the set of future values are exchangeable. In many instances it is of interest to calculate the number R of M future X's that lie in some set [e.g., (t, ∞)] [17]. If X represents survival time, then we might be interested in the fraction that survive until time t. Such a calculation is also easily made since the survival function is

$$\Pr[X \geqslant t \mid \alpha] = e^{-\alpha t} = \theta,$$

and then

$$\Pr[R = r \mid M] = \int \binom{M}{r} \theta^r (1 - \theta)^{M-r} p(\alpha \mid x^{(N)})\, d\alpha = \binom{M}{r} (s + \gamma)^{d+\delta} \sum_{j=0}^{M-r} \binom{M-r}{j} (-1)^j \times [s + \gamma + t(r + j)]^{-(d+\delta)}.$$

It can also be shown that as M grows, $RM^{-1} \to \theta$, where θ is a random variable whose distribution can be obtained from the distribution of α; that is, $-t^{-1}\log\theta = \alpha$ has the posterior distribution* for α whose density is

$$p(\alpha \mid x^{(N)}) \propto \alpha^{d+\delta-1} e^{-\alpha(s+\gamma)}.$$

In cases where little is known a priori about α, it is often suggested that $\gamma = \delta = 0$, which results in the improper prior density that yields a uniform density for $\log\alpha$. The fiducial* approach of Fisher [6] and the structural* approach of Fraser [7] will yield results which are equivalent to making such an assumption in the Bayesian approach.

OTHER MODALITIES

There are several modes that have been suggested for scaling future values which depend on relative likelihood, maximum likelihood* and sufficiency* concepts. Fisher [6] introduced a procedure for scaling the plausibility of future values $X_{(M)}$ after observing $X^{(N)} = x^{(N)}$. His method assumes that $(X^{(N)}, X_{(M)})$ is a set of independently and identically distributed random variables given X. Basically, one computes the maximized relative likelihood function separately for $X^{(N)}$ and $X^{(M)}$ and then jointly so that

$$\sup_\alpha L(\alpha \mid x^{(N)}) / L(\alpha \mid x^{(N)}) = R_N$$

$$\sup_\alpha L(\alpha \mid x_{(M)}) / L(\alpha \mid x_{(M)}) = R_M$$

$$\sup_\alpha L(\alpha \mid x^{(N)}, x_{(M)}) / L(\alpha \mid x^{(N)}, x_{(M)}) = R_{N+M}.$$

Then

$$\frac{R_{N+M}}{R_N R_M} = \frac{L(\hat{\alpha}_{N+M} \mid x^{(N)}, x_{(M)})}{L(\hat{\alpha}_N \mid x^{(N)}) L(\hat{\alpha}_M \mid x_{(M)})} = RL(x_{(M)} \mid x^{(N)})$$

is used to scale the plausibility of values of $x_{(M)}$ for observed $x^{(N)}$. In other words, this ratio scales the values of $x_{(M)}$ according to how likely (in some sense) they make the hypothesis that all the observations are from the same population with parameter α, which is assumed to be true. The usual likelihood function, in contradistinction, scales values of α according to how probable they make a given set of observables. For elaborations of this method, see Kalbfleisch [24]. A proposal by Lejeune and Faulkenberry [28] which can be used to scale future values is to calculate

$$\text{MLP}(x_{(M)} \mid x^{(N)}) = \sup_{\alpha} \prod_{i=1}^{N+M} f(x_i \mid \alpha)$$
$$= f(x^{(N)} \mid \hat{\alpha})\, g(x_{(M)} \mid \hat{\alpha})$$

and scale $x_{(M)}$ according to the function above. They further suggest that a maximum likelihood predicting density be computed as

$$\hat{f}(x_{(M)} \mid x^{(N)})$$
$$= k(x^{(N)}) f(x^{(N)} \mid \hat{\alpha})\, g(x_{(M)} \mid \hat{\alpha})$$

by introducing $k(x^{(N)})$ as the normalizing constant. This is more appropriate than merely inserting the maximum likelihood estimator $\hat{\alpha}$ in $f(x_{(M)} \mid \hat{\alpha}(x^{(N)}))$ as a predicting density estimator. A similar type of scaling function is obtained when $S_N = S(X^{(N)})$ and $S_{N+M} = S(X^{(N)}, X_{(M)})$ are sufficient for α based on the observed set and the total set, respectively. By the properties of sufficiency one obtains the conditional probability function of S_N given $S_{N+M} = s_{N+M}$,

$$f(s_N \mid s_{N+M}) = \text{prlk}(x_{(M)} \mid x^{(N)}),$$

to be independent of α. The above is then used to scale values of $x_{(M)}$ given $x^{(N)}$ [26]. This could also be normalized with respect to $x^{(N)}$ to provide a predicting probability function.

For the simple exponential example given previously, we obtain the following scaling functions for x_{N+1}:

$$RL(x_{N+1} \mid x^{(N)}) \propto x_{N+1}/(N\bar{x} + x_{N+1})^{N+1},$$

$$\text{MLP}(x_{N+1} \mid x^{(N)}) \propto (N\bar{x} + x_{N+1})^{-(N+1)},$$

and

$$\text{prlk}(x_{N+1} \mid x^{(N)}) \propto (N\bar{x} + x_{N+1})^{-N}.$$

CUSTOMARY APPLICATIONS

In sample surveys, the problem is to estimate some function of a finite number of observables—part observed and part unobserved. Clearly, then, this is a prediction problem, even if the function is sometimes misdesignated as a parameter. Direct prediction problems, as such, abound in multivariate regression* [9], time series*, growth curves* [11, 27], and a variety of other special topics where the modeling clearly anticipates the need for prediction. A few less direct areas will be discussed in some detail.

PROBABILITY "ESTIMATION" AND PREDICTIVE DISTRIBUTIONS

One immediate application is to the so-called density estimation* problem or the estimation of the distribution function. Clearly in the Bayesian mode, the predictive distribution (density), which is the expectation of the sampling distribution (density) over the posterior distribution of the parameters, is, for squared error loss, the optimal estimator of the sampling distribution (density). Other loss functions will lead to other estimates (see [17]). Hence probability estimation and whatever is derived from it is contained in this approach—so that even such problems as "goodness of fit" can be managed in this way [20, 12].

A short list of some predictive distributions based on conjugate priors is given below. A longer list appears in Aitchison and Dunsmore [1].

Let $X_1, \ldots, X_n$ be i.i.d. with distribution function $f(x \mid \alpha)$. The first is the probability function, the second is the conjugate prior, and the third the predictive probability function for the next observation denoted by the random variable Z, for $D = (x_1, \ldots, x_n)$. The translated exponential appeared in earlier sections.

Bernoulli: $f(x \mid \alpha) = \alpha^x (1-\alpha)^{1-x}$,

$$x = 0, 1, \quad 0 \leqslant \alpha \leqslant 1,$$

$$p(\alpha \mid \beta_1, \beta_2) \propto \alpha^{\beta_1 - 1}(1-\alpha)^{\beta_2 - 1},$$
$$\beta_1 > 0, \quad \beta_2 > 0,$$

$$f(z \mid r, \beta_1, \beta_2)$$
$$= \frac{B(\beta_1 + r + z, \beta_2 + M - r + 1 - z)}{B(\beta_1 + r, \beta_2 + n - r)},$$

$$r = \sum x_i, \quad z = 0, 1.$$

For $\beta_1 = \beta_2 = 1$, the predictive probability function is the one which results from assuming that α is a priori uniform.

$$\text{Poisson:}\quad f(x \mid \alpha) = \frac{\alpha^x e^{-\alpha}}{x!},$$

$$x = 0, 1, \ldots, \qquad \alpha > 0$$

$$p(\alpha \mid \beta_1, \beta_2) \propto \alpha^{\beta_1 - 1} e^{-\alpha\beta_2},$$

$$\beta_1 > 0, \quad \beta_2 > 0$$

$$f(z \mid r, \beta_1, \beta_2) = \binom{z + \beta_1 + r - 1}{\beta_1 + r - 1}\left(\frac{1}{\beta_2 + n + 1}\right)^z \times \left(\frac{\beta_2 + n}{\beta_2 + n + 1}\right)^{\beta_1 + r}$$

for $r = \sum x_i$ and $z = 0, 1, \ldots$.

The predictive probability function above reduces to the one obtainable from a "noninformative" prior on α when $\beta_1 \to 0$ and $\beta_2 \to 0$.

Normal:

$$f(x \mid \mu, \tau) = \sqrt{\frac{\tau}{2\pi}}\, \exp\left[-\tfrac{1}{2}\tau(x - \mu)^2\right],$$

$$\tau > 0$$

$$p(\mu, \tau) = p(\mu \mid \tau)p(\tau)$$

$$p(\mu \mid \tau, \nu, \beta_1) = \sqrt{\frac{\nu\tau}{2\pi}}\, \exp\left[-\tfrac{1}{2}\nu\tau(\mu - \beta_1)^2\right],$$

$$\nu > 0$$

$$p(\tau \mid \gamma, \beta_2) = \frac{\left(\frac{1}{2}\gamma\right)^{\beta_2/2} \tau^{(\beta_2/2) - 1} e^{-(1/2)\gamma\tau}}{\Gamma(\beta_2/2)},$$

$$\gamma > 0, \quad \beta_2 > 0$$

$$f(z \mid \theta_1, \theta_2, \theta_3) = \frac{\left[1 + (\theta_2\theta_3)^{-1}(z - \theta_1)^2\right]^{-(\theta_2 + 1)/2}}{B\left(\frac{1}{2}, \frac{1}{2}\theta_2\right)\sqrt{\theta_2\theta_3}},$$

where

$$\theta_1 = (\nu + n)^{-1}(\nu\beta_1 + n\bar{x}), \quad \theta_2 = n + \beta_2 - 1,$$

$$\theta_3 = \left[\gamma + (n - 1)s^2 + \nu n(\nu + n)^{-1}(\bar{x} - \beta_1)^2\right] \times (\nu + n - 1)/\left[(\nu + n)\theta_2\right].$$

If we let $\gamma \to 0$, $\nu \to 0$, $\beta_1 \to 0$, or $\theta_1 \to \bar{x}$, $\theta_2 \to n - 1$, $\theta_3 \to (1 + 1/n)s^2$, the resulting predictive density is the one associated with a "noninformative" prior on μ and τ.

CLASSIFICATION

Classification* problems are essentially prediction or, in point of time, retrodiction problems. For the sake of simplicity, consider two populations π_1 and π_2 with training samples $D_1 = (x_{11}, \ldots, x_{1n_1})$ and $D_2 = (x_{21}, \ldots, x_{2n_2})$, where x_{ij} are possibly p-dimensional vectors and a new random variable Z, which will be observed by $Z = z$, and has known prior probability p_i of originating from π_i, $i = 1, 2$ and $p_1 + p_2 = 1$. Assume that π_i is specified by density $f_i(\cdot \mid \alpha_i)$ indexed by α_i and $p(\alpha)$ is the assumed prior probability function for the entire set of distinct parameters $\alpha = \alpha_1 \cup \alpha_2$, which allows for the possibility that some parameters are common to both populations. For $D = (D_1, D_2)$, the posterior probability for the origin of z is

$$\Pr[\pi_i \mid z] \propto p_i f(z \mid D, \pi_i),$$

where the predictive density

$$f(z \mid D, \pi_i) = \int f_i(z \mid \alpha_i) p(\alpha \mid D)\, d\theta.$$

Classification of Z may be made to that population which maximizes the posterior probability if there is no differential cost. For multivariate normal applications, see Geisser [8, 10, 18].

In many cases the two (or more) populations are centered at different points and the ability to discriminate between them depends on how far apart they are. An interesting case occurs when discriminating between monozygotic (identical) and dizygotic (nonidentical) twins. The difference between a pair of twins in each variable under consideration and in each type of twin can reasonably be assumed to have expectation zero. Hence the potential for discriminating between the two types is entirely vested in how they differentially vary about the known expectation zero or more generally how the covariance matrices differ. A very simple example of this occurs where only a single normal variate is at issue. There we can assume that the difference between a pair for each type, say π_1 for the monozygotic and π_2 for the dizygotic twins, is such that $\pi_1 \sim N(0, 2\sigma_1^2)$ and $\pi_2 \sim N(0, 2\sigma_2^2)$. Since, in general, there is no identification for a first and second twin, for convenience we transform to $Z = (1/2)(X - Y)^2$, where $(X - Y)^2$ represents the squared difference. This results in $Z \sim \sigma_i^2 \chi^2(1)$, for $i = 1, 2$, where $\chi^2(1)$ is a chi-squared variate with 1 degree of freedom. We shall illustrate this with a set of data, by calculating the predictive density of Z, reflecting a future twin pair whose zygosis is to be determined. For the sake of this example we shall assume a "noninformative" prior on σ_1 and σ_2 to be $p(\sigma_1, \sigma_2) \propto 1/\sigma_1\sigma_2$. For training samples of size n_1 and n_2 we can calculate

$$f(z \mid D_i, \pi_i) = \frac{\Gamma\left(\frac{n_i + 1}{2}\right)(zS_i)^{-1/2}\left(1 + \frac{z}{S_i}\right)^{-(n_i+1)/2}}{\Gamma(1/2)\Gamma(n_i/2)},$$

where $S_i = \sum_{j=1}^{N_i} z_{ij}$ and $z_{ij} = \frac{1}{2}(x_{ij} - y_{ij})^2$. Hence the posterior probability that a new pair belongs to π_i is

$$P_i \propto p_i f(z \mid D_i, \pi_i),$$

where p_i is the prior probability of the pair belonging to π_i.

Random samples of 10 dermal ridge count differences were obtained from each of a set of male monozygotic and male dizygotic twins from the data reported by Lamy et al. [25] (see Table 1).

A new male twin pair is to be classified as to zygosis. It is assumed that the new pair arose at random from the population of male twin pairs whose prior probability is 30/65 of monozygosis, reflecting the frequency of identical twins in the population of male twin pairs. The twin pair difference is 12. Hence we obtain $P_1 = 0.57$, which makes it more likely that the pair is monozygotic.

Table 1 Ridge Count Difference between Male Twin Pairs

Monozygotes:	12, 3, 3, 16, 11, 6, 1, 6, 8, 8
Dizygotes:	74, 43, 12, 19, 1, 8, 6, 35, 1, 35

MODEL SELECTION

Let M_i be a model, $i = 1, \ldots, k$, which specifies the probability function for a set of observations D to be $f(D \mid M_i, \alpha_i)$ indexed by unknown α_i. Then if p_i is the prior probability of model M_i, the posterior probability of M_i is

$$\Pr[M_i \mid D] \propto p_i f(D \mid M_i),$$

where

$$f(D \mid M_i) = \int f(D \mid M_i, \alpha_i) p(\alpha)\, d\theta$$

is the predictive (marginal) density of the observation set D, $p_1 + \cdots + p_k = 1$ and α represents the set of distinct parameters of $(\alpha_1, \ldots, \alpha_n)$. Again, in the absence of other considerations, the maximum $\Pr[M_i \mid D]$ could be used to select the most appropriate model. For variations on this theme, see Geisser and Eddy [19].

PROBLEMS OF COMPARISON

The comparison of certain attributes of groups or populations comprises a major portion of the statistical enterprise. Current practice often dictates that certain location parameters be made the focus of comparison. For example, in a situation which posits two normal populations $X_1 \sim N(\mu_1, \sigma^2)$ and $X_2 \sim N(\mu_2, \sigma^2)$, the focus may be on inferential statements about $\eta = \mu_1 - \mu_2$, based on samples $D_i = (x_{i1}, \ldots, x_{iN_i})$, $i = 1, 2$. In this case, a Bayesian would base his or her inference on the posterior distribution $P(\eta \mid D_1, D_2)$.

A predictive comparison which includes this as a limiting case would focus on $Z = \bar{Z}_1 - \bar{Z}_2$, where $\bar{Z}_i = M_i^{-1}[Z_{i1} + \cdots + Z_{iM_i}]$, $i = 1, 2$, and display the predictive distribution and density $F(z \mid D_1, D_2)$ and $f(z \mid D_1, D_2)$, respectively. Notice that for $M_1 = M_2 = 1$, we are comparing the distribution of the difference of two observations, one drawn from each population. As M_i grows $Z \rightarrow \eta$, so the former parametric analysis is the limiting case of the latter, but it is quite likely that interest would be focused on a finite number of future values unless only a normative evaluation were at issue. At any rate the predictive comparison is richer and more informative. A variety of comparison problems can be handled from the predictive point of view, in particular optimal ranking and selection problems [12].

REGULATION AND OPTIMIZATION

In problems of regulation—where a series of N trials or experiments are made indexed by $t \in T$, resulting in (t_i, x_i), $i = 1, \ldots, N$—the object is to produce a value in a set X_0 by appropriate choice of t. Closely allied to this is the optimization problem, which requires selecting t to yield an optimal but unknown future value for x (e.g., a minimum or a maximum). If the future experiment had already been performed and x observed but the index t was unknown and required identification, then a calibration* problem results. In all these cases the key to the solution within a Bayesian framework is the predictive distribution of a future X (see [1]).

MODEL CRITICISM

In cases where alternative models are not available, a predictive analysis may also be useful in criticizing an entertained model [2]. Suppose that within a Bayesian framework the model consists of observable X and parameter set θ structured as follows:

$$p(x, \alpha) = f(x \mid \alpha) p(\alpha).$$

By computing the marginal predictive probability function

$$p(x) = \int p(x, \alpha)\, d\alpha,$$

there exists the potential to assess the credibility of the entertained model for an observed set $X = x$. A simple predictive significance test can be defined by calculating

$$\Pr\{p(X) < p(x)\}.$$

If the value above is small enough, a critical reassessment of the model may be in order. Although this procedure allows questioning the model as a whole, it may reject a model merely because of one or a few spurious observations. These potential offenders may be pinpointed by calculating conditional predictive diagnostics based on the predictive density

$$f(x_j \mid x_{(j)}),$$

where $x_{(j)}$ represents all the observations except for x_j. Those x_j which yield relatively small values could indicate precisely where the difficulty lies [16].

INFLUENTIAL OBSERVATIONS

Other methods particularly useful for regression analysis in characterizing and detecting the influence of observations, singly or in sets, on prediction have been developed by Johnson and Geisser [22, 23]. This method compares the predictive probability functions of future observations, f and $f_{(i)}$, with and without the set of observations whose relative influence on prediction is to be determined. Although other measures of influence may be used, a particular scalar measure of the effect that is found useful is the Kullback–Leibler divergence measure

$$I_{(i)} = E_{f_{(i)}}\left[\log(f_{(i)} \mid f)\right].$$

Each observation (or subset of fixed size) is then ranked according to $I_{(i)}$ to determine its relative effect on the predictive distribution. Once influential observations* have been identified, it is up to the practitioner to decide what action, if any, to take with respect to them. For a detailed analysis of such a situation, see Johnson and Geisser [23].

We illustrate this procedure for a multiple linear regression situation where

$$\mathbf{Y} = \mathbf{X}\boldsymbol{\beta} + \mathbf{e}, \qquad \mathbf{e} \sim N(\mathbf{0}, \sigma^2\mathbf{I})$$

$$\mathbf{Y}' = (Y_1, \ldots, Y_N), \qquad \mathbf{e}' = (e_1, \ldots, e_N)$$

$$\mathbf{x}_j' = (x_{i1}, \ldots, x_{ip}), \qquad \boldsymbol{\beta}' = (\beta_1, \ldots, \beta_p)$$

and

$$\mathbf{X} = \begin{bmatrix} x_{11} & x_{12} & \cdots & x_{1p} \\ \vdots & \vdots & & \vdots \\ x_{N1} & x_{N2} & \cdots & x_{np} \end{bmatrix} = \begin{bmatrix} \mathbf{x}_1' \\ \vdots \\ \mathbf{x}_N' \end{bmatrix}$$

with assumed "noninformative" prior density for β and σ^2 to be $p(\beta, \sigma^2) \propto 1/\sigma^2$. In this case we examine the prediction of Z, an $m \times 1$ future vector to be observed, for a given W, an $m \times p$ matrix, where

$$\mathbf{Z} = \mathbf{W}\boldsymbol{\beta} + \mathbf{e}^*, \qquad \mathbf{e}^* \sim N(\mathbf{0}, \sigma^2\mathbf{I})$$

for the observed $\mathbf{Y} = \mathbf{y}$ with and without the ith observation y_i. Consequently,

$$f(\mathbf{z}) = \int f(\mathbf{z} \mid \mathbf{W}, \beta, \sigma^2) p(\beta, \sigma^2 \mid \mathbf{x}, \mathbf{y})\, d\beta\, d\sigma^2$$

$$\propto \left[1 + \frac{\mathbf{A}'\mathbf{B}^{-1}\mathbf{A}}{(N-p)s^2} \right]^{-(N-p+1)/2},$$

where $\mathbf{A} = \mathbf{z} - \mathbf{W}\hat{\boldsymbol{\beta}}$, $\mathbf{B} = \mathbf{I} + \mathbf{W}(\mathbf{X}'\mathbf{X})^{-1}\mathbf{W}'$, $\hat{\boldsymbol{\beta}} = (\mathbf{X}'\mathbf{X})^{-1}\mathbf{X}'\mathbf{y}$ and $(N-p)s^2 = (\mathbf{y} - \mathbf{X}\hat{\boldsymbol{\beta}})'(\mathbf{y} - \mathbf{X}\hat{\boldsymbol{\beta}})$. A similar expression is derived when y_i and the corresponding x_i' are deleted to form $f_{(i)}$.

In cases where no particular $\mathbf{W}$ is at issue, it has been found useful to let $\mathbf{W} = \mathbf{X}$, that is, to ascertain the effect of predicting back on the original set of independent variables, as indicative of an overall canonical assessment. Although in this case $I_{(i)}$ cannot be determined explicitly, an excellent approximation is given by

$$\begin{aligned} 2\hat{I}_{(i)} = {} & \frac{(N-p-2)v_i t_i^2 (N-p-4)}{2(1-v_i)(N-p-3)} \\ & + \frac{v_i(N-p-2)}{2(1-v_i)(N-p-3)} \\ & - \log\left(1 + \frac{v_i}{2(1-v_i)}\right) \\ & + N\left[\frac{N-p-2}{N-p-3}(1-t_i^2) \right. \\ & \left. - \log \frac{N-p-2}{N-p-3}(1-t_i^2) - 1 \right], \end{aligned}$$

where

$$v_i = \mathbf{x}_i'(\mathbf{X}'\mathbf{X})^{-1}\mathbf{x}_i,$$

$$t_i^2 = \frac{(\mathbf{x}_i'\hat{\boldsymbol{\beta}} - y_i)^2}{(1-v_i)(N-p)s^2}.$$

This methodology is illustrated with a set of data given by Aitchison and Dunsmore [1, p. 182]. In Table 2 a canonical assessment of

Table 2 Water Contents (Percentages by Weight) of 16 Soil Specimens Determined by Two Methods, and Associated Influence Measures

Serial Number of Specimen	On-Site Method, y	Laboratory Methods, x	$2\hat{I}_{(i)}$	$\frac{v_i}{1-v_i}$	t_i^2
1	23.7	35.3	0.17	0.16	0.14
2	20.2	27.6	0.05	0.07	0.01
3	24.5	36.2	0.15	0.18	0.13
4	15.8	21.6	0.05	0.09	0.02
5	29.2	39.8	0.08	0.30	0.00
6	17.8	24.1	0.04	0.07	0.03
7	10.1	16.1	0.06	0.19	0.02
8	19.0	27.5	0.05	0.07	0.01
9	24.3	33.1	0.06	0.12	0.01
10	10.6	12.8	0.75	0.30	0.24
11	15.2	23.1	0.04	0.08	0.04
12	11.4	19.6	0.19	0.12	0.17
13	19.7	26.1	0.03	0.07	0.05
14	12.7	19.3	0.06	0.12	0.01
15	12.6	18.8	0.06	0.13	0.00
16	31.8	39.8	1.55	0.33	0.33

the influence of individual observations on the effect of predicting from the simple linear regression is presented. Clearly, specimen 16 is the most influential according to $\hat{I}_{(i)}$ (i.e., in changing the predictive distribution). However, by looking at the next column, which is a measure of how distantly from the center the observation was taken, it is clear that this is also the most distantly observed value. The last column is a measure of the lack of fit to the model of the observation. We note here that $13t_i^2/(1 - t_i^2)$ is an F variate with 1 and 13 degrees of freedom, so that

$$\Pr[F(1, 13) < 5.85] \doteq 0.03,$$

where the observed value of $t_{16}^2/(1 - t_{16}^2)$ is 0.45 and therefore certainly fits the model. Hence the influence of specimen 16 is entirely due to it being distantly observed.

REMARKS

During the second quarter of the twentieth century there was a change in the orientation of theoretical statistics from an analysis of functions of finite observables (observabilism, predictivism) to one on unknown and unknowable parameters. An unknowable parameter is one whose exact value can never be ascertained from a finite number of observations or measurements no matter how large. The analysis of observables such as estimating the total or average response of a finite population from a sample and similar problems had been the major focus of attention of statisticians up until that time.

The introduction of mathematically defined models indexed by parameters and Fisher's clear distinction of parameters* and statistics initiated the stress on parametric inference. Possibly because of the attractive mathematics, hypothesis testing* and the estimation* of parameters, although speedily outgrowing their potential applicability, completely absorbed the attention of mathematical statisticians until rather recently. It has become clear from the work of de Finetti that for the most part, parametric analysis as such can be viewed as a special or limiting case of the predictivistic or observabilistic approach [17]. In most statistical applications there are two basic types of models. The first is the error (or measurement) model, for example, $X = \theta + e$, where θ is a true value, say a physical entity such as the weight of a rock—real and observable but imperfectly measured. Although θ here is generally viewed as a parameter, it is in an extended sense an observable entity (not an index of a population except modeled as such for convenience). In this instance, it will not matter whether we ascribe this to observable or parametric inference. But this model occurs far less frequently than one wherein a sample of units drawn from some population is measured with respect to some attribute or response to an agent or stimulus and these units inherently vary in their response, which has nothing to do with measurement error. Here inference about hypothetical parameters is meaningful only in certain special circumstances and even so is merely a limiting case of a predictive inference. In such cases inference (or decision) may be made for a single future observation or several of them jointly or functions of one or more of them depending on the purposes of the investigation. There are special circumstances where the limiting value of the function of the observable, which serves to define a "parameter," may be of interest. When the predictive distribution of a function of M future observables is analytically difficult or too complex to obtain exactly for moderate or large size, the distribution of the limiting value of the function may serve as a convenient approximation for the distribution in the finite case. Sometimes a normative entity is desirable for evaluative and comparative purposes especially when no particular fixed number of observations is necessarily of critical interest. Such a case might rule out all $1 < M < \infty$ and one would restrict one's attention to the case $M = 1$ or $M \to \infty$; the latter of course yields the parametric case. But even in such a situation it is clearly more informative to present a wide spectrum of values for M, if the calculations are not prohibitive.

In this predictivistic framework a statistical model indexed by parameters is introduced not because it is necessarily the "true" one. Rather, it often serves as an adequate approximation, given what is theoretically assumed and empirically known about whatever the underlying process is that generates the observables. Hence the paramount issue is not the fictive parameters of a convenient and approximate formulation represented by the parametric model but the potential observables. Some current research is oriented toward a complete predictive approach which involves dispensing entirely with the hypothetical parametric model (see [4]).

In summary, almost all areas of statistical application can be effectively managed by a predictivistic approach. For implementing predictivism from a variety of modes, see Geisser [15].

References

[1] Aitchison, J. and Dunsmore, I. R. (1975). *Statistical Prediction Analysis*. Cambridge University Press, New York. (The first text to deal adequately with statistical prediction.)

[2] Box, G. E. P. (1980). *J. R. Statist. Soc. A*, **143**, 383–430. (A discussion of predictive model criticism.)

[3] Butler, R. and Rothman, E. D. (1980). *J. Amer. Statist. Ass.*, **75**, 372, 881–889.

[4] Cifarelli, D. M. and Regazzini, E. (1981). *Exchangeability in Probability and Statistics*, G. Koch and F. Spizzichino, eds. North-Holland, Amsterdam, pp. 185–205. (A discussion on the complete predictive approach.)

[5] de Finetti, B. (1937). La prévision: ses lois logiques, ses sources subjectives. *Ann. Inst. Henri Poincaré*, **7**, 1–68. (The pioneer work on subjectivism.)

[6] Fisher, R. A. (1956). *Statistical Methods and Scientific Inference*. Hafner, New York.

[7] Fraser, D. A. S. (1968). *The Structure of Inference*. Wiley, New York.

[8] Geisser, S. (1964). *J. R. Statist. Soc. B*, **27**, 69–76. (The first Bayesian approach to multivariate normal classification.)

[9] Geisser, S. (1965). *Ann. Math. Statist.*, **36**, 150–159. (Predictive distributions are obtained for observations from the multivariate general linear model.)

[10] Geisser, S. (1966). *Multivariate Analysis*, P. Krishnaiah, ed. Academic Press, New York, pp. 149–163. (The development of Bayesian sequential and joint classification of several observations.)

[11] Geisser, S. (1970). *Sankhyā A*, **32**, 53–64. (Predictive distributions are obtained for observations from the generalized growth curve model.)

[12] Geisser, S. (1971). In *Foundations of Statistical Inference*, B. P. Godambe and D. A. Sprott, eds., Holt, Rinehart and Winston, Toronto, pp. 456–469. (An early introduction to the use of the predictive distribution in statistical analysis.)

[13] Geisser, S. (1974). *Biometrika*, **61**, 101–107. (The application of predictive sample reuse procedures for the random effect and mixed models in ANOVA.)

[14] Geisser, S. (1975). *J. Amer. Statist. Ass.*, **70**, 350, 320–328. (A general development of predictive sample reuse procedures with applications.)

[15] Geisser, S. (1980). In *Bayesian Analysis in Econometrics and Statistics*, A. Zellner, ed. North-Holland, Amsterdam, pp. 363–381. (A primer on predictivism.)

[16] Geisser, S. (1980). *J. R. Statist. Soc. A*, **143**, 416–417. (Introduction of conditional diagnostics.)

[17] Geisser, S. (1982). *Biometrics* (Supp.), **38**, 1, 75–93. (A general discussion of the superiority of prediction to estimation.)

[18] Geisser, S. and Desu, M. (1968). *Biometrika*, **55**, 519–524. (Bayesian classification for multivariate normal observations with zero-mean and uniform covariance matrix. Useful in twin classification.)

[19] Geisser, S. and Eddy, W. F. (1979). *J. Amer. Statist. Ass.*, **74**, 153–160. (Model selection techniques are derived whose goal is optimal prediction.)

[20] Guttman, I. (1966). *J. R. Statist. Soc. B*, **29**, 83–100. (The use of the predictive distribution to estimate the sampling distribution and goodness-of-fit tests based on this idea.)

[21] Jeffreys, H. (1939). *Theory of Probability*. Clarendon Press, Oxford, England. (A pioneer neo-Bayesian book.)

[22] Johnson, W., and Geisser, S. (1982). *Essays in Honor of C. R. Rao*, G. Kallianpur, P. R. Krishnaiah, and J. K. Ghosh, eds. North-Holland, Amsterdam, pp. 343–358. (The introduction of the notion of the influence of sets of observations on future predictions.)

[23] Johnson, W., and Geisser, S. (1983). *J. Amer. Statist. Ass.*, **78**, 137–144. (The application of predictive influence functions to problems of regression.)

[24] Kalbfleisch, J. D. (1971). *Foundations of Statistical Inference*. Holt, Rinehart and Winston, New York, pp. 378–392. (A detailed examination of likelihood methods for prediction.)

[25] Lamy, M. et al. (1957). *Ann. Hum. Genet. Lond.*, **21**, 374–396.

[26] Lauritzen, S. L. (1974). *Scand. J. Statist.*, **1**, 128–134. (A definition of predictive likelihood using sufficiency.)

[27] Lee, J. C. and Geisser, S. (1972). *Sankhyā A*, **34**, 393–412. (Conditional predictive distributions are obtained from the generalized growth curve model.)

[28] Lejeune, M. and Faulkenberry, G. D. (1982). *J. Amer. Statist. Ass.*, **77**, 379, 654–659.

[29] Pearson, K. (1907). *Philos. Mag., 6th Series*, **6**, 365–378.

[30] Wilks, S. S. (1962). *Mathematical Statistics*. Wiley, New York.

SEYMOUR GEISSER

(BAYESIAN INFERENCE
DENSITY ESTIMATION
SPECIFICATION, PREDICTOR)

PREDICTIVE DISTRIBUTION

PREDICTIVE DISTRIBUTION *See* PREDICTIVE ANALYSIS

PREDICTIVE VALIDITY *See* GROUP TESTING

PREDICTOR SPECIFICATION *See* SPECIFICATION, PREDICTOR

PREDOMINANT ITEM

An obsolete term for mode.

PREEMPTIVE DISCIPLINE

Consider a queueing* system in which customers are served in order of arrival. If this discipline is changed so that by virtue of status an arriving customer is served before some others already waiting for service, we then have a priority discipline; *see* PRIORITY QUEUE. Thus it may be more costly to delay some customers than others. Furthermore, if the arriving customer has a higher status than any other customer in the system, he or she may displace a customer being served and hence gain immediate service; the displaced customer returns to the queue. This is called a preemptive (priority) discipline.

Preemptive disciplines were first suggested in connection with delays arising when a single repairperson tends a number of unreliable machines. Phipps [8] noted the possible advantage of interrupting especially long repair periods, but he offered no analysis. Here an interrupted repair job is continued from the point at which it was last interrupted—a preemptive resume discipline. The successful analysis of queues with priority disciplines requires rules governing reentry to service of interrupted customers. The earliest analysis was offered by Stephan [10] and White and Christie [11] for the preemptive resume discipline.

The preemptive resume discipline may be contrasted with a preemptive repeat discipline, wherein service of interrupted customers begins anew. Several cases can occur. For example, in the repeat-identical (also called repeat without resampling) discipline a returning customer needs the same amount of service as requested on his or her previous entry. Alternatively, the repeat-different (also called repeat with resampling) discipline occurs if requested service times at each return are independent. When customer interruptions occur at event times of a Poisson process*, the analysis of the preemptive disciplines so far mentioned may be unified by the notion of customer completion time, which is the total time spent by the customer with the server. This idea was independently discovered by several workers around 1962; see refs. 4 (p. 235) and 6 (p. 55). Many of the tractable problems concerning queues with simple preemptive disciplines had been solved by the mid-1960s. Accounts of this work may be found in refs. 2 and 6, while ref. 9 (pp. 95–104, 210) concentrates on the case where service and interarrival times are exponentially distributed.

Pure preemptive priority disciplines can be modified in many ways by not allowing immediate preemption. Thus in any of the cases above, preemption may be allowed only if the lower-priority customer receiving service has just begun service. This is an example of a discretionary priority [1]. The schemes described so far envision any customer belonging to exactly one of a finite number p of priority classes and hence his or her status may be measured by a positive integer $\leqslant p$, with p denoting least status. Modifications exist whereby it is sensible to allow a continuum of priority classes. For example, arriving customers may purchase entry to a certain class [7, p. 135]; or a customer's status may increase with time spent in the system [7, p. 126]. This could occur through increasing urgency to ensure that service is completed before a preset deadline [5]. These are examples of dynamic priorities.

Historically, machine repair problems have been a significant motivation for the analysis of priority queueing systems, but in the mid-1960s new developments were inspired by the desire to model computer time-sharing systems where customers demand quanta of central processing unit (CPU) time. We shall mention a couple of examples of preemptive disciplines arising in this area of work.

In a last come, first served system an arriving job gains immediate access to the CPU by displacing the present job at the end of the current time quantum. The displaced job is sent to join a queue of previously displaced jobs, which are served in turn when the CPU is otherwise free. A modification is the forward–backward system, in which the CPU selects the job which so far has received least service. In the event of ties, several jobs share the CPU at a proportionately reduced rate of processing. Thus new jobs always preempt existing ones. Finally, multilevel systems permit the CPU to subject jobs to different priority regimes according to their accumulated processing times. See ref. 7 (Chap. 4) for an account of these systems.

Further Reading

The following list is a selection of more recent contributions to the theory and application of preemptive disciplines.

Adira, I. and Domb, I. (1982). *Operat. Res.*, **30**, 97–115. (A model having p priority classes arriving in separate Poisson processes. An arriving customer of class i can preempt a customer of class j who is receiving service only if $j - i$ exceeds a fixed number.)

Brandwajn, A. (1982). *Operat. Res.*, **30**, 74–81. (A finite difference equation analysis of a two-class system where arrival- and service-time rates of low-priority customers depend on their numbers in the system.)

Halfin, S. and Segal, M. A. (1972). *SIAM J. Appl. Math.*, **23**, 369–379. (A system with two classes and several servers. Any customer arriving when all servers are attending to high-priority customers are turned away, but preempted customers queue until service is complete. Recursive formulae are found for the steady-state distribution and moments of the time required to process all waiting preempted customers.)

Hluchyj, M. G., Tsao, C. D., and Boorstyn, R. R. (1983). *Bell Sys. Tech. J.*, **62**, 3225–3245. (A preemptive priority model describing a packet communications system where message sources compete for access to a single transmission channel.)

Jaibi, M. R. (1980). *Ann. Inst. Henri Poincaré*, **16**, 211–223. (A modern analysis of a two-class preemptive resume system. The author uses marked point processes and stochastic differential equations.)

Kodama, M. (1976). *Operat. Res.*, **24**, 500–515. (A repairperson problem in which a system comprises two subsystems, one with units connected in parallel and the other with units in series. Failed series units preempt those being repaired from the parallel unit subsystem.)

Miller, D. R. (1981). *Operat. Res.*, **29**, 945–958. (The author uses Neuts' matrix-geometric methods (next reference) to obtain steady-state distributions of a simple queue having two priority classes. Both preemptive and nonpreemptive disciplines are considered.)

Neuts, M. F. (1981). *Matrix-Geometric Solutions in Stochastic Models—An Algorithmic Approach*. Johns Hopkins University Press, Baltimore, MD. (On pp. 298–300 the author analyses the model of Halfin and Segal as a queue in a Markovian random environment.)

References

The standard, and only, reference book in English on priority disciplines is that by Jaiswal [6]. Cohen's book [2] and Kleinrock's book [7] are useful general references. Reference 3 is in Russian.

[1] Avi-Itzhak, B., Brosh, I., and Naor, P. (1964). *Z. Angew. Math. Mech.*, **44**, 235–242.

[2] Cohen, J. W. (1982). *The Single Server Queue*, 2nd ed. North-Holland, Amsterdam.

[3] Danielyan, E., Dimitrou, B., Gnedenko, B., Klimov, G., and Matveev, V. (1974). *Prioritetnye Sistemy Obslyzhivaniya*. Moscow University, Moscow.

[4] Gaver, D. P. (1965). In *Proc. Symp. Congestion Theory*, W. L. Smith and W. E. Wilkinson, eds. University of North Carolina, Chapel Hill, NC, pp. 228–252.

[5] Jackson, J. R. (1960). *Naval Res. Logist. Quart.*, **7**, 235–269.

[6] Jaiswal, N. K. (1968). *Priority Queues*. Academic Press, New York.

[7] Kleinrock, L. (1976). *Queueing Systems*, Vol. II: *Computer Applications*. Wiley, New York.

[8] Phipps, T.E. (1956). *Operat. Res.*, **4**, 76–85.

[9] Srivastava, H. M. and Kashyap, B. R. K. (1982). *Special Functions in Queueing Theory: And Related Stochastic Processes*. Academic Press, New York.

[10] Stephan, F. F. (1958). *Operat. Res.*, **6**, 399–418.

[11] White, H. and Christie, L. S. (1958). *Operat. Res.*, **6**, 79–95.

(PRIORITY QUEUE
QUEUEING THEORY)

ANTHONY G. PAKES

PREEMPTIVE PRIORITY *See* PREEMPTIVE DISCIPLINE

PREFERENCE FUNCTIONS

Consider a statistical problem where the data are the observed value of a real- or vector-valued random variable X with known density function $f_X(x;\theta)$. Based on the observed value x of X, one wants to draw conclusions about the unknown quantity θ, which is contained in some set Θ. Using the frequency* approach to statistical inference, the result of the statistical analysis is ordinarily given by means of a point estimate, a confidence interval, or the result of a statistical test. Both estimates and tests represent a somewhat crude summary of what the data have to tell, and one may argue that confidence intervals* are somewhat misleading. This is because a confidence interval puts all values inside the interval on an equal footing, and they are sharply set apart from the values outside the interval. However, most statisticians would agree that values near the limits of a confidence interval, whether they are inside or outside, are about equally good or bad.

The concept of preference function is introduced to permit a smoother representation of the knowledge about θ given by the data than by the methods alluded to above.

Definition 1. A function which to each x associates a real-valued function on the parameter space Θ is called a *preference function*.

As a generic notation for preference functions we use $\pi(\cdot;\cdot)$. Thus $\pi(\cdot;x)$ is the function on the parameter space for $X = x$, and $\pi(\theta;x)$ is the value at θ of this function.

With Definition 1, tests, point estimates, and confidence intervals may be considered as special cases of preference functions. For example, a confidence interval may be represented by $\pi(\theta;x) = 1$ for θ in the interval and $\pi(\theta;x) = 0$ for θ outside the interval. Here the values 1 and 0 are chosen arbitrarily; we could have used any values A and B with $A \neq B$.

However, there exist more interesting preference functions that have been used for a long time in statistics. The most notable one is the likelihood* function

$$\pi_L(\theta;x) = \frac{f_X(x;\theta)}{\sup_\theta f_X(x;\theta)}.$$

Apart from being the basic starting point for estimation and testing procedures, its value as a final product of the inferential process has also been advocated by many statisticians.

Another well-known quantity, which may also be considered as a preference function, is the Bayesian posterior* density

$$\pi_B(\theta;x) = \frac{f_X(x;\theta)p(\theta)}{\int f_X(x;\theta)p(\theta)\,d\theta},$$

where $p(\theta)$ is the prior density.

Finally, the plausibility function*

$$\pi_P(\theta;x) = \frac{f_X(x;\theta)}{\sup_x f_X(x;\theta)},$$

introduced by Barndorff-Nielsen [1], may also be considered as a preference function.

PERFORMANCE CHARACTERISTICS

Definition 2. For a given preference function and an observed value x, the point θ_1 is said to be *preferable* to θ_2 if $\pi(\theta_1; x) > \pi(\theta_2; x)$. We call $\pi(\theta; x)$ the *preferability* of θ.

This definition indicates how a preference function is used in practice. Points in the parameter space having relatively high values $\pi(\theta; x)$ are considered to be more reasonable candidates for the true value, and more reasonable the larger $\pi(\theta; x)$ is.

One simple performance criterion of a preference function is its ability to give high values to the true parameter value, when compared to some other value. Thus one wants

$$P_{\theta_1}[\pi(\theta_1, X) < \pi(\theta_2, X)]$$

to be small when $\theta_2 \neq \theta_1$. This may be considered as the risk corresponding to the loss function

$$L(\theta_1, \pi(\cdot; x)) = \begin{cases} 1 & \text{when } \pi(\theta_1; x) < \pi(\theta_2; x) \\ 0 & \text{when } \pi(\theta_1; x) \geqslant \pi(\theta_2; x). \end{cases}$$

An unbiasedness* concept is given by the following:

Definition 3. A preference function is said to be *unbiased* if

$$P_\theta[\pi(\theta; X) \geqslant \pi(\theta'; X)] \geqslant P_\theta[\pi(\theta'; X) \geqslant \pi(\theta; X)]$$

for all $\theta' \neq \theta$.

Thus unbiasedness means that for any given alternative to the true value, the probability of a correct ordering is not smaller than the probability of an incorrect one. The condition for unbiasedness may also be written

$$P_\theta[\pi(\theta; X) \geqslant \pi(\theta'; X)] \geqslant \tfrac{1}{2} + \tfrac{1}{2} P_\theta[\pi(\theta; X) = \pi(\theta'; X)].$$

The second expression to the right disappears in most "continuous" problems.

ONE-PARAMETER EXPONENTIAL FAMILY OPTIMALITY

Consider the one-parameter exponential family

$$f_X(x; \theta) = C(\theta)h(x)\exp(Q(\theta)T(x)),$$

where $Q(\cdot)$ is an increasing function of θ. In the following everything is formulated in terms of the sufficient statistic* $T = T(X)$. To avoid technical details, it will be assumed that $P_\theta(T = t) = 0$ for all θ and t and that $P_{\theta_1}[\pi(\theta_1; T) = \pi(\theta_2; T)] = 0$ for all $\theta_1 \neq \theta_2$. For this family, there exists a preference function, the *acceptability function*, which has certain optimality properties. It is defined by

$$\pi_A(\theta; t) = \begin{cases} 2P_\theta(T \geqslant t) & \text{if } \theta \leqslant \hat{\theta} \\ 2P_\theta(T \leqslant t) & \text{if } \theta \geqslant \hat{\theta}, \end{cases}$$

where $\hat{\theta}$ is the value of θ which makes the observed value of T the median of its distribution. Thus $\hat{\theta}$ is found by the equation

$$P_{\hat{\theta}}(T \geqslant t) = P_{\hat{\theta}}(T \leqslant t).$$

Under the assumptions above, the following optimality property of π_A can be shown [2]:

Theorem. The acceptability function $\pi_A(\theta; t)$ maximizes for any θ_1 and θ_2

$$\min\{P_{\theta_1}[\pi(\theta_1; T) \geqslant \pi(\theta_2; T)], P_{\theta_2}[\pi(\theta_2; T) \geqslant \pi(\theta_1; T)]\}.$$

Moreover,

$$P_{\theta_1}[\pi_A(\theta_1; T) \geqslant \pi(\theta_2; T)] = P_{\theta_2}[\pi_A(\theta_2; T) \geqslant \pi(\theta_1; T)].$$

This theorem shows that when limiting attention to two given values of θ, the acceptability function has a certain minimax property. Also, it achieves a certain balance between the choice of the two values in that the probability of correctly preferring θ_1 to θ_2 equals the probability of correctly preferring θ_2 to θ_1.

It is also easily seen that the acceptability function is unbiased. From its definition it is seen to be twice the P-values* (observed significance probabilities) one would get by systematically taking each value of θ as the null hypothesis and testing against the estimated value $\hat{\theta}$ as the alternative. Note, however, that although π_A is unbiased according to the definition above, it is not equal to the preference function one would obtain using the observed significance levels of the standard unbiased tests.

Example. Let $X_1, \ldots, X_n$ be independent $N(0, \sigma^2)$, let $T = \sum_{i=1}^{n} X_i^2$, and let G_n be the cumulative chi-square distribution* with n degrees of freedom. The value $\hat{\theta}$ needed to calculate the acceptability function is $\hat{\sigma}^2 = T/m_n$, where $G_n(m_n) = \frac{1}{2}$. Then

$$\pi_A(\sigma^2; t) = \begin{cases} 2(1 - G_n(t/\sigma^2)) & \text{when } \sigma^2 \leqslant \hat{\sigma}^2 \\ 2G_n(t/\sigma^2) & \text{when } \sigma^2 \geqslant \hat{\sigma}^2. \end{cases}$$

The preference functions corresponding to the likelihood and the Bayesian posterior density using a noninformative prior* are both of the form

$$\pi_b(\sigma^2; t) = \text{constant}\, \sigma^{-b} \exp(-t/(2\sigma^2)),$$

where $b = n$ for the likelihood, and $b = n + 2$ for the Bayesian posterior density. One finds that

$$P_{\sigma^2}\left[\pi_b(\sigma^2; T) \geqslant \pi_b(\sigma'^2; T)\right] = \begin{cases} 1 - G_n\left(\dfrac{2b \ln(\sigma/\sigma')}{\sigma^2/\sigma'^2 - 1}\right) & \text{if } \sigma' < \sigma \\ G_n\left(\dfrac{2b \ln(\sigma'/\sigma)}{\sigma'^2/\sigma^2 - 1}\right) & \text{if } \sigma' > \sigma. \end{cases}$$

An example of the three preference functions is shown in Fig. 1, taken from Spjøtvoll [2]. It is seen that the Bayesian curve gives higher preferability to smaller values. This corresponds to the fact that it is, in this situation, highly biased. From the expressions above it is found that the probability of a correct ordering of σ relative to σ' tends to $1 - G_n(b)$ and $G_n(b)$ as σ' tends σ from below and above, respectively. Since $b = n$, $n + 2$ are both larger than the median m_n,

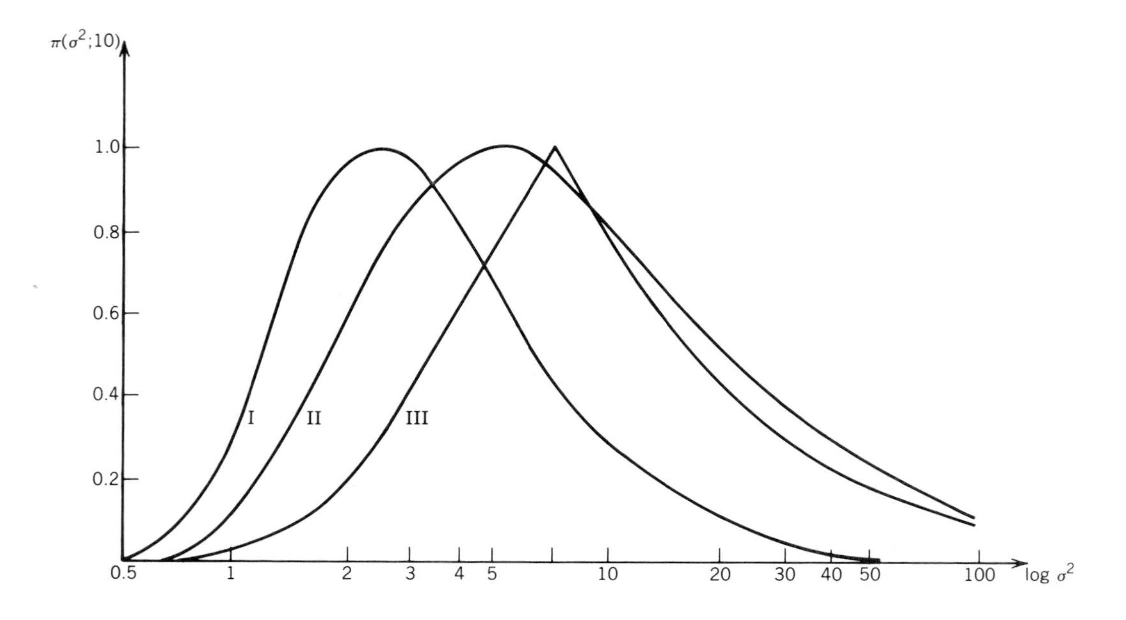

Figure 1 Bayesian posterior density (I), likelihood (II), and acceptability (III) for an unknown variance; example with $n = 2$, $t = 10$. The first two curves are normed to have maximum equal to 1.

both the Bayesian curve and the likelihood curves are biased. Actually, with the case $n = 2$ as in Fig. 1, the probability of ordering σ correctly relative to a smaller, but very close value σ', is $1 - G_2(4) = 0.14$ for the Bayesian curve.

References

[1] Barndorff-Nielsen, O. (1976). *J. R. Statist. Soc. B*, **38**, 103–131.

[2] Spjøtvoll, E. (1983). In *Festschrift for Erich L. Lehmann*, P. J. Bickel, K. Doksum, and J. L. Hodges, Jr., eds. Wadsworth, Belmont, Calif., pp. 409–432. (Introduces the concept of preference functions and gives some basic properties. Motivates the acceptability function.)

(LIKELIHOOD
PLAUSIBILITY FUNCTION
PRIOR DISTRIBUTIONS
POSTERIOR DISTRIBUTIONS)

E. SPJØTVOLL

PREFERENCE MAPPING *See* GRAPHICAL REPRESENTATION OF DATA

PREFERENCES *See* BAYESIAN INFERENCE; DECISION THEORY

PREGIBON'S LINK TEST *See* LINK TESTS

PREMATURE AGGREGATION *See* HALF-NORMAL PLOTS

PREVAILING ITEM

An obsolete term for mode.

PREVISION

Prevision is a concept introduced by B. de Finetti to establish the notion of expectation as the fundamental element of the subjectivist theory of probability. Your prevision, $P(X)$, for the random quantity X may be thought of, intuitively, as the "fair price" for the random gain X [i.e., so that you are indifferent between the random gain X and the certain gain $P(X)$]. (Prevision, as a subjectivist concept, always expresses the judgment of a specified individual, here termed "you.") To make this precise, $P(X)$ must be defined in terms of an operational criterion. One such criterion is as follows. For a general random quantity X, your *prevision* for $X, P(X)$, is the value $\bar{x}$ that you would choose if you were subsequently to suffer a penalty L given by $L = k(X - \bar{x})^2$, where k is a constant defining the units of loss. Your previsions $\bar{x}_1, \ldots, \bar{x}_m$ for the random quantities $X_1, \ldots, X_m$ are said to be *coherent* if there is no other choice $x_1^*, \ldots, x_m^*$ which would lead to a reduction in your overall penalty for all possible combinations of values of $X_1, \ldots, X_m$. (Thus, if you are not coherent, you are a "sure loser.") Coherence is here used in the same intuitive sense as for the axiomatic development of inference and decision making (*see* COHERENCE), namely as a criterion to identify intrinsically contradictory beliefs.

For any random quantities X and Y, coherence imposes the following requirements on your specifications: (1) $\inf X \leqslant P(X) \leqslant \sup X$, and (2) $P(X + Y) = P(X) + P(Y)$. These properties reflect the "fair price" interpretation. (Notice that prevision is finitely additive but not necessarily countably additive.) [(1) and (2) are special cases of the general result that a specification $P(Y_i) = \bar{y}_i$, $i = 1, \ldots, m$, is coherent if and only if the point $(\bar{y}_1, \ldots, \bar{y}_m)$ lies in the closed convex hull of the set of possible values of the random vector $(Y_1, \ldots, Y_m)$ in m-dimensional Euclidean space.]

Prevision can be considered to be expectation, but defined directly as a primitive quantity, rather than in terms of a prespecified probability distribution. For any event H, if you also use H to denote the indicator function of the event (i.e., $H = 1$ if H is true, $H = 0$ if false), then you may interpret $P(H)$ as the probability of H. In this development there is no distinction between probability and expectation, and the notation is chosen to emphasize this. If you have specified a full probability distribution for X, your prevision for X will agree with the usual definition of expectation (at least if X is bounded; there are subtleties for unbounded X). The essential difference arises when you do not want to assess probabilities for all the possible outcomes, in which case $P(X)$, being assigned directly, is well defined, as opposed to your expectation for X, which is defined only through the associated probability distribution for X. Thus you may consider "average" properties without exhaustively considering some limiting partition of possibilities, so that you are free to choose the level of detail at which you wish to describe your beliefs.

Using H interchangeably as the event and the indicator function of your event, your prevision of the random quantity X, conditional on the event H, $P(X \mid H)$, is defined exactly as for $P(X)$, but with the penalty L replaced by the "called off" penalty HL. Coherence imposes the following requirement: (3) $P(HX) = P(H)P(X \mid H)$.

The essential reference on prevision is de Finetti [1, 2], where all the foregoing material is developed. Much of the importance of prevision lies in the depth and thoroughness of the treatment which de Finetti provides in these two volumes. Some statistical implications of this formulation are developed in Goldstein [3] as follows. In the usual Bayesian formulation, your beliefs are collected into a prior probability distribution which is modified by Bayes' theorem in the light of new information to yield a posterior* probability distribution. When working with prevision rather than probability, it is shown that your beliefs should be collected into a prior inner product space, which is modified by a linear transformation, in the light of new information, to yield a posterior inner product space. The relationship between the conditional probability approach and the linear transformation approach is discussed in detail in Goldstein [5]. These developments rely on the condition that if $P_T(X)$ is the actual value that you will assign for $P(X)$ at the future time T, then your current previsions should satisfy the relation: (4) $P(P_T(X)) = P(X)$. [This result would not, in general, be true if $P_T(X)$ were replaced by conditional prevision.] A coherence argument for (4) is given in Goldstein [4].

References

[1] de Finetti, B. (1974). *Theory of Probability*, Vol. 1. Wiley, London.

[2] de Finetti, B. (1975). *Theory of Probability*, Vol. 2. Wiley, London.

[3] Goldstein, M. (1981). *J. R. Statist. Soc. B*, **43**, 105–130.

[4] Goldstein, M. (1983). *J. Amer. Statist. Ass.*, **78**, 817–819.

[5] Goldstein, M. (1984). In *Bayesian Econometrics and Statistics*, Vol. 3 (in honour of B. de Finetti), Prem. K. Goel and A. Zellner, eds. North-Holland, Amsterdam (in press)

(COHERENCE
CONDITIONAL PROBABILITY AND
EXPECTATION
EXPECTED VALUE
SUBJECTIVE PROBABILITIES)

MICHAEL GOLDSTEIN

PRIESTLEY'S TEST FOR HARMONIC COMPONENT(S)

The basic problem may be described as follows. We are given a sample of observations from a time series, X_t, which we suspect may consist of two distinct components, a "signal" Z_t and a "noise" term Y_t. Thus we may write

$$X_t = Z_t + Y_t \qquad (t = 0, \pm 1, \pm 2, \dots),$$

where the signal Z_t consists of a finite sum of periodic terms with different frequencies, that is,

$$Z_t = \sum_{i=1}^{K} A_i \cos(\omega_i t + \phi_i),$$

and the noise* Y_t has a continuous power spectrum, $h_y(\omega)$, say. Here K, $\{A_i\}$, $\{\omega_i\}$, $i = 1, \dots, K$, are all unknown parameters, the $\{\phi_i\}$ may be taken as independent random variables each uniformly distributed over $(-\pi, \pi)$, and $h_y(\omega)$ is an unknown function. Given a set of observations on X_t, say from $t = 1$ to N, the basic problem is to test whether or not a signal is present, that is, to test the null hypothesis $H_0: A_i = 0$, all i, against the alternative hypothesis $H_i: A_i > 0$, for some values of i. In an engineering context we might think of Z_t as representing, for example, a transmitted radio signal (so that X_t then represents the received signal plus noise generated in the transmission medium and receiver). However, in a completely different context Z_t might represent, for example, a "seasonal component" in an economic time series, or indeed any type of strictly periodic component in a general time series*.

Early studies of this problem were based on an extension of the technique of harmonic analysis. In effect, one performs a (numerical) harmonic analysis of the observed series $\{X_t\}$, and then examines the amplitudes of the various trigonometric terms to see which, if any, appear exceptionally large. This procedure reduces to examining the ordinates of a function called the *periodogram*, defined by

$$I_x(\omega) = \frac{2}{N} \left| \sum_{t=1}^{N} X_t e^{-i\omega t} \right|^2,$$

which is usually evaluated at the points $\omega_p = 2\pi p/N$, $p = 0, 1, \ldots, [N/2]$. If the $\{Y_t\}$ are independent zero-mean normal variables (so that Y_t has a *white* or *uniform* spectral density function), then under H_0 the $\{X_t\}$ are similarly zero-mean normal variables, and it is easily shown (see Priestley [8, Chap. 6]) that the $\{I(\omega_p)\}$ are then independent, with $\{I(\omega_p)/\sigma_x^2\}$ having an exponential distribution* with mean 2. Writing $I_p = I(\omega_p)$, it then follows that, under H_0,

$$\Pr\left[I_p/\sigma_x^2 \leqslant z \right] = 1 - \exp(-z/2). \quad (1)$$

On the other hand, it may be shown that when H_1 holds, the periodogram ordinates will be large (relative to their expectations under H_0) when ω_p falls near one of the "true" frequencies, $\{\omega_i\}$, but at other points the ordinates will remain relatively small. The basis of the technique is thus to "search" for exceptionally large periodogram ordinates, their presence indicating the existence of a signal, and their locations indicating the true frequencies of signal components.

However, it is possible that a "large" periodogram ordinate may arise purely from sampling fluctuations even when no signal is present, and we therefore require a suitable test to determine whether certain ordinates are *significantly* large. Using the properties noted above, it follows that under H_0 (and with the $\{Y_t\}$ independent normal variables),

$$\Pr\left[I_p/\sigma_x^2 > z \right] = \exp(-z/2),$$

and Schuster [9] suggested that this distribution could be used to test the significance of a given ordinate (with σ_x^2 replaced by the sample variance of the $\{X_t\}$). However, since we do not know (a priori) the true values of the $\{\omega_i\}$, we would not know which ordinates to select for testing. The natural procedure is to start by testing the largest ordinate, and it is easy to show that under the same assumptions as above,

$$\Pr\left[\max(I_p)/\sigma_x^2 > z \right] = 1 - \{1 - \exp(-z/2)\}^{[N/2]}.$$

This distribution (with σ_x^2 again replaced by its sample estimate) forms the basis of the large sample test proposed by Walker [10] for the maximum periodogram ordinate. Fisher [3], in a celebrated paper, derived an *exact test* for $\max(I_p)$, based on the statistic

$$g = \frac{\max(I_p)}{\sum_{p=1}^{[N/2]} I_p}.$$

(This is, in effect, a "studentized" version of Walker's test statistic, the denominator in Fisher's g-statistic being approximately proportional to the sample variance.) Fisher showed that (for N odd) the exact distribution of g under H_0 is given by

$$\Pr[g > z] = n(1-z)^{n-1} - \frac{n(n-1)}{2}(1-2z)^{n-1} + \cdots + (-1)^a \frac{n!}{a!(n-a)!}(1-az)^{n-1},$$

where $n = [N/2]$ and a is the largest integer less than $(1/z)$.

All the results above are based on the assumptions (a) that H_0 holds and (b) that the process $\{Y_t\}$ is a sequence of independent normal variables (i.e., is *white noise*). When we allow Y_t to have an arbitrary unspecified spectral density function, $h_y(\omega)$, the problem becomes much more formidable, and we may note, for example, that the null distribution of Fisher's g given above is no longer valid. Whittle [11] made an ingenious attempt to reduce the problem to the previous case by *scaling* the periodogram ordinates by $h_y(\omega_p)$, and using the well-known result [8, Chap. 8], that, under general conditions, $\{I_p/h_y(\omega_p)\}$, the scaled ordinates, behave asymptotically (under H_0) like the ordinates of a white noise process. If the function $h_y(\omega)$ were known a priori, this would provide a natural way of dealing with the more general version of the problem. However, in general $h_y(\omega)$ would be an unknown function, and this is the point at which certain fundamental difficulties arise. Since the process $\{Y_t\}$ is unobservable, we cannot estimate $h_y(\omega)$ directly, and if we attempt to estimate this function from the observations $\{X_t\}$, the possible presence of the signal harmonic components will produce spurious peaks in the standard form of "windowed" spectral estimates (see Priestley [8, Chap 7]). The presence of these peaks will considerably reduce the power of a test based on Fisher's g-statistic computed from the scaled periodogram ordinates. On the other hand, if we remove such peaks, then we are assuming, in effect, that they are caused by harmonic components *before* the existence of such components has been established by the test procedure—possibly leading to the conclusion that harmonic components exist even when there is no signal present.

The crucial difficulty presented by the general form of the problem is that if the noise spectral density function, $h_y(\omega)$, becomes too "narrow," the noise process Y_t will look just like a "signal," and there is then no way in which we can disentangle the two components. For example, if $h_y(\omega)$ takes the form of a sufficiently narrow peak centered on frequency ω^* (together with its image in the negative frequency axis), Y_t will be virtually indistinguishable from a harmonic component with frequency ω^*. Accordingly, to make the problem tractable, we must impose a restriction on the "bandwidth" of $h_y(\omega)$. Now it may be shown that any harmonic components present will produce peaks in the periodogram of X_t whose widths are $O(1/N)$, N being the sample size. Consequently, we now impose the condition that the narrowest peak in $h_y(\omega)$ has bandwidth $\gg O(1/N)$. [This situation is one in which purely asymptotic arguments can produce quite misleading results. We could, of course, argue that since $h_y(\omega)$ is a fixed function, the condition above must always hold for sufficiently large N. However, this is largely irrelevant as far as practical analysis is concerned. In practice we are usually given a fixed number of observations, and the question then is whether the bandwidth condition above holds for the particular sample size with which we are working.] *See also* NARROWBAND PROCESS.

Given the foregoing restriction on the bandwidth of $h_y(\omega)$, Priestley [5, 6] suggested a different approach to the problem of detecting harmonic components. In Priestley's method the emphasis is placed on the behavior of the (sample) autocovariance function, rather than the periodogram. It makes use of the following two properties: (a) the autocovariance function of the signal, $R_z(s)$, consists of a sum of cosine terms with the same frequencies as Z_t, and hence never dies out; and (b) the autocovariance function of the noise, $R_y(s)$, is the Fourier transform of a continuous function and hence tends to zero at both ends of the axis. The autocovariance function of the observed process, $R_x(s)$, say, is the sum of the above two autocovariance functions; hence, when a signal is present, $R_x(s)$ may oscillate with varying amplitude over the initial portion of the function but will settle down to a steady oscillatory form for large s. On the other hand, if there is no signal present, $R_x(s)$ will be identical to the autocovariance function of Y_t, and hence will die out. Thus the feature which indicates the presence of a signal is the *behavior of the "tail" of the autocovariance function*. Furthermore, the rate at which $R_y(s)$ decays to zero depends on the bandwidth of $h_y(\omega)$—the larger the bandwidth, the faster the rate of decay. Hence our previous restriction on the bandwidth of $h_y(\omega)$ (in relation to $1/N$) may be expressed alternatively as a restriction on the rate of decay $R_y(s)$; that is, given N observations, we may assume that there exists an integer $m(\ll N)$ such that $R_y(s) \sim 0$, $|s| > m$. Hence the "tail" of the autocovariance function $R_x(s)$ (i.e., the portion with $|s| > m$) will, in effect, be generated purely by the presence of harmonic components (if any), and the presence of such terms can now be detected by performing a harmonic analysis of this "tail" function.

In practice we have to work with the sample autocovariance function, namely

$$\hat{R}_x(s) = (1/N) \sum_{t=1}^{N-|s|} (X_t - \overline{X})(X_{t+|s|} - \overline{X})$$

($\overline{X}$ denoting the sample mean), and we then construct the function

$$P(\lambda) = \frac{1}{2\pi} \sum_{m<|s|\leqslant n} \hat{R}_x(s)\cos(s\lambda), \quad 0 \leqslant \lambda \leqslant \pi,$$

where n is some integer $\leqslant (N-1)$. Under the null hypothesis H_0 (no signal present) we obviously have $E[P(\lambda)] \sim 0$, all λ. However, if X_t contains a harmonic component with frequency ω_r it may be shown that

$$E[P(\lambda)] \sim \begin{cases} 0 & \text{if } |\lambda - \omega_r| \gg O(1/m) \\ O(n-m) & \text{when } \lambda \sim \omega_r. \end{cases}$$

Thus the presence of harmonic components may be detected by searching for "peaks" in the function $P(\lambda)$. In the expression for $P(\lambda)$ above, the contribution of the first m autocovariances has been completely removed; however, we may consider a more general expression in which the contribution of the first m autocovariances is merely "attenuated," leading to the form

$$P(\lambda) = \frac{1}{2\pi} \sum_{s=-(N-1)}^{(N-1)} \{ w_n^{(1)}(s) - w_m^{(2)}(s) \} \hat{R}_x(s) e^{-i\omega s},$$

where $w_n^{(1)}(s)$ and $w_m^{(2)}(s)$ are two general "covariance lag windows" with suitably adjusted rates of decay. The test is constructed by evaluating $P(\lambda)$ over a grid of points at spacings $2\pi/m'$ $(m' < m)$, and forming the cumulative sums, $\sum_{p=1}^{q} P(2\pi p/m')$. It can be shown [8, Chap. 8] that when suitably standardized, the asymptotic distributional properties of such sums can be derived by analogy with the theory of random walks*, and the test for the absence of "peaks" can then be constructed by using the theory of random walks with absorbing barriers. More precisely, we set

$$J_q = \left\{ \frac{N}{m'\Lambda(n,m)} \right\}^{1/2} \sum_{p=1}^{q} P\left(\frac{2\pi p}{m'} \right),$$

$$q = 0, 1, \ldots, [m'/2],$$

where $\Lambda(n,m)[= N \operatorname{var} P(\lambda)/h_y^2(\omega)]$ is an easily evaluated function of n and m, given the specific forms of $w_n^{(1)}(s)$ and $w_m^{(2)}(s)$. It may then be shown that

$$\lim_{N\to\infty} \Pr\left[\frac{\max_q(J_q)}{\{(2\pi)^{-1}\hat{G}(\pi)\}^{1/2}} \leqslant a \right] = 2\Phi(a) - 1,$$

where $\Phi(x)$ is the standardized normal distribution function, and

$$\hat{G}(\pi) = \frac{1}{4\pi} \left[\sum_{s=-m}^{m} \hat{R}_x^2(s) - 2 \sum_{s=-m+1}^{2m} \hat{R}_x^2(s) \right].$$

The full details of the test are given by Priestley [5, 6], who also discusses the practical implementation of this procedure and describes methods for choosing suitable lag windows, $w_n^{(1)}(s)$ and $w_m^{(2)}(s)$, and appropriate values for the parameters, n and m. Further discussion of this test, together with some numerical illustrations, is given in Priestley [8, Chap. 8]. An application of the $P(\lambda)$-test to the Canadian lynx data is described by Bhansali [2], and an application to tide heights is given in Bhansali [1].

It may be mentioned that Hannan [4] proposed a test which is essentially the same as Whittle's statistic, but introduced an important modification in the estimation of

$h_y(\omega)$. Hannan's procedure is, in effect, to estimate $h_y(\omega_p)$ by smoothing the periodogram ordinates of X_t over a narrow band of frequencies in the neighborhood of ω_p, but excluding the ordinate at frequency ω_p. This approach (which is a special case of the "double window" technique—Priestley [7]) removes the main difficulty associated with Whittle's test, but still requires conditions on the bandwidth of $h_y(\omega)$ for its successful application. The resulting test statistic is closely related to the statistic $P(\lambda)$.

References

[1] Bhansali, R. J. (1977). In *Recent Developments in Statistics*, J. R. Barra et al., eds. North-Holland, Amsterdam, pp. 351–356.

[2] Bhansali, R. J. (1979). *J. R. Statist. Soc. A*, **142**, 199–209.

[3] Fisher, R. A. (1929). *Proc. R. Soc. Lond. A*, **125** 54–59.

[4] Hannan, E. J. (1961). *J. R. Statist. Soc. B*, **23**, 394–404.

[5] Priestley, M. B. (1962). *J. R. Statist. Soc. B*, **24**, 215–233.

[6] Priestley, M. B. (1962). *J. R. Statist. Soc. B*, **24**, 511–529.

[7] Priestley, M. B. (1964). *J. R. Statist. Soc. B*, **26**, 123–132.

[8] Priestley, M. B. (1981). *Spectral Analysis and Time Series*, Vol. I. Academic Press, London.

[9] Schuster, A. (1898). *Terr. Mag. Atmos. Electr.*, **3**, 13–41.

[10] Walker, G. (1914). *Calcutta Ind. Met. Memo*, **21**, 9.

[11] Whittle, P. (1952). *Trab. Estadíst.*, **3**, 43–57.

(NARROWBAND PROCESS
TIME-SERIES)

M. B. PRIESTLEY

PRIMARY SAMPLING UNIT

This is a term used in *multistage sampling*. At the first stage of sampling it is necessary to divide the entire *population* or *universe* into a convenient number of primary sampling units (PSUs). In large-scale *area samples** it is usual to form a few thousand PSUs, from which a few hundred would be selected to constitute the first-stage sample. Thus in area samples of the United States it is not uncommon for the PSUs to be counties.

The second stage of selection in multistage sampling takes place only within those PSUs which form the first-stage sample. Secondary sampling units, for instance census tracts, need to be listed only within those selected PSUs. If the number of selected PSUs is held constant, the consequences of forming large PSUs are that many second-stage units need to be listed and that those selected are scattered over relatively large areas. Conversely, if the PSUs are made small, the task of selecting at the second stage involves forming relatively short lists of second-stage units, and those selected are geographically concentrated.

It is usual, particularly in area sampling, for primary sampling units to have unequal probabilities of selection or of inclusion in a sample. These selection or inclusion probabilities are usually proportional to some measure of size, such as the number of persons or dwellings recorded in the PSU at the latest available census.

There are mathematical and operational advantages in selecting PSUs "with replacement." To select with unequal probabilities with replacement, the PSUs within each (usually geographical) stratum are assigned selection probabilities which sum to unity. A series of *n sample draws* is then made using these probabilities. It is possible for a PSU to be selected more than once, hence the expression "with replacement." (Because the same probabilities are used at each draw, the distribution of the numbers of times each PSU is selected in a sample is multinomial*. Since "sampling with replacement" does not necessarily imply that the same probabilities of selection are used at each draw, this type of sampling is occasionally referred to as "multinomial.")

If sampling is to be without replacement at the first stage, the probabilities of inclusion in the sample are usually chosen to sum to n, an integer, which is the required *sample number* or *sample size*. A procedure then has to be specified for selecting the sample in accordance with these inclusion probabilities. Either this procedure itself will be complicated, or the variance estimation will be complicated, inefficient or biased, or the sample must be limited to two units. Sampling "with replacement" (multinomial sampling) is less efficient but can be distinctly more convenient.

In particular, if the PSUs are selected "with replacement," then regardless of the selection methods used at subsequent stages, it is possible to obtain an unbiased estimator of variance (of the unbiased estimator of mean or total) by comparing the estimates based on the individual PSUs. Since these were selected independently, the estimates based on them are also independent. This applies only to estimation of the variance over all stages of sampling combined. To obtain simple unbiased estimators of the variances at the individual stages, sampling must be "with replacement" at the subsequent stages as well as the first (*see* STRATIFIED MULTISTAGE SAMPLING).

In the event that a PSU is selected twice or three times, it is usual to double or triple the number of second-stage sample units to be selected within it. This may in some cases involve multiple selection of second- or even lower-stage units. If final-stage units are selected more than once, unequal weighting must be used in the sample estimator.

Since area sampling effects a great simplification in the sampling procedure and the resulting *clustered samples** are much cheaper to enumerate than unclustered samples, this technique is widely used in large-scale household surveys such as the Current Population Survey and those of the major private polling organizations in the United States and other developed countries.

Although the term "primary sampling unit" is seldom used outside the context of area sampling, it is not confined to it. Thus if a two-stage sample is used in which the first stage consists of a sample of boxes and the second of cards within those boxes, the boxes would be the PSU's.

Bibliography

Barnett, V. (1974). *Elements of Sampling Theory*. Hodder & Stoughton, London, Chap. 5. (Concise treatment at an intermediate mathematical level.)

Sudman, S. (1976). *Applied Sampling*. Academic Press, New York. Chap. 7. (For survey users with limited statistical backgrounds.)

(AREA SAMPLING
STRATIFIED MULTISTAGE SAMPLING
SURVEY SAMPLING)

K. E. R. BREWER

PRIMITIVE ROOT

In any finite field (Galois field*), there is at least one element in the field, different powers of which give all the nonzero elements of the field. Such an element is called a *primitive root* of the field. For example, if the number of elements in the field is $p = 7$, then 3 is a primitive root since (modulo p)

$$3^1 = 3, \quad 3^2 = 2, \quad 3^3 = 6, \quad 3^4 = 4,$$
$$3^5 = 5, \quad \text{and } 3^6 = 1.$$

(Actually, $X^{p-1} = 1$ when X is any element of the field.)

(CONFOUNDING
FRACTIONAL FACTORIAL DESIGNS
GALOIS FIELDS)

PRINCIPAL COMPONENTS *See* COMPONENT ANALYSIS; MULTIVARIATE ANALYSIS

PRINCIPAL COMPONENTS REGRESSION ANALYSIS

Principal components (PC) regression is often used to solve the problem of multicollinearity* [3–5, 8], especially when deleting independent variables to break up the multicollinearity is not a feasible option [3]. One also uses PC regression when there are more independent variables than observations [5].

Basically, the way that PC regression works is that one first finds the set of orthonormal eigenvectors of the correlation matrix of the independent variables. The matrix of PCs is then the product of the eigenmatrix with the matrix of independent variables. The matrix of PCs is such that the first PC will have the maximum possible variance, the second the maximum possible variance among those uncorrelated with the first, and so on. The set of PCs then becomes the set of new regressor variables. Upon completion of this regression, one transforms back to the original coordinate system.

Let the ordinary least squares* (OLS) model be defined (the vector and matrix dimensions are given in parentheses) as

$$\underset{(n\times 1)}{\mathbf{Y}} = \underset{(1\times 1)(n\times 1)}{\gamma_0 \mathbf{1}} + \underset{(n\times p)(p\times 1)}{\mathbf{X}\boldsymbol{\gamma}} + \underset{(n\times 1)}{\boldsymbol{\epsilon}}. \quad (1)$$

To compute the correlation matrix*, it is useful to standardize the variables:

$$\begin{aligned} x_{ij} &= \left(X_{ij} - \bar{X}_j\right) \Big/ \left[\textstyle\sum_i \left(X_{ij} - \bar{X}_j\right)^2\right]^{1/2} \\ &= \left(X_{ij} - \bar{X}_j\right)/\left(s_j\sqrt{n-1}\right) \\ y_i &= \left(Y_i - \bar{Y}\right) \Big/ \left[\textstyle\sum_i \left(Y_i - \bar{Y}\right)^2\right]^{1/2} \\ &= \left(Y_i - \bar{Y}\right)/\left(s_y\sqrt{n-1}\right). \end{aligned} \quad (2)$$

Then (1) becomes

$$\mathbf{y} = \mathbf{x}\boldsymbol{\beta} + \boldsymbol{\epsilon}, \qquad \beta_j = \gamma_j s_j \sqrt{n-1}\,. \quad (3)$$

The correlation matrix $\mathbf{x}'\mathbf{x}$ has eigenvalues and eigenvectors defined by

$$|\mathbf{x}'\mathbf{x} - \lambda_r \mathbf{I}| = 0 \qquad \text{and} \qquad (\mathbf{x}'\mathbf{x} - \lambda_r \mathbf{I})\mathbf{a}_r = \mathbf{0}, \quad (4)$$

$r = 1, 2, \ldots, p$, where $\mathbf{a}'_r = (a_{1r}, a_{2r}, \ldots, a_{pr})$ with constraints $\mathbf{a}'_r\mathbf{a}_s = \delta_{rs}$, where $\delta_{rs} = 1$ if $r = s$ and is zero otherwise.

$$\begin{aligned} &\boldsymbol{\Lambda} = \operatorname{diag}\left[\lambda_1, \lambda_2, \ldots, \lambda_p\right] \text{ such that} \\ &\lambda_1 \geqslant \lambda_2 \geqslant \cdots \geqslant \lambda_p \geqslant 0. \end{aligned} \quad (5)$$

Then, if $\mathbf{A} = (\mathbf{a}_1, \ldots, \mathbf{a}_p)$,

$$\mathbf{A}'\mathbf{x}'\mathbf{x}\mathbf{A} = \mathbf{Z}'\mathbf{Z} = \boldsymbol{\Lambda} \qquad \text{and} \qquad \mathbf{x}'\mathbf{x} = \mathbf{A}\boldsymbol{\Lambda}\mathbf{A}', \quad (6)$$

since $\mathbf{A}$ is orthogonal, where $\mathbf{Z} = \mathbf{x}\mathbf{A}$ is the PC matrix and $\mathbf{Z}'\mathbf{Z}$ is the new design matrix. Then

$$\operatorname{var}(\mathbf{z}_r) = \mathbf{z}'_r\mathbf{z}_r = \lambda_r, \qquad r = 1, \ldots, p, \quad (7)$$

$$\operatorname{tr}(\mathbf{x}'\mathbf{x}) = p = \sum \mathbf{z}'_r\mathbf{z}_r = \sum \lambda_r$$

implies that λ_i/p represents the proportionate contribution of the ith PC to the total variation of the x's.

McCallum [8] has noted that

$$\mathbf{y} = \mathbf{x}\boldsymbol{\beta} + \boldsymbol{\epsilon} = \mathbf{Z}\mathbf{A}'\boldsymbol{\beta} + \boldsymbol{\epsilon}, \quad (8)$$

which implies that

$$\mathbf{h} = \mathbf{A}'\boldsymbol{\beta} \qquad \text{or} \qquad \boldsymbol{\beta} = \mathbf{h}\mathbf{A}, \quad (9)$$

where $\mathbf{h}$ is the coefficient vector of the regression on the PCs. Hence the transformed coefficients for model (1) are obtained from (9) and (3) as

$$\gamma_i = \sum_{j=1}^{p} a_{ij}h_j / \left(s_i\sqrt{n-1}\right), \qquad i = 1, \ldots, p, \quad (10)$$

where i corresponds to the independent variable number,

j corresponds to the PC number

γ_i is the transformed coefficient, model (1)

a_{ij} are elements of the eigenmatrix

h_j are the coefficients obtained from the regression on the PCs

s_i is the standard deviation of independent variable i

(In ridge regression*, $\mathbf{h}$ is referred to as the vector of *canonical coefficients*.) If a subset of the PCs is selected, some of the h_j are zero. The constant term in the transformed equation can be obtained from

$$\gamma_0 = \bar{Y} - \sum_{i=1}^{p} \gamma_i \bar{X}_i\,. \quad (11)$$

McCallum [8] has also observed that

$$\operatorname{var} \mathbf{b} = \sigma^2(\mathbf{x}'\mathbf{x})^{-1} = \sigma^2(\mathbf{A}\Lambda^{-1}\mathbf{A}'), \quad (12)$$

where $\mathbf{b}$ is the least squares estimator of $\boldsymbol{\beta}$. Therefore, the standard errors of the estimators of the coefficients for model (1) (but not the constant term) are, from (12) and (3),

$$\mathrm{SE}(\gamma_i) = \frac{\sigma}{s_i\sqrt{n-1}} \left(\sum_{j=1}^{p} \frac{a_{ij}^2}{\lambda_j} \right)^{1/2}, \quad i = 1, \ldots, p. \quad (13)$$

When a subset of the PCs is selected, the appropriate a_{ij} and λ_j are deleted from (13).

One does not need to be concerned with the interpretation of the PCs or the regression on the PCs, because this is just an intermediate step and one can transform back to the original coordinate system (as shown above).

Three selection rules for picking PCs are [2, 3, 8]:

1. Keeping all p PCs yields zero bias [because this yields OLS, as shown in (8)].
2. Keeping those PCs with the largest eigenvalues minimizes the variances of the estimators of the coefficients. (The proof is given in Greenberg [3] and Coxe [2, p. 226]. One simply needs to substitute **A** for **Z** and **Z** for **W**.)
3. Keeping those PCs which are highly correlated with the dependent variable tends to minimize the mean square errors of the estimators of the coefficients. (Coxe [2] has given an intuitive argument why this rule holds, but it has yet to be proven.)

The reasons given are: (a) McCallum [8] has shown that one can obtain smaller MSEs by eliminating some of the PCs. (b) Since rule 1 yields zero bias and rule 2 yields minimum coefficient variances, minimum MSE should fall somewhere in between. (c) Selecting those PCs that are strongly correlated with the dependent variable is similar to selecting independent variables by stepwise regression* in OLS; that is, both use F-values to determine which terms to retain. It should be noted that selection rules 2 and 3 do not necessarily yield the same PCs model, and hence the resulting coefficients in model 1 can be quite different, especially when the degree of multicollinearity is severe.

There is no definitive source for PC regression analysis. In going from paper to paper, one must be alert to the different standardizations of the variables. Some use s_j in the denominator, others use $s_j\sqrt{n-1}$, and still others do not standardize at all.

References

[1] Anderson, T. W. (1958). *An Introduction to Multivariate Statistics*. Wiley, New York, pp. 272–287. (A rigorous derivation of PC.)

[2] Coxe, K. (1982). *Amer. Statist. Ass., Proc. Bus. & Econ. Statist. Sec.*, pp. 222–227. (Selection rules for PC regression. Comparison with latent root regression. Long on math symbols, short on words. Good list of references.)

[3] Greenberg, E. (1975). *J. Amer. Statist. Ass.*, **70**, 194–197. [Deals with selection rule (2). Good, fairly technical paper.]

[4] Gunst, R. F. and Mason, R. L. (1980). *Regression Analysis and Its Application*. Marcel Dekker, New York, pp. 317–328. (One of the better treatments of the subject in a textbook.)

[5] Johnston, J. (1972). *Econometric Methods*, 2nd ed. McGraw-Hill, New York, pp. 322–331. (A good, simple derivation of PC regression.)

[6] Karson, M. J. (1982). *Multivariate Statistical Methods*. Iowa State University Press, Ames, IA, pp. 191–224. (Many numerical examples. Excellent list of references.)

[7] Massy, W. F. (1965). *J. Amer. Statist. Ass.*, **60**, 234–256. (Derivation of PC regression. Fairly technical paper and one of the first.)

[8] McCallum, B. T. (1970). *Rev. Econ. Statist.*, **52**, 110–113. (Good paper. Easy to read.)

[9] Morrison, D. F. (1967). *Multivariate Statistical Methods*. McGraw-Hill, New York, pp. 221–258. (A rather simple approach. Long list of references.)

[10] Pidot, G. B., Jr. (1969). *Rev. Econ. Statist.*, **51**, 176–188. (Suggests using a combination of the X's and the PC's.)

(COMPONENT ANALYSIS
DISCRIMINANT ANALYSIS
LATENT ROOT REGRESSION

MULTICOLLINEARITY
MULTIVARIATE ANALYSIS)

KEREN L. COXE

PRINCIPAL MOMENTS

An obsolete term for central moments* (moments about the mean).

PRINCIPLE OF PARSIMONY *See* PARSIMONY, PRINCIPLE OF; SCIENTIFIC METHOD IN STATISTICS

PRINCIPLES OF PROFESSIONAL STATISTICAL PRACTICE

PURPOSE AND SCOPE

1. The statistician's job. The statistician's responsibility, whether as a consultant on a part-time basis, or on regular salary, is to find problems that other people could not be expected to perceive. The statistician's tool is statistical theory (theory of probability). It is the statistician's use of statistical theory that distinguishes him from other experts. Statistical practice is a collaborative venture between statistician and experts in subject matter. Experts who work with a statistician need to understand the principles of professional statistical practice as much as the statistician needs them.

Challenges face statisticians today as never before. The whole world is talking about safety in mechanical and electrical devices (in automobiles, for example); safety in drugs; reliability; due care; pollution; poverty; nutrition; improvement of medical practice; improvement of agricultural practice; improvement in quality of product; breakdown of service; breakdown of equipment; tardy buses, trains and mail; need for greater output in industry and in agriculture; enrichment of jobs.

These problems cannot be understood and cannot even be stated, nor can the effect of any alleged solution be evaluated, without the aid of statistical theory and methods. One cannot even define operationally such adjectives as *reliable*, *safe*, *polluted*, *unemployed*, *on time* (arrivals), *equal* (in size), *round*, *random*, *tired*, *red*, *green*, or any other adjective, for use in business or in government, except in statistical terms. To have meaning for business or legal purposes, a standard (as of safety, or of performance or capability) must be defined in statistical terms.

2. Professional practice. The main contribution of a statistician in any project or in any organization is to find the important problems, the problems that other people cannot be expected to perceive. An example is the 14 points that top management learned in Japan in 1950. The problems that walk in are important, but are not usually *the* important problems [4].

The statistician also has other obligations to the people with whom he works, examples being to help design studies, to construct an audit by which to discover faults in procedure before it is too late to make alterations, and at the end, to evaluate the statistical reliability of the results.

The statistician, by virtue of experience in studies of various kinds, will offer advice on procedures for experiments and surveys, even to forms that are to be filled out, and certainly in the supervision of the study. He knows by experience the dangers of (1) forms that are not clear, (2) procedures that are not clear, (3) contamination, (4) carelessness, and (5) failure of supervision.

Professional practice stems from an expanding body of theory and from principles of application. A professional man aims at recognition and respect for his practice, not for himself alone, but for his colleagues as well. A professional man takes orders, in technical matters, from standards set by his professional colleagues as unseen judges, never from an administrative superior. His usefulness and his profession will suffer impairment if he yields to convenience or to opportunity. A professional man feels an obligation to provide services that his client may never comprehend or appreciate.

A professional statistician will not follow methods that are indefensible, merely to please someone, or support inferences based on such methods. He ranks his reputation and profession as more important than convenient assent to interpretations not warranted by statistical theory. Statisticians can be trusted and respected public servants.

Their careers as expert witnesses will be shattered if they indicate concern over which side of the case the results seem to favor. "As a statistician, I couldn't care less" is the right attitude in a legal case or in any other matter.

LOGICAL BASIS FOR DIVISION OF RESPONSIBILITIES

3. Some limitations in application of statistical theory. Knowledge of statistical theory is necessary but not sufficient for successful operation. Statistical theory does not provide a road map toward effective use of itself. The purpose of this article is to propose some principles of practice, and to explain their meaning in some of the situations that the statistician encounters.

Statisticians have no magic touch by which they may come in at the stage of tabulation and make something of nothing. Neither will their advice, however wise in the early stages of a study, ensure successful execution and conclusion. Many a study launched on the ways of elegant statistical design, later boggled in execution, ends up with results to which the theory of probability can contribute little.

Even though carried off with reasonable conformance to specifications, a study may fail through structural deficiencies in the method of investigation (questionnaire, type of test, technique of interviewing) to provide the information needed. Statisticians may reduce the risk of this kind of failure by pointing out to their clients in the early stages of the study the nature of the contributions that they themselves must put into it. (The word *client* will denote the expert or group of experts in a substantive field.) The limitations of statistical theory serve as signposts to guide a logical division of responsibilities between statistician and client. We accordingly digress for a brief review of the power and the limitations of statistical theory.

We note first that statistical inferences (probabilities, estimates, confidence limits, fiducial limits, etc.), calculated by statistical theory from the results of a study, will relate only to the material, product, people, business establishments, and so on, that the frame was in effect drawn from, and only to the environment of the study, such as the method of investigation, the date, weather, rate of flow and levels of concentration of the components used in tests of a production process, range of voltage, or of other stress specified for the tests (as of electrical equipment).

Empirical investigation consists of observations on material of some kind. The material may be people; it may be pigs, insects, physical material, industrial product, or records of transactions. The aim may be enumerative, which leads to the theory for the sampling of finite populations*. The aim may be analytic, the study of the causes that make the material what it is, and which will produce other material in the future. A definable lot of material may be divisible into identifiable sampling units. A list of these sampling units, with identification of each unit, constitutes a *frame*. In some physical and astronomical investigations, the sampling unit is a unit of time. We need not detour here to describe nests of frames for multistage sampling. The important point is that without a frame there can be neither a complete coverage of a designated lot of material nor a sample of any designated part thereof. Stephan introduced the concept of the frame, but without giving it a name [7].

Objective statistical inferences in respect to the frame are the speciality of the statistician. In contrast, generalization to cover material not included in the frame, nor to ranges, conditions, and methods outside the scope of the experiment, however essential for application of the results of a study, are a matter of judgment and depend on knowledge of the subject matter [5].

For example, the universe in a study of consumer behavior might be all the female homemakers in a certain region. The frame might therefore be census blocks, tracts or other small districts, and the ultimate sampling unit might be a segment of area containing households. The study itself will, of course, reach only the people that can be found at home in the segments selected for the sample. The client, however, must reach generalizations and take action on the product or system of marketing with respect to all female homemakers, whether they be the kind that are almost always at home and easy to reach, or almost never at home and therefore in part omitted from the study. Moreover, the female homemakers on whom the client must take action belong to the future, next year, and the next. The frame only permits study of the past.

For another example, the universe might be the production process that will turn out next week's product. The frame for study might be part of last week's product. The universe might be traffic in future years, as in studies needed as a basis for estimating possible losses or gains as a result of a merger. The frame for this study might be records of last year's shipments.

Statistical theory alone could not decide, in a study of traffic that a railway is making, whether it would be important to show the movement of (for example) potatoes, separately from other agricultural products in the northwestern part of the United States; only he who must use the data can decide.

The frame for a comparison of two medical treatments might be patients or other people in Chicago with a specific ailment. A pathologist might, on his own judgment, without further studies, generalize the results to people that have this ailment anywhere in the world. Statistical theory provides no basis for such generalizations.

No knowledge of statistical theory, however profound, provides by itself a basis for deciding whether a proposed frame would be satisfactory for a study. For example, statistical theory would not tell us, in a study of consumer research, whether to include in the frame people that live in trailer parks. The statistician places the responsibility for the answer to this question where it belongs, with the client: Are trailer parks in your problem? Would they be in it if this were a complete census? If yes, would the cost of including them be worthwhile?

Statistical theory will not of itself originate a substantive problem. Statistical theory cannot generate ideas for a questionnaire or for a test of hardness, nor can it specify what would be acceptable colors of dishes or of a carpet; nor does statistical theory originate ways to teach field workers or inspectors how to do their work properly or what characteristics of workmanship constitute a meaningful measure of performance. This is so in spite of the fact that statistical theory is essential for reliable comparisons of questionnaires and tests.

4. Contributions of statistical theory to subject matter. It is necessary, for statistical reliability of results, that the design of a survey or experiment fit into a theoretical model. Part of the statistician's job is to find a suitable model that he can manage, once the client has presented his case and explained why statistical information is needed and how the results of a study might be used. Statistical practice goes further than merely to try to find a suitable model (theory). Part of the job is to adjust the physical conditions to meet the model selected. Randomness*, for example, in sampling a frame, is not just a word to assume; it must be created by use of random numbers for the selection of sampling units. Recalls to find people at home, or tracing illegible or missing information back to source documents, are specified so as to approach the prescribed probability of selection.

Statistical theory has in this way profoundly influenced the theory of knowledge. It has given form and direction to quantitative studies by molding them into the requirements of the theory of estimation*, and other techniques of inference. The aim is, of course, to yield results that have meaning in terms that people can understand.

The statistician is often effective in assisting the substantive expert to improve accepted methods of interviewing, testing, coding, and other operations. Tests properly designed will show how alternative test or field procedures really perform in practice, so that rational changes or choices may be possible. It sometimes happens, for example, that a time-honored or committee-honored method of investigation, when put to statistically designed tests, shows alarming inherent variances between instruments, between investigators, or even between days for the same investigator. It may show a trend, or heavy intraclass correlation* between units, introduced by the investigators. Once detected, such sources of variation may then be corrected or diminished by new rules, which are of course the responsibility of the substantive expert.

Statistical techniques provide a safe supervisory tool to help to reduce variability in the performance of worker and machine. The effectiveness of statistical controls, in all stages of a survey, for improving supervision, to achieve uniformity, to reduce errors, to gain speed, reduce costs, and to improve quality of performance in other ways is well known. A host of references could be given, but two will suffice [2, 3].

5. Statistical theory as a basis for division of responsibilities. We may now, with this background, see where the statistician's responsibilities lie.

In the first place, his specialized knowledge of statistical theory enables him to see which parts of a problem belong to substantive knowledge (sociology, transportation, chemistry of a manufacturing process, law), and which parts are statistical. The responsibility as logician in a study falls to him by default, and he must accept it. As logician, he will do well to designate, in the planning stages, which decisions will belong to the statistician and which to the substantive expert.

This matter of defining specifically the areas of responsibility is a unique problem faced by the statistician. Business managers and lawyers who engage an expert on corporate finance, or an expert on steam power plants, know pretty well what such experts are able to do and have an idea about how to work with them. It is different with statisticians. Many people are confused between the role of theoretical statisticians (those who use theory to guide their practice) and the popular idea that statisticians are skillful in compiling tables about people or trade, or who prophesy the economic outlook for the coming year and which way the stock market will drift. Others may know little about the contributions that statisticians can make to a study or how to work with them.

Allocation of responsibilities does not mean impervious compartments in which you do this and I'll do that. It means that there is a logical basis for allocation of responsibilities and that it is necessary for everyone involved in a study to know in advance what he or she will be accountable for.

A clear statement of responsibilities will be a joy to the client's lawyer in a legal case, especially at the time of cross-examination. It will show the kind of question that the statistician is answerable for, and what belongs to the substantive experts. Statisticians have something to learn about professional practice from law and medicine.

6. Assistance to the client to understand the relationship. The statistician must direct the client's thoughts toward possible uses of the results of a statistical study of the entire frame without consideration of whether the entire frame will be studied or only a small sample of the sampling units in the frame. Once these matters are cleared, the statistician may outline one or more statistical plans and explain to the client in his own language what they mean in respect to possible choices of frame and environmental conditions, choices of sampling unit, skills, facilities and supervision required. The statistician will urge the client to foresee possible difficulties with definitions or with the proposed method of investigation. He may illustrate with rough calculations possible levels of precision to be expected from one or more statistical plans that appear to be feasible, along with rudimentary examples of the kinds of tables and inferences that might be forthcoming. These early stages are often the hardest part of the job.

The aim in statistical design is to hold accuracy and precision to sensible levels, with an economic balance between all the possible uncertainties that afflict data—built-in deficiencies of definition, errors of response, nonresponse, sampling variation, difficulties in coding, errors in processing, difficulties of interpretation.

Professional statistical practice requires experience, maturity, fortitude, and patience, to protect the client against himself or against his duly appointed experts in subject matter who may have in mind needless but costly tabulations in fine classes (five-year age groups, small areas, fine mileage brackets, fine gradations in voltage, etc.), or unattainable precision in differences between treatments, beyond the capacity of the skills and facilities available, or beyond the meaning inherent in the definitions.

These first steps, which depend heavily on guidance from the statistician, may lead to important modifications of the problem. Advance considerations of cost, and of the limitations of the inferences that appear to be possible, may even lead clients to abandon a study, at least until they feel most confident of the requirements. Protection of a client's bank account, and deliverance from more serious losses from decisions based on inconclusive or misleading results, or from misinterpretation, is one of the statistician's greatest services.

Joint effort does not imply joint responsibility. Divided responsibility for a decision in a statistical survey is as bad as divided responsibility in any venture—it means that no one is responsible.

Although they acquire superficial knowledge of the subject matter, the one thing that statisticians contribute to a problem, and which distinguishes them from other experts, is knowledge and ability in statistical theory.

SUMMARY STATEMENT OF RECIPROCAL OBLIGATIONS AND RESPONSIBILITIES

7. Responsibilities of the client. The client will assume responsibility for those aspects of the problem that are substantive. Specifically, he will stand the ultimate responsibility for:

a. The type of statistical information to be obtained.

b. The methods of test, examination, questionnaire, or interview, by which to elicit the information from any unit selected from the frame.

c. The decision on whether a proposed frame is satisfactory.

d. Approval of the probability model proposed by the statistician (statistical procedures, scope, and limitations of the statistical inferences that may be possible from the results).

e. The decision on the classes and areas of tabulation (as these depend on the uses that the client intends to make of the data); the approximate level of statistical precision or protection that would be desirable in view of the purpose of the investigation, skills and time available, and costs.

The client will make proper arrangements for:

f. The actual work of preparing the frame for sampling, such as serializing and identifying sampling units at the various stages.

g. The selection of the sample according to procedures that the statistician will prescribe, and the preparation of these units for investigation.

h. The actual investigation; the training for this work, and the supervision thereof.

i. The rules for coding; the coding itself.

j. The processing, tabulations and computations, following procedures of estimation that the sampling plans prescribe.

The client or his representative has the obligation to report at once any departure from instructions, to permit the statistician to make a decision between a fresh start or an unbiased adjustment. The client will keep a record of the actual performance.

8. Responsibilities of the statistician. The statistician owes an obligation to his own practice to forestall disappointment on the part of the client, who if he fails to understand at the start that he must exercise his own responsibilities in the planning stages, and in execution, may not realize in the end the fullest possibility of the investigation, or may discover too late that certain information that he had expected to get out of the study was not built into the design. The statistician's responsibility may be summarized as follows:

a. To formulate the client's problem in statistical terms (probability model), subject to approval of the client, so that a proposed statistical investigation may be helpful to the purpose.

b. To lay out a logical division of responsibilities for the client, and for the statistician, suitable to the investigation proposed.

c. To explain to the client the advantages and disadvantages of various frames and possible choices of sampling units, and of one or more feasible statistical plans of sampling or experimentation that seem to be feasible.

d. To explain to the client, in connection with the frame, that any objective inferences that one may draw by statistical theory from the results of an investigation can only refer to the material or system that the frame was drawn from, and only to the methods, levels, types, and ranges of stress presented for study. It is essential that the client understand the limitations of a proposed study, and of the statistical inferences to be drawn therefrom, so that he may have a chance to modify the content before it is too late.

e. To furnish statistical procedures for the investigation—selection, computation of estimates and standard errors, tests, audits, and controls as seem warranted for detection and evaluation of important possible departures from specifications, variances between investigators, nonresponse, and other persistent uncertainties not contained in the standard error; to follow the work and to know when to step in.

f. To assist the client (on request) with statistical methods of supervision, to help him to improve uniformity of performance of investigators, gain speed, reduce errors, reduce costs, and to produce a better record of just what took place.

g. To prepare a report on the statistical reliability of the results.

9. The statistician's report or testimony. The statistician's report or testimony will deal with the statistical reliability of the results. The usual content will cover the following points:

a. A statement to explain what aspects of the study his responsibility included, and what it excluded. It will delimit the scope of the report.

b. A description of the frame, the sampling unit, how defined and identified, the material covered, and a statement in respect to conditions of the survey or experiment that might throw light on the usefulness of the results.

c. A statement concerning the effect of any gap between the frame and the universe for important conclusions likely to be drawn from the results. (A good rule is that the statistician should have before him a rough draft of the client's proposed conclusions.)

d. Evaluation (in an enumerative study) of the margin of uncertainty, for a specified probability level, attributable to random errors of various kinds, including the uncertainty introduced by sampling, and by small independent random variations in judgment, instruments, coding, transcription, and other processing.

e. Evaluation of the possible effects of other relevant sources of variation, examples being differences between investigators, between instruments, between days, between areas.

f. Effect of persistent drift and conditioning of instruments and of investigators; changes in technique.

g. Nonresponse and illegible or missing entries.

h. Failure to select sampling units according to the procedure prescribed.

i. Failure to reach and to cover sampling units that were designated in the sampling table.

j. Inclusion of sampling units not designated for the sample but nevertheless covered and included in the results.

k. Any other important slips and departures from the procedure prescribed.

l. Comparisons with other studies, if any are relevant.

In summary, a statement of statistical reliability attempts to present to readers all information that might help them to form their own opinions concerning the validity of conclusions likely to be drawn from the results.

The aim of a statistical report is to protect clients from seeing merely what they would like to see; to protect them from losses that could come from misuse of the results. A further aim is to forestall unwarranted claims of accuracy that a client's public might otherwise accept.

Any printed description of the statistical procedures that refer to this participation, or any evaluation of the statistical reliability of the results, must be prepared by the statistician as part of the engagement. If a client prints the statistician's report, it will be printed in full.

The statistician has an obligation to institute audits and controls as a part of the statistical procedures for a survey or experiment, to discover any departure from the procedures prescribed.

The statistician's full disclosure and discussion of all the blemishes and blunders that took place in the survey or experiment will build for him a reputation of trust. In a legal case, this disclosure of blunders and blemishes and their possible effects on conclusions to be drawn from the data are a joy to the lawyer.

A statistician does not recommend to a client any specific administrative action or policy. Use of the results that come from a survey or experiment are entirely up to the client. Statisticians, if they were to make recommendations for decision, would cease to be statisticians.

Actually, ways in which the results may throw light on foreseeable problems will be settled in advance, in the design, and there should be little need for a client or for anyone else to reopen a question. However, problems sometimes change character with time (as when a competitor of the client suddenly comes out with a new model), and open up considerations of statistical precision and significance of estimates that were not initially in view.

The statistician may describe in a professional or scientific meeting the statistical methods that he develops in an engagement. He will not publish actual data or substantive results or other information about a client's business without permission. In other words, the statistical methods belong to the statistician: the data to the client.

A statistician may at times perform a useful function by examining and reporting on a study in which he did not participate. A professional statistician will not write an opinion on another's procedures or inferences without adequate time for study and evaluation.

SUPPLEMENTAL REMARKS

10. Necessity for the statistician to keep in touch. A statistician, when he enters into a relationship to participate in a study, accepts certain responsibilities. He does not merely draft instructions and wait to be called. The people whom he serves are not statisticians, and thus cannot always know when they are in trouble. A statistician asks questions and will probe on his own account with the help of proper statistical design, to discover for himself whether the work is proceeding according to the intent of the instructions. He must expect to revise the instructions a number of times in the early stages. He will be accessible by mail, telephone, telegraph, or in person, to answer questions and to listen to suggestions.

He may, of course, arrange consultations with other statisticians on questions of theory or procedure. He may engage another statistician to take over certain duties. He may employ other skills at suitable levels to carry out independent tests or reinvestigation of certain units, to detect difficulties and departures from the prescribed procedure, or errors in transcription or calculation.

It must be firmly understood, however, that consultation or assistance is in no sense a partitioning of responsibility. The obligations of a statistician to his client, once entered into, may not be shared.

11. What is an engagement? Dangers of informal advice. It may seem at first thought that statisticians ought to be willing to give to the world informally and impromptu the benefit of their knowledge and experience, without discussion or agreement concerning participation and relationships. Anyone who has received aid from a doctor of medicine who did his best without a chance to make a more thorough examination can appreciate how important the skills of a professional man can be, even under handicap.

On second thought, most statisticians can recall instances in which informal advice backfired. It is the same in any professional line. A statistician who tries to be helpful and give advice under adverse circumstances is in practice and has a client, whether he intended it so or not; and he will later find himself accountable for the advice. It is important to take special precaution under these circumstances to state the basis of understanding for any statements or recommendations, and to make clear that other conditions and circumstances could well lead to different statements.

12. When do the statistician's obligations come to a close? A statistician should make it clear that his name may be identified with a study only as long as he is active in it and accountable for it. A statistical procedure, contrary to popular parlance, is not installed. One may install new furniture, a new carpet, or a new dean, but not a statistical procedure. Experience shows that a statistical procedure deteriorates rapidly when left completely to nonprofessional administration.

A statistician may draw up plans for a continuing study, such as for the annual inventory of materials in process in a group of manufacturing plants, or for a continuing national survey of consumers. He may nurse the job into running order, and conduct it through several performances. Experience shows, however, that if he steps out and leaves the work in nonstatistical hands, he will shortly find it to be unrecognizable. New people come on the job. They may think that they know better than their predecessor how to do the work; or they may not be aware that there ever were any rules or instructions, and make up their own.

What is even worse, perhaps, is that people that have been on a job a number of years think that they know it so well that they cannot go wrong. This type of fault will be observed, for example, in a national monthly sample in which households are to be revisited a number of times; when left entirely to nonstatistical administration, it will develop faults. Some interviewers will put down their best guesses about the family, on the basis of the preceding month, without an actual interview. They will forget the exact wording of the question, or may think that they have something better. They will become lax about calling back on people not at home. Some of them may suppose that they are following literally the intent of the instructions, when in fact (as shown by a control), through a misunderstanding, they are doing something wrong. Or they may depart wilfully, thinking that they are thereby improving the design, on the supposition that the statistician did not really understand the circumstances. A common example is to substitute an average-looking sampling unit when the sampling unit designated is obviously unusual in some respect [1].

In the weighing and testing of physical product, people will in all sincerity substitute their judgment for the use of random numbers. Administration at the top will fail to rotate areas in the manner specified. Such deterioration may be predicted with confidence unless the statistician specifies statistical controls that provide detective devices and feedback.

13. The single consultation. It is wise to avoid a single consultation with a commercial concern unless satisfactory agenda are prepared in advance and satisfactory arrangements made for absorbing advice offered. This requirement, along with an understanding that there will be a fee for a consultation, acts as a shield against a hapless conference which somebody calls in the hope that something may turn up. It also shows that statisticians, as professional people, although eager to teach and explain statistical methods, are not on the lookout for chances to demonstrate what they might be able to accomplish.

Moreover, what may be intended as a single consultation often ends up with a request for a memorandum. This may be very difficult to write, especially in the absence of adequate time to study the problem. The precautions of informal advice apply here.

14. The statistician's obligation in a muddle. Suppose that fieldwork is under way. Then the statistician discovers that the client or duly appointed representatives have disregarded the instructions for the preparation of the frame, or for the selection of the sample, or that the fieldwork seems to be falling apart. "We went ahead and did so and so before your letter came, confident that you would approve," is a violation of relationship. That the statistician may approve the deed done does not mitigate the violation.

If it appears to the statistician that there is no chance that the study will yield results that he could take professional responsibility for, this fact must be made clear to the client, at a sufficiently high management level. It is a good idea for the statistician to explain at the outset that such situations, while extreme, have been known.

Statisticians should do all in their power to help clients to avoid such a catastrophe. There is always the possibility that a statistician may be partly to blame for not being sufficiently clear nor firm at the outset concerning his principles of participation, or for not being on hand at the right time to ask questions and to keep apprised of what is happening. Unfortunate circumstances may teach the statistician a lesson on participation.

15. Assistance in interpretation of nonprobability samples. It may be humiliating, but statisticians must face the fact that many accepted laws of science have come from theory and experimentation without benefit of formal statistical design. Vaccination for prevention of smallpox is one; John Snow's discovery of the source of cholera in London is another [6]. So is the law $F = ma$ in physics; also Hooke's law, Boyle's law, Mendel's findings, Keppler's laws, Darwin's theory of evolution, the Stefan–Boltzmann law of radiation (first empirical, later established by physical theory). All this only means, as everyone knows, that there may well be a wealth of information in a nonprobability sample.

Perhaps the main contribution that the statistician can make in a nonprobability sample is to advise the experimenter against conclusions based on meaningless statistical calculations. The expert in the subject matter must take the responsibility for the effects of selectivity and confounding* of treatments. The statistician may make a positive contribution by encouraging the expert in the subject matter to draw whatever conclusions he believes to be warranted, but to do so over his own signature, and not to attribute conclusions to statistical calculations or to the kind of help provided by a statistician.

16. A statistician will not agree to use of his name as advisor to a study, nor as a member of an advisory committee, unless this service carries with it explicit responsibilities for certain prescribed phases of the study.

17. A statistician may accept engagements from competitive firms. His aim is not to concentrate on the welfare of a particular client, but to raise the level of service of his profession.

18. A statistician will prescribe in every engagement whatever methods known to him seem to be most efficient and feasible under the circumstances. Thus, he may prescribe for firms that are competitive methods that are similar or even identical word for word in part or in entirety. Put another way, no client has a proprietary right in any procedures or techniques that a statistician prescribes.

19. A statistician will, to the best of his ability, at his request, lend technical assistance to another statistician. In rendering this assistance, he may provide copies of procedures that he has used, along with whatever modification seems advisable. He will not, in doing this, use confidential data.

References

[1] Bureau of the Census (1954). Measurement of Employment and Unemployment. Report of Special Advisory Committee on Employment Statistics, Washington, D.C.

[2] Bureau of the Census (1964). Evaluation and research program of the censuses of population and housing. 1960 Series ER60. No. 1.

[3] Bureau of the Census (1963). The Current Population Survey and Re-interview Program. *Technical Paper No. 6.*

[4] Center for Advanced Engineering Study. (1983). *Management for Quality, Productivity, and Competitive Position.* Massachusetts Institute of Technology, Cambridge, Mass.

[5] Deming, W. E. (1960). *Sample Design in Business Research.* Wiley, New York, Chap. 3.

[6] Hill, A. B. (1953). Observation and experiment. *N. Engl. J. Med.*, **248**, 995–1001.

[7] Stephan, F. F. (1936). Practical problems of sampling procedure. *Amer. Sociol. Rev.*, **1**, 569–580.

Bibliography

See the following works, as well as the references just given, for more information on the topic of principles of professional statistical practice.

Brown, T. H. (1952). The statistician and his conscience. *Amer. Statist.*, **6**(1), 14–18.

Burgess, R. W. (1947). Do we need a "Bureau of Standards" for statistics? *J. Marketing*, **11**, 281–282.

Chambers, S. P. (1965). Statistics and intellectual integrity. *J. R. Statist. Soc. A*, **128**, 1–16.

Court, A. T. (1952). Standards of statistical conduct in business and in government. *Amer. Statist.*, **6**, 6–14.

Deming, W. E. (1954). On the presentation of the results of samples as legal evidence. *J. Amer. Statist. Ass.*, **49**, 814–825.

Deming, W. E. (1954). On the Contributions of Standards of Sampling to Legal Evidence and Accounting. Current Business Studies, Society of Business Advisory Professions, Graduate School of Business Administration, New York University, New York.

Eisenhart, C. (1947). The role of a statistical consultant in a research organization. *Proc. Int. Statist. Conf.*, **3**, 309–313.

Freeman, W. W. K. (1952). Discussion of Theodore Brown's paper [see Brown, 1952]. *Amer. Statist.*, **6**(1), 18–20.

Freeman, W. W. K. (1963). Training of statisticians in diplomacy to maintain their integrity. *Amer. Statist.*, **17**(5), 16–20.

Gordon, R. A. et al. (1962). Measuring Employment and Unemployment Statistics. President's Committee to Appraise Employment and Unemployment Statistics. Superintendent of Documents, Washington, D.C.

Hansen, M. H. (1952). Statistical standards and the Census. *Amer. Statist.*, **6**(1), 7–10.

Hotelling, H. et al. (1948). The teaching of statistics (a report of the Institute of Mathematical Statistics Committee on the teaching of statistics). *Ann. Math. Statist.*, **19**, 95–115.

Jensen, A. et al. (1948). The life and works of A. K. Erlang. *Trans. Danish Acad. Tech. Sci.*, No. 2 (Copenhagen).

Molina, E. C. (1913). Computation formula for the probability of an event happening at least c times in n trials. *Amer. Math. Monthly*, **20**, 190–192.

Molina, E. C. (1925). The theory of probability and some applications to engineering problems. *J. Amer. Inst. Electr. Eng.*, **44**, 1–6.

Morton, J. E. (1952). Standards of statistical conduct in business and government. *Amer. Statist.*, **6**(1), 6–7.

Shewhart, W. A. (1931). *Economic Control of Quality of Manufactured Product*. D. Van Nostrand, New York.

Shewhart, W. A. (1939). *Statistical Method from the Viewpoint of Quality Control*. The Graduate School, Department of Agriculture, Washington, D.C.

Shewhart, W. A. (1958). Nature and origin of standards of quality. *Bell Syst. Tech. J.*, **37**, 1–22.

Society of Business Advisory Professions (1957). Report of Committee on Standards of Probability Sampling for Legal Evidence. Graduate School of Business Administration, New York University, New York.

Zirkle, C. (1954). Citation of fraudulent data. *Science*, **120**, 189–190.

(CONSULTING, STATISTICAL)

W. E. DEMING

PRIOR DENSITY *See* PRIOR DISTRIBUTIONS

PRIOR DISTRIBUTIONS

Prior distributions are probability distributions or probability density functions (PDFs) that summarize information about a random variable or parameter known or assumed at a given point in time, *prior* to obtaining further information from empirical data, and so on. The use of the designation "prior distribution" occurs almost entirely within the context of statistical methodology generally labeled "Bayesian inference"*. Bayes' theorem* and a 1763 paper by Thomas Bayes [2] are generally cited as the beginnings of this methodology. However, Stigler [12] cites evidence that the little-known English mathematician Saunderson may have been the true builder of the foundations of what is now termed Bayesian inference.

Suppose that Y and X are two random variables with PDFs $f_Y(y)$ and $f_X(x)$ (Y and X may be discrete or continuous). Denoting the joint distribution of Y and X by $f_{X,Y}(x, y)$ and conditional distributions of Y given X and X given Y by $f_{Y|X}(y|x)$ and $f_{X|Y}(x|y)$, respectively, it is known by definition that

$$f_{Y|X}(y|x)f_X(x) = f_{X,Y}(x, y) = f_{X|Y}(x|y)f_Y(y).$$

Then, for example,

$$f_{X|Y}(x|y) = \frac{f_{Y|X}(y|x)f_X(x)}{f_Y(y)}.$$

This result is one representation of the so-called "Bayes' theorem," true by definition and not an object of debate among statisticians. $f_X(x)$ tells us what is known about X before or prior to the observation of Y, and is thus termed the *prior distribution of* X. Alternatively, $f_{X|Y}(x|y)$ tells us what is known about X after or posterior to the observation of Y (*see* POSTERIOR DISTRIBUTIONS).

The representation of Bayes' theorem that is associated with Bayesian inference and the typical reference to prior distribution is

$$f_{\Theta|\mathbf{Y}}(\boldsymbol{\theta}|\mathbf{y}) = \frac{f_{\mathbf{Y}|\Theta}(\mathbf{y}|\boldsymbol{\theta})f_\Theta(\boldsymbol{\theta})}{f_{\mathbf{Y}}(\mathbf{y})},$$

where $\mathbf{Y}' = (Y_1, Y_2, \ldots, Y_n)$ is a vector of n independent Y_i representing sample observations, each with probability distribution $f_{Y_i|\Theta}(y_i|\boldsymbol{\theta})$ dependent on the value of m parameters in the vector $\boldsymbol{\Theta}$. For example, for normally distributed Y_i,

$$f_{Y_i|\Theta}(y_i|\boldsymbol{\theta}) = \frac{1}{\sqrt{2\pi}\,\sigma}\exp\left[-\frac{1}{2}\left(\frac{y_i-\mu}{\sigma}\right)^2\right], \quad -\infty < y_i < \infty,$$

where $\boldsymbol{\Theta}' = (\mu, \sigma)$, the mean and standard deviation, and $f_{\mathbf{Y}|\Theta}(\mathbf{y}|\boldsymbol{\theta}) = \prod f_{Y_i|\Theta}(y_i|\boldsymbol{\theta})$. The focus of this representation is inference about the typically unknown parameter vector $\boldsymbol{\Theta}$, where the prior distribution $f_\Theta(\boldsymbol{\theta})$ is modified by the empirical data $\mathbf{y}$ to form the posterior distribution $f_{\Theta|\mathbf{Y}}(\boldsymbol{\theta}|\mathbf{y})$. The prior distribution $f_\Theta(\boldsymbol{\theta})$ is an object of debate among statisticians. Classical statisticians perceive parameters as fixed, even if unknown, and reject the idea of any "distribution" for $\boldsymbol{\Theta}$. They also complain that even if one accepts the concept of such a distribution, its exact nature is almost inevitably subjective, based only on some vague degree-of-belief* probability concept rather than the scientifically sound relative-frequency-in-repeated-experiments probability concept. Alternatively, Bayesian statisticians are comfortable using prior distributions for unknown parameters, even if there is considerable subjectivity involved.

There have been attempts to reduce subjectivity by developing procedures for estimating prior distributions. Martz and Waller [9] discuss associating distribution parameters with characteristics of the distribution that one may feel relatively well known, such as chosen percentiles, the distribution mean, and so on. O'Bryan and Walter [10] and Walter [13] discuss specific prior distribution estimation procedures.

Some commonly encountered categorizations of prior distributions are:

Improper Prior: Distribution does not sum or integrate to 1. Otherwise, the prior is proper. Improper priors are often used as approximations to proper prior distributions which are nearly flat in the range of $\mathbf{\Theta}$ indicated to be relevant by the empirical data $\mathbf{y}$. Akaike [1] and Eaton [5] discuss the interpretation and evaluation of improper priors.

Reference Prior: A prior distribution used as a standard in a particular area of study [3, p. 22].

Locally Uniform Prior: Typically refers to a prior density function for a continuous parameter which is at least approximately constant over a limited relevant parameter range, thus expressing equal probability of the parameter falling in any equal-length subrange.

Noninformative Prior: A prior distribution which expresses little knowledge about the parameter relative to the information expected in a size n sample of empirical data. Locally uniform priors are commonly used as noninformative priors. There is a problem, however, in selecting the metric (i.e., $\mathbf{\Theta}$ itself, $\log \mathbf{\Theta}$, $\mathbf{\Theta}^{-1}$, etc.) in which the prior $f_{\mathbf{\Theta}}(\boldsymbol{\theta})$ is locally uniform to best reflect true noninformativeness. Box and Tiao [3, p. 25] discuss this in depth. Jeffrey's rule [7] is a rule guiding the choice of noninformative prior distributions. *See* JEFFREYS' PRIOR DISTRIBUTION.

Conjugate Prior: A prior distribution $f_{\mathbf{\Theta}}(\boldsymbol{\theta})$ which combines $f_{\mathbf{Y}|\mathbf{\Theta}}(\mathbf{y}|\boldsymbol{\theta})$ to produce a posterior distribution $f_{\mathbf{\Theta}|\mathbf{Y}}(\boldsymbol{\theta}|\mathbf{y})$ which is in the same family as the prior distribution; for example, both are normal distributions. See Martz and Waller [9, p. 226], De Groot [4, p. 164], Raiffa and Schlaifer [11, p. 53], and CONJUGATE FAMILIES OF DISTRIBUTIONS.

Exchangeable Prior: A prior distribution $f_{\mathbf{\Theta}}(\boldsymbol{\theta})$ which is not altered by any permutation in the multiparameter $\mathbf{\Theta}$. Exchangeability* has been examined in the context of parameters from multiple populations and multinomial parameters of a single population. See Lindley and Smith [8] and Eaves [6].

References

[1] Akaike, H. (1980). *J. R. Statist. Soc. B*, **42**, 46–52. (On the interpretation of improper prior distributions as limits of data-dependent proper prior distributions.)

[2] Bayes, T. (1958). *Biometrika*, **45**, 298–315. (Reprint of the famous 1763 paper generally credited with giving birth to Bayesian statistical inference.)

[3] Box, G. E. P. and Tiao, G. C. (1973). *Bayesian Inference in Statistical Analysis*. Addison-Wesley, Reading, Mass. (An excellent general statistical reference on Bayesian inference.)

[4] DeGroot, M. H. (1970). *Optimal Statistical Decisions*. McGraw-Hill, New York.

[5] Eaton, M. L. (1982). In *Statistical Decision Theory and Related Topics III*, Vol. 1, S. Gupta and J. Berger, eds. Academic Press, New York, pp. 1329–1352. (Evaluating improper prior distributions.)

[6] Eaves, D. M. (1980). *J. R. Statist. Soc. B*, **42**, 88–93. (Good reference list for work with exchangeable priors.)

[7] Jeffreys, H. (1961). *Theory of Probability*, 3rd ed. Clarendon Press, Oxford, England.

[8] Lindley, D. V. and Smith, A. F. M. (1972). *J. R. Statist. Soc. B*, **34**, 1–41. (Exchangeable priors in the context of the usual linear statistical model.)

[9] Martz, H. F. and Waller, R. A. (1982). *Bayesian Reliability Analysis*. Wiley, New York. (Chapters 5 and 6 contain good general knowledge about Bayesian inference.)

[10] O'Bryan, T. and Walter, G. (1979). *Sankhyā A*, **41**, 95–108.

[11] Raiffa, H. and Schlaifer, R. (1961). *Applied Statistical Decision Theory*. Harvard University Press, Cambridge, Mass.

[12] Stigler, S. M. (1983). *Amer. Statist.*, **37**, 290–296. (A witty search into who really discovered Bayes' theorem.)

[13] Walter, G. (1981). *Sankhyā A*, **43**, 228–245.

Bibliography

See the following works, as well as the references just given, for more information on the topic of prior distributions.

Lindley, D. V. (1965). *Introduction to Probability and Statistics from a Bayesian Viewpoint, Part 1: Probability; Part 2: Inference*, Cambridge University Press, Cambridge, England.

Savage, L. J. (1954). *The Foundations of Statistics*. Wiley, New York. (General reference with philosophical basis for Bayesian statistical inference.)

(BAYESIAN INFERENCE
BAYES' THEOREM
CONJUGATE FAMILIES OF DISTRIBUTIONS
DEGREES OF BELIEF
POSTERIOR DISTRIBUTIONS
POSTERIOR PROBABILITIES
PRIOR PROBABILITIES)

DAVID J. PACK

PRIORITY QUEUE

As a general background regarding ideas and notation, we refer the reader to QUEUEING THEORY. The theory for priority queueing systems is the branch of queueing theory dealing with systems where high-priority customers get faster through the system at the expense of others. The basic reference seems still to be ref. 4, giving a series of different models, with a variety of priority disciplines, and a solid mathematical treatment. In ref. 5, priority queueing systems of special interest in computer applications are considered.

Since the applicability of queueing theory has often been questioned, we will here give reviews of three papers having specific applications in mind. We are not claiming that these are either the best or the most general in the area. One should also remember that even a complex queueing system is more sympathetic to mathematical modeling than is a series of other real-life systems. Hence queueing theory has been and will still be an experimental area to develop tools which are useful in other areas of applied probability.

CLASSICAL PRIORITY QUEUEING MODELS

Assume that the customers are divided into p different priority classes each having a priority index $1, 2, \ldots, p$. The index 1 corresponds to the highest-priority class and p corresponds to the lowest. For simplicity we assume a single server and an infinite waiting-room capacity. Within each class we have a first-in-first-out (FIFO) queueing discipline. Denoting the interarrival-time distribution and the service-time distribution within the ith class by the symbols A_i and B_i, respectively, we can use the notation $A_1/\cdots/A_p/B_1/\cdots/B_p/1$ for this class of models.

However, we have to specify the queueing discipline between customers from different priority classes. Compare the ith and jth classes, where $1 \leqslant i < j \leqslant p$. In the *queue* all customers from the ith class are standing in front of all customers from the jth class and will hence be served first. If, on the other hand, a customer from the jth class is being served on the arrival of a customer from the ith class, there are different alternatives. The priority discipline is *preemptive** if the current service is immediately interrupted and service is started for the arriving customer. It is *nonpreemptive* if service is continued to completion, and it is *discretionary* if the server may use his or her discretion to decide which of the two former strategies to use in each case. For instance, he or she may use the nonpreemptive discipline if the amount of service received exceeds a certain level [4], or if remaining service time is sufficiently small [3].

APPOINTMENT SYSTEM

Consider the $GI/M/1$ queueing model with infinite waiting-room capacity. The customer arriving at $t = 0$ will find $k - 1$ customers waiting. The latter customers belong to the second priority class, whereas the ones arriving in $[0, \infty)$ belong to the first. The priority discipline is nonpreemptive, whereas within each class we have a FIFO queueing discipline.

As a motivation for studying the present model, consider the following specialization of the arrival pattern above which is realistic when, for example, doctors, dentists, or lawyers are consulted. Let the intervals between possible arrivals have fixed length $1/\lambda$ and let the probability of a customer not turning up be $1 - p$. Customers turn up or not independently of each other. The number of intervals of length $1/\lambda$ between the arrivals of two customers are then geometrically distributed with parameter p. The $k - 1$ customers in the second priority class do not have an appointment but are allowed to queue up, for instance, either before the office is opened in the morning or before it is reopened after the lunch break. One is now interested in:

1. Waiting times for customers from both priority classes not to be too long.
2. The initial busy period (starting with k customers in the system) not to be too short.

Small values of k will satisfy **1**, whereas large values satisfy **2**.

In ref. 1 the transient waiting times for customers belonging to both priority classes are arrived at for the general model. Using these results on the special arrival pattern above, optimal values for k have been tabulated for various values of p, λ and service intensity.

COMPUTER TIME SHARING

In ref. 2 a modification of the so-called *round-robin* priority discipline is treated in an $M/M/1$ queueing model. Each program receives a quantum q of service at a time from a single central processor. If this completes its service requirement, it leaves the system. If not, and there is a new arrival during the service of the quantum, it is given an additional quantum. Otherwise, the program joins the end of the queue to await its next turn. The model is analyzed under the assumption of a constant, nonzero overhead when the processor swaps one program for another. Obviously, during periods of high arrival rates, this algorithm has the effect of reducing the system's swapping activities. On the other hand, during periods of low arrival rates the discipline is similar to the conventional round robin, which automatically gives priority to programs with lesser service time requirements.

Reference 2 arrives at expressions for the mean waiting time in queue as a function of the quanta required and for the mean system cost due to waiting. Numerical comparisons with the conventional round-robin discipline are performed.

TELEPHONE EXCHANGE

In ref. 6 a telephone exchange is considered, handling the calls to and from a minor group of subscribers. Let the latter group belong to the second priority class, whereas people from the rest of the world belong to the first class. The Electronic Selector Bar Operator (ESBO) serves customers from both classes. The model one wished to be able to analyse is a modified version of $GI/G/M/G/1$ with a nonpreemptive priority discipline. However, customers of the higher priority are just allowed to wait a fixed length of time before they hear the busy signal and are lost. This is not the case for customers of lower priority. Second, whereas we have a FIFO queueing discipline within the first class, it is *random* within the second (i.e., all customers waiting have equal chance of being served next).

Unfortunately, there seems no way of arriving at the stationary waiting-time distributions for customers of the two classes by using queueing theory of today. As a first approximation ref. 6 starts out from an $M_1/G/M_2/G/1$ model where Laplace transforms of the distributions above are well known. Furthermore, *in this case* it is possible to get a look behind the "Laplace curtain."

References

[1] Dalen, G. and Natvig, B. (1980). *J. Appl. Prob.*, **17**, 227–234.

[2] Heacox, H. and Purdom, P. (1972). *J. ACM*, **19**, 70–91.

[3] Hokstad, P. (1978). *INFOR*, **16**, 158–170.

[4] Jaiswal, N. K. (1968). *Priority Queues*. Academic Press, New York.

[5] Kleinrock, L. (1976). *Queueing Systems*, Vol. 2: *Computer Applications*. Wiley, New York.

[6] Natvig, B. (1977). On the Waiting-Time for a Priority Queueing Model. *Tech. Rep. No. 3*, Dept. of Mathematical Statistics, University of Trondheim, Trondheim, Norway. Prepared for the Norwegian Telecommunications Administration Research Establishments.

(PREEMPTIVE DISCIPLINE
QUEUEING THEORY)

BENT NATVIG

PRIOR PROBABILITIES

Prior probabilities are probabilities that are applicable at a point in time which is *prior* to one obtaining additional information provided by empirical studies, experiments, surveys, and so on, yet to be executed. In the context of a particular random variable or unknown parameter, prior probabilities follow directly from the choice of a *prior distribution**, as the assumption of a probability distribution or probability density function in any statistical endeavor automatically leads to the ability to determine probabilities of various events of interest. *See* PRIOR DISTRIBUTIONS for a full discussion and references.

Suppose for the sake of illustration that you are concerned with the probability of the event E that a particular person known to you only by name is a college graduate. Knowing nothing else, you might reason that about 1 in 20 people in the general population is a college graduate, and thus your probability for E is $\Pr[E] = 0.05$. This is a prior probability. It is based on your knowledge at this time and is not conditioned on any other event (explicitly). It is a probability that you would use prior to getting any other information about the person. If, for example, you knew the person read *Newsweek* or a similar news magazine, you might feel the chances of the person being a college graduate were more than the 1 in 20 in the general population.

The illustration above suggests an element that is frequently present in discussions of prior probabilities—subjective probability assessment. $\Pr[E] = 0.05$ is a statement of degree of belief*, not an objective statement based on repeated experiments or some scientific fact. Prior probabilities may have some objectivity, particularly when they are the posterior probabilities from previous empirical observations (*see* POSTERIOR DISTRIBUTIONS), but in most cases there is inevitably a significant subjective component. The discipline called applied statistical decision theory* as discussed by Raiffa and Schlaifer [11] and Raiffa [10] is one that particularly illustrates the utilization of subjective prior probabilities.

There is a substantial literature on the quantification of degree-of-belief information in a prior probability. Martz and Waller [8] contains a good bibliographic overview. Hampton et al. [3] and Hogarth [4] are good general technical references. The issue of encoding prior probabilities from one expert source is dealt with in Lichtenstein et al. [6] and Lindley et al. [7]. One may also be concerned with forming the prior probabilities from a number of expert sources into a consensus probability, as discussed by DeGroot [2] and Morris [9]. Adams [1] and Kazakos [5] provide other perspectives on the general problem of quantifying prior probabilities.

References

[1] Adams, E. W. (1976). In *Foundations of Probability Theory, Statistical Inference and Statistical The-*

ories of Science, Vol. 1, W. L. Harper and C. A. Hooker, eds. D. Reidel, Boston, pp. 1–22.

[2] DeGroot, M. H. (1974). *J. Amer. Statist. Ass.*, **69**, 118–121. (Consensus prior probabilities via prior distributions which are mixtures of individual expert priors.)

[3] Hampton, J. M., Moore, P. G., and Thomas, H. (1973). *J. R. Statist. Soc. A*, **136**, 21–42. (General reference on the measurement of subjective probability.)

[4] Hogarth, R. M. (1975). *J. Amer. Statist. Ass.*, **70**, 271–289. (General reference on the measurement of subjective probability.)

[5] Kazakos, D. (1977). *IEEE Trans. Inf. Theory*, **23**, 203–210. (Estimation of prior probabilities using a mixture.)

[6] Lichtenstein, S., Fischhoff, B., and Phillips, L. D. (1977). In *Decision Making and Change in Human Affairs*, H. Jungerman and G. de Zeeuv, eds. D. Reidel, Dordrecht, The Netherlands, pp. 275–324. (Encoding prior probabilities.)

[7] Lindley, D. V., Tversky, A., and Brown, R. V. (1979). *J. R. Statist. Soc. A*, **142**, 146–180. (Reconciling inconsistent prior probability assessments.)

[8] Martz, H. F. and Waller, R. A. (1982). *Bayesian Reliability Analysis*. Wiley, New York. [Section 6.4 gives an overview of prior probability (distribution) assessment.]

[9] Morris, P. A. (1977). *Manag. Sci.*, **23**, 679–693. (Consensus prior probabilities.)

[10] Raiffa, H. (1968). *Decision Analysis, Introductory Lectures on Choices under Uncertainty*. Addison-Wesley, Reading, Mass. (Introduction to simple statistical decision theory.)

[11] Raiffa, H. and Schlaifer, R. (1961). *Applied Statistical Decision Theory*. Harvard University Press, Cambridge, Mass.

(DEGREES OF BELIEF
POSTERIOR DISTRIBUTIONS
POSTERIOR PROBABILITIES
PRIOR DISTRIBUTIONS)

DAVID J. PACK

PROBABILISTIC GRAMMAR *See* LINGUISTICS, STATISTICS IN

PROBABILISTIC NUMBER THEORY

Some questions and results of number theory can easily be expressed in probabilistic terms; for others, proofs can be simplified by probabilistic arguments. Probabilistic number theory can therefore be defined as a branch of number theory in which questions are of interest to number theorists, but the proofs utilize techniques of probability theory. There are two major subfields of probabilistic number theory: (a) distribution of values of arithmetical functions, and (b) metric (measure theoretic) results concerning expansions of real numbers. These are discussed in detail in the first two sections. The present article is special in that it is somewhat outside the expertise of most readers. Therefore, a detailed survey of the subject is not aimed at; rather, an expository introduction is given in which the major trends, but not the fine points, are presented.

DISTRIBUTION OF VALUES OF ARITHMETICAL FUNCTIONS

A function $f(n)$ on the positive integers $n = 1, 2, \ldots,$ is called an *arithmetical function*. In other words, an arithmetical function is a sequence in which the emphasis on the variable (index) n is its arithmetical structure rather than its magnitude.

Let us start with an example. Let $f(n)$ be the number of all (distinct) prime divisors of n. Hence, to determine the value of $f(n)$, one first has to express n as the product of powers of primes, and then the number of factors in this product is $f(n)$. One can hardly find any regularity of this $f(n)$, because it is frequently equal to 1 (at prime numbers), but it also exceeds any prescribed value infinitely many times (e.g., when n is a factorial). Here is where probabilistic number theory comes in: One can prove, by an appeal to the Chebyshev inequality*, that, for "most numbers n," $f(n)$ is approximately $\log\log n$. More precisely, if $A(N)$ denotes the number of positive integers $n \leqslant N$ such that

$$|f(n) - \log\log n| < (\log\log n)^{3/4},$$

then $A(N)/N \to 1$ as $N \to +\infty$. To prove this claim, let us introduce the probability spaces $(\Omega_N, \mathscr{A}_N, P_N)$, where $\Omega_N = \{1, 2, \ldots, N\}$, $\mathscr{A}_N$ is the set of all subsets of Ω_N, and P_N is the uniform distribution on Ω_N. Now the function $f_N(n) = f(n)$, $1 \leqslant n \leqslant N$, is a random variable, which can be represented as $f(n) = \sum e_p(n)$, where the summation is over primes $p \leqslant N$, and $e_p(n) = 1$ if p divides n and $e_p(n) = 0$ otherwise. The expectation and the variance of $f(n)$ can easily be computed by means of this representation. Namely, because

$$P_N(e_p(n) = 1) = [N/p]/N$$

and

$$P_N(e_p(n) = e_q(n) = 1) = [N/(pq)]/N,$$

where $p < q$ are prime numbers and $[y]$ signifies the integer part of y, we have

$$P_N(e_p(n) = 1) = 1/p + O(1/N)$$

and

$$P_N(e_p(n) = e_q(n) = 1) = 1/(pq) + O(1/N).$$

Thus

$$E(f(n)) = \sum_{p \leqslant N} 1/p + O(\pi(N)/N),$$

where $\pi(N)$ is the number of primes not exceeding N. However, from number theory we know that $\pi(N) \sim N/\log N$ (the so-called prime number theorem), which yields

$$\begin{aligned} E(f(n)) &= \sum_{m=1}^{N} \frac{\pi(m) - \pi(m-1)}{m} + O\left(\frac{1}{\log N}\right) \\ &= \sum_{m=1}^{N} \pi(m)\left(\frac{1}{m} - \frac{1}{m+1}\right) + O\left(\frac{1}{\log N}\right) \\ &= \sum_{m=1}^{N} \frac{1}{(m+1)\log m} + O(1) \\ &= \log\log N + O(1). \end{aligned}$$

The variance of $f(n)$ can be computed similarly; it also equals $\log\log N$. Finally, observing that $\log\log N$ is so slowly increasing that $\log\log n \sim \log\log N$ for all n satisfying $\sqrt{N} \leqslant n \leqslant N$, say, the Chebyshev inequality now implies our claim on $A(N)$ concerning the closeness of $f(n)$ to $\log\log n$.

This computation represents well the essence of probabilistic number theory. The question, namely the behavior of $f(n)$, originates from number theory; then the basic idea of applying the Chebyshev inequality is a tool of probability theory, but finally, in computing $E(f(n))$, the application of the prime number theorem is unavoidable. This procedure is generally the case in connection with number-theoretic functions. Only the basic ideas come from probability theory; in the actual computations number-theoretic results are frequently utilized.

Let us now consider a whole family of functions $f(n)$ satisfying

$$f(nm) = f(n) + f(m) \tag{1}$$

for all n and m which are relatively prime (i.e., they do not have any common divisor other than 1). Functions satisfying (1) are *additive arithmetical functions*. If, in addition, the additive function $f(n)$ satisfies $f(p^k) = f(p)$ for all primes p and integers $k \geqslant 1$, then $f(n)$ is *strongly additive*. In what follows there is no difference between additive and strongly additive functions. However, computations are much simpler with strongly additive functions, so we restrict ourselves to these. Now, for such functions $f(n)$, (1) implies that

$$f(n) = \sum f(p) e_p(n), \tag{2}$$

where summation is again over all primes p (when n is restricted to Ω_N, then $p \leqslant N$). One easily gets from (2) that $E(f(n)) \sim A_N$ and $V(f(n)) \sim B_N^2$, where

$$A_N = \sum_{p \leqslant N} \frac{f(p)}{p} \quad \text{and} \quad B_N^2 = \sum_{p \leqslant N} \frac{f^2(p)}{p}. \tag{3}$$

Repeating the argument with the Chebyshev inequality, we get that $f(n) \sim A_N$ for "most n" whenever $B_N \to +\infty$ with N.

More accurate results can also be deduced from (2). Because the indicator variables $e_p(n)$ are almost independent, that is, for arbitrary primes $p_1 < p_2 < \cdots < p_k$,

$$P_N(e_{p_1}(n) = e_{p_2}(n) = \cdots = e_{p_k}(n) = 1) = 1/(p_1 p_2 \cdots p_k) + O(1/N), \tag{4}$$

the sums in (2) and (3) suggest that the distribution of $(f(n) - A_N)/B_N$ converges weakly under conditions similar to sums of independent random variables. Although most results for sums of independent random variables are indeed valid for strongly additive functions, unfortunately these cannot be deduced directly from probability theory because the error term $O(1/N)$ in (4) is more troublesome than it looks at first. Namely, if only two primes in (4) are around $\sqrt{N}$, the major term is already smaller than the error term, and it is not easy to show that one can limit the primes in (2) to remain below powers of N. This is why one is forced to consider

$$f_r(n) = \sum_{p \leqslant r} f(p) e_p(n)$$

in place of $f(n)$, where $r = r(N)$ is some slowly increasing function of N, and then one uses probability theory to determine the asymptotic behavior of $f_r(n)$, and number theory (so-called sieve methods) to show that the difference $f(n) - f_r(n)$ is small compared with B_N. This line of attack proves very successful. A few basic results will be quoted as we now turn to the history of the subject. The references are far from complete, even when one adds all references in the survey papers and books listed later.

The probabilistic theory of arithmetic functions started with the paper of Turàn [43], who gave a new proof, essentially through the Chebyshev inequality (although without reference to probability theory), for the asymptotic relation between the number of prime divisors and $\log \log n$. The result itself had been known in number theory, but it was obtained by analytic methods. Both P. Erdös and M. Kac were influenced by Turàn's proof. They, first independently, and later jointly, contributed at the early stages the most significant results of the distribution theory of additive functions. It was Kac who approached the subject from a purely probabilistic point of view by recognizing in the representation at (2) the sum of almost independent random variables. Kac's early work is well set out in his popular monograph [24]. Erdös [7] and Erdös and Wintner [10] found necessary and sufficient conditions for the weak convergence of $P_N(f(n) < x)$ for arbitrary additive functions $f(n)$. Erdös and Kac [9] established that if the sequence $f(p)$ is bounded (over primes p), then, for the strongly additive function $f(n)$, $P_N(f(n) < A_N + B_N x)$ is asymptotically standard normal, where A_N and B_N are defined at (3). It turned out that the boundedness of $f(p)$ is not necessary for asymptotic normality*. In a series of papers, J. Kubilius developed a probabilistic theory of distributions for the truncated functions $f_r(n)$, in which his only requirement on r is that $r \to +\infty$ with N and $\log r/\log N$ should converge to zero as $N \to +\infty$. If in addition $B_r/B_N \to 1$ and $B_N \to +\infty$ as $N \to +\infty$, then $f(n)$ and $f_r(n)$ have the same asymptotic laws. Later, Kubilius published his work in a book which has had several editions; we quote here the English language edition [30]. His contribution to the theory of distributions of arithmetical functions is immeasurable because he not only published significant papers and books on the subject, but established a school at Vilnius which produced a large number of scientists working on the subject and also made a considerable influence on both individual scientists all over the world and the literature. Some of the work of this school is presented in the survey papers of Galambos [11, 14], Kubilius [31], Manstavicius [34], and Schwarz [41], and in the books of Kubilius [30] and Elliott [6]. In the Kubilius model, to show that $(f(n) - f_r(n))/B_N$ converges to zero ("in probability"), quite deep number theoretic results are employed. This induced investigations aimed at reobtaining at least some of the results of Kubilius by purely probabilistic arguments. Several models have been set up, which to various degrees have achieved the desired results, but none of them would have come close to the generality of the Kubilius model. However, through the models of Novoselov [35], Paul [36], Galambos [12], and Philipp [37], interesting probabilistic proofs are given for one result or another in the distribution theory of additive functions. The model of Galambos has the added advantage that it presents a unified treatment of additive functions when the variable is not limited to the set of successive integers. (Such investigations are extensive in the literature; see Kubilius [30], Kàtai [26, 27], and Indlekofer [21].)

For the truncated function $f_r(n)$ Kubilius was able to use the theory of sums of independent random variables. However, for extending his results to additive functions for which a truncation is not possible, other methods are necessary. The method of characteristic functions* proved fruitful, although it required new deep number-theoretic investigations. Namely, since P_N is the uniform distribution over a finite set, the characteristic function of an arithmetical function $f(n)$ is the arithmetical mean of the function $g(n) = \exp(itf(n))$. Now, if $f(n)$ is additive, then, in view of (1), $g(n)$ satisfies the equation

$$g(mn) = g(m)\,g(n) \qquad (5)$$

for all m and n which are relatively prime. A function satisfying (5) is *multiplicative*. Thus asymptotic mean value theorems were needed for multiplicative functions whose absolute value is bounded by 1. Such results were established by Delange [5] and Halàsz [18], who employed tools of analytic number theory (see also Rényi [40]). These mean value theorems gave new life to the distribution theory of arithmetical functions. In a series of papers, Elliott, C. Ryavec, Delange, Halàsz, and the members of the Vilnius school brought the theory of additive functions to a very advanced level. The two-volume set of books by Elliott [6] is based on this new development.

The distribution theory of multiplicative functions has also been studied in detail. Because multiplicative functions which can take both positive and negative values cannot be related to real-valued additive functions, their distribution theory requires a completely different approach. The combined result of Bakstys [1] and Galambos [13] gives necessary and sufficient conditions for the weak convergence of $P_N(g(n) < x)$ for multiplicative functions $g(n)$, and the weak convergence of the normalized multiplicative function $(g(n) - C_N)/D_N$ is characterized in Levin et al. [33].

For arithmetical functions which are neither additive nor multiplicative, approximation methods have been developed: by additive functions in Galambos [15] and Erdös and Galambos [8], and by so-called Ramanujan sums in a series of papers by W. Schwarz and his collaborators; see the survey [41].

The distribution theory of arithmetical functions has several applications within number theory. For example, for estimating certain sums, see De Koninck and Ivic [4], and for its role in characterizing $\log n$ among all additive functions, see Indlekofer [22]. For how much joy the study of this theory can give, one should read the recollections of Kac [25].

METRIC THEORY OF EXPANSIONS OF REAL NUMBERS

When a sequence of "digits" is assigned to real numbers by a specific algorithm, we speak of an expansion. The best known expansion is the decimal, or more generally, the q-adic expansions, in which, when q is an integer, the digits are independent and identically distributed random variables with respect to Lebesgue measure. Hence, although no regularity of the digits of irrational numbers seems to hold, the law of large numbers implies that these digits are quite regular for almost all x. Other theorems of probability theory are directly applicable as well.

The situation is completely different for most other series expansions. As an example, let us look at the so-called Engel series. For a given real number $0 < x \leqslant 1$, the "digits" d_j are defined by the following algorithm: Every d_j is an integer such that

$$1/d_j < x_j \leqslant 1/(d_j - 1),$$

where $x = x_1$ and $x_{j+1} = d_j x_j - 1$. This algorithm leads to the series representation

$$x = 1/d_1 + 1/(d_1 d_2) + 1/(d_1 d_2 d_3) + \cdots,$$

where

$$2 \leqslant d_1 \leqslant d_2 \leqslant d_3 \leqslant \cdots.$$

The digits d_j are strongly dependent in view of their monotonicity. However, the ratios $d_j/(d_{j-1} - 1)$ can well be approximated by independent and identically distributed random variables, which can be utilized for discovering general properties of the sequence d_j, valid for almost all x.

Similar situations arise in connection with most series expansions. That is, the digits are dependent but after some transformation an approximation by independent random variables becomes possible. This is not obvious, though, in most instances, and the challenge in these problems usually is probabilistic: to extend known theorems of probability theory to some weakly dependent random variables. These investigations did indeed contribute to probability theory itself. Most results on series expansions are contained in the books of Galambos [17] and Vervaat [44], although for some expansions there are newer results as well (see, e.g., Deheuvels [3] concerning the Engel series).

Among the expansions which do not result in a series most attention was paid to continued fraction expansions. This is obtained by repeated application of the transformation $Tx = x - [1/x]$, where, once again, $[y]$ signifies the integer part of y. As far back as in the nineteenth century, Gauss recognized that, if instead of length (i.e., Lebesgue measure), an absolutely continuous measure is assigned to the unit interval whose derivative is $1/\{(\log 2)(1 + x)\}$, then the sequence a_n, defined by the algorithm $a_j = [1/x_j]$, where $x = x_1$ and $Tx_j = x_{j+1}$, is stationary. More recent studies revealed that this sequence is stationary mixing, which fact induced intensive studies of mixing sequences of random variables. This benefited both probability theory and number theory; see Ibragimov and Linnik [20], Billingsley [2], and Philipp [37]. For a more classical approach, see Khintchine [29]. On the speed of convergence of the distribution of Tx_n, see Kuzmin [32] and Szüsz [42]. Galambos [16], Philipp [38], and Iosifescu [23] deal with the stochastic behavior of the largest of the digits a_n. The work of Rényi [39] shows the connection between series expansions and continued fractions.

We conclude by noting that there are several works in probabilistic number theory which do not belong to either topic covered in the preceding two sections. These are, however, individual problems rather than a part of a systematic study. We mention two of these. Kàtai [28] is probably the purest among all papers in probabilistic number theory in that he shows that for certain sequences of integers, the number of divisors of n in the neighborhood of $\sqrt{n}$ is normally distributed, in which argument no number theory is applied. The other area we wish to mention is the proof of the existence of certain types of numbers by showing that their probability is positive. Although most of the results are due to Erdös, the most convenient source for consultation is Halberstam and Roth [19].

References

[1] Bakstys, A. (1968). *Litovsk. Mat. Sb.*, **8**, 5–20.

[2] Billingsley, P. (1965). *Ergodic Theory and Information*. Wiley, New York.

[3] Deheuvels, P. (1982). *C. R. Acad. Sci. Paris*, **295**, 21–24.

[4] De Koninck, J.-M. and Ivic, A. (1980). *Topics in Arithmetical Functions*. North-Holland, Amsterdam.

[5] Delange, H. (1961). *Ann. Sci. École Norm. Sup.*, **78**, 273–304.

[6] Elliott, P. D. T. A. (1979/1980). *Probabilistic Number Theory*, Vols. I and II. Springer-Verlag, Berlin.

[7] Erdös, P. (1938). *J. Lond. Math. Soc.*, **13**, 119–127.

[8] Erdös, P. and Galambos, J. (1974). *Proc. Amer. Math. Soc.*, **46**, 1–8.

[9] Erdös, P. and Kac, M. (1940). *Amer. J. Math.*, **62**, 738–742.

[10] Erdös, P. and Wintner, A. (1939). *Amer. J. Math.*, **61**, 713–721.

[11] Galambos, J. (1970). *Ann. Inst. Henri Poincaré B*, **6**, 281–305.

[12] Galambos, J. (1971). *Z. Wahrscheinl. verw. Geb.*, **18**, 261–270.

[13] Galambos, J. (1971). *Bull. Lond. Math. Soc.*, **3**, 307–312.

[14] Galambos, J. (1971). In *The Theory of Arithmetic Functions*, A. A. Gioia et al., eds. Springer-Verlag, Berlin, pp. 127–139.

[15] Galambos, J. (1973). *Proc. Amer. Math. Soc.*, **39**, 19–25.

[16] Galambos, J. (1974). *Acta Arith.*, **25**, 359–364.

[17] Galambos, J. (1976). *Representations of Real Numbers by Infinite Series*. Springer-Verlag, Berlin.

[18] Halàsz, G. (1968). *Acta Math. Acad. Sci. Hung.*, **19**, 365–403.

[19] Halberstam, H. and Roth, K. F. (1966). *Sequences I*. Oxford University Press, Oxford, England.

[20] Ibragimov, I. A. and Linnik, Yu. V. (1971). *Independent and Stationary Sequences of Random Variables*. Wolters-Noordhoff, Groningen.

[21] Indlekofer, K.-H. (1974). In *Proc. Coll. Bolyai Math. Soc. Debrecen*, B. Gyires, ed. North-Holland, Amsterdam, pp. 111–128.

[22] Indlekofer, K.-H. (1981). *Ill. J. Math.*, **25**, 251–257.

[23] Iosifescu, M. (1978). *Trans. 8th Prague Conf. Inf. Theory*, Vol. A. D. Reidel, Dordrecht, The Netherlands, pp. 27–40.

[24] Kac, M. (1959). *Statistical Independence in Probability, Analysis and Number Theory*. Carus Monographs. Wiley, New York.

[25] Kac, M. (1979). *Probability, Number Theory and Statistical Physics*. MIT Press, Cambridge, Mass.

[26] Kàtai, I. (1968). *Compositio Math.*, **19**, 278–289.

[27] Kàtai, I. (1969). *Acta Math. Acad. Sci. Hung.*, **20**, 69–87.

[28] Kàtai, I. (1977). *Publ. Math. Debrecen*, **24**, 91–96.

[29] Khintchine, A. (1964). *Continued Fractions*. Wolters-Noordhoff, Groningen, The Netherlands.

[30] Kubilius, J. (1964). *Probabilistic Methods in the Theory of Numbers*. Transl. Math. Monogr., Vol. 11. American Mathematical Society, Providence, R.I.

[31] Kubilius, J. (1974). In *Current Problems of Analytic Number Theory*, Izd. Nauka i Techn., Minsk, pp. 81–118.

[32] Kuzmin, R. O. (1928). *Rep. Pol. Acad. Sci. (A)*, 375–380.

[33] Levin, B. V., Timofeev, N. M., and Tuljaganov, S. T. (1973). *Litovsk. Mat. Sb.*, **13**, 87–100.

[34] Manstavicius, E. (1980). *Litovsk. Mat. Sb.*, **20**, 39–52.

[35] Novoselov, E. V. (1964). *Izv. Akad. Nauk SSSR, Ser. Mat.*, **28**, 307–364.

[36] Paul, E. M. (1962). *Sankhyā A*, **24**, 103–114, 209–212; *ibid.*, **25**, 273–280 (1963).

[37] Philipp, W. (1971). *Mem. Amer. Math. Soc.*, **114**, 1–102.

[38] Philipp, W. (1976). *Acta Arith.*, **28**, 379–386.

[39] Rényi, A. (1957). *Acta Math. Acad. Sci. Hung.*, **8**, 477–493.

[40] Rényi, A. (1965). *Publ. Math. Debrecen*, **12**, 323–330.

[41] Schwarz, W. (1976). *Jb. Dtsch. Math.-Ver.*, **78**, 147–167.

[42] Szüsz, P. (1961). *Acta Math. Acad. Sci. Hung.*, **12**, 447–453.

[43] Turàn, P. (1934). *J. Lond. Math. Soc.*, **9**, 274–276.

[44] Vervaat, W. (1972). *Success Epochs in Bernoulli Trials with Applications in Number Theory*. Math. Centre Tracts No. 42, Mathematische Centrum, Amsterdam.

(FIRST-DIGIT PROBLEM
LAWS OF LARGE NUMBERS
PROBABILITY THEORY)

JANOS GALAMBOS

PROBABILISTIC VOTING MODELS

In many nations and at many levels of government, the question of who should hold various political offices is settled by voting. As a consequence, there are many public officials, and many politicians who would like to become public officials, who are concerned about the implications of their choices for the decisions that voters will make. In particular, they are concerned about which voters will vote in the next election and who these people will vote for. Two choices that elected officials and office-seeking politicians make that are especially important in this regard are: Their positions on the leading policy questions of the day and the way in which they allocate their campaign resources. As a consequence, researchers have developed mathematical models of the relation between these choices and the decisions that public officials and office-seeking politicians can expect the voters to make. These models can be used to address questions such as: In a given situation in which campaign resources have already been allocated, are there "best" policy positions that can be taken? If so, what are they? In a given situation in which policy positions have already been selected, is there a "best" campaign resource allocation that can be chosen? If so, what is it? In a given situation in which both policy positions and an allocation of resources can be selected, is there a "best" possible combination? If so, what is it? The answers are useful for social scientists who want to know what policy positions and campaign resource allocations can be expected from public officials and office-seeking politicians. They are also potentially useful for campaign strategists who advise political candidates.

TWO EXAMPLES

Before stating the features that differ in the two examples in this section, the characteristics that they have in common will be covered: There are two candidates for a particular political office. They will be indexed by the elements in the set $C = \{1, 2\}$. There are three voters. They will be indexed by the elements in the set $N = \{1, 2, 3\}$.

Each candidate has to decide on what policy proposals he or she wants to make. The only characteristic of a candidate's policy proposals that matters to the voters is the distribution of income that they expect the policies to lead to. For each vector of policy proposals that a candidate can make, each voter has a unique distribution of income that he or she expects the policies to lead to. These expectations are the same for all three voters. As a consequence, the candidate's decision can be viewed as the choice of an income distribution*. It will be assumed that each candidate can, in particular, choose any $\psi_c \in X = \{(x_1, x_2, x_3) \in R^3 : x_1 + x_2 + x_3 = 1, x_1 \geqslant 0.01, x_2 \geqslant 0.01, x_3 \geqslant 0.01\}$ (by a suitable choice of policy proposals).

Each particular voter cares only about his or her own income. More specifically, (for any given $i \in N$) for each pair $x, y \in X$: x is at least as good as y for i if and only if $x_i \geqslant y_i$. This implies that i prefers x to y (i.e., x is at least as good as y for i, but y is not at least as good as x for i) if and only if $x_i > y_i$. It also implies that i is indifferent between x and y (i.e., x is at least as good as y for i and y is at least as good as x for i) if and only if $x_i = y_i$. Thus i's preferences on X can be *represented* by the *utility function* $U_i(x) = x_i$; for each pair $x, y \in X$, $U_i(x) \geqslant U_i(y)$ if and only if x is at least as good as y for i. In addition, assume that $U_i(x)$ measures the intensity of i's preferences. More specifically, the values assigned by $U_i(x)$ have the following interpretations: $U_i(x)/U_i(y) = \frac{1}{2}$ if and only if i likes x only half as much as he or she likes y; $U_i(x)/U_i(y) = 2$ if and only if i likes x twice as much as he or she likes y; $U_i(x)/U_i(y) = 3$ if and only if i likes x three times as much as he or she likes y; and so on. This property implies that each voter's utility* function is unique up to multiplication by a positive scalar. (It is, accordingly, called a *ratio-scale utility function*.)

In the examples, $P_i^c(\psi_1, \psi_2)$ will be used to denote the probability that the individual indexed by i will vote for candidate c when $c = 1$ chooses ψ_1 and $c = 2$ chooses ψ_2. Thus [at any given $(\psi_1, \psi_2) \in X^2$] the expected vote for a given $c \in C$ can be written as

$$\mathrm{EV}^c(\psi_1, \psi_2) = \sum_{i=1}^{3} P_i^c(\psi_1, \psi_2).$$

Each candidate c is concerned solely about his or her expected plurality.

$$\mathrm{Pl}^c(\psi_1, \psi_2) = \mathrm{EV}^c(\psi_1, \psi_2) - \mathrm{EV}^k(\psi_1, \psi_2)$$

(where k is the index for the other candidate), that is, with his or her expected margin of victory (or, phrased differently, how much he or she expects to win or lose by). Furthermore, each candidate wants to maximize his or her expected plurality; in a candidate's view, the larger the expected margin of victory, the better. This implies that, for any specification of the $P_i^c(\cdot)$ functions, the decisions that the two candidates have to make can be appropriately modeled as a two-person, noncooperative game, $(X, X; \mathrm{Pl}^1, \mathrm{Pl}^2)$, in which (1) the two players are the two candidates, (2) the strategy set for each candidate is X, and (3) the payoff functions are $\mathrm{Pl}^1 : X \times X \to R^1$ and $\mathrm{Pl}^2 : X \times X \to R^1$, respectively. By the definitions of Pl^1 and Pl^2, $\mathrm{Pl}^1(\psi_1, \psi_2) + \mathrm{Pl}^2(\psi_1, \psi_2) = 0 \;\; \forall (\psi_1, \psi_2) \in X^2$. Hence the game is zero-sum. *See* GAME THEORY *or* NASH EQUILIBRIUM.

Example 1. Consider the case in which, for each $i \in N$ and each pair $(\psi_1, \psi_2) \in X^2$,

$$P_i^1(\psi_1, \psi_2) = \begin{cases} 1 & \text{if } U_i(\psi_1) > U_i(\psi_2) \\ \frac{1}{2} & \text{if } U_i(\psi_1) = U_i(\psi_2) \\ 0 & \text{if } U_i(\psi_1) < U_i(\psi_2) \end{cases}$$

and

$$P_i^2(\psi_1, \psi_2) = 1 - P_i^1(\psi_1, \psi_2).$$

The resulting game, $(X, X; \mathrm{Pl}^1, \mathrm{Pl}^2)$, has *no* Nash equilibrium. Since the game is zero-sum, this can also be phrased as: The game has *no* saddle point.

The fact that this game has no Nash equilibrium (or, equivalently, has no saddle point) can be seen quite easily. Choose any $(x, y) \in X^2$. Since the game is zero-sum, there are three possibilities:

$$\text{(a)} \qquad \mathrm{Pl}^1(x, y) = \mathrm{Pl}^2(x, y) = 0,$$

$$\text{(b)} \quad \mathrm{Pl}^1(x, y) > 0 > \mathrm{Pl}^2(x, y), \quad \text{or}$$

$$\text{(c)} \qquad \mathrm{Pl}^1(x, y) < 0 < \mathrm{Pl}^2(x, y).$$

Suppose that **(a)** or **(b)** holds. Identify a voter, i, who gets at least as much income at x as anyone else (i.e., for whom we have $x_i \geqslant x_j$, $\forall\, j \neq i$). Let z be the alternative in which this individual gets $z_i = 0.01$ and the others get $z_j = x_j + \frac{1}{2}(x_i - 0.01)$ (i.e., the others split the decrease in i's income). Since $x_i > 0.01$, we have $U_j(z) = z_j > x_j = U_j(x)$ for each $j \neq i$ and $U_i(z) = z_i < x_i = U_i(x)$. Therefore, $P_j^2(x, z) = 1$ and $P_j^1(x, z) = 0$ for each $j \neq i$ and, in addition, $P_i^2(x, z) = 0$ and $P_i^1(x, z) = 1$. Therefore, $\mathrm{Pl}^2(x, z) = +1 > 0 \geqslant \mathrm{Pl}^2(x, y)$. Therefore, if **(a)** or **(b)** holds, (x, y) is *not* a Nash equilibrium. Similar reasoning applies when **(c)** holds.

Example 1 is a game-theoretic version of the voting paradox*, the fact that when majority rule is used to make collective decisions, a society can find itself in a situation where, for each social choice that could be made, there is a feasible alternative that a majority of the voters prefer (and, hence, any particular choice can be overturned by a majority vote). More specifically, this example illustrates how easy it is for the paradox of voting (and the corresponding absence of equilibrium policies) to occur when the issues that are to be resolved involve the distribution of income in the society (*see also* VOTING PARADOX, Arrow [1] and Sen [30]).

Example 2. Consider the case in which the probabilities that describe the voters' choices satisfy the following version of Luce's axiom of "independence from irrelevant alternatives" (*see* LUCE'S CHOICE AXIOM): For each $i \in N$ and $(\psi_1, \psi_2) \in X^2$,

$$\frac{P_i^1(\psi_1, \psi_2)}{P_i^2(\psi_1, \psi_2)} = \frac{U_i(\psi_1)}{U_i(\psi_2)}; \tag{1}$$

assume also (implicit in Example 1) that each voter is going to vote; that is, for each $i \in N$ and $(\psi_1, \psi_2) \in X^2$,

$$P_i^1(\psi_1, \psi_2) + P_i^2(\psi_1, \psi_2) = 1. \tag{2}$$

Equations (1) and (2) imply that for each $i \in N$ and $(\psi_1, \psi_2) \in X^2$,

$$P_i^1(\psi_1, \psi_2) = \frac{U_i(\psi_1)}{U_i(\psi_1) + U_i(\psi_2)},$$

$$P_i^2(\psi_1, \psi_2) = \frac{U_i(\psi_2)}{U_i(\psi_1) + U_i(\psi_2)}.$$

This time, the resulting game $(X, X; \mathrm{Pl}^1, \mathrm{Pl}^2)$ *does* have a Nash equilibrium. Since the game is zero sum, this can also be phrased as: The game *has* a saddle point (*see* GAME THEORY *or* NASH EQUILIBRIUM).

The fact that this game has a Nash equilibrium (or equivalently, has a saddle point) can be seen quite easily. For each $i \in N$ and $(x, y) \in X^2$, $P_i^1(x, y) = x_i/(x_i + y_i)$. Therefore, each $P_i^1(x, y)$ is a concave function of $x = (x_1, x_2, x_3)$, and each $-P_i^1(x, y)$ is a concave function of $y = (y_1, y_2, y_3)$. This, in turn, implies that $\mathrm{EV}^1(x, y)$ is a concave function of x and $-\mathrm{EV}^1(x, y)$ is a concave function of y. Similarly, $\mathrm{EV}^2(x, y)$ is a concave function of y and $-\mathrm{EV}^2(x, y)$ is a concave function of x. Therefore, candidate 1's payoff function, $\mathrm{Pl}^1(x, y) = \mathrm{EV}^1(x, y) - \mathrm{EV}^2(x, y)$, is a concave function of x and candidate 2's payoff function, $\mathrm{Pl}^2(x, y) = \mathrm{EV}^2(x, y) - \mathrm{EV}^1(x, y)$, is a concave function of y. By a similar argument, $\mathrm{Pl}^1(x, y)$ and $\mathrm{Pl}^2(x, y)$ are continuous functions of (x, y). Finally, from its definition, X is a compact, convex subset of R^3. Hence all of the asssumptions in the premise of one of the theorems that has been labeled "Nash's Theorem" are satisfied (*see* NASH EQUILIBRIUM). Therefore, there is a Nash equilibrium in this example.

Where is (or are) the Nash equilibrium (or equilibria) located? This question can be answered by solving the problem: Find the $x \in X$ that maximize(s) $U_1(x) \cdot U_2(x) \cdot U_3(x) = x_1 \cdot x_2 \cdot x_3$ over the set X. There is a unique x which solves this problem: $x = (\frac{1}{3}, \frac{1}{3}, \frac{1}{3})$. This implies that there is a unique Nash equilibrium in the game: $\psi_1 = \psi_2 = (\frac{1}{3}, \frac{1}{3}, \frac{1}{3})$. (See Theorem 1 in Coughlin and Nitzan [9].)

GENERAL MODEL

Models similar to the two examples given above have been studied in the references listed at the end of this entry. They are all special cases of the following general model:

There are two candidates for a particular political office. They will be indexed by the elements in the set $C = \{1, 2\}$. There is a set of possible "locations," X, for the candidates. The set of possible locations is the same for both of them. A particular $x \in X$ could, for instance, specify a position on the policy issues in the election, an allocation of campaign resources, or both of these. ψ_c will be used to denote a particular location for c. There is an index set, N, for the individuals who can vote in the election. A particular $i \in N$ could be a number (e.g., if the voters are labeled as voters $1, \ldots, n$), a specification of various characteristics that a voter can have (such as location of residence, income, age, etc.), a vector that specifies the "ideal" positions on the various policy issues for a voter, or something else.

For each $i \in N$ and $c \in C$, there is a function

$$P_i^c : X \times X \to [0, 1]$$

that assigns to each $(\psi_1, \psi_2) \in X \times X$ a probability for thc event "a voter randomly drawn from the individuals labeled by $i \in N$ will vote for c if candidate 1 is located at ψ_1, and candidate 2 is located at ψ_2." For each $(\psi_1, \psi_2) \in X^2$, $P_i^0(\psi_1, \psi_2) = 1 - P_i^1(\psi_1, \psi_2) - P_i^2(\psi_1, \psi_2)$ will be the probability for the event "a voter randomly drawn from the individuals labeled by $i \in N$ will not vote (i.e., will abstain from voting), if candidate 1 is located at ψ_1 and candidate 2 is located at ψ_2." These can be objective probabilities or they can be subjective probabilities that are believed by both of the candidates.

There is a probability distribution on N which assigns to each measurable set, $B \subseteq N$, a probability for the event "a voter randomly drawn from the individuals who can vote in the election has an index $i \in B$." These probabilities can also be either objective probabilities or subjective probabilities that are believed by both of the candidates.

Each candidate is concerned solely about his or her (a) expected vote, $\mathrm{EV}^c(\psi_1, \psi_2)$, (b) expected plurality, $\mathrm{Pl}^c(\psi_1, \psi_2)$, or (c) probability of winning $W^c(\psi_1, \psi_2)$.

DETERMINISTIC VOTING MODELS

An important special case for the general model given above is the *deterministic voting model*. This terminology comes from analyses of models in which each candidate wants to maximize his or her expected plurality (as in the examples). When [for a given $i \in N$ and $(\psi_1, \psi_2) \in X^2$] $P_i^1(\psi_1, \psi_2) = P_i^2(\psi_1, \psi_2)$, the expected vote from the individual(s) indexed by "i" is split evenly between the two candidates. When this occurs (since each candidate's objective is to maximize his or her expected plurality), the expected votes corresponding to the index i cancel each other out—and, therefore, have no effect on $\mathrm{Pl}^1(\psi_1, \psi_2)$ or $\mathrm{Pl}^2(\psi_1, \psi_2)$. For instance, in Example 1, if $U_3(\psi_1) = U_3(\psi_2)$, then $P_3^1(\psi_1, \psi_2) - P_3^2(\psi_1, \psi_2) = \frac{1}{2} - \frac{1}{2} = 0$. Therefore, the expected plurality for candidate 1 when only voters 1 and 2 are counted is the same as when voters 1, 2, *and* 3 are counted. That is,

$$\begin{aligned}[P_1^1(\psi_1, \psi_2) + P_2^1(\psi_1, \psi_2)] \\ - [P_1^2(\psi_1, \psi_2) + P_2^2(\psi_1, \psi_2)] \\ = [P_1^1(\psi_1, \psi_2) + P_2^1(\psi_1, \psi_2) + P_3^1(\psi_1, \psi_2)] \\ - [P_1^2(\psi_1, \psi_2) + P_2^2(\psi_1, \psi_2) + P_3^2(\psi_1, \psi_2)].\end{aligned}$$

Similarly, the expected plurality for candidate 2 when only voters 1 and 2 are counted is the same as when voters 1, 2, *and* 3 are counted.

From the preceding observations it is thus clear that at any given (ψ_1, ψ_2), the only voter indices that matter (to expected plurality maximizing candidates) are ones with $P_i^1(\psi_1, \psi_2) \neq P_i^2(\psi_1, \psi_2)$. When, in fact, $P_i^1(\psi_1, \psi_2) = 1$ or $P_i^2(\psi_1, \psi_2) = 1$ at a given $i \in N$ and $(\psi_1, \psi_2) \in X^2$, one of two things must be true: (a) there is one voter with the index i and the candidates believe that his or her decision will be completely determined once they choose the strategies ψ_1 and ψ_2, respectively, *or* (b) there is more than one voter with the index i and the decisions made by all these voters will be completely determined (and the same) once the candidates choose the strategies ψ_1 and ψ_2, respectively. Because of this, any model which satisfies the assumptions of the general voting model given above *and* is such that, at each $(\psi_1, \psi_2) \in X^2$ and $i \in N$ where $P_i^1(\psi_1, \psi_2) \neq P_i^2(\psi_1, \psi_2)$, either (i) $P_i^1(\psi_1, \psi_2) = 1$ or (ii) $P_i^2(\psi_1, \psi_2) = 1$, is called a "deterministic voting model." Example 1, for instance, is such a model, whereas Example 2 is not.

The defining characteristic of a deterministic voting model can be restated as: For each $(\psi_1, \psi_2) \in X^2$ and $i \in N$, either the expected votes corresponding to the index i cancel each other out (and, therefore, have no effect on the candidates' expected pluralities) *or* the decision(s) of the voter(s) corresponding to the index i will be completely determined (and identical when candidate 1 chooses ψ_1 and candidate 2 chooses ψ_2. A third way of stating this characteristic is: For each $(\psi_1, \psi_2) \in X^2$ and $i \in N$, (a) $P_i^1(\psi_1, \psi_2) = 1$, (b) $P_i^2(\psi_1, \psi_2) = 1$, *or* (c) $P_i^1(\psi_1, \psi_2) = P_i^2(\psi_1, \psi_2) = \frac{1}{2}(1 - P_i^0(\psi_1, \psi_2))$.

The deterministic voting models that have received the most attention are ones in which (1) each index corresponds to one voter, (2) each voter, i, has a utility function, $U_i(x)$, and (3a) for each voter, i, and each $(\psi_1, \psi_2) \in X^2$,

$$P_i^1(\psi_1, \psi_2) = \begin{cases} 1 & \text{if } U_i(\psi_1) > U_i(\psi_2), \\ \frac{1}{2} & \text{if } U_i(\psi_1) = U_i(\psi_2), \\ 0 & \text{if } U_i(\psi_1) < U_i(\psi_2), \end{cases}$$

$$P_i^2(\psi_1, \psi_2) = 1 - P_i^1(\psi_1, \psi_2)$$

(as in Example 1) or (3b) for each voter, i, and each $(\psi_1, \psi_2) \in X^2$

$$P_i^0(\psi_1, \psi_2) = \begin{cases} 1 & \text{if } U_i(\psi_1) = U_i(\psi_2), \\ 0 & \text{if } U_i(\psi_1) \neq U_i(\psi_2), \end{cases}$$

$$P_i^1(\psi_1, \psi_2) = \begin{cases} 1 & \text{if } U_i(\psi_1) > U_i^0(\psi_2), \\ 0 & \text{if } U_i(\psi_1) \leqslant U_i(\psi_2), \end{cases}$$

$$P_i^2(\psi_1, \psi_2) = 1 - P_i^1(\psi_1, \psi_2) - P_i^0(\psi_1, \psi_2).$$

An exception is McKelvey [26], where it is assumed that (a) for each index, α, all of the voters who are labeled by this index have the same utility function, $U_\alpha(x)$, and (b) for each index α and each $(\psi_1, \psi_2) \in X^2$,

$$\begin{aligned}\left(P_\alpha^0(\psi_1, \psi_2), P_\alpha^1(\psi_1, \psi_2), P_\alpha^2(\psi_1, \psi_2)\right) \\ \in \{(1,0,0), (0,1,0), (0,0,1)\},\end{aligned}$$

$$P_\alpha^1(\psi_1, \psi_2) = 1 \Rightarrow U_\alpha(\psi_1) > U_\alpha(\psi_2),$$

$$P_\alpha^2(\psi_1, \psi_2) = 1 \Rightarrow U_\alpha(\psi_1) < U_\alpha(\psi_2).$$

This formulation provides a deterministic

voting model in which abstentions can occur more frequently than just when $U_\alpha(\psi_1) = U_\alpha(\psi_2)$.

The seminal work on deterministic voting models was done by Hotelling [21], Downs [16], Black [2], and Davis and Hinich [11, 12]. Surveys of results that have been derived in these and other analyses [using either $\mathrm{Pl}^c(\psi_1,\psi_2)$ or $W^c(\psi_1,\psi_2)$ as the objective functions for the candidates] can be found in Plott [29], Kramer [22], and Mueller [27, 28].

PROBABILISTIC VOTING MODELS

Any model which satisfies the assumptions of the general voting model given above, but is *not* a deterministic voting model, is called a *probabilistic voting model*. Example 2, for instance, is such a model. The terminology reflects the fact that in any such model, there is at least one pair of strategies $(\psi_1,\psi_2) \in X^2$ at which there is at least one index that matters (to expected plurality maximizing candidates) where the candidates' beliefs about the voter(s) corresponding to the index are probabilistic in a nontrivial way; the random variable that describes them is nondegenerate. The defining characteristic for these models can be restated as: There is at least one $(\psi_1,\psi_2) \in X^2$ and $i \in N$ where $P_i^1(\psi_1,\psi_2) \neq P_i^2(\psi_1,\psi_2)$ *and* $0 < P_i^1(\psi_1,\psi_2) < 1$ or $0 < P_i^2(\psi_1,\psi_2) < 1$ (or both).

Any model that satisfies all the assumptions of the general voting model given above will be a probabilistic voting model if there is an index i that corresponds to two or more voters and at least one pair of possible locations $(\psi_1,\psi_2) \in X^2$ such that the choices of the voters corresponding to index i are completely determined when candidate 1 chooses ψ_1 and candidate 2 chooses ψ_2, *but* (a) they will not all make the same choice, and (b) those who will vote are not split evenly between the candidates.

Indeed, almost any deterministic voting model can be converted into a probabilistic voting model of this sort by appropriately regrouping the indices in the deterministic voting model into indices for the probabilistic voting model. For instance, let all three voters in Example 1 have the same index i. Then, for the strategies $\psi_1 = (\frac{1}{3},\frac{1}{3},\frac{1}{3})$ and $\psi_2 = (0,\frac{1}{2},\frac{1}{2})$, we have $P_i^1(\psi_1,\psi_2) = \frac{1}{3}$ and $P_i^2(\psi_1,\psi_2) = \frac{2}{3}$. The resulting model is a probabilistic voting model. This approach is the basis for the models analyzed in McKelvey [26]. A voting model that satisfies all the assumptions of the general voting model given above can, alternatively, be a probabilistic voting model if there is at least one voter whose choice of whether to vote and/or which candidate to vote for (if he or she votes) is probabilistic in nature, as in Example 2. (*See* LUCE'S CHOICE AXIOM for a discussion of theories of probabilistic choice behavior.) This approach is the basis for the models analyzed in Brams [3], Brams and Davis [4], Comaner [5], Coughlin and Nitzan [9, 10], Davis and Hinich [13], Hinich [17], Hinich and Ordeshook [18], Hinich et al. [19, 20], and Lake [23]. A voting model that satisfies all the assumptions given above could be a probabilistic voting model because the candidates are uncertain about the choices that voters will make and use subjective probabilities to summarize their expectations about these choices (*see* DECISION THEORY *and* UTILITY THEORY). This is the basis for the model analyzed in Ledyard [24, 25]. Analyses that apply to models in which any of these three interpretations can arise have been carried out in Coughlin [7, 8].

The most closely scrutinized probabilistic voting models can be grouped into three basic categories. The first consists of models in which the description of the candidates' expectations about the voters' decisions to vote (or abstain) are probabilistic, but the description of their expectations about the choices that the voters make between the candidates are deterministic (i.e., they believe that, for each $(\psi_1,\psi_2) \in X \times X$ and $i \in N$, when candidate 1 chooses ψ_1 and candidate 2 chooses ψ_2, the voters corresponding to i who vote will all vote for the same candidate and the candidate who so benefits is completely determined). This category includes the models in Davis and Hinich [13], Hinich and Ordeshook [18], Hinich et al. [19, 20], and McKelvey [26]. The second category consists of models in which there are no abstentions, but the description of the candidates' expectations about the choices that the voters will make in choosing between them is probabilistic: (a) $P_i^0(\psi_1, \psi_2) = 0$, $\forall(\psi_1, \psi_2) \in X \times X$, $\forall i \in N$ and (b) $\exists i \in N$ and $(\psi_1, \psi_2) \in X \times X$, where $P_i^1(\psi_1, \psi_2) \neq \frac{1}{2}$. This category includes Brams [3], Brams and Davis [4], Hinich [17], Coughlin and Nitzan [9, 10], and Coughlin [8]. The third category consists of models in which the description of the candidates' expectations both about voter abstentions and about the choices that the voters will make in choosing between them are probabilistic. This category includes Denzau and Kats [15], Coughlin [6, 7], and Ledyard [24, 25].

For further discussions, see the references.

References

[1] Arrow, K. (1963). *Social Choice and Individual Values*, 2nd ed. Yale University Press, New Haven, Conn.

[2] Black, D. (1958). *The Theory of Committees and Elections*. Cambridge University Press, Cambridge, England.

[3] Brams, S. (1975). *Game Theory and Politics*. MacMillan, London.

[4] Brams, S. and Davis, M. (1982). *Math. Social Sci.*, **3**, 373–388.

[5] Comaner, W. (1976). *J. Public Econ.*, **5**, 169–178.

[6] Coughlin, P. (1982). *Public Choice*, **39**, 427–433.

[7] Coughlin, P. (1983). *Math. Social Sci.*, **4**, 275–292.

[8] Coughlin, P. (1984). *J. Econ. Theory*, **34**, 1–12.

[9] Coughlin, P. and Nitzan, S. (1981). *J. Public Econ.*, **15**, 113–122.

[10] Coughlin, P. and Nitzan, S. (1981). *J. Econ. Theory*, **24**, 226–240.

[11] Davis, O. and Hinich, M. (1966). In *Mathematical Applications in Political Science*, Vol. II, J. Bernd, ed. Southern Methodist University Press, Dallas, Tex.

[12] Davis, O. and Hinich, M. (1968). *Public Choice*, **5**, 59–72.

[13] Davis, O. and Hinich, M. (1972). In *Probability Models of Collective Decision Making*, R. Niemi and H. Weisberg, eds. Charles E. Merrill, Columbus, Ohio.

[14] Davis, O., DeGroot, M., and Hinich, M. (1972). *Econometrica*, **40**, 147–157.

[15] Denzau, A. and Kats, A. (1977). *Rev. Econ. Stud.*, **44**, 227–233.

[16] Downs, A. (1957). *An Economic Theory of Democracy*. Harper, New York.

[17] Hinich, M. (1977). *J. Econ. Theory*, **16**, 208–219.

[18] Hinich, M. and Ordeshook, P. (1969). *Public Choice*, **7**, 81–106.

[19] Hinich, M., Ledyard, J., and Ordeshook, P. (1972). *J. Econ. Theory*, **4**, 144–153.

[20] Hinich, M., Ledyard, J., and Ordeshook, P. (1973). *J. Politics*, **35**, 154–193.

[21] Hotelling, H. (1929). *Econ. J.*, **39**, 41–57.

[22] Kramer, G. (1977). In *Frontiers of Quantitative Economics*, Vol. III, M. Intriligator, ed. North-Holland, Amsterdam.

[23] Lake, M. (1979). In *Applied Game Theory*, S. Brams, A. Schotter, and G. Schwodiauer, eds. Physica-Verlag, Würzburg, Federal Republic of Germany.

[24] Ledyard, J. (1981). In *Essays in Contemporary Fields of Economics*, G. Horwich and J. Quirk, eds. Purdue University Press, West Lafayette, Ind.

[25] Ledyard, J. (1984). *Public Choice*, **44**, 7–41.

[26] McKelvey, R. (1975). *Econometrica*, **43**, 815–844.

[27] Mueller, D. (1976). *J. Econ. Lit.*, **14**, 395–433.

[28] Mueller, D. (1979). *Public Choice*. Cambridge University Press, Cambridge, England.

[29] Plott, C. (1971). In *Frontiers of Quantitative Economics*, Vol. I, M. Intriligator, ed. North-Holland, Amsterdam.

[30] Sen, A. (1970). *Collective Choice and Social Welfare*. Holden-Day, San Francisco.

(DECISION THEORY
GAME THEORY
LUCE'S CHOICE AXIOM
NASH EQUILIBRIUM
UTILITY THEORY
VOTING PARADOX)

PETER COUGHLIN

PROBABILITY FORECASTING

At its simplest, probability forecasting refers to the process of attaching a numerical probability to an uncertain event. Surprisingly, this most basic requirement for any real-world application of probability theory was for long ignored by most statisticians. The motivation for its development as a discipline in its own right has largely arisen from the needs of the field of *probabilistic weather forecasting*, and much of the theoretical work underpinning it originally appeared in the meteorological literature. This special application also forms a convenient framework on which to hang a discussion of general principles.

In many countries the weather service still issues its weather forecasts in purely verbal terms, either as a categorical prediction ("It will be dry and sunny") or with imprecise indications of uncertainty ("A chance of light showers"). But many decisions can often only be taken rationally when the uncertain nature of the weather is admitted and quantified as precisely as possible. To meet this need, the National Weather Service in the United States has, since 1965, formulated and issued *probability of precipitation* (PoP) forecasts. Such a forecast is a number between 0 and 1, intended to be interpreted as the probability that measurable precipitation ($\geqslant 0.01$ inch) will occur during a specific period (generally 12 hours) at a particular point in the area of concern. These PoP forecasts are now popularly accepted by the American public as meaningful and informative.

The forecasts themselves are usually arrived at intuitively by the weather forecaster, after taking account of all the information available. This information may include, as a decision aid, an "objective" probability produced by a computerized statistical analysis of weather records, but the forecast issued is usually interpreted, in the last analysis, as a personal or subjective probability* statement for that forecaster. Questions thus arise concerning the motivation and integrity of forecasters in forming and stating their subjective probabilities, the empirical validity of the forecasts issued, the extent of the agreement between different forecasters, and the way in which the public should use the forecasts. This article will concentrate on the first two issues. A very readable guide is that of Staël von Holstein [30]. The work of Savage [27] on elicitation of subjective probabilities is particularly elegant.

ENCOURAGING HONESTY [28]

To motivate a forecaster F to divulge honestly his personal probability p for the uncertain event A, we can present him with a *decision problem* having economic consequences to him which depend on his decision* d and the outcome a of A (where $a = 1$ if A occurs, $a = 0$ if not). These consequences may be expressed by a loss function $L(a, d)$. If L is measured on a utility* scale, or is small overall in relation to the forecaster's assets, the theory of coherence* implies that he will choose $d = d_p$, where d_p minimizes his expected loss $pL(1, d) + (1 - p) L(0, d) = L(p, d)$, say.

Suppose now that F is asked to quote a value $q \in [0, 1]$, on the understanding that this will commit him to the decision d_q appropriate if q were his "true" probability. Then he clearly has no incentive to quote any $q \neq p$, and should definitely quote $q = p$ if d_p yields the *unique* minimum of $L(p, d)$. His loss, if he quotes q and $A = a$, is $S(a, q) = L(a, d_q)$, so that his expected loss, if he quotes q but believes $\Pr(A) = p$, is $S(p, q) = L(p, d_q)$, which is minimized if $q = p$.

Any such function $S(a, q)$, for which $S(p, q)$ (defined as $E[S(\tilde{a}, q)]$, where $\tilde{a} = 1$ with probability p, 0 with probability $1 - p$) is minimized in q when $q = p$, is a *proper scoring rule*; it is *strictly proper* if this minimum is unique. Such a rule can always be regarded as arising, as above, from a decision problem, for example that having $d \in [0, 1]$ and $L(a, d) = S(a, d)$. Using any such scoring rule as a loss function encourages honesty. (However, difficulties in their use arise when the forecaster's utility is nonlinear in S, or he is motivated to maximize the probability of getting a bigger score than rival forecasters.) In practice, if the elicitor of F's forecast intends to use it as his own in a decision problem he faces, presenting F with a "share" in his own business will yield a scoring rule which is likely to provide the best motivation for F to concentrate his attention on good assessment where it matters most.

The formula $D(p,q) = S(p,q) - S(p,p)$ $(= L(p,d_q) - L(p,d_p))$ defines a *directed distance* of q from p associated with the decision problem or scoring rule, from which $S(a,q)$ may be effectively recovered as $D(a,q)$. Then $D(p,p) = 0$, and $D(p,q)$ is nondecreasing as q departs from p in either direction [27].

Important strictly proper scoring rules are the quadratic *Brier score* [3], with $S(a,q) = (a-q)^2$, and the *logarithmic score* [12] $S(1,q) = -\log q$, $S(0,q) = -\log(1-q)$. The scoring rule $S(a,q) = |a-q|$ is improper, however, having optimal $q_p = 1$ for $p > \frac{1}{2}$, 0 for $p < \frac{1}{2}$. Murphy and Epstein [23] consider the optimal "hedges" q_p for other scoring rules.

We can introduce an operational *definition* of F's personal probability p as the solution of $d_p = d^*$, where d^* is F's chosen decision in his decision problem, and the solution is supposed existent and unique. Equivalently, we take $p = q$, where q is F's quote under a strictly proper scoring rule. General coherence arguments (see especially ref. 27) imply that this solution should not depend on the specific decision problem. An early approach along these lines was that of de Finetti [7, 8], based on the Brier score.

Multipurpose Forecasts

The users of a weather forecast face many diverse decision problems. A particularly simple one, the *cost-loss* problem [31, 32], is that of the fruit farmer, say, who can take protective measures (decision d_1) to guard her plants against frost (event A), at a cost of C; if she does not protect (d_0), she faces a loss L if there is frost, 0 if not. Her expected loss when $\Pr(A) = p$ is $L(p,d_1) = C$, $L(p,d_0) = pL$, so that $d_p = d_1$ if $p > k = C/L$, $d_p = d_0$ if $p < k$. The associated proper scoring rule $S(a,q) = L(a,d_q)$ is not, however, strictly proper, q being an optimal quote whenever q and p both lie on the same side of k, and thus lead to the same decision.

To motivate F to meet the needs of *all* such decision makers, one might proceed as follows [16, 17]. After F states his forecast q, a value k is generated, independently of A, from a known distribution ν on [0, 1], and he is faced with the cost-loss decision problem having $L = 1$, $C = k$, in which he must use his implied optimal decision (i.e., $d_q = d_1$ if $k < q$, d_0 if $k \geqslant q$). His overall expected loss on following this strategy for quote q when $A = a$ is

$$S(a,q) = \int_0^q k\,d\nu(k) + a\int_q^1 d\nu(k).$$

This formula must therefore yield a proper scoring rule [18], and this will be strictly proper if and only if ν has the whole of [0, 1] as its support (so that p and q have to be on the same side of *any* $k \in [0,1]$).

With no essential change, we can replace $S(1,q)$ above by $S(1,q) - \int_0^1 k\,d\nu(k)$, yielding the standard form

$$S(0,q) = \int_0^q k\,d\nu(k),$$
$$S(1,q) = \int_q^1 (1-k)\,d\nu(k). \qquad (1)$$

It can, in fact, be shown [27–29], under weak conditions, that any proper scoring rule is equivalent to one having the form (1), where ν is an arbitrary measure on [0, 1]. (No additional generality is obtained by allowing L above to vary also, as in [25].) In choosing an "all-purpose" scoring rule to elicit a probability in practice, it makes sense to use one whose associated ν gives strongest weight to values of k regarded as most typical of the decision problems faced by users. Taking ν to be the uniform distribution on [0, 1], $d\nu(k) = dk$, effectively recovers the Brier score [16].

FORECASTING CONTINUOUS AND MULTISTATE QUANTITIES

Many of the ideas above extend to the more general case in which the forecaster has to assess a whole distribution Q for a quantity (or quantities) X. Again we can consider a decision problem with loss function $L(x,d)$, define d_Q to minimize $E_Q[L(X,d)]$, and so construct a proper scoring rule $S(x, Q) = L(x,d_Q)$. One important strictly proper scoring rule is the *logarithmic score* $S(x, Q) = -\log q(x)$, where $q(\cdot)$ is the density of Q with respect to a dominating measure μ. It may be shown [1, 15, 27] that whenever X can take more than two values, this is essentially the *only* proper scoring rule which is *local* [i.e., depends on $q(\cdot)$ only through its value at the point x which materializes]. The associated distance function is

$$D(P, Q) = \int p(x)\log[p(x)/q(x)]\,d\mu(x),$$

the *Kullback–Leibler information distance* of Q from P.

In practice one often confines attention to a finite set $\mathbf{A} = (A_s : s = 1,2,\ldots,k)$ of events determined by X, and their corresponding probability forecasts $\mathbf{q} = (q_s : s = 1,2,\ldots,k)$. Let these be judged by a strictly proper scoring rule $S(a,q)$ when $\mathbf{A} = a$. It may be that S is defined for values of $\mathbf{q}$ that violate the probability axioms. However, considerations of coherence imply [27] that any elicited probabilities should obey the axioms, so that if, for example, A_3 is the disjoint union of A_1 and A_2, then $q_3 = q_1 + q_2$ (see also ref. 14 for coherence of deduced probabilities when the loss function is given by an improper scoring rule).

The Brier score in the context above is now $\sum_{s=1}^{k}(a_s - q_s)^2$ and is strictly proper. It is normally applied in the case that the (A_s) form a partition [e.g., wind-speed rating (A_9 = gale, force 9, etc.)]. Then q is a "multistate" forecast. However, when, as in this example, the various states have a natural ordering, the Brier score appears somewhat unsatisfactory, since no credit is given for assessing high probabilities for states close to that materializing. This end can be accomplished by an analog of the cost-loss decision problem: A suitable loss function would have the property that if d_s is the optimal decision when A_s occurs, then the loss in using d_r instead of d_s increases with the distance between r and s. Suitable random selection of such a decision problem will yield a strictly proper scoring rule which is "sensitive to distance." Epstein [11] used this argument to derive his "ranked probability score."

EMPIRICAL VALIDITY: RELIABILITY

A single noncategorical probability forecast (i.e., not 0 or 1) can never be said to have been "right" or "wrong." But when a forecaster has issued a long string of such forecasts, it becomes possible to apply checks of external validity.

Suppose that of n sequential forecasts, the ith is p_i and the realized outcome of the associated event A_i ("rain on day i") is $a_i(= 0$ or $1)$. Then we can compare the overall average forecast probability $\bar{p}_n = n^{-1}\sum_{i=1}^{n} p_i$ with the overall relative frequency of occurrence $\bar{a}_n = n^{-1}\sum_{i=1}^{n} a_i$. If $\bar{p}_n \simeq \bar{a}_n$, the set of forecasts may be regarded as approximately valid on an overall basis. Murphy and Epstein [22] refer to this property as "unbiasedness in the large."

A more incisive test looks at that subset of occasions i for which the forecast probability p_i was at, or suitably close to, some preassigned value p^*, and compares the observed relative frequency in this subset, $\bar{a}(p^*)$ say, with p^*. If $\bar{a}(p^*) \simeq p^*$ for all p^*, the forecasts have been variously termed "unbiased in the small," "reliable," "valid," or "well calibrated." [Of course, sampling variability will lead to unavoidable departures of $\bar{a}(p^*)$ from p^*, particularly if based on small numbers.] The plot of $\bar{a}(p)$ against p is the forecaster's *reliability diagram* or *calibration curve*.

Consider, for illustration, the following probability forecasts and outcomes:

Outcome, a_i	0	0	1	0	1	0	1
Probability, p_i	0.4	0.6	0.3	0.2	0.6	0.3	0.4

Outcome, a_i	1	1	0	1	0	0	1
Probability, p_i	0.5	0.6	0.2	0.6	0.4	0.3	0.5

Table 1 Calibration Calculations

Probability, p^*	Number of Instances	Number of 1's	Relative Frequency of 1's, $a(p^*)$
0.2	2	0	0
0.3	3	1	0.33
0.4	3	1	0.33
0.5	2	2	1
0.6	4	3	0.75

Table 1 details the calculation of $\bar{a}(p^*)$ for these data. For example, of the three occasions on which a probability forecast of 0.3 was given, just one of the three associated events occurred, yielding $\bar{a}(0.3) = \frac{1}{3}$. The reliability diagram for these forecasts and outcomes is plotted in Fig. 1. The diagonal line corresponds, in the absence of sampling fluctuations, to perfect reliability. With such small numbers, there can be little strong evidence against this.

Figure 2 displays the reliability diagram for a sequence of 2820 12-hour PoP forecasts issued by a single weather forecaster in Chicago during the period 1972–1976. These results, which are typical of experienced forecasters, clearly indicate excellent reliability.

For assessing multistate forecasts ($\mathbf{p}_i$), with $\mathbf{p}_i = (p_{i1}, \ldots, p_{ik})$, one might concentrate on reliability for each state separately, or for all states jointly (by extracting the subset for which $\mathbf{p}_i \simeq \mathbf{p}^*$). The latter ("vector reliability") is more common, and provides a more demanding test. Another possibility, not generally recommended, is "scalar reliability," in which the whole set ($p_{is} : i = 1, \ldots, n; s = 1, \ldots, k$) is regarded as a single sequence, thus ignoring the identity of the state being forecast.

For a continuous quantity X, such as maximum temperature over 12 hours, *credible interval forecasting* involves specifying an interval (l, u) with the interpretation that $\Pr(l < X < u)$ is some fixed preassigned value γ (e.g., 75%). Reliability of a set of such forecasts is equivalent to requiring that a proportion γ of the intervals turn out to

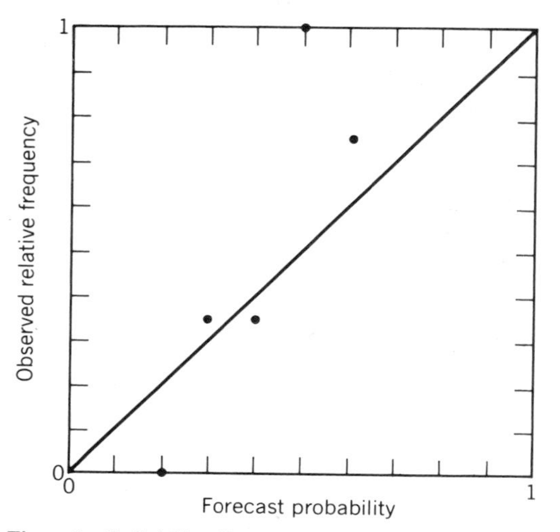

Figure 1 Reliability diagram corresponding to Table 1.

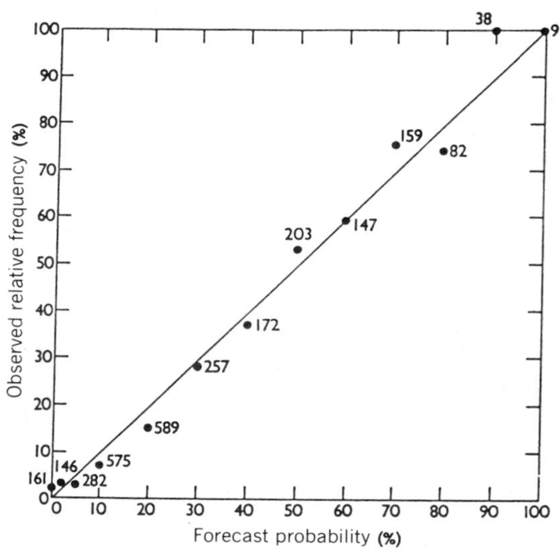

Figure 2 Reliability diagram of a weather forecaster. Reproduced from [24] with the permission of the authors. The number by each point is the number of occasions on which the associated forecast probability was issued.

contain their specified quantity. If one also requires $\Pr(X \leqslant l) = \Pr(X \geqslant u) = 1 - \gamma/2$, say, vector reliability can be applied to this multistate problem.

EMPIRICAL VALIDITY: RESOLUTION

Even when the calibration criterion above is satisfied, the forecasts can be clearly unsatisfactory. Suppose that the weather in fact alternates—dry, wet, dry, wet, ... Consider two forecasters assigning precipitation probabilities. F_1 assigns $p_i = \frac{1}{2}$ always, while F_2's forecast probabilities alternate 0, 1, 0, 1, ... Then both F_1 and F_2 are well calibrated; but F_2's forecasts are perfect, whereas those of F_1 are almost useless.

Now the assignment of probability forecasts to a sequence of events can be considered as comprising two separate tasks, *sorting* and *labeling* [26]. In *sorting*, the sequence of events is divided into disjoint subsequences, such that all the events in any given subsequence are considered (approximately) equally probable. *Labeling* then assigns a numerical value to the common probability in each subsequence. Reliability measures how well the labeling has been carried out but does not address the sorting aspect at all. In the example above, F_1 and F_2 were both completely reliable, but F_2 was clearly the more successful sorter. Another forecaster F_3, who assigns alternating probabilities 1, 0, 1, 0, . . . is also sorting as well as F_2, although the reliability of these forecasts is abominable.

In many ways, good sorting is more important than good labeling: F_3's forecasts are more valuable than those of F_1 once we have obtained enough data to see how to recalibrate them. In effect, ability at sorting depends on substantive (e.g., meteorological) expertise, and ability at labeling on normative expertise (i.e., skill at expressing uncertainty) [34].

The quality of the sorting process, termed *resolution*, might be judged by the forecaster's ability to create a set of subsequences in which the relative frequencies are close to 0 or 1 rather than near the overall relative frequency in the whole sequence. However, it is unreasonable, in general, to expect perfect sorting, which is equivalent to absolutely correct (or absolutely incorrect!) categorical forecasting. When, then, can we say that sorting has been carried out successfully? Curtiss [4] suggests that this has been achieved when within each subsequence of events corresponding to some fixed forecast value, the sequence of actual outcomes exhibits no pattern, but is essentially a "random sequence," or collective in the sense of von Mises [33]. Informally, this means that it should not be possible to extract, from the subsequence, a sub-subsequence in which the relative frequency differs from that in the subsequence. Forecaster F_1 above has clearly failed, and F_2 and F_3 succeeded, by this criterion.

Dawid [5] introduced an extended calibration criterion. Consider a subsequence S_n of $(1, 2, \ldots, n)$, extracted arbitrarily, subject to the constraint that the decision on whether or not i is to be included in S_n should be determined only by the previous outcomes $(a_1, \ldots, a_{i-1})$, not by a_i or any later outcomes. Let $\bar{p}'_n$ be the average forecast probability, and $\bar{a}'_n$ the empirical relative frequency, for the events in S_n. Then we require that $\bar{p}'_n \simeq \bar{a}'_n$. This is justified by showing that if the (p_i) are constructed sequentially as appropriate conditional probabilities from a joint distribution Π for the events $(A_1, A_2, \ldots)$, so that $p_i = \Pi(A_i \mid a_1, a_2, \ldots, a_{i-1})$, then $\bar{p}'_n - \bar{a}'_n \to 0$ except on a set of Π-probability zero. Consequently, if $\bar{p}'_n$ and $\bar{a}'_n$ are not close enough together, there is a suggestion that, in fact, $\bar{p}'_n - \bar{a}'_n \not\to 0$, which serves to discredit Π's probability assignments.

This extended calibration criterion subsumes the earlier one, since the decision to include i in S_n whenever p_i is close enough to some preassigned value p^* depends only on p_i, which in turn depends only on $(a_1, a_2, \ldots, a_{i-1})$. It also allows us to extract a suitable sub-subsequence of this subsequence and still require matching of forecasts and frequencies, as demanded by Curtiss. A sequence of forecasts which satisfies this criterion for every one of a large enough collection of subsequences (e.g., all computable subsequences) is called *completely calibrated*. Such a sequence has both perfect reliability and maximum attainable resolution. It is shown by Dawid [6] that if (p_i^1) and (p_i^2) are each completely calibrated computable forecast sequences, then $p_i^1 - p_i^2 \to 0$ $(i \to \infty)$. Consequently, this criterion of complete calibration is strong enough to exclude all but one limiting assignment of probabilities, as being empirically invalid. Unfortunately, there is no general way of constructing the asymptotically unique valid probability assignments. In practice, one must allow departures from complete calibration ascribable to sampling variation in finite sequences, so that the criterion is not overrestrictive.

Refinement. DeGroot and Fienberg [9, 10] introduce a partial ordering between forecasters, which is closely related to comparative resolution. Each combination of forecast p and observed outcome a can be formally regarded as one observation from a bivariate distribution for two random variables, $\tilde{p}$ and $\tilde{a}$. From a sequence of such forecasts and their outcomes we can build up an empirical bivariate distribution Q for $(\tilde{p}, \tilde{a})$: Q may be regarded as constructed by randomly selecting one of the occasions in the sequence, with equal probability for each, and setting $(\tilde{p}, \tilde{a})$ to the values (p_i, a_i) associated with the selected occasion. Then Q serves as a partial description of the sequence. (To be usable, this empirical distribution may need smoothing or discretisation.) Nature determines the empirical distribution of $\tilde{a}$. The forecaster's input into Q is thus contained in the conditional distributions of $\tilde{p}$, given $\tilde{a} = 0$ and given $\tilde{a} = 1$. (For good resolution, these two distributions should be very different.)

We can now apply concepts from the theory of comparison of experiments [2]. Forecaster A is *sufficient* for, or *more refined than*, forecaster B if B's conditional distribution for $\tilde{p}$ given $\tilde{a}$ can be obtained by first generating $\tilde{p}^*$ from A's conditional distribution and then generating $\tilde{p}$ from some fixed conditional distribution given $\tilde{p}^*$. In other words, B's forecast distributions can be produced by passing those of A through a noisy channel, independently of the actual outcome. Consequently, A's forecasts might be considered more valuable than B's.

ECONOMIC EVALUATION OF FORECASTS

Another approach to assessing the validity of a sequence of forecasts $\mathbf{p} = (p_1, p_2, \ldots, p_n)$ in the light of outcomes $\mathbf{a} = (a_1, a_2, \ldots, a_n)$ is in terms of the economic consequences which would have arisen from using these forecasts to solve a sequence of decision problems. Suppose that before each event A_i is observed, a decision d_i must be made, leading to loss $L(a_i, d_i)$ if $A_i = a_i$ (we suppose the decision space and loss function independent of i, for simplicity). Using the forecasts above corresponds to taking $d_i = d_{p_i}$, the decision optimal under $\Pr(A_i) = p_i$, and thus leads to loss $S(a_i, p_i)$ on the ith event, where S is the proper scoring rule associated with this decision problem. Comparisons between different forecasters might thus be based on their average scores $\bar{S} = n^{-1}\sum_{i=1}^{n} S(a_i, p_i)$. (Since any suitably regular proper scoring rule can be regarded as having arisen from choosing a random cost-loss decision problem, forecaster A will do at least as well as forecaster B for *any* decision problem if he does so for every simple cost-loss problem.)

The average score $\bar{S}$ can be split into components of independent interest. Thus letting E denote expectation under the (perhaps discretized or smoothed) joint empirical distribution Q for $(\tilde{p}, \tilde{a})$ based on the observed forecasts and outcomes, define $\rho(p) = E(\tilde{a} \mid \tilde{p} = p) = Q(\tilde{a} = 1 \mid \tilde{p} = p)$. This is the "recalibrated" version of p, giving the proportion of events occurring when the forecast p is quoted. We introduce $\tilde{r} = \rho(\tilde{p}) = E(\tilde{a} \mid \tilde{p})$. Then $\bar{S}$ can be expressed as

$E[S(\tilde{a}, \tilde{p})]$. But

$$E[S(\tilde{a}, \tilde{p})] = E[E\{S(\tilde{a}, \tilde{p}) \mid \tilde{p}\}]$$
$$= E[S(E(\tilde{a} \mid \tilde{p}), \tilde{p})] = E[S(\tilde{r}, \tilde{p})].$$

Recalling that the directed distance

$$D(p, q) \equiv S(p, q) - S(p, p)(\geqslant 0),$$

we obtain $\bar{S} = S_1 + S_2$, where $S_1 = E[D(\tilde{r}, \tilde{p})]$ and $S_2 = E[S(\tilde{r}, \tilde{r})]$. Note that $S_1 \geqslant 0$, with equality (if S is *strictly* proper) only when $\rho(p) \equiv p$ (almost surely under Q)—that is, when the forecasts are empirically calibrated. Consequently, S_1 can be regarded as a penalty score for poor calibration. On the other hand, S_2 is just the average score accruing to the recalibrated forecasts. Since $S(r, r)$ is concave in r [27], S_2 will tend to be smaller when these are concentrated near the endpoints of [0, 1].

If forecaster A is more refined than forecaster B, A will have a smaller value for S_2 than B for any proper scoring rule [10]. Thus S_2 may be regarded as a penalty for poor resolution. Schervish [28] points out that A will be more refined than B if and only if the recalibrated forecasts of A perform at least as well as those of B in every simple cost-loss problem.

Now let $\pi = E(\tilde{a}) = Q(\tilde{a} = 1)$ denote the empirical relative frequency of 1's in the outcome sequence (called the *climatological probability* in meteorology). Then $E(\tilde{r}) = E[E(\tilde{a} \mid \tilde{p})] = E(\tilde{a}) = \pi$. So $E[S(\tilde{r}, \pi)] = S(E(\tilde{r}), \pi) = S(\pi, \pi)$. Hence

$$S_2 = E[S(\tilde{r}, \tilde{r})]$$
$$= E[S(\tilde{r}, \pi) - D(\tilde{r}, \pi)] = S_0 - S_3,$$

with $S_0 = S(\pi, \pi)$, $S_3 = E[D(\tilde{r}, \pi)] \geqslant 0$. Thus $\bar{S} = S_0 + S_1 - S_3$. In this decomposition, S_0 is the score attained by the least refined well-calibrated forecaster, who always gives probability forecast π; S_3 measures the improvement in resolution over these constant forecasts. Again, S_1 penalizes poor calibration. As an example, let S be the Brier score, so that $S(a, p) = (a - p)^2$, $D(p, q) = (p - q)^2$. Suppose that the probability forecasts are confined to a set $(p_j : j = 1, 2, \ldots, k)$ of values, and that forecast p_j is issued on n_j occasions, of which a proportion ρ_j results in the event occurring. Then

$$n\bar{S} = \textstyle\sum_{j=1}^{k} n_j \{\rho_j(1 - p_j)^2 + (1 - \rho_j)p_j^2\},$$
$$S_0 = \pi(1 - \pi)$$

(with $n\pi = \sum_{j=1}^{k} n_j \rho_j$),

$$nS_1 = \textstyle\sum_{j=1}^{k} n_j(\rho_j - p_j)^2,$$
$$nS_2 = \textstyle\sum_{j=1}^{k} n_j \rho_j(1 - \rho_j)$$
$$nS_3 = \textstyle\sum_{j=1}^{k} n_j(\rho_j - \pi)^2.$$

For this case, the decomposition $\bar{S} = S_1 + S_2$ was first obtained by Sanders [26], and the formula $\bar{S} = S_0 + S_1 - S_3$ by Murphy [21]. The former decomposition was extended, as above, to general proper scoring rules by DeGroot and Fienberg [10]. The extension of the latter decomposition appears to be new.

If a completely calibrated set of forecasts **q** exists, these will be more refined than any other set **p**, and lead to the minimum achievable average loss, for any decision problem. For in any subsequence for which p_i and q_i take on values (approximately) equal, respectively, to some preassigned p^* and q^*, the proportion of events occurring will be q^*. Thus **q** is sufficient for the pair (**p**, **q**) [10], so is more refined than **p**. The average loss on applying the p's will be $L(q^*, d_{p^*}) \geqslant L(q^*, d_{q^*})$, the average loss for the q's.

Extensions of the decompositions above to multistate forecasts are straightforward, but can be based on either a scalar or a vector approach [20]. Indeed, even in the two-state case, scalar partitions will differ from the foregoing essentially vector partitions [19]. The vector partitions appear more relevant in general.

EMPIRICAL EVIDENCE

Many studies of actual forecasting performance have been carried out, investigating reliability, average scores, related "skill scores," and so on. A good review is given by Lichtenstein et al. [13]. The principal finding is that naive forecasters, or experienced experts departing even slightly from their usual tasks, find probability assessment difficult and perform poorly, generally showing overconfidence, with $\rho(p)$ much closer to $\frac{1}{2}$ than p. Feedback sessions, in which penalties arising from applying a proper scoring rule are revealed to the forecaster (or docked from his pay!), can sharpen forecasting ability remarkably. With such practice, U.S. weather forecasters, for example, find themselves well able to discern small differences in probability and are very well calibrated.

References

[1] Bernardo, J. M. (1979). *Ann. Statist.*, **7**, 686–690.

[2] Blackwell, D. (1951). *Proc. 2nd Berkeley Symp. Math. Statist. Prob.* University of California Press, Berkeley, Calif., pp. 93–102.

[3] Brier, G. W. (1950). *Monthly Weather Rev.*, **78**, 1-3.

[4] Curtiss, J. H. (1968). *J. Appl. Meteorol.*, **7**, 3–17.

[5] Dawid, A. P. (1982). *J. Amer. Statist. Ass.*, **77**, 605–613.

[6] Dawid, A. P. (1985). *Ann. Statist.*, **13**, 1251–1274.

[7] de Finetti, B. (1962). In *The Scientist Speculates*, I. J. Good, ed. Heinemann, London, 357–364.

[8] de Finetti, B. (1975). *Theory of Probability* (English transl.). 2 vols. Wiley, New York.

[9] DeGroot, M. H. and Fienberg, S. E. (1982). In *Statistical Decision Theory and Related Topics III*, Vol. 1, S. S. Gupta and J. O. Berger, eds. Academic Press, New York, pp. 291–314.

[10] DeGroot, M. H. and Fienberg, S. E. (1983). *Statistician*, **32**, 12–22.

[11] Epstein, E. S. (1969). *J. Appl. Meteorol.*, **8**, 985–987.

[12] Good, I. J. (1952). *J. R. Statist. Soc. B*, **14**, 107–114.

[13] Lichtenstein, S., Fischhoff, B., and Phillips, L. D. (1982). In *Judgment under Uncertainty: Heuristics and Biases*, D. Kahneman, P. Slovic, and A. Tversky, eds. Cambridge University Press, Cambridge, England.

[14] Lindley, D. V. (1982). *Int. Statist. Rev.*, **50**, 1–26.

[15] McCarthy, J. (1956). *Proc. Natl. Acad. Sci. (USA)*, pp. 654–655.

[16] Murphy, A. H. (1966). *J. Appl. Meteorol.*, **5**, 534–537.

[17] Murphy, A. H. (1969). *J. Appl. Meteorol.*, **8**, 863–873.

[18] Murphy, A. H. (1969). *J. Appl. Meteorol.*, **8**, 989–991.

[19] Murphy, A. H. (1972). *J. Appl. Meteorol.*, **11**, 273–282.

[20] Murphy, A. H. (1972). *J. Appl. Meteorol.*, **11**, 1183–1192.

[21] Murphy, A. H. (1973). *J. Appl. Meteorol.*, **12**, 595–600.

[22] Murphy, A. H. and Epstein, E. S. (1967). *J. Appl. Meteorol.*, **6**, 748–755.

[23] Murphy, A. H. and Epstein, E. S. (1967). *J. Appl. Meteorol.*, **6**, 1002–1004.

[24] Murphy, A. H. and Winkler, R. L. (1977). *Appl. Statist.*, **26**, 41–47.

[25] Pearl, J. (1978). *Int. J. Man–Machine Stud.*, **10**, 175–183.

[26] Sanders, F. (1963). *J. Appl. Meteorol.*, **2**, 191–201.

[27] Savage, L. J. (1971). *J. Amer. Statist. Ass.*, **66**, 783–801.

[28] Schervish, M. J. (1983). A General Method for Comparing Probability Assessors. *Tech. Rep. No. 275*, Dept. of Statistics, Carnegie-Mellon University, Pittsburgh, Pa.

[29] Shuford, E. H., Albert, A., and Massengill, H. E. (1966). *Psychometrika*, **31**, 125–145.

[30] Staël von Holstein, C. A. S. (1970). *Assessment and Evaluation of Subjective Probability Distributions*. Economic Research Institute, Stockholm School of Economics, Stockholm.

[31] Thompson, J. C. (1950). *Monthly Weather Rev.*, **78**, 113–124.

[32] Thompson, J. C. (1952). *Bull. Amer. Meteorol. Soc.*, **33**, 223–226.

[33] von Mises, R. (1957). *Probability, Statistics and Truth*. Allen & Unwin, London.

[34] Winkler, R. L. and Murphy, A. H. (1968). *J. Appl. Meteorol.*, **7**, 751–758.

(METEOROLOGY, STATISTICS IN
SUBJECTIVE PROBABILITIES)

A. P. DAWID

PROBABILITY GENERATING FUNCTIONS *See* GENERATING FUNCTIONS

PROBABILITY, HISTORY OF (OUTLINE)

The history of probability is in essence separate from the history of statistics*, although the statistics of sample data as a mathematical discipline relies on probability as the foundation of quantitative inference. The synthesis of probability with statistics to form mathematical statistics was ultimately achieved at the beginning of the twentieth century, largely due to the English Biometric School*. In this article we adhere to the history of probability proper up to the end of the nineteenth century, with a brief incursion into the twentieth. Thus we shall not mention names as eminent as that of Gauss*, more pertinent to the "theory of errors."

The history of probability theory is traditionally taken to begin in France in 1654, the year of the correspondence between Blaise Pascal* (1623–1662) and Pierre de Fermat* (1601–1665), primarily on the problem of equitable division of stakes in games of chance*. Although there is a "prehistory" of probability, it is due to the enormous prestige of these two workers that later developments crystallized round their work. It is also characteristic of the history that until the nineteenth century its motivation remained, very substantially, games of chance.

The frequently mentioned significant names of the prehistory of probability in the context of gaming are Cardano, Galileo, Pacioli, and Tartaglia in Italy. In a book of Fra Luca Pacioli (ca. 1445–ca. 1517) printed in 1494, there occurs a simple instance of the problem of points* which was later taken up by Tartaglia (Niccolò Fontana, ca. 1500–1557) and Girolamo Cardano (1501–1576), and then solved by Fermat and Pascal. In Cardano's manuscript *Liber de Ludo Aleae* we find (perhaps for the first time) the fundamental notion, in the context of dice games, that the probability of an event in the situation where each of a finite set of possible outcomes has equal weight is the ratio of the favorable cases to the total number of the cases. Galileo Galilei (1564–1642) showed, *inter alia*, a thorough understanding of this principle in reference to a throw of three fair dice.

The keynote of the Fermat–Pascal correspondence is the equitable division of total stake when a game is interrupted before its agreed completion ("problème de partis"), and a solution to the problem of points is produced by an ingenious extension of the sample space to equally likely points. The correspondence contains a numerical instance of the gambler's ruin problem (random walk* with absorbing barriers). Pascal's posthumously published *Traité du triangle arithmetique* is the corresponding treatise; his *Pensées* and portions, disputably written by Pascal, of the *Logique de Port-Royal* (1662) contain elements of the calculus of probabilities also. The next landmark is the publication in 1657 of the first text on probability: *De Ratiociniis in Ludo Aleae* by the Dutch scientist Christiaan Huygens* (1629–1695). Huygens had learned of the problem of points on a visit to Paris in 1655, and his systematization of arguments regarding this problem, and some exercises including the gambler's ruin problem, were inspired directly by the activities of the French mathematicians. His booklet contains a clear allusion to expectation and notions of sampling without and with replacement. Its main function was as a popularizing medium for the calculus of chances until superseded by the very substantial works *Essai d'analyse sur les jeux de hasard* (1708) of Pierre de Montmort* (1678–1719), *Ars Conjectandi* (1713, posthumous) of Jacob (James) Bernoulli* (1654–1705), and *The Doctrine of Chances* (1718) of Abraham De Moivre* (1667–1754). Bernoulli's book, written perhaps about 1690, incorporated Huygens's treatise, with notes, in its first part and the ideas in its fourth part were suggested at least partially by the *Logique de Port-Royal*. In the first part, the gambler's ruin problem is generalized to a totally algebraic setting. In the whole, the general rules of probability calculus are made clear. The most significant achievement, however, is in the fourth part, where it is proved under some restrictions that if $n^{-1}S_n$ is the proportion of successes in n binomial* trials in which the success probability is p, then as $n \to \infty$ for any $\epsilon > 0$,

$$\Pr(|n^{-1}S_n - p| \leqslant \epsilon) \to 1,$$

which has come to be known as Bernoulli's theorem* and is the first version of the weak law of large numbers*. In the preface to *Ars Conjectandi*, James Bernoulli's nephew Nicholas asks that Montmort and De Moivre turn their attention to further applications of probability. One of the results, in part, was the second enlarged edition in 1713 of Montmort's treatise; Montmort had been aware at the time of his first edition of the work of the Bernoullis. Inasmuch as he writes for mathematicians rather than gamblers, using "modern" analysis, this work is historically significant. He worked on the game of *Rencontre* (matching pairs), resolved the problem of points in general terms, and worked with Nicholas Bernoulli (1687–1759) on the problem of duration of play in the gambler's ruin problem. The most important contribution of De Moivre is the refinement of Bernoulli's theorem by obtaining the normal approximation* to the binomial distribution* for general p ("De Moivre's theorem"), thereby establishing the first instance of the central limit theorem*. This work appears in the second and third editions of his *Doctrine*, which also contains the elements of the Poisson approximation to the binomial distribution. The book of the Marquis De Condorcet (1743–1794), *Essai sur l'Application de l'Analyse à la Probabilitè des Décisions Rendues à la Pluralité des Voix* (1785), is important in that it is concerned with social questions and in that it stimulated the fundamental work of Pierre Simon de Laplace* (1749–1827), *Théorie analytique des probabilités* (1812), which is often taken to mark the end of an era. Laplace's book may be regarded as describing the then state of the art not only for probability but for statistics in a systematic mathematical fashion. Perhaps two of Laplace's greatest contributions for probability theory proper are recognition of the mathematical power of transforms (generating functions*, characteristic functions*) and an early general version of the central limit theorem.

The nineteenth century sees work in probability by French mathematicians on the foundations laid by Laplace and the development of a strong probabilistic school in the Russian Empire as the two most subsequently influential streams. (German developments were rather in the direction of statistics.)

Of the French school, a significant landmark was the publication in 1837 of the book by Siméon Poisson* (1781–1840), *Recherches sur la Probabilité des Jugements en Matière Criminelle et en Matière Civile, Précédées des Règles Générales du Calcul des Probabilités* (translated into German in 1841), which, as well as taking up the social context treated by Condorcet and Laplace, is in part a probability text. Poisson considered in particular binomial trials with probability of success which may vary deterministically from trial to trial. This situation gives rise to Poisson's law of large numbers* for the proportion of successes, and a corresponding normal approximation, possibly the first suggestion of a central limit statement for sums of independent but nonidentically distributed random variables. Poisson also arrived at the Poisson distribution as the limit of a negative binomial*, and the probability distribution function makes possibly its first appearance. Augustin Cauchy* (1789–1857) in 1853 discovered the symmetric stable laws* and supplied almost a complete rigorous proof of the central limit theorem by characteristic function methods within the context of least squares, while Jules Bienaymé* (1796–1878) proved by the simple techniques still used today the Bienaymé–Chebyshev inequality*, used it to prove a general weak law of large numbers for sums of i.i.d. random variables, and gave a completely correct statement of the criticality theorem* of branching processes*. The book of Antoine Cournot* (1801–1877), *Exposition de la Théorie des Chances et des Probabilités* (1834) (translated into German in 1849), is largely philosophical and is notable for an early frequentist view of the subject, while that of Joseph Bertrand (1822–1900), entitled *Calcul des Probabilités* (1889), is a significant and sensibly written "modern" textbook on probability and its applications.

The flowering of probability theory in particular (as well as mathematics in general) in the Russian Empire is due in part to the return of the young mathematicians V. Ya. Buniakovsky (1804–1889) and M. V. Ostrogradsky (1801–1861) from Paris, where in the early 1820s both had had contact with Laplace, Cauchy, Fourier, and Poisson. Both popularized probability theory in St. Petersburg, together with A. Yu. Davydov (1823–1885) in Moscow. Buniakovsky wrote the first Russian-language text (heavily influenced by Laplace's *Théorie*) on probability, *Osnovania Matematicheskoi Teorii Veroiatnostei*, and thereby introduced the still-current mathematical terminology into that language. The stage had been set for the founding of the St. Petersburg School by P. L. Chebyshev* (1821–1894), whose probabilistic contributions, with those of his pupils A. A. Markov* (1856–1922) and A. M. Liapunov* (1857–1918), in particular, established it as another center of the subject area and gave rise to much of the direction of modern probability theory. Chebyshev proved rigorously Poisson's weak law of large numbers and popularized the applicability of the Bienaymé–Chebyshev inequality as a tool for proving weak laws. All three worked on rigorous proofs of the central limit theorem for nonidentically distributed independent summands under general conditions, the culmination being Liapunov's theorem*. Markov developed the concept of Markov chains* and thereby gave impetus to the development of stochastic processes*. His textbook *Ischislenie veroiatnostei* (1900, 1908), in its German version in 1912 and in its third (substantially revised) edition of 1913 (and the posthumous edition of 1924), to a large extent laid the foundation for the further development of probability theory.

In regard to stochastic processes, further foundations were laid, again in France, by the *Calcul des probabilités* (1896) of Henri Poincaré (1854–1912), and by Louis Bachelier (1870–1946) through his *Théorie de la speculation* (1900), for the theory of Markov processes.

In England some early impetus for probabilistic work was provided by Augustus de Morgan (1806–1871). *The Logic of Chance* (1876) of John Venn (1834–1923) gave impetus to the study of probability from the standpoint of philosophy, while that of William Whitworth (1840–1905), entitled *Choice and Chance*, of which the first edition appeared in 1867, even now continues to provide exercises on combinatorial probability. Other eminent contributors to the subject were Robert Ellis (1817–1859), Morgan Crofton (1826–1915), and James Glaisher (1848–1928). In the context of statistical mechanics*, important contributors were James Maxwell* (1831–1879) and Ludwig Boltzmann* (1844–1906). For the prehistory of probability see refs. 3, 6, 9, 10, 13, and 15. The Fermat–Pascal correspondence is thoroughly discussed in refs. 3 and 19, while work from this point up to and including Laplace is well covered in the classic ref. 19. Other relevant sources for this period are refs. 1–4, 6, 7, 9, 11–14, and 16.

Unlike the preceding period, the history of probability in the nineteenth century is not adequately represented in monograph form. The most comprehensive is Czuber [2], with the supplement, ref. 21. Modern monograph sources are refs. 5, 8, 9 and 20. For specialized aspects, consult also refs. 1, 13, 17, and 18.

The modern era in the development of probability theory begins with its axiomatization. The first attempts in this direction are due to Sergei Bernstein* (1880–1968), Richard von Mises* (1883–1953), and Emile Borel (1871–1956). A. N. Kolmogorov's *Grundbegriffe der Wahrscheinlichkeitsrechnung* (1933) introduced the now generally accepted axiomatization suitable not only for classical probability theory but also for the theory of stochastic processes*. A few of the significant books which provided an impetus to the development of modern probability are P. Lévy's *Théorie de l'addition des variables aléatoires* (1937); H. Cramér's *Random Variables and Probability Distributions* (1937) and *Mathematical Methods of Statistics* (1946); *Limit Distributions for Sums of Independent Random Variables* (Russian edition 1949; first English edition 1954) by B. V. Gnedenko and A. N. Kolmogorov; W.

Feller's *An Introduction to Probability Theory and Its Applications*, Vol. 1 (1950); and M. Loève's *Probability Theory* (1955).

References

[1] Adams, W. J. (1974). *The Life and Times of the Central Limit Theorem*. Caedmon, New York. (Chapter 2 contains a careful bibliographical analysis of the origin of De Moivre's theorem.)

[2] Czuber, E. (1899). *Jb. Dtsch. Math.-Ver.*, **7**, Part 2, 1–279. (Together with Wölffing [21], the best bibliographical source on probability and statistics in the nineteenth century.)

[3] David, F. N. (1962). *Games, Gods and Gambling*. Charles Griffin, London.

[4] Dupont, P. (1979). *Atti. Accad. Sci. Torino Cl. Sci. Fis. Mat. Natur.*, **113**, 243–261.

[5] Gnedenko, B. V. and Sheynin, O. B. (1978). In *Matematika XIX veka (Mathematics of the nineteenth century)*. Nauka, Moscow, Chap. 4. (Chapter 4, pp. 184–240, deals with the history of probability, beginning with Laplace.)

[6] Gouraud, C. (1848). *Histoire du calcul des probabilités*. A. Durand, Paris.

[7] Hacking, I. (1975). *The Emergence of Probability*. Cambridge University Press, Cambridge, England, Chap. 8.

[8] Heyde, C. C. and Seneta, E. (1977). *I. J. Bienaymé: Statistical Theory Anticipated*. Springer-Verlag, New York. (A modern source on probability and statistics in the nineteenth century.)

[9] Maistrov, L. E. (1967). *Teoriya Veroiatnostei: Istoricheskii Ocherk*. Nauka, Moscow. (English transl.: *Probability Theory. A Historical Sketch*, S. Kotz, trans./ed. Academic Press, New York, 1974.)

[10] Ore, O. (1953). *Cardano, the Gambling Scholar*. Princeton University Press, Princeton, N.J. (This book contains a translation by S. H. Gould of *Liber de Ludo Aleae*.)

[11] Ore, O. (1960). *Amer. Math. Monthly*, **67**, 409–419.

[12] Pearson, K. (1978). *The History of Statistics in the 17th and 18th Centuries*. (Lectures by Karl Pearson 1921–1933, edited by E. S. Pearson.) Charles Griffin, London.

[13] Pearson, E. S. and Kendall, M. G., eds. (1970). *Studies in the History of Statistics*. Charles Griffin, London; Hafner, Darien, Conn. (Vol. II, M. G. Kendall and R. L. Plackett, eds., 1977).

[14] Seneta, E. (1979). In *Interactive Statistics*, D. McNeil, ed. North-Holland, Amsterdam, pp. 225–233.

[15] Sheynin, O. B. (1974). *Arch. History Exact Sci.*, **12**, 97–141.

[16] Sheynin, O. B. (1977). *Arch. History Exact Sci.*, **17**, 201–259.

[17] Sheynin, O. B. (1978). *Arch. History Exact Sci.*, **18**, 245–300.

[18] Stigler, S. M. (1975). *Biometrika*, **62**, 503–517.

[19] Todhunter, I. (1865). *A History of the Mathematical Theory of Probability from the Time of Pascal to That of Laplace*. Cambridge University Press, Cambridge, England. (Reprinted in 1949 and 1961 by Chelsea, New York.)

[20] Ulbricht, K. (1980). *Wahrscheinlichkeitsfunktionen im neunzehnten Jahrhundert*. Minerva, Munich. (Contains a good bibliography and is valuable in presenting a German standpoint, possibly underrepresented in the article.)

[21] Wölffing, E. (1899). *Math. Naturwiss. Ver. Württemberg* (Stuttgart), *Mitt.*, (2), **1**, 76–84. (Supplement to Czuber [2].)

(BERNOULLIS, THE
BERNOULLI'S THEOREM
BINOMIAL DISTRIBUTION
CHEBYSHEV INEQUALITY
CHEBYSHEV, PAFNUTY LVOVICH
COMBINATORICS
CONVERGENCE OF SEQUENCES OF RANDOM VARIABLES
DE MOIVRE, ABRAHAM
FERMAT, PIERRE DE
GAMES OF CHANCE
LAPLACE, PIERRE SIMON DE
LIMIT THEOREM, CENTRAL
MARKOV, ANDREI ANDREEVICH
MOIVRE–LAPLACE THEOREM
MONTMORT, PIERRE RÉMOND DE
PASCAL, BLAISE
POISSON, SIMEON DENIS
PROBLEM OF POINTS
RANDOM WALK
STATISTICS, HISTORY OF)

E. SENETA

PROBABILITY INEQUALITIES FOR SUMS OF BOUNDED RANDOM VARIABLES

If S is a random variable with finite mean and variance, the Bienaymé–Chebyshev inequality states that for $x > 0$,

$$\Pr\left[|S - ES| \geqslant x(\operatorname{var} S)^{1/2}\right] \leqslant x^{-2}. \quad (1)$$

If S is the sum of n independent, identically

distributed random variables, then, by the central limit theorem*, as $n \to \infty$, the probability on the left approaches $2\Phi(-x)$, where $\Phi(x)$ is the standard normal distribution function. For x large, $\Phi(-x)$ behaves as $\text{const.}\, x^{-1}\exp(-x^2/2)$.

S. Bernstein* [2, 3] has shown that under an additional assumption, which is satisfied when the summands of S are uniformly bounded, the upper bound in (1) can be replaced by one which, for n large, behaves as $\exp(-x^2/2)$.

We first discuss Bernstein's inequality* and related results for sums of independent random variables and random vectors, and then extensions of some of these results to certain sums of dependent random variables.

SUMS OF INDEPENDENT RANDOM VARIABLES

Let $X_1, \dots, X_n$ be n independent real-valued random variables with finite means,

$$S = X_1 + \cdots + X_n, \qquad \mu = ES/n. \quad (2)$$

We begin with inequalities that involve no moments other than μ.

Theorem 1. If $0 \leqslant X_i \leqslant 1$, $i = 1, \dots, n$, then for $0 < t < 1 - \mu$,

$$\Pr[S - ES \geqslant nt]$$

$$\leqslant \left\{ \left(\frac{\mu}{\mu + t} \right)^{\mu + t} \left(\frac{1 - \mu}{1 - \mu - t} \right)^{1 - \mu - t} \right\}^n \quad (3)$$

$$\leqslant \exp\{-g(\mu) n t^2\} \quad (4)$$

$$\leqslant \exp\{-2nt^2\}, \quad (5)$$

where

$$g(\mu) = \begin{cases} \dfrac{1}{1 - 2\mu} \ln \dfrac{1 - \mu}{\mu} & \text{for } 0 < \mu < \frac{1}{2}, \quad (6) \\[2ex] \dfrac{1}{2\mu(1 - \mu)} & \text{for } \frac{1}{2} \leqslant \mu < 1. \quad (7) \end{cases}$$

[*Note*: If $t > 1 - \mu$, the probability in (3) is zero.]

The assumption $0 \leqslant X_i \leqslant 1$ has been made to give the bounds a simple form. If, instead, we assume that $a \leqslant X_i \leqslant b$, then μ and t in the statement of Theorem 1 are to be replaced by $(\mu - a)/(b - a)$ and $t/(b - a)$, respectively.

Upper bounds for

$$\Pr[S - ES \leqslant -nt] = \Pr[-S + ES \geqslant nt]$$

and for

$$\Pr[|S - ES| \geqslant nt] = \Pr[S - ES \geqslant nt] + \Pr[S - ES \leqslant -nt]$$

follow from Theorem 1 by appropriate substitutions. Similar remarks apply to the inequalities that follow.

Let

$$\sigma^2 = n^{-1} \operatorname{var} S. \quad (8)$$

Under the conditions of Theorem 1, $\sigma^2 \leqslant \mu(1 - \mu)$, with equality holding only in the case

$$\Pr[X_i = 0] = 1 - \mu, \quad \Pr[X_i = 1] = \mu, \quad i = 1, \dots, n. \quad (9)$$

For this (binomial) case inequality (3) is implicitly contained in Chernoff [5], and the other bounds of Theorem 1, except for (4) with $\mu < \frac{1}{2}$, are due to Okamoto [7]. In the general case Theorem 1 was proved by Hoeffding [6].

If the inequalities of Theorem 1 are written as

$$\Pr[S - ES \geqslant nt] \leqslant A_1 \leqslant A_2 \leqslant A_3,$$

then each A_i is of the form a_i^n, where a_i is a function of μ and t only. Each of these bounds is simpler but cruder than the preceding. We have $A_1 = A_2$ if $t = 1 - 2\mu$ and $A_2 = A_3$ if $\mu = \frac{1}{2}$. If t and μ are fixed as $n \to \infty$, in which case $\Pr[S - ES \geqslant nt] \to 0$ exponentially fast, and if the inequality $A_i < A_{i+1}$ is strict, then the bound A_i is appreciably smaller than A_{i+1} when n is large. On the other hand, if we put $t = yn^{-1/2}$ and hold y and μ fixed, then, as $n \to \infty$,

$$A_1 \to \exp\left(-\frac{y^2}{2\mu(1-\mu)}\right),$$

$$A_2 = \exp(-g(\mu)y^2),$$

$$A_3 = \exp(-2y^2).$$

Note that the limit of A_1 is equal to A_2 for $\frac{1}{2} \leqslant \mu < 1$. When the central limit theorem applies to S,

$$\Pr[S - ES \geqslant n^{1/2}y] \to \Phi(-y/\sigma) \leqslant \Phi\left(-y/\sqrt{\mu(1-\mu)}\right).$$

The following is an extension of bound (5); see Hoeffding [6].

Theorem 2. If $X_1, \ldots, X_n$ are independent and $a_i \leqslant X_i \leqslant b_i$ $(i = 1, \ldots, n)$, then for $t > 0$,

$$\Pr[S - ES \geqslant nt] \leqslant \exp\left\{-2n^2t^2/\textstyle\sum_1^n(b_i - a_i)^2\right\}. \quad (10)$$

We now assume that the X_i have a common mean. For simplicity the mean is taken to be zero.

Theorem 3. If $X_1, \ldots, X_n$ are independent, $\sigma^2 < \infty$, $EX_i = 0$, $X_i \leqslant b$ $(i = 1, \ldots, n)$, then for $0 < t < b$,

$$\Pr[S \geqslant nt] \leqslant \left\{\left(1 + \frac{bt}{\sigma^2}\right)^{-(1+bt/\sigma^2)\sigma^2/(b^2+\sigma^2)} \times \left(1 - \frac{t}{b}\right)^{-(1-t/b)b^2/(b^2+\sigma^2)}\right\}^n \quad (11)$$

$$\leqslant \exp\left\{-\frac{nt}{b}\left[\left(1 + \frac{\sigma^2}{bt}\right) \times \ln\left(1 + \frac{bt}{\sigma^2}\right) - 1\right]\right\} \quad (12)$$

$$\leqslant \exp\left\{-\frac{nt^2}{2(\sigma^2 + bt/3)}\right\}. \quad (13)$$

Here the summands are assumed to be bounded only from above. However, to obtain from Theorem 3 upper bounds for $\Pr[|S| \geqslant t]$, we must assume that they are bounded on both sides.

The bounds in (3) and (11) are related as follows. If the assumptions of Theorem 3 are satisfied and the X_i are also bounded from below, $a \leqslant X_i \leqslant b$ (where $a < 0 < b$), then $\sigma^2 \leqslant -ab$. The bound in (11) is an increasing function of σ^2. If we replace σ^2 by its upper bound $-ab$, we obtain from (11) the inequality which results from (3) when the appropriate substitutions mentioned after the statement of Theorem 1 are made. Note, however, that Theorem 1 does not require that the X_i have a common mean.

The bounds for $\Pr[S \geqslant nt]$ in (11) to (13) are due to Hoeffding [6], Bennett [1], and Bernstein [3], respectively. For proofs of Bernstein's inequality, see also Uspensky [8] and Bennett [1]. For an early version of (13), see Bernstein [2].

Bernstein [3] derived an upper bound for $\Pr[|S| \geqslant nt]$ similar to (13) without assuming that the X_i are bounded. Instead, he assumed that the moments of the X_i satisfy

$$|EX_i^m| \leqslant \tfrac{1}{2}(EX_i^2)m!c^{m-2},$$

$$m = 3, 4, \ldots; \quad i = 1, \ldots, n, \quad (14)$$

with some constant $c > 0$. If $|X_i| \leqslant b$ for all i, then (14) is satisfied with $c = b/3$.

The inequalities of Theorem 3 can be written

$$\Pr[S \geqslant nt] \leqslant B_1 \leqslant B_2 \leqslant B_3.$$

Each B_i is of the form b_i^n, where b_i is a function of t/b and σ/b only. If t, b, and σ are fixed as $n \to \infty$, remarks similar to those concerning Theorem 1 apply. If we put $t = x\sigma n^{-1/2}$ and hold x, b, and σ fixed, then, as $n \to \infty$,

$$B_i \to \exp(-x^2/2), \qquad i = 1, 2, 3.$$

When the central limit theorem applies to S,

$$\Pr[S \geqslant x\sigma n^{1/2}] \to \Phi(-x).$$

Thus in this case the simple bound (13) is nearly as good as the sharper bounds (11) and (12) when n is sufficiently large. For further comparisons between different bounds, see Bennett [1] and Hoeffding [6].

Bernstein's exponential bound (13) has been extended to the cases of sums of independent random vectors taking values in a Euclidean space, a Hilbert space, and a Banach space. See Yurinskiĭ [9], where references to earlier work are given.

SUMS OF DEPENDENT RANDOM VARIABLES

Some of the inequalities discussed above have been extended to certain types of dependent random variables.

(a) Martingales*. Let $X_1, \dots, X_n$ be random variables with finite means, $X_i' = X_i - EX_i$, $S_j' = \sum_1^j X_i'$, and suppose that the sequence $S_1', \dots, S_n'$ is a martingale, that is, $E[X_{j+1}' \mid X_1', \dots, X_j'] = 0$, $j = 1, \dots, n-1$, with probability 1. Bernstein [4] showed that a version of the inequality named for him holds in the present case. Theorems 1 and 2 remain true if the S_j' sequence is a martingale [6, p. 18].

(b) U-statistics*. Let $X_1, \dots, X_n$ be i.i. d., and consider the U-statistic

$$U = \frac{1}{n^{(r)}} \sum\nolimits_{n,r} \phi(X_{i_1}, \dots, X_{i_r}), \qquad n \geqslant r,$$

where ϕ is a measurable function, $n^{(r)} = n(n-1)\dots(n-r+1)$, and the sum $\sum_{n,r}$ is taken over all r-tuples $i_1, \dots, i_r$ of distinct positive integers not exceeding n.

If $0 \leqslant \phi(x_1, \dots, x_r) \leqslant 1$, then the bounds of Theorem 1 with n replaced by $[n/r]$ (the largest integer $\leqslant n/r$) and $\mu = E\phi(X_1, \dots, X_r)$ are upper bounds for $\Pr[U - EU \geqslant t]$. If $\sigma^2 = \operatorname{var}\phi(X_1, \dots, X_r)$ exists and $\phi(x_1, \dots, x_r) \leqslant EU + b$, then the bounds of Theorem 3 with n replaced by $[n/r]$ are upper bounds for $\Pr[U - EU \geqslant t]$.

Similar results hold for k-sample U-statistics and for statistics of the form

$$V = n^{-r} \sum_{i_1=1}^{n} \cdots \sum_{i_r=1}^{n} \phi(X_{i_1}, \dots, X_{i_r}).$$

(c) Sums of finitely dependent random variables. Let the random variables $X_1, \dots, X_n$ be $(r-1)$-dependent; that is, the random vectors $(X_1, \dots, X_i)$ and $(X_j, \dots, X_n)$ are independent if $j - i \geqslant r$. Suppose also that $X_1, \dots, X_n$ are identically distributed. Let $\mu = EX_1$, $S = X_1 + \cdots + X_n$. Under the appropriate boundedness conditions the bounds of Theorems 1 and 3 with n replaced by $[n/r]$, are upper bounds for $\Pr[S - ES \geqslant nt]$.

(d) Sampling from a finite population. Consider a finite population* $\mathscr{P} = \{1, \dots, N\}$ of size N and a real-valued function f defined on $\mathscr{P}$. Let $(Y_1, \dots, Y_n)$ be a simple random sample without replacement from $\mathscr{P}$, $X_i = f(Y_i)$, $S = X_1 + \cdots + X_n$. Let $\mu = EX_1$, $\sigma^2 = N^{-1}\sum_{j=1}^N (f(j) - \mu)^2$. If the appropriate boundedness conditions on $f(j)$ are satisfied, the inequalities of Theorems 1 and 3 hold in this case.

References

[1] Bennett, G. (1962). *J. Amer. Statist. Ass.*, **57**, 33–45.

[2] Bernstein, S. N. (1924). *Učen. Zap. Naŭc.-Issled. Kafedr Ukrainy, Otd. Mat.*, **1**, 30–49 (in Russian). [Reprinted in S. N. Bernstein, *Collected Works*, Vol. IV. Nauka, Moscow, 1964 (in Russian).]

[3] Bernstein, S. N. (1927). *Probability Theory* (in Russian). (Referred to in Uspensky [8].)

[4] Bernstein, S. N. (1937). *Dokl. Akad. Nauk SSSR*, **17**, 275–277 (in Russian). [Reprinted in S. N. Bernstein, *Collected Works*, Vol. IV. Nauka, Moscow, 1964 (in Russian).]

[5] Chernoff, H. (1952). *Ann. Math. Statist.*, **23**, 493–507.

[6] Hoeffding, W. (1963). *J. Amer. Statist. Ass.*, **58**, 13–30.

[7] Okamoto, M. (1958). *Ann. Inst. Statist. Math.*, **10**, 20–35.

[8] Uspensky, J. V. (1937). *Introduction to Mathematical Probability*. McGraw-Hill, New York.

[9] Yurinskiĭ, V. V. (1976). *J. Multivariate Anal.*, **6**, 473–499.

(BERNSTEIN'S INEQUALITY
CHEBYSHEV'S INEQUALITY)

W. Hoeffding

PROBABILITY INTEGRAL TRANSFORMATIONS

The classic probability integral transformation (PIT) theorem can be stated as follows.

Theorem 1. If a random variable X has a continuous distribution function $F(x)$, then the random variable $U = F(X)$ has a uniform distribution on the interval (0, 1), that is, is a $U(0,1)$ random variable (*see* UNIFORM DISTRIBUTIONS).

This result is given in a number of recent books, including refs. 9 (p. 408) and 19 (p. 203). If the distribution function is assumed to be absolutely continuous, so that a density function exists, then the PIT is easily established using the transformation of densities technique.

The earliest use of this result of which we are aware was by R. A. Fisher* [4] in his famous 1930 paper in which he introduced the theory of fiducial* limits or intervals. Although he did not discuss the PIT explicitly as such, his implicit use of it to construct fiducial intervals leaves no doubt that he understood it (*see* FIDUCIAL INFERENCE). Let T be a statistic with continuous distribution function $F_\theta(t)$, and suppose that we wish to estimate θ. For fixed values of t_1, u_1, and u_2 $(0 \leqslant u_1 \leqslant u_2 \leqslant 1)$, put $\Omega_t = \{\theta : u_1 \leqslant F_\theta(t) \leqslant u_2\}$. Then if Ω_t is an interval, it is called a $100(u_2 - u_1)\%$ fiducial interval for θ. This construction was clearly motivated by the PIT.

Fisher [5] again used the PIT in 1932 in the fourth edition of *Statistical Methods for Research Workers*, when he proposed a method for combining tests of significance. Again, he did not discuss the PIT explicitly, and this omission caused considerable confusion among readers (see ref. 21). His combined test of significance is developed as follows. Suppose that $T_1, T_2, \ldots, T_k$ are k test statistics that are independent when the k corresponding null hypotheses hold. Let $p_1, \ldots, p_k$ denote the observed significance levels or p values* of these tests. Then by Theorem 1, the p values are independent, identically distributed (i.i.d.) $U(0,1)$ random variables. If it is desired to test that all the null hypotheses are correct simultaneously against the alternative that at least one is incorrect, then an intuitively appealing test can be made by rejecting when the product of the p-values, say $Q = p_1 \cdots p_k$, is too small, or equivalently, when $-2\ln Q$ is too large. However, since $p_1, \ldots, p_k$ are i.i.d. $U(0,1)$ random variables, it is readily shown that $-2\ln p_1, \ldots, -2\ln p_k$ are i.i.d. $\chi^2(2)$ random variables; therefore, $-2\ln Q = -2\sum \ln p_i$ is a $\chi^2(2k)$ random variable, and the test is easily made by rejecting when $-2\ln Q$ exceeds $\chi^2_\alpha(2k)$.

The principal uses of the PIT in statistics consider a random sample $X_1, \ldots, X_n$ from a parent distribution function $F(\cdot)$, and the i.i.d. $U(0,1)$ random variables $U_1 = F(X_1), \ldots, U_n = F(X_n)$. If $X_{(1)}, \ldots, X_{(n)}$ and $U_{(1)}, \ldots, U_{(n)}$ denote the order statistics* from the samples $X_1, \ldots, X_n$ and $U_1, \ldots, U_n$, respectively, then it follows that $U_{(j)} = F(X_{(j)})$ for $j = 1, \ldots, n$, by the monotonicity of F. These results are fundamental to the development of important fields of statistics such as the distribution theory of order statistics* (see ref. 3) and tolerance regions* theory (see ref. 6).

The PIT result has an important "inverse" which we consider next. If X is a random variable with distribution function $F(x)$—continuous or not—mapping the real number line to the unit interval (0, 1), we define an inverse function, say F^-, that maps the unit interval to the real number line, by $F^-(u) = \inf\{x : F(x) \geqslant u\}$ for $0 < u < 1$. If U is a $U(0,1)$ random variable and if $Y = F^-(U)$, then X and Y are identically distributed random variables [see ref. 19 (p. 203)].

This result also has many important applications in statistics. One use is to generate observations from an arbitrary distribution in simulation* studies by drawing observations on a $U(0,1)$ random variable and transforming them with this result.

By combining the two transformations above, we can transform a random variable with any particular continuous distribution to one with any other distribution. That is, if X has continuous distribution function F and Y has any distribution function G, then Y and the random variable $W = G^-(F(X))$ are identically distributed. If F and G are both continuous, this is sometimes called a *quantile transformation**, since it transforms the pth quantile of the distribution of X to the pth quantile of the distribution of Y for $0 < p < 1$.

Note the generality of these transformation results. They hold for all continuous distributions! We note also a point that will arise later in different contexts. The results hold even if the distribution functions depend on additional variables; those that will be of interest later will be either distribution parameters or random variables that are conditionally fixed.

MULTIVARIATE PROBABILITY INTEGRAL TRANSFORMATION (MPIT)

The PIT is sometimes used to transform the elements of a given sample to a sample from a $U(0, 1)$ distribution. We now consider a result due to Rosenblatt [20] that shows how to transform a set of possibly dependent continuous random variables to a set of i.i.d. $U(0, 1)$ random variables. Let $(X_1, \ldots, X_k)$ be a random vector with an absolutely continuous distribution function $F(x_1, \ldots, x_k)$. Then put $(u_1, \ldots, u_k) = \mathbf{H}(x_1, \ldots, x_k)$, where the transformation $\mathbf{H}$ is defined by

$$
\begin{aligned}
u_1 &= P\{X_1 \leqslant x_1\} \equiv F_1(x_1),\\
u_2 &= P\{X_2 \leqslant x_2 \mid X_1 = x_1\} \equiv F_2(x_2 \mid x_1),\\
&\vdots\\
u_k &= P\{X_k \leqslant x_k \mid X_1 = x_1, \ldots, X_{k-1} = x_{k-1}\}\\
&\equiv F_k(x_k \mid x_1, \ldots, x_{k-1}).
\end{aligned}
$$

For $\mathbf{u} = (u_1, \ldots, u_k)'$, put $S_{\mathbf{u}} = \{(u'_1, \ldots, u'_k)\colon u'_j \leqslant u_j,\ j = 1, \ldots, k\}$, and the distribution function of $\mathbf{U}' \equiv \mathbf{H}(X_1, \ldots, X_k)$ is given by

$$
\begin{aligned}
P(S_{\mathbf{u}}) &= P\{U_j \leqslant u_j, j = 1, \ldots, k\},\\
&= P\{X_1 \leqslant x_1\}P\{X_2 \leqslant x_2 \mid x_1\} \cdots\\
&\quad \times P\{X_k \leqslant x_k \mid x_1, \ldots, x_{k-1}\}\\
&= u_1 u_2 \ldots u_k,
\end{aligned}
$$

and $0 \leqslant u_j \leqslant 1$; $j = 1, \ldots, k$. We have the following result.

Theorem 2. If $(X_1, \ldots, X_k)$ is a random vector with absolutely continuous distribution function $F(x_1, \ldots, x_k)$, then $U_1 = F_1(X_1)$, $U_2 = F_2(X_2 \mid X_1)$, $\ldots$, $U_k = F_k(X_k \mid X_1, \ldots, X_{k-1})$ are i.i.d. $U(0, 1)$ random variables.

This result can be used in problems in statistical inference for multivariate distributions. If $\mathbf{X}_1, \ldots, \mathbf{X}_n$ is a sample of k-variate independent vector observations from an absolutely continuous multivariate distribution, then Theorem 2 shows how these observations can be transformed to nk i.i.d. $U(0, 1)$ random variables. If, for example, the distribution function is known under a simple goodness-of-fit* null hypothesis, then a test for this hypothesis can be constructed by utilizing the i.i.d. $U(0, 1)$ distributions of the nk U's under the null hypothesis.

It should be observed that the random variables $U_1, \ldots, U_k$ in Theorem 2 depend on the order of $X_1, \ldots, X_k$. If these X's are written in a different order, then a set of i.i.d. $U(0, 1)$ random variables will be obtained, but they will, in general, be different random variables. This lack of invariance to permutations of the X's can be important in some contexts.

CONDITIONAL PROBABILITY INTEGRAL TRANSFORMATIONS (CPIT)

When an observed sample is at hand, in order to actually apply either the classic PIT of Theorem 1 or the multivariate PIT of Theorem 2, it is necessary to know the distribution function required exactly. This requirement limits the use of such transformations in many statistical problems, since we often can assume only that the distribution function is a member of some specified class of distribution functions. This fact led Johnson and David [7] in an early pioneering paper to replace the parameters in a distribution function (df) by estimators. They showed that for location-scale classes, the transformed sample values, using this df with estimated parameters, have distributions that do not depend on the values of the parameters.

In this section we give some transformations which can be used when the parent distribution function for a sample is assumed only to be a member of a parametric class of distribution functions. Although the transformations we consider here generalize to multivariate or more complex families of distributions, such as distributions dependent on structural variables, the initial presentation will be for i.i.d. univariate samples.

Let $X_1, \ldots, X_n$ be i.i.d. random variables with common distribution function F assumed to be a member of a parametric class $\mathscr{F} = \{F_\theta : \theta \in \Omega\}$ of continuous distribution functions. Suppose that the family $\mathscr{F}$ admits a ν-component sufficient statistic* T_n, which we assume, without loss of generality, is a symmetric function* of $X_1, \ldots, X_n$, or, equivalently, that T_n is a function of the order statistics of this sample. Let $\tilde{F}_n(x_1, \ldots, x_\alpha)$ denote the conditional distribution function defined by

$$\tilde{F}_n(x_1, \ldots, x_\alpha) = E\{I_{[X_1 \leqslant x_1, \ldots, X_\alpha \leqslant x_\alpha]} \mid T_n\}, \quad \alpha \in \{1, \ldots, n\}.$$

This is an unbiased estimating distribution function for $F(x_1, \ldots, x_\alpha)$; when T_n is complete as well as sufficient, it is the minimum variance unbiased* (MVU) estimating distribution function. It is sometimes called the *Rao-Blackwell* (R-B) *distribution function*. The basic result is stated using these R-B distribution functions.

Theorem 3. If $\tilde{F}_n(x_1, \ldots, x_\alpha)$ is absolutely continuous for an $\alpha \in \{1, \ldots, n\}$, then the α random variables

$$U_1 = \tilde{F}_n(X_1), \quad U_2 = \tilde{F}_n(X_2 \mid X_1), \quad \ldots, \quad U_\alpha = \tilde{F}_n(X_\alpha \mid X_1, \ldots, X_{\alpha-1})$$

are i.i.d. $U(0, 1)$ random variables.

This result is proved in O'Reilly and Quesenberry [10], where it is shown also that α is a maximum when T_n is a minimal sufficient statistic. This theorem is most often useful when $\mathscr{F}$ is a parametric family with, say, ν functionally independent parameters. In many common cases the maximum value of α for which $\tilde{F}_n(x_1, \ldots, x_\alpha)$ is absolutely continuous is then $\alpha = n - \nu$, that is, the sample size reduced by the number of parameters. Even though Theorem 3 gives general transformation results, in particular cases writing out the transformations can itself be a difficult task, so it is worthwhile to consider particular techniques and results that are helpful. The next result is given for this purpose.

A sequence of statistics $(T_n)_{n \geqslant 1}$ is *doubly transitive* if for each n we can compute T_{n+1} from T_n and X_{n+1}; and conversely, if we can compute T_n from T_{n+1} and X_{n+1}. Many important families of distributions have doubly transitive minimal sufficient statistics. When this is the case, the following result from [10] gives the transformations of Theorem 3 in a form that is easier to find in many problems.

Theorem 4. If $\tilde{F}_n(x_1, \ldots, x_\alpha)$ is absolutely continuous for $\alpha \in \{1, \ldots, n\}$, and $(T_n)_{n \geqslant 1}$ is doubly transitive, then

$$U_1 = \tilde{F}_{n-\alpha+1}(X_{n-\alpha+1}), \ldots, U_\alpha = \tilde{F}_n(X_n)$$

is a set of α i.i.d. $U(0, 1)$ random variables.

The R-B distribution functions are known for many classes of distributions, and the transformations of Theorem 4 can be readily obtained explicitly for these classes.

Example: Normal Distributions. Suppose that $X_1, \ldots, X_n$ is a random sample from a $N(\mu, \sigma^2)$ parent distribution. Then $(\bar{X}_n, S_n^2) = (\sum X_j/n, \sum(X_j - \bar{X})^2/n)$ is a doubly transitive minimal sufficient statistic, since one can compute $(\bar{X}_{n+1}, S_{n+1}^2, X_{n+1})$ from $(\bar{X}_n, S_n^2, X_{n+1})$ and conversely. Moreover, the R-B distribution function is given (see ref. 8) for $r > 2$ by

$$\tilde{F}_r(z) = \begin{cases} 0, & z - \bar{X}_r < -(r-1)^{1/2} S_r, \\ 1, & (r-1)^{1/2} S_r < z - \bar{X}_r, \\ G_{r-2}(A_r), & \text{elsewhere,} \end{cases}$$

$$A_r = \frac{(r-2)^{1/2}(z - \bar{X})}{\left[(r-1)S_r^2 - \left(z - \bar{X}_r\right)^2\right]^{1/2}},$$

where G_{r-2} is a Student t-distribution* function with $(r - 2)$ degrees of freedom. Using this result in Theorem 4, we see that the $n - 2$ random variables given by

$$U_{r-2} = G_{r-2}(B_r),$$

$$B_r = \frac{(r-2)^{1/2}(X_r - \bar{X}_r)}{\left[(r-1)S_r^2 - (X_r - \bar{X}_r)^2\right]^{1/2}},$$

for $r = 3, \ldots, n$, are i.i.d. $U(0, 1)$ random variables.

The conditioning approach discussed above can be used in some cases for models more complex than i.i.d. univariate samples. Multivariate normal distributions were considered in ref. 10 and in more detail in refs. 17 and 18. In these papers the transformations were used to establish goodness-of-fit tests for multivariate normality.

The assumption that the original values $X_1, \ldots, X_n$ are identically distributed can also be relaxed in some important models such as regression or design models with normal errors. Some design models have been considered in ref. 15, which gives a multiple samples test of normality, and in ref. 16, which gives a test for a one-way analysis of variance* model.

The independence assumption of the variables $X_1, \ldots, X_n$ may also be relaxed in Theorem 3; however, the relatively simple approach of Theorem 4 is not then available, and finding the transformations is in most cases a formidable task. One place where this approach is potentially helpful is the following. It was pointed out earlier that the MPIT is not a symmetric function of $X_1, \ldots, X_n$, and depends on the order of the variables. The same is true for the CPIT, in general; this has a number of practical consequences. First, when the transformations are used to establish a test of fit, it is possible to obtain different test results from different orderings of the same data. The probability of agreement of tests based on different randomly selected permutations of the sample has been considered in ref. 13, where this probability was found to be high in a large class of cases. Another approach is to replace $(X_1, \ldots, X_\alpha)$ in Theorem 3 by a subset of α order statistics, say $(X_{(1)}, \ldots, X_{(\alpha)})$, where $X_{(1)} \leqslant X_{(2)} \leqslant \cdots \leqslant X_{(n)}$. Then the transformed values will be functions only of the order statistics and will therefore be symmetric functions of the original sample. The task of finding the actual transformation is difficult with this approach, and has been solved only for two rather simple cases of uniform classes [12] and scale parameter exponential classes [11]. The transformations obtained for the exponential distribution* are characterizations of the distribution that are closely related to other characterizing transformations studied in ref. 2.

A second consequence of the order dependence of the CPIT transformations is that it may be exploited to investigate the dependence of a data set on an ordering variable, such as time.

FURTHER REMARKS ON CPITs

The CPIT approach is useful mainly for parametric classes of distributions that admit fixed-dimension sufficient statistics*. Such classes are, under regularity conditions, exponential classes, so that it appears that this method is useful largely just for exponential classes. However, in ref. 12 an extension of the foregoing conditioning approach is considered which gives results for many important families that are not exponential classes because their support is a function of parameters, such as uniform or Pareto distributions*.

The transformations of Theorem 4 possess interesting and useful properties. They have important invariance properties [14]. For a parent class of distribution functions $\mathscr{F}$ that has an associated class of transformations G, say, these transformations will themselves be invariant under the members of G. To illustrate, suppose that $\mathscr{F}$ is a location-scale parametric family of distributions. Then the transformations will be invariant under linear transformations: $g(x_j) = ax_j + b$; $j = 1, \ldots, n$; $a > 0$. This invariance can readily be verified for the transformations of the normal distribution in the preceding section. The CPITs possess similar invariance properties quite generally. This invariance property can sometimes be helpful in the derivation of most powerful invariant tests for separate hypothesis testing* problems (see ref. 14), or, to obtain optimal invariant selection statistics (*see* MODEL CONSTRUCTION: SELECTION OF DISTRIBUTIONS).

If the sufficient statistic T is also complete, then $T = (T_1, \dots, T_\nu)'$ and $U = (U_1, \dots, U_{n-\nu})'$ are independent, since the distribution of U is the same for every parent distribution in the class $\mathscr{F}$, for which T is sufficient and complete [1] (*see also* BASU THEOREMS).

This result has important implications. T contains all the information about the parameters of $\mathscr{F}$, that is, all information on which inferences can be made within the class $\mathscr{F}$. Thus U can be used to make inferences about the class $\mathscr{F}$, perhaps to make goodness-of-fit tests or conduct model selection procedures, and $\mathscr{T}$ can be used to make a parametric test or other inference within $\mathscr{F}$, and the independence of U and T exploited to assess overall error rates.

The remarks above suggest that the transformations are useful largely for investigating the correctness of the parent class $\mathscr{F}$, which will usually be posed under a null hypothesis. It is likely that additional applications of the conditioning approach to transformations and inference described here will be developed.

References

[1] Basu, D. (1955). *Sankhyā*, **15**, 377–380. (The now well-known condition for a statistic to be independent of a complete sufficient statistic was given.)

[2] Csörgö, M. and Seshadri, V. (1971). *Z. Wahrschlein. verw. Geb.*, **18**, 333–339. ("Characterizing" transformations to uniformity are considered for exponential and normal distributions. See also references to other papers in this paper.)

[3] David, H. A. (1981). *Order Statistics*, 2nd ed. Wiley, New York. (A general reference for order statistics.)

[4] Fisher, R. A. (1930). *Proc. Camb. Philos. Soc.*, **26**, Part. 4, 528–535. (Fisher introduced "fiducial" intervals and bounds for a parameter. The distribution function was used as a "pivotal" function, and this was the first use of the PIT.)

[5] Fisher, R. A. (1932). *Statistical Methods for Research Workers*, 4th ed. Oliver & Boyd, Edinburgh. (Fisher implicitly used the PIT to define his "combined test for significance.")

[6] Guttman, I. (1970). *Statistical Tolerance Regions: Classical and Bayesian*. Hafner, Darien, Conn. (A general reference for tolerance regions.)

[7] Johnson, N. L. and David, F. N. (1948). *Biometrika*, **35**, 182–190. (The members of a sample are transformed using a distribution function with parameters replaced by estimates.)

[8] Lieberman, G. J. and Resnikoff, G. J. (1955). *J. Amer. Statist. Ass.*, **50**, 457–516.

[9] Lindgren, B. W. (1968). *Statistical Theory*, 2nd ed. Macmillan, New York.

[10] O'Reilly, F. J. and Quesenberry, C. P. (1973). *Ann. Statist.*, **1**, 79–83. (This paper introduces the important general principle of conditioning a set of RVs on a sufficient statistic and then using this conditional distribution to transform the RVs themselves—the conditional probability integral transformation, CPIT. Results that are helpful for performing these transformations are given for the special case when the original RVs constitute a random sample from an exponential class of distributions.)

[11] O'Reilly, F. J. and Stephens, M. A. (1982). *J. R. Statist. Soc. B*, **44**, 353–360. (Obtain the CPIT for order statistics for the scale parameter exponential distribution.)

[12] Quesenberry, C. P. (1975). *Commun. Statist. A*, **4**, 1149–1155. (This paper extends the CPIT theory for samples to distributions with ranges dependent on parameters.)

[13] Quesenberry, C. P. and Dietz, J. (1983). *J. Statist. Comp. Simul.*, **17**, 125–131. (Studies agreement probabilities for CPIT-Neyman smooth tests made on randomly selected permutations of a sample.)

[14] Quesenberry, C. P. and Starbuck, R. R. (1976). *Commun. Statist. A*, **5**, 507–524. (A theoretical paper that studies optimality properties of CPITs.)

[15] Quesenberry, C. P., Whitaker, T. B., and Dickens, J. W. (1976). *Biometrics*, **32**, 753–759. (Gives a test for normality for multiple samples.)

[16] Quesenberry, C. P., Giesbrecht, F. G., and Burns, J. C. (1983). *Biometrics*, **39**, 735–739. (Consider testing normality in a one-way ANOVA model.)

[17] Rincon-Gallardo, S. and Quesenberry, C. P. (1982). *Commun. Statist. Theor. Meth.*, **11**, 343–358. (An applied paper that considers applying the multivariate normality test of ref. 18.)

[18] Rincon-Gallardo, S., Quesenberry, C. P., and O'Reilly, F. J. (1979). *Ann. Statist.*, **7**, 1052–1057. (Proposes a multiple samples test for multivariate normality.)

[19] Rohatgi, V. K. (1976). *An Introduction to Probability Theory and Mathematical Statistics*. Wiley, New York.

[20] Rosenblatt, M. (1952). *Ann. Math. Statist.*, **23**, 470–472. (Brief but important paper that gives the multivariate probability integral transformation.)

[21] Wallis, W. A. (1942). *Econometrica*, **10**, 229–248. (Interesting historically since it explains why Fisher's combined test of significance works. Some earlier writers had given fallacious accounts.)

(DISTRIBUTION-FREE METHODS
GENERATION OF RANDOM VARIABLES
GOODNESS OF FIT
MULTINORMAL DISTRIBUTION
QUANTILE TRANSFORMATION METHODS
UNIFORM DISTRIBUTIONS)

C. P. QUESENBERRY

PROBABILITY, INVERSE *See* BAYESIAN INFERENCE

PROBABILITY MEASURE

Formally, a probability measure is defined as a measure* μ on an arbitrary space Ω such that $\mu(\Omega) = 1$. We then define an event as a measurable set and the probability of an event E as $\mu(E)$. The space Ω is called "space of elementary events" or "sample space."

In applications, a probability measure is associated with our conceptions of the likelihoods of occurrences of various physical events. Probability measures which are defined on the real line (R^1) result in univariate (one-dimensional) probability distributions, while those on n-dimensional Euclidean space yield multivariate (n-dimensional) probability distributions. *See* MEASURE THEORY IN STATISTICS AND PROBABILITY *and* PROBABILITY THEORY and entries on specific distributions for more details.

Bibliography

Cramér, H. (1946). *Mathematical Methods of Statistics*. Princeton University Press, Princeton, N.J., Chaps. 4 to 9.

Loève, M. (1979). *Probability Theory*, 4th ed. Springer-Verlag, New York. (See especially Chaps. I to III.)

(MEASURE THEORY IN STATISTICS AND PROBABILITY)

PROBABILITY MINIMUM REPLACEMENT SAMPLING DESIGN

We will use the following notation for the parameters of a sampling design:

- *N:* number of sampling units
- *S(i):* size measure associated with sampling unit (i)
- $S(+) = \sum_{i=1}^{N} S(i)$
- *n:* total sample size
- *n(i):* number of times sampling unit (i) is selected

Probability *nonreplacement* (PNR) sample designs are (usually) restricted to sampling frames with $nS(i)/S(+) < 1$, $i = 1, \dots, N$. When this condition holds a *PNR sample design* is one for which

$$\Pr(n(i) = 1) = nS(i)/S(+)$$

and

$$\Pr(n(i) = 0) = 1 - \Pr(n(i) = 1).$$

Probability *minimum replacement* (PMR) sample designs, to be defined below, allow *any* positive size measures $S(i)$. The standardized size measure may be written in terms of two components

$$nS(i)/S(+) = m(i) + f(i),$$

where $m(i)$ is a nonnegative *integer*, while $0 \leqslant f(i) \leqslant 1$. Now, using this notation, a PMR sample design is one for which

$$\Pr\{n(i) = m(i) + 1\} = f(i)$$

and

$$P\{n(i) = m(i)\} = 1 - f(i).$$

[Observe that PNR sample designs are a special case of PMR designs with *all* $m(i)$ equal zero.]

A selection algorithm for a sequential procedure for a PMR sampling and associated unbiased estimation procedures was developed by J. R. Chromy [1].

References

[1] Chromy, J. R. (1981). In *Current Topics in Survey Sampling*, D. Krewski, R. Platek, and J. N. K. Rao, eds. Academic Press, New York, pp. 329–347.

(PROBABILITY PROPORTIONAL TO SIZE (PPS) DESIGNS
SURVEY SAMPLING)

PROBABILITY PLOTTING

Plots and graphs of a wide variety are used in the statistical literature for descriptive purposes. Visual representations have high impact and they provide insight and understanding not readily found by other means. This article will consider methods for the one- and two-sample problems that have plots as their central focus. These methods are useful for informal data analysis and model assessment, while more formal statistical inferences can also be made. Fisher [12] has given an excellent review. *See* GRAPHICAL REPRESENTATION OF DATA *and* GRAPHICAL REPRESENTATION, COMPUTER AIDED for more examples of plots.

CUMULATIVE DISTRIBUTION FUNCTION PLOTS

The cumulative distribution function (CDF) of a random variable X is $F(x) = \Pr(X \leqslant x)$. Many features of the distribution of X are apparent in the CDF, such as likelihood regions, location, scale, percentiles, and skewness. A random sample of size n on X is a collection of independent random variables, each having CDF $F(x)$. The observed, ordered observations for the sample are denoted by $x_1 \leqslant x_2 \leqslant \cdots \leqslant x_n$. The empirical CDF is $F_n(x) =$ (number of observations $\leqslant x)/n$. It is a step function increasing from 0 to 1 with jumps of size $1/n$ at each observation (*see* EDF STATISTICS).

$F_n(x)$ is a natural, nonparametric estimate of $F(x)$ and its graph can be studied informally in making inferences about $F(x)$. A simultaneous $(1-\alpha)100\%$ confidence band for $F(x)$ is

$$F_n(x) - k_\alpha \leqslant F(x) \leqslant F_n(x) + k_\alpha \quad \text{for all } x,$$

where $\Pr(D \leqslant k_\alpha) = 1 - \alpha$ and $D = \sup_x |F_n(x) - F(x)|$, the Kolmogorov–Smirnov statistic*. The confidence band should be plotted with $F_n(x)$ to assess variability. A formal level α test of a hypothesis $H_0: F(x) = F_0(x)$, for a specified CDF $F_0(x)$, can be carried out with the rule: Accept H_0 if $F_0(x)$ lies in the confidence band for all x. The confidence band thus contains all CDFs that would be accepted in this testing problem.

Doksum [9] has improved on this graphical method by developing a confidence band using

$$\sup_{a \leqslant F_n(x) \leqslant b} \left[\frac{|F_n(x) - F(x)|}{\{F(x)(1 - F(x))\}^{1/2}} \right]$$

as a pivot, where $0 < a < b < 1$. The weight function in the denominator serves to equalize the variance of the empirical CDF and the interval (a, b) restricts the scope of the band. The resulting confidence band is slightly wider in the middle but is much narrower in the tails when compared to the first band.

QUANTILE AND PERCENT PLOTS

Suppose that two CDFs $F(x)$ and $G(x)$ are given. These may be empirical or population CDFs. A graphical comparison of the two distributions can be made by plotting the quantiles* of one distribution against those of the other. The quantile of order p for the

CDF $F(x)$ is a number x_p which satisfies $F(x_p) = p$ when $F(x)$ is strictly increasing, but to cover other cases define $F^{-1}(p) = \inf\{x : F(x) \geqslant p\}$ and use $x_p = F^{-1}(p)$. Similarly, let y_p be the quantile of order p for $G(x)$. Then a *quantile plot*, called a *Q-Q plot*, is the graph of (x_p, y_p) for $0 < p < 1$.

If $G(x) = F(x)$, the Q-Q plot will be a 45° line through the origin. If the two distributions differ only in location and scale, $G(x) = F((x - \mu)/\sigma)$ and $y_p = \mu + \sigma x_p$. Thus the Q-Q plot is a straight line with intercept μ and slope σ. This is an important prototype case since the eye can readily perceive straight lines and identify slope and intercept. Departures from linearity can be easily spotted and interpreted.

In a similar manner, a *percent plot*, called a *P-P plot*, is constructed by graphing the points $(F(x), G(x))$ for $-\infty < x < \infty$. This plot will trace a path from $(0, 0)$ to $(1, 1)$. If $G(x) = F(x)$, the plot will be a straight line. Unlike the Q-Q plot, the P-P plot will not be linear in the case of only location-scale differences. Differences in the middles of distributions are usually more apparent for P-P plots than for Q-Q plots. Generalization to multivariate cases is easier with percent plots.

The basic concepts behind these plots can be used in other creative ways. For example, ref. 16 has some good examples of plots of functions $h_1(x_p, y_p)$ versus $h_2(x_p, y_p)$ or just p.

Probability plots are a useful tool in data analysis. One can assess whether a single random sample arises from a distribution with CDF $F_0(x)$ by constructing a Q-Q or a P-P plot of $F_0(x)$ against the empirical CDF $F_n(x)$ of the sample. The Q-Q plot would be represented by points $(F_0^{-1}[(i - 1/2)/n], x_i)$ plotted for $i = 1, \ldots, n$, where $x_1 \leqslant \cdots \leqslant x_n$ are the ordered observations. The first coordinates are quantiles of order $(i - \frac{1}{2})/n$ for $F_0(x)$. Alternatively, quantiles of order $i/(n + 1)$ can be used. If the plot follows a 45° line through the origin, the CDF $F_0(x)$ would informally be accepted. Departures from this line would indicate differences between $F_0(x)$ and the actual CDF. Location or scale differences, heavier tails, existence of outliers*, regions of poor fit, and other peculiarities may be noticed. The Kolmogorov–Smirnov test could be used for a more formal test.

Probability plotting has been widely used for location-scale models, where the observations are assumed to arise from a distribution with CDF $F_0((x - \mu)/\sigma)$, with F_0 known but μ and σ unknown. A Q-Q plot of $F_0(x)$ against $F_n(x)$, as given in the preceding paragraph, will informally test the adequacy of the model. When $F_0(x)$ is the standard normal CDF, this is called a *normal probability plot*. If the location-scale model is correct, the plot should lie along a straight line with intercept μ and slope σ. Departures from linearity would provide insights for choosing an alternative model. *See* GRAPHICAL REPRESENTATION OF DATA for some examples.

Doksum [9] has discussed a formal confidence band that can be graphed and used to test the adequacy of the location-scale model. His procedure is based on a weighted Kolmogorov–Smirnov statistic and has an advantage over the usual Kolmogorov–Smirnov statistic in that μ and σ need not be estimated from the data.

OPTIMAL PLOTTING POSITIONS

With the probability plot for the location-scale model, the slope σ and intercept μ can be crudely estimated from the graph. To improve on this it is natural to use the least-squares* fitted line. Let the plotted points be denoted by (v_i, x_i), where the x_i are the ordered observations and the v_i are called the plotting positions. The basic Q-Q plot uses quantiles from F_0 for the plotting positions. Chernoff and Lieberman [3] considered the problem of determining optimal plotting positions in the sense of minimum mean square error or of unbiased, minimum variance estimates. The resulting optimal plotting positions, given in their paper, are functions of the expectations, variances, and covariances of the order statistics of a sample from $F_0(x)$. An undesirable feature is that the optimal plotting positions need not be in an increasing order.

The order statistics* used in the plot are correlated and have unequal variances, and as a result a weighted least-squares* procedure would be preferable to the ordinary least-squares fit. Cran [4] gave the weighted least-squares solution to the optimal plotting position problem. He also obtained conditions on the weights for the plotting positions to be in increasing order. Barnett [1] noted that the previous results require knowledge of the first and second moments of order statistics, and this information is available only for selected sample sizes and distributions F_0. He suggested an alternative approach which does not require this information and which yields estimates of μ and σ that are nearly optimal.

GAMMA PROBABILITY PLOTS

The gamma probability distribution is widely used in life testing* and reliability* models. It is indexed on three parameters: location, scale, and shape. A graphical method for assessing a gamma probability model consists of making an initial estimate of the shape parameter and then constructing a Q-Q plot to check for linearity as in the preceding section. Reference 17 contains several interesting examples of this approach. It illustrates an important application in assessing the assumption of homogeneity of variance in analysis-of-variance* problems. Under the usual assumptions of independence, normality, and equal variance, the within-cell sample variances should have a gamma distribution* and the gamma plot can be used to spot any discrepancies.

HALF-NORMAL PLOTS*

The half-normal distribution is the distribution of the absolute value of a standard normal random variable; its CDF is $\Phi_0(x) = 2\Phi(x) - 1$ for $x > 0$, where Φ is the standard normal CDF (*see* FOLDED DISTRIBUTIONS). A half-normal probability plot of a sample is a regular Q-Q plot of the quantiles of the absolute values of the data $|x_1| < \cdots < |x_n|$ against the corresponding quantiles of the half-normal distribution, that is, a plot of $(\Phi_0^{-1}[(i - \frac{1}{2})/n], |x_i|)$ for $i = 1, \ldots, n$. If the sample arises from a normal distribution with mean 0 and variance σ^2, this plot should follow a straight line through the origin with slope σ.

Daniel [5] presented an interesting application of this plot in the analysis of 2^p factorial experiments*. The usual set of $2^p - 1$ orthogonal sample contrasts* are used for the half-normal plot. Departures from linearity may indicate heteroscedasticity*, outliers, or real effects. Formal tests for real effects are given; see also Birnbaum [2]. Zahn [18] gave several improvements for Daniel's procedure and corrected the critical limits for the test.

GRAPHICAL ANALYSIS FOR TWO-SAMPLE PROBLEMS

Plots can be used to detect similarities and differences between two distributions. Interest may focus on locations, scales, shapes, tail weight, symmetries, skewness, stochastic ordering, or other features. Assuming two independent samples, let $x_1 \leqslant \cdots \leqslant x_m$ be the ordered sample from a distribution with CDF $F(x)$ and let $F_m(x)$ be the empirical CDF of the sample; let $y_1 \leqslant \cdots \leqslant y_n$ be the ordered sample from a distribution with CDF $G(y)$ and let $G_n(y)$ be the empirical CDF of this sample.

The two samples can be compared with a Q-Q plot of the empiricals F_m versus G_n. If the sample sizes are equal, this is simply a plot of the points (x_i, y_i) for $i = 1, \ldots, n$. For unequal sample sizes, recall that $F_m^{-1}(p) = \inf\{x : F_m(x) \geqslant p\}$ will take the value x_i if $(i-1)/m < p \leqslant i/m$ and similarly for G_n^{-1}. The Q-Q plot is a plot of the points arising from $(F_m^{-1}(p), G_n^{-1}(p))$ for $0 < p < 1$ and this will consist of points (x_i, y_j) for suitable subscripts.

THE SHIFT FUNCTION FOR TWO-SAMPLE PROBLEMS

The Kolmogorov–Smirnov statistic $D' = \sup_x |F_m(x) - G_n(x)|$, or a weighted version of it, can be used to test the hypothesis of identical CDFs for the two-sample problem. This test can be carried out with an additional emphasis on plotting and graphical analysis to provide more information and insight than the basic accept–reject decision. A discussion of this can be found in Doksum and Sievers [9] and Doksum [6, 7, 9].

To motivate this approach, consider the widely used shift model $F(x) = G(x + \Delta)$ for all x, where the Y-population is shifted a uniform amount Δ from the X-population. Comparing populations by a simple shift may often be unrealistic, however. More generally, a shift function $\Delta(x)$ can be defined by the equation $F(x) = G(x + \Delta(x))$ for all x. It follows that $\Delta(x) = G^{-1}(F(x)) - x$ (see Fig. 1). Under general conditions $\Delta(x)$ is the only function for which $X + \Delta(X)$ has the same distribution as Y. We say that $\Delta(X)$ is the amount of *shift* to bring the X variable up to the Y. When X is a control response and Y a treatment response, it may be reasonable to interpret $\Delta(x)$ as the amount that the treatment adds

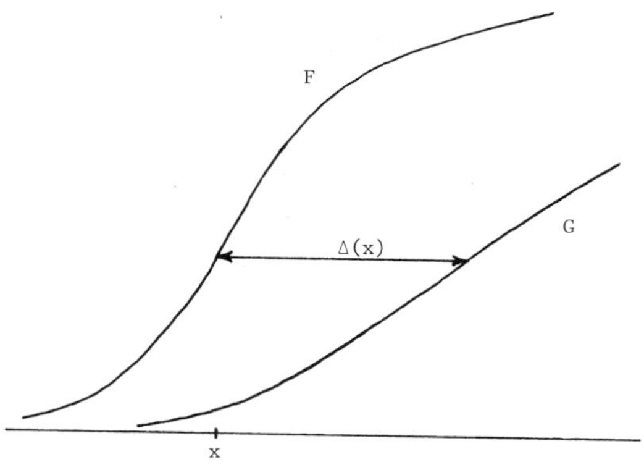

Figure 1 Shift function $\Delta(x)$ between two CDFs F and G.

to a control response x. This effect need not be constant. The set $\{x : \Delta(x) > 0\}$ identifies the part of the population where the treatment has a positive effect. When the pure shift model holds, $\Delta(x) = \Delta$, a constant. When a location-scale model holds, $G(x) = F((x - \mu)/\sigma)$ and $\Delta(x) = \mu + (\sigma - 1)x$, a linear function.

A natural, nonparametric estimate of $\Delta(x)$ is $\hat{\Delta}(x) = G_n^{-1}(F_m(x)) - x$ and this function can be graphed to explore its properties. The plot of $G_n^{-1}(F_m(x))$ at the values x_i is basically the regular Q-Q plot.

A simultaneous confidence band for $\Delta(x)$ can be constructed for more formal inferences. One such band is based on the Kolmogorov–Smirnov statistic D'. With k_α chosen so that $\Pr(D' \leqslant k_\alpha \mid F = G) = 1 - \alpha$, a simultaneous $(1 - \alpha)100\%$ confidence band for $\Delta(x)$ is given by

$$y_{i_1} - x_i \leqslant \Delta(x) < y_{i_2} - x_i \qquad \text{for } x_i \leqslant x < x_{i+1},$$

where $x_0 = -\infty$, $x_{m+1} = \infty$,

$$i_1 = \langle n((i/m) - k_\alpha) \rangle,$$

$$i_2 = \left[n((i/m) + k_\alpha) \right] + 1,$$

$\langle t \rangle$ is the least integer greater than or equal to t, and $[t]$ is the greatest integer less than or equal to t. An example is given in Fig. 2. The confidence band can be used to compare the two distributions in various ways. Does it contain a horizontal line (a pure shift model is accepted), a line through the origin $(0, 0)$ (a pure scale model is accepted), or any straight line (a location-scale model is accepted)? A confidence set will be visible for the set $\{x : \Delta(x) > 0\}$ and similarly other conclusions can be reached, as many in number as desired because of the simultaneous nature of the band.

Doksum [9, 10] gave several other nonparametric bands for $\Delta(x)$ that can be more efficient than the band discussed here. Also, special bands were developed for the location-scale case.

In life testing and reliability studies, multiplicative effects have more direct interest than additive effects. For such cases a scale

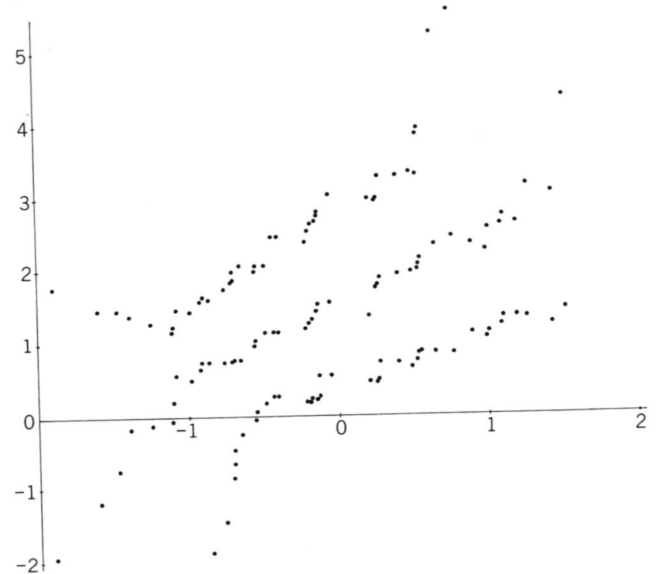

Figure 2 Shift function $\hat{\Delta}(x)$ and an 80% confidence band for simulated samples of sizes $n = m = 50$, where X and Y are $N(0, 1)$ and $N(1, 4)$, respectively.

function $\theta(x) = G^{-1}(F(x))/x$ can be useful. Doksum [8] has developed appropriate graphical analyses and interesting applications for the scale function, including methods for the Weibull* and exponential* models.

PLOTTING FOR SYMMETRY

Suppose that X is a random variable with a continuous CDF $F(x)$. When the distribution is symmetric about a point θ, this point is clearly the measure of location or center. In asymmetric cases, the location is not so clear. Many statistical procedures require symmetric distributions and it is useful to have methods to check on this assumption.

Doksum [7] has discussed some useful graphical methods. He defined a symmetry function by

$$\theta(x) = \left(x + F^{-1}(1 - F(x))\right)/2;$$

it is essentially the only function for which $X - \theta(X)$ and $-(X - \theta(X))$ have the same distributions. For symmetric distributions $\theta(x) = \theta$, a constant. When the distribution is skewed to the right (left), $\theta(x)$ is U-shaped (upside-down U-shaped). He also discussed a location interval $[\underline{\theta}, \bar{\theta}]$, where $\underline{\theta} = \inf_x \theta(x)$ and $\bar{\theta} = \sup_x \theta(x)$, and showed that all reasonable location measures lie in this interval.

The empirical CDF $F_n(x)$ of a sample is used to estimate $\theta(x)$ with

$$\hat{\theta}(x) = \left(x + F_n^{-1}(1 - F_n(x))\right)/2.$$

Using the ordered data $x_1 < \cdots < x_n$, the graph of $\hat{\theta}(x)$ is basically given by a plot of the points $(x_i, (x_i + x_{n+1-i})/2)$ for $i = 1, \ldots, n$. This plot and a simultaneous confidence band should be plotted to assess asymmetry and the location interval.

References

[1] Barnett, V. (1976). *Appl. Statist.*, **25**, 47–50. (A paper on plotting positions.)

[2] Birnbaum, A. (1959). *Technometrics*, **1**, 343–357. (On theory of hypothesis testing for 2^p factorials, half-normal plots.)

[3] Chernoff, H. and Lieberman, G. (1956). *Ann. Math. Statist.*, **27**, 806–818. (Basic paper on estimation of location and scale; theoretical results.)

[4] Cran, G. W. (1975). *J. Amer. Statist. Ass.*, **70**, 229–232. (Use of weighted least-squares line.)

[5] Daniel, C. (1959). *Technometrics*, **1**, 311–341. (Basic reference on half-normal plots for 2^p factorials; readable, examples.)

[6] Doksum, K. (1974). *Ann. Statist.*, **2**, 267–277. (First paper on shift function, theoretical.)

[7] Doksum, K. (1975). *Scand. J. Statist.*, **2**, 11–22. (On plots for asymmetry, theoretical, has examples.)

[8] Doksum, K. (1975). In *Reliability and Fault Tree Analysis*. SIAM, Philadelphia, pp. 427–449. (Focus on plotting with life testing, reliability data; lots of examples and applications.)

[9] Doksum, K. (1977). *Statist. Neerlandica*, **31**, 53–68. (Good survey on a wide range of plotting methods; examples and applications, readable.)

[10] Doksum, K. and Sievers, G. (1976). *Biometrika*, **63**, 421–434. (Basic paper on plotting for shift functions, location/scale model; theory and discussion.)

[11] Fisher, N. I. (1981). *Austral. J. Statist.*, **23**, 352–359.

[12] Fisher, N. I. (1983). *Int. Statist. Rev.*, **51**, 25–58. (Good review of many graphical methods; extensive bibliography.)

[13] Hoaglin, D. C. (1980). *Amer. Statist.*, **34**, 146–149. (Plots for Poisson and other discrete distributions; readable.)

[14] Nair, V. N. (1981). *Biometrika*, **68**, 99–103. (Plots for censored data.)

[15] Schweder, T. and Spjøtvoll, E. (1982). *Biometrika*, **69**, 493–502. (Plots of p values for many tests.)

[16] Wilk, M. B. and Gnanadesikan, R. (1968). *Biometrika*, **55**, 1–17. (Basic paper on Q-Q and P-P plots; lots of examples and applications, readable.)

[17] Wilk, M. B., Gnanadesikan, R., and Huyett, M. J. (1962). *Technometrics*, **4**, 1–15. (Basic paper on gamma probability plots.)

[18] Zahn, D. A. (1975). *Technometrics*, **17**, 189–200. (On modification and corrections for Daniel's half-normal plots.)

Bibliography. (editors' addition)

Harter, H. L. (1984). *Commun. Statist. Theor. Meth.*, **13**, 1613–1633. (Contains 107 references.)

(GRAPHICAL REPRESENTATION, COMPUTER-AIDED
GRAPHICAL REPRESENTATION OF DATA
HALF-NORMAL PLOTS
KOLMOGOROV–SMIRNOV STATISTICS
QUANTILES)

GERALD L. SIEVERS

PROBABILITY PROPORTIONAL TO SIZE (PPS) SAMPLING

USES

PPS sampling is used principally in two contexts.

Multi-Stage Sampling

Samples are often selected in a multistage manner [16, Chap. 10]. For example, in a two-stage national household survey, a sample of areas, the *primary sampling units** (PSUs), may be selected within each of which, subsamples of households, the *subunits*, are drawn. A common procedure is to sample the PSUs with probability proportional to *size*, that is, the number of subunits within each PSU, and to select a fixed number of subunits within each PSU by *simple random sampling** (SRS). This yields a *self-weighting* sample; that is, each subunit has equal probability of selection so that no reweighting is necessary for estimation purposes.

The PPS strategy above has practical advantages over other self-weighting procedures, such as SRS of PSUs, and fixed sampling fractions within PSUs, since it yields a fixed overall sample size (which is usually desirable, given budget constraints) and a fixed number of subunits in each PSU, giving equal work loads to PSU-based interviewers. In terms of sampling precision, the PPS strategy will usually not differ greatly from its competitors [8, p. 310]. A further advantage of the PPS strategy occurs for PSUs (e.g., schools or hospitals) which are of interest for their own sake, when equal subsample sizes may be preferred for a between-PSU analysis [16, p. 418].

For multistage sampling with more than two stages, the use of PPS sampling at every stage except the last, where SRS of fixed size is used, is again self-weighting and the comments above also apply (*see also* STRATIFIED MULTISTAGE SAMPLING).

Single-Stage Designs

For populations of units such as farms or businesses, measures of *size* x_i may be available for each unit i in the population, and x_i may be expected to be approximately proportionately related to a variable y_i to be measured in a survey. For example, x_i may be crop acreage, known for all farms i, and y_i may be wheat yield in a given year. The procedure of sampling units by PPS *may* then permit more efficient estimation of the population total of the y_i than is possible using SRS with the same sample size (see the section "Estimation Theory").

PPS sampling may here be viewed as a limiting form of size-stratified sampling. If the population is divided into strata within which x_i is approximately constant and *if* the within-stratum standard deviation of y_i is proportional to x_i, then optimal *Neyman allocation** for *stratified random sampling* implies that units are selected by PPS.

One problem with PPS sampling is in multiobjective surveys, where, for example, sampling with probability proportional to crop acreage may be more efficient than SRS for estimating wheat yield but may be much less efficient for estimating number of sheep, say. One practical advantage of PPS sampling in this case occurs in area sampling*, where areas may be selected simply using random points on a map [20, p. 207].

ESTIMATION THEORY

For "Single-Stage Designs," let $x_1, \ldots, x_N$ be known size values for the population ($x_i > 0$) and $y_1, \ldots, y_N$ be the values of the variable of interest, where, without loss of generality, $y_1, \ldots, y_n$ are the values observed in the sample. In *with-replacement* PPS sampling (PPSWR) the sample is formed from n independent draws, at each of which the ith unit of the population is selected with probability $p_i = x_i/X$, where $X = x_1 + \cdots + x_N$. In this case the usual estimator [14, 25] of the population total, $Y = y_1 + \cdots + y_N$, is

$$\hat{Y}_{\mathrm{WR}} = \sum_{i=1}^{n} y_i/(np_i).$$

This estimator is unbiased with variance

$$V(\hat{Y}_{\mathrm{WR}}) = n^{-1} \sum_{i=1}^{N} p_i (p_i^{-1} y_i - Y)^2. \quad (1)$$

An unbiased estimator of $V(\hat{Y}_{\mathrm{WR}})$ is

$$\hat{V}(\hat{Y}_{\mathrm{WR}}) = [n(n-1)]^{-1} \sum_{i=1}^{n} (p_i^{-1} y_i - \hat{Y}_{\mathrm{WR}})^2. \quad (2)$$

In PPSWR sampling the same unit may appear more than once in the sample. This is avoided in *without-replacement* PPS sampling (PPSWOR), where the ith unit of the population has probability $\Pi_i = nx_i/X$ of inclusion in the sample (assuming that $nx_{\max} \leqslant X$). Here the usual estimator of Y is the *Horvitz–Thompson estimator** [15]

$$\hat{Y}_{\mathrm{HT}} = \sum_{i=1}^{n} \Pi_i^{-1} y_i .$$

The estimators $\hat{Y}_{\mathrm{HT}}$ and $\hat{Y}_{\mathrm{WR}}$ are formally identical, since $\Pi_i = np_i = nx_i/X$, but are distinguished because they have different distributions under the two sampling schemes. Under PPSWOR, $\hat{Y}_{\mathrm{HT}}$ is still unbiased but now has variance

$$\begin{aligned} V(\hat{Y}_{\mathrm{HT}}) &= \sum_{i=1}^{N} \Pi_i^{-1}(1 - \Pi_i) y_i^2 \\ &\quad + \sum_{i \neq j}^{N} \Pi_i^{-1} \Pi_j^{-1} (\Pi_{ij} - \Pi_i \Pi_j) y_i y_j \\ &= \sum_{i<j}^{N} (\Pi_i \Pi_j - \Pi_{ij})(\Pi_i^{-1} y_i - \Pi_j^{-1} y_j)^2, \end{aligned} \quad (3)$$

where Π_{ij} is the probability that both units i and j are included in the sample. Assuming that $\Pi_{ij} > 0$, $i, j = 1, \ldots, N$, two unbiased estimators of $V(\hat{Y}_{\mathrm{HT}})$ are

$$\begin{aligned} \hat{V}_{\mathrm{HT}}(\hat{Y}_{\mathrm{HT}}) &= \sum_{i=1}^{n} \Pi_i^{-2}(1 - \Pi_i) y_i^2 \\ &\quad + \sum_{i \neq j}^{n} \Pi_i^{-1} \Pi_j^{-1} \Pi_{ij}^{-1} \\ &\qquad \times (\Pi_{ij} - \Pi_i \Pi_j) y_i y_j , \end{aligned}$$

proposed by Horvitz and Thompson [15], and

$$\hat{Y}_{SYG}(\hat{Y}_{HT}) = \sum_{i<j}^{n} \Pi_{ij}^{-1}(\Pi_i \Pi_j - \Pi_{ij}) \times \left(\Pi_i^{-1} y_i - \Pi_j^{-1} y_j\right)^2,$$

proposed by Sen [24] and Yates and Grundy [26]. Of the two estimators, $\hat{V}_{SYG}$ seems to perform better in empirical comparisons [6, p. 8].

In comparing PPSWR and PPSWOR the ratio $V(\hat{Y}_{HT})/V(\hat{Y}_{WR})$ is $(N-n)/(N-1)$ roughly, as in SRS [8, p. 310]. Comparisons between different PPSWOR strategies (see the section "Methods") are given in refs. 2 and 22. The efficiency of the PPS estimators depends on how closely the y_i are proportional to the x_i, as can be seen by examining (1) and (3). Under the regression model, $E(y_i \mid x_i) = \alpha + \beta x_i$, $V(y_i \mid x_i) = \sigma^2 x_i^g$, the PPS strategy may be compared with the use of SRS and the ratio estimator*

$$\hat{Y}_R = X \sum_{i=1}^{n} y_i \Big/ \sum_{i=1}^{n} x_i .$$

The PPS strategy is generally more efficient if $g > 1$ and less efficient if $g < 1$ [8, p. 258]. For example, if $g = 2$ and $\alpha = 0$, then $V_{SRS}(\hat{Y}_R)/V_{PPS}(\hat{Y}_{HT}) \doteq 1 + C_x^2$, where C_x is the coefficient of variation of the x_i. This ratio becomes closer to unity when $\alpha \neq 0$.

For two-stage sampling (see the section "Multistage Sampling") suppose that n PSUs are selected with probability proportional to size, M_i. Under PPSWR an unbiased estimate of the population total, Y, is

$$\hat{Y}_{TS} = M_0 n^{-1} \sum_{i=1}^{n} \bar{y}_i ,$$

where $M_0 = M_1 + \cdots + M_N$ and the $\bar{y}_i$ are the PSU subsample means. The variance of $\hat{Y}_{TS}$ can be expressed as the sum of a between-PSU component of the form of (1) and a within-PSU component V_W [8, p. 307]. An unbiased estimator of $V(\hat{Y}_{TS})$ is

$$\hat{V}(\hat{Y}_{TS}) = \left[n(n-1)\right]^{-1} \sum_{i=1}^{n} (M_0 \bar{y}_i - \hat{Y}_{TS})^2.$$

Under PPSWOR with $\Pi_i = M_i/M_0$ the estimator $\hat{Y}_{TS}$ is again unbiased and its variance is the sum of a between-PSU component of the form of (3) and the same within-PSU component, V_W [8, p. 309]. Unbiased estimation of this variance is relatively complicated, although simplified procedures have been suggested in refs. 10 and 5.

The discussion above is based on the conventional approach of making inferences with respect to sampling distributions induced by the randomized sample design. Alternative discussions of PPS sampling from Bayesian and prediction theory viewpoints are given in refs. 1, 9, and 18.

METHODS

The implementation of PPSWR is in principle straightforward [14], although a simplified procedure for large N has been suggested by Lahiri [17; 20, p. 202].

For PPSWOR (i.e., the selection of a sample of size n with inclusion probabilities given by $\Pi_i = nx_i/X$), many methods are available; 50 are listed in ref. 6. The case $n = 2$ has been especially studied because it corresponds to the common choice of two PSUs per stratum. Methods differ in respect of criteria such as (a) simplicity of implementation, (b) ease of computing the Π_{ij}, (c) $\hat{V}_{SYG}(\hat{Y}_{HT})$ is always nonnegative, and (d) $\hat{Y}_{HT}$ and $\hat{V}_{SYG}(\hat{Y}_{HT})$ are efficient estimators. We assume that $\Pi_i = np_i < 1$ for $i = 1, \ldots, N$, since any unit for which $np_i \geqslant 1$ is included in the sample with certainty and for our present purpose can therefore be excluded from the population.

The simplest procedures to implement are those of *systematic sampling** [19, 12; 20, p. 215], but these suffer because the Π_{ij} may be zero and hence no unbiased estimate of $V(\hat{Y}_{HT})$ may exist.

For $n = 2$ various simple PPSWOR procedures are available for which the Π_{ij} are nonzero and easy to compute. In *Brewer's method* [3] the first unit is selected with probability proportional to $p_i(1-p_i)/(1-2p_i)$, where $p_i = x_i/X$, and the second with probability $p_i/(1-p_j)$, where unit j was drawn first. In *Durbin's method* [10] the first unit is selected with probability p_i and the second with probability

$$p_{i.j} = \frac{p_i\left[(1-2p_i)^{-1} + (1-2p_j)^{-1}\right]}{1 + \sum_{k=1}^{N} p_k(1-2p_k)^{-1}}, \qquad i \neq j,$$

where unit j was drawn first. *Rao's method* [21], a case of *rejective sampling** [13], involves selecting the first unit with probability p_i and the second *with* replacement with probability proportional to $p_i/(1-2p_i)$. If both units are the same, the sample is rejected and the procedure above begun again until a sample of two distinct units is obtained. For all three procedures above, the joint inclusion probabilities are given by $\Pi_{ij} = 2p_ip_{j.i}$. Also in each case $\hat{Y}_{\text{HT}}$ is more efficient than $\hat{Y}_{\text{WR}}$ under PPSWR [3] and $\hat{V}_{\text{SYG}}$ is always nonnegative [21].

For $n > 2$ Rao's method is generalized by Sampford [23], who proposes selecting units $2, 3, \ldots, n$ with replacement with probability proportional to $p_i/(1-np_i)$ and again rejecting samples not consisting of distinct units (*see* REJECTIVE SAMPLING). Brewer's method has also been extended in ref. 4. The computation of the Π_{ij} for these methods, however, does become more complicated as n increases. A straightforward algorithmic approach for general n which obeys at least criteria (a), (b), and (c) above is given by Chao [7].

A method for continuous surveys involving rotation of the PSUs has been suggested by Fellegi [11]. Further methods are reviewed in ref. 6.

References

[1] Basu, D. (1971). In *Foundations of Statistical Inference*, V. P. Godambe and D. A. Sprott, eds. Holt, Rinehart and Winston, Toronto, pp. 203–242.

[2] Bayless, D. L. and Rao, J. N. K. (1970). *J. Amer. Statist. Ass.*, **65**, 1645–1667.

[3] Brewer, K. R. W. (1963). *Aust. J. Statist.*, **5**, 5–13.

[4] Brewer, K. R. W. (1975). *Aust. J. Statist.*, **17**, 166–172.

[5] Brewer, K. R. W. and Hanif, M. (1970). *J. R. Statist. Soc. B*, **32**, 302–311.

[6] Brewer, K. R. W. and Hanif, M. (1983). *Sampling with Unequal Probabilities*. Springer-Verlag, New York. (The most comprehensive reference on methods and estimators.)

[7] Chao, M. T. (1982). *Biometrika*, **69**, 653–656.

[8] Cochran, W. G. (1977). *Sampling Techniques*, 3rd ed. Wiley, New York. (A good reference for the estimation theory.)

[9] Cumberland, W. G. and Royall, R. M. (1981). *J. R. Statist. Soc. B*, **43**, 353–367.

[10] Durbin, J. (1967). *Appl. Statist.*, **16**, 152–164.

[11] Fellegi, I. P. (1963). *J. Amer. Statist. Ass.*, **58**, 183–201.

[12] Grundy, P. M. (1954). *J. R. Statist. Soc. B*, **16**, 236–238.

[13] Hajek, J. (1981). *Sampling from a Finite Population*. Marcel Dekker, New York. (Includes asymptotic results on various PPS schemes not available elsewhere.)

[14] Hansen, M. H. and Hurwitz, W. N. (1943). *Ann. Math. Statist.*, **14**, 333–362. (The original PPS paper.)

[15] Horvitz, D. G. and Thompson, D. J. (1952). *J. Amer. Statist. Ass.*, **47**, 663–685.

[16] Kish, L. (1965). *Survey Sampling*. Wiley, New York, Chap. 7. (Good on the practical context of PPS sampling.)

[17] Lahiri, D. B. (1951). *Bull. Int. Statist. Inst.*, **33**, Bk. 2, 133–140.

[18] Little, R. J. A. (1983). *J. Amer. Statist. Ass.*, **78**, 596–604.

[19] Madow, W. G. (1949). *Ann. Math. Statist.*, **20**, 333–354.

[20] Murthy, M. N. (1967). *Sampling Theory and Methods*. Statistical Publishing Society, Calcutta, Chap. 6.

[21] Rao, J. N. K. (1965). *J. Indian Statist. Ass.*, **3**, 173–180.

[22] Rao, J. N. K. and Bayless, D. L. (1969). *J. Amer. Statist. Ass.*, **64**, 540–559.

[23] Sampford, M. R. (1967). *Biometrika*, **54**, 499–513.

[24] Sen, A. R. (1953). *J. Indian Soc. Agric. Statist.*, **5**, 119–127.

[25] Sukhatme, P. V. and Sukhatme, B. V. (1970). *Sampling Theory of Surveys with Applications*. Asia Publishing House, London, Chap. 2.

[26] Yates, F. and Grundy, P. M. (1953). *J. R. Statist. Soc. B*, **15**, 253–261.

(AREA SAMPLING
HORWITZ–THOMPSON ESTIMATOR
OPTIMAL STRATIFICATION
PRIMARY SAMPLING UNITS

REJECTIVE SAMPLING
REPRESENTATIVE SAMPLING
SIMPLE RANDOM SAMPLING
STRATIFIED MULTISTAGE SAMPLING
STRATIFIED DESIGNS
SYSTEMATIC SAMPLING)

C. J. SKINNER

PROBABILITY SPACES, METRICS AND DISTANCES ON

Metrics and distances (or semidistances) between probability distributions play an important role in statistical inference and in practical applications to study affinities among a given set of populations. A statistical model is specified by a family of probability distributions, usually described by a set of continuous parameters known as *parameter space*. The latter possesses geometrical properties which are induced by the local information content and structures of the distributions. Starting from Fisher's pioneering work [17] in 1925, the study of these geometrical properties has received much attention. In 1945, Rao [24] introduced a Riemannian metric in terms of the Fisher information* matrix over the parameter space of a parametric family of probability distributions, and proposed the geodesic distance induced by the metric as a measure of dissimilarity between two probability distributions. Since then, many statisticians have attempted to construct a geometrical theory in probability spaces and it was only 30 years later that Efron [13] was able to introduce a new affine connection into the geometry of parameter spaces, thereby elucidating the important role of curvature in statistical studies. Significant contributions to Efron's work were made by Reeds [28] and Dawid [11]. The latter has even suggested a geometrical foundation for Efron's approach as well as pointing out the possibility of introducing other affine connections (see also Amari [1, 2]). This recent study has also revived the interest in dissimilarity measures like Rao distance* [25], especially in the closed-form expressions of these distances, for certain families of probability distributions. Some work in these directions was done earlier by Čencov [9, 10]. Atkinson and Mitchell [3], independently of Čencov [9, 10], computed the Rao distances for a number of parametric families of distributions. A unified approach to the construction of distance and dissimilarity measures in probability spaces is given in recent papers by Burbea and Rao [7, 8], and Oller and Cuadras [22] (*see also* J-DIVERGENCES AND RELATED CONCEPTS *and* MEASURES OF SIMILARITY, DISSIMILARITY, AND DISTANCE).

GENERALITIES

We first introduce some notation. Let μ be a σ-finite additive measure, defined on a σ-algebra of the subsets of a measurable space χ. Then, $\mathscr{M} \equiv \mathscr{M}(\chi:\mu)$ stands for the space of all μ-measurable functions on χ, $\mathscr{L} \equiv \mathscr{L}(\chi:\mu)$ designates the space of all $p \in \mathscr{M}$ so that

$$\|p\|_\mu \equiv \int_X |p(x)|d\mu(x) = \int_X |p|d\mu < \infty.$$

By $\mathscr{M}_+ \equiv \mathscr{M}_+(\chi:\mu)$ we denote the set of all $p \in \mathscr{M}$ such that $p(x) \in \mathbb{R}_+ \equiv (0,\infty)$ for μ-almost all $x \in \chi$, and we define $\mathscr{L}_+ \equiv \mathscr{L}_+(\chi:\mu)$ as $\mathscr{L}_+ = \mathscr{M}_+ \cap \mathscr{L}$. We let $\mathscr{P} \equiv \mathscr{P}(\chi:\mu)$ stand for the set of all $p \in \mathscr{L}_+$ with $\|p\|_\mu = 1$. $\mathscr{P}$ is a convex subset of $\mathscr{L}_+$, and $p \in \mathscr{L}_+$ if and only if $p/\|p\|_\mu \in \mathscr{P}$.

In the probability context, a random variable X takes values in the sample space χ according to a probability distribution p assumed to belong to $\mathscr{P}$. If X is continuous, μ will be Lebesgue measure on the Borel sets of a Euclidean sample space χ and, if X is discrete, μ is taken as a counting measure on the sets of a countable sample space χ.

Let $\theta = (\theta_1, \ldots, \theta_n)$ be a set of real continuous parameters belonging to a parameter space Θ, a manifold* embedded in $\mathbb{R}^n$, and let $\mathscr{F}_\Theta = \{p(\cdot \mid \theta) \in \mathscr{L}_+ : \theta \in \Theta\}$ be a parametric family of positive distributions $p = p(\cdot \mid \theta), \theta \in \Theta$, with some regularity properties (see refs. 3, 7, and 12). For example, it is implicitly assumed that

$$\partial_i p \equiv \partial_i p(\cdot \mid \theta) \equiv \partial p(\cdot \mid \theta)/\partial \theta_i$$
$$(p = p(\cdot \mid \theta) \in \mathscr{F}_\Theta, \quad i = 1, \ldots, n)$$

is in $\mathscr{M}$ for every $\theta \in \Theta$, and that for fixed $\theta \in \Theta$, the n functions $\{\partial_i p\}_{i=1}^n$ are linearly independent over χ. We also consider a parametric family of probability distributions $\mathscr{P}_\Theta = \{p(\cdot \mid \theta) \in \mathscr{P} : \theta \in \Theta\}$ which may be viewed as a convex subfamily of $\mathscr{F}_\Theta$.

Let f be a continuous and positive function on $\mathbb{R}_+$ and define

$$ds_f^2(\theta) \equiv \int_\chi \frac{f(p)}{p} [dp]^2 \, d\mu$$
$$(\theta \in \Theta, \quad p = p(\cdot \mid \theta) \in \mathscr{F}_\Theta),$$

where in the integrand, the dependence on $x \in \chi$ and $\theta \in \Theta$ is suppressed and where

$$dp = dp(\cdot \mid \theta) = \sum_{i=1}^n (\partial_i p) \, d\theta_i \, .$$

Here and throughout we freely use the convention of suppressing the dependence on $x \in \chi$ and $\theta \in \Theta$. Thus

$$ds_f^2(\theta) = \sum_{i,j=1}^n g_{ij}^{(f)} \, d\theta_i \, d\theta_j \, , \tag{1}$$

$$g_{ij}^{(f)} = g_{ij}^{(f)}(\theta) = \int_\chi \frac{f(p)}{p} (\partial_i p)(\partial_j p) \, d\mu.$$

It follows that the $n \times n$ matrix $[g_{ij}^{(f)}(\theta)]$ is positive definite for every $\theta \in \Theta$. Hence ds_f^2 gives a Riemannian metric on Θ and

$$g_{ij}^{(f)}(\theta) = E_\theta\big[(f \circ p)(\partial_i \log p)(\partial_j \log p)\big].$$

In the theory of information (*see* INFORMATION THEORY *and* J-DIVERGENCES AND RELATED CONCEPTS) the quantity $-\log p(\cdot \mid \theta)$, for $p(\cdot \mid \theta) \in \mathscr{P}_\Theta$, is known as the amount of *self-information* associated with the state $\theta \in \Theta$. The self-information for the nearby state $\theta + \delta\theta \in \Theta$ is then $-\log p(\cdot \mid \theta + \delta\theta)$. To the first order, the difference between these self-informations is given by

$$d \log p = \sum_{i=1}^n (\partial_i \log p) \, d\theta_i$$

and hence $ds_f^2(\theta)$ measures the weighted average of the square of this first-order difference with the weight $f[p(\cdot \mid \theta)]$. For this reason, the metric ds_f^2 and the matrix $[g_{ij}^{(f)}]$ are called the *f-information metric* and the *f-information matrix*, respectively.

As is well known from differential geometry, $g_{ij}^{(f)}$ $(i, j = 1, \ldots, n)$ is a covariant symmetric tensor* of the second order for all $\theta \in \Theta$, and hence ds_f^2 is invariant under the admissible transformations of the parameters. Let $\theta = \theta(t)$, $t_1 \leqslant t \leqslant t_2$, be a curve in Θ joining the points $\theta^{(1)}, \theta^{(2)} \in \Theta$ with $\theta^{(j)} = \theta(t_j)$ $(j = 1, 2)$. Since $ds_f \equiv (ds_f^2)^{1/2}$ is the line element of the metric ds_f^2, the distance between these points along this curve is

$$\left| \int_{t_1}^{t_2} \frac{ds_f}{dt} \, dt \right| = \left| \int_{t_1}^{t_2} \left\{ \sum_{i,j=1}^n g_{ij}^{(f)}(\theta) \dot{\theta}_i \dot{\theta}_j \right\}^{1/2} dt \right|,$$

where a dot denotes differentiation with respect to the curve parameter t. The geodesic curve joining $\theta^{(1)}$ and $\theta^{(2)}$ such that the distance above is the shortest is called the *f-information geodesic curve* along $\theta^{(1)}$ and $\theta^{(2)}$, while the resulting distance $S_f(\theta^{(1)}, \theta^{(2)})$ is called the *f-information geodesic distance* between $\theta^{(1)}$ and $\theta^{(2)}$. The *f*-information geodesic curve $\theta = \theta(t)$ may be determined from the Euler–Lagrange equations

$$\sum_{i=1}^n g_{ik}^{(f)} \ddot{\theta}_i + \sum_{i,j=1}^n \Gamma_{ijk}^{(f)} \dot{\theta}_i \dot{\theta}_j = 0$$
$$(k = k, \ldots, n) \tag{2}$$

and from the boundary conditions

$$\theta_i(t_j) = \theta_i^{(j)} \qquad (i = 1, \ldots, n; j = 1, 2).$$

Here the quantity $\Gamma_{ijk}^{(f)}$ is given by

$$\Gamma_{ijk}^{(f)} = \frac{1}{2} \left[\partial_i g_{jk}^{(f)} + \partial_j g_{ki}^{(f)} - \partial_k g_{ij}^{(f)} \right], \tag{3}$$

known as the *Christoffel symbol of the first kind* for the metric ds_f^2.

By definition of the *f*-information geodesic curve $\theta = \theta(t)$, its tangent vector $\dot{\theta} = \dot{\theta}(t)$ is of constant length with respect to the metric ds_f^2. Thus

$$\{\dot{s}_f(\theta(t))\}^2 = \sum_{i,j=1}^n g_{ij}^{(f)} \dot{\theta}_i \dot{\theta}_j \equiv \text{const.} \tag{4}$$

The constant may be chosen to be of value 1 when the curve parameter t is chosen to be the arc-length parameter s, $0 \leqslant s \leqslant s_0$ with $s_0 \equiv S_f(\theta^{(1)}, \theta^{(2)})$, $\theta(0) = \theta^{(1)}$, and $\theta(s_0) = \theta^{(2)}$. It is also clear that the f-information geodesic distance S_f on the parameter space Θ is invariant under the admissible transformations of the parameters.

The metric $ds_f^2(\theta)$ may also be regarded as a functional of $p(\cdot\,|\,\theta) \in \mathscr{F}_\Theta$. This functional is convex in $p(\cdot\,|\,\theta) \in \mathscr{F}_\Theta$ if and only if the function $F(x) \equiv x/f(x)$ is concave on $\mathbb{R}_+$. In particular, if f is also a C^2-function on $\mathbb{R}_+$, then this holds if and only if $FF'' \geqslant 2(F')^2$ on $\mathbb{R}_+$. The choice $f(x) = x^{\alpha-1}$ gives the *α-order information metric*

$$ds_\alpha^2(\theta) = E_\theta\left[p^{\alpha-1}(d\log p)^2\right] \qquad (5)$$

with the corresponding *α-order information matrix*

$$g_{ij}^{(\alpha)}(\theta) = E_\theta\left[p^{\alpha-1}(\partial_i \log p)(\partial_j \log p)\right] \qquad (6)$$

and the *α-order information geodesic distance* S_α on Θ. It follows that $ds_\alpha^2(\theta)$ is convex in $p(\cdot\,|\,\theta) \in \mathscr{F}_\Theta$ if and only if $1 \leqslant \alpha \leqslant 2$. We drop the suffix α when $\alpha = 1$. Then ds^2 is known as the *information metric* or the *Fisher amount of information*, while $[g_{ij}]$ is the well-known *information matrix* or *Fisher information* matrix*. The distance S on Θ is the *information geodesic distance* or the *Rao distance** (see refs. 3, 7, and 26), and is invariant under the admissible transformations of the parameters as well as of the random variables. We also note that

$$g_{ij}^{(\alpha)}(\theta) = \int_X p^{\alpha-1}\partial_i\partial_j p\,d\mu - \int_X p^\alpha \partial_i\partial_j \log p\,d\mu.$$

Moreover, for $\alpha \neq 0$,

$$g_{ij}^{(\alpha)}(\theta) = \alpha^{-2}\partial_i\partial_j\int_X p^\alpha\,d\mu - \alpha^{-1}\int_X p^\alpha\partial_i\partial_j\log p\,d\mu.$$

In particular,

$$g_{ij}(\theta) = \partial_i\partial_j\int_X p\,d\mu - \int_X p\partial_i\partial_j\log p\,d\mu$$

and thus

$$g_{ij}(\theta) = -\int_X p\partial_i\partial_j \log p\,d\mu = -E_\theta(\partial_i\partial_j\log p) \qquad (p(\cdot\,|\,\theta) \in \mathscr{P}_\Theta).$$

The metric $ds_f^2(\theta)$ arises as the second-order differential of certain entropy* or divergence functionals along the direction of the tangent space of Θ at $\theta \in \Theta$ [7, 12] (*see also* J-DIVERGENCES AND RELATED CONCEPTS). For example, let $F(\cdot,\cdot)$ be a C^2-function on $\mathbb{R}_+ \times \mathbb{R}_+$ and consider the *F-divergence*

$$D_F(p,q) \equiv \int_X F[p(x), q(x)]\,d\mu(x) \qquad (p, q \in \mathscr{M}_+).$$

We also assume that F satisfies the additional properties: (a) $F(x,\cdot)$ is strictly convex on $\mathbb{R}_+$ for every $x \in \mathbb{R}_+$; (b) $F(x,x) = 0$ for every $x \in \mathbb{R}_+$; and (c) $\partial_y F(x,y)|_{y=x} =$ const. for every $x \in \mathbb{R}_+$.

For $p(\cdot\,|\,\theta^{(1)})$ and $p(\cdot\,|\,\theta^{(2)})$ in $\mathscr{P}_\Theta$ we write

$$\mathscr{D}_F(\theta^{(1)}, \theta^{(2)}) \equiv D_F\left[p(\cdot\,|\,\theta^{(1)}), p(\cdot\,|\,\theta^{(2)})\right] \qquad (\theta^{(1)}, \theta^{(2)} \in \Theta).$$

Then, for $p(\cdot\,|\,\theta) \in \mathscr{F}_\Theta$ and $\theta \in \Theta$,

$$\mathscr{D}_F(\theta,\theta) = 0,$$

$$d\mathscr{D}_F(\theta,\theta) = \int_X \partial_y F(p, y)|_{y=p}(dp)\,d\mu = 0$$

and

$$d^2\mathscr{D}_F(\theta,\theta) = ds_f^2(\theta),$$

where $f(x) = x\partial^2 yF(x,y)|_{y=x}(x \in \mathbb{R}_+)$. It follows that to the second order, infinitesimal displacements

$$\mathscr{D}_F(\theta, \theta + \delta\theta) = \tfrac{1}{2}ds_f^2(\theta).$$

PROPERTIES OF THE INFORMATION METRIC

We describe some further properties of the f-information metric ds_f^2 in the case of the ordinary information metric ds^2 [i.e., when $f(x) \equiv 1$ or when $\alpha = 1$ in (5)] on the parametric space Θ of probability distributions $p(\cdot\,|\,\theta)$ in $\mathscr{P}_\Theta$. A more general discussion may be found in refs. 7 and 8. We shall hereafter omit the summation symbol $\sum$ when the indices are repeated twice and assume that the summation runs from 1 to n. With this summation convention, we have, by virtue of (1), (3), and (6),

$$ds^2 = g_{ij} d\theta_i d\theta_j,$$

$$g_{ij} = E_\theta[(\partial_i \log p)(\partial_j \log p)]$$

$$= -E_\theta[\partial_i \partial_j \log p],$$

$$\Gamma_{ijk} = \tfrac{1}{2}[\partial_i g_{jk} + \partial_j g_{ki} - \partial_k g_{ij}]. \quad (7)$$

The information geodesic curves $\theta = \theta(s)$, where s is the arc-length parameter, are determined, in view of (2), by

$$g_{ij}\ddot{\theta}_i + \Gamma_{ijk}\dot{\theta}_i\dot{\theta}_j = 0 \qquad (k = 1, \ldots, n). \quad (8)$$

Moreover, from (4) we have

$$g_{ij}\dot{\theta}_i\dot{\theta}_k = 1. \quad (9)$$

Thus, for two points $a, b \in \Theta$, or for $p(\cdot \mid a)$, $p(\cdot \mid b) \in \mathscr{P}_\Theta$, the Rao distance $S(a,b)$ is completely determined by (8), a system of n second-order (nonlinear) differential equations, and by the $2n$ boundary conditions $\theta(0) = a$ and $\theta(s_0) = b$ with $s_0 = S(a,b)$. This computation may be facilitated with the aid of the normalization (8).

We denote by $\mathscr{I}$ the Fisher information matrix $[g_{ij}]$, by g^{ij} the elements of its inverse $\mathscr{I}^{-1}$, and the elements of the unit matrix I are denoted by the Kronecker δ_{ij}. $\mathscr{I}^{-1} = [g^{ij}]$ is positive definite and $\mathscr{I}$ is associated with a distribution $p(\cdot \mid \theta) \in \mathscr{P}_\Theta$ of a random variable X. We list the following properties (see Rao [27, p. 323–332] for more details):

1. Let $\mathscr{I}_1$ and $\mathscr{I}_2$ be the information matrices due to two independent random variables X_1 and X_2. Then $\mathscr{I} = \mathscr{I}_1 + \mathscr{I}_2$ is the information matrix due to $X = (X_1, X_2)$ jointly.
2. Let $\mathscr{I}_T$ be the information matrix due to a function T of X. Then $\mathscr{I} - \mathscr{I}_T$ is semi-positive definite.
3. Let $p(\cdot \mid \theta) \in \mathscr{P}_\Theta$ with the corresponding information matrix $\mathscr{I}$. Assume that $f = (f_1, \ldots, f_m)$ is a vector of m statistics (random variables) and define $g(\theta) = (g_1(\theta), \ldots, g_m(\theta))$ by $g_i(\theta) = E(f_i)$ $(i = 1, \ldots, m)$ [i.e., f is an unbiased estimator of $g(\theta)$]. Consider the $m \times m$ and $m \times n$ matrices $V = [V_{ij}]$ and $U = [U_{ij}]$ given by

$$V_{ij} = E_\theta[(f_i - g_i)(f_j - g_j)] \qquad (i, j = 1, \ldots, m),$$

$$U_{ij} = E_\theta[f_i \partial_j \log p] \qquad (i = 1, \ldots, m;\ j = 1, \ldots, n).$$

Then:

(i) The $m \times m$ matrix $V - U\mathscr{I}^{-1}U'$ is semi-positive definite for every $\theta \in \Theta$. The matrix is zero at some $\theta \in \Theta$ if and only if $f = (f_1, \ldots, f_m)$ is of the form $f_i = \lambda_{ik}\partial_k \log p + E_\theta(f_i)$ $(i = 1, \ldots, m)$.

(ii) Suppose, in addition, that

$$\partial_j \int_X f_i(x) p(x \mid \theta)\, d\mu(x) = \int_X f_i(x) \partial_j p(x \mid \theta)\, d\mu(x) \qquad (i = 1, \ldots, m;\ j = 1, \ldots, n).$$

Then U is the Jacobian* matrix $[\partial_j g_i]$ of $g = (g_1, \ldots, g_m)$ with respect to $\theta = (\theta_1, \ldots, \theta_n)$. In particular, when $m = n$ and $g(\theta) = \theta$ (i.e., f is an unbiased estimator of θ), then $V - \mathscr{I}^{-1}$ is semi-positive definite.

The last property constitutes the celebrated *Cramér-Rao* lower bound theorem*, that for any unbiased estimator of θ, its covariance matrix dominates the inverse of the Fisher information matrix.

INFORMATION CONNECTIONS AND CURVATURES

The information metric renders the parameter space Θ as a Riemannian manifold with the metric tensor g_{ij} associated with the distribution $p(\cdot \mid \theta) \in \mathscr{P}_\Theta$. In this context, the Christoffel symbol of the first kind Γ_{ijk} in (7) is called the *first information connection*. As is well known from differential geometry, this natural affine connection induces a parallelism on Θ, the *Levi–Civita parallelism*, which is compatible with the metric tensor g_{ij}, in the sense that the covariant differentiation of the latter vanishes for this connection. Using the summation convention, one introduces the *Christoffel symbol of the second kind* Γ_{ij}^{k} by

$$\Gamma_{ij}^{k} = \Gamma_{ijm} g^{mk}, \qquad (10)$$

also called the *second information connection*. With the aid of this connection, the equation for the information geodesic curves (8) assumes the alternative form

$$\ddot{\theta}_k + \Gamma_{ij}^{k} \dot{\theta}_i \dot{\theta}_j = 0 \qquad (k = 1, \ldots, n). \quad (11)$$

In differential geometry one also considers the *Riemann–Christoffel tensor of the second kind*,

$$R_{ijk}^{l} = \partial_j \Gamma_{ik}^{l} - \partial_k \Gamma_{ij}^{l} + \Gamma_{ik}^{m} \Gamma_{mj}^{l} - \Gamma_{ij}^{m} \Gamma_{mk}^{l}, \qquad (12)$$

and the *Riemann–Christoffel tensor of the first kind*,

$$R_{ijkl} = R_{jkl}^{m} g_{mi}, \qquad (13)$$

also known as the *second information curvature tensor* and the *first information curvature tensor*, respectively. It is worthwhile noticing that

$$R_{ijkl} = -R_{jikl} = -R_{ijlk} = R_{klij},$$
$$R_{ijkl} + R_{iklj} + R_{iljk} = 0,$$

and that the number of distinct nonvanishing components of the tensor R_{ijkl} is $n^2(n^2 - 1)/12$. The latter reduces to 0 when $n = 1$ and to 1 when $n = 2$.

The *mean Gaussian curvature* in the directions of $x = (x_1, \ldots, x_n)$ and $y = (y_1, \ldots, y_n)$ of $\mathbb{R}^n$ is given by

$$\kappa \equiv \kappa(\theta : x, y) = \frac{R_{ijkl} x_i y_j x_k y_l}{(g_{ik} g_{jl} - g_{il} g_{jk}) x_i y_j x_k y_l} \qquad (\theta \in \Theta), \quad (14)$$

also called the *information curvature* in the directions of x and y. It is identically zero if Θ is Euclidean and constant if the space Θ is isotropic (i.e., when κ is independent of the directions x and y), provided that $n > 2$.

Besides the first information connection Γ_{ijk} there are connections leading to parallelisms which differ from the Levi–Civita parallelism. In the context of statistical inference, the choice of such connections should reflect the structure of the distributions in some meaningful manner. Following an idea of Dawid [11], Amari [1] considers the one-parameter family of affine *α-connections* $\overset{\alpha}{\Gamma}_{ijk}$ given by

$$\overset{\alpha}{\Gamma}_{ijk} \equiv \Gamma_{ijk} - \frac{\alpha}{2} T_{ijk} \qquad (\alpha \in \mathbb{R}),$$

where T_{ijk} is the symmetric tensor

$$T_{ijk} \equiv E_\theta\left[(\partial_i \log p)(\partial_j \log p)(\partial_k \log p)\right].$$

The first information connection is the 0-connection. An alternative expression for the α-connection is

$$\overset{\alpha}{\Gamma}_{ijk} = E_\theta\left[(\partial_i \partial_j \log p)(\partial_k \log p)\right] + \frac{1-\alpha}{2} T_{ijk}.$$

The 1-connection was introduced first by Efron [13] and hence is also called the *Efron connection*. The −1-connection, on the other hand, is called the *Dawid connection*, after Dawid [11].

To elucidate the meaningfulness of the α-conditions in statistical problems, we consider two examples suggested by Dawid [11] and described in Amari [1].

Example 1. Consider an exponential family $\mathscr{P}_\Theta$ of distributions $p(\cdot \mid \theta)$ given (using the summation convention) by

$$p(x \mid \theta) = \exp\{T(x) + T_i(x)\theta_i - \psi(\theta)\} \qquad (x \in \chi) \quad (15)$$

with

$$e^{\psi(0)} = \int_\chi e^{T_i(x)\theta_i} e^{T(x)}\, d\mu(x) \qquad (\theta \in \Theta), \quad (16)$$

and specified by the natural free parameters $\theta = (\theta_1, \ldots, \theta_n) \in \Theta$. Here ψ is a C^2-function on Θ, and T and $T_1, \ldots, T_n$ are measurable functions on χ. Under these circumstances

$$\partial_i \log p = T_i(x) - \partial_i \psi(\theta),$$
$$\partial_i \partial_j \log p = -\partial_i \partial_j \psi.$$

Therefore,

$$g_{ij} = \partial_i \partial_j \psi, \tag{17}$$

$$E_\theta[(\partial_i \partial_j \log p)(\partial_k \log p)] = 0, \tag{18}$$

$$\overset{\alpha}{\Gamma}_{ijk} = \frac{1-\alpha}{2} T_{ijk}. \tag{19}$$

Since $\overset{\alpha}{\Gamma}_{ijk}(\theta)$ is identically zero for $\alpha = 1$, the exponential family constitutes an uncurved space with respect to the Efron connection, which hence may also be called the *exponential connection*.

Example 2. Consider a family $\mathscr{P}_\Theta \equiv \mathscr{P}_\Theta(q_1, \ldots, q_{n+1})$ of distributions $p(\cdot \mid \theta)$ given by a mixture of $n+1$ prescribed linearly independent probability distributions on χ,

$$p(x \mid \theta) = q_i(x)\theta_i + q_{n+1}(x)\theta_{n+1} \quad (x \in \chi),$$

where $\theta_{n+1} \equiv 1 - (\theta_1 + \cdots + \theta_n)$, $\theta \in \Theta$, $\Theta = \{\theta = (\theta_1, \ldots, \theta_n) \in \mathbb{R}^n_+ : \theta_{n+1} > 0\}$. In this case

$$\partial_i \log p = p^{-1}(q_i - q_{n+1}),$$

$$\partial_i \partial_j \log p = -(\partial_i \log p)(\partial_j \log p).$$

Therefore,

$$E_\theta[(\partial_i \partial_j \log p)(\partial_k \log p)] = -T_{ijk},$$

$$\overset{\alpha}{\Gamma}_{ijk} = -\frac{1+\alpha}{2} T_{ijk}.$$

Since $\overset{\alpha}{\Gamma}_{ijk}(\theta)$ is identically zero for $\alpha = -1$, this family of mixture distributions constitutes an uncurved space with respect to the Dawid connection, which hence is also called the *mixture connection*.

Once the α-connection $\overset{\alpha}{\Gamma}_{ijk}$ is adopted, the related quantities $\overset{\alpha}{\Gamma}{}^k_{ij}$, $\overset{\alpha}{R}{}^l_{ijk}$, $\overset{\alpha}{R}_{ijkl}$, and $\overset{\alpha}{\kappa}$ are determined by the same rules, (10) and (12) to (14) for determining the corresponding quantities when $\alpha = 0$. For example,

$$\overset{\alpha}{\Gamma}{}^k_{ij} = \overset{\alpha}{\Gamma}_{ijm} g^{mk}$$

and, corresponding to (11), the equation

$$\ddot{\theta}_k + \overset{\alpha}{\Gamma}{}^k_{ij} \dot{\theta}_i \dot{\theta}_j = 0 \qquad (k = 1, \ldots, n)$$

gives the "straight lines" $\theta = \theta(t)$ with respect to the α-connection. When $\alpha = 0$ these straight lines are also the information geodesic curves. This is not necessarily so when $\alpha \neq 0$, for in this case, the α-connection is not compatible with the metric tensor g_{ij}.

The theory of α-connections and their curvatures seems to be particularly applicable in elucidating the structures of the exponential families* as well as of the curved exponential family of distributions. An exponential family may be written in the form (15)–(16) by choosing natural parameters $\theta = (\theta_1, \ldots, \theta_n)$ which are uniquely determined within affine transformations. In this case $(T_1, \ldots, T_n)$ constitutes a sufficient statistic* for the family and has a covariance matrix V equal to $\mathscr{I}$. The corresponding Cramér–Rao lower bound in property 3(i) of the preceding section is always attained. Moreover, the natural parameter space Θ is convex, and, by (17), ψ is convex on Θ. Use of (14) to (19) shows that the α-Riemann–Christoffel curvature tensor of the space is given by

$$\overset{\alpha}{R}_{ijkl} = \frac{1-\alpha^2}{2} \left[T_{jrk} T_{iml} - T_{jrl} T_{imk} \right] g^{mr}.$$

Initially, this formula is valid only for the neutral coordinate system. However, since the formula is given by means of a tensorial equation, its validity does not depend on a particular choice of the coordinates. It follows that for any exponential family $\mathscr{P}_\Theta$,

$$\overset{\alpha}{R}_{ijkl} = (1 - \alpha^2) R_{ijkl},$$

and hence the Efron and the Dawid connections (when $\alpha = 1$ and $\alpha = -1$) render the space Θ as *flat* (or with an *absolute parallelism*).

The curved exponential families can be embedded in the exponential families as subspaces [13, 14]. Thus these families possess various dualistic structures: the Barndorff-Nielsen duality [4] associated with the Legendre transformation, the α – (-α) duality [2] between two kinds of connections, and the α – (-α) duality [2] between two kinds of curvatures. As shown by Amari [2], these dualities are intimately connected and, moreover, the second-order information loss is expressed in terms of the curvatures of the statistical model and the estimator. See Amari [2], Barndorff-Nielsen [4], Dawid [11], Efron [13, 14], and Reeds [28] for a more detailed account. For the general study of connections and curvatures, see the books of Eisenhart [15, 16], Hicks [18], Kobayashi and Nomizi [19], Laugwitz [20], and Schouten [29].

INFORMATIVE GEOMETRY OF SPECIFIC FAMILIES OF DISTRIBUTIONS

An *informative geometry* of distributions $p(\cdot \mid \theta) \in \mathscr{P}_\Theta$ is the geometry associated with the natural affine connection Γ_{ijk} of the information metric ds^2. A description of the informative geometrics of certain well-known families of distributions $\mathscr{P}_\Theta$ may be found in refs. 3, 5, 6, 8, 9, 10, 21, 22, 23, and 25.

References

[1] Amari, S. (1968). *RAAG Mem. No. 4*, pp. 373–418.

[2] Amari, S. (1980). *RAAG Rep. No. 106*, pp. 1–53.

[3] Atkinson, C. and Mitchell, A. F. S. (1981). *Sankhyā*, **43**, 345–365.

[4] Barndorff-Nielsen, O. (1978). *Information and Exponential Families in Statistical Theory*. Wiley, New York.

[5] Bhattacharyya, A. (1943). *Bull. Calcutta Math. Soc.*, **35**, 99–109.

[6] Burbea, J. (1984). Informative Geometry of Probability Spaces. *Tech. Rep.*, Center for Multivariate Analysis, University of Pittsburgh, Pittsburgh, Pa.

[7] Burbea, J. and Rao, C. R. (1982). *J. Multivariate Anal.*, **12**, 575–596.

[8] Burbea, J. and Rao, C. R. (1982). *Prob. Math. Statist.*, **3**, 115–132.

[9] Čencov, N. N. (1965). *Dokl. Akad. Nauk, SSSR*, **164**, 3 (in Russian).

[10] Čencov, N. N. (1972). *Statistical Decision Rules and Optimal Inference*. Nauka, Moscow (in Russian; English translation, 1982, American Mathematical Society, Providence, RI).

[11] Dawid, A. P. (1975). *Ann. Statist.*, **3**, 1231–1234.

[12] Dawid, A. P. (1977). *Ann. Statist.*, **5**, 1249.

[13] Efron, B. (1975). *Ann. Statist.*, **3**, 1189–1217.

[14] Efron, B. (1978). *Ann. Statist.*, **6**, 362–376.

[15] Eisenhart, L. (1926/1960). *Riemannian Geometry*. Princeton University Press, Princeton, N.J.

[16] Eisenhart, L. (1940/1964). *An Introduction to Differential Geometry*. Princeton University Press, Princeton, N.J.

[17] Fisher, R. A. (1925). *Proc. Camb. Philos. Soc.*, **22**, 700–725.

[18] Hicks, N. J. (1965). *Notes on Differential Geometry*. D. Van Nostrand, Princeton, N.J.

[19] Kobayashi, S. and Nomizu, K. (1968). *Foundations of Differential Geometry*, Vol. II. Wiley, New York.

[20] Laugwitz, D. (1965). *Differential and Riemannian Geometry*. Academic Press, New York.

[21] Mahalanobis, P. C. (1936). *Proc. Natl. Inst. Sci. India*, **12**, 49–55.

[22] Oller, J. M. and Cuadras, C. M. (1985). *Sankhyā*, **47**, 75–83.

[23] Pitman, E. J. G. (1979). *Some Basic Theory for Statistical Inference*. Halsted Press, New York.

[24] Rao, C. R. (1945). *Bull. Calcutta Math. Soc.*, **37**, 81–91.

[25] Rao, C. R. (1949). *Sankhyā*, **9**, 246–248.

[26] Rao, C. R. (1962). *J. R. Statist. Soc. B*, **24**, 46–72.

[27] Rao, C. R. (1973). *Linear Statistical Inference and Its Applications*. Wiley, New York.

[28] Reeds, J. (1977). *Ann. Statist.*, **5**, 1234–1238.

[29] Schouten, J. A. (1954). *Ricci-Calculus*. Springer-Verlag, Berlin.

Bibliography

See the following works, as well as the references just given, for more information on the topic of metrics and distances on probability spaces.

Efron, B. (1975). *Ann. Statist.*, **3**, 1189–1217. (An impressive work accompanied by "A discussion on Professor Efron's paper," pp. 1217–1242 of the same issue. A fresh idea was opened up by the introduction of a genuinely new connection—the Efron connection—into the geometry of parameter spaces. This connection is not compatible with the information metric, and thus the resulting Efron's curvature is not the intrinsic Riemannian curvature but rather the curvature of embedding. The latter has the advantage of being amenable to quantitative study in problems of statistical inference. The discussions of C. R. Rao, L. M. LeCam, J. K. Ghosh, J. Pfanzagl, A. P. Dawid, and J. Reeds on this paper, as well as the reply to the discussion by B. Efron, are of particular relevance.)

Fisher, R. A. (1925). *Proc. Camb. Philos. Soc.*, **22**, 700–725. (A classical work, laying out the foundation for the geometry of probability spaces in terms of the Fisher information matrix.)

Rao, C. R. (1945). *Bull. Calcutta Math. Soc.*, **37**, 81–91. (This is one of the first works where differential-geometrical methods, based on the Fisher information metric tensor of the parameter space, have been used to study problems of statistical inference. In particular, Rao distance is introduced as the geodesic distance induced by the metric tensor above.)

(CRAMÉR–RAO LOWER BOUND
DIVERSITY INDICES
EFFICIENCY, SECOND-ORDER
ENTROPY
EXPONENTIAL FAMILIES
FISHER INFORMATION
INFORMATION THEORY AND CODING THEORY
J-DIVERGENCES AND RELATED CONCEPTS
MEASURES OF SIMILARITY, DISSIMILARITY AND DISTANCE
RAO DISTANCE
STATISTICAL CURVATURE
TENSORS)

JACOB BURBEA

PROBABILITY THEORY (OUTLINE)

Probability theory provides a mathematical framework for the description of phenomena whose outcomes are not deterministic but rather are subject to chance*. This framework, historically motivated by a frequency interpretation* of probability, is provided by a formal set of three innocent-looking axioms and the consequences thereof. These are far-reaching and a rich and diverse theory has been developed. It provides a flexible and effective model of physical reality and furnishes, in particular, the mathematical theory needed for the study of the discipline of statistics.

THE FORMAL FRAMEWORK

Elementary (non-measure-theoretic) treatments of probability can readily deal with situations in which the set of distinguishable outcomes of an experiment subject to chance (called elementary events) is finite. This encompasses most simple gambling situations, for example.

Let the set of all elementary events be Ω (called the sample space*) and let $\mathscr{F}$ be the Boolean field of subsets of Ω (meaning just that (a) $A \in \mathscr{F}$ implies that $\bar{A} \in \mathscr{F}$, the bar denoting complementation with respect to Ω and (b) $A_1, A_2 \in \mathscr{F}$ implies that $A_1 \cup A_2 \in \mathscr{F}$). Then the probability P is a set function on $\mathscr{F}$ satisfying the axioms:

A1 $P(A) \geqslant 0$, $A \in \mathscr{F}$.

A2 If $A_1, A_2 \in \mathscr{F}$ and $A_1 \cap A_2 = \emptyset$ (the empty set), then

$$P(A_1 \cup A_2) = P(A_1) + P(A_2).$$

A3 $P(\Omega) = 1$.

These axioms* enable probabilities to be assigned to events in simple games of chance*. For example, in the toss of an ordinary six-sided die, Ω consists of six equiprobable elementary events and for $A \in \mathscr{F}$, $P(A) = n(A)/6$, where $n(A)$ is the number of elementary events in A.

The formulation above can be extended straightforwardly to the case where the number of elementary events in the sample space is at most countable, by extending Axiom **A2** to deal with countable unions of disjoint sets. However, a full measure-theoretic formulation is necessary to deal with general sample spaces Ω and to provide a framework which is rich enough for a comprehensive study of limit results on combinations of events (sets); *see also* MEASURE THEORY IN PROBABILITY AND STATISTICS.

The measure-theoretic formulation has a very similar appearance to the elementary version. We start from a general sample space Ω and let $\mathscr{F}$ be the Borel σ-field of subsets of Ω [meaning that (a) $A \in \mathscr{F}$ implies that $\bar{A} \in \mathscr{F}$ and (b) $A_i \in \mathscr{F}$, $1 \leqslant i < \infty$, implies that $\bigcup_{i=1}^{\infty} A_i \in \mathscr{F}$]. Then as axioms for the probability measure P we have Axioms **A1** and **A3** as before, while

A2 is replaced by:

A2′ If $A_i \in \mathscr{F}$, $1 \leqslant i < \infty$, and $A_i \cap A_j = \varnothing$, $i \neq j$, then

$$P\left(\bigcup_{i=1}^{\infty} A_i\right) = \sum_{i=1}^{\infty} P(A_i).$$

The triplet $(\Omega, \mathscr{F}, P)$ is called a *probability space*. This framework is due to Kolmogorov [14] in 1933 and is very widely accepted, although some authors reject the countable additivity axiom **A2′** and rely on the finite additivity version (**A2**) (e.g., Dubins and Savage [8]). An extension of the Kolmogorov theory in which unbounded measures are allowed and conditional probabilities are taken as the fundamental concept has been provided by Rényi [18, Chap. 2]. For $A, B \in \mathscr{F}$ with $P(B) > 0$, the conditional probability of A given B is defined as $P(A \mid B) = P(A \cap B)/P(B)$.

Suppose that the sample space Ω has points ω. A real-valued function $X(\omega)$ defined on the probability space $(\Omega, \mathscr{F}, P)$ is called a random variable [or, in the language of measure theory, $X(\omega)$ is a *measurable function*] if the set $\{\omega : X(\omega) \leqslant x\}$ belongs to $\mathscr{F}$ for every $x \in \mathbb{R}$ (the real line). Sets of the form $\{\omega : X(\omega) \leqslant x\}$ are usually written as $\{X \leqslant x\}$ for convenience and we shall follow this practice. Random variables and their relationships are the principal objects of study in much of probability and statistics.

IMPORTANT CONCEPTS

Each random variable X has an associated *distribution function* defined for each real x by

$$F(x) = P(X \leqslant x).$$

The function F is nondecreasing, right-continuous, and has at most a countable set of points of discontinuity. It also uniquely specifies the probability measure which X induces on the real line.

Random variables are categorized as having a discrete distribution if the corresponding distribution function is a step function and as having a continuous distribution if it is continuous. Most continuous distributions of practical interest are absolutely continuous, meaning that the distribution function $F(x)$ has a representation of the form

$$F(x) = \int_{-\infty}^{x} f(u)\, du$$

for some $f \geqslant 0$; f is called the *probability density function.*

A relatively small number of families of distributions are widely used in applications of probability theory. Among the most important are the binomial*, which is discrete and for which

$$P(X = r) = \tbinom{n}{r} p^r (1-p)^{n-r},$$

$$0 \leqslant r \leqslant n, \quad 0 < p < 1,$$

and the (unit) normal*, which is absolutely continuous and for which

$$P(X \leqslant x) = (2\pi)^{-1/2} \int_{-\infty}^{x} e^{-(1/2)u^2}\, du,$$

$$-\infty < x < \infty.$$

The binomial random variable represents the numbers of successful outcomes in n trials, repeated under the same conditions, where the probability of success in each individual trial is p. Normal distributions, or close approximations thereto, occur widely in practice, for example in such measurements as heights or weights or examination scores, where a large number of individuals are involved.

Two other particularly important distributions are the Poisson*, for which

$$P(X = r) = e^{-\lambda} \lambda^r / r!,$$

$$r = 0, 1, 2, \ldots, \quad \lambda > 0,$$

and the exponential*, for which

$$P(X \leqslant x) = \lambda \int_0^x e^{-\lambda u}\, du, \qquad x > 0, \quad \lambda > 0;$$

$$= 0 \text{ otherwise}.$$

These commonly occur, respectively, in situations of rare events and as waiting times between phenomena.

A sequence of random variables $\{X_1, X_2, \ldots\}$ defined on the same probability space is called a *stochastic process**. For a finite collection of such random variables $\{X_1, X_2, \ldots, X_m\}$ we can define the joint distribution function

$$F(x_1, x_2, \ldots, x_m) = P(X_1 \leqslant x_1, X_2 \leqslant x_2, \ldots, X_m \leqslant x_m)$$

for $x_i \in \mathbb{R}$, $1 \leqslant i \leqslant m$, and the random variables are said to be *independent** if

$$P(X_1 \leqslant x_1, \ldots, X_m \leqslant x_m) = P(X_1 \leqslant x_1) \cdots P(X_m \leqslant x_m)$$

for all $x_1, \ldots, x_m$. Independence is often an essential basic property and much of probability and statistics is concerned with random sampling, namely independent observations (random variables) with identical distributions.

One step beyond independence is the Markov chain $\{X_1, X_2, \ldots\}$, in which case

$$P(X_m \leqslant x_m \mid X_1 = i_1, \ldots, X_n = i_n) = P(X_m \leqslant x_m \mid X_n = i_n)$$

for any $n < m$ and all $i_1, \ldots, i_n$ and x_m. That is, dependence on a past history involves only the most recent of the observations. Such random sequences with a natural time scale belong to the domain of stochastic processes; *see* MARKOV PROCESSES.

Considerable summary information about random variables and their distributions is contained in quantities called moments. For a random variable X with distribution function F, the rth *moment* ($r > 0$) is defined by

$$EX^r = \int_{-\infty}^{\infty} x^r \, dF(x)$$

provided that $\int_{-\infty}^{\infty} |x|^r \, dF(x) < \infty$. $E|X|^r$ is called the rth *absolute moment*. The operator E is called *expectation* and the first moment ($r = 1$) is termed the *mean*. Moments about the mean $E(X - EX)^r$, $r = 2, 3, \ldots$ are widely used and $E(X - EX)^2$ is termed the *variance** of X, while $(E(X - EX)^2)^{1/2}$ is its *standard deviation**. The mean and variance of X are, respectively, measures of the location and spread of its distribution.

CONVERGENCE

Much of probability theory is concerned with the limit behavior of sequences of random variables or their distributions. The great diversity of possible probabilistic behavior when the sample size is small is often replaced by unique and well-defined probabilistic behavior for large sample sizes. In this lies the great virtue of the limit theory: statistical regularity emerging out of statistical chaos as the sample size increases.

A sequence of random variables $\{X_n\}$ is said to converge *in distribution* (or *in law*) to that of a random variable X (written $X_n \xrightarrow{d} X$) if the sequence of distribution functions $\{F_n(x) = P(X_n \leqslant x)\}$ converges to the distribution function $F(x) = P(X \leqslant x)$ at all points x of continuity of F.

Convergence in distribution does not involve the random variables explicitly, only their distributions. Other commonly used modes of convergence are:

(a) $\{X_n\}$ converges *in probability* to X (written $X_n \xrightarrow{p} X$) if, for every $\epsilon > 0$,

$$\lim_{n \to \infty} P(|X_n - X| > \epsilon) = 0.$$

(b) For $r > 0$, $\{X_n\}$ converges *in the mean of order r* to X (written $X_n \xrightarrow{r} X$) if

$$\lim_{n \to \infty} E|X_n - X|^r = 0.$$

(c) $\{X_n\}$ converges *almost surely* to X (written $X_n \xrightarrow{\text{a.s.}} X$) if for every $\epsilon > 0$,

$$\lim_{n \to \infty} P\Big(\sup_{k \geqslant n} |X_k - X| > \epsilon\Big) = 0.$$

See also CONVERGENCE OF SEQUENCES OF RANDOM VARIABLES. It is not difficult to show that (b) implies (a), which in turn implies convergence in distribution while (c) implies (a). None of the converse implications hold in general.

Convergence in distribution is frequently established by the method of characteristic functions*. The *characteristic function* of a random variable X with distribution function F is defined by

$$f(t) = Ee^{itX} = \int_{-\infty}^{\infty} e^{itx}\, dF(x),$$

where $i = \sqrt{-1}$. If $f_n(t)$ and $f(t)$ are, respectively, the characteristic functions of X_n and X, then $X_n \xrightarrow{d} X$ is equivalent to $f_n(t) \to f(t)$. Characteristic functions are particularly convenient for dealing with sums of independent random variables as a consequence of the multiplicative property

$$Ee^{it(X+Y)} = Ee^{itX}Ee^{itY}$$

for independent X and Y. They have been extensively used in most treatments of convergence in distribution for sums of independent random variables (the classical central limit problem). The basic central limit theorem* (of which there are many generalizations) deals with independent and identically distributed random variables X_i with mean μ and variance σ^2. It gives, upon writing $S_n = \sum_{i=1}^{n} X_i$, that

$$\lim_{n\to\infty} P\left(\sigma^{-1}n^{-1/2}(S_n - n\mu) \leqslant x\right)$$
$$= (2\pi)^{-1/2}\int_{-\infty}^{x} e^{-(1/2)u^2}\, du, \quad (1)$$

the limit being the distribution function of the unit normal law.

Convergence in probability results are often obtained with the aid of Markov's inequality*

$$P(|X_n| > c) \leqslant c^{-r}E|X_n|^r$$

for $r > 0$ and a corresponding convergence in rth mean result. The useful particular case $r = 2$ of Markov's inequality is ordinarily called Chebyshev's inequality*. For example, in the notation of (1),

$$P\left(|n^{-1}S_n - \mu| > \epsilon\right)$$
$$\leqslant \epsilon^{-2}n^{-2}E(S_n - n\mu)^2 \to 0 \quad (2)$$

as $n \to \infty$, which is a version of the weak law of large numbers*, $n^{-1}S_n \xrightarrow{p} \mu$.

Almost sure convergence results rely heavily on the Borel–Cantelli lemmas*. These deal with a sequence of events (sets) $\{E_n\}$ and $E = \bigcap_{r=1}^{\infty}\bigcup_{n=r}^{\infty} E_n$, the set of elements common to infinitely many of the E_n. Then

(a) $\sum P(E_n) < \infty$ implies that $P(E) = 0$.

(b) $\sum P(E_n) = \infty$ and the events E_n are independent implies that $P(E) = 1$.

With the aid of **(a)**, (2) can be strengthened to the strong law of large numbers*,

$$n^{-1}S_n \xrightarrow{\text{a.s.}} \mu \quad (3)$$

as $n \to \infty$.

The strong law of large numbers embodies the idea of a probability as a strong limit of relative frequencies; if we set $X_i = I_i(A)$, the indicator function of the set A at the ith trial, then $\mu = P(A)$. This property is an important requirement of a physically realistic theory.

The law of large numbers, together with the central limit theorem, provides a basis for a fundamental piece of statistical theory. The former yields an estimator of the location parameter μ and the latter enables approximate confidence intervals* for μ to be prescribed and tests of hypotheses* constructed.

OTHER APPROACHES

The discussion above is based on a frequency interpretation of probability, and this provides the richest theory. There are, however, significant limitations to a frequency interpretation in describing uncertainty in some contexts and various other approaches have been suggested. Examples are the development of subjective probability* by de Finetti and Savage (e.g., refs. 7 and 19) and the so-called logical probability of Carnap [3]. The former is the concept of probability used in Bayesian inference* and the latter is concerned with objective assessment of the degree to which evidence supports a hypothesis (*see* BAYESIAN INFERENCE and FOUNDATIONS OF PROBABILITY). A detailed comparative discussion of the principal theories that have been proposed is given in Fine [11].

The emergence of probability theory as a scientific discipline dates from around 1650; a comprehensive sketch of its history is given in PROBABILITY, HISTORY OF (OUTLINE).

THE LITERATURE

Systematic accounts of probability theory have been provided in book form by many authors. These are necessarily at two distinct levels since a complete treatment of the subject has a measure theory prerequisite. Elementary discussions, not requiring measure theory, are given for example in Chung [5], Feller [9], Gnedenko [12], and Parzen [16]. A full treatment can be found in Billingsley [1], Breiman [2], Chow and Teicher [4], Chung [6], Feller [10], Hennequin and Tortrat [13], Loéve [15], and Rényi [17].

References

[1] Billingsley, P. (1979). *Probability and Measure*. Wiley, New York.

[2] Breiman, L. (1968). *Probability*. Addison-Wesley, Reading, Mass.

[3] Carnap, R. (1962). *The Logical Foundations of Probability*, 2nd ed. University of Chicago Press, Chicago.

[4] Chow, Y. S. and Teicher, H. (1978). *Probability Theory. Independence, Interchangeability, Martingales*. Springer-Verlag, New York.

[5] Chung, K. L. (1974). *Elementary Probability Theory with Stochastic Processes*. Springer-Verlag, Berlin.

[6] Chung, K. L. (1974). *A Course in Probability Theory*, 2nd ed. Academic Press, New York.

[7] de Finetti, B. (1974). *Theory of Probability*, Vols. 1 and 2. Wiley, New York.

[8] Dubins, L. E. and Savage, L. J. (1965). *How to Gamble If You Must*. McGraw-Hill, New York.

[9] Feller, W. (1968). *An Introduction to Probability Theory and Its Applications*, Vol. 1, 3rd ed. Wiley, New York.

[10] Feller, W. (1971). *An Introduction to Probability Theory and Its Applications*, Vol. 2, 2nd ed. Wiley, New York.

[11] Fine, T. L. (1973). *Theories of Probability. An Examination of the Foundations*. Academic Press, New York.

[12] Gnedenko, B. V. (1967). *The Theory of Probability*, 4th ed. (Translated from the Russian by B. D. Seckler.) Chelsea, New York.

[13] Hennequin, P. L. and Tortrat, A. (1965). *Théorie des Probabilités et Quelques Applications*. Masson, Paris.

[14] Kolmogorov, A. (1933). *Grundbegriffe der Wahrscheinlichkeitsrechnung*. Springer-Verlag, Berlin. (English trans.: N. Morrison, in *Foundations of the Theory of Probability*. Chelsea, New York, 1956.)

[15] Loève, M. (1977). *Probability Theory*, Vols. 1 and 2, 4th ed. Springer-Verlag, Berlin.

[16] Parzen, E. (1960). *Modern Probability Theory and Its Applications*. Wiley, New York.

[17] Rényi, A. (1970). *Probability Theory*. North-Holland, Amsterdam. (Earlier versions of this book appeared as *Wahrscheinlichkeitsrechnung*, VEB Deutscher Verlag, Berlin, 1962; and *Calcul des probabilités*, Dunod, Paris, 1966.)

[18] Rényi, A. (1970). *Foundations of Probability*. Holden-Day, San Francisco.

[19] Savage, L. J. (1962). *The Foundations of Statistical Inference: A Discussion*. Wiley, New York.

(AXIOMS OF PROBABILITY
BAYESIAN INFERENCE
CHANCE II
CONVERGENCE OF SEQUENCES OF RANDOM VARIABLES
FOUNDATIONS OF PROBABILITY
FREQUENCY INTERPRETATION IN PROBABILITY AND STATISTICS
LAWS OF LARGE NUMBERS
LIMIT THEOREM, CENTRAL
NONADDITIVE PROBABILITIES
PROBABILITY, HISTORY OF (OUTLINE)
SUBJECTIVE PROBABILITIES)

C. C. HEYDE

PROBABLE ERROR

Francis Galton* expressed the values in a normal distribution* as a function of probable error [4], which he described as "the value that one-half of the Errors exceed and the other half fall short of." It is an archaic term for the semi-interquartile range (*see* QUARTILE DEVIATION), for normal variables equal to $0.67449 \times$ (standard deviation*). Probable error as a measure of spread was usually applied to normal variables [6], the probability of obtaining a value within one probable error of the population mean being 0.50, but it was also used for nonnormal data [1].

The term appeared in the early nineteenth century among German astronomers. It was apparently first used explicitly by F. W. Bessel [2] in 1818, was developed subsequently by Gauss*, and was used extensively by Adolphe Quetelet* [5]. Its origin, however, can be traced [3] to Abraham De Moivre's *The Doctrine of Chances* (1756).

Galton [4, pp. 57–58] objected to the name "probable error": "It is astonishing that mathematicians, who are the most precise and perspicacious of men, have not long since revolted against this cumbrous, slipshod, and misleading phrase." He then introduced the term "probable deviation" instead, but it was not adopted by others. Although probable error is still used as a measure occasionally in the physical sciences, it has been replaced in modern statistics by the standard deviation. The term "error" reflects the earlier convention of describing the normal law as the error distribution (*see* LAWS OF ERROR).

References

[1] Aitken, A. C. (1949). *Statistical Mathematics*, 6th ed. Oliver & Boyd, Edinburgh. (See Chapter II.)

[3] Bessel, F. W. (1818). *Über der Ort des Polarsterns*. Berliner Astronomische Jahrbuch für 1818. Berlin.

[3] De Moivre, A. (1756). *The Doctrine of Chances*, 3rd (final) ed. A. Millar, London. (Reprinted by Chelsea, New York, 1967.

[4] Galton, F. (1889). *Natural Inheritance*. Macmillan, London, pp. 57–58.

[5] Quetelet, L. A. (1849). *Letters on the Theory of Probabilities*, O. G. Downes, trans. Layton, London.

[6] Yule, G. U. and Kendall, M. G. (1937). *An Introduction to the Theory of Statistics*, 11th ed. Charles Griffin, London. (See Section 19.10.)

(INTERQUARTILE RANGE
NORMAL DISTRIBUTION
QUARTILE DEVIATION
STANDARD DEVIATION)

PROBABLE ITEM

An obsolete term for mode (*see* MEAN, MEDIAN, MODE, AND SKEWNESS).

PROBIT *See* QUANTAL RESPONSE ANALYSIS

PROBLEM OF POINTS

The "problem of points" is the name given to a simple gaming problem that came into prominence in the Renaissance and was not correctly solved until the seventeenth-century French mathematicians Fermat* and Pascal* considered it. Their famous correspondence about the problem, and Pascal's treatment of it in his *Traité du Triangle Arithmétique*, are important milestones in the theory of probability.

The problem, also known as the "division problem," involves determining how the total stake should be equitably divided when a game is terminated prematurely. Suppose that two players A and B stake equal money on being the first to win n points in a game in which the winner of each point is decided by the toss of a fair coin, heads for A and tails for B. If such a game is interrupted when A still lacks a points and B lacks b, how should the total stake be divided between them?

Particular cases of the problem have been noted in Italian mathematical manuscripts as early as 1380, and the Renaissance mathematicians Pacioli, Tartaglia, and Peverone all made unsuccessful efforts at solutions. It was brought to Blaise Pascal's attention by Antoine Gombaud, Chevalier de Méré, and led to a correspondence with Pierre de Fermat in the late summer of 1654. Unfortunately, not all the letters have survived, but it is possible to reconstruct the sequence of events from the remainder.

Pascal sought Fermat's opinion, and Fermat replied in a letter which, although missing, like Pascal's, must have contained solutions to two gaming problems, one of which was the problem of points, for in his reply on July 29, Pascal expressed his agreement with Fermat's solution to the latter, which had also occurred to him. If A wants a points and B wants b, they argued, then the game must be over in $(a + b - 1)$ further tosses at most, which may occur in 2^{a+b-1} equiprobable ways. Each of these ways may be seen, on examination, to be a win either for A or for B, and on counting them the probability (as we should now say) of A winning is found, and hence the equitable division of the stakes given that the game is not in fact continued. In other words, the actual game has been embedded in a hypothetical game of fixed length which is easy to solve.

In the same letter Pascal offered as an alternative solution an analysis of the possible games in which, working backwards, he uses recursively the idea that if expectations of gain of X and Y units are equally probable, the expectation of gain is $\frac{1}{2}(X + Y)$ units, thereby introducing into probability theory the notion of expectation. Then, having appreciated that with only two players the order of occurrence of heads and tails in the hypothetical game does not matter, he makes use of the fact that the hypothetical game is solved by the partial sum of a binomial distribution* for equal chances. In his contemporary *Traité du Triangle Arithmétique* he proves this result by applying mathematical induction to the method of expectations. The result itself is often attributed to Fermat, and the method of expectations to Huygens*. Not only are they Pascal's but he has also introduced the binomial distribution for equal chances, albeit by reference to the arithmetical triangle rather than algebraically.

The remainder of the Fermat–Pascal correspondence is devoted mostly to the case of three players, which Fermat had suggested could readily be solved by the original "combinatorial" method. Pascal thought that by this Fermat meant assigning each of the possible hypothetical games of fixed length to the appropriate player according to the numbers of points each had won without regard to order, as was correct with two players, and he demurred. Fermat reassured him that he had intended, rather, that each possible *sequence* should be assigned, which leads to the correct solution, and he added for good measure the direct probability solution to an example with three players, in which he simply analyses the possible games in a forward direction (as a modern student would do with the help of a tree diagram) in contrast to Pascal's backward "expectation" solution.

Apart from some minor generalizations introduced by Montmort*, the Bernoulli* brothers, and De Moivre* half a century later, the correspondence between Fermat and Pascal and the account in the latter's *Traité du Triangle Arithmétique* essentially disposed of the problem of points and in doing so created an incipient theory of probability where previously there had only been simple enumerations. The important methodological contributions of operating on expectations and of constructing the binomial distribution for equal chances and applying it to the problem of points were Pascal's alone, however.

Bibliography

The account above relies on the primary sources directly, as follows. For the Fermat–Pascal correspondence, see either *Oeuvres de Fermat*, edited by P. Tannery and C. Henry, Vol. II, *Correspondance* (Gauthier-Villars, Paris, 1894), or B. Pascal, *Oeuvres Complètes*, edited by J. Mesnard, Volume II, (Desclée de Brouwer, Paris, 1970). English translations (which should both be treated with reserve) are available in *A Source Book in Mathematics*, by D. E. Smith (McGraw-Hill, New York, 1929), and *Gods, Games and Gambling*, by F. N. David (Charles Griffin, London, 1962). Pascal's *Traité du Triangle Arithmétique* was originally published by Desprez in Paris in 1665 and is contained in his *Oeuvres* cited above. There is a partial English translation in D. E. Smith (see above) and an even lesser part in *A Source Book in Mathematics*, by D. J. Struik (Harvard University Press, Cambridge, Mass., 1969), but neither includes the treatment of the problem of points. There are many secondary accounts, but we mention only the invaluable *A History of the Mathematical Theory of Probability*, by I. Todhunter (Macmillan, Cambridge, 1865; reprinted by Chelsea, New York, 1965).

(FERMAT, PIERRE DE
GAMES OF CHANCE
PASCAL, BLAISE
PROBABILITY, HISTORY OF (OUTLINE))

A. W. F. EDWARDS

PROBLEM SOLVING IN STATISTICS

A basic approach to problem solving in statistics is to compare sets of numbers: to compare observations with theoretical models, past experience, or other observations in numerical form [9]. Another approach is to make estimations based on models, experience, or observations. Such comparisons and estimations are valid and informative only if the observations and models are appropriately (and measurably) representative of some phenomenon or population. Therefore, a working knowledge of statistics is important in the planning stages of investigations as well as in determining what statistical comparisons and models are appropriate. A common complaint of practicing statisticians is that often statistical consultants are not called in until after the data have been collected [25]. One of the purposes of this article is to stress the full range of the statistical process in order to correct a narrow public image of the discipline.

To provide a systematic framework for a

Table 1 Phases and Steps of Statistical Problem Solving

Phase	Step	Activity
1. Study design	a. Question definition	The question is defined in terms of the need for information.
	b. Alternative strategies development	Alternative strategies are developed and specified for sampling, data collection, and analysis.
	c. Strategy evaluation	The advantages and disadvantages of the feasible alternatives are evaluated.
	d. Strategy selection	A strategy is selected on the basis of costs and the importance of the information to the organization.
2. Data collection	a. Sample design	Sampling procedures are planned on the basis of work done in step 1b and the selection made in step 1d.
	b. Measurement	Observations are chosen and recorded in a form that facilitates analysis
3. Data analysis	a. Statistical analysis	Statistical methods are used for estimating or summarizing.
	b. Reliability assessment	Measures of possible error in results are calculated.
	c. Report generation	The results are reported in a form useful to others.

discussion of problem solving in statistics, a conceptual model has been adapted from information systems literature, including three distinct phases: (1) study design, (2) data collection*, and (3) data analysis. These phases, outlined in Table 1, have been presented as an alternative to an experimental model in a business context [16]; the process model has been revised here to encompass the general procedures required in any application area.

PHASE 1: STUDY DESIGN

The first phase involves determining the optimum strategy to solve the problem. The steps from the beginning of the process until final selection of a strategy are illustrated in Fig. 1. These are important steps since inadequate planning means no data, or worse, meaningless data, to analyze later. The experimenter should consider all the possibilities of misfortune as well as unexpected interactions of the variables, because results are seldom exactly what is anticipated. If a statistical consultant is involved, he or she should take an active part in planning the methodology of the study itself and not wait to help with the analysis of data that may or may not be meaningful [14, 15].

Problem Definition

In the first step of the study design phase the question or problem must be defined in terms of the need for information. Tukey [28] cautions that "finding the question is often more important than finding the answer." This question-finding step can be a very difficult one with none of the tidiness of a well-defined mathematical problem [15, 23]; however, it is crucial to the meaningfulness of the subsequent phases and steps. If the scientist, economist, demographer, or whoever who wants the problem solved has the required statistical skills, the first phase is mainly one of defining the question clearly in order to proceed with an efficient design. "What is the real question I am interested in, and how can it logically be phrased so that it is answerable with data

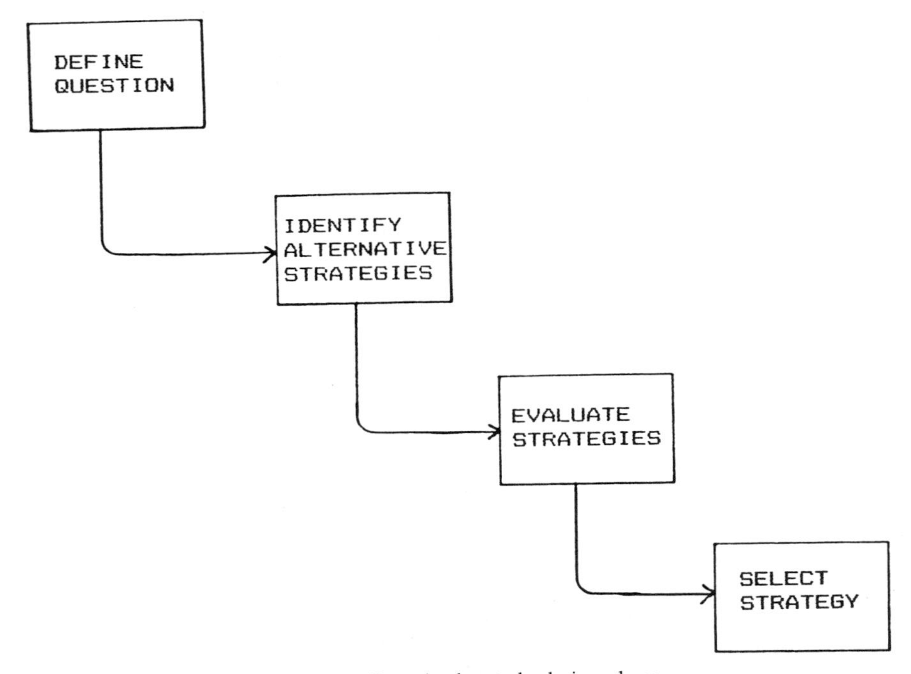

Figure 1 Steps in the study design phase.

that can be obtained with available resources?"

If a statistician is assisting one or more experts in a consulting* situation, the statistician should aid in the problem definition process by discussing the problem and determining the exact question of interest. Consultants have reported that their clients have found this initial problem definition assistance to be the most valuable service provided by the statistician [3, 7, 14].

Several articles [3, 14, 21] specifically focus on the help that statisticians can provide in developing the objectives of a study. Salsburg [23] describes his role in this step as one of asking stupid questions and worrying out loud about everything that could go wrong in the data collection stage. *Statistics: A Guide to the Unknown* [27] provides reasonable examples of statistical designs and analysis to answer such ill-defined, difficult, and interesting questions as: Are police effective? Can people control when they die? In these examples and other real problems, the definition of the logical issues and the experimental design are the main challenge, rather than the choice and execution of analysis techniques.

To be effective in the planning stages of an investigation, the statistician must know enough of the background subject (or ask enough questions) to be sure that the study fits the real problem [13, 24]. Working with a client to determine the logical issues of a problem also may take special interpersonal skills; suggestions on this topic can be found in a discussion of the use of transaction analysis by statistical consultants [3], an identification of consulting roles [14, 25], and a reading list on the training of consultants [2].

Identification of Alternatives

After the problem is well defined, the statistician develops alternative procedures for the experiment, sample survey, or other procedure appropriate for data collection. The value of the information gained in the final phase depends largely on the thoroughness of the data collection and data analysis phases, so the statistician must plan ahead. Failing to allow for dropped test tubes or uncooperative subjects may mean that there will be no usable data for the analysis stage.

Evaluation of Alternatives

For each alternative procedure the statistician should present a cost estimate and an error range indicating the reliability of the answer. The optimum strategy is not necessarily the one that provides the most information. Within any organization cost factors must be weighed against benefits, so the "best" strategy is one that provides adequate information with a justifiable cost and a reasonable amount of time. (A statistician who does not appreciate this fact is unlikely to enjoy a successful career in industry.)

Each feasible alternative must be evaluated. In the final step of phase 1 a strategy is selected on the basis of cost and the importance of the information to the organization. If the statistician has done a good job, the chosen strategy will provide a specified quantity of information at a minimum cost. At the end of phase 1 the project is approved with an understanding of how much time the study will take, how much it will cost, and the quality and type of inferences that can be made from the results. A written statement of the objectives of the study and the time and budget constraints should be prepared by the statistician before the study proceeds to the data collection stage. *See* CONSULTING, STATISTICAL; INDUSTRY, STATISTICS IN; PRINCIPLES OF PROFESSIONAL STATISTICAL PRACTICE.

PHASE 2: DATA COLLECTION

The data collection phase includes two steps: design of the sampling procedure and measurement of observations or collection of data. The time requirements and difficulty level of these steps depend on the problem. The first step consists of planning what will be done in the second step. Time-consuming, complicated observations and measurements will usually require time-consuming, detailed planning in the first step.

Sample Design

The total group to be studied was determined in the problem definition step of the process. The sample may be birds in a field [13], treated water [5], filtered air [21], or earthworms [26], to name a few examples from the literature that discuss the entire problem-solving process. In a quality control* study the target population may be all the machine parts delivered by a particular supplier. Testing millions of machine parts supposedly checked by the original manufacturer would be infeasible for the manufacturer who wishes to use these parts in assembling a product; instead, sampling is used. In a study of the environmental impact of a factory process the population or universe under investigation may be treated water in a lake, filtered air in a plant facility, or shrimp in a bay. Again it would be infeasible to measure impurities in all the air and water or to check every shrimp for contamination. The sampling procedures for the machine parts, air, water, and shrimp are based on probability theory*. The planning for the sample is done in the first step of this data collection phase.

Measurement

The complexity of the planning step varies considerably with the problem. For projects that require complicated data collection procedures, the second step of this phase requires careful planning. The design to select the air, water, and shrimp samples takes considerable time, as to the actual measurement and observation procedures in the second step of this phase. Accountants rely on sampling to audit large organizations. Courts depend on sampling to determine compensation in law suits involving thousands of plaintiffs. Market research surveys always involve sampling. An effective sampling design can ensure a valid audit, appropriate compensation, and invaluable market forecasts, without sacrificing economy (*see* MARKETING, STATISTICS IN).

In the easiest case the sample includes all company computer records pertinent to the question and available at a particular time. In this situation the planning for data collection requires the choice of which file of information is appropriate for the question of interest; this might be the choice of annual, monthly, or end-of-the-day data on a particular date. The next step, the actual data collection, is completed by one command in a computer job that accesses the chosen data file. For example, airlines regularly analyze route schedules and passenger loads. The data source is the company file that is created and updated in the reservation process that goes on every hour of the day in airports and travel agencies around the world. Making this file of passenger information available with one computer command is all that is necessary before proceeding to the analysis phase.

In the other extreme the Bureau of the Census takes years to plan and carry out the data collection for the U.S. census and to make available the results. Demographic studies are used to choose the sample of individuals asked to complete more detailed census forms. The actual collection of data and recording of survey answers involves thousands of full-time and part-time employees of the Bureau of the Census*.

Issues in Data Collection

The role of the statistician in developing sound methodology is stressed in the study design phase. The logical issues of problem solving in the definition of the research question are critical to the value of any study, even though they are not basically a mathematical concern. In the data collection* phase, the role of the statistician as a scientific advisor is just as important. The difficulties of designing the sample and problems with the actual measurement are considerable.

Insofar as statistics is an applied science, justified by the "help it can give in solving a problem . . . [reaching] a decision, on a probabilistic basis" [18], the problem-solving methodology must be repeatable and reproducible [1]. Such repeatability and reproducibility are dependent on the quality of statistical data. Even the most sophisticated multivariate analysis* cannot correct for data that are inconsistent or clearly wrong. Kruskal [17] discusses fascinating problems arising in the public policy application of statistics which are often neglected since they fall in the cracks between disciplinary labels: (1) wrong data (e.g., 1960 census data showing 62 women aged 15 through 19 with 12 or more children) identified by lack of smoothness, logical inconsistencies, or transgressions of general knowledge; (2) ambiguity of classification (when might an unemployed person be said to be in the labor force?); (3) lack of data responsibility (unwillingness to publish the raw data); which relates to (4) confidentiality (can we learn information about specific individuals from a published table of cross classifications?).

Traditionally, the "quality of statistical data" has referred to its accuracy [1], which is often judged by insisting on high precision and low bias. In judging this quality, we must realize that smoothness of repeated observations does not guarantee that the estimate is close to the "truth." When we report "estimate ± something," we imply an emphasis on precision, which does not necessarily imply accuracy. (See Eisenhart [11] for a comprehensive discussion of this distinction.] Therefore, a complete and clear disclosure of the measurement process is necessary for those who will evaluate the results. (*See* CENSUS; MEASUREMENT ERRORS; SAMPLING PLANS; SURVEY SAMPLING; TELEPHONE SURVEYS).

PHASE 3: DATA ANALYSIS

The data analysis phase includes three steps: statistical analysis, reliability assessment, and report generation. The sequence of steps is shown in Fig. 2. The methods employed in this phase are the main concern of statistics.

Statistical Analysis

Articles on specific statistical techniques are a major portion of this encyclopedia, so no attempt at a complete list will be made here. Summary lists of the most frequently used techniques have been compiled for industrial statisticians [22], consultants [19, 20], and statisticians in the federal government [12]. Basic analysis strategies include hypothesis testing* (the approach used in experimen-

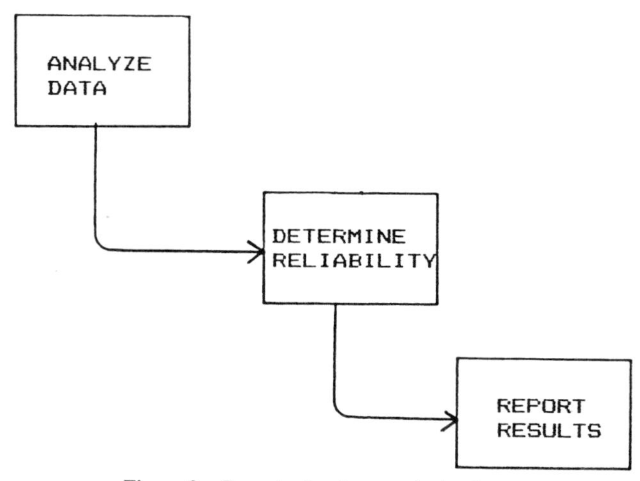

Figure 2 Steps in the data analysis phase.

tation), estimation* techniques, and exploratory data analysis* (for a general understanding of relationships that exist in data). Summaries of statistical applications in various fields are also provided in this encyclopedia; for example, *see* AGRICULTURE; BIOASSAY; BIOSTATISTICS; CHEMISTRY, and so on.

Impact of Computers on Analysis

The availability and accessibility of computers has had a pervasive impact on statistics as complicated statistical procedures come into routine use in analysis of large data sets (*see* COMPUTERS AND STATISTICS). Multivariate techniques of stepwise regression*, discriminant analysis*, and categorical techniques are now feasible as exploratory tools due to the extensive calculating ability of computers. Statistical packages have made data analysis (and misanalysis) widely accessible with multidimensional computer graphics and quick results of data transformation.

In addition to extending the accessibility of established statistical methodology, computers have played a vital role in the development of new statistical techniques. Diaconis and Efron [8] observe that the most commonly used statistical methods were developed when computation was slow and expensive; they point out that many new methods are free of distributional assumptions and are feasible, as a result of the use of computers. Monte Carlo* simulation techniques are easily employed to examine the performance of a new estimator or test. Efron [9] asserts that "computer assisted theory is no less 'mathematical' than the theory of the past; it is just less constrained by the limitations of the human brain."

The empirical Bayes* procedure is an example of a technique feasible only with the advent of computer technology. Empirical Bayes' inference differs from Bayesian inference* in that a prior distribution* is assumed but not known, thus allowing the applied statistician to derive from the data a compromise between classical frequentist and Bayesian estimators. This procedure has been shown to provide accurate estimates in practical multivariate applications and is an example of a procedure whose performance has been assessed without formal proof, but with the aid of a computer.

The computer makes feasible nonparametric estimators involving massive amounts of computation. Consider, for example, the simple problem of determining whether two samples have been drawn from a single population. A frequentist addresses this question with a t-test, which uses only a few calculations but assumes the (common) population is Gaussian. Computers make possible the randomization* approach of deriving all possible ways of partitioning the combined data into two like-sized samples and assessing the likelihood of the original partition among all possible samples. Additionally, computers make possible the use of the jackknife* and bootstrap* procedures to measure accuracy of estimators of population parameters without the Gaussian assumption (*see* SAMPLE REUSE).

Cross-validation, combinations of robustness and efficiency, and analysis of censored data* additionally illustrate the impact of computers on previously impossible data analytic problems. Thus an aid to statistical analysis has now begun to shape the statistical methods themselves.

Reliability Assessment

The second step of the analysis phase is to determine how much error can be expected in the estimations or inferences made from a statistical procedure. Each technique presented in this encyclopedia includes a measure of reliability to indicate how much confidence can be placed in a decision based on the statistical results.

In the final step of the analysis phase, the results must be reported in a form that is intelligible to the user. Stating the results of a statistical investigation in terms of the original objectives of the study is analogous to turning the numerical results of an arithmetic word problem back into words [13]. When complicated statistical procedures have been used, this is a very important step; when controversial issues are involved, it is a very difficult step [22].

Ehrenberg [10] provides tips for effective statistical report writing. Hunter [15] and Salsburg [23] provide motivation. *See also* GRAPHICAL REPRESENTATION OF DATA.

POSTPROCESS RESPONSIBILITIES

At the conclusion of any worthwhile study a decision is made based on the results. The decision may be to market a new drug that has been shown to have no adverse effects, reallocate federal funds to states due to shifts in population, continue with present policies of proven worth, or gather more information for further study. This postprocess decision is usually out of the hands of the statistician. Technically, it should not be the responsibility of the statistician to determine what the implications of a 0.05 significance level are in each of the examples just named or even to set the significance level. But, perhaps, professional responsibilities require some continued interest in the actions taken based on statistical results. Kruskal [17] outlines several problems that affect the value of results of statistical studies for which no one will take responsibility—statistician or client. The ethical guidelines proposed in 1983 have also generated some questions about the responsibility of the statistician as opposed to the responsibility of the organization that acts on the study results. For example, should a statistician allow his or her work to be represented as investigating "cancer cure rates" when success was measured as a cancer victim's failure to die [6]?

CONCLUSIONS

Statistics is a broad problem-solving process used for organizing and thinking about logical issues as well as modeling data to arrive at inferences about the true state of nature. As educators, statisticians should be concerned about teaching statistical problem-solving skills as well as methods. Managers, engineers, and anyone else who requires the services of a statistician should at least learn that the proper time to hire is at the beginning of a statistical study, not after the data are gathered. No amount of probability theory or formula solving with clean data will communicate this knowledge effectively. It is necessary to convey a sense of the importance of structuring real problems so that they can be solved, as well as to teach statistical theory and methods.

References

[1] Bailar, B. A. (1983). *Bull. Int. Statist. Inst.*, 813–834. (A complete discussion of concepts related to the quality of statistical data.)

[2] Baskerville, J. C. (1981). *Amer. Statist.*, **35**, 121–123. (Problem solving is discussed briefly in the context of a course in statistical consulting; bibliography for course included.)

[3] Bishop, T., Petersen, B., and Trayser, D. (1982). *Amer. Statist.*, **36**, 387–390. (Illustrations of types of information that should be obtained during experimental planning stages.)

[4] Boen, J. and Fryd, D. (1978). *Amer. Statist.*, **32**, 58–60. (Interactions between statisticians and clients are explained in terms of transactional analysis; valuable for both parties.)

[5] Box, G. E. P., Hunter, W. G. and Hunter, J. S. (1978). *Statistics for Experimenters*. Wiley, New York. (Chapter 1 is a readable overview of the role of statistics in experimental design; text is oriented toward scientific and engineering problem solving with topics such as "lurking variables" and reducing data models to "white noise.")

[6] Bross, I. D. J. (1983). *Amer. Statist.*, **37**, 12–13. (Examples of statistical results that are misrepresented by policy makers; relevant to issue of boundary of responsibilities of statistician in reporting phase.)

[7] Daniel, C. (1969). *Technometrics*, **11**, 241–245. (Discussion of elements required for successful statistical problem solving in consulting situations.)

[8] Diaconis, P. and Efron, B. (1983). *Science*, Apr., 116–130. (Summary of statistical methods made usable by computers.)

[9] Efron, B. (1979). *SIAM Rev.*, **21**, 460–480. (Very brief summary of standard pre-1950 statistical problem solving, followed by examples of statistical approaches that have changed dramatically due to computers.)

[10] Ehrenberg, A. S. C. (1982). *A Primer in Data Reduction*. Wiley, New York. (Chapters 16 and 17 cover the use of tables and graphs in reports; Chapter 18 includes recommendations on organizations and writing style.)

[11] Eisenhart, C. (1963). *J. Res. Nat. Bur. Stand. C. Eng. Instrum.*, **67**, 161–187. (A discussion of issues relating to precision and accuracy.)

[12] Eldridge, M. D., Wallman, K. K., Wulfsberg, R. M., Bailar, B. A., Bishop, B. A., Bishop, Y. M., Kibler, W. E., Orleans, B. S., Rice, D. P., Schaible, W., Selig, S. M., and Sirken, M. G. (1982). *Amer. Statist.*, **36**, 69–81. (Describes problem solving in a wide variety of positions for statisticians in the federal government.)

[13] Hahn, G. J. (1984). *Technometrics*, **26**, 19–31. (Examples of problem solving in engineering and science with recommendations on how to improve the practice and use of statistics in industry.)

[14] Hunter, W. G. (1981). *Amer. Statist.*, **35**, 72–76. (Discussion of three roles played by statistical consultants; valuable for practicing statisticians and educators in applied statistics programs.)

[15] Hunter, W. G. (1981). *Statistician*, **30**, 107–117. (A very enjoyable series of stories about the reality of applied statistics.)

[16] Jordan, E. W. (1982). *Amer. Statist. Ass., 1982 Proc. Statist. Educ. Sec.* (Examples of problem solving in business with recommendations on how to improve education in statistics for managers.)

[17] Kruskal, W. H. (1981). *J. Amer. Statist. Ass.*, **76**, 1–8. (Discussion of problems, such as data integrity in large data sets, that affect the validity of results of statistical studies, but that are not acknowledged as the responsibility of statisticians.)

[18] Mahalanobis, P. C. (1950). *Sankhyā*, **10**, 195–228. (A delightful essay entitled "Why Statistics?")

[19] Marquardt, D. W. (1979). *Amer. Statist.*, **33**, 102–107. (Very broad view outlined for responsibilities of statisticians in industry.)

[20] Marquardt, D. W. (1981). *Amer. Statist.*, **35**, 216–219. (Argument for total involvement in problem solving by statisticians in industry.)

[21] Price, B. (1982). *Amer. Statist. Ass., 1982 Proc. Phys. Eng. Sci. Sec.* (The need to clarify objectives and specify realistic achievements for any study is discussed within the context of an environmental regulations problem.)

[22] Raiffa, H. (1982). *Amer. Statist.*, **36**, 225–231. (Discussion of the difficulty of maintaining statistical objectivity in controversial application areas of risk analysis for problems such as chemical waste disposal; includes examples that could divest anyone of an image of statistics as a cold science.)

[23] Salsburg, D. S. (1973). *Amer. Statist.*, **27**, 152–154. (Delightful article about how a practicing statistician earns a living.)

[24] Snee, R. D. (1982). *Amer. Statist.*, **36**, 22–87. (Argument for broader view of the field of statistics, including topics that should be added and roles that should be acknowledged.)

[25] Snee, R. D., Boardman, T. J., Hahn, G. J., Hill, W. J., Hocking, R. R., Hunter, W. G., Lawton, W. H., Ott, R. L., and Strawderman, W. E. (1980). *Amer. Statist.*, **34**, 65–75. (Includes list of statistical techniques that are used most often in problem solving by statisticians in industry; discussion of personal characteristics and training necessary for industrial statisticians.)

[26] Sprent, W. G. (1981). *J. R. Statist. Soc. B*, **133**, 139–164. (Difficulties in consulting are discussed; necessity of consultant being involved in projects is stressed by Sprent; wide variety of comments by members follow, including argument about role of statistician in problem solving.)

[27] Tanur, J. M., Mosteller, F., Kruskal, W. H., Link, R. F., Pieters, R. S., Rising, G. R., and Lehmann, E. L., eds. (1978). *Statistics: A Guide to the Unknown*, 2nd ed. Holden-Day, San Francisco. (Series of examples from all fields of applied statistics.)

[28] Tukey, J. W. (1980). *Amer. Statist.*, **34**, 23–25. (An argument for both exploratory and confirmatory methods in data analysis.)

Acknowledgments

We are grateful to G. J. Hahn, W. G. Hunter, and M. M. Whiteside for helpful comments on an early draft of this article. We also thank South-Western Publishing Co. for permission to use copyrighted material, including Figs. 1 and 2, from *Introduction to Business and Economic Statistics* (7th ed.) by C. T. Clark and E. W. Jordan.

(COMPUTERS AND STATISTICS
CONSULTING, STATISTICAL
DATA COLLECTION
DESIGN OF EXPERIMENTS
EXPLORATORY DATA ANALYSIS
HYPOTHESIS TESTING
INDUSTRY, STATISTICS IN
PRINCIPLES OF PROFESSIONAL
STATISTICAL PRACTICE
QUALITY CONTROL, STATISTICS IN
SAMPLING PLANS)

ELEANOR W. JORDAN
DONNA F. STROUP

PROCESSES, DISCRETE

All physical and natural phenomena are subject to environmental and other types of fluctuations or noises. These phenomena are therefore of stochastic nature. A stochastic process* arises when one observes the evolution of such a random phenomenon as a function of time. When the process is observed at discrete time points, one obtains a discrete process. More precisely, a *discrete process* is a sequence X_n, $n \geqslant 1$, of random variables (RVs) defined on an appropriate sample space Ω. A space $\mathscr{X}$ containing all values of the RVs X_n, $n \geqslant 1$, is called the *state space* of the process. In most practical cases $\mathscr{X}$ is a Euclidean space R^m. Consider, for example, the study of a population growth at nonoverlapping generations. Then the population size X_n at generation n is a RV, and X_n, $n \geqslant 1$, is a discrete process. Discrete processes arise in the theories of statistical estimation and optimal filtering. For example, let $\theta(t)$ be a deterministic signal transmitted at time points $t = 1, 2, \dots$, and $\xi(t, \omega)$ be random noises entering the communication channel at these time points. Then the signal received is given by $X(t) = \theta(t) + \xi(t)$. In estimation theory, one needs to determine operations on the signal X that reproduces θ as precisely as possible. We seek an operation that produces a discrete process $\theta(t)$ such that $E[X(t) - \theta(t)] = 0$ and $E[X(t) - \theta(t)]^2$ is minimum. (We will return to this problem after some mathematical preliminaries.)

Let X_t be a discrete process with state space $\mathscr{X}$. Let us first clarify the meaning of the adjective "discrete". We began by assuming that the time parameter set T is discrete (such as $T = 1, 2, \dots$). In defining discrete processes, some authors let t vary continuously while assuming that the state space is discrete. While noting that a countable state space is a particular case of a general state space, we also see that discrete time is not a particular case of continuous time. To be more precise, consider the cases where $T = \{0, 1, 2, \dots\} = N_0$, and $T = R^+$. While R^+ has plenty of limit points, N_0 has ∞ as its only limit point. Consequently, in studying the limit behavior of a discrete-time process one considers the asymptotic behavior at ∞, whereas in the continuous-time process case many new properties such as the local path properties arise. Furthermore, most of the results in the discrete-time case concern the asymptotic behavior at ∞ of functions of the process. Therefore, it is more appropriate to name the discrete-time case as the discrete process; of course, the state space $\mathscr{X}$ can be discrete or continuous.

Let $X = \{X_n(\omega), n \geqslant 1\}$ be a discrete process. It is common to suppress the variable ω in $X_n(\omega)$ and write it only as X_n. If we fix $\omega \in \Omega$, we get a sequence X_n in $\mathscr{X}$, called a *sample path* or *sample sequence*. Thus each sample point in Ω represents a path in the state space $\mathscr{X}$. To keep the presentation simple we shall assume henceforth that $\mathscr{X} = R = R^1$. For every integer $d \geqslant 1$, let us arbitrarily choose time points $n_1 < n_2 < \cdots < n_d$. The joint distribution of the RVs $X_{n_1}, X_{n_2}, \dots, X_{n_d}$ is called a *d-dimensional distribution* of the process $X = \{X_n\}$. The collection of all these finite-dimensional distributions for all possible choices of $n_1 < \cdots < n_d$ and $d \geqslant 1$ governs the probabilistic laws of X. Stochastic processes are classified according to the special properties of their finite-dimensional distributions. We need not distinguish between two processes $X = \{X_n\}$ and $Y = \{Y_n\}$ that have the same finite-dimensional distributions; that is, the multivariate distributions of $X_{n_1}, \dots, X_{n_d}$ and $Y_{n_1}, \dots, Y_{n_d}$ are the same for all $d \geqslant 1$ and all choices of $n_1 < \cdots < n_d \in T$. The researcher is therefore free to choose, from a class of processes having the same finite-dimensional distributions, any process that he or she wishes to work with. In many practical situations the researcher knows the finite-dimensional distributions of a stochastic phenomenon and would like to associate a discrete process with them. Such an association is possible under Kolmogorov's extension theorem [13]. Let $F_{n_1 \dots n_d}(\xi_1, \dots, \xi_d)$ denote the finite-dimensional distributions of the RVs $X_{n_1}, \ldots, X_{n_d}$ of a process X. Then the set of all finite-dimensional distributions satisfies the *consistency condition*

$$\lim_{\xi_k \uparrow \infty} F_{n_1 \ldots n_d}(\xi_1, \ldots, \xi_d) = F_{n_1 \ldots n_{k-1} n_{k+1} \ldots n_d}(\xi_1, \ldots, \xi_{k-1}, \xi_{k+1}, \ldots, \xi_d).$$

Conversely, given a sequence $\{F_n\}$ of n-dimensional distributions satisfying the consistency condition $\lim_{\xi_n \uparrow \infty} F_n(\xi_1, \ldots, \xi_n) = F_{n-1}(\xi_1, \ldots, \xi_{n-1})$, one can find a sample space Ω, an appropriate probability P on Ω, and a process $X = \{X_n\}$ such that $F_n(\xi_1, \ldots, \xi_n) = P\{X_1 \leqslant \xi_1, \ldots, X_n \leqslant \xi_n\}$, $n \geqslant 1$. In place of the finite-dimensional distributions F_n of a process X, one can work with the probabilities $P\{X_1 \in B_1, \ldots, X_n \in B_n\}$, $B_k \in \mathscr{B}$, where $\mathscr{B}$ is the Borel σ-algebra on R. Kolmogorov's extension theorem will transform a process X into a probability measure on $\Omega = R^T$. Treating a discrete process as a probability measure finds nice applications in statistics, for example, in maximum likelihood estimation*. Let $\theta = \theta_t$ be an unknown parameter, and $X(t) = \theta_t + \xi_t$, $t = 0, \pm 1, \ldots,$ where ξ_t is a suitable stationary Gaussian process*. Let μ^θ and μ^0 be the distributions or measures corresponding to the processes X_t and ξ_t, respectively. To obtain the maximum likelihood estimator of θ, one needs to find the Radon–Nikodym* derivative $d\mu^\theta/d\mu^0$, the density function of μ^θ with respect to μ^0 (see Ibragimov and Rozanov [9]).

A discrete process $X = \{X_n\}$ is a *second-order process* if $EX_n^2 < \infty$ for all n. For such a process the functions $\mu(n) = EX_n$ and $\gamma_{ij} = EX_iX_j$ form the *mean sequence* and *correlation matrix* of X, respectively. The covariance matrix of X is defined analogously. Second-order processes can very profitably be studied since, on one hand, the (equivalence class of) second-order RVs form a Hilbert space $L^2(\Omega, \mathscr{T}, P)$ and, on the other hand, for each second-order process with mean sequence $\mu(n)$ ($= 0$, without loss of generality) and covariance matrix $C = [C_{ij}]$, we can find a process $Y = \{y_n\}$, possibly defined on a different probability space, such that Y has the same mean sequence, the same covariance matrix C, and such that the finite-dimensional distributions of Y are all Gaussian (see Ibragimov and Rozanov [9]). Hence, if one has a second-order process to work with, one can assume without much loss of generality that it is a Gaussian process as well. Gaussian processes arise most often in statistical applications. Some of the other useful discrete processes are *processes with uncorrelated RVs*, *processes with orthogonal RVs*, *processes with independent increments*, *Markov chains*, *stationary processes*, *martingales**, and *mixing processes*.

We now present some fairly general examples of discrete processes that arise in statistics and in other applications. Defining a discrete process in a recursive fashion is quite common in statistical estimation theory and learning theory (in psychology).

1. Let us begin with a simple case of T-maze learning in psychology. On each trial a rat is placed at the bottom of a T-maze. The rat makes a choice, at the fork of the T-maze, and proceeds to the end of the left or right arm (response e_0 or e_1). There the rat may or may not find food. Finding food will reinforce the response just made. Consider now a state space $\mathscr{X}$ and a response space $\mathscr{E}$. Corresponding to each response $e \in \mathscr{E}$, there is a contraction or distance diminishing function $f_e: \mathscr{X} \to \mathscr{X}$. If the experiment is in state x_n at trial n, a response e_n is determined and x_{n+1} is given by $x_{n+1} = f_{e_n}(x_n)$. The process x_n, $n \geqslant 1$, is a discrete process roughly describing the learning; for details, see Norman [14] and Cypkin [6].

2. Some statistical estimates are discrete processes defined as functions of other discrete processes. Let θ be a Gaussian signal $N(m, \sigma^2)$. What one observes is $y_k = \theta + \xi_k$, $k \geqslant 1$, where ξ_k, $k \geqslant 1$, are additive noise consisting of independent RVs with a common density $f(x)$. Then the best estimate x_n (in the sense of standard error) is $x_n = E(\theta \mid y_1, \ldots, y_n)$. In principle the estimate x_n can be computed by Bayes' formula. This method is too tedious and does not render itself to practical applications and the study of the solution. For computational purposes one can follow the recursive procedure. For example, $y_{k+1} = \xi_{k+1} + A(y_1, \ldots, y_k)\theta + B(y_1, \ldots, y_k)$. Here A and B are the so-called "controlling" functions. Using such a recursive procedure and appealing to the theorem on normal correlation (stated below) one can obtain the estimates x_n, the tracking errors $e_k = E(\theta - x_k)^2$, and the conditional variance

$$\sigma_k^2 = E\left[(\theta - x_k)^2 | y_1, \ldots, y_k\right].$$

(Further details are given below.)

3. Finally, we give a random walk* model of a collision diffusion process* that arises very commonly in the study of particle transport in dilute gas. In particular, the discrete process we define below approximates the random motion of two elastic particles of identical mass which move in such a way that they exchange velocities when they collide. Let P_1 and P_2 be two particles moving on the same one dimensional lattice such that (a) initially P_1 is an even number of units of distance to the left of P_2; (b) each particle moves independently of each other with steps $+1$ or -1; (c) if at any time n the two particles are not at the same place, then at time $n+1$ each moves $+1$ or -1 step with probability $\frac{1}{2}$; and (d) if at any time n both are at the same point x_0, then at time $n+1$ they move according to $P(\text{both at } x_0 - 1) = \frac{1}{4}$, $P(\text{both at } x_0 + 1) = \frac{1}{4}$, $P(P_1 \text{ at } x_0 - 1, P_2 \text{ at } x_0 + 1) = \frac{1}{2}$. If x_n and y_n, $n \geqslant 0$, are two independent simple random walks with probability $\frac{1}{2}$ for steps of $+1$ or -1 unit, and $x(0) - y(0)$ is a negative even integer, let $\xi_n = \min(x_n, y_n)$ and $\eta_n = \max(x_n, y_n)$. Then the random walk with collision is given by the discrete process $X_n = (\xi_n, \eta_n)$.

As remarked earlier, an important class of results of a discrete process $\{X_n\}$ is the asymptotic behavior as $n \to \infty$. The sequence X_n can converge to a limit X in several senses: convergence* in distribution, in probability, in mean, in mean square, almost sure convergence, and so on. The notions lead to different classification of estimates. For example, let $\xi_1, \ldots, \xi_n$ be independent RVs following the same distribution whose density $f(y, x)$ depends on a parameter x belonging to a class $\mathscr{X}$. Let X_n, $n > 1$, be a process obtained in the procedure of estimating x. Let P_x and E_x be the corresponding probability measure and expectation respectively. An estimate X_n is called *asymptotically unbiased* (respectively, *strongly consistent*, *consistent*, *asymptotically normal*, *strongly asymptotically efficient*, or *asymptotically efficient*) if (as $n \to \infty$), $X_n \to x$ in mean (respectively, almost surely, in probability, $\sqrt{n}(X_n - x) \to N(0, \sigma^2(x))$ in law, $nE_x[X_n - x]^2 \to I^{-1}(x)$, or $\sigma^2(x) = I^{-1}(x)$, where $I(x)$ is the Fisher information*). This classification list clearly indicates the importance of these notions of convergence and the corresponding limit theorems*. (See the annotations in the references at the end of this article.)

Finally, for completeness, we present some statisical results connected with discrete processes that arise as estimates. We will be sketchy for the simple reason that more general results appear elsewhere. Having found the estimate X_n, let X_{n+1} be determined recursively by $X_{n+1} = g(X_n, \xi_{n+1})$, where ξ_{n+1} is the new observation. The first result is basic in the theory of recursive estimation.

(i) Let $\{\xi_n, n \geqslant 1\}$ be independent observations from a population with a common density $f(\xi, x_0)$; $x_0, \xi \in R^1$. Let $m(x)$ and $D(x)$ be the common mean and variance with $x_0 = x$. If $D(x_0) < \infty$ and $m(x)$ is strictly increasing such that $|m(x)|$ increases at most linearly as $|x| \to \infty$, then the recursive estimates defined successively by

$$X_{n+1} - X_n = a_n(\xi_{n+1} - m(x_n)),$$

$$X_0 \equiv \text{constant},$$

with $\sum a_n = \infty$, $\sum a_n^2 < \infty$, give a strongly consistent estimate of x_0. Assume furthermore that $m'(x_0) > 0$ and $2am'(x_0) > 1$.

Then the recursive procedure

$$X_0 = \text{constant},$$
$$X_{n+1} - X_n = (\xi_{n+1} - m(X_n))/(n+1)$$

is asymptotically normal, and for any $k > 0$, $t_i > 0$, $1 \leqslant i \leqslant k$, and $n < n_1 < \cdots < n_k$ such that $\log[n_i/n] \to t_i$ as $n \to \infty$, the joint distribution of $\sqrt{n}(X_n - x_0), \sqrt{n_1}(X_{n_1} - x_0), \ldots, \sqrt{n_k}(X_{n_k} - x_0)$ converges to the finite-dimensional distribution of $X(0), X(t_1), \ldots, X(t_k)$ corresponding to a diffusion process* $X(t)$ satisfying the Ito equation

$$dX(t) = (\tfrac{1}{2} - m'(x_0))X(t)\,dt + \sqrt{D(x_0)}\;d\beta(t),$$

where $\beta(t)$ is a standard Brownian motion*.

(ii) The following theorem on normal correlation is fundamental in the study of recursive estimates or filtering (basic filtering equations are by-products of this theorem); see Liptser and Shiryayev [12]. If $(\theta, X) = ([\theta_1, \ldots, \theta_k], [X_1, \ldots, X_m])$ is a Gaussian vector with $m_\theta = E\theta$, $m_X = EX$, $D_\theta = \mathrm{Cov}(\theta, \theta)$, $D_{\theta X} = \mathrm{Cov}(\theta, X)$, $D_X = \mathrm{Cov}(X, X)$, and if D_X^+ denotes the pseudoinverse matrix of D_X, then the conditional expectation $E[\theta \mid X]$ and the conditional covariance $\mathrm{Cov}(\theta, \theta \mid X)$ are given by

$$E[\theta \mid X] = m_\theta + D_{\theta X} D_X^+ (X - m_X),$$
$$\mathrm{Cov}(\theta, \theta \mid X) = D_\theta - D_{\theta X} D_X^+ D_{\theta X}^*,$$

respectively. In particular, let the RVs $\theta, X_1, \ldots, X_m$ together form a Gaussian vector such that $X_1, \ldots, X_m$ are independent and $\mathrm{Var}(X_k) > 0$. Then

$$E[\theta | X_1, \ldots, X_m] = E\theta + \sum_{i=1}^{m} \frac{\mathrm{Cov}(\theta, X_i)}{\mathrm{Var}(X_i)}(X_i - EX_i).$$

In example 2 presented above, let the additive noises ξ_n [occurring with the signal $\theta \in N(m, \sigma^2)$] be independent unit normal RVs. Then it follows from **(ii)** that $x_n = (m + \sum_{i=1}^n y_i)/(1 + \sigma^2 n)$ and the tracking error $e_n = \sigma^2/(1 + \sigma^2 n)$. To consider the case of y_n defined in terms of the controlling functions A and B, put

$$A_n(y) = A(y_1, \ldots, y_n),$$
$$B_n(y) = B(y_1, \ldots, y_n),$$
$$\Delta x_n = x_{n+1} - x_n, \qquad x_0 = m,$$
$$\Delta\sigma_n = \sigma_{n+1} - \sigma_n, \qquad \sigma_0 = \sigma^2,$$
$$D = 1/\left(1 + \sigma_n A_n^2(y)\right).$$

Applying the theorem on normal correlation, we get

$$\Delta\sigma_n = -DA_n^2(y)\sigma_n^2,$$
$$\Delta x_n = D\sigma_n A_n(y)(y_{n+1} - B_n(y) - A_n(y)x_n).$$

For further details, see Liptser and Shiryayev [12, Chap. 13].

(iii) The following result also plays a central role in obtaining a maximum likelihood estimate*. Let $Y(t)$, $t = 0, \pm 1, \ldots$, be a wide-sense stationary process with spectral representation

$$Y(t) = \int_{-\pi}^{\pi} e^{iut}\left[P_{n-1}(e^{iu})/Q_n(E^{iu})\right] dZ(u),$$

where $dZ(u)$ is an orthogonal random measure with $E(dZ) = 0$, $E|dZ|^2 = du/(2\pi)$, $P_{n-1}(s) = \sum_{k=0}^{n-1} b_k s^k$, $Q_n(s) = \sum_{k=1}^{n} a_k s^k$, $a_n = 1$, $a_k, b_k \in R^1$, and with the roots of $Q_n(s) = 0$ in the unit circle. Then $Y(t) = Y_1(t)$, where $(Y_1(t), \ldots, Y_n(t))$ is an n-dimensional wide-sense stationary process satisfying the recurrent equations,

$$Y_j(t+1) = Y_{j+1}(t) + B_j V(t+1), \qquad j = 1, \ldots, n-1;$$

$$Y_n(t+1) = -\sum_{j=0}^{n-1} a_j Y_{j+1}(t) + B_n V(t+1),$$

$$\text{with } V(t) = \int_{-\pi}^{\pi} e^{iu(t-1)} Z(du),$$

$$E\left[Y_i(\gamma)\overline{V}(t)\right] = 0, \qquad \gamma < t, \quad i = 1, \ldots, n;$$
$$B_1 = b_{n-1},$$
$$B_j = b_{n-1} - \sum_{j=1}^{j-1} B_i a_{n-j+1}, \qquad j = 2, \ldots, n.$$

As an application, consider $X(t) = \theta + Y(t)$, where $-\infty < \theta < \infty$ is an unknown parameter and $Y(t)$ is stationary Gaussian with

$EY(t) = 0$ and spectral density (or power spectrum) $f(u) = (e^{iu} + 1)/(e^{2iu} + e^{iu} + \frac{1}{2})$. By the result above, $Y(t) = Y_1(t)$, where

$$Y_1(t+1) = Y_2(t) + V(t+1),$$

$$Y_2(t+1) = -\tfrac{1}{2}Y_1(t) - Y_2(t) + \tfrac{1}{2}V(t+1).$$

If $m^\theta(t) = E[Y_2(t) \mid \mathscr{A}_t(X)]$, then $m^\theta(t)$ solves

$$m^\theta(t+1) = \frac{\theta - X(t)}{2} m^\theta(t) + \frac{1 + 2k(t)}{2(1+k(t))} \times (X(t+1) - \theta - m^\theta(t)),$$

where

$$k(t+1) = k(t) + \frac{1}{4} - \frac{(1+2k(t))^2}{4(1+k(t))}.$$

Also, $EY_1^2(t) = \frac{12}{5}$, $EY_1(t)Y_2(t) = -\frac{9}{15}$, $EY_2^2(t) = \frac{7}{5}$, $E[X(t) - \theta]^2 = \frac{12}{5}$, and $E[(X(t) - \theta)Y_2(t)] = -\frac{9}{15}$. Now the theorem on normal correlation yields

$$m^\theta(0) = (\theta - X(0))/4, \qquad k(0) = \tfrac{5}{4}.$$

Hence $m^\theta(t) = f(t,x) + g(t)\theta$, where

$$f(t,x) = 4^{-1}\sum_{i=0}^{t-1} X(0)\left[-\frac{3+4k(i)}{2(1+k(i))}\right] + \sum_{i=0}^{t-1}\prod_{j=i+1}^{t-1}\left(-\frac{3+4k(j)}{2(1+k(j))}\right) \times \left[\frac{X(i)}{2} + \frac{1+2k(i)}{2(1+k(i))}X(i+1)\right],$$

$$g(t) = 4^{-1}\prod_{i=0}^{t-1}\left(-\frac{3+4k(i)}{2(k+k(i))}\right) - \sum_{i=0}^{t-1}\prod_{j=i+1}^{t-1}\left(-\frac{3+4k(j)}{2(1+k(i))}\right).$$

For $t \geqslant 1$, set

$$D_t = \frac{5}{12} + \sum_{i=1}^{t} \frac{\{1 + g(i+1)\}^2}{1 + k(i-1)}.$$

The maximum likelihood estimate $\hat{\theta}(t)$ is obtained by maximizing the right side of the Radon–Nikodym derivative

$$\frac{d\mu_X^\theta}{d\mu_X^0} = \exp\left\{\frac{5X(0)\theta}{12} - \frac{25\theta^2}{248} + \sum_{i=1}^{t}[F(i,x)\theta - G(i)\theta^2]\right\},$$

where

$$F(i,x) = \frac{\{1 + g(i-1)\}\{X(i) - f(i-1,x)\}}{1 + k(i-1)},$$

$$G(i) = \frac{\{1 + g(i-1)\}^2}{2\{1 + k(i-1)\}}.$$

Thus

$$\hat{\theta}(t) = D_t^{-1}\left[\frac{5X(0)}{12} + \sum_{i=1}^{t} F(i,x)\right].$$

The books by Cox and Miller [5], Feller [8], Kannan [10], and Karlin and Taylor [11] are of introductory nature and include basic properties and various applications of different types of discrete processes. The first part of Chung [4] is devoted to the theory of discrete Markov chains. For an extensive treatment of theory (and potential theory) of random walks consult Spitzer [16], and for applications, see Barber and Ninham [2]. Even though published in 1953, Doob [7] is a very good advanced level treatise on various types of stochastic processes. Loève [13] and Petrov [15] extensively study various types of limit theorems that are particularly useful in obtaining different types of estimates. Convergence in law is the topic to which Billingsley [3] is devoted; here one can find applications to exchangeable RVs. For detailed statistical study of discrete processes, consult Albert and Gardner [1], Ibragimov and Rozanov [9], and Liptser and Shiryayev [12].

References

[1] Albert, A. E. and Gardner, L. A. (1967). *Stochastic Approximation and Nonlinear Regression*. MIT Press, Cambridge, Mass.

[2] Barber, M. N. and Ninham, B. W. (1970). *Random and Restricted Walks*. Gordon and Breach, New York.

[3] Billingsley, P. (1968). *Convergence of Probability Measures*. Wiley, New York.

[4] Chung, K. L. (1967). *Markov Chains with Stationary Transition Probabilities*, 2nd ed. Springer-Verlag, New York.

[5] Cox, D. R. and Miller, H. D. (1970). *The Theory of Stochastic Processes*. Methuen, London.

[6] Cypkin, Ja. Z. (1973). *Foundations of the Theory of Learning Systems*. Academic Press, New York.

[7] Doob, J. L. (1953). *Stochastic Processes*. Wiley, New York.

[8] Feller, W. (1971). *Introduction to Probability Theory and Its Applications*, Vol. I, 3rd ed. Wiley, New York (Vol. II, 2nd ed., 1966).

[9] Ibragimov, I. A. and Rozanov, Y. A. (1978). *Gaussian Random Processes*. Springer-Verlag, New York.

[10] Kannan, D. (1979). *An Introduction to Stochastic Processes*. North-Holland, New York.

[11] Karlin, S. and Taylor, H. M. (1975). *A First Course in Stochastic Processes*, 2nd ed. Academic Press, New York.

[12] Liptser, R. S. and Shiryayev, A. N. (1977/1978). *Statistics of Random Processes*, Vols. I and II. Springer-Verlag, New York.

[13] Loève, M. (1977/1978). *Probability Theory*, Vols. I and II, 4th ed. Springer-Verlag, New York.

[14] Norman, F. (1972). *Markov Process and Learning Model*. Academic Press, New York.

[15] Petrov, V. V. (1975). *Sums of Independent Random Variables*. Springer-Verlag, New York.

[16] Spitzer, F. (1964). *Principles of Random Walk*. Van Nostrand, Reinhold, New York.

(DIFFUSION PROCESSES
GAUSSIAN PROCESSES
MARKOV PROCESSES
RANDOM WALKS
STOCHASTIC PROCESSES)

D. KANNAN

PROCESSES, EMPIRICAL

If $X_1, X_2, \ldots, X_n$ are independent identically distributed random variables (RVs) with values in a measurable space $(\mathbf{X}, \mathscr{B})$ and with common probability measure P on $\mathbf{X}$, the *empirical measure* or *empirical distribution* $\mathbb{P}_n$ of $(X_1, \ldots, X_n)$ is the measure which puts mass $1/n$ at each X_i, $i = 1, \ldots, n$:

$$\mathbb{P}_n = n^{-1}(\delta_{X_1} + \cdots + \delta_{X_n}), \qquad (1)$$

where $\delta_x(A) = 1$ if $x \in A$, 0 if $x \notin A$, $A \in \mathscr{B}$. Thus $n\mathbb{P}_n(A)$ is simply the number of X_i's in A for any set $A \in \mathscr{B}$. The *empirical process* G_n, defined for each $n \geqslant 1$ by

$$G_n = n^{1/2}(\mathbb{P}_n - P), \qquad (2)$$

may be viewed as a stochastic process* indexed (a) by some class of sets $\mathscr{C} \subset \mathscr{B}$,

$$G_n(C) = n^{1/2}(\mathbb{P}_n(C) - P(C)), \qquad C \in \mathscr{C}, \qquad (3)$$

or (b) by some class of functions $\mathscr{F}$ from $\mathbf{X}$ to the real line R^1,

$$G_n(f) = \int_{\mathbf{X}} f d\{n^{1/2}(\mathbb{P}_n - P)\} = n^{1/2}\int_{\mathbf{X}} f(x)\{\mathbb{P}_n(dx) - P(dx)\}, \qquad f \in \mathscr{F}. \qquad (4)$$

Frequently, in applications of interest the observations $X_1, \ldots, X_n$ are dependent, or nonidentically distributed, or perhaps both. In such cases we will continue to speak of the empirical measure $\mathbb{P}_n$ and empirical process G_n, perhaps with P replaced in (2) by an appropriate average measure.

In the classical case of real-valued random variables, $\mathscr{X} = R^1$, the class of sets $\mathscr{C} = \{(-\infty, x] : x \in R^1\}$ in (3), or the class $\mathscr{F} = \{1_{(-\infty, x]} : x \in R^1\}$ of indicator functions in (4) [where $1_A(x) = 1$ if $x \in A$, 0 if $x \notin A$], yields the usual *empirical distribution function* $\mathbf{F}_n$ given by

$$\mathbf{F}_n(x) = \mathbb{P}_n(-\infty, x] = n^{-1}\{\text{number of } i \leqslant n \text{ with } X_i \leqslant x\}, \qquad (5)$$

and the *empirical process*

$$G_n(x) = n^{1/2}(\mathbf{F}_n(x) - F(x)) \qquad (6)$$

indexed by $x \in R^1$.

The subject of empirical processes is concerned with the large- and small-sample properties of the processes $\mathbb{P}_n$ and G_n, methods for studying these processes, and with the use of these properties and methods to treat systematically the extremely large number of statistics which may be viewed as functions of the empirical measure $\mathbb{P}_n$ or of the empirical process G_n. Much of the motivation for the study of $\mathbb{P}_n, G_n$, and functions thereof comes both historically and in current work from the desirability and attractiveness of nonparametric or distribution-free* statistical methods, methods which have proved to be of interest in a wide variety of problems, ranging from rank* and goodness-of-fit* tests, to density estimation*, clustering and classification*, and survival analysis*. The study of empirical processes also has strong connections with the related probabilistic topics of weak convergence and invariance principles*, as will be seen in the course of this article.

For any fixed set $C \in \mathscr{B}$, $n\mathbb{P}_n(C)$ is simply a binomial RV with parameters n and $P(C)$. Hence, by the classical weak law of large numbers, central limit theorem*, and law of the iterated logarithm*, respectively, as $n \to \infty$,

$$\mathbb{P}_n(C) \xrightarrow{\text{p}} P(C), \tag{7}$$

$$G_n(C) = n^{1/2}(\mathbb{P}_n(C) - P(C)) \xrightarrow{\text{d}} G_P(C) \sim N(0, P(C)(1 - P(C))), \tag{8}$$

and

$$\limsup_{n\to\infty} \frac{|G_n(C)|}{(2\log\log n)^{1/2}} = \limsup_{n\to\infty} \frac{n^{1/2}|\mathbb{P}_n(C) - P(C)|}{(2\log\log n)^{1/2}} = \left[P(C)(1 - P(C))\right]^{1/2} \text{ a.s.} \tag{9}$$

(where we write "$\sim$" for "is distributed as", $N(\mu, \sigma^2)$ denotes the "normal" or Gaussian distribution with mean μ and variance σ^2, "$\xrightarrow{\text{p}}$" denotes convergence in probability, and "$\xrightarrow{\text{d}}$" denotes convergence in distribution or law; *see* CONVERGENCE OF SEQUENCES OF RANDOM VARIABLES). A large part of the theory of empirical processes is concerned with strengthened versions of (7) to (9), versions of these convergence results that hold simultaneously (i.e., uniformly) for all sets C in some collection $\mathscr{C}$. *See* GLIVENKO–CANTELLI THEOREMS for such uniform extensions of (7) and applications thereof. We will concentrate in this article on various uniform analogs of (8) (sometimes called Donsker theorems or functional central limit theorems), and of (9) (which we call Strassen–Finkelstein log-log laws or functional laws of the iterated logarithm). In the same way that Glivenko–Cantelli theorems serve as tools for establishing the consistency of various estimators or statistics, uniform versions of (8) and (9) serve as tools for establishing convergence in distribution (often asymptotic normality*), or laws of the iterated logarithm, respectively, for those estimators and statistics expressible in terms of $\mathbb{P}_n$ or G_n. We have chosen to concentrate this exposition on the large-sample theory* of empirical processes since so little is known concerning finite sample sizes beyond the classical one-dimensional case of real-valued RVs; for useful summaries of finite-sample size results in the one-dimensional case, see Durbin [1], Niederhausen [4] and the references given there.

The article has been divided into the following four sections:

1. The one-dimensional case
2. More general sample spaces and index sets
3. Dependent or nonidentically distributed observations
4. Miscellaneous topics

Many topics have been omitted or are mentioned only briefly. For a nice survey of earlier work and a helpful exposition of weak convergence issues, see Pyke [3]. For a recent comprehensive review of the i.i.d. case, see Gaenssler and Stute [2].

ONE-DIMENSIONAL CASE

Here we focus on the classical empirical distribution function $\mathbf{F}_n$ and empirical process G_n of real-valued random variables given in (5) and (6). In this one-dimensional situation, a significant further simplification is possible by use of the fundamental transformations of nonparametric statistics (the probability integral transformation* and the inverse probability integral transformation or quantile transformation):

$$\text{If } F \text{ is continuous and } X \sim F, \text{ then } F(X) \sim \text{uniform}(0,1); \quad (10)$$

$$\text{If } U \sim \text{uniform }(0,1), \text{ then } F^{-1}(U) \equiv X \sim F \text{ for an arbitrary df } F, \quad (11)$$

where F^{-1} is the left-continuous inverse of F, $F^{-1}(u) = \inf\{x : F(x) \geqslant u\}$. Thus, letting $U_1, U_2, \ldots, U_n$ be independent, identically distributed (i.i.d.) uniform (0, 1) RVs with distribution function $I(t) = t$ on [0, 1], empirical distribution function Γ_n, and corresponding uniform empirical process $\mathbf{U}_n$ defined by

$$\mathbf{U}_n(t) = n^{1/2}(\Gamma_n(t) - t), \qquad 0 \leqslant t \leqslant 1,$$

it follows from (10) and (11) that

$$G_n \circ F^{-1} \stackrel{\mathrm{d}}{=} \mathbf{U}_n \qquad \text{if } F \text{ is continuous}, \quad (12)$$

and

$$G_n \stackrel{\mathrm{d}}{=} \mathbf{U}_n \circ F \qquad \text{for arbitrary } F, \quad (13)$$

where "$\stackrel{\mathrm{d}}{=}$" means equal in distribution or law (so that the two processes are probabilistically equivalent), and "$\circ$" denotes functional composition, $f \circ g(t) = f(g(t))$. By virtue of (12) and (13), we may restrict attention to the uniform empirical process $\mathbf{U}_n$ throughout most of the remainder of this section.

The random function Γ_n is a nondecreasing, right-continuous step function equal to 0 at $t = 0$ and 1 at $t = 1$, which increases by jumps of size $1/n$ at the *order statistics** $0 \leqslant U_{n:1} \leqslant \cdots \leqslant U_{n:n} \leqslant 1$. The random function or process $\mathbf{U}_n$ equals 0 at both $t = 0$ and 1, decreases linearly between successive order statistics with slope $-n^{1/2}$, and jumps upward at the order statistics with jumps of size $n^{-1/2}$. Both Γ_n and $\mathbf{U}_n$ take values in $D = D[0,1]$, the set of functions on [0, 1] which are right continuous and have left limits.

Donsker's Theorem; Weak Convergence of $\mathbf{U}_n$

Convergence in distribution of specific functions of the process $\mathbf{U}_n$ was first treated by Cramér [16], von Mises [65], Kolmogorov [39], and Smirnov [61, 62] in the course of investigations of the now well-known Cramér–von Mises* and Kolmogorov goodness-of-fit statistics. A general unified approach to the large-sample theory of statistics such as these did not emerge until Doob [28] gave his heuristic approach to the Kolmogorov–Smirnov limit theorems. Doob's approach was to note that (a) $\mathbf{U}_n$ is a zero-mean stochastic process on [0, 1] with covariance function

$$\operatorname{cov}[\mathbf{U}_n(s), \mathbf{U}_n(t)] = \min(s,t) - st \qquad \text{for all } 0 \leqslant s, t \leqslant 1; \quad (14)$$

(b) by a simple application of the multivariate central limit theorem*, all the finite-dimensional joint df's of $\mathbf{U}_n$ converge to the corresponding normal df's which are the finite-dimensional joint df's of a mean-zero Gaussian process* $\mathbf{U}$ on [0, 1] with covariance as in (14), called a *Brownian Bridge process*; and (c) hence, for any real-valued "continuous" function g of $\mathbf{U}_n$ it should follow that

$$g(\mathbf{U}_n) \stackrel{\mathrm{d}}{\to} g(\mathbf{U}) \qquad \text{as } n \to \infty \quad (15)$$

in the ordinary sense of convergence in distribution of RVs. For example, for the Kolmogorov statistic $g(u) = \sup_{0 \leqslant t \leqslant 1} |u(t)| \equiv \|u\|$, and Doob [28] showed that the limiting distribution of $g(\mathbf{U}_n) = \|\mathbf{U}_n\| = n^{1/2}\|\Gamma_n - I\|$, obtained earlier by Kolmogorov [39], is exactly that of $g(\mathbf{U}) = \|\mathbf{U}\|$.

A precise formulation of Doob's heuristic approach requires a careful definition of the idea of *weak convergence* of a sequence of stochastic processes*, a notion which extends the more familiar concept of convergence in distribution of random variables or random vector. Donsker [27] succeeded in justifying Doob's [28] heuristic approach, and this in combination with related work on invariance principles by Erdös and Kac [32, 33], Donsker [26], and others led to the development of a general theory of *weak convergence* of stochastic processes (and their associated probability laws) by Prohorov [44] and Skorokhod [59]. This theory has been very clearly presented and further developed in an exemplary monograph by Billingsley [5]; see also Billingsley [6].

Unfortunately for the theory of empirical processes, the space $D = D[0, 1]$ in which $\mathbf{U}_n$ takes its values is inseparable when considered as a metric space with the supremum or uniform metric $\|\cdot\|$ (i.e., $\|f - g\| \equiv \sup_{0 \leqslant t \leqslant 1}|f(t) - g(t)|$), as pointed out by Chibisov [14]; see Billingsley [5, Sec. 18]. This lack of separability creates certain technical difficulties in the weak convergence theory of $\mathbf{U}_n$ and has led to a number of different approaches to the study of its weak convergence: Skorokhod [59] introduced a metric d with which D becomes separable (see Billingsley [5, Sec. 14]), while Dudley [69], Pyke and Shorack [48], and Pyke [47] give different definitions of weak convergence. These difficulties are largely technical in nature, however. Here we follow Pyke and Shorack [48] and Pyke [47] and say that $\mathbf{U}_n \Rightarrow \mathbf{U}$ ("$\mathbf{U}_n$ converges weakly to $\mathbf{U}$") if $g(\mathbf{U}_n) \xrightarrow{d} g(\mathbf{U})$ for all $\|\cdot\|$-continuous real-valued functions g of $\mathbf{U}_n$ for which $g(\mathbf{U}_n)$ $(n \geqslant 1)$ and $g(\mathbf{U})$ are (measurable) RVs. With this definition we have:

Theorem 1. (Donsker [27]). $\mathbf{U}_n \Rightarrow \mathbf{U}$ on $(D, \|\cdot\|)$.

The importance of Theorem 1 for applications in statistics is that the limiting distribution of any statistic that can be expressed as $g(\mathbf{U}_n)$ for some $\|\cdot\|$-continuous measurable function g is that of $g(\mathbf{U})$. For example:

$$\|\mathbf{U}_n\| \xrightarrow{d} \|\mathbf{U}\| \qquad [g(u) = \|u\|]; \quad (16)$$

$$\int_0^1 [\mathbf{U}_n(t)]^2\, dt \xrightarrow{d} \int_0^1 [\mathbf{U}(t)]^2\, dt$$

$$\left[g(u) = \int_0^1 (u(t))^2\, dt \right]; \qquad (17)$$

and

$$n^{1/2}\left(\overline{U}_n - \tfrac{1}{2}\right)$$

$$= -\int_0^1 \mathbf{U}_n(t)\, dt \xrightarrow{d} -\int_0^1 \mathbf{U}(t)\, dt \sim N\left(0, \tfrac{1}{12}\right)$$

$$\left[g(u) = -\int_0^1 u(t)\, dt \right] \qquad (18)$$

as $n \to \infty$. Of course, the distribution function of the RV $g(\mathbf{U})$ must be computed in order to complete the program. For linear functions of $\mathbf{U}_n$ [as in (18)], and hence of $\mathbf{U}$, this is easy: under appropriate integrability conditions, linear functions of the Gaussian process $\mathbf{U}$ have normal distributions with easily computed variances. In general, evaluation of the distribution of $g(\mathbf{U})$ is not an easy task, but well-developed tools are available for quadratic and supremum-type functionals as illustrated by Doob [28] and Darling [24]; see also Sähler [51] and Durbin [30]. More sophisticated applications making essential use of the identities (12) and (13) may be found, for example, in Pyke and Shorack [48] (rank statistics); Shorack [53, 54] (linear combinations of order statistics, quantile*, and spacings processes); and Bolthausen [7], Pollard [43], and Boos [8] (minimum distance estimators* and tests). The weak convergence approach, in combination with the device of almost surely convergent versions of weakly convergent processes (to be discussed in the section "Almost Surely Convergent Constructions; Strong Approximations"), has become a key tool in the modern statistical workshop.

Iterated Logarithm Laws

Following a pattern similar to that outlined above, iterated logarithm laws for specified functions of the process $\mathbf{U}_n$ were established by Smirnov [62], Chung [15], Cassels [12], and others in connection with investigations of particular goodness-of-fit statistics, especially $\|\mathbf{U}_n\|$ and $\|\mathbf{U}_n^+\|$:

$$\limsup_{n\to\infty} \frac{\|\mathbf{U}_n\|}{(2\log\log n)^{1/2}} = \limsup_{n\to\infty} \frac{n^{1/2}\|\Gamma_n - I\|}{(2\log\log n)^{1/2}} = \tfrac{1}{2} \text{ a.s.} \qquad (19)$$

A general law of the iterated logarithm result for the uniform empirical process $\mathbf{U}_n$ comparable to Donsker's theorem emerged in the light of work on almost surely convergent constructions or embeddings of partial sum processes in Brownian motion by Skorokhod [60] and the "invariance principle for the law of the iterated logarithm" by Strassen [63]. Let $\mathscr{U}^0$ be the set of functions on $[0, 1]$ which are absolutely continuous with respect to Lebesgue measure, equal to 0 at 0 and 1, and whose derivatives have L_2-norm no larger than 1; alternatively, $\mathscr{U}^0$ is simply the unit ball of the reproducing kernel Hilbert space with kernel given by the covariance function (14) of the Brownian bridge process $\mathbf{U}$.

Theorem 2. (Finkelstein [34]). With probability 1 every subsequence of

$$\left\{ \frac{\mathbf{U}_n}{(2\log\log n)^{1/2}} : n \geqslant 3 \right\}$$

has a uniformly convergent subsequence, and the set of limit functions is precisely $\mathscr{U}^0$.

In a way completely parallel to the applications of Donsker's theorem given in the preceding section, Finkelstein's theorem yields laws of the iterated logarithm for $\|\cdot\|$-continuous functions g of $\mathbf{U}_n/b_n$, where $b_n = (2\log\log n)^{1/2}$:

$$\limsup_{n\to\infty} g(\mathbf{U}_n/b_n) = \sup\{ g(u) : u \in \mathscr{U}^0\} \text{ a.s.,} \tag{20}$$

where the problem of evaluating the supremum on the right side for specific functions g may be thought of as analogous to the problem of finding the distribution of $g(\mathbf{U})$ in the case of weak convergence. For example, in parallel to (16) to (18), Finkelstein's theorem yields

$$\limsup_{n\to\infty} \|\mathbf{U}_n/b_n\| = \sup\{\|u\| : u \in \mathscr{U}^0\} = \tfrac{1}{2} \text{ a.s.} \tag{21}$$

with equality when $u(t) = \min(t, 1 - t)$, $0 \leqslant t \leqslant 1$;

$$\limsup_{n\to\infty} \int_0^1 [\mathbf{U}_n(t)/b_n]^2\, dt = \sup\left\{ \int_0^1 (u(t))^2\, dt : u \in \mathscr{U}^0 \right\} = 1/\pi^2 \text{ a.s.} \tag{22}$$

with equality when $u(t) = (2^{1/2}/\pi)\sin(\pi t)$, $0 \leqslant t \leqslant 1$; and

$$\limsup_{n\to\infty} \int_0^1 (\mathbf{U}_n(t)/b_n)\, dt = \sup\left\{ \int_0^1 u(t)\, dt : u \in \mathscr{U}^0 \right\} = 1/\sqrt{12}^{\,1/2} \text{ a.s.} \tag{23}$$

with equality when $u(t) = \sqrt{12} \int_0^1 (\min(s,t) - st)\, ds = \sqrt{3}\, t(1-t)$.

Almost Surely Convergent Constructions; Strong Approximations

In Skorokhod's [59] paper the basis for a different and very fruitful approach to weak convergence was already in evidence: that of replacing weak convergence by almost sure (a.s.) convergence. *See also* HUNGARIAN CONSTRUCTIONS OF EMPIRICAL PROCESSES.

Theorem 3. (Skorokhod, Dudley, Wichura). If the processes $\{\mathbb{Z}_n, n \geqslant 0\}$ take values in a metric space $(\mathscr{M}, m)$ and $\mathbb{Z}_n \Rightarrow \mathbb{Z}_0$, then there exists a probability space $(\Omega, \mathscr{A}, P)$ and processes $\{\mathbb{Z}_n^*, n \geqslant 0\}$ defined there such that $\mathbb{Z}_n \overset{d}{=} \mathbb{Z}_n^*$ for all $n \geqslant 0$ and $m(\mathbb{Z}_n^*, \mathbb{Z}_0^*) \to 0$ a.s. as $n \to \infty$.

Skorokhod [59] gave the first version of this result in the case that $(\mathscr{M}, m)$ is complete and separable; Dudley [29] and Wichura [66] proved that the hypotheses of completeness and separability, respectively, could be dropped. See also Billingsley [6, p. 7]. Although the theorem does not tell how to construct the special almost surely convergent $\mathbb{Z}_n^*$ processes, it provides an extremely valuable conceptual tool. For example, in the case of the uniform empirical processes $\{\mathbf{U}_n\}$, the theorem yields the existence of probabilistically equivalent processes $\mathbf{U}_n^*$, $n \geqslant 1$, and a Brownian bridge process $\mathbf{U}^*$ all defined on a common probability space $(\Omega, \mathscr{A}, P)$ such that for each fixed $\omega \in \Omega$ the sequence of functions $\mathbf{U}_n^* = \mathbf{U}_n^*(\cdot, \omega)$ converge uniformly to the continuous function $\mathbf{U}^* = \mathbf{U}^*(\cdot, \omega)$ on $[0, 1]$ as $n \to \infty$; that is,

$$\|\mathbf{U}_n^* - \mathbf{U}^*\| \to 0 \text{ a.s.} \qquad \text{as } n \to \infty. \tag{24}$$

The extreme usefulness of this point of view in dealing with weak convergence problems in statistics was recognized, explained, and advocated by Pyke [45, 46] and has been effectively used to deal with a variety of problems involving two-sample rank statistics, linear combinations of order statistics (*see* L-STATISTICS), spacings*, minimum-distance estimators, and censored data* (the "product limit" estimator) by Pyke and Shorack [48], Pyke [46], Shorack [53–55], and Breslow and Crowley [10], to name only a few outstanding examples. See Pyke [45–47] for excellent expositions of this approach.

When the metric space $\mathscr{M}$ is the real line R^1, so that the $\mathbb{Z}_n$'s are just real-valued RVs, a very explicit construction of the $\mathbb{Z}_n^*$'s is possible using (11): If $F_n(z) \equiv P(\mathbb{Z}_n \leqslant z)$, $n \geqslant 0$, then $F_n \xrightarrow{d} F_0$ implies that $F_n^{-1}(u) \to F_0^{-1}(u)$ for almost all $u \in [0, 1]$ with respect to Lebesgue measure. Hence, if U is a uniform(0, 1) RV, then

$$\mathbb{Z}_n^* \equiv F_n^{-1}(U) \stackrel{d}{=} \mathbb{Z}_n$$

by (11), and

$$\mathbb{Z}_n^* \equiv F_n^{-1}(U) \to F_0^{-1}(U) \equiv \mathbb{Z}_0^* \text{ a.s.}$$

as $n \to \infty$ [45].

The possibility of giving explicit, concrete constructions of the almost surely convergent versions $\mathbf{U}_n^*$ of the uniform empirical process $\mathbf{U}_n$ began to become apparent soon after the appearance of Skorokhod [60] (available in English translation in 1965), and Strassen [63] concerning the embedding of partial sum processes in Brownian motion*. This idea, together with a representation of uniform order statistics as ratios of partial sums of independent exponential RVs, was used by several authors, including Breiman [9], Brillinger [11], and Root [49], to give explicit constructions of the a.s. convergent processes $\{\mathbf{U}_n^*\}$ guaranteed to exist by Skorokhod's [59] theorem. The "closeness" or "rate" of this approximation was studied by Kiefer [37], Rosenkrantz [50], and Sawyer [52].

The method of Skorokhod embedding gave a relatively straightforward and clear construction of versions $\{\mathbf{U}_n^*\}$ converging almost surely. It quickly became apparent, however, that for some purposes the constructed versions $\{\mathbf{U}_n^*\}$ from this embedding method suffered from two inadequacies or deficiencies: The joint distributions of $\mathbf{U}_n^*$ and $\mathbf{U}_{n+1}^*$ (i.e., in n) were not correct; and rates of convergence for specific functions of the $\mathbf{U}_n$ process yielded by the construction were substantially less than those obtainable by other (direct, special) methods ($n^{-1/4}$ or less, rather than $n^{-1/2}$ or a little less). These difficulties were overcome by a Hungarian group of probabilists and statisticians in a remarkable series of papers published in 1975: see Csörgő and Révész [21] and Komlós et al. [40]. By combining the quantile or inverse probability integral transform (11), an ingenious dyadic approximation scheme, and careful analysis, the Hungarian construction* yields uniform empirical processes $\{\mathbf{U}_n^*\}$ which have the correct distributions jointly in n and which are (at least very nearly) as close as possible to a sequence $\{\mathbb{B}_n^*\}$ of Brownian bridge processes (each $B_n^* \stackrel{d}{=} \mathbf{U}$ = Brownian bridge) with the correct joint distributions in n dimensions:

Theorem 4. (Komlós et al. [40]). For the Hungarian construction $\{\mathbf{U}_n^*\}$ of the uniform empirical processes

$$\|\mathbf{U}_n^* - \mathbb{B}_n^*\| = O\left(n^{-1/2}(\log n)^2\right) \text{ a.s.} \quad \text{as } n \to \infty; \qquad (25)$$

that is, there exists a positive constant $M < \infty$ such that

$$\limsup_{n \to \infty} \frac{n^{1/2}}{(\log n)^2} \|\mathbf{U}_n^* - \mathbb{B}_n^*\| \leqslant M \text{ a.s.} \qquad (26)$$

Thus the supremum distance between the constructed uniform empirical processes $\mathbf{U}_n^*$ and the sequence of Brownian bridge processes $\mathbb{B}_n^*$ goes to zero only a little more slowly than $n^{-1/2}$ as $n \to \infty$. This fundamental strong approximation theorem has already proved itself to be of basic importance in a wide and growing range of problems, and has already been generalized and extended in several directions, some of which will be mentioned briefly. The monograph by Csörgő and Révész [23] contains an exposition and a variety of applications.

Other Limit Theorems for $\mathbf{U}_n$

WEIGHTED METRICS. The empirical process $\mathbf{U}_n$ is small near 0 and 1: $\mathbf{U}_n(0) = \mathbf{U}_n(1) = 0$ and $\text{var}[\mathbf{U}_n(t)] = t(1-t)$ for $0 \leqslant t \leqslant 1$. This has led to the introduction of various "weighted" metrics to account for and exploit the small values of $\mathbf{U}_n$ near 0 and 1. Supremum-type weighted metrics $\|\cdot/q\|$ (defined by

$$\|(f-g)/q\| \equiv \sup_{0 \leqslant t \leqslant 1} |(f(t)-g(t))/q(t)|$$

were first introduced by Chibisov [13], who gave conditions for $\mathbf{U}_n \Rightarrow \mathbf{U}$ with respect to $\|\cdot/q\|$: Essentially, q must satisfy

$$q^2(t)/\left[t(1-t)\log\log\left(\{t(1-t)\}^{-1}\right)\right] \to \infty$$

as $t \to 0$ or 1 [e.g., $q(t) = [t(1-t)]^{1/2-\delta}$ with $0 < \delta < \frac{1}{2}$]. This convergence, which strengthens Donsker's theorem, was further investigated by Pyke and Shorack [48], O' Reilly [42], Shorack [56], and Shorack and Wellner [58], and successfully applied to statistical problems by Pyke and Shorack [48], Shorack [53, 54], and subsequently by others. James [36] has established a corresponding weighted version of the Strassen–Finkelstein functional law of the iterated logarithm for $\mathbf{U}_n/b_n$.

The standardized or normalized empirical process $Z_n(t) = \mathbf{U}_n(t)/\sqrt{t(1-t)}$, $0 < t < 1$, has also been thoroughly investigated, largely because it has the appealing feature of having $\text{var}[Z_n(t)] = 1$ for all $0 < t < 1$ and every $n \geqslant 1$. The limit theory of Z_n turns out to be closely linked to the Ornstein–Uhlenbeck process* and the classical work of Darling and Erdös [25] on normalized sums: see Jaeschke [35] and Eicker [31] for distributional limit theorems; and Csáki [19, 20], Shorack [57], and Mason [41] for iterated logarithm-type results.

OSCILLATIONS OF $\mathbf{U}_n$. The oscillation modulus $\omega_n(a)$ of $\mathbf{U}_n$, defined by

$$\begin{aligned}\omega_n(a) &= \sup_{|t-s| \leqslant a} |\mathbf{U}_n(t) - \mathbf{U}_n(s)| \\ &= n^{1/2} \sup_{|t-s| \leqslant a} |\Gamma_n(t) - \Gamma_n(s) - (t-s)|, \\ &\qquad 0 < a \leqslant 1, \end{aligned} \tag{27}$$

arises naturally in many statistical problems including tests for "bumps" of probability and density estimation*. Note that $\omega_n(1)$ is the classical Kuiper goodness-of-fit* statistic (see, e.g., Durbin [30, p. 33]). Cassels [12] established laws of the iterated logarithm for $\omega_n(a)$; Cressie [17, 18] has investigated the limiting distribution of $\omega_n(a)$ for fixed a (which, by Donsker's theorem, is that of $\omega(a) \equiv \sup_{|t-s| \leqslant a} |\mathbf{U}(t) - \mathbf{U}(s)|$); and Stute [64] proved that if $a = a_n \to 0$, $a_n = n^{-\lambda}$, $0 < \lambda < 1$, then

$$\lim_{n \to \infty} \frac{\omega_n(a_n)}{\{2a_n \log(1/a_n)\}^{1/2}} = 1 \text{ a.s.} \tag{28}$$

Stute [64] has exploited this result to obtain several interesting limit theorems for kernel estimates* of density functions. Shorack and Wellner [58] study related oscillation moduli and give weighted-metric convergence theorems related to those of Chibisov [13] and O'Reilly [42].

QUANTILE PROCESSES. An important process closely related to the uniform empirical process $\mathbf{U}_n$ is the uniform quantile process* V_n defined on [0, 1] by

$$V_n(t) = n^{1/2}\left(\Gamma_n^{-1}(t) - t\right), \qquad 0 \leqslant t \leqslant 1, \tag{29}$$

where Γ_n^{-1} is the left-continuous inverse of Γ_n. Γ_n^{-1} and V_n are important for problems involving order statistics since $\Gamma_n^{-1}(i/n) = U_{n:i}$, the ith order statistic of the sample $U_1, \ldots, U_n$ of n i.i.d. uniform (0, 1) RVs. There are many relationships between the processes $\mathbf{U}_n$ and V_n, such as the identity

$$V_n = -\mathbf{U}_n \circ \Gamma_n^{-1} + n^{1/2}\left(\Gamma_n \circ \Gamma_n^{-1} - I\right), \tag{30}$$

which shows that $V_n \Rightarrow V \equiv -\mathbf{U}$, because $\Gamma_n^{-1}(t) \to t$ uniformly in t a.s. and the second term in (30) has supremum norm equal to $n^{-1/2}$; a corresponding functional law of the iterated logarithm for V_n/b_n follows similarly.

For a sample from a general df F on R^1, the *quantile process* $\mathbf{Q}_n$ is defined by

$$\mathbf{Q}_n(t) = n^{1/2}\left(\mathbf{F}_n^{-1}(t) - F^{-1}(t)\right), \quad (31)$$

where $\mathbf{F}_n^{-1}$ denotes the left continuous inverse of the empirical df $\mathbf{F}_n$. By the inverse probability integral transform (11),

$$\mathbf{Q}_n(t) \overset{d}{=} n^{1/2}\left(F^{-1} \circ \Gamma_n^{-1} - F^{-1}\right) = R_n \cdot V_n, \quad (32)$$

where the random difference quotient $R_n \equiv (F^{-1}(\Gamma_n^{-1}) - F^{-1})/(\Gamma_n^{-1} - I)$ can be shown to converge (under appropriate differentiability hypotheses on F^{-1}) to $dF^{-1}/dt = 1/(f \circ F^{-1})$. Thus, at least roughly,

$$\mathbf{Q}_n \Rightarrow \frac{1}{f \circ F^{-1}} V = \frac{-1}{f \circ F^{-1}} \mathbf{U}. \quad (33)$$

For precise formulations of this type of limit theorem, see Shorack [54] and Csörgő and Révész [22], who make use of strong approximation methods together with the deep theorems of Kiefer [38] concerning the process $\mathbf{D}_n = \mathbf{U}_n + V_n$.

MORE GENERAL SAMPLE SPACES AND INDEX SETS

Spurred by questions in many different areas of statistics, the theory of empirical processes has undergone rapid development. The basic theorems of Donsker and Strassen–Finkelstein in one dimension have been generalized to observations X with values in higher-dimensional Euclidean spaces R^k or more general sample spaces; to indexing by classes of sets or functions, and to observations which are dependent or nonidentically distributed. We focus on i.i.d. RVs in higher-dimensional spaces and indexing of these processes by sets and functions; dependent or nonidentically distributed RVs will be discussed in the following section.

A General "Donsker Theorem"

Now, as in the introduction, suppose that $X_1, X_2, \ldots, X_n$ are i.i.d. RVs with values in the measurable space $(\mathscr{X}, \mathscr{B})$, and consider the empirical measures $\mathbb{P}_n$ and empirical process G_n as processes "indexed" by sets C in some class of sets $\mathscr{C} \subset \mathscr{B}$. It turns out that the $\mathscr{C}$-empirical process $\{G_n(C) : C \in \mathscr{C}\}$ will converge weakly only if the class of sets $\mathscr{C}$ is not "too large." The most complete results to date are those of Dudley [70].

Theorem 5. (Dudley [70]). Under measurability and entropy* conditions (satisfied if $\mathscr{C}$ is not "too large"),

$$G_n \Rightarrow G_p \qquad \text{as } n \to \infty,$$

where G_P is a zero-mean Gaussian process indexed by sets $C \in \mathscr{C}$ with continuous sample functions and covariance

$$\operatorname{cov}\left[G_P(A), G_P(B)\right] = P(A \cap B) - P(A)P(B) \quad \text{for all } A, B \in \mathscr{C}.$$

This theorem generalizes and contains as special cases earlier results by Dudley [69], Bickel and Wichura [67], Neuhaus [77], and Straf [85] (all of which dealt with the case $\mathscr{X} = R^k$ and the class $\mathscr{C}$ of lower-left orthants, which yield the usual k-dimensional df $F(x)$ and empirical df $\mathbf{F}_n(x)$, $x \in R^k$) as well as more recent results for convex sets due to Bolthausen [68]. Dudley's results have been used by Pollard [79] to treat chi-square goodness-of-fit* tests with data dependent cells.

If the empirical process G_n is considered as a process indexed by functions f in some class $\mathscr{F}$, $\{G_n(f) : f \in \mathscr{F}\}$, then a "Donsker theorem" will hold if the class $\mathscr{F}$ is not "too large." Roughly speaking, all the functions f in $\mathscr{F}$ must be sufficiently smooth and square integrable (with respect to P). Such a theorem under metric entropy conditions on the class $\mathscr{F}$ was first given by Strassen and Dudley [86] for the case when the sample space $\mathscr{X}$ is a compact metric space such as

$[0,1] \subset R^1$, or $[0,1]^k \subset R^k$. In the case $\mathscr{X} = [0,1]$, the weak convergence of G_n to $G = \{G(f) : f \in \mathscr{F}\}$ holds if the class $\mathscr{F}$ is any of the classes of Lipschitz functions

$$\mathscr{F}_\alpha = \{f : |f(x) - f(y)| \leqslant |x - y|^\alpha \text{ for all } x, y \in [0,1]\}$$

with $\alpha > \frac{1}{2}$; if $\alpha = \frac{1}{2}$, the convergence fails (there are "too many" functions in the class $\mathscr{F}_{1/2}$). Very recently similar (but more difficult) results have been given by Dudley [71] and Pollard [81] without the restriction to compact metric sample spaces $\mathscr{X}$.

Several applications of the properties of empirical processes indexed by functions to problems in statistics have been made: Giné [72] and Wellner [87] use such processes to study test statistics of interest for directional data*; Pollard [80] uses his Donsker theorem to give a central limit theorem for the cluster centers of a clustering method studied earlier in R^1 by Hartigan [73].

General Law of the Iterated Logarithm

In the same way that Dudley's weak convergence theorem in the preceding section generalizes Donsker's theorem, a law of the iterated logarithm for the $\mathscr{C}$-empirical process which generalizes the Strassen–Finkelstein theorem has been proved by Kuelbs and Dudley [76]. We introduce the sets of functions

$$\mathscr{H}^0 = \left\{h \in L^2(\mathscr{X}, \mathscr{B}, P) : \int h\,dP = 0 \text{ and } \int |h|^2\,dP \leqslant 1\right\},$$

$$\mathscr{G}^0_{\mathscr{C}} = \left\{g : \mathscr{C} \to R \text{ defined by } g(C) = \int_C h\,dP,\ C \in \mathscr{C};\ h \in \mathscr{H}^0\right\};$$

$\mathscr{G}^0_{\mathscr{C}}$ is the appropriate analog for the $\mathscr{C}$-empirical process of the set of functions $\mathscr{U}^0$ which arose in the Strassen–Finkelstein theorem.

Theorem 6. (Kuelbs and Dudley [76]). Under the same measurability and entropy conditions as required for weak convergence of the $\mathscr{C}$-empirical process (satisfied if $\mathscr{C}$ is not "too large"), with probability 1 every subsequence of

$$\left\{\frac{G_n}{(2\log\log n)^{1/2}} : n \geqslant 3\right\}$$

restricted to $C \in \mathscr{C}$ has a uniformly convergent subsequence, and the set of limit functions is precisely $\mathscr{G}^0_{\mathscr{C}}$.

This theorem has consequences analogous to those of the Strassen–Finkelstein theorem, and generalizes earlier results for special sample spaces $\mathscr{X}$ and classes $\mathscr{C}$ due to Kiefer [75], Révész [82], Richter [84], and Wichura [88]; it contains the Strassen–Finkelstein theorem as a special case ($\mathscr{X} = [0,1]$, $\mathscr{C} = \{[0,t] : 0 \leqslant t \leqslant 1\}$).

For the $\mathscr{F}$-empirical process (indexed by functions f in some collection $\mathscr{F}$), only partial results are available (see, e.g., Kaufman and Philipp [74]). However, if $\mathscr{F}$ is a class of functions satisfying the hypotheses of Dudley [17] or Pollard [81] sufficient for weak convergence, the following iterated logarithm law should hold: With probability 1 every subsequence of

$$\left\{\frac{G_n(f)}{(2\log\log n)^{1/2}} : n \geqslant 3, f \in \mathscr{F}\right\}$$

has a uniformly (in $f \in \mathscr{F}$) convergent subsequence, and the set of limit functions is

$$\mathscr{G}^0_{\mathscr{F}} = \left\{g : \mathscr{F} \to R^1 \text{ defined by } g(f) = \int fh\,dP, f \in \mathscr{F};\ h \in \mathscr{H}^0\right\}.$$

Almost Surely Convergent Versions; Strong Approximations

In higher-dimensional situations the Skorokhod–Dudley–Wichura theorem continues to guarantee the existence of almost surely convergent versions G_n^* of the empirical process G_n, and this again provides an extremely useful way to treat statistics representable as functions of $\mathbb{P}_n$ and G_n.

Concerning explicit strong approximations much less is known, the best results being those of Philipp and Pinzur [78] (for the case $\mathscr{X} = R^k$, general P, and $\mathscr{C}$ = the lower left orthants) and Révész [82, 83] ($\mathscr{X} = [0,1]^k$, P uniform on $[0,1]^k$, and $\mathscr{C}$ = a class of sets with smooth boundaries).

DEPENDENT OR NONIDENTICALLY DISTRIBUTED OBSERVATIONS

In many cases of practical importance the observations are either nonidentically distributed, or dependent, or both. In comparison to the i.i.d. case treated in the preceding sections, present knowledge of the empirical measures $\mathbb{P}_n$ and corresponding empirical processes G_n is much less complete in these cases. A variety of results are available, however, for the most important case of $\mathscr{X} = R^k$ and $\mathscr{C} = \{(-\infty, x] : x \in R^k\}$, the lower-left orthants.

Independent, Nonidentically Distributed Observations

When the observations $X_1, \ldots, X_n$ have distributions $P_1, \ldots, P_n$ on $\mathscr{X}$, the natural empirical process to consider is

$$G_n = n^{1/2}(\mathbb{P}_n - \overline{P}_n),$$
$$\overline{P}_n = n^{-1}(P_1 + \cdots + P_n).$$

In the case $\mathscr{X} = R^1$ and $\mathscr{C} = \{(-\infty, x] : x \in R^1\}$, sufficient conditions for weak convergence of ("reduced versions" of) G_n have been given by Koul [103], Shorack [121, 122], and Withers [128]. These authors also study the "weighted" or "regression" processes $\mathbf{W}_n = \sum_{i=1}^n c_{ni}(\delta_{X_i} - P_i)/(\sum_i^n c_{ni}^2)^{1/2}$, where the c_{ni}'s are appropriate (regression) constants (see also Hájek [102]); Shorack [122] gives convergence with respect to weighted metrics and convergence theorems for the related quantile processes; Withers [128] allows the observations to be dependent (strong mixing). Interesting inequalities for the limiting distributions of supremum functionals of the process are given by Sen et al. [120] and Rechtschaffen [115]; van Zuijlen [124, 125] gives linear bounds and many useful inequalities.

In the case $\mathscr{X} = R^k$ and $\mathscr{C}$ = the lower-left orthants, conditions ensuring weak convergence of ("reduced" versions of) G_n have been given by Neuhaus [108] and Rüschendorf [119]. Many of the weak convergence theorems above are (naturally) formulated for triangular arrays of RVs with independent RVs in each row.

Although little is known about functional laws of the iterated logarithm analogous to the Strassen–Finkelstein theorem for independent nonidentically distributed observations, a recent inequality due to Bretagnolle [91] makes possible the following extension of the Chung–Smirnov law of the iterated logarithm in the case $\mathscr{X} = R^1$, $\mathscr{C} = \mathscr{O}_1$, and the observations form a *single* independent sequence. Let $\mathbf{F}_n(x) = \mathbb{P}_n(-\infty, x]$, $\overline{F}_n(x) = \overline{P}_n(-\infty, x]$, and $G_n(x) = n^{1/2}(\mathbf{F}_n(x) - \overline{F}_n(x))$ for $x \in R^1$, so $\|G_n\| = n^{1/2}\|\mathbf{F}_n - \overline{F}_n\| = n^{1/2}\sup_x|\mathbf{F}_n(x) - \overline{F}_n(x)|$. Bretagnolle's [91] inequality says that the classical exponential bound of Dvoretzky et al. [97] for the i.i.d. case continues to hold (for arbitrary df's of the observations $F_1, \ldots, F_n$) if their absolute constant is increased by a factor of 4:

$$\Pr(\|G_n\| \geqslant \lambda) = \Pr\left(n^{1/2}\|\mathbf{F}_n - \overline{F}_n\| \geqslant \lambda\right)$$
$$\leqslant 4C\exp(-2\lambda^2)$$

for all $n \geqslant 1$ and all $\lambda > 0$, where C is an absolute constant (weaker inequalities were given earlier by Singh [123] and Devroye [95]). A consequence of Bretagnolle's inequality is that

$$\limsup_{n\to\infty} \frac{\|G_n\|}{(2\log\log n)^{1/2}}$$
$$= \limsup_{n\to\infty} \frac{n^{1/2}\|\mathbf{F}_n - \overline{F}_n\|}{(2\log\log n)^{1/2}} \leqslant \tfrac{1}{2} \text{ a.s.}$$

for independent observations $X_1, X_2, \ldots$ from a completely arbitrary sequence of df's $F_1, F_2, \ldots$.

The results for G_n (and $\mathbf{W}_n$) sketched here have been applied by Koul [104] and Bickel [89] (regression problems), Shorack [121] (linear combinations of order statistics), Sen et al. [120] (strength of fiber bundles), and Gill [101] (censored survival data).

Dependent Observations

Billingsley [90, Sec. 22] proved two different weak convergence or Donsker theorems for the empirical process of a strictly stationary sequence of real-valued random variables with common continuous df F satisfying a weak or ϕ-mixing condition. Billingsley's results have subsequently been extended to other weaker (i.e., less restrictive) mixing conditions by Mehra and Rao [107] (who also consider the regression process $\mathbf{W}_n$ mentioned above and weighted metrics), Gastwirth and Rubin [99] (who introduced a new mixing condition intermediate between weak and strong mixing), and Withers [128]. Puri and Tran [110] provide linear in probability bounds, almost sure nearly linear bounds, and strengthened Glivenko–Cantelli theorems for $\mathbf{F}_n$ under a variety of mixing conditions.

When the (dependent) stationary sequence of observations has values in $\mathscr{X} = R^k$, Donsker theorems for the empirical process have been given by Rüschendorf [118] and Yoshihara [129]. The recent strong approximation results of Philipp and Pinzur [109] apply to strictly stationary R^k-valued observations with common continuous df satisfying a certain strong-mixing property. This strong approximation has, as corollaries, both Donsker (weak convergence) and Strassen–Finkelstein iterated logarithm theorems for the empirical processes of such variables.

An especially interesting Donsker theorem application for the empirical process of mixing variables is to robust location estimators under dependence by Gastwirth and Rubin [100].

Dependent and/or nonidentically distributed observations and the corresponding empirical processes also arise in studies of (a) problems involving finite populations [116, 117]; (b) closely related problems concerning permutation tests* [89, 127]; (c) residuals and "parameter-estimated empirical processes" [92, 96, 106]; (d) Fourier coefficients of an i.i.d. real-valued sample [98]; and (e) the spacings between the points of an i.i.d. sample [93, 94, 105, 111, 112, 114, 121]. An interesting variant on the latter set of problems is Kakutani's method of interval subdivision; see Van Zwet [126] and Pyke [113] for a discussion of Glivenko–Cantelli theorems; analogs of the Donsker theorem and the Strassen–Finkelstein theorem seem to be unknown.

MISCELLANEOUS TOPICS

This section briefly summarizes work concerning (a) censored survival data and the product limit estimator, (b) optimality properties of $\mathbb{P}_n$ as an estimator of P, and (c) large deviation theorems for empirical measures and processes.

Censored Survival Data; The Product Limit Estimator

In many important problems arising in medical or reliability settings, RVs $X_1, \ldots, X_n$ (i.i.d. with common df F) representing "survival times," cannot be observed. Instead, the statistician observes $(Z_1, \delta_1), \ldots, (Z_n, \delta_n)$, where Z_i is the smaller of the lifetime X_i and a censoring time Y_i, $Z_i = \min\{X_i, Y_i\}$, and δ_i equals 1 or 0 according as $Z_i = X_i$ or $Z_i = Y_i$. The statistician's goal is to estimate the df F of the survival times $\{X_i\}$, in spite of the *censoring*.

The nonparametric maximum likelihood estimator of F, the product-limit estimator (or Kaplan–Meier estimator*) $\hat{\mathbf{F}}_n$, was derived by Kaplan and Meier [139]:

$$1 - \hat{\mathbf{F}}_n(t) = \prod_{\{i:\, Z_{n:i} \leqslant t\}} (1 - 1/(n - i + 1))^{\delta_{n:i}},$$

where $Z_{n:1} \leqslant \cdots \leqslant Z_{n:n}$ and $\delta_{n:1}, \ldots, \delta_{n:n}$ denote the corresponding δ's. When there is no censoring, so $Z_i = X_i$ and $\delta_i = 1$ for all $i = 1, \ldots, n$, the product-limit estimator $\hat{\mathbf{F}}_n$ reduces to the usual empirical df $\mathbf{F}_n$.

Study of Donsker or weak convergence theorems for the corresponding empirical process

$$\hat{G}_n = n^{1/2}(\hat{\mathbf{F}}_n - F)$$

was initiated by Efron [134] under the assumption of i.i.d. censoring variables Y_i independent of the X_i (the random censorship model). Efron conjectured the weak convergence of $\hat{G}_n$, and used it in a study of two-sample statistics of interest for censored data. The weak convergence of $\hat{G}_n$ was first proved by Breslow and Crowley [132] under the assumption of i.i.d. censoring variables with common df G by use of a Skorokhod construction and long calculations. Gill [137], following Aalen [130, 131], put the large-sample theory of $\hat{\mathbf{F}}_n$ and $\hat{G}_n$ in its natural setting by using the martingale* theory of counting processes together with a martingale (functional) central limit theorem due to Rebolledo [140] to give a simpler proof of the weak convergence under minimal assumptions on the independent censoring times $\{Y_i\}, \dots, Y_n$. To state the theorem, let $C(t) = \int_0^t (1-F)^{-2}(1-G)^{-1}\,dF$ and set $K(t) = C(t)/(1 + C(t))$.

Theorem 7. (Breslow and Crowley [132]; Gill [137]).

$$\hat{G}_n = n^{1/2}(\hat{\mathbf{F}}_n - F) \Rightarrow (1-F)\cdot \mathbb{B}\circ C$$
$$\stackrel{d}{=} \left(\frac{1-F}{1-K}\right)\cdot \mathbf{U}\circ K \qquad \text{as } n\to\infty$$

where $\mathbb{B}$ denotes standard Brownian motion on $[0,\infty)$.

Gill [138] has given a refined and complete version of this theorem. Aalen [130, 131] and Gill [137, 138] have clarified the extremely important role which counting processes, and their associated martingales, play in the theory of empirical processes in the uncensored as well as the censored case.

Some preliminary iterated logarithm laws for G_n have been established by Földes and Rejtö [135, 136]; iterated logarithm laws also follow from the strong approximations of $\hat{G}_n$ and other related processes provided by Burke et al. [133].

Optimality

Asymptotic minimax theorems demonstrating the asymptotic optimality of the empirical df $\mathbf{F}_n$ in a very large class of estimators of F and with respect to a large class of loss functions were first obtained by Dvoretzky et al. [144] in the i.i.d. case with $\mathscr{X} = R^1$, and by Kiefer and Wolfowitz [145] in the case $\mathscr{X} = R^k$; see also Levit [148]. An interesting representation theorem for the limiting distributions of regular estimates of a df F on $[0,1]$ has been established by Beran [141]. This asserts, roughly speaking, that the limiting process corresponding to any regular estimator of F has a representation as $\mathbf{U}\circ F + W$, where $\mathbf{U}$ is a Brownian bridge process and W is some process on $[0,1]$ independent of $\mathbf{U}$. Hence the empirical df $\mathbf{F}_n$ is an optimal estimator of F in this sense since $G_n = n^{1/2}(\mathbf{F}_n - F) \stackrel{d}{=} \mathbf{U}_n\circ F \Rightarrow \mathbf{U}\circ F$ with $W = 0$ identically.

Motivated by questions in reliability, Kiefer and Wolfowitz [146] showed that the empirical df $\mathbf{F}_n$ remains asymptotically minimax for the problem of estimating a concave (or convex) df (even though $\mathbf{F}_n$ is itself not necessarily concave). Millar [149], using results of LeCam [147], put the earlier asymptotic minimax results in an elegant general setting and gave a geometric sufficient condition in order that the empirical df $\mathbf{F}_n$ be an asymptotically minimax estimator of F in a specified subset of df's. Millar's geometric criterion implies, in particular, that the empirical df is asymptotically minimax for estimating F in the classes of distributions with increasing or decreasing failure rates, or the class of distribution functions with decreasing densities on $[0,\infty)$; also, $\mathbf{F}_n$ is not asymptotically optimal as an estimator of a df symmetric at 0 (the symmetrized empirical df is optimal for this class). Wellner [15] established the asymptotical optimality of the product limit estimator in the case of randomly censored data.

There is a large literature concerning the

power of various tests based on the empirical df and empirical processes; see Chibisov [142, 143] on local alternatives, and Raghavachari [150] concerning the limiting distributions of Kolmogorov statistics under fixed alternatives.

Large Deviations*

Suppose that $X_1, \dots, X_n$ are i.i.d. RVs with values in $\mathscr{X}$, common probability measure P on $\mathscr{X}$, and empirical measures $\mathbb{P}_n$, $n \geqslant 1$, as in the introduction. If Π is a collection of probability measures on $\mathscr{X}$ distant from P, then, by a Glivenko–Cantelli theorem,

$$\Pr(\mathbb{P}_n \in \Pi) \to 0 \qquad \text{as } n \to \infty$$

since $\mathbb{P}_n \to P$ a.s. (in a variety of senses). In fact, this convergence to zero typically occurs exponentially fast as n increases, as demonstrated in problems concerning the Bahadur efficiency* of a variety of test statistics; see Groeneboom et al. [156], Bahadur and Zabell [154], and references therein. The constant appearing in the exponential rate is given by the Kullback–Liebler information of Π relative to P, $K(\Pi, P)$, defined by

$$K(\Pi, P) = \inf_{Q \in \Pi} K(Q, P),$$

$$K(Q, P) = \begin{cases} \int q \log q \, dP, & Q \ll P, \; q \equiv \dfrac{dQ}{dP}, \\ \infty, & \text{otherwise.} \end{cases}$$

Theorem 8. (Groeneboom et al. [156]). If $\mathscr{X}$ is a Hausdorff space, Π is a collection of probability measures on $\mathscr{X}$ satisfying $K(\Pi^0, P) = K(\overline{\Pi}, P) = K(\Pi, P)$, where the interior Π^0 and closure $\overline{\Pi}$ of Π are taken relative to a certain topology τ, then

$$\Pr(\mathbb{P}_n \in \Pi) = \exp(-n[K(\Pi, P) + o(1)]) \qquad \text{as } n \to \infty$$

[i.e., $\lim_{n\to\infty} n^{-1} \log \Pr(\mathbb{P}_n \in \Pi) = K(\Pi, P)$].

Groeneboom et al. [156] give several applications of this general theorem. In the special case of i.i.d. uniform (0, 1) X's and

$$\Pi = \{P : \sup_t (F(t) - t) \geqslant \lambda > 0 \text{ with } F(t) = P(-\infty, t]\},$$

the number $K(\Pi, I)$ has been computed explicitly by Sethuraman [157], Abrahamson [152], Bahadur [153], and Siegmund [158]:

$$\begin{aligned} K(\Pi, I) &= \inf_{0 \leqslant t \leqslant 1-\lambda} \left\{ (\lambda + t) \log\left(\frac{\lambda + t}{t}\right) + (1 - \lambda - t) \log\left(\frac{1 - \lambda - t}{1 - t}\right) \right\} \\ &= (\theta_1 - \theta_2)\lambda + \theta_2 + \log(1 - \theta_2) \equiv g(\lambda), \end{aligned}$$

where $\theta_2 < 0 < \theta_1$ satisfy $\theta_1^{-1} + \theta_2^{-1} = \lambda^{-1}$ and $\theta_1 - \theta_2 = \log[(1 - \theta_2)/(1 - \theta_1)]$. The calculations of Siegmund [158] make the $o(1)$ term explicit in this case.

$$\Pr\left(\sup_{0 \leqslant t \leqslant 1} (\mathbf{F}_n(t) - t) > \lambda\right) \sim h(\lambda) \exp(-n g(\lambda)) \qquad \text{as } n \to \infty,$$

where

$$h(\lambda) \equiv \left\{ \lambda |\theta_2|^{-1} (1 - \theta_2) \times \left[1 + \left(\frac{|\theta_2|}{\theta_1} \right)^3 \left(\frac{1 - \theta_1}{1 - \theta_2} \right) \right] \right\}^{-1/2}.$$

Berk and Jones [155] have some related results.

References

Introduction

[1] Durbin, J. (1973). *Distribution Theory for Tests Based on the Sample Distribution Function*. Reg. Conf. Ser. Appl. Math. No. 9. SIAM, Philadelphia.

[2] Gaenssler, P. and Stute, W. (1979). *Ann. Prob.*, **7**, 193–243.

[3] Niederhausen, H. (1981). *Ann. Statist.*, **9**, 923–944.

[4] Pyke, R. (1972). Jeffrey–Williams Lectures, *Canad. Math. Congr.*, Montreal, pp. 13–43.

The One-Dimensional Case

[5] Billingsley, P. (1968). *Convergence of Probability Measures*. Wiley, New York.

[6] Billingsley, P. (1971). *Weak Convergence of Measures: Applications in Probability*. Reg. Conf. Ser. Appl. Math. No. 5. SIAM, Philadelphia.

[7] Bolthausen, E. (1977). *Metrika*, **24**, 215–227.

[8] Boos, D. D. (1981). *J. Amer. Statist. Ass.*, **76**, 663–670.

[9] Breiman, L. (1968). *Probability*. Addison-Wesley, Reading, Mass.

[10] Breslow, N. and Crowley, J. (1974). *Ann. Statist.*, **2**, 437–453.

[11] Brillinger, D. R. (1969). *Bull. Amer. Math. Soc.*, **75**, 545–547.

[12] Cassels, J. W. S. (1951). *Proc. Camb. Philos. Soc.*, **47**, 55–64.

[13] Chibisov, D. M. (1964). *Select. Transl. Math. Statist. Prob.*, **6**, 147–156.

[14] Chibisov, D. M. (1965). *Theor. Prob. Appl.*, **10**, 421–437.

[15] Chung, K. L. (1949). *Trans. Amer. Math. Soc.*, **67**, 36–50.

[16] Cramér, H. (1928). *Skand. Aktuarietidskr.*, **11**, 141–180.

[17] Cressie, N. (1977). *J. Appl. Prob.*, **14**, 272–283.

[18] Cressie, N. (1980). *Ann. Prob.*, **8**, 828–840.

[19] Csáki, E. (1974). *Limit Theorems of Probability Theory*. Keszthely, Hungary, pp. 47–58.

[20] Csáki, E. (1977). *Zeit. Wahrscheinlichkeitsth. verw. Geb.*, **38**, 147–167.

[21] Csörgő, M. and Révész, P. (1975). *Zeit. Wahrscheinlichkeitsth. verw. Geb.*, **31**, 261–269.

[22] Csörgő, M. and Révész, P. (1978). *Ann. Statist.*, **6**, 882–894.

[23] Csörgő, M. and Révész, P. (1981). *Strong Approximations in Probability and Statistics*. Academic Press, New York.

[24] Darling, D. A. (1957). *Ann. Math. Statist.*, **28**, 823–838.

[25] Darling, D. A. and Erdös, P. (1956). *Duke Math. J.*, **23**, 143–155.

[26] Donsker, M. D. (1951). *Mem. Amer. Math. Soc.*, **6**, 1–12.

[27] Donsker, M. D. (1952). *Ann. Math. Statist.*, **23**, 277–281.

[28] Doob, J. L. (1949). *Ann. Math. Statist.*, **20**, 393–403.

[29] Dudley, R. M. (1968). *Ann. Math. Statist.*, **39**, 1563–1572.

[30] Durbin, J. (1973). *Distribution Theory for Tests Based on the Sample Distribution Function*. Reg. Conf. Ser. Appl. Math. No. 9. SIAM, Philadelphia.

[31] Eicker, F. (1979). *Ann. Statist.*, **7**, 116–138.

[32] Erdös, P. and Kac, M. (1946). *Bull. Amer. Math. Soc.*, **52**, 292–302.

[33] Erdös, P. and Kac, M. (1947). *Bull. Amer. Math. Soc.*, **53**, 1011–1020.

[34] Finkelstein, H. (1971). *Ann. Math. Statist.*, **42**, 607–615.

[35] Jaeschke, D. (1979). *Ann. Statist.*, **7**, 108–115.

[36] James, B. R. (1975). *Ann. Prob.*, **3**, 762–772.

[37] Kiefer, J. (1969). *Zeit. Wahrscheinlichkeitsth. verw. Geb.*, **13**, 321–332.

[38] Kiefer, J. (1970). In *Nonparametric Techniques in Statistical Inference*, M. L. Puri, ed. Cambridge University Press, Cambridge, England, pp. 299–319.

[39] Kolmogorov, A. N. (1933). *Giorn. Ist. Ital. Attuari*, **4**, 83–91.

[40] Komlós, J., Major, P., and Tusnády, G. (1975). *Zeit Wahrscheinlichkeitsth. verw. Geb.*, **32**, 111–131.

[41] Mason, D. M. (1981). *Ann. Prob.*, **9**, 881–884.

[42] O'Reilly, N. E. (1974). *Ann. Prob.*, **2**, 642–645.

[43] Pollard, D. (1980). *Metrika*, **27**, 43–70.

[44] Prohorov, Y. V. (1956). *Theor. Prob. Appl.*, **1**, 157–214.

[45] Pyke, R. (1969). *Lect. Notes Math.*, **89**, 187–200.

[46] Pyke, R. (1970). In *Nonparametric Techniques in Statistical Inference*, M. L. Puri, ed. Cambridge University Press, Cambridge, England, pp. 21–37.

[47] Pyke, R. (1972). Jeffrey–Williams Lectures. *Canad. Math. Congr.*, Montreal, pp. 13–43.

[48] Pyke, R. and Shorack, G. R. (1968). *Ann. Math. Statist.*, **39**, 755–771.

[49] Root, D. H. (1969). *Ann. Math. Statist.*, **40**, 715–718.

[50] Rosenkrantz, W. A. (1967). *Trans. Amer. Math. Soc.*, **129**, 542–552.

[51] Sähler, W. (1968). *Metrika*, **13**, 149–169.

[52] Sawyer, S. (1974). *Rocky Mountain J. Math.*, **4**, 579–596.

[53] Shorack, G. R. (1972). *Ann. Math. Statist.*, **43**, 412–427.

[54] Shorack, G. R. (1972). *Ann. Math. Statist.*, **43**, 1400–1411.

[55] Shorack, G. R. (1974). *Ann. Statist.*, **2**, 661–675.

[56] Shorack, G. R. (1979). *Stoch. Proc. Appl.*, **9**, 95–98.

[57] Shorack, G. R. (1980). *Aust. J. Statist.*, **22**, 50–59.

[58] Shorack, G. R. and Wellner, J. A. (1982). *Ann. Prob.*, **10**, 639–652.

[59] Skorokhod, A. V. (1956). *Theor. Prob. Appl.*, **1**, 261–290.

[60] Skorokhod, A. V. (1961). *Studies in the Theory of Random Processes*. Kiev University Press, Kiev. (English transl.: Addison-Wesley, Reading, Mass., 1965.)

[61] Smirnov, N. V. (1939). *Math. Sb. N.S.*, **6**, 3–26.

[62] Smirnov, N. V. (1944). *Uspekhi Mat. Nauk.*, **10**, 179–206.

[63] Strassen, V. (1964). *Zeit. Wahrscheinlichkeitsth. verw. Geb.*, **3**, 211–226.

[64] Stute, W. (1982). *Ann. Prob.*, **10**, 86–107.

[65] von Mises, R. (1931). *Wahrscheinlichkeitsrechnung*. Franz Deuticke, Leipzig, Germany.

[66] Wichura, M. J. (1970). *Ann. Math. Statist.*, **41**, 284–291.

More General Sample Spaces and Index Sets

[67] Bickel, P. J. and Wichura, M. J. (1971). *Ann. Math. Statist.*, **42**, 1656–1670.

[68] Bolthausen, E. (1978). *Zeit. Wahrscheinlichkeitsth. verw. Geb.*, **43**, 173–181.

[69] Dudley, R. M. (1966). *Ill. J. Math.*, **10**, 109–126.

[70] Dudley, R. M. (1978). *Ann. Prob.*, **6**, 899–929; correction: *Ann. Prob.*, **7**, 909–911.

[71] Dudley, R. M. (1980). Donsker classes of functions. Preprint.

[72] Giné, M. E. (1975). *Ann. Statist.*, **3**, 1243–1266.

[73] Hartigan, J. A. (1978). *Ann. Statist.*, **6**, 117–131.

[74] Kaufman, R. and Philipp, W. (1978). *Ann. Prob.*, **6**, 930–952.

[75] Kiefer, J. (1961). *Pacific J. Math.*, **11**, 649–660.

[76] Kuelbs, J. and Dudley, R. M. (1980). *Ann. Prob.*, **8**, 405–418.

[77] Neuhaus, G. (1971). *Ann. Math. Statist.*, **42**, 1285–1295.

[78] Philipp, W. and Pinzur, L. (1980). *Zeit. Wahrsheinlichkeitsth. verw. Geb.*, **54**, 1–13.

[79] Pollard, D. (1979). *Zeit. Wahrscheinlichkeitsth. verw. Geb.*, **50**, 317–331.

[80] Pollard, D. (1982). *Ann. Prob.*, **10**, 919–926.

[81] Pollard, D. (1982). *J. Aust. Math. Soc. A*, **33**, 235–248.

[82] Révész, P. (1976). *Ann. Prob.*, **5**, 729–743.

[83] Révész, P. (1976). *Lect. Notes Math.*, **566**, 106–126.

[84] Richter, H. (1974). *Manuscripta Math.*, **11**, 291–303.

[85] Straf, M. L. (1971). *Proc. 6th Berkeley Symp. Math. Statist. Prob.*, Vol. 2. University of California Press, Berkeley, Calif., pp. 187–221.

[86] Strassen, V. and Dudley, R. M. (1969). *Lect. Notes Math.*, **89**, 224–231.

[87] Wellner, J. A. (1979). *Ann. Statist.*, **7**, 929–943.

[88] Wichura, M. J. (1973). *Ann. Prob.*, **1**, 272–296.

Dependent or Nonidentically Distributed Observations

[89] Bickel, P. J. (1973). *Ann. Math. Statist.*, **40**, 1–23.

[90] Billingsley, P. (1968). *Convergence of Probability Measures*. Wiley, New York.

[91] Bretagnolle, J. (1980). Communication au *Colloque Int. du CNRS*, **307**, 39–44.

[92] Burke, M. D., Csörgő, M., Csörgő, S., and Révész, P. (1979). *Ann. Prob.*, **7**, 790–810.

[93] Cressie, N. (1979). *Biometrika*, **66**, 619–628.

[94] del Pino, G. E. (1979). *Ann. Statist.*, **7**, 1058–1065.

[95] Devroye, L. P. (1977). *J. Multivariate Anal.*, **7**, 594–597.

[96] Durbin, J. (1973). *Ann. Statist.*, **1**, 279–290.

[97] Dvoretzky, A., Kiefer, J., and Wolfowitz, J. (1956). *Ann. Math. Statist.*, **27**, 642–669.

[98] Freedman, D. and Lane, D. (1980). *Ann. Statist.*, **8**, 1244–1251.

[99] Gastwirth, J. L. and Rubin, H. (1975). *Ann. Statist.*, **3**, 809–824.

[100] Gastwirth, J. L. and Rubin, H. (1975). *Ann. Statist.*, **3**, 1070–1100.

[101] Gill, R. D. (1980). *Censoring and Stochastic Integrals*. Mathematische Centrum, Amsterdam, The Netherlands.

[102] Hájek, J. (1965). *Bernoulli–Bayes–Laplace Seminar*. University of California Press, Berkeley, Calif., pp. 45–60.

[103] Koul, H. L. (1970). *Ann. Math. Statist.*, **41**, 1768–1773.

[104] Koul, H. L. (1971). *Ann. Math. Statist.*, **42**, 466–476.

[105] Koziol, J. A. (1977). *J. R. Statist. Soc. B*, **39**, 333–336.

[106] Loynes, R. M. (1980). *Ann. Statist.*, **8**, 285–298.

[107] Mehra, K. and Rao, M. S. (1975). *Ann. Prob.*, **3**, 979–991.

[108] Neuhaus, G. (1975). *Ann. Statist.*, **3**, 528–531.

[109] Philipp, W. and Pinzur, L. (1980). *Zeit. Wahrscheinlichkeitsth. verw. Geb.*, **54**, 1–13.

[110] Puri, M. L. and Tran, L. T. (1980). *J. Multivariate Anal.*, **10**, 405–425.

[111] Pyke, R. (1965). *J. R. Statist. Soc. B*, **7**, 395–449.

[112] Pyke, R. (1972). *Proc. 6th Berkeley Symp. Math. Statist. Prob.*, Vol. 1. University of California Press, Berkeley, Calif., pp. 417–427.

[113] Pyke, R. (1980). *Ann Prob.*, **8**, 157–163.

[114] Rao, J. S. and Sethuraman, J. (1975). *Ann. Statist.*, **3**, 299–313.

[115] Rechtschaffen, R. (1975). *Ann. Statist.*, **3**, 787–792.

[116] Rosen, B. (1964). *Ark. Math.*, **4**(28), 383–424.

[117] Rosen, B. (1967). *Zeit. Wahrscheinlichkeitsth. verw. Geb.*, **7**, 103–115.

[118] Rüschendorf, L. (1974). *J. Multivariate Anal.*, **4**, 469–478.

[119] Rüschendorf, L. (1976). *Ann. Statist.*, **4**, 912–923.

[120] Sen, P. K., Bhattacharya, B. B., and Suh, M. W. (1973). *Ann. Statist.*, **1**, 297–311.

[121] Shorack, G. R. (1973). *Ann. Statist.*, **1**, 146–152.

[122] Shorack, G. R. (1979). *Statist. Neerlandica*, **33**, 169–189.

[123] Singh, R. S. (1975). *Ann. Prob.*, **3**, 371–374.

[124] van Zuijlen, M. C. A. (1976). *Ann. Statist.*, **4**, 406–408.

[125] van Zuijlen, M. C. A. (1978). *Ann. Prob.*, **6**, 250–266.

[126] Van Zwet, W. R. (1978). *Ann. Prob.*, **6**, 133–137.

[127] Wellner, J. A. (1979). *Ann. Statist.*, **7**, 929–943.

[128] Withers, C. (1976). *Aust. J. Statist.*, **18**, 76–83.

[129] Yoshihara, K. (1975). *Zeit. Wahrscheinlichkeitsth. verw. Geb.*, **33**, 133–138.

Miscellaneous Topics

Censored survival data; the product limit estimator

[130] Aalen, O. O. (1978). *Zeit. Wahrscheinlichkeitsth. verw. Geb.*, **38**, 261–277.

[131] Aalen, O. O. (1978). *Ann. Statist.*, **6**, 701–726.

[132] Breslow, N. and Crowley, J. (1974). *Ann. Statist.*, **2**, 437–453.

[133] Burke, M. D., Csörgő, S., and Horváth, L. (1981). *Zeit. Wahrscheinlichkeitsth. verw. Geb.*, **56**, 87–112.

[134] Efron, B. (1967). *Proc. 5th Berkeley Symp. Math. Statist. Prob.*, Vol. 4. University of California Press, Berkeley, Calif., pp. 831–853.

[135] Földes, A. and Rejtö, L. (1981). *Ann. Statist.*, **9**, 122–129.

[136] Földes, A. and Rejtö, L. (1981). *Zeit. Wahrscheinlichkeitsth. verw. Geb.*, **56**, 75–86.

[137] Gill, R. D. (1980). *Censoring and Stochastic Integrals*. Math. Center Tract No. 124. Mathematical Centre, Amsterdam, The Netherlands.

[138] Gill, R. D. (1983). *Ann. Statist.*, **11**, 49–58.

[139] Kaplan, E. L. and Meier, P. (1958). *J. Amer. Statist. Ass.*, **53**, 457–481.

[140] Rebolledo, R. (1980). *Zeit. Wahrscheinlichkeitsth. verw. Geb.*, **51**, 269–286.

Optimality

[141] Beran, R. (1977). *Ann. Statist.*, **5**, 400–404.

[142] Chibisov, D. M. (1965). *Theor. Prob. Appl.*, **10**, 421–437.

[143] Chibisov, D. M. (1969). *Sankhyā A*, **31**, 241–258.

[144] Dvoretzky, A., Kiefer, J., and Wolfowitz, J. (1956). *Ann. Math. Statist.*, **27**, 642–669.

[145] Kiefer, J. and Wolfowitz, J. (1959). *Ann. Math. Statist.*, **30**, 463–489.

[146] Kiefer, J. and Wolfowitz, J. (1976). *Zeit. Wahrscheinlichkeitsth. verw. Geb.*, **34**, 73–85.

[147] LeCam, L. (1972). *Proc. 6th Berkeley Symp. Math. Statist. Prob.*, Vol. 1. University of California Press, Berkeley, Calif., pp. 245–261.

[148] Levit, B. Ya. (1978). *Theor. Prob. Appl.*, **23**, 371–377.

[149] Millar, P. W. (1979). *Zeit. Wahrscheinlichkeitsth. verw. Geb.*, **48**, 233–252.

[150] Raghavachari, M. (1973). *Ann. Statist.*, **1**, 67–73.

[151] Wellner, J. A. (1982). *Ann. Statist.*, **10**, 595–602.

Large deviations

[152] Abrahamson, I. G. (1967). *Ann. Math. Statist.*, **38**, 1475–1490.

[153] Bahadur, R. R. (1971). *Some Limit Theorems in Statistics*. SIAM, Philadelphia.

[154] Bahadur, R. R. and Zabell, S. L. (1979). *Ann. Prob.*, **7**, 587–621.

[155] Berk, R. H. and Jones, D. H. (1979). *Zeit. Wahrscheinlichkeitsth. verw. Geb.*, **47**, 47–59.

[156] Groeneboom, P., Oosterhoff, J., and Ruymgaart, F. H. (1979). *Ann. Prob.*, **7**, 553–586.

[157] Sethuraman, J. (1964). *Ann. Math. Statist.*, **35**, 1304–1316.

[158] Siegmund, D. O. (1982). *Ann. Prob.*, **10**, 581–588.

Bibliography

Dudley, R. M. (1984). A Course on Empirical Processes. *Lecture Notes in Math.*, **1097**, 2–142. (École d'Été de Probabilités de St.-Flour, 1982.)

Gaenssler, P. (1983). *Empirical Processes*. Inst. Math. Statist., Lecture Notes-Monograph Series (S. S. Gupta, ed.), Vol. 3.

Pollard, D. (1984). *Convergence of Stochastic Processes*. Springer Verlag, New York.

Shorack, G. R. and Wellner, J. A. (1986). *Empirical Processes with Applications to Statistics*. Wiley, New York.

(CONVERGENCE OF SEQUENCES OF RANDOM VARIABLES
GLIVENKO–CANTELLI THEOREMS
HUNGARIAN CONSTRUCTION OF EMPIRICAL PROCESSES
LAW OF THE ITERATED LOGARITHM
STOCHASTIC PROCESSES)

JON A. WELLNER

PROCRUSTES TRANSFORMATION

A procedure used in factor analysis* for transforming a matrix to obtain a matrix approximately of prespecified form, in particular to transform factors so that the factor loading analysis approximates to some desired form (a "target" matrix).

Let **F** denote the $r \times r$ factor loading matrix obtained from data and **H** denote a target matrix, also of order $r \times r$. We seek an $s \times s$ matrix **S** such that $\tilde{\mathbf{H}} = \mathbf{FS}$ "best" matches **H**. "Best" is conventionally defined as meaning "minimizes the sum of squares of differences between corresponding elements of **H** and $\tilde{\mathbf{H}}$." The required **S** is

$$\mathbf{S} = (\mathbf{FF}')^{-1}\mathbf{F}'\mathbf{H}.$$

It represents a regression* of each column of **H** on the corresponding column of **F**. Cliff [2] restricts transforming matrices (**S**) to be orthogonal*. References 1 and 3 to 6 give further details on construction and generalization; see also ref. 7.

References

[1] Browne, M. W. (1967). *Psychometrika*, **32**, 125–132.

[2] Cliff, N. (1966). *Psychometrika*, **31**, 33–42.

[3] Gower, J. C. (1975). *Psychometrika*, **40**, 33–51.

[4] Hurley, J. L. and Cattell, R. B. (1962). *Behav. Sci.*, **7**, 258–262.

[5] Kristof, W. and Wingersky, B. (1971). *Proc. 79th Annu. Conv. Amer. Psychol. Ass.*, pp. 81–90.

[6] Schönemann, P. H. (1966). *Psychometrika*, **31**, 1–10.

[7] Schönemann, P. H. and Carroll, R. H. (1970). *Psychometrika*, **35**, 245–256.

(FACTOR ANALYSIS)

PRODUCER PRICE INDEXES

The Producer Price Indexes (PPI) are calculated and released by the Bureau of Labor Statistics* (BLS). They began publication in 1902 as the Wholesale Price Indexes. They measure the *change* in prices *received* by *producers* for their goods and services. In 1980, BLS published the first installment of these indexes based on a comprehensively revised basis. The January 1986 indexes reflect the final major installment of that revision. The following discussion focuses on the indexes from the revised survey, but the old indexes will be referenced as appropriate (*see also* INDEX NUMBERS).

UNIVERSE AND CLASSIFICATION

Historically, these indexes have measured price changes for selected products of agriculture, manufacturing, and mining. The revision systematically measures the changes in prices received for all products sold by establishments classified in mining or manufacturing according to the Standard Industrial Classification (SIC). The basic building block of the survey is the four-digit industry. Generally following the Census of Manufactures product codes, products of each industry are coded to a seven- or nine-digit level. (The old indexes use an eight-digit product coding scheme that is unique to the PPI.)

There are three families of indexes produced from the same basic data base. The first are the output price indexes for each four-digit industry (493 in mining and manufacturing). These indexes reflect the price change of the output which leaves the particular industry. Sales among establishments within the industry are not included. This so-called net-output approach can also be applied to aggregations of these four-digit industries.

The stage-of-processing price indexes are the most useful summary-level indexes for macroeconomic analysis of inflation. In the old indexes they are calculated for crude, intermediate, and finished goods with appropriate subindexes for specific product areas (food, energy) or end markets. Each product is classified by its degree of fabrication and type of buyers and may be included in more than one stage-of-processing index. The new stage-of-processing indexes will be constructed from aggregates of industries, each industry being classified in total based on the pattern of its purchases and sales as a producer of crude materials, primary manufactures, intermediate manufactures, or finished goods.

The third set consists of price indexes for detailed products. In 1986, there are about 6000 such detailed indexes published each month.

THE INDEXES

Producer Price Indexes are calculated using a modified Laspeyres index of the general form

$$I_1 = \left(\frac{\sum_i P_{1,i} Q_{0,i}}{\sum_i P_{0,i} Q_{0,i}} \right) \times 100$$

where I_1 = index for period 1
$P_{1,i}, P_{0,i}$ = prices for item i in periods 1 and 0 respectively
$Q_{0,i}$ = quantity of item i sold in period 0 (the base period)

These indexes may be viewed as measuring the price change of a constant set of production through time.

PRICES

In order to price the output of a production unit, BLS requests the price which the producer actually received at the time of shipment for the item being priced—the transaction price. The desired price will reflect all applicable discounts and all applicable extras and surcharges. The price is f.o.b. the seller's freight dock and excludes all direct excise taxes and transportation charges. They seek to price all the different types of transactions and types of customers relevant to a particular item. All the conditions of the sale must be identified and held constant through time.

When an item is no longer being sold, it is necessary to replace it in the index. This must be done in such a way that only pure price change is captured by the index, and it remains as close as possible to the concept of pricing a constant set of production through time. This substitution procedure is generally referred to as "quality adjustment." It consists of two steps: (1) identifying all the changes in the specification of the item being priced, and (2) placing a value on each change.

It is the second step which presents the major difficulty. Traditionally, it has been the objective to value specification changes as the sum of the producer's variable costs, fixed costs, and standard markup associated with the item.

Increasingly, there are cases in which the producer cost methodology cannot be applied (e.g., when there is product improvement at reduced cost). Computers, electronic calculators, and other high-technology goods are the prime examples. A technique called hedonic* regression, which has been a research tool for some time, is being investigated for some of these cases. This approach views a particular good as composed of a large number of independent characteristics which are bought in bundles. For example, a computer is composed of certain cycle times, add times, memory size, and so on. A regression equation (in any number of possible functional forms) is estimated with the price of the computer as the dependent variable and the characteristics as the independent variables. The coefficients (or transformations of them) are then implicit prices for the characteristics. The coefficient for memory size may set an implicit value of $100 for each "K" of memory. A new computer with 4K more memory than an old one would thus have a quality increase of $400.

Although its application has been somewhat limited, the hedonic approach has yielded encouraging results. Three major limitations are associated with its application. First, all price-determining characteristics must be specified, which makes it a costly, time-consuming process. Second, it is of no assistance when an entirely new characteristic appears (e.g., a smog-control device on a car). Third, it is necessary to assume that the value of a characteristic is independent of the bundle in which it occurs. This assumption is believed to be approximately true for most cases, but requires evaluation.

USES

Producer Price Indexes are widely used by the general public as indicators of inflation. They are, in conjunction with the Consumer Price Index*, the unemployment rate, and the gross national product*, the chief statistical inputs to national fiscal and monetary policy. They are also heavily relied on by the business and research communities for economic and market analyses and projections. Most major measures of real economic activity (GNP, orders, shipments, inventories, productivity) are derived by using appropriate Consumer and Producer Price Indexes to deflate current dollar aggregates. Finally, increasing numbers of contract sales are using price escalation based on one or more Producer Price Indexes. In 1976, more than $100 billion in contracts had such escalators.

SAMPLING

The universe of interest is the set of all transactions originating in mining and manufacturing establishments. Each transaction is characterized by the industry of the seller and the product sold. It can be portrayed as follows:

Industry/Product	1	2	3	...	10,000
1					
2					
.					
.					
.					
493					

Each four-digit SIC industry (row in the table above) is sampled and estimated independently. A multistage probability-proportional-to-size* sample of transactions is selected for pricing through time. In the first stage, an average of 74 firms in the industry are selected from the unemployment insurance file of all employers in the country. These firms are visited by BLS economists, who use a multistage sampling technique called disaggregation to select an average of five specific items for pricing along with a particular type of transaction if any are sold in more than one way at different prices.

Output price indexes can be constructed for each industry with data from the appropriate row in the matrix above. Product indexes (wherever made) can be constructed from data in the appropriate column. Weights of major product groups and for industries are from the Census of Manufacturers. Within major product groups, weights are developed from the sampling process. Net output weights are developed from the input–output tables of the Bureau of Economic Analysis.

ERRORS

Because the old indexes were not based on probability samples, variances have not been available. Variances are now under development for the revised indexes. A total error profile will also be developed for the program covering such factors of errors in the sampling frame*, nonresponse* bias, and processing errors. Rigorous error control procedures are being installed at all critical stages in the process.

The difficulties associated with quality adjustment may contribute a major portion of the total error in price indexes. This kind of error is unique to estimating price indexes. As a result, the existing literature on errors in surveys is largely not applicable. This topic has received considerable attention from both BLS and academic researchers, but a generally acceptable solution has been elusive.

Bibliography

Archibald, R. B. (1977). *Ann. Econ. Social Meas.*, Winter, 57–72. [Extends Fisher and Schell (below). The direct theoretical basis for the PPI.]

Bureau of Labor Statistics (1982). *BLS Handbook of Methods, Bulletin 2134*, Vol. 1. Bureau of Labor Statistics, Washington, D.C., pp. 43–61. (Official, general description of PPI. Updated editions issued periodically.)

Early, J. F. (1978). *Monthly Labor Rev.*, Apr., 7–15. (General audience description of the theory and methods for the revision.)

Early, J. F. and Sinclair, J. (1981). In *The U.S. National Income and Product Accounts: Selected Topics*, M. Foss, ed. Chicago. (Extensive treatment of actual quality adjustment in the PPI. Illustrative example of hedonic approach.)

Fisher, F. M. and Schell, K. (1972). *The Economic Theory of Price Indexes*. New York. (Rigorous mathematical economic theory underlying modern price indexes.)

Triplett, J. E. (1971). *The Theory of Hedonic Quality Measurement and Its Use in Price Indexes*. Bureau of Labor Statistics, Washington, D.C. (A little dated, but the best introduction to the topic.)

Triplett, J. E. (1975). In P. H. Earl, *Analysis of Inflation*. Lexington Books, Lexington, Mass. (Basic, moderately technical discussion of the problems associated with quality adjustment.)

(BUREAU OF LABOR STATISTICS
CONSUMER PRICE INDEX
GROSS NATIONAL PRODUCT DEFLATOR
HEDONIC INDEX NUMBERS
INDEX NUMBERS
INDEX OF INDUSTRIAL PRODUCTION
PAASCHE–LASPEYRES INDEX NUMBERS)

JOHN F. EARLY

PRODUCTION SCHEDULING *See* LINEAR PROGRAMMING

PRODUCTIVITY MEASUREMENT

Productivity measurement is the estimation of a broad class of ratios of outputs to one input or a combination of associated inputs in particular industries or the entire economy. The aggregate output Q is associated with n inputs (factors) v_i, for $i = 1, 2, \ldots, n$, of production often aggregated into labor (L) and capital (K). The ratios, such as Q/L and Q/K, are *partial factor productivity* indices. The ratio $Q/\sum_{i=1}^{n} \alpha_i v_i$, where the α_i are weights, is referred to as the multifactor or *total factor productivity* (TFP) index. In comparisons over time, the partial index is a measure of the rate of change in total output per unit of a particular input, without separating out the contributions of other factors to the production process. It fails to account for the effects of technological advances, substitution between inputs, scale economies, and changes in input quality and product composition. In contrast, the TFP index is the rate of change in "productive efficiency," or more accurately, in the ratio of total output to a weighted combination of all inputs.

The value of the TFP index must be equal to 1 if *all* inputs, tangible or intangible, and the specification of their combination are correctly accounted for. If the TFP index exceeds unity, as it usually does in the case of industrial economies, it is because some inputs are left out or because the relationship between output and inputs (production function) is incompletely specified. For example, if Q is assumed to be produced by L and K only, and shows an increase of 4% between two points in time, it may turn out that 25% of this increase is explained by increases in the quantities of L and K, and the remaining 75% of the 4%, which is the value of the TFP index, is due to left-out inputs or misspecification. The identification of the misspecification, the left-out inputs, and the decomposition of the TFP index between them is a complex and as yet controversial problem in productivity measurement. The left-out variables have been interpreted to be among the sources and causes of economic growth and have frequently been identified with such factors as the various forms of technological progress, scale economies, research and development, and changes in the nature and quality of inputs, institutions, and availability of relatively cheap inputs, especially energy.

PARTIAL AND TOTAL FACTOR PRODUCTIVITY INDEXES

All productivity measurements before 1942 were of the simple partial factor productivity type, in particular, output per worker, or per hour worked (Q/L). They continue to be published by the statistical agencies of many countries (e.g., ref. 40) and in spite of their evident theoretical limitations, widely used. The reasons are to be found in the general availability of the relevant statistical information and the simplicity of the computational requirements. Furthermore, in interindustry, intercountry, or intertemporal comparisons the average product of labor (output per worker) is an indirect measure of the differences in capital and other resources available for a unit of labor to work with. Thus the partial factor productivity ratios are indicators, however primitive, of differences in technological and economic states, even though the causes of these differences remain unrevealed by them. It is primarily for this latter reason that the main interest in recent years has been in the TFP index.[1]

In 1942, Tinbergen [37] was the first to estimate TFP indexes by introducing a trend function of the form $A(t) = A(0)e^{\lambda t}$ into a log-linear regression of Q on L and K, where $A(t)/A(0)$ was the TFP index in period t relative to period 0. After the mid-1950s several other investigators independently estimated TFP indexes. In particular, in two seminal contributions Solow [32] and Kendrick [18] provided the foundations of two interrelated but conceptually different approaches—one based on the economic concept of the production function and the other on the statistical concept of index numbers*. These are discussed in the following two sections, while the analysis and decomposition of the TFP index are discussed in the final section of this article.

The literature on the subject is large. Recent developments in productivity measurements can be found *inter alia* in refs. 5, 20, 21, 25, and 27. Since 1958, the U.S. Bureau of Labor Statistics* has published five bibliographies on productivity; the latest [39] covers 1976–1978 and contains 1200 publications. Much of this literature is related to the economic theories of capital, production and returns to productive factors (see, e.g., refs. 2, 4, 13, 29, 31, 34, 35, 38, and 42).

THE PRODUCTION FUNCTION APPROACH

In the neoclassical economic theory of production the entire output is explained by the interaction of homogeneous and constant quality productive factors (inputs) as specified by a constant returns to scale (i.e., first-order homogeneous and additive) production function. In a typical two-input case the latter can be written as

$$Q = f(K, L), \tag{1}$$

where Q, K, and L represent output, capital, and labor, respectively. If changes in the quantities of these variables are considered in time, the percentage rate of growth of the output, as derived from (1), can be written as

$$\frac{\dot{Q}}{Q} = E_K \frac{\dot{K}}{K} + E_L \frac{\dot{L}}{L}, \tag{2}$$

where the dotted variables represent time derivatives, and E_K and E_L are defined as

$$\frac{\partial Q}{\partial K}\frac{K}{f} \quad \text{and} \quad \frac{\partial Q}{\partial L}\frac{L}{f}.$$

Because of the assumption of constant returns to scale, E_K and E_L sum to 1 (Euler's identity), and the percentage rate of growth of the output, as shown in (2), is the weighted (linear) combination of the percentage rates of growth of the inputs. Hence the growth rate of the output must lie between, or be equal to, the growth rates of the two inputs.

The empirical evidence is, however, different. In the United States, and in the advanced industrialized countries in general, the growth of labor in the long run has been exceeded by the growth of capital and output, the last two having been approximately equal to each other. In terms of partial productivity indicators, Q/L has grown over time, while Q/K has remained approximately constant. Concomitantly, the capital–labor ratio has increased over time (see e.g., refs. 6, 9, and 32).

To account for this phenomenon, assume (as in ref. 32) that the production function shifts outward as function of time. Then the function can be written as

$$Q(t) = F[L(t), K(t), t], \tag{3}$$

where t designates time and the amounts of the output as well as the constant quality inputs are functions of time. As shown in ref. 26, the production function can take various explicit forms. However, if $F[\cdot]$ is subject to constant returns to scale as in (1), the variable t can be factored out from F and (3) can be restated as

$$Q(t) = A(t) f[L(t), K(t)], \tag{4}$$

where $A(t)$ is a shift factor which determines the cumulative outward shift in the production function itself. $A(t)$ is neutral with respect to, and independent of, the amounts of L and K utilized in the production process. Hence it can be arbitrarily designated as a measure of technological change of a type which is neutral with respect to input combinations and not embodied in (or independent of) the qualities of the inputs themselves. Accordingly, it has been referred to as "neutral disembodied technological change." Analogously to (2), the percentage rate of change of the output over time can be obtained from (4) as

$$\frac{\dot{Q}}{Q} = \frac{\dot{A}}{A} + E_K \frac{\dot{K}}{K} + E_L \frac{\dot{L}}{L}$$

and

$$\frac{\dot{A}}{A} = \frac{\dot{Q}}{Q} - \left(E_K \frac{\dot{K}}{K} + E_L \frac{\dot{L}}{L} \right), \tag{5}$$

where $\dot{A}/A$ represents the percentage rate of growth of the cumulative shift factor, or the difference between the observed growth rate of the output and that part of its growth which is imputable to the rate of growth of the amounts of factors used in production. As a consequence, $\dot{A}/A$ can also be thought of as a residual.

The estimation of the shift factor is based on national income and input statistics requiring the restatement of (5) in value terms. Let the prices of Q, L, and K at time t be P, P^L, and P^K, respectively, and in line with neoclassical purely competitive production theory assume that the prices or returns to factors are determined by the value of their contributions to output (marginal products), so that $P^L = A(t)(\partial f/\partial L)P$ and $P^K = A(t)(\partial f/\partial K)P$. Given the assumption of competition, $PQ = P^L L + P^K K$ at any point in time. We obtain

$$\frac{\dot{A}}{A} = \frac{\dot{Q}}{Q} - \frac{P^L \dot{L} + P^K \dot{K}}{P^L L + P^K K}. \tag{6}$$

In his original 1957 article Solow [32] assumed that $A(t) = A(0)e^{\lambda t}$, where λ represents $[\dot{A}(t)/A(t)]$. In calculation he set $A(0) = 1$ and used the formula

$$A(t) = A(t-1)\left[1 + \frac{\Delta A(t)}{A(t-1)}\right]$$

for $t = 1, 2, \ldots, T$, where Δ is the first difference operator. Based on U.S. time series for 1909–1949 and using a production function of the specific form

$$Q(t) = L(t)^{\alpha} K(t)^{\beta} A(t),$$

where α and β are parameters, he found that $A(1949)/A(1909) = 1.81$; that is, the annual rate of technical progress in the United States over that period was 1.5%.

In fact, by setting $A(0) = 1$, $A(T)$ can be identified as the TFP index. Thus if t is expressed in years and $A(2) = 1.0404$, total productivity increased by 4.04% or at an annual rate $[\dot{A}(t)/A(t)]$ of 2%. It is in this sense that the TFP index is sometimes referred to as a measure of "technical change," one of several alternative descriptions in the literature (see, e.g., ref. 12).

Also, as shown by refs. 15 and 36, we obtain

$$\frac{A(T)}{A(0)} = \frac{Q(T)}{Q(0)} \bigg/ \frac{X(T)}{X(0)}, \tag{7}$$

where $X(T)/X(0)$ is the Divisia input index.

See INDEX NUMBERS *and* DIVISIA INDICES.

THE INDEX NUMBER APPROACH

The TFP index can be defined as an index of output divided by an index of input, an approach implicit in Mills [24], Abramovitz [1], Fabricant [14], and Kendrick [18, 19].

Kendrick's TFP index can be defined as

$$\frac{A(T)}{A(0)} = \frac{Q(T)/Q(0)}{\alpha[L(T)/L(0)] + \beta[K(T)/K(0)]}, \quad (8)$$

where α and β are income shares of labor and capital. The difference between Solow's TFP index and Kendrick's is that the latter is not explicitly based on a formal production function. However, as pointed out by Domar [11], Kendrick's index does imply a production function of the form $Q(t) = A(t)[\alpha L(t) + \beta K(t)]$. Also [22], under certain economic assumptions and for "small" changes in the quantities of the input and output, Kendrick's and Solow's indices are equivalent, and (8) can be rewritten as a Paasche index of $\partial Q/\partial L$ and $\partial Q/\partial K$.

Diewert [10] has proven that under certain economic assumptions, different index number formulas (e.g., those of Laspeyres, Paasche, and Fisher) are consistent with ("exact" for) different functional forms of (3). For example, if (3) takes the form

$$Q(t) = \left[\sum_{i=1}^{2}\sum_{j=1}^{2} a_{ij}L_i(t)K_j(t)\right]^{1/2} \quad (9)$$

where the constants $a_{ij} = a_{ji}$, then, as shown in [10], Fisher's ideal* input index is exact for (9). The problem is of course that the functional form of (3) is unknown. Caves et al. [3] proposed the use of the Törnqvist index for the measurement of input, output, and TFP indexes. Maddala [23] compared empirically four of Diewert's exact indices and concluded that the differences between the underlying four functional forms result in "negligible" differences between the corresponding TFP indices.

ANALYSIS AND DECOMPOSITION OF THE TFP INDEX

It is the output growth not directly attributable to the factors of production which is measured by the TFP index. It has also been referred to as "residual." A large part of the literature on productivity measurement is devoted to attempts to decompose this residual into indices of changes of variables which can account for it. Three approaches to the analysis and decomposition of the residual are described briefly below.

Growth Accounting

Some studies in this vein are Denison [6, 7, 8, 9] and Jorgensen and Griliches [16]. Denison in his implicit framework substituted a set of "explanatory variables," the traditionally defined inputs in the aggregate production, such as (1). He also dropped the assumption of constant returns to scale. His productivity index can be expressed as

$$A(t) = Q(t)/f(x_i\,;\,y_j), \quad (10)$$

where $x_i = x_1, x_2, \ldots, x_n$ and $y_j = y_1, y_2, \ldots, y_m$, are the explanatory variables. Their rates of change over time are Denison's "sources of growth." The x_i are disaggregated and quality-adjusted labor, capital and land, the contributions of which can be estimated by their respective income shares. The y_j are variables which measure scale effects and improvements in resource allocation. The final residual in the growth of output is Denison's productivity index, which he attributes to "advances of knowledge." The method in ref. 16 is similar to Denison's except that the classification of the data and perception of the required quality adjustments are different and Divisia indices are used in the calculations.

Capital Embodied Technological Change

A different concept of technological change than the one presented in Solow's 1957 model is found in refs. 21, 28, 33, and 34, among others. Best known among these is the model based on a production function of the form

$$Q(t) = A(t)L^{\alpha}(t)J^{1-\alpha}(t), \quad (11)$$

where $J(t)$ is the quality-weighted capital

defined by

$$J(t) = \sum_{i=0}^{n} K_i(t)(1 + \lambda)^i, \qquad (12)$$

$K_i(t)$ is the gross amount of capital produced in year i and in use in year t, and λ is the rate of improvement in the productivity of new machines over those produced in the previous year. In this model, the rate of increase in the TFP index is given by

$$\frac{\dot{A}(t)}{A(t)} = \frac{\dot{Q}(t)}{Q(t)} - \alpha \frac{\dot{L}(t)}{L(t)} - (1 - \alpha)\left[\frac{\dot{K}(t)}{K(t)} + \lambda - \lambda\dot{\bar{a}}\right], \qquad (13)$$

where $\bar{a}$ is the average age of capital. Maximum likelihood estimation* of the parameters of this model is discussed in ref. 41.

Best Practice Technique

Salter [30] attributes productivity growth to best practice technique which yields minimum production costs. He decomposes the TFP index into advances in technical knowledge, scale economies, and factor substitution. Salter examines proportionate changes in labor and capital, so as to attribute them to both neutral and nonneutral technical progress as well as to substitution between labor and capital. The discussion of Salter's measures and their application to U.K. and U.S. industries is also contained in ref. 30.

NOTE

1. A survey of partial productivity comparisons between countries can be found in [17]. The weighted mean of the ratio of labor hours worked at two points of time is called the *man-hour index*. The weights are usually the hourly earnings.

References

[1] Abramovitz, M. (1956). *Amer. Econ. Rev.*, **46**, 5–23.

[2] Arrow, K. J. (1962). *Rev. Econ. Stud.*, **29**, 155–173.

[3] Caves, D. W., Christensen, L. R., and Diewert, W. E. (1982). *Econometrica*, **50**, 1343–1414.

[4] Champernowne, D. G. (1961). In *The Theory of Capital*, F. A. Lutz, ed. Macmillan, London.

[5] Cowing, T. G. and Stevenson, R. E. (1981). *Productivity Measurement in Regulated Industries*. Academic Press, New York. (A collection of useful survey and application papers.)

[6] Denison, F. (1962). *The Sources of Economic Growth in the United States and the Alternatives before Us*. Committee for Economic Development, New York.

[7] Denison, F. (1964). *Amer. Econ. Rev.*, **54**, 90–94.

[8] Denison, F. (1967). *Amer. Econ. Rev.*, **57**, 325–332.

[9] Denison, F. (1967). *Why Growth Rates Differ*. The Brookings Institution, Washington, D.C.

[10] Diewert, W. E. (1976). *J. Econometrics*, **4**, 115–145. (Economic index numbers and their use in productivity measurement; intermediate.)

[11] Domar, E. D. (1961). *Econ. J.*, **71**, 709–729. (Concepts and measurement of technical progress; elementary.)

[12] Domar, E. D. (1962). *J. Pol. Econ.*, **70**, 597–608. (Reviews Kendrick's approach to productivity measurement; elementary.)

[13] Eckaus, R. S. and Lefeber, L. (1962). *Amer. Econ. Rev.*, **44**, 113–122.

[14] Fabricant, S. (1959). *Basic Facts on Productivity Change*. National Bureau of Economic Research, New York. (Elementary.)

[15] Hillinger, C. (1970). *Econometrica*, **38**, 773–774. (Defines Divisia form of TFP index; intermediate.)

[16] Jorgenson, D. and Grilliches, Z. (1967). *Rev. Econ. Stud.*, **34**, 249–284. (Concepts, methods and application.)

[17] Kravis, I. B. (1976). *Econ. J.*, **86**, 1–44. (Survey; elementary.)

[18] Kendrick, J. (1961). *Productivity Trends in the United States*. Princeton University Press, Princeton, N.J.

[19] Kendrick, J. (1973). *Postwar Productivity Trends in the United States*. National Bureau of Economic Research, New York.

[20] Kendrick, J. and Vaccara, B. N. eds. (1980). *New Developments in Productivity Measurement and Analysis*. University of Chicago Press, Chicago. (Collection of survey and application papers; elementary.)

[21] Kennedy, C. and Thirlwall, A. P. (1972). *Econ. J.*, **82**, 11–72. (Survey paper; intermediate.)

[22] Kleiman, E., Halevi, N., and Levhari, D. (1966). *Rev. Econ. Statist.*, **48**, 345–347.

[23] Maddala, G. S. (1979). In ref. 27, pp. 309–317.

[24] Mills, F. C. (1952). *Productivity and Economic Progress*. National Bureau of Economic Research, New York. (Elementary.)

[25] Nadiri, M. I. (1970). *J. Econ. Lit.*, **8**, 1137–1177. (Survey; elementary.)

[26] Nadiri, M. I. (1982). In *Handbook of Mathematical Economics*, Vol. 2. K. J. Arrow and M. D. Intriligator, eds. North-Holland, Amsterdam, pp. 431–490. (Economic theory of production; mathematical.)

[27] National Research Council (1979). *Measurement and Interpretation of Productivity*. National Academy of Sciences, Washington, D.C. (Elementary.)

[28] Nelson, R. R. (1964). *Amer. Econ. Rev.*, **54**, 575–606.

[29] Phelps, E. S. (1962). *Quart. J. Econ.*, **76**, 548–567.

[30] Salter, W. E. G. (1960). *Productivity and Technical Change*. Cambridge University Press, Cambridge, England.

[31] Samuelson, P. A. (1962). *Rev. Econ. Stud.*, **29**, 193–206.

[32] Solow, R. (1957). *Rev. Econ. Statist.*, **39**, 312–320. (Reprinted with correction in *Readings in Economic Statistics and Econometrics*, A. Zellner, ed. Little, Brown, Boston, 1968.)

[33] Solow, R. (1959). In *Mathematical Methods in the Social Sciences*, K. Arrow et al., eds. Stanford University Press, Stanford, Calif., pp. 89–104.

[34] Solow, R. (1962). *Amer. Econ. Rev.*, **52**, 76–86.

[35] Solow, R. (1963). *Capital Theory and the Rate of Return*. North-Holland, Amsterdam.

[36] Star, S. and Hall, R. E. (1976). *Econometrica*, **44**, 257–263. (Divisia indices of productivity; intermediate.)

[37] Tinbergen, J. (1942). In L. H. Klassen, et al. *Jan Tinbergen: Selected Papers*, 1959, pp. 182–221; first appeared in 1942 in German, published by North-Holland, Amsterdam.

[38] Tobin, J. (1961). *Amer. Econ. Rev.*, **52**, 26–37.

[39] U.S. Bureau of Labor Statistics (1980). *Productivity: A Selected Annotated Bibliography, 1976–78*. U.S. Government Printing Office, Washington, D.C.

[40] U.S. Bureau of Labor Statistics (1982). *Productivity Measures for Selected Industries, 1954–80*. U.S. Government Printing Office, Washington, D.C.

[41] Wickens, M. R. (1970). *Rev. Econ. Statist.*, **52**, 187–193.

[42] Worswick, G. D. N. (1959). *Oxf. Econ. Pap.*, **11**, 125–142.

(DIVISIA INDICES
FISHER'S IDEAL INDEX NUMBER
INDEX NUMBERS
INDEX OF INDUSTRIAL PRODUCTION
PAASCHE–LASPEYRES INDEX NUMBERS)

NURI T. JAZAIRI
LOUIS LEFEBER

PRODUCT-LIMIT ESTIMATOR *See* KAPLAN–MEIER ESTIMATOR

PRODUCT MOMENT

The expected value of a product of random variables $(X_1, X_2, \ldots, X_m)$ possibly raised to different powers. The usual notation is

$$\mu'_{r_1 r_2 \cdots r_m}(X_1, X_2, \ldots, X_m) = E\left[\prod_{j=1}^{m} X_j^{r_j}\right].$$

(For brevity the symbol $\mu'_{r_1 r_2 \cdots r_m}$ is often used.)

The corresponding *central* product moment is

$$\mu_{r_1 r_2 \cdots r_m} = E\left[\prod_{j=1}^{m} \{X_j - E[X_j]\}^{r_j}\right].$$

To emphasize the distinction between $\mu'_{r_1 r_2 \cdots r_m}$ and $\mu_{r_1 r_2 \cdots r_m}$ the former is sometimes called a *crude* product moment.

Some of the r's may be zero, but at least two should differ from zero.

Another name is *mixed moment*.

Absolute product moments are defined by

$$\nu'_{r_1 r_2 \cdots r_m} = E\left[\prod_{j=1}^{m} |X_j|^{r_j}\right],$$

$$\nu_{r_1 r_2 \cdots r_m} = E\left[\prod_{j=1}^{m} |X_j - E[X_j]|^{r_j}\right].$$

(ABSOLUTE MOMENT
COVARIANCE)

PRODUCT, MULTINOMIAL *See* CONTINGENCY TABLES

PROFILES, CIRCULAR *See* SNOWFLAKES

PROFILES OF DIVERSITY

Diversity is an important concept in ecology and, under various names, it appears in several biological, physical, social, and management sciences*. The common theme is the apportionment of some quantity into a number of well-defined categories. The quantity may be in such a form as abundance, time, resource, investment, or energy. The concern in diversity analysis is about the degree of the apportionment being more diverse or less diverse, expressed variously as diversification or concentration, spread or specialization, monopoly or lack of it, and so on.

As a simple example, consider the apportionment of time spent on two activities, music and mathematics. Suppose that John and Jane apportion their time as follows:

Person	Category	
	Music	Mathematics
John	$\frac{2}{3}$	$\frac{1}{3}$
Jane	$\frac{1}{3}$	$\frac{2}{3}$

If we ask whether John has a different *kind* of specialization/diversification from Jane, the answer is yes: The subject identity matters. Instead, if we ask whether John has a different *degree* of specialization/diversification from Jane, the answer is no: The subject identity does not matter. The *degree* of specialization/diversification, or diversity, is permutation invariant; the identities of the categories are disregarded.

Consider a quantity distributed among a finite set of categories labeled $i = 1, 2, 3, \ldots, s$. Examples of such a quantity are (a) total abundance of an ecological community distributed among s species, and (b) total market sales distributed among s competing firms. Let π_i be the proportion of the quantity found in the ith category, as, for example, the relative abundance of the ith species in an ecological community. We assume for this discussion that $\pi_1 \geqslant \pi_2 \geqslant \cdots \geqslant \pi_s > 0$, and $\pi_1 + \pi_2 + \cdots + \pi_s = 1$. The relative abundance vector, $\boldsymbol{\pi}$, is given by $\boldsymbol{\pi} = (\pi_1, \pi_2, \ldots, \pi_s)$. Since the vast bulk of applications of diversity analysis occurs in the field of ecology (see ref. 1), we will here refer to the categories as *species* and define a *community* C as the pair $C = (s, \boldsymbol{\pi})$.

The concept of *diversity* of a community has been defined by Patil and Taillie [9–12] independent of any particular diversity indices*. In a diverse community, the typical species is relatively rare. Intuitively, a transfer of abundance either from a common to a rare species or from a common to a new, additional species would increase an observer's assessment of the rarity of a typical species. Accordingly, a community $C' = (s', \boldsymbol{\pi}')$ is defined in [10] to be *intrinsically more diverse* than a community $C = (s, \boldsymbol{\pi})$, written $C' \geqslant C$, if C' can be constructed from C by a finite sequence of the following two operations:

1. Introducing a species in this manner: (a) subtract an amount h from the relative abundance π_i of an existing species, where $0 < h < \pi_i$, and (b) add a new species to the community of relative abundance h.
2. Transferring abundance between two existing species in this manner: (a) subtract an amount h from the relative abundance π_i of a species, and (b) add h to the relative abundance π_j of another species, where $0 < h < \pi_i - \pi_j$.

Note that the species' actual labels are ignored in transforming the vector $\boldsymbol{\pi}$ into the vector $\boldsymbol{\pi}'$. It is shown in ref. 12 that C' is intrinsically more diverse than C if and only if every right tail-sum of $\boldsymbol{\pi}'$ is greater than or equal to the corresponding right tail-sum of $\boldsymbol{\pi}$, that is,

$$\sum_{i>k} \pi_i' \geqslant \sum_{i>k} \pi_i, \qquad k = 1, 2, 3, \ldots, s. \quad (1)$$

Since the kth right tail-sum is the combined abundance of those species rarer than the kth ranked species, greater diversity in this sense means a greater amount of rarity.

A diversity measure of a community C is defined in [10, 12] as the *average rarity* of the community, given by

$$\Delta(C) = \sum_{i=1}^{s} \pi_i R(i, \boldsymbol{\pi}). \quad (2)$$

Here $R(i, \boldsymbol{\pi})$ is a numerical measure of rarity associated with the ith species. Different forms for the function R produce different diversity measures. A natural requirement is to use only those forms of R which yield diversity measures with the following properties: (1) $\Delta(C') \geqslant \Delta(C)$ whenever $C' \geqslant C$, and (2) $\Delta(C) = 0$ if C has only one species.

Two general categories of diversity measures are used frequently, based on two types of rarity functions: (1) rarity functions of the *dichotomous* type depend only on the relative abundance of the ith species: $R(i, \boldsymbol{\pi}) = R(\pi_i)$; and (2) rarity functions of the *ranking* type depend only on the (descending) rank of the ith species: $R(i, \boldsymbol{\pi}) = R(i)$.

One dichotomous rarity function suggested in refs. 10 and 12 is

$$R(\pi_i) = (1 - \pi_i^{\beta})/\beta,$$

where β is a constant chosen by the investigator such that $\beta \geqslant -1$. The limiting form $R(\pi_i) = -\log \pi_i$ is used for $\beta = 0$. The result is a parametric family of diversity measures:

$$\Delta_\beta(C) = \left(1 - \sum_{i=1}^{s} \pi_i^{\beta+1}\right) \Big/ \beta. \qquad (3)$$

Three diversity indices* popular in ecological studies are special cases of Δ_β: (a) $\Delta_{-1} = s - 1$, the species richness index, (b) $\Delta_0 = -\sum \pi_i \log \pi_i$, the Shannon index, and (c) $\Delta_1 = 1 - \sum \pi_i^2$, the Simpson index. However, any other values of $\beta \geqslant -1$ also yield diversity indices with the required properties mentioned above. Monotone transformation of Δ_β would preserve the ordering properties required of a diversity index; some have been proposed as indices. The transformation $S_\beta = 1/(1 - \beta\Delta_\beta)^{1/\beta}$ is Hill's [3] family of indices. It has the interpretation of being the numbers equivalent of Δ_β; that is, S_β is the number of species a totally even community must have in order for its value of Δ_β to equal that of the given community. The function $\log S_\beta$ is Rényi's [15] entropy* of order $\beta + 1$, proposed by Pielou [13] as a diversity index. (See refs. 3, 13, and 15 and DIVERSITY INDICES for additional information.)

Another dichotomous rarity function is

$$R(\pi_i) = (1 - \pi_i)\left[1 - (1 - \pi_i)^m\right]/\pi_i,$$

where m is a nonnegative integer. It leads to the family of diversity indices studied in refs. 4 and 16,

$$\Delta_m^{\mathrm{HSG}}(C) = \sum_{i=1}^{s} (1 - \pi_i)\left[1 - (1 - \pi_i)^m\right].$$

The quantity $\Delta_m^{\mathrm{HSG}} + 1$ has the interpretation of being the expected number of species obtained when $m + 1$ individuals are randomly selected from the community C.

One ranking-type rarity function takes the kth ranked species to be the standard against which a species' rarity is assigned the value 1 or 0:

$$R(i) = \begin{cases} 1, & i > k \\ 0, & i \leqslant k. \end{cases}$$

The diversity measure produced, denoted $T_k(C)$, is simply the right tail-sum of the relative abundances:

$$T_k(C) = \sum_{i>k}^{s} \pi_i. \qquad (4)$$

Other ranking-type indices are discussed in ref. 12.

Diversity is a partial ordering in the sense that (a) $C \geqslant C$, (b) $C'' \geqslant C$ whenever $C'' \geqslant C'$ and $C' \geqslant C$, and (c) a given pair of communities need not be comparable. This explains the complaint frequently heard among ecologists (e.g., in ref. 4) that different diversity indices may give different orderings. In ref. 12, for example, a simple illustration involving two communities is given.

The *intrinsic diversity profile* of a community C is a plot of the diversity measure $T_k(C)$ (4) against the ranks $k = 1, 2, \ldots, s$. By plotting the profiles of two communities on one graph, one can tell at a glance whether the communities can be ordered by diversity. By (1), $C' \geqslant C \Leftrightarrow T_k(C') \geqslant T_k(C)$ for all $k = 1, 2, \ldots, s$. Thus the profile of an intrinsically more diverse community is everywhere above or equal to that of a less diverse community (Fig. 1). The profiles will cross if the communities are not intrinsically comparable.

The measure $T_k(C)$ is a family of diversity indices parameterized by k. For larger values of k, T_k is in some sense more sensitive

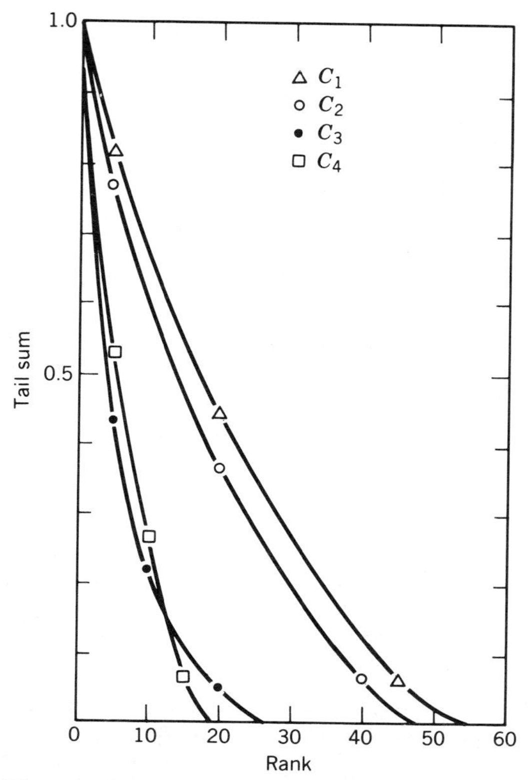

Figure 1 Intrinsic diversity profiles of four communities (data from [11]). It is seen that $C_1 \geqslant C_2$, $C_1 \geqslant C_3$, $C_1 \geqslant C_4$, $C_2 \geqslant C_3$, $C_2 \geqslant C_4$, but that C_3 and C_4 are not intrinsically comparable.

to changes in numbers and abundances of rare species in the community. In ref. 12 and Taillie [20] a formal definition of sensitivity of an index to rare species is provided. In particular, it is shown that the sensitivity of Δ_β to rare species is a strictly decreasing function of β, while that of Δ_m^{HSG} is a strictly increasing function of m.

Other profiles of diversity can be constructed by plotting a parametric diversity index against a parameter which measures sensitivity of the index to rare species. For example, $\Delta_\beta(C)$ can be computed for a given community C using a variety of values of $\beta \geqslant -1$. The plot of Δ_β against β is a type of diversity profile. Similarly, Δ_m^{HSG} may be computed and plotted for a variety of m values. If the diversity profiles of two communities cross, the communities cannot be intrinsically ordered according to diversity.

Additional information on diversity analysis may be found in the references following. Ecological studies using diversity profiles include refs. 2, 11, and 8 (this last study refers to the intrinsic diversity profile as a "k-dominance curve"). In ref. 21 the Δ_β, S_β, and Δ_m^{HSG} families are used to explore relationships between genetic variation and species diversity. In ref. 19 the Δ_β family is applied to forestry. In ref. 17 estimation for diversity index families is studied. In ref. 18 a diversity ordering based on the concept of majorization is proposed. It turns out to be equivalent to the intrinsic diversity ordering (see ref. 12). In ref. 5 the log series distribution is employed as a model of species abundances, and the estimate of the parameter α is used as a diversity index (see also refs. 6 and 7). In ref. 20 it is shown that the parameter α possesses the intrinsic diversity ordering property. Finally, we mention a different approach to diversity analysis taken in ref. 14. *See* RAO'S AXIOMATIZATION OF DIVERSITY.

References

[1] Dennis, B., Patil, G. P., Rossi, O., Stehman, S., and Taillie, C. (1979). In *Ecological Diversity in Theory and Practice*, J. F. Grassle, G. P. Patil, W. Smith, and C. Taillie, eds. International Co-operative Publishing House, Fairland, Md. (Extensive bibliography of applications of diversity in ecology.)

[2] Dennis, B., Patil, G. P., and Rossi, O. (1979). In *Environmental Biomonitoring, Assessment, Prediction, and Management*, J. Cairns, G. P. Patil, and W. E. Waters, eds. International Co-operative Publishing House, Fairland, Md.

[3] Hill, M. O. (1973). *Ecology*, **54**, 427–432.

[4] Hurlbert, S. H. (1971). *Ecology*, **52**, 577–586.

[5] Kempton, R. A. and Taylor, L. R. (1974). *J. Anim. Ecol.*, **43**, 381–399.

[6] Kempton, R. A. and Taylor, L. R. (1976). *Nature (Lond.)*, **262**, 818–820.

[7] Kempton, R. A. and Wedderburn, R. (1978). *Biometrics*, **34**, 25–37.

[8] Lambshead, P. J. D., Platt, H. M. and Shaw, K. M. (1983). *J. Nat. Hist.*, **17**, 859–874.

[9] Patil, G. P. and Taillie, C. (1976). In *Proc. 9th Int. Biom. Conf.*, Vol. II. The Biometric Society, Washington, D.C.

[10] Patil, G. P. and Taillie, C. (1979). In *Ecological Diversity in Theory and Practice*, J. F. Grassle, G. P. Patil, W. Smith, and C. Taillie, eds. International Co-operative Publishing House, Fairland, Md.

[11] Patil, G. P. and Taillie, C. (1979). In *Contemporary Quantitative Ecology and Related Econometrics*, G. P. Patil and M. Rosenzweig, eds. International Co-operative Publishing House, Fairland, Md.

[12] Patil, G. P. and Taillie, C. (1982). *J. Amer. Statist. Ass.*, **77**, 548–561. (The key reference on the intrinsic diversity ordering and various diversity index families. Discussion follows, pp. 561–567.)

[13] Pielou, E. C. (1975). *Ecological Diversity*. Wiley, New York.

[14] Rao, C. R. (1982). *Theor. Popul. Biol.*, **21**, 24–43.

[15] Rényi, A. (1961). *Proc. 6th Berkeley Symp. Math. Statist. Prob.*, Vol. 1. University of California Press, Berkeley, CA, pp. 547–561.

[16] Smith, W. and Grassle, J. F. (1977). *Biometrics*, **33**, 283–292.

[17] Smith, W., Grassle, J. F., and Kravitz, D. (1979). In *Ecological Diversity in Theory and Practice*, J. F. Grassle, G. P. Patil, W. Smith, and C. Taillie, eds. International Co-operative Publishing House, Fairland, Md.

[18] Solomon, D. (1979). In *Ecological Diversity in Theory and Practice*, J. F. Grassle, G. P. Patil, W. Smith, and C. Taillie, eds. International Co-operative Publishing House, Fairland, Md.

[19] Swindel, B. F., Conde, L. F., and Smith, J. E. (1984). *For. Ecol. Manag.*, **7**.

[20] Taillie, C. (1977). *The Mathematical Statistics of Diversity and Abundance*. Ph.D. thesis, Pennsylvania State University.

[21] Taylor, C. E. and Condra, C. (1979). In *Ecological Diversity in Theory and Practice*, J. F. Grassle, G. P. Patil, W. Smith, and C. Taillie, eds. International Co-operative Publishing House, Fairland, Md.

(DIVERSITY INDICES
ECOLOGICAL STATISTICS)

B. Dennis
G. P. Patil

PROGRESSIVE CENSORING SCHEMES

In clinical trials*, survival analysis*, follow-up* studies, and in some problems in reliability*, some incompleteness and/or incompatibility of data may arise due to nonsimultaneous entry of the units into the (experimental) scheme, possible withdrawal of the units from the scheme, and more typically, due to intended termination of the study before all the responses become available. There may be additional complications due to incorporation of interim or repeated analysis schemes for the accumulating data (over the tenure of the study period). For example, in a clinical trial [designed to study the comparative effectiveness or performance of two (or more) drugs (or treatments) relating to a common response] yielding data on the *times to failure* (response) for the different units (subjects) under study, we may encounter a variety of complications due to one or more of the following factors (these call for special attention for a valid and efficient statistical analysis of the clinical trial data):

1. Censoring/truncation. Due to limitations of time, cost of experimentation, and other considerations, the clinical trial may be planned so that after a prefixed period of time, the trial may be terminated and a statistical analysis be performed on the (partial) data accumulated during this tenure of the experiment. This may be termed a *truncated* or *type I censored experiment*. In some other situations, it may be decided to curtail experimentation after a prefixed number (or proportion) of responses becomes available. This may be termed a *type II censoring scheme*. Note that in a type I censoring scheme, the duration of the trial is prefixed, but the number of responses occurring within this period is random. In a type II censoring scheme, this number is fixed in advance, but the duration of the trial is a random variable. In either case, at the termination of the experimentation, there may still be some surviving units whose failure times are not totally known (these are, however, known to be larger than the censoring times). These censored observations introduce some incompleteness in the response data.

2. Staggering entry. The subjects (or units) may not all be available at the beginning of experimentation, so that they may enter (possibly in batches) into the scheme at possibly different points of time. If t_0 is the time point at which the study begins and t^* the time point at which the experimentation is stopped, then for an entry at a time point $t(\geqslant t_0)$, the *exposure time* may be defined as $\max\{0, t^* - t\}$. Thus a *staggering entry plan* for the units results in (possibly) different exposure times for these units (under a given censoring scheme), and this introduces unequal censoring patterns. In most of the cases, this entry plan may be random in nature (e.g., patients arriving in a clinic with a specific disease), so that one would have a *random censoring pattern* for a staggering entry plan.

3. Withdrawal. Some of the subjects may have failures due to causes other than the specific one under study, and hence, their failure times may not throw much light on the particular cause one wants to investigate. Also, due to migration or lack of interest in the participation, other subjects may even drop out of the scheme (during the experimentation). These may be regarded as *withdrawals*. In this setup, withdrawals may be regarded as random, although in some other plans, planned withdrawals may also be considered as a part of the experimentation protocol. For models incorporating various withdrawal patterns (random or nonrandom in nature), some information is lost due to the withdrawal and the statistical analysis scheme for the trial may become more complicated and less efficient, too.

4. Monitoring of clinical trials. In view of the fact that clinical trials involve human subjects, based on medical ethics there may be a need to monitor experimentation from the very beginning: If any significant difference among the treatments (in relation to the responses) is detected at any intermediate stage of the trial, the trial may be stopped at that point and the surviving subjects switched to the better treatment for their benefit. From the statistical point of view, this monitoring of the trial involves interim or repeated statistical analysis on accumulating data, where in a follow-up study, the setup of the classical sequential analysis* may not apply.

For a proper interpretation of progressive censoring schemes (PCSs), all the concepts listed under factors **1** to **4** need to be incorporated. The term PCS has been used in the statistical literature in more than one context. Earlier uses generally refer to a (fixed) single-point censoring scheme when there is a staggering entry plan for the units, so that one essentially has an unequal pattern of censoring (see Nelson [6]). Thus, if we have n $(\geqslant 1)$ units entering into the scheme at time points $t_1, \ldots, t_n$, respectively (where $t_0 \leqslant t_1 \leqslant \cdots \leqslant t_n$), t_0 being the starting point of the study, and if t^* be any (fixed) censoring point, then, at that time point the exposure times for these n units are $t^* - t_1, \ldots, t^* - t_n$, respectively (assuming that $t^* \geqslant t_n$). For these units and under the fixed-point censoring scheme, one has to work with the minimum of the actual failure time and the corresponding exposure time, along with the information whether this corresponds to a failure or a censored event. When $t_1, \ldots, t_n$ are random variables, this may also be termed *random censoring*. The relevance of the term *progressive censoring* lies in the progressive entry of the units resulting in the unequal censoring pattern for them.

In the context of life testing*, the term progressive censoring has also been used for a scheme to further reduce the testing time (see Klein and Basu [4]). For n items under life testing, consider a set $T_1, \ldots, T_m$ of predetermined censoring times. At time T_j, a fixed number c_j $(\geqslant 0)$ of the items are removed from the test, for $i = 1, \ldots, m - 1$; at time T_m either a fixed number c_m are removed from the test or the test is terminated with some units still surviving. Instead of predetermined time points $T_1, \ldots, T_m$, one may also conceive of predetermined numbers $r_1, \ldots, r_m$ such that at the r_jth failure, an additional number c_j of units may be removed and testing be continued, for $j = 1, \ldots, m$, the modification for $j = m$ being the same as in the case of Type I censoring.

In the context of clinical trials with monitoring in mind, a comprehensive definition of progressive censoring schemes, due to Chatterjee and Sen [1] may be given. This extended definition of PCS incorporates statistical monitoring of clinical trials (i.e., factor 4) in a very natural way where various combinations of 1, 2, and 3 may be blended. For simplicity consider first simultaneous entry without any withdrawal. Here n ($\geqslant 1$) units entering into the scheme at a common point of time t_0 are followed through time and their failure times (along with other tagging variables) are recorded in order as they occur. Thus at any time point t ($> t_0$), one has a composite picture of the failures occurring during $(t_0, t]$, the censored observations [having failure times in (t, ∞)], and other design and concomitant variables. In a fixed-point censoring scheme, at some t^* ($> t_0$), where t^* may or may not be random depending on Type II or Type I censoring, based on the picture at t^*, a statistical analysis of the survival data is to be made and conclusions are to be drawn therefrom. In most practical problems, unless the survival distribution is at least roughly known, a fixed-point censoring scheme may lead to considerable loss of efficiency relative to cost. This drawback is particularly felt in exploratory studies. A too early termination of the experiment may lead to an inadequately small set of observations and thus increases the risk of an incorrect decision, while unnecessary prolongation may lead to loss of valuable time with practically no extra gain in efficiency. Also, in a life-testing situation, continuous monitoring is usually adopted. As such, it seems unrealistic to wait until a given time period is over or a given number of failures has occurred, and then to draw inferences from the study. Naturally, one would like to review the results as the experiment progresses, and thereby have the scope for terminating the experiment at an early stage depending on the accumulated evidence from the observations. Such a scheme is termed a *progressive censoring scheme*. Here the censoring is made progressively over time; this flexibility would increase the efficiency relative to cost and would be particularly relevant in long-term studies of an exploratory nature. Although in a PCS, monitoring is done continuously, in many studies (particularly, in nonparametric ones) the picture may change only at the failure points, so that it amounts to reviewing statistically the picture at successive failure points. In this simultaneous entry plan, in a PCS, withdrawals can easily be accommodated by considering the response variable as the minimum of the failure time and the withdrawal time (along with the tagging variable whether it is a failure or censored observation).

PCS may become somewhat more complicated for a staggering entry plan. Here monitoring may commence before all the subjects have entered into the scheme, and failures may also occur among the entries already in hand. Thus if t_0 is the time point at which the study begins, and if $t_1, \dots, t_n$ are the points of entries of the n units in the scheme, then at any time point t (not $\geqslant \max\{t_1, \dots, t_n\}$) we have $n(t)$ entries, where $n(t)$ is nondecreasing in t and $n(t) \leqslant n$, $t \geqslant t_0$; for $t \geqslant \max(t_1, \dots, t_n)$, $n(t) = n$. Further, at the time point t, for the $n(t)$ units having points of entries prior to t, the exposure times are possibly different. This picture is similar to the first interpretation of PCS, sketched earlier, but with sample size $n(t)$ instead of n. The picture in the extended PCS is then obtained by varying t progressively over $t \geqslant t_0$; if at any $t > t_0$, a statistically significant difference among the treatments is detected (based on the picture available at that time), the trial is stopped at that time along with the rejection of the null hypothesis of equality of treatments. Otherwise, the monitoring continues. In this setup, an upper bound for the duration of the trial may also be set in advance. Also, withdrawals may be accommodated in a manner similar to that in the nonstaggering entry plan. For details of this staggering entry PCS, see Majumdar and Sen [5] and Sinha and Sen [8, 9].

Statistical analysis for PCS has opened up a new area of active research. Although technically allied to classical sequential analysis*, lack of independence of the increments over time as well as the lack of stationarity of the related stochastic processes* make it somewhat difficult to apply the classical theory in PCS. Some of these results are discussed in Sen [7] and Tsiatis [10].

References

[1] Chatterjee, S. K. and Sen, P. K. (1973). *Calcutta Statist. Ass. Bull.*, **22**, 13–50.

[2] Elandt-Johnson, R. C. and Johnson, N. L. (1980). *Survival Models and Data Analysis*. Wiley, New York.

[3] Gross, A. J. and Clark, V. A. (1975). *Survival Distributions: Reliability Applications in the Biomedical Sciences*. Wiley, New York.

[4] Klein, J. P. and Basu, A. P. (1981). *Commun. Statist. A*, **10**, 2073–2100.

[5] Majumdar, H. and Sen, P. K. (1978). *Commun. Statist. A*, **7**, 349–371.

[6] Nelson, W. A. (1972). *Technometrics*, **14**, 945–966.

[7] Sen, P. K. (1981). *Sequential Nonparametrics: Invariance Principles and Statistical Inference*. Wiley, New York.

[8] Sinha, A. N. and Sen, P. K. (1982). *Sankhyā B*, **44**, 1–18.

[9] Sinha, A. N. and Sen, P. K. (1983). *Proc. Golden Jubilee Conf. Statist.*, Indian Statistical Institute, Calcutta.

[10] Tsiatis, A. A. (1981). *Biometrika*, **68**, 311–315.

(CENSORED DATA
CENSORING
CLINICAL TRIALS
SURVIVAL ANALYSIS)

P. K. SEN

PROGRESSIVELY CENSORED DATA ANALYSIS

In the entry PROGRESSIVE CENSORING SCHEMES, some statistical problems associated with progressive censoring schemes (PCS) were presented. Valid and efficient statistical analysis procedures of such progressively censored data are considered here.

Consider first a staggered entry but fixed-point censoring scheme (e.g., Nelson [18]), where n items enter at time points $t_1 \leqslant t_2 \leqslant \cdots \leqslant t_n$, respectively, and the study is terminated at a point $t^* \geqslant t_n$. Thus the effective exposure times for the n units are $T_i = t^* - t_i$, $i = 1, \ldots, n$. If X_i stands for the length of life (i.e., failure time) of the ith unit (from the entry point), then this failure occurs during the tenure of the study if $t_i + X_i \leqslant t^*$, and is otherwise censored. Thus, if $X_i^0 = \min(X_i, T_i)$ and $\delta_i = 1$ or 0 according as X_i^0 is equal to X_i or T_i, for $i = 1, \ldots, n$, then the observable random vectors are

$$(X_1^0, \delta_1), \ldots, (X_n^0, \delta_n). \tag{1}$$

Note that if the entry points (t_i) are stochastic, then so are the T_i, so that we have a random censoring scheme; generally, in this context, we assume that T_i are independent and identically distributed random variables with a distribution function G, defined on $[0, \infty)$. For a preplanned entry plan, the T_i are nonstochastic, and we have an unequal censoring plan. In either case, when the distributions of the X_i are of specified forms (involving some unknown parameters), and the same is the case with G, then the likelihood function for the observable random elements in (1) can be used to estimate these unknown parameters and to make suitable tests of significance. These parametric procedures are discussed in detail in books by Elandt-Johnson and Johnson [9], Kalbfleisch and Prentice [13], Gross and Clarke [12], and Lawless [15].

In the nonparametric case (typically in the context of testing the equality of two or more survival functions) Gehan [11], Efron [8], Peto and Peto [19], and others have advocated the use of some simple rank tests, while Breslow and Crowley [2] and others have incorporated the product limit (PL) or the so-called Kaplan–Meier* [14] estimators of the individual survival functions in the formulation of the test procedures. Under random censoring, if in (1) the information contained in the δ_i is ignored, the usual nonparametric tests based on the X_i^0 remain applicable, although they may entail some loss of efficiency due to this sacrifice of information. Sen [24] has studied conditions under which such rank tests are locally optimal. In this context the proportional hazard (PH) model of Cox [5] has been used extensively (*see* COX'S REGRESSION MODEL). Under this model, the hazard function may be quite arbitrary though its dependence on the covariates is of a given parametric form. Thus such models can at best be regarded as quasi-nonparametric. Cox [5, 6] has advocated the use of a *partial likelihood* function* which enables one to employ the classical parametric approach in a general setup. This has been followed up more thoroughly by a host of workers (see Tsiatis [31], Slud [30], Sen [22], Anderson and Gill [1], where other references are cited, and a conference proceedings on survival analysis, edited by Crowley and Johnson [7]).

Let us next consider the second interpretation of PCS, due to Cohen [4] and others. Here also, for Type I or Type II censoring schemes, in the parametric case, the likelihood function can be written explicitly. The usual maximum likelihood* estimators and likelihood ratio tests* may be prescribed in the usual fashion. For exponential*, normal*, lognormal*, and logistic* distributions, such procedures are due to Cohen [4] and Gajjar and Khatri [10], among others. In the nonparametric case, this PCS permits the permutational invariance to generate distribution-free* tests. However, closed expressions for the test statistics may be difficult to prescribe.

Let us proceed to the broad interpretation of PCS, due to Chatterjee and Sen [3], and consider first nonstaggered entry plans. Essentially, we encounter here a time-sequential problem which may be posed as follows. For n items under life testing, corresponding to the actual failure times $X_1, \ldots, X_n$, the observable random vectors are (Z_i, S_i) $i = 1, \ldots, n$, where $Z_1 < \cdots < Z_n$ stand for the order statistics* corresponding to the X_i, and $\mathbf{S}_n = (S_1, \ldots, S_n)$ is the vector of antiranks (i.e., $X_{S_i} = Z_i$, $i = 1, \ldots, n$). As we monitor the experiment from the beginning, we observe the failure points $Z_1, \ldots, Z_n$ in order. Thus at the kth failure point Z_k, the observed data relate to the set

$$\{(Z_1, S_1), \ldots, (Z_k, S_k)\} \quad \text{for } k = 1, \ldots, n. \quad (2)$$

Unfortunately, for different i, the vectors (Z_i, S_i) are neither stochastically independent nor marginally identically distributed. Hence, if one attempts to adapt some sequential procedure based on the partial set in (2), difficulties may arise due to nonhomogeneity and lack of independence of the increments. On the other hand, the PCS relates to looking at the data at each failure point (with a view to rejecting the null hypothesis if the outcome at that point so indicates), so that one essentially encounters a repeated significance testing problem on accumulating data where the increments are neither independent nor homogeneous.

If we denote the likelihood function of the set of observed random variables $\{(Z_1, S_1), \ldots, (Z_k, S_k)\}$ by $p_{nk}(\cdot\,; \boldsymbol{\theta})$, where $\boldsymbol{\theta}$ stands for the parameters associated with the underlying probability law, then the logarithmic derivative of $p_{nk}(\cdot\,; \boldsymbol{\theta})$ (with respect to $\boldsymbol{\theta}$) forms a martingale* array (for every n and $k \leqslant n$); see Sen [21] and Sen and Tsong [25]. Based on this martingale characterization, one may then construct a random function (stochastic process*) from this partial sequence of the derivatives of the log-likelihood function (at the successive failure points). Under fairly general regularity conditions, such a process weakly converges to a Wiener process (in the single-parameter case) or a Bessel process (in the multiparameter case). This weak convergence result enables one to use the distribution theory of suitable functionals (such as the maximum or the maximum absolute value) of the Wiener (or Bessel) process to provide large-sample approximations for the critical levels of the test statistics (in the PCS) based on the partial sequence in (2).

As an illustration, consider the simple model where θ is real valued (e.g., Sen [21]). Suppose that we want to test the null hypothesis $H_0 : \theta = \theta_0$ against one-or two-sided alternatives. Let

$$\lambda_{nk}^0 = (d/d\theta)\log p_{nk}(\cdot,\theta)|_{\theta=\theta_0},$$
$$J_{nk}^0 = E_{\theta_0}(\lambda_{nk}^{02}),$$

$k = 0, 1, \ldots, n$ (where $\lambda_{n0}^0 = J_{n0}^0 = 0$). Also, let

$$t_{nk} = J_{nk}^0 / J_{nn}^0, \qquad W_n(t_{nk}) = \lambda_{nk}^0 \{J_{nn}^0\}^{-1/2},$$

$k = 0, 1, \ldots, n$, and by linear interpolation we complete the definition of $W_n = \{W_n(t) : 0 \leqslant t \leqslant 1\}$. Then under H_0 and some mild regularity conditions, as n increases, $\sup\{W_n(t) : 0 \leqslant t \leqslant 1\}$ [or $\sup\{|W_n(t)| : 0 \leqslant t \leqslant 1\}$] converges in distribution to $D^+ = \sup\{W(t) : 0 \leqslant t \leqslant 1\}$ [or $D = \sup\{|W(t)| : 0 \leqslant t \leqslant 1\}$], where $W = \{W(t) : 0 \leqslant t \leqslant 1\}$ is a standard Wiener process on (0, 1). Hence if D_α^+ (or D_α) is the upper $100\alpha\%$ point of the distribution of D^+ (or D), then we have for large n,

$$P\Big\{\sup_{0 \leqslant t \leqslant 1} W_n(t) > D_\alpha^+ \mid H_0\Big\} \to \alpha,$$
$$P\Big\{\sup_{0 \leqslant t \leqslant 1} |W_n(t)| > D_\alpha \mid H_0\Big\} \to \alpha. \qquad (3)$$

Operationally, at each failure point Z_k, we compute $(J_{nn}^0)^{-1/2}\lambda_{nk}^0$ and if it lies above D_α^+ (for one-sided alternatives) or beyond $\pm D_\alpha$ (for the two-sided case), we stop at that point along with the rejection of the null hypothesis; if not, we continue the study until the next failure and repeat the test at that point. By virtue of (3), this repeated significance testing* scheme has the asymptotic level of significance α. On the other hand, if the λ_{nk}^0 are tested repeatedly against their marginal critical levels, the overall significance level may be much higher than the nominal significance level for each of these marginal tests.

In the nonparametric case, in the context of the simple regression model containing the two-sample model as a special case, Chatterjee and Sen [3] spotted the same martingale property of a suitable version of linear rank statistics, censored at the successive failure points. Let us denote these censored linear rank statistics by T_{nk}, $k = 0, 1, \ldots, n$ (where $T_{n0} = 0$) and the corresponding array of variances (under the null hypothesis H_0 of the equality of all the n distributions) by V_{nk}, $k = 0, \ldots, n$. Defining $t_{nk} = V_{nk}/V_{nn}$, $k = 0, \ldots, n$, and $W_n(t_{nk}) = T_{nk}/V_{nn}^{1/2}$, $k = 0, \ldots, n$, we complete the definition of $W_n = \{W_n(t) : 0 \leqslant t \leqslant 1\}$ by linear interpolation. As in the parametric case (3) holds, so that under a PCS the overall significance level of the repeated significance test based on the T_{nk} will be close to α, for large n. Under H_0 the T_{nk} are jointly distribution-free, so that for small values of n, the exact critical value for this procedure may be obtained by direct enumeration. But these computations become prohibitively laborious, so that for large n, one would naturally use (3) as a handy tool. Majumdar and Sen [17] extended the Chatterjee–Sen [3] results to the case of a multiple regression* model (containing the several sample model as a special case). They were able to show that for the progressively censored rank test statistics, some Bessel process approximations hold. For local alternatives, asymptotic power properties of these rank tests for the simple and multiple regression models were also studied by Chatterjee and Sen [3] and Majumdar and Sen [16]. A detailed discussion of these results is given in Sen [23].

Instead of using linear rank statistics, one may also look at the empirical distributions for the different samples (or more generally, a weighted empirical process) and employ them for testing purposes. Note that the classical Kolmogorov–Smirnov* statistic is usable in a PCS for testing the identity of the two underlying distributions. Sinha and Sen [26, 27] have studied this problem in detail and cited additional references.

Let us finally consider the staggered entry plan. As in the case of random or unequal censoring (treated earlier), here also the entry pattern may or may not be stochastic in nature. Although analysis based on a specific entry pattern may be made using similar martingale characterizations, one might like to adopt some procedure which remains robust against departures from a specified pattern. With this objective in mind, Sen [20] considered a two-dimensional functional central limit theorem* for censored linear rank statistics (where the sample size n is allowed to vary between 1 and N and the failure numbers k between 1 and n). This theorem provides a Brownian sheet approximation for the staggered entry censored linear rank statistics. This result was used by Majumdar and Sen [17] to formulate some (conservative) tests for the two-sample as well as the simple regression model, for the staggered entry PCS; these tests remain insensitive to the entry pattern. Sinha and Sen [28, 29] considered the case of weighted empirical processes* in the staggered entry plans, and developed the tied-down Brownian sheet approximations for these processes. Their tests are also somewhat conservative in nature (due to the fact that the entry pattern is not assumed to be precisely known in advance). Similar Brownian sheet approximations work out well for the Cox [5] proportional hazard model in the staggered entry plan as well as for the parametric likelihood ratio statistics under fairly general regularity conditions.

Thus we may conclude that for the PCS, where independence, homogeneity and simultaneous entry plan may not hold, the basic tool for data analysis is to incorporate martingale theory in the formulation of suitable stochastic processes converging in distribution to appropriate Gaussian ones (in a one- or more-dimensional time parameter), and to employ the distribution theory for these Gaussian processes* to provide large-sample approximations for the specific ones under study. This approach is dealt with elaborately in Sen [23, Chap. 11].

References

[1] Anderson, P. K. and Gill, R. D. (1982). *Ann. Statist.*, **10**, 1100–1120.

[2] Breslow, N. and Crowley, J. (1974). *Ann. Statist.*, **2**, 437–453.

[3] Chatterjee, S. K. and Sen, P. K. (1973). *Calcutta Statist. Ass. Bull.*, **22**, 13–50.

[4] Cohen, A. C. (1963). *Technometrics*, **5**, 327–339.

[5] Cox, D. R. (1972). *J. R. Statist. Soc. B*, **34**, 187–220.

[6] Cox, D. R. (1975). *Biometrika*, **62**, 269–276.

[7] Crowley, J. and Johnson, R. A., eds. (1982). *Survival Analysis*. IMS Lecture Notes–Monographs, Vol. 2. Inst. Math. Statist., Hayward, CA.

[8] Efron, B. (1967). *Proc. 5th Berkeley Symp. Math. Statist. Prob.*, Vol. 4. University of California Press, Berkeley, CA, pp. 831–853.

[9] Elandt-Johnson, R. C. and Johnson, N. L. (1980). *Survival Models and Data Analysis*. Wiley, New York.

[10] Gajjar, A. V. and Khatri, C. G. (1969). *Technometrics*, **11**, 793–803.

[11] Gehan, E. A. (1965). *Biometrika*, **52**, 650–652.

[12] Gross, A. J. and Clark, V. A. (1975). *Survival Distributions: Reliability Applications in the Biomedical Sciences*. Wiley, New York.

[13] Kalbfleisch, J. D. and Prentice, R. L. (1980). *The Statistical Analysis of Failure Time Data*. Wiley, New York.

[14] Kaplan, E. L. and Meier, P. (1958). *J. Amer. Statist. Ass.*, **53**, 457–481.

[15] Lawless, J. F. (1982). *Statistical Models and Methods for Lifetime Data*. Wiley, New York.

[16] Majumdar, H. and Sen, P. K. (1978). *J. Multivariate Anal.*, **8**, 73–95.

[17] Majumdar, H. and Sen, P. K. (1978). *Commun. Statist. A*, **7**, 349–371.

[18] Nelson, W. (1972). *Technometrics*, **14**, 945–965.

[19] Peto, R. and Peto, J. (1972). *J. R. Statist. Soc. A*, **135**, 186–206.

[20] Sen, P. K. (1976). *Ann. Prob.*, **4**, 13–26.

[21] Sen, P. K. (1976). *Ann. Statist.*, **4**, 1247–1257.

[22] Sen, P. K. (1981). *Ann. Statist.*, **9**, 109–121.

[23] Sen, P. K. (1981). *Sequential Nonparametrics: Invariance Principles and Statistical Inference*. Wiley, New York.

[24] Sen, P. K. (1984). *J. Statist. Plan. Infer.*, **9**, 355–366.

[25] Sen, P. K. and Tsong, Y. (1981). *Metrika*, **28**, 165–177.

[26] Sinha, A. N. and Sen, P. K. (1979). *Commun. Statist. A*, **8**, 871–898.

[27] Sinha, A. N. and Sen, P. K. (1979). *Calcutta Statist. Ass. Bull.*, **28**, 57–82.

[28] Sinha, A. N. and Sen, P. K. (1982). *Sankhyā B*, **44**, 1–18.

[29] Sinha, A. N. and Sen, P. K. (1983). *Proc. Golden Jubilee Conf. Statist.*, Indian Statistical Institute, Calcutta.

[30] Slud, E. V. (1982). *Biometrika*, **65**, 547–552.
[31] Tsiatis, A. (1981). *Ann. Statist.*, **9**, 93–108.

(CENSORED DATA
COX'S REGRESSION MODEL
GAUSSIAN PROCESSES
KAPLAN–MEIER ESTIMATOR
LINEAR RANK TESTS
PROGRESSIVE CENSORING SCHEMES
SURVIVAL ANALYSIS)

P. K. SEN

PROJECTION PLOTS

One of the most useful statistical techniques for interpreting bivariate data is the scatter plot*. It can help to identify interesting features of the data such as clusters of observations, outlying points, and possible relationships between two variables. For multivariate data with more than two variables, however, an ordinary scatter plot cannot display the entire data. Finding interesting structure in multivariate data is much more difficult; *see also* MULTIVARIATE GRAPHICS.

There are a variety of ways to modify and enhance scatter plots for use with multivariate data. One possibility is to project the data into two dimensions and then examine a scatter plot constructed using these two dimensions (or equivalently, using these two variables). This is called a *projection plot*.

With multivariate data there are many ways to choose the two dimensions, and projecting into them implies some loss of information from the remaining dimensions. One cannot expect interesting features of the data to show up in all projections, and there may be no single projection that reveals everything important. Some choices—such as looking at scatter plots of all pairs of the original variables—can be made without performing any calculations on the data, while other choices of dimensions can be determined from calculations performed on the specific data set. No single method will always reveal the interesting features of the data, and it is often useful to try several of the projection methods described below.

PROJECTIONS ON PRECHOSEN VARIABLES

Suppose that the analyst has n values (observations) on each of p variables $x_1, x_2, \ldots, x_p$, $p \geqslant 3$, and the goal is to find scatter plots showing interesting features of the data. Consider first the ordinary scatter plot using just variables x_1 and x_2, which ignores the data values on variables $x_3, \ldots, x_p$. Conceptually, the n multivariate observations can be thought of as n points lying in a p-dimensional vector space, with axes corresponding to the variables $x_1, x_2, \ldots, x_p$. The (x_1, x_2) scatter plot corresponds to *projecting* the n points (orthogonally) into the x_1, x_2 plane, which can also be thought of as the plane determined by the dimensions (variables) x_1 and x_2. Other scatter plots will correspond to different projections.

One way to construct a scatter plot that gives information on three variables rather than just two is to change the character that is plotted. Instead of having all plotted characters being the same—such as a *, ·, or + as is typically done—a plot can be constructed in which the location of the character represents the (x_1, x_2) values of an observation as usual, but the character plotted is L if the value of x_3 is in its lowest quartile, M if x_3 is in its middle two quartiles, and U if x_3 is in its upper quartile. Clearly, there are many choices of characters for plotting and many ways of dividing the distribution on the x_3 variable. What is most useful may depend on the specific application.

A related idea which is possible with some plotting devices is to use only one symbol as plotting character, such as a circle, but to make its size vary depending on the value of the x_3 variable. This can permit an impression of perspective on x_3; we view the x_1x_2 plane at a particular value of x_3, and points closest to this plane are plotted with the largest characters, while points farther away (in terms of the x_3 variable) are plotted smaller. For this to work satisfactorily, the character size will generally have to be some nonlinear function of x_3, so that no points are so big as to completely dominate the plot or so small as to be invisible. Typically, iteration is necessary to produce a satisfying and informative plot. Another way of introducing x_3 is to plot the points with different colors, depending on the value of x_3. This idea does not seem to have been used or evaluated very much, possibly because the technology permitting it is still new.

There is no reason to examine only a single scatter plot. One way to construct multiple plots is to look at two variables x_1 and x_2 using different subsets of the observations. Using the ideas behind the preceding three paragraphs, we can first partition the observations using variable x_3, for example, into the four quartiles. Then make four scatter plots of x_1 and x_2 using each of these four subsets; these plots can be arranged on

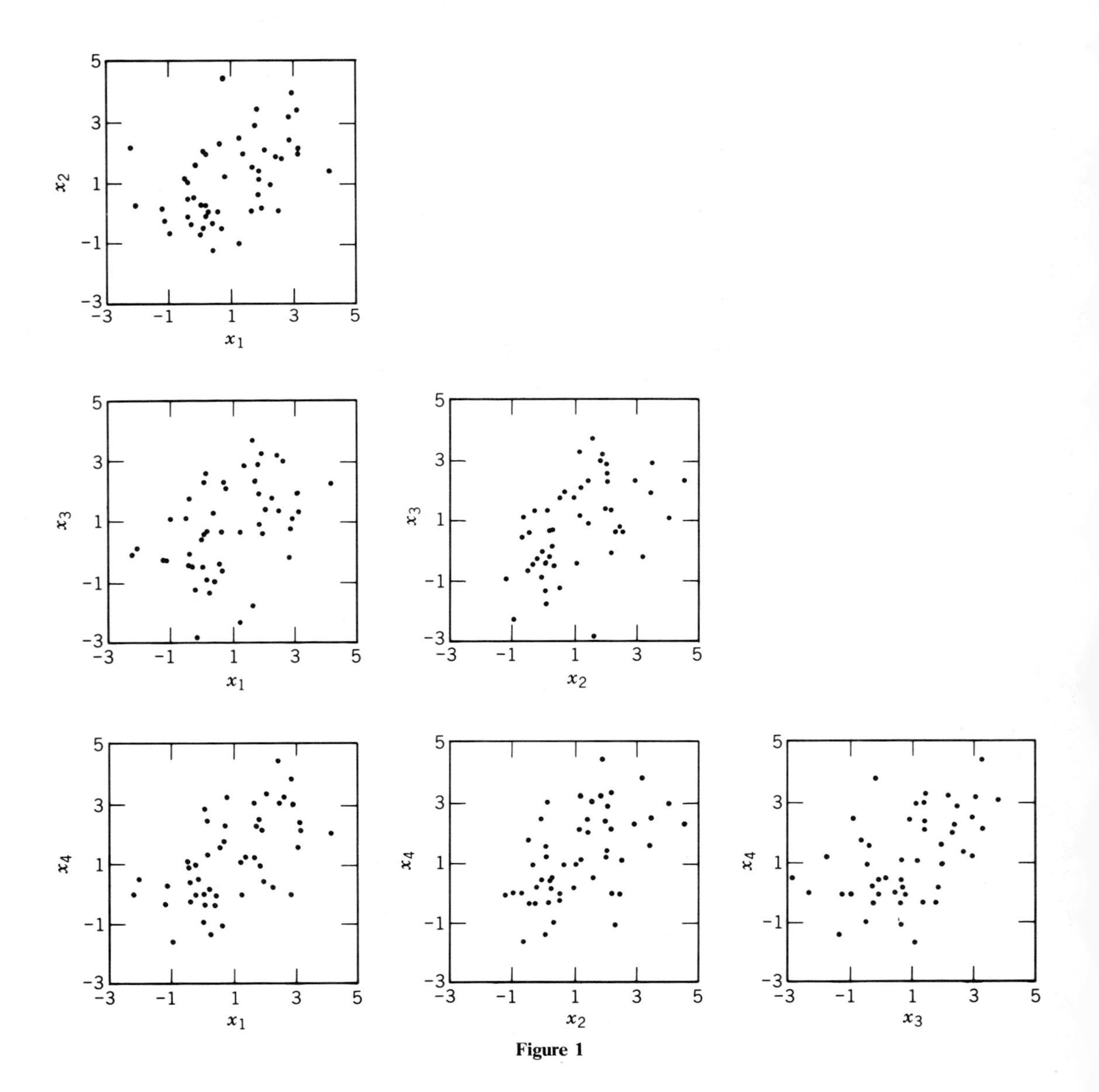

Figure 1

a page from left to right corresponding to the increase in the background variable x_3. A more complicated version of this idea is to partition the observation using two background variables, x_3 and x_4, and then arrange the scatter plots of x_1 and x_2 for the resulting subsets in a rectangular array corresponding to the value on x_3 and x_4. In effect, this is a scatter plot of scatter plots, incorporating four variables in all.

A different approach toward constructing multiple plots is to juxtapose scatter plots involving different variables in some useful way, where now each scatter plot includes different variables but all the observations. Several ways to do this, as well as ways to implement the ideas discussed previously, are described by Tukey and Tukey [7, Chap. 10; Chap. 11, pp. 231–232; Chap. 12, pp. 245–260].

One way to arrange the scatter plots from all pairs of the four variables x_1, x_2, x_3, and x_4 is shown in Fig. 1. For this example there appear to be positive relationships between the pairs of variables. None of these six plots indicates, however, that these 50 observations may consist of two separated groups. Now consider Fig. 2, constructed using $y_1 = (x_1 + x_2 + x_3 + x_4)/2$ and $y_2 = (x_1 + x_2 - x_3 - x_4)/2$. This is an orthogonal projection into two dimensions that do not correspond to any of the original four variables; it does suggest that these data may separate nicely into two groups, those with $y_1 < 2$ and those with $y_1 > 2$.

This example points out that it can be desirable to view many different projections of the data. Simply viewing projections using each variable marginally does not guarantee finding whatever underlying structure that there may be. [The example was constructed to illustrate this point by using two multivariate normal populations with identity covariance matrices, one centered at $(0, 0, 0, 0)$ and the other at $(2, 2, 2, 2)$.]

With more variables the number of scatter plots that could be examined increases greatly. Unfortunately, the chance that a prechosen projection will be sufficiently close to the "right" projection (assuming that there is such a projection that will reveal the structure) decreases dramatically with the number of variables. The result is that when there are a moderate to large number of variables—say, four or more—simply relying on some small number of prechosen projections is not likely to suffice to reveal the structure. Fortunately, calcula-

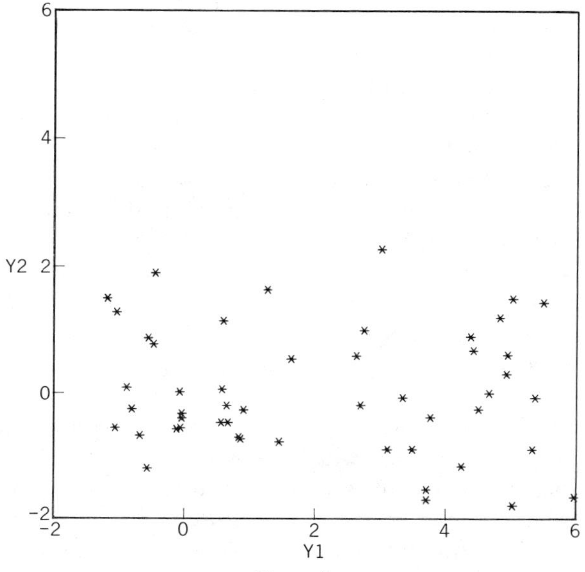

Figure 2

tions are available which can help to find useful projections.

PROJECTIONS FROM MULTIVARIATE METHODS

Calculations from several of the classical multivariate analysis methods give projections which can be useful for displaying the data, although obtaining the projections need not be the only or main reason for performing these analyses. These methods, which are interrelated, include component analysis* (often called principal component analysis), discriminant analysis*, and canonical analysis* (canonical correlations).

Component analysis can be helpful if we are treating the data as one set of observations measured on p variables. Component analysis gives those projections which preserve interpoint distances, based on distances between all pairs of observations using all p variables, as well as possible. Thus a plot of the observations projected on the first two principal components is often used to give a first, overall, two-dimensional picture of multivariate data.

If the data set consists of two or more samples which are assumed Gaussian with equal covariance matrices, and the goal is to find the projections which minimize the error rates of classification, then the appropriate theory is that of linear discriminant analysis. If there are three or more groups, a plot of the data projected on the first two linear discriminant functions can help to understand relationships among the groups.

Canonical analysis is designed for the situation of two sets of continuous variables and one underlying population. Again, projecting the observations on pairs of canonical variates can help to indicate relationships.

The preceding methods can all be obtained from classical multivariate normality assumptions, and all involve linear algebra eigenvector calculations. With recent advances in computational power and interest in methods that are not so closely tied to specific distributional and model assumptions, several approaches have been investigated which can give useful projections for interpreting data.

The basic idea is to define a criterion that measures to what extent a given projection is "interesting," and then to use numerical methods to seek those projections that maximize the criterion. Both parts of the problem can be approached in different ways [6, 3]. In choosing the criterion we do not want a projection in which the observations have a more-or-less uniform scatter, but instead one in which they are clumped (clotted or clustered). There may be several local maxima to the criterion, corresponding to different sorts of structure in the data. These might all be of interest, so it can be desirable to have numerical procedures which find more than a single local maximum. The use of this idea to fit a multiple regression model was given by Friedman and Stuetzle [2].

Work in a similar spirit has been developed for cluster analysis. In clustering, generally some complicated numerical algorithm has been used to group the observations into clusters; the analyst might then need to examine various plots to interpret the results and to study relationships between and within the clusters. But, as with other multivariate data, it is not clear which projections are useful for studying the possible clusters. Gnanadesikan et al. [4] modified some of the component analysis and discriminant analysis calculations in order to produce projection plots which give as uncluttered a view of the observations and clusters as possible, by finding those projections which separate things maximally. Unlike the methods mentioned earlier in this section, these do not require iterative calculations.

Instead of having the computer choose one or a small number of projection plots to examine by maximizing some criterion, a next step is to make even more use of increased computing power by showing *many* projection plots successively in real time, generating movie-like effects. This can now be done on certain specially designed interactive computer graphic systems. A specific projection (involving two dimensions) is shown on a CRT screen; then a third dimension is considered and the data are rotated within this subspace. This enables the user to perceive a spatial three-dimensional picture of the data, in terms of these dimensions. The rotations and dimensions can be controlled by the user. The analyst can see, quickly and interactively, many views of the data. An early such system was developed by Tukey et al. [8]; some recent work and ideas are described by Donoho et al. [1]. Such systems have not yet seen wide use, partly because of their expense; if they do become widely available, they could have a tremendous impact on analysis of multivariate data.

OTHER RELATED METHODS

Several graphical ways to display multivariate data have been devised which do not amount to projection plots. Instead, each multivariate data point is mapped into some appropriate symbol, and the symbols are displayed. Nearby data points should be mapped into symbols that are similar. By comparing the symbols, the analyst may be able to discover interesting relationships among the data points and variables. For Chernoff faces*, the symbol is a cartoon face; for Andrews function plots*, it is a continuous function defined on the interval $(-\pi, \pi)$. Other symbols that have been used include star-shaped figures and trees and castles [5]. Such methods are good for giving general impressions involving all the variables, but it often seems that specific features and details are more apparent in appropriately chosen projection plots. (It can be difficult, of course, to find the appropriate projection plots.) The biplot* is another graphical method, useful for displaying certain kinds of multivariate data.

References

[1] Donoho, D., Huber, P. J., and Thoma, H. M. (1981). In *Computer Science and Statistics: Proceedings of the 13th Symposium on the Interface*, W. F. Eddy, ed. Springer-Verlag, New York, pp. 274–278.

[2] Friedman, J. H. and Stuetzle, W. (1981). *J. Amer. Statist. Ass.*, **76**, 817–823.

[3] Friedman, J. H. and Tukey, J. W. (1974). *IEEE Trans. Computers*, **C-23**, 881–890.

[4] Gnanadesikan, R., Kettenring, J. R., and Landwehr, J. M. (1982). In *Statistics and Probability: Essays in Honor of C. R. Rao*, G. Kallianpur, P. R. Krishnaiah, and J. K. Ghosh, eds. North-Holland, New York, pp. 269–280.

[5] Kleiner, B. and Hartigan, J. A. (1981). *J. Amer. Statist. Ass.*, **76**, 260–269.

[6] Kruskal, J. B. (1969). In *Statistical Computation*, R. C. Milton and J. A. Nelder, eds. Academic Press, New York.

[7] Tukey, P. A. and Tukey, J. W. (1981). In *Interpreting Multivariate Data*, V. Barnett, ed. Wiley, Chichester, England.

[8] Tukey, J. W., Friedman, J. H., and Fisherkeller, M. A. (1976). *Proc. 4th Int. Conf. Stereol.*, Gaithersburg, Md.

(ANDREWS FUNCTION PLOTS
BIPLOT
CHERNOFF FACES
GRAPHICAL REPRESENTATION, COMPUTER-AIDED
GRAPHICAL REPRESENTATION OF DATA
MULTIVARIATE ANALYSIS
MULTIVARIATE GRAPHICS
SCATTER PLOT, TRIPLE)

JAMES M. LANDWEHR

PROPAGATION OF ERROR *See* ERROR ANALYSIS; LAWS OF ERRORS

PROPORTIONAL ALLOCATION

This term is used in *stratified designs** in *finite population sampling**. When a finite population is divided into homogeneous strata and a separate sample is drawn from each stratum, the sample design is said to be *stratified*. For instance, a population of persons might be stratified by sex and age, a particular stratum consisting of females aged 55–59; or a population of retail establishments might be stratified by area, description, and size, a particular stratum consisting of metropolitan hardware stores with annual sales of \$1,000,000–\$1,999,999 at the last retail census.

Proportional allocation is one method which can be used to decide how many *sample units* are to be selected from each stratum, given the *sample number* or *sample size* for the entire survey. In proportional allocation, the sample number for each stratum is chosen to be proportional to the number of population units in each stratum i:

$$n_h = n(N_h/N) \text{ to the nearest integer,}$$

where n_h = sample number for the hth stratum

n = sample number for the whole survey

N_h = number of population units in the hth stratum

N = number of population units in the population.

Thus for the population of persons stratified by age and sex, the number of females aged 55–59 selected in the sample would be proportional to the number of females aged 55–59 in the population as a whole.

Proportional allocation is often (not invariably) appropriate for samples of persons. It is inefficient in circumstances where the population units differ greatly in size or importance, as is the case with surveys of retail establishments. In such circumstances it is preferable to use optimal allocation (*see* OPTIMUM STRATIFICATION).

Proportional allocation is commonly used in small surveys which do not require *multistage sampling*. The corresponding techniques used in multistage surveys are known as *equal probability selection methods* (*epsem* for short).

Bibliography

Sudman, S. (1976). *Applied Sampling*. Academic Press, New York, pp. 112–114. (Indicates why proportional allocation is often preferred to optimum allocation on grounds of simplicity and versatility.)

(NEYMAN ALLOCATION
OPTIMUM STRATIFICATION
PROPORTIONAL SAMPLING
STRATIFIED DESIGNS
SURVEY SAMPLING)

K. R. W. BREWER

PROPORTIONAL HAZARDS MODEL

See COX'S REGRESSION MODEL

PROPORTIONAL REDUCTION IN ERROR (PRE) MEASURES OF ASSOCIATION

Proportional reduction in error measures were developed largely because of the need for measures of association* between two nominal* variables with meaningful interpretations. Unlike measures based upon the chi-square* statistic, which are difficult to interpret, PRE measures have interpretations in terms of probabilities of misclassification which are clear and easily understood.

Suppose that a population is classified on the basis of two nominal attributes, A and B. In contingency table* notation, let p_{ij} be the probability that a member of the population is classified at level i of the (row) variable A and at level j of the (column) variable B, for $i = 1, \dots, r$ and $j = 1, \dots, c$. Then the ith row total $p_{i+} = \sum_j p_{ij}$ is the marginal* probability that an individual falls in class i of attribute A; similarly, p_{+j} is the probability that the individual falls in class j of B.

PRE measures are motivated as follows. Some rule is used to classify a randomly selected member of the population with respect to attribute A, first using only the population information at hand, and again when the member's B class is known as well. The associated PRE measure is the relative reduction in the probability of misclassification:

$$\text{PRE}(\text{A}\,|\,\text{B}) = (P(E) - P(E\,|\,\text{B}))/P(E), \tag{1}$$

where $P(E)$ is the probability of error in the first case and $P(E\,|\,\text{B})$ is the probability of error in the second. Thus (1) is the decrease in the probability of error in predicting (for some prediction rule) a member's A class gained by knowledge of its B class, relative to the probability of error when its B class is unknown.

If it is a priori possible to classify the member on A with certainty [i.e., $P(E) = 0$], then (1) is undefined. Moreover, $0 \leqslant \text{PRE}(A \mid B) \leqslant 1$, with $\text{PRE}(A \mid B) = 0$ when A and B are independent, and $\text{PRE}(A \mid B) = 1$ when knowledge of the member's B class permits errorless prediction of its A class. It can happen that $\text{PRE}(A \mid B) = 0$ even though A and B are not independent. See ref. 1 or 2 for further discussion.

Each classification rule used with (1) yields a PRE measure. One rule (optimal prediction) is to assign the member to the most probable A class, given available information. When the member's B class is unknown, it is assigned to the A class of maximum probability (i.e., to the class m such that $p_{m+} = \max\{p_{1+}, \ldots, p_{r+}\}$). When it is known to fall into B class (column) j, it is assigned to the A class (row) of maximum probability in *column j* (i.e., to the class m such that $p_{mj} = \max\{p_{1j}, \ldots, p_{rj}\}$). The resulting PRE measure is

$$\lambda_{A \mid B} = \left(\sum_{j=1}^{c} p_{mj} - p_{m+} \right) \Big/ (1 - p_{m+}). \quad (2)$$

A second rule (proportional prediction) is to assign the member to an A class according to the appropriate (marginal or conditional) probability distribution for A: The member is assigned to class i with probability p_{i+} when its B class is unknown, and to class i with probability p_{ij}/p_{+j} when its B class is known to be j. This rule yields

$$\tau_{A \mid B} = \frac{\sum_{i=1}^{r} \sum_{j=1}^{c} p_{ij}^2 / p_{+j} - \sum_{i=1}^{r} p_{i+}^2}{1 - \sum_{i=1}^{r} p_{i+}^2}. \quad (3)$$

When a classification of the entire population is unavailable, it is necessary to estimate the measures (2) and (3). For a random sample of size n from a population in which x_{ij} is the number of sample members in A class i and B class j, the analogs to (2) and (3) are

$$\hat{\lambda}_{A \mid B} = \left(\sum_{j=1}^{c} x_{mj} - x_{m+} \right) \Big/ (n - x_{m+}) \quad (4)$$

and (*see* GOODMAN–KRUSKAL TAU AND GAMMA)

$$\hat{\tau}_{A \mid B} = \frac{n \sum_{i=1}^{r} \sum_{j=1}^{c} x_{ij}^2 / x_{+j} - \sum_{i=1}^{r} x_{i+}^2}{n^2 - \sum_{i=1}^{r} x_{i+}^2}. \quad (5)$$

Under the multinomial* sampling model, (4) and (5) are maximum likelihood estimators* of (2) and (3), respectively. Exact sampling distributions of (4) and (5) are not known, but both have asymptotically normal distributions, provided that the sample is not concentrated in a single A class and that (2) and (3) are neither 0 nor 1. Using Gini's definition of total variability in a sample from a categorical variable, Margolin and Light [4, 5] show that, when $\tau_{A \mid B} = 0$,

$$U_{A \mid B}^2 = (n-1)(r-1)\hat{\tau}_{A \mid B} \quad (6)$$

is asymptotically distributed as a chi-square variate with $(r-1)(c-1)$ degrees of freedom.

The PRE measures (2) and (3) treat A and B asymmetrically, with B playing the role of "independent" variable. Versions of (2) to (6) exist which treat A as "independent" and B as "dependent." Symmetric versions of (2) to (5) have also been defined for cases where no natural asymmetry exists between A and B; see ref. 2 or 3.

The measure λ was first proposed by L. Guttman and τ is attributed to W. A. Wallis.

References

[1] Bishop, Y. M. M., Fienberg, S. E., and Holland, P. (1975). *Discrete Multivariate Analysis: Theory and Practice*. MIT Press, Cambridge, Mass., pp. 387–393. (An exposition and derivation of the log-linear model for the analysis of contingency tables, with emphasis on iterative proportional fitting. Contains a chapter on measures of association which includes a concise discussion of the properties of λ and τ.)

[2] Goodman, L. A. and Kruskal, W. H. (1979). *Measures of Associations for Cross Classifications*. Springer-Verlag, New York. (A reprint of four

classical papers, written between 1954 and 1972, on the construction of measures of association for two-way contingency tables. The original and most extensive source of information on λ and τ; these papers contain derivations, sample estimates and related asymptotic calculations, examples, and historical references.)

[3] Liebetrau, A. M. (1983). *Measures of Association*. Sage, Beverly Hills, Calif. (A comprehensive and fairly elementary discussion of a large number of widely used measures of association, including λ and τ.)

[4] Light, R. J. and Margolin, B. H. (1971). *J. Amer. Statist. Ass.*, **66**, 534–544.

[5] Margolin, B. H. and Light, R. J. (1974). *J. Amer. Statist. Ass.*, **69**, 755–764.

(ASSOCIATION, MEASURES OF
CATEGORICAL DATA
CONTINGENCY TABLES
GOODMAN–KRUSKAL TAU AND GAMMA
NOMINAL DATA)

A. M. LIEBETRAU

PROPORTIONAL SAMPLING

This term is closely related to *proportional allocation**. The latter refers strictly speaking to that method of allocating a sample among strata in which the *sample number* for each stratum is chosen to be proportional to the number of *sample units* which each contains. Proportional sampling is a slightly wider concept, encompassing also the actual selection of the sample in accordance with that allocation.

Proportional sampling is commonly used in small surveys which do not require *multistage* sampling*. The corresponding techniques used in multistage surveys are known as *equal probability selection methods* (*epsem* for short).

Bibliography

Sudman, S. (1976). *Applied Sampling*. Academic Press, New York, pp. 112–114. (Indicates why proportional allocation is often preferred to optimum allocation on grounds of simplicity and versatility.)

(PROPORTIONAL ALLOCATION)

K. R. W. BREWER

PROPORTION ESTIMATION IN SURVEYS USING REMOTE SENSING

During 1972–1984 the National Aeronautics and Space Administration (NASA) launched a series of land observatory (LANDSAT) satellites, and developed a remote sensing technology capable of monitoring certain types of earth resources. A major application of LANDSAT remote sensing is the estimation of acreage for agricultural crops. Remotely sensed data consist of radiometric measurements in each of several regions of the electromagnetic spectrum. The sensor system has a radiometer which measures the intensity of radiant energy.

Two radiometers used onboard LANDSAT satellites are the multispectral scanner (MSS) and the thematic mapper (TM). Of the five LANDSAT satellites launched, the first three carried only an MSS sensor and the last two carried both the MSS and TM sensors. Both sensors image a 100-nautical mile wide swath of the earth by viewing the orbital ground track through a mirror, periodically scanning strips perpendicular to the orbital path. The analog signals recorded by the sensors are converted into digital counts onboard the satellite and transmitted to a ground station. Hence the LANDSAT coverage of an area results in a scene consisting of scanlines with a certain number of resolution elements per scanline.

The TM sensor was designed to provide improved spatial and spectral resolution. It consists of six wavelength bands, compared to four for the MSS sensor. It has a smaller size resolution element, approximately 0.2 acre versus 1.1 acres for MSS. The resolution element is called a *pixel*. The TM data transmission rate is 85 megabits per second versus 15 megabits per second for MSS. The digitization counts are from 0 to 255 for TM and from 0 to 63 for MSS.

The first three LANDSAT satellites were launched into circular, near-polar orbits 920 km (570 miles) in altitude. These orbits are "sun-synchronous"; for a given point on the ground, it is the same local time of day when the satellite passes. These satellites pass over a given point on the ground every 18 days. Landsats 4 and 5 were launched into a sun-synchronous, circular, near-polar orbit 705 km (431 miles) in altitude with repetitive coverage every 16 days. Multidate acquisitions over a single location on the ground can be registered with each other and the radiance measurements can be modeled as a function of time. These measurements are influenced by the vegetation, soil type, and atmospheric conditions, and when these are statistically modeled and correlated with ground features, it is feasible to estimate agricultural crop acreages by acquiring and analyzing the remotely sensed data for an area of interest.

Image analysis techniques are used to label spectral classes by crop types. A segment, several square miles in area, is required to delineate discernible patterns and identify possible crop types. A crop can be distinguished from others in a scene by monitoring the temporal development of its fields from planting through harvest and performing multitemporal analysis. This, of course, requires that multitemporal acquisitions of a segment be properly registered since misregistration, especially near field boundaries, can cause measurements to shift from one crop type to another.

STATEMENT OF THE PROBLEM

Let $\mathbf{z}$ be a $p \times 1$ measurement vector for a pixel in an area segment containing a large number of pixels. Suppose that the segment consists of m ground cover classes for which the distribution of $\mathbf{z}$ is given by the probability density function $f_i(\mathbf{z})$, $i = 1, 2, \ldots, m$. Then one may consider that the vectors $\mathbf{z}$ are independent observations sampled from a mixture* population with density function

$$f(\mathbf{z}) = \sum_{i=1}^{m} \lambda_i f_i(\mathbf{z}), \tag{1}$$

where λ_i is the proportion of the segment in ground cover class i satisfying

$$\lambda_i \geqslant 0, \qquad \sum_{i=1}^{m} \lambda_i = 1. \tag{2}$$

The general problem is to estimate the λ_i $(i = 1, 2, \ldots, m)$, given the observed vectors $\mathbf{z}$ for a segment without any knowledge of m or $f_i(\mathbf{z})$. This problem cannot be solved unless one is able to obtain information which would permit estimation, if not a complete determination, of m and $f_i(\mathbf{z})$.

This general mixture problem has been extensively studied. A common approach is to make certain assumptions; one of the following cases is usually considered.

1. The number of classes m is known and $f_i(\mathbf{z})$ is from a known *identifiable* family of distributions, such as multivariate normal.
2. The number of classes m is known, a set of $\mathbf{z}$ observations from each class distribution is available, and the densities $f_i(\mathbf{z})$ are members of some identifiable family, usually known up to a set of parameters.

For a definition of identifiability* and further discussion on this finite mixture problem, see Teicher [7].

In the present context, the number of classes was either fixed or estimated. An initial approach was to develop false color imagery of a segment using its spectral data and have an image analyst, trained in photo-interpretation techniques, identify the potential number of ground cover classes, and label a set of pixels in the segment corresponding to the various classes. There were problems with this approach. First, the analyst was not in a position to identify all ground cover classes in a segment. Clustering the segment data can provide the possible number of classes represented in the data set, but these may not correspond to the ground cover classes in the segment. Second, labeling of pixels was highly labor intensive and time consuming, and hence not enough pixels could be labeled to estimate the class distributions nonparametrically. Generally, an analyst was trained in identifying features for a few major ground cover classes and his or her labeling of pixels was subject to error. These difficulties play a significant role in the approaches to the estimation of crop proportions using LANDSAT data.

CROP PROPORTION ESTIMATION

In this section we survey mainly methods of proportion estimation which were used in practice and evaluated at the NASA Johnson Space Center. These studies were conducted under two major programs, the Large Area Crop Inventory Experiment (LACIE) [6] and the Agriculture and Resources Inventory Surveys Through Aerospace Remote Sensing (AgRISTARS).

The initial approach was to assume that $f_i(\mathbf{z})$ in (1) represents either a parametric family of multivariate normal distributions or mixtures of multivariate normal distributions with mixing weights assumed known. For example, in LACIE, wheat was the class of interest, and therefore there were only two classes, wheat and nonwheat. Since nonwheat may consist of various ground-cover classes in a segment, its spectral distribution was assumed to be a mixture of multivariate normals. A similar model was assumed for wheat to take into account the different varieties, growth stages, and so on. Thus the density functions for the two classes were of the form

$$\begin{aligned} f_1(\mathbf{z}) &= \sum_{k=1}^{m_1} \lambda_{1k} f_{1k}(\mathbf{z}), \\ f_2(\mathbf{z}) &= \sum_{k=1}^{m_2} \lambda_{2k} f_{2k}(\mathbf{z}), \end{aligned} \tag{3}$$

where $f_{ik}(\mathbf{z})$ is the multivariate normal density with mean vector $\boldsymbol{\mu}_{ik}$ and covariance matrix $\boldsymbol{\Sigma}_{ik}$, and λ_{ik} is the relative size of the kth subclass in class i, $i = 1, 2$ and $k = 1, 2, \ldots, m_i$, satisfying

$$\lambda_i = \sum_{k=1}^{m_i} \lambda_{ik}.$$

The number of subclasses m_i and their respective weights λ_{ik} were determined by clustering the data, labeling the clusters as being wheat or nonwheat, and then finding their relative sizes within each class. If the data were not clustered, an image analyst provided a rough estimate of m_i and the subclass weights were assumed equal [i.e., $\lambda_{ik} = 1/m_i$ $(i = 1, 2)$].

Spectral data for the set of pixels labeled by an analyst were used to estimate $f_{ik}(\mathbf{z})$, and hence $f_i(\mathbf{z})$, by replacing the unknown parameters with their corresponding training statistics (i.e., the sample mean vectors and the sample covariance matrices). All pixels in a segment were classified using the maximum likelihood classification* rule (*see* DISCRIMINANT ANALYSIS); a pixel with measurement vector $\mathbf{z}$ is classified as class i if

$$\hat{f}_i(\mathbf{z}) = \max_j \hat{f}_j(\mathbf{z}), \tag{4}$$

where $\hat{f}_i(\mathbf{z})$ is the estimated density function for class i. The class proportion λ_i was estimated by

$$\hat{\lambda}_i = N_i \Big/ \sum_j N_j, \tag{5}$$

where N_i is the number of pixels in the segment classified into the ith class. This estimator is biased, as discussed next, and hence estimation procedures that are relatively unbiased and somewhat insensitive to errors in estimates of class distributions were necessary to obtain reliable crop acreage estimates during LACIE and AgRISTARS.

ESTIMATORS

Let C_1 represent the crop of interest and suppose a pixel either belongs to C_1 or its complement C_2. Suppose that pixels in a segment are classified according to the maximum likelihood classification rule and let R_1 and R_2 be the classification regions corresponding to C_1 and C_2. Then

$$R_i = \left\{\mathbf{z} : \hat{f}_i(\mathbf{z}) = \max_j \hat{f}_j(\mathbf{z})\right\}, \qquad i = 1, 2. \tag{6}$$

Define the conditional probability p_{ij} of a pixel from class j being classified as class i by

$$p_{ij} = \int_{R_i} f_j(\mathbf{z})\, d\mathbf{z}. \tag{7}$$

Suppose that y is the probability that a pixel belongs to C_1 and x is the probability of classifying a pixel as C_1. Then it follows from (7) that

$$x = p_{11}y + p_{12}(1 - y). \tag{8}$$

Define the random variable ϕ by

$$\phi(\mathbf{z}) = \begin{cases} 1, & \text{if } \mathbf{z} \in R_1, \\ 0, & \text{otherwise,} \end{cases} \tag{9}$$

where R_1 is the classification region for C_1. Suppose that the segment has N pixels that are classified as C_1 or C_2. Then the proportion of pixels classified as C_1, given by

$$\hat{y} = \sum_{j=1}^{N} \phi(\mathbf{z}_j)/N, \tag{10}$$

provides an estimate of y. In (10), $\mathbf{z}_j$ denotes the measurement vector for pixel j. This is a biased estimator since $E(\hat{y}) = x$, where x is given in (8). The bias is given by

$$E(\hat{y}) - y = p_{12} - (p_{12} + p_{21})y,$$

where p_{12} and p_{21} are misclassification errors called the *commission* and *omission errors*, respectively. If the commission error p_{12} is large relative to the omission error p_{21}, a strong possibility when the maximum likelihood classifier is used and the true proportion y is small, then the bias may be substantial. If the classification errors are known, an unbiased estimator can be constructed by modifying (10) as follows: From (8),

$$y = (x - p_{12})/(1 - p_{21} - p_{12}), \tag{11}$$

provided that $1 - p_{21} - p_{12} \neq 0$. Replacing x by $\hat{y}$ in (11), an unbiased estimator of y is obtained.

The classification errors are generally unknown and the preceding simple modification is not possible. However, one may consider estimating p_{12} and p_{21} using an independent sample of pixels in the segment, and then estimating y by

$$\hat{\hat{y}} = (\hat{y} - \hat{p}_{12})/(1 - \hat{p}_{21} - \hat{p}_{12}), \tag{12}$$

provided that $1 - \hat{p}_{21} - \hat{p}_{12} \neq 0$, where $\hat{y}$ is given in (10). If $1 - \hat{p}_{21} - \hat{p}_{12} = 0$, then there is no need to modify $\hat{y}$ (i.e., let $\hat{\hat{y}} = \hat{y}$). The estimator $\hat{\hat{y}}$ is then bounded between 0 and 1.

Clearly, for $\hat{\hat{y}}$ to be a reliable and nearly unbiased estimator of y, the estimators $\hat{p}_{21}$ and $\hat{p}_{12}$ should be based on large samples drawn randomly from C_1 and C_2, respectively. In the remote sensing application, it was not possible to select pixels randomly from individual classes C_1 and C_2, since class labels were unknown for the pixels in a segment, and for a large sample pixel labeling by an analyst was too labor intensive and subject to error. There was no apparent way to deal with the latter problem; however, the problem of obtaining random samples was solved by sampling pixels from the segment without having to control sample sizes for C_1 and C_2, as discussed next.

Let n pixels be randomly selected from the segment and be identified for their true class labels. For a pixel with measurement vector $\mathbf{z}$, define the random variable $\eta(\mathbf{z})$ by

$$\eta(\mathbf{z}) = \begin{cases} 1, & \text{if pixel belongs to } C_1, \\ 0, & \text{otherwise.} \end{cases} \tag{13}$$

Clearly, $E[\eta(\mathbf{z})] = y$ and an estimate of y is given by

$$\bar{y} = \sum_{1}^{n} \eta(\mathbf{z}_j)/n, \tag{14}$$

where $\mathbf{z}_j$, $j = 1, 2, \ldots, n$, are the measurement vectors for the n sampled pixels. But $\bar{y}$ is inefficient, as it does not make use of the classification results for the segment. Suppose that S_1 and S_2 are the two strata of pixels classified as C_1 and C_2, respectively. In other words,

$$S_1 = \{u_j : \phi(\mathbf{z}_j) = 1, j = 1, 2, \ldots, N\},$$

$$S_2 = \{u_j : \phi(\mathbf{z}_j) = 0, j = 1, 2, \ldots, N\},$$

where u_j denotes pixel j. Then $\hat{y}$ given in (10) is the proportion of pixels in S_1 and $(1 - \hat{y})$ is the proportion of pixels in S_2. In the sampled pixels, $n_1 = \sum_1^n \phi(\mathbf{z}_j)$ are from S_1 and $n_2 = \sum_1^n [1 - \phi(\mathbf{z}_j)]$ are from S_2. Of the n_1 pixels from S_1, $n_{11} = \sum_1^n \eta(\mathbf{z}_j)\phi(\mathbf{z}_j)$ belong to C_1, and from the n_2 pixels in S_2, $n_{12} = \sum_1^n \eta(\mathbf{z}_j)[1 - \phi(\mathbf{z}_j)]$ belong to C_1. Thus an estimator of y is obtained by

$$\tilde{y} = n_{11}\hat{y}/n_1 + n_{12}(1 - \hat{y})/n_2, \tag{15}$$

provided that $n_i \geq 1$, $i = 1, 2$. Otherwise, let $\tilde{y} = \bar{y}$, where $\bar{y}$ is defined in (14). Since the event that n_1 or n_2 is zero has nonzero probability, $\tilde{y}$ is in general biased. However, as n becomes large, $\tilde{y}$ is asymptotically unbiased.

The estimator in (15) can be rewritten

$$\tilde{y} = \bar{y} + b(\bar{X} - \bar{x}), \tag{16}$$

where $\bar{y}$ is given in (14), $\bar{x} = \sum_i^n \phi(\mathbf{z}_j)/n$, $\bar{X} = \hat{y}$ as defined in (10), and

$$b = \frac{\sum_{j=1}^{n} (\eta(\mathbf{z}_j) - \bar{y})(\phi(\mathbf{z}_j) - \bar{x})}{\sum_{j=1}^{n} (\phi(\mathbf{z}_j) - \bar{x})^2}. \tag{17}$$

If the denominator in b is zero, take $b = 0$. Clearly, b is an estimator of the slope in the linear regression* of $\eta(\mathbf{z})$ onto $\phi(\mathbf{z})$. Thus $\tilde{y}$ can be thought of as a regression estimator in the case of a finite population where the sample mean $\bar{y}$ is adjusted to take into account the auxiliary information of the measurement vector $\mathbf{z}$ based on the correlation between $\phi(\mathbf{z})$ and $\eta(\mathbf{z})$. Properties of this type of estimator are discussed by Cochran [2]. Heydorn [4] discusses the estimators given above in greater detail, noting further extensions of the estimator in (15). Feiveson [3] describes some of these estimators as well as others, particularly mixture decomposition estimators with mixtures of density functions as in (1), marginal distribution functions, or moments.

FEATURE EXTRACTION USING CROP SPECTRAL PROFILE MODELS

The LANDSAT multiband data are highly correlated for both the MSS and TM sensors. A principal component analysis* of MSS data acquired for a segment in a single LANDSAT pass would show that these data are primarily two-dimensional; one corresponds to a measure of vegetation growth and the other to a measure of background reflectance. The two dimensions are called *greenness* and *brightness*, respectively. Recent research has shown that a similar situation holds for the six-band TM data, with the greenness and brightness space augmented by a third dimension related to soil water content. This type of analysis not only reduces the data dimensionality by a factor of 2, but also provides physically interpretable parameters.

From investigations of the greenness and brightness variables, it has been observed that their behavior as a function of time is different for different crops. The temporal behavior of the greenness variable has been approximately modeled by

$$G(t) = G_0 + A(t - T_0)^{\alpha}\exp\left[-\beta(t - T_0)^2\right], \tag{18}$$

where $G(t)$ is the greenness value at time t, G_0 is the bare soil greenness value, T_0 is the date of emergence, α is a crop specific parameter related to the rate of change of greenness in early season ("green up"), and β is another crop specific parameter related to the rate of onset of senescence or ripening of the crop ("green down"). This formulation was extremely important because it related the LANDSAT data to potentially predictable agrophysical parameters such as emergence date, rate of green development, rate of senescence, and total length of growing season.

The model in (18), called the *greenness profile model*, has two inflection points, T_1 and T_2, such that

$$\begin{aligned}\Delta^2 &= (T_2 - T_1)^2 \\ &= 1/(2\beta) + \alpha\left[1 - (1 - 1/\alpha)^{1/2}\right]/(2\beta) \\ &\approx 1/\beta, \end{aligned} \tag{19}$$

and a peak at

$$t = T_p = T_0 + (\alpha/(2\beta))^{1/2} \tag{20}$$

with a maximum greenness value of

$$G_{\max} = G_0 + A(\alpha/(2\beta e))^{\alpha/2}. \tag{21}$$

For corn and soybeans T_1 corresponds closely to the onset of the reproductive phase of these crops and T_2 corresponds to the onset of senescence. Since T_1, T_2, and thus Δ, can be predicted using crop phenology models, this result provides an important feature for labeling crop types. Crop phenology models predict the dates of particular phases in a crop's life cycle. The predictor variables for these models usually include daylength, temperature, and rainfall measurements.

Several empirical studies showed that the variables Δ, T_p, and $G_{\max}$ given in (19) to (21) are good discriminators for crops with different crop phenology and can be used in place of the multitemporal measurement vectors $\mathbf{z}$. Of course, this would not change any of the methods of proportion estimation discussed earlier.

The crop profile model approach to feature extraction was initially proposed by Badhwar [1]. See Houston and Hall [5] for a good review of this approach from a statistical viewpoint. Other references on the use of satellite data for crop surveys are provided in the Bibliography.

References

[1] Badhwar, G. D. (1980). *Photogramm. Eng. Remote Sensing*, **46**, 369–377.

[2] Cochran, W. G. (1977). *Sampling Techniques*. Wiley, New York.

[3] Feiveson, A. H. (1979). *Proc. Tech. Sess., LACIE Symp.*, Vol. 2, NASA-JSC 16015, pp. 643–646. (An excellent review of crop proportion estimation methods in remote sensing from an analytical viewpoint.)

[4] Heydorn, R. P. (1984). *Commun. Statist. Theor. Meth.*, **13**, 2881–2903. (An up-to-date review of crop proportion estimation methods used in practice.)

[5] Houston, A. G. and Hall, F. G. (1984). *Commun. Statist. Theor. Meth.*, **13**, 2857–2880. (Describes LANDSAT MSS and TM data and approaches for relating their temporal behavior to crop growth cycles. Also briefly describes techniques for crop identification and area and yield estimation.)

[6] MacDonald, R. B. and Hall, F. G. (1980). *Science*, **208**, 670–679. (An excellent reference on an approach to global crop forecasting using satellite data.)

[7] Teicher, H. (1961). *Ann. Math. Statist.*, **32**, 244–248.

See the following works, as well as the references just given, for more information on the topic of proportion estimation in crop surveys using remotely sensed data.

Bibliography

Bauer, M. E., ed. (1984). *Remote Sensing Environ.*, **14** (1–3). (A special issue devoted to remote sensing research to inventory and monitor crop, soil, and land resources accomplished by the AgRISTARS program.)

Chhikara, R. S., ed. (1984). *Commun. Statist. Theor. Meth.*, **13**. (A special issue on crop surveys using satellite data.)

Odell, P. L., ed. (1976). *Commun. Statist. A*, **5**. (A special issue on proportion estimation using satellite data.)

(AGRICULTURE, STATISTICS IN
CLASSIFICATION
GEOGRAPHY, STATISTICS IN
METEOROLOGY, STATISTICS IN
SPATIAL DATA ANALYSIS
SPATIAL SAMPLING)

A. Glen Houston
Raj S. Chhikara

PROSPECTIVE STUDIES

In epidemiology, the term "prospective study" usually designates a nonexperimental research design in which all the phenomena under observation occur after the onset of the investigation. This approach is known by a variety of other names, such as follow-up*, incidence, longitudinal*, and cohort studies. Regardless of the title preferred, the prospective approach may be contrasted with other nonexperimental studies that are based on historical information (*see* EPIDEMIOLOGICAL STATISTICS *and* RETROSPECTIVE STUDIES).

The usual plan of prospective research is illustrated in Fig. 1. After initiation of the study, subjects are enrolled according to the level of exposure to the main independent variable. Typically, one group is defined as "exposed" and the comparison group consists of "unexposed" subjects. The group assignment in a prospective study is determined by observations on naturally occurring exposures. The reliance on natural exposures may be contrasted with experimental studies (*see* CLINICAL TRIALS), in which the exposure status is assigned by randomization* (but *see also* HISTORICAL CONTROLS).

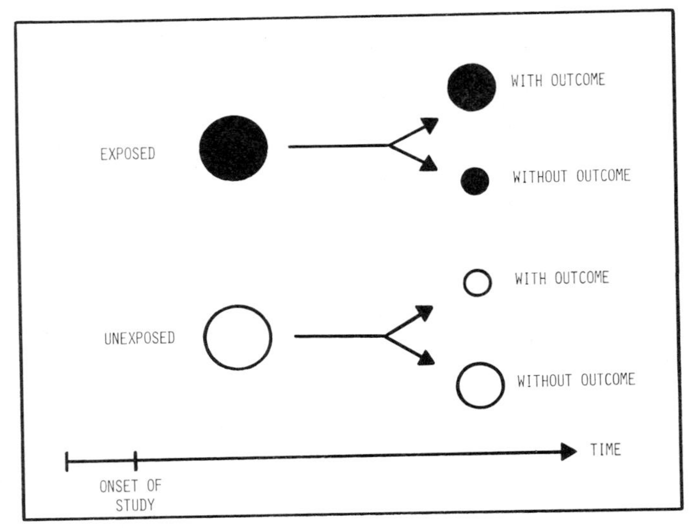

Figure 1 Schematic diagram of a prospective study (shaded areas represent subjects exposed to the antecedent factor; unshaded areas correspond to unexposed subjects).

After the study groups are defined in a prospective investigation, the subjects are followed forward in time for the development of the outcome variable. Then the frequency of the outcome among exposed subjects is compared against the frequency of the outcome among unexposed subjects.

Example. One of the most noteworthy examples of the prospective approach was the investigation of atherosclerotic heart disease in Framingham*, Massachusetts [1]. This study was started in 1948 by the U.S. Public Health Service for the purpose of evaluating possible determinants of heart disease. Framingham was chosen as the site of investigation because it had a stable, cooperative population, with good access to local and referral medical facilities. A sample of 6500 persons between the ages of 30 and 60 years was chosen from a population of about 10,000 persons in that age group. Over 5000 persons without clinical evidence of atherosclerotic heart disease agreed to participate in this research. Each subject was examined at the outset and was reexamined at intervals of two years. The primary outcome was the development of atherosclerotic heart disease. Over the 20 years of follow-up several risk factors were identified, including hypertension, elevated serum cholesterol, and cigarette smoking [6].

THE SELECTION OF SUBJECTS

A variety of factors must be considered in the choice of the study population for prospective research. In particular, the subjects must be accessible, willing to cooperate, and traceable over time. The following sources of subjects are employed commonly in prospective research.

1. **General Population.** A prospective study can be performed in an entire community, as illustrated by the Framingham Study. The choice of a community for investigation may be governed by the stability and demographic features of the population, public support for the study, and the availability of resources over a sufficient period of time to evaluate the outcome.
2. **Special Population.** A prospective study can be conducted among persons

who are affiliated with a particular group. For example, Doll and Hill [2] studied the relationship between cigarette smoking and lung cancer among physicians on the British Medical Register who were living in the United Kingdom. The choice of physicians as a study population offered the advantages of a group that was interested in health research and could be traced through professional organizations.

3. **Clinical Population.** A prospective study can be performed among patients at specific clinical facilities. As an illustration, the Collaborative Perinatal Project was a prospective study of more than 54,000 pregnant women who received prenatal care and delivered at one of 12 cooperating hospitals between 1959 and 1966 [5]. This population was used to evaluate risk factors for perinatal and infant mortality*, as well as for infant morbidity*.

THE MEASUREMENT OF EFFECT

The analysis of a prospective study is based on a comparison of the frequencies of outcomes among exposed and unexposed subjects. The magnitude of effect usually is quantified with one of the following measures.

1. Incidence density ratio (IDR). The IDR is defined as the incidence rate among the exposed divided by the incidence rate among the unexposed. The standard data layout for this type of analysis is portrayed in Table 1. For this analysis, the symbols a and b represent the number of specific outcomes among exposed and unexposed subjects, respectively. The length of observation on each individual, summed over all subjects in the study group are represented by L_1 and L_0, respectively. Thus, the incidence density rate of the outcome among exposed persons is estimated by a/L_1. Similarly, the incidence density of the outcome among unexposed persons is estimated by b/L_0. The IDR is estimated from

$$\widehat{\text{IDR}} = \frac{a/L_1}{b/L_0}.$$

When there is no association between the exposure and outcome of interest, the IDR equals unity. An IDR less than unity implies a reduced incidence of the outcome among exposed subjects, compared with the unexposed. Conversely, an IDR greater than unity indicates that exposed subjects have an elevated incidence of the outcome. A large-sample test of difference from the null value can be constructed from the normal approximation to the binomial distribution* [3]:

$$z = \frac{a - m(L_1/T)}{\left[m(L_1/T)(L_0/T)\right]^{1/2}}.$$

2. Risk ratio (RR). The RR is defined as the risk of the outcome among exposed divided by the risk among unexposed. The standard data layout for this type of analysis is depicted in Table 2. The symbols a and b have the previously defined interpretations. The symbols c and d represent the number of subjects who do not develop the outcome among exposed and unexposed persons, re-

Table 1 Data Layout for Calculation of the Incidence Density Ratio (IDR)

	Exposed Subjects	Unexposed Subjects	Total
Number of Specific Outcome(s)	a	b	$m = a + b$
Population Time of Follow-up	L_1	L_0	$T = L_1 + L_0$

Table 2 Data Layout for Calculation of the Risk Ratio (RR)

	Exposed Subjects	Unexposed Subjects	Total
Number with Outcome	a	b	$m_1 = a + b$
Number without Outcome	c	d	$m_0 = c + d$
Total	$n_1 = a + c$	$n_0 = b + d$	$N = a + b + c + d$

spectively. The risk of the outcome among the exposed is estimated by a/n_1. Similarly, the risk of the outcome among the unexposed is estimated by b/n_0. Thus, the RR is estimated by

$$\widehat{\text{RR}} = \frac{a/n_1}{b/n_0}.$$

The interpretation of the value of the RR is identical to that for the IDR. A large sample test for this measure, based on the hypergeometric distribution*, uses the Mantel–Haenszel* chi-square statistic [4]:

$$\chi^2_{(1)} = \frac{(N-1)(ad-bc)^2}{m_1 m_0 n_1 n_0}.$$

As indicated, the IDR and RR have the same scale of potential values. Although there are no absolute rules for the interpretation of these measures, Table 3 may serve as a rough guideline. For both measures, the absolute lower limit is zero, which results when there are no outcomes among exposed subjects. The upper limit is positive infinity, which results when there are no outcomes among unexposed subjects.

Table 3 Guidelines to the Interpretation of the Incidence Density and Risk Ratios

Value	Effect of Exposure
0–0.3	Strong benefit
0.4–0.5	Moderate benefit
0.6–0.8	Weak benefit
0.9–1.1	No effect
1.2–1.6	Weak hazard
1.7–2.5	Moderate hazard
$\geqslant 2.6$	Strong hazard

RELATIVE MERITS OF PROSPECTIVE STUDIES

When compared with other methods of nonexperimental research, the prospective design offers several advantages. First, incidence rates of the outcome can be measured. Second, there is a clear temporal relationship between the purported cause and effect in prospective research. Third, exposures that are uncommon can be evaluated by selective sampling of exposed persons. Fourth, the amount and quality of information collected is not limited by the availability of historical records or recall of subjects. Fifth, multiple effects of a given exposure can be assessed.

Nevertheless, prospective studies also have important limitations. First, large numbers of subjects may be needed to evaluate rare outcomes. Second, a prospective study may involve years of observation to evaluate chronic processes. A prolonged investigation may suffer from logistical problems, such as turnover of the research staff and variations in the methods used to measure the outcome. Third, it is difficult to follow subjects over protracted time periods and these individuals may change their exposure levels over time. Fourth, prospective research is relatively expensive when compared with alternative nonexperimental methods. Ultimately, the decision to conduct a prospective study is governed by the purpose of the investigation, the nature of the phenomena under observation, and the resources available for the study.

References

[1] Dawber, T. R., Meadors, G. F., and Moore, F. E. (1951). *Amer. J. Public Health*, **41**, 279–286.

[2] Doll, R. and Hill, A. B. (1964). *Brit. Med. J.*, **1**, 1399–1410.

[3] Kleinbaum, D. G., Kupper, L. L., and Morgenstern, H. (1982). *Epidemiologic Research*. Lifetime Learning Publications, Belmont, Calif., p. 286.

[4] Mantel, N. and Haenszel, W. (1959). *J. Natl. Cancer Inst.*, **22**, 719–748.

[5] Niswander, K. and Gordon, M. (1972). *The Women and Their Pregnancies*. National Institutes of Health Publ. 73-379. U.S. Government Printing Office, Washington, D.C.

[6] Truett, J., Cornfield, J., and Kannel, W. (1967). *J. Chronic Dis.*, **20**, 511–524.

Acknowledgments

The preparation of this entry was supported in part by a contract from the National Cancer Institute (N01-CN-61027) to the author, and the manuscript was typed by Ms. Vickie Thomas.

(CLINICAL TRIALS
EPIDEMIOLOGICAL STATISTICS
FOLLOW-UP
FRAMINGHAM: AN EVOLVING LONGITUDINAL STUDY
HISTORICAL CONTROLS
LONGITUDINAL DATA ANALYSIS
PANEL DATA
RETROSPECTIVE STUDIES)

RAYMOND S. GREENBERG

PROXIMITY DATA

Proximity data are numbers representing measures of relatedness of pairs of objects, for example, the judged similarity between political candidates, the confusability of speech sounds, or the number of tourists traveling between different countries. Proximities among a set of n objects are specified in an $n \times n$ matrix whose entry in row i and column j indicates the extent to which objects i and j are related to each other. *Proximity matrices* may be obtained directly, as in the examples above, or derived from rectangular matrices where each of a set of objects under study is described by a set of variables. Some ways of obtaining proximity data are discussed in the next section.

Proximity matrices are often large and complex. Methods for summarizing these data can help to discover important relationships and patterns within the data. Much work has been reported in the past 20 years on the development of models and methods for the representation of structure in proximity data. Most of these representations fall into one of two families: continuous spatial models and discrete network models. *Spatial models* embed objects in some coordinate space so that the metric distances between points represent the observed proximities between the respective objects. *Network models* represent each object as a node in some graph so that the relations among the nodes in the graph reflect the proximities among the objects. Spatial and discrete models that assume the proximity matrices are approximately symmetric are discussed in the sections "Spatial Representations" and "Discrete Representations." Questions about symmetry and some nonsymmetric models are touched on briefly in the section "Nonsymmetric Data."

WAYS OF OBTAINING PROXIMITY DATA

One common procedure for obtaining proximity data is to ask subjects to judge the similarity (or dissimilarity) of the stimulus objects under study. Subjects are presented pairs of stimuli and are asked to rate the proximity of these stimuli on some scale. Although similarity or dissimilarity are most frequently used to elicit the judgments, other words have been used such as association or substitutability. However, specific attributes on which the judgments are to be made are usually not specified to the subjects.

When the stimulus set under study is

large, say 50 to 100 objects, a useful way of obtaining proximity data is to ask subjects to sort the stimuli into mutually exclusive and exhaustive categories; typically, subjects are asked to cluster the items so that stimuli in the same category are more similar to each other than to stimuli in the other categories. A measure of proximity, then, is the number or proportion of times each pair of stimuli is placed into the same category.

Another procedure that requires somewhat less direct judgments about proximity involves measurement of stimulus confusability. Subjects are asked to judge whether two stimuli are the same or different. The number or proportion of times a "same" response is given for two stimuli that are, in fact, different, (i.e., confusability of the stimuli) can be considered to be a proximity measure. Another type of measure for which there is a considerable body of data is the amount of mail, telephone communication, air traffic, and so on, between different locations, people, or organizational units.

It is also possible to compute a derived measure of proximity from a rectangular matrix of profile data where the rows correspond to objects and the columns are variables measured on the objects. For example, a square symmetric matrix of profile proximities between variables could result from computing correlations of the variables over objects. This would measure similarity of the variables. Measures of dissimilarity of the objects could be obtained by computing distances between these objects. A variety of derived proximity measures is described in Anderberg [1, Chaps. 4 and 5] for ratio, interval, ordinal, nominal, and binary variables. Gordon [15, Chap. 2] gives a good summary of ways of computing a proximity matrix from an objects-by-variables matrix and discusses difficulties which can arise in the construction of such proximity measures.

SPATIAL REPRESENTATIONS

A common representation of a matrix of proximities is in terms of a spatial model; objects are represented as points in a multidimensional space and the distance between the points are related in some way to the proximity data. Multidimensional scaling* (MDS) is a family of models and associated methods for constructing such a space. The aim of MDS is to discover the dimensions relevant to the objects under consideration and to determine their coordinates on each dimension. There are several different approaches to MDS. The metric approach (e.g., Torgerson [33]) generally assumes that the relationship between the proximities and distances in space is linear. In nonmetric MDS (e.g., Shepard [28, 29], Kruskal [21, 22]), the relationship need only be monotonic. Two-way MDS assumes that there is only one proximity matrix. Euclidean distances resulting from a two-way MDS analysis are invariant under rigid or orthogonal rotation of the coordinate system. Therefore, the configuration that results from a two-way MDS analysis that assumes Euclidean distances must usually be rotated in order to find the most interpretable dimensions. There are many different computer programs for implementing two-way MDS. One example of such a program that will do both metric and nonmetric MDS is KYST [24]. Other computer programs for two-way scaling are described in MULTIDIMENSIONAL SCALING.

It is possible to have several proximity matrices, for example, one matrix for each of several different subjects or experimental conditions. Three-way MDS procedures will handle these data. The most popular three-way MDS approach, weighted MDS or INDSCAL (INdividual Differences multidimensional SCALing [4, 8, 19]), assumes that there is a set of dimensions common to all individuals (or conditions) but that each individual (or condition) can weight these dimensions differently. Three-way MDS produces a separate space for each individual; these spaces are all related, except that the dimensions are stretched or shrunk according to how salient that dimension is for a particular individual. Unlike two-way MDS, the coordinates from a three-way weighted

MDS analysis are uniquely oriented, since the model is not rotationally invariant. There are also many computer programs that implement the three-way MDS model. One example of a metric weighted MDS program for symmetric data is SINDSCAL [25]. *See also* MULTIDIMENSIONAL SCALING; more general models for three-way MDS are described by Tucker [34] and Harshman [17].

DISCRETE REPRESENTATIONS

Spatial distance models assume that continuous dimensions are the underlying basis for the proximities. Tversky [35] and Tversky and Gati [36] question this assumption. They argue that similarity between objects often should be represented in terms of discrete features. One of the most widely used methods for discrete representation of proximity data is hierarchical cluster analysis*. This produces a set of clusters such that any two clusters are disjoint or one includes the other. The objects being clustered can be represented as terminal nodes of a tree and the distances between objects are the "heights" of the first nonterminal node at which they meet. This leads to a particular kind of metric defined on a tree, the *ultrametric*. *See* HIERARCHICAL CLUSTER ANALYSIS for further details.

Another representation of proximity data in terms of a tree is as a *path-length tree*, also called an additive similarity tree or free tree (see Carroll [5], Sattath and Tversky [27], and Cunningham [13]). While a hierarchical tree has a natural root, path-length trees do not, so it is not necessary to think of them as being hierarchically organized. In a path-length tree, objects are represented as nodes and there are lengths associated with branches; the distance between any two objects is the sum of lengths of the branches that connect the two nodes. A computer program implementing a path-length tree is ADDTREE [27].

Shepard and Arabie [31] describe a nonhierarchical nonspatial model and method for representing data as overlapping clusters. The model assumes that each of the clusters of objects corresponds to a discrete property shared by all and only those objects within that cluster. MAPCLUS is a computer program [2] that uses a mathematical programming* approach for implementing this model. INDCLUS [7] generalizes the Shepard–Arabie model to handle several proximity matrices; it assumes that there is a set of clusters that is common to all individuals, or other data sources, but that each individual can weight these clusters differently. Like MAPCLUS, the INDCLUS computer program uses a mathematical programming procedure to implement the model.

In a hierarchical representation, in order for an object to be represented in more than one cluster, one of the clusters must be a subset of a larger cluster. Overlapping clustering procedures allow an object to be represented in more than one cluster. Another way of classifying an object in more than one way is to assume that the objects are organized into two or more separate hierarchies, not necessarily independent of one another. This can be done by fitting more than one tree to a single proximity matrix; objects are displayed as terminal nodes of one or more different trees (see Carroll [5] and Carroll and Pruzansky [9]). Carroll and Pruzansky [9] describe a hybrid model and fitting procedure for representing structure in proximity data that combines both spatial and discrete representations. They simultaneously fit to the data one or several hierarchical or path-length trees and an N-dimensional space.

Given such a wide variety of possible models and methods, how does one select the correct model for a given set of data? Pruzansky et al. [26] investigated some properties of proximity data that might help determine the appropriate model for a particular data set. They found that the skewness and a property that measures departures from the ultrametric inequality* were useful in distinguishing between a nonmetric MDS model and an additive tree model. However, one should be cautious about trying to find a single correct model for representing proximities. Shepard [30] points out that different methods of analysis may bring out different but equally important aspects of the underlying structure in the data.

NONSYMMETRIC DATA

The models and methods described above assume that the matrix (or matrices) of proximities is symmetric. For many sets of data this is unrealistic. Asymmetries may be due to noise or they may be systematic. If it can be determined that the data are symmetric but the asymmetries are due to noise, it is reasonable to symmetrize the matrix by using some technique for averaging s_{ij} and s_{ji} (e.g., Arabie and Soli [3]). Hubert and Baker [20] describe a method for evaluating the symmetry of a proximity matrix. However, there are data sets where systematic asymmetries might be an inherent characteristic. Several approaches have been described for analyzing nonsymmetric proximity matrices in terms of both spatial and discrete models. For example, if the rows and columns of the matrix are thought of as separate entities, the row and column objects can be displayed separately in the same multidimensional space (e.g., Constantine and Gower [10]). A method of this sort exists for data consisting of several nonsymmetric proximity matrices as well (see DeSarbo and Carroll [14]). Some methods deal with the symmetric and nonsymmetric portions separately [10, 16, 18, 32]. Cunningham [13] uses an extension of a free tree representation to account for nonsymmetric proximities; he introduces bidirectional trees, allowing the length of the branch between node i and node j to be different from the length between node j and node i.

LITERATURE

A book by Coombs [11] contains a general discussion of types of data and types of models as well as the earliest formulation of nonmetric scaling. Carroll and Arabie [6] contains a comprehensive review of the work in MDS through 1979. Since they use a broad definition of MDS, the article includes discussion of discrete models as well as spatial distance models. A monograph on MDS by Kruskal and Wish [23] provides an excellent introduction to spatial distance models for representing proximity data. A book by Coxon [12] is a good introductory text on MDS that includes material about both spatial and discrete representations of proximity data. Additional references are mentioned in HIERARCHICAL CLUSTER ANALYSIS and MULTIDIMENSIONAL SCALING.

References

[1] Anderberg, M. R. (1973). *Cluster Analysis for Applications*. Academic Press, New York.

[2] Arabie, P. and Carroll, J. D. (1980). *Psychometrika*, **45**, 211–235.

[3] Arabie, P. and Soli, S. D. (1980). In *Multidimensional Analysis of Large Data Sets*, R. G. Golledge and J. N. Rayner, eds. University of Minnesota Press, Minneapolis, Minn.

[4] Bloxom, B. (1968). Individual Differences in Multidimensional Scaling. *Res. Bull. 68-45*, Educational Testing Service, Princeton, N.J.

[5] Carroll, J. D. (1976). *Psychometrika*, **41**, 439–463.

[6] Carroll, J. D. and Arabie, P. (1980). *Annu. Rev. Psychol.*, **31**, 607–649.

[7] Carroll, J. D. and Arabie, P. (1983). *Psychometrika*, **48**, 157–169.

[8] Carroll, J. D. and Chang, J. J. (1970). *Psychometrika*, **35**, 283–319.

[9] Carroll, J. D. and Pruzansky, S. (1980). In *Similarity and Choice*, E. D. Lantermann and J. Feger, eds. Hans Huber, Berne, Switzerland, pp. 108–139.

[10] Constantine, A. G. and Gower, J. C. (1978). *Appl. Statist.*, **27**, 297–304.

[11] Coombs, C. H. (1964). *A Theory of Data*. Wiley, New York.

[12] Coxon, A. P. M. (1982). *The User's Guide to Multidimensional Scaling*. Heinemann, Exeter, N.H.

[13] Cunningham, J. P. (1978). *J. Math. Psychol.*, **17**, 165–188.

[14] Desarbo, W. S. and Carroll, J. D. (1981). *Proc. 3rd ORSA/TIMS Special Interest Conf. Market Meas. Anal.*, pp. 157–183.

[15] Gordon, A. D. (1981). *Classification*. Chapman & Hall, London.

[16] Gower, J. C. (1977). In *Recent Developments in Statistics*, J. R. Barra, F. Brodeau, G. Romier, and B. van Cutsem, eds. North-Holland, Amsterdam.

[17] Harshman, R. A. (1972). PARAFAC2: Mathematical and Technical Notes. *Working Paper in Phonetics No. 22*, University of California, Los Angeles, CA, pp. 30–47.

[18] Holman, E. W. (1979). *J. Math. Psychol.*, **20**, 1–15.

[19] Horan, C. B. (1969). *Psychometrika*, **34**, 139–165.

[20] Hubert, L. J. and Baker, F. B. (1979). *Qual. Quant.*, **13**, 77–84.

[21] Kruskal, J. B. (1964). *Psychometrika*, **29**, 1–27.

[22] Kruskal, J. B. (1964). *Psychometrika*, **29**, 115–129.

[23] Kruskal, J. B. and Wish, M. (1978). *Multidimensional Scaling*. Sage, Beverly Hills, Calif.

[24] Kruskal, J. B., Young, F. W., and Seery, J. B. (1973). How to Use KYST 2: A Very Flexible Program to Do Multidimensional Scaling and Unfolding. Bell Laboratories, Murray Hill, N.J.

[25] Pruzansky, S. (1975). How to Use SINDSCAL; A Computer Program for Individual Differences in Multidimensional Scaling. Bell Laboratories, Murray Hill, N.J.

[26] Pruzansky, S., Tversky, A., and Carroll, J. D. (1982). *Psychometrika*, **47**, 3–24.

[27] Sattath, S. and Tversky, A. (1977). *Psychometrika*, **42**, 319–345.

[28] Shepard, R. N. (1962). *Psychometrika*, **27**, 125–140.

[29] Shepard, R. N. (1962). *Psychometrika*, **27**, 219–246.

[30] Shepard, R. N. (1980). *Science*, **210**, 390–398.

[31] Shepard, R. N. and Arabie, P. (1979). *Psychol. Rev.*, **86**, 87–123.

[32] Tobler, W. (1976). *J. Environ. Syst.*, **6**, 271–301.

[33] Torgerson, W. S. (1958). *Theory and Methods of Scaling*. Wiley, New York.

[34] Tucker, L. R. (1972). *Psychometrika*, **37**, 3–27.

[35] Tversky, A. (1977). *Psychol. Rev.*, **84**, 327–352.

[36] Tversky, A. and Gati, I. (1978). In *Cognition and Categorization*, E. Rosch and B. B. Lloyd, eds. Lawrence Erlbaum, Hillsdale, N.J., pp. 79–98.

(HIERARCHICAL CLUSTER ANALYSIS
MEASURES OF SIMILARITY, DISSIMILARITY, AND DISTANCE
MULTIDIMENSIONAL SCALING)

SANDRA PRUZANSKY

PROXIMITY GRAPH *See* GRAPH-THEORETIC CLUSTER ANALYSIS

PROXY VARIABLE

In *stochastic* regression models* of the type

$$y_t = \beta_1 + \beta_2 x_t^* + v_t,$$

where x_t^* is an unobservable true variable and v_t is a stochastic term, x_t^* is sometimes replaced by an *observable* "proxy" variable $\hat{x}_t$ representing x_t^*. Most commonly $\hat{x}_t$ is an estimated or measured (as opposed to "true") value of x_t^*. Formally, $\hat{x}_t = x_t^* + \hat{u}_t$, where $\hat{u}_t$ is a random measurement error. As an example (see, e.g., Judge et al. [1]), in studying the effects of education on income we need to measure the years of education. Instead, the observable proxy variable—years of schooling—is used.

Reference

[1] Judge, G. G., Hill, R. C., Griffith, W., Lütkepohl, H., and Lee, T. C. (1982). *Introduction to the Theory and Practice of Econometrics*. Wiley, New York.

(STOCHASTIC REGRESSION MODELS)

PSEUDO-BAYESIAN INFERENCE *See* QUASI-BAYESIAN INFERENCE

PSEUDO-*F*

This is a statistic (*not* a distribution). More precisely, it is a class of statistics formed by replacing the unbiased estimators $S_i^2 = (n_i - 1)^{-1}\sum_{j=1}^{n_i}(X_{ij} - \bar{X}_i)^2$ of variance σ_i^2 $(i = 1, 2)$ in the variance-ratio* statistic S_2^2/S_1^2 by some other measure of (spread)^2. The class includes, for example, the ratio of squared mean deviations

$$\left\{ n_2^{-1} \sum_{j=1}^{n_2} |X_{2j} - \bar{X}_2| \Big/ \left(n_1^{-1} \sum_{j=1}^{n_1} |X_{1j} - \bar{X}_1| \right) \right\}^2$$

and the ratio

$$\left\{ \frac{a_{n_2}\text{range}(X_{21}, \ldots, X_{2n_2})}{a_{n_1}\text{range}(X_{11}, \ldots, X_{1n_1})} \right\}^2,$$

where the factors a_{n_i} are chosen so that

$$a_{n_i} = K\sigma_i / E\left[\text{range}(X_{i1}, \ldots, X_{in_i})\right]$$

(usually $K = 1$). The value of a_{n_i} depends on the distribution populations. It is commonly supposed, in applications, that these distributions are normal, under which conditions the distribution of the variance ratio (S_2^2/S_1^2) is that of

$$(\sigma_2^2/\sigma_1^2) \times (\text{variable having a } F_{n_1-1, n_2-1} \text{ distribution}).$$

See F-DISTRIBUTION.

PSEUDO-LIKELIHOOD

This is a poor title for a class of procedures in which inferences are based on a derived likelihood for a subset of the parameters. We discuss here marginal, conditional, and profile likelihood; partial likelihood* is a fourth general class of procedures and is discussed in a separate entry.

Suppose that $\mathbf{X} = (X_1, \ldots, X_n)$ is a vector of random variables whose distribution depends on parameters (θ, α). We suppose that θ is of interest and that θ and α may be vectors of parameters. In most cases of interest, α is of very high dimension. We consider the problem of drawing suitable likelihood-based inferences about θ when α is unknown.

MARGINAL, CONDITIONAL, AND PROFILE LIKELIHOODS

The joint likelihood of (θ, α) based on data $\mathbf{X} = \mathbf{x}$ is $L(\theta, \alpha; \mathbf{x}) \propto f_{\mathbf{X}}(\mathbf{x}; \theta, \alpha)$, where $f_{\mathbf{X}}$ represents the joint density function. For convenience, we consider the relative likelihood function

$$R(\theta, \alpha; \mathbf{x}) = L(\theta, \alpha; \mathbf{x}) / L(\hat{\theta}, \hat{\alpha}; \mathbf{x}), \quad (1)$$

which compares each hypothesized (θ, α) with the maximum likelihood* estimate $(\hat{\theta}, \hat{\alpha})$. Likelihood-based inferences on θ, α can be based on (1), but when (θ, α) is of high dimension, the examination and comprehension of the full likelihood surface is difficult. One approach to estimating θ would utilize the *profile* or *maximized relative likelihood* of θ,

$$R_P(\theta; \mathbf{x}) = \sup_{\alpha} R(\theta, \alpha; \mathbf{x}). \quad (2)$$

Note that $R_P(\theta_0; \mathbf{x})$ is the generalized likelihood ratio* statistic for the hypothesis $\theta = \theta_0$, and places an upper bound on the relative likelihood of (θ, α)-pairs with $\theta = \theta_0$.

The profile likelihood can quite generally be used, whereas marginal and conditional likelihoods arise in problems with sufficient structure that the nuisance parameter* α can be eliminated by integration over a subset of the sample space (marginal likelihood) or by conditioning on a particular observed event (conditional likelihood).

Suppose that there exists a one-to-one transformation of $\mathbf{X}$ to U, V such that the joint PDF of U, V factors as

$$f_{U,V}(u, v; \theta, \alpha) = f_{U|V}(u \mid v; \theta, \alpha) f_V(v; \theta), \quad (3)$$

where only the second term depends on θ. The *marginal likelihood* of θ based on V is

$$L_M(\theta; v) \propto f_V(v; \theta). \quad (4)$$

In general, the use of (4) for the estimation of θ will entail a loss of information, namely the information about θ in the first term on the right side of (3). When, as for example through invariance arguments, one can argue that the whole of the information about θ (when α is unknown) is contained in V, the likelihood (4) is highly recommended as the basis of likelihood inference about θ. More generally, the residual term may contain some information about θ, but (3) may still be used for inference on account of its simplicity.

If, on the other hand, the transformation of $\mathbf{X}$ into U, V leads to a joint density which factors as

$$f_{U|V}(u, v; \theta, \alpha) = f_{U|V}(u \mid v; \theta) f_V(v; \theta, \alpha), \tag{5}$$

$$L_C(\theta; u \mid v) \propto f_{U|V}(u \mid v; \theta) \tag{6}$$

is the conditional likelihood of θ based on U for the given V. Again, in certain instances, it may be argued that V contains little or no information about θ when α is unknown so that the conditional likelihood is appropriate for the estimation of θ.

Marginal likelihoods were first introduced within the structural model* by Fraser (see refs. 6 and 7). Conditional likelihoods have been used by many authors. Bartlett [4] suggested their use and extended the argument to allow the conditioning variable V to depend on the parameter θ. Both marginal and conditional likelihoods are discussed and exemplified in refs. 9 and 10.

EXAMPLES AND COMMENTS

Estimation of a Normal Variance

Suppose that duplicate determinations X_{1i}, X_{2i} are made on the ith individual, and that X_{1i}, X_{2i} are independent $N(\mu_i, \sigma^2)$ variates, $i = 1, \ldots, n$. Here $\mu_1, \ldots, \mu_n$ and σ^2 are unknown parameters and our purpose is to estimate σ^2. The profile or maximum relative likelihood of σ^2 is

$$R_P(\sigma^2; \mathbf{x}) \propto \sigma^{-2n} \exp\left\{ -\frac{\sum(x_{1i} - x_{2i})^2}{4\sigma^2} \right\},$$

which is maximized at

$$\hat{\sigma}^2 = (4n)^{-1} \sum (x_{1i} - x_{2i})^2.$$

As $n \to \infty$, $\hat{\sigma}^2$ converges in probability to $\sigma^2/2$. In this example, the profile likelihood is concentrated for large n about an incorrect value of σ^2.

The data for the ith pair can be written $U_i = X_{1i} - X_{2i}$ and $V_i = X_{1i} + X_{2i}$. Note that U_i and V_i are independent and $V_i \sim N(2\mu_i, \sigma^2)$. This suggests that V_i is uninformative about σ^2 when μ_i is unknown and that a marginal likelihood could be based on $U = (U_1, \ldots, U_n)$. Thus

$$L_m(\sigma^2; \mathbf{u}) \propto \sigma^{-n} \exp\left\{ -\frac{\sum(X_{1i} - X_{2i})^2}{4\sigma^2} \right\} \tag{7}$$

and the marginal maximum likelihood estimator is $\hat{\sigma}_M^2 = \sum(X_{1i} - X_{2i})^2/(2n)$, which is consistent. The marginal likelihood takes account of the degrees of freedom lost through the estimation of $\mu_1, \ldots, \mu_n$.

The heuristic argument which suggests the use of $U_1, \ldots, U_n$ for inference about σ^2 can be formalized by group invariance* arguments. The group of transformations $X_{ji} \to X_{ji}^* = X_{ji} + c_i$ ($j = 1, 2$; $i = 1, \ldots, n$), where $-\infty < c_i < \infty$, acts transitively on V_i and leaves U_i invariant. In the parameter space, the induced group acts transitively on $\mu_1, \ldots, \mu_n$ and leaves σ^2 invariant. This suggests that $U_1, \ldots, U_n$ is sufficient for σ^2 when $\mu_1, \ldots, \mu_n$ are completely unknown.

This example illustrates difficulties with the use of the profile likelihood when the dimensionality of the nuisance parameter is large. In other instances, when the nuisance parameter is of small dimension compared to the number of observations, the profile likelihood typically gives a convenient and simple summary.

Estimation of β in the Proportional Hazards Model

The proportional model of Cox (see ref. 5 and COX'S REGRESSION MODEL) with fixed covariates $\mathbf{Z}' = (Z_1, \ldots, Z_p)$ specifies the hazard function for the time to failure T as

$$\lambda(t; \mathbf{Z}) = \lambda_0(t) \exp(\mathbf{Z}'\boldsymbol{\beta}),$$

where $\boldsymbol{\beta}' = (\beta_1, \ldots, \beta_p)$, $\lambda_0(t)$ is a baseline hazard function which is left completely unspecified, and $\boldsymbol{\beta}' = (\beta_1, \ldots, \beta_p)$ is a vector of regression parameters. The model specifies that the rate at which individuals fail at time t is a constant (depending on the measured covariates $\mathbf{Z}$) times a baseline rate.

Consider a sample $T_1, \ldots, T_n$ with corresponding covariates $\mathbf{Z}_1, \ldots, \mathbf{Z}_n$, let $T_{(1)}, \ldots, T_{(n)}$ be the order statistics* and $\mathbf{Z}_{(1)}$,

$\ldots, \mathbf{Z}_{(n)}$ the corresponding covariates. A marginal likelihood can be based on the distribution of the ranks or of the antiranks, $A = ((1), \ldots, (n))$. This gives

$$L_M(\beta; \mathbf{a}) \propto \pi e^{\mathbf{Z}'_{(i)}\beta} \Big/ \sum_{j=i}^{n} e^{\mathbf{Z}'_{(j)}\beta},$$

which is identical to the partial likelihood obtained by Cox (*see* PARTIAL LIKELIHOOD). Group invariance arguments can be used to establish the sufficiency of **A** for the estimation of $\boldsymbol{\beta}$ when $\lambda_0(\cdot)$ is completely unspecified (see ref. 8).

Estimation of a Ratio of Poisson Means

Suppose that X and Y are independent Poisson variates with means α and $\theta\alpha$, respectively. Let $U = X + Y$ and $V = Y$; the joint probability function of U, V factors as

$$\begin{aligned} f_{U,V}(u, v; \theta, \alpha) &= f_u(u; \theta, \alpha) f_{V \mid U}(v; \theta \mid u) \\ &= (\gamma^u e^{-\gamma}/u!)\tbinom{u}{v} \\ &\quad \times \theta^v (1 + \theta)^{-u}, \end{aligned}$$

where $u \geqslant v = 0, 1, 2, \ldots$ and $\gamma = \alpha(1 + \theta)$. A conditional likelihood for θ can be based on the binomial distribution of V, given U, and yields

$$L_C(\theta; v \mid u) \propto \theta^v (1 + \theta)^{-u}, \qquad 0 < \theta < \infty. \tag{8}$$

Further, since the distribution of U depends only on $\gamma = \alpha(1 + \theta)$, it may be argued that no additional information about θ can be extracted from the residual term when α is unknown. In this example, the conditional likelihood (8) is proportional to the profile likelihood.

Estimation of a Common Odds Ratio

Let X_{1i} and X_{2i} be independent Bernoulli variates with parameters p_{1i} and p_{2i}, respectively, where

$$\begin{aligned} \log\{p_{2i}/(1 - p_{2i})\} &= \log\{p_{1i}/(1 - p_{1i})\} + \theta, \\ & \qquad i = 1, \ldots, n. \end{aligned}$$

Here $\exp(\theta)$ represents the odds ratio*, which is assumed common for each i. This model might be entertained when pairs of like individuals are assigned at random to two treatments and the response is binary. A conditional likelihood can be based on the distribution of $\mathbf{V} = (V_1, \ldots, V_n)$ given $\mathbf{U} = (U_1, \ldots, U_n)$, where $V_i = X_{2i}$ and $U_i = X_{1i} + X_{2i}$. If $U_i = 0$ or 2, then V_i is determined, so only discordant pairs (pairs with $U_i = 1$) contribute and

$$\begin{aligned} \Pr\{U_i = 1 \mid V_i = 1\} &= 1 - \Pr\{U_i = 0 \mid V_i = 1\} \\ &= e^\theta/(1 + e^\theta). \end{aligned}$$

If $\lambda = e^\theta/(1 + e^\theta)$, it follows that the conditional likelihood of λ is proportional to

$$\lambda^a (1 - \lambda)^b, \tag{9}$$

where $a(b)$ is the number of discordant pairs with $X_{2i} = 1 (X_{1i} = 1)$.

The conditional likelihood (9) makes no use of information in $\mathbf{U}$ about θ. In ref. 13 it is pointed out that some inferences on θ may be possible given only the value of $\mathbf{U}$. If, for example, almost all the pairs were discordant, this would suggest that θ is not near zero. (If $\theta = 0$, then the probability that a pair is discordant is at most $\frac{1}{2}$, which occurs when $p_{1i} = p_{2i} = \frac{1}{2}$.) It is not clear, however, how this information can be incorporated in the inference about θ, nor that it is desirable to do so.

REMARKS

The problem of assessing the information lost through use of a marginal or conditional likelihood, and of specifying conditions under which a statistic can be said to be sufficient for θ when α is unknown have been considered by several authors. See, for example, refs. 1 to 3 and 13.

Both marginal and conditional likelihoods can be generalized to allow V in (3) or (5) to depend on θ. Extension of the argument to this case are considered in refs. 4, 9, and 10.

Interesting applications of marginal likelihood can be found in refs. 11 and 12.

References

[1] Barnard, G. A. (1963). *J. R. Statist. Soc. B*, **25**, 111–114. (Some aspects of the fiducial argument.)

[2] Barndorff-Nielsen, O. (1976). *Biometrika*, **63**, 567–571. (Nonformation.)

[3] Barndorff-Nielsen, O. (1973). *Biometrika*, **60**, 447–455.

[4] Bartlett, M. S. (1936). *Proc. Camb. Philos. Soc.*, **32**, 560–566. (The information available in small samples.)

[5] Cox, D. R. (1972). *J. R. Statist. Soc. B*, **34**, 187–220. [Regression models and life tables (with discussion).]

[6] Fraser, D. A. S. (1967). *Ann. Math. Statist.*, **38**, 1456–1465. (Data transformations and the linear model.)

[7] Fraser, D. A. S. (1968). *The Structure of Inference*. Wiley, New York.

[8] Kalbfleisch, J. D. and Prentice, R. L. (1973). *Biometrika*, **60**, 267–278. (Marginal likelihoods based on Cox's regression and life model.)

[9] Kalbfleisch, J. D. and Sprott, D. A. (1970). *J. R. Statist. Soc. B*, **32**, 175–208. [Application of likelihood methods to models involving large numbers of parameters (with discussion).]

[10] Kalbfleisch, J. D. and Sprott, D. A. (1974). *Sankhyā A*, **35**, 311–328. (Marginal and conditional likelihood.)

[11] Minder, C. E. and McMillan, I. (1977). *Biometrics*, **33**, 333–341. (Estimation of compartment model parameters using marginal likelihood.)

[12] Minder, C. E. and Whitney, J. B. (1975). *Technometrics*, **17**, 463–471. (A likelihood analysis of the linear calibration problem.)

[13] Sprott, D. A. (1975). *Biometrika*, **62**, 599–605. (Marginal and conditional sufficiency.)

(ANCILLARY STATISTICS
LIKELIHOOD
PARTIAL LIKELIHOOD
QUASI-LIKELIHOOD FUNCTIONS
SUFFICIENT STATISTICS)

JOHN D. KALBFLEISCH

PSEUDO-MEDIAN

The pseudo-median of a distribution F is defined as a median of the distribution of $\frac{1}{2}(X_1 + X_2)$, where X_1 and X_2 are independent random variables, each having the distribution F.

If F is symmetric, the median and pseudo-median coincide. In general, the pseudo-median may differ from the median. Consistent estimators of pseudo-medians are discussed in Hollander and Wolfe [1]. Pseudo-medians should not be confused with quasi-medians*, which are obtained by extension of the *sample* median. *See also* ref. 2.

References

[1] Hollander, M. and Wolfe, D. A. (1973). *Nonparametric Statistical Methods*. Wiley, New York.

[2] Høyland, A. (1965). *Ann. Math. Statist.*, **36**, 174–197.

PSEUDO-RANDOM NUMBER GENERATORS

The need to generate a sequence of numbers $\{X_i\}$ that behaves like a sample of independent uniform $(0, 1)$ random variables arises in several areas, including discrete event simulation* and Monte Carlo* studies. The numbers generated are called *pseudo-random* or *quasi-random* numbers and are produced from generators that yield integers on $[0, m]$, which are then transformed to $[0, 1]$ by division by m.

In theory, a uniform deviate can be transformed to an arbitrary random variable having distribution function $F(\cdot)$. Numerous authors (see Fishman [8], Knuth [18], and Bratley et al. [3]) have compiled algorithms for transformations to commonly required distributions (*see also* GENERATION OF RANDOM VARIABLES).

Several sources of error contaminate the ideal above. The finite word length of the computer can result in errors both in the transformation and in restricting all samples to a finite subset of the rational numbers. However, these are insignificant in comparison to errors resulting from nonuniformity and dependence in the pseudo-random number sequence. The latter source of error is a focal point for research.

Early approaches concentrated on physical devices such as white noise* in electronic circuits and tables of random numbers compiled externally to the computer; see Niederreiter [28] and Jansson [14]. Reproducing results (a necessity both for assuring that the computer representation is an accurate implementation of the model and for gaining efficiencies in the estimation technique through processes known as variance reduction techniques) when using these techniques is problematic and led researchers to consider algorithms for random number generation.

RANDOMNESS

Pseudo-random number generators are usually evaluated on two criteria: their speed of execution and the degree to which the resulting number sequence "appears" random. A generator is acceptable if sequences it produces can pass an assortment of tests for randomness*. The specific tests and passing levels have not been formalized in the literature so there are no canonical measures by which to evaluate a proposed generator (see Knuth [18]).

In practice, researchers seek a generator that yields a sequence $\{X_i\}$ that approximates a k-distributed sequence. The infinite sequence $\{W_i\}$ is *k-distributed* if for numbers $0 \leqslant v_1 < w_1 \leqslant 1, \ldots, 0 \leqslant v_k < w_k \leqslant 1$,

$$\lim_{n \to \infty} \sum_{i=1}^{n} \sum_{j=1}^{k} I(v_j \leqslant W_i < w_j)/n = (w_1 - v_1) \cdots (w_k - v_k),$$

where $I(A) = 1$ if A is true and 0 otherwise. Thus a k-distributed sequence behaves as independent samples from the joint distribution of k independent uniform deviates. A sequence is *∞-distributed* if it is k-distributed for all $k > 0$.

Tests are performed on a given finite sequence to identify deviations from k-distributed uniformity. The tests fall into two categories: empirical and theoretical. The empirical tests are the standard statistical tests one encounters in hypothesis testing* and are based on a sample of $\{X_i\}$. Knuth [18] and Niederreiter [28] catalog many of these; Fishman and Moore [9] demonstrate a comprehensive approach.

Theoretical tests are based on examining the structure that the pseudo-random number generator imparts to the entire sequence $\{X_i\}$. As a result, they test the global properties of the generator; empirical tests evaluate local properties of the generator. The *spectral test** [6, 24] is a theoretical test that is useful for examining the degree to which $\{X_i\}$ is k-distributed. The *lattice test*, developed by Marsaglia [25], measures the lattice structure of k-tuples from $\{X_i\}$. The degree to which the lattice is orthogonal is measured by the ratio of the lengths of the longest to shortest edges, denoted L_k. For the ideal generator $L_k = 1$ for all $k > 1$.

The importance of testing a pseudo-random number generator cannot be overstated. Knuth [18] gives several examples of algorithms which, upon cursory examination, appear to produce highly random sequences that, upon closer examination, are poor approximations to random sequences.

LINEAR CONGRUENTIAL METHODS

Among the most popular and well-studied methods for generating a pseudo-random sequence are the linear congruential generators. These are based on the recurrence scheme

$$Z_{i+1} \equiv aZ_i + c \pmod{m}, \qquad \text{(LCG)}$$

where *seed* Z_0 is an integer $0 \leqslant Z_0 < m$, *multiplier* a is an integer $1 < a < m$, *increment* c is an integer $0 \leqslant c < m$, and *modulus* m is an integer $0 < m$. $\{Z_i\}$ is often referred to as a *linear congruential sequence* or a random number stream, and LCG is called a *mixed congruential generator* when $c > 0$ and a *multiplicative congruential generator* when $c = 0$.

The modulo arithmetic guarantees that each element in $\{Z_i\}$ is an integer in the interval $[0, m-1]$. Consequently, when $\{Z_i\}$ displays sufficiently random behavior, a pseudo-random sequence $\{X_i\}$ can be obtained by normalizing Z_i by the modulus m so that $X_i = Z_i/m$. As a result, the larger the modulus m, the smaller the error that results from the finite word length of the computer. It can also be shown that the sequence $\{Z_i\}$ is periodic in the sense that there exists a $P \leqslant m$ such that $Z_i = Z_{P+i}$ for all $i \geqslant 0$. A large P (and hence a large m) is desirable for achieving apparent randomness of $\{X_i\}$.

The speed of generation is also a consideration in choosing the modulus m. Consider algorithm ALCG for generating a linear congruential sequence according to LCG:

Algorithm ALCG

1. $W \leftarrow aZ_i + c$.
2. $Z_{i+1} \leftarrow W - \lfloor W/M \rfloor$.
3. Return Z_{i+1}.

The notation $\lfloor Y \rfloor$ means the largest integer less than or equal to Y. The division W/m in line 2 (which tends to be a slow operation) can be optimized when $m = z^e$, where z is the base of the computer's arithmetic and e is an integer. When m has this form, division is accomplished through a shift operation which is faster than a divide. As a result, 2^e (10^e) is often chosen as the modulus, where e is the number of bits (digits) in the computer's word. The generators RANDU (formerly used in the IBM Scientific Subroutine Package), with $a = 65{,}539$, $c = 0$, and $m = 2^{31}$ and SUPER-DUPER [25], with $a = 69{,}069$, $c = 0$, and $m = 2^{32}$, are two such examples. One drawback to generators having $m = z^e$ is that the low-order e bits (digits) of $\{Z_i\}$ form a linear congruential sequence with period at most 2^e (10^e). Consequently, the low-order bits (digits) have periods shorter than P and can exhibit extreme departures from randomness.

RELATIONSHIP BETWEEN a, c, Z_0, AND m

Two theorems help to characterize the length of the period P in terms of a, c, Z_0, and m. The first theorem assumes that $c > 0$.

Theorem 1. The linear congruential sequence $\{Z_i\}$ has full period ($P = m$) if and only if:

(a) c is relatively prime to m.
(b) $a - 1$ is a multiple of every prime factor of m.
(c) $a \equiv 1 \pmod 4$ when $m \equiv 0 \pmod 4$.

Example 1. When $m = 2^e$, c is relatively prime to m, and $a \equiv 1 \pmod 4$, LCG generates a sequence with full period 2^e.

A full period is no guarantee that the generator is adequate. Niederreiter [28] presents evidence questioning the use of the following full period generator.

Example 2. The LCG generator on the DEC VAX-11 [7] has $m = 2^{32}$, $a = 69{,}069$, and $c = 1$. Since these satisfy the conditions of Theorem 1, LCG generates a sequence with full period 2^{32}.

When $c = 0$ there is no need to perform the addition in line 1 of ALCG. As a result, LCG with $c = 0$ may execute somewhat faster than one with $c > 0$. Theorem 2 gives the conditions on a, m, and Z_0 to maximize the period of $\{Z_i\}$ when $c = 0$.

Theorem 2. When $c = 0$ the linear congruential sequence $\{Z_i\}$ has maximal period $\lambda(m)$ if:

(a) Z_0 is relatively prime to m.
(b) a is a primitive root modulo m.

The function $\lambda(m)$ is the order of a primitive element modulo m. Knuth [18] defines $\lambda(m)$ explicitly and gives conditions for a to be a primitive root modulo m. In the often encountered case when $p = 2$ and $e > 3$, they simplify to the single requirement that $a \equiv 3$ or 5 (mod 8).

Example 3. The generator with $m = 2^{32}$, $a = 69{,}069$, and $c = 0$ can generate linear congruential sequences with maximal period $\lambda(2^{32}) = 2^{30}$. It achieves this maximal period when Z_0 is relatively prime to m, which in this case is equivalent to Z_0 being odd. When this holds, every element in the sequence $\{Z_i\}$ is odd. Marsaglia [25] favors this multiplier on the basis of its performance on the lattice test. He claims that for $m = 2^{32}$, $L_2 = 1.06$, $L_3 = 1.29$, $L_4 = 1.30$, and $L_5 = 1.25$. However, Niederreiter [28] cautions against using this generator solely on the basis of its performance on the lattice test.

Example 4. The RANDU generator with $m = 2^{31}$, $a = 65{,}539$, and $c = 0$ has maximal period $\lambda(2^{31}) = 2^{29}$. Several authors note deficiencies in the performance of this generator on the spectral and lattice tests [25, 18].

PRIME MODULUS MULTIPLICATIVE GENERATORS

When m is prime and $c = 0$, LCG is a *prime modulus multiplicative generator*. Theorem 2 shows that for these generators the maximal period is $\lambda(m) = m - 1$, an almost full period. Furthermore, since all integers less than m are relatively prime to m, there are no additional restrictions on the seed Z_0.

Example 5. On a 32-bit machine the largest prime that can be directly represented in one word is $2^{31} - 1$. Generators with this modulus have period $2^{31} - 2$. As a result, there are several such generators. These include (a) LLRANDOM [19, 22] with $a = 16{,}807$, used in IMSL [13], APL [15], and SIMPL/I and GPSS [12]; (b) the SIMSCRIPT II simulation language generator for which $a = 630{,}360{,}016$ (Kiviat et al. [17]); and (c) the RANUNI generator [31] and the LLRANDOMII generator [21] with $a = 397{,}204{,}094$.

From an empirical point of view Fishman and Moore [9] study the performance of a multiplicative congruential generator with modulus $2^{31} - 1$ and 16 alternative multipliers; the three multipliers mentioned above are included.

Although a direct implementation of the prime modulus multiplicative congruential generator following ALCG is feasible, some of the efficiencies associated with substituting a shift operation for the division in line 2 when $m = z^e$ can be gained when m is a Mersenne prime. A prime is a Mersenne prime if it can be written in the form $2^e - 1$. An algorithm attributed to Payne et al. [30] is such an implementation.

GENERAL LINEAR CONGRUENTIAL GENERATORS

A generalization of the LCG is obtained by making Z_{i+1} a function not only of Z_i but also of $Z_{i-1}, Z_{i-2}, \ldots, Z_{i-p+1}$, as

$$Z_{i+1} \equiv \sum_{j=1}^{p} a_j Z_{i-j+1} \pmod{m}. \quad \text{(GLCG)}$$

By an extension of Theorem 2, for prime modulus m a set $\{a_j \mid j = 1, \ldots, p\}$ can be found such that the linear congruential sequence generated with GLCG has period $m^p - 1$. The coefficients $\{a_j\}$ must be such that the polynomial

$$f(x) = x^p + a_1 x^{p-1} + a_2 x^{p-2} + \cdots + a_p$$

has a root that is a primitive element of the field with m^p elements. Then $f(x)$ is a primitive polynomial modulo m, which we refer to as condition PPM(m). Knuth [18] gives criteria for testing for primitivity modulo m.

The *additive congruential generators* are the most popular of the GLCG form. These have $a_j = 0$ or 1 for $j = 1, \ldots, p$.

Example 6. The Fibonacci generator with $p = 2$, $a_1 = a_2 = 1$, and large m has been shown to produce a sequence with poor statistical properties and should be avoided; see Bratley et al. [3].

Example 7. A class of generators, which, Knuth [18] reports, produces sequences with good statistical properties but need further testing, has $m \equiv 0 \pmod 2$, $p = 55$, and $a_{24} = a_{55} = 1$, $a_j = 0$ for all other j. These generators have period at least as large as $2^{55} - 1$.

The *linear congruential generators modulo 2* are a special case of GLCG with $m = 2$. This restriction implies that w.l.o.g. the coefficients $\{a_j\}$ can be limited to $\{0, 1\}$. As a result, these generators are also a special case of additive congruential generators. When the $\{a_j\}$ are chosen to satisfy condition PPM(2) then $\{Z_i\}$ is a sequence of bits that is called a *pseudo-noise sequence* or *maximal-length shift register sequence*. Because the $\{a_j\}$ satisfy PPM(2) and $m = 2$ is a prime, the sequence $\{Z_i\}$ has period $2^p - 1$.

Example 8. Because the trinomial $f(x) = x^{55} + x^{24} + 1$ satisfies PPM(2) the generator $Z_i \equiv Z_{i-55} + Z_{i-24} \pmod 2$ has period $2^{55} - 1$. This generator is one of those in the class shown in Example 7.

GENERALIZED FEEDBACK SHIFT REGISTER GENERATORS

Tausworthe [34] suggests that a pseudo-noise sequence $\{Z_i\}$ with period $2^p - 1$ be used as a source of pseudo-random numbers. He examines the properties of the sequence of L-bit numbers $\{X_i\}$ defined by

$$X_i = \sum_{j=1}^{L} 2^{-j} Z_{si+r-dj}$$

$$\text{for } i = 0, \ldots, \lfloor (2^p - 1)/s \rfloor, \qquad \text{(GFSRG)}$$

where $\{Z_i\}$ is a pseudo-noise sequence with period $2^p - 1$, $s \geqslant L$, $d = 1$, $p \leqslant L$, and r is a randomly selected integer on $(0, 2^p - 1)$. This transformation of $\{Z_i\}$ is called the *L-wise decimation of* $\{Z_i\}$. When s and $2^n - 1$ are relatively prime $\{X_i\}$ has period $\lfloor (2^p - 1)/s \rfloor$. In this case the transformation of $\{Z_i\}$ is known as *proper L-wise decimation of* $\{Z_i\}$.

The sequence $\{X_i\}$ is sometimes referred to as a *Tausworthe sequence* and the method of generation as a *Tausworthe generator* or feedback shift register generator. The restriction $s \geqslant L$ is important; Knuth [18] states that for $s = 1$ a sequence with poor statistical properties results.

In practice the sequence $\{Z_i\}$ is usually generated using GLCG with $\sum_j a_j = 2$, so that

$$Z_i \equiv Z_{i-p} + Z_{i-q} \pmod 2.$$

As a result one need only determine integers p and q for which

$$f(x) = x^p + x^q + 1$$

satisfies PPM(2). Primitive trinomials to degree 1000 are given by Zierler and Brillhart [38].

Tausworthe [34] shows that in the limit, as $p \to \infty$, the properly decimated sequence $\{X_i\}$ has the mean, variance, and autocorrelation structure that one expects from independent samples of a uniform deviate. He also shows that for $k \leqslant p/s$ the sequence $\{(X_{i+1}, X_{i+2}, \ldots, X_{i+k})\}$ is approximately uniformly distributed over the unit k-hypercube. However, Tootill et al. [35] have shown that $\{X_i\}$ performs better for some values of p and q than others. They recommend that $s = L = q$, $q\lfloor p/q \rfloor$ be relatively prime to $2^p - 1$ and that $q < p/2$ be neither too small nor too close to $p/2$. They claim if these conditions are satisfied, then GFSRG produces sequences with good runs and uniformity properties.

Example 9. Tootill et al. [36] propose a generator with $p = 607$ and $q = 273$. For $d = 1$, $L = 23$, and $s = 512$, $\{X_i\}$ has a period of $2^{607} - 1$. They claim that for simulations with no fixed dimensionality requirements on the pseudo-random sequence, this generator possesses 26-distributed uniformity.

Lewis and Payne [20] studied sequences $\{X_i\}$ generated by GFSRG with $d = 100p$, $s = 1$, and $L = q$. They suggest that d relatively prime to the period $2^p - 1$ improves the degree to which $\{X_i\}$ approximates a k-distributed sequence for $1 \leqslant k \leqslant \lfloor p/L \rfloor$. Furthermore, they present an algorithm that is both extremely efficient and machine independent. Its greatest drawback is an elaborate initialization process. However, Arvillias and Maritsas [1] suggest changes in the algorithm that simplify initialization when q is an integral power of 2.

In Bright and Enison [4], Enison shows that with proper initialization the Lewis and Payne [20] scheme results in sequences that are 1-distributed. Fushimi and Tezuka [10] have extended this work to give necessary and sufficient conditions on GFSRG for $\{X_i\}$ to be k-distributed.

Example 10. Bright and Enison [4] recommend that p be a Mersenne exponent so that $2^p - 1$ is prime. They propose the generator having $p = 521$, $q = 32$, $L = 64$, $d = 52{,}100$, and $s = 1$. They claim that this generator produces sequences that are 8-distributed.

The attraction of these generators comes from the fact that their implementation can be machine independent. The Lewis and Payne [20] algorithm generates an L-bit integer for any value of s. Thus sequences with extremely large periods can be generated on small-word machines. However, the algorithm requires memory for p L-bit integers and a time-consuming initialization. Although initial results from these generators are encouraging, more empirical and structural study is needed before they are adopted widely [8].

SHUFFLING AND SHIFTING: COMBINATIONS OF GENERATORS

In an attempt to improve the degree to which a pseudo-random sequence approximates a truly random sequence, several authors suggest methods for shuffling the sequence $\{Z_i\}$ or shifting the bits in each Z_i. If $\{V_i\}$ and $\{W_i\}$ are two sequences, preferably generated from independent generators, then the following schemes have been suggested:

(a) $Z_i \equiv (W_i + V_i) \pmod{m}$ [18].

(b) $Z_i = W_k$ where $k = V_i$ [23].

(c) $Z_i = W_k$ where $k = W_i$ [2].

(d) $Z_i = (Y \ll e_1)$ XOR Y, where $Y = W_i$ XOR $(W_i \gg e_2)$ [8].

The XOR indicates the Boolean exclusive-or operation and the $X \ll e$ ($X \gg e$) indicates left (right) shift of X by e bits. In practice, methods **(b)** and **(c)** are performed on segments of the sequence $\{W_i\}$ or $\{V_i\}$. Knuth [18] presents algorithms and recommends strongly the use of method **(c)**. Brown and Solomon [5] show that method **(a)** results in sequences with improved k-distributed properties. They warn, however, that improvements in the k-distribution do not necessarily imply improvements in the marginal distributions of subsets. Fishman [8] reports that the LCG generator SUPER-DUPER ($a = 69{,}069$, $c = 0$, $m = 2^{32}$) uses method **(d)** to lengthen the short period evident in its low-order bits.

Example 11. The combination of three multiplicative congruential generators has been suggested by Wichmann and Hill [37]. Their goal is to use three generators with short period that can be implemented on a 16-bit machine to produce a pseudo-random sequence with long period. Their scheme uses the generators with multipliers 170, 171, and 172, and moduli 30,323, 30,269, and 30,307, respectively. The fractional part of the sum of three uniform deviates is also a uniform deviate; they use this result to combine the three generators. The resulting sequence has period in excess of 2^{44}. They claim that the sequence has good statistical properties, and exhibit a FORTRAN implementation.

LITERATURE

Knuth [18] gives an excellent summary of pseudo-random number generation with emphasis on the congruential generators. Fishman [8], Kennedy and Gentle [16], and Bratley et al. [3] also contain overviews. Bratley et al. [3] go into some detail on feedback shift register generators. Bibliographies can be found in refs. 11, 27, and 29.

References

[1] Arvillias, A. C. and Maritsas, A. E. (1978). *J. ACM*, **25**, 675–686. (A technical article about GFSR generators.)

[2] Bays, C. and Durham, S. D. (1976). *ACM Trans. Math. Software*, **2**, 59–64. (Shifting to improve randomness.)

[3] Bratley, P., Fox, B. L., and Schrage, L. E. (1983). *A Guide to Simulation*. Springer-Verlag, New York. (An introductory guide that focuses on generator implementation. Several FORTRAN programs are given.)

[4] Bright, H. S. and Enison, R. L. (1979). *Computing Surv.*, **11**, 357–370.

[5] Brown, M. and Solomon, H. (1979). *Ann. Statist.*, **7**, 691–695.

[6] Coveyou, R. R. and MacPherson, R. D. (1967). *J. ACM*, **14**, 100–119. (The basis of the spectral test.)

[7] Digital Equipment Corp. (1982). *VAX 11 Runtime Library Ref. Manual*. AA-D036C-TE. Digital Equipment Corp., Maynard, Mass.

[8] Fishman, G. S. (1978). *Principles of Discrete Event Simulation*. Wiley, New York. (An overview of the area with attention to implementation details, empirical and theoretical tests.)

[9] Fishman, G. S. and Moore, L. R. (1982). *J. Amer. Statist. Ass.*, **77**, 129–136. (A comprehensive study of alternative prime modulus LCG.)

[10] Fushimi, M. and Tezuka, S. (1983). *Commun. ACM*, **28**, 516–523. (k-distributivity of GFSRG.)

[11] Hull, T. E. and Dobell, A. R. (1962). *SIAM Rev.*, **4**, 230–254. (Early bibliography.)

[12] IBM Corp. (1981). *PL/I General Purpose Simulation System Program Description/Operations Manual*. SH20-6181-0, IBM Corp., White Plains, N.Y.

[13] International Mathematical and Statistical Libraries, Inc. (1977). *IMSL Library*, Houston, Tex.

[14] Jansson, B. (1966). *Random Number Generators*. Almqvist & Wiksell, Stockholm. (Early approaches.)

[15] Katzan, H., Jr. (1971). *APL User's Guide*. Van Nostrand Reinhold, New York.

[16] Kennedy, W. J. and Gentle, J. E. (1980). *Statistical Computing*. Marcel Dekker, New York. (An overview of the area.)

[17] Kiviat, P. J., Villanueva, R., and Markowitz, H. (1969). *The Simscript II Programming Language*. Prentice-Hall, Englewood Cliffs, N.J.

[18] Knuth, D. (1981). *The Art of Computer Programming*, Vol. 2: *Seminumerical Algorithms*, 2nd ed. Addison-Wesley, Reading, Mass. (A comprehensive guide to LCG, testing, and implementation, less information on GFSRG.)

[19] Learmonth, G. and Lewis, P. A. W. (1973). Naval Postgraduate School Random Number Generator Package LLRANDOM. *Rep. No. NPS55-LW-73061-A*, Naval Postgraduate School, Monterey, Calif.

[20] Lewis, T. G. and Payne, W. H. (1973). *J. ACM*, **20**, 456–468.

[21] Lewis, P. A. W. and Uribe, L. (1981). The New Naval Postgraduate School Random Number Package LLRANDOMII. *Rep. No. NPS55-81-005*, Naval Postgraduate School, Monterey, Calif.

[22] Lewis, P. A. W., Goodman, A. S., and Miller, J. M. (1969). *IBM Syst. J.*, **2**, 136–145.

[23] MacLaren, M. D. and Marsaglia, G. (1965). *J. ACM*, **12**, 83–89.

[24] Marsaglia, G. (1968). *Proc. Natl. Acad. Sci. USA*, **61**, 25–28.

[25] Marsaglia, G. (1972). In *Applications of Number Theory to Numerical Analysis*, S. K. Zaremba, ed. Academic Press, New York, pp. 249–286.

[26] Marsaglia, G. and Bray, T. A. (1968). *Commun. ACM*, **11**, 757–759.

[27] Nance, R. E. and Overstreet, C. (1972). *Computing Rev.*, **13**, 495–508. (Bibliography.)

[28] Niederreiter, H. (1978). *Bull. Amer. Math. Soc.*, **84**, 957–1041. (Monte Carlo technique and number-theoretic aspects of linear congruential generators.)

[29] Page, E. S. (1967). In *The Generation of Pseudo-Random Numbers in Digital Simulation in Operational Research*, S. H. Hollingsdale, ed. American Elsevier, New York, pp. 55–63. (Bibliography.)

[30] Payne, W. H., Rabung, J. R., and Bogyo, T. P. (1969). *Commun. ACM*, **12**, 85–86.

[31] SAS Institute, Inc. (1982). *SAS User's Guide: Basics, 1982 Edition*. SAS Institute, Inc., Cary, N.C.

[32] Sowey, E. R. (1972). *Int. Statist. Rev.*, **40**, 355–371. (Bibliography.)

[33] Sowey, E. R. (1978). *Int. Statist. Rev.*, **46**, 89–102. (Bibliography.)

[34] Tausworthe, R. C. (1965). *Math. Comp.*, **18**, 201–209.

[35] Tootill, J. P. R., Robinson, W. D., and Adams, A. G. (1971). *J. ACM*, **18**, 381–399.

[36] Tootill, J. P. R., Robinson, W. D., and Eagle, D. J. (1973). *J. ACM*, **20**, 469–481.

[37] Wichmann, B. A. and Hill, I. D. (1982). *Appl. Statist.*, **31**, 188–190.

[38] Zierler, N. and Brillhart, J. (1969). *Inf. Control*, **14**, 566–569.

(GENERATION OF RANDOM VARIABLES
MONTE CARLO METHODS
SIMULATION)

MARC-DAVID COHEN

PSEUDO-t

See also PSEUDO-F. The regular Student's t^* is the ratio of a unit normal variable to an independent $\chi/\sqrt{\nu}$ variable (*see* CHI-DISTRIBUTION). It applies (with $\nu = n - 1$) to the distribution of $\sqrt{n}\,(\overline{X} - \mu)/S$, for example, when $S^2 = (n-1)^{-1}\Sigma_{i=1}^{n}(X_i - \overline{X})^2$ and $\overline{X}$

$= n^{-1}\Sigma_{i=1}^{n} X_i$, $X_1, \ldots, X_n$ being independent normal variables with expected value μ and standard deviation σ.

A pseudo-t statistic is obtained by replacing S by some other estimator of σ. This estimator need not be an *unbiased* estimator of σ (indeed, S itself is not unbiased), but it is often felt desirable—perhaps for aesthetic purposes—to arrange that if there is any bias, it is small.

Most commonly, S is replaced by an estimator based on the sample range

$$W = \max(X_1, \ldots, X_n) - \min(X_1, \ldots, X_n).$$

The corresponding pseudo-T is

$$\sqrt{n}\,(\bar{X} - \mu)/(a_n W),$$

where a_n is chosen so that $a_n E[W] = \sigma$ if the population is normally distributed. It is possible to use other denominators, based on the mean deviation* or the interquartile range r, for example. Similar modifications can be used in modifying the denominator of the two-sample t. Approximations to the distribution of pseudo-t (often by a regular t with suitable degrees of freedom, usually fractional) have been developed.

(PSEUDO-F
t-DISTRIBUTION)

PSEUDO-VARIANCE

The $100p\%$ *pseudo-variance* of a continuous cumulative distribution function (CDF) $G(x)$ is the square of the ratio of the difference between its upper and lower $100p\%$ points to the corresponding difference for the standard normal CDF $\Phi(x)$. It is thus equal to

$$v(p) = \left\{ \frac{G^{-1}(1-p) - G^{-1}(p)}{\Phi^{-1}(1-p) - \Phi^{-1}(p)} \right\}^2.$$

For normal distributions, $v(p)$ equals the variance for all p.

For heavy-tailed* symmetric distributions, $v(p)$ increases as p decreases to zero. Andrews et al. [1] noted that the 4.2% pseudo-variance is remarkably stable for distributions in the Pearson system* (see also Pearson and Tukey [3]) and so may serve as a satisfactory index for comparing variability of estimators. Further details can be found in refs. 1 and 2.

References

[1] Andrews, D. F., Bickel, P. J., Hampel, F. R., Huber, P. J., Rogers, W. H., and Tukey, J. W. (1972). *Robust Estimators of Location: Survey and Advances*. Princeton University Press, Princeton, N.J.

[2] Huber, P. J. (1971). *Robust Statistical Procedures*. SIAM, Philadelphia.

[3] Pearson, E. S. and Tukey, J. W. (1965). *Biometrika*, **52**, 533–546.

(ROBUST ESTIMATION
VARIABILITY)

PSEUDO-VARIATES

A term used in the theory of weighted least squares*. Given a linear model

$$Y_u = \sum_{i=1}^{p} \beta_i x_{iu} + \epsilon_u$$

with $\text{var}(\epsilon_u) = w_u^{-1}\sigma^2$, the introduction of *pseudo-variates* $Y_u^* = Y_u\sqrt{w_u}$ and $x_{iu}^* = x_{iu}\sqrt{w_u}$ results in a model

$$Y_u^* = \sum_{i=1}^{p} \beta_i x_{iu}^* + \epsilon_u^*$$

with $\text{var}(\epsilon_u^*) = \sigma^2$. If the ϵ_u's are mutually uncorrelated, so are the ϵ^*'s, and the second model can be analyzed by ordinary least squares. Of course, it is necessary to know the w_u's (at least, their ratios) to apply this method. For more details, see Box et al. [1].

Reference

[1] Box, G. E. P., Hunter, W. G., and Hunter, J. S. (1978). *Statistics for Experimenters*. Wiley, New York.

PSI-FUNCTION *See* DIGAMMA FUNCTION; POLYGAMMA FUNCTION

PSYCHOLOGICAL DECISION MAKING

Decision making is a complex, ill-defined body of knowledge developed and studied by researchers from various disciplines, including mathematics, statistics, economics, political science, management science, philosophy, and psychology. The theory has two interrelated facets, normative and descriptive. The *normative* part of the theory [4, 28, 29] is concerned with optimal rather than actual choices. Its main function is to prescribe which action should be taken, given the structure of the decision problem, the consequences of each action, the objective of the decision maker (DM), and the information available to him. As such, normative decision theory* is a purely deductive discipline. The aim of the *descriptive* part [15, 20, 39] is to account for the DM's values, opinions, and goals, the way in which they are affected by information, perception of the decision task, or the personality of the DM, and the manner by which the DM incorporates them into decisions. As such, descriptive decision theory is a proper branch of psychology.

The distinction between the normative and descriptive parts of the theory is elusive; the two theories are interrelated in most applications. On the one hand, the criteria for optimality or rationality are necessarily based on human judgment and agreement about the criteria is typically not easily reached. On the other hand, in a variety of decision problems, such as most economic investments, goal-oriented DMs try hard to behave rationally. When faced in such problems with obvious errors of judgment (e.g., intransitivity in preference) or calculation, DMs often admit their mistakes and readily reverse their decisions.

TASKS

The experimental tasks employed to study decision behavior fall into two classes. A single-stage decision task is characterized by a triple (E, D, r), where E is a finite nonempty set of possible events, D is a finite nonempty set of possible decisions available to the DM, and r is an outcome or reward uniquely associated with the joint occurrence of a decision and an event. The DM constitutes a "system" that may reside in one of several states, each of which consists of variables assumed to affect decisions. These variables may include the DM's utility function, financial status, or the information he or she has acquired. The latter is often expressed as a subjective probability distribution over the uncertain outcomes associated with each decision. Given that the system is in state s, the DM chooses a course of action d, receives the reward determined by the joint occurrence of the action and the event that obtains, and the task terminates.

A multistage decision task [4] is characterized by a quadruple (E, D, p, r) where E and D are defined as above, p associates with each pair (s, d) a probability distribution on E known as a transformation rule or a transition function, and r is now the outcome associated with the transition from one state of the system to another. At the beginning of each stage of the decision process the system is at state s; the DM makes a decision d, the system moves to state s', selected according to the transformation rule, and the DM receives the reward $r(s, d, s')$ associated with this transition. The process continues for N stages; generally the DM is assumed to maximize some criterion function such as the total subjectively expected reward (or utility*) over all stages or the subjectively expected reward for the last stage only.

Research on multistage decision behavior (multistage betting, inventory control where the demand fluctuates from one stage to another) has been meager [25, 32]. Although human judgment is a continuous, interactive process as individuals cope with their environment, most behavioral decision research has focused on discrete incidents [12]. Because of their simplicity and more fundamental nature, single-stage decision tasks have been the major focus of experimental research. A typical instance is a choice between two gambles:

A: win either $\$a_1$ with probability p_1, $\$a_2$ with probability $p_2, \ldots,$ or $\$a_n$ with probability p_n;

B: win either $\$b_1$ with probability q_1, $\$b_2$ with probability $q_2, \ldots,$ or $\$b_m$ with probability q_m;

where a_i $(i = 1, \ldots, n)$ and b_j $(j = 1, \ldots, m)$ are real numbers, and $\sum p_i = \sum q_j = 1$. A variety of experiments have been conducted in the last 30 years or so, frequently with $m = n = 2$, in which gambles A and B were varied from each other in their expected value, variance, or both. In other tasks, the DMs have been required to rank order several gambles of this type in terms of their attractiveness or riskiness.

Other experimental tasks have been constructed to assess individual utility functions, to measure the subjective probability of both simple and compound events, to measure the perceived riskiness of gambles and relate it to the properties of the gambles, and in particular, to test the implications of theories that have been proposed to describe human decision behavior.

THEORIES

Classifications of psychological theories of decision making depend on the assumptions made about the characteristics of the system. A useful three-way dichotomization is due to Luce and Suppes [23]. The first distinction is whether the theory employs algebraic or probabilistic tools. The second is whether each decision determines a certain outcome (decision making under certainty) or a probability distribution over the outcomes (decision making under risk or uncertainty). The third is whether the theory provides a complete ranking of all available courses of action or merely specifies which one will be selected.

The most intensive study has been of the algebraic theory of maximizing *subjectively expected utility* (SEU). A characteristic of all utility theories, whether under certainty or uncertainty, is that one can assign numerical quantities (*utilities*) to alternatives in such a way that alternative a is chosen from a set of alternatives T if and only if the utility of a is larger than that of any other alternative in T. When it is possible to assign utilities to alternatives, it is said that the DM behaves optimally relative to his utility scale.

When the decision situation is under uncertainty (*see* JUDGMENTS UNDER UNCERTAINTY), then in addition to assigning utilities to alternatives—risky as well as sure—numbers can also be assigned to events. These numbers are called *subjective probabilities** and are interpreted as the DM's evaluation of the likelihood of the event e occurring. These two numerical scales are interlocked [22] as follows: The expected utility of a risky alternative is the sum of the utilities of its component outcomes, each weighted according to the subjective probability of its occurring. Axiomatic systems that justify this representation have been provided by von Neumann and Morgenstern [38], Savage [28], and Krantz et al. [19].

Several experiments have demonstrated that subjects are not always consistent in their choices and that some exhibit regular patterns of inconsistency. Rather than treating these results as errors of judgment, probabilistic choice theories regard them as inherent in the choice mechanism (*see* LUCE'S CHOICE AXIOM AND GENERALIZATIONS). Of the various postulates of these theories, the three that have received much empirical attention are weak, moderate, and strong stochastic transitivity. Probabilistic choice theories that are built on these or related postulates have been proposed and developed by Block and Marschak [2], Restle [27], Luce [21], and Tversky [35].

Another class of theories was developed to study diagnosis, which is conceptualized as the process of revising a subjective probability distribution over a set of events on the basis of some data. These theories have employed Bayes' rule or modified it to yield a psychological theory of revision of opinion. Consider the simplest case of two events e_0

and e_1. One form of Bayes' rule, which yields the posterior odds in favor of e_1 relative to e_0, given the data, is

$$\Omega(e_1 \mid X) = L(X; e_1)\Omega(e_1),$$

where X denotes the data, the posterior odds $\Omega(e_1 \mid X)$ is the ratio of conditional probabilities $p(e_1 \mid X)/p(e_0 \mid X)$, the likelihood ratio $L(X; e_1)$ is the ratio $p(X \mid e_1)/p(X \mid e_0)$, and if $p_0(e_i)$ is the prior probability of event e_i ($i = 0, 1$), then $\Omega(e_1)$, the prior odds, is the ratio $p_0(e_1)/p_0(e_0)$.

FINDINGS

The utility of money or other valuable objects has been assessed either by direct estimation or via some model. Direct estimation techniques have provided some evidence for power utility functions. Indirect measurements of utility, typically conducted via the SEU model, have yielded more variable results. Early studies concluded that it is feasible to measure utility experimentally. Later studies, which were particularly successful in inducing their subjects to be consistent, yielded both linear and nonlinear utility functions. More recent experiments have shown that experimentally derived utility functions depend to a large extent on the values of the objects, the DM's financial status, and in particular, the properties of the utility assessment technique. Based on a series of simple demonstrations, Tversky and Kahneman [37] proposed that what they call a value function is commonly S-shaped, concave about the DM's reference point and convex below it. Consequently, risk-averse decisions are common when DMs must choose between a sure win and a substantial probability of a larger loss, whereas risk-seeking decisions are common when choices between a sure loss and a substantial probability of a larger loss have to be made. For reviews, see Edwards [5, 6], Fishburn [9], Lee [20], and Rapoport and Wallsten [26].

Subjective probabilities may also be measured either by direct psychophysical judgment methods or by inference from the SEU model. Direct psychophysical methods, which usually require the DM to estimate the proportion of one type of element in a display with stimulus elements of two or more types, show that DMs perform the task well and that subjective probability thus measured is roughly equal to observed proportion. Indirect assessment methods [33], which have been used more frequently, have shown more diverse and complex results. There is strong evidence [10] that because of their limited information processing capacity, people do not regard the world in "intuitively probabilistic" terms; rather, they substitute "certainty equivalents" for probability distributions over uncertain outcomes. People do not have satisfactory intuitive notions of the concepts of randomness*, variability, and probability; the subjective probability distributions derived from their decisions are often too "tight"; and probability assessment is highly susceptible to task characteristics such as response mode, payoff, order of information presentation, and feedback.

Several experiments have tested the descriptive power of the SEU model [1, 7 13, 16, 17, 40]. The bulk of the early findings showed that SEU might serve as a first-order approximation to human decisions in experiments involving a choice between a small number of gambles with objective probabilities that are easily discriminable, are quantifiable, and have outcomes that are given in monetary units. In more recent experiments, many of which concentrated on directly testing the axioms of SEU, the proponents of SEU have been greatly outnumbered by its critics [24, 32]. Several experiments showed that utility and subjective probability interact, in violation of SEU theory, or developed counterexamples (e.g., the paradoxes due to Allais and Ellsberg) to the fundamental axioms of the theory. With judicious selection of the experimental design, subjects have been enticed to persistently violate the axioms of SEU, including transitivity [34], the sure-thing principle, and the dominance principle [31].

Consequently, several researchers have proposed replacing SEU with other non-axiomatic, algebraic theories that describe how probabilities and payoffs are integrated into decisions. Prominent among these are dimensional theories that rely on correlational and functional measurement approaches [30], and prospect theory [14]. The *correlational* approach, which uses multiple regression* techniques, postulates that the DM's numerical responses in a choice between gambles constitute some linear combination of the available dimensions of the choice alternatives (i.e., the probabilities and payoffs). *Functional measurement* relies on factorial designs, quantitative response measures, and monotone transformations for rescaling these measures in attempting simultaneously to scale the stimulus attributes and response measures and to determine the combination rule relating the two. *Prospect theory* is a descriptive model proposed to modify SEU theory. It does so by replacing the traditional utility function over absolute amounts by a value function over gains and losses, and the probability function by a decision weighting function. The shapes of both functions are based on generalizations from several psychological studies. Although all these approaches have received some empirical support, they lack the generality and refutability of the axiomatic SEU theory.

In investigating how opinions are revised, a considerable number of studies have compared the DM's numerical estimates of conditional probabilities or posterior odds to $p(e_i \mid X)$ or $\Omega(e_i \mid X)$, respectively. The general but far from unanimous result has been that the estimates were monotonically related to these quantities but were too evenly distributed among the e_i's; i.e., they were conservative relative to the predicted Bayesian values. Assuming the Bayesian model, three alternative explanations of the pervasive conservatism phenomenon have been offered; misperception of the data's diagnostic impact, misaggregation of the data, or response bias. Experimental support of each explanation has been obtained, suggesting again that, as with the SEU theory, the focus of empirical research should not be whether the Bayesian theory is right or wrong, but under what empirical conditions does it hold and in what ways is it systematically violated.

RECENT TRENDS

Recent reviews of the literature on psychological decision making (e.g., Slovic et al. [32]; Einhorn and Hogarth [8]) have noted two general trends. One is a change in orientation of the experimental work. Whereas past studies mainly compared observed and predicted decisions, recent research has focused on the psychological underpinnings of observed decision behavior by attempting to determine how the underlying cognitive processes are modeled by the interaction between the limitations of the DM and the demands of the task. Much of the impetus for this change in orientation can be attributed to demonstrations [15, 36] of several judgmental heuristics (e.g., representativeness, availability, and anchoring), which affect probabilistic judgments in a variety of decision tasks. Other recent studies (e.g., estimation of probabilities via fault trees*, coding processes in risky choice, alternative framings of the same decision problem, and causal schema in probability judgments) have demonstrated that the process of the cognitive representation of the decision task, and the factors that affect it, are of major importance in judgment and choice.

A second and related trend has been a growing interest in studying the process of information search, acquisition, and storage [10, 39], since evaluation and search strategies are interdependent. Concern for how information is acquired has raised questions about the role of memory and attention in decision behavior and has necessitated the use of different research methodologies to account for the dynamics of the process. Both recent trends reflect the growing connection between psychological decision making and more traditional branches of psychology.

References

[1] Becker, G. M. and McClintock, C. G. (1967). *Annu. Rev. Psychol.*, **18**, 239–286. (A review of the literature, 1961–1967.)

[2] Block, H. D. and Marschak, J. (1960). In *Contributions to Probability and Statistics*, I. Olkin, S. Ghurye, W. Hoeffding, W. Madow, and H. Mann, eds. Stanford University Press, Stanford, Calif., pp. 97–132.

[3] Davidson, D., Suppes, P., and Siegel, S. (1957). *Decision Making: An Experimental Approach*. Stanford University Press, Stanford, Calif. (An extensive experimental study.)

[4] DeGroot, M. H. (1970). *Optimal Statistical Decisions*. McGraw-Hill, New York. (A thorough course in the theory and methodology of optimal statistical decisions.)

[5] Edwards, W. (1954). *Psychol. Bull.*, **51**, 380–417. (The first thorough review of the literature up to 1954.)

[6] Edwards, W. (1961). *Annu. Rev. Psychol.*, **12**, 473–498. (A review of the literature 1954–1961.)

[7] Edwards, W. and Tversky, A., eds. (1967). *Decision Making*. Penguin, London. (A collection of early readings, mostly experimental.)

[8] Einhorn, H. J. and Hogarth, R. M. (1981). *Annu. Rev. Psychol.*, **32**, 53–88. (A review of the literature, 1977–1981.)

[9] Fishburn, P. C. (1968). *Manag. Sci.*, **14**, 335–378.

[10] Hogarth, R. M. (1975). *J. Amer. Statist. Ass.*, **70**, 271–294. (An attempt to consider the implications of research on cognitive processes for the practical problem of subjective probability assessments.)

[11] Hogarth, R. M. (1980). *Judgment and Choice: The Psychology of Decision*. Wiley, New York. (A textbook that emphasizes the unstructured, natural way in which people make decisions.)

[12] Hogarth, R. M. (1981). *Psychol. Bull.*, **90**, 197–217. (A discussion of the theoretical and methodological implications of considering judgment as a continuous rather than discrete process.)

[13] Jungerman, H. and de Zeeuw, G., eds. (1977). *Decision Making and Change in Human Affairs*. D. Reidel, Dordrecht, The Netherlands. (A collection of papers on judgment and choice.)

[14] Kahneman, D. and Tversky, A. (1979). *Econometrica*, **47**, 263–291.

[15] Kahneman, D., Slovic, P., and Tversky, A., eds. (1982). *Judgment under Uncertainty: Heuristics and Biases*. Cambridge University Press, Cambridge, England. (A collection of papers on the effects of heuristics and biases of intuitive judgments.)

[16] Kaplan, M. F. and Schwartz, S., eds. (1975). *Human Judgment and Decision Processes*. Academic Press, New York. (A collection of papers on judgment and choice.)

[17] Kaplan, M. F. and Schwartz, S., eds. (1977). *Human Judgment and Decision Processes in Applied Settings*. Academic Press, New York. (A collection of papers on judgment and choices in applied settings.)

[18] Kleinmuntz, B., ed. (1968). *Formal Representations of Human Judgment*. Wiley, New York. (A collection of papers on judgment and choice.)

[19] Krantz, D. H., Luce, R. D., Suppes, P. and Tversky, A. (1971). *Foundations of Measurement*, Vol. I. Academic Press, New York. (An excellent textbook that presents and discusses various axiomatic theories of preference and choice, including SEU, conjoint measurement, and other theories.)

[20] Lee, W. (1971). *Decision Theory and Human Behavior*. Wiley, New York. (An elementary textbook that describes both theoretical and experimental research.)

[21] Luce, R. D. (1959). *Individual Choice Behavior*. Wiley, New York. (A research monograph that presents Luce's probabilistic theory of choice.)

[22] Luce, R. D. (1962). In *Social Science Approaches to Business Behavior*, G. B. Strother, ed. Richard D. Irwin, Homewood, Ill., pp. 141–161.

[23] Luce, R. D. and Suppes, P. (1965). In *Handbook of Mathematical Psychology*, Vol. III, R. D. Luce, R. R. Bush, and E. Galanter, eds. Wiley, New York, pp. 249–410. (An excellent and thorough review of psychological theories of decision making developed prior to 1965.)

[24] March, J. G. (1978). *Bell. J. Econ.*, **9**, 587–608. (A critique of the normative theory of choice.)

[25] Rapoport, A. (1975). In *Utility, Probability, and Human Decision Making*, D. Wendt and C. A. J. Vlek, eds. D. Reidel, Dordrecht, The Netherlands, pp. 349–369. (A review of multistage decision-making experiments.)

[26] Rapoport, A. and Wallsten, T. S. (1972). *Annu. Rev. Psychol.*, **23**, 131–175. (A review of the literature, 1967–1972.)

[27] Restle, F. (1961). *Psychology of Judgment and Choice*. Wiley, New York. (A research monograph.)

[28] Savage, L. J. (1947). *The Foundations of Statistics*. Wiley, New York.

[29] Simon, H. (1957). *Models of Man: Social and Rational*. Wiley, New York. (A collection of papers about decision making and bounded rationality.)

[30] Slovic, P. and Lichtenstein, S. C. (1971). *Organizat. Behav. Hum. Perform.*, **6**, 649–744. (An exten-

sive review and comparative analysis of the Bayesian and regression approaches to information utilization in judgment and choice.)

[31] Slovic, P. and Tversky, A. (1974). *Behav. Sci.*, **19**, 368–373.

[32] Slovic, P., Fischhoff, B., and Lichtenstein, S. C. (1977). Behavioral decision theory. *Annu. Rev. Psychol.*, **28**, 1–39. (A review of the literature, 1972–1977.)

[33] Staël von Holstein, C. A. S. (1970). *Assessment and Evaluation of Subjective Probability Distributions*. The Economic Research Institute at the Stockholm School of Economics, Stockholm.

[34] Tversky, A. (1969). *Psychol. Rev.*, **76**, 31–48.

[35] Tversky, A. (1972). *Psychol. Rev.*, **79**, 281–299.

[36] Tversky, A. and Kahnemann, D. (1974). *Science*, **185**, 1124–1131.

[37] Tversky, A. and Kahnemann, D. (1981). *Science*, **211**, 453–458. (A summary of experimental demonstrations of the effects of the formulation of the decision problem on shifts of preference.)

[38] von Neumann, J. and Morgenstern, O. (1947). *Theory of Games and Economic Behavior*, 2nd ed. Princeton University Press, Princeton, N.J.

[39] Wallsten, T. S., ed. (1980). *Cognitive Processes in Choice and Decision Behavior*. Lawrence Erlbaum, Hillsdale, N.J. (A collection of papers on judgment and choice.)

[40] Wendt, D. and Vlek, C. A. J., eds. (1975). *Utility, Probability and Human Decision Making*. D. Reidel, Dordrecht, The Netherlands. (A collection of papers on decision behavior.)

(DECISION THEORY
JUDGMENTS UNDER UNCERTAINTY
LUCE'S CHOICE AXIOM AND
GENERALIZATIONS
MULTIPLE CRITERIA DECISION MAKING
RISK THEORY
UTILITY THEORY)

AMNON RAPOPORT

PSYCHOLOGICAL SCALING

Psychological scaling refers to measuring peoples' subjective perceptions of stimuli by analyzing their responses according to certain models. Empirical research has demonstrated that the appropriate scaling method depends both on the nature of the stimuli and the nature of the responses.

The origins of psychological scaling reside in psychophysics*, and stem from a classic work by Gustav T. Fechner, the *Elemente der Psychophysik*, 1860 [3], often cited also as marking the formal beginning of experimental psychology. Fechner, trained in physics, discovered the earlier work of a physiologist, Ernst H. Weber, who had experimentally demonstrated that the amount of change in magnitude of a physical stimulus required to produce a "just noticeable difference" in sensation was in a fixed ratio to the initial stimulus magnitude. By assuming that all just noticeable differences of sensation are equal and proportional to the ratio of change in stimulus magnitude to initial stimulus magnitude, Fechner derived the Weber–Fechner equation—that sensation is proportional to the logarithm of stimulus magnitude.

The classical Fechnerian psychophysical methods for gathering and analyzing data were presented and illustrated in detail by the Cornell psychologist, E. B. Titchener [9]. They typically require a "subject" to adjust physical magnitudes to satisfy some criterion of sensation (e.g., to adjust a variable stimulus to be detectably different from a standard stimulus) or to judge whether pairs of fixed stimuli are "the same" or "different." From the results of many such judgments, it is possible to estimate values for the slope and intercept in the Weber–Fechner equation.

S. S. Stevens, a psychologist at Harvard, contributed greatly to psychophysical research, from the 1930s through the 1960s. He postulated an alternative to the Weber–Fechner equation for a broad class of stimuli, whereby the psychological magnitude is related to the physical stimulus magnitude by a power function [7]. Under this formulation, the logarithm of sensation, rather than sensation itself, is proportional to the logarithm of physical stimulus magnitude.

In a series of papers in 1927, the Chicago psychologist L. L. Thurstone developed methods for scaling psychological variables that, unlike those studied in psychophysics, need not be directly coupled with physical stimulus magnitudes. Thurstone's methods could be and have been used to "measure" or "scale" such psychological variables as attitudes, preferences, or aesthetic values. The fundamental assumption underlying Thurstone's methods of psychological scaling is that of a normal distribution of "discriminant processes" associated with stimuli. The distribution is either over occasions, as with a single judge, or over judges. Data may be collected using the method of paired comparisons*, in which pairs of stimuli, X_i, X_j, are presented, and the judge must respond that $X_i > X_j$ or $X_j > X_i$, where "$>$" represents a defined order relation on the subjective scale (e.g., "preferred" or "more favorable"). Alternatively, data may be collected using the method of successive categories, in which each stimulus in a set is assigned by a judge to one of a group of categories that are ordered on the relevant subjective scale. In either case, a subjective continuum is estimated so as to most closely yield simultaneous normal distributions of transformed judgment values for all stimuli. The mean and the standard deviation of a resultant distribution for each stimulus are taken as estimates for the two parameters of the distribution of discriminal processes for that stimulus.

To illustrate the method of successive categories, consider first the following experimental design. A set of stimuli, $X_1, X_2, \ldots, X_n$ is presented to N subjects, randomly selected from a specified subject population. Temporal positions of the X_j are independently randomized for each subject. Subject i is asked to judge each X_j in terms of an explicitly defined attribute, and to classify it into one and only one ordered attribute category k, $k = 1, 2, \ldots, m$. Data are recorded in the form p_{jk}, the cumulative proportion of judgments for the N subjects in which X_j is placed below the upper bound of category k.

The model for relating parameters to data by the method of categorical judgment is detailed by Bock and Jones [1, pp. 214–237]. Estimated parameters include scale values for category boundaries, and means and variances of scale values for the stimuli.

As an example of the use of the method of successive categories, consider data on food preferences collected from a sample of 255 Army enlisted men. Table 1 shows the frequency with which each of 12 food items was rated in each of nine categories on a rating form. (The categories were labeled from "dislike extremely" at the left to "like extremely" on the right.) These frequencies are converted into cumulative proportions (p_{jk}) in Table 2, from which are derived the

Table 1 Frequencies of Ratings over Nine Successive Categories for 12 Food Items

	Successive Category										
Food Item	1	2	3	4	5	6	7	8	9	N	No Response
1. Sweetbreads	24	31	26	29	47	37	32	13	4	243	12
2. Cauliflower	15	6	10	26	27	62	72	30	6	254	1
3. Fresh pineapple	4	1	2	7	16	40	75	77	31	253	2
4. Parsnips	17	30	31	45	47	32	33	11	4	250	5
5. Baked beans	4	5	6	17	27	60	72	56	7	254	1
6. Wieners	0	2	7	9	28	76	91	38	2	253	2
7. Chocolate cake	0	1	2	7	20	33	74	85	32	254	1
8. Salmon loaf	8	9	11	23	36	74	64	25	3	253	2
9. Blueberry pie	2	0	3	14	24	41	70	70	28	252	3
10. Turnips	19	30	33	48	35	40	32	15	1	253	2
11. Liver	21	16	10	15	23	42	71	46	9	253	2
12. Spaghetti	1	1	8	9	24	61	92	46	12	254	1

Source: Ref. 1.

Table 2 Cumulative Proportions, p_{jk}, of Ratings over Nine Successive Categories for 12 Food Items

	Successive Category								
Food Item	1	2	3	4	5	6	7	8	9
1	099	226	333	453	646	798	930	984	1.000
2	059	083	122	224	331	575	858	976	1.000
3	016	020	028	055	119	277	573	877	1.000
4	068	188	312	492	680	808	940	984	1.000
5	015	035	059	126	232	469	752	972	1.000
6	000	008	036	071	182	482	842	992	1.000
7	000	004	012	039	118	248	539	874	1.000
8	032	067	111	202	344	636	889	988	1.000
9	008	008	020	075	171	333	611	889	1.000
10	075	194	324	514	652	810	937	996	1.000
11	083	146	186	245	336	502	783	964	1.000
12	004	008	039	075	169	409	772	953	1.000

Source: Ref. 1.

unit normal deviates (y_{jk}) of Table 3. (An extrapolative method provides estimates of y_{jk} when p_{jk} is less than 0.01 or greater than 0.99.) The values

$$\bar{d}_k = \bar{y}_k - \bar{y}_{k-1}$$

represent scaled category widths. By arbitrarily assigning the center of the middle category (labeled "neither like nor dislike") a scale value of 0, we then estimate scale values of the category boundaries, $\hat{\tau}_k$, as shown in the final row entries of Table 3.

Mean scale values and variances for the stimuli may be estimated by an algebraic solution, or graphically, by determining the

Table 3 Normal Deviates, y_{jk}, Associated with the p_{jk} of Table 2

	Successive Category								
Food Item	1	2	3	4	5	6	7	8	Sum
1	− 1.287	− 0.752	− 0.432	− 0.118	0.374	0.834	1.476	2.144	2.239
2	− 1.563	− 1.385	− 1.165	− 0.759	− 0.437	0.189	1.071	1.977	− 2.072
3	− 2.144	− 2.054	− 1.911	− 1.598	− 1.180	− 0.592	0.184	1.160	− 8.135
4	− 1.491	− 0.885	− 0.490	− 0.020	0.468	0.870	1.555	2.144	2.151
5	− 2.170	− 1.812	− 1.563	− 1.146	− 0.732	− 0.078	0.681	1.911	− 4.909
6	(−2.450)	(−2.071)	− 1.799	− 1.468	− 0.908	− 0.045	1.003	(1.937)	− 5.801
7	(−2.908)	(−2.529)	− 2.257	− 1.762	− 1.185	− 0.681	0.098	1.146	− 10.078
8	− 1.852	− 1.498	− 1.221	− 0.834	− 0.402	0.348	1.221	2.257	− 1.981
9	(−2.705)	(−2.326)	− 2.054	− 1.440	− 0.950	− 0.432	0.282	1.221	− 8.404
10	− 1.440	− 0.863	− 0.456	0.035	0.391	0.878	1.530	(2.464)	2.539
11	− 1.385	− 1.054	− 0.893	− 0.690	− 0.423	0.005	0.782	1.799	− 1.859
12	(−2.413)	(−2.034)	− 1.762	− 1.440	− 0.958	− 0.230	0.745	1.675	− 6.417
$\sum y_{jk}$	− 23.808	− 19.263	− 16.003	− 11.240	− 5.942	1.066	10.628	21.835	− 42.727
$\sum y_{jk}/12$	− 1.984	− 1.605	− 1.333	− 0.937	− 0.495	0.089	0.886	1.820	
d_k		0.379	0.272	0.396	0.442	0.584	0.797	0.934	
$\hat{\tau}_k$	− 1.268	− 0.889	− 0.617	− 0.221	0.221	0.805	1.602	2.536	

Source: Ref. 1.

Table 4 Graphical Estimates for $\hat{\mu}_j$ and $\hat{\sigma}_j$, 12 Food Items

Food Item	$\hat{\mu}_j$	$\hat{\sigma}_j$
1. Sweetbreads	− 0.10	1.13
2. Cauliflower	0.62	1.05
3. Fresh pineapple	1.41	1.02
4. Parsnips	− 0.18	1.00
5. Baked beans	0.88	0.97
6. Wieners	0.82	0.72
7. Chocolate cake	1.46	1.00
8. Salmon loaf	0.52	0.89
9. Blueberry pie	1.25	1.07
10. Turnips	− 0.20	1.06
11. Liver	0.67	1.30
12. Spaghetti	0.99	0.83

Source: Ref. 1.

intercept and slope for linear fits of the graphs of the p_{jk} plotted against the τ_k on normal probability paper. Graphical estimates of mean scale values and standard deviations for the 12 food items are shown in Table 4.

The original Thurstone papers on psychological scaling are reprinted in Thurstone [8]. Further exposition may be found in Edwards [2], Torgerson [10], Bock and Jones [1], and Luce [6].

A distinct concept of scaling, with primary applicability to attitude measurement, is that of Louis Guttman [5], a quantitative social scientist at Hebrew University of Jerusalem. Consider n statements, each representing a level of attitude or belief regarding a selected issue, where each statement might be "accepted" or "rejected" by a particular person. Let the statements be ordered by level of belief, lowest to highest, $1, 2, \ldots, n$. Let i represent the highest ordered statement that a person accepts, $1 < i < n$. Then the set of statements form a Guttman scale if that person accepts all statements $1, 2, \ldots, i$ and rejects all statements $i + 1, i + 2, \ldots, n$. In practice, of course, this condition sometimes might be violated even when, intrinsically, there exists a Guttman scale. This has led to the development of statistical models of scalability (e.g., Goodman [4]).

A form of psychological scaling that has received much attention in recent years is multidimensional scaling*. Related to psychological scaling are theoretical developments in the theory of psychological measurement. *See* MEASUREMENT THEORY.

References

[1] Bock, R. D. and Jones, L. V. (1968). *The Measurement and Prediction of Judgment and Choice*. Holden-Day, San Francisco.

[2] Edwards, A. L. (1957). *Techniques of Attitude Scale Construction*. Appleton-Century-Crofts, New York.

[3] Fechner, G. T. (1966). *Elements of Psychophysics* (English transl.). Holt, Rinehart and Winston, New York.

[4] Goodman, L. A. (1975). *J. Amer. Statist. Ass.*, **70**, 755–768. [Also appears in Goodman, L. A. (1978). *Analyzing Qualitative/Categorical Data*, Abt Books, Cambridge, Mass., pp. 363–401.]

[5] Guttman, L. (1950). In *Measurement and Prediction*, S. A. Stouffer et al. Princeton University Press, Princeton, N.J., pp. 60–90.

[6] Luce, R. D. (1977). *Psychometrika*, **42**, 461–489.

[7] Stevens, S. S. (1961). *Science*, **133**, 80–86.

[8] Thurstone, L. L. (1959). *The Measurement of Values*. University of Chicago Press, Chicago.

[9] Titchener, E. B. (1905). *Experimental Psychology*, Vol. II. Macmillan, New York.

[10] Torgerson, W. S. (1958). *Theory and Methods of Scaling*. Wiley, New York.

(MEASUREMENT THEORY
MULTIDIMENSIONAL SCALING
PAIRED COMPARISONS
PSYCHOLOGY, STATISTICS IN
PSYCHOPHYSICS, STATISTICAL METHODS IN)

LYLE V. JONES

PSYCHOLOGICAL TESTING THEORY

Civil service testing began in China about 3000 years ago, according to DuBois [3]. The Jesuits used written tests in the sixteenth century.

A psychological test is a device for collecting a sample of behavior. A basic problem is how to quantify the behavior. Problems include the optimal design of tests and scoring methods, their evaluation, and the proper interpretation of scores (*see also* FACTOR ANALYSIS).

RELIABILITY

Publishers try to prepare *parallel forms* of a test: forms yielding scores having identical statistical properties. The correlation $\rho_{XX'}$ between scores X and X' on parallel forms is the test *reliability*. By definition, $\mu_X = \mu_{X'}$, $\sigma_X = \sigma_{X'}$, and $\rho_{XV} = \rho_{X'V}$, where V is some external variable.

COMPOSITE SCORES AND TAUTOLOGIES

Frequently, the testee's score X is the sum of his or her scores Y_i on n subtests or questions: $X \equiv \sum_i Y_i$. If $Y_i = 1$ (correct) or 0 (incorrect), then X is the *number-right score*.

Two useful algebraic tautologies follow, expressing properties of X in terms of properties of the Y_i:

$$\sigma_X \equiv \sum_{i=1}^{n} \sigma(Y_i)\rho(X, Y_i), \tag{1}$$

$$\rho_{XV} \equiv \frac{\sum_{i=1}^{n} \sigma(Y_i)\rho(V, Y_i)}{\{\sum_{i=1}^{n}\sum_{j=1}^{n} \sigma(Y_i)\sigma(Y_j)\rho(Y_i, Y_j)\}^{1/2}}. \tag{2}$$

If V is a criterion, ρ_{XV} is called the *test validity*.

A useful lower bound for the test reliability is

$$\alpha \equiv \frac{n}{n-1}\left\{1 - \frac{1}{\sigma_X^2}\sum_{i=1}^{n} \sigma^2(Y_i)\right\} \leqslant \rho_{XX'}. \tag{3}$$

This is *Cronbach's* α, often cited as the "test reliability," whereas it should be called a lower bound. When $Y_i = 0$ or 1, a more easily computed lower bound, called *Kuder–Richardson* formula 21*, is

$$\text{KR}(21) \equiv \frac{n}{n-1}\{1 - \mu_X(n - \mu_X)/(n\sigma_X^2)\} \leqslant \alpha \leqslant \rho_{XX'}. \tag{4}$$

If the Y_i are parallel,

$$\rho_{XX'} \equiv n\rho_{YY'}/\{1 + (n-1)\rho_{YY'}\}; \tag{5}$$

this is the *Spearman–Brown formula*, showing the effect of test length on $\rho_{XX'}$.

For a test containing $n = 50$ questions or *items* scored 0 or 1, typical values might be $\sigma(Y_i) = 0.4583$, $\rho(Y, Y') = 0.14$, $\rho(X, Y_i) = 0.3965$, $\mu_X = 35$. If all Y_i are parallel, we find from (1) that $\sigma_X = 9.085$. From (3) to (5) in this case, $\rho_{XX'} = \alpha = \text{KR}(21) = 0.89$.

ITEM ANALYSIS AND TEST DESIGN

An *item analysis* typically determines the proportion π_i of all testees answering item i correctly, also some measure of item-test correlation such as $\rho(X, Y_i)$. If $Y_i = 0$ or 1, $\mu_X = \sum_i \pi_i$; also $\sigma^2(Y_i) = \pi_i(1 - \pi_i)$. As seen in (5), $\rho_{XX'}$ is maximized by using items with high intercorrelations. In contrast, as seen in (2), the *test validity* is maximized by using items with high $\rho(V, Y_i)$ and low intercorrelations.

TRUE SCORE

If we had scores $X, X', X'', \ldots,$ on parallel test forms for a given testee, we would ordinarily use the mean of these scores in preference to any single score. This establishes an important conclusion: there is an (unobservable) *latent variable* T (called the *true score*) that is more important to us than the *observed score* X. X is here considered a *random variable* for a single testee and also across testees. T is defined as the expectation ($\mathscr{E}$) of X for a fixed testee; T is a random variable across testees.

ERROR OF MEASUREMENT

$$E \equiv X - T \tag{6}$$

is the *error of measurement* in X. The errors of measurement impair the effectiveness of X as a substitute for T. The main purpose of classical test theory is to evaluate the effects of E and to try to determine what the data would be like if the errors were eliminated.

From (6), $\mathscr{E}(E \mid T)$, the expectation of E for fixed T, is zero; the errors are unbiased. Thus in any group of people $\rho_{ET} = 0$. Note that these are not assumptions, they follow from the definition of T. Also, from (6), for a group of people $\sigma_X^2 = \sigma_T^2 + \sigma_E^2$, σ_{XT} (the covariance of X and T) $= \sigma_T^2$, and $\rho_{XT}^2 = 1 - \sigma_E^2/\sigma_X^2$.

CLASSICAL TEST THEORY

The only latent-variable parameter that can be inferred from data up to this point is $\mu_T \equiv \mu_X$. Two assumptions are now made:

$$\rho(E_X, E_Y) = 0, \qquad \rho(E_X, T_Y) = 0, \quad (7)$$

where the subscripts denote two different tests. If we now administer a parallel test form, denoted by a prime [$T' = T$, $\sigma(E' \mid T') = \sigma(E \mid T)$], we can determine the important latent parameters from

$$\begin{aligned} \rho_{XT}^2 (= \rho_{X'T'}^2) &= \rho_{XX'} \\ \sigma_T^2 (= \sigma_{T'}^2) &= \sigma_X^2 \rho_{XX'} \qquad (8) \\ \sigma_E^2 (= \sigma_{E'}^2) &= \sigma_X^2 (1 - \rho_{XX'}). \end{aligned}$$

For the numerical example in the section "Composite Scores and Tautologies," we find from (8) that $\rho_{XT} = 0.944$, $\sigma_T = 8.574$, and $\sigma_E = 3.005$.

For two tests X and Y, $\rho(T_X, T_Y) = \rho_{XY}/\sqrt{\rho_{XX'}\rho_{YY'}}$. This *correction for attenuation* makes it possible to estimate the correlation between true scores on two tests. It follows that $\rho_{XY} \leqslant \rho_{XT}$, in other words, validity $\leqslant \sqrt{\text{reliability}}$.

Classical test theory relies heavily on correlations. This is because the score units of a test are meaningless to most readers. When the score units are familiar, σ_E^2 is a better measure of test effectiveness than $\rho_{XX'}$; σ_E^2 is more stable across groups of testees.

No Gaussian distribution assumptions have been made. Higher-order moments of T, E, and E' can be inferred [7, Chap. 10].

CRITERION-REFERENCED TESTS

Tests are sometimes used only to provide a ranking of testees. When test scores have meaning on an externally fixed scale, however, their accuracy must be evaluated absolutely, not relatively. Correlation coefficients are no longer appropriate for such an evaluation; a part of classical test theory must be replaced (see ref. 1).

ITEM SAMPLING AND MATRIX SAMPLING

In a survey, group means and variances of scores may be important rather than individual scores. In this case, for fixed testing time, it is more efficient to administer a different sample of items to different groups of examinees than to give everyone the same test [11].

If the items are sampled at random from a pool of items and true score is defined as expected performance on the entire pool of items, then the key parameters $\mu_T, \sigma_T^2, \sigma_E^2, \rho_{XT}$ can be estimated from the observed scores without need for parallel forms and without any assumptions. For a given testee, number-right score X has a binomial distribution. An unbiased estimator of $\sigma^2(E \mid T)$ is $X(n - X)/(n - 1)$. Across all testees taking the same test, an unbiased estimator of σ_E^2 is $[\overline{X}(n - \overline{X}) - s_X^2]/(n - 1)$, where s_X^2 is the sample variance of X [7, Chap. 11].

Stratified sampling is desirable in drawing items from a pool to build a test. This implements *domain-referenced testing* [5].

SCORING

Given the polychotomous responses of many testees to a set of test items, how shall a score be assigned to each examinee? A solution to this problem, with no special assumptions made (see [12, Chap. 12]) was obtained by Guttman [4]. The solution is related to the *method of reciprocal averages* [10].

Scoring weights can be assigned empirically so as to maximize test reliability or test validity [7, Chap. 5]. Wilks [13] showed that an unweighted average of item scores will often be almost as good as a weighted average.

When items are scored dichotomously, the number of correct answers (an unweighted average) is the usual test score. Across a group of testees, such scores often approximately follow the beta-binomial distribution [7, Chap. 23; 8].

ITEM RESPONSE THEORY (IRT)

(*See* LATENT STRUCTURE ANALYSIS.) IRT (and test scores) are most useful when all items measure a single dimension of "ability" Θ. IRT provides a statistical model for the probability of a correct response to an item [6]. This probability $P_i(\theta)$ is an increasing function of the testee's θ. If $P_i(-\infty) = 0$ and $P_i(\infty) = 1$, then $P_i(\theta)$ necessarily has the form of a cumulative frequency distribution. If, because of guessing behavior, $P_i(-\infty) = \gamma_i > 0$, then $P_i(\theta)$ is commonly modeled by

$$P_i(\theta) \equiv \Pr(Y_i = 1 \mid \theta) = \gamma_i + (1 - \gamma_i)F\left[\alpha_i(\theta - \beta_i)\right], \quad (9)$$

where α_i, β_i, and γ_i are parameters describing item i, and where F is a cumulative distribution function, usually normal or logistic. (Note: F is *not* related to the frequency distribution of θ in any group of testees.) Since β_i is a property of the items that offsets θ, β_i represents the *item difficulty*. Similarly, α_i represents item *discriminating power*. In a unidimensional test, by definition, all Y_i are independent when θ is fixed (*local independence*).

In IRT a test is evaluated by the *Fisher information* function* of the test score X:

$$I \equiv \frac{\left[\frac{d}{d\theta}\mathscr{E}(X \mid \theta)\right]^2}{\operatorname{var}(X \mid \theta)} \quad (10)$$

[2, Chap. 20], a function of θ, inversely proportional to the square of the length of the asymptotic confidence interval for estimating θ from $X = x$.

When the IRT model holds, the parameters of an item are invariant across groups of testees, except for an arbitrary origin and unit, and a testee's θ is invariant across tests. IRT allows us to predict the properties of all types of test scores from the characteristics of the test items. IRT is useful for test design, for test evaluation, and for optimal scoring.

The Rasch model* is a version of (9) in which items cannot be answered correctly by guessing (all $\gamma_i = 0$), all items are equally discriminating (all $\alpha_i = 1$), and F is logistic*. Under the Rasch model, there is a sufficient statistic (number-right score) for θ, independent of the β_i. Rasch [9] urges that this unique property be required before we can speak of "measurement."

TAILORED TESTING

In tailored testing, the items administered to a testee are chosen from a pool, one at a time, so as to estimate his or her ability θ as accurately as possible. Since different testees take different sets of items, IRT is necessary in order to assign comparable scores. The testee's score is his or her estimated θ, obtained by some standard statistical estimation procedure.

References

[1] Berk, R. A., ed. (1980). *Criterion-Referenced Measurement: The State of the Art*. Johns Hopkins University Press, Baltimore, Md.

[2] Birnbaum, A. (1968). In *Statistical Theories of Mental Test Scores*, F. M. Lord and M. R. Novick, eds. Addison-Wesley, Reading, Mass., Part 5. (The basic ideas of IRT, as developed by a well-known statistical theorist.)

[3] DuBois, P. H. (1970). *A History of Psychological Testing*. Allyn and Bacon, Boston.

[4] Guttman, L. (1941). In *The Prediction of Personal Adjustment*, P. Horst et al., eds. Social Sciences Research Council, New York, pp. 321–344. (Modern statisticians keep reinventing the test-scoring methods derived here.)

[5] Hively, W. (1974). *Domain-Referenced Testing*. Educational Technology Publications, Englewood Cliffs, N.J.

[6] Lord, F. M. (1980). *Applications of Item Response Theory to Practical Testing Problems*. Lawrence Erlbaum, Hillsdale, N.J. (Basic text on IRT.)

[7] Lord, F. M. and Novick, M. R. (1968). *Statistical Theories of Mental Test Scores*. Addison-Wesley, Reading, Mass.

[8] Morrison, D. G. (1978). *Amer. Statist.*, **32**, 23–35.

[9] Rasch, G. (1961). *Proc. 4th Berkeley Symp. Math. Statist. Prob.*, Vol. 4. University of California Press, Berkeley, Calif., pp. 321–323.

[10] Richardson, M. and Kuder, G. F. (1933). *Pers. J.*, **12**, 36–40.

[11] Shoemaker, D. M. (1973). *Principles and Procedures of Multiple Matrix Sampling*. Ballinger, Cambridge, Mass. (The best available book on this subject.)

[12] Torgerson, W. S. (1958). *Theory and Methods of Scaling*. Wiley, New York. (Basic test on assigning numbers to objects or to people.)

[13] Wilks, S. S. (1938). *Psychometrika*, **3**, 23–40.

Bibliography

See the following works, as well as the references just given, for more information on the topic of psychological testing theory.

Bock, R. D. and Aitkin, M. (1981). *Psychometrika*, **46**, 443–459. (Two mathematical statisticians on the frontiers of IRT.)

Cronbach, L. J., Gleser, G. C., Nanda, H., and Rajaratnam, N. (1972). *The Dependability of Behavioral Measurements: Theory of Generalizability for Scores and Profiles*. Wiley, New York. (The authoritative treatise on test reliability.)

Gulliksen, H. (1950). *Theory of Mental Tests*. Wiley, New York. (Very readable basic reference on classical test theory.)

Holland, P. W. and Rubin, D. B., eds. (1982). *Test Equating*. Academic Press, New York. (Mathematical statisticians tackle test score equating.)

Novick, M. R., Jackson, P. H., and Thayer, D. T. (1971). *Psychometrika*, **36**, 261–268. (A Bayesian treatment of classical test theory.)

Popham, W. J., ed. (1971). *Criterion-Referenced Measurement: An Introduction*. Educational Technology Publications, Englewood Cliffs, N.J.

Solomon, H., ed. (1961). *Studies in Item Analysis and Prediction*. Stanford University Press, Stanford, Calif. (An early collection of original work by mathematical statisticians.)

Thorndike, R. L., ed. (1971). *Educational Measurement*, 2nd ed. American Council on Education, Washington, D.C. (Authoritative readable handbook.)

Wright, B. D. and Stone, M. H. (1978). *Best Test Design: A Handbook for Rasch Measurement*. Scientific Press, Palo Alto, Calif.

(FACTOR ANALYSIS
GROUP TESTING
KUDER–RICHARDSON RELIABILITY COEFFICIENTS
LATENT STRUCTURE ANALYSIS
MEASUREMENT STRUCTURES AND STATISTICS
MULTIPLE MATRIX SAMPLING
RASCH MODEL)

FREDERIC M. LORD

PSYCHOMETRICS *See* COMPONENT ANALYSIS; FACTOR ANALYSIS; PSYCHOLOGICAL DECISION MAKING; PSYCHOLOGICAL SCALING; PSYCHOLOGICAL TESTING THEORY

PSYCHOMETRIKA

Psychometrika is a journal whose main concern is the publication of papers on mathematical models and statistical methods for the behavioral sciences, particularly psychology. It was founded in 1935, largely by workers associated with the late L. L. Thurstone at the University of Chicago. Thurstone was the leading mathematical psychologist of the time. The journal has been issued quarterly, dated March, June, September, and December, starting in 1936.

Until recently, the masthead read "A journal devoted to the development of psychology as a quantitative, rational science," and this is an accurate, if somewhat flamboyant, statement of the general purpose of the journal, provided that "psychology" is conceived broadly to include other behavioral sciences. The articles usually have a statistical flavor, recognizing the stochastic nature of behavior, rather than proposing idealized mathematical models. Indeed, current volumes of the journal more often emphasize the development of methods for fitting statistical models to behavioral data rather than the development of the models themselves.

A very wide variety of topics find acceptance in the journal; currently, four or five receive the most attention. One is the fitting of spatial models to behavioral data, particularly data that reflect the psychological "similarity" of stimuli, whether perceptual or conceptual. The topic is frequently called "multidimensional scaling*," and involves fitting one or the other geometrical model to the data, particularly with hope of discovering the latent psychological organization of the stimuli. Seminal papers include Torgerson [27], Shepard [22, 23], Kruskal [13, 14], Carroll and Chang [6], and Takane et al. [26]. These all assume that the appropriate model is an n-dimensional Euclidean one, or some generalization thereof. Recently, there has been an increased interest in fitting tree-like rather than dimensional structures to this kind of data [5, 18, 21]. Clustering methods are also of interest (e.g., Milligan [17]). The model-fitting process is generally not inferential in the classical statistical sense, but there are exceptions [19]. There are generally computer programs that implement the model-fitting process, and they can be fairly sophisticated from the point of view of computational methods. Methodology developed here is widely applied in such diverse fields as marketing* and numerical taxonomy*.

A second active area is what might be called stochastic mental test theory. It involves methods for fitting stochastic models for objective tests (i.e., tests consisting of individually scored items) and/or evaluating the consistency of such mental test data. While older literature focused on the statistical behavior of total scores (e.g., Cronbach [7]), recent interest has been primarily in what is called item response theory (*see* PSYCHOLOGICAL TESTING THEORY). This encompasses a number of stochastic models that relate the probability of getting an item correct to scores on a latent continuum. The primary justification of these models is the desire to locate persons accurately on this latent continuum. An important early work was Lord's [15] monograph, and much of the recent developments are summarized by Lord [16]. Quite a bit of interest is in methods relating to simpler Rasch models* [20] (e.g., Fisher [9]). Again, there is considerable concern with the development of computer programs to implement the data-analysis methods (e.g., Bock and Aiken [3]). An important recent development is the attempt to make the process of objective testing a computer-interactive one, with concomitant increases in testing efficiency. There is wide interest in these methods in such fields as education and sociology* as well as psychology.

Interest also focuses on models for the analysis of covariance* or correlation matrices. This general interest evolves out of an early concern of Thurstone and his students in factor analysis*, but has more recently taken a more general character, embodying more general models for covariance data. The factor analysis model hypothesizes that the covariances among a set of observed variables can be accounted for by a relatively few latent variables. A variety of ad hoc methods for fitting this model to observed data evolved over the years (see, e.g., Harman [10]), but recent interest has focused on methods that are more rigorous and at the same time more flexible. The latter relate primarily to such models and methods as those suggested by Jöreskog [11, 12]. Bentler [2] summarizes some recent generalizations of this literature. It intersects with topics in econometrics* (see Aigner and Goldberger [1]) and is widely used in all the social and behavioral sciences.

A fairly large number of articles are about more traditional types of statistical data analysis, overlapping considerably with the types of papers that might be published in such journals as the *Journal of the American Statistical Association**, *Biometrika**, and the like. To a large extent these have to do with statistical methods or issues of particular concern to psychologists. Some examples are the topics of repeated measures* in analysis of variance [4], principal components [8], measures of correlation [24], canonical* correlation analysis [28], and ordinal statistics [29].

Articles proposing mathematical models for psychological data are also found. These are often models for choice or preference or some other form of human judgment (e.g., Takane [25]), but models for other processes are also welcome.

These are the topics that are the concern of the majority of papers, but the journal is intended to be rather eclectic. Also, interests evolve over time and new emphases emerge. Prospective authors can peruse recent issues to see the most recent trends, but should not hesitate to submit any article that seems to fit the general purpose.

Recently, *Psychometrika* instituted a section called Computational Psychometrics. It is intended to publish a wide variety of things relevant to the computer implementation of psychometric and statistical methods. One subsection is for the publication of reviewed algorithms, either in conjunction with or subsequent to the publication of a regular article. It also publishes reviews and evaluations of widely distributed programs, and more informal notes announcing the availability of programs, equipment, and so on.

The journal also includes a book review section for books that are of special interest to its readers.

Psychometrika employs a blind review process, typically involving three independent reviewers. Instructions for authors were published in the March 1984 issue and are available from the current editor, whose name and address appear inside the back cover of the journal. At the time of writing, the acceptance rate for submitted manuscripts was about 30%. Articles must be submitted in English, but about a third of recent authors come from outside North America.

References

[1] Aigner, D. J. and Goldberger, A. S. (1977). *Latent Variables in Socioeconomic Models*. North-Holland, Amsterdam.

[2] Bentler, P. M. (1983). *Psychometrika*, **49**, 493–517.

[3] Bock, R. D. and Aitkin, M. (1981). *Psychometrika*, **46**, 443–459.

[4] Boik, R. J. (1981). *Psychometrika*, **46**, 241–255.

[5] Carroll, J. D. and Arabie, P. (1983). *Psychometrika*, **48**, 157–170.

[6] Carroll, J. D. and Chang, J. J. (1970). *Psychometrika*, **35**, 283–319.

[7] Cronbach, L. J. (1951). *Psychometrika*, **16**, 297–334.

[8] Eckart, C. T. and Young, G. (1936). *Psychometrika*, **1**, 211–218.

[9] Fischer, G. H. (1983). *Psychometrika*, **48**, 3–26.

[10] Harman, H. H. (1975). *Modern Factor Analysis*. University of Chicago Press, Chicago.

[11] Jöreskog, K. G. (1969). *Psychometrika*, **34**, 183–202.

[12] Jöreskog, K. G. (1978). *Psychometrika*, **43**, 443–479.

[13] Kruskal, J. B. (1964). *Psychometrika*, **29**, 1–27.

[14] Kruskal, J. G. (1964). *Psychometrika*, **29**, 115–129.

[15] Lord, F. M. (1952). *A Theory of Test Scores*. Psychom. Monogr. No. 7.

[16] Lord, F. M. (1980). *Application of Item Response Theory to Practical Testing Problems*. Lawrence Erlbaum, Hillsdale, N.J.

[17] Milligan, G. W. (1981). *Psychometrika*, **46**, 187–199.

[18] Pruzansky, S., Tversky, A., and Carroll, J. D. (1982). *Psychometrika*, **47**, 3–24.

[19] Ramsay, J. O. (1978). *Psychometrika*, **43**, 145–160.

[20] Rasch, G. (1960). *Probabilistic Models for Some Intelligence and Attainment Tests*. Dansmark Paedogogiske Institute, Copenhagen.

[21] Sattath, S. and Tversky, A. (1977). *Psychometrika*, **42**, 319–345.

[22] Shepard, R. N. (1962). *Psychometrika*, **27**, 125–140.

[23] Shepard, R. N. (1962). *Psychometrika*, **27**, 219–246.

[24] Steiger, J. H. and Browne, M. W. (1984). *Psychometrika*, **49**, 11–24.

[25] Takane, Y. (1982). *Psychometrika*, **47**, 225–241.

[26] Takane, Y., Young, F. W., and de Leeuw, J. (1977). *Psychometrika*, **42**, 7–67.

[27] Torgerson, W. S. (1950). *Psychometrika*, **17**, 401–419.

[28] van de Geer, J. P. (1984). *Psychometrika*, **49**, 79–94.

[29] Wackerly, D. D. and Robinson, D. H. (1983). *Psychometrika*, **48**, 183–194.

NORMAN CLIFF

PSYCHOPHYSICS, STATISTICAL METHODS IN

Psychophysics is a subfield of experimental psychology which studies the effects of external stimuli on the various sensory systems (visual, auditory, tactile, etc.) of human or animal organisms. A basic concept is that the energy in a stimulus is coded into a sensory signal interpretable by the organism for some decision making. Psychophysics uses behavioral data to investigate the form of that coding.

DETECTION AND DISCRIMINATION

In a classical experimental paradigm, data are collected to estimate the probability p_{sn} that a stimulus s is detected over a masking background of "noise" n. The symbol s denotes a real number measuring a physical intensity expressed in standard units; n may denote a physical intensity or a (possibly infinite dimensional) vector, such as a spectral density function. For any value of the noise* n, the function $p_n : s \to p_{sn}$ is a *psychometric function*, and often assumed to be continuous and strictly increasing. The detection probabilities p_{sn} may be estimated from frequency data in the framework of some mathematical model. Such a model specifies any psychometric function up to the values of some parameters which, it is hoped, capture some essential features of the sensory mechanisms. An example is the Gaussian model

$$p_{sn} = \frac{1}{\sqrt{2\pi}} \int_{-\infty}^{\gamma(s)-\eta(n)} e^{-z^2/2}\, dz \qquad (1)$$

with γ and η, two real-valued functions depending on the stimulus and the background, respectively. Equation (1) arises as a special case of the model

$$p_{sn} = \Pr[U_s \geqslant V_n], \qquad (2)$$

in which U_s and V_n are two jointly distributed random variables, which may be regarded as symbolizing the excitations evoked in some neural location by the presentation of the stimulus and of masking noise. If U_s and V_n are independent, normally distributed random variables, with expectations $\gamma(s)$ and $\eta(n)$, respectively, and a common variance equal to $\frac{1}{2}$, then (2) reduces to (1). With a different interpretation, this model is also encountered in choice theory (*see* RANDOM UTILITY MODEL). Variants of it are attributed to Thurstone [13, 14].

Other distributional assumptions are also made concerning U_s and V_n. Some researchers suppose that a given sensory mechanism behaves as a counting process. Under appropriate conditions, this leads to the conclusion that U_s and V_n have a Poisson* distribution [2, 3, 5, 9]. For another example, it is sometimes postulated that a sensory mechanism operates as if it were computing the maximum of the excitation registered over a large number of sensory "channels," an assumption leading one to suppose that each of U_s and V_n has an extreme value distribution* (e.g., double exponential) [12, 16, 17]. Typically, the available data do not permit a clear choice between these models. Accordingly, the distributional assumptions are often dictated by considerations of convenience. Of primary interest are the functions $\gamma(s)$ and $\eta(n)$ in (1), which provide a parametrization of the mechanisms responsible for sensory coding. These functions are sometimes regarded as *sensory scales*, measuring the sensations evoked by the stimulus and the noise. In this connection, (1) is a special case of the equation

$$p_{sn} = F[\gamma(s) - \eta(n)], \qquad (3)$$

with F, a continuous, strictly increasing, but otherwise unspecified function. The fact that γ and η, but not F, are of central interest suggests a critical change in the paradigm.

Writing f^{-1} for the inverse of a one-to-one function f, and defining the *sensitivity function* $\xi_\pi(n) = p_n^{-1}(\pi)$ (see Fig. 1), (3) can be rewritten

$$\xi_\pi(n) = \gamma^{-1}[F^{-1}(\pi) + \eta(n)]. \qquad (4)$$

For a fixed value of the detection probabilities $\pi = p_{sn}$, the right member of (4) depends only on the critical functions ν and η. The form of these functions can thus be investigated by estimating $\xi_\pi(n)$ experimentally for various values of the masking background n.

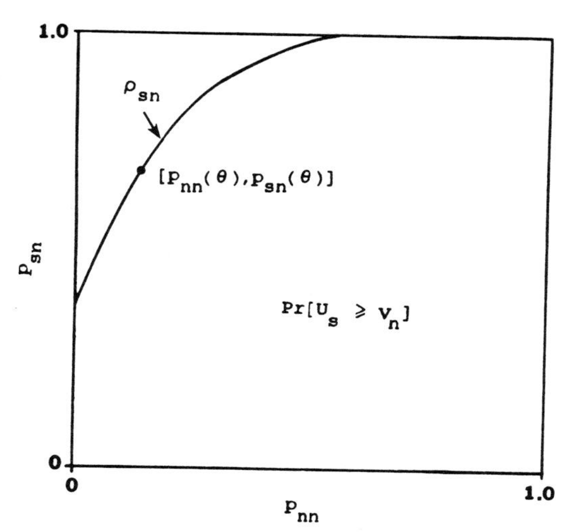

Figure 1 Psychometric function $p_n : s \to p_{sn}$ and sensitivity function $\xi_\pi(n) = p_n^{-1}(\pi)$.

Many data are consistent with the equation

$$\xi_\pi(n) = n^\beta G(\pi), \qquad \beta > 0,$$

which is referred to as the *near-miss-to-Weber's law*. (See ref. 5; we assume here that n is a positive real number measuring the intensity of the background.) This is a special case of (4), in which $\gamma(s) = \log s^{1/\beta}, \eta(n) = \log n$, and $G(\pi) = \exp[\beta F^{-1}(\pi)]$. When $\beta = 1$, one obtains *Weber's law*.

In modern laboratories, experimental estimates of the values $\xi_\pi(n)$ are secured by computerized applications of stochastic approximation* techniques, in particular, the staircase method*. The subject is presented on successive trials with backgrounds n_1, $n_2, \ldots, n_k, \ldots$. The choice of each value n_k depends on the probability π in (4), the previous value n_{k-1}, and the subject's response. Under the assumptions of the stochastic approximation technique used, either n_k or $(1/n_k)\sum_{i=1}^{k} n_k$ tends to $\xi_\pi(n)$. To avoid the subject detecting a trend in the succession of the n_k values, several staircase procedures may be "interleaved."

Psychophysicists also investigate the ability of subjects to discriminate between stimuli in the same sensory continuum. In a psychoacoustic context, for example, p_{ab} may denote the probability of the subject judging a as louder than b. Most of the concepts and methods used in the study of detection also apply to discrimination.

SIGNAL DETECTION THEORY

A difficulty in psychophysics is that any experimental task has cognitive components, this term covering a variety of factors, such as response bias, guessing strategies, motivation, and so on. Consider a detection task in which, in half of the trials, only the masking background is presented. There are thus two kinds of trials: the *sn* or *stimulus trials*, and the *nn* or *noise-alone trials*. Two types of errors can be made: (a) the subject may fail to report a stimulus presented on an *sn* trial; this is a *miss*; (b) the subject may report a detection on a *nn* trial; this is a *false alarm*. The two types of correct responses are called *hit* and *correct rejection*. Various kinds of strategies are available in this task. For example, when the subject is not quite sure that the stimulus was presented on a trial, he or she may nevertheless claim to have detected it. This strategy would be advantageous in a situation in which the misses are much more heavily penalized than the false alarms, for example, in the *payoff matrix* in Table 1, in which the numbers represent monetary values.

If the subject reports a detection on every trial, whether or not the stimulus was presented and these two possibilities are equiprobable, the average gain per trial is 1.5. The opposite strategy of responding "no de-

Table 1 Example of a Payoff Matrix in Signal Detection Theory

		Responses: Yes	Responses: No
Stimulus	Yes	4 Hit	− 4 Miss
	No	− 1 False alarm	3 Correct rejection

tection" on every trial results in an average loss of 0.5. This payoff strategy thus favors a positive guessing strategy over a conservative one. In fact, the data indicate a remarkable tendency of the subjects to alter their strategy according to nonsensory aspects of the task. A problem for the psychophysicist is thus that of disentangling the purely sensory aspects of the task from those resulting from cognitive factors.

The solution of this problem was inspired by some application of statistical decision theory* to communication theory*, in the design of optimal decision procedures for the detection of signal in noise [6, 10, 11, 15]. In psychophysics, however, considerations of optimality play a minor role. Let Θ be a large collection of payoff matrices. For any $\theta \in \Theta$, let $p_{sn}(\theta)$ and $p_{nn}(\theta)$ be the probability of a hit and of a false alarm, respectively. The set $\rho_{sn} = \{[p_{nn}(\theta), p_{sn}(\theta)] \mid \theta \in \Theta\}$ is referred to as a *receiver-operating-characteristic* (ROC) *curve*. Ideally, ρ_{sn} is assumed to be (the graph of) a continuous, increasing, typically concave function mapping the open interval (0, 1) into itself. In principle, the information contained in the ROC curve ρ_{sn} captures the purely sensory aspects of the detection of the stimulus s over the masking background n. This position is rationalized by the standard model of signal detection theory, the assumptions of which are as follows. In the spirit of the model of (2), the effect of the stimulation on an sn trial (respectively, nn trial) is represented by a random variable U_s (respectively, V_n). To each payoff matrix $\theta \in \Theta$ corresponds a real number λ_θ, the *criterion*. The separation of the sensory aspects of the task from the cognitive ones is achieved through the equations

$$p_{sn}(\theta) = \Pr[U_s > \lambda_\theta], \tag{5}$$

$$p_{nn}(\theta) = \Pr[V_n > \lambda_\theta]. \tag{6}$$

This means that a detection is reported on any trial if the sampled value of the induced random variable exceeds the criterion. If the random variables U_s and V_n are independent, then

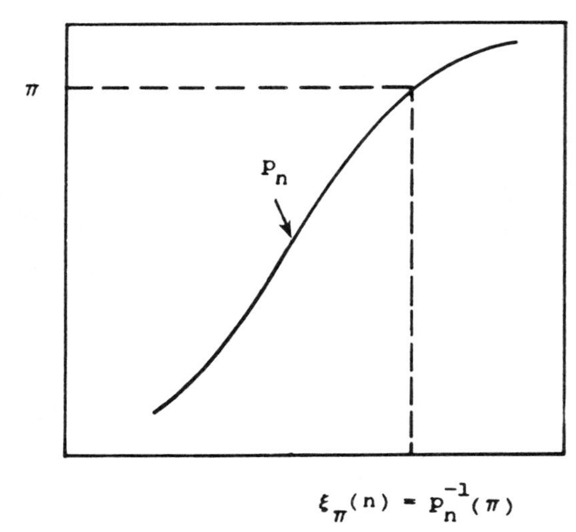

Figure 2 ROC curve ρ_{sn}.

$$\int_0^1 \rho_{sn}(p)\, dp = \Pr[U_s > V_n].$$

Thus the area under the ROC curve provides a measure of the detectability of the stimulus over the masking background, which is, in theory, unimpaired by cognitive factors (see Fig. 2).

In practice, assumptions are made concerning the distributions of the random variables U_s and V_n, which specify the ROC curve up to the values of some parameters. A few experimental points suffice to estimate these parameters and thus to locate the ROC curve. As in the case of (2), a frequent assumption is that U_s and V_n are Gaussian.

In a special case of this theory, the subject strategy, as embodied in (5) and (6), is consistent with an optimal statistical decision procedure based on the computation of likelihood ratios, and maximizing the expected gain.

OTHER PARADIGMS AND CONCEPTS

Many other concepts and methods are used in psychophysics. One important current must be mentioned here.

Some researchers favor experimental methods in which a human subject is required to provide a "direct" estimate of the sensory state evoked by stimulus. In the *magnitude estimation* paradigm advocated by S. S. Stevens and his followers [4, 8], the subject gives a numerical response quantifying the subjective impression of the stimulus intensity. A variation of this method (*cross-modality matching*) has the subject compare the intensity of stimuli from different sensory continua (say, the brightness of a monochromatic light and the loudness of a pure tone), and adjust one of the two intensities so as to equalize the two sensory impressions. Generally, the data are consistent with the hypothesis that the various sensory continua, when the physical intensities are measured in standard ratio scale units, are related to each other and to the number continuum by power laws (see ref. 7).

The *functional measurement* methods of N. H. Anderson are similar to Stevens's "direct" methods, in that the subject is asked to give a quantitative response (e.g., a number on a rating scale). The stimuli, however, vary according to several dimensions (such as the size and the weights of cubical blocks rated for "heaviness"; see ref. 1). The aims of the analysis are to parse out the effects of the stimulus dimensions on the sensory impressions; this is typically, and often successfully, achieved by regression or ANOVA techniques.

References

[1] Anderson, N. H. (1970). *Percept. Psychophys.*, **8**, 153–170.

[2] Green, D. M. and Luce, R. D. (1974). In *Contemporary Developments in Mathematical Psychology*, Vol. 2: *Measurement, Psychophysics, and Neural Information Processing*, D. H. Krantz, R. C. Atkinson, R. D. Luce, and P. Suppes, eds. W. H. Freeman, San Francisco, pp. 372–415.

[3] Luce, R. D. and Green, D. M. (1974). *J. Acoust. Soc. Amer.*, **56**, 1554–1564.

[4] Marks, L. E. (1974). *Sensory Processes*. Academic Press, New York.

[5] McGill, W. J. and Goldberg, J. P. (1968). *J. Acoust. Soc. Amer.*, **44**, 576–581.

[6] Peterson, W. W., Birdsall, T. L., and Fox, W. C. (1954). *Trans. IRE Prof. Group Inf. Theory*, **PGIT-4**, 171–212.

[7] Stevens, S. S. (1966). *Percept. Psychophys.*, **1**, 5–8.

[8] Stevens, S. S. (1975). *Psychophysics: Introduction to Its Perceptual Neural, and Social Prospects*. Wiley, New York.

[9] Strackee, G. and van der Gon, J. J. D. (1962). *Statist. Neerlandica*, **16**, 17–23.

[10] Tanner, W. P., Jr. and Swets, J. A. (1954). *Psychol. Rev.*, **61**, 401–409.

[11] Tanner, W. P., Jr. and Swets, J. A. (1954). *Trans. IRE Prof. Group Inf. Theory*, **PGIT-4**, 213–221.

[12] Thompson, W. A., Jr. and Singh, J. (1967). *Psychometrika*, **32**, 255–264.

[13] Thurstone, L. L. (1927). *Psychol. Rev.*, **34**, 273–286.

[14] Thurstone, L. L. (1927). *Amer. J. Psychol.*, **38**, 368–389.

[15] Van Meter, D. and Middleton, D. (1954). *Trans. IRE Prof. Group Inf. Theory*, **PGIT-4**, 119–141.

[16] Wandell, B. and Luce, R. D. (1978). *J. Math. Psychol.*, **17**, 220–235.

[17] Yellott, J. I., Jr. (1977). *J. Math. Psychol.*, **15**, 109–144.

Bibliography

See the following works, as well as the references just given, for more information on the topic of statistical methods in psychophysics.

Birnbaum, M. H. (1982). In *Social Attitudes and Psychophysical Measurement*, B. Wegener, ed. Lawrence Erlbaum, Hillsdale, N.J., pp. 410–485. (A discussion of the controversies in psychological measurement.)

Egan, J. P. (1975). *Signal Detection Theory and ROC Analysis*. Academic Press, New York. (Contains a discussion of distributional assumptions in signal detection theory.)

Falmagne, J.-C. (1985). *Elements of Psychophysical Theory*. Oxford University Press, New York. (An introductory text for the mathematically minded reader.)

Fechner, G. T. (1966). *Elements of Psychophysics*, D. H. Howes and E. C. Boring, eds., H. E. Adler, trans. Holt, Rinehart and Winston, New York. (Originally published, 1860. The original monograph of the founder of psychophysics. At least of historical interest.)

Gescheider, G. A. (1976). *Psychophysics: Method and Theory*. Lawrence Erlbaum, Hillsdale, N.J. (An elementary, nontechnical introduction.)

Green, D. M. and Swets, J. A. (1974). *Signal Detection Theory and Psychophysics*. Krieger, Huntington, N.Y. (The basic text.)

Levitt, H. (1970). *J. Acoust. Soc. Amer.*, **49**, 467–476. [A paper presenting the application of stochastic approximation methods in psychophysics (in particular, psychoacoustics).]

(PSYCHOLOGICAL SCALING)

JEAN-CLAUDE FALMAGNE

PUBLIC OPINION POLLS

HISTORY

The history of modern public opinion polling began in 1935 with the first publication of the Gallup polls. Researchers, however, have pointed to references to data gathering in the Bible and during the Middle Ages. As early as 1824, newspapers were using informal straw polls to predict elections in the United States. In Europe at the end of the nineteenth century Booth was conducting his survey of the poor working people in London and Le Play did the first family budget surveys. In the United States marketing and advertising research procedures were being developed and used by Scott, Cherington, and Parlin after 1910. Commercial and advertising research continued to grow after World War I and provided the foundation of methods for studying public opinion.

Gallup argued from the beginning that small, carefully selected quota samples* were superior to very large uncontrolled samples for measuring opinion. In 1936, the *Literary Digest* conducted a very large mail survey with more than 10,000,000 persons whose names had been obtained from automobile registrations and telephone directories. The same procedure had been used successfully in predicting the outcomes in several previous elections. Based on responses, the *Literary Digest* predicted a sweeping Landon victory. The much smaller Gallup, Roper, and Crossley polls correctly predicted a Roosevelt victory and established the credibility of public opinion polls among the media and politicians, including President Roosevelt, who used Gallup poll results as a source of information, especially on foreign policy issues [2].

Shortly thereafter the U.S. Department of Agriculture established the Division of Program Surveys to measure farmer attitudes. The first academic survey organization, the National Opinion Research Center, was started at the University of Denver in 1941 and then moved to the University of Chicago.

Opinion polls continued to increase in popularity during World War II. There were many surveys of public morale both in the United States and in other countries. The U.S. Army conducted large and sophisticated studies of the attitudes of soldiers, under the leadership of Samuel Stouffer and his colleagues [11]. After the war, public opinion research organizations were organized in every country in Western Europe. The Program Surveys group left the U.S. Department of Agriculture and under the leadership of Rensis Likert established the Survey Research Center at the University of Michigan. State and regional polls have been established by many universities as well as commercial researchers.

Major questions about the validity of poll results were raised in 1948 after all the polls incorrectly predicted a Dewey victory in the Presidential election. Statisticians pointed out the deficiencies in the quota samples then in use, and improved sampling methods were introduced [10]. The major reason for the error in 1948, however, was that the polls stopped interviewing too soon and missed late changes in opinion. More recently, polls interview through the last day prior to elections.

The concern about poll accuracy soon died down as polls again resumed a series of successful predictions. The use of polls by politicians as an aid in election campaigns grew rapidly, as did their use in the social sciences, program evaluation, and as a source of news for magazines and television in addition to their traditional newspaper audience.

As the fraction of households with telephones rose to the high 80% and then low

90% range and as field costs of face-to-face interviewing rose sharply, more and more polls were conducted on the telephone. The development of computer-assisted telephone interviewing (CATI) accelerated the use of telephone interviewing, which is today the most popular method of data collection*; *see* TELEPHONE SURVEYS, COMPUTER-ASSISTED.

The legality of polls has been questioned by some communities that have attempted to regulate polling, both face-to-face and telephone. There has also been a continuing controversy about whether poll results change the outcome of elections by discouraging voters or changing the voter's intentions. Most recently, this controversy has centered about exit polls that are conducted with voters as they leave the polling place.

USES OF POLLS

Of the many uses of polls, the most important is their use by policymakers in the private and public sectors. Since Franklin D. Roosevelt, U.S. presidents as well as leaders of many other countries have used polls as one source of information about public attitudes. For political leaders, the successful use of poll data is not to promote programs and policies that make the user more popular, but as a guide to public knowledge and attitudes that leaders will need to accept or modify in their policies.

Local government executives and national and local legislative bodies are also potential users of poll data, but much less data is available on a local level, so that informal sources are more likely to be used. While some political figures mistrust polling procedures, almost all of them use polls as an aid to reelection.

Polls are heavily used for business and program applications. It was estimated that of total expenditures of $1.2 billion for polls and surveys in the United States, less than 10% was for public opinion research with the rest for market and program applications [7]. The major uses include new product testing, customer satisfaction with products or services, use of media, testing advertising, and the polls as economic indicators (*see also* MARKETING, STATISTICS IN). The users are businesses, nonprofit organizations such as hospitals and schools, and local and federal administrators of a broad range of national and community programs. Extensive uses of polls are also made by professional and trade associations and other interest groups. These groups survey not only their own members, but also the general public, to determine the status of the group and public attitudes on issues of concern to it.

The use of polls in legal cases and administrative hearings has increased rapidly since 1960. Earlier polls were excluded as hearsay evidence, but recently courts have accepted the idea that public opinion can be measured, and expert witnesses can report on how the results were obtained and what they mean. Polls have been used to determine whether media coverage in a local area has been so intense as to require a change of venue, whether advertising is deceptive and whether there is confusion about a product trademark. A growing but controversial use of polls is as an aid to attorneys in jury selection and in presenting evidence to juries.

Polls and surveys are the most widely used method of collecting data in the social sciences, especially in sociology and political science [12] (*see* DATA COLLECTION *and* SURVEY SAMPLING). Many social scientists conduct secondary analyses of poll data gathered by the federal government or from large survey organizations. Major social science archives of poll data exist and can readily be accessed by researchers [16]. The largest of these in the United States is the Roper Center at the University of Connecticut.

The best known use of polls is as news. When the polls concentrate on issues they perform a service by helping to crystallize public opinion by providing people information on what others think. When the polls are used simply to predict in advance the outcome of elections, their main function is to entertain and to increase viewing and readership.

METHODS OF DATA COLLECTION

There are three major methods of collecting data: telephone, face-to-face, and mail. There are variations on each of these methods and they are also used in combination. The criteria for judging alternative procedures are the quality of the sample that will cooperate, the length and complexity of the questionnaire, and the costs.

Telephone Surveys

In the United States during the 1980s about 94% of all households have telephones. Very large percentages of households have telephones in some European countries and the percentage in other countries is growing rapidly. Thus the major bias caused by the omission of nontelephone households that caused a huge error in the 1936 *Literary Digest* poll is no longer found today. For this reason, telephone interviews are the most widely used method of data gathering for opinion polls. Costs are about half of face-to-face interviewing because no travel is involved, and there are no telephone charges and virtually no interviewer time spent if a respondent is not at home.

Most telephone interviewing is conducted from central locations where a group of interviewers are under close supervision, so that interviewer quality is tightly controlled. It has been found that cooperation rates can be as high and interviews can be as long on the telephone as face to face, and that telephone interviewing is substantially superior to mail questionnaires [5]. The major limitations are questions that require the respondent to look, touch, or taste something. Even here, combined methods may be used. Respondents can be mailed material and then telephoned for their opinions.

A major technological innovation has been the general introduction of computer-assisted telephone interviewing (CATI), now used by most polling organizations. Instead of a paper questionnaire, the interviewer uses a computer terminal on which the proper question appears in sequence. It is possible to individualize questionnaires based on respondent answers to earlier questions as well as to introduce experimental variations. Also, editing of responses is conducted as the interview proceeds. An improper answer is immediately rejected by the computer. Results are available for analysis immediately after the interview is completed.

Mail Surveys

Mail methods are the cheapest to use, even when done carefully. This involves an initial mailing and two follow-up* mailings to those who do not respond, plus a small cash gift [3, 4]. With general population samples only short questionnaires of fewer than four pages obtain reasonably high cooperation. Longer questionnaires that require much writing or following complex instructions are likely to result in very low cooperation, although these may sometimes be possible with special groups of highly educated respondents.

Opinion polls by mail have been successful when conducted at the local level and dealing with important local issues. They are also successful when conducted by groups and associations with their members. In the United States there have been no successful national mail opinion polls, in part because there is no convenient complete mailing list of households or individuals as in some European countries. Combined methods using self-administered questionnaires and telephone are sometimes used. When successful, mail or combined procedures can obtain cooperation rates as high as for personal methods.

Face-to-Face Interviewing

This once major method of polling has become much less popular because of the high cost of locating respondents at home. Nevertheless, it still is the most flexible method and allows the respondent to do tasks that are impossible by mail or telephone. Some very high quality surveys by federal agencies and universities are still conducted face-to-face, although even here follow-up interviews for panel* studies are conducted by telephone where possible.

Some face-to-face polls are conducted in central locations such as shopping malls or recreation areas. When the population is intended to be users of such facilities, this can be a very efficient sampling method (but *see* QUOTA SAMPLING). Users of such facilities differ from the general population in many ways, however, and generalizing from location samples is difficult even if the facilities have been selected using probability procedures [14].

SAMPLING

Almost all current public opinion polls use either strict or modified probability procedures in selecting their samples. The most commonly used procedure on the *telephone* is random digit dialing*. A list of all telephone area codes and working exchanges is available. Exchanges are selected at random and the last four digits are chosen using random numbers. This eliminates the need for directories, which are seriously incomplete in large cities because of unlisted numbers.

Random digit dialing results in many selected numbers being ineligible because they belong to businesses or other institutions or simply are not assigned. Procedures have been developed for substantially reducing ineligible numbers by clustering the sample in banks of numbers known to contain working household numbers [17].

Although strict probability samples would require that the selection of a respondent within the household also use probability methods, most polls use procedures that balance for gender since otherwise a substantial overrepresentation of women would be found. The short time period allowed for most polls means that people who are hard to find at home are likely to be underrepresented (*see also* SURVEYS, HOUSEHOLD).

The *face-to-face sampling* of individuals is a far more complex and costly enterprise since local interviewers must be hired and trained. Area probability samples are always multistage. At the first stage, counties or metropolitan areas are selected. Since these geographic areas vary enormously in size, they are chosen by a procedure called sampling with probability proportional to size*. This gives the largest metropolitan areas a greater chance of selection than smaller counties and results in the largest metropolitan areas such as New York, Los Angeles, and Chicago always being selected. The sample is still self-weighting because at later stages a smaller fraction is taken from the larger locations.

Within selected counties or metropolitan areas, additional selection is made of smaller geographic units in two or three stages, using census data. Selected blocks or segments are then either listed by the polling organization or directories (available for middle-sized cities) are used. Within a block or segment, all households have an equal probability of selection. To save interviewer travel costs, about five households will usually be chosen for each block. Once a household is selected, all members in the household are listed and one is chosen at random for interviewing. As a final step, it is necessary to weight the sample for household size both in phone and area samples since a respondent's probability of selection within the household depends on household size.

As with phone surveys, area samples are faced with serious problems of finding the selected respondent at home; usually, three attempts are made. Phone samples can do many more because there are no travel costs. Modified procedures that reduce the number of callbacks or that eliminate them entirely underrepresent people who are at home less often, introducing a potentially serious bias [6, 9, 13].

By far the simplest sampling procedure is used when there is a list available for a *mail study*. No clustering is required since the only purpose of clustering is to save interviewing costs. Using either hand or computer methods, the desired sample is selected either by pure random selection (*see* SIMPLE RANDOM SAMPLING), for which a different random number is required for each selected unit, or by systematic sampling*, where a random start is made and then every *n*th unit is chosen.

SAMPLE BIASES

Since not all selected respondents can be located or are willing to be interviewed after being located, some sample biases are inevitable. The following are typically found: Rural and small-town respondents are overrepresented since cooperation rates decline as the size of the place increases. The largest metropolitan areas have cooperation rates about 20% lower than in rural areas. There is some tendency to underrepresent both the youngest and oldest respondents for different reasons. The youngest respondents (age 18–35) are difficult to find at home, while the oldest respondents (over 65) are easy to find but more likely to refuse to be interviewed. There are no major social class or racial or ethnic biases. Biases for telephone and face-to-face methods are roughly comparable, but for different reasons. Respondents who cannot be reached because they will not open their doors are also likely to refuse when telephoned.

Since noncooperation occurs before the interview, there is seldom a high correlation* between the topic of the poll and sample biases for personal surveys. Much more serious sample biases are possible, however, for mail polls. Here the respondent can see all the questions before deciding whether to cooperate or not. People who feel strongly about the issues in the mail questionnaire, either pro or con, are more likely to respond than those who are in the middle. Also on mail surveys there is likely to be a clear education bias, with highest levels of cooperation coming from the best educated.

RESPONSE EFFECTS

From the earliest days of polling it was recognized that question wording* could affect responses on attitude questions. This is particularly true when issues are new or not very important so that people do not have well developed opinions.

Also there are often substantial errors in the reporting of behavior caused by memory error and social desirability of the responses. Well-known examples are the substantial overstatement of voting and charitable giving and the understatement of alcoholic consumption. In some cases, changes in the method of administration or wording of the questionnaire can improve results. Thus group or self-administered questionnaires are less likely to obtain overstatements of desirable behavior, and randomized response* methods may increase reporting of undesirable behavior. The use of open rather than closed questions increases the reported quantities of alcohol consumed [15, 1].

However, it is sometimes impossible to obtain useful information from respondents even if they answer questions. As illustrations, although household informants can provide accurate information about the employment status of other family members, parents are very poor informants about the drug-use behavior of their teenaged children. Questions asking about annual household consumption of specific grocery products or even about purchases in a short time period have very large measurement errors* unless the data are collected by diaries. On attitude questions, respondents are unwilling to admit ignorance and may express views on issues they have never heard about or that do not exist.

Although response effects are common, they are not ubiquitous. Where poll questions deal with important issues results are identical, even with substantial question wording changes. For nonthreatening questions, such as ownership or purchase of durable goods, responses are likely to be reasonably accurate, although there is always some memory error.

INTERVIEWER EFFECTS

If the poll is conducted by an interviewer, there is always the possibility that the interviewer will influence the results. In the most direct way, this occurs if the characteristics of the interviewer are related to the topic of the poll. Thus it has been found consistently that black and white respondents give different answers to black and white interviewers on questions about racial issues [8]. No effects are found, however, if the interviewer characteristics are unrelated to the topic.

Interviewer beliefs about the topic may have some effect in special cases. This may occur if the respondent gives an ambiguous answer to an open question. Deliberate interviewer falsification is very rare. It is possible for the interviewer to prompt the respondent with cues that suggest an answer, but the better trained the interviewer, the less likely this is to occur.

Measuring interviewer effects has generally been difficult because it requires randomized assignments. With face-to-face interviewing, randomized assignments are usually impractical because of interviewer travel costs and availability. Computer-assisted telephone interviewing, however, makes randomized assignments and the computation of interviewer components of variance practical.

References

[1] Bradburn, N. M. and Sudman, S. (1979). *Improving Interview Method and Questionnaire Design: Response Effects to Threatening Questions in Survey Research*. Jossey-Bass, San Francisco. (Reports results of experiments for improving reporting of threatening questions.)

[2] Cantril, H. (1967). *The Human Dimension: Experiences in Policy Research*. Rutgers University Press, New Brunswick, N.J. (A memoir by an early academic pollster who was a consultant to President Franklin Roosevelt.)

[3] Dillman, D. A. (1978). *Mail and Telephone Surveys: The Total Design Method*. Wiley, New York. (Gives detailed discussion of procedures for improving cooperation on mail surveys.)

[4] Erdös, P. L. (1970). *Professional Mail Surveys*. McGraw-Hill, New York. (A very useful introduction to successful mail surveys, primarily for business uses.)

[5] Groves, R. M. and Kahn, R. L. (1979). *Surveys by Telephone: A National Comparison with Personal Interviews*. Academic Press, New York. (The best source to date of experiences with phone surveys and comparisons to face-to-face interviewing.)

[6] Hansen, M. H., Hurwitz, W. N., and Madow, W. G. (1953). *Sample Survey Methods and Theory*, 2 Vols. Wiley, New York. (The classic work on applied probability sampling. Still very useful for large-scale surveys.)

[7] Honomichl, J. (1982). *Advertising Age*. June 21, pp. 48, 52.

[8] Hyman, H. (1954). *Interviewing in Social Research*. University of Chicago Press, Chicago. (The classic work on interviewer effects, but tends to exaggerate their importance.)

[9] Kish, L. (1965). *Survey Sampling*. Wiley, New York. (Standard text for courses in applied sampling.)

[10] Mosteller, F. et al. (1949). *The Pre-election Polls of 1948*. Social Science Research Council, New York. (The analysis of deficiencies in 1948 polling procedures that led to improved sampling methods.)

[11] Stouffer, S. et al. (1949). *The American Soldier*. Princeton University Press, Princeton, N.J. (A classic work that shaped the direction of social science after World War II.)

[12] Sudman, S. (1976). *Annu. Rev. Sociol.*, **2**, 107–120.

[13] Sudman, S. (1976). *Applied Sampling*. Academic Press, New York. (An elementary text on sampling for users with limited statistical backgrounds.)

[14] Sudman, S. (1980). *J. Marketing Res.*, **17**, 423–31.

[15] Sudman, S. and Bradburn, N. M. (1974). *Response Effects in Surveys: A Review and Synthesis*. Aldine, Chicago. (A meta analysis of 500 surveys that generalizes types and magnitudes of response effects.)

[16] Sudman, S. and Bradburn, N. M. (1982). *Asking Questions*. Jossey-Bass, San Francisco, pp. 15–16.

[17] Waksberg, J. (1978). *J. Amer. Statist. Ass.*, **73**, 40–46.

(BIAS
DATA COLLECTION
ELECTION PROJECTIONS
PANEL DATA
QUESTION WORDING EFFECT IN SURVEYS
QUOTA SAMPLING
RANDOM DIGIT DIALING
SURVEY SAMPLING
SURVEYS, HOUSEHOLD

TELEPHONE SURVEYS, COMPUTER-ASSISTED)

SEYMOUR SUDMAN

PURCHASING POWER PARITY

The purchasing power parity (PPP) is a price index which measures the price level in country s relative to that in the *base* country r, where the prices compared are expressed in national currency units. It is therefore a rate of exchange between two national currencies, except that while the official exchange rate is the number of units of the currency of country s required to buy one unit of the currency of country r, the PPP is the number of units of the currency of country s required to buy goods and services equivalent to what can be bought with one unit of the currency of country r. While the former provides a conversion factor to compare *nominal* income and output and for the settlement of international transactions, the PPP provides a conversion factor to compare *real* income and output between countries.

BRIEF HISTORY

The early history of PPP comparisons can be traced to comparisons of the real incomes of England, France, and Holland by Sir William Petty [24] in 1665 and 1676 and by Gregory King [17] in 1688. However, major international comparisons of real income began to appear only in this century. Two landmark studies based on consumer goods were made by the U.K. Board of Trade [29–33] between 1907 and 1911 and by the International Labour Office [15] in 1932. In the 1950s the Organization for European Economic Cooperation [11, 12] sponsored pioneering comparisons for gross domestic product and its components. These were followed recently by extensive comparisons sponsored by the United Nations and the World Bank [19, 21, 22] and by the European Economic Community [9]. There are also scores of highly specialized studies by individual scholars and government agencies, most notably by Bergson [5] and by the Joint Economic Committee of the U.S. Congress [34]. A more detailed historical discussion can be found in refs. 18 and 25.

TYPES AND PROPERTIES OF PPP COMPARISONS

There are direct or "full-scale" and indirect or "shortcut" methods of PPP comparisons. In this section we define the direct comparisons and their desired properties, and in the following section we discuss some of the methods used in these comparisons and present an example. In the final section we discuss briefly some of the shortcut methods.

The direct PPP comparisons require the specification of a sample of commodities representative of total consumption or production of the countries compared, the collection of the prices and quantities of these commodities during a specified period, and the application of some index number* formulas. Thus suppose that one observes the *same* set of n commodities $A = (a_1, a_2, \ldots, a_n)$ in $m \geqslant 2$ countries at a given time. Let p_{ij} be the price of a_i in country j expressed in national currency units and q_{ij} the corresponding quantity in physical units. P_{rs} is a weighted average of the ratios (p_{is}/p_{ir}). The weights are functions of the quantities of either or both countries r and s or of all the m countries if $m > 2$. The corresponding quantity index, Q_{rs}, which is a measure of real income in s relative to r, is obtained by dividing P_{rs} into the expenditure ratio V_{rs} $(= \sum_{i=1}^{n} p_{is}q_{is} / \sum_{i=1}^{n} p_{ir}q_{ir})$. If $m = 2$, the comparisons are binary or *bilateral*, and if $m > 2$, they are multicountry or *multilateral*. The methods used in these comparisons are chosen on the basis of certain desired properties which include in particular the following.

1. **Base-Country Invariance.** For $m = 2$, this requires that $P_{rs}P_{sr} = 1$.

2. **Transitivity.** For $m > 2$, this requires that $P_{rs}P_{st} = P_{rt}$ for any three countries r, s, and t among the m countries.
3. **Consistency in Aggregation*.** This requires that the value of the index is the same whether computed directly for the set A or as the weighted sum of the equivalent indices of disjoint subsets of A (*see* LOG-CHANGE INDEX NUMBERS).
4. **Characteristicity.** This requires that the weights used in computing P_{rs} are representative of the patterns of consumption or production of the two countries r and s.

METHODS OF DIRECT PPP COMPARISONS

Laspeyres, Paasche*, and especially Fisher's ideal index* are among the index number formulas widely used in bilateral PPP comparisons. The PPP between two countries r and s calculated by Fisher's ideal index, F_{rs}, is given by

$$F_{rs} = \left[\frac{\sum_i p_{is} q_{ir}}{\sum_i p_{ir} q_{ir}} \cdot \frac{\sum_i p_{is} q_{is}}{\sum_i p_{ir} q_{is}} \right]^{1/2}. \quad (1)$$

This index satisfies properties **1** and **4**, but it fails property **3** and, in multilateral comparisons, it also fails **2**. A multilateral version of (1) which satisfies property **2**, F^*_{rs}, is given by

$$F^*_{rs} = \left[F^2_{rs} \prod_{\substack{t=1 \\ t \neq r,s}}^{m} \frac{F_{rt}}{F_{st}} \right]^{1/m}, \quad (2)$$

where the indices in the brackets are obtained from (1). This index is known as the EKS [9, 18, 22] after the initials of its three authors: Elteto, Koves, and Szulc.

It has been argued in ref. 6 on economic grounds in favour of the Törnqvist bilateral index, H_{rs}, which is given by

$$\ln H_{rs} = \frac{1}{2} \sum_i^m (w_{ir} + w_{is}) \ln(p_{is}/p_{ir}), \quad (3)$$

where $w_{ij} = (p_{ij}q_{ij} / \sum_{i=1}^n p_{ij}q_{ij})$ for $j = r$ and s. This index, like (1), satisfies properties **1** and **4**, but it fails property **3** and, in multilateral comparisons, it also fails property **2**. However, for $m > 2$, ref. 6 provides a multilateral version of (3), H^*_{rs}, which satisfies property **2** and is given by

$$\ln H^*_{rs} = \ln \overline{H}_s - \ln \overline{H}_r, \quad (4)$$

where

$$\ln \overline{H}_i = \frac{1}{m} \sum_{j=1}^{m} \ln H_{ij} \qquad \text{for } i = r, s \quad (5)$$

and the indices H_{ij} in (5) are obtained from (3).

Another method of multilateral comparisons is the Geary–Khamis index (*see* INDEX NUMBERS). This method has been used in refs. 19, 21, and 22, and a version of it called the Gerardi index [10] has been used in ref. 9. Voeller [36, p. 81] lists many index number properties which the Geary–Khamis index satisfies, including properties **1** and **2**. But it fails property **3** and also property **4** unless $m = 2$. The Gerardi index, however, satisfies property **3** in addition to the other properties which the Geary–Khamis index satisfies [36, p. 83]. These two indices are studied in great detail in ref. 14. A serious problem common to all multilateral index number formulas is that for $m > 2$ the PPP between two countries r and s is not independent of the value of m.

A typical example of the main use of PPP, based on [18, pp. 5–10], is shown in Table 1, which is self-explanatory. (The PPPs in this example were calculated by the Geary–Khamis method.)

SHORTCUT METHODS

Part of the economic literature of real income comparisons analyzes the relation between estimated real incomes and living standards in the countries compared, as in refs. 1, 20, 23, 26, 27, 28 and 35, and the relation between nominal and real incomes, as in refs. 2, 3, 8, and 16. Based on economic arguments, the relation between nominal and real income can be estimated by regression methods. Thus David [8] estimated

Table 1 Per Capita Gross Domestic Product, Official Exchange Rates, and Purchasing Power Parities in 1970

				Per Capita GDP			
					In U.S. Dollars Converted		
Country	Currency Unit (1)	Official Exchange Rate to U.S. Dollars (2)	Purchasing Power Parity to U.S. Dollars (3)	In National Currency Units (4)	At Official Exchange Rate (5) = (4) ÷ (2)	At Purchasing Power Parity (6) = (4) ÷ (3)	Price Index (U.S. Prices = 100) (7) = (3) ÷ (2)
France	Franc	5.5540	4.480	16,118	2,902	3,598	81
West Germany	Deutsche Mark	3.6600	3.140	11,272	3,080	3,590	86
India	Rupee	7.5000	2.160	736	98	341	29
Italy	Lira	625.0000	483.000	1,062[a]	1,699	2,199	77
Japan	Yen	360.0000	244.000	721[a]	2,003	2,955	68
United Kingdom	Pound	0.4167	0.308	893	2,143	2,899	74
United States	Dollar	1.0000	1.0000	4,801	4,801	4,801	100

[a] In thousands of national currency units.

from cross-section data for industrial countries the regression equation

$$Y_r/Y_{rs} = 0.559 + 0.441(Y_r/Y^*_{rs}), \quad (6)$$

where Y_r is the per capita income of the base country r (the United States) in its own currency and Y_{rs} and Y^*_{rs} are, respectively, nominal and real per capita income of country s expressed in the currency of country r. The estimated coefficient of (Y_r/Y^*_{rs}) in (6) was interpreted [8, p. 985] to mean that "the 'real' percentage gap between the per capita income of a given country and the United States is only (0.441) four-ninths of the percentage gap indicated by a straight exchange rate conversion expressing all incomes in dollars." Put differently, this also means that the official exchange rate of country s in terms of U.S. dollars is only four-ninths of the corresponding PPP. This method was further developed and applied in ref. 20 and other similar regression methods were applied in refs. 4 and 7. On comparative assessment of many shortcut methods, see ref. 13.

References

[1] Afriat, S. N. (1972). In *International Comparisons of Prices and Output*, D. J. Daly, ed. National Bureau of Economic Research, New York, pp. 1–69. (Economic theory, technical but foundational paper; references.)

[2] Balassa, B. (1964). *J. Pol. Econ.*, **72**, 584–596. (Economic theory of a shortcut method.)

[3] Balassa, B. (1973). *Econ. J.*, **83**, 1258–1267.

[4] Beckermann, W. and Bacon, R. (1966). *Econ. J.*, **76**, 519–536. (Elementary.)

[5] Bergson, A. (1978). *Productivity and the Social System: The USSR and the West*. Harvard University Press, Cambridge, Mass.

[6] Caves, D. W., Christensen, L. R., and Diewert, W. E. (1982). *Econ. J.*, **92**, 73–86. (Intermediate; few references.)

[7] Clague, C. and Tanzi, V. (1972). *Econ. Int.*, **25**, 3–18.

[8] David, P. A. (1972). *Econ. J.*, **82**, 591–605.

[9] European Community Statistical Office (1977). *Comparisons in Real Values of the Aggregates ESA*. European Community, Luxemburg.

[10] Gerardi, D. (1982). *Rev. Income Wealth*, **28**, 381–405.

[11] Gilbert, M. and Kravis, I. (1954). *An International Comparison of National Products*. OEEC, Paris.

[12] Gilbert, M. and Associates (1958). *Comparative National Products and Price Levels*. OEEC, Paris.

[13] Heston, A. (1973). *Rev. Income Wealth*, **19**, 79–104. (Elementary; few references.)

[14] Hill, T. P. (1982). *Multilateral Measurements of Purchasing Power and Real GDP*. Publications of the European Community, Luxemburg. (Elementary; few references.)

[15] International Labour Office (1932). *A Contribution to the Study of International Comparisons of Costs of Living*, Ser. N, No. 17. ILO, Geneva, Switzerland.

[16] Jazairi, N. T. (1979). *Econ. J.*, **89**, 127–130.

[17] King, G. (1688). *Two Tracts*, G. C. Barnett, ed. Johns Hopkins University Press, Baltimore, 1936.

[18] Kravis, I. B. (1984). *J. Econ. Lit.*, **22**, 1–39. (Elementary survey; references.)

[19] Kravis, I. B., Heston, A., and Summers, R. (1978). *International Comparisons*, Johns Hopkins University Press, Baltimore, Md.

[20] Kravis, I. B., Heston, A., and Summers, R. (1978). *Econ. J.*, **88**, 215–242.

[21] Kravis, I. B., Heston, A., and Summers, R. (1982). *World Product and Income*. Johns Hopkins University Press, Baltimore, Md.

[22] Kravis, I. B., Zoltan, K., Heston, A., and Summers, R. (1975). *A System of International Comparisons*. Johns Hopkins University Press, Baltimore, Md.

[23] Marris, R. (1984). *J. Econ. Lit.*, **22**, 40–57. (Elementary, economic theory; few references.)

[24] Petty, W. (1665/1676). In *Economic Writings*, C. H. Hull, ed. Published in 1899 by Cambridge University Press, Cambridge, England.

[25] Ruggles, R. (1967). In *Ten Economic Studies in the Tradition of Irving Fisher*, Wiley, New York, pp. 171–205.

[26] Samuelson, P. (1964). *Rev. Econ. Stat.*, **46**, 145–154. (Economic theory.)

[27] Samuelson, P. (1974). *Econ. J.*, **84**, 595–608. (Economic theory.)

[28] Sen, A. (1979). *J. Econ. Lit.*, **17**, 1–45. (Economic theory, elementary survey; references.)

[29] U.K. Board of Trade (1908). *Cd. 3864*. H. M. Stationery Office, London.

[30] U.K. Board of Trade (1908). *Cd. 4032*. H. M. Stationery Office, London.

[31] U.K. Board of Trade (1909). *Cd. 4512*. H. M. Stationery Office, London.

[32] U.K. Board of Trade (1910). *Cd. 5065*. H. M. Stationery Office, London.

[33] U.K. Board of Trade (1911). *Cd. 5609*. H. M. Stationery Office, London.

[34] U.S. Congress (1982). *USSR: Measures of Economic Growth and Development*. U.S. Government Printing Office, Washington, D.C.

[35] Usher, D. (1968). *The Price Mechanism and the Meaning of National Income Statistics*. Oxford University Press, Oxford. (Theory and application.)

[36] Voeller, J. (1981). *Purchasing Power Parities for International Comparisons*. Verlag, Königstein, German Democratic Republic.

(FISHER'S IDEAL INDEX NUMBER
INDEX NUMBERS
PAASCHE–LASPEYRES INDEX NUMBERS)

NURI T. JAZAIRI
RASESH THAKKAR

PURI'S EXPECTED NORMAL SCORES TEST

When comparing k populations under completely randomized designs (i.e., one-way* layouts), statisticians are often interested in testing for ordered treatment effects among the k groups. *See also* ISOTONIC INFERENCE *and* ORDER-RESTRICTED INFERENCES. To treat such problems, numerous statistical procedures have been devised. Under conditions appropriate for classical analysis, the statistician may select a (parametric) procedure from among those developed by Bartholomew [2] or by Abelson and Tukey [1]. On the other hand, under more general conditions one would select a (nonparametric) procedure from among those (chronologically) developed by Terpstra [25], Jonckheere [18], Chacko [11], Puri [21], Conover [13], Johnson and Mehrotra [17], Tryon and Hettmansperger [27], or Wolf and Naus [30].

DEVELOPMENT OF PURI'S TEST

If θ_i represents some location parameter (e.g., the median) and if it is assumed that $F_i(x) = F(x - \theta_i)$ for some unknown CDF F, the essential problem becomes one of ordered location shift, and the statistician would test

$$H_0 : \theta_1 = \theta_2 = \cdots = \theta_k \qquad (1)$$

against

$$H_1 : \theta_1 \leqslant \theta_2 \leqslant \cdots \leqslant \theta_k \qquad (2)$$

(with at least one inequality strict).

Puri [21] proposed and developed a family of rank procedures for testing against such ordered alternatives. The test statistic devised in Section 2 of his paper possesses several interesting properties. Depending on the type of "constants" selected for his term $E_\nu^{(i,j)}$, the Puri test becomes equivalent to the test developed by Jonckheere [18] for comparing k groups against ordered alternatives (*see also* JONCKHEERE TESTS FOR ORDERED ALTERNATIVES) as well as to the tests developed by Mann and Whitney [19], Hoeffding [16] and Terry [26], and Savage [22] for comparing two groups against one-tailed alternatives. Since the Jonckheere test can be thought of as a k-sample generalization of the one-sided two-sample Mann–Whitney–Wilcoxon* procedure [19, 29], the work of Puri can also be viewed as basically a k-sample extension of a class of procedures devised for the two-sample problem.

DESCRIPTION

Since Puri was more concerned with the mathematical and large-sample properties of the tests than with actual applications (see refs. 20, 10, and 21), Berenson [6] described a modified and simplified version of the powerful Puri expected normal scores procedure, which facilitates its usage when the sample sizes are equal.

To perform the test, consider a one-way layout with k groups and (to avoid ties) n continuous observations per group. Let X_{ij} represent the jth observation in the ith group ($i = 1, 2, \ldots, k$ and $j = 1, 2, \ldots, n$). Replace all the nk observations by their corresponding ranks R_{ij}, such that a rank of 1 is given to the smallest of the X_{ij} and a rank of nk is given to the largest value. Now replace the ranks with ξ_{ij}—their corresponding "expected normal order statistics." First presented in 1938 by Fisher and Yates [14] in an attempt to transform a set of nonnormal data into one of normal form, the ξ_{ij} represent the (expected) Z values for each of the nk order statistics when taken from a standardized normal distribution. These expected normal scores are easily read from tables—a most extensive set of which was published by Harter [15]. *See* FISHER–YATES TESTS *and* NORMAL SCORES TESTS.

The test statistic for the Puri expected normal scores procedure can now be written [6] as

$$V = n \sum_{i=1}^{k} (2i - k - 1)\xi_{i\cdot} , \qquad (3)$$

where $\xi_{i\cdot} = \sum_{j=1}^{n} \xi_{ij}$ and $i = 1, 2, \ldots, k$. The decision rule is to reject the null hypothesis of form (1) in favor of the ordered location alternative of form (2) if

$$V \geqslant Z_{\alpha} \left[\frac{n^3 k(k^2 - 1)}{3(nk - 1)} \sum_{i=1}^{k} \sum_{j=1}^{n} \xi_{ij}^2 \right]^{1/2}, \qquad (4)$$

where Z_{α} is the upper α percent critical value under the standardized normal distribution.

The version of the test above has shown to be useful provided $nk \geqslant 15$ (see ref. 7). A detailed application is demonstrated in ref. 6.

POWER COMPARISONS

Asymptotic power comparisons of the test with some of its competitors under the restriction of equal sample sizes have been given by Bartholomew [3–5], Puri [21], Johnson and Mehrotra [17], and Berenson and Gross [9]. In these studies, two extreme types of ordered location alternatives are considered—the equally spaced linear trend alternative of the form

$$H_1 : \theta_1 < \theta_2 < \cdots < \theta_k \qquad (5)$$

(where $\theta_{i+1} - \theta_i$ is constant for all $i = 1, 2, \ldots, k - 1$) and the end gap alternative of the form

$$H_1 : \theta_1 = \theta_2 = \cdots = \theta_{k-1} < \theta_k . \qquad (6)$$

The results of these studies indicate that Puri's expected normal scores procedure is excellent for testing against equally spaced linear trend alternatives as in (5). These findings have been corroborated by Berenson [7] in an extensive Monte Carlo* study of the small-sample (empirical) powers of the various k sample tests under the restriction of equal n's. Puri's procedure may be highly recommended over its competitors as a test against linear trend alternatives as in (5), especially when the statistician is known to be sampling from (a) normal populations; (b) light-tailed populations of the beta family, regardless of symmetry; (c) heavy-tailed* and moderately skewed populations (such as Rayleigh and Weibull); and (d) heavy-tailed and very skewed populations (such as Erlang and exponential).

A note of caution: Although the Puri test has been highly recommended for testing against equally spaced ordered location alternatives, one might prefer the procedure developed by Tryon and Hettmansperger [27] for such situations. While the Puri test enjoys (slight) power* advantages, the Tryon–Hettmansperger procedure requires only the ranks (R_{ij}) and not a set of normal scores (ξ_{ij}) so that it possesses more "practical power" as defined by Tukey [28]. In addition, the statistician is typically unaware of the form of the parent population(s) being sampled as well as the form that the true ordered location alternative would take [between the extremes of (5) and (6)]. Thus, in general, from ref. 7 the most highly recommended procedure for testing against unspecified ordered location alternatives of form (2) is the Johnson–Mehrotra test [17].

EXTENSIONS

Skillings and Wolfe [23, 24] have extended the Puri test to the two-way layout by developing an "among-blocks" rank test for randomized complete block designs. While they demonstrate that their statistic has a limiting normal distribution, they do not provide a useful formula for practical applications. However, such an expression is presented in ref. 8.

References

[1] Abelson, R. P. and Tukey, J. W. (1963). *Ann. Math. Statist.*, **34**, 1347–1369.

[2] Bartholomew, D. J. (1959). *Biometrika*, **46**, 36–48.

[3] Bartholomew, D. J. (1959). *Biometrika*, **46**, 328–335.

[4] Bartholomew, D. J. (1961). *Biometrika*, **48**, 325–332.

[5] Bartholomew, D. J. (1961). *J. R. Statist. Soc. B*, **23**, 239–272.

[6] Berenson, M. L. (1978). *Educ. Psychol. Meas.*, **38**, 905–912.

[7] Berenson, M. L. (1982). *Psychometrika*, **47**, 265–280, 535–539.

[8] Berenson, M. L. (1982). *Commun. Statist. Theor. Meth.*, **11**, 1681–1693.

[9] Berenson, M. L. and Gross, S. T. (1981). *Commun. Statist. B*, **10**, 405–431.

[10] Bhuchongkul, S. and Puri, M. L. (1965). *Ann. Math. Statist.*, **36**, 198–202.

[11] Chacko, V. J. (1963). *Ann. Math. Statist.*, **34**, 945–956.

[12] Chernoff, H. and Savage, I. R. (1958). *Ann. Math. Statist.*, **29**, 972–994.

[13] Conover, W. J. (1967). *Ann. Math. Statist.*, **38**, 1726–1730.

[14] Fisher, R. A. and Yates, F. (1963). *Statistical Tables for Biological, Agricultural, and Medical Research*, 6th ed. Hafner, New York.

[15] Harter, H. L. (1961). *Biometrika*, **48**, 151–165.

[16] Hoeffding, W. (1951). *Proc. 2nd Berkeley Symp. Math. Statist. Prob.*, University of California Press, Berkeley, pp. 83–92.

[17] Johnson, R. A. and Mehrotra, K. G. (1971). *J. Indian Statist. Ass.*, **9**, 8–23.

[18] Jonckheere, A. R. (1954). *Biometrika*, **41**, 133–145.

[19] Mann, H. B. and Whitney, D. R. (1947). *Ann. Math. Statist.*, **18**, 50–60.

[20] Puri, M. L. (1964). *Ann. Math. Statist.*, **35**, 102–121.

[21] Puri, M. L. (1965). *Commun. Pure Appl. Math.*, **18**, 51–63.

[22] Savage, I. R. (1956). *Ann. Math. Statist.*, **27**, 590–615.

[23] Skillings, J. H. and Wolfe, D. A. (1977). *Commun. Statist. A*, **6**, 1453–1463.

[24] Skillings, J. H. and Wolfe, D. A. (1978). *J. Amer. Statist. Ass.*, **73**, 427–431.

[25] Terpstra, T. J. (1952). *Nederl. Akad. Wet. Indag. Math.*, **55**, 327–333.

[26] Terry, M. E. (1952). *Ann. Math. Statist.*, **23**, 346–366.

[27] Tryon, P. V. and Hettmansperger, T. P. (1973). *Ann. Statist.*, **1**, 1061–1070.

[28] Tukey, J. W. (1959). *Technometrics*, **1**, 31–48.

[29] Wilcoxon, F. (1945). *Biometrics*, **1**, 80–83.

[30] Wolf, E. H. and Naus, J. I. (1973). *J. Amer. Statist. Ass.*, **68**, 994–997.

(BLOCKS, RANDOMIZED COMPLETE
FISHER–YATES TESTS
ISOTONIC INFERENCE
JONCKHEERE TESTS FOR ORDERED ALTERNATIVES
NORMAL SCORES TESTS
ONE-WAY ANALYSIS OF VARIANCE
ORDER-RESTRICTED INFERENCES
PAGE TEST FOR ORDERED ALTERNATIVES
RANK TESTS
WILCOXON-TYPE TESTS)

M. L. BERENSON

PURPOSIVE SAMPLING *See* REPRESENTATIVE SAMPLING

PURPOSIVE SELECTION *See* REPRESENTATIVE SAMPLING

P VALUES

INTRODUCTION AND DEFINITIONS

A P value is frequently used to report the result of a statistical test of significance. A P value is defined as the probability, calculated under the null distribution, of obtaining a test result as extreme as that observed in the sample (in a particular direction). This definition implies that P values are useful when the alternative hypothesis is one-sided because then the relevant direction is determined by the alternative.

As a simple illustration, suppose that we are testing H_0: $\mu = 50$ vs. $H_1 : \mu > 50$ for a normal population with $\sigma = 10$, and a sample of size 36 gives $\bar{x} = 54$. A value of $\bar{X}$ that exceeds the hypothesized μ value of 50 by a significant amount should make us tend to believe that the alternative $\mu > 50$ is true. Therefore, we might find the probability of obtaining a sample $\bar{X}$ as large as 54, the observed value, when $\mu = 50$. This is the P value, calculated as $P = \Pr[\bar{X} \geqslant 54 \mid \mu = 50] = \Pr[Z \geqslant 2.4] = 0.0082$. If we need to make a decision and this probability is regarded as small, we should reject H_0; if this probability is regarded as large, we should not reject H_0. Alternatively, we might simply report the P value.

The more classical method of carrying out a test of significance is to preselect a significance level α, the probability of a type I error, and use α to determine a critical region for the test statistic. In the example of the preceding paragraph, this classical procedure would lead us to find a number c_α such that $\Pr[\bar{X} \geqslant c_\alpha \mid \mu = 50] = \alpha$. For $\alpha = 0.01$, $c_\alpha = 50 + 1.64(10/\sqrt{36}) = 52.73$. The 0.01-level test is to reject H_0 if $\bar{X} \geqslant 52.73$ and accept H_0 otherwise. With $\bar{x} = 54$, we reject H_0 for $\alpha = 0.01$.

Hence the relationship between P values and the classical method is that if $P \leqslant \alpha$, we should reject H_0, and if $P > \alpha$, we should accept H_0. This relationship shows that an important interpretation of a P value is that it is the smallest level at which the observations are significant in a particular direction. This relationship has prompted some other names for a P value, such as the critical level, borderline level, level attained, or observed level of significance. A P value is also sometimes called a probvalue or an associated probability.

ADVANTAGES AND DISADVANTAGES OF *P* VALUES

In well-behaved problems the P value is well-defined, with a specific probability interpretation. It is not arbitrary, as a preselected significance level usually is in practice, and it is objective. An investigator who can report only a P value conveys the maximum amount of information contained in the sample and permits all readers to essentially choose their own level of significance and make their own individual decision about the null hypothesis. This is especially important when there is no justifiable reason for choosing one particular level of significance.

P values are especially useful for test statistics that have a discrete null distribution because in such cases a preselected significance level frequently is not attainable (except by resorting to randomized tests*). This phenomenon occurs primarily with distribution-free* or nonparametric test statistics, where P values are widely used.

A disadvantage of the P value is that its meaning is sometimes misinterpreted. In a single experiment, the P value does indeed measure the degree of agreement or disagreement in a particular direction between the observations and the null hypothesis, but this interpretation cannot be extended to another experiment (or to another sample size). A lengthy discussion of possible misinterpretations is given in Gibbons and Pratt [1].

Another disadvantage of P values is that they do not take power* (or the probability of a type II error) into account. This criticism, of course, applies equally to classical significance test procedures where sample size is chosen arbitrarily as opposed to being chosen in a manner that would guarantee a preselected value for the probability of each type of error. When P values are used with distribution-free tests, where alternative distributions are not known in general, this criticism is inappropriate.

Perhaps the most persuasive argument against an *exclusive* use of P values is the problem of defining a P value when the appropriate rejection region is in both tails of the null distribution. One approach to a solution here is to simply report a one-tailed P value that reflects the direction in which the sample statistic differs from its expected value under H_0. Another approach is to double the one-tailed P value, but this solution can only be recommended when the null distribution is symmetric (or nearly so), because then it corresponds to the classical procedure that divides the rejection region into two parts of equal probability. Many other possibilities for defining P values for a two-tailed rejection region have been advanced, but most of them can lead to absurdities or at least anomalies; see Gibbons and Pratt [1] for a complete exposition and references.

SUMMARY

P values provide a useful alternative approach to classical inference. They are especially appropriate for distribution-free tests (or any test statistic that has a discrete null distribution) with appropriate critical region in only one tail. They are also useful for one-tailed classical tests when the probability of a type II error is not taken into account, either because there is to be no specific critical region or because of practical considerations. Most modern elementary statistical methods books discuss P values and use them to some extent.

Reference

[1] Gibbons, J. D. and Pratt, J. W. (1975). *Amer. Statist.*, **29**, 20–25; *ibid.*, **31**, 99 (1977).

(DISTRIBUTION-FREE METHODS
HYPOTHESIS TESTING
INFERENCE, STATISTICAL, I, II
RANDOMIZED TESTS
SIGNIFICANCE TESTS, HISTORY)

JEAN D. GIBBONS

PYRAMID SCHEME MODELS

Promoters of chain letters and related pyramid business ventures who promise potential participants that they can earn a substantial return on their investment primarily by recruiting new participants are subject to prosecution under various laws [5] concerning fraud and "misrepresentation," as the pool of potential participants is exhausted before most actual participants recover their initial fee much less make a profit.

One approach to demonstrating this relies on the geometric progression. For example, assuming that each participant obtains two new recruits every month, after n months the total number of participants would be 3^n. After a year and a half ($n = 18$), this number (about 387×10^6) would be larger than the U.S. population, so the process must stop. This approach has been used successfully in some cases, e.g., *State of Connecticut v. Bull Investment Group* [32 Conn., Sup. 279 (1975)]; however, it was rejected as being too far removed from the real world in *Ger-ro-Mar, Inc. v. FTC* [518 F.2d, 33 (1975)]. Furthermore, promoters have modified their business ventures by limiting the total number of active participants or the number of persons a participant may recruit, unless he or she rejoins, in an effort to avoid prosecution.

Stochastic models of such pyramid processes were developed to provide more realistic models of their growth and ultimate extinction. Two models emphasizing how much a participant's potential earnings depends on the time of entry will be described.

A QUOTA PYRAMID SCHEME

The "Golden Book of Values" case ([6] and *State of Connecticut v. Bull Investment Group* [32 Conn., Sup. 279 (1975)]) offered dealerships in the development of a book of discounts to be sold to the public. After paying \$2500 for a dealership, a participant could earn about \$100 for obtaining a merchant to advertise a special discount in the book and could recruit other dealers, receiving \$900 for each one. Thus the recruitment process provided the largest share of the potential earnings described in the brochure. The total number of dealerships sold in an area was limited to N (270 in this case).

To model the number of recruits a participant may make, consider the kth entrant and his or her chance of recruiting the next $(k+1)$st participant. Assuming that the current k participants have an equal chance, the kth entrant has probability $1/k$ of recruiting the $(k+1)$st entrant. Similarly, the kth entrant has probability $1/i$ of recruiting the $(i+1)$st entrant for $i = k, \ldots, N$. Thus the number of recruits is expressible as

$$S_k = \sum_{i=1}^{N-1} X_i, \qquad (1)$$

where X_i is a Bernoulli random variable taking the values 1 and 0 with probabilities $1/i$ and $1 - 1/i$, respectively.

The misrepresentation in the promotional brochure can be demonstrated by the following consequences of (1), derived in refs. 6 and 3.

1. The expected fraction of all participants who recruit *no* one is *one-half*.
2. Participants who enter after the process reaches 37% of its ultimate size cannot expect to recruit more than *one* person.
3. The probability of a participant recruiting sufficiently many new entrants to recoup his or her initial fee, $P[S_k \geqslant 3]$, decreases rapidly with k. Indeed, for k equal to one-third of the ultimate size, $P[S_k \geqslant 3]$ is slightly less than 0.1. Moreover, a Lorenz curve* of recruits by time of entry [7] shows that 33% of all participants are expected to be recruited by the first 10%, while participants entering during the last half of the process collectively are expected to recruit only 15% of the total.
4. For large N (i.e., implicitly assuming that the pyramid develops [6]) the expected fraction of all participants who obtain exactly r recruits tends to $(1/2)^{r+1}$.

This last result was also obtained by Meir and Moon [9], who observed that the graph describing the process (i.e., which connects each participant to his or her recruiter), forms a recursive tree, and participants with exactly r recruits at the nth stage correspond to nodes of degree $r+1$. Hence the result follows from Na and Rapoport [10]. For finite n, an exact expression has been obtained [4].

A PYRAMID LIMITING THE NUMBER OF RECRUITS

A chain letter [2] was circulated with six names on it. A purchaser invests $\$2x$, paying x to the person selling him the letter and $\$x$ to the person at the top of the letter. The purchaser then sells two letters, with his name at the bottom. The promoters note that the purchaser recoups the purchase price when he sells two letters and after his recruits sell their letters and the process continues, he will ultimately receive $\$64x$. Moreover, they asserted that once participants earn money they will rejoin, so the pool of potential buyers is continually replenished.

To describe the chain after the kth letter is sold, denote the number of participants who have sold 0, 1, or 2 letters by U_k, V_k, and W_k, respectively. Assuming that the $U_k + V_k$ persons who can sell the $(k+1)$st letter are equally likely to be successful, the process is a Markov chain.

As $U_k + V_k + W_k$ increase by one at each step and $U_{k+1} - U_k = W_{k+1} - W_k$, $W_k = U_k - c$, and $V_k = k + 2c - 2U_k$. Thus the process is described by the Markov chain $\{U_k\}$ with $U_0 = c$.

Assuming that all participants who sell their two letters rejoin, the mathematical model implies that the process stops before $2n$ letters are sold in a population of n potential participants. Since the number of persons selling the letter must be less than n for the process to continue, $U_k + V_k < n$, while $V_k > 0$, that is, $k + c - U_k < n$, and $2U_k < k + c$, or $U_k < (k+c)/2$. In order for $k + c - U_k < n$ and $U_k < (k+c)/2$ both to hold, $k < 2n - 2c$. Thus the process cannot continue past $2n - 2c$ steps, and consequently only a very small fraction of the participants can possibly gain large amounts of money.

The argument above already implies that most purchasers cannot earn the planned returns. Using the diffusion approximation [1, 2, 11] to the limiting Markov chain, it can be shown that:

1. At the time the process stops a total of approximately $1.62n$ letters will have been sold.
2. At this stopping point, all n persons are behind by either $\$x$ or $\$2x$, depending on whether they have 1 or 2 letters to sell and approximately 61.8% of them are in the second category.

FURTHER DEVELOPMENTS

Multilevel pyramid schemes [5] pay a commission to a participant when his or her descendants (i.e., recruits and their subsequent recruits) are successful. Using a Polya–Eggenberger model [8], one can show [3] that a person joining at the one-third point of the process has only a 30% chance of having three or more descendants (and even fewer direct recruits), so most participants should expect to lose money.

Although the stochastic model described assumes that all current participants are equally likely to recruit the next one, the basic conclusion that most investors lose money remains valid when recruiting ability varies among participants. To further study the robustness of the results implied by these models, more research is needed to model the growth of the process in real time and the fact that unsuccessful participants often drop out. The lifetime of the process is important, as the decision for *Ger-ro-Mar, Inc. v. FTC* [518 F.2d, 33 (1975)] rejected the "geometric progression" argument, in part because the company had existed for nine years and attracted only a few thousand participants.

References

[1] Arnold, L. (1974). *Stochastic Differential Equations: Theory and Applications*. Wiley, New York.

[2] Bhattacharya, P. K. and Gastwirth, J. L. (1983). In *Recent Advances in Statistics; Papers in Honor of Herman Chernoff*, M. H. Rizvi, J. S. Rustagi, and D. Siegmund, eds. Academic Press, New York.

[3] Bhattacharya, P. K. and Gastwirth, J. L. (1983). (To appear in a special issue of *Operations Research*, N. Singpurwalla, ed.)

[4] Dondajewski, M. and Szymanski, J. (1982). *Bull. Acad. Pol. Sci., Sér. Sci. Math.*, **30**(5, 6).

[5] Feore, J. (1974). *Georgetown Law J.*, 1257–1293.

[6] Gastwirth, J. L. (1977). *Amer. Statist.*, **31**, 79–82.

[7] Hammick, (1984). Statistical Properties of Pyramid Schemes and Related Stochastic Processes. (Ph.D. dissertation to be submitted to the George Washington University.)

[8] Johnson, N. L. and Kotz, S. (1977). *Urn Models and Their Application*. Wiley, New York.

[9] Meir, A. and Moon, J. W. (1978). *Utilitas Math.*, **14**, 313–333.

[10] Na, H. S. and Rapoport, A. (1970). *Math. Biosci.*, **6**, 313–329.

[11] Strook, D. W. and Varadhan, S. R. S. (1969). *Commun. Pure Appl. Math.*, **22**, 479–530.

Bibliography

See the following works, as well as the references just given, for more information on the topic of pyramid scheme models.

Bailey, N. T. J. (1957). *The Mathematical Theory of Epidemics*. Charles Griffin, London.

Feller, W. (1957). *An Introduction to Probability Theory and Its Applications, Vol. 1*, 2nd ed. Wiley, New York.

Hodges, J. L., Jr. and LeCam, L. (1960). *Ann. Math. Statist.*, **31**, 737–740.

Percus, O. E. (1983). *Tech. Rep.*, Courant Institute of Mathematical Sciences, New York.

(MARKOV PROCESSES)

J. L. GASTWIRTH

Q

Q-Q PLOTS *See* GRAPHICAL REPRESENTATION OF DATA

QR ALGORITHM

The QR algorithm is the preferred method for calculating all eigenvalues* and (optionally) eigenvectors of a nonsparse matrix. A typical use of eigensystems in statistics is in principal components analysis* [2, 3, 6], in which some or all of the eigensystem of a correlation matrix $\mathbf{X}^T\mathbf{X}$ are computed in attempting to determine a few uncorrelated variables which have associated with them the bulk of the variance of the sample. A related application is the calculation of the singular value decomposition of a matrix $\mathbf{X}$ (*see* LINEAR ALGEBRA, COMPUTATIONAL; ref. 9 or 14). The latter may be used to find the least-squares* solution of the overdetermined system $\mathbf{Xb} = \mathbf{y}$, and hence to solve linear regression* problems. It is especially useful when $\mathbf{X}$ is rank deficient or nearly so. The singular value problem is equivalent to an eigenvalue problem—the singular values of $\mathbf{X}$ are the square roots of the eigenvalues of $\mathbf{X}^T\mathbf{X}$, and the right singular vectors of $\mathbf{X}$ are the eigenvectors of $\mathbf{X}^T\mathbf{X}$. (The left singular vectors of $\mathbf{X}$ are the eigenvectors of $\mathbf{X}\mathbf{X}^T$.) The most widely used algorithm for calculating the singular value decomposition is a variant of the QR algorithm. Because of the equivalence just noted, the principal components problem for $\mathbf{X}^T\mathbf{X}$ can be viewed as a problem of calculating singular values and singular vectors of $\mathbf{X}$. This point of view has considerable merit: If the singular vectors of $\mathbf{X}$ are computed directly, the computed principal components will be more accurate than if $\mathbf{X}^T\mathbf{X}$ is formed and its eigenvectors are then computed [14].

AVAILABILITY OF PROGRAMS

High-quality software for the QR algorithm is widely available. Several ALGOL procedures are given in the Handbook [17]. FORTRAN programs which are mostly translations of the procedures in ref. 17 are given in the EISPACK Guide [13]. FORTRAN programs which use the QR algorithm to calculate the singular value decomposition are given by Businger and Golub [1], Lawson and Hanson [9], and the LINPACK User's Guide [4].

HISTORY

The QR algorithm was introduced independently by Francis [5] and Kublanovskaya [8] in 1961. It is an orthogonal variant of Rutishauser's (1955) LR algorithm [12], itself an outgrowth of the quotient-difference algorithm [11]. Today's implementations of the QR algorithm are not fundamentally different from the algorithms described by Francis [5]. Golub [7] was the first to suggest (in 1968) that the QR algorithm be used to calculate the singular value decomposition. High-quality software soon followed [1, 17].

OTHER ALGORITHMS FOR THE EIGENVALUE PROBLEM

The QR algorithm has supplanted the earlier Jacobi method [17, p. 202]. The slicing (Sturm sequence, bisection) method [17, p. 249] is superior to the QR algorithm when only a few eigenvalues are required. For large, sparse matrices other methods are preferred. In this rapidly evolving area Lanczos methods seem to be the best available at present. Subspace iteration is also widely used. See ref. 10.

DESCRIPTION OF THE QR ALGORITHM

This section describes the algorithm but does not attempt to explain why it works. For a detailed explanation, see ref. 15, in which it is shown that the QR algorithm is just a sophisticated extension of the venerable power method.

Let $\mathbf{A}$ be an n by n real, symmetric, and positive semidefinite matrix. These assumptions simplify the analysis and are valid in typical statistical applications. (Every correlation matrix $\mathbf{X}^T\mathbf{X}$ has these properties.) However, most of the theory can be extended to the nonsymmetric case. The QR algorithm is based on the QR matrix decomposition: Every matrix can be expressed as a product, $\mathbf{A} = \mathbf{QR}$, where $\mathbf{Q}$ is orthogonal and $\mathbf{R}$ is upper triangular [14] (*see* LINEAR ALGEBRA, COMPUTATIONAL). If $\mathbf{A}$ is nonsingular, then $\mathbf{Q}$ and $\mathbf{R}$ can be specified uniquely if we require that the entries on the main diagonal of $\mathbf{R}$ be positive. The basic (unshifted) QR algorithm generates a sequence of matrices $(\mathbf{A}_m)$ as follows: Given $\mathbf{A}_{m-1}$, let $\mathbf{Q}_m$ and $\mathbf{R}_m$ be the orthogonal and upper triangular matrices, respectively, such that $\mathbf{A}_{m-1} = \mathbf{Q}_m\mathbf{R}_m$. Then let $\mathbf{A}_m = \mathbf{R}_m\mathbf{Q}_m$. In summary,

$$\mathbf{A}_{m-1} = \mathbf{Q}_m\mathbf{R}_m, \qquad \mathbf{R}_m\mathbf{Q}_m = \mathbf{A}_m. \quad (1)$$

The algorithm starts with $\mathbf{A}_0 = \mathbf{A}$. The matrices in the sequence $(\mathbf{A}_m)$ are orthogonally similar since $\mathbf{A}_m = \mathbf{Q}_m^T\mathbf{A}_{m-1}\mathbf{A}_m$, so they all have the same eigenvalues as $\mathbf{A}$. Under most conditions [10, 15, 16] the matrices will converge to diagonal form, exposing the eigenvalues on the main diagonal.

The eigenvectors can be found by accumulating the transforming matrices $\mathbf{Q}_m$. Letting $\hat{\mathbf{Q}}_m = \mathbf{Q}_1\mathbf{Q}_2 \cdots \mathbf{Q}_m$, we have $\mathbf{A}_m = \hat{\mathbf{Q}}_m^T\mathbf{A}\hat{\mathbf{Q}}_m$. For m sufficiently large, $\mathbf{A}_m$ will be effectively diagonal, from which it follows that the columns of $\hat{\mathbf{Q}}_m$ form a complete orthonormal set of eigenvectors of $\mathbf{A}$.

The basic QR algorithm is too inefficient to be competitive. The two reasons are that (a) the cost of each QR step in (1) is high, and (b) the rate of convergence of $(\mathbf{A}_m)$ is usually slow. The arithmetic cost of the QR decomposition is about $\frac{2}{3}n^3$ operations. (By an *operation* we mean one floating-point multiplication and one floating-point addition.) The cost of the second step in (1) is about $\frac{1}{6}n^3$ operations. These costs can be decreased dramatically by first reducing $\mathbf{A}$ to tridiagonal form by an orthogonal similarity transformation. This can be done in about $\frac{2}{3}n^3$ operations using elementary reflectors (Householder transformations), and it only has to be done once because the tridiagonal form is preserved by the QR algorithm [10]. The cost of one iteration of the QR algorithm for a tridiagonal matrix is of order n rather than n^3.

The problem of slow convergence is solved by making *shifts of origin*. The shifted

QR algorithm has the form

$$\begin{aligned} \mathbf{A}_{m-1} - \sigma_{m-1}\mathbf{I} &= \mathbf{Q}_m\mathbf{R}_m, \\ \mathbf{R}_m\mathbf{Q}_m + \sigma_{m-1}\mathbf{I} &= \mathbf{A}_m, \end{aligned} \tag{2}$$

where the *shift* σ_{m-1} is chosen to approximate an eigenvalue of $\mathbf{A}$. Let $a_{ij}^{(k)}$ denote the (i, j) entry of $\mathbf{A}_k$. Since $a_{nn}^{(m)}$ approaches an eigenvalue as $m \to \infty$, a good choice is $\sigma_m = a_{nn}^{(m)}$, the *Rayleigh quotient shift* [10]. A better choice is the *Wilkinson shift*, which takes σ_m to be that eigenvalue of the lower right hand 2 by 2 submatrix of $\mathbf{A}_m$ which is closer to $a_{nn}^{(m)}$. With this shift the algorithm is guaranteed to converge [10]. With either shift convergence to the nth eigenvalue is of third order (cubic), where it is only of first order (linear) in the unshifted case. Numerous other shifting strategies have been proposed [10]. The eigenvalue $\lambda_n \approx a_{nn}^{(m)}$ can be regarded as found as soon as $|a_{n,n-1}^{(m)}|$ is sufficiently small. At that point the problem is *deflated* by deletion of the last row and column of $\mathbf{A}_m$, and the search continues with the focus shifted to the next eigenvalue. With the Wilkinson shift the average number of QR steps required per eigenvalue is less than two.

Example. We will calculate the eigenvalues of the 2 by 2 matrix

$$\mathbf{A} = \begin{bmatrix} 8 & 2 \\ 2 & 5 \end{bmatrix},$$

which are easily seen to be $\lambda_1 = 9$ and $\lambda_2 = 4$. We consider the unshifted algorithm first; $\mathbf{A} = \mathbf{A}_0 = \mathbf{QR}$, where

$$\mathbf{Q} = \frac{1}{\sqrt{68}}\begin{bmatrix} 8 & -2 \\ 2 & 8 \end{bmatrix}, \quad \mathbf{R} = \frac{1}{\sqrt{68}}\begin{bmatrix} 68 & 26 \\ 0 & 36 \end{bmatrix}.$$

$\mathbf{Q}$ is orthogonal and $\mathbf{R}$ is upper triangular. Then

$$\mathbf{A}_1 = \mathbf{RQ} = \frac{1}{68}\begin{bmatrix} 596 & 72 \\ 72 & 288 \end{bmatrix}$$

$$\doteq \begin{bmatrix} 8.7647 & 1.0588 \\ 1.0588 & 4.2353 \end{bmatrix}.$$

Notice that $\mathbf{A}_1$ is closer to diagonal form than $\mathbf{A}_0$ is, in the sense that its off-diagonal entries are closer to zero, also that the main diagonal entries of $\mathbf{A}_1$ are closer to the eigenvalues. On subsequent iterations the main diagonal entries of $\mathbf{A}_i$ give better and better estimates of the eigenvalues until after 10 iterations they agree with the true eigenvalues to seven decimal places. The rate of convergence is governed by the ratio of the eigenvalues, which is $\frac{4}{9}$ in this case. If this ratio were closer to 1, then more iterations would be required. If it were closer to 0, fewer iterations would be needed. We now consider the shifted algorithm. The objective of shifting is to shift the eigenvalues in such a way that the ratio of the shifted eigenvalues is closer to zero. The Rayleigh quotient shift of $\mathbf{A}$ is $\sigma_0 = 5$. Then the eigenvalues of $\mathbf{A} - \sigma_0\mathbf{I}$ are $\lambda_1 - 5 = 4$ and $\lambda_2 - 5 = -1$. The new ratio is $-\frac{1}{4}$, which is closer to 0 than $\frac{4}{9}$ is. $\mathbf{A} - \sigma_0\mathbf{I} = \mathbf{QR}$, where

$$\mathbf{Q} = \frac{1}{\sqrt{13}}\begin{bmatrix} 3 & 2 \\ 2 & -3 \end{bmatrix}, \qquad \mathbf{R} = \frac{1}{\sqrt{13}}\begin{bmatrix} 13 & 6 \\ 0 & 4 \end{bmatrix}.$$

Then

$$\mathbf{A}_1 = \mathbf{RQ} + \sigma_0\mathbf{I} = \frac{1}{13}\begin{bmatrix} 51 & 8 \\ 8 & -12 \end{bmatrix} + \begin{bmatrix} 5 & 0 \\ 0 & 5 \end{bmatrix}$$

$$\doteq \begin{bmatrix} 8.9231 & 0.6154 \\ 0.6154 & 4.0769 \end{bmatrix}.$$

Notice that the main diagonal entries of $\mathbf{A}_1$ are quite close to the eigenvalues. On the next step the Rayleigh quotient shift is $\sigma_1 = 4.0769$. The eigenvalues of $\mathbf{A}_1 - \sigma_1\mathbf{I}$ are $\lambda_1 - \sigma_1 = 4.9231$ and $\lambda_2 - \sigma_1 = -0.0769$, whose ratio is very close to zero. Therefore, $\mathbf{A}_2$ is expected to be substantially better than $\mathbf{A}_1$. In fact,

$$\mathbf{A}_2 \doteq \begin{bmatrix} 8.999981 & 0.009766 \\ 0.009766 & 4.000019 \end{bmatrix}.$$

On the next step the shift is $\sigma_2 = 4.000019$, which gives an even better ratio, and

$$\mathbf{A}_3 \doteq \begin{bmatrix} 9.000000 & 3.7 \times 10^{-7} \\ 3.7 \times 10^{-7} & 4.000000 \end{bmatrix}.$$

It is evident from this example that shifting introduces a positive feedback mechanism which accelerates convergence dramatically. The Wilkinson shift is even better than the Rayleigh quotient shift; it chooses the shift based on a 2 by 2 submatrix, which would be the whole matrix in this case. The shift so chosen would be exactly an eigenvalue, and the algorithm would converge in one iteration.

NOTES ON IMPLEMENTATIONS

Of the many ways of carrying out the computations there are two basic variants. The *explicit* QR algorithm performs each QR step in a straightforward manner, essentially as shown in (2). The *implicit* QR algorithm obtains the same result by a different sequence of operations [10, 14]. In both cases the transforming matrix $\mathbf{Q}_m$ is built up as a product of elementary reflectors (Householder transformations) or plane rotators (Givens transformations).

Real nonsymmetric matrices may have complex eigenvalues, which always occur in conjugate pairs. It is therefore desirable to use complex shifts. A variant of the implicit QR algorithm called the *double* QR algorithm effectively takes two QR steps at once, using complex conjugate shifts χ and $\bar{\chi}$. Not only is the resulting matrix real, but the entire double step is carried out in real arithmetic. The double QR algorithm converges not to diagonal form, but to block upper triangular form, with a 2 by 2 block on the main diagonal for each complex conjugate pair of eigenvalues.

Another variant of the implicit QR algorithm is used to find the singular values of $\mathbf{X}$. First $\mathbf{X}$ is reduced by elementary reflectors to an upper bidiagonal matrix $\mathbf{Y}$. Then $\mathbf{Y}^T\mathbf{Y}$ is tridiagonal, and the square roots of the eigenvalues of $\mathbf{Y}^T\mathbf{Y}$ are the singular values of $\mathbf{Y}$ (and of $\mathbf{X}$). $\mathbf{Y}$ is further reduced to diagonal form (exposing the singular values on the main diagonal) by a sequence of transformations which is equivalent to the QR algorithm applied to $\mathbf{Y}^T\mathbf{Y}$. Thus the algorithm effectively calculates the eigenvalues of $\mathbf{Y}^T\mathbf{Y}$ without ever forming $\mathbf{Y}^T\mathbf{Y}$.

References

[1] Businger, P. A. and Golub, G. H. (1969). *Numer. Math.*, **7**, 269–276. (A FORTRAN program to calculate the singular value decomposition.)

[2] Chambers, J. M. (1977). *Computational Methods for Data Analysis*. Wiley, New York.

[3] Dempster, A. P. (1969). *Elements of Continuous Multivariate Analysis*. Addison-Wesley, Reading, Mass.

[4] Dongarra, J. J. et al. (1979). *LINPACK Users' Guide*. SIAM, Philadelphia. (Contains high-quality FORTRAN programs for systems of linear equations and least-squares problems. Includes a singular value decomposition program.)

[5] Francis, J. G. F. (1961/1962). *Computer J.*, **4**, 265–271, 332–345.

[6] Gnanadesikan, R. (1977). *Methods of Statistical Data Analysis of Multivariate Observations*. Wiley, New York.

[7] Golub, G. H. (1968). *Apl. Mat.*, **13**, 44–51.

[8] Kublanovskaya, V. N. (1961). *Ž. Vyčisl. Mat. Mat. Fiz.*, **1**, 555–570; *USSR Com. Math. Math. Phys.*, **3**, 637–657.

[9] Lawson, C. L. and Hanson, R. J. (1974). *Solving Least Squares Problems*. Prentice-Hall, Englewood Cliffs, N.J.

[10] Parlett, B. N. (1980). *The Symmetric Eigenvalue Problem*. Prentice-Hall, Englewood Cliffs, N.J. (Discusses many recent innovations.)

[11] Rutishauser, H. (1954). *Zeit. angew. Math. Phys.*, **5**, 233–251.

[12] Rutishauser, H. (1955). *C. R. Acad. Sci. Paris*, **240**, 34–36.

[13] Smith, B. T. et al. (1976). *Matrix Eigensystem Routines—EISPACK Guide*, 2nd ed. Springer-Verlag, New York. (Contains FORTRAN listings which are mostly translations of the ALGOL procedures given in ref. 17.)

[14] Stewart, G. W. (1973). *Introduction to Matrix Computations*. Academic Press, New York. (The leading textbook on numerical linear algebra.)

[15] Watkins, D. S. (1982). *SIAM Rev.*, **24**, 427–440. (Tells what the QR algorithm is and why it works.)

[16] Wilkinson, J. H. (1965). *The Algebraic Eigenvalue Problem*. Clarendon Press, Oxford. (A rich mine of information. Difficult to penetrate.)

[17] Wilkinson, J. H. and Reinsch, C. (1971). *Handbook for Automatic Computation*, Vol. 2: *Linear Algebra*. Springer-Verlag, New York. (Contains ALGOL procedures for the QR and other algorithms.)

(EIGENVALUE
EIGENVECTOR
LINEAR ALGEBRA, COMPUTATIONAL)

D. S. WATKINS

QR FACTORIZATION *See* LINEAR ALGEBRA, COMPUTATIONAL

QUADE TEST FOR GENERAL ALTERNATIVES *See* WILCOXON-TYPE TESTS FOR ORDERED ALTERNATIVES IN RANDOMIZED BLOCKS

QUADRANT DEPENDENCE *See* DEPENDENCE, CONCEPTS OF

QUADRATIC FORMS

Let $y_1, \ldots, y_n$ be n real random variables and a_{ij}, $i, j = 1, \ldots, n$ be n^2 given real numbers. Put $\mathbf{y} = (y_1, \ldots, y_n)'$ and let $\mathbf{A}$ be the $n \times n$ matrix whose (i, j)th element is $(a_{ij} + a_{ji})/2$. Then $Q = \sum_1^n \sum_1^n a_{ij} y_i y_j = \mathbf{y}'\mathbf{A}\mathbf{y}$ is a *quadratic form* in $y_1, \ldots, y_n$. If the y_i's are independently distributed, then the rank of $\mathbf{A}$ is defined as the degrees of freedom (d.f.) of Q. If $\mathbf{A}$ is positive definite or semipositive definite, Q is a *definite form*; it is an *indefinite form* if some eigenvalues* of $\mathbf{A}$ are positive and some eigenvalues of $\mathbf{A}$ negative. Q is a *central form* if $E(y_j) = 0$ for all j and a *noncentral form* if $E(y_j) \neq 0$ for some j. If the y_j's are $p \times 1$ column vectors $\mathbf{y}_j$, one has then a p-dimensional multivariate quadratic form $\mathbf{Q} = \mathbf{YAY}'$, where $\mathbf{Y} = (\mathbf{y}_1, \ldots, \mathbf{y}_n)$ is the $p \times n$ matrix with jth column $\mathbf{y}_j$.

Quadratic forms arise very often in many statistical procedures (see ref. 26), most important, in regression analysis and in analysis of variance* and covariance models (*see* GENERAL LINEAR MODEL). This follows from the fact that all sums of squares can be expressed as definite quadratic forms. The sums of squares divided by their respective degrees of freedom are the *mean squares*. As a simple illustration, consider a one-way fixed-effects* model:

$$y_{ij} = \mu + \alpha_i + e_{ij},$$
$$i = 1, \ldots, k, \qquad j = 1, \ldots, n_i,$$

where the parameters α_i, $i = 1, \ldots, k$, satisfy the restriction $\sum_1^k n_i \alpha_i = 0$. Put $n = \sum_1^k n_i$,

$$\bar{y}_i = \left(\sum_{j=1}^{n_i} y_{ij}\right) \Big/ n_i, \qquad \bar{y} = \left(\sum_{i=1}^{k} n_i \bar{y}_i\right) \Big/ n,$$
$$\mathbf{y}' = (\mathbf{y}'_1, \ldots, \mathbf{y}'_k),$$
$$\text{with } \mathbf{y}_i = (y_{i1}, \ldots, y_{in_i})'.$$

Denote by $\mathbf{J}_n$ the $n \times n$ matrix of 1's and by

$$\sum_1^k \oplus \mathbf{A}_i = \operatorname{diag}(\mathbf{A}_1, \ldots, \mathbf{A}_k)$$

the direct sum of matrices $\mathbf{A}_1, \ldots, \mathbf{A}_k$. Then the sums of squares S_μ, S_α, and S_e for $\mu, (\alpha_1, \ldots, \alpha_k)$ and $(e_{ij}, i = 1, \ldots, k, j = 1, \ldots, n_i)$ are given, respectively, by

$$S_\mu = n\bar{y}^2 = \mathbf{y}'\mathbf{B}_1\mathbf{y},$$
$$S_\alpha = \sum_1^k n_i(\bar{y}_i - \bar{y})^2 = \mathbf{y}'\mathbf{B}_2\mathbf{y},$$
$$S_e = \sum_1^k \sum_1^{n_i} (y_{ij} - \bar{y}_i)^2 = \mathbf{y}'\mathbf{B}_3\mathbf{y},$$

where

$$\mathbf{B}_1 = \frac{1}{n}\mathbf{J}_n, \qquad \mathbf{B}_2 = \mathbf{I}_n - \left\{\sum_1^k \oplus \left(\frac{1}{n_i}\mathbf{J}_{n_i}\right)\right\},$$
$$B_3 = \mathbf{I}_n - \left\{\sum_1^k \oplus \left(\frac{1}{n_i}\mathbf{J}_{n_i}\right)\right\}.$$

The degrees of freedoms for S_μ, S_α, and S_e are given, respectively, by $\operatorname{rank}\mathbf{B}_1 = 1$, $\operatorname{rank}\mathbf{B}_2 = k - 1$, and $\operatorname{rank}\mathbf{B}_3 = n - k$.

SAMPLING DISTRIBUTIONS OF QUADRATIC FORMS FROM NORMAL UNIVERSES

Fisher [15] noted that the sampling distributions of sample variances from normal universes are scaled chi-square distributions*. Cochran [6] showed that when sampling from normal universes, the distribution and independence of quadratic forms are related to their degrees of freedom*, a result referred to later as *Cochran's theorem*. After this prior work, at least 150 papers have been published concerning distribution theories of quadratic forms; most of these publications have been listed in Johnson and Kotz [26], Anderson et al. [2], Scarowsky [39], Khatri [30, 31], Jensen [25], Rice [37], Ruben [38], Anderson and Styan [1], Lize van der Merwe and Crowther [33], and Tan [49]. For a brief summary, let $x_1, \ldots, x_n$ be a random sample from a normal universe with mean μ and variance σ^2; further, let

"$\sim$" denote "distributed as" and $\chi_f^2(\delta^2)$ a noncentral chi-square distribution* with f d.f. and noncentrality δ^2. Then the following give some of the important results concerning the distributions of quadratic forms $Q = \mathbf{x}'\mathbf{A}\mathbf{x}$ and $Q_i = \mathbf{x}'\mathbf{A}_i\mathbf{x}$, $i = 1, \ldots, k$, from normal universes. Extensions to singular normal distributions can be found in Good [16], Shanbhag [40], Styan [42], Rao and Mitra [36], Tan [45], and Khatri [27, 30]. Multivariate extensions can be found in Hogg [22] and Khatri [27–31].

1. $Q \sim \sigma^2\chi_f^2(\delta^2)$, where $f = \text{rank}\,\mathbf{A}$ and $\delta^2 = \mu^2(\mathbf{1}_n'\mathbf{A}\mathbf{1}_n)/\sigma^2$, iff $\mathbf{A}^2 = \mathbf{A}$.
2. If $\mathbf{A}^2 = \mathbf{A}$, then $Q \sim \sigma^2\chi_f^2(\delta^2)$ iff the universe is normal.
3. $Q_1, \ldots, Q_k$ $(k \geqslant 2)$ are independently distributed of one another iff $\mathbf{A}_i\mathbf{A}_j = \mathbf{0}$ for all $i \neq j$.
4. Suppose that $\mathbf{A} = \sum_1^k \mathbf{A}_i$ and consider the following conditions (*see also* IDEMPOTENT MATRIX):
   (a) $f = \sum_1^k f_i$, $f_i = \text{rank}\,\mathbf{A}_i$, $i = 1, \ldots, k$.
   (b) $Q \sim \sigma^2\chi_f^2(\delta^2)$.
   (c) $Q_i \sim \sigma^2\chi_{f_i}^2(\delta_i^2)$, $\delta_i^2 = \mu^2(\mathbf{1}_n'\mathbf{A}_i\mathbf{1}_n)/\sigma^2$, $i = 1, \ldots, k$.
   (d) $Q_1, \ldots, Q_k$ $(k \geqslant 2)$ are independently distributed of one another.

   Then (a) and (b) imply (c) and (d); and any two conditions of (b), (c), and (d) imply all the conditions.
5. Let $\lambda_1 > \lambda_2 > \cdots > \lambda_l$ be all the distinct eigenvalues of $\mathbf{A}$ with multiplicities $r_1, r_2, \ldots, r_l$, respectively, $(\sum_1^l r_i = n)$. Then $Q \sim \sum_1^l \lambda_i \chi_{r_i}^2(\Delta_i^2)$, a linear combination of independently distributed noncentral chi-square variables $(\chi_{r_i}^2(\Delta_i^2))$, where

$$\Delta_i^2 = \mu^2(\mathbf{1}_n'\mathbf{E}_i\mathbf{1}_n)/\sigma^2,$$

$$\mathbf{E}_i = \prod_{j \neq i} \left\{ \frac{1}{(\lambda_i - \lambda_j)} (\mathbf{I}_n - \lambda_j \mathbf{A}) \right\}, \quad i = 1, \ldots, l.$$

As a simple illustration for the applications of the results above, consider again the one-way fixed-effects model as given above. Assume now that the e_{ij}'s are distributed independently as normal with expectation 0 and variance σ^2. Then, with $\boldsymbol{\alpha} = (\alpha_1, \ldots, \alpha_k)'$, $\mathbf{y} \sim N_n(\mathbf{u}, \sigma^2\mathbf{I}_n)$, where

$$\mathbf{u} = \mu\mathbf{1}_n + \left(\sum_1^k \oplus \mathbf{1}_{n_i}\right)\boldsymbol{\alpha},$$

$\mathbf{1}_n$ being a $n \times 1$ column of 1's. Now, $\mathbf{u}'\mathbf{B}_3\mathbf{u} = 0$, $\mathbf{u}'\mathbf{B}_2\mathbf{u} = \sum_1^k n_i\alpha_i^2$, and $\mathbf{B}_2\mathbf{B}_3 = \mathbf{0}$. Further, $\mathbf{B}_2^2 = \mathbf{B}_2$ and $\mathbf{B}_3^2 = \mathbf{B}_3$. Thus S_α and S_e are distributed independently of each other; $S_e \sim \sigma^2\chi_{(n-k)}^2(0) = \sigma^2\chi_{n-k}^2$, and $S_\alpha \sim \sigma^2\chi_{(k-1)}^2(\delta_2^2)$, where $\delta_2^2 = (\sum_1^k n_i\alpha_i^2)/\sigma^2$. It follows that

$$(n-k)S_\alpha/\{(k-1)S_e\} \sim F(k-1, n-k; \delta_2^2),$$

a noncentral F distribution* with $(k-1, n-k)$ d.f. and noncentrality parameter δ_2^2.

SERIES EXPANSIONS AND COMPUTATIONS OF DISTRIBUTION FUNCTIONS

Various series expansions of quadratic forms have been discussed in Johnson and Kotz [26], including the power series expansion, the chi-square and the noncentral chi-square expansion, and the Laguerre polynomial expansion. The latter two are the most commonly used expansions for computing the distribution functions of definite quadratic forms. Depending on the moments of the first term, Laguerre series* expansions may further be classified as one moment series, two moments series, and three moments series, leading to different approximations of the distribution functions of definite quadratic forms [56]. Alternative competitive approximations of definite quadratic forms have been the Wilson–Hilferty approximation* [34] and the beta-chi-square approximation [33]. In connection with analysis of variance models, the two moments Laguerre series have been used extensively for deriving approximations to the F-ratio from normal and nonnormal universes for two basic reasons (see refs. 47, 48, 52, and 54): (a) the first term of the series provides a basis for the Satterthwaite approximation (see ref. 52); and (b) it appears to be more convenient mathematically and computationally than other Laguerre polynomial expansion approximations.

As a simple example, consider two positive-definite quadratic forms Q_1 and Q_2 with $EQ_i > 0$, and put $\rho_i = \text{var}(Q_i)/EQ_i$ and $b_i = (EQ_i)/\rho_i$, $i = 1, 2$. Then the two moments series leads to the following approximation for $P_r\{Q_1 \leqslant zQ_2\}$:

$$P_r\{Q_1 \leqslant zQ_2\}$$

$$\simeq I_w(b_1, b_2) + \sum_{i=3}^{k}\sum_{u=0}^{i}\binom{i}{u}(-1)^u C_{iu}(b_1, b_2; w)$$

$$+ \sum_{i=1}^{k-1}\sum_{j=1}^{k-i} d_{ij} D_{ij}(b_1, b_2; w),$$

where

$$C_{iu}(b_1, b_2; w) = d_{i0} I_w(b_1 + u, b_2) + d_{0i} I_w(b_1, b_2 + u),$$

$$D_{iu}(b_1, b_2; w) = \sum_{s=0}^{i}\sum_{t=0}^{j}\binom{i}{s}\binom{j}{t}(-1)^{s+t} I_w(b_1 + s, b_2 + t),$$

where $k \geqslant 1$ is a given integer ($\sum_{i=3}^{k}$ is defined as 0 if $k < 3$), $w = \rho_2 z/(\rho_2 z + \rho_1)$, and

$$I_w(f_1, f_2) = \int_0^w x^{f_1 - 1}(1 - x)^{f_2 - 1}\, dx / B(f_1, f_2),$$

the incomplete beta* integral; the coefficient d_{ij} is the expected value of the product of Laguerre polynomials $L_i^{(b_1)}(Q_1/\rho_1)$ and $L_j^{(b_2)}(Q_2/\rho_2)$ and has been expressed as simple functions of cumulants and mixed cumulants of Q_1 and Q_2 in Tan and Cheng [51]. Formulas for mixed cumulants and cumulants of quadratic forms and bilinear forms up to order 4 have been given in Tan and Cheng [50]; using these formulas d_{ij} can readily be computed. As shown in refs. 47, 48, 51, and 54, the approximation above would appear to work well for definite forms; however, its application to indefinite forms in general is quite difficult (for distribution theories, see Johnson and Kotz [26] and Khatri [30, 31]). Press [35] and Harville [18, 19] have expressed the distributions of the difference of chi-square variables in terms of special functions. Realizing the difficulty in using Laguerre expansion approximations, Bayne and Tan [5] used the Pearson system* to fit distributions of differences of chi-square variables.

Multivariate Extension of Series Expansions

Series expansions for multivariate quadratic forms have been derived and discussed in Crowther [7], Hayakawa [20, 21], Khatri [28, 29], and Krishnaiah and Walker [32]. These expansions involve zonal polynomials* and Laguerre polynomials in matrix argument; computations for these series are quite difficult and further complicated by the slow convergence of series involving zonal polynomials.

Computation by Numerical Methods

Computations of the distribution functions of quadratic forms using numerical methods have been developed by Davis [9] and Sheil and O'Muircheartaigh [41] for positive-definite forms and by Davis [13] and Rice [37] for general forms. Earlier works along this line for definite and indefinite quadratic forms were given in Grad and Solomon [17] and Imhof [23].

QUADRATIC FORMS FROM NONNORMAL UNIVERSES

Quadratic forms from nonnormal universes have been considered by Atiqullah [3, 4], Subrahmanian [43, 44], Tiku [55, 57, 58], Ito [24], Davis [12, 14], Tan [46], and Tan and Wong [53]. David and Johnson [8] have derived mixed cumulants up to order 4 of between groups sums of squares and within groups sums of squares in one-way fixed-effects models. See Tan and Cheng [50] for mixed cumulants of bilinear and quadratic forms. Davis [11] and Tan [49] have shown that if the cumulants are finite, sums of squares in analysis-of-variance models can be expanded formally in terms of products of Laguerre polynomials and gamma densities. This suggests that Laguerre polynomial expansions can be used to approximate joint distributions of quadratic forms, especially definite forms, from nonnormal universes.

Using Laguerre polynomial expansions with $k = 4$, Tiku [55, 57] derived approximations to the null and nonnull distributions of the F-ratio in one-way fixed-effects models and evaluated the effects of departure from normality of the F-test. Similar approximations with $m = 4$ to the null and nonnull sampling distribution of the F-ratio in one-way random effects models were derived by Tan and Wong [54]. A summary of the robustness of these tests has been given in Tan [49]. Some multivariate extensions of these approximations have been given in Davis [12, 14].

References

[1] Anderson, T. W. and Styan, G. H. P. (1982). In *Statistics and Probability: Essays in Honor of C. R. Rao*, G. Kallianpur, P. R. Krishnaiah, and J. K. Ghosh, eds. North-Holland, Amsterdam, pp. 1–23.

[2] Anderson, T. W., Das Gupta, S. and Styan, G. P. H. (1972). *A Bibliography of Multivariate Statistical Analysis*. Oliver & Boyd, Edinburgh.

[3] Atiqullah, M. (1962). *J. R. Statist. Soc. B*, **24**, 140–147.

[4] Atiqullah, M. (1962). *Biometrika*, **49**, 83–91.

[5] Bayne, C. K. and Tan, W. Y. (1981). *Commun. Statist. A*, **10**, 2315–2326.

[6] Cochran, W. G. (1934). *Proc. Camb. Philos. Soc.*, **30**, 178–191.

[7] Crowther, C. G. (1975). *S. Afr. Statist. J.*, **9**, 27–36.

[8] David, F. N. and Johnson, N. L. (1951). *Biometrika*, **38**, 43–57.

[9] Davis, R. B. (1973). *Biometrika*, **60**, 415–417.

[10] Davis, A. W. (1976). *Biometrika*, **63**, 661–670.

[11] Davis, A. W. (1977). *J. Amer. Statist. Ass.*, **72**, 212–214.

[12] Davis, A. W. (1980). *Biometrika*, **67**, 419–427.

[13] Davis, R. B. (1980). *Appl. Statist.*, **29**, 323–333.

[14] Davis, A. W. (1982). *J. Amer. Statist. Ass.*, **77**, 896–900.

[15] Fisher, R. A. (1925). *Metron*, **5**, 90–104.

[16] Good, J. J. (1969). *Biometrika*, **56**, 215–216; correction: *Biometrika*, **57**, 225 (1970).

[17] Grad, A. and Solomon, H. (1955). *Ann. Math. Statist.*, **26**, 464–477.

[18] Harville, D. A. (1971). *Ann. Math. Statist.*, **42**, 809–811.

[19] Harville, D. A. (1971). *ARL71-0131 No. 7071*. Air Force Systems Command, U.S. Air Force, Wright-Patterson Air Force Base, Ohio.

[20] Hayakawa, T. (1966). *Ann. Inst. Statist. Math.*, **18**, 191–201.

[21] Hayakawa, T. (1973). *Ann. Inst. Statist. Math.*, **25**, 205–230.

[22] Hogg, R. V. (1963). *Ann. Math. Statist.*, **34**, 935–939.

[23] Imhof, J. P. (1961). *Biometrika*, **48**, 419–426.

[24] Ito, K. (1980). In *Handbook of Statistics, Vol. 1*, P. R. Krishnaiah, ed. North-Holland, Amsterdam, pp. 199–236.

[25] Jensen, D. R. (1982). *SIAM J. Appl. Math.*, **32**, 297–301.

[26] Johnson, N. L. and Kotz, S. (1970). *Distributions in Statistics*, Vol. 2: *Continuous Univariate Distributions*. Wiley, New York.

[27] Khatri, C. G. (1968). *Sankhyā A*, **30**, 267–280.

[28] Khatri, C. G. (1971). *J. Multivariate Anal.*, **1**, 199–214.

[29] Khatri, C. G. (1977). *S. Afr. Statist. J.*, **11**, 167–179.

[30] Khatri, C. G. (1980). In *Handbook of Statistics*, Vol. 1, P. R. Krishnaiah, ed. North-Holland, Amsterdam, pp. 443–469.

[31] Khatri, C. G. (1982). In *Statistics and Probability: Essays in Honor of C. R. Rao*, G. Kallianpur, P. R. Krishnaiah, and J. K. Ghosh, eds. North-Holland, Amsterdam, pp. 411–417.

[32] Krishnaiah, P. R. and Walker, V. B. (1973). *Commun. Statist.*, **1**, 371–380.

[33] Merwe, L. van der and Crowther, N. A. S. (1982). *S. Afr. Statist. J.*, **16**, 133–145.

[34] Mudholkar, G. S. and Trevidi, M. C. (1981). *J. Amer. Statist. Ass.*, **76**, 479–485.

[35] Press, S. J. (1966). *Ann. Math. Statist.*, **37**, 480–487.

[36] Rao, C. R. and Mitra, S. K. (1971). *Generalized Inverse of Matrices and its Applications*. Wiley, New York.

[37] Rice, S. O. (1980). *SIAM J. Sci. Statist. Comput.*, **1**, 438–448.

[38] Ruben, H. (1978). *Sankhyā A*, **40**, 156–173.

[39] Scarowsky, I. (1973). M.S. thesis, McGill University, Montreal.

[40] Shanbhag, D. N. (1968). *Biometrika*, **55**, 593–595.

[41] Sheil, J. and O'Muircheartaigh, I. (1977). *Appl. Statist.*, **26**, 92–98.

[42] Styan, G. P. H. (1970). *Biometrika*, **57**, 567–572.

[43] Subrahmanian, K. (1966). *Sankhyā A*, **28**, 389–406.

[44] Subrahmanian, K. (1968). *Sankhyā A*, **30**, 411–432.

[45] Tan, W. Y. (1977). *Canad. J. Statist.*, **5**, 241–250.

[46] Tan, W. Y. (1980). *S. Afr. Statist. J.*, **14**, 47–59.

[47] Tan, W. Y. (1982). *Commun. Statist. Theor. Meth.*, **11**, 731–751.

[48] Tan, W. Y. (1982). *J. Statist. Comp. Simul.*, **16**, 35–55.

[49] Tan, W. Y. (1982). *Commun. Statist. Theor. Meth.*, **11**, 2482–2513.

[50] Tan, W. Y. and Cheng, S. S. (1981). *Commun. Statist. A*, **10**, 283–298.

[51] Tan, W. Y. and Cheng, S. S. (1984). *Sankhyā B*, **46**, 188–200.

[52] Tan, W. Y. and Wong, S. P. (1977). *J. Amer. Statist. Ass.*, **72**, 875–881.

[53] Tan, W. Y. and Wong, S. P. (1977). *Sankhyā B*, **30**, 245–257.

[54] Tan, W. Y. and Wong, S. P. (1980). *J. Amer. Statist. Ass.*, **75**, 655–662.

[55] Tiku, M. L. (1964). *Biometrika*, **51**, 83–95.

[56] Tiku, M. L. (1965). *Biometrika*, **52**, 415–427.

[57] Tiku, M. L. (1972). *J. Amer. Statist. Ass.*, **66**, 913–916.

[58] Tiku, M. L. (1975). In *Applied Statistics*, R. P. Gupta, ed. American Elsevier, New York.

(CHI-SQUARE DISTRIBUTION
GENERAL LINEAR MODEL
IDEMPOTENT MATRIX
MULTINORMAL DISTRIBUTION
WISHART DISTRIBUTION)

W. Y. TAN

QUADRATIC VARIANCE FUNCTIONS

See NATURAL EXPONENTIAL FAMILIES

QUADRAT SAMPLING

A quadrat is a small-sample plot used by ecologists investigating populations of sessile or sedentary organisms, for example plants in a meadow, or benthic organisms on shallow sea or lake floors. A quadrat is usually (but not necessarily) square and is often (but not always) 1 meter square in size.

Quadrat sampling is usually done for one of two reasons: (1) to estimate the size of a biological population in a defined area, and (2) to examine the spatial pattern of a population of organisms so that biological conclusions may be drawn.

The size of a population may be defined as the number of individuals comprising it, the total biomass of all the individuals comprising it, or (in the case of vegetatively spreading plants) as the percentage cover of the plants. Whichever of these measures of population size is used, sampling with quadrats as the sampling units is a convenient estimation method. Quadrats can be used for simple random sampling*, stratified sampling, systematic sampling*, or other methods. With organisms that are difficult or laborious to count, the best procedure may be double sampling* with eyeball estimates of the quantities in some of the quadrats as auxiliary variable.

The spatial pattern of a species population is usually studied by recording the quantity (number of individuals or biomass) of the species in each quadrat of a grid, or block, of contiguous quadrats. Various ways of interpreting such data have been devised [2, 3]. They aim to answer such questions as: Are the organisms arranged in clumps? If so, what are the means and variances of (1) the numbers of individuals (or the biomass) per clump, and (2) the areas of the clumps? Do the clumps have abrupt boundaries or do they merge into each other? Is there evidence for hierarchical clumping, or for long-distance trends in population density? Is the pattern isotropic? And the like.

Ripley [3] gives examples of one- and two-dimensional spectral analyses applied to data from rows, and grids, of quadrats respectively.

The most direct method of studying the spatial pattern of a biological population is to construct an isopleth map showing how population density varies throughout an area [1]. Quadrat sampling is used to provide the data for such maps. The quadrats (usually circular in shape) are systematically arranged with their centers at the mesh points of a square grid. Quadrat size may be so chosen that adjacent quadrats are overlapping, or tangent to each other, or separated by gaps.

References

[1] Pielou, E. C. (1974). *Population and Community Ecology*. Gordon and Breach, New York.

[2] Pielou, E. C. (1977). *Mathematical Ecology*. Wiley, New York.

[3] Ripley, B. D. (1981). *Spatial Statistics*. Wiley, New York.

(ECOLOGICAL STATISTICS
LINE INTERCEPT SAMPLING
LINE INTERSECT SAMPLING
LINE TRANSECT SAMPLING
SPATIAL DATA ANALYSIS)

E. C. PIELOU

QUADRATURE *See* n-DIMENSIONAL QUADRATURE; NUMERICAL INTEGRATION

QUADRIGAMMA FUNCTION

The second derivative of the digamma* or psi function $[\Psi(x) = d \log \Gamma(x)/dx]$. It is customarily denoted by

$$\Psi''(x) = \frac{d^3 \log \Gamma(x)}{dx^3}.$$

QUADRINOMIAL DISTRIBUTION

A multinomial distribution* with four cells. It is used in the theory of 2×2 tables*.

QUADRINORMAL DISTRIBUTION

A distribution with cumulative distribution function

$$F_X(x) = \frac{3}{\sigma\sqrt{(2\pi)}} \int_{-\infty}^{x} t^{3/2} \exp\left(-\tfrac{1}{2} y^2 (t\sigma^2)^{-1}\right) dt\, dy.$$

The distribution is symmetrical about zero; its variance is $\frac{1}{2}\sigma^2$. It belongs to a class of modified normal distributions constructed by Romanowskiĭ [1, 2].

References

[1] Romanowskiĭ, M. (1964). *Bull. Géod.*, **73**, 195–216.

[2] Romanowskiĭ, M. (1965). *Metrologia*, **4**(2), 84–86.

(EQUINORMAL DISTRIBUTION
LINEO-NORMAL DISTRIBUTION
MODIFIED NORMAL DISTRIBUTIONS
RADICO-NORMAL DISTRIBUTION)

QUALITATIVE DATA *See* CATEGORICAL DATA; NOMINAL DATA; ORDINAL DATA

QUALITY CONTROL, STATISTICAL

Statistical quality control refers to the statistical procedures used to control the quality of the output of some production process, including the output of services.

Quality of output is characterized in various ways, frequently by the percentage or fraction of items that do not individually conform to product specifications. (Actually, comparison may be made with "test limits" that allow for measurement errors* and lie outside the specification limits; see ref. 22.) For some products (e.g., sheets of material) quality is characterized by the number of nonconformities per "unit area of opportunity," say, per square foot, or per 100 unit areas. Sometimes the quality of a lot or process is characterized by its mean quality or occasionally by a percentile.

In early writings and in some current ones, the terms *defect* and *defective* are used in the general sense of nonconforming. In the more restricted sense now approved, these terms relate to a product that does "not satisfy intended normal, or reasonably foreseeable usage requirements" [2].

Statistical quality control is principally concerned with control charts* and sampling inspection plans (*see* SAMPLING PLANS).

Control charts are used in deciding whether variations in the quality of the output of a production process can be viewed as coming from a statistically stable population. They are used to study past operations and to monitor current operations. If a chart suggests the existence of special causes of variation, it is expected that the process can be adjusted to attain a state of statistical control at an acceptable level. Sampling inspection plans provide an economic inspection procedure with known risks of passing nonconforming output.

Control charts and sampling inspection plans will be discussed in turn. A concluding section presents some pertinent history and indicates current trends. The statistical design of experiments* may be used in the study of control procedures and in the establishment of an inspection system as well as in the design of the product itself, but this aspect will not be discussed.

CONTROL CHARTS

The principles and considerations involved in process quality control were presented in a memorandum prepared in May 1924 by Walter E. Shewhart of the Bell Telephone Laboratories. His thoughts were later developed in several articles in the *Bell System Technical Journal* (see particularly refs. 54 and 55) and subsequently published in 1931 in his book *The Economic Control of Quality of Manufactured Product* [56]. Lectures given at the Graduate School of the U.S. Department of Agriculture, *Statistical Method from the Viewpoint of Quality Control*, were published in 1939 [57]. The charts he developed for the study of a process are today known as *Shewhart control charts*.

Shewhart Control Charts

Such a chart consists of a plot of some sample statistic—a mean*, range*, or sample proportion—for segments of output known as *rational subgroups*. Time is a rational basis for subgrouping of output, so rational sub-

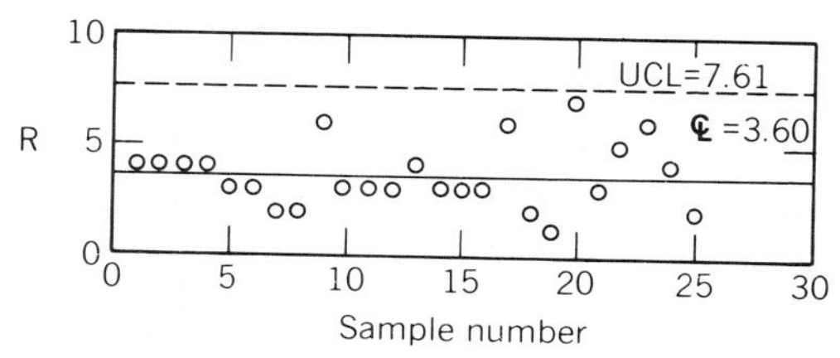

Figure 1 Shewhart R-chart (redrawn from Johnson and Leone [31]).

groups might be small portions of output occurring at intervals of, say, 2 hours, a day, or a week; they might also consist of portions of the output of different operators or different machines. Figure 1 shows a Shewhart range or R-chart.

The assumption underlying control chart analysis is that variation in the quality of output is due in part to chance causes described in ASQC Standard A1 [2] as "generally numerous and individually of relatively small importance . . . which are not feasible to detect or identify" and in part to special or "assignable" causes that are feasible to detect and identify [2]. If in the analysis of a given segment of output, subgroups are appropriately chosen, the variation within subgroups will be representative of that due to chance causes, while variation due to assignable causes will tend to show up in differences between subgroups. Every Shewhart control chart has an upper or lower control limit or both; these represent the limits of variation in the plotted statistic that may reasonably be expected if chance forces are alone at work. If a point exceeds these limits, it is taken as a signal of the presence of an assignable cause. The R-chart of Fig. 1 has only an upper limit, but there may be cases in which it is desirable for an R-chart to have a lower limit as well.

Determination of a State of Statistical Control

If output for a past period consists of units of product, the quality characteristics of which have been individually measured, and if samples of equal size are available for a set of rational subgroups, these data may be effectively studied to determine the existence of a state of statistical control by plotting sample averages on an $\bar{X}$-chart and sample ranges on an R-chart. (With modern computer facilities available, the more efficient sample standard deviation chart or s-chart may be preferred to the R-chart. Our discussion will be in terms of the R-chart.) If there are no points outside estimated $3\sigma_R$ control limits on the R-chart, and no long runs above or below the central line or other evidence of nonrandom variation (see Fig. 1), the average range ($\bar{R}$) can be used to set up estimated $3\sigma_{\bar{X}}$ control limits on the $\bar{X}$-chart falling at $\bar{\bar{X}} \pm 3\bar{R}/(d_2\sqrt{n})$. See ref. 17 (Chap. 21). (The variation between the sample averages on the $\bar{X}$-chart cannot be used to determine the control limits for this chart since that variability may be contaminated by the presence of assignable causes that it is the function of the $\bar{X}$-chart to detect.) If the points on the $\bar{X}$-chart also all fall within the control limits and there are no long runs above or below the central line on the chart or other evidence of nonrandom variation, the analysis will suggest that assignable causes have not been active in the period studied.

To make a positive statement that control exists, the data studied should be reasonably extensive. Shewart stated [57, p. 46] that it has "been observed that a person would seldom if ever be justified in concluding that a state of statistical control of a given repetitive operation or production process has been reached until he has obtained, under presumably the same essential conditions, a sequence of not less than twenty-five samples of four that satisfied Criterion I (i.e., all points fall within 3σ control limits)." This criterion continues to be commonly accepted. For a theoretical study of the operating characteristics of $\bar{X}$-charts used to study past data, see ref. 34.

The use of a pair of $\bar{X}$- and R-charts to determine a state of statistical control is similar to running a one-way analysis of variance*. The control charts have the advantage, however, of retaining the order of the subgroups and offering a graphical presentation. With the availability of modern computer facilities, both analyses could be incorporated in a single computer program. For attributes data, lack of control in past output may be detected by a number-of-nonconforming-units chart (an np-chart), a sample-proportion-nonconforming chart (a p-chart*), or a chart for the sample number of nonconformities (a c-chart*). See refs. 3 and 17. An np-chart for k subgroups is similar to the analysis of a $2 \times k$ contingency table*, showing the numbers of conforming and nonconforming items for each subgroup. See ref. 17 (Chap. 28, Sec. 3).

Analysis of Data from a Controlled Process

When a process is in a state of statistical control, certain meaningful statistical analyses can be carried out. Without control, these analyses may have little meaning and may even be misleading. For example, with a controlled process, a histogram* of the variation in product quality will show process capabilities with respect to given specifications and may indicate the gain from possible recentering of the process. Also, the hypothesis of normal variability in output quality can be tested. In addition, the attainment of a state of statistical control in the manufacture of piece parts will permit random assembly and the computation of statistical limits for the assembled unit.

Attaining and Maintaining a State of Statistical Control

If study of past production data reveals that a process is in a state of statistical control, the central lines and control limits of the charts that were used may be extended for use in monitoring current production. If study of past data does not show control, investigation of the process may reveal assignable causes that can be eliminated, and control charts for current use can be set up with tentative central lines and control limits that can be revised as experience with the process develops. Sometimes standard values for the central lines and control limits are used.

After a process has been currently brought into a state of statistical control, the effectiveness of a control chart in detecting departures from the state of control becomes a major interest. This has led to proposals for modifying the usual Shewhart control chart when employed for purposes of current control by using other than 3σ control limits and sample sizes other than 4 or 5 (see e.g., ref. 15). Interest in effectiveness has also led to the introduction of "warning limits" on a Shewhart chart, say at $\pm 2\sigma$ (see refs. 46 and 68) and to development of geometric moving-average* control charts (see ref. 50), cumulative sum* (CUSUM) control charts (see refs. 45 and 48), combinations of Shewhart and CUSUM charts (see ref. 38), and moving sum* (MOSUM) and moving sum of squares (MOSUM-SQ) controls (see refs. 6 and 7).

The use of sums and averages can be very effective. A CUSUM chart, for example, will detect a sudden small but persistent change in a process mean more quickly [i.e., the average run length (ARL) following the change will be shorter] than in the case of an ordinary Shewhart chart and will facilitate locating the time of occurrence of the change. If deviations from a designated central value are cumulated, a V-mask can be used not only to detect changes but also to estimate the new level of operations. Furthermore, cusum charts readily lend themselves to the use of computers. [For textbook discussions, see refs. 17 (Chap. 22) and 31 (Sec. 10.4).] Geometric moving averages weight past observations in accordance with their distance in time from the most recent observation and MOSUMs limit the number of past observations included in a sum, making both procedures more sensitive to a current shift in the process mean.

Broader studies that pertain to the overall cost-effectiveness of control charts used for current control have led to consideration of prior probabilities* of occurrence of assignable causes and to the inclusion of the frequency of sampling as a design parameter. (See refs. 15, 11, and 18 and the 15 other references cited in ref. 18.) Adaptive control charts have been developed that anticipate changes in the process and permit early adjustments (see ref. 9). For a list of a variety of special control charts, see ref. 2. Acceptance control charts are described below, following the discussion of sampling inspection. For a review of recent developments in control charts used to monitor a process, see ref. 21, and for a review and literature survey of the economic design of control charts, see ref. 42.

Two-sided cusum charts that sum deviations from a target or reference value and use a V-mask based on a known value of the process standard deviation, derived, say, from a previously constructed R-chart or s-chart, have been likened to sequential sampling* plans in reverse [30]. The discussion of MOSUMs in ref. 6 runs entirely in terms of sequential testing of the hypothesis of a constant process mean.

Other Uses of Control Charts

Shewhart charts can be used for purposes other than control of product quality. In general, when data arise out of repetitious operations, say, for example, in making a series of measurements, a Shewhart chart can be used in deciding whether statistically stable conditions exist. See, for example, ERROR ANALYSIS.

SAMPLING INSPECTION PLANS

In the years when Shewhart was developing his ideas on control charts, Harold F. Dodge and Harry G. Romig were working at the Bell Telephone Laboratories on sampling inspection plans. Their historic paper on "A method of sampling inspection" appeared in the *Bell System Technical Journal* of October 1929. Work continued in subsequent years, culminating in their *Sampling Inspection Tables—Single and Double Sampling* [14], published in book form in 1944.

The discussion that follows seeks to develop an understanding of the principles of sampling inspection as applied to the acceptance and rejection of lots. For a discussion of sampling inspection applied to the output of a process not normally involving the formation of lots, see ref. 17 (Chap. 17). The initial discussion here pertains to *attributes inspection*, in which interest centers on the percent of the items in a lot or process that do not conform to product specifications. With minor modifications it can be related to *sampling inspection*, in which interest centers on the number of nonconformities per unit area of opportunity. A brief section is presented on life testing* and a concluding section discusses the sampling of bulk* material. In the two latter cases interest centers on the mean quality of a lot or process. Reference 62 is an example of the case in which interest centers on a specified percentile. *See also* ACCEPTANCE SAMPLING; INSPECTION SAMPLING; SAMPLING PLANS.

Single-Sampling Plans and Schemes for Attributes Inspection of a Series of Lots

A *single-sampling plan* for attributes inspection calls for taking a random sample of specified size n from each lot submitted for inspection, testing the sample items in accordance with a prescribed test method, and noting the number of items that do not conform to product specifications. If for a given lot the number of nonconforming items in a sample is less than or equal to a specified acceptance number c, the lot is accepted; otherwise, it is rejected. Except for the rectifying inspection plans discussed below, the disposition of a rejected lot is not part of the sampling inspection plan.

OC CURVE. The operating characteristic or OC curve of a sampling inspection plan gives the probability of lot acceptance as a function of submitted product quality. As explained in ACCEPTANCE SAMPLING, two types of OC curves are distinguished. A type A OC curve gives the probability of accepting a particular lot as a function of the quality of that lot. A type B OC curve relates to the process producing the product being inspected and gives the proportion of lots that can be expected to be accepted in a continuing series as a function of the process average. Figure 2 depicts a type B OC curve.

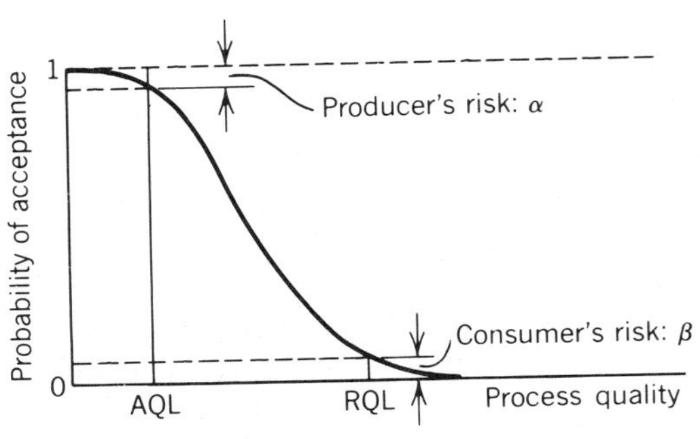

Figure 2 Type B operating characteristic (OC) curve for a sampling inspection plan (redrawn from Johnson and Leone [31]).

OC FEATURES OF SPECIAL INTEREST: THE AQL AND LQ. A feature of the OC curve of a sampling inspection plan that is of special interest is the *acceptable quality level** (AQL). For attributes inspection the AQL is defined as the maximum percent nonconforming or maximum number of nonconformities per unit that for purposes of acceptance sampling can be considered acceptable as a process average (see ref. 2). The naming of an AQL carries with it the implication that the sampling plan will have a high probability of lot acceptance, say 0.95 or higher, for submitted product of AQL quality or better. A given sampling plan will be acceptable to a producer if he or she believes the process capabilities are such that the process average will be at least as good as the AQL. If a producer agrees to a plan with a stated AQL, the consumer will expect that most of the product that he or she will receive will be of AQL quality or better.

In addition to the AQL, a feature of a sampling plan that the consumer will be particularly interested in is the probability of accepting an especially poor lot that might be occasionally turned out by the producer.

He may thus wish to name a *limiting quality* (LQ) and seek a plan for which the probability of lot acceptance is low, say less than 0.10, for lots of LQ quality or worse. For relatively large lots an LQ with probability of acceptance β will be approximately the same as a process level with a β probability of lot acceptance. This is the *rejectable quality level* (RQL; see Fig. 2).

For a given lot size a sampling inspection plan could be tailor-made to accord with the producer's and consumer's interests regarding the AQL and LQ [see ref. 17 (Chap. 7, Sec. 2.3)]. Some modification may be necessary if an unsatisfactorily large sample size is needed to meet the desired goals. An alternative procedure would be to abstract a mutually satisfactory individual sampling plan from some table of sampling schemes (see below) or other systematized table. It will be noted that "producer" and "consumer" could be different departments of the same company.

EFFECT OF A LOT SAMPLING INSPECTION PLAN. If a single-sampling plan is applied to the inspection of the product turned out by a process in a state of statistical control and if the quality of the lots from the process follows a binomial distribution*, the number of nonconforming items in the uninspected portion of a lot is independent of the number of nonconforming items in the sample (see ref. 43). For this case, the distribution of the number of nonconforming items in the uninspected portion of a lot will be the same whether a lot has been accepted or rejected. Thus in this case the sampling plan has no direct effect on the quality of the lots submitted for inspection.

If the sampling plan is applied, however, to lots from processes of different quality levels, it will accept more lots from a process of good quality and reject more lots from a process of poor quality. Hence, if a producer's process is of poor quality, the percentage of lots rejected will be high and the producer will be under pressure to improve it. Thus with every sampling plan for inspecting a continuing series of lots, the producer will be under pressure to turn out product quality in the neighborhood of the AQL, where the probability of lot acceptance is high.

SAMPLING INSPECTION SCHEMES. Awareness of the pressure of lot rejection noted above has led to the development of sampling *schemes* in which a switch to a tightened inspection plan (say, a plan with a smaller acceptance number) is required when there is evidence of deterioration in process quality, indicated, say, by the rejection of two of the last five lots submitted for inspection. This switch to tightened inspection significantly increases the percentage of poor lots rejected and thus augments the pressure on the producer to turn out product of AQL quality or better. [It is suggested by Hamaker in ref. 26 (p. 275) that this extra pressure may possibly be excessive and thus undesirable. He has also personally indicated to the writer that in practice, when rejections tend to run high on normal inspection, some feedback to the supplier is desirable to help them locate the source of the deterioration of the process. If this does not lead to improvement, tightened inspection can be imposed.] Most schemes also provide for discontinuance of inspection if a producer has been on tightened inspection for a specified period of time. Further incentive to produce good quality is usually provided in these sampling schemes by providing for a shift to a smaller sample size and hence less costly inspection when there is evidence that the process is running at AQL quality or better. Sampling inspection schemes can only be applied, of course, to inspection of a continuing series of lots and are usually referred to as AQL schemes, since this is a key element in their application.

RECTIFYING INSPECTION PLANS. Whereas sampling inspection schemes seek long-range AQL quality assurance through application of economic pressures and incentives, there are attributes sampling inspection procedures, referred to as *rectifying inspection plans*, that seek assurance on average outgoing quality* (AOQ) by formally incorporating a provision for 100% inspection of rejected lots as part of the sampling plan. The worst AOQ for various levels of incoming quality is the *average outgoing quality limit** (AOQL), a key figure in an attributes rectifying inspection plan. Such plans are suitable only for a continuing series of lots and, of course, cannot be applied if tests are destructive.

Procedures for Cutting Average Sample Size

Considerable effort has gone into developing sampling inspection procedures that will allow smaller samples to be used on the average without changing sampling risks. These procedures include double sampling*, multiple sampling, and item-by-item sequential sampling*. Thus a double-sampling plan that on the first sample permits decisions to accept, reject, or take a second sample may have an average sample size or *average sample number** (ASN) that for levels of incoming quality in the neighborhood of the AQL is less than the sample size of a single-sampling plan with approximately the same OC curve. Multiple-sampling plans that allow the taking of several samples before a decision is required may have even smaller ASNs. Unit sequential sampling that allows a decision to be made as each sample item is tested and has an OC curve that matches a comparable single-sampling plan at two points, say at the AQL and LQ, will have the smallest ASN at these points (see ref. 65). If the product quality characteristic has a specific distributional form (e.g., normal), a variables sampling inspection plan (one, for example, based on the sample mean and sample standard deviation) can be used to attain assurance regarding the percent nonconforming at a considerable saving in sample size. However, the reduction in cost resulting from a reduction in average sample size obtained by the foregoing procedures may be partly offset by greater administrative costs or in the case of variables sampling, by higher measurement costs. There is also the problem of matching various procedures so that they will have approximately the same OC curve. Furthermore, a given variables plan can be applied only to a single quality characteristic. In conclusion, an ASQC standard has been written for "skip-lot" sampling* that will reduce average inspection costs.

Minimum-Cost Designs

There may be cases, say in application to within-company operations, in which it is feasible to derive minimum-cost sampling inspection plans and schemes. Reference 24 presents a mathematical discussion of Bayesian* sampling plans that consider the expected distribution of lot quality, the cost of inspection, and the cost of accepting a nonconforming item without inspection. Interesting comments on process curves, "discovery sampling," the minimax* principle, and other economic theories of sampling inspection will be found in ref. 25. In some cases the study of costs may suggest that the only economic alternatives are 100% inspection or zero inspection (see refs. 12 and 24). In accepting such alternatives it must be remembered that routine 100% inspection can be boring and hence far from perfect. For zero lot inspection there will also be a need for some checks on the process, to assure that it is maintained at a satisfactory level.

Special References

References 24 and 53 and the bibliographies they contain provide a comprehensive survey of sampling inspection by attributes. The reader will also find ref. 26 interesting. Reference 13 is a review paper by H. F. Dodge.

Tables

Systems of sampling schemes for attributes inspection, indexed by AQLs, lot-size ranges, and inspection levels, and allowing for single, double, and multiple sampling, are provided by U.S. MIL-STD-105D, its

U.S. civilian counterpart ANSI Z1.4, and by the international standard ISO 2859. Similar systems for sampling inspection by variables for percent nonconforming, allowing for single sampling only, are provided by U.S. MIL-STD 414 and its civilian counterparts ANSI Z1.9 and ISO 3951 (*see* INTERNATIONAL STANDARDIZATION: APPLICATION OF STATISTICS). The Dodge–Romig sampling inspection tables contain sampling plans indexed by LTPDs (= LQs) and AOQLs and by process levels for which the plans yield minimum average total inspection. A 1983 supplement to ISO 2859 offers single, double, and multiple sampling plans indexed by LQs and by inspection levels and lot size ranges that are compatible with those of MIL-STD-105D. All these tables give OC curves for individual plans and ANSI Z1.4 gives scheme OC curves. Reference 27 gives a small set of sampling plans indexed by indifference quality ($= p_{0.50}$ or *point of control*) and lot sizes, and refs. 60 and 61 give the slope of the tangent of the OC curve at the point of inflection for a number of sampling plans. Reference 24 contains a large set of tabular data useful for deriving sampling plans when the probability of lot acceptance can be assumed to be based on a Poisson distribution* with parameter λ and λ can be assumed to have a prior gamma distribution*, and ref. 53 contains a large collection of tables pertaining to acceptance sampling.

Acceptance Control Charts

An acceptance control chart (see ref. 20) is a cross between a control chart and a sampling inspection plan. It is intended for use in situations in which process capabilities are well within specification limits and there is little interest in variations in process levels that do not lead to product falling outside these limits. The limits on an acceptance control chart are based on the product specification limits and are intended to detect assignable causes that will lead to the production of an unacceptable fraction of nonconforming product. A succession of sample points within these limits does not necessarily mean that the process is in a state of statistical control as defined above. Acceptance control charts may be viewed as a form of repetitive application of a variables sampling inspection plan with known, constant σ. They should be accompanied by an R-chart or an s-chart for continued validation of the assumption of constant σ.

Life Testing*

When the quality characteristic of an item that is of interest is its life, a sampling inspection plan will probably seek to give assurance regarding mean life. If items are used in sequence, mean life will become the mean time between failures (MTFB). Reference 63 is a sampling standard suitable for controlling the MTBF when the distribution of life is exponential*. A more general type of life distribution which includes the exponential is the Weibull distribution*, which has been studied in depth. See, for example refs. 28, 31, 40, and 66. Weibull probability paper is described in ref. 44. Life testing is a very important part of reliability* engineering.

Sampling Inspection of Bulk Material

Sampling inspection of bulk material takes a special form since physical averaging can be substituted for arithmetic averaging of separate test results. It usually consists of random or systematic selection of increments of the material (generally when it is moving on a conveyor belt or being loaded or unloaded), compositing these increments to form a single gross sample, and then reducing this gross sample to a laboratory sample from which test units are drawn for analysis. Pilot studies of how interincrement variances relate to increment size and how the subsampling variances (called *reduction variances*) relate to the size of the composite sample are found useful in designing a sampling plan and in setting up rough confidence limits for the mean of a lot or process. In inspection of a series of lots, the use of two composites instead of one or the taking of parallel subsamples will provide routine checks on the basic variances obtained in a preliminary pilot study. See ref. 16 and for composite sampling in particular, see refs. 10 and 51. JIS M8100, *General Rules for Methods of Sampling Bulk Materials*, is a well-received Japanese standard and ref. 23 is a comprehensive study of sampling particulate material.

HISTORY

American Developments

Work on statistical quality control was pioneered in the Bell Telephone Laboratories in the late 1920s and early 1930s under the leadership of Walter E. Shewhart, Harold F. Dodge, and Harry G. Romig. Shewhart and Dodge also fostered interest in statistical quality control in the American Society for Testing and Materials. This led to the publication in the 1930s of material on statistical methods that in revised form is now issued as the *ASTM Manual on Presentation of Data and Control Chart Analysis* (STP 15D). Under Shewhart's guidance, Leslie Simon started work on statistical quality control at the Picatinny Arsenal in 1934, and Simon's book *An Engineer's Manual of Statistical Methods* appeared in 1941.

The big boom in statistical quality control in the United States came during and immediately following World War II (*see also* MILITARY STATISTICS). Bell Telephone personnel were sent to Washington to develop a sampling inspection program for Army Ordnance, and the Statistical Research Group, Columbia University, prepared a manual on sampling inspection for the U.S. Navy. The principles of the latter were subsequently expounded in *Sampling Inspection*, edited by H. A. Freeman, M. Friedman, F. Mosteller, and W. A. Wallis, and published in 1947 by McGraw-Hill. (These Army and Navy sampling tables were later amalgamated and issued in 1950 as the Department of Defense MIL-STD-105A.) A significant contribution of the Columbia research group was the work by A. Wald* on sequential sampling. The Statistical Research Group Report No. 255, *Sequential Analysis of Statistical Data: Applications*, was published by Columbia University Press in loose-leaf form in 1945 and Wald published his book on sequential analysis* in 1947 [64]. Almost a year before America's entry into the war, the U.S. War Department induced the American Standards Association to initiate the preparation of instructional material that subsequently became the American War Standards Z1.1-1941, *Guide for Quality Control*; Z1.2-1941, *Control Chart Method of Analyzing Data*; and Z1.3-1942, *Control Chart Method of Controlling Quality during Production*. This material was used in training courses given in numerous cities during the war. In 1944, publication of the journal *Industrial Quality Control* was begun under the auspices of the University of Buffalo (*see* QUALITY TECHNOLOGY, JOURNAL OF).

The interest in statistical quality control aroused during World War II led to the formation in 1946 of the American Society for Quality Control. The society took over publication of *Industrial Quality Control* and subsequently replaced it by the two journals *Quality Progress* and the *Journal of Quality Technology**. In 1959, it initiated with the American Statistical Association* the joint publication of *Technometrics**. ASQC technical conferences are held annually. It has many local chapters and its membership as of March 31, 1982, was 37,254.

The year 1946 also saw the establishment of ASTM Committee E-11 on Quality Control of Materials under the chairmanship of Harold Dodge. This subsequently became the current ASTM Committee E-11 on Statistical Methods. Besides giving advice to ASTM technical committees regarding statistical matters, Committee E-11 has issued standards on indication of significant figures in specified limiting values (E29), on probability sampling of materials (E105, E122, and E141), on the use of the terms *precision* and *accuracy* as applied to measurement of a property of a material (E177), on dealing with outlying observations (E178), and on conducting an interlaboratory study to determine the precision of a test method (E691). The committee also has responsibility for maintaining ASTM Special Technical Publication 15D, noted above.

The idea of using variables sampling involving the sample mean and sample standard deviation or sample range to give assurance regarding the percent nonconforming had been considered by the Columbia Research Group, but the Group was disbanded in 1945 before it was possible to publish tables of variables plans. The Office of Naval Research, however, undertook in 1948 to support this work at Stanford University. The book by Bowker and Goode on *Sampling Inspection by Variables* was published in 1952 and the basic paper by Lieberman and Resnikoff on "Sampling plans for inspection by variables," which subsequently became the Department of Defense Standard MIL-STD-414, was published in the *Journal of the American Statistical Association** in 1955. The paper by A. J. Duncan on the economic design of $\bar{X}$-charts, published in 1956 (see ref. 15), was also supported under Stanford University's contract with the Office of Naval Research.

In 1950, the American Statistical Association sponsored a symposium on acceptance sampling [1] in which Paul Peach of the University of North Carolina presented a paper on "Statistical developments in acceptance sampling by attributes prior to 1941," Edwin G. Olds of Carnegie Institute of Technology presented a paper on "War-time developments in acceptance sampling by attributes," J. H. Curtis of the National Bureau of Standards* presented a paper on "Lot quality measured by average or variability," and W. Allen Wallis of Stanford University presented a paper on "Lot quality measurement by proportion defective." The proceedings include considerable discussion by others in attendance at the symposium, as well as closing remarks by the chairman, John W. Tukey, of Princeton University.

A more recent American development has been the organization in 1974 of the American National Standards Institute (ANSI) Z-1 Committee on Quality Assurance, with a Subcommittee on Statistical Methods to serve in an advisory capacity regarding quality control standards. ASQC serves as secretariat for the committee.

British Developments

In England, statistical analyses relating to the variation in quality of output were undertaken as early as the 1920s by Bernard Dudding of the General Electric Co.'s research laboratories at Wembley. The need to study small-sample variations in means against a probability background was emphasized in his 1929 company paper. Since reference is made in this paper to articles by Shewhart and others in Vols. 1–6 of the *Bell System Technical Journal* (see, e.g., refs. 54 and 55), it is not clear whether the application of sampling theory as a control for mean values occurred to Dudding before he came across these earlier papers. In 1925, L. H. C. Tippett, a statistician for the British cotton industry, published a paper [58] on the means of sample ranges from a normal distribution which contained results that subsequently had significant use in statistical quality control.

In 1931, Egon S. Pearson* of University College, London, visited the United States and spent some days with Shewhart. As a result, Shewhart was invited to come to England in May 1932 to give three lectures in University College on *The Role of Statistical Method in Industrial Standardization*. Following this, things moved rapidly in England. Pearson gave a paper entitled "A survey of the uses of statistical method in the control and standardization of the quality of manufactured products" at the December 1932 meeting of the Royal Statistical Society*, and the next year the society established an Industrial and Agricultural Research Section. Following Shewhart's lectures, the British Standards Institute set up a small committee on Statistical Methods in Standardization and Specification, including Egon

Pearson, Dudding, and representatives from various British Industries. An outcome was BS 600-1935, *The Application of Statistical Methods to Industrial Standardization and Quality Control*, which bore Egon Pearson's name. Another "guide" subsequently issued was BS 2564 (1955), *Control Chart Technique when Manufacturing to a Specification*, by B. P. Dudding and W. J. Jennett (originally prepared for the General Electric Co. in 1944). An excellent review of these early developments in Great Britain is contained in ref. 49.

In the 1950s and 1960s interest developed in Great Britain in improved procedures for detecting process changes. A particularly significant contribution was E. S. Page's article on cusum charts [45] that appeared in 1954. (See also his 1963 paper [48].) In 1959, G. A. Barnard [5] proposed the use of a V-mask for a two-sided cusum scheme, and Kenneth W. Kemp in a 1961 paper [33] discussed the "Average run length* of the cumulative sum chart when a V-mask is used." A review paper by Page [47] was published in 1961. In 1955, Page published a paper on the use of warning limits (see ref. 46).

Also in the 1950s and 1960s, interest was manifested in applying Bayesian procedures to sampling inspection. In 1954, G. A. Barnard presented a paper at a meeting of the Royal Statistical Society entitled "Sampling inspection and statistical decisions" [4]. In February 1960 at a meeting at Imperial College in London, further papers on Bayesian aspects of sampling inspection were presented by A. Hald of the University of Copenhagen and by G. B. Wetherill and D. R. Cox of the University of London. These papers were published in the August 1960 issue of *Technometrics* together with the discussion (pp. 275–372) that followed.

Other Developments

From the United States and Great Britain interest in statistical quality control spread to other countries. W. Edwards Deming's lectures on statistical quality control in Japan following World War II led to the development there of one of the finest quality control systems in the world. Prominent in this development was the Japanese Union of Scientists and Engineers. For an account of the Japanese development, see ref. 32.

American and British demand for European goods was influential in developing an interest in statistical quality control in Europe. H. C. Hamaker, who played a prominent roll in the development of statistical quality control procedures in the Philips Electric Co. at Eindhoven (see ref. 25), has suggested that it was the success that these methods had in Philips's English factories during World War II that provided the main stimulus for their introduction at Eindhoven. A European Organization for Quality Control (EOQC) was established in 1956 and became active in promoting statistical quality control on the continent. In 1965, it published *A Glossary of Terms Used in Quality Control* with their equivalents (in the 1972 edition) in 14 European languages (see ref. 19).

Further international activity has taken the form of standards sponsored by the International Organization for Standardization (ISO) Technical Committee 69 on Applications of Statistical Methods. The United States participates in this through the American National Standards Institute (ANSI), and the United Kingdom through the British Standards Institute. Special international cooperation has taken the form of quality control agreements sponsored by the ABCA Group, made up of representatives of the American, British, Canadian, and Australian armies. (New Zealand participates through Australia.) American, British, and Canadian military representatives constituted the ABC committee that was responsible for the 1963 revision of MIL-STD-105 known as MIL-STD-105D.

Interest in statistical quality control has thus become worldwide, a great quantity of literature on the subject exists, and the application of statistical quality control techniques is widespread.

Recent Trends

Prominent in recent years has been the interest in the dynamic aspects of sampling inspection schemes. Koyama's 1979 paper [35] studies the switching characteristics of MIL-STD-105D. Hald in ref. 24 devotes a whole chapter to switching rules. ANSI/ASQC Z1.4 (the civilian version of MIL-STD-105D) gives data on the operating characteristics of its sampling schemes for the case in which the producer makes no change in his or her process and there is no discontinuance of inspection. These hypothetical oper ating characteristics are useful in comparing various schemes. However, they do not generally give a realistic representation of what would happen in practice, since when the process is running at levels significantly worse than the AQL, the resulting high rate of rejections will usually lead to attempts to improve it.

A very recent development has been the mounting interest in the consideration of costs in the determination of inspection procedures. Prominent in this has been W. E. Deming's book *Quality, Productivity and Competitive Position* [12]. The book was published in 1983, but a preliminary version was available for use in his courses and seminars for some time before that. The considerable discussion given to Bayesian sampling plans in A. Hald's 1981 book on sampling inspection plans [24] has also heightened the interest in relative costs.

There has been an increased interest in lot LQs of sampling inspection plans. Whereas ISO Std 2859 (the international counterpart of MIL-STD-105D) is intended for inspection of a series of lots and is indexed by AQLs, a 1983 supplement is indexed by LQs and is intended for inspection of lots in isolation. In 1978 [52] Schilling discussed "lot-sensitive" sampling plans.

Interest in CUSUM procedures has continued strong. In 1976, Lucas [37] discussed the design and use of V-mask control schemes, and in 1982 [38], a combination of Shewhart and CUSUM schemes. In 1982, Lucas and Crosier [39] discussed the use of a "fast initial response" for a CUSUM scheme. At the 1982 annual conference of the American Statistical Association, Donald Marquardt noted the loss of information in going from continuous to attribute data and recommended the retention of continuous data in that form with the use of CUSUM techniques for process control; these techniques can easily be computerized and the computer results can readily be understood by operating personnel. ISO Technical Committee 69 is preparing a standard for cusum charts.

Improved methods for reporting quality assurance audit results have been presented by Hoadley in his paper "The quality measurement plan (QMP)*" [29]. In the QMP, confidence intervals are computed from both current and past data and are derived from a new Bayesian approach to the empirical Bayes* problem for Poisson observations.

Computer graphics are being used increasingly. For a prominent example of the use of J. Tukey's "box and whisker plots" (see ref. 59 and EXPLORATORY DATA ANALYSIS) in the reporting of process quality, see Hoadley [29]. Video tapes of lectures on statistical quality control are currently being sold or briefly rented by the publishers of Deming's book [12] for training courses using his book as a text. Such presentations will probably become an important form of instruction in this field.

References

[1] *Acceptance Sampling—A Symposium* (1950). American Statistical Association, Washington, D.C.

[2] American Society for Quality Control (1978). Definitions, Symbols, Formulas and Tables for Quality Control. *ANSI/ASQC Standard A1-1978.*

[3] American Society for Testing and Materials (1976). ASTM Manual on Presentation of Data and Control Chart Analysis. *STP 15D.*

[4] Barnard, G. A. (1954). *J. R. Statist. Soc. B*, **16**, 151–165.

[5] Barnard, G. A. (1959). *J. R. Statist. Soc. B*, **21**, 239–257; discussion, 257–271.

[6] Bauer, P. and Hackl, P. (1978). *Technometrics*, **20**, 431–436.

[7] Bauer, P. and Hackl, P. (1980). *Technometrics*, **22**, 1–7.

[8] Bowker, A. H. and Goode, H. P. (1952). *Sampling Inspection by Variables*. McGraw-Hill, New York.

[9] Box, G. E. P. and Jenkins, G. M. (1970). *Time Series Analysis, Forecasting and Control*. Holden-Day, San Francisco.

[10] Brown, G. H. and Fisher, J. I. (1972). *Technometrics*, **14**, 663–668.

[11] Chiu, W. K. and Wetherill, G. N. (1974). *J. Quality Tech.*, **6**, 63–69.

[12] Deming, W. E. (1983). *Quality, Productivity and Competitive Position*. MIT Center for Advanced Engineering Study, Cambridge, Mass.

[13] Dodge, H. F. (1969). *J. Quality Tech.*, **1**. Part I, Apr., 77–88; Part II, July, 155–162; Part III, Oct., 225–232; **2**. Part IV, Jan. 1970, 1–8.

[14] Dodge, H. F. and Romig, H. G. (1959). *Sampling Inspection Tables—Single and Double Sampling.* Wiley, New York. (First edition in 1944.)

[15] Duncan, A. J. (1956). *J. Amer. Statist. Ass.*, **51**, 228–242.

[16] Duncan, A. J. (1962). *Technometrics*, **4**, 319–344.

[17] Duncan, A. J. (1974). *Quality Control and Industrial Statistics*, 4th ed. Richard D. Irwin, Homewood, Ill.

[18] Duncan, A. J. (1978). *Technometrics*, **20**, 235–243.

[19] European Organization for Quality Control—EOQC (1972). *Glossary of Terms Used in Quality Control with Their Equivalents in Arabic, Bulgarian, Czech, Dutch, French, German, Italian, Norwegian, Polish, Portuguese, Romanian, Russian, Spanish and Swedish.* EOQC, Rotterdam, The Netherlands.

[20] Freund, R. A. (1957). *Ind. Quality Control*, **14**(4), 13–23.

[21] Gibra, I. N. (1975). *J. Quality Tech.*, **7**, 183–192.

[22] Grubbs, F. E. and Coon, H. F. (1954). *Ind. Quality Control*, **10**(5), 15–20.

[23] Gy, P. M. (1979). *Sampling of Particulate Materials: Theory and Practice*. Elsevier, New York.

[24] Hald, A. (1981). *Statistical Theory of Sampling Inspection by Attributes*. Academic Press, New York.

[25] Hamaker, H. C. (1958). *Appl. Statist.*, **7**, 149–159.

[26] Hamaker, H. C. (1960). *Bull. Int. Statist. Inst.*, **37**, 265–281.

[27] Hamaker, H. C., Taudin Chanot, J. J. M., and Willemze, F. G. (1950). *Philips Tech. Rev.*, **11**, 362–370.

[28] Harter, H. L. and Moore, A. H. (1976). *IEEE Trans. Rel.*, **R-25**, 100–104.

[29] Hoadley, B. (1981). *Bell Syst. Tech. J.*, **60**, 215–273.

[30] Johnson, N. L. (1961). *J. Amer. Statist. Ass.*, **56**, 835–840.

[31] Johnson, N. L. and Leone, F. C. (1977). *Statistics and Experiemntal Design in Engineering and the Physical Sciences*, 2nd ed. Wiley, New York. (This discusses a variety of uses of cusum charts.)

[32] Juran, J. M. (1975). *Rep. Statist. Appl. Res.*, **22**, 66–72. Union of Japanese Scientists and Engineers, Tokyo.

[33] Kemp, K. W. (1961). *J. R. Statist. Soc. B*, **23**, 149–153.

[34] King, E. P. (1952). *Ann. Math. Statist.*, **23**, 384–394.

[35] Koyama, T. (1979). *Technometrics*, **21**, 9–19.

[36] Lieberman, G. J. and Resnikoff, G. J. (1955). *J. Amer. Statist. Ass.*, **50**, 457–516.

[37] Lucas, J. M. (1976). *J. Quality Tech.*, **8**, 1–8.

[38] Lucas, J. M. (1982). *J. Quality Tech.*, **14**, 51–59.

[39] Lucas, J. M. and Crosier, R. B. (1982). *Technometrics*, **24**, 199–205.

[40] Mann, N. R., Schafer, R. E., and Singpurwalla, N. D. (1974). *Methods for Statistical Analysis of Reliability and Life Data*. Wiley, New York.

[41] Marquardt, D. W. (1982). *Amer. Statist. Ass., Proc. Sec. Statist. Educ.*, pp. 21–27.

[42] Montgomery, D. C. (1980). *J. Quality Tech.*, **12**, 75–89.

[43] Mood, A. M. (1943). *Ann. Math. Statist.*, **13**, 415–425.

[44] Nelson, L. S. (1967). *Ind. Quality Control*, **24**, 452.

[45] Page, E. S. (1954). *Biometrika*, **41**, 100–114.

[46] Page, E. S. (1955). *Biometrika*, **42**, 243–257.

[47] Page, E. S. (1961). *Technometrics*, **3**, 1–9.

[48] Page, E. S. (1963). *Technometrics*, **5**, 307–315.

[49] Pearson, E. S. (1973). *Statistician*, **22**, 165–179.

[50] Roberts, S. W. (1959). *Technometrics*, **1**, 239–250.

[51] Rohde, C. A. (1976). *Biometrics*, **32**, 273–282.

[52] Schilling, E. G. (1978). *J. Quality Tech.*, **10**, 47–51.

[53] Schilling, E. G. (1982). *Acceptance Sampling in Quality Control*. Marcel Dekker, New York.

[54] Shewhart, W. A. (1926). *Bell Syst. Tech. J.*, **5**, 593–606.

[55] Shewhart, W. A. (1927). *Bell Syst. Tech. J.*, **6**, 722–735.

[56] Shewhart, W. A. (1931). *Economic Control of Quality of Manufactured Product*. D. Van Nostrand, New York.

[57] Shewhart, W. A. with the editorial assistance of W. E. Deming. (1939). *Statistical Method from the Viewpoint of Quality Control*. Graduate School, U.S. Dept. of Agriculture, Washington, D.C.

[58] Tippett, L. H. C. (1925). *Biometrika*, **17**, 364–387.

[59] Tukey, J. W. (1977). *Exploratory Data Analysis*. Addison-Wesley, Reading, Mass.

[60] U.S. Dept. of the Army, Chemical Corps Engineering Agency (1953). A Method of Discrimination for Single and Double Sampling OC Curves Utilizing the Tangent at the Point of Inflection. *ENASR No. PR-7*.

[61] U.S. Dept. of the Army, Chemical Corps Engineering Agency (1954). A Method of Fitting and Comparing Variables and Attributes Operating Characteristic Curves Using the Inflection Tangent with Tables of Inflection Tangents for Variables Sampling OC Curves. *ENASR No. PR-12*.

[62] U.S. Dept. of Defense (1973). Maintainability Verification/Demonstration/Evaluation. *MIL-STD-471A*.

[63] U.S. Dept. of Defense (1977). Reliability Design Qualification and Production Acceptance Tests: Exponential Distribution. *MIL-STD-781C*.

[64] Wald, A. (1947). *Sequential Analysis*. Wiley, New York.

[65] Wald, A. and Wolfowitz, J. (1948). *Ann. Math. Statist.*, **29**, 326–329.

[66] Weibull, W. (1951). *J. Appl. Mech.*, **18**, 293–297.

Bibliography

See the following works, as well as the references just given, for more information on the topic of statistical quality control.

American Society for Quality Control. *Publications Catalogue*, 230 West Wells Street, Milwaukee, WI 53202.

Ishikawa, K. (1982). *Guide to Quality Control*, 2nd rev. ed. Asian Productivity Organization, Tokyo. [May be ordered through the American Society for Quality Control (see address above). The book, a translation from the Japanese, is a summary of articles and exercises originally published in the magazine *Quality Control for the Foreman*.]

Juran, J. M., Gryna, F. M., and Bingham, R. S., eds. (1974). *Quality Control Handbook*, 3rd ed. McGraw-Hill, New York. (A wide coverage.)

Ott, E. R. (1975). *Process Quality Control*. McGraw-Hill, New York. (Contains many examples.)

(ACCEPTANCE SAMPLING
AVERAGE OUTGOING QUALITY
AVERAGE OUTGOING QUALITY LIMIT (AOQL)
AVERAGE RUN LENGTH
BULK SAMPLING
CONTROL CHARTS
CUMULATIVE SUM CONTROL CHARTS
DODGE–ROMIG LOT TOLERANCE TABLES
EXPLORATORY DATA ANALYSIS
INSPECTION SAMPLING
INTERNATIONAL STANDARDIZATION: APPLICATION OF STATISTICS
MILITARY STATISTICS
MOVING SUMS (MOSUM)
SAMPLING PLANS)

ACHESON J. DUNCAN

QUALITY MEASUREMENT PLAN (QMP)

An important function of quality assurance is to measure the outgoing quality of a product or service. In the Western Electric Company (1983) this is done with the Quality Measurement Plan (QMP), a system for analyzing and reporting attribute quality *audit* results.

A quality audit is a highly structured system of inspections done continually on a sampling basis. Sampled product is inspected and defects are assessed for failure to meet the engineering requirements. The results during a *rating period* (eight per year) are aggregated and compared to a *quality standard*. The quality standard is a target value for defects per unit, which reflects a trade-off between manufacturing cost, operating costs, and customer need. Specific examples of audits are (a) a functional test audit for digital hybrid integrated circuits and (b) a workmanship audit for spring-type modular cords.

QMP replaces Shewhart-type control charts* such as the p-chart, the c-chart, and the T-rate, all described in ref. 4. *Box and whisker plots* are used to graphically display the posterior distribution* of *true current quality* (*see* EXPLORATORY DATA ANALYSIS). The posterior distribution* is computed from both current and past data and is derived from a Bayesian approach to the empirical Bayes* problem for Poisson observations.

The idea for QMP evolved from the work of Efron and Morris on empirical Bayes estimation [1], in which they illustrate the method with baseball batting averages. There is a clear analogy between traditional quality assurance attribute measures and batting averages. QMP advances the empirical Bayes method in two directions. The first is to consider a given product type as a random selection from a large population of product types. This leads to a Bayes estimate of the empirical Bayes *shrinkage weight* rather than the usual maximum likelihood estimate*. The second is to provide an empirical Bayes interval estimate rather than just a point estimate.

The complete documentation of QMP is given in ref. 2, where the rationale, mathematical derivations, dynamics, operating characteristics, and many examples are discussed. This article is a summary of material in both refs. 2 and 3.

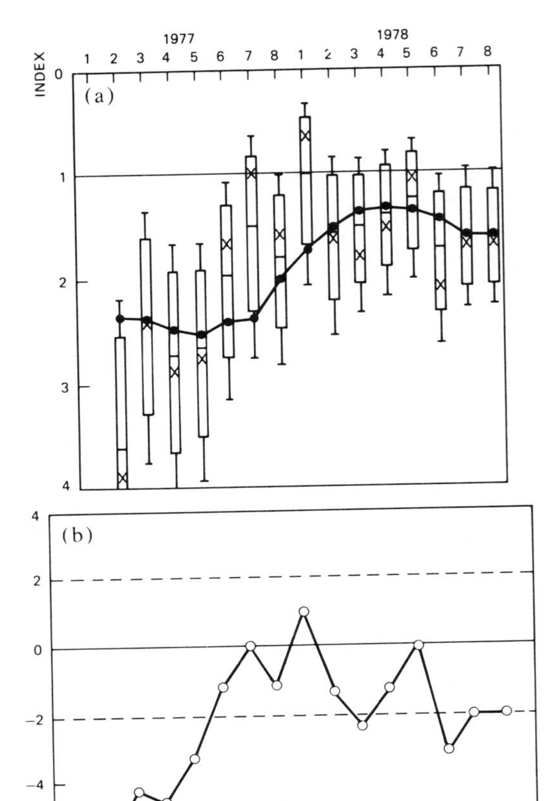

Figure 1 (From ref. 2. Reprinted with permission of the *Bell System Technical Journal*, Copyright AT&T, 1981.)

QMP ANALYSIS

As an example, consider Fig. 1. This is a comparison of the QMP reporting format (a) with the old *T-rate* reporting format (b), which is based on the Shewhart control chart [4]. Each year is divided into eight periods. In Fig. 1*b*, the *T*-rate is plotted each period and measures the difference between the observed and standard defect rates in units of sampling standard deviation (given standard quality). The idea is that if the *T*-rate is, for example, less than -2 or -3, the hypothesis of standard quality is rejected.

The *T*-rate is simple but has problems; for example, it does not measure quality. A *T*-rate of -6 does not mean that quality is twice as bad as when the *T*-rate is -3. The *T*-rate is only a measure of statistical evidence with respect to the hypothesis of standard quality. QMP provides a complete statistical inference of the true quality process.

Under QMP, a box and whisker plot (Fig. 1*a*) is plotted each period. The box plot is a graphical representation* of the posterior distribution of current true quality on a *quality index scale*. "One" is the standard value on the quality index scale. "Two" means twice as many defects as expected under the standard. The posterior probability that the true quality index is larger than the top whisker is 0.99. The top of the box, the bottom of the box, and the bottom whisker correspond to probabilities of 0.95, 0.05, and 0.01, respectively.

The heavy dot is a Bayes estimate of the long-run *process average*, the X is the observed value in the current sample, and the dash is the Bayes estimate of the current true quality index and is called the *best measure* of current quality. This is an empirical Bayes estimate—a shrinkage toward the process average. The process averages (dots) are joined to show trends.

Although the T-rate chart and the QMP chart sometimes convey similar messages, there are differences. The QMP chart provides a measure of quality; the T-rate chart does not. For example, in period 6, 1978, both charts imply that the quality is substandard, but the QMP chart also implies that the true quality index is somewhere between 1 and 3. Comparing period 6, 1977, with period 4, 1978, reveals similar T-rates but QMP box plots with different messages.

The QMP chart is a modern control chart for defect rates. However, it can be used for management reporting rather than as part of a shop process control system. This is the way it is used at present in the Western Electric Company.

STATISTICAL FOUNDATIONS OF QMP

For rating period t [$t = 1, \ldots, T$(current period)], the data from the audit are of the form

n_t = audit sample size

x_t = defects observed in the audit sample

e_t = expected defects in the audit sample when the quality standard is met

$= sn_t$,

where s represents the standard defects per unit. In practice, defectives or weighted defects are sometimes used as the quality measure. These cases can be treated by the theory presented herein via a transformation [2, p. 229].

We measure the defect rate with the *quality index*

$$I_t = x_t / e_t,$$

which is the defect rate expressed as a multiple of the standard defect rate. So $I_t = 2$ means that we observed twice as many defects as expected.

The formulas used for computing the QMP box plots shown in Fig. 1*a* were derived from a Bayesian analysis of a particular statistical model. The assumptions of the model are:

1. x_t is the observed value of a random variable X_t, whose sampling distribution is Poisson* with mean $= n_t\lambda_t$, where λ_t is the true defect rate per unit. For convenience, we reparameterize λ_t on a quality index scale as

   $$\theta_t = \text{true quality index} = \lambda_t / s.$$

   So the standard value of θ_t is 1.
2. θ_t, $t = 1, \ldots, T$, is a random process (or random sample) from a gamma distribution* with

   $$\theta = \text{process average}$$

   $$\gamma^2 = \text{process variance},$$

   which are *unknown*. This assumption makes this a parametric empirical Bayes model.
3. θ and γ^2 have a joint prior distribution*. The physical interpretation of this prior is that each product type has its own value of θ and γ^2, and these vary at random across product types according to the foregoing prior distribution. This assumption makes this a Bayes empirical Bayes model. It is now a full Bayesian model, specifying the joint distribution of all variables. The quality rating in QMP is based on the conditional or posterior distribution of current quality (θ_T) given all the data, $\mathbf{x} = (x_1, \ldots, x_T)$.

POSTERIOR DISTRIBUTION OF CURRENT QUALITY

The formulas for the posterior mean and variance of current quality are given in ref. 2. Here we provide the essence of them. The posterior mean of current quality (θ_T) is

$$\hat{\theta}_T = \hat{\omega}_T \hat{\theta} + (1 - \hat{\omega}_T) I_T,$$

where

$\hat{\theta}$ = Bayes estimate of the process average (θ),

$\hat{\omega}_T$ = Bayes estimate of the shrinkage weight,

$\omega_T = [\theta/e_T]/[(\theta/e_T) + \gamma^2]$.

This is a weighted average of the estimated process average $\hat{\theta}$ and the sample index I_T. It is the dynamics of the shrinkage weight ($\hat{\omega}_T$) that causes the Bayes estimate to work so well. For any t, the sampling variance of I_t (under the Poisson assumption) is θ_t/e_t. The expected value of this is $E[\theta_t/e_t] = \theta/e_t$. So the shrinkage weight (ω_T) is of the form

$$\frac{\text{sampling variance}}{(\text{sampling variance}) + (\text{process variance})}.$$

If the process is stable, the process variance is relatively small and the weight is mostly on the process average; but if the process is unstable, the process variance is relatively large and the weight is mostly on the current sample index. The reverse is true of the sampling variance. If it is large (e.g., small expectancy), the current data are weak and the weight is mostly on the process average; but if the sampling variance is small (e.g., large expectancy), the weight is mostly on the current sample index. In other words, ω_T is monotonically increasing with the ratio of sampling variance to process variance.

The posterior variance of current quality (θ_T) is

$$\begin{aligned} V_T = {} & (1 - \hat{\omega}_T)\hat{\theta}_T/e_T \\ & + \hat{\omega}_T^2(\text{posterior variance of the process average}) \\ & + (\hat{\theta} - I_T)^2(\text{posterior variance of the shrinkage weight}). \end{aligned}$$

If the process average and variance are known, the posterior variance of θ_T can easily be shown to be $(1 - \omega_T)\theta_T/e_T$, which is estimated by the first term in V_T. But since the process average and variance are unknown, the posterior variance has two additional terms. One contains the posterior variance of the process average and the other the posterior variance of the shrinkage weight.

The first term dominates. A large $\hat{\omega}_T$ (relatively stable process), a small $\hat{\omega}_T$ (good current quality), and a large e_T (large audit) all tend to make the posterior variance of θ_T small. If $\hat{\omega}_T$ is small, the second term is negligible. This is because the past data are not used much, so the uncertainty about the process average is irrelevant. If the current sample index is far from the process average, the third term can be important. This is because outlying observations add to our uncertainty.

If the process average and variance were known, the posterior distribution would be gamma, so we approximate the posterior distribution with a gamma fitted by the method of moments*. The parameters of the fitted gamma are α = shape parameter = $\hat{\theta}_T^2/V_T$, τ = scale parameter = $V_T/\hat{\theta}_T$.

QMP BOX AND WHISKER PLOT

For the box and whisker plots shown in Fig. 1*a*, the top whisker, top of box, bottom of box, and bottom whisker are the 1st, 5th, 95th, and 99th percentiles of the approximate posterior distribution of current quality, respectively. So a posteriori, there is a 95% chance that the current quality is worse than the top of the box.

EXCEPTION REPORTING

For QMP, there are two kinds of exceptions: *Below normal* means that the top whisker is worse than the standard (1); that is, there is more than a 0.99 posterior probability that current true quality is worse than standard. *Alert* means that the top of the box is worse than standard but the product is *not* below normal; that is, the posterior probability that current quality is worse than standard is

Table 1

1977 period	7703	7704	7705	7706	7707	7708
t	1	2	3	4	5	6
Number of defects (x_t)	17	20	19	12	7	11
Quality index (I_t)	2.4	2.9	2.7	1.7	1.0	1.6

between 0.95 and 0.99. Products that meet these conditions are highlighted in an exception report.

NUMERICAL EXAMPLE

To illustrate the QMP calculations, consider the eighth period of 1977 in Fig. 1*a*. Six periods of data ($T = 6$) are used—period 3 through period 8. The *sample size* (n_t) in each period is 500 and the *standard defects per unit* (s) is 0.014, so the *expected number of defects* in each sample is $e_t = sn_t = (0.014)(500) = 7$.

The *number of defects* (x_t) observed in each period is given in Table 1 together with the *quality index*, $I_t = x_t/e_t = x_t/7$.

The posterior distribution analysis described in the section "Posterior Distribution of Current Quality" is simplified because of the equal sample sizes. The Bayes estimate ($\hat{\theta}$) of the *process average* is approximately the average of the I_t's, $(2.4 + 2.9 + 2.7 + 1.7 + 1.0 + 1.6)/6 = 2.0$.

The estimated *sampling variance* for the current period is $\hat{\theta}/e_6 = 2.0/7 = 0.28$. The *process variance* is approximated by the *total variance* of the I_t's minus the *average sampling variance*. The total variance is thus $\sum_{t=1}^{6}(I_t - \hat{\theta})^2/5 = 0.54$ and the average sampling variance is 0.28, because the sample sizes do not vary. So the process variance is $0.54 - 0.28 = 0.26$. The Bayes estimate of the *shrinkage weight* is $\hat{\omega}_6 = 0.28/(0.28 + 0.26) = 0.52$. The *posterior mean of current quality* is $\hat{\omega}_T\hat{\theta} + (1 - \hat{\omega}_T)T_T = (0.52)(2.0) + (0.48)(1.6) = 1.8$.

The dominant term in the *posterior variance of current quality* is $(1 - \hat{\omega}_T)\hat{\theta}_T/e_T = (0.48)(1.8)/7 = 0.123$. The second term is $\hat{\omega}_T^2$(posterior variance of the process average) $\approx \hat{\omega}_T^2$(total variance)$/T = (0.52)^2(0.54)/6 = 0.024$. The *posterior variance of the shrinkage weight* can be shown to be 0.019; hence the third term is $(2.0 - 1.6)^2(0.019) = 0.003$. So the posterior variance (V_T) of current quality is $0.123 + 0.024 + 0.003 = 0.15$.

The approximate *posterior distribution of current quality* is a gamma distribution with *shape parameter*

$$\alpha = \hat{\theta}_T^2/V_T = (1.8)^2/0.15 = 21.6$$

and *scale parameter*

$$\tau = V_T/\hat{\theta}_T = 0.15/1.8 = 0.083.$$

The *box and whisker plot* is constructed from the 1st, 5th, 95th, and 99th percentiles of this gamma distribution, which are 1.01, 1.2, 2.5, and 2.8, respectively.

The quality in period 8, 1977, is *below normal*, because the top whisker is below the standard (1). The corresponding *T-rate* shown in Fig. 1*b* is given by

$$T\text{-rate} = \frac{e_t - x_t}{\sqrt{e_t}} = \frac{7 - 11}{\sqrt{7}} = -1.5.$$

This is well within the control limit of -2.0. So with QMP, we reach a different conclusion, because of the poor quality history.

References

[1] Efron, B. and Morris, C. (1977). *Sci. Amer.*, **236**(5), 119–127.

[2] Hoadley, B. (1981). *Bell Syst. Tech. J.*, **60**, 215–271.

[3] Hoadley, B. (1981). *Proc. 31st Electron. Components Conf.*, pp. 499–505.

[4] Small, B. B. (Chairman of Writing Committee) (1958). *Western Electric Statistical Quality Control Handbook.* Western Electric, Indianapolis, Ind. (Sixth printing by Mack Printing Co., Easton, Pa., 1982.)

(EMPIRICAL BAYES THEORY
EXPLORATORY DATA ANALYSIS
NOTCHED BOX-AND-WHISKER PLOT
POSTERIOR DISTRIBUTIONS
PROCESS AVERAGE
QUALITY CONTROL, STATISTICAL
SHRINKAGE ESTIMATORS)

BRUCE HOADLEY

QUALITY TECHNOLOGY, JOURNAL OF

The American Society for Quality Control* (ASQC) had by 1965 grown to the point where its only publication, *Industrial Quality Control**, was unable to continue to serve adequately both as the monthly news magazine and as an outlet for technical papers. In consequence of this, a first proposal was made to the Executive Committee of the ASQC by Lloyd S. Nelson on February 11, 1965, to split *Industrial Quality Control* into a monthly magazine, *Quality Progress*, and a technical quarterly, the *Journal of Quality Technology*.

Subsequent proposals to the board of directors of the ASQC over the next few years resulted in final approval for *Quality Progress*, which began publication in January 1968, and for the *Journal of Quality Technology*, which began publication in January 1969.

The *Journal of Quality Technology* exists to serve the members of the ASQC by publishing papers that will teach applications of statistics in the quality sciences. It also publishes new results that have direct application. It is described on the front cover as "A Quarterly Journal of Methods, Applications and Related Topics." The inside front cover contains the statement: "The objective of the *Journal of Quality Technology* is to contribute to the technical advancement of the

TABLE OF CONTENTS

Vol. 12, No. 2, April 1980

ARTICLES

Figure 1 Contents page from the *Journal of Quality Technology*, **12** (Apr. 1980).

field of quality technology by publishing papers that emphasize the practical applicability of new techniques, instructive examples of the operation of existing techniques, and results of historical researches. Expository, review, and tutorial papers are also acceptable if they are written in a style suitable for practicing engineers."

Four issues per year are published (January, April, July, and October). An issue contains four to six main articles and four departments, each under its own editor. The departments are *Reviews of Standards and Specifications*, *Computer Programs*, *Technical Aids*, and *Book Reviews*. Some idea of the type of material published can be gotten from the table of contents for April 1980, shown in Fig. 1.

Approximately 40% of the manuscripts submitted are eventually accepted. The editor manages an editorial review board of about a dozen members who serve as anonymous referees for manuscripts. The only paid member of the journal staff is the editorial assistant to the editor. The circulation is about 15,000.

Editors of the journal have been Lloyd S. Nelson (founding editor, 1969–1970), H. Alan Lasater (1971–1973), Lawrence D. Romboski (1974–1976), John S. Ramberg (1977–1979), Harrison M. Wadsworth (1980–1982), Peter R. Nelson (Lloyd Nelson's son), and Samuel S. Shapiro (1986–).

LLOYD S. NELSON

QUANDT–RAMSEY (MGF) ESTIMATOR

The MGF estimator was proposed by Quandt and Ramsey [8] to estimate the parameters of normal mixtures and of switching regression models. The estimator is based on the minimization of a sum of squared differences between population and sample values of the moment generating function* (hence the name "MGF").

Consider first the normal mixtures case. Here we have a random sample of size n from a normal mixture model; y_i $(i = 1, \dots, n)$ is drawn from $N(\mu_1, \sigma_1^2)$ with probability λ, and from $N(\mu_2, \sigma_2^2)$ with probability $(1 - \lambda)$. The vector of parameters to be estimated is $\gamma = (\mu_1, \sigma_1^2, \mu_2, \sigma_2^2, \lambda)'$. The parameters may be estimated consistently by the method of moments* [2] or by maximum likelihood* [6]. In the latter case it is necessary to find a local maximum of the likelihood function, since the likelihood function is unbounded. The MGF estimator is an alternative consistent* estimator.

Suppose that we pick $Q \geqslant 5$ values $\theta_1, \theta_2, \dots, \theta_Q$. For a given θ_j, the population moment generating function is

$$G(\gamma, \theta_j) = \lambda \exp\left(\theta_j \mu_1 + \theta_j^2 \sigma_1^2/2\right) + (1 - \lambda)\exp\left(\theta_j \mu_2 + \theta_j^2 \sigma_2^2/2\right), \quad (1)$$

while the sample moment generating function is

$$\bar{z}_j = \frac{1}{n} \sum_{i=1}^{n} \exp(\theta_j y_i). \quad (2)$$

The MGF estimator is defined as the value of γ that minimizes the criterion

$$\sum_{j=1}^{Q} \left[\bar{z}_j - G(\gamma, \theta_j)\right]^2. \quad (3)$$

The MGF estimator is consistent and asymptotically normal. It is not asymptotically efficient. As noted by Johnson [5], it amounts to the method of moments applied to the exponentiated data. This raises the obvious questions of how many moments to use (i.e., choice of Q) and which moments to use (i.e., choice of $\theta_1, \dots, \theta_Q$, given Q).

An improved version of the MGF estimator, suggested by Schmidt [9], is based on the observation that the $\bar{z}_j$ are correlated with each other and have unequal variances. Thus more efficient estimates can be arrived at by minimizing a generalized sum of squares, rather than the ordinary sum of squares as in (3). Define

$$\bar{\mathbf{z}} = (\bar{z}_1, \dots, \bar{z}_Q)' \quad \text{and}$$

$$\mathbf{G}(\gamma, \boldsymbol{\theta}) = \left[G(\gamma, \theta_1), \dots, G(\gamma, \theta_Q)\right]',$$

and let Ω be the $Q \times Q$ covariance matrix of $\bar{\mathbf{z}}$; it has typical element

$$\Omega_{ij} = G(\gamma, \theta_i + \theta_j) - G(\gamma, \theta_i)G(\gamma, \theta_j). \quad (4)$$

Then the improved MGF estimator is defined as the value of γ that minimizes the criterion

$$[\bar{\mathbf{z}} - \mathbf{G}(\gamma, \boldsymbol{\theta})]'\boldsymbol{\Omega}^{-1}[\bar{\mathbf{z}} - \mathbf{G}(\gamma, \boldsymbol{\theta})]. \quad (5)$$

It is consistent and asymptotically normal. For a given set of θ's, the improved MGF estimator is asymptotically more efficient than the Quandt–Ramsey MGF estimator.

The choice of Q (how many moments to use) is now difficult because, for a given set of θ's, adding one more θ cannot decrease the asymptotic efficiency of the estimates. Thus the choice of Q must depend on the small-sample properties of the estimates, which are known. Schmidt [9] evaluates the asymptotic covariance matrices of the estimates for various values of Q and finds that the efficiencies compare favorably with the efficiency of the MLE if Q is fairly large (perhaps 15). He also compares the asymptotic efficiency of the improved MGF estimator to that of the "optimal" method of moments estimator, using the same number of moments (but not having exponentiated the data); there is little reason to prefer the MGF estimator.

In the switching regression model, the means of the two normal distributions are nonconstant; they depend on the values of some explanatory variables. Let $\mathbf{x}_i$ be the ith observation on a vector of explanatory variables. Then y_i is drawn from $N(\mathbf{x}_i'\boldsymbol{\beta}_1, \sigma_1^2)$ with probability λ, and from $N(\mathbf{x}_i'\boldsymbol{\beta}_2, \sigma_2^2)$ with probability $(1 - \lambda)$. The vector of unknown parameters becomes $\gamma = (\boldsymbol{\beta}_1', \sigma_1^2, \boldsymbol{\beta}_2', \sigma_2^2, \lambda)'$. The switching regression model is an extension of the normal mixture model to the regression case. This changes some of the algebra but little of the substance of the preceding discussion.

A variety of other estimators are similar in nature to the MGF estimator. Brockwell and Brown [1] consider estimation of the parameters of the stable distribution using negative moments and compare this to estimation using the empirical MGF. Press [7] and Feuerverger and McDunnough [3] consider estimation based on the empirical characteristic function rather than the MGF. All of these can be considered as moments estimators, after some transformation of the data. A general treatment of such estimators can be found in Hansen [4].

References

[1] Brockwell, P. J. and Brown, B. M. (1981). *J. Amer. Statist. Ass.*, **76**, 626–631.

[2] Day, N. E. (1969). *Biometrika*, **56**, 453–474. (Considers method of moments estimator for the normal mixture case.)

[3] Feuerverger, A. and McDunnough, P. (1981). *J. Amer. Statist. Ass.*, **76**, 379–387.

[4] Hansen, L. P. (1982). *Econometrica*, **50**, 1029–1054. (A general treatment of method of moments estimators.)

[5] Johnson, N. L. (1978). *J. Amer. Statist. Ass.*, **73**, 750. (A comment on Quandt and Ramsey [8].)

[6] Kiefer, N. M. (1978). *Econometrica*, **46**, 427–434. (A proof of consistency of the MLE in the normal mixture model.)

[7] Press, S. J. (1972). *J. Amer. Statist. Ass.*, **67**, 842–846. (Proposes estimates of the stable distribution using the empirical characteristic function.)

[8] Quandt, R. E. and Ramsey, J. B. (1978). *J. Amer. Statist. Ass.*, **73**, 730–737. (Article that proposes the MGF estimator.)

[9] Schmidt, P. (1982). *Econometrica*, **50**, 501–516. (Proposes a more efficient version of the MGF estimator.)

(ASYMPTOTIC NORMALITY
ESTIMATION, POINT
GENERATING FUNCTIONS
MIXTURE DISTRIBUTIONS)

PETER SCHMIDT

QUANGLE

A quangle (quality control angle chart) [1, 3, 5] is a diagram for representing a series of numbers by directions (rather than lengths, as on a conventional graph). It consists of a chain of equal straight-line segments, one for

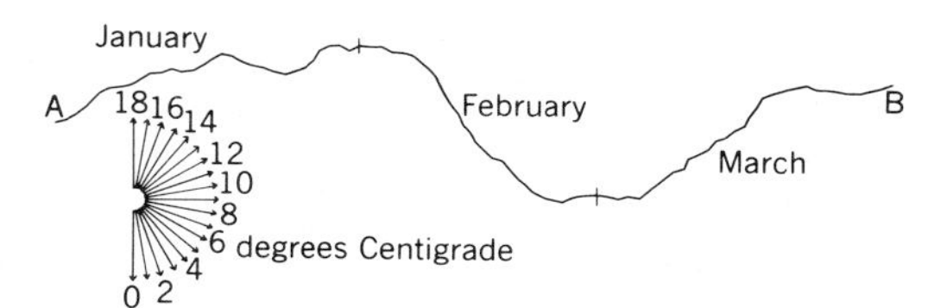

Figure 1 Quangle of the maximum daily temperatures in London for the first three months of 1983.

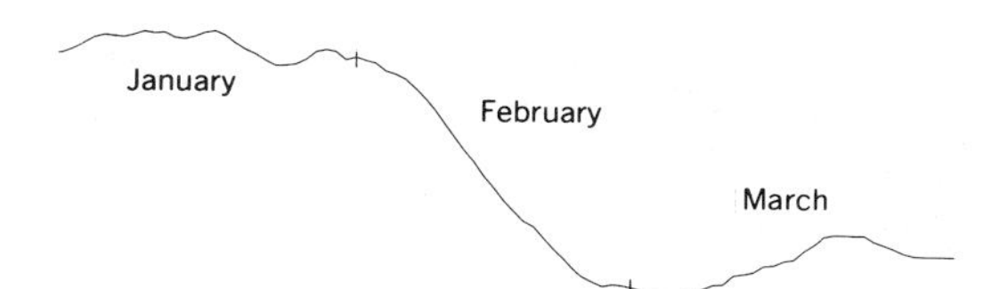

Figure 2 CUSUM of the same data, taking the mean temperature for January as target value.

each term of the series, and each pointing in the direction corresponding to that term.

Figure 1 shows a quangle of the maximum temperature in London for each day of the first three months of 1983. It consists of 90 segments, starting at A and finishing at B. The orientation has been chosen to that the general direction is across the paper. The "horizontal" direction represents 9°C. (The average for the three months was 9.28°C.) The value chosen for the horizontal is called the *neutral value*. The angular scale is 10° of angle for 1°C. In accordance with the usual convention, a change in direction anticlockwise corresponds to an increase in temperature.

The quangle shows that on the whole, February was a cooler month than January or March, with a particularly cold spell from February 9 to 15, when the maximum temperature reached only 2 or 3°C. There was a warm fortnight from March 5 to 19, with maxima around 12°C, followed by a rather cooler week. It is important to choose an appropriate angular scale, so that relevant changes can be seen clearly but minor fluctuations are not exaggerated.

Any series of numbers can be plotted as a quangle. The general shape of the plot will show at a glance whether the data are on the whole uniform or whether there is a sudden (or gradual) change in the mean. If the data form a stationary time series*, the quangle will tend to go in a constant direction, with fluctuations about this direction corresponding to the standard deviation. A sudden change in the underlying mean corresponds to a bend in the quangle, whereas a gradual change gives rise to a steady curve.

In industrial quality control* a CUSUM (cumulative sum control chart*) [4] is often used to indicate when the mean value of a process has changed. If the system is then brought "under control" again by altering the mean, a new CUSUM is started. However, in certain situations where it is not possible to correct the process immediately, it is preferable to have a chart that will monitor successive changes, responding equally sensitively to each. A CUSUM changes direction less for a given change in the mean as it gets steeper, whereas in a quangle the change in direction is always proportional to the change in mean. As long as the process is in control, the two charts are virtually indistinguishable, and so can be used interchangeably. Figures 2 and 3 show two CUSUMS of the temperature data used in Fig. 1. In Fig. 2 the target value is the mean temperature for January (10.46°C), whereas Fig. 3 uses the mean for February (5.75°C). The scale has been chosen to make them directly comparable to the quangle: that is, the ratio of the units in the y and x directions is equal to the angular scale in radians.

The January section of Fig. 2 is very similar to that in Fig. 1. The February section is longer (because the CUSUM takes equal

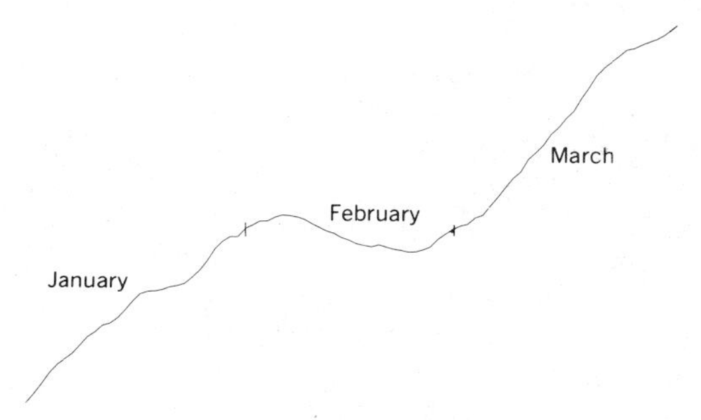

Figure 3 Another CUSUM, taking the mean temperature for February as target value.

intervals in the x-direction). In Fig. 3, the February section is similar to the quangle (but rotated through about 35°), while the January and March sections are distorted. In summary, a CUSUM indicates well when a process departs from a preassigned target value, but a quangle gives a representative picture of the whole series. It is therefore more appropriate for use in examining a series retrospectively for change points. The target value affects the shape of the CUSUM. The neutral value affects the orientation, but not the shape, of a quangle.

For small amounts of data it is easy to draw a quangle on squared paper, using a 360° protractor. For routine use, it is better to write a computer program for use with a graph plotter.

Finding the appropriate angular scale may require some experimentation. A reasonable rule is to take about 30° to correspond to the minimum change that needs to be detected. For retrospective use, a scale of 30° to 1 standard deviation will probably show up salient features.

The theoretical properties of the quangle involve the statistics of directional data* [2]. If the data follow a von Mises distribution (*see* DIRECTIONAL DISTRIBUTIONS), the direction of the quangle is a consistent estimator of the mean.

Acknowledgment

I would like to thank the London Weather Centre for permission to use their temperature data, and my colleague S. G. Thompson for the computer program that drew the quangle.

References

[1] Lear, E. (1877). "The Quangle Wangle's Hat." [In *Laughable Lyrics*, a fourth book of nonsense poems, songs, botany, music, etc. (Edward Lear is responsible only for the name.)]

[2] Mardia, K. V. (1972). *Statistics of Directional Data*. Academic Press, London.

[3] North, W. R. S. (1982). *Appl. Statist.*, **31**, 155–158. (A fuller account of the quangle and its properties.)

[4] Page, E. S. (1961). *Technometrics*, **3**, 1–9. (A discussion of the CUSUM and other control schemes.)

[5] Thompson, S. G. (1983). *Statist. Med.* **2**, 147–153.

(CUMULATIVE SUM CONTROL CHARTS
DIRECTIONAL DATA ANALYSIS
QUALITY CONTROL, STATISTICAL)

W. R. S. NORTH

QUANTAL DATA *See* BIOASSAY, STATISTICAL METHODS IN; CATEGORICAL DATA; NOMINAL DATA; QUANTAL RESPONSE ANALYSIS

QUANTAL RESPONSE ANALYSIS

Quantal response analysis is concerned with the relationship between one or more dependent or response variables on the one hand and a set of independent or explanatory variables on the other, in the situation in which the response is assessed in terms of a discrete, qualitative, or categorical* variable. The simplest form of quantal variable involves a dichotomy, with two response categories which may be assigned the conventional labels 0 and 1, respectively. More generally, response may be recorded in terms of a polychotomy with several classes or categories, which may or may not reflect some implied underlying ordering (*see also* NOMINAL DATA).

APPLICATIONS

In the past, the main applications of quantal response analysis have been in biometry and medical statistics, including pharmacology, toxicology, occupational medicine, and follow-up* studies of groups of patients with particular characteristics, conditions, or treatment regimes (*see* BIOSTATISTICS *and* STATISTICS IN MEDICINE). Quantal models have commonly been concerned with the death of a human being or other biological organism or the appearance of a particular symptom, in response to a stimulus such as a drug, a poison, a period of exposure to a hazardous environment, or merely the passage of time. Many biometric applications involve just one response variable and just one stimulus, although some consideration has been given both to multiple quantal responses and mixtures of several stimuli administered jointly. *See also* BIOASSAY, STATISTICAL METHODS IN.

More recently, there have been an increasing number and range of economic applications, in a variety of contexts which include labor force participation, choice of occupation, migration*, transportation, housing, education, legislation, and criminology. Many of the variables of interest in this field are either naturally expressed in a discrete or categorical form or are recorded in this way. Because of the complexities of the situation, large numbers of explanatory variables are commonly used; in addition, the growing availability and sophistication of survey data has led to a sharp increase in the raw material for such studies. Both developments reflect the growing use and power of modern computing facilities.

There is an extensive and growing literature on the applications of quantal response analysis in biometry; for example, see refs. 6, 8, and 13. Economic analysis is less well served in this context, although Daganzo [7] has dealt with one specific application—transport model choice—and Amemiya [1] has produced an excellent survey of the use of qualitative response models in both econometrics* and biometry.

FRAMEWORK FOR QUANTAL RESPONSE MODELS

A general framework for the formulation of quantal response models in biometry has been provided by Ashford [3], who proposed that a biological organism should be regarded as a system comprising several subsystems, $S_1, S_2, \ldots, S_k$. When a stimulus consisting of d components, $D_1, D_2, \ldots, D_d$, is applied at level $\mathbf{z} = (z_1, z_2, \ldots, z_d)$, the organism manifests a reaction $\mathbf{y} = (y_1, y_2, \ldots, y_k)$, where y_i, $i = 1, 2, \ldots, k$, refers to the underlying response of subsystem S_i; y_i may in principle be expressed in quantitative terms, but is not necessarily observable in this form. For example, human subjects are assessed in terms of several distinct respiratory symptoms, each of which corresponds to the quantal response of a particular subsystem. The same framework may also be employed in economic analysis, although the interpretation of the components reflects the field of application.

The underlying stimulus-response relation may then be represented in terms of an additive model of the form

$$\mathbf{y} = \mathbf{x}(\mathbf{z}) + \boldsymbol{\epsilon}. \tag{1}$$

In this expression, $\mathbf{x}(\mathbf{z})$ is the vector *effects function* of the stimulus $\mathbf{z}$. The random vector $\boldsymbol{\epsilon}$ represents the "tolerance" or "susceptibility" of the individual subject and is distributed with joint cumulative distribution function (CDF) $T(\boldsymbol{\epsilon})$ and joint probability density function (PDF) $t(\boldsymbol{\epsilon})$ in the population of subjects. In general, the tolerance components (ϵ_i, ϵ_j), $(i, j = 1, 2, \ldots, k;\ i \neq j)$, will be positively correlated, since they reflect different characteristics of the same subject.

RELATION BETWEEN QUANTITATIVE AND QUANTAL RESPONSES

The value of the underlying reaction $\mathbf{y}$ will determine the particular quantal or semi-quantal category of the observed system response. Suppose that the k-variate reaction space is divided into mutually exclusive and exhaustive regions corresponding to the various categories. In the simplest situation, just one subsystem S is involved; for such a subsystem quantal response, the two regions correspond to the subdivision of the y-axis at a point which, without loss of generality, may be taken as zero. The probability of subsystem quantal response may then be expressed as

$$\begin{aligned} P(\mathbf{z}) &= P\{y \geqslant 0\} \\ &= P\{x(\mathbf{z}) + \epsilon \geqslant 0\} \\ &= 1 - T(-x(\mathbf{z})), \end{aligned} \tag{2}$$

where $T(\epsilon)$ is the (marginal) CDF of ϵ.

If we make the plausible assumption that $T(\epsilon)$ is symmetrical, (2) may be expressed in the form

$$P(\mathbf{z}) = T(x(\mathbf{z})). \qquad (3)$$

The form of the right-hand side of (3) emphasizes that, on the basis of a quantal response alone [i.e., using only the information contained in $P(\mathbf{z})$], it is not possible to determine the forms of both the tolerance distribution and the effects function without ambiguity, since for any given $P(\mathbf{z})$ and specified marginal CDF $T(\epsilon)$, the form of $x(\mathbf{z})$ is uniquely determined.

A semiquantal response involving s ordered categories $C_1, C_2, \ldots, C_s$ may also be represented in terms of the response of a single subsystem S. The boundaries of the various categories may be defined as the points $\eta_1, \eta_2, \ldots, \eta_{s-1}$ on the y-axis, where $\eta_1 < \eta_2 < \cdots < \eta_{s-1}$, the rth response category C_r corresponding to all points y in the interval (η_{r-1}, η_r), where η_0 and η_s are taken as $-\infty$ and $+\infty$, respectively. Without loss of generality, any one of the η_i may be set as equal to zero. The probability that the response falls within category C_r is then

$$\begin{aligned} P_r(\mathbf{z}) &= P\{\eta_{r-1} \leqslant y_1 < \eta_r\} \\ &= P\{\eta_{r-1} - x(\mathbf{z}) \leqslant \epsilon \leqslant \eta_r - x(\mathbf{z})\} \\ &= T\{\eta_r - x(\mathbf{z})\} - T\{\eta_{r-1} - x(\mathbf{z})\}. \end{aligned} \qquad (4)$$

Quantal and semiquantal responses defined in relation to a complete system may also be represented in terms of the underlying reaction of more than one subsystem. Thus a "weakest link" type of system might involve the quantal response of any one of k constituent subsystems. The corresponding probability is

$$\begin{aligned} P(\mathbf{z}) &= P\{\text{any } y_i > 0\} \\ &= 1 - P\{\text{all } y_i \leqslant 0\} \\ &= 1 - T(-x_1(\mathbf{z}), \\ &\qquad -x_2(\mathbf{z}), \ldots, -x_k(\mathbf{z})). \end{aligned} \qquad (5)$$

In general, if the quantal response corresponds to a region R in the underlying reaction space, the corresponding probability is

$$P(\mathbf{z}) = \int\int \cdots \int_R t(\mathbf{x}(\mathbf{z}) - \boldsymbol{\epsilon})\, d\epsilon_1\, d\epsilon_2 \cdots d\epsilon_k. \qquad (6)$$

Semiquantal responses may similarly be defined in terms of a mutually exclusive and exhaustive set of regions. Such categories will not in general reflect any intrinsic ordering of the various response variables, although, for example, regions defined by a set of parallel hyperplanes would in some sense imply an underlying gradation.

THE TOLERANCE DISTRIBUTION: PROBITS AND LOGITS

Various algebraic forms have been proposed for the distribution of the tolerances $\boldsymbol{\epsilon}$ in (1). Turning first to the marginal distribution of a particular component, say ϵ, the following types of distribution have been employed in the analysis of quantal data.

Linear Probability Model

$$T(\epsilon) = \begin{cases} 0, & \epsilon \leqslant 0, \\ \epsilon, & 0 < \epsilon \leqslant 1, \\ 1, & \epsilon > 1. \end{cases} \qquad (7)$$

This model restricts the range of variation of the tolerance component, which may well be unrealistic in many applications. This deficiency must be balanced against the advantage of computational simplicity. The model has been found useful in the preliminary analysis of economic data.

Probit Model

The scale used for the measurement of the underlying response is arbitrary. Given that the mean of the tolerance distribution* has already been set equal to zero, the variance may be equated to unity and the tolerance may be assumed to be a standardized variable. The most widely used tolerance distribution of this type is the standardized normal,

$$T(\epsilon) = \Phi(\epsilon) = \int_{-\infty}^{\epsilon} (1/\sqrt{2\pi})\epsilon^{-(1/2)t^2}\, dt, \qquad (8)$$

which forms the basis of probit analysis (*see also* BIOASSAY, STATISTICAL METHODS IN). The *probit* of the proportion p is defined in terms of the inverse of Φ as

$$p^* = \Phi^{-1}(p) + 5. \qquad (9)$$

The factor 5 was originally included in (9) for convenience of manual computation, to avoid the occurrence of negative values of the probit. In the more modern usage of the term "probit," this factor is excluded from the definition (*see* NORMAL EQUIVALENT DEVIATE).

Logistic Model

An alternative "smooth" standardized distribution to the normal is the logistic distribution*, employed in the form

$$T(\epsilon) = L(\lambda, \epsilon) = \epsilon^{-\lambda\epsilon}/(1 + \epsilon^{-\lambda\epsilon}). \qquad (10)$$

Taking $\lambda = \pi/\sqrt{3}$, the variance of the tolerance distribution may be set equal to unity. However, some authors consider that closer overall agreement with the standardized normal distribution may be obtained by setting $\lambda = 1.6$. The inverse of the relation $L(1, \epsilon)$ defines the *logit** transformation (or the logarithm of the *odds ratio*; *see* ODDS RATIO ESTIMATORS)

$$l = \ln[p/(1-p)]. \qquad (11)$$

Cox [6, p. 33] has suggested a modified form of (11) which avoids infinite values of the logit corresponding to $p = 0$ and $p = 1$.

The logit and probit transformations are similar for values of p that lie toward the center of the range and differ substantially only for values close to zero or unity. In practical applications, extremely large numbers of observations are required to differentiate between the two transformations [5]. The logit transformation has the advantage of greater ease of computation, which must be set against the fact that the probit, by virtue of its derivation from the normal distribution, is closer to the mainstream of statistical analysis.

A two-parameter class of tolerance distributions has been proposed [14] which includes both the normal and logistic distributions as special cases,

$$t(m_1, m_2, \epsilon) = \frac{\exp(m_1\epsilon)\{1 + \exp\epsilon\}^{-(m_1+m_2)}}{\beta(m_1, m_2)}, \qquad (12)$$

where β represents the beta function. The logistic model is given by $m_1 = m_2 = 1$, while (12) converges to a normal distribution as $m_1, m_2 \to \infty$. Other limiting special cases are the extreme minimal value ($m_1 = 1$, $m_2 \to \infty$) and the extreme maximum value ($m_1 \to \infty$, $m_2 = 1$).

MULTIVARIATE TOLERANCE DISTRIBUTIONS

The assumption of a standardized multivariate normal distribution for the joint distribution of several tolerance components corresponds to multivariate probit analysis [2]. This has the advantage of standardized normal univariate marginal distributions, as used in probit analysis, coupled with a representation of the association between pairs of tolerance components in terms of the correlation coefficient. Representations based on a multivariate logistic distribution have also been proposed. Morimune [11] compared the Ashford–Sowden bivariate probit model with the Nerlove–Press [12] multivariate logit model in an application concerned with home ownership and number of rooms in house, and concluded that the probit was to be preferred.

ORDERED RESPONSE VARIABLES: RANKITS

In circumstances such as sensory tests for degrees of preference, the response is measured by a variable that can be placed in rank order, but cannot be assigned a quantitative value (*see* ORDINAL DATA). The (subjective) ranks may be replaced by the corresponding *rankits* [10]. The rankit* of the rth in rank order from a sample of n observations is defined as the expected value of the rth-order statistic* of a random sample of size n from a standardized normal distribution. This transformation is in a sense a generalization of the probit transformation and produces a response variable that may be regarded as quantitative in form.

MIXED DISCRETE-CONTINUOUS RESPONSE VARIABLES: TOBITS

Response may be recorded in terms of a variable that is capable of being measured continuously over only part of the real line. For example, household expenditures on luxury goods are zero at low income levels. Individuals available for employment may be working full-time or for a certain number of hours part-time or may be unemployed during a particular week. The general "two-limit" Tobit model envisages an underlying quantitative response which is only measurable in categorical form at the lower and upper limits of the range of variation. If one or other of the extreme categories is not present, the Tobit model is called "right-truncated" or "left-truncated," as appropriate. Tobit models invariably involve a standardized normal distribution of tolerances and the probabilities of obtaining one or other of the categorical responses may be calculated on the basis of (2) and (8); *see also* ECONOMETRICS.

THE EFFECTS FUNCTION

The effects function $\mathbf{x}(\mathbf{z})$ in (1) represents the average contribution of the stimulus $\mathbf{z}$ to the response $\mathbf{y}$ of the subject. In general, this function and its individual components will include parameters $\boldsymbol{\theta} = (\theta_1, \theta_2, \ldots, \theta_s)$ which must be estimated. The form of $\mathbf{x}(\mathbf{z})$ and its components $x_i(\mathbf{z})$ reflects the particular application. In most but not all applications it is reasonable to suppose that any given component of the effects function will be a monotonically increasing function of the level of any particular component of the stimulus, for fixed levels of any other components. Usually, in the absence of any stimulus all components of the effects function will be zero. Linear functions of the form $\mathbf{z}'\boldsymbol{\theta}$ of the components z_i of the stimulus $\mathbf{z}$ have been commonly used, conforming to the classical general linear model*. Nonlinear functional forms may, however, be appropriate in certain circumstances.

The simplest situation, commonly used in the biological assay of a single drug, involves a single subsystem S for which the effects function is expressed in terms of the logarithm of the dosage of a single drug. The underlying dosage–response relation then takes the form

$$y = \theta_1 + \theta_2 \ln z_1 + \epsilon, \tag{13}$$

where ϵ is a standardized variable and the corresponding probability of a quantal response is, from eq. (3),

$$P(z_1) = T(\theta_1 + \theta_2 \ln z_1). \tag{14}$$

Both the response y and the stimulus z may be subject to errors of observation. Suppose that the observed value of y (which forms the basis of the observed quantal response) is denoted $y' = y + \phi_y$ and that the observed stimulus $(\ln z_1)' = \ln z_1 + \phi_z$. The underlying dosage–response relation (13) may then be expressed in the form

$$y' = \theta_1 + \theta_2 \ln z_1 + \epsilon', \tag{15}$$

where $\epsilon' = \epsilon_1 + \phi_y - \theta_2 \phi_z$. Given that ϕ_y and ϕ_z follow independent normal distributions with zero mean and variances σ_y^2 and σ_z^2, respectively, ϵ' is itself normally distributed with zero mean and variance $(1 + \sigma_y^2 + \theta_z^2 \sigma_z^2)$. This means that for a standardized normal distribution of tolerances ϵ the corresponding observed dosage–response relation is

$$\begin{aligned} y'' &= y' \Big/ \sqrt{(1 + \sigma_y^2 + \theta q_z^2 \sigma_z^2)} \\ &= (\theta_1 + \theta_2 \ln z) \Big/ \sqrt{(1 + \sigma_y^2 + \theta_z^2 \sigma_z^2)} + \epsilon. \end{aligned} \tag{16}$$

Comparison of (13) and (16) with (14) shows that the effect of errors of observation is to reduce the slope of the quantal response relationship between the inverse of the CDF of the tolerance distribution and the log dose, but the position corresponding to a 50% response, the LD50 (*see* BIOASSAY, STATISTICAL METHODS IN), is unchanged.

In bioassay more complex forms of the effects function have been proposed to represent stimuli with several components, corresponding to mixtures of drugs or poisons, which are becoming increasingly important in both pharmacology and toxicology (see, e.g., ref. 9). An approach based on a simplified representation of the underlying mode of biological action of the stimulus has been put forward by Ashford (see ref. 3), presenting models based on drug-receptor theory that permit both monotonic and nonmonotonic dosage–response relationships for a single-component stimulus and a simple representation of the phenomena of synergism and antagonism.

The majority of examples in economics arise from the representation of events whose outcome is determined by the decisions of a customer, a product, or some other economic unit. There is an implicit assumption that rational decisions are made which optimize some inherent utility [15]. An effects function $\mathbf{x}'\boldsymbol{\theta}$ is commonly thought to be appropriate.

ESTIMATION AND HYPOTHESIS TESTING

The main objective of the statistical analysis of quantal response data is the assessment of the parameters $\boldsymbol{\theta}$ contained within the effects function, although in certain applications further parameters relating to the tolerance distribution are also of interest, including the correlation coefficients associated with a multivariate normal distribution of tolerances. As far as the raw data are concerned, a distinction is often made between structures in which only one or a few observations correspond to any given level of stimulus and situations in which many observations are involved. If possible, the number of observations at any stimulus level should be sufficient for large-sample theory* to apply in order to minimize the effects of sampling variations on the observed proportions of quantal or semiquantal responses.

The most commonly used estimation procedure is maximum likelihood* (ML), which may be applied to both data structures. Suppose that for any particular value z_i of the stimulus or independent variable, there are n_i observations, of which r_i manifest the quantal response of some subsystem S. (If there is only one observation at any particular stimulus level, r_i will be zero or unity.) Assuming that the observations are independent and applying the binomial theorem on the basis of (3), the log likelihood may be written (apart from a constant term) in the form

$$L = \sum_i \{ r_i \ln T(x(\boldsymbol{\theta}, \mathbf{z}_i)) + (n_i - r_i) \ln(1 - T(x(\boldsymbol{\theta}, \mathbf{z}_i))) \}. \tag{17}$$

The ML estimator of $\boldsymbol{\theta}$ is given by the solution of the simultaneous equations

$$\frac{\partial L}{\partial \theta_k} = \sum_i \frac{\{ r_i - n_i T(x(\boldsymbol{\theta}, \mathbf{z}_i)) \}}{T(x(\boldsymbol{\theta}, \mathbf{z}_i)) \{1 - T(x(\boldsymbol{\theta}, \mathbf{z}_i))\}} \times t(x(\boldsymbol{\theta}, \mathbf{z}_i)) \frac{\partial x(\boldsymbol{\theta}, \mathbf{z}_i)}{\partial \theta_k} = 0, \tag{18}$$

where $k = 1, 2, \ldots, s$.

In general, explicit algebraic solutions of (18) do not exist and the ML estimators must be obtained by an iterative method. For the probit or logit models, L is globally concave, which implies that the ML estimator is unique if it is bounded; hence any iterative procedure that is guaranteed to converge to a stationary point must also converge to the global maximum. Various methods of iteration have been used. Powell's method has the advantage of involving the calculation of the function L, but not of its derivatives, in contrast to the Newton–Raphson method* and the method of scoring, which both involve the first derivatives.

See ITERATED MAXIMUM LIKELIHOOD ESTIMATES.

For a semiquantal response, the log likelihood involves the application of the multinomial distribution* in conjunction with expressions of the form (4) in a generalization of (17). For the Tobit model, the log likelihood involves the sum of components corresponding to the discrete and continuous elements. The log likelihoods corresponding to multiple subsystem and system quantal responses (which reflect the state of several subsystems) may also be defined in terms of the binomial or multinomial distributions.

Under certain general conditions the ML estimator is consistent and asymptotically normal. The large-sample variance–covariance matrix of the ML estimators may be calculated as the inverse of the matrix $[E\{-\partial^2 L/\partial\theta_k\partial\theta_l\}]$, where $k, l = 1, 2, \ldots, s$. Tests of hypotheses about the parameters $\boldsymbol{\theta}$ may be based on the ML estimators using Wold's test and the likelihood ratio test*.

Other forms of estimation have been proposed, largely in an attempt to reduce the computational difficulties associated with the ML method. The most widely used is the modified minimum logit χ^2 method [4] (*see* MINIMUM CHI-SQUARE).

References

[1] Amemiya, T. (1981). *J. Econ. Lit.*, **19**, 1483–1536.

[2] Ashford, J. R. (1970). *Biometrics*, **26**, 535–546.

[3] Ashford, J. R. (1981). *Biometrics*, **37**, 457–474.

[4] Berkson, J. (1955). *J. Amer. Statist. Ass.*, **50**, 130–162.

[5] Chambers, E. A. and Cox, D. R. (1967). *Biometrika*, **54**, 573–578.

[6] Cox, D. R. (1970). *The Analysis of Binary Data*. Methuen, London.

[7] Daganzo, C. (1979). *Multinomial Probit*. Academic Press, New York.

[8] Finney, D. J. (1971). *Probit Analysis*, 3rd ed. Cambridge University Press, Cambridge, England.

[9] Hewlett, P. S. and Plackett, R. L. (1979). *The Interpretation of Quantal Response in Biology*. Edward Arnold, London.

[10] Ipsen, J. and Jerne, N. K. (1944). *Acta Pathol.*, **21**, 343–361.

[11] Morimune, K. (1979). *Econometrica*, **47**, 957–976.

[12] Nerlove, M. and Press. S. J. (1973). Univariate and Multivariate Log-Linear and Logistic Models. *Mimeograph No. R-1306-EDA/NIH*, Rand Corporation, Santa Monica, Calif.

[13] Plackett, R. L. (1974). *The Analysis of Categorical Data*. Charles Griffin, London.

[14] Prentice, R. L. (1976). *Biometrics*, **32**, 761–768.

[15] Thurstone, L. (1927). *Psychol. Rev.*, **34**, 273–286.

(BIOASSAY, STATISTICAL METHODS IN
ECONOMETRICS
ITERATED MAXIMUM LIKELIHOOD ESTIMATES
NORMAL EQUIVALENT DEVIATE
PHARMACEUTICAL INDUSTRY, STATISTICS IN
QUANTIT ANALYSIS
RANKIT)

J. R. ASHFORD

QUANTILE ESTIMATION

Consider the usual framework in which a random variable X has a cumulative distribution function $F(x)$ and probability density (or mass) function $f(x)$. If a random sample of size n were obtained from this distribution, the mean, standard deviation, and other useful parameters could be estimated. Often, specific quantiles* or theoretical percentage points of a distribution are of interest to examine. For example, to determine the cholesterol level below which 95% of clinically observed values occur, one would estimate the 0.95 quantile, also known as the 95th percentile of the distribution. Commonly used quantiles have special names: the 0.50 quantile is the *median*, the 0.25 quantile is the *lower quartile*. If the median is estimated by a sample median, these median estimates are known as *least absolute deviation* (LAD) *estimates*. Their use is discussed in MEDIAN ESTIMATES AND SIGN TESTS*.

Placing the discussion in a mathematical framework, the pth population quantile of X is the value T_p given by $F(T_p) = p (0 < p < 1)$. If X is not continuous, T_p is defined by $p(X < T_p) \leqslant p \leqslant p(X \leqslant T_p)$. This gives T_p

uniquely unless $F(T_p) = p$, in which case T_p lies in an interval. The median may be estimated directly; *see* MEDIAN ESTIMATION, INVERSE.

DISTRIBUTION-FREE ESTIMATORS

Even before the era of computers, quantiles could easily be estimated from one or two order statistics*, $X_{1:n}, X_{2:n}, \ldots, X_{n:n}$ from a random sample $X_1, X_2, \ldots, X_n$ obtained from a continuous random variable. Many such estimators have been presented; one of the simplest is the following. Let

$$k = \begin{cases} np & \text{if } np \text{ is an integer,} \\ [np] + 1 & \text{if } np \text{ is not an integer;} \end{cases}$$

then the pth sample quantile is

$$Q_1(p) = X_{k:n}.$$

If $f_X(x)$ is differentiable in the neighborhood of T_p and $f_x(T_p) \neq 0$, then the distribution of $Q_1(p)$ tends to a normal distribution with mean T_p and variance

$$n^{-1}p(1-p)f^2(T_p) \text{ as } n \to \infty.$$

An alternative estimator of T_p can be written as

$$Q_2(p) = (1-g)X_{j:n} + gX_{j+1:n},$$

where $j = [p(n+1)]$ and $g = p(n+1) - j$. A special case of $Q_2(p)$ is the sample median

$$Q_2(0.5) = \begin{cases} X_{(n+1)/2:n} & \text{if } n \text{ is odd,} \\ (X_{n/2:n} + X_{(n/2)+1:n})/2 & \text{if } n \text{ is even.} \end{cases}$$

Schmeiser [24] presents a straightforward technique for estimation using grouped data*.

It has long been known that binomial probabilities can be combined to form confidence intervals* for quantiles. The well-known sign test* employs this principle.

For a continuous distribution, pairs of order statistics can be used to construct a distribution-free confidence interval for the pth population quantile, that is, [4],

$$\begin{aligned} P(X_{j:n} \leqslant T_p \leqslant X_{k:n}) &= I_p(j, n-j+1) - I_p(k, n-k+1) \\ &= \sum_{i=j}^{k-1} \binom{n}{i} p^i (1-p)^{n-i}, \end{aligned}$$

where $I_p(a,b)$ is the incomplete beta function*. A confidence interval with coefficient $\geqslant 1-\alpha$ can be constructed by choosing k and j such that

$$\sum_{i=j}^{k-1} \binom{n}{i} p^i (1-p)^{n-i} \geqslant 1 - \alpha.$$

Choice of j and k is arbitrary, but usually one tries to make $k - j$ as small as possible. For the case of the median, this procedure results in taking $k = n - j + 1$ and gives confidence

$$2^{-n} \sum_{i=j}^{n-j} \binom{n}{i}.$$

This result is due to Thompson [27]; see also Scheffé and Tukey [23]. Breth [3] generalizes the technique so that more than one pair of order statistics may be used to obtain simultaneous confidence intervals* for various quantiles. Wilks [30] gives corresponding results for a confidence interval for quantiles based on a simple random sample from a finite population. Procedures for stratified samples are considered by McCarthy [16], Loynes [13], and Sedransk and Meyer [25].

Many refinements exist to improve the estimates. Krewski [10] and Reiss and Ruschendorf [18] have provided details on construction of interval estimates with tighter bounds than the simpler methods presented.

OTHER DISTRIBUTION-FREE METHODS

Linear Combinations of Order Statistics

Kaigh and Lachenbruch [9] propose an estimator using linear combinations of order statistics (*see* L-STATISTICS). For a fixed integer k, $1 \leqslant k \leqslant n$, consider a simple random

Table 1

Limit	Sign Test	Harrell–Davis	Kaigh–Lachenbruch	Bootstrap
Upper	184	190.6	189.7	189.3
Lower	220	219.1	219.8	220.7

sample (without replacement) from the complete sample $X_1, \ldots, X_n$ and denote the ordered observations in the subsample by

$$Y_{1;k:n}, \ldots, Y_{k;k:n}.$$

The estimate of T_p based on the observations in the subsample is $Y_{j;k,n}$, where $j = [(k+1)p]$. Define as the *total sample estimator* of T_p the average over all possible subsamples of size k of the subsample estimator. This estimator is of the form

$$Q_3(p) = \sum_{i=1}^{k} w_i X_{i:n},$$

where

$$w_i = \binom{i-1}{r-1}\binom{n-i}{k-r} \Big/ \binom{n}{k} \qquad \text{and}$$

$$r = \left[(k+1)p\right].$$

The choice of k is somewhat arbitrary but should be made so that little weight is given to the extreme order statistics. The choice of k that minimizes the mean squared error* requires knowledge of the underlying distribution.

Harrell and Davis [7] propose the estimator

$$Q_4(p) = \sum_{i=1}^{n} u_i X_{i:n},$$

where

$$u_i = I_{i/n}\{p(n+1), (1-p)(n+1)\} - I_{(i-1)/n}\{p(n+1), (1-p)(n+1)\}.$$

Both $Q_3(p)$ and $Q_4(p)$ have smaller mean squared errors than either $Q_1(p)$ or $Q_2(p)$ for a wide variety of distributions. In addition, both are asymptotically normally distributed and jackknife* estimates of their variance can be computed. This leads to confidence intervals of the form

$$Q_j(p) \pm Z_{1-\alpha/2} s_j(p), \qquad j = 3, 4,$$

where $s_j(p)$ is the square root of the jackknife variance estimate and $Z_{1-\alpha/2}$ is the appropriate normal distribution percentile. Kaigh [8] suggests replacement of $Z_{1-\alpha/2}$ with $t_{n-k,1-\alpha/2}$ for $Q_3(p)$.

Bootstrap* Method

Efron [5] gives the details of point and interval bootstrap estimators of the median. Appropriate modification would lead to estimators of other quantiles. In Steinberg [26], interval estimates are obtained for the median using the above-mentioned techniques. On a particular data set, the upper and lower 99% confidence limits based on a sample of size 51 are as in Table 1.

DISTRIBUTION-SPECIFIC QUANTILE ESTIMATORS

General Estimators

Kubat and Epstein [11] propose linear asymptotically unbiased estimators of T_p based on two or three order statistics selected in a neighborhood of the sample quantile. The method requires knowledge of the form of the distribution F, but the location and scale parameters are unknown.

The Normal Distribution

A minimum variance unbiased estimator* for the pth quantile of the normal distribution* is given by

$$Q_5(p) = \overline{X} + ksZ_p,$$

where

$$k = \left(\frac{n-1}{2}\right)^{1/2} \Gamma\left(\frac{n-1}{2}\right) \Gamma\left(\frac{n}{2}\right).$$

Confidence intervals for quantiles from the normal distribution can be constructed using the noncentral t-distribution*. The interval is defined by

$$P\{X + u_{\alpha/2}s < \mu + Z_p\sigma \leqslant x + u_{1-\alpha/2}s\} = \gamma,$$

where the values of u are based on the noncentral t-distribution. Tables allowing computation of the u are given by Owen [17].

The Exponential Distribution

For the single-parameter exponential distribution* with probability density function (PDF) $\theta^{-1}e^{-x/\theta}$; $\theta > 0$; $x > 0$, the pth quantile is $-\theta \ln(1-p) = v_p\theta$ (say). Robertson [19] has shown that the estimator with minimum mean square error in predicted distribution functions is given by $K\overline{X}$, where

$$K = n\left[e^{v_p/(n+1)} - 1\right] / \left[2 - e^{v_p/(n+1)}\right].$$

For the two-parameter exponential distribution with PDF $\theta^{-1}e^{-(x-\lambda)/\theta}$; $x > \lambda$; θ, $\lambda > 0$, the pth quantile is $\lambda + v_p\theta$. An estimator of the pth quantile is

$$Q_5(p) = X_{1:n} + \left(V_p + \frac{1}{n}\right)\left(\overline{X} - X_{1:n}\right).$$

Rukhin and Strawderman [20] show that $Q_5(p)$ is inadmissible for most practical situations and develop an improved estimator

$$Q_6(p) = Q_5(p) - 2U_n/(n+1),$$

$$U_n = (b - 1 - n^{-1})\left(\overline{X} - X_{1:n}\right) - (bn-1)X_{1:n}.$$

Other Distributions

For the Weibull distribution* Lawless [12], Mann and Fertig [14, 15], and Schafer and Angus [22] have proposed relevant estimators. Angus and Schafer [2] and Eubank [6] discuss estimation of the quantiles from the logistic distribution*, Ali et al. [1] from a double-exponential distribution, Umbach et al. [28] from the Pareto distribution*, and Lawless [12] and Mann and Fertig [15] from the extreme-value distribution*. Wheeler [29] estimates parameters of Johnson* curves.

References

[1] Ali, M. M., Umbach, D., and Hassanein, K. M. (1981). *Commun. Statist. A*, **10**, 1921–1932.

[2] Angus, J. E. and Schafer, R. E. (1979). *Commun. Statist. A*, **8**, 1271–1284.

[3] Breth, M. (1980). *Aust. J. Statist.*, **22**, 207–211.

[4] David, H. A. (1981). *Order Statistics*, 2nd ed. Wiley, New York.

[5] Efron, B. (1981). *J. Amer. Statist. Ass.*, **76**, 312–319.

[6] Eubank, R. L. (1981). *Scand. Actu. J.*, 229–236.

[7] Harrell, F. E., Jr. and Davis, C. E. (1982). *Biometrika*, **69**, 635–640.

[8] Kaigh, W. D. (1983). *Commun. Statist. Theor. Meth*, **12**, 2427–2443.

[9] Kaigh, W. D. and Lachenbruch, P. A. (1982). *Commun. Statist. Theor. Meth.*, **11**, 2217–2238.

[10] Krewski, D. (1976). *J. Amer. Statist. Ass.*, **71**, 420–422.

[11] Kubat, P. and Epstein, B. (1980). *Technometrics*, **22**, 575–581.

[12] Lawless, J. F. (1975). *Technometrics*, **17**, 255–261.

[13] Loynes, R. M. (1966). *J. R. Statist. Soc. B*, **28**, 497–512.

[14] Mann, N. R. and Fertig, K. W. (1975). *Technometrics*, **17**, 361–368.

[15] Mann, N. R. and Fertig, K. W. (1977). *Technometrics*, **19**, 87–93.

[16] McCarthy, P. J. (1965). *J. Amer. Statist. Ass.*, **60**, 772–783.

[17] Owen, D. B. (1968). *Technometrics*, **10**, 445–478.

[18] Reiss, R. D. and Ruschendorf, L. (1976). *J. Amer. Statist. Ass.*, **71**, 940–944.

[19] Robertson, C. A. (1977). *J. Amer. Statist. Ass.*, **72**, 162–164.

[20] Rukhin, A. L. and Strawderman, W. E. (1982). *J. Amer. Statist. Ass.*, **77**, 159–162.

[21] Sarhan, A. E. and Greenberg, B. G., eds. (1962). *Contributions to Order Statistics*. Wiley, New York, pp. 383–390.

[22] Schafer, R. E. and Angus, J. E. (1979). *Technometrics*, **21**, 367–370.

[23] Scheffé, H. and Tukey, J. W. (1945). *Ann. Math. Statist.*, **16**, 187–192.

[24] Schmeiser, B. W. (1977). *Commun. Statist. B*, **6**, 221–234.

[25] Sedransk, J. and Meyer, J. (1978). *J. R. Statist. Soc. B*, **40**, 239–252.

[26] Steinberg, S. M. (1983). Confidence Intervals for Functions of Quantiles Using Linear Combinations of Order Statistics. Unpublished Ph.D. dissertation, University of North Carolina, Chapel Hill, N.C.

[27] Thompson, W. R. (1936). *Ann. Math. Statist.*, **7**, 122–128.

[28] Umbach, D., Ali, M. M., and Hassanein, K. M. (1981). *Commun. Statist. A*, **10**, 1933–1941.

[29] Wheeler, R. E. (1980). *Biometrika*, **67**, 725–728.

[30] Wilks, S. S. (1962). *Mathematical Statistics*. Wiley, New York.

(*L*-STATISTICS
MEDIAN ESTIMATES AND SIGN TESTS
MEDIAN ESTIMATION, INVERSE
ORDER STATISTICS
QUANTILES)

C. E. DAVIS
S. M. STEINBERG

QUANTILE PROCESSES

Let X be a real-valued random variable (RV) with *distribution function F*, defined to be right-continuous on the real line $\mathbb{R}$; that is, if the underlying probability space is $(\Omega, \mathscr{A}, P)$, then

$$F(x) := P\{\omega \in \Omega : X(\omega) \leqslant x\}, \qquad x \in \mathbb{R}, \tag{1}$$

where := denotes "is defined as."

Let Q be the *quantile function* of the RV X, defined to be the left-continuous inverse of F,

$$Q(y) := \inf x\{x \in \mathbb{R} : F(x) \geqslant y\}, \qquad 0 < y < 1, \tag{2}$$
$$Q(0) := Q(0+), \qquad Q(1) := Q(1-).$$

Q has the fundamental property that for $x \in \mathbb{R}$ and $0 < y < 1$,

$$F(x) \geqslant y \quad \text{if and only if} \quad Q(y) \leqslant x, \tag{3}$$

and that

$$F(Q(y)-) \leqslant y \leqslant F(Q(y)). \tag{4}$$

Consequently, if F is a continuous distribution function, then

$$F(Q(y)) = y \quad \text{and}$$
$$Q(y) = \inf\{x \in \mathbb{R} : F(x) = y\}, \quad 0 < y < 1. \tag{5}$$

Hence the so-called *probability integral transformation** $F : X \rightarrow F(X)$ yields

$$\text{If } F \text{ is } continuous, \text{ then } F(X) \text{ is a uniform } (0, 1) \text{ RV}. \tag{6}$$

Also, if U is a uniform (0, 1) RV, then by (3) we have

$$P\{Q(U) \leqslant x\} = P\{U \leqslant F(x)\} = F(x)$$
$$= P\{X \leqslant x\}, \qquad x \in \mathbb{R}, \tag{7}$$

for an arbitrary distribution function F of a RV X; that is, the so-called *inverse probability integral transformation* or *quantile transformation* $Q : U \rightarrow Q(U)$ yields

$$\text{If } U \text{ is a uniform } (0, 1) \text{ RV, then } Q(U) \stackrel{\mathscr{D}}{=} X \tag{8}$$

for any RV X having an *arbitrary* distribution function F, where "$\stackrel{\mathscr{D}}{=}$" means equality in distribution. In this particular example it means that $P\{Q(U) \leqslant x\} = P\{X \leqslant x\}$, $x \in \mathbb{R}$ [see (7)].

Given a *random sample* $X_1, X_2, \ldots, X_n$, $n \geqslant 1$, on a RV X with distribution function F and quantile function Q, one of the basic problems of statistics is to estimate the latter functions on the basis of the said random sample.

Let $F_n(x)$, $x \in \mathbb{R}$, be the proportion of those observations of a given random sample of size $n \geqslant 1$ which are less than or equal to x. F_n is called the *empirical distribution function* of $X_1, \ldots, X_n$ (*see* EDF STATISTICS). An *equivalent* way of defining F_n is in terms of the *order statistics** $X_{1:n} \leqslant X_{2:n} \leqslant \cdots \leqslant X_{n:n}$ of a random sample $X_1, \ldots, X_n$, as follows:

$$F_n(x) := \begin{cases} 0 & \text{if } X_{1:n} > x, \\ k/n & \text{if } X_{k:n} \leqslant x < X_{k:n} \\ & \quad (k = 1, \dots, n-1), \\ 1 & \text{if } X_{n:n} \leqslant x, \quad x \in \mathbb{R}. \end{cases} \tag{9}$$

The *empirical quantile function* Q_n of $X_1, \dots, X_n$ is defined to be the left-continuous inverse of the right-continuously defined F_n, that is,

$$Q_n(y) := \inf\{x \in \mathbb{R} : F_n(x) \geqslant y\} = X_{k:n} \quad \text{if } (k-1)/n < y \leqslant k/n \quad (k = 1, \dots, n). \tag{10}$$

From a *probabilistic* point of view, the distribution function F view of the distribution of a RV X is equivalent to its view via the quantile function Q. From a *statistical* point of view, F and Q represent two natural, complementary views of a distribution. F_n is a picture of (statistic for) F, while Q_n is a picture of (statistic for) Q. They are two different pictures (empirics) of the distribution of a RV X, which complement each other, featuring the shape of a distribution from two different angles. Both have their much intertwined but nevertheless distinct distribution theories. To facilitate the description, let $U_1, U_2, \dots, U_n$, $n \geqslant 1$, be independent, identically distributed (i.i.d.) uniform (0, 1) RVs with distribution function $G(y) = y$ on [0, 1], empirical distribution function $G_n(y)$, and quantile function $U_n(y)$. Then by (3),

$$\{F_n(x);\, x \in \mathbb{R},\, n \geqslant 1\} \overset{\mathscr{D}}{=} \{G_n(F(x));\, x \in \mathbb{R},\, n \geqslant 1\} \tag{11}$$

if F is *arbitrary*, by (5) applied to (11);

$$\{F_n(Q(y));\, 0 < y < 1,\, n \geqslant 1\} \overset{\mathscr{D}}{=} \{G_n(y);\, 0 < y < 1,\, n \geqslant 1\} \tag{12}$$

if F is *continuous*, by (8);

$$\{Q_n(y);\, 0 < y < 1,\, n \geqslant 1\} \overset{\mathscr{D}}{=} \{Q(U_n(y));\, 0 < y < 1,\, n \geqslant 1\} \tag{13}$$

if F is *arbitrary*, and by (5) applied to (13);

$$\{F(Q_n(y));\, 0 < y < 1,\, n \geqslant 1\} \overset{\mathscr{D}}{=} \{U_n(y);\, 0 < y < 1,\, n \geqslant 1\} \tag{14}$$

if F is *continuous*.

The empirical process β_n is defined by

$$\{\beta_n(x) := n^{1/2}(F_n(x) - F(x));\; x \in \mathbb{R},\, n \geqslant 1\}, \tag{15}$$

while the *uniform empirical process* α_n is defined by

$$\{\alpha_n(y) := n^{1/2}(G_n(y) - y);\; 0 \leqslant y \leqslant 1,\, n \geqslant 1\}. \tag{16}$$

The *uniform quantile process* u_n is defined by

$$\{u_n(y) := n^{1/2}(U_n(y) - y);\; 0 \leqslant y \leqslant 1,\, n \geqslant 1\}, \tag{17}$$

and the *general quantile process* q_n by

$$\{q_n(y) := n^{1/2}(Q_n(y) - Q(y));\; 0 < y < 1,\, n \geqslant 1\}. \tag{18}$$

By (11), (15), and (16) we have

$$\{\beta_n(x);\, x \in \mathbb{R},\, n \geqslant 1\} \overset{\mathscr{D}}{=} \{\alpha_n(F(x));\, x \in \mathbb{R},\, n \geqslant 1\} \tag{19}$$

if F is *arbitrary*, and by (12), (15), and (16),

$$\{\beta_n(Q(y));\, 0 < y < 1,\, n \geqslant 1\} \overset{\mathscr{D}}{=} \{\alpha_n(y);\, 0 < y < 1,\, n \geqslant 1\} \tag{20}$$

if F is *continuous*.

By virtue of (19) and (20), when studying the empirical process β_n, we may restrict attention to the uniform empirical process α_n. According to (20), the distribution of the empirical process β_n does not depend on the form of F as long as the latter is continuous. This is the so-called *distribution-free property of* β_n for continuous F (see Doob [27]). According to (19), the distribution of the empirical process β_n does depend on the form of F if the latter is discrete, but it can be still expressed in terms of the distribution of $\alpha_n(F)$. *See also* PROCESSES, EMPIRICAL.

Unfortunately, there are no simple relationships between u_n and q_n like those of (19) and (20) for α_n and β_n. By (13) and (18) we have

$$\{q_n(y);\, 0 < y < 1, n \geqslant 1\} \stackrel{\mathscr{D}}{=} \{n^{1/2}(Q(U_n(y)) - Q(y));\; 0 < y < 1, n \geqslant 1\} \tag{21}$$

if F is *arbitrary*, and by (5), (14), and (17),

$$\{u_n(y);\, 0 < y < 1, n \geqslant 1\} \stackrel{\mathscr{D}}{=} \{n^{1/2}(F(Q_n(y)) - F(Q(y)));\; 0 < y < 1, n \geqslant 1\} \tag{22}$$

if F is *continuous*. Let F now be an absolutely continuous distribution function on $\mathbb{R}$, let $f := F'$ be its density function (with respect to Lebesgue measure), and assume that $f(\cdot) > 0$ on the support of F. Then any one of (21) and (22) implies, if $a \wedge b$ and $a \vee b$ denote the minimum and maximum of a and b, respectively,

$$\{q_n(y);\, 0 < y < 1, n \geqslant 1\} \stackrel{\mathscr{D}}{=} \{u_n(y)/f(Q(\theta_{y,n}));\, 0 < y < 1,\; U_n(y) \wedge y < \theta_{y,n} < U_n(y) \vee y,\; n \geqslant 1\}. \tag{23}$$

Clearly then, even if F is continuous and also differentiable, the distribution of q_n depends on F through $f(Q)$, the so-called *density-quantile function* of F, or, equivalently, through $Q' = 1/f(Q)$, the so-called *quantile-density function* of F. The latter two terminologies are due to Parzen [49]. Tukey [69] calls Q the *representing function*, and Q' the *sparsity function*.

Since $u_n(k/n) = -\alpha_n(U_{k:n})$, where $U_{1:n} \leqslant U_{2:n} \leqslant \cdots \leqslant U_{n:n}$ are the order statistics of a uniform (0, 1) random sample $U_1, \ldots, U_n$, it is reasonable to expect that the asymptotic distribution theory of α_n and u_n should be the same. Due to (23), we have

$$\{f(Q(y))q_n(y);\, 0 < y < 1, n \geqslant 1\} \stackrel{\mathscr{D}}{=} \left\{u_n(y)\frac{f(Q(y))}{f(Q(\theta_{y,n}))};\, 0 < y < 1,\; U_n(y) \wedge y < \theta_{y,n} < U_n(Y) \vee y,\; n \geqslant 1\right\}. \tag{24}$$

This, in turn, suggests that $f(Q)q_n$ should have the same kind of asymptotic theory as u_n if f is "nice." Hence we introduce ρ_n, the *re-normed general quantile process*, by defining it to be

$$\{\rho_n(y) := f(Q(y))q_n(y);\, 0 < y < 1, n \geqslant 1\}, \tag{25}$$

and attempt to summarize here the intertwined distribution theories of α_n, u_n, and ρ_n.

First we define several Gaussian processes* which play a basic role in approximating the said empirical processes.

Wiener Process (Brownian Motion*). A real-valued separable Gaussian process $\{W(t);\, t \geqslant 0\}$ with continuous sample paths is called a Wiener process if $EW(t) = 0$ and $EW(s)W(t) = s \wedge t$, $s, t \geqslant 0$.

Brownian Bridge.

$$\{B(s);\, 0 \leqslant s \leqslant 1\} := \{W(s) - sW(1);\, 0 \leqslant s \leqslant 1\};$$

hence $EB(s) = 0$ and

$$EB(s)B(y) = s \wedge y - sy,\; 0 \leqslant s, y \leqslant 1.$$

Kiefer Process*.

$$\{K(y,t);\, 0 \leqslant y \leqslant 1, t \geqslant 0\} := \{W(y,t) - y(W(1,t);\; 0 \leqslant y \leqslant 1, t \geqslant 0\},$$

where $\{W(y,t);\, y \geqslant 0,\, t \geqslant 0\}$ is a real valued two-time parameter separable Gaussian process with $EW(y,t)=0$ and $EW(y,t)W(s,l)=(y\wedge s)(t\wedge l)$ $(y,s,t,l\geqslant 0)$. Hence $EK(y,t)=0$ and

$$EK(y,t)K(s,l)=(t\wedge l)(y\wedge s-ys)\quad(0\leqslant y,s\leqslant 1;\ t,l\geqslant 0).$$

A Kiefer process at integer-valued arguments $t=n$ can be viewed as the partial sum process of a sequence of *independent* Brownian bridges $\{B_i(y);\ 0\leqslant y\leqslant 1\}_{i=1}^{\infty}$:

$$\{K(y,n);\ 0\leqslant y\leqslant 1,\ n=1,2,\ldots\}=\left\{\sum_{i=1}^{n}B_i(y);\ 0\leqslant y\leqslant 1,\ n=1,2,\ldots\right\},$$

and

$$\{B_n(y);\ 0\leqslant y\leqslant 1\}=\{K(y,n)-K(y,n-1);\ 0\leqslant y\leqslant 1\}\quad(n=1,2,\ldots)$$

is a sequence of *independent* Brownian bridges.

For proof of existence and further properties of these Gaussian processes, we refer to Csörgő and Révész [18, Chap. 1].

The theory of weak convergence of α_n to a Brownian bridge B was initiated by Doob [27], formalized by Donsker [26], Prohorov [54], Skorohod [65], and summarized and further developed in Billingsley [5] and Breiman [6]. See also Dudley [28], Wichura [71], Durbin [29, 30], Burke et al. [8], and for a review and further references, *see* PROCESSES, EMPIRICAL and GLIVENKO–CANTELLI THEOREMS, Serfling [57], and Gaenssler and Stute [33]. To facilitate our approach to describing u_n and ρ_n, we must say a few words on strong approximation of α_n.

The first strong approximation result of α_n by a sequence of Brownian bridges B_n was proved by Brillinger [7], using the Skorohod embedding* [66] based results of Strassen [67] for partial sums of i.i.d. RVs. A fundamental breakthrough for α_n is

Theorem 1 (Komlós et al. [41]). For α_n there exists a probability space with a sequence of Brownian bridges B_n and a Kiefer process $K(y,t)$ such that

$$\sup_{0\leqslant y\leqslant 1}|\alpha_n(y)-B_n(y)|\overset{\text{a.s.}}{=}O(n^{-1/2}\log n)\tag{26}$$

and

$$n^{-1/2}\sup_{1\leqslant k\leqslant n}\ \sup_{0\leqslant y\leqslant 1}|k^{1/2}\alpha_k(y)-K(y,k)|\overset{\text{a.s.}}{=}O(n^{-1/2}\log^2 n),\tag{27}$$

that is, there exists a positive constant $M<\infty$ such that on the said probability space

$$\limsup_{n\to\infty} n^{-1/2}\sup_{0\leqslant y\leqslant 1}|\alpha_n(y)-B_n(y)|/\log n\leqslant M\tag{28}$$

and

$$\limsup_{n\to\infty}\ \sup_{1\leqslant k\leqslant n}\ \sup_{0\leqslant y\leqslant 1}|k^{1/2}\alpha_k(y)-K(y,k)|/\log^2 n\leqslant M\tag{29}$$

with probability 1.

It follows by (20) that Theorem 1 holds true for β_n if F is continuous, while (19),

$$\sup_{x\in\mathbb{R}}|\alpha_n(F(x))-B_n(F(x))|\leqslant\sup_{0\leqslant y\leqslant 1}|\alpha_n(y)-B_n(y)|,$$

and

$$\sup_{x\in\mathbb{R}}|k^{1/2}\alpha_k(F(x))-B_n(F(x))|\leqslant\sup_{0\leqslant y\leqslant 1}|k^{1/2}\alpha_k(y)-K(y,k)|$$

together imply that Theorem 1 also holds true for β_n if F is arbitrary.

This fundamental theorem has proved to be of basic importance in a wide and growing range of practical and theoretical problems alike. The monographs by Csörgő and Révész [18], and Csörgő [13] contain an exposition, further developments, and a variety of applications. Recent developments and applications are also reviewed in a paper by Csörgő and Hall [14]. *See also* PROCESSES, EMPIRICAL and references therein. As to quantile processes, Theorem 1 has inspired

Theorem 2 (Csörgő and Révész [15, 16]). For u_n there exists a probability space with a sequence of Brownian bridges B_n such that

$$\sup_{0 \leqslant y \leqslant 1} |u_n(y) - B_n(y)| \overset{\text{a.s.}}{=} O(n^{-1/2}\log n), \tag{30}$$

and on the probability space of Theorem 1 with the Kiefer process of (27), we have

$$n^{-1/2} \sup_{1 \leqslant k \leqslant n} \sup_{0 \leqslant y \leqslant 1} |k^{1/2}u_k(y) - K(y,k)| \overset{\text{a.s.}}{=} 2^{-1/4}\left(n^{-1/4}(\log\log n)^{1/4}(\log n)^{1/2}\right). \tag{31}$$

Due to Bártfai [4] and/or to the Erdös–Rényi [32] theorem, the $O(\cdot)$ rate of convergence of (26) and (30) is best possible (see Komlós et al. [41], or Csörgő and Révész [18, Theorem 4.4.2]). The result (31) is closely tied to that of Kiefer [40] [see (40)], which implies that the Kiefer process of (27) cannot be nearer to u_n than given in (31). A direct examination of this problem for the sake of producing better rates of convergence, with a necessarily differently constructed Kiefer process, would be desirable.

The idea behind the next theorem is to compare the uniform quantile process u_n to the quantile process ρ_n, where ρ_n is based on a random sample $X_1, \ldots, X_n$ on a continuous F, and u_n is now based on the general uniform (0, 1) random sample $U_1 := F(X_1)$, $\ldots, U_n := F(X_n)$, $n \geqslant 1$.

Theorem 3 (Csörgő and Révész [16]). Let $X_1, X_2, \ldots,$ be i.i.d. RVs with a continuous distribution function F and assume that

(a) F is twice differentiable on (a,b), where $a = \sup\{x : F(x) = 0\}$, $b = \inf\{x : F(x) = 1\}$, $-\infty \leqslant a < b \leqslant +\infty$,
(b) $F'(x) = f(x) > 0$ on (a,b),
(c) for some $\gamma > 0$ we have

$$\sup_{0<y<1}\left[y(1-y)\frac{|f'(Q(y))|}{f^2(Q(y))}\right] \leqslant \gamma.$$

Then, with $\delta_n = 25n^{-1}\log\log n$,

$$\sup_{\delta_n \leqslant y \leqslant 1-\delta_n} |\rho_n(y) - u_n(y)| \overset{\text{a.s.}}{=} O(n^{-1/2}\log\log n).$$

If, in addition to (a), (b) and (c), we also assume that

(d)

$$A = \lim_{x \downarrow a} f(x) < \infty,$$

$$B = \lim_{x \uparrow b} f(x) < \infty,$$

(e) one of
  (e, α) $A \wedge B > 0$,
  (e, β) if $A = 0$ (respectively, $B = 0$), then f is nondecreasing (respectively, nonincreasing) on an interval to the right of a (respectively, to the left of b),

then if (e, α) obtains

$$\sup_{0 \leqslant y \leqslant 1} |\rho_n(y) - u_n(y)| \overset{\text{a.s.}}{=} O(n^{-1/2}\log\log n),$$

and if (e, β) obtains

$$\sup_{0 \leqslant y \leqslant 1} |\rho_n(y) - u_n(y)| \overset{\text{a.s.}}{=} \begin{cases} O(n^{-1/2}\log\log n) & \text{if } \gamma < 1 \\ O\left(n^{-1/2}(\log\log n)^2\right) & \text{if } \gamma = 1 \\ O\left(n^{-1/2}(\log\log n)^{\gamma}(\log n)^{(1+\epsilon)(\gamma-1)}\right) & \text{if } \gamma > 1, \end{cases}$$

where $\epsilon > 0$ is arbitrary and γ is as in (c).

This theorem is a strong invariance principle for ρ_n in terms of u_n. For further developments, see Csörgő et al. [20]. The latter results are reviewed in Csörgő [13, Sec. 5.4]. When combined with (22) and Theorem 2, it results in a strong invariance principle for ρ_n in terms of the Brownian bridges B_n and the Kiefer process $K(y, t)$ of Theorem 2 (for a discussion of these results, we refer the reader to Csörgő and Révész [18, Chaps. 4, 5] and Csörgő [13, Chap. 3]). It can be used to construct confidence bands for the quantile function Q (see Csörgő and Révész [19] and Csörgő [13, Chap. 4]), for example: under conditions (a) to (c) of Theorem 3 we have

$$\lim_{n\to\infty} P\{ Q_n(y - n^{-1/2}c) \leqslant Q(y) \leqslant Q_n(y + n^{-1/2}c); \epsilon_n \leqslant y \leqslant 1 - \epsilon_n\}$$
$$= P\Big\{ \sup_{0\leqslant y\leqslant 1} |B(y)| \leqslant c\Big\}$$
$$= 1 - \sum_{k\neq 0} (-1)^{k+1} e^{-2k^2c^2}, \qquad c > 0, \tag{32}$$

where $\{\epsilon_n\}$ is such that $n^{1/2}\epsilon_n \to \infty$ $(n\to\infty)$, and $B(\cdot)$ is a Brownian bridge.

Notice that the confidence band of (32) for $Q(\cdot)$ on the interval $[\epsilon_n, 1-\epsilon_n]$ does not require the estimation of the unknown density quantile function $f(Q(\cdot))$.

A word about condition (c) of Theorem 3. In the literature of nonparametric statistics it is customary to define the so-called *score function* (see, e.g., Hajek and Šidák [35, p. 19]):

$$J(y) = -f'(Q(y))/f(Q(y))$$
$$= -\frac{d}{dy} f(Q(y)), \qquad 0 < y < 1.$$

Hence the said condition (c) can be viewed as a condition on the score function J and the quantile-density function $Q' = 1/f(Q)$. For a review of estimates of Q' (respectively, J), see Csörgő [13, Sec. 4.1 and Chap. 10]. For examples and an excellent discussion of tail monotonicity assumptions of extreme-value theory as related to condition (c), see Parzen [49, Sec. 9] and Parzen [53]. Theorem 3 as well as the statement of (32) were recently extended to quantile-quantile plots (*see* Q-Q PLOTS) by Aly and Bleuer [1], and to the quantile process of the product-limit estimator (*see* KAPLAN–MEIER ESTIMATOR) by Aly et al. [1, 2] (for a preview of the latter results, see Csörgő [13, Chap. 8] and references therein).

The modern theory of sample quantiles was initiated by Bahadur [3], who, in terms of our notation, studied the following representation of the yth sample quantile $Q_n(y)$:

$$Q_n(y) = Q(y) + \frac{1 - F_n(Q(y)) - (1-y)}{f(Q(y))} + R_n(y), \tag{33}$$

as a stochastic process* in n for fixed $y \in (0,1)$.

For an excellent review of sample quantiles $Q_n(y)$ vs. $Q(y)$ for $y\in(0,1)$ fixed, see Serfling [57, Secs. 2.3.1 to 2.6.6].

On assuming (a) and (b) of Theorem 3, and continuing along the lines of (24) and (25) we get

$$\rho_n(y)/n^{1/2} = \frac{u_n(y)}{n^{1/2}} + \frac{u_n(y)}{n^{1/2}}\left[\frac{f(Q(y))}{f(Q(\theta_{y,n}))} - 1\right]$$
$$= \frac{u_n(y)}{n^{1/2}} + \frac{u_n(y)}{n^{1/2}} \frac{y - \theta_{y,n}}{f(Q(\theta_{y,n}))} \frac{f'(Q(\delta_{y,n}))}{f(Q(\delta_{y,n}))}$$
$$:= n^{-1/2}u_n(y) + n^{-1/2}\epsilon_n(y), \tag{34}$$

where $\theta_{y,n}\wedge y < \delta_{y,n} < \theta_{y,n}\vee y$ with $\theta_{y,n}\in (U_n(y)\wedge y, U_n(y)\vee y)$.

Consequently, by (33) and (34), we get

$$f(Q(y))R_n(y) = n^{-1/2}u_n(y) + n^{-1/2}\alpha_n(y) + n^{-1/2}\epsilon_n(y) \tag{35}$$

with

$$n^{-1/2}\epsilon_n(y) = n^{-1/2}(\rho_n(y) - u_n(y)). \tag{36}$$

Let $R_n^*(y) := n^{-1/2}(u_n(y) + \alpha_n(y))$. Then by (35) and (36), we get

$$f(Q(y))R_n(y) - R_n^*(y) = n^{-1/2}\epsilon_n(y)$$
$$= n^{-1/2}(\rho_n(y) - u_n(y)), \qquad y\in(0,1). \tag{37}$$

Hence, *given the conditions of Theorem* 3, we have (see Csörgő and Révész [16, 18], and Csörgő [13])

$$\sup_{0\leqslant y\leqslant 1} |f(Q(y))R_n(y) - R_n^*(y)|$$

$$= \begin{cases} O(n^{-1}\log\log n) \\ \qquad \text{if } (\mathrm{e},\alpha) \text{ obtains, or } \gamma < 1, \\ O\big(n^{-1}(\log\log n)^2\big) \qquad \text{if } \gamma = 1, \\ O\big(n^{-1}(\log\log n)(\log n^{(1+\epsilon)(\gamma-1)})\big) \\ \qquad \text{if } \gamma > 1, \end{cases} \tag{38}$$

where $\epsilon > 0$ is arbitrary and γ is as in (c).

The latter result is a strong invariance statement for the two-parameter stochastic process $\{f(Q(y))R_n(y);\ 0 < y < 1,\ n = 1, 2, \dots\}$ in terms of $\{R_n^*(y);\ 0 < y < 1,\ n = 1, 2,, \dots,\}$, provided, of course, that one can prove an almost sure rate of convergence to zero of the latter remainder term so that, via (39), it should be inherited by the former. The final answer concerning R_n^* was given by Kiefer [40]:

$$\limsup_{n\to\infty} \sup_{0\leqslant y\leqslant 1} \Big[|R_n^*(y)|n^{3/4}(\log n)^{-1/2} \times (\log\log n)^{-1/4}\Big] \overset{\text{a.s.}}{=} 2^{-1/4}. \tag{40}$$

This, in turn, when combined with (38), gives [16]

$$\limsup_{n\to\infty} \sup_{0\leqslant y\leqslant 1} \Big[|f(Q(y))R_n(y)|n^{3/4} \times (\log n)^{-1/2}(\log\log n)^{-1/4}\Big] \overset{\text{a.s.}}{=} 2^{-1/4} \tag{41}$$

under the conditions of Theorem 3, and thus extends Kiefer's [40] theory of deviations between the sample quantile and empirical processes (for a detailed discussion, see Csörgő [13, Chap. 6]). Shorack [62] showed that (40) itself can be also deduced from strong invariance considerations.

From $\sup_{0\leqslant y\leqslant 1} n^{-1/2}|u_n(y)| \overset{\text{a.s.}}{\to} 0$ $(n\to\infty)$, an obvious Glivenko–Cantelli theorem* for the uniform quantile process u_n, by Theorem 3 we can also conclude that

$$\sup_{0\leqslant y\leqslant 1} n^{-1/2}|\rho_n(y)| = \sup_{0\leqslant y\leqslant 1} f(Q(y))|Q_n(y) - Q(y)| \overset{\text{a.s.}}{\to} 0 \qquad (n\to\infty),$$

a Glivenko–Cantelli theorem for ρ_n under the conditions of Theorem 3. Mason [44] proved: *Let $\nu > 0$ be fixed, and assume that Q is continuous. Then we have*

$$\limsup_{n\to\infty} \sup_{0\leqslant y\leqslant 1} (y(1-y))^{\nu}|Q_n(y) - Q(y)| \overset{\text{a.s.}}{=} \begin{cases} 0 & \text{if } \int_0^1 |Q(y)|^{1/\nu}\,dy < \infty, \\ \infty & \text{if } \int_0^1 |Q(y)|^{1/\nu}\,dy = \infty. \end{cases}$$

The latter is a very nice Glivenko–Cantelli theorem for the quantile process, replacing the tail conditions of Theorem 3 by the moment conditions

$$\int_0^1 |Q(y)|^{1/\nu}\,dy = \int_{-\infty}^{\infty} |x|^{1/\nu}\,dF(x) < \infty.$$

Mason [44] also discusses the relationship between his moment condition and those of Theorem 3.

For applications of Theorem 3, see Csörgő [13, Chaps. 7 to 10] and references therein to topics on quadratic forms* of the quantile process of the product-limit estimator, nearest-neighbor empirical density, and score functions. For related results on nearest-neighbor density function estimators, we refer the reader to Moore and Yackel [47] and references therein.

With the sequence of Brownian bridges B_n of Theorem 2 and under the conditions of Theorem 3, we have

$$\sup_{0\leqslant y\leqslant 1} |\rho_n(y) - B_n(y)| = o_P(1) \qquad (n\to\infty), \tag{42}$$

and hence the weak convergence of $\rho_n(\cdot)$ to a Brownian bridge $B(\cdot)$ in Skorohod's $D[0,1]$ space also follows. For a description of the latter space, we refer the reader to Billingsley [5, Chap. 3].

On the other hand, Theorems 2 and 3 were proved having mainly strong approximations in mind. It is of independent interest to see if (42) could be true under milder conditions of $f = F'$ than those of Theorem 3. Toward this end we first note that an improved construction for approximating the uniform quantile process u_n (see Csörgő et al. [23, Paper 1, Sec. 1]) leads to the following common generalization of (26) and (30) of Theorems 1 and 2, respectively:

Theorem 4 (Csörgő et al. [22]). For u_n and α_n there exists a probability space with a sequence of Brownian bridges B_n such that for every $0 < \lambda < \infty$ as $n \to \infty$,

$$\sup_{\lambda/n \leqslant y \leqslant 1-\lambda/n} \frac{n^{\nu}|(-1)u_n(y) - B_n(y)|}{(y(1-y))^{(1/2)-\nu}} = \begin{cases} O_P(\log n) & \text{when } \nu = \frac{1}{2}, \\ O_P(1) & \text{when } 0 \leqslant \nu < \frac{1}{2}, \end{cases} \tag{43}$$

and

$$\sup_{\lambda/n \leqslant y \leqslant 1-\lambda/n} \frac{n^{\nu}|\alpha_n(y) - B_n(y)|}{(y(1-y))^{(1/2)-\nu}} = \begin{cases} O_P(\log n) & \text{when } \nu = \frac{1}{4}, \\ O_P(1) & \text{when } 0 \leqslant \nu < \frac{1}{4}. \end{cases} \tag{44}$$

For the far-reaching implications of this theorem, see the two papers in Csörgő et al. [23] and the first four papers in Csörgő et al. [24]. Here we illustrate only its *immediate* usefulness in studying ρ_n.

Let p be any such *positive* function on (0, 1) for which

$$\lim_{s \downarrow 0} p(s) = \lim_{s \uparrow 1} p(s) = \infty, \tag{45}$$

where by calling p *positive* on (0, 1) we mean that

$$\inf_{c \leqslant s \leqslant 1-c} p(s) > 0 \qquad \text{for all } 0 < c < \tfrac{1}{2}.$$

Then by Theorem 4 we get

Theorem 5 (Csörgő et al. [22]). On the probability space of Theorem 4 we have, as $n \to \infty$,

$$\sup_{1/(n+1) \leqslant y \leqslant n/(n+1)} \frac{|(-1)u_n(y) - B_n(y)|}{\left[(y(1-y))^{1/2} p(y)\right]} = o_P(1), \tag{46}$$

$$\sup_{1/(n+1) \leqslant y \leqslant n/(n+1)} \frac{|\alpha_n(y) - B_n(y)|}{\left[(y(1-y))^{1/2} p(y)\right]} = o_P(1), \tag{47}$$

and

$$\sup_{0<y<1} \frac{|\alpha_n(y) - \overline{B}_n(y)|}{\left[(y(1-y))^{1/2} p(y)\right]} = o_P(1), \tag{48}$$

with any function p as in (44), where

$$\overline{B}_n(y) = \begin{cases} B_n(y) & \text{for } y \in \left[\dfrac{1}{(n+1)}, \dfrac{n}{(n+1)}\right] \\ 0 & \text{elsewhere.} \end{cases} \tag{49}$$

Let q be any positive function defined on (0, 1), nondecreasing in a neighborhood of zero and nonincreasing in a neighborhood of 1. Such a function q will be called an *Erdös–Feller–Kolmogorov–Petrovski* (E-F-K-P) *upper class function* of a Brownian bridge $\{B(s);\ 0 \leqslant s \leqslant 1\}$ if and only if

$$\limsup_{s \downarrow 0} |B(s)|/q(s) = \limsup_{s \uparrow 1} |B(s)|/q(s) \overset{\text{a.s.}}{=} \beta \tag{50}$$

for some constant $0 \leqslant \beta < \infty$.

An E-F-K-P upper class function q of a Brownian bridge will be called a *Chibisov–O'Reilly function* if $\beta = 0$ in (50).

Any E-F-K-P upper class function q of a Brownian bridge has the property that

$$\lim_{s \downarrow 0} q(s)/s^{1/2} = \lim_{s \uparrow 1} q(s)/(1-s)^{1/2} = \infty. \tag{51}$$

For characterizations of E-F-K-P upper

class and Chibisov–O'Reilly functions in terms of integrals, we refer the reader to Csörgő et al. [22, Prop. 3.1, Theorems 3.3, 3.4].

If, in addition to p being a function as in (45), it is also such that

$$w(y) := (y(1-y))^{1/2} p(y), \qquad 0 < y < 1, \tag{52}$$

is an E-F-K-P upper class, or a Chibisov–O'Reilly function of a Brownian bridge, then by (48) and some further thought we have

Theorem 6 (Csörgő et al. [22, Theorems 4.2.1, 4.2.2]). Let w be a positive function on $(0, 1)$ such that it is nondecreasing in a neighborhood of zero and nonincreasing in a neighborhood of 1. On the probability space of Theorem 4 we have, as $n \to \infty$,

$$\sup_{0<y<1} |\alpha_n(y) - B(y)|/w(y) = \begin{cases} O_P(1) & \text{if and only if } w \text{ is an E-F-K-P upper class function on } (0,1), \\ o_P(1) & \text{if and only if } w \text{ is a Chibisov–O'Reilly function on } (0,1). \end{cases} \tag{53}$$

The second statement of (53) was first proved by Chibisov [12], assuming some regularity conditions on w, and then [48] by O'Reilly, assuming only the continuity of w, which is not assumed in Theorem 6.

By Csörgő et al. [22, Lemma 4.2.2] we have

$$\sup_{1/(n+1) \leqslant y \leqslant n/(n+1)} |B(y)|/w(y) \xrightarrow{\mathscr{D}} \sup_{0<y<1} |B(y)|/w(y) \tag{54}$$

for any Browian bridge *whenever w is an E-F-K-P upper class function*. Hence by (46), (48), and (54) we have

$$\sup_{1/(n+1) \leqslant y \leqslant n/(n+1)} |u_n(y)|/w(y) \xrightarrow{\mathscr{D}} \sup_{0<y<1} |B(y)|/w(y) \tag{55}$$

and

$$\sup_{0<y<1} |\alpha_n(y)|/w(y) \xrightarrow{\mathscr{D}} \sup_{0<y<1} |B(y)|/w(y), \tag{56}$$

whenever w is an E-F-K-P upper class function. In fact, (56) holds if and only if w is an E-F-K-P upper class function [see Csörgő et al. [22, Theorem 4.2.3]. Equation (55) can also be thus extended, *provided*, for example, that we redefine u_n to be zero on the outside of the interval $[1/(n+1),\ n/(n+1)]$ (see Csörgő et al. [22, Sec. 4.3]). *Let $\bar{u}_n$ be the latter redefined uniform quantile process. Then* (53) *and* (56) *hold true if we replace α_n by $(-1)\bar{u}_n$ in them.*

From the second statement of (53), weak convergence of α_n/w on $D[0,1]$ follows if w is a Chibisov–O'Reilly function. The second statement of (53), however, does not imply (56). Consequently, as far as convergence in distribution of supfunctionals of α_n/w (and also those of u_n/w and ρ_n/w, of course) is concerned, a Chibisov–O'Reilly type theorem like that of (53) is far from being optimal. It excludes all those E-F-K-P upper class functions w from the game of playing (56) which are not necessarily Chibisov–O'Reilly functions.

To illustrate the promised immediate usefulness of these results in studying ρ_n, we need one more result from Csörgő et al. [22, Lemma 4.2.1]:

For any Chibisov–O'Reilly function w on $(0,1)$ there exists an E-F-K-P upper class function w^* on $(0,1)$ such that with $g(y) := w(y)/w^*(y)$, we have (57)

$$\lim_{s \downarrow 0} g(y) = \lim_{s \uparrow 1} g(y) = \infty.$$

From now on, let F be an absolutely continuous distribution function with density function $f = F'$ that is positive on (a, b), the support of F, where a and b are defined as in (a) of Theorem 3. Given a random sample $X_1, \ldots, X_n$ on F, define ρ_n in terms of these RVs and u_n in terms of the F induced uniform $(0,1)$ RV $U_1 = F(X_1)$,

$\ldots, U_n = F(X_n)$. Then, *à la* (24),

$$\rho_n(y) = u_n(y) + u_n(y)\left[\frac{f(Q(y))}{f(Q(\theta_{y,n}))} - 1\right]$$
$$:= u_n(y) + u_n(y)\eta_n(y),$$
$$y \in (0,1), \quad \theta_{y,n} \in I_{n,y}, \quad n \geqslant 1, \quad (58)$$

where $I_{n,y} := (U_n(y) \wedge y, U_n(y) \vee y)$.

Corollary 1. *Let F be an absolutely continuous distribution function with density function $f = F'$ that is positive on (a, b), the support of F. Assume, as $n \to \infty$, that for some E-F-K-P upper class function w on* $(0, 1)$,

$$\sup_{1/(n+1) \leqslant y \leqslant n/(n+1)} \sup_{\theta_{y,n} \in I_{n,y}} w(y)|\eta_n(y)| \xrightarrow{P} 0, \quad (59)$$

or that

for any given $0 < \delta < 1$ and $\epsilon > 0$ there exist $0 < c < 1$ and n_0 such that for some E-F-K-P upper class function w on $(0, 1)$,

$$\text{(a)}\ P\left\{\sup_{1/(n+1) \leqslant y \leqslant c} \sup_{\theta_{y,n} \in I_{n,y}} W(y)|\eta_n(y)| > \epsilon\right\} \leqslant \delta$$

and (symmetrically) (60)

$$\text{(b)}\ P\left\{\sup_{1-c \leqslant y \leqslant n/(n+1)} \sup_{\theta_{y,n} \in I_{n,y}} W(y)|\eta_n(y)| > \epsilon\right\} \leqslant \delta$$

for all $n \geqslant n_0$ if f is continuous.

Then on the probability space of Theorem 4, *we have*

$$\sup_{1/(n+1) \leqslant y \leqslant n/(n+1)} |\rho_n(y) - u_n(y)| = o_P(1) \quad (n \to \infty), \quad (61)$$

and

$$\sup_{1/(n+1) \leqslant y \leqslant n/(n+1)} |\rho_n(y) - B_n(y)| = o_P(1) \quad (n \to \infty). \quad (62)$$

Given (58), the proofs of (61) and (62) are immediate through (43) and (55). Naturally, purely analytic conditions on $f(Q)$ are preferable to the *in probability* conditions of (59) and (60). Several such sets of analytic conditions can be fashioned in terms of only the conditions (d) and (e) of Theorem 3, or in terms of regularly varying $f(Q)$ functions, which are sufficient for (59) and/or (60) to hold true. For hints as to how to go about this, see Csörgő et al. [20] and Csörgő [13, Chap. 5]. If we redefine u_n (and hence also ρ_n) to be equal to zero on $[0, 1/(n+1))$ and $(n/(n+1), 1]$, then (62) will look like (42) and it can be put in terms of weak convergence on $D[0,1]$. For an earlier result, see Stute [68].

For further use in what follows we quote here Wellner [70, Lemma 2] based on the result of Csörgő [13, Theorem 1.5.1, eq. (2.8)]. *Under the conditions* (a) *to* (c) *of Theorem* 3, *we have*

$$\lim_{\epsilon \to \infty} \limsup_{n \to \infty} P\{A_n > \epsilon\} = 0, \quad (63)$$

where

$$A_n = \sup_{1/(n+1) \leqslant y \leqslant n/(n+1)} \sup_{\theta_{y,n} \in I_{n,y}} |\eta_n(y)|.$$

We note that (63) implies (60), with any E-F-K-P upper class function w for which we have $\lim_{s\downarrow 0} w(s) = \lim_{s \uparrow 1} w(s) = 0$, and hence, *under the conditions* (a) *to* (c) *of Theorem* 3, *we have* (61) *and* (62) *on the probability space of Theorem* 4. (The latter is a restatement, on a different probability space, of Csörgő [13, Theorem 2.1].)

Given Theorem 6 in terms of $\bar{u}_n$, by (57) and (63) we have

Corollary 2. *Let F be a continuous distribution function satisfying conditions* (a) *to* (c) *of Theorem* 3. *Then on the probability space of Theorem* 4 *we have, as $n \to \infty$,*

$$\sup_{1/(n+1) \leqslant y \leqslant n/(n+1)} \frac{|\rho_n(y) - u_n(y)|}{w(y)} = o_P(1) \quad (64)$$

and

$$\sup_{1/(n+1) \leqslant y \leqslant n/(n+1)} \frac{|\rho_n(y) - B_n(y)|}{w(y)} = o_P(1) \quad (65)$$

with any Chibisov–O'Reilly function w on $(0, 1)$.

The latter corollary can be restated under milder conditions, fashioned after Csörgő [13, Theorem 5.3.1]:

Corollary 3. *Let F be an absolutely continuous distribution function with density function $f = F'$ that is positive on (a, b), the support of F. Assume that for some function g as in* (57),

$$\sup_{1/(n+1) \leqslant y \leqslant n/(n+1)} \ \sup_{\theta_{y,n} \in I_{n,y}} |\eta_n(y)|/g(y) \xrightarrow{P} 0 \qquad (n \to \infty), \quad (66)$$

or that

for any given $0 < \delta < 1$ and $\epsilon > 0$ there exist $0 < c < 1$ and n_0 such that

$$P\left\{ \sup_{1/(n+1) \leqslant y \leqslant c} \ \sup_{\theta_{y,n} \in I_{n,y}} |\eta_n(y)|/g(y) > \epsilon \right\} \leqslant \delta \quad (67)$$

and (symmetrically)

$$P\left\{ \sup_{1-c \leqslant y \leqslant n/(n+1)} \ \sup_{\theta_{y,n} \in I_{n,y}} |\eta_n(y)|/g(y) > \epsilon \right\} \leqslant \delta$$

for all $n \geqslant n_0$ if f is continuous.

Then on the probability space of Theorem 4 *we have* (64) *and* (65).

We note that (63) implies (67) with any function g as in (57), and hence we get Corollary 2 again. On replacing the g function of Csörgő [13, Theorem 5.3.1] and that of Csörgő et al. [20] by a g function as in (57), several sets of analytic conditions which are sufficient for (66) and/or (67) can be fashioned in terms of only the conditions (d) and (e) of Theorem 3, or in terms of regularly varying $f(Q)$ functions, along the same lines as in the two references just quoted. If w on (0, 1) is such a Chibisov–O'Reilly function that $w(s)/\{s(1-s)\}^{1/2}$ $\nearrow$ near zero and 1, then w^* can be taken to be $(y(1-y)\log\log[1/\{y(1-y)\}])^{1/2}$ in (57), and then Corollary 3 reduces to Theorem 5.3.1 in ref. 13 and the corresponding results of Csörgő et al. [20] hold true in their original form.

If we redefine u_n (and hence ρ_n) to equal zero on $[0, 1/(n+1))$ and $(n/(n+1), 1]$, then (65) of Corollary 2 will read as

$$\sup_{0<y<1} |\rho_n(y) - B_n(y)|/w(y) = o_P(1), \qquad n \to \infty \quad (68)$$

with any Chibisov–O'Reilly function w, and a similar statement holds also under the conditions of Corollary 3. Hence we have also weak convergence of ρ_n/w on $D[0, 1]$. (For earlier results we refer to Shorack [58, 59, 63], and for related results to Pyke and Shorack [55].)

For entirely invariance-based complete proofs and extensions of the quoted Chibisov [12]–O'Reilly [48] and E-F-K-P upper class theorems for α_n/w and u_n/w, see Csörgő et al. [22, Secs. 3, 4.2, 4.3, 4.6, 5]. In Sec. 4.4 of ref. 22 the Darling and Erdős [25]-based results of Eicker [31] and Jaeschke [39] on the asymptotic distribution of the supremum of standardized empirical and quantile processes $\alpha_n(y)/(y(1-y))^{1/2}$ and $u_n(y)/(y(1-y))^{1/2}$ are also proven, entirely by invariance principles*. For similar results on $\rho_n/(y(1-y))^{1/2}$ under the conditions (a) to (c) of Theorem 3, see Csörgő and Révész [18, Theorem 5.5.1], or Csörgő and Révész [17]. Using the just quoted paper of Csörgő et al. [22, Theorem 4], the asymptotic distribution of Rényi-type statistics are derived, which unify and generalize results of Rényi [56] and Csáki [9] (see also Csörgő et al. [24, Papers 3, 4]), while in ref. 22 (Sec. 4.6) the necessary and sufficient condition is obtained for an invariance principle for weighted increments of the uniform quantile and empirical processes (see also Mason [45]). For iterated logarithm-type results for $\alpha_n(y)/(y(1-y))^{1/2}$, see Csáki [10, 11], Shorack [61], and Mason [43].

Quantile and moment techniques were successfully combined with the Chibisov–O'Reilly theorems in the monograph of Csörgő et al. [21] to provide a unified asymptotic theory for mean residual life (see also Hall and Wellner [36, 37]), total time on test, scaled total time on test, empirical Lorenz, and empirical concentration (see also Goldie [34]) processes.

For further applications of and developments on quantile processes, see Shorack [60], Parzen [49–53], Mason [46], Horváth [38], LaRiccia and Mason [42], and the monograph of Shorack and Wellner [64]. In my recent work on quantiles [13] I found Wellner [70] a very useful reference. I wish to thank David M. Mason for his advice while this exposition went through several revisions.

References

[1] Aly, E.-E. A. A. and Bleuer, S. (1983). *Tech. Rep. Ser. Lab. Res. Statist. Prob., No. 16*, Carleton University/University of Ottawa, Canada.

[2] Aly, E.-E. A. A., Csörgő, M., and Horváth, L. (1985). *J. Multivariate Anal.*, **16**, 185–210.

[3] Bahadur, R. R. (1966). *Ann. Math. Statist.*, **37**, 577–580.

[4] Bártfai, P. (1966). *Studia Sci. Math. Hung.*, **1**, 161–168.

[5] Billingsley, P. (1968). *Convergence of Probability Measures*. Wiley, New York.

[6] Breiman, L. (1968). *Probability*. Addison-Wesley, Reading, Mass.

[7] Brillinger, D. L. (1969). *Bull. Amer. Math. Soc.*, **75**, 545–547.

[8] Burke, M. D., Csörgő, M., Csörgő, S., and Révész, P. (1979). *Ann. Prob.*, **7**, 790–810.

[9] Csáki, E. (1974). English translation in: *Select. Transl. Math. Statist. Prob.*, **15**, 229–317 (1981).

[10] Csáki, E. (1975). *Limit Theorems in Probability Theory (Coll. Math. Soc. J. Bolyai 11)*, P. Révész, ed. North-Holland, New York, pp. 47–58.

[11] Csáki, E. (1977). *Z. Wahrscheinl. verw. Geb.*, **38**, 147–167.

[12] Chibisov, D. (1964). *Select. Transl. Math. Statist. Prob.*, **6**, 147–156.

[13] Csörgő, M. (1983). *Quantile Processes with Statistical Applications*. CBMS Reg. Conf. Ser. Appl. Math. 42. SIAM, Philadelphia.

[14] Csörgő, S. and Hall, P. (1984). *Aust. J. Statist.*, **26**, 189–218.

[15] Csörgő, M. and Révész, P. (1975). *Limit Theorems in Probability Theory (Coll. Math. Soc. J. Bolyai 11)*, P. Révész, ed. North-Holland, New York, pp. 59–71.

[16] Csörgő, M. and Révész, P. (1978). *Ann. Statist.*, **6**, 882–894.

[17] Csörgő, M. and Révész, P. (1979). *Optimizing Methods in Statistics (Proc. Int. Conf.)*, I. S. Rustagi, ed. Academic Press, New York, pp. 125–140.

[18] Csörgő, M. and Révész, P. (1981). *Strong Approximations in Probability and Statistics*. Akadémiai Kiadó, Budapest/Academic Press, New York.

[19] Csörgő, M. and Révész, P. (1981). *Two Approaches to Constructing Simultaneous Confidence Bounds for Quantiles*. Carleton Math. Ser. No. 176, Carleton University, Ottawa.

[20] Csörgő, M., Csörgő, S., Horváth, L., and Révész, P. (1982). To appear in *Proc. 7th Conf. Prob. Theory*, Brasov, Aug. 29–Sept. 4, 1982.

[21] Csörgő, M., Csörgő, S., Horváth, L., and Mason, D. M. (1983). *An Asymptotic Theory for Empirical Reliability and Concentration Processes*. Book manuscript in progress.

[22] Csörgő, M., Csörgő, S., Horváth, L., and Mason, D. M. (1983). Weighted empirical processes, Paper No. 1 in M. Csörgő et al. [23].

[23] Csörgő, M., Csörgő, S., Horváth, L., and Mason, D. M. (1984). *Tech. Rep. Ser. Lab. Res. Statist. Prob., No. 24*, Carleton University/University of Ottawa, Canada.

[24] Csörgő, M., Csörgő, S., Horváth, L., and Mason, D. M. (1984). *Tech. Rep. Ser. Lab. Res. Statist. Prob., No. 25*, Carleton University/University of Ottawa, Canada.

[25] Darling, D. A. and Erdős, P. (1956). *Duke Math. J.*, **23**, 143–145.

[26] Donsker, M. (1952). *Ann. Math. Statist.*, **23**, 277–283.

[27] Doob, J. L. (1949). *Ann. Math. Statist.*, **20**, 393–403.

[28] Dudley, R. M. (1968). *Ann. Math. Statist.*, **39**, 1563–1572.

[29] Durbin, J. (1973a). *Ann. Statist.*, **1**, 279–290.

[30] Durbin, J. (1973b). *Distribution Theory for Tests Based on the Sample Distribution Function*. CBMS Reg. Conf. Ser. Appl. Math. 9. SIAM, Philadelphia.

[31] Eicker, F. (1979). *Ann. Statist.*, **7**, 116–138.

[32] Erdős, P. and Rényi, A. (1970). *J. Anal. Math.*, **23**, 103–111.

[33] Gaenssler, P. and Stute, W. (1979). *Ann. Prob.*, **7**, 193–243.

[34] Goldie, C. M. (1977). *Adv. Appl. Probability*, **9**, 765–791.

[35] Hájek, J. and Šidák, Z. (1967). *Theory of Rank Tests*. Academic Press, New York.

[36] Hall, W. J. and Wellner, J. A. (1979). Estimation of Mean Residual Life. Unpublished manuscript.

[37] Hall, W. J. and Wellner, J. A. (1981). In *Statistics and Related Topics*, M. Csörgő, D. A. Dawson, J. N. K. Rao, and A. K. M. E. Saleh, eds. North-Holland, Amsterdam, pp. 169–184.

[38] Horváth, L. (1983). Candidatus dissertation, The Hungarian Academy of Sciences, Budapest.

[39] Jaeschke, D. (1979). *Ann. Statist.*, **7**, 108–115.

[40] Kiefer, J. (1970). *Nonparametric Techniques in Statistical Inference*, M. L. Puri, ed. Cambridge University Press, Cambridge, England, pp. 299–319.

[41] Komlós, J., Major, P., and Tusnády, G. (1975). *Z. Wahrscheinl. verw. Geb.*, **32**, 111–131.

[42] LaRiccia, V. N. and Mason, D. M. (1983). Optimal goodness-of-fit tests for location/scale families of distributions based on the sum of squares of L-statistics (in press).

[43] Mason, D. M. (1981). *Ann. Prob.*, **9**, 881–884.

[44] Mason, D. M. (1982). *Z. Wahrscheinl. verw. Geb.*, **59**, 505–513.

[45] Mason, D. M. (1983). *Stochastic Processes Appl.*, **15**, 99–109.

[46] Mason, D. M. (1984). *Ann. Prob.*, **12**, 243–255.

[47] Moore, D. S. and Yackel, J. W. (1977). *Ann. Statist.*, **5**, 143–154.

[48] O'Reilly, N. (1974). *Ann. Prob.*, **2**, 642–651.

[49] Parzen, E. (1979). *J. Amer. Statist. Ass.*, **74**, 105–131.

[50] Parzen, E. (1979). In *Robust Estimation Workshop Proceedings*, R. Launer and G. Wilkinson, eds. Academic Press, New York, pp. 237–258.

[51] Parzen, E. (1979). Density-Quantile Estimation Approach to Statistical Data Modeling. *Tech. Rep. No. A-5*, Statistical Institute, Texas A&M University, College Station, Tex.

[52] Parzen, E. (1979). Statistical Science, Statistical Data Modeling, and Statistical Education. *Tech. Rep. No. A-6*, Statistical Institute, Texas A&M University, College Station, Tex.

[53] Parzen, E. (1980). Quantile Functions, Convergence in Quantile, and Extreme Value Distribution Theory. *Tech. Rep. No. B-3*, Statistical Institute, Texas A&M University, College Station, Tex.

[54] Prohorov, Yu. V. (1956). *Teor. Verojatn. Primen.*, **1**, 177–238.

[55] Pyke, R. and Shorack, G. R. (1968). *Ann. Math. Statist.*, **39**, 755–771.

[56] Rényi, A. (1953). *Acta Math. Acad. Sci. Hung.*, **4**, 191–231.

[57] Serfling, R. J. (1980). *Approximation Theorems of Mathematical Statistics*. Wiley, New York.

[58] Shorack, G. R. (1972). *Ann. Math. Statist.*, **43**, 412–427.

[59] Shorack, G. R. (1972). *Ann. Math. Statist.*, **43**, 1400–1411.

[60] Shorack, G. R. (1979). *Stoch. Processes Appl.*, **9**, 95–98.

[61] Shorack, G. R. (1980). *Aust. J. Statist.*, **22**, 50–59.

[62] Shorack, G. R. (1982). *Z. Wahrscheinl. verw. Geb.*, **61**, 369–373.

[63] Shorack, G. R. (1982). IMS *Bull.*, **11**, Abstr. 82t-2, 60.

[64] Shorack, G. R. and Wellner, J. A. (1983). Book manuscript in progress.

[65] Skorohod, A. V. (1956). *Teor. Verojatn. Primen.*, **1**, 289–319.

[66] Skorohod, A. V. (1961). *Studies in the Theory of Random Processes*. Addison-Wesley, Reading, Mass.

[67] Strassen, V. (1965). *Proc. 5th Berkeley Symp. Math. Statist. Prob.*, Vol. 2, University of California Press, Berkeley, Calif., pp. 315–344.

[68] Stute, W. (1982). *Ann. Prob.*, **10**, 86–107.

[69] Tukey, J. W. (1965). *Proc. Nat. Acad. Sci.*, **53**, 127–134.

[70] Wellner, J. A. (1978). *Z. Wahrscheinl. verw. Geb.*, **45**, 73–88.

[71] Wichura, M. J. (1970). *Ann. Math. Statist.*, **41**, 284–291.

(BROWNIAN MOTION
GAUSSIAN PROCESSES
GLIVENKO–CANTELLI THEOREMS
KIEFER PROCESS
PROCESSES, EMPIRICAL
QUANTILES
QUANTILE TRANSFORMATION METHOD)

MIKLÓS CSÖRGŐ

QUANTILES

Quantiles play a fundamental role in statistics, although their use is often disguised by notational and other artifices. They are the critical values we use in hypothesis testing* and interval estimation and often are the characteristics of a distribution we wish most to estimate. Sample quantiles are utilized in numerous inferential settings and, recently, have received increased attention as a useful tool in data modeling.

Historically, the use of sample quantiles in statistics dates back, at least, to Quetelet* [40], who considered the use of the semi-interquartile range as an estimator of the probable error* for a distribution. Subsequent papers by Galton* and Edgeworth (see, e.g., Galton [17] and Edgeworth [11,

12], and references therein) discussed the use of other quantiles, such as the median, in various estimation settings. Sheppard [45] and then Pearson [38] studied the problem of optimal quantile selection for the estimation of the mean and standard deviation of the normal distribution by linear functions of subsets of the sample quantiles. Pearson's paper also contained most of the details involved in the derivation of the asymptotic distribution of a sample quantile. The large-sample behavior of a sample quantile was later investigated by Smirnoff [47], who gave a rigorous derivation of its limiting distribution. Smirnoff's results were generalized in a landmark paper by Mosteller [33], which, along with work by Ogawa [34], generated considerable interest in quantiles as estimation tools in location and scale parameter models. In more recent years quantiles have been utilized in a variety of problems of both classical and robust statistical inference, and have played an important part in the work of Tukey [50] and Parzen [36] on exploratory data analysis* and nonparametric data modeling.

In this article the focus will be on the role of quantiles in various areas of statistics both as parameters of interest as well as means to other ends. We begin by defining the notion of population and sample quantiles.

Let F be a distribution function (DF) for a random variable X and define the associated *quantile function* (QF) by

$$Q(u) = F^{-1}(u) = \inf\{x : F(x) \geqslant u\}, \qquad 0 < u < 1. \quad (1)$$

Thus, for a fixed p in $(0, 1)$, the pth *population quantile* for X is $Q(p)$. It follows from definition (1) that knowledge of Q is equivalent to knowledge of F. Further relationships between F and Q are

(a) $FQ(u) \geqslant u$ with equality when F is continuous.

(b) $QF(x) \leqslant x$ with equality when F is continuous and strictly increasing.

(c) $F(x) \geqslant u$ if and only if $Q(u) \leqslant x$.

Another important property of the QF which follows easily from relationship (c) is that if U has a uniform distribution on $[0, 1]$, then $Q(U)$ and X have identical distributions. This fact provides one of the basic tools in many areas of statistical analysis. For example, in statistical simulation* it has the consequences that a random sample from the uniform distribution may be used in conjunction with Q, to obtain a random sample from X.

The sample analog of Q is obtained by use of the *empirical distribution function* (EDF*). Let $X_{1:n}, X_{2:n}, \ldots, X_{n:n}$ denote the order statistics* for a random sample of size n from a distribution F; then the usual empirical estimator of F is

$$\tilde{F}(x) = \begin{cases} 0, & x < X_{1:n}, \\ j/n, & X_{j:n} \leqslant x < X_{j+1:n}, \\ & \quad j = 1, \ldots, n-1, \\ 1, & x \geqslant X_{n:n}. \end{cases} \quad (2)$$

Replacing F with $\tilde{F}$ in (1) gives the *sample* or *empirical quantile function* (EQF)

$$\tilde{Q}(u) = X_{j:n}, \qquad \frac{j-1}{n} < u \leqslant \frac{j}{n}, \qquad j = 1, \ldots, n. \quad (3)$$

Thus the fundamental sample statistics $\tilde{Q}$, $\tilde{F}$, and the order statistics are all closely related. From (2) and (3), knowledge of any one implies knowledge of the other two.

The previous discussions apply to both continuous and discrete random variables. However, in subsequent work it will be assumed that F is continuous and admits a density $f = F'$. In this case we also define the *density-quantile function* (DQF)

$$fQ(u) = f(Q(u)), \qquad 0 \leqslant u \leqslant 1.$$

Differentiation of $FQ(u) = u$ reveals that Q and fQ are related by $Q'(u) = 1/fQ(u)$. Table 1 contains DF's, QF's, densities and DQF's for several common continuous distributions.

Table 1 Distribution, Quantile, Density, and Density-Quantile Functions for Selected Probability Laws

Probability Law	Distribution Function	Quantile Function	Density Function	Density-Quantile Function
Normal	$\Phi(x) = \int_{-\infty}^{x} \phi(x)\,dx$	$\Phi^{-1}(u)$	$\phi(x) = (2\pi)^{-1/2} e^{-x^2/2}$	$\phi\Phi^{-1}(u) = (2\pi)^{-1/2} e^{-\lvert\Phi^{-1}(u)\rvert^2/2}$
Lognormal	$\Phi(\log x)$	$e^{\Phi^{-1}(u)}$	$\frac{1}{x}\phi(\log x)$	$\phi\Phi^{-1}(u) e^{-\Phi^{-1}(u)}$
Exponential	$1 - e^{-x}, \quad x > 0$	$-\log(1-u)$	e^{-x}	$1-u$
Weibull	$1 - \exp(-x^c) \quad c, x > 0$	$c\{-\log(1-u)\}^{1-1/c}$	$cx^{c-1}e^{-x^c}$	$c(1-u)[-\log(1-u)]^{1-1/c}$
Extreme value	$\exp(-e^{-x})$	$-\log\log\frac{1}{u}$	$e^{-x}e^{-e^{-x}}$	$-u\log(u)$
Logistic	$\{1 + e^{-x}\}^{-1}$	$\log\frac{u}{1-u}$	$e^{-x}(1+e^{-x})^{-2}$	$u(1-u)$
Pareto	$1 - (1+x)^{-\nu}, \quad \nu, x > 0$	$(1-u)^{-1/\nu} - 1$	$\nu(1+x)^{-(\nu+1)}$	$\nu(1-u)^{1+1/\nu}$
Cauchy	$0.5 + \pi^{-1}\arctan x$	$\tan[\pi(u-0.5)]$	$[\pi(1+x^2)]^{-1}$	$\pi^{-1}\sin^2 \pi u$
Uniform	$x, \quad 0 \leqslant x \leqslant 1$	u	1	1
Reciprocal of a uniform	$1 - \frac{1}{x+1}, \quad x > 0$	$(1-u)^{-1} - 1$	$(x+1)^{-2}$	$(1-u)^2$
Double exponential	$\begin{cases} \frac{1}{2}e^{x}, & x < 0 \\ 1 - \frac{1}{2}e^{-x}, & x > 0 \end{cases}$	$\begin{cases} \log 2u, & u < 0.5 \\ -\log 2(1-u), & u > 0.5 \end{cases}$	$\frac{1}{2}e^{-\lvert x\rvert}$	$\begin{cases} u, & u < 0.5 \\ 1-u, & u > 0.5 \end{cases}$

SAMPLE QUANTILES: ASYMPTOTIC PROPERTIES AND NONPARAMETRIC INFERENCE

Sample quantiles provide nonparametric estimators of their population counterparts that are optimal in the sense that, for any fixed p in $(0, 1)$ no other translation equivariate asymptotically median unbiased estimator* is asymptotically more concentrated about $Q(p)$ than is $\tilde{Q}(p)$; similar properties hold for tests about $Q(p)$ based on $\tilde{Q}(p)$ [39].

Several alternatives to $\tilde{Q}$ as defined in (3) also have useful properties. This estimator duplicity stems, in part, from the discreteness of $\tilde{F}$, which entails that for $\tilde{F}(X_{j-1:n}) \leqslant p \leqslant \tilde{F}(X_{j:n})$ any value between $X_{j-1:n}$ and $X_{j:n}$ can, intuitively, act as the pth sample quantile. Thus one could consider combining both $X_{j-1:n}$ and $X_{j:n}$ or, more generally, several order statistics in the neighborhood of $X_{j:n}$ to obtain an estimator of $Q(p)$. Such considerations have led to the usual definition for the sample median, which agrees with $\tilde{Q}$ (0.5) only when n is odd, and have prompted several authors to propose linearized versions of $\tilde{Q}$ (see, e.g., Parzen [36]). Estimators of $Q(p)$ that utilize local smoothing of the order statistics near $X_{j:n}$ and appear to have good small-sample properties have been suggested by Kaigh and Lachenbruch [26], Kaigh [25], and Harrell and Davis [20]. Reiss [41] has considered the use of quasi-quantiles and shown them to be superior to sample quantiles when compared on the basis of deficiency rather than efficiency (*see* EFFICIENCY, SECOND-ORDER).

For $0 < p < 1$ it is well known (see Serfling [44]) that for fQ positive and continuous near p, $\tilde{Q}(p)$ is asymptotically normally distributed with mean $Q(p)$ and variance $p(1-p)/\{nfQ(p)^2\}$. An extension of this result to k quantiles, for fixed $k \geqslant 1$, can be found in Mosteller [33] and Walker [51], with the case of k growing with n treated by Ikeda and Matsunawa [23]. Necessary and sufficient conditions for the existence of moments for sample quantiles and for the convergence of these moments to those of the limiting distribution are provided by Bickel [4]. For a discussion of the asymptotic properties of $\tilde{Q}(p)$ for certain types of dependent samples, see Sen [43] and Babu and Singh [1].

From the asymptotic distribution of $\tilde{Q}(p)$ an asymptotic $100(1-\alpha)\%$ confidence interval for $Q(p)$ is given by

$$\tilde{Q}(p) \pm \Phi^{-1}(\alpha/2)\sqrt{p(1-p)/\{nfQ(p)^2\}}\ ,$$

which, unfortunately, requires knowledge of $fQ(p)$. This difficulty can be resolved by using instead the interval $(\tilde{Q}(k_1/n), \tilde{Q}(k_2/n))$, where k_1 and k_2 are integers chosen so that

$$k_1 \simeq np - \Phi^{-1}(\alpha/2)\sqrt{np(1-p)}$$

and

$$k_2 \simeq np + \Phi^{-1}(\alpha/2)\sqrt{np(1-p)}\ .$$

The latter interval is asymptotically equivalent to the former but utilizes the asymptotic relationship between $\tilde{Q}(k_i/n)$ and $\tilde{Q}(p)$ to estimate $fQ(p)$ (see Serfling [44, p. 103]). An alternative but similar approach is given in Walker [51]. For an exact confidence interval based on order statistics, see Wilks [55, p. 329]. Interval estimates obtained by bootstrapping and jackknifing have been proposed by Harrel and Davis [20] and Kaigh [25].

In testing hypotheses about $Q(p)$ the most widely known procedure is probably the quantile test, based on the fact that if H_0: $Q(p) = Q_0(p)$ is true, then the number of sample quantiles below or equal to $Q_0(p)$ will be binomial with parameters n and p. As a result, the binomial distribution may be utilized to obtain an exact test, or the normal approximation to the binomial for an approximate test, of H_0. The quantile test, as well as several other tests concerning the median, can be found in standard texts such as Conover [8].

From a data modeling perspective, what is of interest is not $Q(p)$ for some particular p, but rather the entire function $Q(\cdot)$, as its knowledge is equivalent to knowing the data's underlying probability law. Thus we now consider the construction of nonparametric estimators $\hat{Q}(\cdot)$ that are random functions or stochastic processes* on (0, 1) (this is the quantile domain analog of nonparametric probability distribution and density estimation*). The natural estimator of $Q(\cdot)$ is $\tilde{Q}(\cdot)$, whose asymptotic distribution theory, when considered as a stochastic process, has been studied by Shorack [46], Csörgő and Révész [10], Csörgő [9], Mason [32], and others. From their work it follows that when fQ is positive and differentiable on [0, 1] and satisfies certain other regularity conditions near 0 and 1, $\sqrt{n}\, fQ(u)\{\tilde{Q}(u) - Q(u)\}$ converges in distribution to a Brownian bridge process on (0, 1), that is, a zero-mean normal process with covariance kernel $K(u,v) = u - uv$, $u \leqslant v$ (analogous results for a linearized version of $\tilde{Q}$ and for the case of randomly censored data can be found in Bickel [4], Sander [42], and Csörgő [9]). Tests in this setting are of the goodness-of-fit* variety. The asymptotic distribution of many classical statistics, such as $\sup_{0<u<1} f_0 Q_0(u) \,|\, \tilde{Q}(u) - Q_0(u)|$, are available under the null hypothesis $H_0: Q(\cdot) = Q_0(\cdot)$, for specified Q_0, from Csörgő and Révész [10, p. 171] and Csörgő [9, Chap. 7]. Consequently, such statistics can be utilized to conduct quantile based goodness-of-fit tests. Another goodness-of-fit procedure that can be naturally formulated in the quantile domain is the Shapiro–Wilk* test for normality (see Csörgő and Révész [10, pp. 202–212] and Csörgő [9]). *See* also QUANTILE PROCESSES.

Several procedures are available for constructing smooth estimators of Q formed from suitably rich function classes. The Tukey lambda distribution and its generalizations as well as g-and-h distributions are examples of curves derived specifically for this purpose (*see* LAMBDA DISTRIBUTIONS *and* G-AND-H DISTRIBUTIONS). Other techniques, developed by Parzen [36], utilize certain analogies with time-series* analysis to provide estimators for fQ and Q as well as goodness-of-fit tests.

Another important asymptotic result is the Bahadur representation for sample quantiles, which describes the relationship between the $\tilde{F}$ and $\tilde{Q}$ processes. One statement of this result is that for fQ positive and differentiable at p, with probability 1,

$$n^{1/2}(\tilde{Q}(p) - Q(p)) = \frac{n^{1/2}(p - \tilde{F}Q(p))}{fQ(p)} + O(n^{-1/4}(\log n)^{3/4}).$$

This has the immediate consequence that both $n^{1/2}(\tilde{Q}(p) - Q(p))$ and also $n^{1/2}(p - \tilde{F}Q(p))/fQ(p)$ have identical asymptotic distributions. The Bahadur representation may also be used to obtain a law of the iterated logarithm* for sample quantiles, namely, with probability 1 [2]

$$\overline{\lim_{n\to\infty}} \pm \frac{n^{1/2}[\tilde{Q}(p) - Q(p)]}{(2\log\log n)^{1/2}} = \frac{[p(1-p)]^{1/2}}{fQ(p)}.$$

More general results and references may be found in Kiefer [27], Csörgő and Révész [10], and Csörgő [9]. The case of ϕ-mixing random variables is treated by Sen [43] and Babu and Singh [1].

PARAMETER ESTIMATION

We consider the use of quantile-based estimators in parametric models of the form $F(x) = F_0(x;\boldsymbol{\theta})$, where F_0 is a known distributional form and $\boldsymbol{\theta}$ is a vector of unknown parameters. An important special case that we will focus on initially is the location and scale parameter model where

$$F(x) = F_0\left(\frac{x-\mu}{\sigma}\right)$$

for μ and σ unknown location and scale parameters. In this instance $Q(u) = \mu + \sigma Q_0(u)$, where Q_0 is the QF for F_0.

The problem of location parameter estimation for symmetric distributions has been a subject of extensive study. Several quick estimators of μ that are useful for data from symmetric distributions are based on *symmetric quantile averages* of the form

$$\tilde{\mu}(p) = [\tilde{Q}(p) + \tilde{Q}(1-p)]/2, \quad 0 < p \leqslant 0.5.$$

One example is the Tukey trimean* $\{\tilde{\mu}(0.5) + \tilde{\mu}(\frac{1}{4})\}/2$; another is the estimator suggested by Gastwirth [18], $0.4\tilde{\mu}(0.5) + 0.6\tilde{\mu}(\frac{1}{3})$, which was found to be nearly 80% as efficient (asymptotically) as the best estimators for the Cauchy, double exponential, logistic, and normal distributions. For references and a general discussion of the robustness* and efficiency properties of symmetric quantile averages, see Brown [6].

General classes of estimators for μ are also conveniently (and usefully) formulated in the quantile domain. For example, if ψ is an odd function, an M-estimator* of μ is a solution to $\int_0^1 \psi(\tilde{Q}(u) - \hat{\mu})\,du = 0$. Similarly, an R-estimator satisfies

$$\int_0^1 J[u - \tilde{F}(2\hat{\mu} - \tilde{Q}(u))]\,du = 0,$$

with J an odd function on $[-1, 1]$, whereas an L-estimator* can be written explicitly as $\int_0^1 h(\tilde{Q}(u))\,dM(u)$ for some function h and some signed measure M on $(0, 1)$. See Huber [22] and Fernholz [16] for further discussion of these estimators.

Asymptotically efficient quantile-based estimators of both μ and σ that are applicable to general F_0 (not necessarily symmetric) have been given by Parzen [36, 37]. Using results from the preceding section it can be seen that, asymptotically, location and scale parameter estimation can be considered as a regression analysis problem for the quantile process via the model

$$f_0 Q_0(u)\,\tilde{Q}(u) = \mu f_0 Q_0(u) + \sigma f_0 Q_0(u)\,Q_0(u) + \sigma_B B(u), \qquad u \in [0, 1], \tag{4}$$

where $\sigma_B = \sigma/\sqrt{n}$ and $B(\cdot)$ is a Brownian bridge process. Under appropriate restrictions on $f_0 Q_0$ and the product $f_0 Q_0 \cdot Q_0$, continuous-time regression techniques can be utilized to obtain asymptotically efficient estimators for μ and σ,

$$\begin{bmatrix} \hat{\mu} \\ \hat{\sigma} \end{bmatrix} = \mathbf{A}^{-1} \begin{bmatrix} \int_0^1 W_\mu(u)\tilde{Q}(u)\,du \\ \int_0^1 W_\sigma(u)\tilde{Q}(u)\,du \end{bmatrix}, \tag{5}$$

where $\mathbf{A}$ is the usual Fisher information* matrix, $W_\mu(u) = -(f_0 Q_0)''(u) f_0 Q_0(u)$ and $W_\sigma(u) = -[f_0 Q_0(u) Q_0(u)]'' f_0 Q_0(u)$.

Many estimators based on quantiles and order statistics have strong ties to model (4) and the estimators in (5). For instance, L-estimation of location and scale (*see* L-STATISTICS) can be motivated from (5) through consideration of alternative weight functions in place of W_μ and W_σ. By using an analog of model (4) that holds for left and right censored data*, estimators similar to those given by Weiss [52] and Weiss and Wolfowitz [53] can be obtained. Through sampling from model (4) at a set of $k < n$ points $U = \{u_1, \ldots, u_k\}$ which satisfy $0 < u_1 < \cdots < u_k < 1$, "observations" $f_0 Q_0(u_i)\tilde{Q}(u_i)$ can be obtained that asymptotically have means $\mu f_0 Q_0(u_i) + \sigma f_0 Q_0(u_i) Q_0(u_i)$, $i = 1, \ldots, k$, and variance–covariance matrix consisting of the elements $\sigma_B^2 u_i(1 - u_j)$, $i \leqslant j$, $i, j = 1, \ldots, k$. Thus generalized least squares may be utilized to obtain asymptotically best linear unbiased estimators of μ and σ. Since their derivation by Ogawa [34], an extensive literature has developed on these latter estimators and the associated problem of optimal selection for the *spacing*, U (*see* OPTIMAL SPACING PROBLEMS).

The estimation of a particular quantile $Q(p)$, say, is often of interest in parametric settings such as the location and scale parameter model. As $Q(p) = \mu + \sigma Q_0(p)$, we see that to estimate $Q(p)$ for this model, it suffices to estimate μ and σ. This may be accomplished, for instance, using the estimators in (5) or maximum likelihood estimators. Alternatives that have good asymptotic efficiency properties and provide computational savings by using appropriate subsets of the sample quantiles have been suggested by Kubat and Epstein [30], Eubank [14], and

Koutrouvelis [29]. Estimators for extreme quantiles have been studied by Weissman [54] and Boos [5].

For the estimation of a parameter vector $\boldsymbol{\theta}$, not necessarily of the location/scale variety, LaRiccia [31] has proposed a minimum quantile distance approach based on the distance measure

$$D(\boldsymbol{\theta}) = \int_0^1 W(u;\boldsymbol{\theta})[\tilde{Q}(u) - Q_0(u;\boldsymbol{\theta})]^2\,du, \tag{6}$$

where $W(u;\boldsymbol{\theta})$ is some specified weight function and $Q_0(u;\boldsymbol{\theta})$ is the QF for $F_0(x;\boldsymbol{\theta})$. Under certain restrictions, the estimator obtained by minimizing (6) as a function of $\boldsymbol{\theta}$ is asymptotically normal. An optimal weight function has also been provided for single-parameter situations that, in the special case of location or scale parameter estimation, results in the estimator obtained by Parzen from model (4). Unlike minimum distance* procedures based on F_n (see Parr and Schucany [35]) the robustness properties of *minimum quantile distance estimators*, such as those obtained from (6), have as yet to be extensively investigated. Nevertheless, this approach seems promising and is intuitively appealing since quantile-based methods are closely related to regression techniques [as exemplified by model (4)] and are directly related to various data-oriented diagnostics such as Q-Q plots* (*see* GRAPHICAL REPRESENTATION OF DATA). For the extension of these estimators to randomly censored data, see Eubank and LaRiccia [15].

DESCRIPTIVE STATISTICS AND EXPLORATORY DATA ANALYSIS

Many of the diagnostic measures and tabular summaries utilized in descriptive and exploratory data analysis (EDA) can be conveniently formulated in terms of sample quantiles. For example, the five-, seven-, and nine-number data summaries that are a basic tool in EDA are all, essentially, collections of symmetrically chosen sample quantiles. This point is illustrated by the five-number summary* proposed by Tukey [50], which (for large n) is equivalent to the use of $\tilde{Q}(0.5)$ (the median), $\tilde{Q}(0.25)$, and $\tilde{Q}(0.75)$ (the quartiles), and the extremes

$$\tilde{Q}\left(\frac{1}{n+1}\right) \quad \text{and} \quad \tilde{Q}\left(\frac{n}{n+1}\right).$$

Similarly, a seven-number summary suggested by Parzen [36] consists of the median and quartiles as well as the eighths, $\tilde{Q}(0.125)$ and $\tilde{Q}(0.875)$, and the sixteenths, $\tilde{Q}(0.0625)$ and $\tilde{Q}(0.9375)$. Such data summaries are frequently utilized to obtain a transformation which gives a data set an approximately symmetric or normal distribution (see Tukey [50] and Emerson and Stoto [13]). The transformation is first applied to the summary and, using various diagnostic measures, is checked for the desired properties. If the diagnostics indicate that the transformation is satisfactory, it is then applied to the entire data set. Thus, in this case, models for the data are developed by modeling $\tilde{Q}$.

Symmetric quantile averages are frequently utilized with data summaries to provide measures of centrality as well as diagnostics. Familiar examples are the median, $\tilde{\mu}(0.5)$, and the midrange*, $\tilde{\mu}(1/[n+1])$. Measures of spread are often constructed from the midspreads* $\tilde{\sigma}(p) = \tilde{Q}(1-p) - \tilde{Q}(p)$, $0 < p < 0.5$, as exemplified by the sample range* $\tilde{\sigma}(1/[n+1])$ and the interquartile range* $\tilde{\sigma}(0.25)$. When the data are approximately normal, $\tilde{\sigma}(p)/[\Phi^{-1}(1-p) - \Phi^{-1}(p)]$ provides an estimator of the population standard deviation. A special case of this is the pseudostandard deviation, $\tilde{\sigma}(0.25)/1.35$, discussed by Koopmans [28, p. 63]. Various diagnostic measures based on $\tilde{\mu}(p)$ and $\tilde{\sigma}(p)$, including measures of skewness and tail length, may be found in Parzen [36, Sec. 11].

A useful graphical tool proposed by Parzen [36, 37] is the quantile box plot. This is a graph of a linearized version of Q upon which p boxes have been superimposed with coordinates $(p, \tilde{Q}(p))$, $(p, \tilde{Q}(1-p))$, $(1-p, \tilde{Q}(p))$, and $(1-p, \tilde{Q}(1-p))$ for $p = \frac{1}{4}, \frac{1}{8}$, and $\frac{1}{16}$. A horizontal line is drawn across the quartile box ($p = \frac{1}{4}$) at the median $\tilde{Q}(0.5)$ to

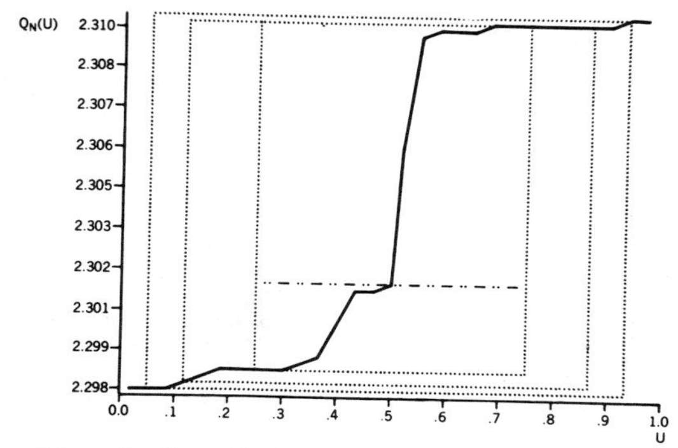

Figure 1 Quantile box plot for the Rayleigh data.

aid in the visual assessment of symmetry. A vertical line with length $\tilde{\sigma}(0.25)/\sqrt{n}$ is also frequently placed at the median to provide an approximate confidence interval for $Q(0.5)$ as well as an indication of the size of the data set from which the plot derives. When examining the plot one looks for sharp rises (infinite slopes) which, when occurring inside the quartile box, indicate the presence of two or more modes and suggest the presence of outliers* otherwise. Flat intervals (0 slopes) correspond to probability masses and, consequently, are indicative of discrete random variables.

An illustration of a quantile box plot is provided in Fig. 1 for the Rayleigh data [50, p. 49] of 15 weights of standard volumes of nitrogen obtained from air and other sources. The sharp rise in $\tilde{Q}$ indicates possible bimodality. It was Rayleigh's recognition of this characteristic which led him to the discovery of argon.

CONDITIONAL QUANTILES

When the relationship between two random variables X and Y is being studied, it is frequently of interest to estimate the conditional quantiles of Y for a given value or values of X. An important special case occurs when X and Y satisfy a linear model. In this case one possible definition of an empirical quantile function has been given by Bassett and Koenker [3]. Procedures for inference about a conditional quantile, assuming a linear model, have been developed by Steinhorst and Bowden [48] and Kabe [24], under the assumption of normal errors. An alternative nonparametric approach suggested by Hogg [21] allows the error distribution to depend on the independent variable. A parametric alternative to Hogg's procedure is provided by Griffiths and Willcox [19]. Consistent estimators of conditional quantiles in general have been derived by Stone [49] under very mild conditions. An alternative is suggested by Parzen in the discussion of Stone's paper. If the conditional quantile function can be assumed monotone in X, strongly consistent estimators presented in Casady and Cryer [7] can be utilized.

References

[1] Babu, J. G. and Singh, K. (1978). *J. Multivariate Anal.*, **8**, 532–549.

[2] Bahadur, R. R. (1966). *Ann. Math. Statist.*, **37**, 577–580.

[3] Bassett, G. and Koenker, R. (1982). *J. Amer. Statist. Ass.*, **77**, 407–415.

[4] Bickel, P. J. (1967). *Proc. 5th Berkeley Symp. Math. Statist. Prob.*, Vol. 1. University of California Press, Berkeley, Calif., pp. 575–591.

[5] Boos, D. D. (1984). *Technometrics*, **26**, 33–39.

[6] Brown, B. M. (1981). *Biometrika*, **68**, 235–242.

[7] Casady, R. J. and Cryer, J. D. (1976). *Ann. Statist.*, **4**, 532–541.

[8] Conover, W. J. (1971). *Practical Nonparametric Statistics*. Wiley, New York.

[9] Csörgö, M. (1983). *Quantile Processes with Statistical Applications*. SIAM, Philadelphia,

[10] Csörgö, M. and Révész, P. (1981). *Strong Approximations in Probability and Statistics*. Academic Press, New York.

[11] Edgeworth, F. Y. (1886). *Philos. Mag., 5th Ser.*, **22**, 371–383.

[12] Edgeworth, F. Y. (1893). *Philos. Mag., 5th Ser.*, **36**, 98–111.

[13] Emerson, J. D. and Stoto, M. A. (1982). *J. Amer. Statist. Ass.*, **77**, 103–108.

[14] Eubank, R. L. (1981). *Ann. Statist.*, **9**, 494–500.

[15] Eubank, R. L. and LaRiccia, V. N. (1984). *J. Multivariate Anal.*, **14**, 348–359.

[16] Fernholz, L. T. (1983). von Mises Calculus for Differentiable Statistical Functionals. Lect. Notes Statist., **19**. Springer-Verlag, New York.

[17] Galton, F. (1889). *Natural Inheritance*. Macmillan, New York.

[18] Gastwirth, J. L. (1966). *J. Amer. Statist. Ass.*, **61**, 929–948.

[19] Griffiths, D. and Willcox, M. (1978). *J. Amer. Statist. Ass.*, **73**, 496–498.

[20] Harrel, F. E. and Davis, D. E. (1982). *Biometrika*, **69**, 635–640.

[21] Hogg, R. V. (1975). *J. Amer. Statist. Ass.*, **70**, 56–59.

[22] Huber, P. J. (1981). *Robust Statistics*. Wiley, New York.

[23] Ikeda, S. and Matsunawa, T. (1972). *Ann. Inst. Statist. Math.*, **24**, 33–52.

[24] Kabe, D. G. (1976). *J. Amer. Statist. Ass.*, **71**, 417–419.

[25] Kaigh, W. D. (1983). *Commun. Statist. Theor. Meth.*, **12**, 2427–2443.

[26] Kaigh, W. D. and Lachenbruch, P. A. (1982). *Commun. Statist. Theor. Meth.*, **11**, 2217–2238.

[27] Kiefer, J. (1970). *Proc. Conf. Nonparametric Tech. Statist. Inference*, M. L. Puri, ed., pp. 349–357.

[28] Koopmans, L. H. (1981). *An Introduction to Contemporary Statistics*. Duxbury, Boston.

[29] Koutrouvelis, I. A. (1981). *Commun. Statist. A*, **10**, 189–201.

[30] Kubat, P. and Epstein, B. (1980). *Technometrics*, **4**, 575–581.

[31] LaRiccia, V. N. (1982). *Ann. Statist.*, **10**, 621–624.

[32] Mason, D. M. (1984). *Ann. Prob.*, **12**, 243–255.

[33] Mosteller, F. (1946). *Ann. Math. Statist.*, **17**, 377–408.

[34] Ogawa, J. (1951). *Osaka Math. J.*, **3**, 175–213.

[35] Parr, W. C. and Schucany, W. R. (1980). *J. Amer. Statist. Ass.*, **75**, 616–624.

[36] Parzen, E. (1979). *J. Amer. Statist. Ass.*, **74**, 105–121.

[37] Parzen, E. (1979). In *Robustness in Statistics*, R. Launer and G. Wilkinson, eds. Academic Press, New York, pp. 237–258.

[38] Pearson, K. (1920). *Biometrika*, **13**, 113–132.

[39] Pfanzagl, J. (1975). In *Statistical Methods in Biometry*, W. J. Ziegler, ed. Birkhauser Verlag, Basel, Switzerland, pp. 111–126.

[40] Quetelet, A. (1846). *Lettres à S.A.R. le Duc Régnant de Saxe-Cobourg et Gotha, sur la Theorie des Probabilitiés Appliquée aux Sciences Morales et Politiques*. M. Hayes, Brussels, Belgium.

[41] Reiss, R. D. (1980). *Ann. Statist.*, **8**, 87–105.

[42] Sander, J. M. (1975). *Tech. Rep. No. 11*, Stanford University, Stanford, Calif.

[43] Sen, P. K. (1972). *J. Multivariate Anal.*, **2**, 77–95.

[44] Serfling, R. J. (1980). *Approximation Theorems of Mathematical Statistics*. Wiley, New York.

[45] Sheppard, W. F. (1899). *Philos. Trans. R. Soc. Lond.* (A), **192**, 101–167.

[46] Shorack, G. R. (1972). *Ann. Math. Statist.*, **43**, 1400–1411.

[47] Smirnoff, N. V. (1935). *Metron*, **12**, 59–81.

[48] Steinhorst, R. K. and Bowden, D. C. (1971). *J. Amer. Statist. Ass.*, **66**, 851–854.

[49] Stone, C. J. (1977). *Ann. Statist.*, **5**, 595–645.

[50] Tukey, J. W. (1977). *Exploratory Data Analysis*. Addison-Wesley, Reading, Mass.

[51] Walker, A. M. (1968). *J. R. Statist. Soc. B.*, **30**, 570–575.

[52] Weiss, L. (1964). *Naval Res. Logist. Quart.*, **11**, 125–134.

[53] Weiss, L. and Wolfowitz, J. (1970). *Z. Wahrscheinl. verw. Geb.*, **16**, 134–150.

[54] Weissman, I. (1978). *J. Amer. Statist. Ass.*, **73**, 812–815.

[55] Wilks, S. S. (1962). *Mathematical Statistics*. Wiley, New York.

Acknowledgment

Research sponsored by Office of Naval Research contract N00014-82-K-0209.

(DENSITY ESTIMATION
EDF STATISTICS
EXPLORATORY DATA ANALYSIS
LARGE-SAMPLE THEORY
ORDER STATISTICS
Q-Q PLOTS
QUANTILE PROCESSES
QUANTILE TRANSFORMATION METHODS)

RANDALL L. EUBANK

QUANTILE TRANSFORMATION METHODS

Let ξ be uniformly distributed on [0, 1] and, for a distribution function $F(x)$ on the real line (assumed left continuous), define the inverse $F^{-1}(t) = \sup(x : F(x) \leqslant t)$. Then $F^{-1}(\xi)$ has distribution function $F(x)$.

Now suppose that $F(x)$ and $G(x)$ are two distribution functions on the real line. If we are given a random variable Y of distribution G, it is useful to be able to construct a random variable X of distribution F such that the difference $|X - Y|$ is as small as possible. Closeness of random variables themselves is a stronger property than closeness of their distributions and leads to a variety of very powerful approximation results related to the central limit theorem*.

An effective approach to the problem above is to define ξ, which is uniformly distributed on [0, 1], by $\xi = G(Y)$ and then set $X = F^{-1}(\xi)$. This construction is a *quantile transformation*.

The quantile transformation possesses an important optimality property. Suppose that $\int |x|\,dF(x) < \infty$ and $\int |x|\,dG(x) < \infty$, and let $\hat{X}, \hat{Y}$ be any pair of random variables with marginal distributions F, G, while X and Y are as defined above. Then [5] for any convex function f on the real line,

$$\begin{aligned}\inf Ef(\hat{X} - \hat{Y}) &= Ef(X - Y) \\ &= \int_0^1 f\big(F^{-1}(x) - G^{-1}(x)\big)\,dx.\end{aligned}$$

This approach was introduced by Bártfai [1] and later strikingly exploited by Komlós et al. [3, 4]. In most applications the role of $F(x)$ is played by $F_n(x) = P(S_n < n^{1/2}x)$, where S_n is a sum of n independent and identically distributed random variables (i.i.d. RVs) with mean zero and variance unity, while $G(x)$ is $\Phi(x)$, the distribution function of the unit normal law. Bounds are sought on $|S_k - T_k|$, where $T_1, T_2, \ldots$ is a sequence of partial sums of i.i.d. unit normal RVs. The method produces results which are improvements over those obtained via the use of the Skorokhod representation (*see* INVARIANCE PRINCIPLES AND FUNCTIONAL LIMIT THEOREMS *and* PROCESSES, EMPIRICAL) at the expense of stronger assumptions on F, including the existence of higher moments.

For example, if $ES_1 = 0$, $ES_1^2 = 1$, and $ES_1^4 < \infty$, Strassen [6] showed, using the Skorokhod representation, that a version of S_n, T_n could be found such that

$$|S_n - T_n| = O\big(n^{1/4}(\log n)^{1/2}(\log\log n)^{1/4}\big) \text{ a.s.}$$

and that, no matter what assumptions were made of F, a result better than $O(n^{1/4})$ could not be achieved. However, Komlós et al. [3, 4] showed, using the quantile transformation method, that if $ES_1 = 0$, $ES_1^2 = 1$, and $E\exp(tS_1)$ exists in a neighborhood of $t = 0$, there is a version of S_n, T_n such that

$$|S_n - T_n| = O(\log n) \text{ a.s.}$$

The quantile transformation method has also been usefully employed to provide bounds on the approximation of the empirical distribution function (process) of a random sample by a Brownian bridge process.

Substantial accounts giving results which have been obtained using the quantile transformation method in the light of those provided by other techniques can be found in the review article of Major [5] and, very comprehensively, in the monograph of Csörgő and Révész [2].

References

[1] Bártfai, P. (1970). *Studia Sci. Math. Hung.*, **5**, 41–49.

[2] Csörgő, M. and Révész, P. (1981). *Strong Approximations in Probability and Statistics*. Academic Press, New York.

[3] Komlós, J., Major, P., and Tusnády, G. (1975). *Zeit. Wahrscheinl. verw. Geb.*, **32**, 111–131.

[4] Komlós, J., Major, P., and Tusnády, G. (1976). *Zeit. Wahrscheinl. verw. Geb.*, **34**, 33–58.

[5] Major, P. (1978). *J. Multivariate Anal.*, **8**, 487–517.

[6] Strassen, V. (1967). *Proc. 5th Berkeley Symp. Math. Statist. Prob.*, Vol. 2. University of California Press, Berkeley, Calif., pp. 315–343.

(ASYMPTOTIC NORMALITY
INVARIANCE PRINCIPLES AND
FUNCTIONAL LIMIT THEOREMS
LIMIT THEOREM, CENTRAL
PROCESSES, EMPIRICAL
QUANTILE PROCESSES
QUANTILES)

C. C. HEYDE

QUANTIT ANALYSIS

Quantit analysis [2] is a parametric method for analyzing quantal response assays (*see* QUANTAL RESPONSE ANALYSIS). In comparison with similar methods based on two-parameter (location and scale) families of tolerance distributions* such as the normal (probit analysis), logistic (logit analysis), and uniform (linit analysis) distributions, quantit analysis is based on a richer three-parameter distribution (*see* OMEGA DISTRIBUTION) which includes a shape parameter in addition to the location and scale parameters. The cumulative distribution function $F(x)$ and the probability density function $f(x)$ of the omega distribution with shape parameter v are characterized by

$$F[x(p)] = p,$$

$$f[x(p)] = 1 - |2p - 1|^{v+1},$$

and

$$x(p) = \int_{1/2}^{p} \{f[x(z)]\}^{-1}\,dz,$$

where $0 < p < 1$ and $v > -1$. The purpose of quantit analysis is to provide improved estimates of extreme effective dosages (e.g., ED95 and ED99, those that cure 95% and 99%, respectively). Most methods, such as the previously mentioned ones based on a two-parameter family, yield similar estimates for ED50, the median effective dosage, while the estimates of extreme dosages may differ drastically.

Let c different dosages of a preparation be applied to c groups of subjects, and suppose that s_i responses are produced among n_i subjects at an appropriately transformed dosage value denoted by x_i $(i = 1, \ldots, c)$. Quantit analysis, like other quantal response assay methods, is based on the conditional probability given by

$$P(s_i \mid x_i, n_i) = \binom{n_i}{s_i} P_i^{s_i}(1 - P_i)^{n_i - s_i},$$

where

$$P_i = \int_{-\infty}^{\alpha + \beta x_i} f(t)\,dt = F(\alpha + \beta x_i)$$

and α and β are the location and scale parameters, respectively. The probability density function $f(t)$ is termed the *tolerance distribution*. A symmetric tolerance distribution will reasonably describe the data if an appropriate transformation of the raw dosage values is used. A relatively rich collection of transformations such as the one proposed by Box and Cox* [1] is essential for this purpose. The intent of utilizing the omega distribution is to provide a class of symmetric tolerance distributions* with a substantial variety of shapes.

The likelihood function corresponding to quantit analysis is given by

$$L = \prod_{i=1}^{c} \binom{n_i}{s_i} P_i^{s_i}(1 - P_i)^{n_i - s_i},$$

where $P_i = F(\alpha + \beta x_i)$, and the tolerance distribution is given by

$$f(\alpha + \beta x_i) = 1 - |2P_i - 1|^{v+1}.$$

In addition,

$$\alpha + \beta x_i = h_v(P_i) = \int_{1/2}^{P_i} (1 - |2z - 1|^{v+1})^{-1}\,dz$$

is the *quantit* of P_i and is essential in determining point and interval estimates of $y_P = \text{ED}100P$, the transformed dosage value at which $100P$ percent of the subjects respond.

While a complete description of the computations for quantit analysis is given elsewhere [2, 3], the main features will be described. The method for obtaining maximum likelihood* (ML) estimates $(\hat{\alpha}, \hat{\beta}, \hat{v})$ of the parameters (α, β, v) involves a Newton–Raphson* procedure for determining $\hat{\alpha}$ and $\hat{\beta}$ combined with an efficient search procedure for finding $\hat{v}$. Since $v = 20$ yields an omega distribution which is very similar to the uniform distribution (i.e., as $v \to \infty$), the search procedure for $\hat{v}$ is confined to all the two-decimal-place numbers from -0.99 through 20.00. The reason for this combined Newton–Raphson and search procedure is that a complete Newton–Raphson procedure for determining the ML estimates $(\hat{\alpha}, \hat{\beta}, \hat{v})$ for the unrestricted parameter space (i.e., $v > -1$) failed to converge for roughly a third of the examples given by Copenhaver and Mielke [2]. An efficient procedure for determining the quantit of P, $h_v(P)$, is required, since this result must be calculated repeatedly in the computations for quantit analysis. As a consequence, a closed expression for $h_v(P)$ is given in an example of a more general result by Magnus et al. [3]. In particular,

$$h_v(P) = \frac{\delta}{2} \sum_{j=0}^{\infty} \frac{x^{j(v+1)+1}}{j(v+1)+1},$$

where

$$x = |2P - 1|, \qquad \delta = \begin{cases} 1, & P > \frac{1}{2}, \\ 0, & P = \frac{1}{2}, \\ -1, & P < \frac{1}{2}, \end{cases}$$

and

$$\begin{aligned} \sum_{j=0}^{\infty} \frac{x^{j(v+1)+1}}{j(v+1)+1} &= -\frac{d}{k} \sum_{m=1}^{[(k-1)/2]} \cos\frac{2\pi m d}{k} \\ &\quad \times \ln\left(1 - 2x^{1/d}\cos\frac{2\pi m}{k} + x^{2/d}\right) \\ &\quad + \frac{2d}{k} \sum_{m=1}^{[(k-1)/2]} \sin\frac{2\pi m d}{k} \\ &\quad \times \tan^{-1}\frac{x^{1/d}\sin(2\pi m/k)}{1 - x^{1/d}\cos(2\pi m/k)} \\ &\quad - \frac{d}{k}\ln(1 - x^{1/d}) \\ &\quad + \frac{d\{1 + (-1)^k\}}{2k}\ln(1 + x^{1/d}) \\ &\quad - \sum_{j=-[(d-1)/k]}^{-1} \frac{x^{j(v+1)+1}}{j(v+1)+1}, \end{aligned}$$

$v + 1 = k/d$, k and d are positive integers, and $[c]$ denotes the greatest integer $\leqslant c$. This closed expression for $h_v(P)$ reduces the computation time of quantit analysis by at least one-half and often much more (e.g., nine-tenths) for specific examples. The invariance property of ML estimators yields the ML estimator of $y_P = \text{ED}100P$ given by

$$\hat{y}_P = [h_v(P) - \hat{\alpha}]/\hat{\beta}.$$

An approximate propagation of error estimator of the variance of $\hat{y}_P$ is given by

$$\begin{aligned} \widehat{\text{var}}(\hat{y}_P) &\doteq \big[\widehat{\text{var}}(\hat{\alpha}) + \hat{y}_P^2\, \widehat{\text{var}}(\hat{\beta}) \\ &\quad + 2\hat{y}_P\, \widehat{\text{cov}}(\hat{\alpha}, \hat{\beta})\big]/\hat{\beta}^2, \end{aligned}$$

where $\widehat{\text{var}}(\hat{\alpha})$, $\widehat{\text{var}}(\hat{\beta})$, and $\widehat{\text{cov}}(\hat{\alpha}, \hat{\beta})$ are estimates obtained from the inverse of the estimated information matrix [2].

The examples of Copenhaver and Mielke [2] are all based on a logarithmic transformation* of the raw dosage values. As mentioned previously, a richer class of transformations would be preferred to achieve the goal of having a symmetric tolerance distribution. However, these examples do indicate that the interval estimates of ED99 using quantit analysis are larger (smaller) than corresponding results using either probit or logit analysis when $\hat{v}$ is less (greater) than 1.

References

[1] Box, G. E. P. and Cox, D. R. (1964). *J. R. Statist. Soc. B*, **26**, 211–246. (Describes transformations for attaining symmetry.)

[2] Copenhaver, T. W. and Mielke, P. W. (1977). *Biometrics*, **33**, 175–186. (Introduces quantit analysis.)

[3] Magnus, A., Mielke, P. W., and Copenhaver, T. W. (1977). *Biometrics*, **33**, 221–223. (The computational efficiency of quantit analysis depends on this procedure for summing an infinite series.)

(BIOASSAY, STATISTICAL METHODS IN
OMEGA DISTRIBUTION
QUANTAL RESPONSE ANALYSIS
TOLERANCE DISTRIBUTIONS)

PAUL W. MIELKE, JR.

QUANTUM HUNTING

A data set consisting of N positive observations X_j is said to be *quantal* with *quantum* q when

$$X_j = a + M_j q + e_j \qquad (j = 1, 2, \ldots, N),$$

where $a \geqslant 0$, $q > 0$, and the M's and e's are independent random variables, with the M's all positive integers, and the e's all (say) Gaussian with expectation zero and variance σ^2. Normally, σ is required to be small. The data set is then said to display *quanticity*. The bias term a is often absent, as we shall assume here, except in one or two parenthetic comments. Quantum hunting is the activity in which one tests a data set for quanticity and tries to estimate q and σ, and a if present.

The first investigation in this field was made by Richard von Mises* [6], who analyzed the series of atomic weights of the chemical elements before the discovery of isotopes. That data set is strongly recommended as an excellent test example for those commencing such work. The first modern approach is that of Broadbent [1]. The more recent technique presented here was introduced in refs. 2 and 3, and makes use of a functional statistic called the *cosine quantogram*, defined by

$$\phi(\tau) = \sqrt{2/N} \sum_{j=1}^{N} \cos(2\pi X_j \tau),$$

where N is the number of observations in the data set and τ the reciprocal of the current value of the trial quantum q. The *sine quantogram* ψ is defined similarly, with "sin" instead of "cos." When there is no bias, then for $q = 0$ both quantograms will largely consist of noisy oscillations, while the cosine quantogram alone will respond with a positive peak at $\hat{\tau}$ near to $\tau = 1/q$ if a true quantum $q > 0$ is present. If the bias is not zero, however, then the quantal response will be divided between the sine and the cosine quantograms, and by comparing these two responses it is possible both to detect and measure the effect at q and to estimate the amount a of bias. When there is a risk that one may be in this situation, the experimenter should also plot the *modular quantogram* $\frac{1}{2}(\phi^2 + \psi^2)$, together with the *polar quantogram* in which the current point on the plot has coordinates $(\phi(\tau), \psi(\tau))$. The four quantograms taken together often give much greater insight than the cosine quantogram alone.

Typically (in the null case), the cosine quantogram has the fixed large value $\sqrt{2N}$ at $\tau = 0$ and then falls smoothly away from this, while for larger τ it displays what are close to random oscillations in the manner of a standardized stationary Gaussian time series whose ordinates have mean zero and variance 1. (In principle [4], the covariance function for this time series can be estimated from the behavior of the sine and cosine quantograms near $\tau = 0$, but this is scarcely a practicable procedure and is virtually never used.) The sine quantogram starts with the value zero at $\tau = 0$; otherwise, its oscillatory behavior is very similar.

If a quantal effect with quantum q is present (and $a = 0$), then the cosine quantogram will have a peak at $\hat{\tau}$ near $\tau = 1/q$, of height H, where approximately we shall have

$$\mathbf{E}(H) = \sqrt{2N} \exp(-2\pi^2\sigma^2/q^2).$$

(This follows from the formula for $\mathbf{E} \cos Z$, where Z is Gaussian.) When the ratio σ/q is

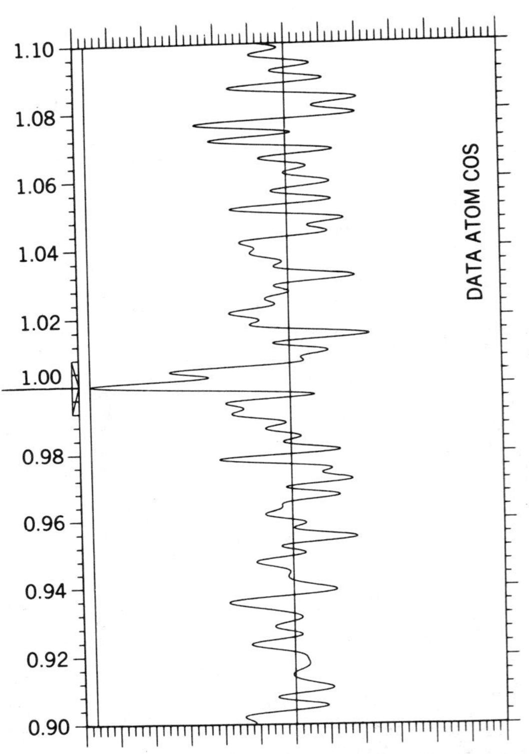

Figure 1 Cosine quantogram (CQG) for the atomic weights of 82 chemical elements. The prominent peak occurs at $\hat{\tau} = 1.000$ ($\hat{q} = 1.00$), $H = 5.67$. (Subsidiary peaks of heights 4.48 and 3.61 were found at $\tau = 2.001$ and 3.001, respectively.) The horizontal span of the diagram here ranges from $q = 0.909$ ($\tau = 1.1$) on the right to $q = 1.11$ ($\tau = 0.9$) on the left. The vertical span ranges from $\phi = -6$ (bottom) to $\phi = +6$ (top). In fact, no search range is really needed in this example because the suspected quantum $q = 1.0$ is here known in advance. The vertical arrow points to the location of the highest peak seen in this search range, and the horizontal cursor enables one to read off its height. The rest of the plot is noise, though the "forked" nature of the main peak may perhaps contain some significant information (compare this with Fig. 4).

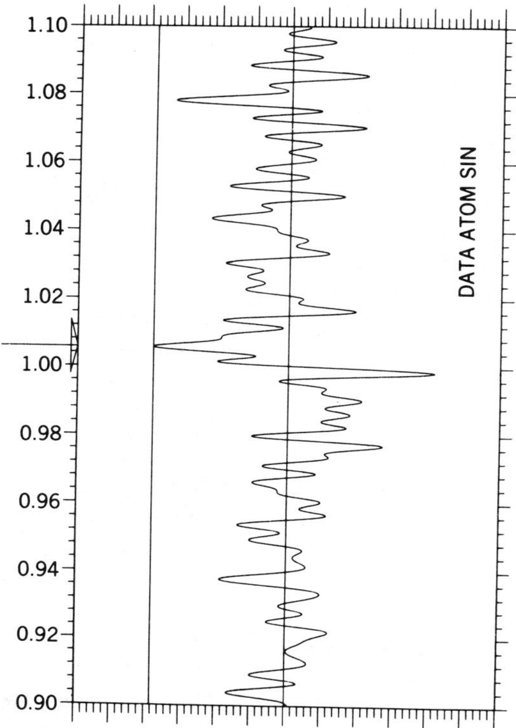

Figure 2 Sine quantogram corresponding to the CQG of Fig. 1. There is here a much smaller peak ($H = 3.83$) at $\hat{\tau} = 1.006$ ($\hat{q} = 0.995$), with subsidiary peaks as before at $\hat{\tau} = 2.005$ and 3.009. This indicates the existence of a slight a-effect.

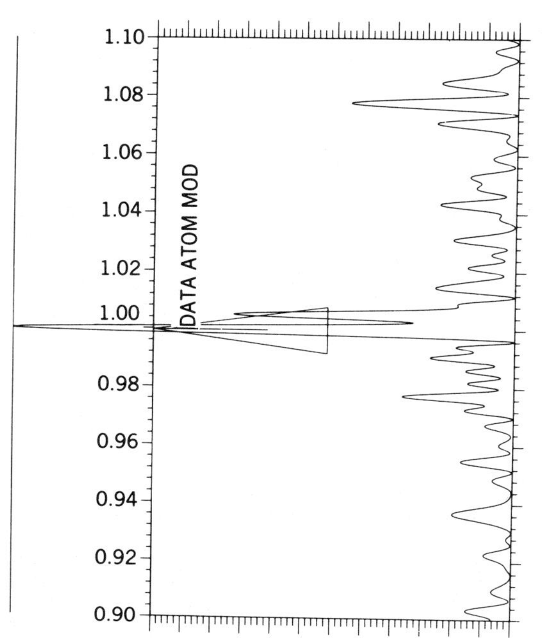

Figure 3 Modular quantogram corresponding to the CQG of Fig. 1. The peak of height $H = 16.71$ is off the scale, and $\hat{\tau} = 1.000$ ($\hat{q} = 1.00$). Subsidiary peaks at $\hat{\tau} = 2.000$ and 3.000 have heights 11.3 and 7.1, respectively.

too large in relation to the sample size N, then the value of $\mathbf{E}(H)$ is less than 2.0 and we cannot expect to detect the quanticity even if it is present; the "noise" will be too great. But if a significant quantal peak *is* found, this last relation can be used to estimate the value $\hat{\sigma}$ of σ, and that value for σ should, of course, be an acceptable one in the light of the origin of the data; this provides a very useful secondary check on the practical significance of the analysis.

Figures 1 to 4 show all four quantograms for the atomic weights of 82 chemical elements.

In the cosine quantogram shown in Fig. 1, the massive peak at a τ-value corresponding to $q = 1.0$ is of height 5.7 and so scarcely needs testing for significance (i.e., it is too big to be accepted as a standard Gaussian ordinate). But in marginal cases, and always when q is not known in advance, a careful test is needed; we explain how this is done below. First, however, we comment on various points of interpretation. As in the analysis of time series, one must be careful about harmonics. It is clear that there will be secondary peaks (of progressively less height) at τ-values corresponding to $q/2, q/3, \ldots$. There may also be secondary peaks at other rational multiples of q, for example, at the τ-value corresponding to $2q$ if the M-values are predominantly even. Because $\phi(0)$ has the large value $\sqrt{2N}$, and because ϕ oscillates between $\pm\sqrt{2N}$ like an almost periodic function for large τ, there will always be large positive ordinates in the cosine quantogram (for quantal data or not) when τ is near to zero, and occasionally (but surely) when τ is large. Thus no meaningful quantum hunting is possible until the provider of the data is firmly committed to a fixed *search range* $[q_0, q_1]$ (excluding zero and infinity) within which it seems sure that the

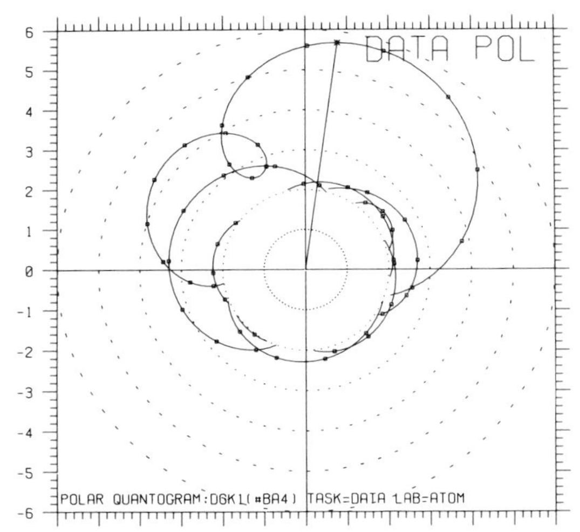

Figure 4 Polar quantogram corresponding to the CQG of Fig. 1. Here for $0.9 < \tau < 1.1$ we plot $\phi(\tau)$ as the vertical coordinate and $\psi(\tau)$ as the horizontal coordinate (but increasing to the *left*). The portions of the spiral for which $\sqrt{(\phi^2 + \psi^2)} < 2$ are omitted. The asterisk joined by a radius to the origin marks the maximum peak on the CQG. "Pips" on the spiral indicate points at τ-increments of 0.000555 (five times the plotting interval).

quantum, if present, will be found. In the von Mises example we can take $q_0 = q_1 = 1$, so in the plots we make $q_0 < q_1$ and both close to unity. Normally, however, the user must be warned that if the search range is too narrow, the quantum may be missed, while if it is too large, false quanta may be found. From the search range we compute $\tau_0 = 1/q_1$ and $\tau_1 = 1/q_0$, and the quantogram to be studied is then that defined above, *but now confined to the interval* $[\tau_0, \tau_1]$. To be sure of picking up the highest peak within this interval, the curve must be plotted sufficiently finely. The quantogram consists of many superimposed oscillations, the τ-wavelength of the average one being about $1/X_{av}$ (where X_{av} is the average of the X-values), and ideally there should be (say) about 10 plotted points in a τ-interval of that length. Experimentation with an artificially small search range and various scales of plotting will soon reveal the plotting density necessary to pick up all the peaks, and this practice is strongly recommended. In later follow-up studies it will often be possible to manage with coarser (quicker!) plotting. If the data values are quoted, say, to the nearest 0.1, then obviously a fictitious quantum of value 0.1 will inevitably be signalled at $\tau = 10.0$. (This provides a useful test of the program.) If a proportionately large number of data values are very close together on the X-scale, the common value to which they approximate may appear as a fictitious quantum. If two different quanta are present, so that

$$X_j = M_j' q' + M_j'' q'' + e_j,$$

then the quantum signals may be eroded (possibly with changes of sign) and a fictitious quantum may appear at their highest common (approximate) factor; the analysis of such situations is very difficult.

We now come to the question of significance tests*. (This is almost trivial when $q_0 = q_1$.) It is scarcely practicable, when $q_0 < q_1$, to make a theoretically based test using the exceedance distribution for a section of a time series having a largely unknown covariance function. One must then resort to data-based simulations, to be constructed from independent random corruptions of the data planned in such a way that (a) the original quantal effect is eroded, and (b) relevant ancillaries (controlling the *propensity* to produce fictitious quantal effects) are retained. A useful procedure is the one detailed below. Here we suppose that a highest peak of height H has been observed at $\hat{\tau} = 1/\hat{q}$ and is to be tested.

1. Arrange the data values in increasing order of size as $X_1, X_2, \ldots, X_N$.
2. Put $Y_j = X_{j+1} - X_1$, for $j = 1, 2, \ldots, N-1$.
3. For consecutive cells of width $\hat{q}$ on the Y-scale, starting at zero, reassign all the data values found in each cell to values randomly and uniformly distributed within that cell, thus getting $Y_1^*, Y_2^*, \ldots, Y_{N-1}^*$.

4. Now put $X_1^* = X_1$ and $X_j^* = X_1 + \lambda Y_{j-1}^*$ for $j \geqslant 2$. Choose the coefficient λ so that $X_1^*, X_2^*, \ldots, X_N^*$ have the same sum of squares as $X_1, X_2, \ldots, X_N$.
5. Draw the quantogram for the starred data, and record the height H^* of the highest peak in the interval $[\tau_0, \tau_1]$, proceeding exactly as in the analysis of the real data.
6. Repeat steps **3** to **5** a large number of times (at least 100), and then assess the significance of the observed peak H against the data-based simulation set of peaks H^*.

The rationale of this procedure is that in asymptotic theory for the null case $\sum X_j^2$ largely controls the distribution of the height of the maximum peak found in a given search interval. This, and X_1, are the "relevant ancillaries." The computation of the factor λ in step **4** will give no trouble. A quadratic equation has to be solved; it is the root near to unity that is to be taken.

Preliminary accounts of these procedures were set out in refs. 2 and 3, but those are now somewhat out of date, and to that extent the present account is intended to replace them.

Reference 1 describes the earlier and quite different technique devised by S. R. Broadbent, which together with the one presented here has been much used in studies of archaeological measurements. These suggest (at a significance level about 1%) that a unit of about 5.4 (2.7) feet may have been employed in constructing the diameters (radii) of the neolithic stone circles of western Britain. The corresponding estimate for σ (when X = circle diameter) came out to be 1.5 feet. The search range for q used in the analysis of the diameters was the range from 2 to 10 feet.

Now 2.7 is about the length of a pace, and $1\frac{1}{2}$ feet is about the size of a marker stone, so although not necessarily accepting all the claims [5] of Alexander Thom concerning the "megalithic yard," we may feel that the case is strong enough to justify at least a more extensive survey.

References

[1] Broadbent, S. R. (1956). *Biometrika*, **43**, 32–44. [See also *ibid.*, **42**, 45–57 (1955).] (Broadbent laid the foundations on which all subsequent work has been done; his approach makes use of a quite different functional statistic.)

[2] Kendall, D. G. (1974). *Philos. Trans. R. Soc. Lond.* (A), **276**, 231–266. (Some errors are corrected in ref. 3.)

[3] Kendall, D. G. (1977). *Proceedings of the Symposium to Honour Jerzy Neyman*, R. Bartoszyński et al., eds. PWN, Polish Scientific Publishers, Warsaw, pp. 111–159. (Both refs. 2 and 3 are now somewhat out of date, and the present article should replace them, but ref. 3 is still useful for background and supplementary comments.)

[4] Kent, J. T. (1975). *J. Appl. Prob.*, **12**, 515–523. [See also S. Csörgő, *J. Appl. Prob.*, **17**, 440–447.]

[5] Thom, A. (1967). *Megalithic Sites in Britain*. Clarendon Press, Oxford. (This will serve as an introduction to a large and still-growing literature which goes well beyond the relatively simple matter of the analysis of the diameters of *circles*, treated in Thom's book in the context of other more speculative questions for which the status of the data is much more open to question. Even for circles there is some residual anxiety: "a circle is a circle is a circle"—or is it a reerected "circle"?)

[6] von Mises, R. (1918). *Phys. Zeit.*, **19**, 490–500. (This paper contains the first mention of what is now known as the "von Mises distribution.")

(TIME SERIES)

D. G. KENDALL

QUANTUM MECHANICS AND PROBABILITY: AN OVERVIEW

HISTORICAL SURVEY

Quantum mechanics arose from a need to reconcile certain new experimental observations in physics with the (incorrect) predictions of classical mechanics (Newton's laws of motion) and classical electromagnetic theory (Maxwell's equations). These experiments—the photoelectric effect, the blackbody radiation spectrum, the characteristic line spectra of the chemical elements, x-rays, and Compton scattering—seemed to indicate that electromagnetic radiation manifests a particle-like aspect when interacting with microscopic systems. This was puzzling because physicists had thought that the question of whether light (known to be a form of electromagnetic radiation) was composed of particles (Newton's theory) or was a wave in a continuous medium (Huygens's theory) had been settled definitely in favor of waves by the experiments of Young, Fraunhofer, and their successors [8]. The fact that Maxwell's theory of electromagnetism predicted waves which propagated at the same speed as light, and that Hertz had generated such waves and measured their velocity to be that of light, was regarded as the final proof of the wave nature of light. Hence it was a shock to the world of physics when Planck discovered that the observed electromagnetic radiation spectrum of a blackbody could be reproduced theoretically only if thermodynamics were combined with an additional hypothesis, namely that light must be emitted and absorbed in packets of definite energy [17]

$$E = h\nu, \tag{1}$$

where h is Planck's constant and ν is the light frequency. Einstein's subsequent application of this *quantum hypothesis* to explain the photoelectric effect [9] introduced yet another discordant gap into what had appeared to be a seamless whole. *See also* PHYSICS, STATISTICS IN (EARLY HISTORY).

A second advance came in 1924 when de Broglie [4] suggested a symmetry between radiation and matter. He proposed that since light (whose wave aspect is revealed through diffraction and interference) sometimes behaves like particles, we might expect massive particles (such as electrons or atoms) sometimes to behave like waves. The notion that matter has a wavelike aspect bore immediate fruit by explaining the old Bohr–Sommerfeld empirical quantization rules as well as the precise quantization of the energies of atoms. The reason the wave theory can accomplish this task is that if matter obeys a wave equation analogous to that describing the vibrations of a stretched string, the boundary conditions appropriate to a confined matter wave would necessarily permit only certain wavelengths, namely those that fit naturally and smoothly into the confinement region. Because of the Planck–Einstein relation (1) between energy and vibration frequency, this quantization of wavelengths convincingly implies quantization of energy levels. Thus it was not long before Schrödinger proposed a wave equation to describe the motion of particles [18], and Davisson and Germer [3] and G. P. Thomson [21] independently verified that electrons diffract from crystals (thereby demonstrating experimentally the wave aspect of particles).

The new wave mechanics was an immense success. It predicted correctly the spectrum of hydrogen, the properties of diatomic molecules, and the binding energy of the helium atom. In the space of a decade the pioneers of quantum mechanics developed a theoretical framework for quantum mechanics virtually identical to that used today, and generated an enormous number of solutions to specific problems in atomic, nuclear, and solid-state physics.

UNDERLYING IDEAS OF QUANTUM MECHANICS

The development of quantum mechanics was motivated by several experimental facts:

1. At the microscopic level, all events seem to be local. That is, when a photon is absorbed by a metal, it is absorbed at a point, ejecting only one electron, rather than over a widespread area and transferring its energy and momentum to many electrons simultaneously. Similarly, when an electron is absorbed, its charge appears *in toto* at the point of absorption, rather than a little here and a little somewhere else.

2. Events at the atomic scale of distances exhibit interference and diffraction—phenomena which at the macroscopic level have always been thought to be characteristic of waves.
3. Measurements of microscopic systems obey the Heisenberg uncertainty principle [14]. We can easily convince ourselves that this is characteristic of wave-like phenomena, because in order to determine the wavelength of a pulse of length L, we must count how many oscillations are contained in the pulse. But this is always uncertain by ~ 1 wave, because it is hard to define the beginning and end of a pulse of finite duration (see Fig. 1). If we identify the pulse with a particle, the particle's position is determined only to a precision of $\sim L$; hence the uncertainty in wave-number multiplied by that of position is always of order unity or greater. The de Broglie relation between a particle's wavelength and momentum,
$$\lambda = h/p, \tag{2}$$
then leads to the standard form of the Heisenberg uncertainty relation,
$$\Delta x \Delta p \gtrsim h/(4\pi). \tag{3}$$
(The additional factor of 4π results from considering pulse shapes which yield a minimum lower bound on the uncertainty product.)
4. Because Planck's constant h is very small, the de Broglie waves characterizing macroscopic objects are extremely short—far smaller even than nuclear dimensions. Our senses are so coarse that they average over many wave crests, thus destroying all interference phenomena we might otherwise observe. This is why quantum mechanics plays no role in our normal experience.

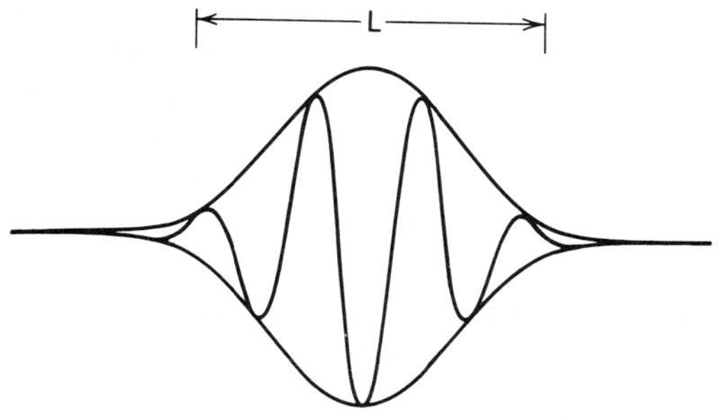

Figure 1 Pulse of finite duration.

The central innovation in quantum mechanics was the introduction of a *wavefunction*, that is, a complex-valued function of coordinates and time, whose evolution in time describes the dynamics of the system to which it corresponds. The time evolution of the wavefunction is determined by a first-order partial differential equation
$$\frac{\partial \psi}{\partial t} = -i\left(\frac{2\pi}{h}\right)H\psi, \tag{4}$$
where i is the square root of -1 and H is a differential operator which can depend on coordinates, gradient operators, and even time (the time dependence of H occurs with systems perturbed by external forces that vary with time). The operator H is a representation of the classical Hamiltonian of the system, which is constructed according to certain rules: The chief of these is that canonical momenta p are replaced by derivatives with respect to the corresponding canonical coordinates q via the identification
$$p \rightarrow -i\frac{h}{2\pi}\frac{\partial}{\partial q}. \tag{5}$$
(This rule must be applied with discretion, however, because in curvilinear coordinates, or when describing rigid-body rotation, its naive application leads to ambiguities [15].) For the sake of completeness, note that it is possible to regard the wavefunctions as abstract vectors in a Hilbert space, and the Hamiltonian as an abstract operator on that space. The conversion from an abstract vector to a wavefunction which depends on coordinates is then a form of unitary transformation from one representation to another. This is the procedure followed in modern quantum mechanics texts.

One of the most useful signposts in the unfamiliar terrain of quantum physics was the *correspondence principle*, which required that in the limit of macroscopic systems, quantum dynamics must lead to classical mechanics (Newton's laws). For this to happen, it was necessary to reinterpret the idea of a particle's position $x(t)$ in terms of an expectation value; that is, only the expectation $\langle x(t)\rangle$, defined in terms of the wavefunction $\psi(x,t)$ by

$$\langle x(t)\rangle = \int dx\, \psi^*(x,t) x \psi(x,t), \tag{6}$$

has physical significance. If we take the Hamiltonian of a particle moving along one spatial axis under the influence of the potential $V(x)$ to be

$$H = -\left(\frac{h}{2\pi}\right)^2 \frac{\partial^2}{\partial x^2} + V(x), \tag{7}$$

then it is easy to see, by differentiating (6) with respect to time and then applying (4) and its complex conjugate, that

$$\frac{d}{dt}\langle x(t)\rangle = i\int dx\, \psi^*(x,t)(Hx - xH)\psi(x,t) \tag{8}$$

and hence that

$$\frac{d^2}{dt^2}\langle x(t)\rangle = -\int dx\, \psi^*(x,t) \times \left[H(Hx - xH) - (Hx - xH)\right] \times \psi(x,t) \tag{9}$$

or

$$\frac{d^2}{dt^2}\langle x(t)\rangle = \frac{-1}{m}\int dx\, \psi^*(x,t) \frac{\partial V(x)}{\partial x} \psi(x,t). \tag{10}$$

Now, the force acting on a particle is just the negative gradient of its potential energy. In general, the right side of (10) involves the expectation of all moments of x. However, for macroscopic objects the wavelength is typically so small that the quantum fluctuations of position and momentum are also small in comparison with their expected values. Hence it is permissible to replace $\langle -V'(x)\rangle$ by $-V'(\langle x\rangle)$, thereby obtaining Newton's law of motion. (This result is known as *Ehrenfest's theorem.*) To summarize, when we interpret the position of a particle in terms of (6), we obtain the classical equations of motion in the limit that fluctuations are ignored. (This is, of course, the sense in which any stochastic process yields a deterministic equation of motion.) The physical content of this mathematical formalism is simply that our classical picture of the trajectory of a particle acted on by an external force requires simultaneously measuring both the position and the momentum of the particle at successive (short) time intervals. For macroscopic objects the measurement process introduces only negligibly small perturbations of the values of these quantities relative to their mean values, so Newtonian physics adequately approximates the dynamics of such objects.

Like all successful physical theories, wave mechanics immediately generated more problems than it solved. The chief of these was how to interpret the "wavefunction" associated with the particles, which we obtain by solving the Schrödinger wave equation. It would be an overstatement to suggest that the interpretation proposed by Bohr, Born, and others (the so-called Copenhagen school of quantum mechanics) [2, 19] and which today is the predominant way of looking at things, was immediately accepted by acclamation. Einstein steadfastly refused to accept the probabilistic interpretation of quantum mechanics all his life, a refusal based on his intuition that "God does not play dice!" Various authors have attempted to find logical paradoxes in the Copenhagen interpretation [10], or to find mechanistic theories of a "guiding wave" that would retain the wave equation for particles while reintroducing determinism at the microscopic level [5]. All such attempts have ended in failure.

PROBABILISTIC INTERPRETATION

The idea that certain physical phenomena must be treated in a probabilistic fashion was certainly not new in the 1920s and 1930s, and certainly not unique to quantum mechanics. For example, by about 1900 all of thermodynamics had been placed on a statistical basis through the efforts of Maxwell*, Boltzmann*, and Gibbs. There are two crucial differences between the use of probability in quantum mechanics and in classical physics, and the exploration of these differences will comprise the remainder of this section.

To appreciate the role of probability in quantum physics, we first examine how it is employed in classical physics. Here probability is applicable to systems whose final states are highly unstable with regard to small variations in the initial conditions. Thus a coin toss or the fall of a die is generally regarded as a random event because minute variations in the initial conditions (e.g., at the level of environmental noise) will considerably alter the outcome. As long as the variability of the outcomes with respect to small variations in the initial conditions is sufficiently chaotic (even though entirely deterministic) in the sense that standard statistical tests for the absence of correlation* are satisfied (thereby rendering prediction of the outcome effectively impossible), we are entitled to regard such events as random. The same justification can be invoked for statistical methods in classical thermodynamics or fluid mechanics. Even though the microscopic constituents are supposed to obey deterministic and time-reversal-invariant laws of motion, large aggregates of such constituents become enormously sensitive to small variations in initial state; hence the "equal likelihood" hypothesis for the statistical ensembles of thermal physics makes good sense.

The probabilistic interpretation of quantum mechanics has an entirely different conceptual foundation. The mathematical objects that we interpret as probability distributions are all that can *ever* be known, in principle, about a system. That is, the entire goal of quantum dynamics is to calculate the probability distribution of the outcomes of a particular experiment [1]. The fact that we deal with probabilities is not the reflection of lack of detailed information about the initial state; there simply is *no* other information to be had. (The Copenhagen interpretation contains certain subtleties and logical difficulties; the very notion of "experiment" presupposes an external "observer" who performs measurements using macroscopic instruments. This dichotomy between the quantum system being observed and the classical external observer raises profound questions about epistemology, the nature of "consciousness," and similar philosophical concerns [11]. We shall return to these problems in the next section.)

The second way that quantum and classical ideas of probability differ is in the composition of probabilities. Here we encounter the problems that early workers in the field found paradoxical and confusing, because it is at this point in the theory that the question of "wave–particle duality" enters. As Feynman has emphasized [13], the central axiom of quantum mechanics is that to each physical process, representing the transition of a system from an initial state A to a final state Z, there corresponds a complex number, the *probability amplitude*. The probability that the system makes the transition in question is the squared modulus of the probability amplitude. The law of composition of probabilities is then as follows:

> If there is more than one way that a system can make a transition from initial state A to final state Z, say by passing through intermediate states B, C, D, and so on, then as long as we make no measurement to determine *which* of the intermediate states the system goes through, the total amplitude for the transition A to Z is the sum of the individual amplitudes involving the various intermediate states. The corresponding probability is the squared modulus of the total amplitude. Conversely, if we determine by some measurement that the system has gone through, say, intermediate state D, then the total amplitude is simply the partial amplitude corresponding only to that intermediate state, and the corresponding probability of the squared modulus of that partial amplitude.

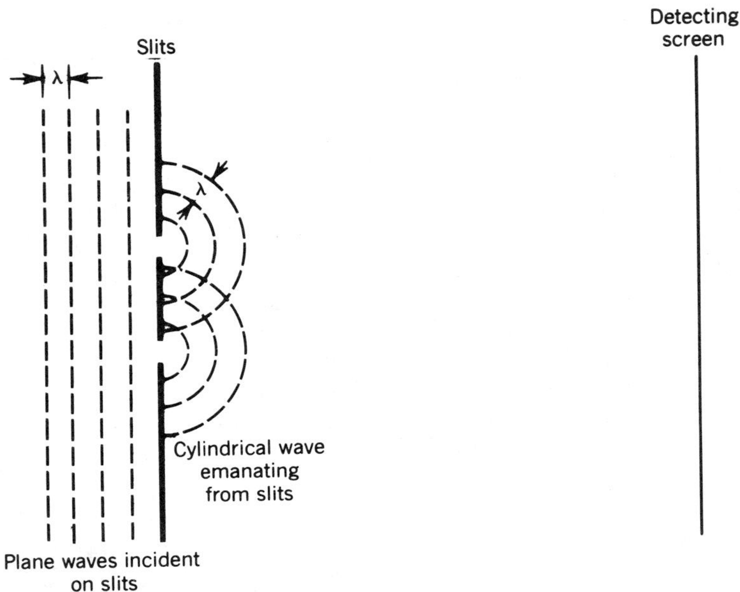

Figure 2 Electron beam incident on two narrow parallel slits.

To see what this law means in a specific example, consider the apparatus shown in Fig. 2, where electrons from a point source fall on a fluorescent screen after passing through two narrow parallel slits in an opaque screen (the slits are of width a and are separated by a distance d). Clearly, an electron emitted by filament A can pass through either slit B or slit C and end up somewhere on the detecting screen Z. The amplitude $\psi(x, y, z, t)$ for an electron to propagate from slit B or C to screen Z is obtained by solving the free-particle Schrödinger equation

$$\frac{\partial \psi}{\partial t} = i \frac{h}{4m\pi} \nabla^2 \psi \tag{11}$$

and is given for $\theta = \arctan(x/z)$ by

$$\phi(\theta) = \text{const}.\exp[-ih^2k^2t/(8m\pi^2)]|\cos\theta|^{1/2} \times \cos\left(\frac{kd}{2}\sin\theta\right)\sin\left(\frac{ka}{2}\sin\theta\right)\Big/\sin\theta, \tag{12}$$

where x is the vertical distance along the screen. If we make no attempt to determine whether the electron has gone through slit B or slit C, we must add the amplitudes and square them, thereby obtaining a classical two-slit interference pattern (Fig. 3). By direct experiment, namely, by adjusting the electron current to be so low that only one electron at a time is in the apparatus, we see that the observed interference pattern is a probability distribution—the individual electrons fall randomly, but after enough have impinged on screen Z, the interference pattern emerges. We also see from this experiment that the interference of the amplitudes corresponding to the two slits is entirely characteristic of electrons taken one at a time. That is, it is not some sort of collective effect depending on the mutual interaction of swarms of electrons. This buildup of the interference pattern from individual events in a probabilistic manner was actually first proposed for photons from a weak light source as early as 1909 [20], and represented one of the cornerstones of the probabilistic interpretation. (The experiment actually failed to see the effect because the photographic emulsions available in 1909 were too "slow." Rather, it was used to place an upper limit on Planck's constant h.)

What happens if we try to determine which of the two slits the electron passed through? Suppose that we cover first slit B, then C: we obtain patterns like those in Fig.

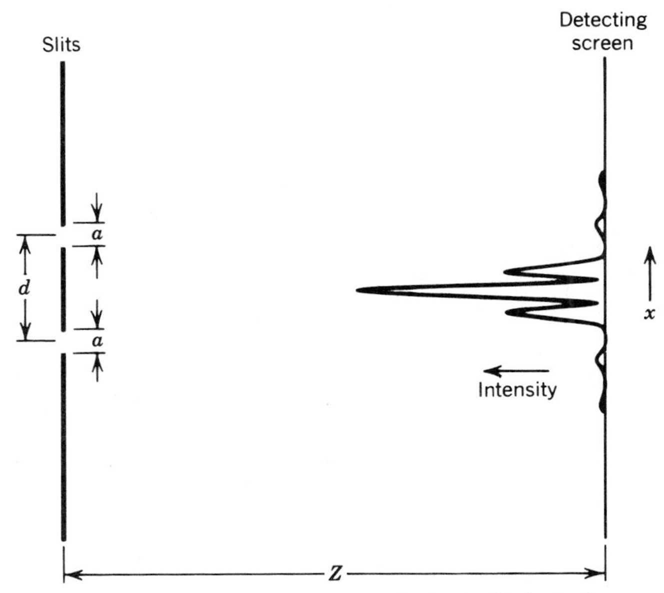

Figure 3 Two-slit interference pattern obtained with both slits open.

4*a*, given by the squares of each of the corresponding amplitudes. Their sum, Fig. 4*b*, exhibits no interference pattern. But wait a moment: What if instead of blocking off the electrons (so that we guaranteed which slit they had gone through) we had merely placed a light behind one of the slits, so that if an electron went through it would scatter some light and we would see a flash? This situation can be analyzed precisely, but to avoid excessive detail we simply state the result in qualitative terms: If the light is sufficiently intense that there is a high probability that an electron passing through that slit will scatter a photon, and if the wavelength of the photon is sufficiently short that we can resolve which of the two slits the electron passed through (this means that the wavelength must be somewhat smaller than the spacing of the slits), then a random amount of momentum will be transferred to that electron [7]. This will effectively average the two-slit interference pattern over a distance large compared with its oscillations, thereby causing it to look precisely like the sum of the two individual slit patterns and exhibiting no interference. In other words, quantum mechanics predicts that if we try to tell, by any means whatsoever, which slit the electron goes through, we destroy the two-slit interference pattern. Moreover, because the interference between the two slits averages to zero when integrated along the plane of the fluorescent screen, the total probability remains unity. That is, the electron goes *somewhere*.

There is one further situation that we must consider when speaking of the quantum law for composition of probabilities, because we shall need the idea it embodies in the next section. Suppose that it were possible to determine which intermediate state a system passed through, even long after the experiment is finished. For example, electrons possess intrinsic spin and therefore (for reasons that go beyond the scope of this article) can be polarized either "up" or "down" with respect to some definite direction. What if slit B contained a polarizing window, so that it permitted only "up" electrons to pass through it, and what if slit C were similarly equipped and passed only "down" electrons? By looking at the individual electrons reaching Z to see whether they are polarized "up" or "down," we can tell immediately whether they went

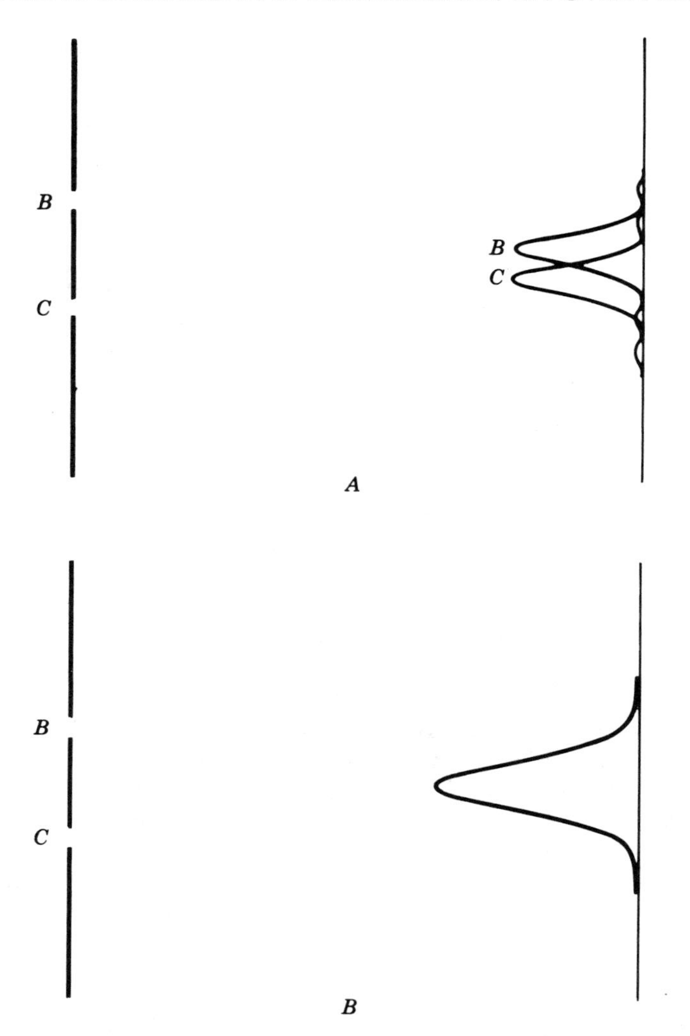

Figure 4 (*a*) Results from one slit at a time; (*b*) the sum of the intensity patterns from slits B and C taken separately. This is also what is observed if any attempt is made to determine which slit the electrons pass through.

through slit B or C. The answer is that the resulting pattern exhibits no interference, because this is essentially the same experiment as having closed off one slit and then the other. Even if we do not actually measure the polarization at Z, there is no interference pattern.

To summarize, an essential difference between classical ideas of probability and the interpretation of the experimental results of quantum physics is the law of composition of probabilities of disjoint trials. Classically, an electron can go through slit B or slit C, and the resulting probability is the sum of the probabilities of the experiments conducted independently (since the slits do not interact with each other). We find that the data compel us to add the *amplitudes* of disjoint trials and then to compute the total probability as the squared modulus of the sum. For macroscopic systems, as emphasized above, the wavelength associated with the quantum interference effects becomes small even compared with nuclear dimensions. Thus any random disturbance of the system will average the quantum interference pattern over many oscillations, washing it out. This is how quantum mechanical laws of probability approach classical ones in macroscopic systems.

THE ROLE OF THE OBSERVER

The chief concern of physics is the analysis of dynamical processes, that is, processes that evolve in time. The mathematical description of quantum phenomena implied by (4) includes dynamics in the following way. We are given an initial amplitude ψ_0 whose squared modulus is the probability distribution of the system variables at time $t = 0$ (the beginning of the experiment). Time evolution then proceeds according to the Schrödinger equation, which is of first order in the time variable, so that the initial condition specifies completely the subsequent probability distribution of the system. As long as we are content to regard the wavefunction as a probability amplitude and to ascribe no further physical significance to it, we encounter no problems of interpretation. If we wish to ascribe an independent physical reality to the wavefunction of a particle or of a complex system, all sorts of philosophical difficulties arise.

To make this discussion precise, it is important to define clearly what we mean by "observation." Implicit in the definition is the idea of "observables," by which we mean numerical quantities associated with quantum systems, and to which we give the names "energy," "position," "charge," "mass," "momentum," or "intrinsic angular momentum" (spin), or the vector component of spin along a particular direction. There is by no means universal agreement as to why nature singles out these entities as the appropriate observables. For the most part observables are the extension to the microworld of concepts which apply to macroscopic physics, although some (such as "isospin," "hypercharge," "flavor," or "color") are picturesquely named abstractions with no macroscopic counterparts. With each of these observables we associate a Hermitian linear operator defined on the linear vector space of wavefunctions, whose expectation value [in the sense of (6)] is what we actually measure. Some measurements always yield a definite value of the observable (i.e., with no statistical fluctuation). For this to happen, it is necessary that the wavefunction representing the state of the system at that time be an eigenfunction of the operator representing the observable. To state this more mathematically, if

$$\langle(\mathscr{O} - \langle\mathscr{O}\rangle)^2\rangle \equiv \langle\mathscr{O}^2\rangle - \langle\mathscr{O}\rangle^2 = 0 \quad (13)$$

then by virtue of the Cauchy–Bunyakovsky–Schwartz inequality [i.e., equality holds in (13) if and only if the Hilbert-space vectors ψ and $\mathscr{O}\psi$ are proportional to each other], ψ is an eigenvector of $\mathscr{O}$.

Now quantum mechanics as described above gives a perfectly unequivocal description of the evolution of a wavefunction as it interacts with an experimental apparatus. Consider the following experiment. Electrons from an unpolarized source are incident on a polarization "filter" whose function is to produce a spatial or temporal separation between those electrons with spin "up" and those with spin "down." [Such a filter might be a Stern–Gerlach apparatus, for example; or more schematically, a spin-dependent potential barrier of the form $V(x) = (1 + 2s_z)V_0\theta((a/2)^2 - x^2)$, where V_0 can be as large as we wish.] Clearly, with such an apparatus we can be sure that all the electrons that emerge are in an eigenstate of the z-component of their spin operator, and thereby possess a definite value for this observable. Now, to calculate the results of an experiment where we send a beam of particles of unknown (random) spin orientations into a polarization filter, we first construct a wave packet which is a superposition of plane waves of all wavelengths (representing solutions of the Schrödinger equation in the absence of interaction). This packet is to be fairly sharply peaked about some definite momentum, so that it represents a fairly monochromatic beam. It will propagate at the (group) velocity

$$v_g = \frac{d\omega(k)}{dk} \quad (14)$$

and, because the phase velocity depends on wavelength, it will eventually disperse (we shall return to this point) and broaden. The packet will also have to represent particles with a definite spin orientation with respect to the axis singled out by the filter. The condition that the beam be unpolarized (or that it have some definite proportion with spins up and the remainder with spins down) is put in at the end when we calculate probabilities. We calculate what happens to the particles with spins up and multiply by the fraction with that orientation, and add the result to the corresponding quantity for the spin-down particles, since as mentioned in the section "Probabilistic Interpretation," such probabilities are disjoint. The evolution of the wave packet now proceeds as follows. A packet representing an "up" particle experiences no interaction and so propagates through the apparatus undisturbed. A "down" packet, conversely, gives rise to two packets which, a long time after the incident packet entered the apparatus, look like a transmitted packet (perhaps with some time delay built in) with a very small amplitude (the transmission coefficient is assumed to be small) and a reflected packet moving in the opposite direction from the original, with an amplitude nearly the same as the incident packet. We can easily see what happens to a packet representing a superposition of up and down particles (this is not the same as a probability distribution of orientations in the beam, but rather represents particles with a definite orientation along some direction which does not coincide with the axis of the filter): the up component propagates undisturbed while the down component is almost completely reflected. Hence after a long time a packet which is a superposition of up and down is separated into two disjoint packets which have no spatial overlap, each of which is essentially pure in spin orientation with respect to the preferred axis of the filter. Thus when light is incident on a polaroid, we can be sure that all the photons that pass through have a definite linear polarization.

When we observe a photon hitting a photographic emulsion, we do not observe either its wavefunction or its *probability* of arriving in a particular spot. We see rather that it has either arrived, or not arrived, at that spot. Thus our act of observing the photon has apparently caused its wavefunction to "collapse" to a state with a definite value of its position (or at least, to a superposition of states whose position observables are confined to a small interval), in the sense that subsequently that photon's state will evolve from the initial condition that at the time of observation it *definitely* appeared where it exposed the silver grain in the photograph. The mysterious aspect of the observation process is how the probability gets converted into a definite outcome. For example, should the observer be included in the total wavefunction of the system? It is sometimes said that the wavefunction is in some sense a measure of our knowledge of the system, and that it is not therefore surprising that a packet should disperse, since this simply represents the expected decay of knowledge with time. Alternatively, one might consider the wavefunction to have an independent physical existence, perhaps something like a field strength. In this view there is a pronounced asymmetry between the observer and that which is observed. The interaction between them must be highly nonlinear and essentially discontinuous in order to produce the "collapse" of the wavefunction. But because quantum mechanics is an inherently *linear* theory, the process of observation must be somehow "special." Some authors [16, 23] have gone so far as to suggest that the special role of the observer is the result of (or the cause of—they are not specific) consciousness and/or self-awareness. Others [6, 12, 22] have proposed that each observation that converts a probability into an outcome is accompanied by a bifurcation in the physical universe, so that alternate worlds are generated in one of which, for example, a photon went through the apparatus, whereas in the other it was reflected. Given the rate of events on the microscopic scale, the rate of universe generation must dwarf the term "astronomical."

All evidence presently available supports

the idea that observations could as well be conducted by simple counting apparatus as by sentient beings. If we want to suppose that the image on a photograph has no existence (i.e., that the wavefunction of photons *cum* reduced silver grains has not collapsed) until a self-aware human being has looked at it, we venture onto the very thin ice of solipsism. One or another of the mystical viewpoints of the Berkelian variety described above may, after all, be correct. However, it is as reasonable to suppose that they arise out of semantic confusion between the words "intelligence," "observation," and "knowledge" and therefore represent logic as circular as St. Anselm's "proof" of the existence of God.

Most physicists today support the view that the wavefunction is no more than a probability measure with no independent physical significance, whose sole predictive power is restricted to the distribution of outcomes of large ensembles of identical experiments. The deeper question of how probabilities become actualities is one that we ignore mainly because we possess no objective way to investigate the question at present, and there is at least some suspicion that the question itself may be as meaningless as the paradox of the Spanish barber. That nature should behave in the odd way implicit in the probabilistic interpretation of quantum mechanics defies our intuition. But as we have seen in other instances, our everyday experiences generate intuitions that have little relevance to the unfamiliar realms now accessible to our exploration.

References

[1] Bohm, D. (1951). *Quantum Theory*. Prentice-Hall, Englewood Cliffs, N.J., p. 81ff.

[2] Cropper, W. H. (1970). *The Quantum Physicists*. Oxford University Press, London, p. 120ff.

[3] Davisson, C. J. and Germer, L. H. (1927). *Phys. Rev.*, **30**, 705.

[4] de Broglie, L. (1924). *Philos. Mag., 6th Ser.*, **47**, 446.

[5] de Broglie, L. (1960). *Nonlinear Wave Mechanics—A Causal Interpretation*. Elsevier, Amsterdam.

[6] DeWitt, B. (1970). *Phys. Today*, **23**(Sept.), 30.

[7] Dicke, R. H. (1981). *Amer. J. Phys.*, **49**, 925. (In this article Dicke analyzes the consequences of using photons to determine a particle's position in considerable detail and resolves several of the apparently paradoxical consequences of Heisenberg's original analysis.)

[8] Drude, P. (1900). *The Theory of Optics*. (Reprint of the English translation by C. R. Mann and R. A. Millikan of the German edition, Dover, New York, p. 124ff.)

[9] Einstein, A. (1905). *Ann. Phys.*, **17**, 132.

[10] Einstein, A., Podolsky, B., and Rosen, N. (1935). *Phys. Rev.*, **47**, 777.

[11] Epstein, Paul S. (1945). *Amer. J. Phys.*, **13**, 127.

[12] Everett, H., III. (1957). *Rev. Mod. Phys.*, **29**, 454.

[13] Feynman, R. P. (1961). *The Theory of Fundamental Processes*. W. A. Benjamin, Menlo Park, Calif.

[14] Heisenberg, W. (1927). *Zeit. Phys.*, **43**, 1972.

[15] Kemble, E. C. (1937). *The Fundamental Principles of Quantum Mechanics*. McGraw-Hill, New York, p. 293ff.

[16] Peierls, Sir R. (1979). *Surprises in Theoretical Physics*. Princeton University Press, Princeton, N. J., pp. 23–34.

[17] Planck, M. (1900). *Verh. Deut. Phys. Ges.*, **2**, 237; *Ann. Phys.*, **4**, 553 (1901).

[18] Schrödinger, E. (1926). *Ann. Phys.*, **79**, 361, 489, 784; *ibid.*, **80**, 437 (1926); *ibid.*, **81**, 109 (1926).

[19] Stapp, H. P. (1972). *Amer. J. Phys.*, **40**, 1098. (Stapp's article is one of the most literate and complete expositions of the Copenhagen interpretation to be found in the general literature. It is intended for readers who already have considerable familiarity with the formalism of quantum mechanics and who wish to explore its philosophical basis.)

[20] Taylor, G. I. (1909). *Proc. Camb. Philos. Soc.*, **15**, 114. [Modern variants of this experiment can be found in Merli, P. G., Missiroli, G. F., and Pozzi, G. (1976); *Amer. J. Phys.*, **44**, 306.]

[21] Thomson, G. P. and Reid, A. (1927). *Nature*, **119**, 890.

[22] Wheeler, J. A. (1957). *Rev. Mod. Phys.*, **29**, 463.

[23] Wigner, E. P. (1962). In *The Scientist Speculates*, I. J. Good, ed. Basic Books, New York, p. 284.

(PHYSICS, STATISTICS IN (EARLY HISTORY)
QUANTUM MECHANICS: STATISTICAL INTERPRETATION
QUANTUM PHYSICS AND FUNCTIONAL INTEGRATION)

J. V. Noble

QUANTUM MECHANICS: STATISTICAL INTERPRETATION

The statistical interpretation of quantum mechanics arose from the discussion of scattering experiments. In a typical experimental setup of electron scattering we can distinguish the following parts:

1. An electron source, emitting electrons under specified, reproducible conditions (including perhaps focusing, shielding, etc.)
2. A target (scattering center)
3. A screen (detector) where the electrons are registered after the scattering process

One may run the experiment at sufficiently low intensity for a long time so that one can be sure that the behavior of a single electron is studied (i.e., the interaction between different electrons in the beam is negligible). The source may then be regarded as a preparation procedure for a statistical ensemble of electrons. Under optimal conditions this ensemble may be mathematically described by a Schrödinger wavefunction. In the language of quantum mechanics, one says that the source prepares a *state* of an electron. Operationally, the statement that an electron is in a specific state means that the electron is a member of a specific statistical ensemble, the ensemble prepared by a specified source.

The screen provides the opportunity for each electron in the ensemble to produce one observable event, such as the blackening of a grain in a photographic emulsion covering the screen or a scintillation flash originating from a small area on the screen. In our example the different possible events correspond to different distinguishable small surface cells on the screen. The general characteristic feature of such an observation procedure (here the screen) is that it supplies a set of mutually exclusive possible events which may only be caused by the system under observation (here the electron) and, in the ideal case, each individual in the ensemble will have to produce one such event.

If we call an event a *measuring result* and parametrize the event p_i by a number a_i (or several numbers $a_i^{(1)} \cdots a_i^{(n)}$) called the "measured value(s)," we are lead to the quantum mechanical concept of an *observable*. It is the same as the "observation procedure" mentioned above, except that equivalent procedures are not distinguished; for example, in our example it is irrelevant whether the screen is a photographic plate or a scintillation screen, the elementary events caused by an electron being essentially in one-to-one correspondence. On the other hand, the concept of observable involves a specific assignment of measured values to events. For instance, we could assign to each surface cell of the screen the Cartesian x-coordinate of its center as the measured value and thereby obtain an observable called X; or we might choose the y-coordinate and obtain the observable Y; or we might use other parametrizations of the events. The screen then obviously provides a "simultaneous measuring instrument" for the observables X and Y (and all other possible labellings of the possible events).

The target may—for conceptual simplicity—be taken together with the source as a part of the "state preparing procedure" or alternatively together with the screen as part of the "observation procedure." Or we may give it an independent significance as an "operation" that changes the state (i.e., we consider it as a transformation in the set of states).

The fundamental experience in experiments of the sort described is that one is not able to construct a source so that all individual systems in the ensemble prepared behave in the same way (i.e., produce the same event, "hit the screen at the same position"). Given the source (plus target), only the probability for the different events is determined. The question to be answered by the theory is: Given a state s and a potential event p, what is the probability $w(p;s)$ for this event to happen? Experimentally, it is, of course, given by the ratio between the total number of cases in which the specified event has been produced and the total number of systems prepared.

One important property of the set of all states of a system can be immediately abstracted from the operational definition. If it is possible to prepare the states s_1 and s_2, then one can also prepare any mixture between the two with weights λ and $1 - \lambda$, respectively ($0 < \lambda < 1$). This mixture we denote by $s = \lambda s_1 + (1 - \lambda)s_2$, and since there is no physical interaction between any individuals in the ensemble, the probabilities for any event in the states s, s_1, and s_2 must be related by

$$w(p; s) = \lambda w(p; s_1) + (1 - \lambda)w(p; s_2).$$

Mathematically, this means that the set $\mathscr{S}$ of all states is a convex set that may be canonically embedded in a real linear space. While the mixing of states is always possible, the purification has its limitations. There are states that cannot be obtained as the mixtures of any other states; these are the *pure states* or the extremal points of the convex body $\mathscr{S}$. All other states may be ultimately expressed as convex combinations (mixtures) of pure states.

In the mathematical formalism of quantum mechanics a pure state s is represented by a vector Ψ of length 1 in a Hilbert space $\mathscr{H}$, a potential event (experimentally corresponding to a detector which gives a signal when the event is produced) by an orthogonal projection operator P acting on $\mathscr{H}$. The probability w is given by

$$w(p; s) = \|P\Psi\|^2, \tag{1}$$

the square of the length of the projected vector.

There have been many investigations of whether this specific mathematical structure can be understood as a consequence of simple operational principles [1]. If one focuses on the structure of the set of events, one has to show that they form an "orthocomplemented, orthomodular, and semimodular lattice." This is the approach of *quantum logic*, where the events (respectively, the detectors that signal them) are regarded as "questions" or logical "propositions" [2]. If one focuses on the set of states, one has to show that they form the base of a "self-dual, facially homogeneous cone" in a real linear space [3]. If one focuses on the (pure) operations, one has to show that they form a "Baer semigroup" [4]. Any one of these three structural statements is essentially equivalent to the mathematical formalism of quantum mechanics. Concerning the underlying principles, we have to be content here with a few remarks.

Obviously, the events (propositions) may be partially ordered. We may call a proposition p_1 finer than p_2 if

$$w(p_1; s) \leqslant w(p_2; s)$$

for all states s. It is seen from (1) that in quantum mechanics the finest propositions ("elementary events") correspond to projectors on one-dimensional subspaces of $\mathscr{H}$; that is, they can, like the pure states, be characterized by unit vectors in $\mathscr{H}$ and (1) becomes in that case

$$w(p_\Phi; s_\Psi) = |(\Phi, \Psi)|^2.$$

Thus in quantum mechanics there is a one-to-one mapping between pure states and elementary events and the probability function w is symmetric with respect to the interchange:

$$w(p_\Phi; s_\Psi) = w(p_\Psi; s_\Phi).$$

This property, together with some maximality assumptions about the sets of states and events, implies that state space is the base of a self-dual cone. The existence of sufficiently many pure operations brings the homogeneity, and after this there remain only a few possibilities:

1. The situation of classical mechanics, where state space is a simplex (i.e., every mixed state has a unique decomposition into pure states)
2. The quantum mechanical structure with a Hilbert space over either the real, the complex, or the quaternionic field

3. A few exceptional cases when there are only two or three mutually exclusive events

The quantum theoretical description of nature has raised two (not entirely unconnected) epistemological questions: Are the fundamental laws really indeterministic, or is the statistical description only due to our ignorance? This is the question of the possible existence of "hidden variables" which may have so far escaped our observation. The other, more fundamental, question concerns the meaning of physical reality. The notions of "state" and "event" as used in quantum theory are not properties of the "system" (the electron in our example). Rather, the state is an attribute of the system plus the source and the event an attribute of the system plus the detector. If one wanted to assume, for example, that a single particle always has some position in space and the role of the screen is merely to reveal this position to us, we would immediately get into contradictions with experimental facts. If one wanted to assume that the Schrödinger wavefunction is a full description of the state of each individual particle, in the sense that each particle, if far away from any other matter, has to have a definite wavefunction (which may be unknown to us), one would get a contradiction to the generalized Einstein–Podolsky–Rosen phenomenon which has recently been tested experimentally [5]. One may, of course, consider the particle together with the source and the screen as one physical system. But then, if one wishes to discuss microscopic states of this big system and events caused by it, one meets the same problem again. One has to bring preparation and observation procedures for the big system into the picture without which the notions of microscopic state and event are not defined. Thus one is always forced to make some cut, mentally dividing the physical world into an observed system and the rest of the universe, while the interplay between these two parts cannot be ignored in the very definition of microscopic state and event of the "system." If, on the other hand, one is content with a coarser description of the big system, the indeterminacy of events already results from our ignorance of the precise conditions. One can, then, also maintain that an event—such as a flash from the scintillations screen—has either happened or not happened, irrespective of observation. The finality of the event is then tied to the irreversibility as it is known in thermodynamics. This irreversibility is usually interpreted, however, not as a fundamental one but as a practical one, due to the high complexity of the system.

References

[1] The following books give an account of such efforts up to 1970: G. Mackey, *The Mathematical Foundations of Quantum Mechanics*, Benjamin, New York, 1963; J. M. Jauch: *Foundations of Quantum Mechanics*, Addison-Wesley, Reading, Mass., 1968; G. Ludwig, Deutung des Begriffs physikalische Theorie und axiomatische Grundlegung der Hilbert-Raum-Struktur der Quantenmechanik, *Lect. Notes Phys.*, **4**, Springer-Verlag, Heidelberg, 1970.

[2] This approach was started by G. Birkoff and J. von Neumann, "The logic of quantum mechanics," *Ann. Math.*, **37**, 823–843 (1936). Concerning the role of semimodularity, see J. M. Jauch and C. Piron, "On the structure of quantal proposition systems," *Helv. Phys. Acta*, **42**, 842–848 (1969).

[3] E. M. Alfsen and F. W. Schultz: "State space of Jordan algebras," *Acta Math.*, **140**, 155–190 (1978); H. Araki, *On the Characterization of the State Space of Quantum Mechanics*, RIMS Publications, Kyoto, 1979; E. B. Vinberg, "The structure of the group of automorphisms of a homogeneous convex cone," *Trans. Mosc. Math. Soc.*, 63–93 (1965); J. Bellisard and B. Jochum: "Homogeneous self dual cones versus Jordan algebras," *Ann. Inst. Fourier*, Vol. XXVIII, Fascicule 1, 1978.

[4] J. C. T. Pool, "Baer*-semigroups and the logic of quantum mechanics," *Commun. Math. Phys.*, **9**, 118–141 (1968).

[5] J. S. Bell, *Physics*, **1**, 195 (1964); J. F. Clauser, M. A. Horne, A. Shimony, and R. A. Holt, *Phys. Rev. Lett.*, **23**, 880 (1969).

(QUANTUM MECHANICS AND PROBABILITY)

RUDOLPH HAAG

QUANTUM PHYSICS AND FUNCTIONAL INTEGRATION

Methods based on integration in function space are a powerful tool in the understanding of quantum dynamics, especially in quantum field theory, where they not only give a useful formulation of perturbation theory but also provide one of the few tools that go beyond perturbation theory.

Since the time-dependent Schrödinger equation (in units with $\hbar = 1$)

$$i\dot{\psi}_t(x) = -(2m)^{-1}\Delta\psi_t(x) + V(x)\psi_t(x) \quad (1)$$

is similar to the diffusion equation

$$-\dot{\phi}_t = -\Delta\phi_t(x), \quad (2)$$

for which Wiener invented functional integration [11], it is not surprising that functional integration can be used to study (1). Of course, (1) is solved by $\psi_t = e^{-itH}\psi_0$; $H = -(2m)^{-1}\Delta + V$, while (2) is solved by $\phi_t = e^{-2tH_0}\phi_0$; $H_0 = -\frac{1}{2}\Delta$. Feynman's initial formal framework of "path integrals in quantum mechanics [3]" involved the study of e^{-itH}. Although there has been some partially successful rigorous study of Feynman "integrals" (one must extend the notion of measure in all these approaches) [1, 4, 5], most of the rigorous work and an increasing fraction of the heuristic literature has studied the semigroup e^{-tH} rather than e^{-itH}.

At first sight it seems surprising that the study of e^{-tH} would be a suitable substitute for the study of e^{-itH}, but the semigroup even has some real advantages. (1) If one wants to understand the lowest eigenvalue*, E (*ground state energy*), and corresponding eigenvector (*ground state*) of H, the operator e^{-tH} is often more useful than e^{-itH}; for example,

$$E = -t^{-1}\ln\|e^{-tH}\|.$$

(2) If one views the parameter t as $1/(kT)$, with k equal to Boltzmann's constant and T the temperature, then the semigroup is the basic object of quantum statistical mechanics. (3) In quantum field theory, the formal continuation from it to t should replace Lorentz invariance by invariance under a Euclidean group of rotations and various hyperbolic equations by better-behaved elliptic equations. This passage to Euclidean invariance is often called *analytic continuation to the Euclidean region* in the theoretical physics literature.

In units with $m = 1$, the basic formula for e^{-tH} is the *Feynman–Kac formula*, named after fundamental contributions by their authors [3, 6]:

$$(e^{-tH}f)(x) = E_x\left(e^{-\int_0^t V(b(s))\,ds}f(b(t))\right). \quad (3)$$

In this formula, b is Brownian motion* and E_x denotes expectation with respect to paths starting at x, that is, $b(s)$ is an R^ν-valued Gaussian process* with mean x and covariance:

$$E_x((b_i(s) - x_i)(b_j(t) - x_j)) = \delta_{ij}\min(t,s).$$

For detailed conditions on f, V for (3) to hold and several proofs, see ref. 8. An especially useful way of looking at (3) is in terms of the Lie–Trotter formula:

$$\exp(-tH) = \lim_{n\to\infty}\left[\exp(-tH_0/n)\exp(-tV/n)\right]^n$$

and Riemann sum approximations to $\int_0^t V(b(s))\,ds$.

While (3) is very useful to study H, it can also be turned around and used to study Brownian motion; for example [see ref. 8 (pp. 58–60)], one can compute the distribution of the Lebesgue measure of $\{s \leqslant t \mid b(s) > 0\}$ by using (3) with V the indicator function of $\{x \mid x > 0\}$.

The differential operator

$$H(a) \equiv \tfrac{1}{2}(-i\nabla - a)^2 + V \quad (4)$$

is the Hamiltonian operator for particles (with $\hbar = 1$, $m = 1$, $e/(mc) = 1$) moving in a magnetic field $B = \text{curl}\,(a)$ and potential V. There is an analog of (3) for $H(a)$, sometimes called the *Feynman–Kac–Ito formula*:

$$(e^{-tH}f)(x) = E_x\left(e^{-iF(b)}e^{-\int_0^t V(b(s))\,ds}f(b(t))\right) \quad (5)$$

with

$$F(b) = \int_0^t a(b(s)) \cdot db - \frac{1}{2} \int_0^t (\operatorname{div} a)(b(s))\, ds. \quad (6)$$

In (6), $\int a \cdot db$ is an Ito stochastic integral* (*see also* BROWNIAN MOTION); if the Stratonovich integral is used instead, the second term in (6) will not be present. Since F is real-valued, (5) immediately implies the *diamagnetic inequality* of Nelson and Simon:

$$|(e^{-tH(a)}f)(x)| \leqslant (e^{-tH(a=0)}|f|)(x),$$

which has been very useful in the study of operators of the form (4).

As an intermediate situation to describing quantum field theory, one can consider $P(\phi)_1$-processes. Let $h = -\frac{1}{2}d^2/dx^2 + v(x)$ on $L^2(R, dx)$. Suppose that e is the ground-state energy and $\Omega(x)$ the ground state. Let $P_t(x, y)$ be the integral kernel of the operator $\Omega \exp[-t(h - e)]\Omega^{-1}$. $P_t(x, y)$ has the semigroup property; it is positive since $\Omega(x) \geqslant 0$, and $\int P_t(x, y)\, dx = 1$ since e^{-th} is symmetric and $e^{-th}\Omega = \Omega$. Thus $P_t(x, y)$ defines a Markov process* with invariant measure $\Omega^2\, dx$. The Markov process $q(t)$ with this invariant measure as initial distribution and transition kernel $P_t(x, y)$ is called the $P(\phi)_1$-*process* [since V is often a polynomial $P(x)$]; we use E to denote the corresponding expectation. The expectations $E^{(v)}$ and $E^{(v+w)}$ associated to h and $h + w$ are related by

$$E^{(v+w)}(F) = \lim_{t \to \infty} Z_t^{-1} E^{(v)} \left\{ F \exp\left[-\int_{-t}^{t} w(q(s))\, ds \right] \right\} \quad (7)$$

with

$$Z_t = E^{(v)} \exp\left[-\int_{-t}^{t} w(q(s)) \right] ds.$$

This sets up an analogy to statistical mechanics which is often useful.

The $P(\phi)_1$-process with $v(x) = \frac{1}{2}x^2$ is of particular interest since it is a Gaussian process. Its covariance is

$$E(q(s)q(t)) = \tfrac{1}{2}\exp(-|t - s|). \quad (8)$$

Up to changes of scale of q and/or t, it is the unique stationary, Markov-Gaussian process, called the *Ornstein–Uhlenbeck* velocity process*, or occasionally the *oscillator process*, since $-\frac{1}{2}d^2/dx^2 + \frac{1}{2}x^2$ is the Hamiltonian of a harmonic oscillator.

Quantum field theories analytically continued to the Euclidean region are analogs of $P(\phi)_1$-processes with the time, s, now a multidimensional variable. In generalizing (8), it is important to realize that the right side of (8) is the integral kernel of the operator $(-d^2/dt^2 + 1)^{-1}$. Let $G_0(x - y)$ be the integral kernel of the operator $(-\Delta + 1)^{-1}$ on $L^2(R^\nu, d^\nu x)$. The generalized Gaussian process $\phi(x)$ with mean zero and covariance

$$E(\phi(x)\phi(y)) = G_0(x - y)$$

is called the *free Euclidean field* or occasionally the *free Markov field* (since it has a kind of multitime Markov property). If $(-\Delta + 1)^{-1}$ is replaced by $(-\Delta + m^2)^{-1}$, the phrase "of mass m" is added to "free Euclidean field." This process is the analytic continuation to the Euclidean region of quantum field theory describing noninteracting spinless particles.

The process above is only "generalized" [i.e., $\phi(x)$ must be smeared in (x), since G_0 is singular at $x = y$, if $\nu \geqslant 2$]. For $\nu = 2$, the singularity is only logarithmic, but it is a power singularity if $\nu \geqslant 3$. The natural analog of (7) is to try to construct an expectation by

$$\lim_{\Lambda \to R^\nu} Z(\Lambda)^{-1} \int F(\phi) \exp\left[-\int_\Lambda w(\phi(x))\, d^\nu x \right], \quad (9)$$

where

$$Z(\Lambda) = \int \exp\left[-\int_\Lambda w(\phi(x))\, d^\nu x \right].$$

Such a construction would yield models of quantum fields. In $\nu = 2$ and 3 ($\nu = 4$ is the physical case), this program, called *constructive quantum field theory*, has been successful, due to the efforts of many mathematical physicists, most notably J. Glimm and A. Jaffe.

In (9), there are two general problems. The limit $\Lambda \to R^\nu$ is not trivial to control; it has been controlled by making an analogy to the corresponding limit in statistical mechanics and extending various ideas from that discipline.

The other problem is special to quantum field theory. When $\nu = 1$, the oscillator process is supported on continuous functions, but for $\nu \geq 2$, the corresponding free field is supported on distributions that are signed measures with probability zero! Thus the typical choice $w(\phi(x)) = \phi(x)^4$ is meaningless since $\phi(x)$ does not have powers. This is the celebrated problem of *ultraviolet divergences*, which occurred in all the earliest attempts to study quantum field theory. It is solved by *renormalization theory*. For example, let $\phi_f(x) = \int f(x-y)\phi(y)\,d^\nu y$, which yields a nice process if $f \in C_0^\infty$. We want to let $f \to \delta$. One lets w depend on f and in (9) lets $w(\phi)$ be replaced by $w_f(\phi_f)$ and then also takes a limit as $f \to \delta$. For example, in $\nu = 2$ dimensions, one can take

$$w_f(\phi(x)) = \phi_f(x)^4 - 6E(\phi^2(x))\phi_f^2(x) \quad (10)$$

and control the limit in (9) if first $f \to \delta$ and then $\Lambda \to \infty$. In (10), $E(-)$ denotes expectation with respect to the free Euclidean field, so

$$E(\phi_f^2(x)) = \int f(x)G_0(x-y)f(y)\,dx\,dy \to \infty$$

as $f \to \delta$. The resulting theory is often denoted as the ϕ_2^4 theory. There have been developed ϕ_3^4 theories and $P(\phi)_2$ for a large class of polynomials P. Fermion theories and certain gauge models have been constructed in two and three dimensions, but this goes beyond the scope of this article.

An especially useful device in the theory is the *lattice approximation*, where x is replaced by a discrete variable and $-\Delta$ by a finite difference* operator. One realizes that these models are lattice systems* with unbounded spins and one uses intuition from that subject. As the lattice spacing goes to zero, one can hope to recover the continuum theory.

Recently, steepest descent ideas have come into use in functional integration pictures of quantum theory. A major role is played by functions minimizing an exponent. The quantity in the exponent is usually called *Euclidean action* and the minimizing functions are *instantons*.

Glossary

Fermion Theories: Field theories describing particles that obey the Pauli exclusion principle (also called Fermi–Dirac* statistics, hence "fermions").

Gauge Models (also called "Yang–Mills" or "non-Abelian gauge" theories): These are the current popular quantum field models which describe both weak interactions (Weinberg–Salam theories) and strong interactions (quantum chromodynamics).

Perturbation Theory: Expansion in physical coupling constants of the basic quantities of a quantum field theory which is especially useful in quantum electrodynamics, where the natural coupling constant is small (about 1/137).

Quantum Field Theory: The only successful way of synthesizing quantum mechanics and special relativity.

References

[1] Albeverio, S. and Hoegh-Krohn, R. (1976). Mathematical Theory of Feynman Path Integrals. *Lect. Notes Math.*, **523**, Springer-Verlag, New York. (This reference and refs. 2 and 5 constitute three attempts at mathematical formulations of a path integral view of e^{-itH}.)

[2] DeWitt-Morette, C., Maheshwari, A., and Nelson, B. (1979). *Phys. Rep.*, **50**, 255–372.

[3] Feynman, R. P. (1948). *Rev. Mod. Phys.*, **20**, 367–387. (This reference and ref. 6 are two classics with a wealth of insight.)

[4] Feynman, R. P. and Hibbs, A. (1965). *Quantum Mechanics and Path Integrals*. McGraw-Hill, New York. (The main reference on the formal path integral for studying e^{-itH}.)

[5] Fujiwara, D. (1974). *Proc. Japan Acad.*, **50**, 566–569; *ibid.*, **50**, 699–701; *ibid.*, **54**, 62–66 (1978).

[6] Kac, M. (1950). *Proc. 2nd Berkeley Symp. Math. Statist. Prob.* University of California Press, Berkeley, Calif., pp. 189–215.

[7] Nelson, E. (1960). *Phys. Rev.*, **150**, 1079–1085. (Another interface of stochastic processes and quantum theory.)

[8] Simon, B. (1974). *The $P(\phi)_2$ Euclidean (Quantum) Field Theory*. Princeton University Press, Princeton, N. J. (This reference and ref. 10 cover the details of Euclidean constructive quantum field theory.)

[9] Simon, B. (1979). *Functional Integration and Quantum Physics*. Academic Press, New York. (A general discussion of Brownian motion and e^{-tH}.)

[10] Velo, G. and Wightman, A. S., eds. (1973). *Constructive Quantum Field Theory*. Springer-Verlag, Berlin.

[11] Wiener, N. (1923). *J. Math. Phys.*, **2**, 131–174.

(LATTICE SYSTEMS
QUANTUM MECHANICS AND PROBABILITY
QUANTUM MECHANICS: STATISTICAL INTERPRETATION
STOCHASTIC INTEGRALS)

BARRY SIMON

QUARTIC EXPONENTIAL DISTRIBUTION

The quartic exponential distribution has a probability density function of the form

$$f(x) \propto \exp\left[-\left(\alpha_1 x + \alpha_2 x^2 + \alpha_3 x^3 + \alpha_4 x^4\right)\right],$$
$$\alpha_4 > 0, \quad -\infty < x < \infty.$$

It belongs to the general exponential family*. Fisher [2] discussed a general "polynomial" exponential family, but the fourth-degree case seems to be first explicitly considered by O'Toole [4] and subsequently by Aroian [1].

Matz [3] studied the quartic exponential family in detail, with special emphasis on maximum likelihood* estimation of parameters. Application to bimodal data seems the principal use of this distribution.

References

[1] Aroian, L. A. (1948). *Ann. Math. Statist.*, **19**, 589–592.

[2] Fisher, R. A. (1921). *Phil. Trans. R. Soc. Lond. A*, **222**, 309–368.

[3] Matz, A. W. (1978). *Technometrics*, **20**, 475–484.

[4] O'Toole, A. L. (1933). *Ann. Math. Statist.*, **4**, 1–29, 79–93.

(EXPONENTIAL DISTRIBUTION
EXPONENTIAL FAMILIES)

QUARTILE *See* QUANTILES

QUARTILE DEVIATION

This is one-half of the interquartile* distance. It is also called the *semi-interquartile range*.

(PROBABLE ERROR
QUARTILE DEVIATION, COEFFICIENT OF)

QUARTILE DEVIATION, COEFFICIENT OF

The quartile deviation* is an appropriate measure of variation when the median is used as the measure of central tendency. It is defined as

$$Q = \tfrac{1}{2}(Q_3 - Q_1)$$

where Q_1 and Q_3 are the first and third quartiles*, respectively.

The *relative variation* is measured by expressing the quartile deviation as the percentage of the midpoint between Q_1 and Q_3. This percentage is denoted by

$$V_Q = \frac{(Q_3 - Q_1)/2}{(Q_3 + Q_1)/2} \times 100$$
$$= \frac{Q_3 - Q_1}{Q_3 + Q_1} \times 100$$

and is called the coefficient of quartile deviation. It is an analog of the coefficient of variation*.

QUARTIMAX

An algorithm for *orthogonal* rotation occasionally used in factor analysis*. The quartimax method rotates the factors in such a manner as to accomplish for a given variable one and only one major loading on a given factor. An apparently undesirable property of this method is a tendency toward producing a general factor with all or most of the variables having high loadings. Compare with Varimax*. See, e.g., Rummel [1] for details of the computations.

Reference

[1] Rummel, R. J. (1970). *Applied Factor Analysis*. Northwestern University Press, Evanston, Ill.

(FACTOR ANALYSIS
VARIMAX)

QUASI-BAYESIAN INFERENCE

Quasi-Bayesian (or pseudo- or semi-Bayesian) inference refers to a combination of, or compromise between Bayesian* and classical procedures sometimes used in the analysis of hierarchical models.

In the hierarchical model, multiple levels or stages of uncertainty are generated in the following way. First, a likelihood function, which is conditional on values of unknown parameters, represents the data generated by some random process. A Bayesian statistician then puts a prior distribution* on these unknown parameters, and this distribution itself is conditional on the values of other parameters called hyperparameters. If these hyperparameters are unknown, a distribution (sometimes called a hyperprior) may be given for their values, and so on.

In quasi-Bayesian inference, the hyperparameters at some stage are estimated by non-Bayesian methods, such as maximum likelihood estimation*, thus eliminating a hyperprior at that stage. Although this practice is not viewed as Bayesian, it has been termed part of a "Bayes/non-Bayes compromise" in ref. 1.

For example, suppose that $(n_1, \ldots, n_K)$ represent a sample from a K-category multinomial distribution*, with $\sum_{i=1}^K n_i = n$. Let p_i represent the probability attached to cell i, with $\sum_{i=1}^K p_i = 1$. A Bayesian statistician would represent uncertainty about $(p_1, \ldots, p_K)$ in a prior distribution; for example, consider a (symmetrical) Dirichlet* prior distribution having density

$$f(p_i, \ldots, p_k \mid \alpha) = \frac{\Gamma(K\alpha)}{\{\Gamma(\alpha)\}^K} \prod_{i=1}^K p_i^{\alpha-1},$$

where $\alpha > 0$ is the hyperparameter. A Bayes estimate of p_i, under squared-error loss, would be $E(p_i \mid \alpha) = (n_i + \alpha)/(n + K\alpha)$. In order to test $H_0: p_i = 1/K$, $i = 1, \ldots, K$, the Bayes factor would be given by

$$F(\alpha) = \frac{K^n \Gamma(K\alpha)\left[\prod_{i=1}^K \Gamma(n_i + \alpha)\right]}{[\Gamma(\alpha)]^K \Gamma(n + K\alpha)}.$$

The hyperparameter α is seen to play an important role in inference.

If α is uncertain, a full Bayesian analysis would require that a hyperprior, $f(\alpha)$, be specified. A quasi-Bayesian approach might use maximum likelihood estimation to circumvent the need for $f(\alpha)$. In other words, define $\alpha_{\max}$ as the value of α that maximizes $F(\alpha)$, and let $F_{\max} = F(\alpha_{\max})$. Then a quasi-Bayesian estimate of p_i would be

$$(n_i + \alpha_{\max})/(n + K\alpha_{\max}).$$

Since $F_{\max}$ tends to overestimate the Bayes factor against H_0, alternative interpretations of $F_{\max}$ may be employed; see ref. 1.

Reference

[1] Good, I. J. and Crook, J. F. (1974). *J. Amer. Statist. Ass.*, **69**, 711–720.

(BAYESIAN INFERENCE
EMPIRICAL BAYES METHODS
ESTIMATION, POINT)

R. L. TRADER

QUASIBINOMIAL DISTRIBUTIONS

The simplest form for the probability function of a quasibinomial variate r is

$$\Pr(r=k)=\binom{n}{k}p(p+k\phi)^{k-1} \times(1-p-k\phi)^{n-k}$$
$$(k=0,1,\ldots,n;\ 0<p<1),\quad (1)$$

where n is a positive integer and $p+n\phi \leqslant 1$. It was introduced by Consul [3] in connection with a two-urn model. A modified urn model due to Consul and Mittal [4] led to the more general four-parameter quasibinomial form

$$\Pr(r=k) = \binom{n}{k}\frac{ab}{(a+b)(a+k\theta)}\left(\frac{a+k\theta}{a+b+n\theta}\right)^k \times\left(\frac{b+n\theta-k\theta}{a+b+n\theta}\right)^{n-k-1}$$
$$(k=0,1,\ldots,n),\quad (2)$$

which reduces to the binomial form when $\theta=0$.

An identity due to Jensen [7] is needed to derive expressions for the moments. For the mean,

$$E(r)=na/(a+b),$$

and for the second factorial moment,

$$E\{r(r-1)\}=\frac{n(n-1)a}{a+b}\times\left[1-\frac{b\Psi(n,a,b)}{a+b+n\theta}\right],$$

where

$$\Psi(n,a,b) = \sum_{s=0}^{n-2}\frac{(n-2)!}{(n-2-s)!}\left(\frac{\theta}{a+b+n\theta}\right)^s.$$

SOME PROPERTIES

1. There are interesting limiting cases. For example, if n is large, a and θ small with $na=c$, $n\theta=d$, then the quasibinomial in the limit becomes the generalized Poisson distribution*. Under certain restrictions, another limiting form produces the normal distribution.
2. The sum of quasibinomial variates in general is not a quasibinomial variate.
3. Properties of the binomial distribution, such as the location of the largest probability, expression of the partial sum of probabilities in integral form, and so on, are not readily available for the quasibinomial distribution.
4. An optimal asymptotic test for a quasibinomial distribution has been given by Fazal [5].

See also Janardan [6] and Korwar [9].

LINKS WITH LAGRANGE'S EXPANSION

In 1770, Lagrange [10] gave the formula

$$f(t)=f(a)+\sum_{k=1}^{\infty}(u^k/k!) \times D^{k-1}[f'(t)\{\phi(t)\}^k]_{t=a} \quad (3)$$

for a root of the equation $t=a+u\phi(t)$ under certain restrictions. In the early part of the present century, Jensen [7] exploited this formula, together with a second basic formulation [see his expression (5)], to produce numerous identities, for the most part involving positive components and hence often translatable into distributional forms. A well-known example is

$$\exp(at)=\sum_{k=0}^{\infty}\frac{a(a+k\beta)^{k-1}}{k!}u^k$$
$$[t=u\exp(\beta t)].\quad (4)$$

Similarly, from a convolution* point of view, Jensen used $\{\exp(at)\}*\{\exp(bt)\}$ to derive

$$(a+b)(a+b+n\beta)^{n-1} = \sum_{k=0}^{n}\binom{n}{k}a(a+k\beta)^{k-1} \times b(b+n\beta-k\beta)^{n-k-1},\quad (5)$$

which relates to a quasibinomial [8].

Surprisingly enough, some of these formulas were given nearly a century earlier by Abel [1] and Cauchy [2], the latter considering expansions such as the binomial form

$$[f(x,z)+f(z,x)]^n F(x,z)$$

and using the calculus of residues. For example,

$$[(x+a+n)^n-(x+a)^n]/a$$
$$=\sum_{k=0}^{n-1}\binom{n}{k}(a+n-k)^{n-k-1}(x+k)^k. \qquad (6)$$

For generalizations, see Riordan [11].

LINKS WITH AN OPERATIONAL FORM OF LAGRANGE

The function $t\exp(-\beta t)$ has a powerful interpretation in terms of finite differences*. For if $t=\ln(1+\Delta)$, where $\Delta f(x)=f(x+1)-f(x)$, $E\{f(x)\}=f(x+1)$, then

$$[t\exp(-\beta t)]^k \equiv D^k E^{-k\beta}, \qquad (7)$$

the operators being commutative and distributive. Also recall that

$$f(x+h)\equiv E^h f(x)\equiv e^{hD}\{f(x)\}.$$

Jensen [7] uses Lagrange expansions as operators; for example, from (4) he produces

$$\Phi(x+a)=\sum_{k=0}^{\infty}\frac{a(a+k\beta)^{k-1}}{k!}$$
$$\times\Phi^{(k)}(x-k\beta), \qquad (8)$$

leading to a form similar to Consul's expression (2) for a quasibinomial distribution when $\Phi(x)=x^n$, $x=1-p$, $a=p$.

Again using the Lagrange expansion for $t=v(1+t)^\beta$ and the function $(1+t)^a$, where $t=\Delta$, quasidistributions may be constructed; for example,

$$\binom{a+\beta+nt}{n}=\binom{\beta+nt}{n}$$
$$+a\sum_{k=1}^{n}\frac{1}{a+kt}\binom{a+kt}{k}$$
$$\times\binom{\beta+nt-kt}{n-k}.$$

To derive the quasihypergeometric probability function (see Johnson and Kotz [8, p. 84, eq. (70)], filter out the coefficients of v^n in the convolution

$$(1+t)^a*(1+t)^b=(1+t)^{a+b}$$
$$[t=v(1+t)^\beta],$$

using

$$(1+t)^a=\sum_{k=0}^{\infty}\frac{a}{a+k\beta}\binom{a+k\beta}{k}v^k. \qquad (9)$$

Varieties of this kind are legion—the inverse problem of discovering the correct Lagrange form is another matter.

Note that the validity of the expansions set up from the operational approach has to be scrutinized.

References

[1] Abel, N. H. (1826). *J. Math.*, **1**, 159–160. (Before the age of 18, proved the binomial expansion for nonrational index.)

[2] Cauchy, A. L. (1826). *Exer. Math.*, first year, 44–55.

[3] Consul, P. C. (1974). *Sankhyā B*, **36**, 391–399.

[4] Consul, P. C. and Mittal, S. P. (1975). *Biom. Z.*, **17**, 67–75.

[5] Fazal, S. S. (1976). *Biom. Z.*, **18**, 619–622.

[6] Janardan, K. G. (1974). *SDSW*, **3**, 359–364.

[7] Jensen, J. L. W. V. (1902). *Acta Math.*, **26**, 307–318. [Systematic uses of operational calculus are attributed to Oliver Heaviside (1850–1925). It has been used to derive Stirling's formula for $\ln\Gamma(\cdot)$.]

[8] Johnson, N. L. and Kotz, S. (1982). *Int. Statist. Rev.*, **50**(1), 71–101. (Gives a comprehensive view of recent developments in discrete distributions.)

[9] Korwar, R. M. (1977). *Commun. Statist. A*, **6**, 1409–1415.

[10] Lagrange, J. L. (1770). *Oeuvres*, **2**, 25.

[11] Riordan, J. (1968). *Combinatorial Identities*. Wiley, New York. (Chapter 6 treats the difference and differential operators Δ and D.)

(BINOMIAL DISTRIBUTION
GENERALIZED HYPERGEOMETRIC DISTRIBUTION
HYPERGEOMETRIC DISTRIBUTION
LAGRANGE AND RELATED PROBABILITY DISTRIBUTIONS
LAGRANGE EXPANSIONS

MULTIVARIATE HYPERGEOMETRIC DISTRIBUTIONS
POWER SERIES DISTRIBUTIONS)

L. R. SHENTON

QUASI-EXPERIMENT *See* OBSERVATIONAL STUDIES

QUASI-FACTORIAL DESIGNS

The name *quasi-factorial* design was given by Yates [4] to a class of designs that bear some resemblance to *factorial* designs*. These designs are used for comparing a large number of treatments in blocks, where there are more treatments than plots in a block, so that they are *incomplete block designs**. The word *quasi* comes from a Latin word meaning "as if," but a common present use, and dictionary definition, is "in appearance only." To make matters more confusing, the sets into which the treatments of the design are divided are known not as quasi-factors but as *pseudo-factors*. The word *pseudo* comes from a Greek word meaning "false," and its modern use is, in practice and in the dictionary, "deceptively resembling." Perhaps it is just as well that for most practical purposes, the name "quasi-factorial design" has almost disappeared from the modern statistical literature; instead, this type of design is known as a *lattice design**.

The idea underlying the name "quasi-factorial design" is that it shall be possible to establish a correspondence between the treatments of the design and the treatment combinations of a factorial set, and to make use of this correspondence to get a simpler form of statistical analysis for the design [2]. In the present computer age, ease of statistical analysis is no longer the problem it once was. Thus although quasi-factorial designs remain valuable in their own right, the original quasi-factorial concept has outlived its usefulness.

The simplest, and commonest, quasi-factorial or lattice designs are those for n^2 treatments in blocks of n plots each; such designs are *resolvable**, that is, a number of blocks, here n, form a complete replicate. If a fully orthogonal set of $n \times n$ *Latin squares** exists, it is possible to have $(n + 1)$ replicates of the treatments in the design, which then forms a *balanced* lattice. This is the same as a *balanced incomplete block* design*, and standard methods of analysis of such designs are well known. With fewer replicates the design is not balanced, and indeed may not come into the category of *partially balanced* incomplete block designs*. However, the use of pseudo-factors renders a quasi-factorial design analyzable, and a general recursive algorithm for analysis of variance*, which includes these designs, was described by Wilkinson [3] and implemented in the general statistical program GENSTAT [1]. With n^2 treatments and only two replicates it is sensible to regard the treatments as an $n \times n$ pseudo-factorial arrangement; however, with more than two replicates it is still possible to implement the algorithm, by having further pseudo-factors whose levels correspond to the treatments of a Latin square.

References

[1] Alvey, N. G. et al. (1977). Genstat, a General Statistical Program. Rothamsted Experimental Station, Harpenden, England.

[2] Kempthorne, O. (1952). *Design and Analysis of Experiments*. Wiley, New York.

[3] Wilkinson, G. N. (1970). *Biometrika*, **57**, 19–46.

[4] Yates, F. (1936). *J. Agric. Sci.*, **26**, 424–455. (The paper that introduced these designs, and the leading reference.)

(LATTICE DESIGNS)

G. H. FREEMAN

QUASI-INDEPENDENCE

HISTORICAL REMARKS

The concept of quasi-independence (QI) arose in response to a dispute between Har-

Table 1 Relationship between Radial Asymmetry and Locular Composition in Staphylea (Series A)

	Locular Composition	Coefficient of Radial Symmetry								
		$j=1$ 0.00	2 0.47	3 0.82	4 0.94	5 1.25	6 1.41	7 1.63	8 1.70	9 1.89
$i=1$	3 even 0 odd	462	—	—	130	—	—	2	—	1
2	2 even 1 odd	—	614	138	—	21	14	—	1	—
3	1 even 2 odd	—	443	95	—	22	8	—	5	—
4	0 even 3 odd	103	—	—	35	—	—	1	—	0

ris (e.g., ref. 16) and Pearson (e.g., ref. 19) on how to test for independence in *incomplete* contingency tables*. A classic set of data used by both Harris and Pearson is reproduced in Table 1. A valid model for these data must recognize that 18 of the 36 cell counts are zero a priori (e.g., a "locular composition" value of "3 even 0 odd" cannot occur when the "coefficient of radial symmetry" is 0.47, 0.82, 1.25, 1.41, or 1.70). Contingency tables with "structural zeros" are found in genetics* (e.g., certain chromosome combinations are lethal), the social sciences (e.g., a cross-classification of birth order and sibship size produces a "triangular" table), epidemiology (e.g., some diseases are sex specific), psychology (e.g., data on dyad formation, since subjects do not form dyads with themselves), and other areas. See refs. 2, 6, and 10 for examples. With the partial exception of Watson [20], correct methods for dealing with such data did not appear until the 1960s in the work of Caussinus [3] and Goodman [8–10].

Another context where the QI concept arose was in the analysis of mobility tables [8, 10]. Table 2 is a classic 3×3 mobility table used by Goodman. Although this table is complete in the sense that all cell frequencies are positive, it can be worthwhile to *superimpose* an incomplete table on it in order to simplify the analysis of association. For example, the $(1, 1)$ cell in Table 2 is relatively large, and it is natural to ask what association remains after the row–column affinity in this cell is eliminated.

Table 2 Cross-Classification of a Sample of British Males According to Subject's Status and Father's Status

	Father's Status	Subject's Status		
		$j=1$ Upper	2 Middle	3 Lower
$i=1$	Upper	588	395	159
2	Middle	349	714	447
3	Lower	114	320	411

THE QI MODEL

Let p_{ij} denote the population proportion in cell (i, j) of an $I \times J$ contingency table. Rows and columns are *independent* when

$$p_{ij} = \alpha_i \beta_j \quad (\text{for } i = 1, \ldots, I; j = 1, \ldots, J) \quad (1)$$

for sets of positive constants α_i and β_j. When these constants are normalized so that $\sum_i \alpha_i = \sum_j \beta_j = 1$, $\alpha_i = p_{i\cdot}$ and $\beta_j = p_{\cdot j}$ are the row and column *marginal proportions*, respectively. The definition of independence implies that $p_{ij} > 0$ for all i and j, so this model cannot be directly relevant when some of the $p_{ij} = 0$ (as is the case in incomplete tables).

Let U denote the set of deleted cells and S the set of nondeleted cells. In analyzing incomplete tables, U is the set of cells having structural zeros*. As in Caussinus [3] or Goodman [10], quasi-independence is defined by

$$p_{ij} = \alpha_i \beta_j \qquad [\text{for cells } (i, j) \text{ in } S] \quad (2)$$

for sets of positive constants α_i and β_j. It is no longer generally true that these constants correspond to marginal proportions, even after normalization. Because of this, maximum likelihood estimates* often do not exist in closed form. The QI model says nothing about the p_{ij} for cells (i, j) in U; these can be zero (as for incomplete tables) or positive (as for complete tables where some entries are blanked out).

Two other forms are often used when the QI model is applied to complete tables. The first consists of (2) plus the relationship

$$p_{ij} = \alpha_i \beta_j \gamma_k \qquad [\text{for cells } (i, j) \text{ in } U], \quad (3)$$

where k indexes cells in the set U ($k = 1, \ldots, K$, for K deleted cells) and γ_k is a parameter specific to the kth deleted call. The QI model applied to complete tables is a special model for *interaction**, since the γ_k depend on both i and j values. When $\gamma_k > 1$ for all k, the QI model is equivalent to a *latent class* model, which is a special kind of finite mixture model for categorical data*. The latent class model has $K + 1$ "latent classes." The proportions π_k in these classes (or mixing weights) are given by the relationships

$$\pi_k = p_{ij} - \alpha_i \beta_j$$

$$[\text{for } k = 1, \ldots, K, \text{ and } (i, j) \text{ in } U], \quad (4a)$$

$$\pi_{K+1} = 1 - \sum_{k=1}^{K} \pi_k . \quad (4b)$$

Much early work on the QI model was prompted by its affinity to the latent class model (see refs. 4 and 8 for discussion and references). The alternative versions of the model are useful both for computation and for summarization of results. For example, either the γ_k or the π_k values may be used to assess the importance of the kth deleted cell in the overall relationship (see ref. 11).

ESTIMATION

Maximum likelihood estimation is discussed in refs. 1 to 3, 10, and 14, among others. Other estimation methods have not been emphasized in the literature dealing with QI or related models, although weighted least-squares* methods can be used. When the observed frequencies arise from either *independent Poisson* or *multinomial* sampling schemes, the likelihood equations are

$$\sum_j \delta_{ij}(x_{ij} - \hat{m}_{ij}) = 0, \quad (5a)$$

$$\sum_i \delta_{ij}(x_{ij} - \hat{m}_{ij}) = 0, \quad (5b)$$

where $\delta_{ij} = 1$ if cell (i, j) is in S and $\delta_{ij} = 0$ otherwise, x_{ij} is the observed frequency in cell (i, j), and $\hat{m}_{ij}$ is the maximum likelihood estimate of the corresponding expected frequency. The observed values of the sufficient statistics* are

$$x_{i\cdot}^* = \sum_j \delta_{ij} x_{ij}, \qquad x_{\cdot j}^* = \sum_i \delta_{ij} x_{ij}.$$

Various numerical methods have been used to solve these equations. Early treatments (e.g., refs. 1, 3, 8, and 10) emphasized variants of *iterative proportional fitting**. Newton–Raphson methods*, extensions of Newton's elementary (one-dimensional) method, and the EM algorithm (*see* MISSING INFORMATION PRINCIPLE) have also been used. A comparison of algorithms applied to two-way tables is provided in ref. 18. Most of the computer programs cited in the entry CONTINGENCY TABLES can deal with QI models or related models for incomplete tables.

DEGREES OF FREEDOM

Much of the technical literature on QI models deals with determining the degrees of freedom for chi-square* statistics. Correct rules can be found in ref. 10 or 14. To illustrate the source of the difficulty, consider Table 2, where the (1, 1) cell is deleted. A chi-square statistic for the independence model applied to a 3×3 table has $(3 - 1)(3 - 1) = 4$ degrees of freedom (d.f.). (The value of Pearson's chi-squared statistic is 505.5, so independence is decisively rejected.) Since one cell is deleted, the QI model has $4 - 1 = 3$ d.f. [Pearson's chi-squared is 137.6 for this QI model; removal of the (1, 1) cell provides a dramatic reduction in chi-squared.]

Now consider Table 1, where 18 cells are zero a priori. The logic above would indicate that d.f. $= (4 - 1)(9 - 1) - 18 = 6$, but the correct number is 7. Rearrange the entries in Table 1 so that row levels are 1, 4, 2, 3 and column levels are 1, 4, 7, 9, 2, 3, 5, 6, 8. When this is done, there is a 2×4 complete subtable in the upper left and a 2×5 complete subtable in the lower right. No rows or columns are shared by the two subtables. Table 1 can thus be broken up into two *separable* subtables, as recognized by Pearson [19]. The QI model here means that independence holds in each subtable; d.f. $= (2 - 1)(4 - 1) + (2 - 1)(5 - 1) = 7$. (We obtain Pearson chi-squared statistics 1.4 and 6.1 for these subtables, and the sum 7.5 is a chi-square statistic on 7 d.f. The QI model cannot be rejected for Table 1.) Rules for d.f. calculations applied to separable tables, triangular tables, and other special tables commonly encountered in practice are provided in refs. 2, 6, and 10. If Newton–Raphson estimation methods are used, many difficulties associated with d.f. calculations can be avoided. If a QI model is overparameterized (giving incorrect d.f.), this fact will appear as a rank problem in the iterative matrix inversions associated with this procedure.

REMARKS

Recent research has emphasized the relationship between QI and many other models. Fienberg and Larntz [7] show how Bradley–Terry models* for paired-comparisons* experiments are equivalent to QI models when the data are arranged appropriately in a contingency table format. Larntz and Weisberg [17] show how QI and related models can be used for psychological data on dyad formation. Goodman [12] uses QI models to test the scalability of a set of three or more dichotomous items. Clogg [4] links QI to latent structure* models.

QI models are especially attractive for square contingency tables where there is a one-to-one correspondence between row and column categories. Let $S = \{(i, j) : i = j\}$ denote the set of deleted cells; S refers to diagonal entries (consistent responses). For $I = J = 3$, the QI model in (2) is equivalent to the following condition on the expected frequencies: $(m_{12}m_{23}m_{31}/(m_{21}m_{32}m_{13})) = 1$. For this case QI is equivalent to quasi-symmetry* (QS). For $I = J > 3$, QI implies QS. Because QI is nested within QS it often forms a natural baseline model that can be used to partition chi-squared statistics whenever QS models are relevant. Many special models for two-way tables based on both QI and QS can be found in ref. 11. Extensions of QI to multidimensional contingency tables* are taken up in refs. 1, 2, and 5, among others.

Any log-linear model for a complete multiway contingency table can be considered as a *quasi-log-linear model* for a subset S of the cells in the complete table. Let the log-linear model be denoted as

$$\boldsymbol{\nu} = \mathbf{X}\boldsymbol{\lambda}, \tag{6}$$

where $\boldsymbol{\nu}$ is a t-dimensional vector of logarithms of expected frequencies (t is the number of cells), $\mathbf{X}$ (of order $t \times s$) is the design matrix* (assumed to be of full column rank), and $\boldsymbol{\lambda}$ is an s-dimensional vector of parameters. A quasi-log-linear model results when (6) is posited for only a subset S of the t cells. This means that some elements in $\boldsymbol{\nu}$, the corresponding *rows* of $\mathbf{X}$, and possibly some elements of $\boldsymbol{\lambda}$ are deleted. Problems in d.f. calculations (or in estimability*) arise because deleting rows of $\mathbf{X}$ will often create rank problems if some columns of $\mathbf{X}$ (corresponding to elements of $\boldsymbol{\lambda}$) are not also deleted.

Quasi-log-linear models have many uses in the social sciences [2–5, 11, 15]. Such models are also used occasionally to detect cell "outliers" [2]. Nested sequences of such models can be used to partition association between (among) discrete variables, as in Goodman [10, 11, 13]. Often a major problem is determining whether a zero cell frequency is, in fact, zero a priori or merely the result of sparse sampling relative to small (but nonzero) cell probabilities.

References

[1] Bishop, Y. M. M. and Fienberg, S. E. (1969). *Biometrics*, **22**, 119–128. (Extends quasi-independence to multiway tables.)

[2] Bishop, Y. M. M., Fienberg, S. E., and Holland, P. W. (1975). *Discrete Multivariate Analysis: Theory and Practice*. MIT Press, Cambridge, Mass. (Chapter 5 presents intermediate-level survey of models for incomplete tables, including QI model.)

[3] Caussinus, H. (1965). *Ann. Fac. Sci. Univ. Toulouse*, **29**, 77–182. [Proposes an iterative proportional fitting algorithm for QI model. Also develops many related models (e.g., model of quasi-symmetry).]

[4] Clogg, C. C. (1981). *Amer. J. Sociol.*, **86**, 836–868.

[5] Clogg, C. C. (1982). *J. Amer. Statist. Ass.*, **77**, 803–815. (Uses QI and related models for analysis of ordinal data.)

[6] Fienberg, S. E. (1981). *The Analysis of Cross-Classified Categorical Data*, 2nd ed. MIT Press, Cambridge, Mass. (Chapter 9 gives a nontechnical survey of models for incomplete tables.)

[7] Fienberg, S. E. and Larntz, K. (1976). *Biometrika*, **63**, 245–254.

[8] Goodman, L. A. (1961). *J. Amer. Statist. Ass.*, **56**, 841–868. (Links QI model to classical "mover-stayer" model.)

[9] Goodman, L. A. (1965). *Amer. J. Sociol.*, **70**, 564–585. (Develops QI model for mobility tables.)

[10] Goodman, L. A. (1968). *J. Amer. Statist. Ass.*, **63**, 1091–1131. (Best survey article. Compares various algorithms, shows use of model for partitioning association, and deals with d.f. calculations for many types of tables.)

[11] Goodman, L. A. (1972). *Proc. 6th Berkeley Symp. Math. Statist. Prob.*, Vol. 1, University of California Press, Berkeley, Calif., pp. 646–696. (Develops many special models for 2-way tables, most built on quasi-independence model.)

[12] Goodman, L. A. (1975). *J. Amer. Statist. Ass.*, **70**, 755–768. (Applied quasi-independence to analyze item scalability.)

[13] Goodman, L. A. (1979). *Biometrics*, **35**, 651–656.

[14] Haberman, S. J. (1974). *Analysis of Frequency Data*. University of Chicago Press, Chicago. (Chapter 7 provides technical survey of quasi-independence and other models for incomplete tables; a key theoretical work.)

[15] Haberman, S. J. (1979). *The Analysis of Qualitative Data*, Vol. 2: *New Developments*. Academic Press, New York. (Chapter 7 gives a nontechnical survey; the Newton–Raphson algorithm is emphasized and methods for ordinal data are discussed.)

[16] Harris, J. A. (1927). *J. Amer. Statist. Ass.*, **22**, 460–472. (One of several papers by author criticizing Pearson's chi-square tests applied to incomplete tables.)

[17] Larntz, K. and Weisberg, S. (1976). *J. Amer. Statist. Ass.*, **71**, 455–461.

[18] Morgan, B. J. T. and Titterington, D. M. (1977). *Biometrika*, **64**, 265–269.

[19] Pearson, K. (1930). *J. Amer. Statist. Ass.*, **25**, 320–323. (One of several papers defending Pearson's methods for chi-square tests.)

[20] Watson, G. S. (1956). *Biometrics*, **12**, 47–50.

Acknowledgment

The preparation of this entry was supported in part by Grants SES-7823759 and SES-8303838 from the Division of Social and Economic Sciences, National Science Foundation.

(CATEGORICAL DATA
CHI-SQUARE TESTS
CONTINGENCY TABLES
ITERATIVE PROPORTIONAL FITTING
LATENT STRUCTURE ANALYSIS
QUASI-SYMMETRY)

CLIFFORD C. CLOGG

QUASI-LIKELIHOOD FUNCTIONS

The term *quasi-likelihood* is used in a number of related senses. First, it describes a method of estimation that is applicable under second-moment assumptions rather than full distributional assumptions. Second, it describes a function which, when maximized, produces the estimator described above and a consistent estimate of its variance. Third, ratios of this maximized function may be used for significance testing* or for model selection in much the same way that likelihood-ratio* statistics are used. The unifying theme is that second-moment assumptions rather than full distributional assumptions are used throughout. Reliance on low-order moment assumptions is often appealing in applications where more detailed assumptions would often be suspect and difficult to check.

The following example, which is analyzed in greater detail toward the end of this article, is introduced here to motivate the general discussion in the sections that follow.

Example: Regression with Constant Coefficient of Variation. Suppose that the random variables $Y_1, \ldots, Y_n$ are uncorrelated, that $E(Y_i) = \mu_i < \infty$, and that $\text{var}(Y_i) = \sigma^2\mu_i^2$, so that the coefficient of variation*, σ, rather than the variance, is constant over all observations. Suppose further that inference is required for (β_0, β_1), where

$$\log(\mu_i) = \beta_0 + \beta_1(x_i - \bar{x}), \qquad i = 1, \ldots, n \tag{1}$$

and $x_1, \ldots, x_n$ are known constants.

The principal features of this example that we wish to emphasize are:

1. The relation (1) between $\boldsymbol{\mu} = E(\mathbf{Y})$ and $\boldsymbol{\beta} = (\beta_0, \beta_1)$ is not linear in $\boldsymbol{\beta}$.
2. The covariance matrix of $\mathbf{Y}$ is given by
$$\text{cov}(\mathbf{Y}) = \sigma^2\mathbf{V}(\boldsymbol{\mu}), \tag{2}$$
where $\mathbf{V}(\cdot)$ is a matrix of known functions and σ^2 is known as the *dispersion* parameter*.
3. The model is specified entirely in terms of first and second moments of $\mathbf{Y}$ so that it is not possible to write down the likelihood function.

We require a method of estimation that is reliable under the second moment assumptions just described.

LEAST SQUARES

To simplify the notation we consider arbitrary nonlinear regression* models $\boldsymbol{\mu} = \boldsymbol{\mu}(\boldsymbol{\beta})$, where $\boldsymbol{\beta}$ is a p-dimensional unknown parameter vector. It is assumed that the model is identifiable in the sense that if $\boldsymbol{\beta} \neq \boldsymbol{\beta}'$, then $\boldsymbol{\mu}(\boldsymbol{\beta}) \neq \boldsymbol{\mu}(\boldsymbol{\beta}')$. Special cases include (1) above and generalized linear models* [3, 4]. In the special case where $\mathbf{V}(\cdot)$ is a matrix of constants, the method of nonlinear weighted least squares minimizes the weighted sum of squares

$$X^2(\boldsymbol{\beta}) = [\mathbf{Y} - \boldsymbol{\mu}(\boldsymbol{\beta})]^T\mathbf{W}[\mathbf{Y} - \boldsymbol{\mu}(\boldsymbol{\beta})] \tag{3}$$

where $\mathbf{W} = \mathbf{V}^{-1}$. The estimating equations may be written

$$\mathbf{D}^T\mathbf{W}[\mathbf{Y} - \boldsymbol{\mu}(\hat{\boldsymbol{\beta}})] = \mathbf{0}, \tag{4}$$

where $\mathbf{D} = \{\partial\mu_i/\partial\beta_j\}$ is order $n \times p$ and is a function of $\boldsymbol{\beta}$.

To extend the method of weighted least squares* to the case where $\mathbf{V}(\cdot)$ is not constant, we may choose to generalize either (3) or (4) in the obvious way, simply taking $\mathbf{W}$ to be a function of $\boldsymbol{\mu}(\boldsymbol{\beta})$. Unfortunately, (3) and (4) are no longer equivalent in the sense that the solution $\hat{\boldsymbol{\beta}}$ to (4) no longer minimizes (3). To emphasize this distinction, we refer to (4) as the *quasi-likelihood equations*: they have the appealing geometrical interpretation that the residual vector $\mathbf{Y} - \boldsymbol{\mu}(\hat{\boldsymbol{\beta}})$ is orthogonal to the columns of $\mathbf{D}$ with respect to the inner product matrix $\mathbf{W}$, both evaluated at $\hat{\boldsymbol{\beta}}$. We take the view that the appropriate generalization of least squares is based on (4) and not on (3) [2, 5].

On the other hand, the method of minimum chi-square* [1] is based on (3). Thus it seems appropriate to refer to the minimizing value $\tilde{\boldsymbol{\beta}}$ as the "minimum chi-square estimate" of $\boldsymbol{\beta}$. The two methods coincide only if $\mathbf{V}$ is constant.

QUASI-LIKELIHOOD FUNCTIONS

The estimation equations (4) may be written in the form $\mathbf{U}(\hat{\boldsymbol{\beta}}) = \mathbf{0}$, where

$$\mathbf{U}(\boldsymbol{\beta}) = \mathbf{D}^T\mathbf{W}[\mathbf{Y} - \boldsymbol{\mu}(\boldsymbol{\beta})].$$

Both $\mathbf{D}$ and $\mathbf{W}$ are treated as functions of $\boldsymbol{\beta}$. It may be verified that

$$E\{\mathbf{U}(\boldsymbol{\beta})\} = \mathbf{0},$$

$$E\{\mathbf{U}'(\boldsymbol{\beta})\} = -\mathbf{D}^T\mathbf{W}\mathbf{D},$$

$$\text{cov}\{\mathbf{U}(\boldsymbol{\beta})\} = \sigma^2\mathbf{D}^T\mathbf{W}\mathbf{D},$$

so that, apart from the factor σ^2 above,

$\mathbf{U}(\boldsymbol{\beta})$ behaves like the derivative of a log-likelihood. Consistency* and asymptotic normality* of $\hat{\boldsymbol{\beta}}$ follow from the three properties listed above.

In the majority of statistical problems it is possible to construct a function $l(\boldsymbol{\beta})$ such that

$$\frac{\partial l}{\partial \boldsymbol{\beta}} = \mathbf{U}(\boldsymbol{\beta}). \qquad (5)$$

The function $l(\boldsymbol{\beta})$ is then known as a log-quasi-likelihood and can be used in much the same way as an ordinary log-likelihood, for example, in the construction of likelihood-ratio tests*. Furthermore, $\hat{\boldsymbol{\beta}}$ maximizes $l(\boldsymbol{\beta})$ and

$$\operatorname{cov}(\hat{\boldsymbol{\beta}}) \simeq \sigma^2(\mathbf{D}^T\mathbf{W}\mathbf{D})^{-1},$$

which is proportional to the inverse of the expected second derivative matrix of $l(\boldsymbol{\beta})$.

In general, for arbitrary covariance matrices $\mathbf{V}(\boldsymbol{\mu})$, $\mathbf{U}'(\boldsymbol{\beta})$ is not symmetric and no solution to (5) exists. This has no effect on the statistical properties of $\hat{\boldsymbol{\beta}}$, but there may be other statistical implications that have not been investigated. It does seem paradoxical, however, that the least-squares equations (4) cannot be formulated as a minimization problem. Note that (5) has a solution if $\mathbf{V}(\boldsymbol{\mu}) = \mathbf{A}\mathbf{C}(\boldsymbol{\mu})\mathbf{A}^T$, where $\mathbf{C}(\cdot)$ is diagonal and $\mathbf{A}$ is constant, or if $\boldsymbol{\beta}$ is scalar, but these conditions do not appear to be necessary.

ASYMPTOTIC PROPERTIES

The discussion here refers to the solution $\hat{\boldsymbol{\beta}}$ to (4) in the limit as $n \to \infty$. The principal results are

$$n^{1/2}(\hat{\boldsymbol{\beta}} - \boldsymbol{\beta}) \xrightarrow{D} N_p\left\{\mathbf{0}, n\sigma^2(\mathbf{D}^T\mathbf{W}\mathbf{D})^{-1}\right\} \qquad (6)$$

and $\operatorname{plim}\{X^2(\boldsymbol{\beta})/(n-p)\} = \sigma^2$, where N_p refers to the p-variate normal distribution. Apart from the dispersion factor σ^2, this is exactly the kind of result that one would expect from maximizing a log-likelihood. The results above are independent of whether or not (5) has a solution.

In contrast, the minimum chi-square estimator $\tilde{\boldsymbol{\beta}}$ based on (3) does not have similar desirable properties in the same limit. In fact, $\tilde{\boldsymbol{\beta}}$ is generally inconsistent for $\boldsymbol{\beta}$. However, other limits such as $\mu_i \to \infty$ may be considered in which $\tilde{\boldsymbol{\beta}}$ is consistent [1].

In applications it is frequently the case that we need to consider a sequence of nested hypotheses, say $H_0: \boldsymbol{\beta} \in \omega_0$, $H_1: \boldsymbol{\beta} \in \omega_1, \ldots$, where ω_0 has dimension q, ω_1 has dimension p, and $\omega_0 \subset \omega_1 \subset \cdots$. It is then very convenient, if a solution to (5) exists, to use the quasi-likelihood ratio

$$\Lambda_{01} = 2l(\hat{\boldsymbol{\beta}}_1) - 2l(\hat{\boldsymbol{\beta}}_0)$$

as a test statistic, where $\hat{\boldsymbol{\beta}}_j$ is the estimate of $\boldsymbol{\beta}$ under H_j. The approximate null distribution of Λ_{01} is $\sigma^2\chi^2_{p-q}$. If $l(\boldsymbol{\beta})$ does not exist or if it cannot be computed explicitly, score tests based on $\mathbf{U}(\hat{\boldsymbol{\beta}}_0)$ may be used instead of Λ_{01}. The principal advantage of Λ is that test statistics for a sequence of nested models are additive (*see* PARTITION OF CHI-SQUARE).

Example (Continued). Continuing with model (1), we suppose for definiteness that $n = 10$, $\mathbf{Y} = (0.53, 0.38, 1.97, 1.65, 1.27, 1.34, 0.05, 0.45, 1.88, 0.19)$ and $x_i = i/2$. The quasi-likelihood function is $l = -\sum\{y_i/\mu_i + \log(\mu_i)\}$, where μ_i satisfies (1). The estimating equations (4) may be written

$$\sum(y_i - \hat{\mu}_i)/\hat{\mu}_i = 0 \qquad \text{and}$$

$$\sum(x_i - \bar{x})(y_i - \hat{\mu}_i)/\hat{\mu}_i = 0$$

and give values $\hat{\boldsymbol{\beta}} = (-0.034, -0.078)$, with covariance matrix $\operatorname{diag}\{\sigma^2/n, \sigma^2/\sum(x_i - \bar{x})^2\}$ estimated as $\operatorname{diag}\{0.066, 0.032\}$ with $s^2 = X^2(\hat{\boldsymbol{\beta}})/8 = 0.66$. The quasi-likelihood-ratio statistic for testing $H_0: \beta_1 = 0$ is $\{2l(\hat{\boldsymbol{\beta}}) - 2l(\hat{\boldsymbol{\beta}}_0)\}/s^2 = 0.15$, so that there is no evidence of any relationship between X and Y.

The minimum chi-square estimates based on (3) are $\tilde{\boldsymbol{\beta}} = (0.384, 0.031)$ and $X^2(\tilde{\boldsymbol{\beta}}) = 3.36$, as opposed to $X^2(\hat{\boldsymbol{\beta}}) = 5.30$. It is evident that $\tilde{\boldsymbol{\beta}}$ and $\hat{\boldsymbol{\beta}}$ are not estimating the same quantity, in agreement with earlier claims concerning the consistency of $\tilde{\boldsymbol{\beta}}$.

An alternative procedure, and one that seems appealing if the data are symmetrically distributed or nearly so, is to assume normality and independence and to use maximum likelihood. The resulting estimators are consistent and in this case the values are

$$\boldsymbol{\beta}^* = (-0.030, -0.001),$$

$$\hat{\sigma}^2 = X^2(\boldsymbol{\beta}^*)/10 = 0.51.$$

The asymptotic variance of β_1^* is given by $\sigma^2(1+2\sigma^2)^{-1}/\sum(x_i-\bar{x})^2$, which is less than the asymptotic variance of $\hat{\beta}_1$ by a factor of $(1+2\sigma^2)^{-1}$. However, this estimate of variance, based on the second derivative matrix of the log-likelihood, is heavily dependent on the assumptions of normality and independence. If the normality assumption is dropped, it can be shown that, asymptotically in n,

$$\operatorname{var}(\beta_1^*) = \sigma^2\left[1+2\sigma\gamma_1+\sigma^2(\gamma_2+2)\right] \times (1+2\sigma^2)^{-2} \Big/ \sum(x_i-\bar{x})^2, \tag{7}$$

where γ_1 and γ_2 are the skewness and kurtosis of Y_i, assumed constant for each i.

In fact, these data were generated from a distribution with $\sigma^2 = 1$, $\beta_0 = \beta_1 = 0$, $\gamma_1 = 2$, $\gamma_2 = 6$, so that $\operatorname{var}(\beta_1^*) = 13/[9\sum(x_i-\bar{x})^2]$ or more than four times the apparent variance. Furthermore, although β_1^* appears to be three times as efficient* as $\hat{\beta}_1$, it is, in fact, 44% less efficient by (7). Efficiency is, of course, only a minor consideration. A more serious objection to β_1^* is that the true variance (7) is difficult to estimate and naive application of maximum likelihood theory is likely to lead to grossly misleading inferences.

References

[1] Berkson, J. (1980). *Ann. Statist.*, **8**, 457–487. [Recommends the method of minimum chi-square, of which (3) is a special case, in the context of discrete data and appears to consider the limit $\mu_i \to \infty$ rather than $n \to \infty$.]

[2] McCullagh, P. (1983). *Ann. Statist.*, **11**, 59–67. [Establishes asymptotic properties of quasi-likelihood statistics assuming a solution to (5) exists.]

[3] McCullagh, P. and Nelder, J. A. (1983). *Generalized Linear Models*. Chapman & Hall, London. (Emphasizes applications.)

[4] Nelder, J. A. and Wedderburn, R. W. M. (1972). *J. R. Statist. Soc. A*, **135**, 370–384. (Demonstrates the connection between least squares and maximum likelihood and discusses implications for computing.)

[5] Wedderburn, R. W. M. (1974). *Biometrika*, **61**, 439–477. (Introduces the term "quasi-likelihood" and considers independent observations satisfying a model of the generalized linear type.)

(ESTIMATION, POINT
GENERALIZED LINEAR MODELS
LEAST SQUARES
MAXIMUM LIKELIHOOD ESTIMATION
MINIMUM CHI-SQUARE)

P. McCullagh

QUASI-LINEARIZATION *See* NEWTON–RAPHSON METHODS

QUASI-MEDIANS

These concepts were introduced by Hodges and Lehmann [1]. Let $x_{(1)} \leqslant \cdots \leqslant x_{(n)}$ be the sample observations ordered from smallest to largest. Then for $i = 0, 1, 2, \ldots,$

$$\bar{\theta}_i = \begin{cases} \dfrac{x_{(k+1-i)} + x_{(k+1+i)}}{2} & \text{if } n = 2k+1, \\[2ex] \dfrac{x_{(k-i)} + x_{(k+1+i)}}{2} & \text{if } n = 2k. \end{cases}$$

form a sequence of quasi-medians. (Each $\bar{\theta}_i$ is the average of two symmetrically ordered x observations; the sample median is equal to $\bar{\theta}_0$).

Hodges and Lehmann computed approximations to the variance $\bar{\theta}_i$. Quasi-medians were used by Hollander and Wolfe [2] as estimators associated with sign statistics.

References

[1] Hodges, J. L. and Lehmann, E. H. (1967). *J. Amer. Statist. Ass.*, **62**, 926–931.

[2] Hollander, M. and Wolfe, D. A. (1973). *Nonparametric Statistical Methods*. Wiley, New York.

(PSEUDO-MEDIAN)

QUASI-NEWTON METHODS *See* NEWTON–RAPHSON METHODS

QUASI-RANDOM NUMBERS *See* QUASI-RANDOM SEQUENCES

QUASI-RANDOM SAMPLING

Quasi-random sampling is a special case of systematic sampling*. From a list whose elements are in random order, a sample is obtained by choosing every kth element. The first element in the sample is selected randomly from the first k elements in the list.

The term "quasi-random sampling" developed in the fields of sociology* and economics. Currently it is seldom used. A review of the development of this topic can be found in ref. 1. An early example of the use of quasi-random sampling is ref. 2.

References

[1] Buckland, W. R. (1951). *J. R. Statist. Soc. B*, **13**, 208–215.

[2] Hilton, J. (1924). *J. R. Statist. Soc.*, **87**, 544–570.

(SYSTEMATIC SAMPLING)

STEPHEN G. NASH

QUASI-RANDOM SEQUENCES

The concept of a quasi-random sequence arose in the application of number-theoretic techniques to Monte Carlo methods*. To estimate $\int f$ over the s-dimensional unit box $I^s = [0,1]^s$, a sequence of random points $\{x_i\} \subset I^s$ is generated, and the approximation

$$\int f \simeq \frac{1}{N} \sum_{n=1}^{N} f(x_n)$$

is used. By the strong law of large numbers* and the central limit theorem*, this converges almost surely, with expected integration error $O(N^{-1/2})$. This error does not depend explicitly on the dimension s (although the implied constant may), making this technique attractive for large-dimensional problems.

Although the sequence $\{x_n\}$ can be generated using a pseudo-random number generator*, deterministic sequences designed for Monte Carlo integration can be more efficient. Such sequences are called *quasi-random*. Unlike a pseudo-random sequence, which imitates the behavior of a truly random sample, a quasi-random sequence will have only a particular random characteristic. For this application, the sequence should be more evenly distributed over I^s.

More precisely, for a set $E \subset [0,1]$, let $A(E;N)$ count the number of $x_n \in E$. Define the *discrepancy* D_N^* of the N numbers $x_1, \ldots, x_N$ in $[0,1]$ by

$$D_N^* = \sup_{0 \leqslant t \leqslant 1} \left| \frac{A([0,t);N)}{N} - t \right|.$$

If f is of bounded variation $V(f)$ on $[0,1]$, then [5]

$$\left| \frac{1}{N} \sum_{n=1}^{N} f(x_n) - \int_0^1 f(x)\,dx \right| \leqslant V(f) D_N^*.$$

This formula separates the influences of the function f and the sequence $\{x_n\}$ on the effectiveness of the Monte Carlo scheme.

In higher dimensions, a formal analog of this result also holds [4]. In this case, the function f should be of bounded variation in the sense of Hardy and Krause (the variations of f and its restrictions to coordinate subspaces must all be finite). Related bounds can be obtained for some integrals over more general regions [7].

These results suggest using *low-discrepancy* or *quasi-random* sequences for Monte Carlo integration. With some such sequences, the error can be reduced to $O(N^{-1}(\log N)^{s-1})$. In one dimension, it can be shown that $D_N^* \geqslant 1/(2N)$ for any N numbers in I, and that $D_N^* = 1/(2N)$ for the sequence

$$1/(2N), 3/(2N), \ldots, (2N-1)/(2N).$$

However, to use this sequence, N must be prespecified. Instead, in practice it is preferable to use portions of a single infinite sequence whose subsequences have low discrepancy for any N. For example, if α is an irrational number that has a continued-fraction expansion with uniformly bounded partial quotients (such as the square root of a prime number), then the sequence of fractional parts $\{\alpha\}, \{2\alpha\}, \ldots$ satisfies $D_N^* = O(N^{-1}\log N)$. Sequences with even lower discrepancies can also be constructed [1, 8]. For quasi-random sequences in higher dimensions, see refs. 2, 3, and 9.

The techniques above do not exploit regularity in the function f, beyond its bounded variation. If the function is periodic of period 1 in each variable, and if its Fourier coefficients decay sufficiently rapidly, then an approximation of the form

$$\int f(x)\,dx \simeq \frac{1}{N}\sum_{n=1}^{N} f\left(\frac{n}{N}g\right)$$

can be used, where $N \geqslant 2$ is a fixed integer and $g \in Z^s$ is an s-dimensional lattice point. The lattice point g is chosen to reduce the error in the approximation above [12, 13]. In dimension $s = 2$, g can be optimally selected [11]; tables exist for $3 \leqslant s \leqslant 10$ [6].

A similar technique is based on using multiples of a point $\alpha = (\alpha_1, \ldots, \alpha_s)$ for which $1, \alpha_1, \ldots, \alpha_s$ are linearly independent over the rationals. The function f must again be periodic. An example is $\alpha = (a, a^2, \ldots, a^s)$, where $a = p^{1/(s+1)}$ and p is prime [7].

If the function f is not periodic, it may be replaced by a related periodic function which has the same integral. Alternatives are to change variables or to add polynomials to the integrand (see refs. 10 and 11). These techniques can increase the cost of the method, however.

References

[1] Faure, H. (1981). *Bull. Soc. Math. Fr.*, **109**, 143–182 (in French).

[2] Halton, J. H. (1960). *Numer. Math.*, **2**, 84–90.

[3] Hammersley, J. M. (1960). *Ann. New York Acad. Sci.*, **86**, 844–874.

[4] Hlawka, E. (1961). *Ann. Mat. Pura Appl.*, **54**, 325–333 (in German).

[5] Koksma, J. F. (1942–43). *Mathematica B (Zutphen)*, **11**, 7–11 (in German).

[6] Maisonneuve, D. (1972). In ref. 13, pp. 121–201 (in French).

[7] Niederreiter, H. (1973). In *Diophantine Approximation and Its Applications*, C. F. Osgood, ed. Academic Press, New York, pp. 129–199.

[8] van der Corput, J. G. (1935). *Nederl. Akad. Wet. Proc.*, **38**, 813–821, 1058–1066 (in Dutch).

[9] Warnock, T. T. (1972). In ref. 13, pp. 319–343.

[10] Zaremba, S. K. (1970). *Aequationes Math.*, **4**, 11–22.

[11] Zaremba, S. K. (1970). *Fibonacci Quart.*, **8**, 185–198.

[12] Zaremba, S. K. (1972). In ref. 13, pp. 39–119 (in French).

[13] Zaremba, S. K., ed. (1972). *Applications of Number Theory to Numerical Analysis*. Academic Press, New York.

Bibliography

See the following works, as well as the references just given, for more information on the topic of quasi-random sequences.

Kuipers, L. and Niederreiter, H. (1974). *Uniform Distribution of Sequences*. Wiley, New York. (Discusses the concepts of discrepancy and equi-distribution of sequences.)

Niederreiter, H. (1978). *Bull. Amer. Math. Soc.*, **84**, 957–1041. (Extensive survey of the theory of quasi- and pseudo-Monte Carlo methods.)

(GENERATION OF RANDOM VARIABLES
MONTE CARLO METHODS
PSEUDO-RANDOM NUMBER GENERATORS)

STEPHEN G. NASH

QUASI-RANGES *See* RANGES

QUASI-SYMMETRY

QUASI-SYMMETRY, MARGINAL SYMMETRY, AND COMPLETE SYMMETRY

This entry should be read in conjunction with the entry MARGINAL SYMMETRY, henceforth referred to as MS; much of the practical interest in quasi-symmetry lies in its role as a background structure against which to test the null hypothesis of marginal symmetry.

Both marginal symmetry and quasi-symmetry are properties of the distribution of a t-dimensional Response $(R_1, \ldots, R_t)$

measured on a randomly selected Individual, each R_i taking one of the values $1, \ldots, r$. The probability that $R_1 = j_1, \ldots, R_t = j_t$ is denoted by $\rho(j_1, \ldots, j_t) = \rho(\mathbf{j})$. The marginal probability that $R_i = j$ is denoted by ρ_j^i and the hypothesis H_2 of marginal symmetry is

$$H_2 : \rho_j^i \text{ is constant with respect to } i.$$

A fuller description of the following examples, together with sample data, is given in MS.

Example 1. Individual = voter in a mayoral election being contested by three candidates, $t = 2$, $r = 3$; R_1 and R_2 are the voting preferences one month before and one week before the election.

Example 2. Individual = boy, $t = 3$, $r = 2$; R_1, R_2, and R_3 are each binary responses with levels 1 = considered inactive, 2 = considered active; R_1 is a self-assessment and R_2 and R_3 are assessments by the boy's parent and teacher.

Quasi-symmetry is easier to appreciate if we first consider the hypothesis H_0 of *complete symmetry*. In Example 1, and in general when $t = 2$, H_0 simply means that the matrix $(\rho(j_1, j_2))$ is symmetric, that is,

$$H_0 : \rho(j_1, j_2) = \rho(j_2, j_1).$$

In Example 2 H_0 is defined by

$$H_0 : \rho(1,1,2) = \rho(1,2,1) = \rho(2,1,1) \quad \text{and} \quad \rho(1,2,2) = \rho(2,1,2) = \rho(2,2,1).$$

In general, H_0 is defined by

$$H_0 : \rho(j_1, \ldots, j_t) \text{ is constant with respect to permutations of } (j_1, \ldots, j_t).$$

It is obvious that $H_0 \Rightarrow H_2$; that is, marginal symmetry is more general than complete symmetry.

Quasi-symmetry, denoted by H_1, is also more general than H_0. It can be defined by the symmetry of the log-linear interactions, but the best working definition of H_1 says that $\rho(j_1, \ldots, j_t)$ is the product of a function of $j_1, \ldots,$ a function of j_t, and a completely symmetric function of $j_1, \ldots, j_t$; that is,

$$H_1 : \rho(j_1, \ldots, j_t) = \lambda_{j_1}^1 \ldots \lambda_{j_t}^t \nu(j_1, \ldots, j_t),$$

where ν is completely symmetric. In the case $t = 2$, H_1 says that $\rho(j_1, j_2) = \lambda_{j_1}^1 \lambda_{j_2}^2 \nu(j_1, j_2)$. Further, H_1 is expressible through the cross-product ratio conditions

$$\frac{\rho(j, j')\rho(j', j'')\rho(j'', j)}{\rho(j, j'')\rho(j'', j')\rho(j', j)} = 1. \qquad (1)$$

Since $H_0 \Rightarrow H_1$ and $H_0 \Rightarrow H_2$ it is immediate that $H_0 \Rightarrow H_1 \wedge H_2$, the latter hypothesis denoting the conjunction of H_1 and H_2. Not immediate, but more important, is that $H_1 \wedge H_2 \Rightarrow H_0$. In other words, complete symmetry is the conjunction of quasi-symmetry and marginal symmetry. A proof is given by Darroch and Speed [6]. None of the symmetry properties involve the "diagonal" probabilities $\rho(j, \ldots, j)$.

It follows from $H_1 \wedge H_2 \Leftrightarrow H_0$ that, assuming that quasi-symmetry holds, testing marginal symmetry is equivalent to testing complete symmetry. In this way the problem of testing a linear hypothesis, namely marginal symmetry, can be handled using log-linear methodology, because quasi-symmetry is log-linear and complete symmetry is log-linear (and linear).

Proofs of results stated here, and other details, are given in Darroch [5].

TESTS OF MARGINAL SYMMETRY ASSUMING QUASI-SYMMETRY

Denote by $y(\mathbf{j})$ the frequency of response $\mathbf{j}$ in a sample of s individuals. The probability distribution of the $y(\mathbf{j})$ is multinomial and underlies the various test statistics that have been used for testing H_2: marginal symmetry assuming H_1: quasi-symmetry, that is, for testing H_0: complete symmetry against H_1. All test statistics are approximately chi-square on H_0 and have $(t-1)(r-1)$ degrees of freedom.

The likelihood-ratio test* is possibly the easiest to perform given the wide availability of computing systems for performing likelihood-ratio tests of log-linear hypotheses (e.g., GLIM*). The hypothesis H_1 should first be tested against the alternative hypothesis H of no restriction, and this test can also be performed as a likelihood-ratio test. A good discussion of this test and of other aspects of quasi-symmetry is given by Bishop et al. [3, pp. 286–293].

Darroch [5] constructed the "conditional likelihood score" statistic for testing H_0 against H_1. It has the theoretical advantage over other tests of being expressible in closed form. It looks at the magnitude of the differences

$$d_j^i = y_j^i - y_j^{\cdot},$$

where y_j^i is the marginal frequency of $R_i = j$ and $y_j^{\cdot}$ is the average over i. Let $\mathbf{d}$ denote the vector of the $(t-1)(r-1)$ d_j^is, $i = 1, \dots, t-1$, $j = 1, \dots, r-1$. The conditional likelihood score statistic takes the form

$$\mathbf{d}'\mathbf{V}_d^{-1}\mathbf{d}.$$

It closely resembles the adjusted Wald statistic

$$\mathbf{d}'\tilde{\mathbf{\Sigma}}_d^{-1}\mathbf{d},$$

for testing H_2 against H, discussed in MS. Both statistics involve the univariate marginal frequencies y_j^i and the bivariate marginal frequencies $y_{jj'}^{ii'}$, $i \neq i'$. The typical elements of $\mathbf{V}_d$ and $\tilde{\mathbf{\Sigma}}_d$ are

$$\mathbf{V}_d : \operatorname{cov}\left(d_j^i, d_{j'}^{i'}\right) = \left(\delta_{ii'} - t^{-1}\right)\left(\delta_{jj'} y_j^{\cdot} - y_{jj'}^{\cdot\cdot}\right), \tag{2}$$

$$\begin{aligned} \tilde{\mathbf{\Sigma}}_d &: \operatorname{cov}\left(d_j^i, d_{j'}^{i'}\right) \\ &= (1-\delta_{ii'}) y_{jj'}^{ii'} - (1-t^{-1})\left(y_{jj'}^{i\cdot} + y_{jj'}^{\cdot i'} - y_{jj'}^{\cdot\cdot}\right) \\ &\quad + \delta_{jj'}\left\{\delta_{ii'} y_j^i - t^{-1}\left(y_j^i + y_j^{i'} - y_j^{\cdot}\right)\right\}. \end{aligned} \tag{3}$$

The averages $y_{jj'}^{i\cdot}$ and $y_{jj'}^{\cdot\cdot}$ are calculated over $i' : i' \neq i$ and over $(i, i') : i' \neq i$. When $t = 2$, (2) and (3) are identical. In other words, we get the same answer in testing H_2, whether or not we test and accept H_1 first.

Example 1 (See MS). The likelihood ratio test statistic for testing H_1: quasi-symmetry against H is 31.65, which, on one degree of freedom, is very highly significant. Thus quasi-symmetry is not acceptable. This can be seen directly [recall (1)] from the cross-product ratio

$$\frac{17 \times 22 \times 15}{72 \times 35 \times 32} = 0.07,$$

which is very different from 1. The likelihood ratio test statistic for testing H_0 against H_1, that is, for testing H_2 assuming H_1, is 16.63. The conditional likelihood score statistic* is 16.29, identical to the adjusted Wald statistic for testing H_2 against H, which was reported in MS. All these tests are on 2 degrees of freedom.

Example 2 (See MS). The likelihood ratio test statistic for testing H_1 against H is 2.73, which, on two degrees of freedom, shows that H_1 is very acceptable. The likelihood ratio test statistic for testing H_2 assuming H_1 is 18.84, which, on two degrees of freedom, makes H_2: marginal symmetry unacceptable. The conditional likelihood score statistic calculated using (2) is 19.19. The adjusted Wald statistic for testing H_2 against H is 16.29, the same value as in Example 1. (See MS for a discussion of this coincidence.)

MATCHED PAIRS AND t-PLETS

The individual may comprise a matched pair* ($t = 2$) or, in general, a matched t-plet of individuals, on the ith of which is measured the response R_i.

Example 3 (See MS). $t = 2$, $r = 3$, Individual = matched (case, control) pair, the case and control being people under 30 with and without heart disease. The response is level of smoking. The likelihood ratio test statistic for testing H_1 against H is 0.34, which, on one degree of freedom, shows that H_1 is acceptable. The likelihood ratio test statistic for testing H_2 assuming H_1 is 16.51, while the conditional likelihood score statistic is 15.52. The latter value is the same as the value of the adjusted maximum likelihood statistic, reported in MS.

The matching of the individuals in a (case, control) pair, or more generally in a t-plet, can be modeled as follows. The population of individuals is divided into S homogeneous strata. In Example 3, for instance, each stratum is homogeneous with respect to sex, age, occupation, and place of residence. In the hth stratum let π_{h1j} and π_{h2j} be the probabilities of response j for the case and control. The probability of a matched pair from a randomly selected stratum exhibiting response (j_1, j_2) is then

$$\rho(j_1, j_2) = S^{-1} \sum_{h=1}^{S} \pi_{h1j_1} \pi_{h2j_2}.$$

In MS it was shown that H_2 is equivalent to G_2 where

$$G_2 : \pi_{\cdot 1j} = \pi_{\cdot 2j},$$

with "." denoting average. G_2 may thus be described as average (over strata) disease, exposure independence.

The hypotheses H_0 and H_1 are related by implication to two hypotheses which feature prominently in the analysis of stratified frequencies. These are, in the language of case-control studies, G_0: within stratum (disease, exposure) independence and G_1: no (stratum, disease, exposure) interaction. They are defined by

$$G_0 : \pi_{hij} = \phi_{hj},$$
$$G_1 : \pi_{hij} = \theta_{hi} \phi_{hj} \psi_{ij}.$$

Therefore, under G_0,

$$\rho(j_1, j_2) = S^{-1} \sum_{h=1}^{S} \phi_{hj_1} \phi_{hj_2}. \qquad (4)$$

Property (4) implies the complete symmetry of $(\rho(j_1, j_2))$. Thus G_0 implies H_0. Similarly, G_1 implies H_1, because, under G_1,

$$\rho(j_1, j_2) = \psi_{1j_1} \psi_{2j_2} \nu(j_1, j_2),$$

where

$$\nu(j_1, j_2) = S^{-1} \sum_{h=1}^{S} \theta_{h1} \theta_{h2} \phi_{hj_1} \phi_{hj_2}$$

and is completely symmetric. Thus there is a close relationship between no interaction and quasi-symmetry which helps to give credence to the latter. McCullagh [10] discusses other aspects of the relationship; see the section "Other Aspects of Quasi-Symmetry."

Under G_0 the matrix $(\rho(j_1, j_2))$ given by (4) is seen to be nonnegative definite. This implies, for instance, that

$$\frac{\rho(j, j)\rho(j', j')}{\rho(j, j')\rho(j', j)} \geqslant 1$$

and this inequality also holds under G_1. The amount by which the left side is greater than 1 can be shown to reflect the effectiveness of the matching.

Example 3 (Continued). The sample versions of these inequalities are not all satisfied since

$$\frac{13 \times 5}{16 \times 6} < 1.$$

However, this discrepancy is not significant.

CONNECTIONS WITH THE MANTEL–HAENSZEL TEST AND COCHRAN'S Q-TEST

The Mantel–Haenszel statistic* is designed for the following problem. There are s fixed strata, rather than a random sample of strata. There are t treatments to be compared. Some individuals from each stratum are given treatment i, $i = 1, \ldots, t$. A response, taking one of the values $1, \ldots, r$, is measured on each individual. The problem is to test for treatment–response independence after allowing for strata. A very common application is to the retrospective study in which $t = 2$, the two "treatments" being the possession or nonpossession of a disease and the "response" being the exposure variable.

Birch [1, 2] proved that the Mantel–Haenszel statistic is the conditional likelihood score statistic CLS$(G_0 \mid G_1)$ for testing G_0 against G_1. It turns out that when there is only one individual per (stratum, treatment), the Mantel–Haenszel statistic, which is expressed as a function of an $s \times t \times r$ table

(x_{hij}), is also expressible as a function of the r^t table $(y(\mathbf{j}))$. Darroch [5] found that when this is done, $\text{CLS}(G_0 \mid G_1) = \text{CLS}(H_0 \mid H_1)$, where $\text{CLS}(H_0 \mid H_1) = \mathbf{d}'\tilde{\boldsymbol{\Sigma}}_d^{-1}\mathbf{d}$, the conditional likelihood score statistic that we had earlier. In other words, when there is only one individual per (stratum, treatment), the Mantel–Haenszel test can be interpreted as a test of marginal symmetry assuming quasi-symmetry. Earlier, Mantel and Haenszel [9] and Mantel and Byar [8] had shown, mainly through numerical examples, that there must be strong connections between the Mantel–Haenszel test and tests of marginal symmetry.

Cochran's [4] Q-test is a special case of the Mantel–Haenszel test, specialized to the case of matched t-plets and a binary response ($r = 2$). Thus it too may be interpreted as a test of marginal symmetry assuming quasi-symmetry (*see* COCHRAN'S Q-STATISTIC).

OTHER ASPECTS OF QUASI-SYMMETRY

Referring to the section "Matched Pairs and t-Plets," let

$$\rho_h(j_1, j_2) = \pi_{h1j_1}\pi_{h2j_2},$$

the probability that a matched pair from stratum h exhibits response (j_1, j_2). Then, under G_1, $\rho_h(j_1, j_2)/\rho_h(j_2, j_1)$ is independent of h and has the same value as $\rho(j_1, j_2)/\rho(j_2, j_1)$ under H_1, namely the ψ cross-product ratio

$$\frac{\psi_{1j_1}\psi_{2j_2}}{\psi_{1j_2}\psi_{2j_1}}.$$

McCullagh [10] notes this fact, its implications, and its generalization to $t > 2$. He also discusses the fact that the transition matrix of a reversible Markov chain is quasi-symmetric.

Fienberg and Larntz [7] discuss the relationship between the Bradley–Terry model* for paired-comparison* probabilities and quasi-symmetry.

References

[1] Birch, M. W. (1964). *J. R. Statist. Soc. B*, **26**, 313–324.

[2] Birch, M. W. (1965). *J. R. Statist. Soc. B*, **27**, 111–124.

[3] Bishop, Y. M. M., Fienberg, S. E., and Holland, P. W. (1975). *Discrete Multivariate Analysis: Theory and Practice*. MIT Press, Cambridge, Mass.

[4] Cochran, W. G. (1950). *Biometrika*, **37**, 256–266.

[5] Darroch, J. N. (1981). *Int. Statist. Rev.*, **49**, 285–307.

[6] Darroch, J. N. and Speed, T. P. (1979). Multiplicative and Additive Models and Interactions. *Res. Rep. 49*, Department of Theoretical Statistics, Aarhus University, Aarhus, Denmark.

[7] Fienberg, S. E. and Larntz, K. (1976). *Biometrika*, **63**, 245–254.

[8] Mantel, N. and Byar, D. P. (1978). *Commun. Statist. A*, **7**, 953–976.

[9] Mantel, N. and Haenszel, W. (1959). *J. Natl. Canad. Inst.*, **22**, 719–748.

[10] McCullagh, P. (1982). *Biometrika*, **69**, 303–308.

Bibliography

See the following works, as well as the references just given, for more information on the topic of quasi-symmetry.

Breslow, N. E. and Day, N. E. (1980). *Statistical Methods in Cancer Research*. IARC Sci. Publ. No. 32. IARC, Lyon, France.

Haberman, S. (1979). *Analysis of Qualitative Data*, Vol. 2. Academic Press, New York.

Plackett, R. L. (1981). *Analysis of Categorical Data*, 2nd ed. Charles Griffin, London.

(CONTINGENCY TABLES
MANTEL–HAENSZEL STATISTIC
MARGINAL SYMMETRY)

J. N. DARROCH

QUENOUILLE'S ESTIMATOR

Quenouille [6] proposed and developed [7] a method to reduce estimation bias*. This work can be considered classical in that it stimulated considerable useful work on unbiased and almost-unbiased point estimation, some of which is discussed here. It also led to new variance estimation techniques, such as jackknifing* and bootstrapping*, for complex sampling designs [3]. If, for a given series of observations $x_1, x_2, \dots, x_n$, we have a function $t_n(x_1, x_2, \dots, x_n)$ which provides an estimate of an unknown parameter θ, then for a wide variety of statistics it is true that

$$E(t_n - \theta) = \frac{a_1}{n} + \frac{a_2}{n^2} + \frac{a_3}{n^3} + \cdots$$

and if $t'_n = nt_n - (n-1)t_{n-1}$, then

$$E(t'_n) = \theta - \frac{a_2}{n^2} - \frac{a_2 + a_3}{n^3} - \cdots$$

Similarly,

$$t''_n = \left[n^2 t'_n - (n-1)^2 t'_{n-1} \right] \Big/ \left[n^2 - (n-1)^2 \right]$$

is biased to order $1/n^3$, and so on.

It is possible to calculate similar statistics from any subset of the observations to achieve corrections for bias. Durbin [2] studies the idea above in terms of ratio estimators* of the form $r = y/x$, where the regression of y on x is linear ($y = \alpha + \beta x + u$) and where x is normally distributed with $E(x) = 1$, $V(x) = h$ of order $O(n^{-1})$, and $V(u) = \delta$ of $O(n^{-1})$. He splits the sample n into two halves which yield the ratio estimators $r_1 = y_1/x_1$ and $r_2 = y_2/x_2$, where $y = \frac{1}{2}(y_1 + y_2)$ and $x = \frac{1}{2}(x_1 + x_2)$. Quenouille's estimator is

$$t = 2r - \tfrac{1}{2}(r_1 + r_2).$$

Durbin shows that

$$V(r) = \alpha^2(h + 8h^2 + 69h^3) + \delta(1 + 3h + 15h^2 + 105h^3)$$

and

$$V(t) = \alpha^2(h + 4h^2 + 12h^3) + \delta(1 + 2h + 8h^2 + 108h^3).$$

For sufficiently large n, t has both a smaller bias and variance than r.

If the regression of y on x is as before, but now x has the gamma distribution* with density $x^{m-1}\exp(-x)/\Gamma(m)$, the difference between the mean square errors* of t and r is

$$M(r) - M(t) = \frac{\alpha^2(m-16)}{m(m-1)(m-2)^2(m-4)} + \frac{\delta(m-10)}{(m-1)(m-2)^2(m-4)}.$$

The following conclusions can be drawn from this: If $m > 16$, then $\text{MSE}(t) < \text{MSE}(r)$; if $m \leqslant 10$, then $\text{MSE}(r) < \text{MSE}(t)$; if $V(x) < \frac{1}{4}$, then $\text{MSE}(t) < \text{MSE}(r)$.

Rao [8] derives the variance of t for general g to order $O(n^{-3})$ for a linear relationship between y and x and for x normally distributed. This variance of t is

$$V(t) = \alpha^2\left[h + \frac{2g}{g-1}h^2 - 6g\frac{4g^2 - 14g + 11}{(g-1)^3}h^3 \right] + \delta\left[1 + \frac{g}{g-1}h - 2g\frac{2g^2 + 9g - 8}{(g-1)^3}h^2 - 3g\frac{28g^4 - 149g^3 + 260g^2 - 173g + 32}{(g-1)^5} \right].$$

Both the bias and variance of t are decreasing functions of g.

Rao and Webster [11] showed that the bias and variance of the generalization t_Q of t,

$$t_Q = gr - (g-1)\bar{r}_g,$$

where

$$\bar{r}_g = g^{-1}\sum_{j=1}^{g} r_j,$$

are decreasing functions of g. Using $g = n$, they showed that there is little difference in efficiency between t_Q and the Tin [12] estimator

$$t_{T_1} = r\left[1 + \theta_1\left(\frac{s_{xy}}{\bar{x}\bar{y}} - \frac{s_x^2}{\bar{x}^2} \right) \right],$$

where $\theta_1 = (1/n - 1/N)$.

Hutchison [4] conducted Monte Carlo comparisons of some ratio estimators. He compared the version of the Quenouille esti-

Table 1 Ratio of Mean Square Errors of Estimators for s Strata to that of r from 1000 Replicates of n Pairs (x, y), where $x = zx_p$, z is Poisson with Mean μ, x_p is $\chi^2_{(m)}/m$, $m = 20$, and $y = \chi^2_{(z)}$, given x

		$s = 1$			$s = 30$		
		μ			μ		
n	Estimator	10.0	5.0	2.5	10.0	5.0	2.5
6	t_{HR}	1.052	1.090	1.088	0.817	0.818	1.335
	t_M	0.989	0.997	1.077	0.763	0.760	1.167
	t_Q	0.978	0.981	0.996	0.753	0.752	1.131
	t_B	0.976	0.976	0.989	0.767	0.777	1.073
	t_{T1}	0.976	0.976	0.990	0.762	0.768	1.086
12	t_{HR}	1.113	1.079	1.051	0.977	1.065	1.085
	t_M	0.993	0.998	0.993	0.873	0.979	0.972
	t_Q	0.987	0.993	0.989	0.865	0.973	0.967
	t_B	0.987	0.991	0.989	0.868	0.968	0.965
	t_{T1}	0.987	0.991	0.989	0.867	0.969	0.965

mator

$$t_Q = wr - (w-1)\bar{r}_g$$

where

$$w = g\{1 - (n-m)/N\},$$

$g = mn$, and $g = n$ (so that $m = 1$), with the Tin estimator* t_{T_1}, the Beale estimator t_B, the Hartley–Ross estimator t_{HR}, and with Mickey's estimator t_M.

The comparisons are made for two models. The first is $E(y \mid x) = \beta x$, where x has a lognormal distribution with parameters μ and σ^2. The second is $y = \beta x + e$, where β is a constant and $\epsilon \sim N(0, kx^{\gamma})$ for given x, where k and γ are constants. All estimators considered are unbiased for this model. The second model is basically the same except that $x = zx_p$, where z has a Poisson distribution with mean μ and $x_p \sim \chi^2_{(m)}/m$ (Table 1). He concludes that t_{B_1} and t_{T_1} are generally most efficient, closely followed by Quenouille's estimator t_Q.

Rao and Beegle [10] compare eight ratio estimators: t_Q, t_B, t_M, t_{HR}, r, t_{T_1}, t_{T_3} (*see* TIN ESTIMATORS), and the Pascual estimator* t_P. They assume two models.

The first model is $y_i = 5(x_i + e_i)$, where $x_i \sim N(10, 4)$ and $e_i \sim N(0, 1)$, so that the correlation* between y_i and x_i is $\rho = 0.89$ and the coefficient of variation* of x is $C_x = 0.2$. They call this the Lauh and Williams model. The second model is $y_i = \alpha + \beta x_i + e_i$, where x_i and y_i have a bivariate normal distribution* with $\mu_x = 5$, $\sigma_x^2 = 45$, $\bar{y} = 15$, $\sigma_y^2 = 500$ and $\rho = 0.4, 0.6$, or 0.8, and where $e_i \sim N[0, \sigma_y^2(1-\rho^2)]$, independent of x_i. They call this *Tin's model*.

Results for the two models are shown in Tables 2 and 3. The authors consider variance comparisons meaningful for the Lauh and Williams model but favor the interquartile range* for the Tin model. This range is the distance between the upper and lower quartile points and hence contains one-half of the 1000 computed values of an

Table 2 Ratio of Variance Estimator to That of r for 1000 Replicates using Lauh and Williams Model [10]

Estimator	$n = 6$	$n = 12$
t_{HR}	0.990	0.995
t_M	0.990	0.995
t_Q	0.990	0.995
t_B	0.990	0.995
t_{T1}	0.990	0.995
t_{T3}	0.990	0.995
t_P	0.990	0.995

Table 3 Interquartile Ranges of Estimators for 1000 Replicates using Tin's Model [10]

	n				
Estimator	4	6	10	20	50
r	2.8	2.3	1.8	1.07	0.72
t_{HR}	4.0	3.2	2.5	1.63	1.05
t_{M}	2.6	1.7	1.4	0.96	0.68
t_{Q}	2.7	1.9	1.4	0.93	0.67
t_{B}	1.7	1.6	1.4	0.96	0.68
t_{T1}	2.1	1.7	1.4	0.94	0.68
t_{T3}	4.0	3.0	2.2	1.24	0.77
t_{P}	4.0	3.1	2.2	1.26	0.77

estimator. The authors conclude that T_{Q}, t_{T_1}, and t_{B} perform quite efficiently, especially if almost unbiased rather than unbiased estimators are satisfactory.

DeGraft-Johnson and Sedransk [1] and Rao [9] compare two-phase versions of the Quenouille estimate to other two-phase ratio estimators. The Quenouille estimator with $g = 2$ did not perform as well generally as Tin and Beale estimators in the first study. Two Quenouille-type estimators with $g = n$ performed as well as other ratio estimators in the second study.

Miller [5] concludes that the Quenouille estimator is usually one of the most efficient ratio estimators, requires more computation than competitors, but has an easily computed estimate of its variability.

References

[1] deGraft-Johnson, K. T. and Sedransk, J. (1974). *Ann. Inst. Statist. Math.*, **26**, 339–350.

[2] Durbin, J. (1959). *Biometrika*, **46**, 477–480.

[3] Efron, B. (1982). CBMS-NSF Regional Conference Series, *Appl. Math. No. 38*, SIAM, Philadelphia.

[4] Hutchison, M. C. (1971). *Biometrika*, **58**, 313–321.

[5] Miller, R. G. (1974). *Biometrika*, **61**, 1–15.

[6] Quenouille, M. H. (1949). *J. R. Statist. Soc. B*, **11**, 68–84.

[7] Quenouille, M. H. (1956). *Biometrika*, **43**, 353–360.

[8] Rao, J. N. K. (1965). *Biometrika*, **52**, 647–649.

[9] Rao, P. S. R. S. (1981). *J. Amer. Statist. Ass.*, **76**, 434–442.

[10] Rao, J. N. K. and Beegle, L. D. (1967). *Sankhyā B*, **29**, 47–56.

[11] Rao, J. N. K. and Webster, J. T. (1966). *Biometrika*, **53**, 571–577.

[12] Tin, M. (1965). *J. Amer. Statist. Ass.*, **60**, 294–307.

(MICKEY'S UNBIASED RATIO AND REGRESSION ESTIMATORS
PASCUAL'S ESTIMATOR
RATIO ESTIMATORS
TIN ESTIMATORS)

H. T. SCHREUDER

QUENOUILLE'S TEST

In its simplest form Quenouille's test is a significance test* for autoregressive time-series* models. An observed time series $\{y_t : t = 1, \ldots, n\}$ is an autoregression of order p if

$$y_t - \phi_1 y_{t-1} - \cdots - \phi_p y_{t-p} = a_t,$$

where $\{a_t\}$ is a series of independent, identically distributed variates and the constants $\phi_1, \ldots, \phi_p$ are the autoregressive coefficients. Quenouille [4] showed that, defining r_k to be the kth sample correlation

$$r_k = \sum_{t=k+1}^{n} (y_t - \bar{y})(y_{t-k} - \bar{y}) \Big/ \sum_{t=1}^{n} (y_t - \bar{y})^2,$$

and writing $\phi_0 = -1$, the quantities

$$R_k = \sum_{i=0}^{p} \sum_{j=0}^{p} \phi_i \phi_j r_{k-i-j}, \qquad k > p,$$

are normally distributed with mean zero and variance $n^{-1}[\mathrm{var}(a_t)/\mathrm{var}(y_t)]^2$ in large samples. This result is still true when the ϕ_j are replaced by estimated values obtained from least-squares* fitting of the autoregressive model, so the R_k may be used to test the fit of the model, either individually or via the statistic

$$Q_m = n\left[\frac{\mathrm{var}(y_t)}{\mathrm{var}(a_t)}\right]^2 \sum_{k=1}^{m} R_{p+k}^2,$$

which has an asymptotic χ_m^2 distribution

when the fitted model is true. This is Quenouille's test.

As an example, Quenouille's test will be applied to series F of Box and Jenkins [1]. The series consists of 70 consecutive yields from a batch chemical process, and inspection of the partial correlations* of the series suggests that a first- or second-order autoregression may be an appropriate model. For a first-order model calculation of R_2 leads to a test statistic $Q_1 = 2.42$, which corresponds to a significance level of 12% for a χ_1^2 distribution. Although not highly significant, this value suggests that it may be worthwhile to consider a model of higher order. When a second-order model is fitted there is no doubt about the fit; none of the first dozen R_k statistics is significant and the overall goodness-of-fit* statistic Q_{15}, based on R_3 through R_{17}, has the value 8.55, which is not at all significant and indicates no inadequacy of the model.

The quantities R_k are seen to be linear combinations of the sample correlations, weighted so as to make them independent. Another interpretation has been given by Quenouille [5]; R_k is asymptotically equivalent to the partial correlation between y_t and y_{t-k} with the intervening variables y_{t-i}, $i = 1, \ldots, k-1$, held constant. This interpretation means that Quenouille's test is particularly apt for autoregressive models, since these are commonly identified and their order estimated by examination of the partial correlation function of the observed series. In view of this point and the fact that its better-known rival, the portmanteau test* for residual correlation, does not share its advantage of consisting of independent components, Quenouille's test deserves to be more widely known and used than it is at present.

Quenouille's test can be extended to cover a wider range of time-series* models, although with some loss of convenience. For autoregressive-integrated moving-average (ARIMA)* processes, a test can be based on the R_k [6], although these are no longer independent. Alternatively, one may generalize the partial correlations by defining statistics $\hat{d}_i$, which for an ARIMA process are independent of each other and of the estimated parameters and are such that the test statistic $n\sum \hat{d}_i^2$ has an asymptotic χ^2 distribution if the fitted model is true [2]. Hosking [3] has also derived an extension of Quenouille's test to multivariate ARIMA processes.

References

[1] Box, G. E. P. and Jenkins, G. M. (1970). *Time Series Analysis Forecasting and Control.* Holden-Day, San Francisco.

[2] Hosking, J. R. M. (1980). *J. R. Statist. Soc. B*, **42**, 170–181.

[3] Hosking, J. R. M. (1981). *J. R. Statist. Soc. B*, **43**, 219–230.

[4] Quenouille, M. H. (1947). *J. R. Statist. Soc. A*, **110**, 123–129.

[5] Quenouille, M. H. (1949). *J. R. Statist. Soc. B*, **11**, 68–84.

[6] Walker, A. M. (1950). *J. R. Statist. Soc. B*, **12**, 102–107.

(AUTOREGRESSIVE-INTEGRATED MOVING-AVERAGE (ARIMA) MODELS
PORTMANTEAU TEST
TIME SERIES)

J. R. M. HOSKING

QUESTIONNAIRES *See* QUESTION-WORDING EFFECTS IN SURVEYS; PUBLIC OPINION POLLS

QUESTION-WORDING EFFECTS IN SURVEYS

The data collection* instrument in survey research (*see* SURVEY SAMPLING) is the questionnaire, a document that contains the set of questions for which responses are required. The questionnaire may be administered by interviewers, either face to face with respondents or by telephone, or it may be a self-administered form on which respondents record their answers, as in mail surveys. In interview surveys, interviewers are usually instructed to adhere rigidly to the questionnaire, keeping to the designated order of questions and asking them exactly as specified. If a respondent fails to provide a satisfactory answer, the interviewer is permitted only to follow up with standard probes or to repeat the question. Whichever mode of data collection is used, the questionnaire is a carefully constructed standardized instrument that aims to collect data that can be meaningfully aggregated for statistical analyses. Standardization is needed to yield comparable data from different respondents, for it has been well documented that survey responses can be sensitive to variations in the precise wording, format, and context of the questions asked.

Much of the research on question wording, format, and context deals with opinion questions. This research has mainly employed a split-ballot technique in which alternative forms of a question are administered to comparable samples of respondents. Numerous split-ballot experiments were conducted during the 1940s and 1950s, with the results being reported in the *Public Opinion Quarterly* and similar journals, in the collection of papers by Cantril [5], and in the classic *The Art of Asking Questions* by Payne [12]. After this spate of activity, interest in such experiments waned for a time but has now revived. The results of a recent wide-ranging program of research on opinion questions are reported by Schuman and Presser [16]. *See also* PUBLIC OPINION POLLS.

Rugg [13] conducted a well-known early split-ballot experiment in which he compared responses to the questions "Do you think the United States should allow public speeches against democracy?" and "Do you think the United States should forbid public speeches against democracy?" While only 25% of respondents to the first question would allow such speeches, 46% of respondents to the second question would not forbid them (excluding the "don't knows"). The approximately 20% difference between these percentages, which has been replicated by Schuman and Presser [16], illustrates how responses to opinion questions can sometimes be affected by what might appear to be slight wording changes.

In addition to wording changes per se, research on opinion questions has examined the effects of various aspects of question format. One issue concerns the balance built into the question between the alternative opinions offered to the respondent. Thus the question "Do you favor X?" is unbalanced, "Do you favor or oppose X?" has a token balance, while "Do you favor X or Y?" is balanced. Experimental comparisons between the token and unbalanced forms have generally found only small differences in response distributions, but the balanced form has sometimes given rise to large differences. The inclusion of the Y alternative in the balanced form may, however, effectively change the question so that these differences cannot necessarily be attributed to format variation [8]. A related issue is that of an acquiescence response set, which denotes a tendency for respondents to give "agree" rather than "disagree" responses regardless of content. There is some evidence to support the existence of this set [16].

Another group of question format issues concerns the alternative response options offered to respondents. An initial consideration is whether respondents should be supplied with a list of alternative responses (a closed question) or whether they should be left to compose their own answers (an open question). Experiments have demonstrated that open and closed questions can produce markedly different response distributions. Moreover, although constructing the alternatives for the closed question to correspond to the open question responses reduces the differences, it does not remove them entirely [14, 16]. When, as is often the case, a closed question is used, decisions then need to be made on whether to offer a "don't know" alternative explicitly or to accept such an answer only when it is given spontaneously, whether to offer a middle neutral alternative or not, and in which order to present the alternatives.

As some experimental studies have shown,

many respondents who are ignorant about the issue involved in a closed question will nevertheless choose one of the alternatives offered [3, 15]. Thus, in the experiment conducted by Bishop et al. [3], for instance, respondents were asked: "Some people say that the 1975 Public Affairs Act should be repealed. Do you agree or disagree with this idea?" Although the act was entirely fictitious, 16% of the respondents agreed and 18% disagreed with its repeal. One way that can be used to screen out some uninformed respondents from expressing an opinion is to make an explicit offer of a "don't know" answer in the alternatives presented. A more forceful way is to use a preliminary filter question of the form "Do you have an opinion on ... ?," with the opinion question then being asked only of those answering "yes" to the filter. Another factor that influences the extent to which respondents without opinions are screened out is the instruction given to interviewers on how readily they are to accept "don't know" answers; as a study by Smith [18] indicates, differences in interviewer instructions can lead to an appreciable difference in the proportions of "don't know" responses.

The explicit offer of a middle alternative may be justified on the grounds that some respondents may hold a considered neutral view on the topic under study, and they need the offer of that alternative in order to be able to indicate their position. The argument against offering a neutral—or "don't know"—alternative is that there is a risk that many respondents may choose it as a way to avoid deciding on their preference. As several experiments have shown, when the neutral alternative is offered, the proportion of respondents choosing it may increase substantially [11, 16].

The presentation order of the alternatives with closed questions may affect the responses, and the nature of this order effect may depend on whether the alternatives are presented in written form or orally. Some experiments have shown a primacy effect, favoring alternatives at the top of the list, when the alternatives are presented in written form [2], while some have shown a recency effect, favoring later alternatives, when they are presented orally [16].

Yet another issue examined with opinion questions is the effect of reversing the order of two adjacent questions. In many cases the change of order has had no discernible effect for either question, but there are a few examples where the response distributions have been affected. A situation where an order effect has been found on several occasions is when the two questions comprise a general one and a specific one on the same topic. In this case the response distribution for the specific question is as a rule unaffected by the question order, but that for the general question may differ according to whether the question comes before or after the specific question [10, 17, 20]. In the Kalton et al. [10] experiment, for example, respondents were asked a general question, "Do you think that driving standards generally are lower than they used to be, or higher than they used to be, or about the same?" and an equivalent specific question about "the driving standards among younger drivers." The order of these questions was varied for two random halves of the sample. The responses to the specific question were unchanged by the question order: both when it was asked first and when it was asked after the general question, 35% of respondents said that the driving standards of younger drivers were lower than they used to be. The question order did, however, affect the responses to the general question: 34% of respondents said that general standards were lower when the general question was asked first, but only 27% gave this answer when the general question followed the specific question. Further analysis revealed that this question order effect applied only with respondents aged 45 and over: among these respondents, 38% said general driving standards were lower when the general question was asked first, compared with only 26% when the question was asked second.

The demonstrated sensitivity of the distributions of opinion responses to the precise wording, format, and context of the questions has led experienced survey researchers to interpret these distributions with considerable caution. When the overall distribution of opinion is of major interest, the researcher can seek protection against question artifacts by using several questions that differ in wording and format. However, partly because of the instability of overall distributions, researchers usually place little emphasis on them. Rather, they concentrate their analyses mainly on examining the associations between the opinions and other responses, for example, examining the variation in the distributions of opinions for respondents in different age groups or with different levels of education. This type of analysis is justified under the assumption that although overall distributions may be affected by questioning effects, associations between opinions and other responses will be unaffected. Although this assumption of "form-resistant correlation" [16] may often hold as a reasonable approximation, there are nevertheless examples to demonstrate that it does not always do so.

While questioning effects with factual questions differ from those with opinion questions, there are nevertheless some similarities. Responses to factual questions can be equally affected by wording changes, but such effects are usually more easily accounted for. Responses to factual questions may also be affected by format, as for instance by the presentation order of response alternatives. There is also evidence that factual responses can be affected by other items on the questionnaire. One illustration of this effect comes from a survey in which respondents were asked to report their readership of a sizable number of periodicals. The presentation order of the different types of periodical was varied for different parts of the sample, with a resulting variation in reported readership levels [1]. The largest variation occurred with the weekly publications: when they appeared last in the presentation order, their reported level of readership was only three-fourths of what it was when they appeared first. Another illustration comes from an examination by Gibson et al. [7] of the effects of including supplements on the responses to the core items in three major surveys conducted by the U.S. Bureau of the Census*. In the National Crime Survey Cities Sample, a random half of the sample was asked a lengthy series of attitude questions about crime and the police before being asked the core crime victimization questions. This half sample reported victimization rates for personal crimes and for property crimes that were greater than the rates reported by the half sample asked only the victimization questions by around 20% and 13%, respectively (see also Cowan et al. [6]). A particularly disturbing feature of this effect of question context on responses is that it may invalidate comparisons between surveys: it is well recognized that comparison of survey results is hazardous unless the identical question is used, but these findings indicate that even repetition of the identical question may not yield comparable responses if the question is asked in a different context.

Two issues of especial concern with factual questions are obtaining information about sensitive matters and minimizing memory errors. Several approaches are available for attempting to elicit accurate reports on sensitive matters, of which the randomized response* technique is a recent addition. The use of records, diaries, and aided recall methods can sometimes be useful for reducing memory errors. Cannell and his colleagues have experimented with several procedures for dealing with problems of both memory errors and of sensitive questions, including lengthening the questions, including instructions on the questionnaire to inform respondents of what is expected of them in answering the questions, the use of feedback to tell them how they are performing, and securing their commitment to respond conscientiously. The evidence from the various experiments using these techniques suggests that each of them leads to an improvement in reporting [4]. While further research is needed, these techniques hold promise for improving the quality of survey responses.

There is a considerable, widely scattered literature on questioning effects. Introductions to this literature are provided by Kalton et al. [10], Kalton and Schuman [9], Schuman and Presser [16], Sudman [19], and Sudman and Bradburn [20].

References

[1] Belson, W. A. (1962). *Studies in Readership*. Business Publications, London.

[2] Belson, W. A. (1966). *J. Advert. Res.*, **6**(4), 30–37.

[3] Bishop, G. F., Oldendick, R. W., Tuchfarber, A. J., and Bennett, S. E. (1980). *Public Opinion Quart.*, **44**, 198–209.

[4] Cannell, C. F., Miller, P. V., and Oksenberg, L. (1981). In *Sociological Methodology, 1981*, S. Leinhardt, ed. Jossey-Bass, San Francisco.

[5] Cantril, H., ed. (1944). *Gauging Public Opinion*. Princeton University Press, Princeton, N.J.

[6] Cowan, C. D., Murphy, L. R., and Wiener, J. (1978). *Proc. Sec. Surv. Res. Meth., Amer. Statist. Ass.*, pp. 277–282.

[7] Gibson, C. O., Shapiro, G. M., Murphy, L. R., and Stanko, G. J. (1978). *Proc. Sec. Surv. Res. Meth., Amer. Statist. Ass.*, pp. 251–256.

[8] Hedges, B. M. (1979). *Statistician*, **28**, 83–99.

[9] Kalton, G. and Schuman, H. (1982). *J. R. Statist. Soc. A*, **145**, 42–73.

[10] Kalton, G., Collins, M., and Brooks, L. (1978). *Appl. Statist.*, **27**, 149–161.

[11] Kalton, G., Roberts, J., and Holt, D. (1980). *Statistician*, **29**, 65–78.

[12] Payne, S. L. (1951). *The Art of Asking Questions*. Princeton University Press, Princeton, N.J.

[13] Rugg, D. (1941). *Public Opinion Quart.*, **5**, 91–92.

[14] Schuman, H. and Presser, S. (1979). *Amer. Sociol. Rev.*, **44**, 692–712.

[15] Schuman, H. and Presser, S. (1980). *Amer. J. Sociol.*, **85**, 1214–1225.

[16] Schuman, H. and Presser, S. (1981). *Questions and Answers in Attitude Surveys*. Academic Press, New York.

[17] Schuman, H., Presser, S., and Ludwig, J. (1981). *Public Opinion Quart.*, **45**, 216–223.

[18] Smith, T. W. (1982). *Public Opinion Quart.*, **46**, 54–68.

[19] Sudman, S. (1980). *Statistician*, **29**, 237–273.

[20] Sudman, S. and Bradburn, N. M. (1982). *Asking Questions*. Jossey-Bass, San Francisco.

(PUBLIC OPINION POLLS
SURVEY SAMPLING)

GRAHAM KALTON

QUETELET, ADOLPHE

Born: February 22, 1796, in Ghent, Belgium.

Died: February 17, 1874, in Brussels, Belgium.

Contributed to: descriptive statistics, demography, vital statistics, statistics in the social sciences.

Adolphe Quetelet was one of the nineteenth century's most influential social statisticians. He was born Lambert Adolphe Jacques Quetelet in Ghent, Belgium on February 22, 1796. He received a doctorate of science in 1819 from the University of Ghent, with a dissertation on conic sections. From 1819 on he taught mathematics in Brussels, founded and directed the Royal Observatory, and he dominated Belgian science for a half century, from the mid-1820s to his death in 1874.

Early in 1824 Quetelet spent three months in Paris, where he studied astronomy and probability and learned what he could about the running of an observatory. Upon his return to Brussels he convinced the government to found an observatory, and for most of his life he operated from this base, giving particular attention to its meteorological functions. But one science was not enough to contain his energy and interests, and from about 1830 he became heavily involved in statistics and sociology*.

Quetelet's international reputation was made in 1835 with the publication of a treatise where he coined the term "social physics" [7, 8, 12]. This work is today best known for the introduction of that now-famous character, the "average man" ("l'homme moyen"). The average man began as a simple way of summarizing some characteristic of a population (usually a national population), but he took on a life of his own, and in some of Quetelet's later work he is presented as an ideal type, as if nature were shooting at the average man as a target and deviations from this target were errors. The concept was criticized by Cournot* and others, for example on the grounds that an individual average in all dimensions might not even be biologically feasible (the average of a set of right triangles may not be a right triangle).

In 1846 he published a book [9, 10] on probability and social science in the form of a series of letters to two German princes he had tutored (one of them, Albert, had married Queen Victoria of England in 1840). That book exerted a significant influence on social statistics by demonstrating that as diverse a collection of human measurements as the heights of French conscripts and the chest circumferences of Scottish soldiers could be taken as approximately normally distributed. This gave a further dimension to the idea of an average man—deviations from the average were normal, just as errors of observation deviated normally from their mean. Quetelet devised a scheme for fitting normal distributions (actually, symmetric binomial distributions with $n = 999$) to grouped data* that was essentially equivalent to the use of normal probability paper [14]. The appearance of the normal curve in such areas so far from astronomy and geodesy had a powerful influence on Francis Galton's* thinking, and reading of this work of Quetelet may have inspired James Clark Maxwell* in formulating his kinetic theory of gases.

Quetelet made few technical contributions to statistics, although in 1852 he did anticipate some later work on the use of runs* for testing independence. Quetelet derived the expected numbers of runs of different lengths both for independent and for simple Markov sequences, and compared them with rainfall records, concluding that there was strong evidence of persistence in rainy or dry weather [11, 14]. In his earlier 1835 treatise, Quetelet was led to a forerunner of a measure of association in 2×2 tables, although the measure was neither developed nor expressed algebraically [2]. In other work (e.g., ref. 9) Quetelet gave much attention to classifying sources of variation as due to accidental, periodic, or constant causes, in a sort of informal precursor to the analysis of variance* or the decomposition of time series*.

Quetelet was a prolific writer and editor. He wrote a dozen books, founded and wrote much of the material for several journals, and still found time to fill the pages of the official publications of the Belgian Académie Royale des Sciences. In addition, he carried on an immense correspondence with scientists and others all over Europe [1, 15]. He was a highly successful entrepreneur of science who was instrumental in the founding of the Statistical Society of London, the International Statistical Congresses, and the Statistical Section of the British Association for the Advancement of Science, not to mention several Belgian bureaus and commissions, and similar activities in meteorology. He was the first foreign member of the American Statistical Association*. The historian of science George Sarton has called him the "patriarch of statistics" [13].

References

[1] Diamond, M. and Stone, M. (1981). *J. R. Statist. Soc. A*, **144**, 66–79, 176–214, 332–351. (Presents manuscript material on the relationship between Quetelet and Florence Nightingale*.)

[2] Goodman, L. and Kruskal, W. H. (1959). *J. Amer. Statist. Ass.*, **54**, 123–163.

[3] Hankins, F. H. (1908). *Adolphe Quetelet as Statistician*. Longman, New York.

[4] Landau, D. and Lazarsfeld, P. F. (1978). *International Encyclopedia of Statistics*, Vol. 2, pp. 824–834. (Reprinted from *The International Encyclopedia of the Social Sciences*, with minor additions.)

[5] Lazarsfeld, P. F. (1961). *Isis*, **52**, 277–333. (An excellent discussion of Quetelet's role in the quantification of sociology.)

[6] Lottin, J. (1912). *Quetelet, Statisticien et Sociologue*. Alcan, Paris/Institut Supérieur de Philosophie, Louvain.

[7] Quetelet, A. (1835). *Sur l'homme et lè développement de ses facultés, ou Essai de physique sociale*, 2 vols. Bachelier, Paris.

[8] Quetelet, A. (1842). *A Treatise on Man and the Development of his Faculties*. W. & R. Chambers, Edinburgh. (Reprinted by Burt Franklin, 1968.)

[9] Quetelet, A. (1846). *Lettres à S.A.R. le Duc Régnant de Saxe-Cobourg et Gotha, sur la Théorie des Probabilités, appliquée aux Sciences Morales et Politiques*. Hayez, Brussels.

[10] Quetelet, A. (1849). Letters Addressed to H.R.H. the Grand Duke of Saxe Coburg and Gotha, on the Theory of Probabilities as Applied to the Moral and Political Sciences. Layton, London. (A mediocre translation of ref. 9.)

[11] Quetelet, A. (1852). *Bulletins de l'Académie Royale de Belgique*, **19**, 303–317.

[12] Quetelet, A. (1869). *Physique Sociale, ou Essai sur le Développement des Facultés de l'Homme*, 2 vols. Muquardt, Brussels. (An expanded second edition of ref. 7.)

[13] Sarton, G. (1935). *Isis*, **23**, 6–24.

[14] Stigler, S. M. (1975). *Bull. Int. Statist. Inst.*, **46**, 332–340.

[15] Wellens-De Donder, L. (1966). *Mémoires de l'Académie royale de Belgique*, **37**(2), 1–299. (A catalog of an extensive archive of Quetelet's correspondence.)

STEPHEN M. STIGLER

QUETELET INDEX

The Quetelet index is the ratio

$$1000 \times \text{weight}/(\text{height})^2$$

with weight measured in kilograms and height in centimeters. It is used as an index of build; the greater the index, the sturdier the build.

QUEUEING THEORY

Queueing theory is concerned with the modeling of situations where customers wait for service and congestion occurs through randomness in service times and patterns of arrival. Examples include customer queues at checkout counters in a supermarket and queues of jobs in a network of time-sharing computers.

The simplest queueing model comprises three elements, the first of which is a point process describing arrival instants of customers at a given place and the most common assumption is that customers arrive singly at event times of a renewal* process. Second, customers require a period of service from one of k servers; it is easiest to assume that service times are independent, identically distributed and independent of the arrival process. Finally, there is a queue discipline specifying how customers queue in order of arrival within an infinite waiting room and move from the head of the queue to the first available server. Servers are idle only if no customer awaits service. All other situations are best regarded as modifications of this simple model.

Analysis of the queueing model may comprise determination of the distributions of the queue length at any time, the waiting time before commencement of service of a customer, the duration of periods of continual work by the server—the busy periods—and the idle periods during which the server waits for further customers. Determination of finite time distributions is difficult but often it is possible to determine their large time limits, the so-called equilibrium theory.

Further discussion is facilitated by Kendall's [38] notation $A/B/k$ for the simple queueing model; A denotes the type of interarrival time distribution, B the service-time distribution, and k the number of servers. For special purposes this notation is sometimes augmented. For example, an $A/B/k$ queue having a finite waiting room of size N is denoted $A/B/k/N$. In particular, M denotes the negative exponential* or Markov distribution, E_j is the j-fold convolution* of M, or Erlang distribution*, and D denotes the deterministic distribution. The general k-server queueing model is denoted by $GI/G/k$. Here, however, we shall concentrate mainly on the more tractable single-server case and refer the reader to the companion entry MULTISERVER QUEUES for an account of the complications engendered by several servers. Also, *see* NETWORKS OF QUEUES for an account of the properties of interconnected queues.

The earliest papers on queueing theory were written by A. K. Erlang [7] during the period 1909–1929, to solve congestion problems arising in teletraffic contexts. He discussed the equilibrium theory of the $M/M/k$, $M/D/k$, and $M/E_j/1$ systems with finite waiting rooms. Other workers extended and consolidated Erlang's work, and this is summarized by Fry [27]. F. Pollaczek [59] and A. Khinchin [40] independently worked on the equilibrium theory of $M/G/1$, obtaining expressions (called Pollaczek–Khinchin formulae*) for the Laplace–Stieltjes transform and expectation of the equilibrium waiting-time distribution in terms of the arrival rate and service-time distribution [41; 44, p. 177]. Subsequently, Pollaczek (1934) considered $M/G/k$, and E. Volberg (1939) obtained equilibrium and some time-dependent properties of $E_j/G/1$. In 1942, E. Borel obtained the distribution of the number of customers served during a typical busy period of $M/D/1$. See refs. 37, 38, and 65 (p. 20), for more details.

The modern phase of queueing theory was marked by the publication by D. G. Kendall [37] in 1951 of a paper which acquainted a wide circle of mathematicians with these early contributions. More important, by using the concept of embedded processes*, Kendall showed how the equilibrium theory of $M/G/1$ could be derived from Markov chain theory (*see* MARKOV PROCESSES), then recently expounded in the first edition of Feller's famous text [25]. The queue-length process $\{Q(t): t \geqslant 0\}$ is a birth-and-death process* for $M/M/1$ but is not even Markovian otherwise. Roughly, this is because both the current queue length and elapsed service time of the customer receiving service are required to decouple the past and future evolution of $\{Q(t)\}$. This difficulty vanishes if queue lengths are inspected at successive departure epochs—if these are denoted by $\{D_n\}$ then $\{Q(D_n+)\}$ is a Markov chain. Kendall also generalized Borel's work by showing that customers served during a busy period of $M/G/1$ can be grouped into generations of a Galton–Watson process*. Consequently, the class of distributions of the numbers of customers served during a busy period essentially coincides with the class of Lagrange distributions* [58].

SUBSEQUENT DEVELOPMENT OF THE SIMPLE QUEUE MODEL

The theory of the $GI/G/1$ system developed rapidly during the five years following Kendall's paper [37]. Lindley [53] obtained the fundamental relation $W_{n+1} = (W_n + U_n)^+$, where W_n is the time spent waiting for service by the nth customer, who arrives at T_n and demands service lasting S_n, and $U_n = S_{n+1} - (T_{n+1} - T_n)$. Thus $\{W_n\}$ is a random walk* on the positive real line with a reflecting barrier at the origin. Lindley exploited this observation to show that an equilibrium waiting-time distribution exists if $EU_n < 0$ but not if $EU_n \geqslant 0$; then waiting times can be arbitrarily long. These results were later extended to the $GI/G/k$ system [42, 78]. Lindley derived an integral equation* for the equilibrium waiting-time distribution and solved it for $M/G/1$ and $D/E_j/1$. This equation, of Wiener–Hopf type, was treated by Smith [67]. The deep connection between $\{W_n\}$ and the random walk $\{V_n\}$, where $V_0 = 0$ and $V_n = \sum_{j=1}^{n} U_j$ $(n \geqslant 1)$, was exposed by Spitzer [68], who showed that if $W_0 = 0$, then W_n and $\max_{0 \leqslant j \leqslant n} V_j$ have the same distribution. He also obtained the identity

$$\sum_{n=0}^{\infty} t^n E(e^{-\theta W_n}) = \exp\left[\sum_{n=1}^{\infty} (t^n/n) E(e^{-\theta V_n^+})\right],$$

independently discovered using more analytical methods by Pollaczek [60], and now called the *Pollaczek–Spitzer identity*. See ref. 43 for an elegant algebraic derivation and illuminating discussion of related matters.

The virtual waiting time $W(t)$ $(t \geqslant 0)$ is

the total work load facing the server at time t. This concept was introduced by Takács [72] and analyzed for the $M/G/1$ system, where $\{W(t)\}$ is a Markov process. It is not Markovian for other $GI/G/1$ systems, but the waiting times constitute an embedded Markov chain, since $W_n = W(T_n^-)$. Non-Markovian processes can be rendered Markovian by adding supplementary variables. For example, if $E(t)$ is the expended service time of the customer, if any, receiving service at time t, then the stochastic process* having state $(Q(t), E(t))$ if $Q(t) > 0$ and (0) otherwise is Markovian for the $M/G/1$ system. The idea was first introduced by Cox [17] and later exploited by others [34, 35].

Properties of $GI/G/1$ can be obtained by numerous methods. The use of transform and generating function* techniques is described by Takács in ref. 73 and combinatorial methods in ref. 74. Prabhu [62] describes the use of Wiener–Hopf and fluctuation theory and Cohen [13] the use of complex variable methods, originally due to Pollaczek, and regenerative processes in ref. 12. Frequently, these analytical techniques yield derived distributions in forms having limited practical utility, for example, as an integral transform*. Although a transform can be used to obtain moments and tail behavior, explicit inversion is not often possible (but see ref. 28 on numerical inversion). Recognizing these problems, some researchers have sought explicit expressions for derived distributions, but often they occur as infinite series containing high-order convolutions.

Several approximation procedures have been devised to circumvent these problems. For example, it is often possible to obtain bounds for the distributions of interest [45, 48, 71]. Again, transient distributions may be adequately approximated by the equilibrium distribution if some estimate of the rate of convergence of the former is available. In most practical cases convergence is geometrically fast [76]. Finally, if the arrival- and service-time distributions of a system can be closely approximated by those of a second, computationally tractable system, then we can expect that derived distributions of the first system will be closely approximated by the corresponding and easily computed quantities of the second system. Neuts [56] has shown that this approach can be powerfully implemented with phase-type distributions*.

Standard statistical methods can be used to estimate parameters of queueing models. For example, if the continuous-time queue-length process is Markovian, then likelihood methods provide point estimates and confidence intervals from continuously observed records of queue lengths [1, 18, 30]. Frequently, ad hoc methods must be devised to handle incomplete observations and/or the non-Markovian nature of many derived queueing processes. Inference procedures for point processes [19] will yield the characteristics of the arrival process from observations of successive arrival times. Again, the method of moments* can be used to construct approximate estimates of the service-time distribution of an $M/G/1$ system from observations on waiting times [18, 30]. The regenerative structure of many queueing systems can be exploited to obtain estimates by simulation*. For example, the lengths of successive busy periods of a $GI/G/1$ system, and the numbers of customers served therein, are serially independent. Estimates of the limiting expected waiting time can be obtained by observing successive busy periods [21, 64].

The stochastic processes of interest in the $GI/G/1$ system are functionals of the random walk $\{V_n\}$. When $EU_n \simeq 0$, the central limit theorem* shows that $\{V_n\}$ can be approximated by a Brownian motion* process. It is plausible that the distribution of the queueing functional of interest is approximated by that of the same functional of the Brownian motion, which in most cases can be computed. The resolution of this question is encompassed by the theory of functional limit and heavy traffic theorems. They provide useful approximations for much more general systems than $GI/G/1$ and with only weak dependence assumptions [5, 32, 77]. Related to this are continuity theorems [39] which show that distributions of functionals, like waiting times, are insensitive to small perturbations of service-time and interarrival-time distributions. These theorems show that Neuts' work [56] effectively solves a wide range of queueing problems.

VARIANTS OF THE SIMPLE QUEUEING MODEL

The $GI/G/k$ system can be generalized in a myriad ways, and we mention only a few. Thus independence assumptions can be weakened while retaining some qualitative features and approximation procedures of the simple system [5, 49, 54]. General arrival processes can be modeled with point processes* theory [26, 63], which also facilitates understanding of the process of departures [22]. Arrival and service rates may vary with the state of the system [49, 55], particularly if control policies are implemented [20, 75]. Variations of the queue discipline include finite waiting rooms and bounds on waiting times [13] and arrivals and servicing of batches of customers [11, 13, 65]. Again, arriving customers may not join the queue with a probability depending on queue length (balking) or may leave after waiting for some unpredictable period (reneging) [70]. Many of these variants can usefully be modeled by discrete-state Markov processes [15, Chaps. 3, 4; 24, 69]. Arriving customers may be allocated to priority classes giving high-priority customers earlier access to the server than previously arrived lower-priority customers, even to the extent of interrupting the service of such customers [33; 45, Chap. 3].

These modifications mainly concern individual service facilities, but frequently individual queues are part of a larger network where customers leaving one service facility may queue for service at another. Although such networks were first studied in the 1950s, the desire to understand computer and communication networks, for example, has recently sparked intense research [16, 46, 51]. The equilibrium theory of networks in which service and interarrival distributions are negative-exponential can be elucidated by concepts of reversibility of Markov processes [36]. Many results so obtained extend to certain classes of networks allowing more general distributions [9]. A recent dynamical approach portrays a network as a system of interacting point processes and makes extensive use of martingale* theory [4, 6]. Diffusion approximations and other weak convergence techniques are also efficacious [8, 52]. *See also* NETWORKS OF QUEUES.

CONCLUDING REMARKS

Queueing theory has been a very popular area of research over the past 40 years, with some 50 published books and a journal literature numbering about 4000 papers. Yet critics have contended that much of this literature deals with unimportant variations of the basic theme, has little theoretical interest or practical value, and that the literature which is useful is only the elementary part of the theory [10; 31, p. 280; 50, pp. vii, 27; 61, p. 243; 66]. For example, Lee [50] shows how complicated problems of congestion at airline terminals can be resolved with quite simple queueing models. There are many mathematical investigations of systems where arrival- and service-time distributions vary in response to queue lengths, waiting times, and so on [49]. However, little of this work is tailored to real situations and the state dependencies analyzed arise more from the analyst's imagination than through empirical investigation.

Most researchers stand aside from such criticism, but on occasion rebuttals have been published. Thus refs. 3 and 47 are the first of several rejoinders to ref. 10, the most vigorous attack on the subject. Points made include:

1. The assertion of little applicability is simply false; many successful applications are either not reported in the literature or are reported in journals not normally read by the critics [47].
2. The literature contains a much higher proportion of practically motivated papers than critics are willing to admit.

Bhat [3] asserts that 65% of the 700-odd references in ref. 65 are application oriented.

3. Kolesar [47] asserts that much of the theoretical literature is relevant in practice and he cites several papers, including ref. 51, used by himself.

4. Prabhu [61] observes that the criticisms above can be leveled at other theoretical disciplines and that it is not realistic to suspend research until all extant theory is utilized.

The structures of queueing theory also arise in dam theory* and risk theory* [62]. These structures are largely amenable to analysis based on Markov chains and sums of random variables, and hence queueing theory has been a very attractive subject for exploration by probabilists [31, p. 281]. The development of queueing theory has also been closely related to other topics, for example, point processes*, semi-Markov processes*, and functional limit theorems. Problems arising in queueing theory have inspired progress in these areas, and reciprocally, it has been enriched by general developments in these subjects. There is no sign that this symbiotic relationship is weakening, and hence queueing research is likely to remain fertile for some time to come. Queueing theory first arose in response to problems of telephony, and the need to measure the performance of modern communication systems has inspired a resurgence of research activity [45, 46]. Indeed, Kleinrock [44] states that queueing theory is "one of the few tools we have for analyzing the performance of computer systems" and he further asserts that the successful application of queueing theory to this problem is a significant reason for its continuing popularity among researchers.

References

Together refs. 23, 65, and 66 give a bibliography complete up to about 1966, but there is no complete listing of the thousands of articles written thereafter. Papers of more recent years listed below have been chosen largely for their bibliographic value. References 15 and 57 contain (almost) complete lists of English language books on queues and are updated by the following list. See also the reference list for MULTISERVER QUEUES.

[1] Basawa, I. V. and Prakasa Rao, B. L. S. (1980). *Statistical Inference for Stochastic Processes*. Academic Press, New York.

[2] Bhat, U. N. (1969). *Manag. Sci.*, **15**, B-280–B-294. (Useful historical survey.)

[3] Bhat, U. N. (1978). *Interfaces*, **8**, 27–28. (Rebuts ref. 10.)

[4] Boel, R. (1981). In *Stochastic Systems: The Mathematics of Filtering and Identification and Applications*, M. Hazewinkel and J. C. Willems, eds. D. Reidel, Dordrecht, The Netherlands, pp. 141–167. (Surveys networks.)

[5] Borovkov, A. A. (1976). *Stochastic Processes in Queueing Theory*. Springer-Verlag, Berlin. (Highly mathematical, stressing generality and weak convergence.)

[6] Bremaud, P. (1981). *Point Processes and Queues: Martingale Dynamics*. Springer-Verlag, Berlin.

[7] Brockmeyer, E., Halstrom, H. L., and Jensen, A. (1948). *The Life and Works of A. K. Erlang*. Transactions 2. Danish Academy of Technical Sciences, pp. 1–277. (Reprinted in 1960 in Acta Polytech. Scand. Appl. Maths. Comp. Machinery No. 6.)

[8] Bruell, S. C. and Balbo, G. (1980). *Computational Algorithms for Closed Queueing Networks*. North-Holland, New York.

[9] Burman, D. Y. (1981). *Adv. Appl. Prob.*, **13**, 846–859. (Surveys and extends insensitivity results.)

[10] Byrd, J. (1978). *Interfaces*, **8**, 22–26. (Castigates queueing theory.)

[11] Chaudhry, M. L. and Templeton, J. G. C. (1983). *A First Course in Bulk Queues*. Wiley, New York.

[12] Cohen, J. W. (1976). On Regenerative Processes in Queueing Theory. *Lect. Notes Econ. Math. Syst.*, **121**. Springer-Verlag, Berlin.

[13] Cohen, J. W. (1982). *The Single Server Queue*, 2nd ed. North-Holland, Amsterdam. (Encyclopediac in scope.)

[14] Cohen, J. W. and Boxma, O. J. (1983). *Boundary Value Problems in Queueing System Analysis*. North-Holland, Amsterdam. (Analytical treatment of two-dimensional random walk with applications to some two-server queues.)

[15] Cooper, R. B. (1981). *Introduction to Queueing Theory*, 2nd ed. North-Holland, New York. (Good introductory text containing a useful guide to journals that publish queueing theory papers.)

[16] Courtois, P. J. (1977). *Decomposability: Queueing and Computer System Applications*. Academic Press, New York.

[17] Cox, D. R. (1955). *Prob. Camb. Philos. Soc.*, **51**, 433–441.

[18] Cox, D. R. (1965). In *Proc. Symp. Congestion Theory*, W. L. Smith and W. E. Wilkinson, eds. The University of North Carolina Press, Chapel Hill, N.C., pp. 289–316.

[19] Cox, D. R. and Lewis, P. A. W. (1966). *The Statistical Analysis of Series of Events*. Methuen, London.

[20] Crabill, T. B. (1977). *Operat. Res.*, **25**, 219–232. (Bibliography on optimal control of queues.)

[21] Crane, M. A. and Lemoine, A. J. (1977). An Introduction to the Regenerative Method for Simulation Analysis. *Lect. Notes Control Inf. Sci.*, **4**. Springer-Verlag, Berlin.

[22] Daley, D. J. (1976). *Adv. Appl. Prob.*, **8**, 395–415. (Surveys the output process.)

[23] Doig, A. (1957). *Biometrika*, **44**, 490–514. (An extensive bibliography.)

[24] Doorn, E. van (1980). Stochastic Monotonicity and Queueing Applications of Birth–Death Processes. *Lect. Notes Statist.*, **4**. Springer-Verlag, Berlin.

[25] Feller, W. (1968). *An Introduction to Probability Theory and Its Applications*, Vol. I, 3rd ed. Wiley, New York.

[26] Franken, P., König, D., Arndt, U., and Schmidt, V. (1982). *Queues and Point Processes*. Wiley, Chichester, England. (Mathematically advanced.)

[27] Fry, T. C. (1928). *Probability and Its Engineering Uses*. Van Nostrand, New York. (Surveys the earliest work.)

[28] Gaver, D. P., Jr. (1966). *Operat. Res.*, **14**, 444–459.

[29] Gelenbe, E., Labetoulle, J., Marie, R., Metivier, M., Pujolle, G., and Stewart, W. (1980). *Réseaux de files d'attente*. Editions Homme et Techniques, Suresnes, France.

[30] Harris, C. (1974). In Mathematical Methods in Queueing Theory, A. B. Clarke, ed. *Lect. Notes Econ. Math. Syst.*, **98**. Springer-Verlag, Berlin, pp. 157–184. (Surveys statistical inference for queues.)

[31] Hedye, C. C. (1981). *Aust. J. Statist.*, **23**, 273–286.

[32] Iglehart, D. L. (1973). *Adv. Appl. Prob.*, **5**, 570–594. (Surveys weak convergence theory.)

[33] Jaiswal, N. K. (1968). *Priority Queues*. Academic Press, New York.

[34] Keilson, J. and Kooharian, A. (1960). *Ann. Math. Statist.*, **31**, 104–112.

[35] Keilson, J. and Kooharian, A. (1962). *Ann. Math. Statist.*, **33**, 767–791.

[36] Kelly, F. P. (1979). *Reversibility and Stochastic Networks*. Wiley, Chichester, England.

[37] Kendall, D. G. (1951). *J. R. Statist. Soc. B*, **13**, 151–185. (A seminal paper.)

[38] Kendall, D. G. (1953). *Ann. Math. Statist.*, **24**, 338–354.

[39] Kennedy, D. P. (1977). *Bull. ISI*, **47**, Book 2, 353–365. (Survey of continuity theorems.)

[40] Khinchin, A. (1932). *Mat. Sb.*, **39**, 73–84.

[41] Khintchine, A. Y. (1969). *Mathematical Methods in the Theory of Queueing*, 2nd ed. Charles Griffin, London. (Translation of the 1955 Russian original. This booklet collects the contributions of one of the pioneers and was also important to the development of point process theory.)

[42] Kiefer, J. and Wolfowitz, J. (1955). *Trans. Amer. Math. Soc.*, **78**, 1–18.

[43] Kingman, J. F. C. (1966). *J. Appl. Prob.*, **3**, 285–326. (Beautiful exposition of $GI/G/1$ waiting time theory.)

[44] Kleinrock, L. (1975). *Queueing Systems*, Vol. 1: *Theory*. Wiley, New York. (Excellent introduction to Markov queues, $M/G/1$, $GI/M/1$ and their variants.)

[45] Kleinrock, L. (1976). *Queueing Systems*, Vol. 2: *Computer Applications*. Wiley, New York. (Continues with $GI/G/1$ theory and networks.)

[46] Kobayashi, H. and Konheim, A. (1977). *IEEE. Trans. Commun.*, **COM-25**, 2–29. (Survey of queueing models for computer networks.)

[47] Kolesar, P. (1979). *Interfaces*, **9**, 77–82. (Rebuts ref. 10.)

[48] Köllerström, J. (1976). *Math. Proc. Camb. Philos. Soc.*, **80**, 521–525.

[49] König, D., Rykov, V. V., and Schmidt, V. (1983). *J. Sov. Math.*, **21**, 938–994. (Surveys queues with various kinds of dependencies.)

[50] Lee, A. M. (1966). *Applied Queueing Theory*. St. Martin's Press, New York. (Interesting case studies using elementary theory.)

[51] Lemoine, A. J. (1977). *Manag. Sci.*, **24**, 464–484. (Surveys network equilibrium theory.)

[52] Lemoine, A. J. (1978). *Manag. Sci.*, **24**, 1175–1193. (Surveys application of weak convergence theory to networks.)

[53] Lindley, D. V. (1952). *Proc. Camb. Philos. Soc.*, **48**, 277–289.

[54] Loynes, R. M. (1962). *Proc. Camb. Philos. Soc.*, **58**, 497–520.

[55] Minh, Do Le. (1980). *Math. Operat. Res.*, **5**, 147–159. (State-dependent queues.)

[56] Neuts, M. F. (1981). *Matrix-Geometric Solutions in Stochastic Models—An Algorithmic Approach*. Johns Hopkins University Press, Baltimore.

[57] Newell, G. F. (1982). *Applications of Queueing Theory*, 2nd ed. Chapman & Hall, London. (Concerned with approximate methods and contains a useful list of applications to road traffic situations.)

[58] Pakes, A. G. and Speed, T. P. (1977). *SIAM J. Appl. Math.*, **32**, 745–754.

[59] Pollaczek, F. (1930). *Math. Z.*, **32**, 64–100, 729–750.

[60] Pollaczek, F. (1952). *C. R. Acad. Sci. Paris*, **234**, 2334–2336.

[61] Prabhu, N. U. (1975). *Stoch. Processes Appl.*, **3**, 223–258.

[62] Prabhu, N. U. (1980). *Stochastic Storage Processes—Queues, Insurance Risk, and Dams*. Springer-Verlag, Berlin.

[63] Rolski, T. (1981). Stationary Random Processes Associated with Point Processes. *Lect. Notes Statist.*, **5**. Springer-Verlag, Berlin.

[64] Rubinstein, R. Y. (1981). *Simulation and the Monte Carlo Method*. Wiley, New York.

[65] Saaty, T. L. (1961). *Elements of Queueing Theory*. McGraw-Hill, New York. (Useful general text.)

[66] Saaty, T. L. (1966). *Naval Res. Logist. Quart.*, **13**, 447–476. (Extensive bibliography.)

[67] Smith, W. L. (1953). *Proc. Camb. Philos. Soc.*, **49**, 449–461.

[68] Spitzer, F. (1956). *Trans. Amer. Math. Soc.*, **82**, 323–339.

[69] Srivastava, H. M. and Kashyap, B. R. K. (1982). *Special Functions in Queueing Theory: And Related Stochastic Processes*. Academic Press, New York.

[70] Stanford, B. E. (1979). *Math. Operat. Res.*, **4**, 162–178.

[71] Stoyan, D. and Daley, D. J. (1982). *Comparison Methods for Queues and Other Stochastic Models*. Wiley, New York.

[72] Takács, L. (1955). *Acta Math. Acad. Sci. Hung.*, **6**, 101–129.

[73] Takács, L. (1962). *Introduction to the Theory of Queues*. Oxford University Press, New York.

[74] Takács, L. (1968). *Combinatorial Methods in the Theory of Stochastic Processes*. Wiley, New York.

[75] Teghem, J. (1982). In *Operations Research in Progress*, G. Feichtinger and P. Kall, eds. D. Reidel, Dordrecht, The Netherlands, pp. 333–345. (Surveys optimal control for queues.)

[76] Tweedie, R. L. (1983). In *Probability, Statistics and Analysis*, J. F. C. Kingman and G. E. H. Reuter, eds. Cambridge University Press, Cambridge, England, pp. 260–276. (Surveys rates of convergence to equilibrium distributions.)

[77] Whitt, W. (1974). In Mathematical Methods in Queueing Theory, A. B. Clarke, ed. *Lect. Notes Econ. Math. Syst.*, **98**. Springer-Verlag, New York, pp. 307–350. (Surveys heavy traffic results.)

[78] Whitt, W. (1982). *Math. Operat. Res.*, **7**, 88–94.

(BIRTH-AND-DEATH PROCESSES
DAM THEORY
EMBEDDED PROCESSES
MARKOV PROCESSES
MULTISERVER QUEUES
NETWORKS OF QUEUES
PHASE-TYPE DISTRIBUTIONS
RANDOM WALK
RENEWAL THEORY
RISK THEORY
STOCHASTIC PROCESSES, POINT)

ANTHONY G. PAKES

QUINCUNX

An arrangement of five objects (e.g., spots on a die) in a square or rectangle with one at each corner and one in the middle is called a quincunx. In statistics the term is used to refer to a probability device (see Fig. 1) designed to demonstrate mechanically the generation of normal or binomial curves. This device was originated by Francis Galton* [1, 2] and consists of a glass-enclosed board with rows of equally spaced nails or pins. These rows are arranged so that each successive row is shifted with each nail directly below the midpoint of two adjacent nails in the row above. Except for those on the boundary of the formation, each nail is the center of a quincunx of five nails. Metal balls (shot) are poured through a funnel or hopper and directed at the middle nail in the top row so that each ball has a 50 : 50 chance of darting to the right or left. The same condition holds for all successive lower

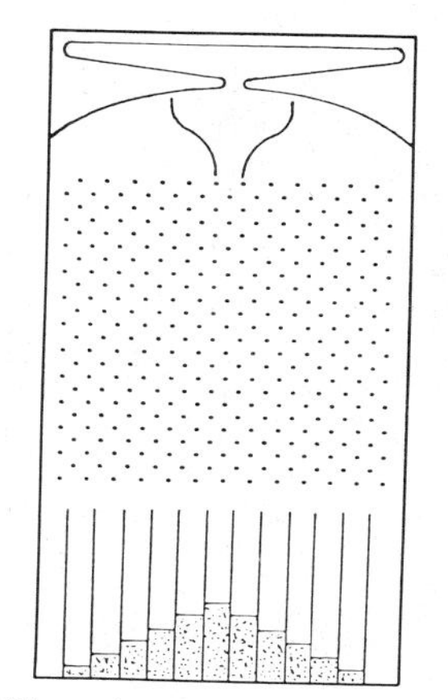

Figure 1 From Pearson [6].

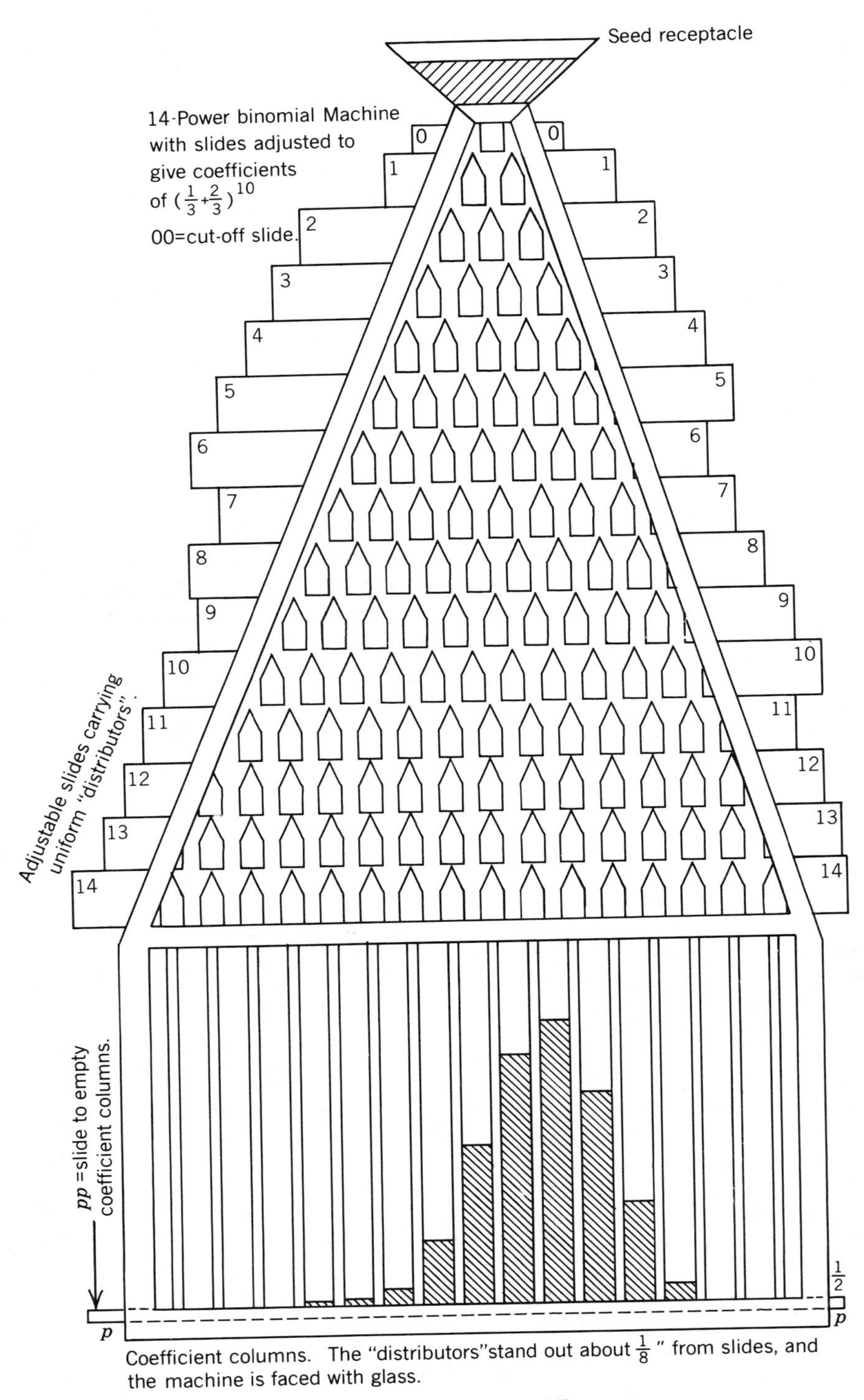

Figure 2 From Pearson [4].

nails that a ball may approach. In effect, each ball is a unit in an independent random walk* with a probability of 0.50 of moving one step to the right or left for each lower nail it approaches. The final row, instead of nails, consists of a set of vertical columns into which all balls drop forming a histogram* of columns of balls. The theoretical resulting histogram is binomial* with $p = \frac{1}{2}$ and n equal to the number of rows (of pins plus collector column row).

If the number of balls is large, the resulting observed histogram will fairly accurately represent the binomial histogram. If the number of rows is reasonably large (20 or 30 or so), this empirical histogram will serve as a good representation of the normal* curve since a normal probability distribution closely approximates a symmetric binomial probability distribution. Further, the resulting observed "normal curve" can be considered as a supportive demonstration of the physical validity of Hagen's derivation of the normal probability distribution, based on the idea that a normal random variable is a sum of many small independent elementary "errors"; *see* LAWS OF ERROR, II. Galton also used a quincunx-type device to illustrate the phenomenon of revision (i.e., regression toward the mean) in biological populations. [A number of authors use Galton's *Natural Inheritance* as the original source of the quincunx. This appears to be the proper source for this name; however, Galton used a similar device 12 years earlier (1877) and referred to its use some three years before this. See also ref. 5.]

Modifications of Galton's original device include replacing the pins with wedges and the balls with other items, such as seed or sand. Each wedge is placed to divide the stream of sand or seed into two streams. Karl Pearson* [4] constructed such a device with the position of the wedges adjustable so that each wedge will split a stream into two parts in the proportion $p : q$, with p having arbitrary values according to the placement of the wedges (see Fig. 2). Varying choices for p produce varying skewed binomial distributions and, consequently, a wide variety of distribution shapes can be physically generated by such a device.

Recently, a computer program has been developed for a microcomputer with video display facility for simulating a quincunx [3]. The program also provides a slow-mode option to enable the viewer to follow the drop of individual balls and, thereby, to study the pattern of runs to the left or right.

References

[1] Galton, F. (1877). *Nature (Lond.)*, **15**, 492–495, 512–514, 532–533.

[2] Galton, F. (1889). *Natural Inheritance*. Macmillan, London, p. 63.

[3] Hilsenrath, J. and Field, B. (1983). *The Mathematics Teacher*, **76**, 571–573.

[4] Pearson, K. (1895). *Philos. Trans. R. Soc. Lond.* (A), **186**, 343–414 + 9 plates.

[5] Pearson, K. (1930). *The Life, Letters and Labours of Francis Galton*, Vol. IIIA. Cambridge University Press, London, England.

[6] Pearson, E. S. (1969). Some Historical Reflections Traced through the Development of the Use of Frequency Curves. *Tech. Rep. No. 38* (THEMIS). Department of Statistics, Southern Methodist University, Dallas, Tex.

(BINOMIAL DISTRIBUTION
GALTON, FRANCIS
LAWS OF ERROR, II: THE GAUSSIAN DISTRIBUTION
NORMAL DISTRIBUTION)

HARRY O. POSTEN

QUOTA SAMPLING

Quota sampling is a method of *judgmental selection* in which specified numbers (quotas) of particular types of *population units* are required in the final or realized sample. For example, a quota-sampling design for a survey of shoppers in a particular shopping center might specify that the sample include 50 white males, 150 white females, 25 nonwhite males, and 75 nonwhite females, but give no further instructions on how these quotas were to be filled. Quota sampling is therefore fundamentally different from *prob-*

ability sampling, in which each population unit has a known nonzero probability of inclusion in the sample. Quota sampling is used in situations where probability sampling is impracticable, unduly costly, or considered to be unnecessary.

Estimates based on quota sampling lack the usual criterion of acceptability which applies to most estimates based on probability samples, namely that they are actually or nearly *unbiased over repeated sampling*. A sample *estimator* is said to be unbiased in this sense if its design-based *expectation* is equal to the value which would have been obtained if the entire population had been enumerated. This means that if the same selection and estimation procedures are repeated many times, the average of the resulting estimates is close to that population value.

The extent of the validity that can be ascribed to an estimate based on a quota sample in fact depends on the extent to which an implicit *population model* underlying the quota-sampling design accurately describes the population being sampled. To illustrate, suppose that the objective of sampling is to estimate the mean over the entire population of the item Y_i ($i = 1, 2, \dots, N$), where N is the number of units in the population. This population mean will be denoted by $\bar{Y}$ and defined by

$$\bar{Y} = N^{-1} \sum_{i=1}^{N} Y_i. \tag{1}$$

Suppose further that a sample of n distinct units is selected from this population (the method of selection not as yet specified) and that the values y_i ($i = 1, 2, \dots, n$) are observed on these n sample units. (Note that the y_i values are simply a subset of the Y_i values. If the first unit selected in the sample happened to be the 47th population unit, then y_1 would be numerically the same as Y_{47}.) Finally, suppose that the sample estimator $\bar{y}$ of the population mean $\bar{Y}$ is

$$\bar{y} = n^{-1} \sum_{i=1}^{n} y_i. \tag{2}$$

The validity of the estimator $\bar{y}$ can be established in one of two ways.

1. If the n sample units were selected using *simple random sampling* without replacement**, a form of probability sampling in which each population unit has the same probability of inclusion in the sample, the estimator $\bar{y}$ is unbiased over repeated sampling and its (design based) variance is

$$\text{var}_d(\bar{y}) = \frac{N-n}{Nn(N-1)} \sum_{i=1}^{N} \left(Y_i - \bar{Y}\right)^2. \tag{3}$$

2. Regardless of the manner in which the sample was selected, if the Y_i values were generated by the stochastic process*

$$Y_i = \mu + \epsilon_i, \tag{4}$$

where μ is a constant and the ϵ_i are independent random variables with mean zero and variance σ^2, then the expectation of both $\bar{Y}$ and $\bar{y}$ under that stochastic process is μ and the (model based) variance of $\bar{y}$ considered as an estimator (not of μ but) of $\bar{Y}$ is

$$\text{var}_m(\bar{y}) = \left[(N-n)/(Nn)\right]\sigma^2. \tag{5}$$

The best unbiased estimator of σ^2 obtainable from the whole population is

$$\hat{\sigma}^2 = (N-1)^{-1} \sum_{i=1}^{N} \left(Y_i - \bar{Y}\right)^2. \tag{6}$$

Comparing (5) and (6) with (3), the model under which the population is generated by the stochastic process (4) can be used to validate both the estimator (2) and its variance formula (3) even though the sample is not a probability sample.

Estimates of this kind based on quota samples are therefore validated to the extent that, for each type of unit for which a separate quote is specified, the population can be regarded as generated by a separate stochastic process of the type (4).

An example of a situation where quota sampling has been used successfully is the following. The Australian Bureau of Statistics collects its international migration* statistics using Incoming and Outgoing Passenger Cards which are completed by the passengers themselves. Some years ago the cards were redesigned in a fashion that was intended to be easier to follow, and it was considered desirable to test out the new design in a realistic situation.

Most international passengers arrive and depart by air, and it was necessary to conduct the test at airports in a situation where passengers were continually arriving and departing. It was not possible to select probability samples in these circumstances. Instead, quotas were set for persons arriving from and departing to various overseas airports in accordance with statistics of passenger movements.

Quota sampling was especially suitable for this test because:

1. Probability sampling was impracticable.
2. The statistics were required for internal decision making rather than for public use.
3. The effect observed was very clear cut (in favor of the new format).

Usually, however, statistical offices tend to avoid the use of quota sampling on the grounds that the assumptions on which it is based may not be reliable [i.e., the model (4) may not describe reality sufficiently well].

Dangers in the use of quota sampling arise chiefly from the element of subjective choice allowed in the selection of the sample in the field. Interviewers tend to choose, if they can, potential respondents whom it would be pleasant and not threatening to interview. In the shopping center situation mentioned above, the following types of shopper tend to be avoided by most interviewers:

Aggressive-looking shoppers
Down-at-heel shoppers
Harrassed-looking shoppers
Impressive-looking shoppers
Mothers with several small children
Shoppers laden down with parcels

The reader can no doubt add to this list.

The element of subjective choice in quota sampling tends to invalidate the model (4) in that it breaks any population, or any subpopulation for which a quota is set, into recognizable and relevant subsets with different values of μ and σ^2.

An important special case of quota sampling is probability sampling with quotas, a term used frequently in household-based surveys. Probability sampling is used down to the level of street block, and quotas (e.g., on the numbers of respondents in various age–sex cells) are set either within the individual blocks or over a number of blocks. The chief problem here is that the respondents tend to be chosen from those who spend most time at home. As a result, characteristics which are strongly correlated with time spent at home, such as extent of television viewing, can be quite poorly measured. For many practical purposes, however, probability sampling with quotas provides an acceptable compromise between full probability sampling, which can be extremely expensive, and pure quota sampling, which is unduly hazardous.

Bibliography

Barnett, V. D. (1974). *Elements of Sampling Theory*. English Universities Press, London, pp. 102–103.

Conway, F. (1967). *Sampling: An Introduction for Social Scientists*. Minerva Ser. No. 18. Allen & Unwin, London, pp. 133–134.

Deming, W. E. (1950). *Some Theory of Sampling*. Wiley, New York, pp. 11–14.

Deming, W. E. (1960). *Sample Design in Business Research*. Wiley, New York, pp. 31–33.

Huff, D. (1954). *How to Lie With Statistics*. Gollancz, London, pp. 11–26. (An amusing and instructive account of "samples with a built-in bias," not limited to quota sampling.)

Stephan, F. F. and McCarthy, P. J. (1958). *Sampling Opinions*. Wiley, New York. (Several relevant passages can be found using the index.)

Stuart, A. (1962). *Basic Ideas of Scientific Sampling*. Griffin's Statist. Monogr. Courses, No. 4, London, pp.

10–11. (Explains some of the dangers of nonrandom sampling.)

Sudman, S. (1966). *J. Amer. Statist. Ass.*, **61**, 749–771.

Sukhatme, P. V. (1954). *Sampling Theory of Surveys with Applications*. Indian Society of Agricultural Statistics, New Delhi/Iowa State University Press, Ames, Iowa, p. 10.

Tull, D. S. and Albaum, G. S. (1973). *Survey Research, A Decisional Approach*. Intext Educational Publications, New York, pp. 38–39.

(PROPORTIONAL ALLOCATION
SIMPLE RANDOM SAMPLING
SURVEY SAMPLING)

K. R. W. BREWER

QUOTIENT METHOD

Let m be a smooth function of time, that is, a function which can be eliminated or greatly reduced by forming finite differences*. If an observation $x_t = m_t + u_t$ (u_t a random function), we use the variate difference method; if $x_t = m_t + m_t u_t$, the quotient method is appropriate.

The quotient method is a variant of the variate difference method* and has been developed by Strecker [6, 7, 9]. The model of the variate difference method is additive; here the model is multiplicative.

Let $x_t > 0$ ($t = 1, 2, 3, \ldots, N$) be the observations of a time series*; m_t is a smooth systematic part, u_t a random variable, and

$$x_t = m_t + m_t u_t = m_t(1 + u_t).$$

Taking logarithms, we have

$$\log x_t = \log m_t + \log(1 + u_t) \simeq \log m_t + u_t.$$

The approximation holds only for small values of u_t.

We define differences recursively:

$$\Delta \log x_t = \log x_{t+1} - \log x_t,$$

$$\Delta^k \log x_t = \Delta(\Delta^{k-1} \log x_t).$$

The taking of differences is a linear operation. We have

$$\Delta^k \log x_t = \Delta^k \log m_t + \Delta^k \log(1 + u_t)$$

$$\simeq \Delta^k \log m_t + \Delta^k u_t.$$

Now $\log m_t$ may be approximated by a polynomial of order $K = k_0 - 1$, say (recall that m_t is a smooth function). The properly reduced variances of the observations of order $k_0, k_0 + 1, k_0 + 2, \ldots,$ must be equal apart from random fluctuations.

The variances of the difference series are computed as follows. For the original observations,

$$V_0 = \sum_{t=1}^{N} (\log x_t - \overline{\log x})^2 / (N - 1),$$

$$\overline{\log x} = \sum_{t=1}^{N} \log x_t / N,$$

and for the difference series,

$$V_k = \sum_{t=1}^{N-k} (\Delta^k \log x_t)^2 / [(N - k)_{2k}C_k],$$

where ${}_{2k}C_k$ denotes the binomial coefficient $\binom{2k}{k}$. Using the formulae given in Tintner [8, p. 51ff] we get tests for the differences $V_{k+1} - V_k$. These yield a large-sample test for the estimation of k_0.

A more general test is due to T. W. Anderson [1]. We test

$$(V_r - V_q)(N - q)/V_q.$$

The quantity

$$\sqrt{\frac{N-q}{2\left[\dfrac{\binom{4r}{r}}{\binom{2r}{r}^2} + \dfrac{\binom{4q}{2q}}{\binom{2q}{q}^2} - 2\dfrac{\binom{2q+2r}{q+r}}{\binom{2q}{q}\binom{2r}{r}}\right]}}$$

$$\times (V_r - V_q)/V_q$$

has for large samples a normal distribution with mean zero and variance 1.

For a small-sample test based on a circular distribution, see Rao and Tintner [4]. For a consideration of the multiple-choice problem, see Rao and Tintner [5]. See also Tintner et al. [9].

By testing successively $V_0 - V_1, V_1 - V_2, \ldots,$ we estimate k_0. We may then use V_{k_0} as an approximation to $\sigma^2 = Eu_t^2$. $K = k_0 - 1$ is the degree of a polynomial which approximately represents $\log m_t$. Now let

$k_0 = 2n$ for k_0 even and $k_0 = 2n + 1$ for k_0 odd. A smoothing formula that minimizes

$$E\{\log(1 + u_t) + \log f_t\}^2$$

results as follows:

$$\log f_t = b_1 \Delta^{2n}(u_{t-n}) + b_2 \Delta^{2n+2}(u_{t-n-1}) + \cdots,$$

where the b_i are constants. Hence

$$\log m_t \approx \log m_t' = \log x_t + \log f_t,$$

$$\log m_t' = \sum_{s=1}^{m} g_{mn}(s)(\log x_{t+s} + \log x_{t-s}) + g_{mn}(0) \log x_t .$$

g_{mn} are optimum weights; these are tabulated [8, pp. 100 ff.]. n depends on the degree of the polynomial and m represents the desired accuracy.

Instead of fitting a polynomial to the whole series of observations, we can use moving averages*, which are more flexible; see Fuller [9]. Strecker [7, 9] suggests a nonparametric test for the choice between the additive variate difference method and the multiplicative quotient method. A more general model is the Box and Cox [2] transformation*; see Zarembka [10]. Transform the variable x as follows:

$$x^{(\lambda)} = (x^{\lambda} - 1)/\lambda,$$

so that $\lim_{\lambda \to 0} x^{(\lambda)} = \log_e x$. Then $\lambda = 0$ corresponds to the quotient method and $\lambda = 1$ to the variate difference method.

References

[1] Anderson, T. W. (1971). *Time Series Analysis*. Wiley, New York.

[2] Box, G. E. P. and Cox, D. R. (1964). *J. R. Statist. Soc. B*, **26**, 211–252.

[3] Fuller, W. A. (1976). *Introduction to Statistical Time Series*. Wiley, New York.

[4] Rao, J. N. K. and Tintner, G. (1962). *Sankhyā A*, **24**, 385–394.

[5] Rao, J. N. K. and Tintner, G. (1963). *Aust. J. Statist.*, **5**, 106–116.

[6] Strecker, H. (1949). *Mitteilungsbl. Math. Statist.*, **1**, 115–130.

[7] Strecker, H. (1970). *Metrika*, **16**, 130–187; **17**, 257–259 (1971).

[8] Tintner, G. (1940). *Variate Difference Method*. Principia Press, Bloomington, Ind.

[9] Tintner, G., Rao, J. N. K., Strecker, H. (1978). *New Results in the Variate Difference Method*. Vandenhoeck and Ruprecht, Göttingen, Federal Republic of Germany.

[10] Zarembka, P. (1974). In *Frontiers of Econometrics*, P. Zarembka, ed. Academic Press, New York, p. 81.

(MOVING AVERAGES
TIME SERIES
VARIATE DIFFERENCE METHOD)

GERHARD TINTNER
HEINRICH STRECKER

R

RADEMACHER FUNCTIONS

Let $x = \sum_{m=1}^{\infty} e_m/2^m$ be the dyadic expansion of the real number x, $0 \leqslant x \leqslant 1$, where $e_m = e_m(x)$ is either 0 or 1. If we set $r_m(x) = 1 - 2e_m(x)$ $(m = 1, 2, 3, \ldots)$, then the $\{r_m(x)\}$ form a system of functions, orthonormal over the interval $(0, 1)$; these are the *Rademacher functions*, as usually defined at present. It is easy to visualize these functions (see Fig. 1).

The only values taken by the functions $r_m(x)$ are $+1$ and -1; occasionally, one also sets $r_m(x) = 0$, if x is a dyadic rational, where the $r_m(x)$ are discontinuous. One observes that if one "rounds off" the corners of the graph, then $r_1(x)$ resembles $\sin 2\pi x$, $r_2(x)$ behaves like $\sin 4\pi x$, and generally, $r_m(x)$ mimics the behavior of $\sin(2\pi \cdot 2^{m-1}x)$; the precise relation between these functions is $r_m(x) = \operatorname{sgn} \sin 2^m \pi x$.

These functions were defined by Hans Rademacher in Part VI of his paper [5] on series of general orthogonal functions. For the sake of comparison, one should observe that Rademacher's $\psi_m(x)$ are actually our $-r_m(x)$. The change of sign made here conforms to contemporary usage and, although it is only a trivial modification, it makes the system more convenient.

Rademacher introduced these functions to prove that his two estimates for the Lebesgue functions $L_n(x)$ of general orthonormal systems, namely $L_n(x) = O(n^{1/2}(\log n)^{3/2+\epsilon})$ unconditionally and $L_n(x) = L_n = O(n^{1/2})$ if the $L_n(x)$ are independent of x, are essentially the best possible. Indeed, Rademacher showed that for the $\{r_m(x)\}$, just as for the orthonormal system $\{\cos 2\pi mx, \sin 2\pi mx\}$ of trigonometric functions, the Lebesgue functions $L_n(x)$ are independent of x, so that they reduce to Lebesgue constants L_n.

Parenthetically, besides the mentioned systems of trigonometric and of Rademacher functions, it seems that only one other system of orthogonal functions is known for which the Lebesgue functions reduce to constants, namely that of the Haar functions. In the case of the $\{r_m(x)\}$, Rademacher showed that for $n \to \infty$, the Lebesgue constants satisfy the asymptotic equality $L_n \simeq (2n/\pi)^{1/2}$; this proves, in particular, that for *arbitrary* orthogonal systems, $L_n = O(n^{1/2})$ is the best possible. For *specific* systems one may, of course, do better; indeed, for the aforementioned system of Haar functions, one has $L_n(x) = 1$, and $L_n(x)$ is independent not only of x, but also of n.

The original reason for the consideration

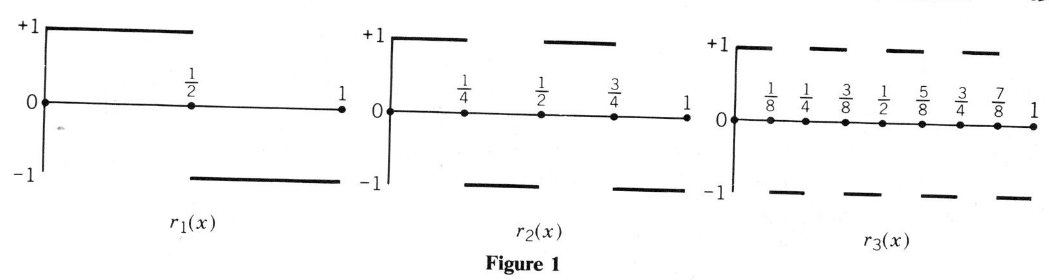

Figure 1

of the system $\{r_m(x)\}$ had been the construction of a counterexample, and since then the system has proven useful for the construction of many other counterexamples, but the interest and general relevance of the Rademacher functions far outweigh that particular use.

As pointed out in ref. 5, the system $\{r_m(x)\}$ is not complete; indeed, the functions $f_{mk}(x) = r_m(x)r_k(x)$, $m \neq k$, are not identically zero but are orthogonal over $(0, 1)$ to all $r_j(x)$. The system is, however, easily completed; the resulting complete system is known as the system of *Walsh functions* $\{w_m(x)\}$.

The system $\{r_m(x)\}$ has many remarkable properties, some of which are shared by other lacunary systems [see ref. 4 (Chap. VII)]; these are orthonormal systems $\{\phi_m(x)\}$ such that if $\sum_{m=1}^{\infty} c_m^2$ converges and if the series $\sum_{m=1}^{\infty} c_m\phi_m(x)$ converges (strongly) in L^2, then it represents a function $f(x) \in L^p$ for some $p > 2$, the order of lacunarity. The Rademacher system $\{r_m(x)\}$ is lacunary to all orders p, $1 \leqslant p < \infty$.

Among its properties are the following:

1. Let c_m $(m = 1, 2, \ldots)$ be real numbers and denote the formal series $\sum_{m=1}^{\infty} c_m r_m(x)$ by $S(x)$, regardless of convergence. If $\sum_{m=1}^{\infty} c_m^2$ converges, then $S(x)$ converges a.e. (almost everywhere) on $(0, 1)$ and represents there a function $f(x)$. The series $S(x)$ is called the R-F (Rademacher–Fourier) series of $f(x)$ and the c_m are the R-F coefficients of $f(x)$. One recalls that the same condition suffices to ensure the convergence a.e. of Haar function series, while for arbitrary orthonormal systems one needs the stronger [best possible; see ref. 4 (p. 162)] condition $\sum_{m=1}^{\infty} c_m^2 \log^2 m < \infty$ for convergence a.e. If, on the other hand, $\sum_{m=1}^{\infty} c_m^2$ diverges, then $S(x)$ diverges a.e.
2. The condition $\sum_{m=1}^{\infty} |c_m| < \infty$ is necessary and sufficient for the convergence of $\sum_{m=1}^{\infty} c_m r_m(x)$ everywhere on $(0, 1)$.
3. If $\sum_{m=1}^{\infty} |c_m|$ diverges, while $\lim_{m\to\infty} c_m = 0$, then, given reals $a \leqslant b$ (not necessarily finite), each subinterval of $(0, 1)$ contains a set $\{\xi\}$ of the power of the continuum such that the partial sums $s_n(\xi)$ of $S(\xi)$ satisfy $\liminf_{n\to\infty} s_n(\xi) = a$, $\limsup_{n\to\infty} s_n(x) = b$.
4. If $f(x) \in L(0, 1)$ and also $f(x) \in L^p(0, 1)$ and if we denote the p-norm of $f(x)$ by $\|f\|_p = \{\int_0^1 |f(x)|^p\, dx\}^{1/p}$, then, if $\{c_m\}$ are the R-F coefficients of $f(x)$, the two inequalities

$$\frac{2p-2}{3p-2} \sum_{m=1}^{\infty} c_m^2 \leqslant \|f(x)\|_p^2 \leqslant \left(1 + \frac{p}{2}\right) \sum_{m=1}^{\infty} c_m^2$$

hold. From the first inequality follows the convergence of $\sum_{m=1}^{\infty} c_m^2$ if $f(x) \in L^p$ for some $p > 1$, and then the R-F series of $f(x)$ converges at least a.e. The second inequality shows that the convergence of $\sum_{m=1}^{\infty} c_m^2$ implies that $f \in L^p$ for $1 \leqslant p < \infty$.

In analogy with the Riesz–Fischer theorem* we also have:

5. (a) If $\sum_{m=1}^{\infty} c_m^2$ converges, then there exists a continuous function $f(x)$ that has the R-F coefficients c_m.
   (b) If $\lim_{m\to\infty} c_m = 0$, then there exists

a Lebesgue integrable function $f(x)$, with the R-F coefficients c_m.

(c) In either case, $c_m = \int_0^1 f(x) r_m(x)\,dx$.

To be able to state some converse results, because of the incompleteness of the system $\{r_m(x)\}$ we also need the following concept. We say that $f(x)$ *belongs* to the (generally incomplete) orthogonal set $\{\phi_m(x)\}$ if $f(x)$ is orthogonal to every function $g(x)$ that is itself orthogonal to all functions $\phi_m(x)$.

6. (a) If $f(x)$ belongs to $\{r_m(x)\}$ and $f(x) = \sum_{m=1}^{\infty} c_m r_m(x)$ formally, then $\sum_{m=1}^{\infty} c_m^2$ converges and $f(x)$ is represented by its R-F series at least a.e.

(b) If, furthermore, $f(x)$ is bounded and measurable, then $\sum_{m=1}^{\infty} |c_m|$ converges and [see (2)] $f(x)$ is represented by its R-F series everywhere on $(0, 1)$.

These results are due to many authors, among them, in addition to Rademacher, Steinhaus, Kaczmarz, Paley, Zygmund, Kolmogorov, and Menchov; for an attribution of discoveries, see refs. 3 and 4.

The completion $\{w_m(x)\}$ of the system $\{r_m(x)\}$ was discovered independently by Rademacher [6], by Kaczmarz [3], and by Walsh [7] and earlier was sometimes referred to as the *Rademacher–Kaczmarz* or *Walsh–Kaczmarz system*. For details on the history of the contributions to the theory of the Rademacher and Walsh functions, see ref. 1.

The importance of the system $\{r_m(x)\}$ of Rademacher functions for statistics comes from the fact that it may be considered as a model for the tossing of a fair coin (i.e., for a random variable* $X(x)$ with $\Pr[X(x) = +1] = \Pr[X(x) = -1] = \frac{1}{2}$). To see that, let us consider the set $E_j \subset [0, 1]$ on which $r_j(x) = \delta_j$, where $\delta_j =$ either $+1$ or -1. From the definition of the $r_j(x)$ (see also Fig. 1), it follows that its measure $\mu(E_j) = \frac{1}{2}$. More generally, if for $j = 1, 2, 3, \ldots$ we denote by E_j the set on which $r_j(x) = \delta_j$, where the sequence $\delta_1, \delta_2, \ldots$ consists only of $+1$ or -1, then for $j \neq k$, $\mu(E_j \cap E_k) = \mu(r_j(x) = \delta_j$ and $r_k(x) = \delta_k) = 2^{-2}$ and

$$\mu\left(\bigcap_{j=1}^{m} E_j\right) = \mu(r_1(x) = \delta_1, r_2(x) = \delta_2, \ldots, r_m(x) = \delta_m)$$
$$= 2^{-m} = \prod_{j=1}^{m} \mu(E_j)$$
$$= \prod_{j=1}^{m} \mu(r_j(x) = \delta_j). \qquad (1)$$

If, on the other hand, we interpret the event E_j as the outcome of the jth toss of a fair coin, then the probability of m consecutive tosses with preassigned outcomes is 2^{-m} and, from the statistical independence* of the tosses it follows that

$$\Pr\left[\bigcap_{j=1}^{m} E_j\right] = 2^{-m} = \prod_{j=1}^{m} \Pr[E_j]. \qquad (2)$$

Comparison of (1) and (2) justifies our previous claim that the system $\{r_m(x)\}$ may be interpreted as a sample space for the tossing of a fair coin, that is, for a random variable $X(x)$, that may take only one of the two values $+1$ or -1, each with equal probability $\frac{1}{2}$.

Using the $\{r_m(x)\}$, one can express with ease many probabilistic concepts. So, for example,

$$\Pr[X = +1 \text{ exactly } k \text{ times in a sample of } n \text{ values}] = \mu\left\{\sum_{m=1}^{n} r_m(x) = 2k - n\right\}.$$

Next, the weak law of large numbers* for X is equivalent to the statement that for every $\epsilon > 0$,

$$\lim_{n\to\infty} \mu\left\{\left|\sum_{m=1}^{n} r_m(x)\right| > \epsilon n\right\} = 0.$$

Similarly, the normal* (Gauss) law may be formulated as follows:

$$\lim_{n\to\infty} \mu\left\{a < n^{-1/2} \sum_{m=1}^{n} r_m(x) < b\right\} = (2\pi)^{-1} \int_a^b e^{-z^2/2}\,dz.$$

As a last example, let us consider the random variable $Z_n = X_1 + \cdots + X_n$, where for $1 \leqslant m \leqslant n$, each $X_m = X_m(x)$ is a random variable with $\Pr[X_m = +1] = \Pr[X_m = -1] = \frac{1}{2}$; also set $Z = \lim_{n\to\infty} Z_n$. Then the expected value*

$$E(Z_n) = \int_0^1 \left(\sum_{m=1}^{n} r_m(x) \right) dx$$

$$= \sum_{m=1}^{n} \int_0^1 r_m(x)\, dx = 0,$$

whence $E(Z) = \lim_{n\to\infty} E(Z_n) = 0$. It is convenient to remark that one cannot write directly $E(Z) = \int_0^1 (\sum_{m=1}^{\infty} r_m(x))\, dx$, because the series diverges.

There exist many other important and fascinating applications of the Rademacher functions in analysis (e.g., the proof of Vieta's trigonometric identity), in number theory (e.g., the Borel normality* of almost all real numbers), in probability theory*, and so on. Many of these topics are beautifully discussed in the book [2] by M. Kac.

References

[1] Grosswald, E. (1981). Analytic Number Theory, *Lect. Notes Math.*, **899**, Springer-Verlag, New York, pp. 1–9.

[2] Kac, M. (1959). *Statistical Independence in Probability, Analysis and Number Theory*. Carus Math. Monogr. 12, Math. Ass. Amer.; distributed by Wiley, New York. (A beautifully written, most pleasant book to read.)

[3] Kaczmarz, S. (1929). *C. R. I-er Congr. Math. Pays Slaves*, Warsaw, pp. 189–192.

[4] Kaczmarz, S. and Steinhaus, H. (1935). *Theorie der Orthogonalreihen*. Monogr. Mat. VI. Warsaw. (Reprinted, Chelsea, New York, 1951.) (A very clearly written, comprehensive presentation.)

[5] Rademacher, H. (1922). *Math. Ann.*, **87**, 112–138.

[6] Rademacher, H. (1922). Unpublished. (Long-lost manuscript, found posthumously and dated by its author January 1922.)

[7] Walsh, J. (1923). *Amer. J. Math.*, **55**, 5–24.

(ORTHOGONAL EXPANSIONS
RIESZ–FISCHER THEOREM)

EMIL GROSSWALD

RADIAL ERROR

If X and Y are two random variables (RVs) measured along Cartesian coordinate axes, the distribution of the radial distance $R = (X^2 + Y^2)^{1/2}$ has numerous applications. For example, if (X, Y) is the impact point of a missile aimed at a target at the origin, the miss distance is called the *radial error*. If the lethal radius of a missile is r, the probability of destroying a target is $P[R \leqslant r] = F(r)$, the cumulative distribution function (CDF) of R. The median of R, obtained by solving $F(r) = 0.5$, is called the *circular probable error**.

The RV R is clearly the radial distance of a point undergoing a random walk* [12]. Other applications of R are in the fields of signal detection [10], meteorology* [2], and astronomy [14], among others.

Assuming X and Y to be normally and independently distributed with zero means and standard deviations σ_X and σ_Y such that $c = (\sigma_X/\sigma_Y) \leqslant 1$, Chew and Boyce [3] give the following probability density function (PDF) and CDF of R.

$$f(r) = \{r/(\sigma_X\sigma_Y)\}\exp(-ar^2)I_0(br^2), \qquad 0 \leqslant r < \infty \quad (1)$$

$$F(r) = \frac{1}{2a\sigma_X\sigma_Y} \sum_{n=0}^{\infty} \left(\frac{b}{2a}\right)^{2n} \left(\frac{1}{n!}\right)^2 \gamma(2n+1, ar^2), \quad (2)$$

$a = (1 + c^2)/(4\sigma_X^2)$, $b = (1 - c^2)/(4\sigma_X^2)$; also $I_0(br^2) = \sum_{j=0}^{\infty} (j!)^{-2}(\frac{1}{2}br^2)^{2j}$ is the modified Bessel function* of the first kind and zero order, and $\gamma(a, x) = \int_0^x e^{-t}t^{a-1}\, dt$ is the incomplete gamma function*. Values of the mean, median, mode, and standard deviation of R are tabulated in Chew and Boyce [3] for $c = 0.0(0.1)1.0$. Tables of $F(r)$ by Harter [9] have been extended by Lowe [11] and DiDonato and Jarnagin [4] to the case where the means of X and Y are not zero.

Without explicitly using its PDF, Scheuer [13] gives the nth moment of R in the case where X and Y have zero means but nonzero covariance σ_{XY}.

$$E(R^n) = (2\sigma_1^2)^{(1/2)n}\Gamma(\tfrac{1}{2}n + 1) \times F(-\tfrac{1}{2}n, \tfrac{1}{2}; 1; k^2) \qquad (3)$$

where

$$\sigma_1^2 = \tfrac{1}{2}\left\{(\sigma_X^2 + \sigma_Y^2) + \left[(\sigma_X^2 - \sigma_Y^2) + 4\sigma_{XY}^2\right]^{1/2}\right\},$$

$$\sigma_2^2 = \tfrac{1}{2}\left\{(\sigma_X^2 + \sigma_Y^2) - \left[(\sigma_X^2 - \sigma_Y^2) + 4\sigma_{XY}^2\right]^{1/2}\right\},$$

$$k^2 = (\sigma_1^2 - \sigma_2^2)/\sigma_1^2, \qquad \Gamma(a) = \gamma(a, \infty)$$

$$F(a, b; c; z) = \frac{\Gamma(c)}{\Gamma(a)\Gamma(b)} \sum_{n=0}^{\infty} \frac{\Gamma(a+n)\Gamma(b+n)}{\Gamma(c+n)n!} z^n$$

is the hypergeometric function [1, p. 556]. Explicit formulas for the first three moments and a recurrence formula are also given.

Edmundson [7] gives the following PDF and CDF of R in n dimensions, assuming independence, zero means, and common variance σ^2.

$$f(r) = \left[\Gamma(\tfrac{1}{2}n)2^{(n-2)/2}\sigma^n\right]^{-1} r^{n-1} e^{-r^2/(2\sigma^2)}, \qquad (4)$$

$$F(r) = 2P\left(\tfrac{1}{2}n, (2\sigma^2)^{-1} r^2\right), \qquad (5)$$

where $P(a, x) = \gamma(a, x)/\Gamma(a)$ is the incomplete gamma function ratio [1, p. 260]. The cases $n = 2$ and 3 are called the Rayleigh* and Maxwell distributions*, respectively. The former is sometimes incorrectly referred to as the circular normal distribution*.

Surveys of more general so-called target coverage* problems are given in Guenther and Terragno [8] and Eckler [6]. $F(r)$ is the probability of a missile destroying a target only if the damage function $d(R)$ is a "cookie-cutter" function [i.e., $d(R) = 1$ for $R \leqslant r$ and zero for $R > r$]. More realistically, $d(R)$ is a continuously decreasing function of R.

The estimation problem is considered in Eckler [6] and, in the case of truncated observations, in Dyer [5].

References

[1] Abramowitz, M. and Stegun, I. A., eds. (1972). *Handbook of Mathematical Functions with Formulas, Graphs, and Mathematical Table*. Wiley, New York.

[2] Brooks, C. E. P. and Carruthers, N. (1953). *Handbook of Statistical Methods in Meteorology*. H. M. Stationery Office, London.

[3] Chew, V. and Boyce, R. (1962). *Technometrics*, **4**, 138–140.

[4] DiDonato, A. R. and Jarnagin, M. P. (1962). *Math. Comp.*, **16**, 347–355.

[5] Dyer, D. D. (1974). *Operat. Res.*, **22**, 197–201.

[6] Eckler, A. R. (1969). *Technometrics*, **11**, 561–589. (Contains 60 references.)

[7] Edmundson, H. P. (1961). *Operat. Res.*, **9**, 8–21.

[8] Guenther, W. C. and Terragno, P. J. (1964). *Ann. Math. Statist.*, **35**, 232–260. (Contains 58 references.)

[9] Harter, H. L. (1960). *J. Amer. Statist. Ass.*, **55**, 723–731.

[10] Helstrom, C. W. (1968). *Statistical Theory of Signal Detection*, 2nd ed. Pergamon Press, Elmsford, N.Y.

[11] Lowe, J. R. (1960). *J. R. Statist. Soc. B*, **22**, 177–187.

[12] Papoulis, A. (1965). *Probability, Random Variables, and Stochastic Processes*. McGraw-Hill, New York.

[13] Scheuer, E. M. (1962). *J. Amer. Statist. Ass.*, **57**, 187–190; corrigenda: **60**, 1251 (1965).

[14] Trumpler, R. J. and Weaver, H. F. (1962). *Statistical Astronomy*. Dover, New York.

(BIVARIATE NORMAL DISTRIBUTION
CIRCULAR NORMAL DISTRIBUTION
TARGET COVERAGE)

V. CHEW

RADICO-NORMAL DISTRIBUTION

A distribution with cumulative distribution function (CDF)

$$F_X(x) = \frac{3}{\sigma^2\sqrt{2\pi}} \int_{-\infty}^{x} \exp\left[-\tfrac{1}{2} y^2 (t\sigma^2)^{-1}\right] dt\, dy.$$

The distribution is symmetrical about zero; its variance is $3\sigma^2/5$. It belongs to a class of modified normal distributions* constructed by Romanowskiĭ [1, 2].

References

[1] Romanowskiĭ, M. (1964). *Bull. Géod.*, **73**, 195–216.

[2] Romanowskiĭ, M. (1965). *Metrologia*, **4**(2), 84–86.

(EQUINORMAL DISTRIBUTION
LINEO-NORMAL DISTRIBUTION
MODIFIED NORMAL DISTRIBUTIONS
QUADRI-NORMAL DISTRIBUTION)

RADON–NIKODYM THEOREM

THE THEOREM

A precise statement of the theorem is as follows: Let $(X, \mathscr{S}, \mu)$ be a σ-finite measure space and ν a σ-finite, possibly signed measure on $(X, \mathscr{S})$. If ν is absolutely continuous with respect to μ, then there is a finite-valued, measurable function f uniquely determined almost everywhere (μ) such that

$$\int_X g\,d\nu = \int_X gf\,d\mu$$

whenever either integral is well defined, and in particular for every E in $\mathscr{S}$ we have $\nu(E) = \int_E f\,d\mu$. The function f is called the *Radon–Nikodym derivative* of ν with respect to μ and is denoted by $f = d\nu/d\mu$.

Thus the Radon–Nikodym theorem enables us to express integration with respect to a measure ν which is absolutely continuous to μ, in terms of integration with respect to μ via their Radon–Nikodym derivative. Its proof can be found in any book on measure theory or real analysis. Like most "derivatives," the Radon–Nikodym derivative can be expressed at almost every point as the limit of the ratios of the two measures over sets that contain the point and belong to a refining sequence of partitions (see Hewitt and Stromberg [2, p. 373]). The theorem is not necessarily valid for measures that are not σ-finite (see Zaanen [3, Sec. 33]).

ν is *absolutely continuous* with respect to μ if all sets of μ-measure zero have also ν-measure zero. In contrast to absolute continuity* is the notion of singularity: μ and ν are singular if they live on disjoint sets [i.e., for some E in $\mathscr{S}$, $\mu(E) = 0$ and $\nu(E^c) = 0$, or $|\nu|(E^c) = 0$ in the signed case]. The *Lebesgue decomposition theorem* says that ν can always be decomposed into two terms, ν_a and ν_s ($\nu = \nu_a + \nu_s$), of which one is absolutely continuous and the other singular to μ, so that

$$\int_X g\,d\nu = \int_X gf\,d\nu + \int_{E_0} g\,d\nu,$$

where $f = d\nu_a/d\mu$ and $\mu(E_0) = 0$, and in particular $\nu(E) = \int_E f\,d\mu + \nu(E \cap E_0)$.

The Radon–Nikodym theorem has applications to analysis, probability, and statistics.

APPLICATIONS

Probability Density Functions

Absolutely continuous distribution functions $F(x)$ have densities $f(x)$: $F(x) = \int_{-\infty}^{x} f(t)\,dt$ and $F'(x) = f(x)$ for almost every x. Similarly, each absolutely continuous function is the indefinite integral of its derivative.

Change of Variable of Integration

The real variable of integration can be changed by an absolutely continuous function h as follows:

$$\int_{h(a)}^{h(b)} g(x)\,dx = \int_a^b g(h(t))h'(t)\,dt,$$

whenever either integral is well defined.

Local Times

The local time* of a function or of a random process or field plays an important role in the study of its smoothness properties (for a comprehensive survey, see Geman and Horowitz [1]). It is well defined as the Radon–Nikodym derivative of its occupation measure as follows. Let f be a (Borel measurable) function defined on the interval I, and ν the occupation measure of f defined for every Borel subset B of the real line as the "time" the function f spends at values in B [i.e., $\nu(B) = \mu\{x \in I : f(x) \in B\}$, where μ is Lebesgue measure]. If the occupation measure ν is absolutely continuous with respect to Lebesgue measure, its Radon–Nikodym derivative $\alpha(x) = (d\nu/d\mu)(x)$ is called the *local time* or *occupation density* of the function f and carries the interpretation that $\alpha(x)$ is the time spent by the function f at the level x.

Conditional Probability and Expectation*

Conditional probabilities and expectations are defined in the general (nondiscrete) case as Radon–Nikodym derivatives. To be specific, let us fix two random variables X and Y with $\mathscr{E}|X| < \infty$ and define $\mathscr{E}(X \mid Y = y)$. Notice that when Y is continuous, $P(Y = y) = 0$ for all y, and the conditional expectation cannot be defined in the elementary fashion. It can be defined though, at least for almost every value y of Y, by means of the Radon–Nikodym theorem via the relationship

$$\mathscr{E}[X 1_B(Y)] = \int_B \mathscr{E}[X \mid Y = y]\, dF_Y(y)$$

for all Borel sets B. Indeed, the left-hand side defines a finite signed measure on the Borel sets, which is absolutely continuous with respect to the distribution F_Y of Y, and thus by the Radon–Nikodym theorem there is a function of y, defined uniquely for almost every y (F_Y) and denoted by $\mathscr{E}[X \mid Y = y]$, which satisfies the foregoing relationship.

Likelihood Ratios*

The Neyman–Pearson* test of two hypotheses is based on the likelihood ratio, which is closely related to Radon–Nikodym derivatives. Let P_0 and P_1 be the probability measures associated with the two hypotheses. The likelihood ratio $L = (dP_1/d\mu)/(dP_0/d\mu)$ is the ratio of the Radon–Nikodym derivatives, where μ is a probability measure with respect to which both P_0 and P_1 are absolutely continuous [such as $(P_0 + P_1)/2$]. An alternative expression is via the Lebesgue decomposition of P_1 with respect to P_0: $L = \infty$ on the (P_0-null) event where the singular part of P_1 lives, and outside this event $L = dP_{1a}/dP_0$, the Radon–Nikodym derivative of the absolutely continuous part of P_1. If a finite sample is available to the statistician, the probabilities P_0 and P_1 are defined on a finite-dimensional Euclidean space and typically μ is Lebesgue or counting measure and L the ratio of two probability densities. If a continuous sample (such as the sample of a time series over an interval), is available to the statistician, the probabilities P_0 and P_1 are defined on infinite-dimensional function spaces, typically no natural dominating measure μ exists, and one works with the alternative expression via the Lebesgue decomposition; in such cases L is quite difficult to compute and has been calculated only in few cases, such as when under both hypotheses the observed continuous random process is Gaussian, or Poisson, or more generally infinitely divisible.

References

[1] Geman, D. and Horowitz, J. (1980). *Ann. Prob.*, **8**, 1–67.

[2] Hewitt, E. and Stromberg, K. (1965). *Real and Abstract Analysis*. Springer-Verlag, New York.

[3] Zaanen, A. C. (1967). *Integration*. North-Holland, Amsterdam.

(ABSOLUTE CONTINUITY
MEASURE THEORY IN PROBABILITY
AND STATISTICS)

S. CAMBANIS

RADON TRANSFORMATION

The Radon transform of a two-variable function $f(x, y)$ is defined as

$$R(f) = p(\mu \mid \theta) = \int_{-\infty}^{\infty} f(u\cos\theta - v\sin\theta,\; u\sin\theta + v\cos\theta)\, dv.$$

If $f(x, y)$ is the probability density function (PDF) of random variables X, Y then $p(\mu \mid \theta)$ is the PDF of $U = X\cos\theta + Y\sin\theta$, that is, the projection of the point (X, Y) on the line $Y/X = \tan\theta$.

The transform is useful in the analysis of spatial data* and signal processing. Reference 1 contains a discussion of estimation problems, and a bibliography.

Reference

[1] Pattey, P. A. and Lindgren, A. G. (1981). *IEEE Trans. Acoust. Speech Signal Process.*, **29**, 994–1002.

(SPATIAL DATA ANALYSIS
SPATIAL PROCESS)

RAIKOV'S THEOREM

Raikov's theorem is concerned with the following factorization problem of the Poisson* characteristic function*

$$\phi(t) = \exp\left[\lambda(e^{it} - 1)\right].$$

Assume that $\phi_1(t)$ and $\phi_2(t)$ are two characteristic functions satisfying

$$\phi(t) = \phi_1(t)\phi_2(t). \qquad (1)$$

What can be said about the two factors $\phi_1(t)$ and $\phi_2(t)$?

One immediately finds the following two solutions of (1):

$$\text{(a)} \quad \phi_1(t) = e^{ict}, \qquad \phi_2(t) = \phi(t)e^{-ict}, \qquad c \text{ real};$$

and

$$\text{(b)} \quad \phi_1(t) = \exp\left[\lambda_1(e^{it} - 1) + ict\right], \qquad \lambda_1 > 0, \quad c \text{ real},$$

$$\phi_2(t) = \exp\left[\lambda_2(e^{it} - 1) - ict\right], \qquad \lambda_2 > 0,$$

with $\lambda_1 + \lambda_2 = \lambda$.

Now, Raikov's theorem states that (a) and (b) represent *all* solutions of (1). It thus implies that if a Poisson random variable X can be decomposed as $X = Y + Z$, where $Y \geqslant 0$ and $Z \geqslant 0$ are independent random variables (neither of which is identically zero), then both Y and Z are Poisson [note that the assumption $Y \geqslant 0$, $Z \geqslant 0$ implies that $c = 0$ in both (a) and (b)].

Raikov's theorem is of interest in its own right. It also has implications in applied fields, such as superpositions of independent renewal* processes, and decompositions (e.g., by rarefactions) of Poisson processes*, which in turn play important roles in reliability theory*, telecommunications, nuclear physics, and other fields.

Raikov's original work appeared in [3]. Lukács [2] discusses in detail the problem of factorization of characteristic functions. Some basic results on renewal processes, in which Raikov's theorem plays an implicit role, can be found in Galambos and Kotz [1, Chap. 4].

References

[1] Galambos, J. and Kotz, S. (1978). *Characterizations of Probability Distributions*. Springer-Verlag, Berlin.

[2] Lukács, E. (1970). *Characteristic Functions*. Charles Griffin, London.

[3] Raikov, D. A. (1938). *Izv. Akad. Nauk SSSR, Ser. Mat.*, **2**, 91–124.

(CHARACTERISTIC FUNCTIONS
POISSON PROCESSES)

JANOS GALAMBOS

RAMSEY'S PRIOR

An experimenter attempting to estimate a potency curve by running a quantal-response* experiment may be unwilling to specify the functional form for the curve or be unable to run more than one experimental unit per dose level. In this situation Ramsey [6] proposed looking at the potency curve as a cumulative distribution function (CDF) and then employing a Bayesian nonparametric prior. The prior* he chose is essentially the Dirichlet process prior of Ferguson [4]. The posterior distribution on the potency curve is a mixture* of Dirichlet processes which Antoniak [1] showed to be computationally intractable. However, Ramsey obtained the mode of the posterior*, which he then used as his estimate of the potency curve. His estimator is a smoothed version of the isotonic regression* estimator.

Using his estimate of the potency curve, Ramsey showed that one may estimate any effective dose. Disch [3] derived the prior and posterior distributions of any effective dose assuming Ramsey's prior on the potency curve. The posterior is difficult to calculate exactly, so two approximations were developed.

Ramsey also addressed the question of optimal experimental design if his prior is to be used in the analysis of the data. He showed by example that an optimal design seems to assign one experimental unit per dose level. Kuo [5] has done a theoretical study tending to corroborate this choice of design, although for a different optimality criterion than Ramsey used. She also showed that for her special case (the Dirichlet process* prior had parameter uniform over [0, 1]), the optimal design did not pick the design doses uniformly spaced but shifted somewhat toward the prior estimate of the ED50.

Ramsey's model and prior will now be described. A dose, x, is administered to an experimental unit which then either responds with probability $P(x)$ or fails to respond. The function P is the potency curve and is assumed to be an increasing function. Observational doses $x_1, x_2, \ldots, x_M$ are chosen, and n_i experimental units are assigned to dose x_i with the result that s_i units independently respond, $i = 1, 2, \ldots, M$. The joint likelihood for $P(x_1), P(x_2), \ldots, P(x_m)$ is proportional to

$$\prod_{i=1}^{M} \left[P(x_i) \right]^{s_i} \left[1 - P(x_i) \right]^{n_i - s_i}. \qquad (1)$$

Ramsey viewed P as a CDF and essentially assigned a Dirichlet process prior to P where the parameter of the process is βP^*. The function P^* is the experimenter's prior guess at the potency curve, and the constant β determines the degree of smoothing in the posterior estimate where for $\beta = 0$ the posterior estimate is the isotonic regressor and for $\beta \to \infty$ the posterior estimate is P^*. Alternatively, β may be thought of as the experimenter's strength of belief in the accuracy of P^* measured in number of pieces of data at a given dose level. The Dirichlet process prior implies in particular that the joint density of $P(x_1), P(x_2), \ldots, P(x_M)$ is proportional to

$$\prod_{i=1}^{M+1} \left[P(x_i) - P(x_{i-1}) \right]^{\beta(P^*(x_i) - P^*(x_{i-1}))} \qquad (2)$$

when integrating with respect to

$$\prod_{i=1}^{M} dP(x_i) \Big/ \prod_{i=1}^{M+1} \left[P(x_i) - P(x_{i-1}) \right]$$

as Ramsey does. Note that $P(x_0) = P^*(x_0) = 0$ and $P(x_{M+1}) = P^*(x_{M+1}) = 1$. A consequence of (2) is that $P(x_i)$ has a beta distribution. In fact, as Ferguson [4] showed, $P(x)$ has a beta distribution regardless of the value of x. Taking the product of (1) and (2) and setting the partial derivatives with respect to $P(x_i)$, $i = 1, 2, \ldots, M$, to zero yields the following system of equations:

$$\frac{n_i}{P(x_i)\left[1 - P(x_i)\right]} \left[\frac{s_i}{n_i} - P(x_i) \right] = \beta \left[\frac{P^*(x_{i+1}) - P^*(x_i)}{P(x_{i+1}) - P(x_i)} - \frac{P^*(x_i) - P^*(x_{i-1})}{P(x_i) - P(x_{i-1})} \right].$$

This system may be solved to get the posterior mode using a constrained Newton–Raphson method* and a special matrix inversion method applicable to tridiagonal matrices*. If $x_i \leqslant x \leqslant x_{i+1}$, then the modal estimate is $P(x)$, where

$$P(x) - P(x_i) = \frac{P^*(x) - P^*(x_i)}{P^*(x_{i+1}) - P^*(x_i)} \left[P(x_{i+1}) - P(x_i) \right],$$

so that the posterior mode has the same shape as the prior guess, P^*, between x_i and x_{i+1}.

Disch [3] observed that Ramsey's prior implies a prior on the ED(p), where, for example, ED(0.5) = ED50. Let $F_{P(x)}(p)$ be the prior CDF of $P(x)$. Since Ramsey's prior implies that $P(x)$ has a priori a beta distribution*, $F_{P(x)}(p) = I_p(a,b)$, where $a = \beta P^*(x)$, $b = \beta(1 - P^*(x))$, and

$$I_p(a,b) = \{\Gamma(a+b)/[\Gamma(a)\Gamma(b)]\} \times \int_0^1 x^{a-1}(1-x)^{b-1}\,dx$$

is the incomplete beta function. Hence the prior CDF of ED(p) is

$$\begin{aligned}\Pr\{\mathrm{ED}(p) \leqslant x\} &= 1 - \Pr\{P(x) < p\}\\ &= 1 - I_p(a,b)\\ &= I_{1-p}[\beta\{1 - P^*(x)\},\\ &\qquad \beta P^*(x)].\end{aligned}$$

The posterior CDF may be approximated by replacing P^* with the modal estimate of Ramsey with β and $\beta + \min_i(D_i)$, where $D_i = n_i$ + (number of successes at dose levels less than x_i) + (number of failures at dose levels greater than x_i) [3].

Investigating the implied priors for several effective doses can be a useful tool in properly assessing Ramsey's prior. However, Ramsey showed that his modal estimator can still provide an improvement to isotonic regression even when P^* is badly specified, provided that β is moderate.

References

[1] Antoniak, C. (1974). *Ann. Statist.*, **2**, 1152–1174.

[2] Ayer, M., Brunk, H. D., Ewing, G. M., Reid, W. T., and Silverman, E. (1955). *Ann. Math. Statist.*, **26**, 641–647.

[3] Disch, D. (1981). *Biometrics*, **37**, 713–722. (Extends Ramsey's results to the distribution of effective doses.)

[4] Ferguson, T. S. (1973). *Ann. Statist.*, **1**, 209–230. (The basic introduction to Dirichlet processes.)

[5] Kuo, L. (1983). *Ann. Statist.*, **11**, 886–895. (Discusses the experimental design question.)

[6] Ramsey, F. L. (1972). *Biometrics*, **28**, 841–858. (Introduces Ramsey's prior and has several examples.)

(ISOTONIC REGRESSION
QUANTAL RESPONSE ANALYSIS
TOLERANCE DISTRIBUTION)

D. DISCH

RAMSEY THEORY *See* GRAPH THEORY

RANDOM BALANCE DESIGNS

In the statistical literature, the word "balanced" is used in several contexts (*see* BALANCING IN EXPERIMENTAL DESIGN; GENERAL BALANCE); Preece [2] wrote an interesting article about the muddle in terminology surrounding it. Random balanced designs is another such class of balanced designs. These, initially introduced by Satterthwaite [3, 4] for industrial experiments where the effects of several input variables are to be studied, have not gained much popularity and are not even known to some statisticians.

Satterthwaite [4] distinguishes between exact balance and random balance in connection with a factorial experiment* treatment structure. Consider a $s_1 \times s_2 \times \cdots \times s_n$ factorial experiment in n factors $A_1, A_2, \ldots, A_n$, where A_i is at s_i levels. The factor A_j is *exactly balanced* with respect to the factor A_i if the distribution of the levels of A_j is the same for each level of the factor A_i. This is clearly the case for a complete factorial with all treatment combinations. There are several known fractional factorials meeting the exact balance criterion.

The factor A_j is *randomly balanced* with respect to the factor A_i if the random sampling process used to select the s_j levels of factor A_j is the same for each level of factor A_i. The factor A_j is *randomly unbalanced* with respect to the factor A_i if the random sampling process used to select the s_j levels of factor A_j is not the same for each level of factor A_i.

Consider a 3^2 experiment with factors A and B, where the levels of a are represented by a_0, a_1, a_2 and the levels of B by b_0, b_1, b_2. Then the nine treatment combinations a_0b_0, a_0b_1, a_0b_2, a_1b_0, a_1b_1, a_1b_2, a_2b_0, a_2b_1, and

a_2b_2 provide exact balance for factor B with respect to factor A. Alternatively, suppose that we choose two of the three levels of B by simple random sampling at each level of A and obtain the six treatment combinations a_0b_1, a_0b_2, a_1b_0, a_1b_2, a_2b_0, and a_2b_1; then factor B is randomly balanced with respect to factor A. Instead, if one chooses levels b_0, b_1, with each of levels a_0, a_2 of A and levels b_1, b_2 with level a_1 of A, getting the treatment combinations a_0b_0, a_0b_1, a_1b_1, a_1b_2, a_2b_0, and a_2b_1, then factor B is randomly unbalanced with respect to factor A.

No restrictions are put on the random process of selecting the levels of a factor A_j that is randomly balanced with respect to the factor A_i. However, if the correlation coefficient between two independent variables exceeds a prechosen value, then at least one of these two variables is eliminated from the experiment, to avoid singularities in the information matrix. The situation is analogous to the problem of multicollinearities* in regression analysis.

Note that the terms "balance" and "unbalance" are closely connected with the underlying model used in the analysis. A set of treatment combinations providing a balanced design under one model need not be a balanced design with another model.

Any type of desired analysis using data analytic tools, multiple regression*, or ANOVA methods can be carried on all or on a subset of the factors for a random balance design. A numerical illustration of a random balanced design with a synthesized example was given by Budne [1].

References

[1] Budne, T. A. (1959). *Technometrics*, **1**, 139–155. (Analysis of random balanced designs is illustrated.)

[2] Preece, D. A. (1982). *Utilitas Math.*, **21C**, 85–186. (Review article on the concept of balance in experimental designs.)

[3] Satterthwaite, F. E. (1957). *J. Am. Statist. Ass.*, **52**, 379–380. (Random balance designs introduced.)

[4] Satterthwaite, F. E. (1959). *Technometrics*, **1**, 111–137. (Detailed treatment of random balance designs given.)

(BALANCING IN EXPERIMENTAL DESIGN
GENERAL BALANCE
UNBALANCEDNESS OF DESIGN, MEASURES OF)

DAMARAJU RAGHAVARAO

RANDOM DETERMINANT *See* RANDOM MATRIX

RANDOM-DIGIT DIALING, SAMPLING METHODS

Random-digit dialing is a sampling methodology that uses as a frame all possible telephone numbers. In the United States, a 10-digit telephone number is comprised of a three-digit area code, a three-digit central office code, and a four-digit suffix. There are presently 34,000 active area code/central office code combinations in use in the United States (based on information updated monthly by AT&T Long Lines Division). 340,000,000 possible 10-digit telephone numbers exist when a four-digit suffix in the range 0000 to 9999 is added to all 34,000 distinct six-digit area code/central office code combinations.

Simple random-digit dialing involves selecting area code/central office code combinations at random from a list of combinations in use within the geographic area of interest, and then selecting four-digit random numbers as suffixes. Simple random-digit dialing provides complete coverage of all residential telephone numbers but at the cost of a sample largely comprised of nonworking and nonhousehold numbers. Seventy-five percent of sample numbers will be business, government, pay phone, and unassigned numbers. The total survey costs are quite high with simple random-digit dialing, since three-fourths of the initial call attempts will be made to nonworking or nonhousehold numbers.

A variation of simple random-digit dialing which reduces unproductive calls has been described by Waksberg [4]. This form of random-digit dialing involves two-stage sampling. As a first step, all possible choices for the seventh and eighth digits are added to all six-digit area code/central office code combinations obtained from AT&T. These eight-digit numbers are treated as primary sampling units* (PSUs). A random two-digit suffix is selected and appended to a randomly selected eight-digit number. This 10-digit number is then dialed. If the dialed number is a residential number, an interview is attempted and the PSU is retained in sample. Additional random two-digit suffixes are selected and dialed with the same eight-digit PSU until a fixed number of residential phone numbers are contacted within that PSU. If the original number dialed was not a working number or was not residential, the PSU is eliminated from the sample and no further calls are made to that PSU. The foregoing steps are repeated until a predesignated number of PSUs are selected for inclusion in the sample. The total sample size for such a survey is the product of the predesignated number of PSUs and the number of residential interviews desired per PSU.

With this design, all residential telephone numbers have the same final probability of selection, despite the fact that the first-stage and second-stage probabilities are unknown. In addition, the PSUs are selected with probabilities proportionate to size* (PPS), based on the actual number of residential phone numbers within the PSU.

Since a fairly high proportion of 100-phone-number blocks contain no residential phone numbers (generally unassigned blocks), this two-stage design reduces the number of call attempts to these blocks since they would be rejected after the first stage.

Overall, the two-stage Waksberg approach increases the contact rate for residential numbers from about 25% to close to 60%. Thus substantially reduced costs are associated with this method when compared to simple random-digit dialing.

Both simple random-digit dialing and Waksberg random-digit dialing are alternatives to methods collectively known as *directory sampling*. One form of directory sampling, simple random directory dialing, uses as a frame all listed telephone numbers in current telephone directories for the geographic area of interest. Frame errors with such a methodology are significant. Numbers unlisted by customer request and numbers unlisted due to the time lag of directory publishing together account for about 22% of total U.S. household telephone numbers. This problem of unlisted numbers is most severe in large cities, where over 40% of household numbers are unlisted. Households with unlisted numbers move more frequently, are more urban, are younger, and have less education than do listed households.

A second form of directory sampling, directory "$N + 1$" dialing, involves the addition or subtraction of a constant from the four-digit suffix of each number in a directory sample. Although this methodology reduces the unlisted number frame error, it does not fully overcome the disadvantage associated with directory samples. Most significantly, new central office codes are missed due to the time lag involved between telephone company assignment of a central office code and publication of the telephone directory.

A more complete description of the two forms of random-digit dialing and of alternative forms of directory-based telephone samples can be found in Frey [2]. A detailed comparison of random-digit-dialing telephone surveys and personal interview surveys can be found in Groves and Kahns [3]. Blankenship [1] provides an introduction to telephone survey methodology.

References

[1] Blankenship, A. B. (1977). *Professional Telephone Surveys*. McGraw-Hill, New York.

[2] Frey, J. H. (1983). *Survey Research by Telephone*. Sage, Beverly Hills, Calif.

[3] Groves, R. M. and Kahn, R. L. (1979). *Surveys by Telephone: A National Comparison with Personal Interviews*. Academic Press, New York.

[4] Waksberg, J. (1978). *J. Amer. Statist. Ass.*, **73**, 40–46.

(SURVEY SAMPLING
TELEPHONE SURVEYS, COMPUTER-ASSISTED)

KENNETH H. GROSS

RANDOM EFFECTS MODEL *See* VARIANCE COMPONENTS

RANDOM FIELDS

Random fields are simply stochastic processes* whose "time" parameter ranges over some space more complicated than the usual real line. Usually, the parameter space is either all or part of Euclidean N-space (yielding an "N dimensional random field"), but it may, on occasion, be something far more complex. As is well known, the theory of stochastic processes is vast, and so, *a fortiori*, must be the theory of random fields. Consequently, it will be impossible in this brief article to give anything other than the most condensed of outlines of this theory. Indeed, we shall do little more than indicate the various areas of current interest in random fields, while providing a directory of introductions to these areas.

The article is divided into four sections. We begin with a brief historical survey, describe some applications, and then discuss, in turn, various choices for the parameter space of random fields, different types of probabilistic structure, and finally, some sample path properties. We shall often refer to MARKOV RANDOM FIELDS (MRFs) to avoid unnecessary overlap of material. Indeed, the two articles should be read in conjunction to obtain a more complete picture of the area.

SOME HISTORY AND APPLICATIONS

The history of continuous-parameter random fields dates back at least to the beginning of the twentieth century, at which time there was considerable interest in the mathematical theory of turbulence. It was not, however, until the middle of the century that serious advances were made. In 1952, Longuet-Higgins [19], a Cambridge applied mathematician, wrote the first of a series of path-breaking papers in which he studied a number of sample path properties of Gaussian random fields. His motivation came from the study of ocean surfaces, which he modeled as either two-dimensional (when time was fixed) or three-dimensional (when time was allowed to vary) random fields. Consequently, the sample path properties to which he devoted most attention were related to level-crossing behavior and the structure of local maxima (see below for more on this), these subjects being of obvious oceanograpic interest.

Independently of Longuet-Higgins but at about the same time, interest in random fields started developing in the Soviet Union, and in 1957 and 1961 Yaglom published two important papers developing the mathematical basis of continuous parameter random fields, considering, for example, their spectral* theory. By the late 1960s considerable effort was being devoted to studying the sample path properties of random fields, particularly by Belyaev and members of his group, culminating in a collection of papers [7, 8].

More recently, this work was consolidated and expanded in a series of papers by Adler and co-workers. These developments, as well as an overview of previous results and the general theory of continuous parameter random fields, are described in detail in Adler [1].

At the same time that the theoretical advances described above were occurring, random fields were being used as mathematical models in a number of different areas. One of the most fruitful applications has been to the modeling of the microscopic roughness of macroscopically smooth surfaces, such as steel, plastic, and so on. A recent review of the rather extensive literature of this area is given in Archard et al. [5], with more sophisticated models given in Adler and Firman [3].

Continuous parameter random fields have also been applied to such diverse areas as forestry [23], geomorphology [20], geology* [16], turbulence [6, 21, 25], and seismology [29]. The book by Mandelbrot [22] discusses a wide variety of further applications of random fields exhibiting fractal* properties, while a number of more pedestrian applications are listed in Adler [1].

CHOICES OF THE PARAMETER SET

There are essentially three different choices of parameter set over which a random field can be defined. The choice that one would initially expect to yield the most mathematically tractable models is that of a finite or countable subset of Euclidean N-space, such as a simple lattice or graph. These are discussed in some detail in the companion entry on MRFs, so we shall say no more on them here, other than to note that despite the apparent simplicity of these parameter spaces, the difficulties inherent in the analysis of the resulting fields are formidable.

The second class of random field, which we discuss in some detail below, involves processes $X(\mathbf{t})$ for which $\mathbf{t} = (t_1, t_2, \ldots, t_N)$ is allowed to vary in a continuous fashion over some subset of N-space. Here, when $N = 1$, we have a simple continuous-time stochastic process*; when $N = 2$, a random surface; and when $N = 3$, a spatial process such as local pressure or speed in fluid turbulence. These random fields, defined on parameter sets of intermediate difficulty, are those which have been studied with the greatest degree of success.

The widest class of fields is the so-called *generalized fields*, which are defined not on points in N-space but on functions, themselves defined on N-space. Since the class of index functions may be a family of delta functions, all ordinary fields, whether they be indexed either by a lattice or continuously on N-space, are clearly special cases of generalized fields. The level of generality afforded by this distinctly more complex parameter space is generally not of great significance for simple applications of random fields, so we shall not discuss this case further. However, it is important to note that these fields arise in a natural fashion in physics and are central to any discussion of Markov-type properties for random fields (*see* MRF).

STRUCTURAL PROPERTIES OF RANDOM FIELDS

Once the parameter set of a random field has been decided on, further study, be it theoretical or applied, will fall into one of a number of essentially disjoint areas depending on the probabilistic structure of the field. For example, as discussed in detail in the entry MRF, the assumption of any type of Markov structure immediately imposes a very distinct probabilistic structure on the distributions associated with the field that dictates specific directions of investigation and exploitation.

A generally milder assumption than Markovianess is that the field satisfies some type of martingale* property, which is, as usual, characterized by considering expectations of the field conditioned on the "past." However, since the parameter set of a random field is not totally ordered, there is no unique way to define the "past," and thus no unique way in which to define the martingale property. The result of this has been a proliferation of different types of multiparameter martingales, originating with the work of Wong and Zakai [31]. There has been substantial interest in this area over the past decade, with the French school of probabilists having made substantial contributions, many of which have been published in the Springer series *Lecture Notes in Mathematics* either as proceedings of the Strasbourg seminars (e.g., vols. 784, 850, 920, 921) or as special topics volumes (e.g., vols. 851, 863).

A class of stochastic processes on the line about which a great deal is known is the class of processes with *independent increments*. Random fields possessing an analogous property are reasonably easy to define but considerably more difficult to study. For example, only recently have the most basic sample path properties of right continuity and left limits (suitably defined) been established. It seems that it is going to be extremely difficult to obtain as broad a picture of these fields as has been possible for their one-dimensional counterparts, primarily because one of the most important tools used in the study of the latter, namely *stopping times*, is not really available in the multiparameter case.

Indeed, the fact that stopping times cannot be defined easily in the multiparameter setting (consider trying to generalize to N-space the simplest of all stopping times, the first hitting time) has been a serious impediment to creating a theory in this setting as rich as the one-dimensional theory. Some progress in this direction has been made via the notion, in two dimensions, of *stopping lines*, but this tool is not as powerful as its one-dimensional counterpart. Once again, the French school of probabilists has been active in this area, although the original idea is due to Wong and Zakai [31] with important contributions by Merzbach (e.g., ref. 24).

Moving away from structural properties based on some form of dependence on the past, we come to the properties of *stationarity*, *isotropy*, and *self-similarity**. Each of these properties describes some form of statistical homogeneity of the process. Stationarity requires that the field behave (statistically) in the same fashion in every region of its domain of definition. Isotropy requires that the joint distribution of the field at two points be a function only of the distance between them. Both of these requirements can be expressed in terms of the spectral representation* of the field, details of which can be found in Adler [1]. It is worth noting that properties such as stationarity are essentially no more difficult to study in the N-parameter case than in the one-parameter case, since their definition and study are in no way connected with ordering properties of the parameter space. The property of self-similarity is somewhat more complex than the other two just described. Basically, it describes the fact that after the parameter set has been rescaled in some fashion, the original field can be recovered (statistically) by a simple linear transformation of the field values. This property (which is related to the famous "renormalization group" of statistical physics) and its implications for the possible distributions of a random field are discussed in some detail in Dobrushin and Sinai [14] and the references listed therein.

The last structural property that we shall consider is that of *Gaussianity*. It is in the case of Gaussian random fields that there has been the most success in developing, for fields, an analog of many of the results known about one-dimensional processes. There are two reasons for this. The first is the comparatively simple and tractable form of the multivariate normal distribution, which allows many problems to be solved for Gaussian processes that are completely intractable in general. The other reason is that, even on the real line, many of the aspects of Gaussian processes that have been of central interest are not connected to the total ordering of the line, and so are far easier to extend to multiparameter situations than are those of, for example, Markov processes* or martingales. (For relationships between two-parameter martingale theory and Gaussian fields, see Bromley and Kallianpur [10].) Since the theory is so rich in the Gaussian case, we shall now devote some time to considering it in detail.

GAUSSIAN RANDOM FIELDS

Let us assume now that a Gaussian random field $X(\mathbf{t})$ is defined continuously on N-space and has mean $E\{X(\mathbf{t})\} = 0$. Then all the statistical properties of X depend only on the covariance function $R(\mathbf{s}, \mathbf{t}) = E\{X(\mathbf{s}) \cdot X(\mathbf{t})\}$ of X. For example, whether or not the field will have continuous sample paths can be read off in a reasonably straightforward fashion from the behavior of the covariance function at the origin. A little reflection shows why this must be the case. If $R(\mathbf{t})$ decays slowly in a neighborhood of the origin, close values of the field will be highly correlated, and thus highly dependent, leading to continuous sample paths. Rapid decay will lead to the opposite phenomenon. Indeed, the behavior of the covariance near the origin also determines such things as Hausdorff dimension* properties of the sample paths and the properties of local time*. Many of these phenomena are not, in principle, more difficult to study in the random field situation than they are for simple stochastic processes on the line, although they occasionally involve more sophisticated calculations.

One problem that is significantly more difficult in the multiparameter case is that of *level crossings*. The level crossings of a continuous process in univariate time are simply those times t at which the process takes the value u, where u is some predetermined value [i.e., $\{t : X(t) = u\}$]. Knowledge of the number of level crossings of various levels is a useful tool in many applications of stochastic processes, and one of the most useful results available is the so-called Rice formula, which states that for stationary Gaussian processes* the mean number of level crossings in time $[0, T]$ is given by

$$\lambda^{1/2}(\sigma\pi)^{-1/2}\exp\left(-\tfrac{1}{2}u^2/\sigma^2\right)$$

where λ and σ are parameters of the process. (For details, see Cramér and Leadbetter [13].) It has long been a problem of some considerable interest to obtain a generalization of the Rice formula for random fields. However, the generalization of the notion of level crossings to the multiparameter situation involves uncountable sets of the form $\{\mathbf{t} : X(\mathbf{t}) = u\}$ which, when $N = 2$ form families of contour lines in the plane, when $N = 3$ form surfaces in three-dimensional space, and in general define $(N-1)$-dimensional manifolds*. Clearly, then, not only is it not trivial to generalize Rice's formula, but it is not even clear how best to generalize the notion of level crossings itself. Let it suffice to say here that this can be done, by employing a number of concepts from integral geometry and differential topology, and an interesting and useful theory of multidimensional level crossings can be developed. For details see Adler [1].

Another problem of both theoretical interest and immediate applicability is the behavior of Gaussian fields in the neighborhood of high maxima. Early work in this area was done by Longuet-Higgins, who was interested in the shape of wave crests, and whose results, together with more recent advances, have been put to substantial use in studying the contact between two surfaces modeled as random fields. Both Lindgren [18] and Nosko [26–28] made significant contributions to this problem, with some more recent results due to Wilson and Adler [30]. Essentially all these authors have studied various aspects of the fact that in the vicinity of a high maximum (or low minimum) a Gaussian field exhibits almost deterministic behavior, so that its sample paths, appropriately normalized, look just like its covariance function in a neighborhood of the origin.

Problems related to the distribution of the global maximum of Gaussian fields over large sets have also been studied in considerable detail, and a number of asymptotic results giving values for $\Pr[\sup(X(\mathbf{t}) : \mathbf{t} \in S) > x]$, for large x, have been obtained (e.g., Hasofer [17], Bickel and Rosenblatt [9]). Nonasymptotic results are much harder to find, although good bounds to the foregoing excursion probability are known in some special cases (e.g., Goodman [15], Cabaña and Wschebor [11, 12], Adler [2]).

Finally, note that some of the results on level crossings and sample path behavior have been extended from the Gaussian case to classes of processes that exhibit, in some form, a near-Gaussian behavior. An example of such a field is given by the so-called "χ^2 field," which can be defined as the sum of a number of squared, zero-mean, independent, Gaussian fields. Although this field often exhibits highly non-Gaussian behavior, the fact that it is defined via Gaussian fields makes it analytically tractable.

Further Reading

The reader interested in pursuing this subject further would be best advised to begin with the three books listed in the references. Adler [1] deals with continuous parameter fields, Dobrushin and Sinai [14] with lattice indexed and generalized fields, while Mandelbrot [22] has something to say about nearly everything.

References

[1] Adler, R. J. (1981). *The Geometry of Random Fields*. Wiley, Chichester, England.

[2] Adler, R. J. (1984). *Ann. Prob.*, **12**, 436–444.

[3] Adler, R. J. and Firman, D. (1981). *Philos. Trans. R. Soc.*, **303**, 433–462.

[4] Adler, R. J., Monrad, D., Scissors, R. H., and Wilson, R. (1983). *Stoch. Proc. Appl.*, **15**, 3–30.

[5] Archard, J. F. et al. (1975). In *The Mechanics of Contact between Deformable Bodies*, A. D. de Pater and J. J. Kalker, eds. Delft University Press, Delft, The Netherlands, pp. 282–303.

[6] Batchelor, G. K. (1953). *The Theory of Homogeneous Turbulence*. Cambridge Monogr. Mech. Appl. Math. Cambridge University Press, New York.

[7] Belyaev, Yu. K., ed. (1972). *Bursts of Random Fields*. Moscow University Press, Moscow (in Russian).

[8] Belyaev, Yu. K. (1972). *Proc. 6th Berkeley Symp. Math. Statist. Prob.*, Vol. 2. University of California Press, Berkeley, pp. 1–17.

[9] Bickel, P. and Rosenblatt, M. (1972). In *Multivariate Analysis III*, P. R. Krishnaiah, ed. Academic Press, New York, pp. 3–15.

[10] Bromley, C. and Kallianpur, G. (1980). *Appl. Math. Optim.*, **6**, 361–376.

[11] Cabaña, E. M. and Wschebor, M. (1981). *J. Appl. Prob.*, **18**, 536–541.

[12] Cabaña, E. M. and Wschebor, M. (1982). *Ann. Prob.*, **10**, 289–302.

[13] Cramér, H. and Leadbetter, M. R. (1967). *Stationary and Related Stochastic Processes*. Wiley, New York.

[14] Dobrushin, R. L. and Sinai, Ya. G. (1980). *Multicomponent Random Systems*. Marcel Dekker, New York.

[15] Goodman, V. (1976). *Ann. Prob.*, **4**, 977–982.

[16] Harbaugh, J. W. and Preston, F. W. (1968). In *Spatial Analysis: A Reader in Statistical Geography*, B. J. L. Berry and S. F. Marble, eds. Prentice-Hall, Englewood Cliffs, N.J., pp. 218–238.

[17] Hasofer, A. M. (1978). *Adv. Appl. Prob. Suppl.*, **10**, 14–21.

[18] Lindgren, G. (1972). *Ark. Math.*, **10**, 195–218.

[19] Longuet-Higgins, M. S. (1952). *J. Mar. Res.*, **11**, 245–266.

[20] Mandelbrot, B. B. (1975a). *Proc. Natl. Acad. Sci. USA*, **72**, 3825–3828.

[21] Mandelbrot, B. B. (1975b). *J. Fluid Mech.*, **72**, 401–416.

[22] Mandelbrot, B. B. (1982). *The Fractal Geometry of Nature*. W. H. Freeman, San Francisco.

[23] Matérn, B. (1960). *Commun. Swed. For. Res. Inst.*, **49**, 1–144.

[24] Merzbach, E. (1980). *Stoch. Proc. Appl.*, **10**, 49–63.

[25] North, G. R. and Cahalan, R. F. (1981). *J. Atmos. Sci.*, **38**, 504–513.

[26] Nosko, V. P. (1969). *Sov. Math. Dokl.*, **10**, 1481–1484.

[27] Nosko, V. P. (1969). *Proc. USSR–Japan Symp. Prob.* (Harbarovsk, 1969), Novosibirsk, pp. 216–222 (in Russian).

[28] Nosko, V. P. (1970). *Vestnik Mosk. Univ. Ser. I Mat. Meh.*, pp. 18–22 (in Russian).

[29] Robinson, E. A. (1967). *Statistical Communication and Detection*. Charles Griffin, London.

[30] Wilson, R. J. and Adler, R. J. (1982). *Adv. Appl. Prob.*, **14**, 543–565.

[31] Wong, E. and Zakai, M. (1974). *Z. Wahrscheinl. verw. Geb.*, **29**, 109–122.

(FRACTALS
MARKOV RANDOM FIELDS
STOCHASTIC PROCESSES)

ROBERT J. ADLER

RANDOM GRAPHS

A random graph is a pair $(\mathscr{G}, P)$ where $\mathscr{G}$ is a family of graphs and P is a probability distribution over $\mathscr{G}$ (*see* GRAPH THEORY). Two extensively studied cases are:

1. **The Edge Probability Model.** $\mathscr{G}$ is the set of *all* graphs on n-vertices where any particular m-edge graph occurs with probability $p^m(1-p)^{\binom{n}{2}-m}$. Such a random graph is often denoted by $K_{n,p}$ to indicate selection of edges from the complete graph K_n, with each edge chosen independently with probability p.

2. **The Equiprobable Model.** $\mathscr{G}$ is a particular class of graphs on n vertices and m edges (e.g., trees, regular graphs) where each graph occurs with equal probability.

Intuitively, the edge probability model is readily identified with applications, say in chemistry or biology, where edges are introduced ("evolve") independently with probability p. Equivalently, for engineering and communications applications, the edges in the edge probability model may be considered to be independently deleted (through "failure") with probability $(1 - p)$. The alternative equiprobable model has a more obvious tie to enumeration problems for particular classes of graphs in combinatorial graph theory.

Applications of probabilistic methods in combinatorics* were started by Szekeres, Turán, Szele, and Erdös in the 1930s and 1940s (see Erdös [5]), while the theory of random graphs was initiated by Erdös and Rényi in the 1960s in their seminal paper on the evolution of random graphs [7] followed by a sequence of fundamental contributions (for the entire collection, see ref. 6 or 22). Since then several hundred papers have been published in various areas of random graph theory. See the books by Bollobás [1, 3], Palmer [21], Erdös and Spencer [8], and Sachkov [23] as well as review papers [2, 10, 12, 24].

ALMOST CERTAIN PROPERTIES OF RANDOM GRAPHS

The principal asymptotic results for the edge probability model of random graphs are essentially equivalent to those of the equiprobable model over the class of *all* n-vertex m-edge graphs if we take p and $m/\binom{n}{2}$ to be asymptotically of the same order. For convenience we state results pertaining to the class of all n-vertex graphs as properties of the random graph $K_{n,p}$. Erdös and Rényi's fundamental result for random graphs is that for many so-called monotone structural properties A (such as connectivity) of $K_{n,p}$, there exists a threshold function $t_A(n)$ for property A where, with $p = p(n)$, $t_A = t(n)$,

$$\lim_{n\to\infty} \Pr\{K_{n,p} \text{ has property A}\} = \begin{cases} 0 & \text{if } p/t_A \to 0, \\ 1 & \text{if } p/t_A \to \infty, \end{cases} \quad \text{as } n \to \infty.$$

The notion of random graph *evolution* relates to the changes in the typical structure of $K_{n,p}$ as the edge occurrence probability p increases. By "typical structure" we mean properties possessed almost surely (a.s.), that is, with probability 1 as $n \to \infty$.

Consider how the typical structure of $K_{n,p}$ changes during its evolution. For very sparse graphs, characterized by the property that $pn^{3/2} \to 0$ as $n \to \infty$, $K_{n,p}$ a.s. consists only of disjoint edges and isolated vertices. When $pn^{3/2} \to c$, where c is a positive constant, trees on at most three vertices appear with positive probability. For $pn \to 0$, $K_{n,p}$ is a.s. a union of disjoint (isolated) trees. For $pn \to c$, where $0 < c < 1$, cycles are present in $K_{n,p}$ with a positive probability and $K_{n,p}$ is a.s. a union of disjoint trees and components containing exactly one cycle. The structure of the random graph changes dramatically when $pn = c$ as c passes through the value 1. For $c > 1$, $K_{n,p}$ a.s. consists of one gigantic component of order comparable to n, along with a number of isolated trees and relatively small components with exactly one cycle. This epoch of the evolution is often called *phase transition* (e.g., the gel point of polymer chemistry). When p increases further, the gigantic component subsumes the smaller components. As p reaches the order of magnitude $\log n/n$, only isolated vertices remain outside the large component. Finally, as $pn - \log n \to \infty$, the random graph becomes a.s. connected. At this same point the graph is a.s. asymmetric, and for even n a.s. has a perfect matching. More specifically, for $pn - \log n - (r-1)\log\log n \to \infty$, the graph a.s. has minimum degree at least r and is r-connected, and is Hamiltonian (see [16]) for $r = 2$. Proceeding to somewhat denser graphs where $pn^{2/(r-1)} \to \infty$, the graph a.s.

has a complete subgraph on r vertices. Note that all of these important transitions occur in the region of relatively sparse edge density, in that all occur with the edge probability asymptotically smaller than any fixed constant c. See the fundamental paper of Erdös and Rényi [7], where the notion of evolution was introduced and the majority of basic facts about the evolutionary process were proved.

NUMERICAL CHARACTERISTICS OF RANDOM GRAPHS

The theory of random graphs is not only concerned with dichotomous graph properties. The exact and/or approximate distributions of numerical graph characteristics constituting random variables defined over random graphs have been investigated. The number of vertices of a given degree, the numbers of independent and dominating sets of vertices, and the number of subgraphs of a given type are examples of such random variables which have been investigated. For typical results dealing with subgraphs, see Karoński [13].

The asymptotic distribution of extremal and/or rank statistics associated with random graphs, such as the maximal and minimal degree, have been investigated extensively. Often, the extremal characteristic asymptotically has its probability distribution concentrated at only one or two discrete values.

In particular, for the "relatively dense" random graphs characterized by a constant edge probability $p = c$ for any $0 < c < 1$, we trivially obtain that the random graph $K_{n,p}$ a.s. has radius 2, diameter 2, longest path length $n - 1$, smallest cycle size 3, and largest cycle size n as $n \to \infty$. More interesting is the result that the largest clique size and largest independent set size a.s. assume one of at most two values in the neighborhood of $2 \log n / \log(1/p)$ for largest clique size and $2 \log n / \log(1/(1-p))$ for largest independent set size (see Matula [17] and Bollobás and Erdös [4]). The chromatic number of $K_{n,p}$ with $b = 1/(1-p)$ is known to a.s. fall in the interval

$$\left[\frac{(\frac{1}{2} - \epsilon)n}{\log_b n}, \frac{(1+\epsilon)n}{\log_b n} \right]$$

for any $0 < \epsilon < \frac{1}{2}$, with an open conjecture that the result is a.s. in a narrow interval about the lower limit $n/(2 \log_b n)$ (see Grimmett and McDiarmid [11]).

RANDOM TREES

A *random tree* is understood to be a tree picked at random from a given family of trees T, such as the family of all n^{n-2} labeled trees on n vertices, as well as families of plane and binary trees. Characteristics studied include the number of vertices of a given degree, the distance of a given vertex from the root (*altitude*), the number of vertices at altitude k (width), the *height* of a rooted tree (the maximal altitude of its vertices), and the diameter of the random tree. Metric characteristics related to the length of paths between specified vertices in a random tree have also been investigated. Details on these topics and more general structures, such as random forests (collections of trees), random hypertrees (connected hypergraphs without cycles), and k-dimensional trees, are contained in Moon's monograph [19] and a section devoted to trees in Karoński's review paper [12].

RANDOM REGULAR GRAPHS AND DIGRAPHS

Besides trees, the equiprobable model has been applied to random regular graphs (all vertex degrees equal) and out-regular digraphs (all vertex out-degrees equal). The asymptotic enumeration of the number of graphs in the family of all d-regular graphs (all vertices of degree d) on n vertices has provided a procedure to analyze the average properties of this important family of graphs. See Bollobás's review paper [2].

A mapping $f: V \to V$ of an n-element set into itself may be interpreted as a digraph with out-degree 1 at each vertex. Mappings without fixed points correspond to digraphs without loops. One-to-one mappings (permutations) correspond to regular digraphs with both in-degree and out-degree 1. Mappings correspond to digraphs where each component contains at most one cycle, and permutations yield components each of which is a directed cycle. A *random mapping* and a *random permutation* then may each be defined by the process of uniform selection from the appropriate set of corresponding digraphs as per the equiprobable model. Issues such as connectedness and the cycle-length distribution of random mappings have been investigated. See the monograph by Kolchin [15], or the review paper by Mutafchiev [20]. Generalizing this approach, random digraphs of out-degree d corresponding to random d-multiple-valued mappings have also been investigated using the equiprobable model.

RANDOM TOURNAMENTS

A tournament between n players can be represented by a directed graph where an arc (i, j) denotes that player i defeated player j. If the outcomes of the tournament's $\binom{n}{2}$ matches are chosen at random, the resulting digraph is called a *random tournament*. Random tournaments can serve as a model for statistical multiple-comparison* methods. A wide range of problems and results on random tournaments is presented in the monograph by Moon [18].

PERCOLATION ON LATTICES

In a regular lattice suppose that we color edges (or vertices) white or black independently with probability p and $q = 1 - p$, respectively. Assume that a fluid spreads through this lattice, with the white edges (or vertices) interpreted as "open" and black as "closed." With this interpretation, percolation theory* is concerned with the structural properties of such random lattices. See the monograph by Kesten [14].

STATISTICAL GRAPHS

Statistical graph models deal with the situation when the structure of the initial graph (population graph) is unknown or partially unknown, but we are able to randomly observe its subgraphs (sample graphs). Generally speaking, statistical graph theory is concerned with the estimation of various characteristics of the population graph based on knowledge of its sample graphs. A comprehensive survey is provided by Frank [9].

References

[1] Bollobás, B. (1979). *Graph Theory: An Introductory Course*. Springer-Verlag, New York.

[2] Bollobás, B. (1981). *Combinatorics (Swansea, 1981) Lond. Math. Soc. Lect. Notes Ser.* **52**, Cambridge University Press, Cambridge, England, pp. 80–102.

[3] Bollobás, B. (1985). *Random Graphs*. Academic Press, London.

[4] Bollobás, B. and Erdös, P. (1976). *Math. Proc. Camb. Philos. Soc.*, **80**, 419–427.

[5] Erdös, P. (1985). *Random Graphs '83, Ann. Discrete Math.*, **28**, M. Karoński and A. Ruciński, eds. North Holland, Amsterdam.

[6] Erdös, P. (1973). *The Art of Counting*. MIT Press, Cambridge, Mass. (Selected writings, J. Spencer, ed.).

[7] Erdös, P. and Rényi, A. (1960). *Publ. Math. Inst. Hung. Acad. Sci.*, **5**, 17–61.

[8] Erdös, P. and Spencer, J. (1974). *Probabilistic Methods in Combinatorics*. Academic Press, New York.

[9] Frank, O. (1980). In *Sociological Methodology, 1981*, S. Leinhardt, ed. Jossey-Bass, San Francisco.

[10] Grimmett, G. R. (1983). *Selected Topics in Graph Theory 2*, L. Beineke and R. Wilson, eds. Academic Press, London, pp. 201–235.

[11] Grimmett, G. R. and McDiarmid, C. J. H. (1975). *Math. Proc. Camb. Philos. Soc.*, **77**, 313–324.

[12] Karoński, M. (1982). *J. Graph Theory*, **6**, 349–389.

[13] Karoński, M. (1984). *Balanced Subgraphs of Large Random Graphs*. Adam Mickiewicz University Press, Poznań, Poland.

[14] Kesten, H. (1982). *Percolation Theory for Mathematicians*. Birkhauser Boston, Cambridge, Mass.

[15] Kolchin, V. F. (1984). *Random Mappings*. Izd. Nauka, Moscow (in Russian).

[16] Komlós, J. and Szemeredi, E. (1972). *Discrete Math.*, **43**, 55–64.

[17] Matula, D. W. (1972). *Notices Amer. Math. Soc.*, **19**, A-382. [See also ref. 21 (pp. 75–80).]

[18] Moon, J. W. (1968). *Topics on Tournaments*. Holt, Rinehart, and Winston, New York.

[19] Moon, J. W. (1970). *Counting Labelled Trees*. Canad. Math. Congress, Montreal, Quebec, Canada.

[20] Mutafchiev, L. R. (1984). *Proc. 13th Spring Conf. Union Bulg. Math.*, pp. 57–80.

[21] Palmer, E. M. (1985). *Graphical Evolution: An Introduction to the Theory of Random Graphs*. Wiley, New York.

[22] Rényi, A. (1976). *Selected Papers*, P. Turán, ed. Akadémiai Kiadó, Budapest.

[23] Sachkov, V. N. (1978). *Probabilistic Methods in Combinatorial Analysis*. Izd. Nauka, Moscow, (in Russian).

[24] Stepanov, V. N. (1973). *Voprosy Kibernetiki*. Izd. Nauka, Moscow, pp. 164–185. (In Russian.)

(GRAPH-THEORETIC CLUSTER ANALYSIS
GRAPH THEORY
PERCOLATION THEORY)

MICHAL KAROŃSKI
DAVID W. MATULA

RANDOMIZATION—I

We discuss issues involved in arguments for and against randomization in the design of surveys* and comparative experiments and the use of randomization tests* in the latter. For brevity we focus on full-response surveys of fixed size obtained by random sampling without replacement from a finite population and comparing two treatments by a design related to the survey framework. In both cases the observations are supposed free of measurement error.

Let $U = \{u_f : f \in F\}$ be an indexed set of N distinct units. The indexing set F is the sampling frame and a subset J of it is a sample. When x is a real-valued function on U, we write x_f for $x(u_f)$, $\bar{x}(J)$ for the mean of the J-sample x-values, and $\bar{x}$ for $\bar{x}(F)$, the population mean. The J-sample variance of x is $s_x^2(J) = (n-1)^{-1}\sum_J[x_f - \bar{x}(J)]^2$, where $n > 1$ is the size of J. As J runs equiprobably through $\mathscr{F}_n$, the set of all samples of size n, $E\{s_x^2(J)\}$ is the population variance $\sigma_x^2 = [(N-1)/N]s_x^2(F)$, $E\{\bar{x}(J)\}$ is $\bar{x}$, and $\mathrm{var}\{\bar{x}(J)\}$ is $[(N-n)/(Nn)]\sigma_x^2$. If J is chosen at random from $\mathscr{F}_n$, $\bar{x}(J)$ is an unbiased estimate of $\bar{x}$ and $[(N-n)/(Nn)]s_x^2(J)$ is an unbiased estimate of its variance.

The aim of a survey is usually seen as the estimation of the population quantities $\bar{x}$, $N\bar{x}$ from sample x-values. The preceding random sampling procedure seems to achieve this because the $\bar{x}(J)$ obtained then locates $\bar{x}$ with an estimated standard error $[(N-n)/(Nn)]^{1/2}s_x(J)$. This view is questioned by Basu [2, 3, 4], on the grounds that the likelihood principle* entails that the way the sample x-values are obtained plays no role in inference from them to the population quantity $\bar{x}$. Before considering this objection, we discuss a comparative experiment open to the same criticism.

Suppose that we wish to compare two treatments, A and B, when only one of them can be applied to a given unit and do so by applying A to the u_j with j in the sample J and B to the u_k with k in the sample $K = F - J$. Let the individual potential effects of A and B on u_f be x_f and y_f. When J is chosen at random from $\mathscr{F}_n$, the mean effects $\bar{x}$, $\bar{y}$ are located by $\bar{x}(J)$, $\bar{y}(K)$ and their estimated standard errors. The mean treatment difference $\Delta = \bar{x} - \bar{y}$ is located by $\Delta(J) = \bar{x}(J) - \bar{y}(K)$, with a standard error that may be conservatively approximated by the sum of those of $\bar{x}(J)$ and $\bar{y}(K)$, because

$$\mathrm{var}\{\Delta(J)\} \leqslant \left[\left(\frac{N-n}{Nn}\right)^{1/2}\sigma_x + \left(\frac{n}{N(N-n)}\right)^{1/2}\sigma_y\right]^2,$$

equality holding when $\bar{x}(J)$, $\bar{y}(K)$ are perfectly negatively correlated, for instance when $x_f - y_f$ does not depend on f. Thus randomized design, with the A-treatment group determined by the random choice of J from $\mathscr{F}_n$, gives an estimate of the error in taking the mean treatment difference between the samples J and K to be the differential treatment effect in the population U. The provision of such estimates of error was seen by Fisher* [14] as one of the principal aims of experimental randomization, and he argued they could not be obtained without it. Against this, Harville [17] asserts that randomization is no basis for inference because we know what sample J was realized when the experiment is carried out and just how it was obtained is irrelevant from the standpoint of inference conditional on that realization. This position is similar to that reached about surveys by Basu, whose argument we now consider.

If J in $\mathscr{F}_n$ is the sample chosen in a survey, the data comprise the function $x_J = \{(j, x_j) : j \in J\}$ with domain J. Let $D(x) = \{x_J : J \in \mathscr{F}_n\}$ be the set of possible data sets for samples from $\mathscr{F}_n$ when x is the population function. Let $\mathscr{X}$ be a finite set of feasible population functions x, and let $D(\mathscr{X}) = \cup\{D(x) : x \in \mathscr{X}\}$. Consider a sampling procedure Π over $\mathscr{F}_n$ yielding J in $\mathscr{F}_n$ with probability $\pi(J) > 0$. If ξ in $D(\mathscr{X})$ has domain J, then conditional on x in $\mathscr{X}$, $p(\xi \mid x)$, the probability that the observed data set is ξ, is $\pi(J)$ for $\xi = x_J$ and 0 otherwise. For a given ξ with domain J, the likelihood function on $\mathscr{X}$ [i.e., $L(x \mid \xi) = p(\xi \mid x)$] is $\pi(J)$ when $x_J = \xi$ and 0 otherwise. If X_ξ is the number of x in $\mathscr{X}$ with $x_J = \xi$, the normalized likelihood $\bar{L}(x \mid \xi)$ is X_ξ^{-1} when $x_J = \xi$, 0 otherwise. It does not depend on the sampling procedure Π. If one accepts the likelihood principle, this means that Π is irrelevant to inference about x from ξ. All that is relevant is the observed data set x_J. Basu [2] concludes that there is little, if any, use for randomization in survey design. Others, like Kempthorne [22], see this conclusion as so opposed to intuition that it casts doubt on the likelihood principle.

A similar argument applies to the comparative experiment. Conditional on potential A and B treatment effects x and y, the probability $p\{(\xi, \eta) \mid (x, y)\}$ that the observed A and B data sets are ξ and η, with domains J and $K = F - J$, is $\pi(J)$ when $\xi = x_J$, $\eta = y_K$, and 0 otherwise. The likelihood principle again suggests that randomization in design has no role in the analysis of such an experiment. This view is shared by Bayesians, who, accepting the likelihood principle, take the posterior density to be the normalized restriction of the prior density to the (x, y) with $x_J = \xi$, $y_K = \eta$.

Fisher's arguments for randomization in refs. 14 and 15 have been defended and clarified by Greenberg [16], White [39], Easterling [9], and Kempthorne in the works cited below, but the issues remain controversial. Kruskal [29] sees Fisher's own exposition as confusing. But it is arguable that Fisher's critics, not understanding the practical motivation behind randomization, have been misled by what he himself saw as their mistaken emphasis on the deductive reasoning of pure mathematics instead of the inductive procedures of the practical scientist, [6, p. 256], a point also made by Kempthorne [26].

From a practical viewpoint, the effect one is talking about is $\Delta(J)$, not Δ, and the principal question of the comparative experiment is not inference about Δ, regarded as an unknown parameter, but whether $\Delta(J)$ is more appropriately said to be a treatment effect than a J effect or a general combined treatment—J effect, which might, as a special case, include a "treatment × J" interaction*. If it is, the *observed* $\Delta(J)$ is the treatment effect of the experiment and it is *its* magnitude and direction that have contextual meaning. If the A-treatment group consisted of the n oldest units, it might be said the treatments were confounded with age and that $\Delta(J)$ is then a treatment–age effect. If the units are plots, the treatments manures and x, y crop yields, the practical question is whether the observed difference $\Delta(J)$ can be attributed to the manures rather than to something else (e.g., greater soil fertility within A-plots). In practice, randomization in design is not directed to inference from $\Delta(J)$ to Δ but to such possibilities of misnomenclature [13]. It has to do with obtaining samples J, K which hopefully will not exhibit features that might invalidate talking about the treatment-J effect $\Delta(J)$ as if it were only a treatment effect.

But it is often difficult to specify these features beyond the fact that even after blocking*, there is a possibility of confounding* with a few variables whose practical importance was not foreseen. Thus there is usually a relatively small unknown collection C of samples in $\mathscr{F}_n$ that one wishes to avoid. If their number $m \ll \binom{N}{n}$ and J is chosen at random from $\mathscr{F}_n$, the probability that one of them is obtained is correspondingly small [i.e., $m/\binom{N}{n}$] whatever C may be. This is the practical purpose of randomization in design. Whether it can be achieved in other ways is a moot point. When analyzing the experiment one will still need to enquire if randomization did avoid unwanted samples. Cornfield [8] is a good example of such an enquiry.

Similar remarks apply to surveys. It is true that once J is known just how it was obtained does not alter the fact that all one knows about x is x_J, but it may have a bearing on whether x_J meets the aims of practical enquiry. An example of a potential design flaw that was quickly remedied because the practical aims would not have been met is noted in Brewer [7].

In the comparative experiment, $\bar{x}$ is known when $y = x$ and $\Delta(J)$ is then $N[\bar{x}(J) - \bar{x}]/(N - n)$. Following Fisher [14] the distribution of this $\Delta(J)$ as J runs through $\mathscr{F}_n$ has been used in the so-called *randomization test**, whereby the hypothesis of no treatment effect, $y = x$, is rejected when $|\Delta(J)|$ is improbably large. The rationale seems to be that it is preferable to hold $y \neq x$ than concede that design randomization has led to an improbably extreme outcome, but the grounds for this preference remain obscure. Easterling [9] suggests that a descriptive interpretation might be more appropriate, and Finch [10] argues that such tests are rudimentary descriptive procedures that may be applied even in the absence of these design randomizations. Interest in randomization procedures was stimulated by the recovery from them of ANOVA* normal theory tests in general settings (Welch [38], Pitman [31], Kempthorne [18, 19]). These ANOVA tests and a descriptive interpretation of standard error have also been recovered in a descriptive setting by Finch [11, 12]. This calls into question the claim of Fisher [14] and Barbacki and Fisher [1] that systematic designs such as those suggested by Student [35, 36] are ruled out because randomization is necessary to obtain valid estimates of error. Not only is such a claim difficult to sustain with inference conditional on the sample used but, more important from a practical viewpoint, these error estimates have a descriptive interpretation that does not depend on the way the sample was obtained.

In summary, randomization is an easily reportable, reproducible, and unambiguous way of collecting data to meet specific aims of practical enquiry. But the analysis of data so collected is primarily descriptive, not inferential. For example, not randomizing in the comparative experiment might result in x_J, y_K not meaning what we want them to mean in the performed experiment. But once it is performed, the practical question, in the absence of confounding, is whether the data set (x_J, y_K) is better described as one noisy constant on F or two noisy constants, one on J and one on K, and how sensitive this description is to data perturbation. This question may be investigated by the two-sample descriptive procedure of Finch [11, 12].

References

[1] Barbacki, S. and Fisher, R. A. (1936). *Ann. Eugen. (Lond.)*, **7**, 189–193.

[2] Basu, D. (1969). *Sankhyā A*, **31**, 441–454.

[3] Basu, D. (1971). In *Waterloo Symposium on Foundations of Statistical Inference*. Holt, Rinehart and Winston, Toronto, pp. 203–234.

[4] Basu, D. (1978). In *Survey Sampling and Measurement*, N. K. Namboodiri, ed. Academic Press, New York, pp. 267–339.

[5] Basu, D. (1980). *J. Amer. Statist. Ass.*, **75**, 575–595.

[6] Box, J. F. (1978). *R. A. Fisher: The Life of a Scientist*. Wiley, New York.

[7] Brewer, K. R. W. (1981). *Aust. J. Statist.*, **23**, 139–148.

[8] Cornfield, J. (1971). *J. Amer. Medical Ass.*, **217**, 1676–1687. (A detailed post hoc examination of the extent to which randomization had been successful in a large clinical trial.)

[9] Easterling, R. G. (1975). *Commun. Statist.*, **4**, 723–735.

[10] Finch, P. D. (1979). *Biometrika*, **66**, 195–208. (Introduces procedures for assessing descriptions and argues that randomization tests are a rudimentary version of them.)

[11] Finch, P. D. (1980). *Biometrika*, **67**, 539–550. (Gives a descriptive interpretation of the usual ANOVA procedures that does not depend on the assumption of normality.)

[12] Finch, P. D. (1981). *Aust. J. Statist.*, **23**, 296–299. (Gives a descriptive interpretation of standard error.)

[13] Finch, P. D. (1982). *Aust. J. Statist.*, **24**, 146–147.

[14] Fisher, R. A. (1926). *J. Min. Agric.*, **33**, 503–513.

[15] Fisher, R. A. (1935). *The Design of Experiments.* Oliver & Boyd, Edinburgh.

[16] Greenberg, B. G. (1951). *Biometrics*, **7**, 309–322.

[17] Harville, D. A. (1975). *Amer. Statist.*, **29**, 27–31.

[18] Kempthorne, O. (1952). *The Design and Analysis of Experiments.* Wiley, New York. (One of the few texts to include a detailed discussion of randomization in design.)

[19] Kempthorne, O. (1955). *J. Amer. Statist. Ass.*, **50**, 946–967.

[20] Kempthorne, O. (1966). *J. Amer. Statist. Ass.*, **61**, 11–34.

[21] Kempthorne, O. (1969). In *New Developments in Survey Sampling*, N. L. Johnson and H. Smith, Jr., eds.

[22] Kempthorne, O. (1971). Contribution to Rao [32].

[23] Kempthorne, O. (1975). In *A Survey of Statistical Design and Linear Models*, J. N. Srivastava, ed. North-Holland, Amsterdam, pp. 303–331.

[24] Kempthorne, O. (1977). *J. Statist. Plann. Inf.*, **1**, 1–25.

[25] Kempthorne, O. (1978). In *Contributions to Survey Sampling and Applied Statistics: Papers in Honor of H. O. Hartley*, H. A. David, ed. Academic Press, New York, pp. 11–28.

[26] Kempthorne, O. (1980). Contribution to Basu [5].

[27] Kempthorne, O. and Doerfler, T. E. (1969). *Biometrika*, **56**, 231–248.

[28] Kempthorne, O. and Folks, J. L. (1971). *Probability, Statistics, and Data Analysis*. Iowa State University Press, Ames, Iowa.

[29] Kruskal, W. (1980). *J. Amer. Statist. Ass.*, **75**, 1019–1029.

[30] Pearson, E. S. (1939). *Biometrika*, **31**, 159–179.

[31] Pitman, E. J. G. (1937). *Biometrika*, **29**, 322–335.

[32] Rao, C. R. (1971). In *Waterloo Symposium on Foundations of Statistical Inference*. Holt, Rinehart and Winston, Toronto, pp. 177–190.

[33] Smith, T. M. F. (1976). *J. R. Statist. Soc. A*, **139**, 183–204. (A useful review of the theory of surveys.)

[34] Student (1923). *Biometrika*, **15**, 271–293.

[35] Student (1936). *J. R. Statist. Soc. Suppl.*, **3**, 115–122.

[36] Student (1936). *Nature (Lond.)*, **88**, 971.

[37] Thornett, M. L. (1982). *Aust. J. Statist.*, **24**, 137–145.

[38] Welch, B. L. (1937). *Biometrika*, **29**, 21–52.

[39] White, R. F. (1975). *Biometrics*, **31**, 552–572.

(ANALYSIS OF VARIANCE
BAYESIAN INFERENCE
CONFOUNDING
DESIGN OF EXPERIMENTS
LIKELIHOOD PRINCIPLE)

PETER D. FINCH

RANDOMIZATION—II

Consider a given population of experimental units (e.g., plots of land, children of age 6, trees, mice). These units—children, trees, and mice—alter over time, and the dynamics over time is influenced by various factors (e.g., nutrition and environment). Suppose that we wish to modify the dynamics as, for instance, that we wish children of age 6 to grow more rapidly than they have in the past. Suppose also that we have acts of intervention, such as defined supplementations of diet or instantaneous or semicontinuous or continuous treatment with drugs of various sorts. In attempting to assess the effects of modes of intervention, we have to call on all the scientific opinion that is available. We shall find that the general system we are looking at is so complex and the background of partial knowledge is so weak that we are forced to do a comparative experiment. Name the various interventions we wish to compare with the word "treatments." Then we have to apply each treatment to a subset of our population of units. Necessarily, each unit can receive only one of the treatments. Suppose that we have 30 units and wish to compare six treatments, with equal attention to them. Then we have to select five units to receive any one of the treatments. This is obvious, and the only design problem in performing the comparative experiment is to have a process of assignment of treatments to units.

The principle of randomization in its simplest form is merely that we partition the 30 units into six subsets, each of five units, at random, and then assign each one of the treatments to one of the subsets. The implementation of this prescription can be done in various ways which will not be discussed. Having decided on the assignment of treatments to units, we then perform the study with the resulting treatment–unit combinations, and observe the chosen outcome variables at the end of the experimental period. The design so induced is called a *completely randomized design*. If we surmise that subsets of the original 30 units would behave differently, we would consider a different scheme of treatment allocation. The randomized block design for t treatments is based on the surmise that in the totality of available units, there are disjoint subsets of t units, such that variability within subsets in the outcome variables under the same treatment would be small. If we surmise that there are subsets of size k (e.g., if we are experimenting with mice that come to us in litters of, say, $4 = k$), we will adopt a scheme of random allocation that utilizes the surmise. We shall then have, if $k < t$, some sort of randomized incomplete block design, and if $k > t$, some sort of randomized super-complete randomized block design. Suppose that we have t^2 units on which we wish to compare t treatments, and the t^2 units may be classified into a $t \times t$ two-way array with one unit in each cell of the array. Then we impose on the two-way array a random $t \times t$ Latin square*, and associate each treatment with one letter of the Latin square. This yields a (randomized) Latin square design. If we classify the units by two factors, one nested by the other, we obtain a split-plot design*.

The ideas in randomized experiments correspond closely to those used in finite sampling of a finite population*. So, for example, the completely randomized design is analogous to simple random sampling, the randomized block design to stratified sampling, and the randomized Latin square design to deep sampling. There is, however, one very critical difference between randomization in pure sampling and randomization in comparative experiments. In the latter, we wish to compare treatments, whereas in the former, there is, in a certain sense, only one treatment. Also, in basic sampling, there is no experimental period and no dynamics over time.

In both sampling* and experimental design*, use is made of surmises about the variability among units in the population of units, that is, of a process of informal empirical Bayesian* thinking. In the case of the randomized block design, the Latin square design, and the split-plot design, for instance, the surmises are categorical, and the randomization imposes a classificatory structure that is used to restrict the randomization. The places of the units in this classificatory structure is "concomitant information."

We may extend the ideation to a more general situation in which the concomitant information is not solely classificatory. A classical procedure, on which there has much controversy over the past five decades, is to use an idea of balance*. One attempts to choose a partition of the units into treatment subsets that are closely alike with respect to the concomitant variables. In the Latin square case controversy arose on the use of what are called systematic squares, the idea being that the subsets of units which receive the different treatments should appear to be balanced relative to each other with respect to position in the square. The classical example of this is the Knut-Vik square*. A simpler example occurs with experimentation with units that occur on the time line, as with successive experimental periods of a chemical reactor. Consider experimentation "on a line." Then concomitant information on the unit is the position of the unit on the line. Let X_{1j}, $j = 1, 2, \ldots, r$, be the positions of the units that receive treatment 1, with $X_{2j}, \ldots, X_{tj}$ having corresponding meanings. Then a natural way to make the assignment is to require $\text{ave}_{(j)}\{X_{ij}\}$ to be as nearly the same as possible for the various values of i. This could result in the units of some treatments being closer together than the units with other treatments. So one might then also require $\text{ave}_j(X_{ij}^2)$ to be as constant as possible. This would give a partition of the units that is somewhat balanced with respect to position on the line. One would then assign the subsets of the partition at random to the treatments. This process is thought to be desirable because it would lead to treatment differences with smaller error than would occur without the balancing. It had very great appeal before Fisher [6] and Yates [14] wrote about it. Their criticisms arise from a simple question: Given that one has used a systematic design of the type outlined, how is one to assess the statistical significance of whatever differences in treatment means one observes, and how is one to obtain standard errors for treatment differences? One would, according to the Fisher [6] discussion, use a conventional linear model with treatment effects entering additively and with a conventional linear form in the concomitants or covariates: that is, one would use the ordinary analysis-of-covariance* process. As Fisher [6] describes, such a procedure will cause an overestimation of the variance under randomization of differences of treatment means. This would have the consequence that even in the presence of actual treatment effects, one would conclude that there is no evidence for treatment effects.

Fisher [6] and Yates [14] wished to achieve (a) unbiased estimation of treatment effects and unbiased estimation of the error of estimated treatment effects; and (b) validity of the test of significance of the null hypothesis of no treatment effects or of any hypothesized set of treatment effects. To examine these, we have to envisage a population of repetitions. Fisher* showed in the case of the Latin square design that if one's population of repetitions is the totality of possible Latin squares, then with fixed (but unknown) unit values and no treatment effects, the expectation over the randomization set of the mean square for treatments is equal to the expectation of the residual mean square if one uses the conventional analysis of variance. This property was called "unbiasedness of design" by Yates and holds for the common designs. If additivity holds, in that the observation obtained is equal to a unit effect plus a treatment effect plus purely random measurement error, then one obtains unbiased estimation of error of estimated treatment effects for the standard designs by conventional least squares (for proofs, see Kempthorne [7]. To address tests, we have to decide what inferential process we shall follow when we have obtained the results of the experiment. The elementary and natural process that is almost uniformly presented in teaching texts is to apply the model $\mathbf{y}_i = \mathbf{x}_i'\boldsymbol{\beta} + \boldsymbol{\tau}_{t(i)} + \mathbf{e}_i$, with the error vector $\mathbf{e}$ being multivariate normal* with null mean vector and variance matrix equal to $\sigma^2\mathbf{I}$, with the associated tests. This is altered, say, for a split-plot design with respect to error by assuming that the whole of the error consists of two parts, one part being common to several observations and the other unique to each observation. This error assumption is patently unreasonable in most experimental contexts. It amounts to assuming that the units we use are a random sample from some population of units that we could have used. Because we are interested in comparing the several treatments, we shall and should attempt to obtain a set of units that are as alike as possible. Then, clearly, the set of units we use will be a correlated subset of any "whole" population that is available.

The way out of the difficulty was given by Fisher, namely, the use of randomization tests. Suppose that we have the results of the experiment. Then if there are no treatment effects, our actual data set is the same as we would get with any of the different plans we might have used. We may then superimpose each of the different plans on the observed data set, compute for each such imposition the value of treatment-determined criterion, and obtain the randomization distribution of the criterion. We place the observed value of the criterion in this randomization distribution and calculate the probability under randomization that we would have obtained a criterion value equal to or greater than that actually obtained. This probability is then called the *significance level* (SL) of the test of the null hypotheses of no treatment differences. The property of this process that is thought, by some, to be very valuable is: $\Pr(\mathrm{SL} \leqslant \alpha \mid \text{null hypotheses}) = \alpha$, if α is an achievable level. This ideation was exposited by Kempthorne [7-10]. If we wish to examine the nonnull hypotheses that treatment j gives a result greater than treatment 1 by δ_j ($j = 2, 3, \ldots, t$), then we merely substract δ_j from the observations under treatment j and make a randomization test of the null hypothesis on the adjusted data. The idea of sampling the randomization distribution was given first by Eden and Yates [3], was examined mathematically by Dwass [2], and exemplified for complex multivariate responses by Cox and Kempthorne [1]. It is now a routine way of testing hypotheses on treatment effects. Edgington [4] describes randomization tests for simple experiments.

The next step in ideation counters the problems that are associated with the fact that almost any randomization scheme will produce plans that are unreasonable from an intuitive viewpoint; for example, with six units "on a line" and two treatments A and B, a possible plan with the completely randomized design is AAABBB, which is "unreasonable." The new idea is that one considers, say, the completely randomized design, which gives, say, M possible plans before treatment randomization. One looks at each of the M plans, perhaps intuitively or perhaps with computations, using ideas of distance between subsets, and one rejects some of the plans as "unreasonable." Suppose that we delete these plans from the test and are left with M_1 plans as "reasonable." Then we pick at random one these M_1 plans and assign treatments at random to the treatment symbols in the chosen plan. After obtaining the response data, we use the subset of M_1 plans for randomization tests on treatment differences. Ordinarily, one would like M_1 to be at least 100 (if this is possible) in order to enable the experiment to give significance at a level no greater than 0.01, if there are in fact large treatment effects.

The generality of the prescription is clear. The undoubted obscurity in the whole randomization choice of plan and then test evaluation is the choice of criterion that we are to use to measure deviation from the null hypothesis of no treatment differences. The commonly used process is to apply ordinary linear model theory with independent and homoscedastic error to obtain a criterion. Such a process can be applied regardless of the nature of the concomitant variation.

Another aspect of experiment analysis is estimation of treatment effects. This requires assumption of a model for treatment effects, and the natural one is to assume additivity, as explained earlier. Let $\delta_j^i = 1$ if treatment j falls on unit i and 0 otherwise. Then the observed mean of r observations on treatment j is

$$\frac{1}{r}\sum_i \delta_j^i (u_i + t_j) = t_j + \frac{1}{r}\sum_i \delta_j^i u_i .$$

If $\Pr(\delta_j^i = 1) = 1/t$, as occurs with randomization of any equi-replicate plan with respect to treatments, the expectation of this is $t_j + \bar{u}$, $\bar{u}$ being the average of unit values. So we then have unbiasedness of treatment means in the sense that the average for treatment j under repetitions, of which we have only one, is what we would observe if we placed it on all the units. Under the additivity assumption the difference of treatment means is

$$t_j - t_{j'} + \frac{1}{r}\left(\sum_i \delta_j^i u_i - \sum_i \delta_{j'}^i u_i\right).$$

In the case of the simple balanced designs, we find that variances of estimated treatment differences are given by the use of the simple linear model under special circumstances, as exposited by Kempthorne [7]. There are, however, unresolved obscurities in that to obtain intervals of uncertainty for a parameter θ, given an estimate $\hat{\theta}$ and a so-called standard error of $\hat{\theta}$ (i.e., estimated standard deviation of $\hat{\theta}$), $\mathrm{SE}(\hat{\theta})$, we need to know as a fact that $(\hat{\theta} - \theta)/\text{SE}(\hat{\theta})$ is a pivotal quantity with a known distribution (e.g., a t-distribution). From this we should infer that interval estimation is just a particular form of a significance test*—an inversion thereof. We accomplish our aim by the use of the randomization test* procedure. The process that is often followed is to use randomization, then to verify that the randomization distribution of a criterion is like the distribution of that criterion under a normal law (or infinite model) model assumption in what are thought to be critical respects, and finally to use the normal law theory of error as an approximation to the actual randomization distribution of criterion (see Pitman [11] and Welch [12]). The idea of incorporating design random variables, incorporating the randomization that is used, to lead to derived models and expected values of mean squares in ordinary analysis of variance* has been described by Kempthorne [7] and by Wilk and Kempthorne [13] for basic designs.

The modern view held by supporters of randomization ideas is that the randomization of design and the use of randomization tests on the preselected class of possible designs should be viewed as an integral whole. Any relation to conventional analysis of variance associated with linear models is, in general, fortuitous.

Other modes of inference (e.g., Bayesian*) will not be discussed. There appears to be no justification for randomization ideas in such inferences.

If one follows the ideas of randomization and randomization test inference, it is clear that the inferences obtained may be associated unambiguously with the population of experimental units that is used in the experiment. If we have a large population of experimental units and select (e.g., at random) a subset of that population, then extension of the statistical test from the set of units that is used to the totality of available units in the whole population can be justified by the randomization test argument, using a simple conditional argument. So one can use randomization arguments to justify a (partial) inference to the set of experimental units that is actually used or sampled. Extension of such an inference to some larger population [e.g., from an experiment on 6-year-old children in 1950 to all 6-year-old children of the United States in 1950, or to the (nonexistent in 1950) population of 6-year-old-children in the United States in 1990], is a matter of substantive judgment, which can be aided by a statistical formulation but cannot be justified by substantive reasoning. It follows that randomization with randomization inference is not a panacea.

The ideas of randomization tests, based on the experiment randomization used, were difficult to apply in most real experimental situations, because of the necessity of considering each of the possible realizations. Analysis of variance was easy computationally, with easy significance testing if one used the assumption that errors are independent, homoscedastic and Gaussian, that is, "normal law theory" or "infinite model theory" (Kempthorne, [7]). On the basis of the work of Pitman [11] and Welch [12], it seemed that randomization distributions of test statistics can be well approximated by such normal law theory. This was the justification of normal law theory used by Kempthorne [7]. With the advent of high-speed computation, the need to call on such approximation decreased and exact randomization testing of null hypotheses could be done. However, even today, a more complete inference to give exact confidence intervals requires massive computation, as do considerations of power of tests. Curiously, it is informal experience that the approximation is good in many circumstances, but it can be very bad, in that the set of achievable levels by the randomization test can be very small and bounded away from zero. In the case of the 3×3 Latin square design, the only achievable levels are 1 and $\frac{1}{2}$.

The ideas of experiment randomization and associated randomization tests must be clearly distinguished from testing techniques based merely on random relabeling of observations, as exposited by Finch [5]. Such techniques question whether an observed labeling is significant within a chosen class of relabelings. Such procedures may give a significance level that is the same as would result from a randomized experiment with a set of possible plans the same as the set allowed for in the population of relabelings.

The ideas of randomized design and analysis are aimed, clearly, at the proving of causation* by intervention, in contrast to random relabeling, which may be termed purely data analytic.

The distinction between randomization tests and permutation tests* is important. The latter are based on the assumption of random sampling, an assumption that is often patently false or unverifiable, even though necessary to make an attack on the substantive problem being addressed.

Acknowledgment

This article is Journal Paper No J-11257 of the Iowa Agriculture and Home Economics Experiment Station, Ames, Iowa. Project 890.

References

[1] Cox, D. F. and Kempthorne, O. (1963). *Biometrics*, **19**, 307–317.

[2] Dwass, M. (1957). *Ann. Math. Statist.*, **28**, 181–187.

[3] Eden T. and Yates, F. (1933). *J. Agric. Sci.*, **23**, 6–16. (The first Monte Carlo examination of a randomization distribution.)

[4] Edgington, E. S. (1980). *Randomization Tests*. Marcel Dekker, New York. (Exposition of computational procedures.)

[5] Finch, P. D. *Biometrika*, **66**, 195–208.

[6] Fisher, R. A. (1935). *The Design of Experiments*. Oliver & Boyd, Edinburgh. (The basic ideas of randomization, but incomplete because of reversion to infinite model theory.)

[7] Kempthorne, O. (1952). *The Design and Analysis of Experiments*. Wiley, New York. (Reprinted by Krieger, Huntington, N.Y.) (Theory of randomization and infinite model analysis of experiments.)

[8] Kempthorne, O. (1955). *J. Amer. Statist. Ass.*, **50**, 946–967. (Exposition of inference based solely on randomization.)

[9] Kempthorne, O. (1975). In *A Survey of Statistical Design and Linear Models*. North-Holland, Amsterdam. (Randomization ideas and obscurities in Fisher's writings.)

[10] Kempthorne, O. (1977). *J. Statist. Plann. Inf.*, **1**, 1–25. (General basic discussion.)

[11] Pitman, E. J. G. (1937). *Biometrika*, **29**, 322–335. (Approximation of randomization distributions.)

[12] Welch, B. L. (1937). *Biometrika*, **29**, 21–52. (Approximation of randomization distributions.)

[13] Wilk, M. B. and Kempthorne, O. (1956). *Ann. Math. Statist.*, **27**, 950–985. (Linear models based on randomization ideas.)

[14] Yates, F. (1933). *Empire J. Experimental Agriculture*, **1**, 235–244.

(ANALYSIS OF VARIANCE
DESIGN OF EXPERIMENTS
INFERENCE, STATISTICAL
RANDOMIZATION TESTS)

OSCAR KEMPTHORNE

RANDOMIZATION, CONSTRAINED

When, as in agricultural experiments, the experimental units have a spatial, or even temporal, layout, there is always the possibility that ordinary randomization* may produce a plan where the pattern of the treatments accidentally corresponds closely to the underlying spatial pattern of the plots. For example, if there were 12 plots side by side in a long strip and four treatments to be replicated three times each, ordinary randomization might give a plan such as

$$\text{A A A B B B C C C D D D.} \tag{1}$$

The chance of obtaining such a plan is 1/15,400, which is not high, but neither is it negligibly small. The sheer number of experiments that are conducted each year make it certain that plans like (1) will be used from time to time if ordinary randomization is properly used.

However, many experimenters will be unhappy about using plans like (1), for several reasons. The first is the fear that, even in the absence of treatment effects, the yields might exhibit a pattern corresponding to the linear layout. This pattern may consist of either or both of two parts: (a) a fixed trend, such as a fertility trend along the length of the experimental area; and (b) inequality of correlations, so that, for example, yields on neighboring plots are more highly correlated than those on nonadjacent plots. For simplicity, we shall refer to both of these as *plot pattern*. Although part of the purpose of randomization is to allow for our ignorance about point (b), by ensuring equality of correlations *on the average*, many people will find plan (1) too extreme, too far from average, for randomization-based inference to be applicable.

The second reason is a similar worry about competition or interference between neighboring treatments. If one variety of wheat germinates early, it may steal nutrients from neighboring plots, or grow tall and shade its neighbors. In fungicide and insecticide experiments, a treatment giving poor control can severely affect its neighbors, as pests or disease may multiply in the poorly controlled plot and then spread to neighboring plots. If plan (1) is used and treatment C provides poor control, then treatment A has an advantage over B and D. Such treatment interference is a quite different problem from plot pattern, although the two are often confused.

A third reason is more pragmatic, but nonetheless important. Randomization must be *seen* to be random. No matter how objectively the randomization is carried out, a plan like (1) will always cause some people to think it systematic and unfair. Suppose that plan (1) is used in a variety trial, and variety D fails to perform well enough to gain inclusion in the national list of recommended varieties. The breeder of variety D may then complain that his variety was unfairly allotted to the worst part of the field; and it may be hard to refute this. Even a statistician's colleague—a scientist or a field worker—faced with a plan like (1) may question the statistician's judgment, and so may decide to alter the plan, or other aspects of the experiment, without consulting or informing the statistician, possibly destroying months of careful planning.

A plan like (1) is also a poor guard against crop damage by disease or wildlife, which tends to occur in patches. If two or three of the D plots in plan (1) have their crop destroyed by disease, it is impossible to tell whether this is an unfortunate accident or whether variety D is particularly susceptible. If there are reasons to believe the former, then it is legitimate to use missing value techniques (*see* INCOMPLETE DATA) or to fit an extra model term for disease presence, but in either case the comparison of D with the remaining varieties is either impossible (if all three plots are lost), or has very low precision.

SOLUTIONS

The most obvious solution to the problem of undesirably patterned plans like (1) is simply to reject them and rerandomize. Since this destroys the exact justification for the form of the analysis of variance, this is not satisfactory unless the proportion of plans to be rejected is very small. Moreover, these must be specified in advance, for it is easy to find undesirable patterns in almost any plan if it is scrutinized long enough; and post hoc rejection raises doubts about objectivity and fairness. Until they are required to specify them in advance, many experimenters do not realize just how many plans they would reject. For example, in a trial in which each of eight people measured each of four sows twice, the experimenter initially requested that all plans be rejected which entailed any person measuring the same sow twice in succession. With ordinary randomization, this would have entailed rejecting over 99.98% of plans.

In many cases blocking* provides a better solution. For example, plan (1) could be avoided by dividing the strip into four blocks of three plots each, and insisting that each block contain every treatment. If blocks are chosen carefully, and if the subsequent analysis is modified to take account of the blocks, many of the problems of bad patterns simply disappear. However, block-

Table 1. Plan for $2^4 \times 4$ Factorial Experiment on Beans[a]

2*np*	0*dp*	1	3*dn*	0*nk*	2*dk*	1*dnpk*	3*pk*
0*dnp*	2*p*	3*d*	1*n*	2*dnk*	0*k*	3*npk*	1*dpk*
1*d*	3*n*	2*dnp*	0*p*	3*dpk*	1*npk*	2*k*	0*dnk*
3	1*dn*	0*np*	2*dp*	1*pk*	3*dnpk*	0*dk*	2*nk*
0*pk*	2*dnpk*	3*nk*	1*dk*	2	0*dn*	3*dp*	1*np*
3*dnk*	1*k*	0*dpk*	2*npk*	1*dnp*	3*p*	0*n*	2*d*
1*nk*	3*dk*	2*pk*	0*dnpk*	3*np*	1*dp*	2*dn*	0
2*dpk*	0*npk*	1*dnk*	3*k*	0*d*	2*n*	1*p*	3*dnp*

[a] 0, 1, 2, 3, borax at four levels; *d*, farmyard manure; *n*, nitrochalk; *p*, superphosphate; *k*, muriate of potash.

ing is not always sufficient. Table 1 shows a plan for a $2^4 \times 4$ factorial experiment*. The square layout was blocked into eight rows and eight columns, and ordinary randomization was used on a design confounding* high-order interactions* with rows and with columns. Nevertheless, the main effect of potash (levels indicated by presence or absence of the symbol *k*) was completely confounded with the contrast between pairs of diagonally opposite corners of the square. Moreover, blocking may not be desirable if there are few plots. In a trial comparing two fungicides on eight large plots in a 2×4 layout, the only way to avoid plans such as (2) to (4) by blocking is to consider both the two rows and the four columns as blocking systems, which reduces the number of degrees of freedom for error from 6 to 2; this is quite inadequate for meaningful conclusions.

```
A A A A
B B B B      (2)

A A B B
A A B B      (3)

A B B A
A B B A      (4)
```

A third suggestion is to fit one or more covariates to account for positional effects. This is a reasonable solution if suspected plot pattern is the basic objection to certain plans, but is not without difficulty. Some treatment effects may be more confounded with the covariate than others, so that effects are estimated with different precision. There are problems of randomization theory in the presence of covariates (see Cox [8, 9]). Modern computers make short work of the arithmetical problems of the analysis of such nonorthogonal designs, but their programming needs care, and there may be difficulties in *interpreting* such an analysis to a nonstatistical client. However, a major problem in using a covariate is that it may give a poor representation of the plot pattern: for example, a linear covariate for plot number would not be helpful if the plot pattern had a curved smooth trend or a clumpy arrangment. Recent work [18] on nearest-neighbor models attempts to take account of general, but unknown, underlying smooth trends, but the theory is not yet sufficiently worked out for any general methods to be recommended. A more serious case against covariates is that they are inadequate to deal with the other three objections to bad patterns: they are (a) insufficient to ensure that the experiment is seen to be fair; (b) incapable of preventing patches of disease coinciding with patches of a single treatment; and (c) quite inappropriate for the problem of treatment interference.

What is required is a method of randomization which ensures that bad plans cannot occur, while guaranteeing that the usual analysis of variance is still valid. *Constrained randomization* is the term used for such methods. Constrained randomization is often used in conjunction with blocking; the experiment is analyzed in the normal way appropriate to the given block structure, which usually permits straightforward interpretation of results.

THEORY AND METHODS

The main justification for constrained randomization is that randomization dictates, or at least validates, the analysis (see ref. 1). This is true whether randomization is done by randomly choosing a plan from a set of plans, by randomly labeling the treatments, by randomly labeling the plots, or a combination of these methods (see ref. 16). For simplicity, we concentrate on the first method.

Fisher* [12] argued that the justification for the usual analysis of a completely randomized design is that, if a plan is picked at random from among all those with the correct replications of the given treatments, the probability that a specified pair of plots are both allotted the same treatment is independent of the pair of plots. But the same is true of several *incomplete* randomization schemes, such as the set (5) of plans for three treatments in a linear strip of six plots. Therefore, this incomplete randomization scheme validates the same analysis of variance as complete randomization. Moreover, the highly undesirable plan AABBCC can never occur. For *strong* validity [14], that is, to ensure that all normalized treatment contrasts are estimated with the same variance, one must also randomly allocate the actual treatments to the letters A to C.

ABBACC ABACCB ABCBAC
ABCCBA AABCBC (5)

Similarly, Fisher [12] showed that the analysis of variance of a complete block design* depends only on the fact that the probability (over the randomization) of a pair of plots in different blocks having the same treatment is independent of the pair of plots. Likewise, for a Latin square design, all that is necessary is a constant probability of two given plots in different rows and columns receiving the same treatment: this may be achieved by taking as the randomization set one or more complete sets of mutually orthogonal Latin squares* [11]. Fisher's justification of these observations was brief, and they have been rediscovered by many subsequent authors.

A randomization set is said to be *valid* if it has the required property of constant probability on suitable pairs of plots. Methods that have been found for generating valid randomization sets include the use of combinatorial objects, such as balanced incomplete block designs* and Latin squares*, the theory of *D*-optimal designs, and 2-transitive permutation groups. These methods need specialist knowledge, so we shall not go into details here, but present some results (*see also* ORTHOGONAL DESIGNS).

Unfortunately, there is no single recipe for constrained randomization. The undesirable plans vary from experiment to experiment; they depend on the shape and size of the experimental area and on the nature of the treatment structure, whether factorial or not, whether quantitative or qualitative, and, if quantitative, on the precise levels. The experimenter must specify which plans would be undesirable; the statistician can then test the known methods for generating valid randomization sets to see if there is a set which omits all the undesirable plans. If there is not, the experimenter may have to weaken or modify the criteria of desirability.

EXAMPLES

No Blocking

Youden [22] showed that a valid randomization set for t treatments equireplicated in tr plots with no blocking is equivalent to a resolvable balanced incomplete-block design (BIBD) for tr treatments in blocks of size r. The treatments of the BIBD correspond to

Table 2

		Replicate		
I	II	III	IV	V
14	13	15	16	12
23	26	24	25	35
56	45	36	34	46

the original plots, and each replicate of the BIBD gives a plan in the randomization set: each block of the replicate gives the positions for a single treatment in the original design. Table 2 shows a resolvable BIBD for six treatments in blocks of size two: this yields the randomization set (5), with the plans in the same order as the replicates.

Complete Blocks

Suppose that a 3×3 factorial experiment is to be conducted in linear blocks of size 9. Within any block, it is undesirable for any three consecutive plots to receive the same level of either treatment factor (called A and B, say) or of either component of their interaction, AB or AB^2. It is also undesirable for the four plots at either end of the block to contain all three occurrences of any level of A, B, AB, or AB^2. A valid randomization scheme that avoids these patterns consists of randomly allocating A and B to two different rows of Table 3, and then randomizing the levels of each factor independently.

Latin Squares

Freeman [13] introduced quasi-complete Latin squares to balance the effects of treatment interference in two dimensions, assuming an east neighbor to have the same effect as a west neighbor, and similarly for north and south. Each treatment must have each other treatment as a row neighbor exactly twice, and similarly for columns. Ordinary randomization of Latin squares would destroy this neighbor balance, so a constrained randomization scheme is needed, all of whose plans are quasi-complete Latin squares. Bailey [4] showed how this could be done when the number of treatments is a power of an odd prime number. For example, for seven treatments the sequence (6) may be used.

$$0 \quad 6 \quad 1 \quad 5 \quad 2 \quad 4 \quad 3 \qquad (6)$$

Choose a random permutation of the form

$$x \to ax + b \qquad \text{modulo } 7;$$

apply this to the sequence (6) and write the result along the top border of the square. Repeat the process, this time writing the result down the left-hand border of the square. Now the entry in any plot is obtained by adding the two corresponding border entries modulo 7. For example, if we choose the permutations $x \to 2x + 1$ and $x \to 5x + 4$, respectively, we obtain the layout in Table 4.

Table 3

0	1	0	2	0	2	2	1	1
0	0	1	1	2	0	2	1	2
0	1	1	0	2	2	1	2	0
0	1	2	1	1	2	0	0	2

Table 4

	1	6	3	4	5	2	0
4	5	3	0	1	2	6	4
6	0	5	2	3	4	1	6
2	3	1	5	6	0	4	2
1	2	0	4	5	6	3	1
0	1	6	3	4	5	2	0
3	4	2	6	0	1	5	3
5	6	4	1	2	3	0	5

Other Shapes

If there are three treatments replicated three times in a square layout, it is unfortunate if one treatment occurs entirely in one row, in one column, or in one corner, as in plans (7) (a) to (c), respectively.

(a)	(b)	(c)	
A A A	A B A	B A A	
B C B	C B C	C B B	(7)
C C B	A B C	C C A	

Table 5

A A B	A B A	A B C	A C C
C A B	C C B	A C A	C B B
C B C	A C B	B B C	B A A

Two-way blocking would avoid these but would seriously reduce the number of error degrees of freedom. A valid constrained randomization set that avoids both these problems is given in Table 5: one plan should be chosen at random, and then the treatments should be randomly allocated to the letters A to C.

The 2×4 layout is discussed in ref. 5.

HISTORY

Yates [20] first drew attention to the problem of bad patterns, citing an experiment done at Rothamsted Experimental Station [19], whose plan is given in Table 1. The solution he suggested was constrained randomization, which he called *restricted randomization*; this name is still used in the United Kingdom. Grundy and Healy [14] found a valid restricted randomization scheme for Yates's particular example; they also gave the scheme in Table 3 for a 3×3 experiment in linear blocks of size 9. They showed the importance of 2-transitive permutation groups, and how such groups could be combined for a simple crossed layout. Dyke [10] extended their ideas to deal with linear blocks of 16 plots.

Independently of the British work, Youden [21, 22] considered the problem of bad patterns in linear layouts, and advocated constrained randomization. He showed how to obtain valid randomization sets from resolvable balanced incomplete block designs. He and Kempthorne [15] used this method to find good restricted randomization schemes for three and four treatments in linear arrays of six and eight plots, respectively.

More recently, White and Welch [17] have introduced D-optimal design theory into constrained randomization, while Bailey [2] has extended the use of permutation groups, giving a catalog of valid restricted randomization schemes for several common situations. The theory in ref. 7 shows how 2-transitive permutation groups may be combined to give valid restricted randomization schemes for more complicated block structures, such as those described in NESTING AND CROSSING IN DESIGN.

References

[1] Bailey, R. A. (1981). *J. R. Statist. Soc. A*, **144**, 214–223.

[2] Bailey, R. A. (1983). *Biometrika*, **70**, 183–198. (Includes a catalog of restricted randomization schemes for linear blocks of up to 32 plots.)

[3] Bailey, R. A. (1983). In *Recent Trends in Statistics*, S. Heiler, ed. Vandenhoeck und Ruprecht, Göttingen, pp. 9–26. (Gives more detail of the technical arguments.)

[4] Bailey, R. A. (1984). *J. R. Statist. Soc. B*, **46**, 323–334. (Randomization of quasi-complete Latin squares.)

[5] Bailey, R. A. (1985). *Int. Statist. Rev.*, **53**, 171–182. (Restricted randomization in place of losing degrees of freedom by blocking.)

[6] Bailey, R. A. (1985). *Statistics and Decisions, Suppl.*, **2**, 237–248. (Problems when randomization is constrained to preserve neighbor-balance.)

[7] Bailey, R. A., Praeger, C. E., Rowley, C. A., and Speed, T. P. (1983). *Proc. Lond. Math. Soc.*, **47**, 69–82.

[8] Cox, D. R. (1956). *Ann. Math. Statist.*, **27**, 1144–1151. (Randomization in the presence of covariates.)

[9] Cox, D. R. (1982). In *Statistics and Probability: Essays in Honor of C. R. Rao*, G. Kallianpur, P. R. Krishnaiah, and J. K. Ghosh, eds. North-Holland, Amsterdam, pp. 197–202.

[10] Dyke, G. V. (1964). *J. Agric. Sci.*, **62**, 215–217. (Restricted randomization for linear blocks of 16 plots.)

[11] Fisher, R. A. (1935). *J. R. Statist. Soc. Suppl.*, **2**, 154–157.

[12] Fisher, R. A. (1966). *The Design of Experiments*. Oliver & Boyd, Edinburgh (1st ed., 1935).

[13] Freeman, G. H. (1979). *J. R. Statist. Soc. B*, **41**, 253–262.

[14] Grundy, P. M. and Healy, M. J. R. (1950). *J. R. Statist. Soc. B*, **12**, 286–291. (The first real progress in restricted randomization, including the introduction of permutation groups.)

[15] Kempthorne, O. (1961). In *Analysis of Variance Procedures*, ARL 149, by O. Kempthorne, G. Zyskind, S. Addelman, T. N. Throckmorton and R. F. White. Aeronautical Res. Lab., Office of Aerospace Res., U.S. Air Force, Wright-Patterson Air Force Base, Ohio, pp. 190–202.

[16] Preece, D. A., Bailey, R. A., and Patterson, H. D. (1978). *Aust. J. Statist.*, **20**, 111–125.

[17] White, L. V. and Welch, W. J. (1981). *J. R. Statist. Soc. B*, **43**, 167–172. (Restricted randomization using *D*-optimal design theory.)

[18] Wilkinson, G. N., Eckert, S. R., Hancock, T. W., and Mayo, O. (1983). *J. R. Statist. Soc. B*, **45**, 151–178.

[19] Yates, F. (1937). The Design and Analysis of Factorial Experiments, *Imp. Bur. Soil Sci. Tech. Commun.*, **35**.

[20] Yates, F. (1948). *J. R. Statist. Soc. A*, **111**, 204–205. (Introduction of the problem.)

[21] Youden, W. J. (1964). *Technometrics*, **6**, 103–104.

[22] Youden, W. J. (1972). *Technometrics*, **14**, 13–22. (Independent introduction of the problem, first published in 1956.)

(ANALYSIS OF VARIANCE
BLOCKING
BLOCKS, BALANCED INCOMPLETE
CONCOMITANT VARIABLES
CONFOUNDING
DESIGN OF EXPERIMENTS
FACTORIAL EXPERIMENTS
INCOMPLETE DATA
INTERACTION
LATIN SQUARES
NESTING AND CROSSING IN DESIGN
OPTIMAL DESIGN OF EXPERIMENTS
RANDOMIZATION)

R. A. BAILEY

RANDOMIZATION TESTS

A randomization test is a permutation test* which is based on randomization* (random assignment), where the test is carried out in the following manner. A test statistic is computed for the experimental data (measurements or observations), then the data are permuted (divided or rearranged) repeatedly in a manner consistent with the random assignment procedure, and the test statistic is computed for each of the resulting data permutations. These data permutations, including the one representing the obtained results, constitute the reference set for determining significance. The proportion of data permutations in the reference set with test statistic values greater than or equal to (or, for certain test statistics, less than or equal to) the value for the experimentally obtained results is the *P*-value* (significance or probability value). Determining significance on the basis of a distribution of test statistics generated by permuting the data is characteristic of all permutation tests; it is when the basis for permuting the data is random assignment that a permutation test is called a randomization test.

The preceding definition is broad enough to include procedures called randomization tests that depend on random sampling as well as randomization. The modern conception of a randomization test is a permutation test that is based on randomization alone, where it does not matter how the sample is selected. It is this narrower conception of randomization tests that will be the concern of this article.

The null hypothesis for a randomization test is that the measurement for each experimental unit (e.g., a subject or a plot of land) is the same under one assignment to treatments as under any alternative assignment. Thus, under the null hypothesis, assignment of experimental units to treatments randomly divides the measurements among the treatments. Each data permutation in the reference set represents the results that, if the null hypothesis is true, would have been obtained for a particular assignment. (The validity of the foregoing procedure depends on all possible assignments being equally probable; situations with unequally probable assignments require modification of the foregoing procedure and will not be dealt with here.)

Example. Suppose that five subjects have been assigned randomly to treatments A and B, with two subjects for A and three for B, and that these results were obtained: A: 1, 2; B: 3, 5, 9. A two-tailed (double-tailed) t test

Table 1

A	B	$\lvert t \rvert$
1, 2	3, 5, 9	1.81
1, 3	2, 5, 9	1.22
1, 5	2, 3, 9	0.52
1, 9	2, 3, 5	0.52
2, 3	1, 5, 9	0.83
2, 5	1, 3, 9	0.25
2, 9	1, 3, 5	0.83
3, 5	1, 2, 9	0.00
3, 9	1, 2, 5	1.22
5, 9	1, 2, 3	3.00

gives a t of 1.81, and the randomization test procedure, rather than a t table, is used to determine the significance of t. The data are divided into two parts with two measurements for A and three for B, in all $\binom{5}{2} = 10$ ways. The reference set of 10 data permutations is shown in Table 1, with the absolute value of t for each data permutation. Only the first (obtained) and last data permutations have $\lvert t \rvert$'s as large as 1.81, the obtained value, so the P-value is 2/10, or 0.20.

If the null hypothesis was true, random assignment of the subjects to the treatment conditions was, in effect, random assignment of the measurements to the treatment conditions. There were 10 ways the measurements could have been "assigned," and those are the 10 data permutations above, which provide the distribution of t for determining the significance of the obtained t. Each data permutation represents the results associated with a particular assignment, so the same reference set would be used even if a different test statistic was computed. For example, if σ_A^2/σ_B^2 was the test statistic, that variance ratio would be computed for each of the data permutations above to provide a distribution of values for determining the significance of the variance ratio obtained.

If a repeated-measurement* design was employed, with each of five subjects taking treatments A and B in an order determined randomly and independently for each subject, the $2^5 = 32$ possible assignments to treatments would be associated with 32 data permutations which differed only in the order of the measurements within pairs.

Just as the reference set of data permutations is independent of the test statistics, so is the null hypothesis. A difference between means may be used as a test statistic, but the null hypothesis does not refer to a difference between means. The null hypothesis, no matter what test statistic is used, is that there is no differential effect of the treatments for any of the subjects. (Exceptions, such as a directional null hypothesis [6, pp. 137–138], are possible but seldom used.) Thus the alternative hypothesis* is that the measurement of at least one subject would have been different under one of the other treatment conditions. Inferences about means must be based on nonstatistical considerations; the randomization test does not justify them.

CONTRIBUTIONS OF FISHER AND PITMAN

The test usually regarded as being the first randomization test was published by Fisher* in 1935 [13, pp. 43–47]. The hypothetical experiment and test were described as follows. Random samples of 15 seeds were taken from each of two populations to test the null hypothesis of identity of the populations with respect to the size of plants the seeds would produce under the same conditions. As the seeds could not be grown under identical conditions, to prevent population differences from being confounded with soil and environmental differences, the seeds in the samples were assigned randomly to the pots in which they were to be planted. Pots with similar soil and locations were paired, and for each of the 15 pairs of pots it was randomly determined which of a pair of seeds would be planted in each pot. After the plants reached a certain age, the heights were measured and paired off as they would be for a paired t test. Fisher, however, carried out a randomization test instead. He generated a reference set of 2^{15} data permutations by switching the signs of the differences for all 15 pairs and computing the difference between totals for the two types of seeds as the test statistic. (Switching the sign of a difference has the same effect on the test statistic as switching the two measurements on which the difference depends.) The P-value was the proportion of those data permutations with as large a difference between totals as the difference obtained. The procedure for determining significance was definitely novel, and this test seems to be the first permutation test of any kind, not just the first randomization test. As this test is based on random sampling as well as random assignment, however, it is not a test of the modern kind, where the nature of the sampling is irrelevant.

Fisher described another test in 1935 [13, pp. 11–25] that is at times cited as a randomization test but which, although based on randomization, did not explicitly involve permuting of data. In a hypothetical experiment, a lady was told that she would be presented with eight cups of tea in a random order and that there would be four cups with milk added to tea and four cups with tea added to milk before being mixed. The lady's task was to distinguish between the two types of preparation on the basis of taste. (If eight presentation times were designated in advance, as they should be, the eight times would be the experimental units and the responses "milk first" and "tea first" would be the "measurements" for the eight experimental units.) She was hypothesized to identify correctly all eight cups, and Fisher determined the P-value to be $1/70$, there being only one way out of $\binom{8}{4} = 70$ divisions of the cups into the two categories that would correctly match the way they were presented. Unlike the other test, this test did not involve a new procedure for significance determination; there was no explicit permuting of data. Nevertheless, the demonstration of how randomization could provide the basis for a valid test for a single subject influenced the development of single-subject randomization tests.

In a series of articles in 1937 and 1938 [25–27], Pitman* provided a theoretical basis for experimental and nonexperimental applications of permutation tests. Some investigators believe Pitman's contributions to randomization tests to be more important than Fisher's. It is certainly true that whereas Fisher dealt with *randomization* at length, he barely touched on randomization *tests*. Pitman was thorough and clear in his treatment of randomization tests; Fisher was neither. The principal contribution of Pitman was demonstrating that randomization tests could be based on randomization alone, for nonrandomly selected experimental units. As random selection from populations is uncommon in experimentation, a statistical test that is valid in the absence of random sampling is very useful.

RANDOMIZATION AND RANDOMIZATION TESTS

Randomization and randomization tests tend to be similarly evaluated, and what is nominally a criticism of randomization tests frequently is a criticism of the use of randomization in an experiment. In a series of papers, headed by a paper by Basu [1], debating the value of randomization tests, the principal disagreement concerned the relative advantages of systematic and random assignment of experimental units to treatments. Similarly, criticisms of the application of randomization tests to single-subject experimental data [17; 18, pp. 328–329] stem primarily from the belief that decisions to introduce or withdraw experimental treatments should be made during the course of the experiment on the basis of a subject's responses. It has been pointed out [9, 10] that although such criticisms seem to be directed toward randomization tests, they are in fact criticisms of a widespread philosophy of experimentation which requires randomization in order for *any* statistical test to be acceptable.

ROBUSTNESS* OF PARAMETRIC TESTS

Many studies have been performed to assess the robustness* (insensitivity to violations of assumptions) of parametric tests when populations are nonnormal and variances are heterogeneous. (See Bradley [2, pp. 24–43] for a critical evaluation of those studies.) For the typical experiment, in which there is randomization but not random sampling, these studies have little relevance. The important consideration is the robustness of the tests in the absence of random sampling.

Building on work begun by Pitman, several investigators have investigated null and nonnull distributions of parametric test statistics (primarily F) under randomization. Kempthorne [19] has been the principal contributor to this field of study, which is called randomization analysis. (Despite the similarity of names, randomization *analysis* and randomization *tests* have different objectives and employ different approaches to statistical testing.) It is frequently asserted that randomization analysis has demonstrated that parametric tests give essentially the same P-values as randomization tests and consequently are valid for nonrandom samples, given experimental randomization. That is a misconception; parametric tests have *not* been demonstrated to be robust in the absence of random sampling.

Parametric tests are based on infinite, continuous distributions and randomization tests on finite, discrete distributions of test statistics, so a parametric distribution can never exactly correspond to a randomization test distribution. Parametric and randomization test P-values are very similar for some data sets and very different for others, and there is no way of knowing the degree of correspondence without determining both P-values for a given set of data.

The experimenter who is concerned about the validity of a particular application of a parametric test can perform a randomization test instead of approximating the randomization test P-value by use of the parametric test. For many years, the only serious objection raised against using randomization tests instead of parametric tests for analysis of experimental data has been the impracticality of performing the large amount of computation required for a randomization test, but the availability of high-speed computers makes the objection obsolete.

RANK TESTS

Rank tests* were produced in great numbers immediately after the writings of Fisher and Pitman on experimental and nonexperimental permutation tests. Rank tests are permutation tests applied to ranks and are usually represented as requiring the assumption of random sampling. When rank tests are applied to ranks of experimental data from a randomized experiment, however, the tests become randomization tests and do not require the assumption of random sampling. In his book on rank tests, Lehmann [22] discusses the tests in terms of both the random sampling and the randomization models.

RESTRICTED NULL HYPOTHESES

The reference set for a randomization test represents outcomes that, under the null hypothesis, would be associated with alternative assignments. A reference set consisting of data permutations associated with all possible assignments can be used to test the *general* null hypothesis of no effect of any treatment manipulation. To test *restricted* null hypotheses, however, that refer to some but not all treatment manipulations, a reference *subset* with data permutations for only some of the possible assignments (randomizations) must be used. The following examples of the use of reference subsets to extend the applicability of randomization tests to complex experimental designs employ procedures used by Edgington [6, 7].

Consider the random assignment of six subjects to treatments A, B, and C with two subjects per treatment, where the following results are obtained: A: 3, 5; B: 7, 9; C: 6, 4. A planned comparison of A and B cannot use the reference set of data permutations associated with all 90 possible assignments. The restricted null hypothesis of no difference between A and B does not refer to C; consequently, under the null hypothesis, depending on the effect of C, the subjects providing 6 and 4 may or may not have provided those values if they had been assigned to other treatments. There is thus no basis for constructing data permutations for those random assignments. A procedure that generates only data permutations representing results dictated by the restricted null hypothesis is to divide the four A and B measurements between A and B in all $\binom{4}{2} = 6$ ways to generate a reference subset of six data permutations. This procedure provides data permutations for that subset of randomizations where the same two subjects are always assigned to C.

Another example of a restricted null hypothesis is the hypothesis of no effect of factor A in a completely randomized factorial experiment* with factors A and B. Factors A and B each have two levels, and both factors are experimentally manipulated. Eight subjects are randomly assigned to the four cells, with two subjects per cell. For a test of factor A alone, data permutations for only 36 of the 2520 possible assignments can be determined under the restricted null hypothesis, namely, those generated by permuting the obtained data for the four subjects at one level of B over both levels of A in conjunction with permuting the data for the four subjects at the other level of B over both levels of A. Those 36 data permutations constitute the reference subset for determining the significance of the obtained F (*see* F-TESTS).

The preceding two examples illustrate an approach that can be used in generating a reference subset for any restricted null hypothesis, which is to hold fixed the results of all treatment manipulations not specified in the null hypothesis and permute the data only with respect to the specified treatment manipulations.

For none but the simplest of experiments is there only a general null hypothesis to test. The applicability of randomization tests is thus greatly extended through the use of reference subsets to test restricted null hypotheses.

PERMUTATION GROUPS

After noting that reference sets based on all assignments frequently would be too large to be practical, Chung and Fraser [3] proposed the use of reference subsets to reduce the number of data permutations to a practical level. An example will illustrate their approach.

Three subjects are assigned at random to 5-, 10-, and 15-gram drug dosages, with one subject per dosage level. The test statistic is the product-moment coefficient of correlation* (r) between dosage and response magnitude. The experimental results are these: 5-g dosage, 62; 10-g dosage, 78; 15-g dosage, 83. The reference set based on all assignments would consist of the six possible pairings of the drug dosages and responses. A reference subset of three pairings, however, will be used here to determine significance. The reference subset consists of the data permutation obtained and the two data permutations shown below, with their correlation coefficients:

Dosage (g)			
5	10	15	r
62	78	83	+ 0.96
78	83	62	− 0.73
83	62	78	− 0.23

The procedure for deriving the last two data permutations from the first (obtained) data permutation is this: Move the first measurement in the sequence of three measurements to the end to produce a new data permutation, apply the same procedure to that data permutation to produce another one, and continue until no new data permutations are produced. This procedure produces only three data permutations because applying the procedure to the third data permutation produces the first. Starting the permuting operations with any of the three data permutations will result in the same subset of data permutations. Thus the operations form what Chung and Fraser call a *permutation group*, which is defined as a collection of operations that will result in the same set of data permutations, no matter which member of the reference set (or subset) is the one that is initially permuted.

The significance is based on the foregoing reference subset of three data permutations. As the correlation +0.96 obtained is the largest correlation, the P-value is $\frac{1}{3}$.

DATA-PERMUTING AND RANDOMIZATION-REFERRAL TESTS

From the earliest times, randomization tests have been based on permutation groups, but until the example of Chung and Fraser the permutation groups were ones that generated a reference set of data permutations for all randomizations. The demonstration that randomization tests can be based on permutation groups that provide a reference subset stimulated a study of the role of permutation groups [11], which will be outlined in this section.

The study proceeded from a division of randomization tests into two classes: (1) *data-permuting* tests, which do not require knowledge of which experimental unit provided a given measurement, and (2) *randomization-referral* tests, which do require such knowledge. All of the examples considered up to this point, including the drug dosage example, employed data-permuting tests, and, in fact, virtually all randomization tests are of this kind. An additional example of a data-permuting test, consequently, is unnecessary. It is necessary, however, to explain the randomization-referral test.

The randomization-referral test involves these steps to produce a reference set (or subset) of data permutations: (1) Before the experiment, partition the entire set of randomizations (possible assignments) into subsets such that each randomization belongs to one and only one subset. (The performance of a "null" partition that leaves the set unpartitioned, is an acceptable first step.) (2) After the experiment, on the basis of the obtained data permutation and the obtained randomization (the actual assignment), transform the randomizations in the subset containing the obtained randomization into data permutations to form the reference subset. For instance, before performing the drug-dosage experiment, the set of six possible randomizations, showing the order in which subjects a, b, and c could be assigned to the drug dosage levels, is divided into two subsets: {abc, acb} and {bac, bca, cab, cba}. Suppose that the measurements for the data permutation obtained, (62, 78, 83), were obtained from subjects a, c, and b, respectively. Then abc, the other randomization in the subset containing acb, is transformed into 62, 83, 78, by substituting 62 for a, 78 for c, and 83 for b. As the data permutation obtained provides the larger correlation coefficient of the two, the P-value is $\frac{1}{2}$. If any of the four randomizations in the other subset had been the obtained randomization, there would have been four data permutations constituting the reference subset for determining significance. The rationale is simple: when a distribution of all randomizations is divided in advance into subsets (or left undivided as a "subset" of itself) the probability of the obtained randomization providing a test statistic value that is one of the k largest of the n values for its subset is no greater than k/n, so the determination of the P-value as k/n is valid.

The conceptual simplicity of a randomization-referral test makes it easy to assess its validity. This is convenient because the validity of data-permuting tests can then be demonstrated by showing that there is a valid randomization-referral test that is equivalent in the sense that it necessarily gives the same reference set and, consequently, the same P-value. The use of permutation groups, as defined by Chung and Fraser, for data-permuting tests ensures that there is an equivalent randomization-referral test. An example will show what is meant.

Let us consider a randomization-referral test that will necessarily give the same reference subset as the data-permuting procedure in the drug-dosage example. The randomization subsets are generated by applying to each of the six randomizations the same permutation group that was applied to the data permutations in the data-permuting test. That is, for each randomization, the first letter is moved repeatedly to the end until no new randomizations are produced. When that is done, the randomizations fall into two subsets, within each of which the randomizations are related to each other in the same way as the data permutations are interrelated within the reference set: any of the randomizations would generate the others in the subset by applying the permutation group to it. The two subsets are {abc, bca, cab} and {acb, cba, bac}. Now, it can be seen that no matter which randomization provided 62, 78, 83, the other randomizations in its subset would be transformed into 78, 83, 62, and 83, 62, 78, resulting in the same reference set as was provided by the data-permuting procedure. In general, if a randomization-referral test with the randomizations partitioned by applying the same permutation group to the entire set of randomizations as is applied to the data permutations in a data-permuting test is valid, the data-permuting test is valid.

In the study, application of the foregoing approach to the specific data-permuting procedures given in the section on restricted null hypotheses provided a rigorous demonstration of their validity. Also, it was shown that for any test of a restricted null hypothesis, the procedure of holding fixed the results of all treatment manipulations not specified in a restricted null hypothesis and permuting the data only with respect to the specified treatment manipulations is a prescription for an appropriate permutation group.

To sum up, the study of the role of permutation groups resulted in a comprehensive rationale for randomization tests, based on three key concepts: the permutation group, the data-permuting test, and the randomization-referral test. The practical and theoretical consequences are far-reaching.

RANDOMIZATION TESTS IN THE COMPUTER AGE

Access to computers makes it possible to carry out computations on thousands of data permutations quickly and inexpensively. Even with high-speed computers, however, it is frequently necessary to use a reference subset. For example, for a completely randomized design, with 10 subjects per treatment for three treatments, there are over 5 trillion possible assignments.

The use of permutation groups to generate systematic (nonrandom) reference subsets is less practical than forming random reference subsets. A random procedure that is completely valid was discovered by several investigators, the first being Dwass [4]. The data are permuted randomly n times, and the reference set consists of the n data permutations plus the obtained data permutation. Green [15, pp. 172–173; 16] has provided efficient algorithms for random permuting.

The validity of a random permuting procedure is harder to assess than the validity of a data-permuting test because we must be sure that (1) the random procedure provides a random sample of data permutations from the relevant data-permuting reference set (or subset), and (2) the data-permuting test for the sampled reference set would be valid.

Until randomization tests were developed for restricted null hypotheses, the uses of randomization tests were limited, but now parametric tests, no matter how complex, can be transformed into distribution-free* tests by determining significance by the randomization test procedure. Randomization tests also can be custom-made to meet special experimental requirements.

Conventional experimental designs can be combined to provide more complex and efficient experiments. For example, 10 subjects may be paired and assigned randomly within the five pairs to two treatments, while

10 other subjects, for whom there is no good basis for pairing, are randomly and equally divided between the two treatments, without pairing. The test statistic is the difference between means, computed over all 20 subjects, and the reference set is derived by permuting the data for the five pairs of subjects in all $2^5 = 32$ ways, in conjunction with permuting the data for the other 10 subjects in all $\binom{10}{5} = 252$ ways.

Two areas in which randomization tests have been strongly recommended in recent years are behavior therapy single-subject research [5; 7, Chap. 10; 8–10, 12, 23, 29] and weather modification* research [14, 24, 28]. These areas have several problems in common that militate against the use of parametric tests: nonrandom sampling, constraints on randomization, and lack of independence of observations. Each area can profit from studying tests used in the other area. Tests that were developed for other purposes also have relevance to single-subject research and weather modification. Zerbe and Walker [30] devised randomization tests of differences between growth curves* which have potential for single-subject experimentation, and Klauber's [20, 21] space-time clustering randomization tests for determining whether certain events that occur close together in space are also close together in time would appear to have uses in weather modification experimentation.

The most complete and up-to-date guide to the application of randomization tests is a recent book by Edgington [7]. It discusses the logic of randomization tests, presents computer programs for commonly used tests, and describes ways to develop new tests and programs.

Acknowledgment

This work was supported by a Killam Resident Fellowship from the University of Calgary.

References

[1] Basu, D. (1980). *J. Amer. Statist. Ass.*, **75**, 575–582. (Followed by comments by five other contributors, plus a rejoinder.)

[2] Bradley, J. V. (1968). *Distribution-Free Statistical Tests*. Prentice-Hall, Englewood Cliffs, N.J.

[3] Chung, J. H. and Fraser, D. A. S. (1958). *J. Amer. Statist. Ass.*, **53**, 729–735.

[4] Dwass, M. (1957). *Ann. Math. Statist.*, **28**, 181–187.

[5] Edgington, E. S. (1967). *J. Psychol.*, **65**, 195–199.

[6] Edgington, E. S. (1969). *Statistical Inference: The Distribution-Free Approach*. McGraw-Hill, New York. (Chapter on randomization tests comprises one-third of book; substantial portions of Chapters 6 and 7 also deal with randomization tests.)

[7] Edgington, E. S. (1980). *Randomization Tests*. Marcel Dekker, New York. (Practical guide for development and application of randomization tests; 11 FORTRAN computer programs; many numerical examples.)

[8] Edgington, E. S. (1980). *J. Educ. Statist.*, **5**, 235–251. (Validity of single-subject randomization tests.)

[9] Edgington, E. S. (1980). *J. Educ. Statist.*, **5**, 261–267.

[10] Edgington, E. S. (1983). *Contemp. Psychol.*, **28**, 64–65.

[11] Edgington, E. S. (1983). *J. Educ. Statist.*, **8**, 121–135.

[12] Edgington, E. S. (1984). In *Progress in Behavior Modification*, Vol. XVI, M. Hersen, R. M. Eisler, and P. M. Miller, eds. Academic Press, New York. (Single-subject randomization tests, including rank tests; randomization; statistical and experimental independence.)

[13] Fisher, R. A. (1951). *The Design of Experiments*, 6th ed. Hafner, New York. (Similar to the first edition, which is harder to find; a classic in experimental design; first randomization test; discusses randomization in depth.)

[14] Gabriel, K. R. (1979). *Commun. Statist. A*, **8**, 975–1015.

[15] Green, B. F. (1963). *Digital Computers in Research*. McGraw-Hill, New York.

[16] Green, B. F. (1977). *Amer. Statist.*, **31**, 37–39.

[17] Kazdin, A. E. (1980). *J. Educ. Statist.*, **5**, 253–260.

[18] Kazdin, A. E. (1982). *Single-Case Research Designs: Methods for Clinical and Applied Settings*. Oxford University Press, New York.

[19] Kempthorne, O. (1952). *Design and Analysis of Experiments*. Wiley, New York. (A widely known book employing randomization analysis of many complex statistical procedures.)

[20] Klauber, M. R. (1971). *Biometrics*, **27**, 129–142.

[21] Klauber, M. R. (1975). *Biometrics*, **31**, 719–726.

[22] Lehmann, E. L. (1975). *Statistical Methods Based on Ranks*. Holden-Day, San Francisco.

[23] Levin, J. R., Marascuilo, L. A., and Hubert, L. J. (1978). In *Single Subject Research: Strategies for Evaluating Change*, T. R. Kratochwill, ed. Academic Press, New York.

[24] Miller, A. J., Shaw, D. E., and Veitch, L. G. (1979). *Commun. Statist. A*, **8**, 1017–1047.

[25] Pitman, E. J. G. (1937). *J. R. Statist. Soc. B*, **4**, 119–130.

[26] Pitman, E. J. G. (1937). *J. R. Statist. Soc. B*, **4**, 225–232.

[27] Pitman, E. J. G. (1938). *Biometrika*, **29**, 322–335.

[28] Tukey, J. W., Brillinger, D. R., and Jones, L. V. (1978). *The Management of Weather Resources*, Vol. II: *The Role of Statistics in Weather Resources Management*. Department of Commerce, U.S. Government Printing Office, Washington, D.C. (Discusses many aspects of randomization and randomization tests in weather modification research.)

[29] Wampold, B. E. and Furlong, M. J. (1981). *J. Behav. Assess.*, **3**, 329–341.

[30] Zerbe, Z. O. and Walker, S. H. (1977). *Biometrics*, **33**, 653–657.

(DESIGN AND ANALYSIS OF EXPERIMENTS
EXPERIMENTAL DESIGN
FISHER, R. A.
PERMUTATION TESTS
RANDOMIZATION
RANDOMIZATION, CONSTRAINED)

EUGENE S. EDGINGTON

RANDOMIZED BLOCKS *See* BLOCKS, RANDOMIZED COMPLETE

RANDOMIZED GAMMA DISTRIBUTION

The randomized gamma distribution is a kind of compound gamma distribution*, and is also closely connected with the noncentral chi-square distribution*. Denote by $g(x;\theta,s)$ the probability density function (PDF) $[1/\{\theta\Gamma(s)\}](x/\theta)^{s-1}\exp(-x/\theta)$ of the gamma distribution*. Let $X_1, X_2, \ldots$ be independent random variables from the gamma distribution with PDF $g(x,1,1)$ (i.e., exponential distribution* with mean 1). Then $\sum_{i=1}^{k} X_i$ has a gamma distribution with PDF $g(x;1,k)$ for $k = 1, 2, \ldots$. If K is a Poisson random variable with mean μ, then for fixed ρ $(= 0, 1, \ldots)$, $\sum_{i=1}^{\rho+K+1} X_i$ is distributed as a randomized gamma distribution. On the other hand, if $X_1, X_2, \ldots$ are independent random variables from the gamma distribution with PDF $g(x;2,\frac{1}{2})$ (i.e., chi-square distribution with 1 degree of freedom), and if K is a Poisson random variable with mean $\delta/2$, then $\sum_{i=1}^{n+2K} X_i$ is distributed as the noncentral chi-square distribution with n degrees of freedom and noncentrality parameter δ.

The definition of randomized gamma distribution can be generalized to the case where ρ (> -1) is a *real* number [1]. For fixed real $\rho > -1$ consider the gamma distribution with PDF $g(x;1,\rho+K+1)$. Taking the parameter K as an integer-valued random variable subject to the Poisson distribution* with mean μ we get the randomized gamma distribution with PDF:

$$w(x;\ \mu,\rho) = (x/\mu)^{\rho/2}\{\exp(-\mu - x)\} I_\rho\left(2\sqrt{\mu x}\right),$$

$x > 0$, $\mu > 0$, $\rho > -1$, where $I_\nu(z)$ is the modified Bessel function* of the first kind given by

$$I_\nu(z) = \sum_{m=0}^{\infty} \frac{1}{m!\,\Gamma(m+\nu+1)} (z/2)^{2m+\nu}.$$

Mean: $\mu + \rho + 1$

Variance: $2\mu + \rho + 1$

Moments about the mean:

$$\mu_3 = 2(3\mu + \rho + 1)$$

$$\mu_4 = 3(4\mu^2 + 4\mu\rho + 12\mu + \rho^2 + 4\rho + 3)$$

$$\mu_5 = 4(30\mu^2 + 25\mu\rho + 55\mu + 5\rho^2 + 16\rho + 11)$$

νth moment about zero:

$$\mu^\nu + \textstyle\sum_{i=1}^{\nu}\binom{\nu}{i}(\rho + \nu - i + 1)\cdots(\rho+\nu)\mu^{\nu-i}$$

Moment generating function:

$$(1-t)^{-\rho-1}\exp\{\mu t/(1-t)\},\ t < 1$$

Convolution formula: If X_1 and X_2 are independent random variables with PDFs $w(x;\ \mu,\rho)$ and $g(x;1,s)$, respectively, the sum $X = X_1 + X_2$ has a randomized gamma distribution with PDF $w(x;\ \mu,\rho+s)$—the convolution formula $w(x;\ \mu,\rho) * g(x;1,s) = w(x;\ \mu,\rho+s)$ given by Feller [1] holds.

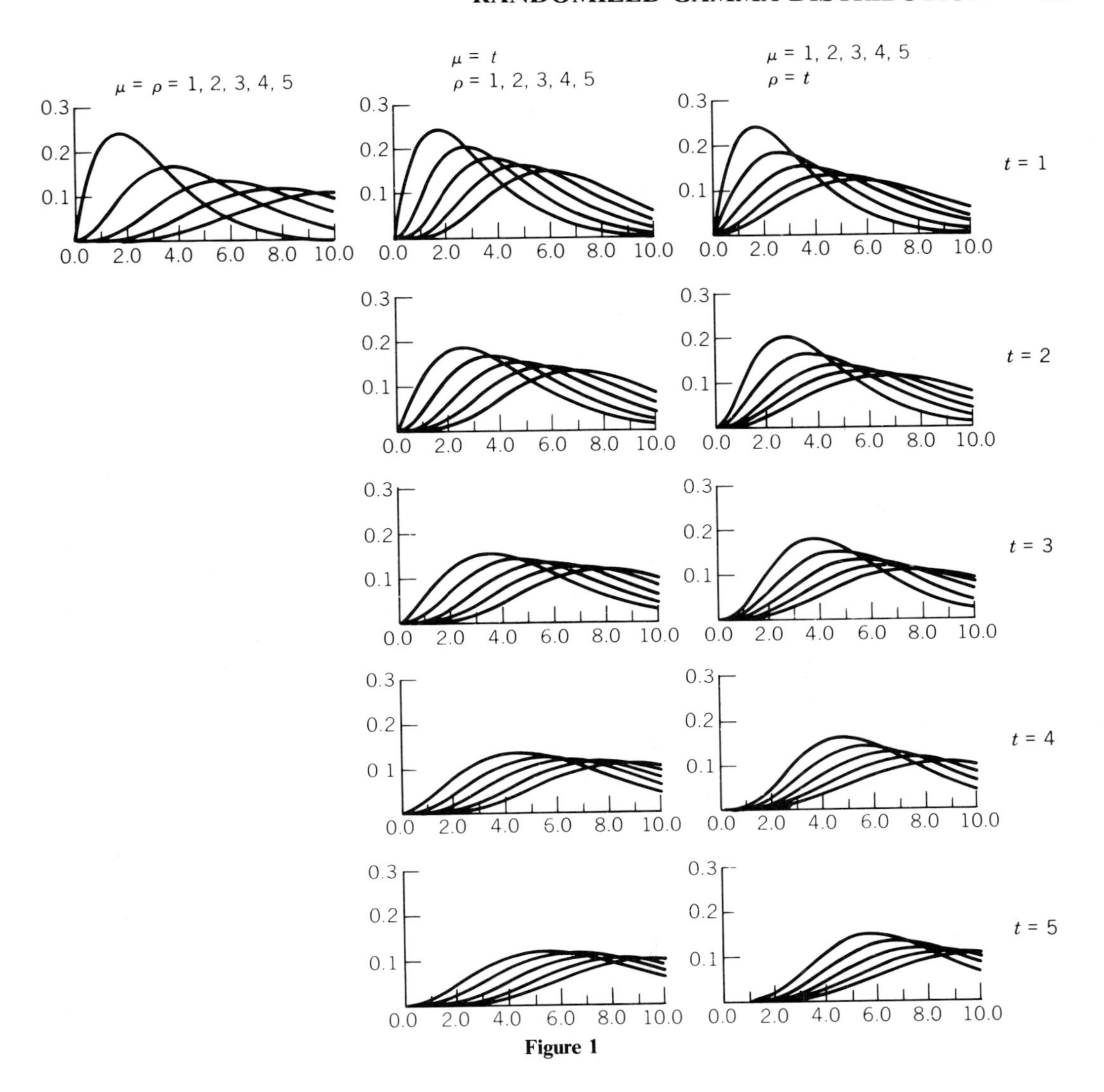

Figure 1

Denote by $f(x;\ \beta, \theta, \lambda)$ the PDF

$$f(x;\ \beta, \theta, \lambda) = Cx^{(\lambda-1)/2}\{\exp(-\theta x)\} I_{\lambda-1}(\beta\sqrt{x}),\quad x \geqslant 0,$$

$$C = (2/\beta)^{\lambda-1}\theta^{\lambda}\exp\{-\beta^2/(4\theta)\},$$

$\theta > 0$, $\lambda > 0$, $\beta > 0$, for type I Bessel function distributions. Then the randomized gamma distribution with PDF $w(x;\ \mu, \rho)$ and the noncentral chi-square distribution with n degrees of freedom and noncentrality parameter δ are the special cases of the type I Bessel function distributions obtained with the parameter values $\beta = 2\sqrt{\mu}$, $\theta = 1$, $\lambda = \rho + 1$ and $\beta = \sqrt{\delta}$, $\theta = \frac{1}{2}$, $\lambda = n/2$, respectively. For additional information, see refs. 2 to 5.

Figure 1 shows the graphs of the PDFs $w(x;\ \mu, \rho)$. In drawing the figure we used a program in ref. 2.

References

[1] Feller, W. (1966). *An Introduction to Probability Theory and Its Applications*, Vol. 2. Wiley, New York, Chap. II, Sec. 7.

[2] Hirano, K., Kuboki, H., Aki, S., and Kuribayashi, A. (1983). *Comput. Sci. Monogr., No. 19*, Inst. Statist. Math., Tokyo, Appendix.

[3] Kotz, S. and Srinivasan, R. (1969). *Ann. Inst. Statist. Math., Tokyo*, **21**, 201–210.

[4] McNolty, F. (1967). *Sankhyā B*, **29**, 235–248.

[5] Springer, M. D. (1979). *The Algebra of Random Variables*. Wiley, New York, Chap. 9.

KATUOMI HIRANO

RANDOMIZED MODEL *See* RANDOMIZATION—I; RANDOMIZATION—II

RANDOMIZED RESPONSE

Description

Randomized response is a method used in statistical surveys whose goal is to reduce or eliminate response errors which occur when respondents are queried about sensitive or highly personal matters. First introduced by Warner [40], a randomized response design requests information in an indirect manner. By using a randomizing device, the respondent selects a question on a probability basis from two or more questions at least one of which is sensitive. The respondent answers the question selected without revealing to the interviewer which question was chosen.

Since the type and ranges of the responses are the same for each question, no respondent can be classified with certainty *a posteriori* with respect to the sensitive characteristic. The privacy of the respondent is thereby protected but the information obtained from a sample of respondents is sufficient, with knowledge of the probability distribution(s) used in the design, to compute unbiased estimates of parameters associated with the sensitive characteristic, such as the proportion with that attribute. In most applications it is assumed that each respondent truthfully answers the question selected by the randomizing device. In view of the protection provided the respondents, randomized response offers unusual opportunities for eliciting accurate information on sensitive subjects, including even illegal and socially deviant behavior. It has no applicability to the individual respondent except in a probability sense and is used to describe only population groups.

CATEGORICAL DATA DESIGNS

Theoretical development of randomized response designs was relatively rapid, especially during the first several years after the initial Warner design. The Warner design and three alternate classes of designs highlight this early period.

The *original Warner design* developed a method for estimating the proportion π_A of persons with a sensitive attribute, A, without requiring the individual respondent to report his actual classification, whether it be A or not-A, to the interviewer. The respondent is provided with a randomizing device in order to choose one of two statements of the form:

I belong to group A (selected with probability p).

I do *not* belong to group A (selected with probability $1 - p$).

Without revealing to the interviewer which statement has been chosen, the respondent answers "yes" or "no" according to the statement selected. The responses to either question divide the sample into two mutually exclusive and complementary classes. With a random sample of n respondents, π_A is estimated by

$$\hat{\pi}_A = \frac{\hat{\lambda} - (1 - p)}{2p - 1}, \qquad p \neq \tfrac{1}{2},$$

where $\hat{\lambda}$ is the observed proportion of "yes" answers. This is an unbiased estimate if all respondents answer the selected statement truthfully, with variance given by

$$\operatorname{var}(\hat{\pi}_A) = \frac{\pi_A(1 - \pi_A)}{n} + \frac{p(1 - p)}{n(2p - 1)^2}.$$

The second term on the right-hand side represents the addition to the variance attributable to the uncertainty of the question selected. Had a direct question been asked, its response bias need not be too great for the resulting mean square error to exceed $\text{var}(\hat{\pi}_A)$.

Abul-Ela et al. [3] extended Warner's design to the trichotomous case to estimate the proportions of three related, mutually exclusive groups, one or two of which are sensitive, and showed that the extension is easily made to the multichotomous case. The basis for this design was to use an extra independent sample for each additional parameter in the multinomial case. Eriksson [12] and Bourke [7] have shown, however, how multinomial proportions can be estimated with only one sample using a different kind of randomizing device (see below).

The *unrelated question design* is attributed to Walt R. Simmons [23], who suggested that confidence of respondents in the anonymity provided by the randomized response might be increased, and hence their cooperation and the truthfulness of their responses, if one of the two questions referred to a nonsensitive, innocuous attribute, say Y, unrelated to the sensitive attribute A. Two such questions (in statement form) might be:

(A) I had an induced abortion last year.

(Y) I was born in the month of April.

The theoretical framework for this unrelated question randomized response design was developed extensively by Greenberg et al. [18], including recommendations on the choice of parameters. Abul-Ela [2] and Horvitz et al. [23] provided some earlier, but limited results for this design.

If π_Y, the proportion in the population with the nonsensitive attribute Y, is known in advance, only one sample is required to estimate the proportion with the sensitive attribute π_A. The estimator of π_A given π_Y, and its variance are

$$\hat{\pi}_A \mid \pi_Y = \frac{\hat{\lambda} - (1-p)\pi_Y}{p}$$

$$\text{var}(\hat{\pi}_A \mid \pi_Y) = \frac{\lambda(1-\lambda)}{np^2},$$

where $\hat{\lambda}$ is the proportion of "yes" answers as before, and λ is the probability of a "yes" response. If π_Y is not known, the design can be altered to permit estimation not only of the proportion with the sensitive characteristic, but π_Y as well. This requires two independent samples of sizes n_1 and n_2 to be selected with different probabilities, p_1 and p_2, of choosing the sensitive question. Moors [36] showed that for optimally allocated n_1 and n_2, it is best to choose $p_2 = 0$ in order to reduce the sampling variance of $\hat{\pi}_A$. That is, the randomizing device is used in the first sample only, while the second sample is asked the unrelated question directly to estimate π_Y. Moors further showed that with optimal choice of n_1 and n_2, and $p_2 = 0$, the unrelated question design would be more efficient than the Warner design for $p_1 > \frac{1}{2}$, regardless of the choice of π_Y. Dowling and Shachtman [11] proved that $\hat{\pi}_A$ in the unrelated question design has less variance than $\hat{\pi}_A$ in the original Warner design, for all π_A and π_Y, provided that p [or the $\max(p_1, p_2)$ in the two-sample case] is greater than approximately one-third.

Folsom et al. [14] evaluated an alternative two-sample design which consists of using two nonsensitive alternate questions Y_1 and Y_2, in conjunction with the sensitive question A. One sample is used to estimate π_{Y_1} in a direct question with the randomizing device choosing between A and Y_2. In the second sample the roles of Y_1 and Y_2 are reversed. Both samples are used to estimate π_A. In practice, two samples are never required inasmuch as Y can be selected such that π_Y is known or incorporated into the randomizing device as described below.

Contamination (permutation) designs are discussed by Boruch [6], who suggested that error be introduced into classificatory data during the interview for inquiries of sensitive attributes. He proposed to accomplish this by presenting the respondent with a single question (or statement) and instructing him or her to lie or tell the truth depending on the outcome of a randomizing device.

A different version of Boruch's contamination design had been proposed by Warner [41] to reduce the loss in efficiency with respondents who do not belong to the sensitive class. This loss may be reduced by keeping the proportion of false positives low.

Multiple trials designs seek to reduce additional variance introduced by the random selection of questions (or response sets) by repeated trials with each respondent. The use of two trials per respondent has been discussed by Horvitz et al. [23], Gould et al. [17], Liu and Chow [30], Chow and Liu [9] and Greenberg et al. [21]. Ordinarily, doubling the number of observations will reduce the variance in random samples by one-half. The gain in efficiency is somewhat less with two trials per respondent (pooled data), however, because of the correlation between the two responses.

QUANTITATIVE DATA DESIGNS

Randomized response designs need not be restricted to qualitative data. Greenberg et al. [20] extended the theory appropriate to the unrelated question randomized response design to estimating the mean and variance of the distribution of a quantitative measure, whether discrete or continuous. Assuming two independent samples of size n_1 and n_2, unbiased estimators for the means of the sensitive and nonsensitive distributions, μ_A and μ_Y, respectively, are

$$\hat{\mu}_A = \left[(1-p_2)\bar{Z}_1 - (1-p_1)\bar{Z}_2\right]/(p_1-p_2),$$

$$\hat{\mu}_Y = \left(p_2\bar{Z}_1 - p_1\bar{Z}_2\right)/(p_2-p_1),$$

where p_1 and p_2 are the probabilities with which the sensitive question is selected in the two samples, $p_1 \neq p_2$, and $\bar{Z}_1$ and $\bar{Z}_2$ are the sample means computed from the responses in the two samples. The variance of $\hat{\mu}_A$ is given by

$$V(\hat{\mu}_A) = \frac{1}{(p_2-p_1)^2}\left[(1-p_2)^2 V(\bar{Z}_1) + (1-p_1)^2 V(\bar{Z}_2)\right],$$

where the variance of $\bar{Z}_i$ for $i = 1, 2$ is

$$V(\bar{Z}_i) = \left[\sigma_Y^2 + p_i(\sigma_A^2 - \sigma_Y^2) + p_i(1-p_i)(\mu_A - \mu_Y)^2\right]/n_i.$$

From the standpoint of both efficiency and privacy of the respondent the nonsensitive variable Y should be chosen with mean and variance as close as possible to the corresponding moments for the sensitive variable.

Poole [37] used a contamination design to estimate the distribution function of a continuous variable. The respondent is asked to multiply the true response to the variable of interest by a random number and to tell only the result to the interviewer. This randomized information, together with known properties of the distribution of the random multiplier, is sufficient to estimate the distribution of the sensitive variable.

Eriksson [12] discussed in some detail a design for collecting sensitive data on a discrete quantitative variable. It is essentially the quantitative counterpart to his design, discussed below, for estimating a multinomial distribution*. He suggests a deck with two types of cards for the randomizing device; the first type of card asks the sensitive question and the second type of card asks the respondent to state a designated value for Y as shown on the card. For example,

> *Card-Type 1:* How many abortions have you ever had?
>
> *Card-Type 2:* State the numerical value shown on this card. (Cards will have values 0, 1, 2, . . . , m).

With a random sample of n respondents, an unbiased estimator of μ_A is

$$\hat{\mu}_A = \left[\bar{Z} - (1-p)\mu_Y\right]/p,$$

where $\bar{Z}$ is the mean response for all n respondents, μ_Y the mean of numerical values shown on card-type 2, and p the proportion of cards of the first type in the deck. The variance of $\hat{\mu}_A$ reduces to $\sigma_A^2/(np^2)$ when $\mu_A = \mu_Y$ and $\sigma_A^2 = \sigma_Y^2$.

THE LINEAR RANDOMIZED RESPONSE MODEL

The various randomized response designs discussed above have been developed without any apparent unifying framework. Warner [41] has supplied that framework with his formulation of a general linear randomized response model. Let X be a random vector of q elements, some of which are sensitive measures, with $E(X) = \pi$ or some linear function of π without observing sample values of X. The observations reported by the ith respondent in a random sample of size n are represented by the r-element column vector Y_i defined by the matrix product $Y_i = T_i X_i$, where T_i is an observation from an $r \times q$ random matrix of known distribution. The actual values of T_i and X_i (observed value of X) remain unknown to the interviewer and the researcher. Since $E(Y_i) = \tau_i \pi$, the model may be written as $Y = \tau\pi + U$, where $E(U) = 0$ and $E(UU')$ is the covariance matrix. Thus the linear randomized response model may be interpreted as an application of the generalized linear regression* model with weighted least-squares* used to estimate the unknown π. The randomized response designs discussed above are all special cases of this general model.

COMPARISON OF ALTERNATIVE RANDOMIZED RESPONSE DESIGNS

Under the best of design conditions, with p of the order of 0.7, the estimated proportion with the sensitive characteristic will have about twice the variability of the estimate obtained with direct questioning. This ratio can be reduced by schemes offering less privacy protection but possibly at the expense of introducing bias.

Greenberg et al. [21] have compared six different randomized response designs for estimating the proportion belonging to a sensitive class by computing design effects for π_A ranging from 0.05 to 0.40, p from 0.50 to 0.90, and sample sizes 100, 400, and 800 for each design. The design effect was computed as the ratio of the variance for the particular randomized response design to the mean square error of the estimated proportion obtained by a direct question with the same number of respondents. It was assumed that a proportion T_A of respondents with the sensitive attribute A would answer the direct question truthfully. It was also assumed that respondents not in the sensitive group would answer the direct question truthfully, as would all respondents for each of the randomized response designs. The designs compared included the Warner design and unrelated question design, an extended contamination design, and the Chow and Liu [9] multiple trials designs.

The comparisons indicate rather conclusively that randomized response designs should be given consideration over direct questioning whenever $T_A < 1.00$, and, depending on π_A, p and T_A, with samples as small as 100. Except for the Warner design, all the designs compared were more efficient than direct questioning for $n \geqslant 400$ and $\pi_A \geqslant 0.10$, $p \geqslant 0.70$ and $T_A < 0.90$. None of the designs was found to be uniformly superior, although the unrelated question designs were generally efficient, and the multiple trials design slightly better when equivalent to 10 trials per respondent and $\pi_A \geqslant 0.20$.

RANDOMIZING DEVICES

Randomized response requires a mechanical device or rule of procedure that selects at random with known probability the question to be answered by the respondent. This process must be unbiased, completely adaptable to use in the field, and easily understood and acceptable to the respondent with assurance of the protection of privacy in its use. The device suggested by Warner [40] in the original paper was a spinner with an arrow pointer similar to the type found in games that involve a chance element. This device is not too satisfactory, because the spinner must be perfectly horizontal to be unbiased and confusion results when the arrow rests on the line separating two segments.

In early field trials of randomized response reported by Horvitz et al. [23], use was made instead of decks of specially printed cards. An identical pack of 50 cards with each card containing one of two alternate questions was used by each interviewer.

The respondent was asked to mix the cards and select one for reply. This is not a completely satisfactory randomizing device since interviewers can lose cards from the deck. Further, respondents may not shuffle the cards adequately to generate the expected probabilities for selecting the alternate questions.

Cards are also used by Eriksson [12] and Bourke [7] in an ingenious manner with the unrelated question to estimate $k > 2$ mutually exclusive classes with $(k - 1)$ sensitive groups and requiring only one sample. For example, Bourke suggests cards on which each has three (or more) statements listed in a numbered, permuted sequence. After selecting a card at random, the respondent replies by stating the number associated with his or her category on that card. Another variation using special cards for quantitative discrete variables by Eriksson was described in the foregoing text. All card designs, however, have the problem of adequate shuffling to reproduce the design probabilities.

To overcome the sampling difficulties with cards, Abernathy et al. [1] developed a randomizing device that consisted of a sealed, transparent, plastic box approximately the size of a pack of cigarettes. Inside are 50 colored beads, tested to be electrostatically free. When the box is shaken, beads can roll down a small track to a window at the end of the track which displays one and only one bead. On the front surface of the box, there is a printed card with each color bead in front of the question to be answered. The respondent is advised to inspect the box on the obverse side to see that there are at least two colors of beads inside. The respondent is also encouraged to shake the box several times to assure that each color can appear in the window. This device fulfills all the criteria mentioned except that sets of boxes are not always easily accessible to investigators.

The plastic box can also be used as a randomizing device with Π_Y built into the design. Using three colors of beads, say red, white, and blue, the respondent answers the sensitive question if the red bead appears, and is instructed to reply "yes" when the white one shows, and "no" when the bead is blue. The value of Π_Y is the proportion of white beads to the combined total number of white and blue beads.

A variation of the sealed box is the sampler developed by Chow and Liu [9] and discussed by Liu et al. [32] and Liu and Chow [31]. They constructed a volumetrical flask with a thin narrow neck and inside the spherical bottle are six red and six white balls. The respondent is told that the red balls refer to the sensitive category and the white ones to the nonsensitive category. The respondent is then asked to shake the device thoroughly before turning it upside down, permitting exactly five balls to move into the neck of the bottle, which is frosted on one side so that the interviewer cannot observe the result of the trial. Without mentioning color, respondents who belong to the sensitive class report the number of red balls in the neck of the bottle and respondents who do not belong to the sensitive class report the number of white balls. This yields an estimate with variance roughly equivalent to the variance of the Warner estimate based upon five trials per respondent. As with the plastic box, surveyors may not find this special bottle to be easily accessible.

After the unrelated question technique was developed, the Warner restriction that $p \neq \frac{1}{2}$ could be dropped. This opened up the possibility of using a coin as the randomizing device. The coin apparently works well except for the disadvantage that since the probability of selecting the sensitive question is only 50%, sample size must be rather large (i.e., four times greater than for $p = 1$) if the sampling variance is to be kept within reasonable bounds.

A simple variation of the coin toss is to use two coins. The respondent is told to answer the unrelated question if both coins are heads, but to answer the sensitive question otherwise. This procedure reduces variance by more than one-half over a single coin but may also decrease confidence in the technique and ability to follow instructions.

A randomizing procedure that can be

used in telephone* interviews as well as mail questionnaires is to instruct the respondent to answer a specific question according to his or her birth month. For example, if the respondent was born in any one of the months from January through August, the first question is answered; if the month of birth was September through December, the second question is answered. Respondent's month of birth should not be used where the date might be known to the survey designer or interviewer, and the unrelated question must also avoid the question of birth. In the latter case, a good substitute for the unrelated question is the month of birth of the respondent's mother.

RESPONDENT HAZARD

An honest, affirmative response to a direct question which is truly sensitive results in classification of the respondent as a member of stigmatizing group A with certainty. The purpose of the randomized response technique is to reduce the hazard to the respondent who belongs to group A as being perceived as such a member and redistributing some of the jeopardy to the usually larger number of not-A (i.e., $\bar{A}$) group members who also reply in the affirmative.

A "yes" response by a randomly selected respondent to the question chosen by a randomizing device in a randomized response design increases the prior probability* Π_A (that the respondent belongs to the sensitive group) to a posterior probability* greater than Π_A but less than 1. The exact posterior probability or hazard for a given design can be computed using Bayes' theorem* and specific parameter values for p, Π_A, and Π_Y. The challenge to the statistician is to choose parameter values p and Π_Y such that the hazard to the respondent, whether from group A or $\bar{A}$, is at a level low enough to induce an accurate response, yet at the same time minimize the added variance associated with the use of a randomizing device to select the question to be answered.

Lanke [28] measured the "risk of suspicion" for those respondents who reply "yes" in an unrelated question design, where Π_Y is known, as

$$P(A \mid \text{Yes}) = \frac{\Pi_A p + \Pi_A(1-p)\Pi_Y}{\Pi_A p + (1-p)\Pi_Y}.$$

He recommended that this risk should be bounded and Y chosen so that $\Pi_Y = 1$, resulting in the maximum number of respondents replying "yes" for a given p. Loynes [34] came to the same conclusion.

Anderson [4] pointed out that there is also a risk of suspicion if the response is "no," and he added to $P(A \mid \text{Yes})$ the value of $P(A \mid \text{No}) = 1 - P(\bar{A} \mid \text{No})$. Leysieffer [29] had previously generalized this by saying that any respondent is in jeopardy if $P(A \mid \text{Response}) > \Pi_A$, and comparisons of efficiency in design must be for a given level of respondent jeopardy. Leysieffer also concluded that among all unrelated question designs with the same level of jeopardy, $\Pi_Y = 1$ has the smallest variance.

Greenberg et al. [22] have addressed this problem by pointing out that too much reduction in hazard for A respondents endangers the cooperation of $\bar{A}$ respondents of which there are $(1 - \Pi_A)/\Pi_A$ times more of them. They defined the benefit B_A to the A respondents as $(1 - H_A)$, where

$$H_A = P(\text{A is perceived as A} \mid \text{Yes}),$$

and the corresponding benefit $B_{\bar{A}}$ to the $\bar{A}$ respondents as $(-H_{\bar{A}})$, where

$$H_{\bar{A}} = P(\bar{A} \text{ is perceived as A} \mid \text{Yes}).$$

Inasmuch as B_A is a true reduction in hazard or gain and $B_{\bar{A}}$ is really a loss, they maximized the ratio $B_A/B_{\bar{A}}$. This ratio is a monotonically decreasing function of $(1-p)\Pi_Y$. This means that to maximize the real benefit for both A and $\bar{A}$ respondents as well as to reduce the sampling error of $\hat{\Pi}_A$, it is better to select an unrelated question with a low frequency, usually in the order of magnitude of Π_A, and to choose as high a value for p as will be acceptable to the respondents. (A value of p in the range 0.50 to 0.75 has proved to be acceptable in many field trials.) These criteria for selecting p and

Π_Y had previously been recommended by Greenberg et al. [18] and Moors [36] as the preferred strategy for randomized response unrelated question designs without formal consideration of respondent hazard.

VALIDATION AND COMPARISON STUDIES

Optimal validation tests of randomized response require knowledge of the actual attribute or behavioral category for each person selected for the test. As a minimum, validation can also be tested by having knowledge of the actual value of the population parameter to be estimated by the randomized response sample. When validation data are not available, an indication of the efficacy of the technique may be obtained by comparison of the randomized response-derived estimates with estimates derived from other methods. An indication of efficacy may also be obtained by using the two-sample unrelated question design to estimate π_Y when that population parameter was actually known beforehand, such as proportion born in a given month. Finally, confidence that the procedure was properly administered and accepted by respondents may also be gained through analysis of the consistency and plausibility of randomized response estimates across demographic and socioeconomic subgroups of the sample.

The validity of the technique has been tested on a variety of characteristics considered to be subject to inaccurate responses in sample surveys, including illegitimate births, persons charged with drunken driving, persons registered to vote, number of college courses failed, and number of arrests in a specified time period (see, e.g., Horvitz et al. [23], Locander et al. [33], Lamb and Stern [27] and Tracy and Fox [39]).

Comparison tests have been reported for variables such as illicit drug usage, abortion incidence and prevalence, alcoholic beverage drinking and driving behavior, and child abuse (see, e.g., Brown and Harding [8], Goodstadt and Gruson [16], Zdep et al. [43], Krotki and Fox [26], I-Cheng et al. [24], Gerstel et al. [15], Barth and Sandler [5], Zdep and Rhodes [42], and Fidler and Kleinknect [13]).

Overall, although conflicting results have been realized in some of the validation and comparison tests, the randomized response technique has generally yielded a greater frequency of reporting of sensitive behavior, and its estimates are closer to actual values than those obtained by direct questioning.

APPLICATIONS

Reports in the literature on experience with randomized response designs have increased considerably as awareness of the technique has become widespread. One of the earliest studies, reported by Abernathy et al. [1], provided estimates of the incidence and prevalence of induced abortion in urban North Carolina. Both single-sample and two-sample unrelated-question designs were used. Women in the same sample were also queried about use of the contraceptive pill; with the same type of plastic box randomizing device, Greenberg et al. [19] reported the findings of this component of the study as well as the results of using the randomizing device with a value for Π_Y built in to obtain information on emotional problems. Experience with randomized response collection of quantitative data on number of abortions and also on annual income was reported by Greenberg et al. [20].

Other applications reported in the literature include a study of organized crime in Illinois which used the Warner design with $p = 0.20$ to estimate the proportions of the adult population in metropolitan areas of the state that engaged in illegal betting and the proportion which have had contact with organized crime reported by IIT Research Institute [25]; a study that used the two-alternate-questions design with a coin toss to estimate the annual incidence of at-fault automobile accidents among licensed drivers who drink alcoholic beverages reported by Folsom et al. [14]; a study of the extent of deliberate concealment of deaths in household surveys of vital events in the Philippines reported by Madigan et al. [35]; and a study of abortion in a national survey reported by Shimizu and Bonham [38].

An excellent bibliography of randomized response, including additional applications of the technique, may be found in Daniel [10].

References

[1] Abernathy, J. R., Greenberg, B. G., and Horvitz, D. G. (1970). *Demography*, **7**, 19–29. (Early application of randomized response. Excellent interview procedures.)

[2] Abul-Ela, A. A. (1966). Unpublished Ph.D. thesis, University of North Carolina, Chapel Hill, N.C.

[3] Abul-Ela, A. A., Greenberg, B. G., and Horvitz, D. G. (1967). *J. Amer. Statist. Ass.*, **62**, 990–1008.

[4] Anderson, H. (1976). *Int. Statist. Rev.*, **44**, 213–217.

[5] Barth, J. T. and Sandler, H. M. (1976). *J. Stud. Alcohol*, **37**, 690–693.

[6] Boruch, R. F. (1972). *Social Sci. Res.*, **1**, 403–414.

[7] Bourke, P. D. (1974). Errors in Surveys Research Project. *Rep. No. 74*, Institute of Statistics, University of Stockholm, Sweden.

[8] Brown, G. H. and Harding, F. D. (1973). *Tech. Rep. 73-9*, Human Resources Research Organization, Alexandria, Va.

[9] Chow, L. P. and Liu, P. T. (1973). A New Randomized Response Technique: The Multiple Answer Model. Department of Population Dynamics, School of Hygiene and Public Health, Johns Hopkins University, Baltimore, Md. (mimeo).

[10] Daniel, W. W. (1979). Collecting Sensitive Data by Randomized Response: An Annotated Bibliography. *Res. Monogr. No. 85*. College of Business Administration, Georgia State University, Atlanta, Ga. (Thorough coverage, well annotated.)

[11] Dowling, T. A. and Shachtman, R. (1975). *J. Amer. Statist. Ass.*, **70**, 84–87.

[12] Eriksson, S. A. (1973). *Int. Statist. Rev.*, **41**, 101–113.

[13] Fidler, D. S. and Kleinknect, R. E. (1977). *Psychol. Bull.*, **84**, 1045–1049.

[14] Folsom, R. E., Greenberg, B. G., Horvitz, D. G., and Abernathy, J. R. (1973). *J. Amer. Statist. Ass.*, **68**, 525–530.

[15] Gerstel, E. K., Moore, P., Folsom, R. E., and King, D. A. (1970). Mecklenburg County Drinking—Driving Attitude Survey. Report Prepared for U.S. Department of Transportation, Research Triangle Institute, Research Triangle Park, N.C.

[16] Goodstadt, M. W. and Gruson, V. (1975). *J. Amer. Statist. Ass.*, **70**, 814–818.

[17] Gould, A. L., Shah, B. V., and Abernathy, J. R. (1969). *Amer. Statist. Ass., Proc. Social Statist. Sec.*, pp. 351–359. (Suggests respondent behavior model to explain response distribution.)

[18] Greenberg, B. G., Abul-Ela, A. A., Simmons, W. R., and Horvitz, D. G. (1969). *J. Amer. Statist. Ass.*, **64**, 520–539.

[19] Greenberg, B. G., Abernathy, J. R., and Horvitz, D. G. (1970). *Milbank Mem. Fund Quart.*, **48**, 39–55.

[20] Greenberg, B. G., Kuebler, R. R., Jr., Abernathy, J. R., and Horvitz, D. G. (1971). *J. Amer. Statist. Ass.*, **66**, 243–250.

[21] Greenberg, B. G., Horvitz, D. G., and Abernathy, J. R. (1974). *Reliability and Biometry, Statistical Analysis of Lifelength*, F. Prochan and R. J. Serfling, eds. SIAM, Philadelphia, pp. 787–815.

[22] Greenberg, B. G., Kuebler, R. R., Abernathy, J. R., and Horvitz, D. G. (1977). *J. Statist. Plann. Inf.*, **1**, 53–60.

[23] Horvitz, D. G., Shah, B. V., and Simmons, W. R. (1967). *Amer. Statist. Ass., Proc. Social Statist. Sec.*, pp. 65–72.

[24] I-Cheng, C., Chow, L. P., and Rider, R. V. (1972). *Stud. Family Plann.*, **3**, 265–269.

[25] IIT Research Institute and the Chicago Crime Commission (1971). A Study of Organized Crime in Illinois. Report prepared for the Illinois Law Enforcement Commission, Chicago.

[26] Krotki, K. J. and Fox, B. (1974). *Amer. Statist. Ass., Proc. Social Statist. Sec.*, pp. 367–371.

[27] Lamb, C. W., Jr., and Stern, D. E., Jr. (1978). *J. Marketing Res.*, **15**, 616–621.

[28] Lanke, J. (1976). *Int. Statist. Rev.*, **44**, 197–203.

[29] Leysieffer, F. W. (1975). Respondent Jeopardy in Randomized Response Procedures. *FSU Statist. Rep. M338, ONR Tech. Rep. No. 93*, Department of Statistics, Florida State University, Tallahassee, Fla.

[30] Liu, P. T. and Chow, L. P. (1976). *Biometrics*, **32**, 607–618.

[31] Liu, P. T. and Chow, L. P. (1976). *J. Amer. Statist. Ass.*, **71**, 72–73.

[32] Liu, P. T., Chow, L. P., and Mosley, W. H. (1975). *J. Amer. Statist. Ass.*, **70**, 329–332.

[33] Locander, W., Sudman, S., and Bradburn, N. (1974). *Amer. Statist. Ass., Proc. Social Statist. Sec.*, pp. 21–27. (See also *J. Amer. Statist. Ass.*, **71**, 269–275, 1976.)

[34] Loynes, R. M. (1976). *J. Amer. Statist. Ass.*, **71**, 924–928.

[35] Madigan, F. C., Abernathy, J. R., Herrin, A. N., and Tan, C. (1976). *Popul. Stud.*, **30**, 295–303.

[36] Moors, J. J. A. (1971). *J. Amer. Statist. Ass.*, **66**, 627–629.

[37] Poole, W. K. (1974). *J. Amer. Statist. Ass.*, **69**, 1002–1005.

[38] Shimizu, I. M. and Bonham, G. S. (1978). *J. Amer. Statist. Ass.*, **73**, 35–39.

[39] Tracy, P. E. and Fox, J. A. (1981). *Amer. Sociol. Rev.*, **46**, 187–200.

[40] Warner, S. L. (1965). *J. Amer. Statist. Ass.*, **60**, 63–69. (Original paper introducing randomized response.)

[41] Warner, S. L. (1971). *J. Amer. Statist. Ass.*, **66**, 884–888.

[42] Zdep, S. M. and Rhodes, I. N. (1976). *Public Opinion Quart.*, **40**, 531–537.

[43] Zdep, S. M., Rhodes, I. N., Schwarz, R. M., and Kilkenny, M. J. (1979). *Public Opinion Quart.*, **43**, 544–549.

(SURVEY SAMPLING)

B. G. Greenberg
J. R. Abernathy
D. G. Horvitz

RANDOMIZED TESTS

Statistical inference* procedures are generally based on the outcome of a random experiment or an observable random variable. In a randomized procedure, the inference is also based on the outcome of some additional random experiment which is unrelated to the original random experiment. A randomized test is one that rejects with a specified probability for each possible outcome X. An ordinary (nonrandomized) test is a special case of a randomized test where the specified rejection probability is always equal to 1 or zero. In symbols, a randomized test is frequently denoted by $\phi(x)$, the critical function, $0 \leqslant \phi(x) \leqslant 1$, which represents the probability of rejection of the null hypothesis for various outcomes x. In a nonrandomized test, $\phi(x)$ takes on only the values 0 and 1.

As a numerical example, consider a binomial test at level 0.10 with null hypothesis $H_0: p \geqslant 0.4$ and alternative $H_1: p < 0.4$ for p the probability of success in the population. The random variable on which the test should be based is the number of successes X in a sample of n and the appropriate rejection region is small values of X. For $n = 6$, say, we should reject for $X \leqslant c$, where c is chosen such that

$$\Pr[X \leqslant c] = \sum_{k=0}^{c} \binom{6}{k} (0.4)^k (0.6)^{6-k} = 0.10. \tag{1}$$

The left tail cumulative binomial probabilities for $n = 6$, $p = 0.4$ are $\Pr[X \leqslant 0] = 0.0467$, $\Pr[X \leqslant 1] = 0.2333$. It is clear that there is no integer c which satisfies (1); hence a nonrandomized test at level 0.10 would reject for $X = 0$ only and the actual level is 0.0467. A randomized test at level 0.10 is to reject always when $X \leqslant 0$ and reject with probability $\phi(1)$ when $X = 1$, where $\phi(1)$ is chosen such that

$$\Pr[X \leqslant 0] + \phi(1)\Pr[X = 1] = 0.10$$

$$0.0467 + 0.1866\phi(1) = 0.10 \text{ or } \phi(1) = 0.2856.$$

If we observe $X = 1$, a separate randomized procedure is performed to see whether or not rejection of H_0 is appropriate.

The probability of rejection of H_0 by this randomized test for any p is equal to

$$\Pr[p] = (1-p)^6 + 0.2856\left[6p(1-p)^5\right]; \tag{2}$$

this value is always greater than $(1-p)^6$, the probability of rejection by the nonrandomized test, for any $p = 0, 1$. Therefore, the power of this randomized test for any $0 < p < 0.4$ is always larger than that of the nonrandomized test, and the probability of a type I error never exceeds the given level 0.10 for any $p \geqslant 0.4$.

A right-tail randomized test should be used for $H_0: p \leqslant 0.4$ and $H_1: p > 0.4$. If the alternative is two-sided, as in $H_0: p = 0.4$, $H_1: p \neq 0.4$, a two-tailed randomized test should be used. This would be of the form: reject always if $X < c_1$ or $X > c_2$, reject with probability ϕ_1 if $X = c_1$, reject with probability ϕ_2 if $X = c_2$, and do not reject otherwise. For a test at exact level α, c_1, c_2, ϕ_1, and ϕ_2

must be chosen such that

$$\Pr[X < c_1] + \phi_1\Pr[X = c_1] + \Pr[X > c_2] + \phi_2\Pr[X = c_2] = \alpha, \qquad (3)$$

where $c_1 < c_2$, $0 \leqslant \phi_1, \phi_2 \leqslant 1$ or $\phi_1 + \phi_2 \leqslant 1$ if $c_1 = c_2$.

Randomized tests are probably seldom used in applications, even in the one-sided case where they are uniquely defined. In practice, most researchers take the conservative approach and use a nonrandomized test with a critical value that makes the exact level as close to α as possible without exceeding α. Randomized tests are important in theory though, especially for test statistics with a discrete null sampling distribution when the choice of exact levels is extremely limited. In comparing the power of two tests, for example, the exact level of the two tests must be the same for the comparison to be unbiased. Further, a randomized test is usually the optimum test for a statistic with a discrete null sampling distribution.

Few textbooks discuss randomized tests. Notable exceptions are Lehmann [1] and Pratt and Gibbons [2].

References

[1] Lehmann, E. L. (1959). *Testing Statistical Hypotheses*. Wiley, New York. (Graduate-level text and reference book on inference.)

[2] Pratt, J. W. and Gibbons, J. D. (1981). *Concepts of Nonparametric Theory*. Springer-Verlag, New York. (Graduate-level text and reference book on nonparametric statistical inference.)

(HYPOTHESIS TESTING
NEYMAN–PEARSON LEMMA)

JEAN DICKINSON GIBBONS

RANDOM MATRICES

Random matrices occur in many contexts, ranging from dynamical systems in physics and biology through to multivariate statistics and the Wishart* matrix.

For example, in quantum mechanics* the energy levels of a system are described by eigenvalues of an operator called the Hamiltonian, which is represented as a random matrix (e.g., Mehta [9]). In mathematical biology, model ecosystems are often discussed in terms of a square (random) community matrix $\mathbf{A} = \mathbf{B} - \mathbf{I}$, with $\mathbf{I}$ the identity and $\mathbf{B}$ having elements which are independent and identically distributed (i.i.d.) with zero mean and variance σ^2 (e.g., May [8, pp. 62–65] and Cohen and Newman [1, Sec. 4]). The eigenvalues* of $\mathbf{A}$ determine the stability consequences of various biological assumptions as the size of $\mathbf{A}$ increases.

Another standard application comes from demography* and concerns the evolution of an age-structured population which is modeled by the equation

$$\mathbf{Y}_{t+1} = \mathbf{X}_{t+1}\mathbf{Y}_t, \qquad t \geqslant 0,$$

where $\mathbf{Y}_k$ is the (column) vector of the number of individuals in each age class at time k and $\{\mathbf{X}_k\}$ is a sequence of random matrices of vital rates. Here the behavior of the system is determined by that of the backwards matrix products $\mathbf{X}_t\mathbf{X}_{t-1} \cdots \mathbf{X}_1$ (e.g., Tuljapurkar and Orzack [11], Heyde and Cohen [6]).

Quite a substantial theory has been developed which elucidates the examples above and a range of similar ones. Most of this theory is concerned with investigating either

1. The spectrum of an $n \times n$ random matrix as n increases, or
2. The asymptotic behavior of a product of t random matrices (of fixed size) as t increases.

We shall illustrate this theory with some particular results, first relating to problem 1.

Theorem 1 (Geman [4]). Let $\{v_{ij},\ i \geqslant 1,\ j \geqslant 1\}$ be i.i.d. random variables with $Ev_{11} = 0$ and $E|v_{11}|^n \leqslant n^{\alpha n}$ for all $n \geqslant 2$ and some α. Then, if $\mathbf{V}_n = (v_{ij})_{1 \leqslant i,j \leqslant n}$, the maximum eigenvalue of $n^{-1}\mathbf{V}_n\mathbf{V}_n'$, $\lambda_{\max}(n)$, satisfies

$$\lim_{n\to\infty} \lambda_{\max}(n) = 2^2 Ev_{11}^2 \text{ a.s.} \quad \text{(almost surely)}.$$

Next, let $\mathbf{A}(t)$, $t \geqslant 1$, be random $n \times n$ matrices with elements $(\mathbf{A}(t))_{ij}$, $1 \leqslant i, j \leqslant n$, and let $x(0)$ be a nonzero vector in $\mathbb{R}^n$. Define

$$\mathbf{x}(t) = \mathbf{A}(t)\mathbf{A}(t-1)\cdots\mathbf{A}(1)\mathbf{x}(\mathbf{0}), \qquad t \geqslant 1,$$

and let $\|\mathbf{x}\|$ denote $(\sum_{j=1}^{n} x_j^2)^{1/2}$ if $\mathbf{x}' = (x_1, \ldots, x_n)$. Also, write $\log^+ u$ for $\max(0, \log u)$.

Theorem 2 (Furstenberg and Kesten [3]). If $\{\mathbf{A}(t),\ t \geqslant 1\}$ is a stationary sequence of random matrices and

$$E\left(\log^+ \max_i \sum_{j=1}^{n} |\mathbf{A}(1)|_{ij}\right) < \infty,$$

then $\lim_{t\to\infty} \log\|\mathbf{x}(t)\|$ exists a.s. and is finite.

In general the limit in Theorem 2 may be random but under more specific assumptions much more explicit results can sometimes be obtained, as in the following theorem.

Theorem 3 (Cohen and Newman [1]). Suppose that $\{\mathbf{A}(1)_{ij},\ 1 \leqslant j, j \leqslant n\}$ are i.i.d. $N(0, s^2)$ variables. Then

$$\begin{aligned}\lim_{t\to\infty} & t^{-1}\log\|\mathbf{x}(t)\| \\ &= \tfrac{1}{2}\left(\log s^2 + \log 2 + \psi(n/2)\right) \\ &= \log\lambda \text{ a.s.,}\end{aligned}$$

say, independent of $\mathbf{x}(0)$, where ψ is the digamma function*. Moreover, for any $\mathbf{x}(0) \neq 0$,

$$t^{-1/2}(\log\|\mathbf{x}(t)\| - t\log\lambda)$$

converges in distribution to $N(0, \sigma^2)$ with $\sigma^2 = \frac{1}{4}\psi'(n/2)$.

Now suppose that the $\mathbf{A}(t)$ are independent while the elements of $\boldsymbol{\beta}_t = (\beta_1^{(t)}(\mathbf{x}(t)), \ldots, \beta_n^{(t)}(\mathbf{x}(t)))$ are measurable functions of the elements of $x(t)$. Girko [5] has given sufficient conditions for the convergence of $\boldsymbol{\beta}_t - E\boldsymbol{\beta}_t$ to infinitely divisible* laws.

Rather surprisingly, it can happen that the product $\mathbf{A}(t)\mathbf{A}(t-1)\cdots\mathbf{A}(1)$ of i.i.d. nonnegative matrices of size n can converge in distribution, *without normalization*, as $t \to \infty$, to a distribution not concentrated on the zero matrix. Kesten and Spitzer [7] provide various equivalent sufficient conditions for this behavior. It should be noted that this behavior is impossible for $n = 1$ since $\log\{\mathbf{A}(t)\ldots\mathbf{A}(1)\}$ is then a sum of i.i.d. random variables and it is well known that such a sum cannot converge in distribution unless all the summands $\log \mathbf{A}(i)$ are zero with probability 1.

Finally, mention needs to be made of random matrices occurring in multivariate analysis*. Here the emphasis is usually on distribution theory rather than asymptotic behavior. However, for situations where multivariate normality cannot realistically be assumed, asymptotic expansions* can provide useful information. For example, Edgeworth-type expansions have been used to investigate the asymptotic joint distributions of certain functions of the eigenvalues of the sample covariance matrix, correlation matrix, and canonical correlation matrix by Fang and Krishnaiah [2].

Now let $\mathbf{u}_i$ have the $N_p(\boldsymbol{\mu}_i, \boldsymbol{\Sigma})$ distribution, $i = 1, 2, \ldots, k$, namely p-dimensional normal with mean vector $\boldsymbol{\mu}_i$ and covariance matrix $\boldsymbol{\Sigma}$. Then the most ubiquitous of the random matrices studied in multivariate analysis is $\mathbf{S} = \sum_{j=1}^{k} \mathbf{u}_j\mathbf{u}_j'$. This has a Wishart distribution on k degrees of freedom, $W_p(k, \boldsymbol{\Sigma}, \mathbf{M})$, where $\mathbf{M}'$ is a $p \times k$ matrix with $\boldsymbol{\mu}_1, \boldsymbol{\mu}_2, \ldots, \boldsymbol{\mu}_k$ as its columns (e.g., Rao [10, Chap. 8]). For more details, *see* MULTIVARIATE ANALYSIS *and* WISHART DISTRIBUTION.

References

[1] Cohen, J. E. and Newman, C. M. (1984). *Ann. Prob.*, **12**, 283–310.

[2] Fang, C. and Krishnaiah, P. R. (1982). *J. Multivariate Anal.*, **12**, 39–63.

[3] Furstenberg, H. and Kesten, H. (1960). *Ann. Math. Statist.*, **31**, 457–469.

[4] Geman, S. (1980). *Ann. Prob.*, **8**, 252–261.

[5] Girko, V. L. (1982). *Theor. Prob. Appl.*, **27**, 837–844.

[6] Heyde, C. C. and Cohen, J. E. (1985). *Theor. Popul. Biol.*, **27**, 120–153.

[7] Kesten, H. and Spitzer, F. (1984). *Z. Wahrscheinl. verw. Geb.*, **67**, 363–386.

[8] May, R. M. (1974). *Stability and Complexity in Model Ecosystems*. Princeton University Press, Princeton, N.J.

[9] Mehta, M. L. (1967). *Random Matrices and the Statistical Theory of Energy Levels*. Academic Press, New York.

[10] Rao, C. R. (1973). *Linear Statistical Inference and Its Applications*, 2nd ed. Wiley, New York.

[11] Tuljapurkar, S. D. and Orzack, S. H. (1980). *Theor. Popul. Biol.*, **18**, 314–342.

(MATRIX-VALUED DISTRIBUTIONS
MULTIVARIATE ANALYSIS
WISHART DISTRIBUTION)

C. C. HEYDE

RANDOMNESS AND PROBABILITY—COMPLEXITY OF DESCRIPTION

The problem of randomness remained controversial for a long time after the introduction of the axioms* for probability theory* in Kolmogorov [13]. (The survey by Fine [9] is a good introduction to several alternative approaches to the foundations* of probability theory.) We outline the solution offered by algorithmic information theory*, its applications, and its history.

Given a measurable space $(\Omega, \mathscr{E})$, an experiment with outcome $\omega \in \Omega$, and a probability distribution μ over the events in $\mathscr{E}$, a statistician's task is to test the hypothesis that the distribution of possible outcomes of the experiment was μ. To the degree to which μ is accepted, we can consider ω *random* with respect to μ. (Some properties of μ can often be taken for granted. In statistics, ω is generally a large independent sample, but this is not true in other applications: for example, testing pseudo-random sequences, or inductive inference*.)

The following example from Levin [21] points out the principal difficulty. Suppose that in some country, the share of votes for the ruling party in 30 consecutive elections formed a sequence $0.99x_i$ where for every even number i, x_i is the ith digit of $\pi = 3.1415\ldots$. Although many of us would feel that these elections were not fair, to prove this mathematically turns out to be surprisingly difficult.

In fair elections, every sequence ω of n digits has approximately the probability $P_n(\omega) = 10^{-n}$ to appear as the actual sequence of last digits. Let us fix n. If the government guarantees the fairness, it must agree for any function $t(\omega)$ with $\sum_\omega P_n(\omega) t(\omega) \leqslant 1$ (a payoff function, a *martingale**), to the following *bet*: we pay 1 dollar in advance and receive $t(X)$ dollars for the actual sequence X of last digits. Let $t_0(\omega)$ be $10^{n/2}$ for all sequences ω whose even digits are given by π, 0 otherwise. If we could propose the bet t_0, these elections would cost the government $10^{n/2} - 1$ dollars. Unfortunately, we must propose the bet *before* the elections, and it is unlikely that we would come up exactly with the martingale t_0. We need a new idea to proceed.

Let us introduce the *complexity* of a mathematical object as its length of definition (in some formal sense to be specified below). Notice that t_0 is a very *simple* martingale, and that since the number of short strings of symbols is small, so is the number of martingales having a short definition. Therefore, we can afford to make *all* simple bets in advance and still win by a wide margin (for n large enough).

It turns out below that the principle saying that on a random outcome, all sufficiently simple martingales take small values, can be replaced by the more elegant principle saying that *a random outcome* itself *is not too simple*.

FORMAL RESULTS

We assume here the sample space Ω to be discrete: we identify it with the set of natural numbers. Some additional technical problems arise in the case when Ω is continuous (e.g., the set of infinite sequences of natural numbers); but the results are analogous, and even more spectacular, because with nontrivial sets of probability 0, a sharp distinction occurs between random and nonrandom sequences.

Binary strings will be identified with the natural numbers they denote. A *universal computer* will be used to *interpret* a description. We assume some familiarity with the theory of computations (see, e.g., Yasuhara [38]). Let us consider a computer (e.g., Turing machine) T with binary strings p as programs, natural numbers x, and $T(p,x)$ as data and output (if defined). We suppose that if $T(p,x)$ is defined, then $T(q,x)$ is not defined for any prefix q of p. Such machines are called *self-delimiting* (SD).

The *conditional complexity* $K_T(x \mid y)$ of the number x with respect to the number y is the length of the shortest program p for which $T(p, y) = x$. The *invariance theorem* asserts that the function $K_T(x \mid y)$ depends only weakly on the machine T: there is a SD computer U with the property that for any SD machine T a constant c_T exists with $K_U(x \mid y) \leqslant K_T(x \mid y) + c_T$. We fix U and put $K(x \mid y) = K_U(x \mid y)$ and $K(x) = K(x \mid 0)$. [The same quantity is denoted by $I(x)$ in ALGORITHMIC INFORMATION THEORY.] The function $K(x \mid y)$ is not computable. We can compute a nonincreasing, convergent sequence of approximations to it (it is *semicomputable* from above), but will not know how far to go in this sequence for some prescribed accuracy.

In what follows we consider only probability measures μ which are *computable*: there is a program computing $\mu(\omega)$ for any outcome $\omega \in \Omega$ to any desired degree of accuracy. All our logarithms have base 2. A statement $A(\ldots O(1) \ldots)$ means that there is a constant c for which $A(\ldots c \ldots)$ holds.

Theorem 1. For any computable function μ with $\sum_\omega \mu(\omega) \leqslant 1$, we have

$$K(\omega) \leqslant -\log \mu(x) + K(\mu) + O(1),$$

where $K(\mu)$ is defined in the obvious way.

Put $d(\omega \mid \mu) = -\log \mu(\omega) - K(\omega)$.

Theorem 2. For any y, $\sum_\omega 2^{-K(\omega \mid y)} < 1$. Hence $2^{d(\omega \mid \mu)}$ is a martingale with respect to μ. By Markov's inequality*, for any $m > 0$,

$$\mu\{\omega : K(\omega) < -\log \mu(\omega) - m\} < 2^{-m}.$$

Applying Theorem 1 to the election example with $\mu = P_n$, we get $K(x \mid n) \leqslant n \log 10 + O(1)$ for all n-digit strings x (these can be coded with natural numbers). Theorem 2 says that this estimate is sharp for most sequences x. (We used conditional complexity, since n is a parameter in P_n.) If every other digit of x comes from π, then $K(x \mid n) \leqslant (n \log 10)/2 + c$ with some constant c.

Theorems 1 and 2 say that with large probability, the complexity $K(\omega)$ of a random outcome ω is close to its upper bound $-\log \mu(\omega) + K(\mu)$. This law occupies a distinguished place among the "laws of probability," because if the outcome ω violates *any* law of probability, the complexity falls far below the upper bound. Indeed, if ω_0 does not satisfy some law of probability, then for some large number m, there is a computable martingale $t(\omega)$ of complexity $< m/2$, with $t(\omega_0) > 2^m$. Then Theorem 1 can be applied to $v(\omega) = \mu(\omega)t(\omega)$, and we get

$$\begin{aligned} K(\omega) &\leqslant -\log \mu(\omega) - m + K(v) + O(1) \\ &\leqslant -\log \mu(\omega) - m/2 + K(\mu) + O(1). \end{aligned}$$

A more general theorem says that the martingale $2^{d(\omega \mid \mu)}$ is *maximal* (up to a multiplicative constant) among all martingales that are semicomputable (from below). Hence the quantity $-\log \mu(\omega) - K(\omega)$ is a *universal test of randomness*; it can be used to measure the *deficiency of randomness* in the outcome ω with respect to the distribution μ.

APPLICATIONS

Algorithmic information theory (AIT) justifies the intuition of random sequences as nonstandard analysis justifies infinitely small quantities. Any statement of classical probability theory is provable without the notion of randomness, but some of them are easier to find using this notion. Due to the incomputability of the universal randomness test, only its approximations can be used in practice. *Pseudo-random sequences* are ones whose Kolmogorov complexity is very low but which withstand all easily computable randomness tests. Such sequences play an important role in several areas of computer science. Their existence can be proved using some difficult unproven (but plausible) assumptions of computation theory. See ref. 8 and the sources listed under "Bibliography."

Information theory*

Since with large probability, $K(\omega)$ is close to $-\log \mu(\omega)$, the *entropy** $-\sum_\omega \mu(\omega)\log \mu(\omega)$ of the distribution μ is close to the *average complexity* $\sum_\omega \mu(\omega)K(\omega)$. The complexity $K(x)$ of an object x can indeed be interpreted as the distribution-free definition of *information content**. The conditional complexity $K(x \mid y)$ obeys to identities analogous to the information-theoretical identities for conditional entropy, but these identities are less trivial to prove in AIT.

Inductive Inference

The incomputable "distribution" $M(\omega) = 2^{-K(\omega)}$ has the remarkable property that, tested by the quantity $d(\omega \mid M)$, all outcomes ω are "random" with respect to it. This and other considerations suggest that it represents well the intuitive concept of *a priori probability**. The martingale $2^{d(\omega \mid \mu)}$ is the *likelihood ratio** between the hypothesis μ and the a priori hypothesis M.

Conditional a priori probability as a general inductive inference formula (see, e.g., Solomonoff [31] for justification) can be viewed as a mathematical form of "Occam's razor": the advice to predict by the simplest rule fitting the data. Since a priori probability is incomputable, finding maximally efficient approximations can be considered the main open problem of inductive inference. Sometimes, even a simple approximation gives nontrivial results (see Barzdin' [1]).

Logic

Some theorems of mathematical logic (in particular, Gödel's theorem) have a strong quantitative form in AIT, with new philosophical implications (see Levin [18, 21] and Chaitin [3, 4]). Levin based a new system of intuitionistic analysis on his independence principle (see below) in Levin [19, 21].

HISTORY OF THE PROBLEM

P. S. Laplace* thought that the set of all "regular" sequences has small probability (see Laplace [16]). R. von Mises* called an infinite binary sequence a *Kollektiv* if the relative frequencies converge in any subsequence selected according to some (nonanticipating) "rule" (see von Mises [36]). As pointed out by A. Wald*, Mises's definitions are sound if a countable set of possible rules is fixed. A. Church proposed to understand "rule" as "recursive function." J. Ville proved that a Kollektiv can violate, for example, the law of the iterated logarithm* (see Ville [34]). He proposed to introduce a countable set $\mathscr{C}$ of martingales, and to call a sequence x random if any function from $\mathscr{C}$ has an upper bound on the segments of x.

For the solution of the problem of inductive inference, R. J. Solomonoff introduced complexity and a priori probability in Solomonoff [30] and proved the invariance theorem. A. N. Kolmogorov independently introduced complexity as a measure of individual information content and randomness, and proved the invariance theorem (see Kolmogorov [14, 15]). P. Martin-Löf defined randomness for infinite sequences. His concept is essentially equivalent to the one suggested by Ville if $\mathscr{C}$ is the set of semicomputable martingales (see Schnorr [26]). The incomputability properties of $K(x)$ have noteworthy philosophical implications (see Chaitin [3, 4], Bennett and Gardner [2]).

L. A. Levin defined the a priori probability M as a maximal (to within a multiplicative constant) semicomputable measure. He introduced *monotonic complexity* and characterized random sequences by the behavior of the complexity of their segments (see Levin [17, 23]). Theorems 1 and 2 are special cases of these theorems for discrete probability distributions. In Levin [18] and Gács [10], the information-theoretical properties of the self-delimiting complexity (a special case of the monotonic complexity) are exactly described. C. P. Schnorr discovered independently a part of Levin [17], G. J. Chaitin a part of Levin [19], and Gács [10] (see Schnorr [27] and Chaitin [5]). Related results were proved in Willis [37].

In Levin [20, 21], Levin defined the deficiency of randomness $d(\omega \mid \mu)$ in a uniform manner for all (computable or incomputable) measures μ. He proved that all outcomes are random with respect to the a priori probability M. In Levin [18], he proved the *law of information conservation*, stating that the information $I(\alpha : \beta)$ in a sequence α about a sequence β cannot be significantly increased by any algorithmic processing of α (even using random number generators*). In its present form, this law follows from a so-called law of randomness conservation via the definition of information as $I(\alpha : \beta) = d((\alpha, \beta) \mid M \times M)$. Levin suggested the independence principle, saying that any sequence α arising in *nature* contains only finite information $I(\alpha : \beta)$ about any sequence β defined by mathematical means. With this principle, he showed that the use of more powerful notions of definability in Martin-Löf's test does not lead to fewer random sequences among those arising in nature.

References

At this time (1986), no easily readable, up-to-date overview of AIT is available. The most recent, fairly comprehensive work, that of Levin [21], can be recommended only to devoted readers. The work by Levin and Zvonkin [23] is comprehensive and readable but not quite up to date. The surveys by Schnorr [26, 27] and Chaitin [7] can be used to complement it.

AIT created many interesting problems of its own; see, for example, Chaitin [5, 6], Gács [10–12], Levin [22], Loveland [24], Solovay [32], and Schnorr [26, 28], and the technically difficult results in Solovay [33] and V'iugin [35].

[1] Barzdin', Ya. M. and Freivald (1972). *Sov. Math. Dokl.*, **206**, 1224–1228.

[2] Bennett, C. and Gardner, M. (1979). *Sci. Am.*, **241**(5), 20–34.

[3] Chaitin, G. J. (1974). *J. ACM*, **21**, 403–424.

[4] Chaitin, G. J. (1975). *Sci. Amer.*, **232**(5), 47–52.

[5] Chaitin, G. J. (1975). *J. ACM*, **22**, 329–340.

[6] Chaitin, G. J. (1976). *Comput. Math. Appl.*, **2**, 233–245.

[7] Chaitin, G. J. (1977). *IBM J. Res. Dev.*, **21**, 350–359.

[8] Daley, R. P. (1975). *Math. Syst. Theory*, **9**(1), 83–94.

[9] Fine, T. (1973). *Theories of Probability*. Academic Press, New York.

[10] Gács, P. (1974). *Sov. Math. Dokl.*, **15**, 1477–1480.

[11] Gács, P. (1980). *Zeit. Math. Logik Grundlag. Math.*, **26**, 385–394.

[12] Gács, P. (1983). *Theor. Computer Sci.*, **22**, 71–93.

[13] Kolmogorov, A. N. (1956). *Foundations of the Theory of Probability*. Chelsea, New York.

[14] Kolmogorov, A. N. (1965). *Prob. Inf. Transm.*, **1**, 4–7.

[15] Kolmogorov, A. N. (1968). *IEEE Trans. Inf. Theor.*, **IT-14**, 662–664.

[16] Laplace, P. S. (1819). *A Philosophical Essay on Probabilities*. Dover, New York, pp. 16–17.

[17] Levin, L. A. (1973). *Sov. Math. Dokl.*, **14**, 1413–1416.

[18] Levin, L. A. (1974). *Prob. Inf. Transm.*, **10**, 206–210.

[19] Levin, L. A. (1976). *Sov. Math. Dokl.*, **17**, 601–605.

[20] Levin, L. A. (1976). *Sov. Math. Dokl.*, **17**, 337–340.

[21] Levin, L. A. (1984). *Inf. Control*, **61**, 15–36.

[22] Levin, L. A. and V'iugin, V. V. (1977). *Lect. Notes Computer Sci.*, **53**, 359–364 (Proc. 1977 MFCS Conf.). Springer-Verlag, Berlin.

[23] Levin, L. A. and Zvonkin, A. K. (1970). *Russ. Math. Surv.*, **25**, 83–124.

[24] Loveland, D. W. (1969). *Inf. Control*, **15**, 510–526.

[25] Martin-Löf, P. (1966). *Inf. Control*, **9**, 602–619.

[26] Schnorr, C. P. (1971). Zufälligkeit und Wahrscheinlichkeit. *Lect. Notes Math.*, **218**. Springer-Verlag, New York.

[27] Schnorr, C. P. (1973). *J. Comput. Syst. Sci.*, **7**, 376–388.

[28] Schnorr, C. P. (1975). Basic Problems in Methodology and Linguistics. In *Proc. 5th Int. Congr. Logic Methods Philos. Sci.*, Butts and Hintikka, eds. Reidel, Dordrecht, pp. 193–211.

[29] Schnorr, C. P. and Fuchs, P. (1977). *J. Symb. Logic*, **42**, 329–340.

[30] Solomonoff, R. J. (1964). *Inf. Control*, **7**, 1–22 II., *ibid.*, 224–254.

[31] Solomonoff, R. J. (1978). *IEEE Trans. Inf. Theory*, **IT-24**, 422–432.

[32] Solovay, R. (1975). Unpublished manuscript.

[33] Solovay, R. (1977). *Non-classical Logic, Model Theory and Computability*, A. I. Arruda et al., eds. North-Holland, Amsterdam, pp. 283–307.

[34] Ville, J. L. (1939). *Étude critique de la notion de collectif*. Gauthier-Villars, Paris.

[35] V'iugin, V. V. (1976). *Sov. Math. Dokl.*, **229**, 1090–1094.

[36] von Mises, R. and Geiringer, H. (1964). *The Mathematical Theory of Probability and Statistics*. Academic Press, New York.

[37] Willis, D. G. (1970). *J. ACM*, **17**, 241–259.

[38] Yasuhara, A. (1971). *Recursive Function Theory and Logic*. Academic Press, New York.

Bibliography

Blum, M. and Micali, S. (1984). *SIAM J. Comput.*, **13**, 850–864.

Goldreich, O., Goldwasser, S. and Micali, S. (1984). In *Proc. 25th IEEE Symp. Foundations Comput. Sci.*

Yao, A. C. (1982). In *Proc. 23rd IEEE Symp. Foundations Comput. Sci.*

(ALGORITHMIC INFORMATION THEORY
FOUNDATIONS OF PROBABILITY)

PETER GÁCS

RANDOMNESS, TESTS OF

Randomness tests are tests of the null hypothesis that a set of observations or measurements can be considered random. They are an important part of statistical methodology because it is frequently important to examine for nonrandomness data that occur in an ordered sequence or series. Randomness tests are also useful to check on the validity of the assumption of randomness which is inherent in most statistical procedures.

A test for the null hypothesis of randomness is difficult because there are many different kinds of nonrandomness (e.g., a tendency to cluster or bunch, a tendency to mix or alternate direction, a serial correlation* or autocorrelation, a trend*, a cyclical or seasonal pattern of movement, etc.). Because of this, the alternative hypothesis is usually quite general, sometimes stating simply nonrandomness. The tests covered here are used primarily in this general situation; they include tests based on runs* in a dichotomous sequence, on runs up and down in an ordinal measured sequence, on successive differences*, on times of occurrence, and chi-square goodness of fit* tests. References for other tests are also given.

The conclusion from one of these randomness tests is usually simply a rejection or nonrejection of the null hypothesis. However, when using a test that is known to be particularly sensitive to a specific kind or indication of nonrandomness, the investigator may wish to conclude that the sequence does or does not exhibit that specific kind of nonrandomness. When the alternative is more specific (e.g., states an upward or downward trend), other tests may be more appropriate (*see* TREND TESTS), although the general tests covered here can also be used for such specific alternatives.

Tests for the randomness of a set of numbers generated by a computer are also important. Most of the procedures covered in this entry can also be used for this situation, although many others have been developed specifically for such an application. Knuth [37] discusses many tests of this kind.

TESTS BASED ON RUNS OF DICHOTOMOUS VARIABLES

For an ordered sequence of dichotomous observations, a *run* is defined as a succession of one or more identical symbols which are preceded and followed by the other symbol or no symbol at all. For example, the ordered sequence of nine symbols, S S F S S F F S F, has three runs of S's, the first and second of length 2, the third of length 1, and three runs of F's, of lengths 1, 2, and 1, respectively. Both the total number of runs and their lengths provide clues to lack of randomness because too few runs, too many runs, a run of excessive length, and so on, rarely occur in a truly random sequence. Of course, the number of runs and their lengths are highly interrelated, so that only one of these characteristics needs to be considered. The total number of runs is the characteristic commonly used to test the null hypothesis of randomness. This is actually a test for nonrandomness of the appearance of S and F in the sequence.

Suppose that there are m S's and n F's in an ordered sequence of $m + n = N$ dichotomous symbols. There are $\binom{N}{m} = N!/m!n!$ distinguishable arrangements, and each is equally likely under the null hypothesis of randomness. The null distribution of the total number of runs, R, is easily derived by combinatorial arguments [31, 39, 58, 63] as

$$P(R = r) = \begin{cases} \dfrac{2\binom{m-1}{k-1}\binom{n-1}{k-1}}{\binom{m+n}{m}} & \text{for } r = 2k \\[2ex] \dfrac{\binom{m-1}{k-1}\binom{n-1}{k} + \binom{m-1}{k}\binom{n-1}{k-1}}{\binom{m+n}{m}} & \text{for } r = 2k+1 \end{cases} \tag{1}$$

for $k = 1, 2, \ldots, N/2$. If m and n approach infinity while m/n approaches a constant, Wald and Wolfowitz [63] show that the null distribution of

$$Z = \frac{R - 1 - 2mn/N}{\sqrt{\dfrac{2mn(2mn - N)}{N^2(N-1)}}} \tag{2}$$

is approximately standard normal. A continuity correction* of ± 0.5 can be incorporated in the numerator of (2).

The alternative A_+: tendency to mix, is supported if there are too many runs and hence a right-tail rejection region (or P-value) is appropriate; the alternative A_-: tendency to cluster, calls for a left-tail rejection region. A two-tailed rejection region should be used for the alternative A: nonrandomness. Tables of the exact distribution of R for $m \leqslant n \leqslant 20$ are given in Swed and Eisenhart [61] and reproduced in some textbooks on applied nonparametric or distribution-free statistics (e.g., Bradley [8], Daniel [14], and Gibbons [22]).

For an ordered sequence of N observations $X_1, X_2, \ldots, X_N$ that are measured on at least an ordinal scale, this same runs test can be applied by transforming the observations into two kinds of symbols. David [15] suggests this test for detecting serial correlation in time-series* data by comparing each observation with the magnitude of the median M of the sequence (i.e., X_i is replaced by an S if $X_i - M \geqslant 0$ and by an F if $X_i - M < 0$). The distribution of R is given by (1); this test is frequently called *runs above and below the median*.

Goodman [23] proposes another number of runs test, based on the statistic

$$S = NR/(2mn) - 1, \tag{3}$$

where R, m, and n are defined as before and X_i is replaced by an S if $X_i - k \geqslant 0$ and by an F if $X_i - k < 0$ for k any appropriate constant. The asymptotic distribution of $Z = \sqrt{N}\,S$ is approximately standard normal. Granger [25] shows that this test is quite powerful for detecting serial correlation* in nonstationary time series with $N \geqslant 100$, $m \geqslant N/3$ and $n \geqslant N/3$, for any k.

O'Brien [45] suggests a different test of randomness for a dichotomous sequence; it is a slight modification of the criterion given by Dixon [17] for the hypothesis that two mutually independent random samples are drawn from the same population. O'Brien's test criterion for m S's and n F's is

$$C^2 = \sum_{i=1}^{n+1} \left(W_i - \overline{W}\right)^2 / m^2, \tag{4}$$

where W_i = number of S's between the $(i-1)$th F and the ith F, $i = 2, 3, \ldots, n$,

W_1 = number of S's prior to the first F,

W_{n+1} = number of S's subsequent to the nth F,

$$\overline{W} = \sum_{i=1}^{n+1} W_i/(n+1) = m/(n+1).$$

Small values of C^2 suggest a systematic arrangement and large values suggest clustering. Dixon [17, p. 286] gives a table of critical values of C^2 for $n \leqslant m \leqslant 10$ and shows that $mn^2C^2/(2N)$ has approximately the chi-square distribution with $n/2$ degrees of freedom. O'Brien finds this test more sensitive to multiple clustering than the number of runs test or the rank tests* to detect trend. It is, of course, basically a test on the lengths of the runs of the S's.

Another test of randomness for a dichotomous sequence, given in Moore [40], is called the *group test*, according to David [15]. This test is sequential and therefore uses fewer data in general than do the nonsequential tests mentioned earlier. The alternative H_1 is dependence of the kind found in a simple Markov chain. Here we specify both α and β, the probabilities of a type I and type II error, respectively, and then take observations one at a time. At each stage we calculate a quantity L and follow the following stopping rule:

If $L > (1-\beta)/\alpha$, accept H_1.
If $L < \alpha/(1-\beta)$, accept H_0.
If $\beta(1-\alpha) \leqslant L \leqslant (1-\beta)/\alpha$, take more observations.

The L is a function of α, β, the number of runs at that stage, and the conditional probability of an S given that it is preceded by an S. For observations measured on at least an ordinal scale, the dichotomy may be effected by comparing each observation with the median of all observations collected at that stage.

A different use of the runs test based on (1) or (2) arises when the data consist of two mutually independent random samples of sizes m and n, say $X_1, X_2, \ldots, X_m$ and $Y_1, Y_2, \ldots, Y_n$, and the null hypothesis is $F_X = F_Y$. If the $N = m + n$ observations are pooled and arranged from smallest to largest while keeping track of which observation comes from which sample, the resulting pattern of X's and Y's should be well mixed if the samples come from identical populations; hence R is defined here as the number of runs of X's and Y's. For the alternative that the populations are not identical, a left-tail critical region for R should be used. Raoul and Sathe [52] give a computer program for this technique, frequently called the *Wald–Wolfowitz runs test* [63]. This test is consistent against a variety of differences between populations, but is not particularly sensitive to any specific kind of difference between the populations (like location) and is not very powerful for specific alternatives (see Gibbons [20]).

TESTS BASED ON RUNS UP AND DOWN

For an ordered sequence of data measured on at least an ordinal scale, a sequence of two types of symbols can be generated by noting the magnitude of each observation relative to that of the observation immediately preceding it in the sequence. If the preceding value is smaller, a "run up" is started; if it is larger, a "run down" is started. For example, the ordered sequence $X_1, X_2, \ldots, X_N$ can be transformed into a sequence of $(N-1)$ plus and minus signs by looking at the signs of the successive differences $X_{i+1} - X_i$ for $i = 1, 2, \ldots, N-1$. The null hypothesis of randomness can be tested using the total number of runs (whether up or down) statistic, say V. The null distribution of V is derived in Kermack and McKendrick [33] and Moore and Wallis [43], and studied further in Olmstead [46]. The most convenient table for $N \leqslant 25$ is given in Edgington [18]; this table is reproduced in Bradley [8] and Gibbons [22]. For large samples the standard normal distribution can be used with the random variable

$$Z = \frac{V - (2N-1)/3}{\sqrt{\dfrac{16N-29}{90}}}; \qquad (5)$$

a continuity correction of ± 0.5 can be incorporated in the numerator of (5).

If the alternative to randomness is trend or gradual oscillation, the appropriate rejection region is left tail; for the alternative rapid oscillation, the appropriate rejection region is right tail.

Methods for computing the asymptotic power of this test are given in Levene [38] and tables are given for alternatives of a constant location shift from one observation to the next and for normal and uniform populations.

An equivalent procedure is called the *turning points* test for randomness based on the number of reversals of direction of movement in the series. The turning points statistic is always one less than R because a reversal cannot occur at X_N.

TESTS BASED ON SUCCESSIVE DIFFERENCES

In an ordered sequence of N observations measured on at least an ordinal scale, X_1, $X_2, \dots, X_N$, a test of randomness can be based on the squares of the differences of successive observations with the ratio statistic

$$\eta = \frac{\sum_{i=1}^{N-1}(X_{i+1} - X_i)^2}{\sum_{i=1}^{N}(X_i - \bar{X})^2} \tag{6}$$

investigated by von Neumann [62]. Young [67] gives critical values of $|1 - \eta/2|$ for $8 \leqslant N \leqslant 25$, and Hart [27] gives critical values of η for $4 \leqslant N \leqslant 60$, while Hart and von Neumann [28] give tail probabilities, in each case for normal samples. For large N (say $N \geqslant 20$) and a normal population the statistic

$$Z = \left(1 - \frac{\eta}{2}\right) \Big/ \sqrt{(N-2)/(N^2-1)}$$

is approximately standard normal [67]. A trend alternative calls for a right-tail rejection region and a short oscillation alternative calls for left-tail. Kleijnen et al. [34] investigate the power of this procedure in testing for independence in simulation subruns and recommend that the number of subruns be at least 100.

Bartels [2] gives a distribution-free rank version of the statistic (6) for ordinal scale data as

$$\begin{aligned} \text{RVN} &= \frac{\sum_{i=1}^{N-1}(R_{i+1} - R_i)^2}{\sum_{i=1}^{N}(R_i - \bar{R})^2} \\ &= \frac{S}{\sum_{i=1}^{N}(R_i - \bar{R})^2} = \frac{12S}{N(N^2-1)} \end{aligned} \tag{7}$$

where R_i is the rank of X_i in the ordered sequence. This reference gives tables of critical values of S for $N \leqslant 10$ and of RVN for $10 \leqslant N \leqslant 100$ and shows that

$$Z = \frac{\text{RVN} - 2}{\sqrt{4/N}}$$

is asymptotically standard normal. Monte Carlo* comparisons of power against the alternative of first-order autoregression for several different distributions and $N = 10$, 25, 50 show that this test is always far superior to the runs up and down tests, and frequently superior to the normal theory von Neumann test. Hence this procedure may be used more specifically as a test for serial correlation (*see* SERIAL CORRELATION).

A simpler procedure that is also distribution free is the difference sign test proposed by Wallis and Moore [65], which is simply the number c of positive signs among the $X_{i+1} - X_i$. They give the null distribution; it approaches the normal distribution with mean $(N-1)/2$ and variance $(N+1)/12$. This procedure is appropriate only with a trend alternative.

TESTS BASED ON TIMES OF OCCURRENCE

If we are observing the occurrence or nonoccurrence of some phenomenon over a fixed period of time, the time of each occurrence can be noted and then transformed to points on the interval (0, 1). A test for the randomness of these points may be of interest.

Bartholomew [3] discusses the test statistic

$$\frac{1}{2}\sum_{i=1}^{N+1}\left|X_i - \frac{1}{N+1}\right|,$$

where $X_1, X_2, \dots, X_N$ are successive intervals between the observed points, $X_1 = 0$, and $X_{N+1} = 1$ − largest observation. This statistic is then a comparison between each observed interval and the interval expected with uniform spacing, $1/(N+1)$, and hence gives a test for randomness of spacing with large values indicating nonrandomness. The moments and asymptotic distribution of this test are developed in Sherman [55] and it is sometimes called the *Sherman statistic*.

Bartholomew [4] examines the properties of a test proposed in Cox [11] that is based on the sum of the times of occurrence. Bartholomew shows a simple relation of this test statistic to the one-sided Kolmogorov–Smirnov* goodness-of-fit test and gives some power calculations.

CHI-SQUARE TESTS FOR RANDOMNESS

Chi-square goodness-of-fit* tests can sometimes be used to test for randomness in interval scale data by testing the null hypothesis that subsequences of a certain size n have a constant probability θ of a certain dichotomous characteristic, say, Success (S) and Failure (F), and hence the number of S's in each subsequence follows the binomial distribution with parameters n and θ. For example, a sequence of N observations could be divided into k subsequences, each of size n, where $kn = N$. Suppose that a success is defined as the event that an observation is an even number. Under the null hypothesis that the observations are independent, identically distributed random variables, the expected number of S's in each subsequence is

$$e_i = N\binom{n}{i}(0.5)^n$$

for $i = 1, 2, \ldots, k$. These expected frequencies can be compared with the respective observed frequencies of success f_i using the ordinary chi-square goodness-of-fit test. A success could be defined in other ways, as, for example, a final digit greater than or equal to 5, and so on.

Wallis and Moore [65] suggest using a chi-square goodness-of-fit test with classes defined as runs of length 1, runs of length 2, and runs of length 3 or more, as a test for randomness. The respective expected frequencies for these classes are $5(N-3)/12$, $11(N-4)/60$, and $(4N-21)/60$.

ADDITIONAL NOTES

Other general tests for randomness have been proposed, some based on the lengths of runs. These are discussed in Kermack and McKendrick [32, 33], Mood [39], Mosteller [44], Olmstead [46], David [15], Bateman [7], Goodman [23], Burr and Cane [10], David and Barton [16], and Barton [5]. Mood [39] gives a history of runs tests and extends the theory of runs to sequences consisting of more than two kinds of symbols, as do Wallis and Roberts [66]. Some discussion of power functions appears in David [15], Bateman [7], Barton and David [6], and Levene [38], but power* cannot be calculated without giving a specific alternative.

APPLICATIONS

A major area of application or randomness tests is in quality control. If nonrandomness is detected and can then be explained, the cause can perhaps be corrected promptly to restore the process to randomness. Applications in quality control are given in Mosteller [44], Olmstead [47–50], and Prairie et al. [51], and in many quality control* textbooks.

If we have data that give measurements on at least an ordinal scale as well as times of occurrence, we have time-series data in the traditional sense. This is another important use of tests of randomness, especially when the alternative is trend (*see* TREND TESTS) or serial correlation (*see* SERIAL CORRELATION). Alam [1] gives a class of unbiased tests for this situation. In addition to those mentioned specifically here, the best known are those based on the Spearman rank correlation* coefficient (Daniels test), the Kendall tau* coefficient (Mann test), runs up and down (turning points test), and the rank serial correlation test. Specific procedures with time-series applications are given in Wallis and Moore [65], Moore and Wallis [43], Wald and Wolfowitz [64], Olmstead [47], Foster and Stuart [19], Cox and Stuart [13], Goodman and Grunfield [24], Sen [54], Knoke [35, 36], Gupta and Govindarajulu [26], and in many books. Stuart [59, 60] compares the asymptotic relative efficiencies of several of these tests relative to normal regression alternatives. Knoke [36] gives a table of power comparisons for both normal and nonnormal distributions.

Trend tests might be called tests of randomness of occurrence over time. Tests of randomness of occurrence over space are also sometimes of interest. One group of such tests is based on distance from nearest neighbor; these have applications in statistical ecology for studying the pattern of plant distributions. Some tests based on distance and proposed by others are discussed in Holgate [30]. Other tests are proposed and discussed in Holgate [29]. Pearson [50] compares four tests based on Kolmogorov–Smirnov and von Mises statistics. Brown and Rothery [9] give tests based on squared distances. In a recent survey, Ripley [53] compares seven randomness tests for spatial* point patterns that are all based on some function of this distance.

Stephens [57] describes tests for randomness of directions in the form of points on the circumference of a unit circle, gives tables, and some results for power. These tests have application in the study of wind directions and migratory habits of birds. A similar study for the three-dimensional problem is given in Stephens [56]. Cox and Lewis [12] give several applications of randomness in series of events.

EXAMPLE

An illustration of these procedures is provided for the data given in Bartels [2, p. 42] (see Table 1) on changes in stock levels from 1968–1969 to 1977–1978 after deflating by the Australian gross domestic product (GDP) price index (base period 1966–1967). We will test the null hypothesis of randomness against the alternative of a trend using three of the procedures given here.

The signs of successive differences of the deflated stock changes are

$$-, -, -, -, +, -, -, +, -,$$

so that the runs up and down test statistic is $V = 5$. The left-tail P-value is 0.2427, so we cannot reject the null hypothesis of randomness by this procedure.

The ranks of the deflated stock changes are

$$9, 6, 5, 3, 1, 10, 7, 4, 8, 2,$$

Table 1[a]

Year	Change in Stocks (millions of dollars)	GDP Index (1966–1967 = 100)	Deflated Stock Change
1968–1969	561	106.2	528
1969–1970	386	110.9	348
1970–1971	309	117.2	264
1971–1972	−25	125.5	−20
1972–1973	−227	136.3	−167
1973–1974	893	155.4	575
1974–1975	757	184.5	410
1975–1976	−8	211.8	−4
1976–1977	1009	234.9	430
1977–1978	−309	254.1	−122

[a] Source: Ref. [2].

and the numerator of the rank von Neumann test statistic from (7) is $S = 169$, which is again not significant.

The median of the 10 deflated stock changes is between 264 and 348. If we let B denote an observation below the median and A above, the sequence of changes is

$$A, A, B, B, B, A, A, B, A, B,$$

which has $U = 6$ runs for $m = 5$, $n = 5$. The left-tail P-value then is $P > 0.5$ so we cannot reject the null hypothesis of randomness.

In this example, all three procedures illustrated lead to the same conclusion. This will not always be the case because each procedure is sensitive to randomness in a different way.

On the other hand, the Wallis and Moore [65] test designed specifically for trend gives $c = 2$ with a one-tailed P-value of 0.013, and hence we would conclude a downward trend with this test.

References

[1] Alam, K. (1974). *J. Amer. Statist. Ass.*, **69**, 738–739.

[2] Bartels, R. (1982). *J. Amer. Statist. Ass.*, **77**, 40–46.

[3] Bartholomew, D. J. (1954). *Biometrika*, **41**, 556–558.

[4] Bartholomew, D. J. (1956). *J. R. Statist. Soc. B*, **18**, 234–239.

[5] Barton, D. E. (1967). *Technometrics*, **9**, 682–694.

[6] Barton, D. E. and David, F. N. (1958). *Biometrika*, **45**, 253–256.

[7] Bateman, G. (1948). *Biometrika*, **35**, 97–112.

[8] Bradley, J. V. (1968). *Distribution-Free Statistical Tests*. Prentice-Hall, Englewood Cliffs, N.J. (Elementary; Chapters 11 and 12 cover runs tests and runs up and down tests, respectively.)

[9] Brown, D. and Rothery, P. (1978). *Biometrika*, **65**, 115–122.

[10] Burr, E. J. and Cane, G. (1961). *Biometrika*, **48**, 461–465.

[11] Cox, D. R. (1955). *J. R. Statist. Soc. B*, **17**, 129–157.

[12] Cox, D. R. and Lewis, P. A. W. (1966). *The Statistical Analysis of Series of Events*. Methuen, London.

[13] Cox, D. R. and Stuart, A. (1955). *Biometrika*, **42**, 80–95.

[14] Daniel, W. W. (1978). *Applied Nonparametric Statistics*. Houghton Mifflin, Boston. (Elementary; Section 2.3 covers number of runs tests.)

[15] David, F. N. (1947). *Biometrika*, **34**, 335–339.

[16] David, F. N. and Barton, D. E. (1962). *Combinatorial Chance*. Charles Griffin, London.

[17] Dixon, W. J. (1940). *Ann. Math. Statist.*, **11**, 199–204.

[18] Edgington, E. S. (1961). *J. Amer. Statist. Ass.*, **56**, 156–159.

[19] Foster, F. G. and Stuart, A. (1954). *J. R. Statist. Soc. B*, **16**, 1–22.

[20] Gibbons, J. D. (1964). *J. R. Statist. Soc. B*, **26**, 293–304.

[21] Gibbons, J. D. (1985). *Nonparametric Statistical Inference*, 2nd ed. Marcel Dekker, New York. (Intermediate level; mostly theory; runs tests are covered in Chapters 3 and 7.)

[22] Gibbons, J. D. (1985). *Nonparametric Methods for Quantitative Analysis*, 2nd ed. American Sciences Press, Columbus, Ohio. (Elementary; applied approach; many numerical examples; runs tests are covered in Chapter 8.)

[23] Goodman, L. A. (1958). *Biometrika*, **45**, 181–197.

[24] Goodman, L. A. and Grunfield, Y. (1961). *J. Amer. Statist. Ass.*, **56**, 11–26.

[25] Granger, G. W. J. (1963). *J. Amer. Statist. Ass.*, **58**, 728–736.

[26] Gupta, S. S. and Govindarajulu, Z. (1980). *Biometrika*, **67**, 375–380.

[27] Hart, B. I. (1942). *Ann. Math. Statist.*, **13**, 445–447.

[28] Hart, B. I. and von Neumann, J. (1942). *Ann. Math. Statist.*, **13**, 207–214.

[29] Holgate, P. (1965). *J. Ecol.*, **53**, 261–266.

[30] Holgate, P. (1965). *Biometrika*, **52**, 345–353.

[31] Ising, E. (1925). *Zeit. Phys.*, **31**, 253–258.

[32] Kermack, W. O. and McKendrick, A. G. (1937). *Proc. R. Soc. Edinb.*, **57**, 332–376.

[33] Kermack, W. O. and McKendrick, A. G. (1937). *Proc. R. Soc. Edinb.*, **57**, 228–240.

[34] Kleijnen, J. P. C., van der Ven, R. and Sanders, B. (1982). *Eur. J. Operat. Res.*, **9**, 92–93.

[35] Knoke, J. D. (1975). *Biometrika*, **62**, 571–576.

[36] Knoke, J. D. (1977). *Biometrika*, **64**, 523–529.

[37] Knuth, D. E. (1968). *The Art of Computer Programming*, Vol. II. Addison-Wesley, Reading, Mass.

[38] Levene, H. (1952). *Ann. Math. Statist.*, **22**, 34–56.

[39] Mood, A. M. (1940). *Ann. Math. Statist.*, **11**, 367–392.

[40] Moore, P. G. (1953). *Biometrika*, **40**, 111–114.

[41] Moore, P. G. (1955). *J. Amer. Statist. Ass.*, **50**, 434–465; corrigenda: *ibid.*, **51**, 651 (1956).

[42] Moore, P. G. (1958). *Biometrika*, **45**, 89–95; corrigenda: *ibid.*, **46**, 279 (1959).

[43] Moore, G. H. and Wallis, W. A. (1943). *J. Amer. Statist. Ass.*, **38**, 153–164.

[44] Mosteller, F. (1941). *Ann. Math. Statist.*, **12**, 228–232.

[45] O'Brien, P. C. (1976). *Biometrics*, **32**, 391–401.

[46] Olmstead, P. S. (1946). *Ann. Math. Statist.*, **17**, 24–33.

[47] Olmstead, P. S. (1952). *Ind. Quality Control*, **9**(3), 32–38.

[48] Olmstead, P. S. (1958). *Bell Syst. Tech. J.*, **37**, 55–82.

[49] Olmstead, P. S. (1958). Statistical quality control in research and development. *Trans. 13th Midwest Quality Control Conf.*, Kansas City, Mo., Nov. 6–7, 1958.

[50] Pearson, E. S. (1963). *Biometrika*, **50**, 315–326.

[51] Prairie, R. R., Zimmer, W. J., and Brookhouse, J. K. (1962). *Technometrics*, **4**, 177–185.

[52] Raoul, A. and Sathe, P. T. (1975). *J. Quality Tech.*, **7**, 196–199.

[53] Ripley, B. D. (1979). *J. R. Statist. Soc. B*, **41**, 368–374.

[54] Sen, P. K. (1965). *J. Amer. Statist. Ass.*, **60**, 134–147.

[55] Sherman, B. (1950). *Ann. Math. Statist.*, **21**, 339–361.

[56] Stephens, M. A. (1964). *J. Amer. Statist. Ass.*, **59**, 160–167.

[57] Stephens, M. A. (1969). *J. Amer. Statist. Ass.*, **64**, 280–289.

[58] Stevens, W. L. (1939). *Ann. Eugen. (Lond.)*, **9**, 10–17.

[59] Stuart, A. (1954). *J. Amer. Statist. Ass.*, **49**, 147–157.

[60] Stuart, A. (1956). *J. Amer. Statist. Ass.*, **51**, 285–287.

[61] Swed, F. S. and Eisenhart, C. (1943). *Ann. Math. Statist.*, **14**, 66–87.

[62] von Neumann, J. (1941). *Ann. Math. Statist.*, **12**, 367–395.

[63] Wald, A. and Wolfowitz, J. (1940). *Ann. Math. Statist.*, **11**, 147–162.

[64] Wald, A. and Wolfowitz, J. (1943). *Ann. Math. Statist.*, **14**, 378–388.

[65] Wallis, W. A. and Moore, G. H. (1941). *J. Amer. Statist. Ass.*, **36**, 401–409.

[66] Wallis, W. A. and Roberts, H. V. (1956). *Statistics: A New Approach*. Free Press, Glencoe, Ill. (Elementary; runs tests are discussed in Chapter 18.)

[67] Young, L. C. (1941). *Ann. Math. Statist.*, **12**, 293–300.

(GOODNESS-OF-FIT
RUNS
RUN TESTS, MULTIDIMENSIONAL
SERIAL CORRELATION
SUCCESSIVE DIFFERENCES
TREND TESTS)

JEAN DICKINSON GIBBONS

RANDOM NOISE *See* NOISE

RANDOM NUMBER GENERATION *See* GENERATION OF RANDOM VARIABLES

RANDOM SAMPLING, SIMPLE *See* SIMPLE RANDOM SAMPLING

RANDOM SETS OF POINTS

Probability theory starts with (real) random variables taking their values in R, and then proceeds to consider complex random variables, random vectors, random matrices*, random (real) functions, and so on, in a natural sequence of generalizations. The concept of a *random set* is more subtle and has to be approached obliquely via suitable random functions. However, there is a natural link with a familiar object; a point process (supposed "unmarked") is obviously a random set of a sort, and in the familiar simple cases we can describe it adequately by means of the "counting random variables"

$$N(T) = \text{number of random points in } T,$$

where T is a "test set": we write $\mathscr{T}$ for the class of test sets in use. This approach fits neatly into the classical theory of (real) stochastic processes*, although it should be noted that the customary "time parameter" t has now been replaced by the test set T. But in many important cases it is not very helpful. For example, it is no good if we wish to describe the set of zeros of Brownian motion*; on R it is natural to take the test sets to be open intervals, and then $N(T)$ will almost surely be zero or infinite, for each T.

The fruitful idea lying at the root of the

whole theory is due to Choquet [3] and is connected with capacity theory. We continue to employ test sets, but now, instead of thinking of them as sampling frames, we think of them as targets which may or may not be hit by the random set. The idea was elaborated about 10 years later, independently by Davidson [4] and Kendall [7], and by Matheron [8]. We now give a necessarily brief sketch.

A test set T will be called a *trap*, and to start with it is sufficient to suppose that the collection of traps consists of enough nonempty sets T to cover the ambient space X in which the random set is to live. We now write f for any 0-1-valued function of T, and we associate with an arbitrary subset E of X a function f_E defined as follows:

$$f_E(T) = 1\ (0) \text{ when } T \text{ does (does not) hit } E.$$

Now it may happen that the union of E and all the traps disjoint with E is the whole of X. When this is so, we shall call E, *$\mathscr{T}$-closed* (this is not a topological closure!).

We next look at some special 0-1 functions of T, called *strong incidence functions* (SIFs). An f is to be a SIF when and only when it satisfies the following condition:

SIF: no trap T such that $f(T) = 1$ can be covered by a (finite or infinite) collection of traps for which f is zero.

We shall also need the weaker definition WIF for a *weak* incidence function; this differs from SIF only in that the collection of traps referred to there must now be finite collections. Obviously, a SIF is a WIF, but not conversely. It is then a basic theorem [7] that f can be expressed in the form f_E if and only if it is a SIF, and that while there can be many solutions E to the equation $f = f_E$, there is one and only one that is $\mathscr{T}$-closed. We have thus succeeded in identifying the concept of a SIF with that of a $\mathscr{T}$-closed set, so that to build a theory of random $\mathscr{T}$-closed sets we have only to consider the class of 0-1-valued T-parametered stochastic processes whose realizations are almost surely SIFs. In practice the choice of the trapping system $\mathscr{T}$ will be made so that the random sets we wish to discuss are all $\mathscr{T}$-closed (and there will usually be a variety of acceptable trapping systems, from which we select one that is convenient for our purposes.).

There is a hidden obstacle here because of the arbitrary unions permitted in the SIF definition. In practice, we have to start with the concept of a random WIF, for the measurability problems are minimal there, and then pass from that to the concept of a random SIF. It is at this point that the theories in refs. 7 and 8 begin to diverge. For the most part the theory of Matheron is directed to the very important special case when the ambient space is R^k for some finite k. The simplest version of the theory in ref. 7 covers the case when X is second-countable locally compact Hausdorff (which of course allows X to be R^k, or indeed any reasonable manifold). This situation is also dealt with by Matheron. What makes these cases specially simple is that we have some countability present; we can choose the trapping system to be a countable base of open sets or (if X is R^n) the set of open balls with rational radii and centers. In either case the $\mathscr{T}$-closed sets will be exactly the closed sets in the ordinary topological sense. If for some reason we wish to consider a random open set, we apply the theory to its complement. A more general solution, given in ref. 7, discards all overt reference to a topology for X and instead imposes two extra set-theoretic conditions on the trapping system; these distil from the simpler topological case described above what is essential for the construction of a (measurable!) random SIF to go through. We omit the details.

When working with ordinary (real) random variables the absolutely basic tool is the use of distribution functions and characteristic functions* to characterize laws of distribution. This is more than a manipulative tool; it enables us to identify two apparently distinct theoretical constructions as operationally equivalent. The extreme case of this remark is that for *practical* purposes two real stochastic processes (Y_t) and (Z_t) are the same when, for each m and each $(t_1, t_2, \ldots, t_m)$, the vectors $(Y_{t_1}, Y_{t_2}, \ldots, Y_{t_m})$ and $(Z_{t_1}, Z_{t_2}, \ldots, Z_{t_m})$ have the same m-dimensional distribution function. Thus we can speak of the collection of such "finite-dimensional distributions" as the "name" of the process. Two processes with the same name specify the same real situation, although one may have much better analytical properties than the other. So we need to know how we are to specify the name in the case of a random set.

The solution in ref. 7 to this problem again exploits a fundamental invention of Choquet: the *avoidance function*. To simplify the notation, let us write U for any finite union of traps. Then, given a model for a random $\mathcal{T}$-closed set, E say, we define the avoidance function A by

$$A(U) = \Pr(E \text{ does not hit } U).$$

Because the random set will ultimately have been defined in terms of a random WIF g, we observe that one can also write

$$A(U) = \Pr\Big[\, g(T_j) = 0 \text{ for each } T_j \text{ which is a member of the finite collection whose union is } U \Big].$$

Notice that, precisely because g is a WIF, the condition within the brackets is unchanged if we respecify U by representing it in a different way as a finite union of traps.

Now such functions A have some beautiful analytical properties. In particular, we always have $A(\varnothing) = 1$ (where $\varnothing$ is the empty set), and furthermore for any traps $T_1, T_2, \ldots, T_n$, and any finite union V of (perhaps other) traps, we have

$$A(V) - \sum_i A(V \cup T_i) + \sum_{i<j} A(V \cup T_i \cup T_j) + \cdots + (-)^n A(V \cup T_1 \cup \cdots \cup T_n) \geqslant 0.$$

(This follows from a classical inclusion-exclusion* argument.) In addition, the function $A(\cdot)$ will be "continuous from below" in the sense that $A(V)$ converges (downward) to $A(U)$ when V swells up to U as limit. These three conditions characterize an avoidance function associated with a random $\mathcal{T}$-closed set, and there is a good sense in which two constructions of a random $\mathcal{T}$-closed set can be regarded as equivalent for practical purposes when and only when they have the same A. So A is the sought-for "name," and we now have the basic prerequisites for the random-set calculus.

Turning now to the applications, these arise in many biological, geological, and technological contexts wherever the morphological structure (e.g., of composite materials) is of importance. For these, reference should be made to refs. 1, 5, 6, 8, 9, and 11. There are also other quite different applications to control theory*, but here further sophistication is needed to allow the introduction of the essential "filtering" concepts.

It seems appropriate to give a little more detail in the classical paradigm for stochastic geometry: the theory of random sets of lines. A *sensed* line L in R^2 can be identified relative to a given origin and frame of axes by quoting the pair (p, ϕ), where ϕ specifies the direction of the line in $[0, 2\pi)$ $(= S^1)$, and P measures its (signed!) lateral displacement from the parallel and similarly sensed line through the origin. In this way we have identified the set of all sensed lines with the cylinder $C = R \times S^1$.

But now suppose that the lines are not sensed. We can continue to use the coordinates above, but now we must remember that $(p, \phi + \pi)$ is to be identified with $(-p, \phi)$. A diagram does not much help here, but readers will be able to convince themselves that we can represent all the lines, each of them once only, if we agree to restrict ϕ to the half-open interval $[0, \pi)$. Actually, it is a little more convenient to use the closed interval $[0, \pi]$ and to remember that we are now left with the identification of edges, $(p, \pi) = (-p, 0)$. J. F. Adams pointed out that the need to make the final "twisted" identification tells us that topologically the natural home for the family of all unsensed lines is an (open) Möbius band M. So random sets of unsensed lines can be identified with borel sets in M. In particular, a Poisson field of unsensed lines in R^2 is just a Poisson point process on M, and so on.

We close with two last references. An alternative approach to random sets will be found in ref. 2; it is very different from that described here. On the historical side the earliest reference to random sets known to me is a paper by Herbert Robbins [10], which, however, does not go into the questions we have been examining.

References

[1] Bartlett, M. S. (1964). The spectral analysis of two-dimensional point processes. *Biometrika*, **51**, 299–311.

[2] Carter, D. S. and Prenter, P. M. (1972). Exponential spaces and counting processes. *Z. Wahrscheinl. verw. Geb.*, **21**, 1–19.

[3] Choquet, G. (1955). Theory of capacities. *Ann. Inst. Fourier*, **5**, 131–295.

[4] Davidson, R. (1967). *Some Geometry and Analysis in Probability Theory*. Smith's Prize Essay, Cambridge, England. (See ref. 5.)

[5] Harding, E. F. and Kendall, D. G., eds. (1974). *Stochastic Geometry*. Wiley, London.

[6] Kendall, D. G. (1974). An introduction to stochastic geometry, pp. 3–9 in ref. 5.

[7] Kendall, D. G. (1974). Foundations of a theory of random sets, pp. 322–376 in ref. 5.

[8] Matheron, G. (1975). *Random Sets and Integral Geometry*. Wiley, New York.

[9] Ripley, B. D. (1981). *Spatial Statistics*. Wiley, New York.

[10] Robbins, H. E. (1944). On the measure of a random set. *Ann. Math. Statist.*, **15**, 70–74.

[11] Tweedie, R. L., ed. (1978). Spatial patterns and processes. *Suppl. Adv. Appl. Prob.*, pp. 1–143.

(GEOMETRY IN STATISTICS
RANDOM TESSELLATIONS
STOCHASTIC PROCESSES)

D. G. KENDALL

RANDOM SUM DISTRIBUTIONS

Let X_i, $i = 1, 2, \ldots$ be independent and identically distributed random variables (i.i.d. RVs) with characteristic function $\phi(u)$ and write $S_n = \sum_{i=1}^{n} X_i$. Let N be a nonnegative integer-valued random variable which is independent of the sequence $\{X_i\}$. Then S_N has characteristic function*

$$Ee^{iuS_N} = \sum_{n=0}^{\infty} E(e^{iuS_n} \mid N = n)P(N = n)$$
$$= \sum_{n=0}^{\infty} \phi^n(u)P(N = n)$$

and

$$ES_N = ENEX_1,$$
$$\operatorname{var} S_N = (EN)\operatorname{var} X_1 + (\operatorname{var} N)(EX_1)^2$$

when the various moments exist. This random sum model has found quite widespread use. For example, an application in entomology interprets N as the number of female insects in a region and X_i as the number of eggs laid by female i, so that S_N is the total number of eggs laid in the region.

In the case where the distribution of N depends on a parameter λ, a discussion of the asymptotic properties of S_N as $\lambda \to \infty$ has been given by Robbins [6].

When N depends on the X_i, it is not possible in general to obtain the distribution of S_N explicitly. However, substantial research relating to random sums has been conducted in two important contexts. These are in the theory of optimal stopping* and in the study of the preservation of classical limit theory (laws of large numbers* and central limit theorem*) under random time changes.

The general theory of optimal stopping is concerned with a finite-mean stochastic process* $\{x_n\}$ and stopping rules t. That is, t is a positive integer-valued random variable such that the event $\{t = n\}$ depends only on the values $x_1, \ldots, x_n$. The theory deals with the attainability of $V = \sup Ex_t$, where the supremum is taken over the class of all possible stopping rules t for which Ex_t exists. A particular rule is said to be *optimal* if $Ex_t = V$.

Many problems within this framework are concerned with random sums. For example, if the X_i, $i = 1, 2, \ldots$, are i.i.d. with zero mean and unit variance and $S_n = \sum_{i=1}^{n} X_i$, then it is known that there is an optimal stopping rule for the process $\{n^{-1}S_n\}$ [e.g., ref. 2, (Theorem 4.11)]. However, an exact description of such an optimal stopping rule is not available, nor is a precise value for V in this context. A comprehensive discussion of optimal stopping* is provided in Chow et al. [2].

Randomly stopped sums have some important applications in statistics, for example in sequential analysis*. For Wald's sequential probability ratio test we have Y_1, $Y_2, \ldots,$ a sequence of i.i.d. RVs with PDF $p(y \mid \theta)$, and we define

$$X_i = \log\left[p(Y_i \mid \theta_1)/p(Y_i \mid \theta_0) \right]$$

and $S_k = \sum_{i=1}^{k} X_i$. Then, to test $H_0: \theta = \theta_0$ vs. $H_1: \theta = \theta_1$, we use the rule to stop at $t = n$ if $S_n \leqslant a$ or $S_n \geqslant b$ for appropriately assigned levels a and b.

In the context of the preservation of the classical limit theorem under random time changes we first consider the laws of large numbers. Here we take S_n as a sum of n independent random variables, while N_n is any nonnegative integer-valued random variable defined on the same probability space. Then, using $\xrightarrow{\text{a.s.}}$ and $\xrightarrow{\text{p}}$ to denote almost sure convergence and convergence in probability, respectively,

$$n^{-1}S_n \xrightarrow{\text{a.s.}} 0 \text{ and } N_n \xrightarrow{\text{a.s.}} \infty$$
$$\text{imply } N_n^{-1}S_{N_n} \xrightarrow{\text{a.s.}} 0,$$

and

$$n^{-1}S_n \xrightarrow{\text{a.s.}} 0 \text{ and } N_n \xrightarrow{\text{p}} \infty$$
$$\text{imply } N_n^{-1}S_{N_n} \xrightarrow{\text{p}} 0,$$

but

$$n^{-1}S_n \xrightarrow{\text{p}} 0 \text{ and } N_n \xrightarrow{\text{a.s.}} \infty$$
$$\text{do not imply } N_n^{-1}S_{N_n} \xrightarrow{\text{p}} 0$$

as $n \to \infty$ [5, Theorem 10.1]. More detailed assumptions on the behavior of N_n as $n \to \infty$ can be imposed to repair this last result. Indeed, if $n^{-1}S_n \xrightarrow{\text{p}} 0$ and for any $\epsilon > 0$ there exists

$$0 < a = a(\epsilon) < b = b(\epsilon) < \infty$$

and a sequence $\{f(n),\ n \geqslant 1\}$ with $f(n)\uparrow\infty$ as $n \to \infty$ such that

$$P(af(n) < N_n < bf(n)) \geqslant 1 - \epsilon \qquad (1)$$

for $n > n_0(\epsilon)$, then $N_n^{-1}S_{N_n} \xrightarrow{\text{p}} 0$ as $n \to \infty$ [5, Theorem 10.2]. It should be noted that the condition (1) is weaker than $(f(n))^{-1}N_n \xrightarrow{\text{p}} \lambda$ (> 0 a.s.). These results can be extended to more general sums, and in particular to martingales, without difficulty.

As an illustration of the foregoing result based on (1), let $Z_0 = 1, Z_1, Z_2, \ldots$ denote a supercritical Galton–Watson* branching process* with nondegenerate offspring distribution and let $1 < m = EZ_1 < \infty$. This branching process evolves in such a way that

$$Z_{n+1} = Z_n^{(1)} + Z_n^{(2)} + \cdots + Z_n^{(Z_n)},$$

where the $Z_n^{(i)}$ are i.i.d., each with the offspring distribution. Then, identifying Z_n with N_n and $Z_{n+1} - mZ_n$ with S_{N_n}, we obtain consistency of $Z_n^{-1}Z_{n+1}$ as an estimator of m, it being known that (1) always holds in this case.

For the central limit theorem, comprehensive results are available for the case of i.i.d. RVs. These have found important application in contexts such as renewal theory* and queueing theory*. As usual, write S_n for the sum of n i.i.d. RVs, here with zero mean and unit variance. If $n^{-1}N_n \xrightarrow{\text{p}} \lambda$, where λ is a positive random variable, then $N_n^{-1/2}S_{N_n}$ converges in distribution to the unit normal law (e.g., Billingsley [1, Theorem 17.2], where a stronger functional central limit result is established). Sharp results on the rate of this convergence in the case where λ is a constant have been provided by Landers and Rogge [3], and for nonuniform convergence, [4]. For some generalizations to dependent variables, see Billingsley [1, Theorem 17.1].

References

[1] Billingsley, P. (1968). *Convergence of Probability Measures*. Wiley, New York.

[2] Chow, Y. S., Robbins, H., and Siegmund, D. (1971). *Great Expectations: The Theory of Optimal Stopping*. Houghton Mifflin, Boston.

[3] Landers, D. and Rogge, L. (1976). *Z. Wahrscheinl. verw. Geb.*, **36**, 269–283.

[4] Landers, D. and Rogge, L. (1978). *Metrika*, **25**, 95–114.

[5] Révész, P. (1968). *The Laws of Large Numbers*. Academic Press, New York.

[6] Robbins, H. (1948). *Bull. Amer. Math. Soc.*, **54**, 1151–1161.

(OPTIMAL STOPPING RULES
SEQUENTIAL ANALYSIS
SEQUENTIAL ESTIMATION)

C. C. HEYDE

RANDOM TESSELLATIONS

For the purposes of this entry, a *tessellation* is an arrangement of (convex) polygons fitting together without overlapping so as to cover the plane R^2 (or sometimes $X \subset R^2$); or (convex) polyhedra similarly covering R^3; or, indeed, (convex) polytopes similarly covering R^d ($d = 1, 2, \dots$). [Geometry: d-dimensional Euclidean space R^d contains flats of varying dimensions: 0-flat = point, 1-flat = line, 2-flat = plane, 3-flat = solid space, ..., $(d-1)$-flat = hyperplane, d-flat = R^d itself. (s, d) below always relates to "s-flats in R^d."]

Everyone is familiar with deterministic tessellations, such as the hexagonal planar or the cubical spatial. However, we do not live in a deterministic world, so that many real-life tessellations are "random" (e.g., the granular structure of metals) both in space and in plane section. Hence the need for a theory of random tessellations, quite apart from their intrinsic interest.

Random tessellations naturally lie in the province of geometric probability theory*. In particular, the following specific examples are based on the invariant geometric measures and the Poisson fields of geometric objects discussed in that entry.

We first discuss certain natural specific random planar tessellations, which comprise random polygons.

LINE-GENERATED RANDOM TESSELLATION

Suppose that L is an arbitrary (directed) line in R^2, and that $\mathbf{P}(0,1)$ is a Poisson point process of constant intensity $2\tau/\pi$ on L, with "particles" $\{a_i\}$. Independently, suppose that $\{\theta_i\}$ are independent, identically distributed (i.i.d.) random variables with density $\frac{1}{2}\sin\theta$ $(0 \leqslant \theta < \pi)$. Write L_i for the line through a_i making an angle θ_i with L. Then $\{L_i\}$ is $\mathbf{P}(1,2)$, the standard Poisson line process in R^2 of intensity τ, with the following basic properties:

1. It is homogeneous and isotropic in R^2 [i.e., stochastically invariant under Euclidean motions (of the coordinate frame)].
2. For any bounded convex domain K of R^2, the number of lines of $\mathbf{P}(1,2)$ intersecting it has a Poisson distribution* with mean value $\tau S(K)/\pi$ (S = perimeter).

In an obvious way (Fig. 1), $\mathbf{P}(1,2)$ determines a random tessellation $\mathscr{L}$ of R^2.

For a polygon, write N, S, A for the number of its vertices (or sides), its perimeter, and its area, respectively. $\mathbf{P}(1,2)$ being homogeneous in R^2, it may be anticipated that the values of (N, S, A) for the almost surely countable number of members of $\mathscr{L}$ conform to some trivariate probability distribution, and similarly for more general translation-invariant polygon characteristics Z. In

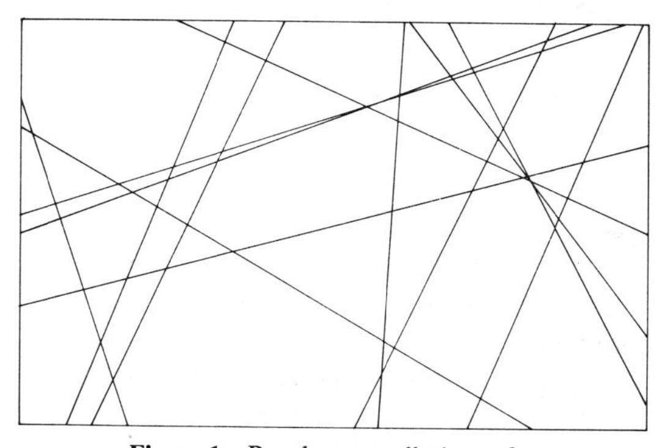

Figure 1 Random tessellation $\mathscr{L}$.

fact, if $F_{Z,r}$ is the empiric distribution function (df) of Z for those polygons of $\mathscr{L} \subset$ the disk center 0 radius r, then there exists a df F_Z such that, for any given Z-value Z', almost surely $F_{Z,r}(Z') \to F_Z(Z')$ as $r \to \infty$. A key tool in the proof of such properties is Wiener's multiparameter ergodic theorem [36]. Distributions, moments, and so on, corresponding to such df's F_Z are naturally called "ergodic," a term often omitted in practice. Loosely speaking, the ergodic distribution of Z for $\mathscr{L}$ is the distribution of Z for a "uniform random" member of $\mathscr{L}$.

Ergodic Properties of $\mathscr{L}$

For further details, see Miles [16, 21] and Solomon [30, Chap. 3].

1. The inradius distribution is exponential (τ).
2. The distribution of width in a given direction has density

$$(8\tau^2 x/\pi^3)K_{-1}(2\tau x/\pi)$$

($x \geqslant 0$) [14, p. 183].

3. The conditional distribution of S given N is $\Gamma(N-2, \tau/\pi)$ [~sum of $N-2$ independent exponential (τ/π) random variables].
4. $\Pr[N=3] = 2 - (\pi^2/6) = 0.355066$

$$\Pr[N=4] = -\tfrac{1}{3} - \tfrac{7}{36}\pi^2 + 4\int_0^{\pi/2} x^2 \cot x \, dx = 0.381466.$$

(Pr[$N = 4$]: from refs. 34 and 35.)

5. $$E[N] = 4, \qquad E[N^2] = (\pi^2 + 24)/2;$$
$$E[S] = 2\pi/\tau, \qquad E[S^2] = \pi^2(\pi^2+4)/(2\tau^2);$$
$$E[A] = \pi/\tau^2, \qquad E[A^2] = \pi^4/(2\tau^4),$$
$$E[A^3] = 4\pi^7/(7\tau^6); \quad E[SA] = \pi^4/(2\tau^3),$$
$$E[AN] = \pi^3/(2\tau^2), \quad E[NS] = \pi(\pi^2+8)/(2\tau).$$

Other ergodic distributional properties of $\mathscr{L}$ have been estimated by Monte Carlo methods* [4].

Many properties extend to the *anisotropic case*, in which the $\{\theta_i\}$ have probability element $\propto \sin\theta \, d\Theta(\theta)$; the (arbitrary) df Θ then governs the orientation distribution of $\mathscr{L}$ [16]. This model has immediate relevance to the location-time trajectories of a collection of vehicles, each moving with its own constant velocity, along a highway [30, Chap. 3; 32]. If the lines are randomly and independently "thickened," the thickness w of a line with orientation θ having an arbitrary conditional distribution depending on θ, then the polygonal interstices between the thick lines have the same ergodic distributions as before [14, p. 167; 16]! Although Kallenberg [10] disproved Davidson's conjecture that a homogeneous line process with almost surely no parallel lines is necessarily a mixture of anisotropic **P**(1, 2), for all practical purposes the conjecture applies, severely restricting the range of line-generated homogeneous tessellations.

VORONOI TESSELLATION

Suppose that **P**(0, 2) is a Poisson point process of point "particles" with constant intensity ρ in R^2. Those points of R^2 closer to a given particle of **P**(0, 2) than to any other particle constitute a convex polygon, and the aggregate of such polygons determines the Voronoi (sometimes Dirichlet, or Thiessen) random tessellation $\mathscr{V}$ relative to **P**(0, 2) as "base" (Fig. 2). The multiplicity of description reflects the repeated discoveries and manifold applications of this model. Each common side of adjoining polygons is part of the perpendicular bisector of the corresponding particles. Each vertex is a common vertex of three adjoining polygons, and is the circumcenter of the three corresponding particles. No ergodic distributions of $\mathscr{V}$ are known, but mean values are

$$E[N] = 6, \qquad E[S] = 4\rho^{-1/2},$$
$$E[A] = \rho^{-1}.$$

Gilbert [5] numerically evaluated an integral to find $E[A^2] = 1.280\rho^{-2}$, and various other ergodic distributional quantities have been estimated in a number of Monte Carlo stud-

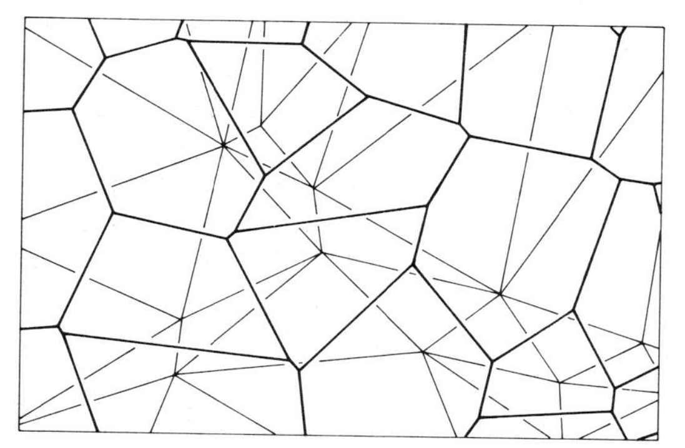

Figure 2 Random tessellations $\mathscr{V}$ (continuous thick line) and $\mathscr{D}$ (triangles, dashed thin line).

ies, the most extensive being by Hinde and Miles [8]. Green and Sibson [7] have devised an efficient computer algorithm to generate $\mathscr{V}$ relative to given (finite) particle aggregates. Again by computer, Tanemura and Hasegawa [33] iterated the Voronoi operation by replacing each particle of the "old" point process by a "new" particle at the old vertex centroid. Their "limit" random tessellation is similar to a regular hexagonal lattice, but without its preferred 120° directions; its difference from $\mathscr{V}$ is strikingly illustrated on p. 482 of their article. Solomon and Stephens [31] have fitted Pearson* and generalized gamma* distributions to the limited moment information available for both $\mathscr{L}$ and $\mathscr{V}$.

DELAUNAY TESSELLATION

Constructing the line segments joining neighboring particles of $\mathscr{V}$ yields the triangular Delaunay random tessellation $\mathscr{D}$ of R^2 (Fig. 2), in a certain sense "dual" to $\mathscr{V}$. Its full ergodic distribution is known [17]: representing a triangle by its circumradius r and the directions θ_1, θ_2, θ_3 relative to the circumcenter of the particle vertices on its circumference ($0 \leqslant \theta_1, \theta_2, \theta_3 < 2\pi$), r is ergodically independent of $(\theta_1, \theta_2, \theta_3)$ with densities

$$2(\pi\rho)^2 r^3 \exp(-\pi\rho r^2) \qquad (r \geqslant 0),$$

$$(24\pi^2)^{-1}|\sin(\theta_2 - \theta_3) + \sin(\theta_3 - \theta_1) + \sin(\theta_1 - \theta_2)|.$$

For further details, see Kendall [11], and for a statistical application, see Mardia et al. [12].

GENERALIZED VORONOI TESSELLATIONS

A point $x \in R^2$ has (almost surely) not only a nearest particle of $\mathbf{P}(0,2)$, but also a set of n nearest particles ($n = 2, 3, \ldots$). The set of x having these n particles as nearest n (in some order) is a convex polygon, and the aggregate of such polygons constitutes the generalized Voronoi random tessellation $\mathscr{V}_n$ of R^2 ($n = 2, 3, \ldots$) [17, 24]. Note that $\mathscr{V}_1 \equiv \mathscr{V}$. R. J. Maillardet has developed a computer algorithm for generating them up to $n \sim 500$—for illustrations, see Miles and Maillardet [24, p. 110]. Unlike any of the tessellations already considered, including $\mathscr{V}$, almost surely they contain no triangles. Two ergodic mean values of $\{\mathscr{V}_n\}$ ($n = 1, 2, \ldots$) are essentially geometrical, demanding simply that (say) the underlying point process be homogeneous and ergodic, with particles almost surely in sufficiently general position; they are

$$E_n[N] = 6, \qquad E_n[A] = 1/[(2n-1)\rho]$$

[the latter was derived by Miles [17] for $\mathbf{P}(0,2)$ base, its full generality subsequently being demonstrated by Maillardet].

The following geometric identity due to Sibson [28] demands only that the particles be in general position (i.e., there is no need even for randomness). Suppose that y is a particle and that its associated $\mathscr{V}_1$ polygon T has N sides. Writing y_i for a neighboring particle and T_i for the polygon of $\mathscr{V}_2$ corresponding to y and y_i ($i = 1, \ldots, N$), we have $T \subset \cup_{i=1}^{N} T_i$ and

$$|T|y = \sum_{i=1}^{N} |T \cap T_i| y_i \qquad (|\cdots| = \text{area}).$$

This result, which generalizes to all dimensions, has application in spatial interpolation and smoothing.

Another interesting $\mathbf{P}(0,2)$-based random tessellation is "Gilbert's radiating segments model" [3], for which $E[N]=4$ but little else is known.

GENERAL HOMOGENEOUS PLANAR RANDOM TESSELLATIONS

Knowledge of the foregoing *specific* random tessellations is relatively complete. When it comes to *general* random tessellations, present knowledge is rather limited and fragmentary—therein lies a prime area for further research.

The first result in this direction [15] was the relation $E[N]=2\chi/(\chi-2)$, where χ is the (ergodic) mean number of sides of the tessellation meeting of each vertex ($\chi=4,3$ above). Ambartzumian [1] considered homogeneous and isotropic random tessellations $\mathscr{T}$—he calls them "random mosaics"—as a special case of random fields* of line segments. His approach is through consideration of the intersection of an arbitrary transect line L with $\mathscr{T}$. As above, write $\{a_i,\theta_i\}$ for the successive intersection points of L with the segment sides of $\mathscr{T}$ and the accompanying angles. He proved that if $\{a_i\}$ is independent of $\{\theta_i\}$, and the θ_i mutually independent, then the vertices of $\mathscr{T}$ can only be of T and X types. Moreover, if there are no T vertices, then $\mathscr{T}$ is determined by a τ-mixture of $\mathbf{P}(1,2)$. Later [2] he showed how the ergodic distribution of $(a_{i+1}-a_i, \theta_i, \theta_{i+1})$, which does not depend on i, permits the determination of the tessellation side-length ergodic distribution, together with $E[A]$, $E[A^2]$, and $E[AS]$; these are stereological* results. Miles [22, Sec. 3.4.6] sketched an "ergodic program" for general homogeneous random tessellations, based on the property that the ergodic df's F are related to the df's G for the single random polygon containing an arbitrary point a of R^2 (e.g., the origin) by

$$F(dA,dZ)=G(dZ,dZ)/A\int A^{-1}G(dA),$$

which has the heuristic meaning that "the chance that a falls in a given polygon $\propto$ the area of that polygon." Cowan [3] partly filled in this picture, demonstrating in particular that certain edge effects on the boundary of the increasing polygon sampling circle are asymptotically negligible when homogeneity prevails. He also generalized Matschinski's result [15].

RANDOM TESSELLATIONS IN HIGHER DIMENSIONS AND OTHER SPACES

Poisson planes $\mathbf{P}(2,3)$ tessellate R^3 into random convex polyhedra, while Poisson points $\mathbf{P}(0,3)$ form the basis of the Voronoi polyhedral and dual Delaunay tetrahedral tessellations. These polyhedra are "simple" in the sense that almost surely each vertex of each polyhedron lies in three polyhedron faces. Basic properties are given in Miles [20]. Results for the corresponding polytopal tessellations in R^d generated by $\mathbf{P}(d-1,d)$ and $\mathbf{P}(0,d)$, and properties of Poisson flats $\mathbf{P}(s,d)$, are to be found in Miles [18, 22]. An extensive account of the fundamental role of Poisson flats in the theory of random sets*, together with the tessellation determined by $\mathbf{P}(d-1,d)$, is given by Matheron [14]. General flat processes are considered by Kallenberg [9]; in particular, he demonstrated that homogeneous Poisson flat processes are ergodic (= metrically transitive).

In an obvious notation, s-flat sections of $\mathscr{L}(d)$ simply yield another $\mathscr{L}(s)$, but s-flat sections of $\mathscr{V}(d)$ yield new tessellations $\mathscr{V}(s,d)$ $(0<s<d)$. $\{\mathscr{V}(3,d)\}$ $(d=3,4,\ldots)$ is a class of polyhedral tessellations of R^3, with the mean number of plane faces decreasing from 15.54 to 13.99 as d increases from 3 to ∞ [20, 23]; they are "normal" inasmuch as each edge, vertex belongs to 3, 4 polyhedra, respectively. Extending beyond our original definition, random tessellations on the surface of a sphere [19] and the hyperbolic plane [27] have also been considered.

CONCLUDING REMARKS

Over the years there have been many empiric studies of naturally occurring irregular tessellations, such as those contained in several papers by Marvin and Matzke [13]; see especially their references. However, the stochastic theory of random tessellations has a relatively brief history, initiated by the pathbreaking approach to $\mathscr{L}$ of Santaló [26] and Goudsmit [6].

Here are some instances of the practical occurrence of random tessellations: $\mathscr{L}$-like ones in the fibrous structure of thin paper sheets, in geological faulting and fragmentation, and as a representation of vehicular traffic along a road (see above); planar $\mathscr{V}$-like ones in mudcrack formations, crystal growth, and all manner of territorial situations (a particle is the "nucleus" of a polygonal "cell"); spatial $\mathscr{V}$-like ones in foam and froth, crystal growth, and cellular and granular structure.

The tessellation $\mathscr{V}$ has recently found useful application in the statistical analysis of point process data [25, Chap. 8]. Many possible additional applications are discussed by Sibson [29]. Future statistical work may demand the development of more diverse classes of specific random tessellations.

References

[1] Ambartzumian, R. V. (1972). Random fields of segments and random mosaics on the plane. *Proc. 6th Berkeley Symp. Math. Statist. Prob.*, Vol. 3, University of California Press, Berkeley, Calif., pp. 369–381.

[2] Ambartzumian, R. V. (1974). In *Stochastic Geometry*, E. F. Harding and D. G. Kendall, eds. Wiley, London, pp. 176–191.

[3] Cowan, R. (1978). The use of the ergodic theorems in random geometry. *Suppl. Adv. Appl. Prob.*, **10**, 47–57.

[4] Crain, I. K. and Miles, R. E. (1976). Monte Carlo estimates of the distributions of the random polygons determined by random lines in the plane. *J. Statist. Comput. Simul.*, **4**, 293–325.

[5] Gilbert, E. N. (1962). Random subdivisions of space into crystals. *Ann. Math. Statist.*, **33**, 958–972.

[6] Goudsmit, S. (1945). Random distribution of lines in a plane. *Rev. Mod. Phys.*, **17**, 321–322.

[7] Green, P. J. and Sibson, R. (1978). Computing Dirichlet tessellations in the plane. *Comput. J.*, **21**, 168–173.

[8] Hinde, A. L. and Miles, R. E. (1980). Monte Carlo estimates of the distributions of the random polygons of the Voronoi tessellation with respect to a Poisson process. *J. Statist. Comput. Simul.*, **10**, 205–223.

[9] Kallenberg, O. (1976). On the structure of stationary flat processes. *Z. Wahrscheinl. verw. Geb.*, **37**, 157–174.

[10] Kallenberg, O. (1977). A counterexample to R. Davidson's conjecture on line processes. *Math. Proc. Camb. Philos. Soc.*, **82**, 301–307.

[11] Kendall, D. G. (1983). The shape of Poisson–Delaunay triangles. In *Studies in Probability and Related Topics in Honour of Octav Onicescu*, M. C. Demetrescu and M. Iosifescu, eds. Nagard, Montreal, pp. 321–330.

[12] Mardia, K. V., Edwards, R., and Puri, M. L. (1977). Analysis of central place theory. *Bull. Int. Statist. Inst.*, **47**(2), 93–110.

[13] Marvin, J. W. and Matzke, E. B. (1939). *Amer. J. Bot.*, **26**, 100–103, 280–288, 288–295, 487–504.

[14] Matheron, G. (1975). *Random Sets and Integral Geometry*. Wiley, New York.

[15] Matschinski, M. (1954). Considérations statistiques sur les polygones et les polyèdres. *Publ. Inst. Statist. Univ. Paris*, **3**, 179–201.

[16] Miles, R. E. (1964). Random polygons determined by random lines in a plane. *Proc. Natl. Acad. Sci. (USA)*, **52**, 901–907; II, 1157–1160.

[17] Miles, R. E. (1970). On the homogeneous planar Poisson point process. *Math. Biosci.*, **6**, 85–127.

[18] Miles, R. E. (1971). Poisson flats in Euclidean spaces: Part II: Homogeneous Poisson flats and the complementary theorem. *Adv. Appl. Prob.*, **3**, 1–43.

[19] Miles, R. E. (1971). Random points, sets and tessellations on the surface of a sphere. *Sankhyā A*, **33**, 145–174.

[20] Miles, R. E. (1972). The random division of space. *Suppl. Adv. Appl. Prob.*, **4**, 243–266.

[21] Miles, R. E. (1973). The various aggregates of random polygons determined by random lines in a plane. *Adv. Math.*, **10**, 256–290.

[22] Miles, R. E. (1974). In *Stochastic Geometry*, E. F. Harding and D. G. Kendall, eds. Wiley, London, pp. 202–227.

[23] Miles, R. E. (1984). Sectional Voronoi tessellations. *Rev. Unión Mat. Argentina*, **29**, 310–327.

[24] Miles, R. E. and Maillardet, R. J. (1982). The basic structures of Voronoi and generalized Voronoi polygons. In *Essays in Statistical Science*, J. Gani and E. J. Hannan, eds., *Appl. Prob. Trust. J. Appl. Prob.*, **19A**, 97–111.

[25] Ripley, B. D. (1981). *Spatial Statistics*. Wiley, New York.

[26] Santaló, L. A. (1941). Valor medio del numero de partes en que una figura convexa es dividida por *n* rectas arbitrarias. *Rev. Union Mat. Arg.*, **7**, 33–37.

[27] Santaló, L. A. and Yañez, I. (1972). Averages for polygons formed by random lines in Euclidean and hyperbolic planes. *J. Appl. Prob.*, **9**, 140–157.

[28] Sibson, R. (1980). A vector identity for the Dirichlet tessellation. *Math. Proc. Camb. Philos. Soc.*, **87**, 151–155.

[29] Sibson, R. (1980). The Dirichlet tessellation as an aid in data analysis. *Scand. J. Statist.*, **7**, 14–20.

[30] Solomon, H. (1978). *Geometric Probability*. CBMS-NSF Reg. Conf. Ser. Appl. Math. 28. SIAM, Philadelphia.

[31] Solomon, H. and Stephens, M. A. (1980). Approximations to densities in geometric probability. *J. Appl. Prob.*, **17**, 145–153.

[32] Solomon, H. and Wang, P. C. C. (1972). Nonhomogeneous Poisson fields of random lines, with applications to traffic flow. *Proc. 6th Berkeley Symp. Math. Statist. Prob.*, Vol. 3. University of California Press, Berkeley, Calif., pp. 383–400.

[33] Tanemura, M. and Hasegawa, M. (1980). Geometrical models of territory: I. Models for synchronous and asynchronous settlement of territories. *J. Theor. Biol.*, **82**, 477–496.

[34] Tanner, J. C. (1983). The proportion of quadrilaterals formed by random lines in a plane. *J. Appl. Prob.*, **20**, 400–404.

[35] Tanner, J. C. (1983). Polygons formed by random lines in a plane: some further results. *J. Appl. Prob.*, **20**, 778–787.

[36] Wiener, N. (1939). The ergodic theorem. *Duke Math. J.*, **5**, 1–18.

(GEOMETRICAL PROBABILITY THEORY
RANDOM SETS OF POINTS)

R. E. MILES

RANDOM UTILITY MODELS

Utility theory* is concerned with structures of preferences or indifferences on a set of choice alternatives. Its origins go back at least to the eighteenth century, when Jeremy Bentham and others began the study of what has been called algebraic utility theory. Much of the development has been carried out in economics. In this work a subject's preferences or indifferences are nonprobabilistic. A landmark in this tradition is the celebrated theorem of von Neumann and Morgenstern [7], giving a set of sufficient conditions for the existence of a (real-valued) utility function on the set of choice alternatives. *See* UTILITY THEORY.

In a *probabilistic* utility model, on the other hand, it is supposed that when presented with a subset of alternatives A from a universe Ω, the subject will choose item $x \in A$ with some probability $p_x(A)$. Probabilistic models have been developed mainly in psychology, partly in response to the observed probabilistic choice behavior of subjects in psychophysical experiments. The subclass of *random utility models* has played a prominent role in the theory. A set of choice probabilities $\{p_x(A) : x \in A, A \subseteq \Omega\}$ is a random utility model if there are random variables $\{U_x : x \in \Omega\}$ such that

$$p_x(A) = \Pr\Big(U_x = \max_{y \in A}\{U_y\}\Big), \quad x \in A \subseteq \Omega.$$

It is usually assumed that the maximizing variable is unique with probability 1.

The underlying idea is that a subject's "utility" for stimulus x is a random quantity U_x, and that he or she makes a choice according to which utility variable is, on that realization of $\{U_x\}$, the largest. The classical embodiment of this idea is L. L. Thurstone's law of comparative judgment [6], according to which the U_x's are identically distributed apart from location shifts. Thurstone described five cases for the distributional form of the U_x; the best known is his case V model, wherein the U_x are independent normally distributed variables with a common variance. This form of the Thurstone model has been widely applied in the scaling of psychometric stimuli and in educational* testing. In the latter the "alternatives" might be tasks T_i and respondents R_j, with "R_j preferred to T_i" taken to mean that R_j is successful at task T_i. The model is also related to probit analysis. Note that the location parameters in Thurstone's models may be viewed loosely as the "utilities" of the choice alternatives. Thus the important notion of a unidimensional utility scale can be retained in a natural way. Conversely, the Thurstone model will not be appropriate in multidimensional situations, such as commonly arise in market research (preferences among cars, soft drinks, etc.). *See also* THURSTONE'S THEORY OF COMPARATIVE JUDGMENTS.

The properties of the general random utility model have for the most part been studied by mathematical psychologists. As in the economic literature for algebraic utility theory, the emphasis has been on the development of axiomatic theory; except in connection with the Thurstone model, there seems to have been little empirical work or application. We conclude by summarizing some of the results; see, for example, Luce and Suppes (1965).

The class of random utility models is a very broad one but does not cover all choice situations. Consider, for instance, the guest at a restaurant who, out of deference to his host, will not choose item x, the most expensive on the menu. The introduction of an even more expensive item will then increase the chance of his choosing x. This example violates the *regularity* condition

$$p_x(A) \geqslant p_x(B) \qquad \text{whenever } A \subseteq B.$$

It is easy to see that any random utility model satisfies regularity. On the other hand, there are some random utility models which violate the rather mild *weak stochastic transitivity* condition

$$p(x;y) \geqslant \tfrac{1}{2}, \qquad p(y;z) \geqslant \tfrac{1}{2} \Rightarrow p(x,z) \geqslant \tfrac{1}{2},$$

where $p(u;v)$ denotes $p_u(\{u,v\})$. For example, we can choose three random variables U_x, U_y, U_z (which may even be independent) such that

$$P(U_x > U_y > U_z) = 0.3$$

$$P(U_y > U_z > U_x) = 0.3$$

$$P(U_z > U_x > U_y) = 0.4.$$

Then $p(x;y) = 0.7$ and $p(y;z) = 0.6$ but $p(z;x) = 0.7$. See Blyth [1] for some interesting examples of this kind.

There has been considerable interest in the relationship between the Thurstone form of the independent random utility model and Luce's choice axiom*; see Yellott [8]. Yellott also characterizes the condition that a vector of independent variables $\{U_x : x \in \Omega\}$ is equivalent to another independent vector $\{V_x\}$, in the sense that the two collections of utility models obtained by varying the location parameters are the same and thus $\{U_x\}$ and $\{V_x\}$ are experimentally indistinguishable. For $|\Omega| \geqslant 3$, the equivalence condition is essentially that for each $x \in \Omega$ there are constants $a_x > 0$, b_x such that $(a_x U_x + b_x)$ has the same distribution as V_x. Strauss [4] gives the corresponding result for arbitrary dependent random vectors U, V. Note that the random utility model with dependencies among the variables is a natural way to model similarities between the choice alternatives [5]. Falmagne [2] derives the necessary and sufficient conditions on the choice probabilities $\{p_x(A)\}$ for them to be a random utility model.

References

[1] Blyth, C. (1972). *J. Amer. Statist. Ass.*, **67**, 366–373.

[2] Falmagne, J. C. (1978). *J. Math. Psychol.*, **18**, 52–72.

[3] Luce, R. D. and Suppes, P. (1965). In *Handbook of Mathematical Psychology*, Vol. 3, R. D. Luce, R. R. Bush, and E. Galanter, eds. Wiley, New York.

[4] Strauss, D. J. (1979). *J. Math. Psychol.*, **20**, 35–52.

[5] Strauss, D. J. (1981). *Br. J. Math. Statist. Psychol.*, **34**, 50–61.

[6] Thurstone, L. L. (1927). *Psychol. Rev.*, **34**, 273–286.

[7] von Neumann, J. and Morgenstern, O. (1947). *Theory of Games and Economic Behavior*, 2nd ed. Princeton University Press, Princeton, N.J.

[8] Yellott, J. I. (1977). *J. Math. Psychol.*, **15**, 109–144.

(DECISION THEORY
LUCE'S CHOICE AXIOM AND GENERALIZATIONS
THURSTONE'S THEORY OF COMPARATIVE JUDGMENTS
UTILITY THEORY)

DAVID J. STRAUSS

RANDOM VARIABLE *See* PROBABILITY THEORY

RANDOM VARIABLES, COMPUTER-GENERATED *See* GENERATION OF RANDOM VARIABLES

RANDOM VECTOR *See* PROBABILITY THEORY

RANDOM WALKS

By a *random walk* we will mean a sum of random variables. Some authors restrict the term to mean discrete sums, so that the time variable is an integer, but in this article our terminology will include limiting diffusion processes* which are continuous in time. Problems that are naturally phrased in random walk terminology were discussed by Pascal and Fermat (see Maistrov [15]) in their correspondence related to the theory of gambling*. The first use of the terminology "random walk" appeared in a note to *Nature* in 1905 by Karl Pearson [22], in the form of a query: "A man starts from a point 0 and walks l yards in a straight line: he then turns through any angle whatever and walks another l yards in a straight line. He then repeats this process n times. I require the probability that after these n stretches he is at a distance between r and $r + dr$ from his starting point 0." Even the random walk in this specific formulation had a precedent in earlier literature, as Lord Rayleigh (J. W. R. Strutt) was quick to point out in the same volume of *Nature* [24–25]. Since this early statement, the random walk has suggested important research directions in probability theory* and has played a significant role as a mathematical model for many scientific phenomena. Some of the former are summarized in the monograph by Spitzer [27], and some of the latter are discussed in a recent introductory article [28].

Pearson's random walk is one in which the increment of change in any step along an axis can take on any value between a minimum and maximum. Problems related to gambling* in which discrete units of money change hands suggest the study of random walks on lattice structures and, most recently, random walks defined on more abstract mathematical structures such as groups have been analyzed [13]. Additionally, one can introduce time as a variable in different ways for modeling processes that are continuous in time. For example, consider a random walk on an infinite one-dimensional lattice with sites separated by Δx, and let the time between successive steps be Δt. Provided that the mean step size is zero and the variance is finite, it follows that when Δx, $\Delta t \to 0$ in such a way that $(\Delta x)^2/\Delta t = D =$ constant, the probability density function (PDF) of displacement satisfies

$$\frac{\partial P}{\partial t} = D \frac{\partial^2 P}{\partial x^2}. \tag{1}$$

Aside from its importance in biological and physical applications, the continuum limit is useful as an approximation in many statistical problems because properties of partial differential equations are frequently easier to analyze than those of partial difference equations. An interesting illustration of this notion is provided in a monograph of Khintchin [14], in which many useful central limit* results are obtained by analyzing continuous rather than discrete processes. The solution to (1) for a diffusing particle initially at $x = 0$ is

$$P(x,t) = (4\pi Dt)^{-1/2}\exp\left[-x^2/(4Dt)\right], \tag{2}$$

which is seen to be a normal distribution with mean equal to zero and variance equal to $2Dt$. The time t therefore plays the same role as n, the number of steps of a discrete random walk or the number of terms in a discrete sum. Equation (2) can also be obtained by passing to the continuum limit of the solution to the discrete analog of (1), which is a heuristic way of establishing the relation between the sum of independent random variables and the diffusion limit.

We note that the problems mentioned so far involve sums of independent random variables. There are important and particularly difficult problems posed by sums of dependent random variables which have so far eluded complete solution. These will be mentioned later.

In the study of random walks with independent steps a central role is played by the characteristic function*, or the appropriate analog for functions in continuous time. For example, for a random walk on a D-dimensional Euclidean lattice, let $p(j_1, j_2, \ldots j_D)$ be the probability that the transition in a single step is equal to j (note here the assumption of a homogeneous lattice). A convenient definition of the generating function is

$$\lambda(\boldsymbol{\Theta}) = \sum_{\{\mathbf{j}\}} p(\mathbf{j})\exp(i\mathbf{j} \cdot \boldsymbol{\Theta}), \tag{3}$$

where $\mathbf{j} \cdot \boldsymbol{\Theta} = j_1\Theta_1 + j_2\Theta_2 + \cdots + j_D\Theta_D$ is a dot product. The probability that a random walker starting from $\mathbf{r} = \mathbf{0}$ is at site $\mathbf{r}$ at step n is

$$P_n(\mathbf{r}) = \frac{1}{(2\pi)^D} \int_{-\pi}^{\pi} \cdots \int_{-\pi}^{\pi} \lambda^n(\boldsymbol{\Theta})\exp(-i\mathbf{r} \cdot \boldsymbol{\Theta}) \, d\Theta_1 d\Theta_2 \cdots d\Theta_D. \tag{4}$$

The central limit theorem together with correction terms can be obtained from this representation when n is large and the appropriate moments of single-step transition probabilities are finite. A second function important in many analyses is the probability that a random walker reaches r for the first time at step n. This function is sometimes called the *first-passage-time probability*. It will be denoted by $F_n(\mathbf{r})$ and is related to $P_n(\mathbf{r})$ by

$$\begin{aligned} P_n(\mathbf{r}) &= \sum_{j=0}^{n} F_j(\mathbf{r}) P_{n-j}(\mathbf{0}), && \mathbf{r} \neq \mathbf{0} \\ &= \delta_{n,0}\,\delta_{\mathbf{r},\mathbf{0}} + \sum_{j=1}^{n} F_j(\mathbf{0}) P_{n-j}(\mathbf{0}), && \mathbf{r} = \mathbf{0}, \end{aligned} \tag{5}$$

where $\delta_{n,0}$ is a Kronecker delta*. While the $F_j(\mathbf{r})$ are not easily related to the $P_j(\mathbf{r})$ directly, the generating functions* with respect to the time variable,

$$\begin{aligned} F_n(\mathbf{r}; z) &= \sum_{n=0}^{\infty} F_n(\mathbf{r}) z^n, \\ P_n(\mathbf{r}; z) &= \sum_{n=0}^{\infty} P_n(\mathbf{r}) z^n \end{aligned} \tag{6}$$

are related by

$$\begin{aligned} F(\mathbf{r}; z) &= P(\mathbf{r}; z)/P(\mathbf{0}; z), && \mathbf{r} \neq \mathbf{0} \\ &= (P(\mathbf{0}; z) - 1)/P(\mathbf{0}; z), && \mathbf{r} = \mathbf{0} \end{aligned} \tag{7}$$

Some interesting properties follow from these identities. A famous problem posed by Pólya in 1921 asks for the probability that a random walker initially at the origin will return to that point at some future step [23]. The probability of such return is, from (6), equal to $F(\mathbf{0}; 1)$. Thus, by (7), when $P(\mathbf{0}; 1)$ is infinite the random walker is certain to return to the origin (i.e., it does so with probability equal to 1). An explicit expression for $P(\mathbf{0}; 1)$ can be derived from (4) and (6):

$$P(\mathbf{0}; 1) = \frac{1}{(2\pi)^D} \int_{-\pi}^{\pi} \cdots \int_{-\pi}^{\pi} \frac{d\Theta_1 \cdots d\Theta_D}{1 - \lambda(\boldsymbol{\Theta})}, \tag{8}$$

so that the question can be answered by studying properties of $\lambda(\boldsymbol{\Theta})$ in different dimensions. When the single-step variances are finite (i.e., $\partial^2\lambda/\partial\Theta_i^2|_{\boldsymbol{\Theta}=\mathbf{0}} < \infty$, $i = 1, 2, \ldots, D$), the return probabilities are equal to 1 in $D = 1$ and two dimensions, and < 1 in higher dimensions. This is proved by noting that the integrand is singular when $\boldsymbol{\Theta} = \mathbf{0}$, so that $P(\mathbf{0}; 1)$ is finite when the singularity is integrable. The argument proceeds by observing that when the variances of the single-step transition probabilities are finite, $1 - \lambda(\boldsymbol{\Theta})$ is a quadratic form of $\boldsymbol{\Theta}$ in a neighborhood of $\boldsymbol{\Theta} = \mathbf{0}$. The numerator, transformed to local spherical coordinates, gives a factor Θ^{D-1} which cancels the singularity when $D \geqslant 3$. Random walks that return to the origin with probability equal to 1 are known as *recurrent*; otherwise, they are termed *transient*. This terminology also appears in the theory of Markov chains, with a similar meaning. Even in one and two dimensions the average time to return to the origin is infinite. When transition probabilities are in the domain of attraction* of stable laws, random walks can be transient even in one dimension. Asymptotic properties of the $F_n(\mathbf{r})$ can be deduced from analytic properties of $P(\mathbf{r}; z)$ regarded as a function of z, which, in turn, in most cases of interest can be related to the behavior of $\lambda(\mathbf{\Theta})$ in a neighborhood of $\mathbf{\Theta} = \mathbf{0}$.

More elaborate versions of these classic results have been investigated in recent years [27]. These, for example, replace the notion of recurrence to a point by recurrence to a given set. For this more complicated problem not only the time to reach the set is of interest, but also the site at which it is reached. The time to reach a given set is sometimes termed the *hitting probability* and the time at which the set is reached is known as the *hitting time*. All of these problems can be subsumed under the heading of first-passage-time problems. Let the random walker initially be at a point belonging to a set S. Let R_n be the random walker's position at step n. The first passage time for leaving S is defined by the relation

$$T = \min_n \{ n \mid R_n \notin S \}. \tag{9}$$

Much is known about properties of the first passage time, T, particularly in one dimension [12], where the geometry of S is usually a connected interval. In higher dimensions the many possible configurations of S render any general analysis of first passage times quite difficult, although particular symmetric configurations sometimes allow a solution. In one dimension, when S is a connected interval whose boundary consists of its two end points the use of Wald's identity [31] allows one to generate identities and approximations for the moments of first passage time and the place at which the random walk leaves S. When the second moment of step size is finite, an elegant analysis by Erdös and Kac [8] allows one to approximate the PDF of first passage time. The use of these statistics is crucial to sequential analysis* and it was in the context of that subject that Wald devised his famous identity [31, 32]; *see also* FUNDAMENTAL IDENTITY OF SEQUENTIAL ANALYSIS.

The notion of a random walk with boundaries is suggested by many applications. For example, if the size of a queue is modeled as a random walk, and queue size is limited, say, by a finite waiting room, the maximum queue size can be regarded either as an absorbing or a reflecting barrier. In the first case the random walk is terminated when the barrier is reached, and in the second, the random walk continues after the barrier is reached, but the first event thereafter must lead to a decrease in queue size. Different rules are possible for the behavior of the random walk after contact with a reflecting barrier. The effects of barriers can be included in the mathematical statement of the random walk problem as boundary conditions to the equations that describe the evolution of the system. The boundary condition corresponding to an absorbing barrier sets the PDF of the random walk equal to zero at the barrier. Thus, if a diffusion process whose evolution is governed by (1) is to include an absorbing barrier at $x = 0$, one would need to solve (1) subject to the boundary condition $P(0, t) = 0$. Although there are different possibilities for the definition of a reflecting barrier, the most common usage is to set the gradient of the PDF in a direction normal to the barrier equal to 0. In the case of simple diffusion [exemplified by Equation (1)] a reflecting barrier at $x = 0$ would be incorporated by the boundary condition

$$\left. \frac{\partial P}{\partial x} \right|_{x=0} = 0. \tag{10}$$

The inclusion of any barrier changes the nature of the PDF. As an example, consider (1) with an absorbing barrier at $x = 0$, and the random walker initially at $x = x_0$. The resulting PDF is found to be

$$P(x,t) = \frac{1}{4\pi Dt} \left\{ \exp\left[-\frac{(x - x_0)^2}{4Dt} \right] - \exp\left[-\frac{(x + x_0)^2}{4Dt} \right] \right\}, \tag{11}$$

which has the property that

$$\lim_{t\to\infty} \int_{-\infty}^{\infty} P(x,t)\,dx = 0, \qquad (12)$$

in contrast to the barrier-free case, in which the integral = 1 for all time. Sequential analysis can be regarded as the study of random walks in the presence of certain absorbing barriers, which is the point of view adopted in the pioneering studies by Wald [31].

A property of lattice random walks that has attracted attention recently is that of the number of distinct sites visited during the course of an n-step walk. We denote this quantity by S_n. It is not too difficult to find asymptotic expressions for $E(S_n)$ in different dimensions, but the calculation of higher moments presents a much more challenging problem. One can find $E(S_n)$ by observing that

$$E(S_n) = \sum_{j=0}^{n} \sum_{\mathbf{r}} F_j(\mathbf{r}), \qquad (13)$$

which allows us to use (7) together with a Tauberian theorem* to infer $E(S_n)$ for large n [17]. No such representation can be given for higher moments. One finds, specifically, that for zero-mean and finite-variance random walks,

$$E(S_n) \sim \begin{cases} n^{1/2} & \text{one-dimensional} \\ n/\ln n & \text{two-dimensional} \\ n & \text{three-dimensional} \end{cases} \qquad (14)$$

asymptotically, where "$\sim$" means "asymptotic to" and where each term on the right side of this equation is to be multiplied by a constant calculated from the actual transition probabilities. The basic difficulty in the analysis of S_n is that it is not a Markovian random variable. Jain and Pruitt, by a sophisticated analysis, have determined the asymptotic distribution of S_n in three dimensions [11]. Typical further properties that have attracted research effort include the occupancy of a set by an n-step walk, that is, the fraction of time spent by the random walker in the set [5], and the spans, of a random walk. These are the dimensions of the smallest box with sides parallel to the coordinate axes entirely enclosing the random walk [4]. Another generalization useful in problems that arise in chemical physics is to keep the discrete lattice structure but allow the time between steps to be continuously variable [18]. The literature, both pure and applied, abounds with calculations of different random walk functionals that are too numerous to summarize with any completeness here.

APPLICATIONS

From the generality of our definition of random walks one might guess that many applications of the basic model have been made. This is indeed the case, and random walk methodology is central to a variety of subject areas in statistics and the natural sciences. Pearson's original query related to a theory of biological migration that he had developed. At about the same time Bachelier developed a diffusion model for speculation in the stock market [1]. Because even the enumeration of these applications would lead to too large an article, we mention only several representative ones.

A major application of random walks in statistics is to sequential analysis*, in which results of an experiment conducted sequentially in time are summarized in terms of a probability ratio, which in turn can be represented as a random walk in a bounded region of the plane. The random walk continues as long as it remains within the region. Termination with acceptance of the appropriate hypothesis occurs when any part of the boundary is reached [32]. A related application is that of CUSUM charts. The statistical questions of interest are related to the determination of the boundaries, the probability of termination at one or another section of the boundary, and the determination of properties of the first passage times for leaving the region of interest. All of these questions lead to difficult analytical problems. The one general result of Wald mentioned earlier provides approximations for the relevant parameters but is valid only in one dimension (although some limited generalizations to higher dimensions have been made). There is not much known about random walks in regions of a general shape because of the mathematical complexity of the related calculations, although some results are available for specific curvilinear boundaries.

Any Markov process* can be regarded as a random walk, and the many applications of Markov methods lead to corresponding problems about random walks. For example, queueing theory* has led to some elegant investigations on properties of one-dimensional random walks. A large class of problems relates to the probability of reaching a half-line and first-passage-time problems for that event [20]. These would arise, for example, if one were interested in the maximum queue size during a given interval of time. An interesting early application of these ideas is to the theory of insurance, in which one is interested in the initial amount of capital necessary to insure avoidance of bankruptcy with predetermined probability [26]. This is a sophisticated generalization of the classical gambler's ruin problem. Similar questions arise in inventory theory* and dam theory* [19].

Many applications of random walks in the physical sciences have been made which in turn have generated some profoundly difficult problems in the theory of random walks. An excellent example of this is provided by the self-avoiding walk. Random walks are often used to model configurational properties of long-chain molecules [8]. Since no two atoms can occupy the same site, one is led to study random walks forbidden to intersect themselves. The mathematical difficulty inherent in these models is that the walks are not Markovian. Few results about self-avoiding walks are known rigorously, although an enormous amount is known by simulation* and from approximate theories. As an example of results that differ from standard random walk results, the mean square end-to-end distance of a zero-mean two-dimensional self-avoiding walk is asymptotically proportional to $n^{3/2}$ in contrast to n for the unrestricted walk. The corresponding result in three dimensions is $n^{6/5}$. In four and higher dimensions $E(R_n^2) \sim n$ for both unrestricted and self-avoiding walks. The limiting probability density function for the end-to-end distance is distinctly nonnormal in two dimensions, and is probably normal in four or more dimensions, but these results have not been proved rigorously [6].

Another class of problems suggested by applications in both the biological and physical sciences is that of random walks in random environments. The simplest example of this class is a random walk on a lattice in which the transition probabilities associated with each lattice site are themselves random variables. For example, random walks on lattices in which each point is absorbing with probability $c < 1$ are of current interest in chemical applications [34]. The techniques applicable to homogeneous random walks are not useful for random environment problems (e.g., generating or characteristic functions). An example of a random environment problem is the diffusion process representing population growth* in a time-dependent random environment [30]. If the population size at time t is $X(t)$, then a typical model has the form

$$dX = f(X)\,dt + g(X)\,dW, \qquad (15)$$

where dW is $W(t+h) - W(t)$, where $W(t)$ is a representation of Brownian motion* and h tends to 0. Many problems of this kind can be analyzed by the Ito calculus and other methods used for solving stochastic differential equations.

There is a considerable literature on random walk models of stock price behavior dating back to the work of Bachelier [1, 3, 9, 21]. One line of research in this area is the determination of the distribution of such quantities as the logarithm of the ratio of stock price on two successive days. It is known that the distribution of this parameter is highly peaked and heavy-tailed compared with a normal distribution. Mandelbrot was the first to suggest that these variables have a stable law* distribution [3, 9, 16]. However, it is not certain at this time whether the distribution is even stationary, and over intervals where stationarity exists, it is not known whether there are more appropriate distributions than stable laws. Whether characterization of stock market behavior as a random walk will lead to useful predictive information is also unclear.

Investigators have found that the locomotion of many microorganisms can be described in terms of random walks or diffusion processes. This finding has been used to try to distinguish mechanisms by which microorganisms respond to chemically favorable substances ("chemotaxis"). The observations show that bacterial motion takes place along approximately straight line segments randomly interrupted by periods during which the organism rotates in place, then chooses a new direction of motion. Thus the organism performs a type of Pearson random walk. Two different phenomenological responses to a chemical attractant have been identified. In one the organism tends to turn in the general direction of the attractant, the lengths of the straight-line segments being uncorrelated with turn angles. In the second the random times spent in the straight-line segments tend to be longer on average when motion is toward the attracting source. Other responses are undoubtedly possible. These experimental investigations suggest further research into developing statistical tests to distinguish between different random walk models. A good reference to this work is ref. 20.

Other areas in which random walks play an important role include the theory of the Kolmogorov–Smirnov statistics* [7] and crystallography [29]. In the first of these, problems related to first passage times to curved boundaries are a focus for current research, while in the second, approximations to the PDF of various forms of the Pearson random walk are of interest. Recently, many models have been suggested for the study of multiparticle random walks in which there is a nontrivial interaction between the different walkers [28]. This class of problems is a simplified form of questions that arise in statistical mechanics, where one is interested in determining the equilibrium states of interacting particles. Progress in this area is difficult and results are mainly available for one-dimensional systems, using techniques that cannot be generalized to the more realistic three-dimensional case.

References

[1] Bachelier, L. (1900). *Théorie de la speculation*. Gauthier-Villars, Paris.

[2] Baxter, G. (1961). *J. Anal. Math.*, **9**, 31–70.

[3] Cootner, P. H. (1964). *The Random Character of Stock Market Prices*. MIT Press, Cambridge, Mass. (Contains reprints of significant studies prior to 1964. An English translation of Bachelier's thesis is included.)

[4] Daniels, H. E. (1941). *Proc. Camb. Philos. Soc.*, **37**, 244–251.

[5] Darling, D. A. and Kac, M. (1957). *Trans. Amer. Math. Soc.*, **84**, 444–458.

[6] Domb, C., Gillis, J. and Wilmers, G. (1965). *Proc. Phys. Soc.*, **85**, 625–645.

[7] Durbin, J. (1973). *Distribution Theory for Tests Based on the Sample Distribution Function*. SIAM, Philadelphia. (A compact account of distribution related to Kolmogorov–Smirnov tests and their generalizations.)

[8] Erdös, P. and Kac, M. (1946). *Bull. Amer. Soc.*, **52**, 292–302. (The genesis of the idea of an invariance principle, developed so beautifully later by Donsker.)

[9] Fama, E. F. (1965). *J. Bus.*, **38**, 34–105.

[10] Hermans, J. J., ed. (1978). *Polymer Solution Properties*: Part II. *Hydrodynamics and Light Scattering*. Dowden, Hutchinson & Ross, Stroudsburg, Pa. (A reprinting of classic papers in polymer physics, many of which relate to the application of random walk theory to the characterization of polymer configurations.)

[11] Jain, N. C. and Pruitt, W. E. (1971). *J. Anal. Math.*, **24**, 369–393.

[12] Kemperman, J. H. B. (1961). *The Passage Problem for a Stationary Markov Chain*. University of Chicago Press, Chicago. (A tightly written and highly theoretical account of one-dimensional first-passage-time problems. Exact rather than approximate results are emphasized.)

[13] Kesten, H. (1959). *Trans. Amer. Math. Soc.*, **92**, 336–354.

[14] Khintchin, A. (1948). *Asymptotische Gesetze der Wahrscheinlichkeitsrechnung*. (Reprinted by

Chelsea, New York.) (Although 50 years old, a monograph still worth reading for derivations of central limit theorems that emphasize the relation to diffusion processes rather than the use of characteristic functions.)

[15] Maistrov, L. E. (1974). *Probability Theory. A Historical Sketch*, S. Kotz, trans. Academic Press, New York. (An interesting account of some of the early problems in probability considered by Cardano, Tartaglia, Fermat, Pascal, and others.)

[16] Mandelbrot, B. (1963). *J. Bus.*, **36**, 394–419. (Also reprinted in ref. 3.)

[17] Montroll, E. W. (1964). *Proc. Symp. Appl. Math. Amer. Math. Soc.*, **16**, 193–208.

[18] Montroll, E. W. and Weiss, G. H. (1965). *J. Math. Phys.*, **6**, 167–181. (Written for physicists, but contains a number of applications of Tauberian methods to random walk problems.)

[19] Moran, P. A. P. (1959). *The Theory of Storage*. Wiley, New York.

[20] Nossal, R. (1980). In *Biological Growth and Spread. Mathematical Theories and Applications*, W. Jager and H. Röst, eds. Springer-Verlag, New York, pp. 410–440.

[21] Osborne, M. F. M. (1959). *J. Operat. Res. Amer.*, **7**, 145–173.

[22] Pearson, K. (1950). *Nature (Lond.)*, **72**, 294.

[23] Polya, G. (1921). *Math. Ann.*, **84**, 149–160.

[24] Rayleigh, Lord (J. W. R. Strutt) (1905). *Nature (Lond.)*, **72**, 318.

[25] Rayleigh, Lord (J. W. R. Strutt) (1880). *Philos. Mag., 5th Ser.*, **10**, 73–78.

[26] Seal, H. (1978). *Survival Probabilities: The Goal of Risk Theory*. Wiley-Interscience, New York.

[27] Spitzer, F. (1975). *Principles of Random Walk*, 2nd ed. Springer-Verlag, New York. (A classic monograph on properties of lattice random walks. Diffusion processes are not covered and few applications are discussed.)

[28] Spitzer, F. (1977). *Bull. Amer. Math. Soc.*, **83**, 880–890.

[29] Srinivisan, R. and Parthasarathy, S. (1976). *Some Statistical Applications in X-Ray Crystallography*. Pergamon Press, London. (The Pearson random walk and its generalizations are applied as an aid to determining crystal structure from x-ray data.)

[30] Turelli, M. (1977). *Theor. Popul. Biol.*, **12**, 140–178. (Discussion of the use of stochastic differential equation techniques to ecological models.)

[31] Wald, A. (1944). *Ann. Math. Statist.*, **15**, 283–294.

[32] Wald, A. (1947). *Sequential Analysis*. Wiley, New York.

[33] Weiss, G. H. (1983). *Amer. Sci.*, **71**, 65–71.

[34] Weiss, G. H. and Rubin, R. J. (1983). *Adv. Chem. Phys.*, **52**, 363–504. (A review of analytical techniques for random walks with an account of many applications, mainly in chemical physics. Over 400 references are included.)

(CUMULATIVE SUM CONTROL CHARTS
DIFFUSION PROCESSES
KOLMOGOROV–SMIRNOV STATISTICS
MARKOV PROCESSES
QUEUEING THEORY
RISK THEORY
SEQUENTIAL ANALYSIS
STABLE DISTRIBUTIONS)

GEORGE H. WEISS

RANGE-PRESERVING ESTIMATORS

An estimator is said to be range preserving if its values are confined to the range of what it is to estimate. The property of being range preserving is an essential property of an estimator, a *sine qua non*. Other properties, such as unbiasedness, may be desirable in some situations, but an unbiased estimator that is not range preserving should be ruled out as an estimator. [We are not speaking of uses of estimators for purposes other than estimation (e.g., as test statistics).]

Suppose that the observation vector $\mathbf{X}$ takes values in a space $\mathscr{X}$ and that the probability distribution P of $\mathbf{X}$ is known to belong to a family $\mathscr{P}$ of distributions. It is desired to estimate a function $\theta(P)$ defined for $P \in \mathscr{P}$. To fix ideas, suppose that $\theta(P)$ takes values in a Euclidean space $\mathbb{R}^k$, $k \geqslant 1$.

Following Hoeffding [3], the set

$$\Theta = \{\theta(P) : P \in \mathscr{P}\}$$

will be called the *prior range* of $\theta(P)$. The *posterior range*, Θ_x, of $\theta(P)$ is, informally, the least set in which $\theta(P)$ is known to lie after an observation x from $\mathscr{X}$ has been made. For example, if $\mathscr{P}$ is the family of the uniform distributions P_η on the interval $(\eta, \eta + 1)$, $\eta \in \mathbb{R}^1$, then $\Theta = \mathbb{R}^1$, $\Theta_x = [x - 1, x]$. In many common estimation problems the posterior range coincides with the prior range.

The estimator $t(\mathbf{x})$ of $\theta(P)$ is said to be

range preserving if $t(\mathbf{x}) \in \Theta_x$ with P probability 1, for all $P \in \mathscr{P}$. Some types of estimators are range preserving by definition. Suppose that for each $\theta' \in \Theta$ there is a unique $P_{\theta'} \in \mathscr{P}$ such that $\theta(P_{\theta'}) = \theta'$. If a maximum likelihood* estimator (defined as a value in Θ that maximizes the likelihood) exists, it is range preserving. Similarly, Bayes estimators are range-preserving. On the other hand, roots of the maximum likelihood equations may fail to preserve the range (e.g., a sample of $n = 2$ observations from a mixture of two known probability densities with unknown mixture parameter). The same is true of moment estimators; thus Rider [6] has shown that in the case of a mixture of two exponential distributions the moment estimator is not range preserving.

Few general results on range-preserving estimators are available. The fact that standard unbiased estimators of (positive) variance components often can take negative values has attracted considerable attention. LaMotte [4] characterized linear combinations of variance components for which there exist unbiased, nonnegative quadratic estimators. He showed that the "error" component in ANOVA models is the only single component that can be so estimated. Pukelsheim [5] proved analogous results on the estimation of linear combinations of variance–covariance components. Hartung [1] derived nonnegative minimum-biased invariant estimators in variance component* models.

Hoeffding [3] gave some necessary conditions for the existence of unbiased range-preserving estimators, including the following. The prior range Θ of $\theta(P)$, $P \in \mathscr{P}$, is assumed to be a subset of a Euclidean space $\mathbb{R}^k$. The convex hull of Θ is denoted $C\Theta$. A supporting hyperplane $H = H_c$ of $C\Theta$ at the point $\theta(P_0)$ is given by a $c \neq 0$ in $\mathbb{R}^k$ such that

$$(c, \theta(P)) \geqslant (c, \theta(P_0)) \qquad \text{for all } P \in \mathscr{P}.$$

Proposition 1. Suppose that there is a $P^0 \in \mathscr{P}$ such that $C\Theta$ has a supporting hyperplane H at $\theta(P^0)$. Then if the estimator $t(x)$ is both range preserving and unbiased, we have

$$t(x) \in \Theta \cap H \text{ with } P^0 \text{ probability 1.}$$

Example 1. Let

$$\mathscr{X} = \{0, 1, \ldots, n\},$$
$$\mathscr{P} = \{P_u : 0 \leqslant u \leqslant 1\},$$
$$P_u(x) = \binom{n}{x} u^x (1-u)^{n-x},$$

$x \in \mathscr{X}$; $\theta(P_u) = u$. The posterior range of u given x is $(0, 1)$ if $1 \leqslant x \leqslant n-1$; $[0, 1)$ if $x = 0$; $(0, 1]$ if $x = n$. The unbiased estimator $t(x) = x/n$ is range preserving. In accordance with Proposition 1, $t(x) = 0$ with P_0 probability 1 and $t(x) = 1$ with P_1 probability 1.

Proposition 2. Let P^0 and H satisfy the conditions of Proposition 1. If there is a distribution $P' \in \mathscr{P}$ dominated by P^0 and satisfying

$$\theta(P') \notin H,$$

then no range-preserving estimator of $\theta(P)$ is unbiased.

Example 2. Let X_1 and X_2 be independent normal with common unknown mean μ and variance 1. Denote by P_μ the corresponding distribution of (X_1, X_2), and consider estimating $\theta(P_\mu) = \mu^2$. By Proposition 2 with $P^0 = P_0$, $P' = P_\mu$ ($\mu \neq 0$), H the point 0, no range-preserving estimator of μ^2 is unbiased. (This can be seen directly from the fact that the only unbiased estimate of μ^2 is X_1X_2, and the probability of $X_1X_2 < 0$ is positive.)

For the case of a sample from a multinomial distribution*, Hoeffding [2] gave a necessary and sufficient condition for the (unique) unbiased estimator to be range preserving.

References

[1] Hartung, J. (1981). *Ann. Statist.*, **9**, 278–292.

[2] Hoeffding, W. (1984). *J. Amer. Statist. Ass.*, **79**, 712–714.

[3] Hoeffding, W. (1983). *Unbiased Range-Preserving Estimators. A Festschrift for Erich L. Lehmann.* Wadsworth, Belmont, Calif., pp. 249–260.

[4] LaMotte, L. R. (1973). *J. Amer. Statist. Ass.*, **68**, 728–730.

[5] Pukelsheim, F. (1981). *Ann. Statist.*, **9**, 293–299.

[6] Rider, P. R. (1961). *Ann. Math. Statist.*, **32**, 143–147.

W. HOEFFDING

RANGES

The *range* of a statistical distribution or random variable X is $b - a$, where $[a,b]$ is the support of X. The range is infinite if a or b is infinite.

The *range* of an ordered set of data

$$x_{(1)} \leqslant x_{(2)} \leqslant \cdots \leqslant x_{(n)}$$

is $w = x_{(n)} - x_{(1)}$. The data set is of most interest when it comprises realizations of order statistics* $X_{(1)}, \ldots, X_{(n)}$ from a sample of size n based on a parent distribution having cumulative distribution function* (CDF) $F(\cdot)$ and probability density function (PDF) $f(\cdot)$, with $W = X_{(n)} - X_{(1)}$.

Other statistics related to the range are:

1. *Quasi-ranges* $W_{(r)} = X_{(n-r+1)} - X_{(r)}$, $r \leqslant [\frac{1}{2}n]$.
2. *Studentized ranges**, discussed in the entry of that name.
3. The *mean range* $W_{n,k}$ of k sample ranges, each based on samples of size n.
4. *Range ratios* (see David [7, Secs. 5.4, 7.7]).
5. The *bivariate range* R. If (X_i, Y_i) for $i = 1, \ldots, n$ are n target impact points, say, then R is the maximum distance between any two of them, that is,

$$R = \max_{i \neq j} \left[(X_i - X_j)^2 + (Y_i - Y_j)^2 \right]^{1/2}.$$

Cacoullos and DeCicco [3] give approximations to the distribution and percentage points of R when the data come from a bivariate normal distribution.

BASIC PROPERTIES

In most applications F is continuous. The CDF of W is then given by

$$G(w) = n \int_{-\infty}^{\infty} f(x) \left[F(x+w) - F(x) \right]^{n-1} dx, \qquad w > 0 \quad (1)$$

with PDF

$$g(w) = n(n-1) \int_{-\infty}^{\infty} [F(x+w) - F(x)]^{n-2} \times f(x) f(x+w) dx, \qquad w > 0. \quad (2)$$

The moments can be derived from those of $X_{(1)}$ and $X_{(n)}$; for example,

$$E(W) = E(X_{(n)}) - E(X_{(1)}),$$

$$\operatorname{var}(W) = \operatorname{var}(X_{(n)}) - 2\operatorname{cov}(X_{(1)}, X_{(n)}) + \operatorname{var}(X_{(1)}).$$

The PDF of the quasi-range $W_{(r)}$ is given by

$$g_r(w) = \frac{n!}{[(r-1)!]^2 (n-2r)!} \times \int_{-\infty}^{\infty} f(x) f(x+w) \times \left[F(x)(1 - F(x+w)) \right]^{r-1} \times \left[F(x+w) - F(x) \right]^{n-2r} dx. \quad (3)$$

In most applications F is rectangular or normal. Studies of W and of $W_{(r)}$ have been made, however, for cases where F has a gamma* [13] or a logistic distribution* [14].

Example 1. X is rectangular* on the interval $0 \leqslant x \leqslant 1$. Then $W_{(r)}$ has a beta distribution* with PDF

$$g_r(w) = \frac{1}{B(n-2r+1, 2r)} w^{n-2r} \times (1-w)^{2r-1}, \qquad 0 \leqslant w \leqslant 1,$$

the PDF of the range being the special case $r = 1$. Further,

$$E\left(W_{(r)}{}^k\right) = \frac{(n-2r+k)!\, n!}{(n-2r)!\,(n+k)!}, \qquad k = 1, 2, 3, \ldots,$$

and in particular,

$$E(W) = \frac{n-1}{n+1},$$

$$\operatorname{var}(W) = \frac{2(n-1)}{(n+2)(n+1)^2}.$$

Example 2. X is normally distributed with variance σ^2. A summary of results appears in Patel and Read [2], Secs. 8.4, 8.5], including expressions for the PDFs of W and $W_{(r)}$; explicit expressions for the distribution of W are given for the cases $n = 2$, 3, and 4, and for moments of W for $n = 2$, 3, 4, and 5. Some approximations to the distributions of W and of $W_{(r)}$ are given; see also David [7, Sec. 7.3].

Approximations to the distribution of W when the parent distribution is normal are based on chi* or chi-square*. The most accurate and adequate is given [4, 7, 24] by

$$W/\sigma = (\chi_\nu^2/c)^\alpha,$$

where ν, c, and α are derived by equating the first three moments of W/σ with those of the right-hand side. These approximations were also developed for the distribution of the mean range $W_{n,k}$ under normality assumptions; see Bland et al. [2] for exact distributions of $W_{2,2}$, $W_{2,3}$, $W_{3,2}$, and $W_{2,4}$.

A listing of available tables of percent points, cumulative probabilities and moment values of W, $W_{(r)}$, the ratio of independent ranges W_1/W_2, and of studentized ranges appears in Patel and Read [21, Table 8.1]. Thus Owen [20, pp. 139–140] gives percent points and moments of W for $n \leqslant 100$ to three decimal places; Harter [10] gives these, and CDF and PDF values to at least six places, as well as percent points and CDF and PDF values for the ratio W_1/W_2 with samples up to size 15. Harter [11] tabulates percent points, CDF, and moment values of $W_{(r)}$ to at least five places. See also Pearson and Hartley [23].

Barnard [1] gives a FORTRAN computer program for the probability integral of W under normality assumptions; see also el Lozy [9].

RANGES IN INFERENCE

The range W, quasi-range $W_{(r)}$, and mean range W are used primarily as measures of dispersion. The use of the range in place of the standard deviation σ in examining stability of variation in industrial processes was suggested in 1927 by Student [27]. In 1925, Tippett [28] studied the estimation of σ by W and fitted Pearson-type curves to the CDF of W under normality by the method of moments*. The exact derivation of the CDF of W was given by Hartley [12], but in the meantime the use of the range to measure variation in control charts* had been proposed by Pearson and Haines [22] as an alternative to the standard deviation. This occurred in 1933, only two years after Shewhart [26] had introduced control charts as a practical tool in quality control*. The impetus to measure variation by the range comes, of course, from its greater ease of computation. Since the efficiency* of W as an estimator of σ in normal populations is at least 0.91 when $n \leqslant 7$ and at least 0.85 when $n \leqslant 10$ (see David [7, Sec. 7.3]), the range is still used in computing control limits for $\bar{x}$-charts, and is still used together with the mean range to control for variability in R-charts (*see* CONTROL CHARTS) because of its ease of calculation, but also because it compares favorably with the sample standard deviation s in small samples in nonnormal populations [6]. Table 1 shows for $n \leqslant 30$ the

Table 1

n	2	3	4	5	6	7
d_2	1.128	1.693	2.059	2.326	2.534	2.704
d_3	0.853	0.888	0.880	0.864	0.848	0.833
Eff.	1	0.992	0.975	0.955	0.933	0.911
n	8	10	12	15	20	30
d_2	2.847	3.078	3.258	3.472	3.735	4.086
d_3	0.820	0.797	0.778	0.756	0.729	0.693
Eff.	0.890	0.850	0.814	0.766	0.700	—

values d_2 and d_3 such that in normal samples $E(W) = d_2\sigma$ and s.d.$(W) = d_3\sigma$, as well as the efficiency (eff.) of W relative to that of the minimum variance unbiased estimator*

$$\left\{\Gamma\left[\tfrac{1}{2}(n-1)\right]/\left[\sqrt{2}\Gamma\left(\tfrac{1}{2}n\right)\right]\right\}$$

$$\times\left[\sum(X_i - X)^2\right]^{1/2} \quad \text{of } \sigma.$$

Among quasi-ranges in providing estimators of σ in normal samples (Cadwell [5]), $W = W_{(1)}$ is most efficient for $n \leqslant 17$ and $W_{(2)}$ for $18 \leqslant n \leqslant 30$; $W_{(3)}$ becomes preferable for $n \geqslant 31$. Dixon [8] determined subsets of quasi-ranges $W_{(i)}$, not necessarily consecutive, such that $k'\sum W_{(i)}$ is unbiased and most efficient in estimating σ among such subsets.

For a discussion of a quick test using the mean range in a one-way analysis of variance, see David [7, Sec. 7.7] and Sheesley [25]. Nelson [18, 19] summarizes the use of W and of $W_{n,k}$ in normal samples to estimate σ and test for heterogeneity of variance. McDonald [15–17] has constructed subset selection procedures (choosing a subset of k populations so that, based on available samples, the "best" population is included in the subset with a given probability) for uniform and normal populations, in which the selection is based on range or quasi-range statistics.

References

[1] Barnard, J. (1978). *Appl. Statist.*, **27**, 197–198.

[2] Bland, R. P., Gilbert, R. D., Kapadia, C. H., and Owen, D. B. (1966). *Biometrika*, **53**, 245–248.

[3] Cacoullos, T. and DeCicco, H. (1967). *Technometrics*, **9**, 476–480.

[4] Cadwell, J. H. (1953). *Biometrika*, **40**, 336–346.

[5] Cadwell, J. H. (1953). *Ann. Math. Statist.*, **24**, 603–613.

[6] David, H. A. (1962). In *Contributions to Order Statistics*, A. E. Sarhan and B. G. Greenberg, eds. Wiley, New York, pp. 94–128.

[7] David, H. A. (1981). *Order Statistics*, 2nd ed. Wiley, New York. (A standard work, well written and presented. See Chapter 7 in particular.)

[8] Dixon, W. J. (1957). *Ann. Math. Statist.*, **28**, 806–809.

[9] el Lozy, M. (1982). *Appl. Statist.*, **31**, 99.

[10] Harter, H. L. (1969). *Order Statistics and Their Use in Testing and Estimation*, Vol. 1. Aerospace Research Laboratories, U.S. Air Force, Dayton, Ohio.

[11] Harter, H. L. (1969). *Order Statistics and Their Use in Testing and Estimation*, Vol. 2. Aerospace Research Laboratories, U.S. Air Force, Dayton, Ohio.

[12] Hartley, H. O. (1942). *Biometrika*, **32**, 334–348.

[13] Lee, K. R. and Kapadia, C. H. (1982). *Commun. Statist. Simul. Comp.*, **11**, 175–195.

[14] Malik, H. J. (1980). *Commun. Statist. A*, **14**, 1527–1534.

[15] McDonald, G. C. (1976). *Technometrics*, **18**, 343–349.

[16] McDonald, G. C. (1977). *Commun. Statist. A*, **6**, 1055–1079.

[17] McDonald, G. C. (1978). *Sankhyā B*, **40**, 163–191.

[18] Nelson, L. S. (1975). *J. Quality Tech.*, **7**, 46–48.

[19] Nelson, L. S. (1975). *J. Quality Tech.*, **7**, 99–100.

[20] Owen, D. B. (1962). *Handbook of Statistical Tables*. Addison-Wesley, Reading, Mass.

[21] Patel, J. K. and Read, C. B. (1982). *Handbook of the Normal Distribution*. Marcel Dekker, New York. (See Chapter 8 for results on the range, quasi-ranges, mean range, studentized ranges, and range ratios.)

[22] Pearson, E. S. and Haines, J. (1935). *J. R. Statist. Soc. Suppl.*, **2**, 83–98.

[23] Pearson, E. S. and Hartley, H. O. (1966). *Biometrika Tables for Statisticians*, Vol. 1. Cambridge University Press, London.

[24] Pillai, K. C. S. (1950). *Ann. Math. Statist.*, **21**, 100–105.

[25] Sheesley, J. H. (1981). *J. Quality Tech.*, **13**, 184–185.

[26] Shewhart, W. A. (1931). *Economic Control of Quality of Manufactured Product*. D. Van Nostrand, New York.

[27] Student (1927). *Biometrika*, **19**, 151–164.

[28] Tippett, L. H. C. (1925). *Biometrika*, **17**, 364–387.

(CONTROL CHARTS
L-STATISTICS
MIDRANGES
ORDER STATISTICS
SHORTCUT METHODS)

CAMPBELL B. READ

RANK ANALYSIS OF COVARIANCE

See RANK ORDER STATISTICS

RANKED SET SAMPLING

Ranked set sampling was first suggested by McIntyre [4], without the supporting mathematical theory, as a method for estimating mean pasture yields. He found that the yield of several small plots could be ranked fairly accurately with negligible cost, while exact measurement of the yield of each was time consuming. To take advantage of this ability, he proposed ranked set sampling, a procedure that improves the efficiency of the sample mean as an estimator of the population mean μ in situations where the characteristic of interest is difficult or expensive to measure, but may be readily "judgment ordered." This term was introduced later by Dell and Clutter [2, 3] to mean ranking of the observations by eye or some other relatively cheap method not requiring actual measurement. Meanwhile, Takahasi and Wakimoto [13] supplied the theory missing from McIntyre's intuitively appealing suggestion and explored the extent to which the efficiency is improved when the judgment ordering can be performed perfectly. Dell and Clutter go on to show that even if the sample cannot be ranked perfectly by judgment, the procedure may still produce improved precision in the estimator. This is important for their applications in forestry, since accurate ranking is usually not possible.

ESTIMATION OF MEAN

MTW Procedure

The ranked set sampling procedure, as originally described by McIntyre and Takahasi and Wakimoto, consists of drawing n random samples from the population, each of size n, and (judgment) ranking each of them. Then the smallest observation from the first sample is chosen for measurement, as is the second smallest observation from the second sample. The process continues in this way until the largest observation from the nth sample is measured, for a total of n measured observations, one from each order class. The entire cycle is repeated m times until a total of mn^2 elements have been drawn from the population but only mn have been measured. These mn observations are the ranked set sample and the selection process described will be referred to as the *MTW procedure*.

For example, suppose that 3 is the maximum number of observations which can be ordered by eye, but enough time (or money) is available for six exact measurements to be made. Then choose $n = 3$ and $m = 2$. The sampling scheme can be diagrammed as follows:

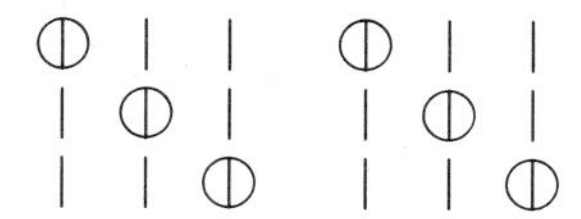

Each row represents a (judgment) ordered sample, and the circled observations indicate those to be chosen for measurement.

Let $X_{[j]i}$ denote the jth judgment order statistic* from the ith cycle. Then

$$\hat{\mu}_{\text{MTW}} = \frac{1}{mn} \sum_{i=1}^{m} \sum_{j=1}^{n} X_{[j]i}$$

can be shown [3, 13] to be an unbiased estimator of the population mean μ. Furthermore, its variance is never greater than that of a sample mean from a random sample of size mn. The relative efficiency* (RE) is

$$\text{RE}(\overline{X}, \hat{\mu}_{\text{MTW}}) = \frac{\text{var}\,\overline{X}}{\text{var}\,\hat{\mu}_{\text{MTW}}}$$

$$= \left(1 - \frac{1}{n\sigma^2} \sum_{j=1}^{n} \tau_j^2\right)^{-1} \geqslant 1,$$

where σ^2 is the population variance and $\tau_i = E[X_{[j]i} - \mu]$. These properties remain true when the ranking is imperfect. If judgment ranking is no better than random, $\text{RE}(\overline{X}, \hat{\mu}_{\text{MTW}}) = 1$.

The amount of improvement in precision depends on the underlying distribution. When ranking is perfect, so that $X_{[j]i}$ is the jth order statistic,

$$\text{RE}(\overline{X}, \hat{\mu}_{\text{MTW}}) \leqslant \tfrac{1}{2}(n+1),$$

Table 1 RE($\overline{X}$, $\hat{\mu}_{\text{MTW}}$) for Given n

	Distribution			
	2	3	4	5
Uniform	1.50	2.00	2.50	3.00
Normal	1.47	1.91	2.35	2.77
Exponential	1.33	1.64	1.92	2.19

with equality achieved if and only if X is uniform [13]. However, other unimodal distributions show RE close to this maximum value [3], as shown in Table 1.

As an example of the computation process, suppose that the values of the circled observations from the sampling scheme diagrammed above are chosen from a normal population and are as given in Table 2. Then $\hat{\mu}_{\text{MTW}} = 19.3$ and, if the ranking was done accurately,

$$\text{var}(\hat{\mu}_{MTW}) = (1/1.91)(\sigma^2/6);$$

that is, the variance of the estimator is nearly halved over that of the sample mean from a simple random sample.

It may be possible to judgment order X by ranking on a concomitant* Y; that is, in the first sample of size n, the X associated with the smallest Y is measured, in the second sample, the X associated with the second smallest Y is measured, and so on. As in stratified sampling, the strength of the correlation* between X and Y determines the amount gained by this procedure. If the regression of X on Y is linear (e.g., bivariate normal), where $\rho(X, Y) = \rho$, then

$$\text{RE}(\overline{X}, \hat{\mu}_{\text{MTW}}) = \left(1 - \frac{\rho^2}{n\sigma^2} \sum_{j=1}^{n} \tau_j^2\right)^{-1}.$$

The variance of $\hat{\mu}_{\text{MTW}}$ can be decreased by an increase in either m, the number of cycles, or n, the size of each sample. However, accurate judgment ordering of a large number of observations would be difficult in most experimental situations, so an increase in sample size is typically implemented by increasing m rather than n.

Table 2 Values of $X_{[j]i}$ Where $n = 3$, $m = 2$

	j		
i	1	2	3
1	15.6	15.1	26.2
2	15.8	19.9	23.4

Takahasi [11] has modeled the judgment ordering process for $n = 2$ by assuming that the elements can be accurately ordered when and only when $|X_{[1]i} - X_{[2]i}| \geqslant \delta$. (This assumption is conservative since correct ordering can occur by chance when $|X_{[1]i} - X_{[2]i}| \leqslant \delta$.) He found by looking at examples that for distributions that are highly concentrated, judgment ordering and thus ranked set sampling will not be very effective.

Other Procedures

A generalization of the MTW procedure proposed by Yanagawa and Shirahata [15] provides a method for estimating the population mean when N samples of size n are chosen from the population and N (one from each sample) are measured. If $N \neq mn$, the best scheme is not necessarily to measure one from each order class.

Their method, called the *YS procedure*, differs by the manner in which an observation is chosen for measurement after ranking of each sample. The jth largest observation from the ith sample is chosen for measurement with probability p_{ij}, where $\sum_{j=1}^{n} p_{ij} = 1$ for each i. Then $P = \{p_{ij};\ i = 1, \ldots, N;\ j = 1, \ldots, n\}$ is called the *selective probability matrix*. They show that:

(a) The arithmetic mean of these N measured observations, $\hat{\mu}_{\text{YS}}$, is an unbiased estimator of μ if and only if $\sum_{i=1}^{n} p_{ij} = N/n$ for $j = 1, \ldots, n$. Denote those P satisfying this condition by $\underline{P}$.

(b) $\text{var}(\hat{\mu}_{\text{YS}}) \leqslant \text{var}(\overline{X}_N)$ for $P \in \underline{P}$.

Furthermore, they give a procedure for finding P that minimizes $\text{var}(\hat{\mu}_{\text{YS}})$ over $\underline{P}$.

When $N = mn$, we have $\hat{\mu}_{\text{YS}} = \hat{\mu}_{\text{MTW}}$ by choosing $p_{ij} = 1$ for $i = j, j + n, \ldots, j + (m - 1)n$. The MTW estimator is shown to be optimal for this case.

The YS procedure was improved still further by Yanagawa and Chen [14]. They modified the selective probability matrix to an $l \times n$ matrix, where $N = 2l$ [or $(l + 1) \times n$ if $N = 2l + 1$]. After judgment ordering the n observations in each sample, the jth smallest element from the ith sample is paired with the $(n + 1 - j)$th element from the $(N + 1 - i)$th sample and is called pair j. Then pair j is selected for measurement from the ith [and therefore $(N + 1 - i)$th] sample with probability p_{ij}, with $\sum_{j=1}^{n} p_{ij} = 1$ for $i = 1, \ldots, l$. This is referred to as the *MG procedure*.

The arithmetic mean $\hat{\mu}_{MG}$ of these $N = 2l$ measured observations is unbiased if

$$\sum_{i=1}^{l} (p_{ij} + p_{i,m+1-j}) = N/m$$

for $j = 1, \ldots, n$. The optimal unbiased estimator $\hat{\mu}_{MG}$ has variance smaller than the optimal estimator from the YS procedure, unless $N = m$, in which case they are identical.

ESTIMATION OF OTHER PARAMETERS

Variance

The estimator

$$\hat{\sigma}^2 = \sum_{i=1}^{m} \sum_{j=1}^{n} (X_{[j]i} - \hat{\mu}_{MTW})^2/(mn - 1)$$

is an asymptotically (as either m or n increases) unbiased estimator of σ^2 [9]. Further, if mn is large enough, it has greater precision than the usual variance estimator s^2 based on a sample of size mn. Examples show that "large enough" is quite small (between 3 and 8) for all distributions considered. The gain from ranked set sampling for estimation of variance is modest compared with that for the mean, however, and it is not likely that the procedure would be worthwhile for estimation of variance alone.

Correlation Coefficient

Let (X, Y) denote a bivariate normal random vector. Suppose that X is difficult to either measure or rank, but a sample from the Y population can at least be easily ordered. For example, consider the problem of a psychologist who wishes to estimate the correlation* between an individual (X) and a group (Y) intelligence test. The scores from group tests are readily available from students' records or can easily be administered. Individual tests are extremely costly since they require a large amount of time and professional expertise to administer and score.

Suppose that a ranked set sample of Y's and their concomitant X's are selected and measured in order to estimate their correlation ρ. If the sampling of the Y's is done according to the MTW procedure, there is no improvement in the precision of any estimator of ρ over Pearson's correlation coefficient r from a random sample of the same size. If, however, the extreme order statistics* from the X sample and their concomitant Y's are measured, giving a selective probability matrix of the form

$$p = \begin{bmatrix} p_1 & 0 & \cdots & 0 & q_1 \\ p_2 & 0 & \cdots & 0 & q_2 \\ \vdots & & & & \\ p_N & 0 & \cdots & 0 & q_N \end{bmatrix},$$

(where $p_i + q_i = 1$ for $i = 1, \ldots, N$), then a considerable improvement in estimation of ρ can be achieved.

RELATED METHODS

Ranked set sampling has some similarities to stratified sampling, where the population of each of the n order statistics is a stratum. Although the "stratification" is done after the samples are selected, there is an advantage over post-stratification, since the number chosen from each stratum can be controlled.

Mosteller [5] investigated methods for choosing the best subset of order statistics from a sample for estimating the mean, variance, or correlation coefficient from normal populations. For this procedure, exact measurements must be made on only the smaller subset, although one must be able to rank the larger sample accurately. O'Connell and David [6] have a similar scheme that is useful when a difficult-to-measure variable has a concomitant that is more easily measured, or at least ranked. In this method, optimally spaced ordered concomitants indicate which subset of the sample should be measured to produce the best estimate of the mean of a symmetrical population.

Barnett et al. [1] discuss a procedure that makes use of concomitant variables to estimate ρ in a bivariate normal* population. Their asymptotically efficient estimator is computed entirely from the concomitants of the ordered observations of one of the variables, say X, and does not require measurement of the Y's. Thus their estimator would be valuable when one of the variables is impossible to measure, but can be ranked accurately.

Nomination sampling* is a procedure for estimating population quantities when exact measurements of only one order statistic from each of several samples are known. This method is not necessarily used because measurement is expensive, but perhaps because only a subset of the values is available (e.g., censoring of the sample may have occurred).

References

[1] Barnett, V., Green, P. J., and Robinson, A. (1976). *Biometrika*, **63**, 323–328.

[2] David, H. A. and Levine, D. N. (1972). *Biometrics*, **28**, 553–555. (This is an appendix to Dell and Clutter's paper, which provides some insight into how much damage is done to the estimator by inaccurate ordering.)

[3] Dell, T. R. and Clutter, J. L. (1972). *Biometrics*, **28**, 545–553. (This is the best explanation of the basics of ranked set sampling.)

[4] McIntyre, G. A. (1952). *Aust. J. Agric. Res.*, **3**, 385–390.

[5] Mosteller, F. (1946). *Ann. Math. Statist.*, **17**, 377–407.

[6] O'Connell, M. J. and David, H. A. (1976). *Essays in Probability and Statistics in Honor of J. Ogawa*. Shinko, Tsusho Company, Tokyo, pp. 451–466. (Discusses use of concomitants in double sampling.)

[7] Shirahata, S. (1977). *Bull. Math. Statist.*, **17**, 33–47.

[8] Stokes, S. L. (1977). *Commun. Statist. A*, **6**, 1207–1211.

[9] Stokes, S. L. (1980). *Biometrics*, **36**, 35–42.

[10] Stokes, S. L. (1980). *J. Amer. Statist. Ass.*, **75**, 989–995.

[11] Takahasi, K. (1969). *Ann. Inst. Statist. Math.*, **21**, 249–255.

[12] Takahasi, K. (1970). *Ann. Inst. Statist. Math.*, **22**, 421–428.

[13] Takahasi, K. and Wakimoto, K. (1968). *Ann. Inst. Statist. Math.*, **20**, 1–31.

[14] Yanagawa, T. and Chen, S. (1980). *J. Statist. Plann. Inf.*, **4**, 33–44.

[15] Yanagawa, T. and Shirahata, S. (1976). *Aust. J. Statist.*, **18**, 45–52.

(CONCOMITANT VARIABLES
ORDER STATISTICS)

S. LYNNE STOKES

RANK ESTIMATORS, SEQUENTIAL
See SEQUENTIAL RANK ESTIMATORS

RANKING PROCEDURES

A statistical ranking procedure uses sample data to order or rank members of a family of k populations with respect to the relative magnitudes of some parameter in such a way that we maintain control over the probability that the ranking asserted to be correct is indeed correct.

Ranking procedures are one approach in multiple decision theory* (see Gupta and Huang [13]), a field greatly influenced by Abraham Wald*, where we use a simple (zero–one) loss function and our risk is an incorrect decision. A related approach in multiple decision theory is selection procedures (*see* SELECTION PROCEDURES). The two kinds of problems are frequently linked together under the heading of ranking and selection procedures.

Selection procedures differ from ranking

procedures in that they seek to select (identify) certain ones of the k populations as having certain parameter characteristics and say nothing specific (except by elimination) about the remaining populations. For example, a selection procedure might have the goal of selecting only the one population with the largest parameter value, or selecting a subset of populations that contains the one with the largest parameter. In such cases we maintain control over the probability that the selection is correct.

It is also possible to combine the approaches of ranking and selection procedures in the sense that we may want to select the best t populations for $t < k$ and rank these t with respect to their parameter values. The complete ranking problem is analogous to the above, where $t = k$. Procedures that combine the subset selection approach pioneered by Gupta [12] with ranking are also available; these are covered extensively in Chapter 19 of Gupta and Panchapakesan [15], which is a comprehensive discussion of various aspects of selection and ranking. Other important references here are Huang and Panchapakesan [17] and Gupta and Miescke [14]. Dudewicz and Koo [8] give a comprehensive categorized bibliography of selection and ranking procedures.

The problem of a complete ranking was first mentioned in Bechhofer [1] and tables for applying the procedure to rank $k = 3$ normal populations according to their variances were given in Bechhofer and Sobel [3]. The procedures were slow to develop for $k > 3$ because of the mathematical complexity of computations involved. To date, procedures have been developed for ranking normal means (with variances known or unknown), and for ranking normal variances (with means known or unknown), using the indifference zone approach. Researchers are currently working on solutions for other distributions, other parameters, and with other approaches. Related confidence procedures for estimating certain parameter values are also being developed. Confidence procedures directly analogous to the complete ranking problem are only those that simultaneously estimate all of the parameter values.

THE COMPLETE RANKING PROBLEM

Suppose that we have k independent populations $\pi_1, \pi_2, \ldots, \pi_k$, where the cumulative distribution function (CDF) of π_i is $G(x; \theta_i)$ for $i = 1, 2, \ldots, k$, where $G(x; \theta)$ is a stochastically increasing function of θ [i.e., $G(x; \theta') \geqslant G(x; \theta'')$ for $\theta' < \theta''$ for all x]. The problem is to use parameter estimates computed from sample data to order or rank the populations with respect to their true θ values. Denote the true ordered θ values by $\theta_{[1]} \leqslant \theta_{[2]} \leqslant \cdots \leqslant \theta_{[k]}$. The ranking procedure used should guarantee that the probability is at least some specified value P^* that the ranking asserted from the estimates is correct whenever the true parameter configuration $\boldsymbol{\theta} = (\theta_1, \theta_2, \ldots, \theta_k)$ lies in a certain specified subset of the parameter space, called the preference zone (PZ). We require $P^* > 1/k!$ since the probability that a random ordering of the k populations (not based on sample data) is correct is equal to $1/k!$. The remaining portion of the parameter space is called the *indifference zone* (IZ), and hence this formulation is called the indifference zone approach to ranking procedures.

If the parameter of interest is a location parameter, the preference zone is where the differences of successive θ values are at least some specified value δ^*, that is, the set

$$\mathrm{PZ} = \{\boldsymbol{\theta} : \theta_{[i]} - \theta_{[i-1]} \geqslant \delta^*, i = 2, 3, \ldots, k\}.$$

The preference zone for the scale parameter problem, where θ denotes a population variance, is where the ratios of successive θ values are at least some specified value δ^*, that is, the set

$$\mathrm{PZ} = \{\boldsymbol{\theta} : \theta_{[i]}/\theta_{[i-1]} \geqslant \delta^*, i = 2, 3, \ldots, k\}.$$

The configuration of θ values in the preference zone for which the probability of a correct ranking (PCR) is a minimum is called the *least favorable configuration*

(LFC). If we can determine a common sample size n required from each population such that the PCR is equal to the specified value P^* when the parameters are in the least favorable configuration*, we know that the PCR is at least P^* for any configuration in the preference zone for that n value and hence we have a conservative solution for n. The least favorable configuration for the location problem is

$$\text{LFC} = \{\boldsymbol{\theta} : \theta_{[i]} - \theta_{[i-1]} = \delta^*, \quad i = 2, 3, \ldots, k\},$$

and for the scale problem is

$$\text{LFC} = \{\boldsymbol{\theta} : \theta_{[i]}/\theta_{[i-1]} = \delta^*, \quad i = 2, 3, \ldots, k\}.$$

To solve the complete ranking problem for any distribution and any parameter, the requirements are a procedure for using the sample data to order the populations, and an expression (or set of tables) that relates the values of n, δ^*, and P^*. Such a solution will enable a person designing an experiment to determine the minimum common sample size n required to assert that the ranking resulting from sample data is correct with probability at least P^* for all parameter configurations in the preference zone specified by δ^*. Alternatively, the sample size n may be fixed, either because the data have already been obtained or because n is restricted by cost or other practical limitations. In this case, the same solution can be used to construct a graph of the pairs (δ^*, P^*) that are satisfied for the fixed value of n.

ONE-STAGE PROCEDURES FOR THE NORMAL MEANS PROBLEM

A special case of the location problem arises when the sample observations from π_i are normally distributed with unknown mean μ_i and known variance σ^2 which is common for each of the k populations. The parameter is μ_i. For a one-stage procedure, a common sample size n is taken from each population and the ordering asserted is the same as that of the $\bar{X}_i$, the mean of the sample from π_i, for $i = 1, 2, \ldots, k$. The investigator specifies P^* and δ^* and determines the required sample size as a function of τ from

$$n = (\tau\sigma/\delta^*)^2. \tag{1}$$

This procedure was given in Bechhofer [1]; tables for τ when $k = 3$ can be computed from tables of the bivariate normal distribution with $\rho = -0.5$, but the computation when $k \geqslant 4$ is much more difficult. Carroll and Gupta [7] give an approximate solution for τ based on numerical integration*, or Gibbons et al. [11, Table P.1] can be used. Alternatively, if n is outside the control of the investigator, the same τ value for any chosen P^* can be used to compute the (δ^*, P^*) pairs that are satisfied by the selection procedure with

$$\delta^* = \tau\sigma/\sqrt{n}\,. \tag{2}$$

Another special case arises for normal populations with known but unequal variances. The same tables can be used if the sample sizes for the k populations are chosen so that

$$\frac{\sigma_1^2}{n_1} = \frac{\sigma_2^2}{n_2} = \cdots = \frac{\sigma_k^2}{n_k}\,. \tag{3}$$

The value n determined from (1) is used to set the total of all sample sizes as $N = kn$ and then find the respective individual sample sizes from

$$n_i = \frac{N\sigma_i^2}{\sum \sigma_i^2}\,. \tag{4}$$

TWO-STAGE PROCEDURES FOR THE NORMAL MEANS PROBLEM

The more typical location problem in practice is normal distributions with unknown variances. Then a single-stage procedure will not satisfy the requirements of the problem. A two-stage procedure is required, where an initial sample is taken from each population in order to obtain some useful information about the variances before asserting a ranking.

If the unknown variances are assumed common, the procedure is to take an initial sample of n_0 observations from each population and compute a pooled sample variance as

$$s^2 = \frac{s_1^2 + s_2^2 + \cdots + s_k^2}{k}, \qquad (5)$$

where each sample variance is the usual unbiased estimator of the population variance. In the second stage, the investigator specifies δ^* and P^* and takes an additional $n - n_0$ observations from each of the k populations, where n is given as a function of h by

$$n = \max\left[n_0, \left\{2(sh/\delta^*)^2\right\}^+ \right] \qquad (6)$$

and $\{c\}^+$ means the smallest integer equal to or greater than c. Then the mean of all n observations from both stages is calculated for each sample, and the populations are ranked according to these sample values. This procedure was given by Bechhofer et al. [2].

Dunnett and Sobel [9] give tables of h for $k = 3$. Freeman et al. [10] give additional tables, or Table P.2 of Gibbons et al. [11] can be used for $P^* = 0.95$ and $k = 3, 4, 5, 6$, $n_0 = 10(10)100, 200, 500$.

If the unknown variances are not assumed equal, a similar but more complex two-stage procedure can be used. An initial sample of n_0 observations is taken from each population and used to compute $s_1^2, s_2^2, \ldots, s_k^2$. Then $n_i - n_0$ additional observations are taken from each population where

$$n_i = \max\left[n_0 + 1, \left\{(s_i h/\delta^*)^2\right\}^+ \right]. \qquad (7)$$

This solution and tables of h were given by Beirlant et al. [4]. The final step is to compute a weighted mean of all the observations in the ith sample, $X_{i1}, X_{i2}, \ldots, X_{in_i}$, as

$$\overline{X}_i = \sum_{j=1}^{n_0} a_i X_{ij} + \sum_{j=n_0+1}^{n_i} b_i X_{ij}, \qquad (8)$$

where a_i and b_i are the solutions to the system

$$n_0 a_i^2 + (n_i - n_0) b_i^2 = (\delta^*/s_i h)^2,$$
$$n_0 a_i + (n_i - n_0) b_i = 1, \qquad (9)$$

for $i = 1, 2, \ldots, k$. Note that the s_i in (9) are computed from only the initial n_0 observations. The populations are then ranked according to the values of these weighted means.

ONE-STAGE PROCEDURES FOR THE NORMAL VARIANCES PROBLEM

A special case of the scale parameter problem arises when the sample observations from π_i are normally distributed with mean μ_i (known or unknown) and unknown variance σ_i^2. The scale parameter is σ_i^2 and the appropriate estimator is the sample variance

$$s_i^2 = \sum_{j=1}^{n} \left(X_{ij} - \overline{X}_i\right)^2/(n-1)$$

if μ_i is unknown, and

$$V_i = \sum_{j=1}^{n} (X_{ij} - \mu_i)^2/n$$

if μ_i is known, where $X_{i1}, X_{i2}, \ldots, X_{in}$ is the sample of n observations from π_i. The populations are ranked according to the values of these estimators. Bechhofer and Sobel [3] considered this problem for $\mu_1, \mu_2, \ldots, \mu_k$ known and $k = 2$ and 3, and gave some tables. Schafer [18] used Monte Carlo methods* to evaluate the probability of a correct complete ranking for this problem, and these tables are readily available in Schafer and Rutemiller [19], as well as in Gibbons et al. [11, Table P.3] for $k = 3, 4, \ldots, 8$. Bishop and Dudewicz [6] and Bishop [5] give a general solution that can be used for any distribution.

NUMERICAL EXAMPLE

An example where a complete ranking with respect to means is needed is provided by Holt [16]. Four helmet types, No Helmet, PASGT A, PASGT B, and Standard M-1, are to be ranked with respect to integrated sum mean values of induced muscle stress based on a sample of $n = 144$ observations on each type. Low muscle stress, as measured by electromyography, is an important characteristic of a good infantry helmet. The data, assumed to be normally distributed with a common variance, are given as follows:

Helmet	Mean ($n = 144$)
None	9,949
PASGT A	10,745
PASGT B	10,539
Standard	10,785
$s = 1,196.91$	

Holt set $P^* = 0.90$ and $\delta^* = 170$ for these $k = 4$ populations and assumed a common $\sigma = 1196.91$ to compute n as $n = 335$ using $\tau = 2.599$ and (1). He then concluded that since only $n = 144$ observations were taken and 335 are required, the order of the sample means does not reflect the order of the population means with any confidence. In fact, the order observed for $n = 144$ could be asserted to be the correct ranking with $\delta^* = 170$ for a P^* of only about 0.68. This example shows how sample size determination can be used effectively in a statistical analysis even when the n is determined in advance or by other considerations.

Note that, in general, large sample sizes are required for any reasonable P^* value even for moderate k and typical δ^* values. This is because the goal of a complete ranking is asking a great deal from the data.

References

[1] Bechhofer, R. E. (1954). *Ann. Math. Statist.*, **25**, 16–39.

[2] Bechhofer, R. E., Dunnett, C. W., and Sobel, M. (1954). *Biometrika*, **41**, 170–176.

[3] Bechhofer, R. E. and Sobel, M. (1954). *Ann. Math. Statist.*, **25**, 273–289.

[4] Beirlant, J., Dudewicz, E. J., and van der Meulen, E. C. (1981). Complete statistical ranking, with tables and applications. *Mededelingen uit het Wiskundig Instituut Katholieke Universiteit*, Leuven, Belgium, No. 145, December.

[5] Bishop, T. A. (1978). *Technometrics*, **20**, 495–496.

[6] Bishop, T. A. and Dudewicz, E. J. (1977). *IEEE Trans. Reliab.*, **R-26**, 362–365.

[7] Carroll, R. J. and Gupta, S. S. (1977). *J. Statist. Comput. Simul.*, **4**, 145–157.

[8] Dudewicz, E. J. and Koo, J. O. (1982). *The Complete Categorized Guide to Statistical Selection and Ranking Procedures*. American Sciences Press, Columbus, Ohio.

[9] Dunnett, C. W. and Sobel, M. (1954). *Biometrika*, **41**, 153–169.

[10] Freeman, H., Kuzmack, A., and Maurice, R. J. (1967). *Biometrika*, **54**, 305–308.

[11] Gibbons, J. D., Olkin, I., and Sobel, M. (1977). *Selecting and Ordering Populations: A New Statistical Methodology*. Wiley, New York. (Elementary and applied; extensive tables and references; Chapter 12 covers ranking procedures.)

[12] Gupta, S. S. (1956). On a Decision Rule for a Problem in Ranking Means. Ph.D. dissertation, *Inst. Statist. Mimeo Ser. No. 150*, University of North Carolina, Chapel Hill.

[13] Gupta, S. S. and Huang, D. Y. (1981). *Multiple Statistical Decision Theory*. Springer-Verlag, New York.

[14] Gupta, S. S. and Miescke, K.-J. (1981). *Sankhyā B*, **43**, 1–17.

[15] Gupta, S. S. and Panchapakesan, S. (1979). *Multiple Decision Procedures: Theory and Methodology of Selecting and Ranking Populations*. Wiley, New York. (A comprehensive theoretical treatment of ranking and selection; extensive references.)

[16] Holt, W. R. (1980). Personal communication.

[17] Huang, D. Y. and Panchapakesan, S. (1978). *J. Chinese Statist. Ass.*, **16**, 5801–5810.

[18] Schafer, R. E. (1974). In *Reliability and Biometry: Statistical Analysis of Lifelength*, F. Proschan and R. J. Serfling, eds., SIAM, Philadelphia, pp. 597–617.

[19] Schafer, R. E. and Rutemiller, H. C. (1975). *Technometrics*, **17**, 327–331.

(LEAST FAVORABLE CONFIGURATION
MULTIPLE COMPARISONS
SELECTION PROCEDURES)

JEAN DICKINSON GIBBONS

RANK INTERACTION

To introduce the concept of rank interaction, let us take a very simple example. Suppose that compared to "no treatment," some treatment increases the level of a particular variate in women, whereas it decreases the level of the same variate in men. In that case, obviously an interaction in the usual sense is present, but for the purposes of this article, we may even say that a rank interaction is present, as the effect of the treatment on women is the opposite of that on men. Rank interaction was first formally dealt with in the paper of De Kroon and Van der Laan [3].

Let X_{ij} be a set of random variables with continuous cumulative distribution functions $F(x + \theta_{ij})$, with

$$\theta_{ij} = \mu + \alpha_i + \beta_j + \gamma_{ij},$$

$$\sum_i \alpha_i = \sum_j \beta_j = \sum_i \gamma_{ij} = \sum_j \gamma_{ij} = 0,$$

i referring to the classes generated by factor A ($i = 1, 2, \ldots, I$) and j to the classes generated by factor B ($j = 1, 2, \ldots, J$).

Suppose that for every pair (i, j), a random sample of m (> 1) observations is taken from the random variable X_{ij}, m having the same value for each cell (i, j). If we want to investigate whether factor A has any effect, we can test the null hypothesis H_0 that factor A has no effect at all, against the alternative H_1 that H_0 does not hold. More formally,

$$H_0 : \alpha_1 + \gamma_{1j} = \alpha_2 + \gamma_{2j} = \cdots = \alpha_I + \gamma_{Ij} \quad \text{for } j = 1, 2, \ldots, J,$$

$$H_1 : H_0 \text{ is not true},$$

which implies that at least one of the equalities stated under H_0 is violated.

Now a test statistic T will be defined that can be written as the sum of two statistics T_1 and T_2, the first one being sensitive to deviations from zero of the α_i's and the second to deviations from zero of the γ_{ij}'s, but only if the last-mentioned deviations imply that the rankings of $\alpha_1 + \gamma_{1j}$, $\alpha_2 + \gamma_{2j}$, $\ldots, \alpha_I + \gamma_{Ij}$ are not identical for different values of j, thus satisfying the property given in the example. In other words, T_2 will not be sensitive to all deviations from zero of the γ_{ij}'s, but only to "rank interaction $A^*(B)$," which can be defined as follows.

Definition. Rank interaction $A^*(B)$ (discordance) exists if the rankings of the coordinates of the vector $(\theta_{1j}, \theta_{2j}, \ldots, \theta_{Ij})$ are not identical for different values of j. If these rankings *are* identical, rank interaction $A^*(B)$ is said not to exist (concordance).

This definition is not symmetric with respect to A and B. In other words, rank interaction of type $A^*(B)$ does not imply rank interaction of type $B^*(A)$, and vice versa.

Now the test statistic T is defined by

$$T = \sum_{j=1}^{J} K_j,$$

where K_j denotes the Kruskal–Wallis statistic applied within class j of factor B, using the classification generated by factor A. As the observations in different classes of B are statistically independent, it is obvious that T is the sum of J independent Kruskal–Wallis test* statistics.

Defining T_1 to be the Friedman statistic for the case of J blocks of rankings over I classes, each class containing m observations, the statistic T_2 is defined by

$$T = T_1 + T_2,$$

T_1 being sensitive to differences between the α_i's, and T_2 being sensitive to differences between the ranking vectors of

$$(\alpha_1 + \gamma_{1j}, \alpha_2 + \gamma_{2j}, \ldots, \alpha_I + \gamma_{Ij}), \quad j = 1, \ldots, J.$$

Rewriting H_0 as

$$H_0 : \{\alpha_1 = \alpha_2 = \cdots = \alpha_I = 0\} \cap \{\gamma_{ij} = 0 \text{ for all } i \text{ and } j\},$$

it becomes clear that T_1 is sensitive to violation of the first component of H_0, namely $\{\alpha_1 = \alpha_2 = \cdots = \alpha_I = 0\}$, whereas T_2 is sensitive to violation (of a certain kind) of the second component of H_0.

A very simple numerical example may elucidate the procedure. Suppose that we have two blocks and two treatments as well, with three observations per cell. Let the rankings be as follows:

	Treatment 1	Treatment 2
Block 1	1 3 2	6 4 5
Block 2	5 4 6	3 1 2

Obviously, the Friedman statistic is equal to zero in this case, whereas within each block the Kruskal–Wallis statistic reaches its maximum. The combination of both (two-sided) Kruskal–Wallis statistics gives a significant result at the 5% level. Hence H_0 should be rejected at the 5% confidence level. As can be calculated very quickly, the values for T, T_1, and T_2 are 7.7, 0, and 7.7, respectively.

Depending on the problem that one wants to consider in practice (i.e., which alternative to the null hypothesis is thought to be relevant), we can choose one of the three statistics T, T_1, or T_2. Clearly, T serves as an omnibus test, whereas T_1 will be used if one is mainly concerned with differences between α_i's, and T_2 will be used if one aims at proving the presence of rank interaction. The three test statistics have one-sided critical regions: large values lead to rejection of H_0.

One can also use both statistics T_1 and T_2 simultaneously, each at confidence level $\alpha/2$, rejecting H_0 if at least one of the outcomes of T_1 or T_2 is larger than the corresponding critical value. It can be proved that, under H_0, the statistics T_1 and T_2 are asymptotically ($m \to \infty$) independent and distributed as chi-square statistics, with $I - 1$ and $(I - 1)(J - 1)$ degrees of freedom, respectively.

The procedures proposed can easily be generalized to the case where the number of observations per cell, say m_{ij}, may differ, provided that the scheme is connected and Kruskal–Wallis can be applied for each class of B. If the design happens to be orthogonal, the generalization is straightforward. For arbitrary values of m_{ij}, a very nice and natural generalization of Friedman's test statistic may be found. It is possible to adapt the proposed procedure using aligned ranks that one can test against interaction in the traditional sense rather than against rank interaction [2].

In De Kroon and Van der Laan [2, 3] many details are discussed and a great number of computer results with respect to the exact distribution are given. For general information about permutation tests*, see Conover [1] or Lehmann [4].

References

[1] Conover, W. J. (1971). *Practical Nonparametric Statistics*. Wiley, New York.

[2] De Kroon, J. P. M. and Van der Laan, P. (1981). *Statist. Neerlandica*, **35**, 189–213.

[3] De Kroon, J. P. M. and Van der Laan, P. (1983). *Statist. Neerlandica*, **37**, 1–14.

[4] Lehmann, E. L. (1975). *Nonparametric Statistical Methods Based on Ranks*. McGraw-Hill, New York.

(INTERACTION
KRUSKAL–WALLIS TEST
RANK TESTS)

J. P. M. De Kroon

RANKIT

The expected value of a unit normal order statistic. *See* NORMAL SCORES TESTS; QUANTAL RESPONSE ANALYSIS.

RANK LIKELIHOOD

Rank likelihoods or, more fully, marginal likelihoods of ranks, have been used in models for the statistical analysis of various problems, ranging from the analysis of paired comparisons* to randomly censored* survival data. Ranks, rather than the original values of observations, if available, prove useful in statistical analysis if there are no strong prior beliefs about the distributional properties of the observations or the distributional properties of the observations are difficult to assess, as can be the case with censored data.

Let $W_1, \ldots, W_n$ be independent random

variables, with probability density functions (PDFs) $f_j(w;\boldsymbol{\theta})$ $(j = 1, \ldots, n)$ known except for the vector parameter $\boldsymbol{\theta}$. Suppose that $G(\cdot)$ is an unknown monotone-increasing differentiable function and we can observe Y_j's $(j = 1, \ldots, n)$ with $G(Y_j) = W_j$ $(j = 1, \ldots, n)$. In such a case the ranks of the Y_j's are marginally sufficient for $\boldsymbol{\theta}$ (see Kalbfleisch [5]), or the ranks provide the basis for invariant tests for $\boldsymbol{\theta}$ (see Cox and Hinkley [4, Sec. 5.3]). The model implies that for observations $y_1, \ldots, y_n$ we are willing to assume a parametric model after some arbitrary transformation.

The rank likelihood for the Y_j's is then given by

$$\Pr(Y_j \text{ has rank } r_j, j = 1, \ldots, n) = \Pr(W_j \text{ has rank } r_j, j = 1, \ldots, n),$$

and the latter probability can be found for observed ranks $r_1, \ldots, r_n$ and given values of $\boldsymbol{\theta}$. Now

$$\Pr(W_j \text{ has rank } r_j, j = 1, \ldots, n) = \int_A \prod_{j=1}^{n} \{ f_{(j)}(u_j;\boldsymbol{\theta})\, du_j \}, \tag{1}$$

where $A = \{(u_1, \ldots, u_n) : u_1 < \cdots < u_n\}$ and (j) is the index of the $Y_1, \ldots, Y_n$, which is the jth smallest in the ordered sample of Y's (i.e., $r_{(j)} = j$).

AN IMPORTANT EXACT RESULT

In one particularly important case, (1) can be found analytically. Savage [13] shows that if Lehmann alternatives* are considered for the CDFs of the W_j's, that is,

$$F_j(w;\boldsymbol{\theta}) = w^{\delta_j} \quad (\delta_j > 0,\ 0 \leqslant w \leqslant 1),$$

with $\boldsymbol{\theta} = (\delta_1, \ldots, \delta_n)$, then

$$\Pr(W_1 < \cdots < W_n) = \left(\prod_{j=1}^{n} \delta_j \right) \left[\prod_{j=1}^{n} \left(\sum_{i=1}^{j} \delta_i \right) \right]^{-1}. \tag{2}$$

Putting $n = 2$, in (2), essentially gives the Bradley and Terry [1] model for paired comparisons*, and $n = 3$ gives a model for triple comparisons. The value of (1) can also be found if the PDFs of the W_j's are given by $f_j(w;\boldsymbol{\theta}) = \exp\{w - \theta_j - \exp(w - \theta_j)\}$, where $\exp[w - \exp(w)]$ is an extreme value* density and $\boldsymbol{\theta} = (\theta_1, \ldots, \theta_n)$. We then have

$$\Pr(W_1 < \cdots < W_n) = \exp\left(-\sum_{j=1}^{n} \theta_j \right) \times \left\{ \prod_{j=1}^{n} \left[\sum_{i=j}^{n} \exp(-\theta_i) \right] \right\}^{-1}. \tag{3}$$

If $\delta_j = \exp(-\theta_j)$, then, with these W_j's, $\Pr(W_n < \cdots < W_1)$ is given by the right-hand side of (2). If the θ_j's are related by a linear model, $\theta_j = \mathbf{x}_j^T \cdot \boldsymbol{\beta}$, where $\mathbf{x}_j$ is a vector of known explanatory variables, then (3) gives Cox's* [3] proportional hazards regression model, for survival times with no censoring, except that $\boldsymbol{\beta}$ has been replaced by $-\boldsymbol{\beta}$.

Kalbfleisch and Prentice [6] define the rank vector $\mathbf{r}$ for a sample that contains randomly right censored observed survival times and obtain the rank likelihood

$$f(\mathbf{r};\boldsymbol{\beta}) = \exp\left(-\sum_{j=1}^{k} \mathbf{x}_{(j)}^T \cdot \boldsymbol{\beta} \right) \times \left\{ \prod_{j=1}^{k} \left[\sum_{i \in R(t_{(j)})} \exp(-\mathbf{x}_i^T \cdot \boldsymbol{\beta}) \right] \right\}^{-1}, \tag{4}$$

where for the k $(\leqslant n)$ distinct uncensored survival times $t_{(1)} < \cdots < t_{(k)}$, $R(t_{(j)})$ is the set of all indices i so that $t_i \geqslant t_{(j)}$, $j = 1, \ldots, k$. Here $\mathbf{x}_{(j)}$ refers to the explanatory variable of the uncensored observation with survival time ranked j among $t_{(1)} < \cdots < t_{(k)}$. The rank likelihood (4) is identical to the partial likelihood* of Cox [3], except for the changed sign of $\boldsymbol{\beta}$. This identity occurs provided that there are no ties in the data and there are no time-dependent covariates in Cox's [3] regression model. The hazard function for Y_j, where Y_j is observed and $G(Y_j) = W_j$, is given by

$$\lambda(y;\mathbf{x}) = G'(y)\exp[G(y)]e^{-\mathbf{x}_j^T \cdot \boldsymbol{\beta}},$$

and since $G(y)$ is unspecified, so also is $G'(y)\exp[G(y)]$, giving the proportional hazards condition.

LINEAR REGRESSION* MODELS AND TEST STATISTICS

The choice $f_j(w;\boldsymbol{\theta}) = \exp[w - \theta_j - \exp(w - \theta_j)]$ of $f_j(w;\boldsymbol{\theta})$ belongs to the location shift family of choices of $f_j(w;\boldsymbol{\theta})$ specified by $f_j(w;\boldsymbol{\theta}) = f_e(w - \theta_j)$, where $f_e(w)$ is a known density, such as standard normal, double exponential, logistic, or an extreme value density as given above. When $\theta_j = \mathbf{x}_j^T \cdot \boldsymbol{\beta}$ the model for the W_j's is the linear model $W_j = \mathbf{x}_j^T \cdot \boldsymbol{\beta} + e_j$, where the $e_1, \ldots, e_n$ are independent with the completely known PDF $f_e(w)$. Therefore, the model for the responses $Y_1, \ldots, Y_n$, is that after some unknown monotone transformation, the $g(Y_j) = W_j$ follow the linear model $W_j = \mathbf{x}_j^T \cdot \boldsymbol{\beta} + e_j$. In general, the rank likelihood (1) cannot be calculated for arbitrary choices of $f_e(w)$, an exception being (4), of course.

If for this case (1) is denoted by $f(\mathbf{r};\boldsymbol{\beta})$ and $l(\mathbf{r};\boldsymbol{\beta}) = \log f(\mathbf{r};\boldsymbol{\beta})$, then tests of $\boldsymbol{\beta} = \mathbf{0}$ are based on the score statistic* derived from $l(\mathbf{r};\boldsymbol{\beta})$, and an estimate of the variance is given by the observed information at $\boldsymbol{\beta} = \mathbf{0}$; note that Cox and Hinkley [4, Sec. 6.3] discuss a similar approach but consider variances under randomization* distributions. The resulting statistic has an approximate χ^2 distribution, with degrees of freedom equal to the dimension of $\boldsymbol{\beta}$, if the hypothesis $\boldsymbol{\beta} = 0$ is true (see Prentice [12]).

The vector-valued score statistic $l'(\mathbf{r};\mathbf{0}) = \mathbf{X}^T\mathbf{a}$, where $\mathbf{X}^T = (\mathbf{x}_1, \ldots, \mathbf{x}_n)$ and $(\mathbf{a})_j = E\{g(W_{r_j:n})\}$, where $W_{1:n} < \cdots < W_{n:n}$ are the order statistics of a sample of size n from the population with density $f_e(w)$ and $g(w) = -f_e'(w)/f_e(w)$. For the two-sample problem, where $\mathbf{x}_j$ is a scalar taking the value 0 or 1 according to which sample the observation belongs, the statistic $l'(\mathbf{r};\mathbf{0})$ is a two-sample rank statistic, being (a) the Wilcoxon* statistic if $f_e(w)$ is the logistic density, (b) the normal scores* statistic if $f_e(w)$ is the normal density, or (c) the log-rank* or Savage's rank statistic if $f_e(w) = \exp[w - \exp(w)]$, an extreme value density.

Applying these ideas to case (a), we find the score statistic to be

$$\sum_{i=1}^{n}\left(2\frac{r_i}{N+1} - 1\right),$$

where $r_1, \ldots, r_n$ are the ranks of one of the two samples of combined size N. We note this statistic is just $2(N+1)^{-1}W - n$, where W is the Mann–Whitney–Wilcoxon* statistic. The estimate of variance is the reciprocal of the observed information at $\beta = 0$, that is, the reciprocal of

$$\frac{2}{N^2 - 1}\sum_{i=1}^{n} r_i(N + 1 - r_i) - \frac{8}{(N+1)^2(N-1)}\sum\sum_{r_i < r_j} r_i(N + 1 - r_j).$$

Evidence from Latta [8] suggests that better approximations for P-values* result from using variance estimates based on observed information than from randomization distributions.

FURTHER INFERENCE

The rank likelihood $f(\mathbf{r};\boldsymbol{\beta})$ can be used to estimate and find confidence intervals for $\boldsymbol{\beta}$. To do this, without extensive numerical computation to evaluate $f(\mathbf{r};\boldsymbol{\beta})$ at several values of $\boldsymbol{\beta}$, it is found that the Taylor series approximation

$$\log f(\mathbf{r};\boldsymbol{\beta}) = \text{const.} + \boldsymbol{\beta}^T \cdot l'(\mathbf{r};0) + \tfrac{1}{2}\boldsymbol{\beta}^T l''(\mathbf{r};0)\boldsymbol{\beta} \qquad (5)$$

is adequate for many choices of $\mathbf{r}$ and design matrix $\mathbf{X}$; for Bayesian applications, see Brooks [2] (two-sample problem), Pettitt [10] (regression model), and Pettitt [9] (change-point problems).

If $f_e(w)$ is chosen to be normal, the approximation (5) is surprisingly good, and reasonably so for the logistic choice of $f_e(w)$.

As an example using the normal density for $f_e(w)$, consider the two-sample problem with the ordered combined sample consisting of ABBAABAABB; see Brooks [2, Table 2]. With $x = 0$ for an A, $x = 1$ for a B, β represents the difference between the B and A populations in the regression model. In this case the approximation (5) gives

$$\log f(\mathbf{r}; \beta) = \text{const.} + \beta(1.01) - \tfrac{1}{2}\beta^2(2.30), \quad (6)$$

so that treating (6) as an exact log likelihood gives the estimate (1.01/2.30) or 0.44 of β with standard error $(2.30)^{-1/2}$ or 0.66. From a Bayesian point of view, if a locally uniform prior for β is assumed, then, for these data, β has a posterior distribution which is approximated by a normal distribution with mean 0.44 and standard deviation 0.66. This posterior distribution can be used to find the predictive probability that a typical A element is less than a typical B element. This is found to be 0.39 using the approximate posterior distribution*. Using the Mann–Whitney–Wilcoxon statistic, this probability is estimated to be 0.40. Pettitt [10] extends these ideas to the general linear model* and Pettitt [11] considers applications to censored data using the logistic density choice of $f_e(w)$.

References

[1] Bradley, R. A. and Terry, M. E. (1952). *Biometrika*, **39**, 324–345. (Paired comparison methods.)

[2] Brooks, R. J. (1978). *J. R. Statist. Soc. B*, **40**, 50–57.

[3] Cox, D. R. (1972). *J. R. Statist. Soc. B*, **34**, 187–220.

[4] Cox, D. R. and Hinkley, D. V. (1974). *Theoretical Statistics*. Chapman & Hall, London.

[5] Kalbfleisch, J. D. (1978). *J. Amer. Statist. Ass.*, **73**, 167–170.

[6] Kalbfleisch, J. D. and Prentice, R. L. (1973). *Biometrika*, **60**, 267–278.

[7] Kalbfleisch, J. D. and Prentice, R. L. (1980). *The Statistical Analysis of Failure Time Data*. Wiley, New York, Chap. 6.

[8] Latta, R. B. (1981). *J. Amer. Statist. Ass.*, **76**, 713–719.

[9] Pettitt, A. N. (1981). *Biometrika*, **68**, 443–450.

[10] Pettitt, A. N. (1982). *J. R. Statist. Soc. B*, **44**, 234–243.

[11] Pettitt, A. N. (1983). *Biometrika*, **70**, 121–132.

[12] Prentice, R. L. (1978). *Biometrika*, **65**, 167–179.

[13] Savage, I. R. (1956). *Ann. Math. Statist.*, **27**, 590–615.

(COX'S REGRESSION MODEL
DISTRIBUTION-FREE METHODS
PARTIAL LIKELIHOOD
SCORE STATISTICS
SURVIVAL ANALYSIS)

A. N. PETTITT

RANK OF MATRIX

The rank $r(\mathbf{A})$ of a matrix $\mathbf{A}$ is the maximum number of independent rows or columns of $\mathbf{A}$. If $\mathbf{A}$ is a $m \times n$ matrix and $r(\mathbf{A}) = \min(m, n)$, $\mathbf{A}$ is said to be of *full rank*. [$r(\mathbf{A})$ *cannot* exceed this value.] Otherwise, $\mathbf{A}$ is said to be *singular*.

RANK ORDER STATISTICS

Nonparametric methods, compared to their parametric counterparts, require, generally, less stringent regularity conditions for their validity, without necessarily compromising on their efficiency. In the usual hypothesis-testing* problems, this validity robustness of nonparametric procedures is largely due to the fact that for such problems the ranks (or signed ranks) of the observations are usually the maximal invariants. These rank tests are also applicable when the data are measured only on an ordinal scale (where a parametric model may not be very reasonable and the nonparametric ones may have distinctly more appeal). The efficiency robustness of nonparametric methods rests on skillful (efficient) choice of statistics which are functions of these maximal invariants (i.e., ranks). Since, typically, ranking is made with reference to the order statistics*, such rank statistics are termed *rank order statistics*.

The use of rank order statistics is not confined to the hypothesis-testing problems where the ranks are the maximal invariants. By now, in nonparametric (and robust) statistical inference (covering both the univariate and multivariate models), these rank order statistics play a fundamental role in a broad class of problems of genuine practical interest. This entry deals with this composite picture. SIGNED RANK STATISTICS are discussed separately. In this entry there will be no coverage of signed rank statistics. Also, we suggest reference to DISTRIBUTION-FREE METHODS *and* RANK TESTS, where classical nonparametric testing procedures have been discussed at a more expository level. This would help the reader to understand the current entry (which is a bit more theoretical in character).

LINEAR RANK STATISTICS (LRS)

Let $X_1, \dots, X_n$ be independent random variables with continuous distribution functions $F_1, \dots, F_n$, respectively. Let R_i be the *rank* of X_i among $X_1, \dots, X_n$, for $i = 1, \dots, n$ [we neglect the ties among the observations (with probability 1), by virtue of the assumed continuity of the F_i]. We may define formally the R_i by denoting the ordered random variables corresponding to $X_1, \dots, X_n$ by $Z_1 < \cdots < Z_n$, so that $X_i = Z_{R_i}$, $i = 1, \dots, n$, and $(R_1, \dots, R_n)$ represent a permutation of $(1, \dots, n)$. Let $a_n(1), \dots, a_n(n)$ be a set of *scores* (real numbers) and let $c_1, \dots, c_n$ be n given numbers (not all equal). Then a linear rank* statistic is typically defined as

$$L_n = \sum_{i=1}^{n} (c_i - \bar{c}_n) a_n(R_i), \quad \text{where } \bar{c}_n = n^{-1} \sum_{i=1}^{n} c_i . \tag{1}$$

L_n is typically used to test for the null hypothesis $H_0: F_1 = \cdots = F_n = F$ (unknown) against suitable alternatives, and the choice of the c_i may be dictated by these alternatives. We illustrate this with the following.

(I) Two-Sample Rank Statistics

Suppose that $X_1, \dots, X_{n_1}$ have a common distribution F and $X_{n_1+1}, \dots, X_n$ (where $n = n_1 + n_2$) have a common distribution G, and we want to test for the equality of F and G. Here n_1 and n_2 are both positive integers, and one may be interested in a shift [i.e., $G(x) = F(x + a)$ for some $a \neq 0$] or scatter [i.e., $G(x) = F(x/a)$ for some $a \neq 1$] alternative. In such a case, one may naturally choose $c_1 = \cdots = c_{n_1} = 0$ and $c_{n_1+1} = \cdots = c_n = 1$. Then L_n in (1) reduces to a conventional two-sample rank statistic. In this setup, the scores are usually taken as

$$a_n(i) = E[\phi(U_{ni})] \quad \text{or} \quad \phi(i/(n+1)), \quad i = 1, \dots, n, \tag{2}$$

where $\phi = \{\phi(u), 0 < u < 1\}$ is some *score-generating function* and $U_{n1} < \cdots < U_{nn}$ are the ordered random variables of a sample of size n from the uniform $(0, 1)$ distribution. The following are the notable cases:

1. **Wilcoxon Scores.** $\phi(u) = u$ [i.e., $a_n(i) = i/(n+1)$, $i = 1, \dots, n$];
2. **Log-rank Scores.** $\phi(u) = -1 - \log(1 - u), 0 < u < 1$;
3. **Normal Scores.** $\phi(u) = \Phi^{-1}(u)$, $0 < u < 1$, where $\Phi(\cdot)$ is the standard normal distribution function;
4. **Median Scores.** $\phi(u) = \operatorname{sign}(u - \frac{1}{2})$, $0 < u < 1$.

The corresponding L_n are termed the Wilcoxon–Mann–Whitney statistic, log-rank* (Savage) statistic, normal scores* statistic, and the Brown–Mood* median statistic, respectively. All of these are suitable for the shift model (or "stochastically larger" alternatives). We shall discuss their desired properties later. For the two-sample scale problem, the commonly adopted scores are (a) Klotz's scores: $\phi(u) = [\Phi^{-1}(u)]^2 - 1$, $0 < u < 1$; (b) Ansari–Bradley* scores: $\phi(u) = \frac{1}{2} - |u - \frac{1}{2}|$, $0 < u < 1$; (c) and Mood's score: $\phi(u) = (u - \frac{1}{2})^2$, $0 < u < 1$. These rank statistics are discussed in detail in Puri and Sen [29, Chap. 3].

(II) Regression Rank Statistics

For the independent random variables X_i, $i \geqslant 1$, one may conceive of the simple regression model: $X_i = \alpha + \beta d_i + e_i$, $i \geqslant 1$, where the e_i are identically distributed, and the null hypothesis H_0 relates to no regression (i.e., $\beta = 0$) against one- or two-sided alternatives. If these d_i are specified, one may then choose $c_i = d_i$ and define the linear rank statistic L_n as in (1). On the other hand, in some situations, we may not have the precise knowledge of these d_i, but they are known to satisfy some ordering (i.e., $d_1 \leqslant \cdots \leqslant d_n$). In such a case, one may choose the c_i rather arbitrarily but satisfying the same ordering. The scores for this model may also be chosen in the same way as in the two-sample problem.

We may note that for either of the models above, under H_0, the vector $(R_1, \ldots, R_n)$ takes on each permutation of $(1, \ldots, n)$ with the common probability $(n!)^{-1}$ (whenever F is continuous), so that the distribution of L_n, under H_0, does not depend on F. Thus L_n provides a distribution-free test for H_0. This explains the validity robustness of L_n. In this context, the choice of the scores plays a vital role, and we discuss that in detail later. The moments of L_n under H_0 can easily be computed as $E[L_n \mid H_0] = 0$ and $E[L_n^2 \mid H_0] = A_n^2 C_n^2$, where

$$C_n^2 = \sum_{i=1}^{n} (c_i - \bar{c}_n)^2,$$

$$A_n^2 = (n-1)^{-1} \sum_{i=1}^{n} \left[a_n(i) - \bar{a}_n \right]^2;$$

$$\bar{a}_n = n^{-1} \sum_{i=1}^{n} a_n(i).$$

The exact null distribution of L_n depends on the c_i, $a_n(i)$, and n in a very involved manner, and the enumeration may become prohibitively laborious as n increases. For this reason, it becomes necessary to approximate this exact null distribution of L_n by some simple ones. Under appropriate regularity conditions on the c_i and the scores $a_n(i)$, for large n, $L_n/(C_n A_n)$ has closely the standard normal distribution (under H_0) and this result has been discussed in PERMUTATIONAL CENTRAL LIMIT THEOREMS. In this context, we may remark that although the X_i are independent, their ranks R_{ni} are not so, and hence the classical central limit theorems may not apply for the L_n. In a very elegant fashion, Hájek [17] approximated the L_n by a sum of independent random variables on which the classical central limit theorem applies neatly. Alternatively, as in Sen and Ghosh [39], one may consider the martingale characterizations of linear rank statistics (under H_0) and incorporate the martingale* central limit theorems* for the asymptotic normality* of L_n [37, Chap. 4].

The distribution of L_n, when H_0 may not hold, in general, depends on the underlying distribution. If we consider the simple regression model

$$F_i(x) = F(x - \alpha - \beta c_i),$$

$$i = 1, \ldots, n, \quad -\infty < x < \infty, \quad (3)$$

where F is assumed to have an absolutely continuous density function f (with a first derivative f' almost everywhere), and if we consider the scores $a_n(i; f) = E\phi_f(U_{ni})$, $i = 1, \ldots, n$, where

$$\phi_f(u) = -f'\big(F^{-1}(u)\big)/f\big(F^{-1}(u)\big),$$

$$0 < u < 1, \quad (4)$$

then the corresponding linear rank statistic, denoted by $L_{n,f}$, provides a *locally most powerful* (LMP) rank test for $\beta = 0$ against $\beta < 0$ (or > 0). Thus with the optimal choice of the score generating function in (4), the corresponding L_n is a LMP rank test statistic. This result gives us some justifications for the choice of the particular scores discussed earlier. In particular, if F is normal, then (4) leads us to the normal scores, while for the logistic distribution, we have the Wilcoxon scores relating to (4). The median scores correspond to the case where F is a double-exponential distribution. For exponential distributions* differing in the scale parameter, the optimal scores are the log-rank scores. A very nice account of this theory is given in Hájek and Šidák [20, pp. 64–71]. The case of scale alternatives is also considered there. This LMP character of $L_{n,f}$ is among the rank-based tests, but it extends to a wider class under an asymptotic setup to be discussed later.

When H_0 does not hold, the uniform permutation distribution* of the vector of ranks may not hold, and as a result, neither the martingale property nor the permutation central limit theorem*, referred to earlier, may be tenable. In this case, the exact distribution of L_n, even for small n, may be dependent on $F_1, \ldots, F_n$ and the c_i in such an involved manner that it may be very difficult to evaluate it. For this reason, one is again interested in providing suitable approximations to such nonnull distributions. Specifically, one may like to know whether for some suitable normalizing constants μ_n and σ_n (depending on the F_i, the c_i and the score function), $(L_n - \mu_n)/\sigma_n$ has closely a normal distribution, and, if so, what can be said about the margin of error for such normal approximations? This has been an active area of fruitful research during the past three decades with outstanding contributions by Hoeffding [22], Chernoff and Savage [10], Hájek [18, 19], Pyke and Shorack [30], and others. Hoeffding was the first to incorporate a projection of such a statistic as a sum of independent random variables plus a remainder term (converging to 0 at a faster rate), and used the classical central limit theorem. Hájek [18] used the same idea for local alternatives and obtained the asymptotic normality under very mild conditions on the score function; in his 1968 paper [19], he has a more general result where the "contiguity*" condition on the density functions has been waived. For the two-sample case, Chernoff and Savage [10] used the novel expansion based on the empirical distributions, and Pyke and Shorack [30] looked into the problem through the weak convergence of some related empirical processes*. For local alternatives, contiguity along with the martingale* property (under H_0) provides an easy avenue for the asymptotic normality results under mild regularity conditions. These are all discussed in Sen [37, Chap. 4]. *See also* ASYMPTOTIC NORMALITY. It appears that for local (contiguous) alternatives when F has an absolutely continuous density function with a finite Fisher information*, the LMP rank test statistics, discussed earlier, are asymptotically most powerful [18].

LRS VECTOR CASE

In the same setup as in (1), we consider now the case where the c_i are q-vectors, for some $q \geqslant 1$. We write $\mathbf{c}_i = (c_{i1}, \ldots, c_{iq})'$, $i = 1, \ldots, n$ and $\bar{\mathbf{c}}_n = n^{-1}\sum_{i=1}^{n} \mathbf{c}_i$. Replacing the $\mathbf{c}_i$ and $\bar{\mathbf{c}}_n$ in (1) by $\mathbf{c}_i$ and $\bar{\mathbf{c}}_n$, respectively, we obtain a q-vector of LRS, denoted by $\mathbf{L}_n$. These LRS arise typically in a several sample problem and/or a multiple regression model.

(III) Several Samples Model

Consider k ($\geqslant 2$) independent samples of sizes $n_1, \ldots, n_k$, respectively, from distributions $F_1, \ldots, F_k$, all assumed to be continuous. We want to test for the equality of $F_1, \ldots, F_k$. We may set $n = n_1 + \cdots + n_k$, $q = k - 1$ and let $\mathbf{c}^{(1)} = \mathbf{0}$, $\mathbf{c}^{(2)} = (1, 0, \ldots, 0)'$, $\mathbf{c}^{(3)} = (0, 1, \ldots, 0)', \ldots, \mathbf{c}^{(k)} = (0, \ldots, 0, 1)'$. Then, for the n_j observations from the jth sample, we let $\mathbf{c}_i = c^{(j)}$, so that the jth component of $\mathbf{L}_n$ reduces to the sum of the rank scores for the $(j+1)$th sample observations, adjusted from the average scores, for $j = 1, \ldots, q$, where ranking is made with respect to the combined sample observations. The reason for choosing $q = k - 1$ (not k) is that the $q \times q$ matrix

$$\mathbf{C}_n = \sum_{i=1}^{n} (\mathbf{c}_i - \bar{\mathbf{c}}_n)(\mathbf{c}_i - \bar{\mathbf{c}}_n)' \tag{5}$$

is positive definite (p.d.), while, for $q = k$, this would have resulted in a $k \times k$ matrix of rank $k - 1$. However, one may work with generalized inverses* and eliminate this arbitrariness.

(IV) Multiple Regression* Model

Consider the linear model pertaining to independent $X_1, \ldots, X_n$, where $X_i = \alpha + \boldsymbol{\beta}'\mathbf{c}_i + e_i$, $i = 1, \ldots, n$, and, as before, the e_i are identically distributed. The null hypothesis to be tested is $H_0: \boldsymbol{\beta} = \mathbf{0}$, against $\boldsymbol{\beta} \neq \mathbf{0}$.

Note that under H_0, the X_i remain identically distributed and independent random variables. The several sample location problem is then a special case of this model, where we define the $\mathbf{c}_i$ as in (III) and where the jth component of $\boldsymbol{\beta}$ stands for the difference of locations of the $(j+1)$th and the first populations, $j = 1, \ldots, q$. Here also, the scores $a_n(i)$ may be defined as in (2), with the score generating functions as in the case of the simple regression model.

As in the simple regression model, here also under H_0, $\mathbf{L}_n$ is a distribution-free statistic (q-vector) and $E[\mathbf{L}_n \mid H_0] = \mathbf{0}$ and $E[\mathbf{L}_n\mathbf{L}_n' \mid H_0] = A_n^2 \cdot \mathbf{C}_n$. As such, as an appropriate test statistic, one usually takes a quadratic form* $\mathscr{L}_n = A_n^{-2}\mathbf{L}_n\mathbf{C}_n^{-}\mathbf{L}_n$, where $\mathbf{C}_n^{-}$ is a generalized inverse of $\mathbf{C}_n$. One nice property of $\mathscr{L}_n$ is that it remains invariant under $\mathbf{c}_i \to \mathbf{d}_i = \mathbf{B}\mathbf{c}_i$, $\mathbf{B}$ nonsingular, so that the choice of any particular set of $\mathbf{c}_i$ is not very crucial. For the several-sample problem, using the Wilcoxon scores, one arrives in this manner at the classical Kruskal–Wallis statistic. Other multisample rank order test statistics are obtainable by using the appropriate scores. Under H_0, for small sample sizes, the exact distribution of $\mathscr{L}_n$ can be obtained by direct enumeration of the permutation distribution of the ranks; again, this process may become prohibitively laborious as the sample sizes increase. However, the permutational central limit theorem applies to the vector $\mathbf{L}_n$ as well [17], and hence the Cochran theorem (on quadratic forms) ensure that for large sample sizes, under H_0, $\mathscr{L}_n$ has closely a chi-square distribution* with q degrees of freedom when $\mathbf{C}_n$ is of rank q. The martingale property of $\{\mathbf{L}_n; n \geqslant 1\}$, under H_0 remains intact, and deeper asymptotic results on $\mathbf{L}_n$ follow on parallel lines [37, Chap. 4]. For the multiple regression model, the score function (4) still leads to some locally optimal rank test, where local optimality is interpreted in the light of locally maximin power, most stringency, or locally best average power of such rank tests [20]. Study of the asymptotic multinormality of $\mathbf{L}_n$, when H_0 may not hold, based on the Crameŕ–Wold device*, reduces to the case of simple regression models, and hence no new technique is needed. Local asymptotic power and optimality (for contiguous alternatives) have also been studied by a host of workers (in the finite Fisher information case) under the same setup as in the simple regression model. Details of these can be found in Hájek and Šidák [20] and Puri and Sen [29], among other places.

(V) Rank Order Statistics for Restricted Alternatives

In the multiparameter case, the null hypothesis is often tested against some restricted alternatives. For example, in the multisample location model, the equality of the locations may be tested against an alternative that the locations (or the distributions) are ordered. In the multiple regression model, similarly, the null hypothesis $H_0: \boldsymbol{\beta} = \mathbf{0}$ may be tested against an orthant alternative $\boldsymbol{\beta} \geqslant \mathbf{0}$ (with at least one strict inequality sign) or an ordered alternative $\beta_1 \leqslant \cdots \leqslant \beta_q$ (or $\beta_1 \geqslant \cdots \geqslant \beta_q$), with at least one strict inequality sign being true. In such a case, the rank order statistics on which the tests are based may be somewhat different in nature. Often, an ad hoc procedure is employed in combining the elements of $\mathbf{L}_n$ into a single scalar statistic. For example, if

$$\mathbf{b} = \left(-\frac{q-1}{2}, \ldots, \frac{q-1}{2}\right),$$

then one can take the inner product $\mathbf{b}'\mathbf{L}_n$ as a suitable test statistic for the ordered alternative problem. Alternatively, one may consider the usual rank correlation* between the elements of $\mathbf{L}_n$ and $(1, \ldots, q)$, and consider that as a test statistic for the same problem. Some detailed accounts of such ad hoc tests are given in Puri and Sen [29] and Hollander and Wolfe [22], among others. Another possibility, initiated in Sen [38], is to use the union–intersection (UI-) principle* of Roy [30] in conjunction with the directional LMP rank tests to construct appropriate UI-LMP rank tests for such restricted alternatives. The class of restricted alternatives may be conceived of as the set-theoretic union of various specific (directional) ones in such a way that for each of these specific alternatives, LMP rank tests statistics can be obtained in a convenient manner. Then one may consider the union of these (LMP rank test) critical regions as the critical region for the overall test. It turns out that such UI-LMP test statistics are still appropriate quadratic forms in the elements of $\mathbf{L}_n$. However, the particular form of these statistics depends on the specific realization of $\mathbf{L}_n$ and the quadratic form may have a discriminant with rank equal to an integer less than or equal to q. Although such UI-LMP rank tests may be genuinely distribution-free (under H_0), their distributions are generally more complicated; in the asymptotic case, they are expressible as weighted combinations of chi distributions with varying degrees of freedom (i.e., chi-bar distributions in the terminology of Barlow et al. [3]). Boyd and Sen [6, 7] have considered such statistics for some simple linear models; other references are cited in these papers.

(VI) Aligned Rank Order Statistics

In many problems, the null hypothesis H_0 may fail to ensure the identity of the distributions $F_1, \dots, F_n$. For example, if we have $X_i = \theta + \boldsymbol{\beta}'\mathbf{c}_i + e_i$, $i = 1, \dots, n$, where the e_i are independent with the common distribution (F), then, if we write $\boldsymbol{\beta}' = (\boldsymbol{\beta}_1', \boldsymbol{\beta}_2')$, $\mathbf{c}_i' = (\mathbf{c}_{i(1)}', \mathbf{c}_{i(2)}')$, so that $X_i = \theta + \boldsymbol{\beta}_1'\mathbf{c}_{i(1)} + \boldsymbol{\beta}_2'\mathbf{c}_{i(2)} + e_i$, $i \geqslant 1$, under the null hypothesis $H_0: \boldsymbol{\beta}_1 = \mathbf{0}$ (against $\boldsymbol{\beta}_1 \neq \mathbf{0}$) with $\boldsymbol{\beta}_2$ as a nuisance parameter, the X_i may not all have the same distribution. As such, the ranking may not be very meaningful or may not give the desired information for formulating the test statistic in a meaningful way. In such a case, if we have an estimator of the nuisance parameter (i.e., $\hat{\boldsymbol{\beta}}_{2,n}$ of $\boldsymbol{\beta}_2$) and consider the aligned observations (or residuals) by substituting this estimator (i.e., taking $\hat{X}_i = X_i - \hat{\boldsymbol{\beta}}_{2,n}'\mathbf{c}_{i(2)}$, $i = 1, \dots, n$), then a rank order statistic based on these aligned observations is termed an *aligned rank order statistic*. In the simplest case of blocked experiments, the block effects are usually the nuisance parameters when one is interested in the treatment effects. The residuals obtained by using suitable estimates of the block effects can be incorporated in the construction of some aligned rank order statistics which are again permutationally (conditionally) distribution-free under the null hypothesis of no treatment effects. For some detailed study of these aligned rank statistics, we may refer to Sen [35] and Puri and Sen [29, Chap. 7]. In a general linear model, for the subhypothesis-testing problem, such aligned rank order statistics are very useful; they may not be distribution-free (even in a permutational or conditional setup), but they are robust and asymptotically distribution-free. In this setup, the asymptotic theory rests very much on some asymptotic linearity results on linear rank statistics, mostly due to Jurečková [25, 26]. If the X_i are independent and identically distributed random variables with a distribution F, and if we define $X_i(\mathbf{b}) = X_i - \mathbf{b}'\mathbf{c}_i$, $i = 1, \dots, n$, where the $\mathbf{c}_i$ are given q-vectors, then for $L_n(\mathbf{b})$ based on these aligned observations, defined as in (1), we have the following: for every finite and positive K, as $n \to \infty$,

$$\sup\left\{\frac{|\mathbf{L}_n(\mathbf{b}) - \mathbf{L}_n(\mathbf{0}) + \mathbf{C}_n\mathbf{b}\gamma|}{n^{1/2}} : \|\mathbf{b}\| < \frac{K}{n^{1/2}}\right\} \to 0$$

in probability, where $\mathbf{C}_n$ is defined by (5) and γ is a suitable positive constant, depending on the distribution and the score function. Based on this basic linearity result, aligned rank order statistics have been considered by Sen and Puri [40] and Adichie [2] to construct some asymptotically distribution-free rank tests for subhypotheses in linear models. For the particular case of $q = 1$ [i.e., $X_i(b) = X_i - bc_i$, $i = 1, \dots, n$], if the score function ϕ is nondecreasing, the statistic $L_n(b)$ is nonincreasing in b. This fact has been tacitly used by Adichie [1] to obtain rank order estimators of the regression coefficient. For the general linear model, under additional regularity conditions, such rank estimators based on aligned rank statistics have been studied by Jurečková [26]. For the particular case of single- or two-sample models, such rank estimators have been considered earlier by Hodges and Lehmann [21] and Sen [33], among others.

(VII) Bilinear Rank Order Statistics

In the context of testing the null hypothesis H_0 of independence of two variates (X, Y) based on a sample $(X_1, Y_1), \dots, (X_n, Y_n)$, one considers a rank order statistic of the form

$$M_n = \sum_{i=1}^{n} \left[a_n(R_i) - \bar{a}_n \right] \left[b_n(Q_i) - \bar{b}_n \right], \tag{6}$$

where $R_i(Q_i)$ is the rank of $X_i(Y_i)$ among $X_1, \dots, X_n$ $(Y_1, \dots, Y_n)$, for $i = 1, \dots, n$, and the scores $a_n(i)$, $b_n(i)$ are defined as in (2). If $Y_{n:1} < \cdots < Y_{n:n}$ be the ordered values of the Y_i and if the corresponding X_i has the rank R_i^*, $1 \leqslant i \leqslant n$, then M_n can be written equivalently as

$$\sum_{i=1}^{n} \left[a_n(R_i^*) - \bar{a}_n \right] \left[b_n(i) - \bar{b}_n \right],$$

and under H_0, this has the same distribution as L_n in (1). This product-sum form is a special case of the bilinear form

$$M_n^* = \sum_{i=1}^{n} h_n(R_i^*, i), \tag{7}$$

where $\{h_n(i, j),\ 1 \leqslant i \leqslant n,\ 1 \leqslant j \leqslant n\}$ is defined suitably. This may be termed a *bilinear rank statistic*. Under H_0, the distribution of M_n^* is again generated by the $n!$ equally likely realizations of $(R_1^*, \dots, R_n^*)$. Permutational central limit theorems for M_n^* were considered by Hoeffding [23] and Motoo [27]. For the LMP property of such statistics, we refer the reader to Hájek and Šidák [20].

(VIII) Adjustments for Ties and Grouped Data

So far, we have considered the case where ties among the observations are neglected, with probability 1. In practice, due to a rounding-off procedure or interval measurements, one may have ties or even data recorded on interval scales. For tied observations, one can distribute the total ranks equally among them (called the *midranks*) and then work with the statistics L_n in (1) with R_i replaced by the corresponding midranks for the tied observations (and without any change for the untied ones). This may be termed a *type I adjusted* (for ties) rank order statistic. Alternatively, one may consider the scores for all the tied observations averaged and redefine L_n in (1) with these adjusted scores—this may be termed a *type II* adjusted rank order statistic. For Wilcoxon scores, either type will lead to the same rank order statistic, while for nonlinear ϕ, in general, they may not agree totally. Similar adjustments are necessary to define M_n in (6) when there are ties among the X_i or the Y_i or both. As here, ranking is made separately for each coordinate; the procedure for ties adjustments needs to be applied for each coordinate. Further, in (7), one may define the R_i^* in a similar manner for ties, although it becomes a bit more complicated. Often, data are collected on ordinal scale or on ordered intervals. For such a case, some asymptotically optimal rank order statistics were studied by Sen [34] and Ghosh [14, 15], among others, *see* RANK TESTS, GROUP. These statistics are only conditionally distribution-free (under H_0), and the same picture prevails in the vector case.

(IX) Multivariate Rank Order Statistics

If $\mathbf{X}_1, \dots, \mathbf{X}_n$ are stochastic p-vectors (for some $p \geqslant 1$) with distributions $F_1, \dots, F_n$, all defined on the p-dimensional Euclidean space E^p, then to test for the null hypothesis $H_0: F_1 = \cdots = F_n = F$ (unknown), one may use suitable rank order statistics for each of the p coordinates, and combine these into a single statistic for the overall test. If $\mathbf{X}_i = (X_{i1}, \dots, X_{ip})'$, $i = 1, \dots, n$, then, for the jth coordinate, we may consider the n observations $X_{1j}, \dots, X_{nj}$ and as before (1), define the rank of X_{ij} among these n observations by R_{ij}, for $i = 1, \dots, n$ and $j = 1, \dots, p$. The rank order statistic L_{nj} can be defined as in (1), where the R_i are replaced by R_{ij}, for different j; the scores need not be the same. The vector $\mathbf{L}_n = (L_{n1}, \ldots, L_{np})'$ is then used in the construction of appropriate test statistics. There is, however, a basic difference between the univariate and the multivariate cases. Though, for each j $(= 1, \ldots, p)$, marginally, L_{nj} is a distribution-free statistic under H_0, the joint distribution of $\mathbf{L}_n$, generally, depends on the unknown F (when H_0 holds). Thus, in the multivariate case, $\mathbf{L}_n$ is not (generally) genuinely distribution-free (under H_0). Chatterjee and Sen [8] have formulated a rank-permutation principle which renders these statistics as permutationally (conditionally) distribution-free and yields conditionally (permutationally) distribution-free tests for H_0. We let $\mathbf{R}_i = (R_{i1}, \ldots, R_{ip})$, for $i = 1, \ldots, n$ and consider the *rank-collection matrix* $\mathbb{R} = (\mathbf{R}_1, \ldots, \mathbf{R}_n)$, each row of which is a permutation of the numbers $(1, \ldots, n)$. We permute the columns of $\mathbb{R}$ in such a way that the first row is in the natural order, and denote the resulting matrix as $\mathbb{R}^*$, the *reduced rank-collection matrix*. Let $S(\mathbb{R}^*)$ be the set of $n!$ rank collection matrices which are reducible to a common $\mathbb{R}^*$ by such column permutations only. Then, under H_0, the conditional distribution of $\mathbb{R}$ over the set $S(\mathbb{R}^*)$ is uniform, with each element having the same conditional probability $(n!)^{-1}$.

With respect to this conditional probability law, the vector $\mathbf{L}_n$ is also conditionally (permutationally) distribution-free, and hence a test based on $\mathbf{L}_n$ is also. If we define $\mathbf{V}_n = ((v_{njj'}))$ by letting

$$v_{njj'} = n^{-1}\sum_{i=1}^{n} a_{nj}(R_{ij})a_{nj'}(R_{ij'}) - \left(n^{-1}\sum_{i=1}^{n} a_{nj}(i)\right)\left(n^{-1}\sum_{i=1}^{n} a_{nj'}(i)\right) \quad (8)$$

for $j, j' = 1, \ldots, p$, then, under the conditional probability law, $\mathbf{L}_n$ has a null mean vector and dispersion matrix $n(n-1)^{-1}C_n^2 \cdot \mathbf{V}_n$, where $C_n^2 = \sum_{i=1}^{n}(c_i - \bar{c}_n)^2$. Thus one may use $\mathscr{L}_n = (n-1)n^{-1}C_n^{-2}(\mathbf{L}_n'\mathbf{V}_n^{-}\mathbf{L}_n)$ as a test statistic for testing H_0. The exact conditional (permutational) distribution of $\mathscr{L}_n$ may be obtained by direct enumeration when n is small, while for large n, multivariate permutational central limit theorems may be used to conclude that under H_0, the permutational distribution of $\mathscr{L}_n$ is closely a central chi-square distribution with p degrees of freedom when the rank of $\mathbf{V}_n$ is p (in probability). The theory extends directly to the case where the $\mathbf{c}_i$ $(= (c_{i1}, \ldots, c_{iq})')$ are q-vectors (for some $q \geqslant 1$), so that the $\mathbf{L}_{nj}$ $[= (L_{nj1}, \ldots, L_{njq})']$ are also q-vectors. In such a case, under the same permutational (conditional) setup, the covariance of L_{njk} and $L_{nj'k'}$ is given by $v_{njj'}C_{nkk'}$, where the $v_{njj'}$ are defined by (8) and $\mathbf{C}_n = ((C_{nkk'}))$ is defined by (5), for $j, j' = 1, \ldots, p$ and $k, k' = 1, \ldots, q$. The $pq \times pq$ matrix with the elements $v_{njj'}C_{nkk'}$ is formally written as $\mathbf{V}_n \otimes \mathbf{C}_n$, where $\otimes$ stands for the Kronecker product. Thus if we roll out the $p \times q$ matrix $\mathbf{L}_n$ into a pq-vector $\mathbf{L}_n^*$, we may consider the test statistic as $(n-1)^{-1}n\{(\mathbf{L}_n^*)'(\mathbf{V}_n \otimes \mathbf{C}_n)^{-}(\mathbf{L}_n^*)\}$. This extension covers the multivariate several sample situation with the specific choice of the $\mathbf{c}_i$ as in (III) and the general multivariate linear model which extends (IV) to the p-variate case. We may refer to Puri and Sen [29, Chap. 5], where these multivariate multisample rank order statistics have been studied in greater detail.

The test statistic $\mathscr{L}_n$ is designed primarily to test for the identity of all n distributions $F_1, \ldots, F_n$. As in the univariate case [see (VI)], in the context of subhypothesis testing, one may have a null hypothesis that does not ensure the identity of all these distributions, and hence the permutational (conditional) distribution-freeness discussed earlier may not be tenable in such a case. As in (VI), here also we may use coordinate-wise aligned rank order statistics and then construct a suitable quadratic form in these aligned statistics as a test statistic. This theory has been studied in detail in Sen and Puri [40]. These aligned tests are generally only asymptotically distribution-free. Further, the statistic $\mathscr{L}_n$ (or its counterpart based on aligned rank statistics) is designed primarily for testing against a global alternative.

If one is interested in testing a null hypothesis against some restricted alternatives (i.e., orthant or ordered ones), then one may again use some ad hoc procedures or may appeal to the UI-principle [see (V)]. These UI-rank order statistics are suitable quadratic forms in $\mathbf{L}_n^*$, but their forms depend on the particular forms of these alternatives as well as the actual structure of $\mathbf{L}_n^*$. For some specific restricted alternative hypotheses testing problems, in the general multivariate case, these rank order statistics have been studied in detail by Chinchilli and Sen [11, 12], where other references are also cited.

The multivariate rank order statistics described above are all based on the coordinatewise (or aligned) rankings. There are, however, some other possibilities. One may use a scalar function $y = g(\mathbf{x})$, so that the $\mathbf{X}_i$ are reduced to some univariate Y_i for $i = 1, \ldots, n$, and then one can use the classical univariate rank order statistics on these Y_i to test for suitable hypotheses; these tests may be genuinely distribution-free and may be easy to apply. However, the choice of a suitable $g(\cdot)$ remains an open problem. Further, this dimensional reduction of data (i.e., from the p to the univariate case) invariably leads to some loss of information that may not be recovered even by efficient choice of scores. Finally, we may remark that unlike the parametric tests (based on the maximum likelihood estimators in general linear models), in the multivariate case, the rank order tests are not invariant under nonsingular transformations on the observation vectors.

(X) Rank Order Statistics for the Analysis-of-Covariance* Problem

The analysis-of-covariance model can be regarded as a special case of the multivariate analysis of variance* model, where only one coordinate relates to the primary variate and the rest to the covariates. The vector of concomitant variates is assumed to have the same distribution for all i $(= 1, \ldots, n)$, while the (marginal and conditional) distributions of the primary variables may differ from the different observations. Quade [30] considered the simplest case of the Wilcoxon scores and suggested the use of the usual parametric analysis with the $\mathbf{X}_i$ being replaced by the $\mathbf{R}_i$. Puri and Sen [28] formulated the problem in a more general framework and considered the following covariate-adjusted rank order statistics. Define the L_{nj} and $v_{njj'}$ as in the multivariate case [see (8)], and let V_{nij} be the cofactor of v_{nij} in V_n, for $i, j = 1, \ldots, p$. Then let

$$L_n^0 = \sum_{j=1}^{p} (V_{n1j}/V_{n11}) L_{nj} \qquad \text{and}$$

$$v_n^* = |V_{n11}|/|V_n|. \tag{9}$$

If the first variate is the primary one and the rest covariates, L_n^0 is the covariate-adjusted rank order statistic, and the corresponding test statistic for testing the hypothesis of randomness* is

$$\mathscr{L}_n^0 = (n-1)n^{-1} C_n^{-2} v_n^* (L_n^0)^2. \tag{10}$$

Here also we may appeal to the rank permutation principle for ensuring the conditional (permutational) distribution-free structure of $\mathscr{L}_n^0$, while for large sample sizes, the null hypothesis distribution of $\mathscr{L}_n^0$ can be adequately approximated by a chi-square distribution with 1 degree of freedom. Here, also, the $\mathbf{c}_i$ may be taken as q-vectors, so that in (9), $\mathbf{L}_n^0$ will be a q-vector, too. In (10), we need to replace $C_n^{-2}(L_n^0)^2$ by $(\mathbf{L}_n^0)'\mathbf{C}_n^{-}(\mathbf{L}_n^0)$, and the asymptotic distribution (under H_0) will be chi square with q degrees of freedom. This covers the case of rank order tests for the analysis-of-covariance problem in the one-way layout. For two-way layouts see Puri and Sen [29, Chap. 7].

(XI) Mixed Rank Order Statistics

As in (VII), consider the bivariate model. In the context of nonparametric tests for regression (of X on Y) with stochastic predictors, Ghosh and Sen [16] have considered some *mixed rank statistics*, which may be outlined as follows. Define the ranks R_i and the scores $a_n(i)$, as in (6). Then consider the statistic

$$T_n = \sum_{i=1}^{n} \left[a_n(R_i) - \bar{a}_n \right] b(Y_i), \qquad (11)$$

where $b(\cdot)$ is some suitable function. For linear regression, one may take $b(x) = x$. Note that under the hypothesis (H_0) of no regression of X on Y, the ranks R_i have the same permutation distribution as in (6), so that given $Y_1, \ldots, Y_n$, T_n is conditionally (permutationally) distribution-free. Various properties of such mixed rank statistics (where only partial sets of ranks are used) were studied by Ghosh and Sen [16]. These mixed rank statistics are very similar to some other statistics, known as *induced order statistics* or *concomitants of order statistics* (see Bhattacharya [5] and David and Galambos [13]). Let $X_{n,1} < \cdots < X_{n,n}$ be the order statistics corresponding to $X_1, \ldots, X_n$. Let $Y_{n(i)} = Y_k$ when $X_{n,i} = X_k$, for $k = 1, \ldots, n$. Then the $Y_{n(i)}$ are termed the *concomitants* of order statistics. Consider a linear function of these concomitants,

$$S_n = \sum_{i=1}^{n} a_{ni} b(Y_{n(i)}), \qquad (12)$$

where the a_{ni} are nonstochastic constants. Note that, by definition, $X_i = X_{n,R_i}$ for $i = 1, \ldots, n$. Thus $Y_{n(R_i)} = Y_i$, for $i = 1, \ldots, n$, and hence S_n in (12) can be written as $\sum_{i=1}^{n} a_{nR_i} b(Y_i)$. As a result, whenever the a_{ni} are expressible as $a_n(i)$, $i = 1, \ldots, n$, T_n and S_n are very much related to each other. Such mixed rank statistics are also very useful in testing for specified forms of regression functions.

(XII) Censored Rank Order Statistics

In many situations, the observations may be censored from the left or right or both. In this respect, *see* CENSORING *and* PROGRESSIVE CENSORING SCHEMES. We may rewrite L_n in (1) as $\sum_{i=1}^{n}(c_{S_i} - \bar{c}_n)a_n(i)$, where the S_i are the *antiranks* (i.e., $R_{S_i} = S_{R_i} = i$, for $i = 1, \ldots, n$). In the context of right censoring, for some r $(1 \leqslant r \leqslant n)$, one observes $(S_1, \ldots, S_r)$, while the remaining $n - r$ antiranks are not individually observable. If we define

$$L_{nr} = \sum_{i=1}^{n} (c_{S_i} - \bar{c}_n)\left[a_n(i) - a_n^*(r) \right], \qquad (13)$$

where

$$a_n^*(r) = (n - r)^{-1} \sum_{j=r+1}^{n} a_n(j), \qquad 0 \leqslant r \leqslant n - 1;$$

$$a_n^*(n) = 0, \qquad (14)$$

then L_{nr} is termed a (right) censored linear rank statistic, where censoring is made at the rth order statistic $X_{n,r}$. In the context of a type II censoring plan, r is prefixed, and under the null hypothesis (of the identity of all the n distributions), L_{nr} is a genuinely distribution-free statistic. In the context of a type I censoring scheme, r is itself a positive integer-valued random variable, and L_{nr} is not genuinely distribution-free; however, given r, L_{nr} is conditionally distribution-free [8]. The particular construction in (13) is motivated by some projection results, and Basu et al. [4] have established some locally most powerful (rank) test structures for such statistics. In the context of progressive censoring, one sequentially observes $L_{nr} : 0 \leqslant r \leqslant n$. The joint distribution of these censored rank statistics (at various r) does not depend on the underlying distribution when the null hypothesis holds, and hence a test based on these is also genuinely distribution-free. This fact, together with some suitable Brownian motion* approximations for such sequences, has been utilized by Chatterjee and Sen [9] in the formulation of suitable time-sequential rank tests in clinical trials*. Various generalizations of this procedure are listed in Sen [37, Chap. 11].

(XIII) Sequential Rank Order Statistics

When observations are available sequentially, at each stage one has to recompute the ranks of the available observations. This is often quite tedious, and hence sometimes only the rank of the last observation (among the older ones) is computed at each stage. This is termed *sequential ranking*. With respect to the model for which L_n in (1) is properly defined, we then have the following related statistic:

$$L_n^* = \sum_{i=1}^{n} (c_i - \bar{c}_n)\left[a_i(R_{ii}) - \bar{a}_i\right], \qquad n \geq 1, \tag{15}$$

where the c_i, $a_i(j)$, $j \leq i$, $i \geq 1$, $\bar{a}_i$, and $\bar{c}_n$ are defined as before, and R_{ii} is the rank of X_i among $X_1, \ldots, X_i$, for $i \geq 1$. Note that under the null hypothesis H_0 (that all the n distributions are the same), the R_{ii} are independent and each R_{ii} can take on the values $1, \ldots, i$ with the same probability i^{-1}, $i \geq 1$, so that L_n^* is a distribution-free statistic. In fact, under H_0, $L_{n+1}^* - L_n^*$ is independent of L_n^*. (Recall that $L_{n+1} - L_n$ is not generally independent of L_n, although there is a martingale structure holding under H_0.) When H_0 does not hold, this independent increment property of L_n^* may not hold, and its distribution theory becomes more complicated. Under the null as well as local (contiguous) alternatives, L_n and L_n^* behave quite similarly [37, Chap. 4].

(XIV) Sequential Censored Rank Order Statistics

In the context of clinical trials with a staggered entry plan, one may encounter a two-dimensional array of sequential censored rank statistics, namely $\{L_{kq};\ 0 \leq q \leq k,\ 1 \leq k \leq n\}$, where the L_{kq} are defined as in (15) for the sample size k and censoring number q. The triangular array, under the null hypothesis of identical distributions, form a genuinely distribution-free set of statistics. Asymptotic theory (including permutational limit theorems) for such an array was developed in Sen [36]. Compared to the asymptotic normality of the L_n here, we have some Brownian sheet approximations for suitable (two-dimensional time-parameter) stochastic processes* constructed from these arrays.

References

[1] Adichie, J. N. (1967). *Ann. Math. Statist.*, **38**, 894–904.

[2] Adichie, J. N. (1978). *Ann. Statist.*, **6**, 1012–1026.

[3] Barlow, R. E., Bartholomew, D. J., Bremner, J. M., and Brunk, H. D. (1972). *Statistical Inference under Order Restrictions*. Wiley, New York.

[4] Basu, A. P., Ghosh, J. K., and Sen, P. K. (1983). *J. R. Statist. Soc. B*, **45**, 384–390.

[5] Bhattacharya, P. K. (1974). *Ann. Statist.*, **2**, 1034–1039.

[6] Boyd, M. N. and Sen, P. K. (1983). *Commun. Statist. Theor. Meth.*, **12**, 1737–1754.

[7] Boyd, M. N. and Sen, P. K. (1984). *Commun. Statist. Theor. Meth.*, **13**, 285–303.

[8] Chatterjee, S. K. and Sen, P. K. (1964). *Calcutta Statist. Ass. Bull.*, **13**, 18–58.

[9] Chatterjee, S. K. and Sen, P. K. (1973). *Calcutta Statist. Ass. Bull.*, **22**, 13–50.

[10] Chernoff, H. and Savage, I. R. (1958). *Ann. Math. Statist.*, **29**, 972–994.

[11] Chinchilli, V. M. and Sen, P. K. (1981). *Sankhyā B*, **43**, 135–151.

[12] Chinchilli, V. M. and Sen, P. K. (1981). *Sankhyā B*, **43**, 152–171.

[13] David, H. A. and Galambos, J. (1974). *J. Appl. Prob.*, **11**, 762–770.

[14] Ghosh, M. (1973). *Ann. Inst. Statist. Math.*, **25**, 91–107.

[15] Ghosh, M. (1973). *Ann. Inst. Statist. Math.*, **25**, 108–122.

[16] Ghosh, M. and Sen, P. K. (1971). *Ann. Math. Statist.*, **42**, 650–661.

[17] Hájek, J. (1961). *Ann. Math. Statist.*, **32**, 506–523.

[18] Hájek, J. (1962). *Ann. Math. Statist.*, **33**, 1124–1147.

[19] Hájek, J. (1968). *Ann. Math. Statist.*, **39**, 325–346.

[20] Hájek, J. and Šidák, Z. (1967). *Theory of Rank Tests*. Academic Press, New York.

[21] Hodges, J. L., Jr. and Lehmann, E. L. (1963). *Ann. Math. Statist.*, **34**, 598–611.

[22] Hoeffding, W. (1948). *Ann. Math. Statist.*, **19**, 293–325.

[23] Hoeffding, W. (1951). *Ann. Math. Statist.*, **22**, 558–566.

[24] Hollander, M. and Wolfe, D. A. (1973). *Nonparametric Statistical Methods*. Wiley, New York.

[25] Jurečková, J. (1969). *Ann. Math. Statist.*, **40**, 1889–1900.

[26] Jurečková, J. (1971). *Ann. Math. Statist.*, **42**, 1328–1338.

[27] Motoo, M. (1957). *Ann. Inst. Statist. Math.*, **8**, 145–154.

[28] Puri, M. L. and Sen, P. K. (1969). *Ann. Math. Statist.*, **40**, 610–618.

[29] Puri, M. L. and Sen, P. K. (1971). *Nonparametric Methods in Multivariate Analysis*. Wiley, New York.

[30] Pyke, R. and Shorack, G. R. (1968). *Ann. Math. Statist.*, **39**, 755–771.

[31] Quade, D. (1967). *J. Amer. Statist. Ass.*, **62**, 1187–1200.

[32] Roy, S. N. (1953). *Ann. Math. Statist.*, **24**, 220–238.

[33] Sen, P. K. (1963). *Biometrics*, **19**, 532–552.

[34] Sen, P. K. (1967). *Ann. Math. Statist.*, **38**, 1229–1239.

[35] Sen, P. K. (1968). *Ann. Math. Statist.*, **39**, 1115–1124.

[36] Sen, P. K. (1976). *Ann. Prob.*, **3**, 13–26.

[37] Sen, P. K. (1981). *Sequential Nonparametrics: Invariance Principles and Statistical Inference*. Wiley, New York.

[38] Sen, P. K. (1982). *Coll. Mat. Soc. Janos Bolyai*, **32**, 843–858.

[39] Sen, P. K. and Ghosh, M. (1972). *Sankhyā A*, **34**, 335–348.

[40] Sen, P. K. and Puri, M. L. (1971). *Z. Wahrscheinl. verw. Geb.*, **39**, 175–186.

(BROWNIAN MOTION
LINEAR RANK TESTS
ORDER STATISTICS
PERMUTATIONAL CENTRAL LIMIT THEOREMS
RANK TESTS
RANKED SET SAMPLING
SEQUENTIAL RANK ESTIMATORS
SIGNED RANK STATISTICS
UNION-INTERSECTION PRINCIPLE)

P. K. SEN

RANK SUM TESTS *See* MANN–WHITNEY–WILCOXON STATISTIC; RANK TESTS

RANK TESTS (EXCLUDING GROUP RANK TESTS)

Rank tests are statistical tests based on a function of the rank order statistics* of a set of observations. Hence they provide a method for testing hypotheses when data are measured only on an ordinal scale. Tests based on ranks are usually distribution-free (*see* DISTRIBUTION-FREE TESTS), and therefore these methods are frequently applied to data that are collected as measurements on an interval or ratio scale but then transformed to ranks because the assumptions needed to apply other tests cannot be justified. Many rank tests have corresponding confidence procedures for interval estimation of certain parameters.

Rank tests are usually some function of the positive integers assigned to the observations in such a way that the smallest observation is denoted by 1, the next smallest by 2, and so on. The most familiar rank tests are of this type, including those covered in this entry. However, the idea of rank tests is easily generalized to other types of assignments of integers, and to assignments of constants other than integers (e.g., normal scores*; *see* NORMAL SCORES TESTS). An interesting geometric interpretation of rank tests that serves to illustrate and unify the various kinds is given in Cook and Seiford [5].

These tests are widely applicable; are based on very minimal assumptions, frequently only that the population is continuous so that a unique assignment can be made; and are simple to perform. Further, the reduction in efficiency over what might be obtained using data measured on a scale higher than ordinal is usually quite small.

The literature on rank tests is quite extensive, and new applications, developments, and studies of their properties are reported each year. Most of the current performance studies are based on Monte Carlo* methods. All books on distribution-free and/or nonparametric statistics contain extensive treatment of rank tests. These include Noether [20], Bradley [1], Gibbons [8, 9], Hollander and Wolfe [14], Marascuilo and McSweeney [19], Daniel [6], Randles and Wolfe [22], Conover [4], and Pratt and Gibbons [21]. Books devoted exclusively to rank tests include Hájek and Šidák [11], Hájek [10], and Lehmann [17]. Most of the recent elementary textbooks on general statistical methods include at least one chapter that discusses some of the best known rank tests. Some of these books integrate a coverage of rank tests with the presentation of classical tests. A general discussion of statistical inference based on ranks is given in Hettmansperger and McKean [13] and the references given therein.

The rank tests discussed here are the primary ones for ungrouped data in the one-sample and paired-sample location problem, the general two-sample problem, the k-sample location problem, and independence. These tests include the Wilcoxon signed-rank test, the two-sample linear rank test and the Mann–Whitney–Wilcoxon* test, the Kruskal–Wallis test*, and the Spearman and Kendall rank correlation coefficients. Rank tests for grouped data are covered in RANK TESTS—GROUPED DATA. Rank tests for treatment-control experiments are covered in STEEL STATISTICS.

ONE-SAMPLE (OR PAIRED-SAMPLE) RANK TESTS FOR LOCATION

Assume that $X_1, X_2, \ldots, X_N$ are a random sample of N observations measured on at least an ordinal scale and drawn from a population that is symmetric about its median M and continuous at M. The null hypothesis is $H_0: M = M_0$, a specified number.

The procedure is to take the differences $X_i - M_0$ and order their absolute values from smallest to largest while keeping track of the original sign of $X_i - M_0$. These ordered absolute values are denoted by $Z_{1:N}$, $Z_{2:N}, \ldots, Z_{N:N}$. A rank test statistic is any statistic of the form

$$T_+ = \sum_{i=1}^{N} a_i Z_i, \tag{1}$$

where

$$Z_i = \begin{cases} 1 & \text{if } Z_{i:N} \text{ corresponds to an } X_j - M_0 > 0 \\ 0 & \text{if } Z_{i:N} \text{ corresponds to an } X_j - M_0 < 0, \end{cases}$$

and the a_i are given constants, called *weights* or *scores*.

The most common rank test statistic uses $a_i = i$ in (1). This is the Wilcoxon signed-rank test* proposed by Wilcoxon [23]. Then large values of T_+ lead to rejection of H_0 in favor of the alternative $M > M_0$, and small values favor the alternative $M < M_0$. The exact distribution of T_+ is tabled extensively in Harter and Owen [12, Table II].

If we have a random sample of pairs $(X_1, Y_1), (X_2, Y_2), \ldots, (X_N, Y_N)$ and the distribution of differences $D = X - Y$ is symmetric about its median M_D and continuous at M_D, the same test can be carried out on the differences $X_i - Y_i - M_0$.

Other signed rank statistics use functions of the normal scores as weights in (1).

TWO-SAMPLE RANK TESTS

Assume that $X_1, X_2, \ldots, X_m$ and $Y_1, Y_2, \ldots, Y_n$ are mutually independent random samples of observations measured on at least an ordinal scale and drawn from continuous populations F_X and F_Y, respectively. The null hypothesis in the general two-sample problem is $H_0: F_X(u) = F_Y(u)$ for all u. A general class of test statistics is based on some function of the pattern of arrangement of X's and Y's in the pooled array of the $m + n = N$ sample observations, denoted by $Z_{1:N}, Z_{2:N}, \ldots, Z_{N:N}$. The class known as linear rank statistics has the test statistic

$$T_X = \sum_{i=1}^{N} a_i Z_i, \tag{2}$$

where

$$Z_i = \begin{cases} 1 & \text{if } Z_{i:N} \text{ corresponds to an } X \\ 0 & \text{if } Z_{i:N} \text{ corresponds to a } Y, \end{cases}$$

and the a_i are given weights or scores. The statistic T_X is linear in the indicator variables, and no similar restriction is implied for the constants.

Linear rank statistics are distribution-free for any set of constants, and an appropriate choice of the constants makes the test sensitive to various kinds of differences between F_X and F_Y.

The most common two-sample rank test

uses $a_i = i$ in (2) so that T_X is simply the sum of the X ranks in the pooled array. This is the Mann–Whitney–Wilcoxon test proposed by Wilcoxon [23] and Mann and Whitney [18]. This test is especially appropriate for the location model where $F_Y(u) = F_X(u - M_X + M_Y)$ (*see* LOCATION TESTS), and then the null hypothesis may be written as $H_0: M_X = M_Y$ in terms of the respective medians M_X and M_Y. The alternative $M_X > M_Y$ is suggested by a large value of T_X and $M_X < M_Y$ by a small value. The exact distribution of T_X for these weights is tabled extensively in Buckle et al. [2].

Linear rank statistics are also available for the scale model where $F_Y(u) = F_X(u\sigma_X/\sigma_Y)$ or $F_{Y-M_Y}(u) = F_{X-M_X}(u\sigma_X/\sigma_Y)$ for scale parameters σ_X and σ_Y, respectively, and H_0: $\sigma_X/\sigma_Y = 1$ (*see* SCALE TESTS).

Other two-sample linear rank statistics that are well known use functions of the normal scores as weights.

The basic idea of two-sample rank statistics can also be extended to the case of k mutually independent random samples that are measured on at least an ordinal scale. These observations are pooled and arranged from smallest to largest, and assigned scores according to their relative position in the array. The test statistic is then some function of the sum of the scores for each of the k samples. The most common test of this type is the Kruskal–Wallis [16] test. Other tests use functions of the normal scores as weights.

Rank tests are also available for k related samples measured on at least an ordinal scale. The best known of this type is the Friedman [7] test (*see* FRIEDMAN'S CHI-SQUARE TEST).

RANK TESTS FOR INDEPENDENCE

Assume that $(X_1, Y_1), (X_2, Y_2), \dots, (X_N, Y_N)$ are a random sample of pairs measured on at least an ordinal scale and drawn from a bivariate distribution which is continuous. The null hypothesis is that X and Y are independent random variables so that the product of the marginal distributions equals the bivariate distribution.

The procedure here is to assume without loss of generality that the pairs are listed so that $X_1 < X_2 < \cdots < X_N$, and hence would have ranks $1, 2, \dots, N$, respectively. Then we note the corresponding order of the set of Y observations. A rank test statistic is some function of this arrangement of the Y ranks. The most common test statistics here are the Spearman rank correlation coefficient*, which is based on the sum of squares of the differences of the corresponding X and Y ranks, and the Kendall tau* statistic, which is a function of the minimum number of inversions required in the Y set to make its ranks also appear in natural order, as do the X ranks. These statistics are fully covered in Kendall [15]. These tests are applicable primarily for alternatives of correlation, trend, or regression.

References

[1] Bradley, J. V. (1968). *Distribution-Free Statistical Tests*. Prentice-Hall, Englewood Cliffs, N.J.

[2] Buckle, N., Kraft, C. H., and van Eeden, C. (1969). *Tables Prolongées de la Distribution de Wilcoxon–Mann–Whitney*. Presse de l'Université de Montréal, Montreal.

[3] Chernoff, H. and Savage, I. R. (1958). *Ann. Math. Statist.*, **29**, 972–994.

[4] Conover, W. J. (1980). *Practical Nonparametric Statistics*. Wiley, New York.

[5] Cook, W. D. and Seiford, L. M. (1983). *Amer. Statist.*, **37**, 307–311.

[6] Daniel, W. (1978). *Applied Nonparametric Statistics*. Houghton Mifflin, Boston.

[7] Friedman, M. (1937). *J. Amer. Statist. Ass.*, **32**, 675–701.

[8] Gibbons, J. D. (1985). *Nonparametric Methods for Quantitative Analysis*, 2nd ed. American Sciences Press, Columbus, Ohio.

[9] Gibbons, J. D. (1985). *Nonparametric Statistical Inference*, 2nd ed. Marcel Dekker, New York.

[10] Hájek, J. (1969). *Nonparametric Statistics*. Holden-Day, San Francisco.

[11] Hájek, J. and Šidák, Z. (1967). *Theory of Rank Tests*. Academic Press, New York.

[12] Harter, H. L. and Owen, D. B., eds. (1972). *Selected Tables in Mathematical Statistics*, Vol. 1. Markham, Chicago.

[13] Hettmansperger, T. P. and McKean, J. W. (1978). *Psychometrika*, **43**, 69–79.

[14] Hollander, M. and Wolfe, D. A. (1973). *Nonparametric Statistical Methods*. Wiley, New York.

[15] Kendall, M. G. (1962). *Rank Correlation Methods*. Hafner, New York.

[16] Kruskal, W. H. and Wallis, W. A. (1952). *J. Amer. Statist. Ass.*, **47**, 583–621; errata: *ibid.*, **48**, 905–911 (1953).

[17] Lehmann, E. L. (1975). *Nonparametrics: Statistical Methods Based on Ranks*. Holden-Day, San Francisco.

[18] Mann, H. B. and Whitney, D. R. (1947). *Ann. Math. Statist.*, **18**, 50–60.

[19] Marascuilo, L. A. and McSweeney, M. (1977). *Nonparametric and Distribution-Free Methods for the Social Sciences*. Brooks/Cole, Monterey, Calif.

[20] Noether, G. E. (1967). *Elements of Nonparametric Statistics*. Wiley, New York.

[21] Pratt, J. W. and Gibbons, J. D. (1981). *Concepts of Nonparametric Theory*. Springer-Verlag, New York.

[22] Randles, R. H. and Wolfe, D. A. (1979). *Introduction to the Theory of Nonparametric Statistics*. Wiley, New York.

[23] Wilcoxon, F. (1945). *Biometrics*, **1**, 80–83.

(DISTRIBUTION-FREE METHODS
FRIEDMAN'S CHI-SQUARE TEST
LOCATION TESTS
NORMAL SCORES TESTS
RANK TESTS, GROUPED DATA
SCALE TESTS
SLIPPAGE TESTS
STEEL STATISTICS)

JEAN DICKINSON GIBBONS

RANK TESTS, GROUPED DATA

Generally, statistical theory is addressed to individually known observations in a sample but in practice, even though the parent distribution is continuous, it is common for the sample to be grouped. This results in so-called grouped data*, where usual nonparametric methods (for continuous variables) are not strictly applicable. Therefore, technically, one would seek statistical methodology to estimate or test a hypothesis regarding the concerned parameters based on group data. The "median test" [5] for the location problem and the "Westenberg test" [10] for the scale problem may be regarded as tests based on grouped data. For the two-sample problem, Gastwirth [2] heuristically proposed a class of group rank tests which are asymptotically most powerful and showed its relationship with "quick estimators" of location and scale parameters based on selected order statistics* (see Sarhan and Greenberg [8]). Saleh and Dionne [7] considered Lehmann alternatives* and obtained the locally most powerful grouped rank test. Further, they suggested how one may choose the optimum group limits via percentiles, so that the test is optimum. Later, Dionne [1] considered the location, scale and joint location, and scale group tests. For simple regression models, Sen [9] obtained the asymptotically most powerful group rank test for the regression parameter. Ghosh [3, 4] and Saleh [6] extended the study to the multiple regression* model.

The specific methodology for the group rank test is now outlined for the two-sample location and scale problems. For the group rank test for the regression problem, see Sen [9], Ghosh [3, 4], and Saleh [6].

GROUP RANK TEST PROCEDURE (TWO-SAMPLE LOCATION AND SCALE PROBLEMS)

ASSUMPTIONS. $X_1, \dots, X_m$ and $Y_1, Y_2, \dots, Y_n$ are mutually independent random samples of observations that are each measured on at least an ordinal scale and drawn from continuous populations F_X and F_Y, respectively.

TEST PROCEDURE. F_X and F_Y are populations from which the sample observations have been drawn. The null hypothesis is $F_Y(u) = F_X(u)$ for all u. The $m + n = N$ observations are pooled and ordered from the smallest to the largest value, say, $W_{(1)} \leqslant \cdots \leqslant W_{(N)}$, where $1, 2, \dots, N$ are the ranks of the ordered observations. Let $0 = \lambda_0 < \lambda_1 < \cdots < \lambda_k < \lambda_{k+1} = 1$ be any k fractiles and let $\gamma_j = [N\lambda_j] + 1$, $j = 1, \ldots, k$, where $[N\lambda_j]$ denotes the largest integer not exceeding $N\lambda_j$. Then the ordered values of the observations with ranks $\gamma_1, \gamma_2, \ldots, \gamma_k$ are

$$W_{(\gamma_1)} < W_{(\gamma_2)} < \cdots < W_{(\gamma_k)},$$

grouping the pooled sample into $(k + 1)$ groups. Let the frequencies of X-observations in the $k + 1$ groups $(-\infty, W_{(\gamma_1)}), (W_{(\gamma_1)}, W_{(\gamma_2)}), \ldots, (W_{(\gamma_k)}, \infty)$ be $M_1, M_2, \ldots, M_{k+1}$, respectively. Further, let $F_X^{-1}(\lambda_j)$ and $f_X(F_X^{-1}(\lambda_j))$ be the quantile function and density-quantile functions of the distribution F_X evaluated at λ_j, $j = 1, \ldots, k$.

Test Statistic for Location Problem

Let M_X and M_Y be the medians of F_X and F_Y, respectively. The null hypothesis is $F_Y(u) = F_X(u)$ for all u or $M_X = M_Y$ under the additional assumption $F_Y(u) = F_X(u - M_X + M_Y)$. Then, if $g_j = f_X(F_X^{-1}(\lambda_j))$, the test statistic is

$$T_X^{(k)} = \sum_{j=0}^{k} \left[\frac{g_{j+1} - g_j}{\lambda_{j+1} - \lambda_j} \right] M_{j+1},$$

which is the weighted sum of the frequencies of X-observations in the $(k + 1)$ random intervals. Since the sample is large, the quantity

$$Z = T_X^{(k)} \Big/ \sqrt{\sum_{j=0}^{k} \frac{[g_{j+1} - g_j]^2}{\lambda_{j+1} - \lambda_j}}$$

can be treated as an approximately standard normal variate if the null hypothesis is valid. In that case the appropriate rejection regions are as follows:

Alternative	Rejection Region
$M_X > M_Y$	$Z > Z_\alpha$
$M_X < M_Y$	$Z < Z_\alpha$
$M_X \neq M_Y$	$\lvert Z\rvert < Z_{\alpha/2}$

where α $(0 < \alpha < 1)$ is the level of significance of the test and Z_α is the corresponding critical value. As for the choice of weights, we may use, for example, the Wilcoxon score or the normal scores*, which are

$$a_{j+1} = -\frac{\lambda_{j+1}(1 - \lambda_{j+1}) - \lambda_j(1 - \lambda_j)}{\lambda_{j+1} - \lambda_j}, \quad j = 0, \ldots, k$$

or

$$a_{j+1} = -\frac{\phi(\Phi^{-1}(\lambda_{j+1})) - \phi(\Phi^{-1}(\lambda_j))}{\lambda_{j+1} - \lambda_j}, \quad j = 0, 1, \ldots, k.$$

Now the question of choosing appropriate λ_j $(j = 1, \ldots, k)$ for these tests arises. They are:

Wilcoxon's group rank test:
$\lambda_j = j/(k + 1)$, $j = 1, \ldots, k$.

Normal group rank test:

$$\lambda_j = \Phi\left[\sqrt{3}\,\Phi^{-1}(j/(k + 1))\right], \quad j = 1, \ldots, k.$$

Example: Normal Group Rank Test. For $k = 5$, the optimum spacings are

$$\lambda_1 = 0.074, \quad \lambda_2 = 0.255, \quad \lambda_3 = 0.500,$$
$$\lambda_4 = 0.745, \quad \lambda_5 = 0.926.$$

These may be found in Sarhan and Greenberg [8, Table 10 E.1]. The weights are then computed to be -6.2002, -2.6190, -0.6115, 0.6831, -1.1527, and -4.5820. The value of

$$K_1 = \sum_{j=0}^{k} \frac{[\phi(\Phi^{-1}(\lambda_{j+1})) - \phi(\Phi^{-1}(\lambda_j))]^2}{\lambda_{j+1} - \lambda_j} = 0.9420.$$

Using the weights and the K_1-value, one can compute the Z-score once the group frequencies M_{j+1} are known from the grouped data.

Test Statistic for Scale Problem

Let σ_X and σ_Y be the scale parameters for the distributions F_X and F_Y, respectively. The null hypothesis is $\sigma_Y = \sigma_X$ with additional assumption

$$F_Y(u) = F_X\left(u \frac{\sigma_X}{\sigma_Y}\right) \quad \text{for all } u.$$

Then, if $h_j = F_X^{-1}(\lambda_j) f_X(F_X^{-1}(\lambda_j))$, the test statistic is

$$S_X^{(k)} = \sum_{j=0}^{k} \left[\frac{h_{j+1} - h_j}{\lambda_{j+1} - \lambda_j} \right] M_{j+1},$$

which is the weighted sum of the frequencies of X-observations in the $(k+1)$ random intervals. Note the difference between the weights for location and scale problem. Again we can treat (for large samples)

$$Z = S_X^{(k)} \Big/ \sqrt{\sum_{j=0}^{k} \frac{(h_{j+1} - h_j)^2}{\lambda_{j+1} - \lambda_j}}$$

as an approximate standard normal variable if the null hypothesis is valid. We reject the null hypothesis whenever $Z > Z_\alpha$. In this case, the rejection region corresponds to $\sigma_X > \sigma_Y$. The optimum spacings for the normal score test corresponding to the problem of estimating σ for the normal case may be found in Sarhan and Greenberg [8].

Test Statistics for Lehmann Alternatives

Let $F_Y(u) = [F_X(u)]^k$ for some $k > 1$; this is called a Lehmann alternative, which states that random variables Y are distributed as the largest of k of the X random variables. For this case, the test statistic for testing the null hypothesis $k = 1$ is

$$S_X^{(k)} = \sum_{j=0}^{k} \alpha_{j+1} M_{j+1},$$

where $\alpha_{j+1} = 1 - (\lambda_{j+1} - \lambda_j)^{-1}\{(1 - \lambda_{j+1}) \ln(1 - \lambda_{j+1}) - (1 - \lambda_j)\ln(1 - \lambda_j)\}$, $j = 0, \ldots, k$. For large samples, the quantity

$$Z = S_X^{(k)} \Big/ \sqrt{\sum_{j=1}^{k} \frac{(u_j e^{-u_j} - u_{j-1} e^{-u_{j-1}})^2}{e^{-u_{j-1}} - e^{-u_j}}}$$

$$\text{with } u_j = \ln(1 - \lambda_j), \quad j = 1, \ldots, k,$$

can be treated as approximately a standard normal variate if the null hypothesis is valid. In this case, the appropriate rejection region is $Z > Z_\alpha$ for $k > 1$. The optimum spacings are discussed in Saleh and Dionne [7]. The nearly best spacings are defined by

$$\lambda_j = 1 - (1 - j/(k+1))^3, \qquad j = 1, \ldots, k.$$

References

[1] Dionne, J.-P. (1978). Some Contributions to Two Sample Nonparametric Tests for Location and Scale under Multiple Censoring. Ph.D. thesis, Carleton University, Ottawa. (Gives details of group rank statistics.)

[2] Gastwirth, J. L. (1966). *J. Amer. Statist. Ass.*, **61**, 929–948. (Discusses robust procedures and puts forth the idea of the group rank test and its connection with ABLUE of location and scale parameters.)

[3] Ghosh, M. (1973). *Ann. Inst. Statist. Math.*, **25**, 91–107. (Discusses group rank tests for multiple regression model.)

[4] Ghosh, M. (1973). *Ann. Inst. Statist. Math.*, **25**, 108–122. (Discusses group rank tests for multiple regression model.)

[5] Mood, A. M. (1950). *Introduction to the Theory of Statistics*. McGraw-Hill, New York. (Theoretical; no examples, or exercises.)

[6] Saleh, A. K. Md. E. (1969). *J. Statist. Res.*, **3**, 1–17. (Discusses group rank tests for multiple regression model.)

[7] Saleh, A. K. Md. E. and Dionne, J.-P. (1977). *Commun. Statist. A*, **6**, 1213–1221. (Discusses basic theory of group rank test and ARE under Lehmann alternatives.)

[8] Sarhan, E. A. and Greenberg, B. G. (1962). *Contributions to Order Statistics*. Wiley, New York.

[9] Sen, P. K. (1967). *Ann. Math. Statist.*, **38**, 1229–1239. (First extension of regression test group data.)

[10] Westenberg, J. (1948). *Proc. Kon. Ned. Akad. Wet.*, **51**, 252–261.

(RANK TESTS (EXCLUDING GROUP RANK TESTS))

A. K. Md. Eshanes Saleh

RAO–BLACKWELLIZATION *See* MINIMUM VARIANCE UNBIASED ESTIMATOR; RAO–BLACKWELL THEOREM

RAO–BLACKWELL THEOREM

In its simplest form this theorem provides a means of reducing the variance of an unbiased estimator T of a parameter θ in the presence of a sufficient statistic* S.

Theorem 1. Let $X_1, X_2, \ldots, X_n$ be a random sample from a distribution having discrete or continuous probability density function $f(x;\theta)$, $\theta \in$ parameter space Θ. Let $S = S(X_1, \ldots, X_n)$ be a sufficient statistic, and $T = T(X_1, \ldots, X_n)$ be an unbiased estimator (UE) of θ. Define

$$T' = E(T \mid S).$$

Then:

(a) T' is a statistic, and is a function of S.
(b) $E[T' \mid \theta] = \theta$ (i.e., $E[T' \mid \theta]$ is unbiased).
(c) $\mathrm{Var}[T' \mid \theta] \leqslant \mathrm{var}[T \mid \theta]$ (1)

for *all* $\theta \in \Theta$, with strict inequality for *some* $\theta \in \Theta$, unless $T' = T$ almost surely.

This form of the theorem is sometimes stated for T as a UE of a function $\tau(\theta)$ or with S replaced by a set of jointly sufficient statistics $S_1, \ldots, S_k$ (see, e.g., Mood et al. [3, Sec. 5.2]). The most important application of the theorem is as a stepping-stone toward identifying uniformly minimum variance UEs; *see* LEHMANN–SCHEFFÉ THEOREM *and* MINIMUM VARIANCE UNBIASED ESTIMATION.

Theorem 1 is a special case of a more general version. Let $L(\theta, d)$ be a *convex loss function* in d when θ is estimated by $\delta = \delta(X_1, \ldots, X_n)$, and δ is observed to have the real value d, that is,

$$L[\theta, \gamma d_1 + (1-\gamma) d_2] \leqslant \gamma L(\theta, d_1) + (1-\gamma) L(\theta, d_2) \quad (2)$$

for any $a < d_1 < d_2 < b$ and $0 < \gamma < 1$. $L(\theta, \delta)$ is *strictly convex* if (2) holds with strict inequality for all such values of d_1, d_2, and γ.

Theorem 2. Let $X_1, \ldots, X_n$ be a random sample from a member of a family $\mathscr{P}$ of distributions indexed by θ, $\theta \in \Theta$, and let S be sufficient for $\mathscr{P}_\theta$. Let δ be an estimator of $\tau(\theta)$, with a strictly convex loss function $L(\theta, d)$ in d; further, suppose that the risk

$$R(\theta, \delta) = EL(\theta, \delta) < \infty.$$

Let $\eta(S) = E[\delta \mid S]$. Then the risk of the estimator $\eta(S)$ satisfies

$$R(\theta, \eta) < R(\theta, \delta) \quad (3)$$

unless $\delta = \eta$ almost surely.

In (3), $<$ is replaced by $\leqslant$ if $L(\theta, d)$ is convex but not strictly convex. Theorem 2 holds when δ and τ are vector valued. See Lehmann [2, Secs. 1.6, 4.5]; the theorem does not hold without the convexity property of the loss function L.

The Rao–Blackwell theorem was first proved by Rao [4] for estimators with squared error loss and independently by Blackwell [1] in the form of Theorem 1. See also Rao [5, Sec. 5a.2] for further discussion and other early references.

References

[1] Blackwell, D. (1947). *Ann. Math. Statist.*, **18**, 105–110.

[2] Lehmann, E. L. (1983). *Theory of Point Estimation*. Wiley, New York.

[3] Mood, A. M., Graybill, F. A., and Boes, D. C. (1974). *Introduction to the Theory of Statistics*, 3rd ed. McGraw-Hill, New York.

[4] Rao, C. R. (1945). *Bull. Calcutta Math. Soc.*, **37**, 81–91.

[5] Rao, C. R. (1973). *Linear Statistical Inference and Its Applications*, 2nd ed. Wiley, New York.

(ESTIMATION, POINT
LEHMANN–SCHEFFÉ THEOREM
MINIMUM VARIANCE UNBIASED ESTIMATION
SUFFICIENT STATISTICS
UNBIASEDNESS)

CAMPBELL B. READ

RAO'S AXIOMATIZATION OF DIVERSITY MEASURES

Statistical measures such as standard deviation*, mean deviation*, range*, and Gini's coefficient of concentration* are introduced to study the variability of a quantitative characteristic of individuals in a population. For a long time no such general measures were proposed for studying the variability of a qualitative characteristic (attribute) such as the eye color of individuals, which has a finite number of alternatives (categories). Strangely enough, the concept of entropy* in information theory provided certain functionals in the space of multinomial distributions*, which have been accepted by biologists as measures of qualitative variation and termed *measures of diversity*. Let $\underline{p} = (p_1, \ldots, p_k)'$ be the vector of the relative frequencies of k categories of an attribute in a population. Then some of the entropy functions used as diversity measures are (see Burbea and Rao [1] and Rao [5])

$$H_S(\underline{p}) = -\sum p_i \log p_i \qquad \text{(Shannon)}, \quad (1)$$

$$H_\alpha(\underline{p}) = \left(1 - \sum p_i^\alpha\right)/(\alpha - 1),$$
$$\alpha > 0, \quad \alpha \neq 1 \qquad \text{(Havrda and Charvát)}, \quad (2)$$

$$H_R(\underline{p}) = \log \sum p_i^\alpha/(1 - \alpha),$$
$$\alpha > 0, \quad \alpha \neq 1 \qquad \text{(Rényi)}, \quad (3)$$

$$H_\gamma(p) = \left[1 - \left(\sum p_i^{1/\gamma}\right)^\gamma\right] \Big/ \left[1 - 2^{\gamma - 1}\right],$$
$$\gamma > 0 \qquad \text{(-entropy)}, \quad (4)$$

$$H_p(p) = -\sum p_i \log p_i - \sum (1 - p_i) \log (1 - p_i)$$
$$\text{(paired Shannon entropy)}. \quad (5)$$

All these measures have the property that they attain the maximum value when $p_i = 1/k$ for all i (maximum diversity) and the minimum value zero only when one p_i has the value unity and the rest are zero, which appear to be logical requirements for any diversity measure. Are there other natural conditions associated with the concept of diversity that a measure of diversity should satisfy? This problem is discussed in relation to possible uses of a diversity measure. Uses of diversity measures are discussed by Pielou [4].

AXIOMS OF DIVERSITY

Let $\mathscr{P}$ be a convex set of probability distributions and H a real-valued functional defined on $\mathscr{P}$. To characterize H as a measure of diversity of a distribution, we consider the following axioms.

C_0 $H(P) = -J_0(P) \geqslant 0 \ \forall P \in \mathscr{P}$, and $= 0$ if P is degenerate.

C_1 If $P_1, P_2 \in \mathscr{P}$ and $\lambda_1, \lambda_2 \in R^+$, $\lambda_1 + \lambda_2 = 1$, then

$$J_1(P_1, P_2 : \lambda_1, \lambda_2) = \lambda_1 J_0(P_1) + \lambda_2 J_0(P_2) - J_0(\lambda_1 P_1 + \lambda_2 P_2) \geqslant 0$$

(i.e., J_0 is a convex functional on $\mathscr{P}$). (We define in general for $P_1, P_2, \ldots \in \mathscr{P}$ and $\lambda_1, \lambda_2, \ldots \in R^+$, $\lambda_1 + \lambda_2 + \cdots = 1$,

$$J_1(\{P_i\} : \{\lambda_i\}) = \sum \lambda_i J_0(P_i) - J_0\left(\sum \lambda_i P_i\right)$$

and call it the first-order Jensen difference [5, 6].)

C_2 Let $P_{11}, P_{12}, P_{21}, P_{22} \in \mathscr{P}$ and $\lambda_1, \lambda_2 \in R^+$, $\lambda_1 + \lambda_2 = 1$; $\mu_1, \mu_2 \in R^+$, $\mu_1 + \mu_2 = 1$. Further, let $P_{i\cdot} = \mu_1 P_{i1} + \mu_2 P_{i2}$, $P_{\cdot j} = \lambda_1 P_{1j} + \lambda_2 P_{2j}$. Then

$$J_2(\{P_{ij}\} : \{\lambda_i \mu_j\}) = \lambda_1 J_1(\{P_{1j}\} : \{\mu_j\}) + \lambda_2 J_1(\{P_{2j}\} : \{\mu_j\}) - J_1(\{P_{\cdot j}\} : \{\mu_j\}) \geqslant 0,$$

(i.e., J_1 defined on $\mathscr{P}^2$ is convex). (J_2 defined on $\mathscr{P}^4$ is called the second-order Jensen difference [6].)

C_i We can recursively define J_3 from J_2, J_4 from J_3, and so on, where in general J_i is defined on $\mathscr{P}^{2^i}$. We write the condition $J_i \geqslant 0$, which is equivalent to saying that J_{i-1} is convex on $\mathscr{P}^{2^{i-1}}$.

We call H an *ith-order* diversity measure if the conditions $C_0, \ldots, C_i$ are satisfied. If C_i is satisfied for all i, then H is called a *perfect* diversity measure. We consider applications of diversity measures of different orders.

Consider a set of populations characterized by probability distributions $\{P_i\}$ with a priori probabilities $\{\lambda_i\}$, $i = 1, \ldots, k$. If H is a first-order diversity measure, then we have the decomposition

$$H\left(\sum_1^k \lambda_i P_i\right) = \sum_1^k \lambda_i H(P_i) + J_1(\{P_i\} : \{\lambda_i\}),$$

$$\text{i.e.,} \quad T = W + B,$$

where W and B are nonnegative. T is the total diversity in the overall population (mixture of populations) and W is the average diversity within individual populations. The excess of T over W is B, which may be interpreted as a measure of diversity between populations. The ratio $G = B/T$ is the index of diversity* between populations introduced by Lewontin [3], which is used extensively in genetic research (see Rao [5]). The convexity condition C_1 meets the intuitive requirement that diversity is possibly increased by mixing. Thus any first-order diversity measure enables apportionment of diversity (APDIV) as between and within populations. Indeed, with such a measure, APDIV analysis can be carried out for any hierarchically classified set of populations to determine the proportion of diversity at any level of classification, as shown in Rao and Boudreau [7].

Let us consider populations with probability distributions $\{P_{ij}\}$ indexed by the levels $i = 1, \ldots, p$ of a factor C and levels $j = 1, \ldots, q$ of a factor D. Let $\{\lambda_i \mu_j\}$ be a priori probabilities associated with $\{P_{ij}\}$. The marginal distribution and the a priori probabilities associated with the levels of C are

$$\left\{P_{i\cdot} = \sum \mu_j P_{ij}\right\} \quad \text{and} \quad \{\lambda_i\}.$$

Similarly, $\{P_{\cdot j}\}$ and $\{\mu_j\}$ are defined and finally $P_{\cdot\cdot} = \sum\sum \lambda_i \mu_j P_{ij}$ represents the grand population (mixture* of populations). If H satisfies C_0 and C_1, we have the basic decomposition

$$H(P_{\cdot\cdot}) = \sum\sum \lambda_i \mu_j H(P_{ij}) + J_1(\{P_{ij}\} : \{\lambda_i \mu_j\}), \qquad (6)$$

$$\text{i.e.,} \quad T = W + B,$$

where B is the diversity between cells (populations indexed by i, j). If H satisfies the conditions C_0, C_1, and C_2, then B admits the decompositions

$$J_1(\{P_{ij}\} : \{\lambda_i \mu_j\}) = J_1(\{P_{i\cdot}\} : \{\lambda_i\}) + J_1(\{P_{\cdot j}\} : \{\mu_j\}) + J_2(\{P_{ij}\} : \{\lambda_i \mu_j\}),$$

$$\text{i.e.,} \quad J_1(C, D) = J_1(C) + J_1(D) + J_2(CD), \qquad (7)$$

where all the terms are nonnegative. $J_1(C)$ and $J_2(D)$ represent the main effects of the factors C and D, respectively, while $J_2(CD) \geqslant 0$ may be interpreted as interaction* between the factors C and D. Thus we have a generalization of ANOVA for two-way classification with balanced data in terms of a general diversity measure satisfying the axioms C_0, C_1, and C_2, which we call ANODIV (*analysis of diversity*). The analysis can be set out in a tabular form as in the case of ANOVA (see Table 1).

The extension of ANODIV to three-way or multiply classified data is done in the same way as in ANOVA. If H also satisfies the condition C_3 and the populations in a three-way classification are characterized by $\{P_{ijk}\} : \{\lambda_i \mu_j \nu_k\}$, then we have the following decomposition into main effects and first- and second-order interactions:

$$\begin{aligned} H(P_{\cdots}) - \sum\sum\sum \lambda_i \mu_j \nu_k H(P_{ijk}) &= J_1(C, D, E) \\ &= J_1(C) + J_1(D) + J_1(E) \\ &\quad + J_2(CD) + J_2(DE) + J_2(CE) + J_3(CDE), \end{aligned} \qquad (8)$$

which provides the formula for computing the three-factor interaction $J_3(CDE)$ based on the expressions for main effects and two-

Table 1 ANODIV for Two-Way Balanced Classification

C (main effect)	$J_1(\{P_{i\cdot}\} : \{\lambda_i\}) = J_1(C)$
D (main effect)	$J_1(\{P_{\cdot j}\} : \{\mu_j\}) = J_1(D)$
CD (interaction)	$J_2(\{P_{ij}\} : \{\lambda_i \mu_j\}) = J_2(CD)$
Between cells	$J_1(\{P_{ij}\} : \{\lambda_i \mu_j\}) = J_1(C, D)$
Within cells	$\sum\sum \lambda_i \mu_j H(P_{ij})$
Total	$H(P_{\cdot\cdot})$

factor interactions. It is interesting to note that an alternative expression for $J_3(CDE)$ is

$$J_3(CDE) = J_1(C,D,E) - J_1(C,D) - J_1(C,E) - J_1(D,E) + J_1(C) + J_1(D) + J_1(E), \quad (9)$$

as in the case of ANOVA.

Similarly, a four-way classified data set can be analyzed in terms of main effects and interactions provided that H also satisfies the condition C_4. The higher-order interactions are obtained recursively from the expressions for lower-order interactions such as (8) and (9). The ANODIV is a very general technique applicable to any type of data provided that a suitable measure of diversity of a population can be defined.

Do there exist diversity measures satisfying one or more of the conditions C_i, $i = 0, 1, 2, \ldots$? Burbea and Rao [1] have shown that in the case of multinomial distributions, the Shannon entropy (1) and paired Shannon entropy (5) satisfy the conditions C_0, C_1, and C_2 but not $C_3, C_4, \ldots$, so that one can do two-way ANODIV using these diversity measures. The Havrda and Charvát entropy (2) satisfies C_0, C_1, C_2 for α only in the range $(1,2]$ when $k \geqslant 3$ and for α in the range $[1,2] \cup [3,11/3]$ when $k = 2$ and C_3, $C_4, \ldots$ do not hold for any α except when $\alpha = 2$. Is there a perfect diversity measure that enables ANODIV to be carried out for multiply classified data of any order? Rao [6] introduced a measure, called *quadratic entropy*,

$$H(P) = \int_{\Omega\times\Omega} d(X_1, X_2)dP(X_1)dP(X_2), \quad P \in \mathscr{P}, \quad (10)$$

where P is a probability measure defined on a probability space $(\Omega, \mathscr{B})$ and $d(X_1, X_2)$ is a measure of difference between any two points X_1 and X_2 in Ω. It is shown that if $d(X_1, X_2)$ is a conditionally negative definite function, that is,

$$\sum_1^n \sum_1^n d(X_i, X_j)a_i a_j \leqslant 0, \quad \forall n \text{ and } \forall a_i \notin \sum a_i = 0, \quad (11)$$

then (10) satisfies C_i for all i. For instance, in the case of a univariate distribution, the choice of the function $d(X_1, X_2) = (X_1 - X_2)^2$, which is conditionally negative definite, leads to the diversity measure

$$H(P) = 2\int (X - \mu)^2 dP(X) = 2V(X),$$

which is a multiple of the variance of the distribution. Since $V(X)$ is a perfect diversity measure, it can be used for ANODIV of multiply classified data of any order, which is the basis of the ANOVA technique. However, there are other functionals besides $V(X)$ which lead to perfect diversity measures such as the one obtained by choosing $d(X_1, X_2) = |X_1 - X_2|$. In a recent paper, Lau [2] has shown that a perfect diversity measure must have the representation (10), thus characterizing Rao's quadratic entropy.

References

[1] Burbea, J. and Rao, C. R. (1982). *IEEE Trans. Inf. Theory*, **IT-28**, 489–495, 961–963.

[2] Lau, Ka-Sing (1986). *Sankhyā A*, **45** (in press).

[3] Lewontin, R. C. (1972). *Evolut. Biol.*, **6**, 381–398.

[4] Pielou, E. C. (1975). *Ecological Diversity*. Wiley, New York.

[5] Rao, C. R. (1982). *Theor. Popul. Biol.*, **21**, 24–43.

[6] Rao, C. R. (1982). *Sankhyā A*, **44**, 1–22.

[7] Rao, C. R. and Boudreau, R. (1984). *Human Population Genetics: Pittsburgh Symposium*. van Nostrand Reinhold, pp. 277–296.

(DIVERSITY INDICES)

C. Radhakrishna Rao

RAO SCORING TEST *See* SCORE STATISTICS

RARE-EVENT RISK ANALYSIS

Rare events are considered to be those which occur so infrequently that direct observation is improbable. Risk resulting from these events is particularly important in two areas of concern. The first aspect addresses the "zero-infinity dilemma" [1] for risks with very low probability and very high consequences, such as those associated with nuclear power plant meltdowns. The second involves very low probabilities of occurrence of events where the numbers of people exposed are high, but whose measurements are masked by spontaneous occurrences, uncontrolled variables, conflicting risks, synergistic and antagonistic processes with other threats, and so on. These kinds of events are associated with the problem of detecting cancer in animals and human populations for substances whose potency is not very high.

Risk, the potential for harm, is some function of the probability of an event and the magnitude of the consequences associated with possible outcomes. When the function is multiplicative, the expected value of risk results, a measure of central tendency. Such a measure has little meaning for an event so rare that it cannot be observed. There are a number of approaches to gaining information on the "meaning" of rare events that can be useful. Some of these will be considered together with their limitations.

DEFINITIONS

A *rare event* may be defined as

$$np < 0.01 \text{ per year}$$

that is, once in a 100 years on the average, or

$$np < NP$$

where n is the number of trials, p the probability of event occurrence in a test population, N the total number of trials occurring in the parent population, and P the number of spontaneous and competing events occurring in the parent population. Both of the definitions above are arbitrary. The first refers to the zero–infinity dilemma and our inability to acquire historical data for something that occurs less often than 100 years. Other values may be used, but should describe the difficulty of observing the events. The second is the problem in measuring potency (the problem of measuring a signal embedded in noise*). An event that is not rare we will term an *ordinary* event.

If one cannot get historical data about systems under study because the events are rare, what can be done? In terms of directly obtainable information, little will be demonstrated. However, probabilities and belief are intertwined at all levels. Belief often replaces hard information, even if the most objective probability experts would argue against this procedure.

PROBABILITY AND BELIEF

All probability estimates involve degrees of belief and are by definition subjective. The three classical approaches to estimating probability in increasing order in terms of the degrees of belief required in each case are:

A priori information (logical approach)
Likelihood of occurrence (frequentist approach)
Subjective estimates (behavioral approach)

For rare events it has been traditional to use a combination of these approaches which form two other approaches.

Modeled Estimates: This comprises a study of the behavior of similar systems for which data are available, which with reasoned modification, is used as a model for the system under analysis (e.g., the estimate of rupture of steam boilers in general to provide an estimate of the probability of rupture of nuclear reactor boilers). Here the belief structure involves the confidence one has in comparing such systems (e.g., does radiation damage increase failures in boilers?).

System Structuring: The failure of systems may be rare because of redundancy, so analysis of the failure probability of component parts and their interconnection is used to synthesize an estimate of

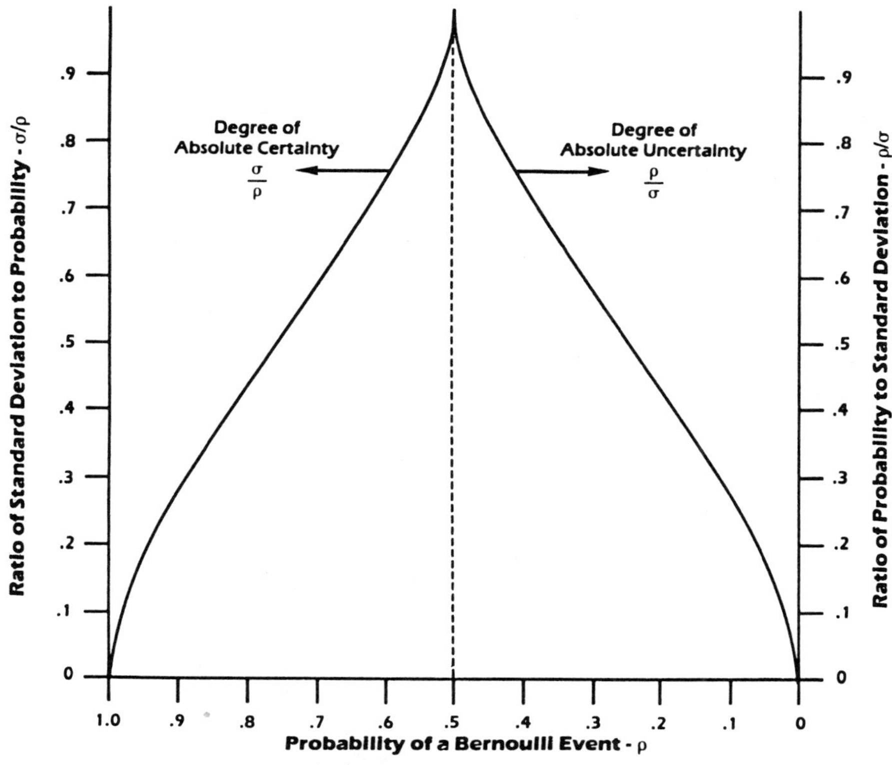

Figure 1 Variability of knowledge of a Bernoulli event as a function of its probability.

system behavior (e.g., event trees and fault trees in nuclear reactors). The belief structures involve the degree of knowledge about individual component behavior, how components behave in a system, and the degree to which all important system combinations can be ascertained. Such systems are always open-ended since the combinational possibilities are astronomical in number.

Realizing the limitations of these approaches, the concept of putting error ranges on the estimates of probability has evolved. The object is to estimate the confidence one has in his or her estimate. This represents a "degree of belief" about one's "degree of belief." Without a keen understanding of implications and limitations of such approaches, they have often been applied erroneously, such as in the Reactor Safety Study [5].

RARE-EVENT BEHAVIOR IN BINOMIAL FORM

Even if a binomial process is known to exist, the information content for rare events is limited. The coefficient of variation* for a binomial distribution* is

$$\text{C.V.} = \sigma/\bar{x} = \frac{\sqrt{np(1-p)}}{np} \simeq \frac{\sqrt{np}}{np} .$$

The significance is that np is a small number for rare events and the square root of a small number is always a larger number. The form of the distribution of uncertainty is shown in Fig. 1 for $n = 1$. The coefficient of variation is inverted at $p = 0.5$ for symmetry. On the left, the coefficient of variation tending toward zero is a measure of certainty. On the left, the coefficient of variation tending toward zero is a measure of certainty. On the right, the inverse is a measure of uncertainty. At a value of $p = 0.1$ the standard deviation is three times the mean value, already a very uncertain situation. As n is increased, the C.V. improves only by the square root of n. The same curve holds for the Poisson distribution*.

The very nature of the binomial and Poisson distributions indicates wide ranges of uncertainty for values of p that are very small. Thus expected value becomes most meaningful at values of $p = 0.5$ since both p and $(1 - p)$ are at their maximum values, but has little meaning at low or high values of p.

SUBJECTIVE BEHAVIOR TOWARD RARE EVENTS

Even if one knew exactly the probability of a rare event, (e.g., 1 in 1 million chance of occurrence in the next year), what does it mean?

One can say that it is unlikely to occur or that it is less likely to occur than an event with a probability of 1 in 100,000 over the next year. It is possible but unlikely that the first event will occur but not the second. All one can do is compare a known but small probability with that of familiar events with meaningful probabilities (i.e., benchmark events). We can say that the estimated probability is higher or lower than a benchmark, but suggesting that it is 10 times lower (for a 10-times-lower probability estimate) begins to challenge meaning. The use of a benchmark provides means for making comparative risk estimates as opposed to absolute risk levels.

ABSOLUTE RISK VS. RELATIVE RISK

For a go/no-go type of decision, one would like to have a meaningful absolute risk estimate. For selection of one of a set of alternatives, only relative risk estimates are required. As will be seen, relative risk evaluations can be quite useful in decision making.

Absolute Risk: an estimate of the likelihood of an event with a specific consequence (type III).

Relative Risk: an estimate of the relative likelihood of an event in terms of the likelihood of other events of a similar magnitude (type I) or the comparison of event magnitudes for events with the same likelihood (type II).

Absolute risk estimates may or may not be useful for decision making, depending on where the risk estimates and their ranges of uncertainty lie. Decisions are always made against some reference or set of references. Benchmarks are one form of reference that do not necessarily imply acceptability. They are risks of a similar nature that people have experienced, which provide a reference to real conditions. However, if the results of analysis show that the range of uncertainty in estimates of probability of occurrence encompass reasonable benchmarks, resolution of the decision by probabilistic methods is unlikely. If the benchmarks fall outside the range of uncertainty, probabilistic approaches can be effective.

As an example, the worst estimate of risk for high-level radioactive waste disposal seems to be lower than benchmarks that seem to be in the acceptable range [4]. If the bands of uncertainty of probability estimates encompass the range of acceptable risk levels, the decision cannot meaningfully be based on probability estimates. Figure 2 illustrates this problem using nuclear accidents and high-level radioactive waste disposal as examples. The scale on the left is a measure of absolute risk in terms of the probability of the number of fatalities that might occur in a year. Some benchmarks are shown on the right, including worldwide fallout from nuclear weapons already committed, planned releases from the nuclear fuel cycle for 10,000 GWe years of operation (the maximum production possible from available uranium resources without breeding), 1% of natural radiation background, and radon and radiation from undisturbed ura-

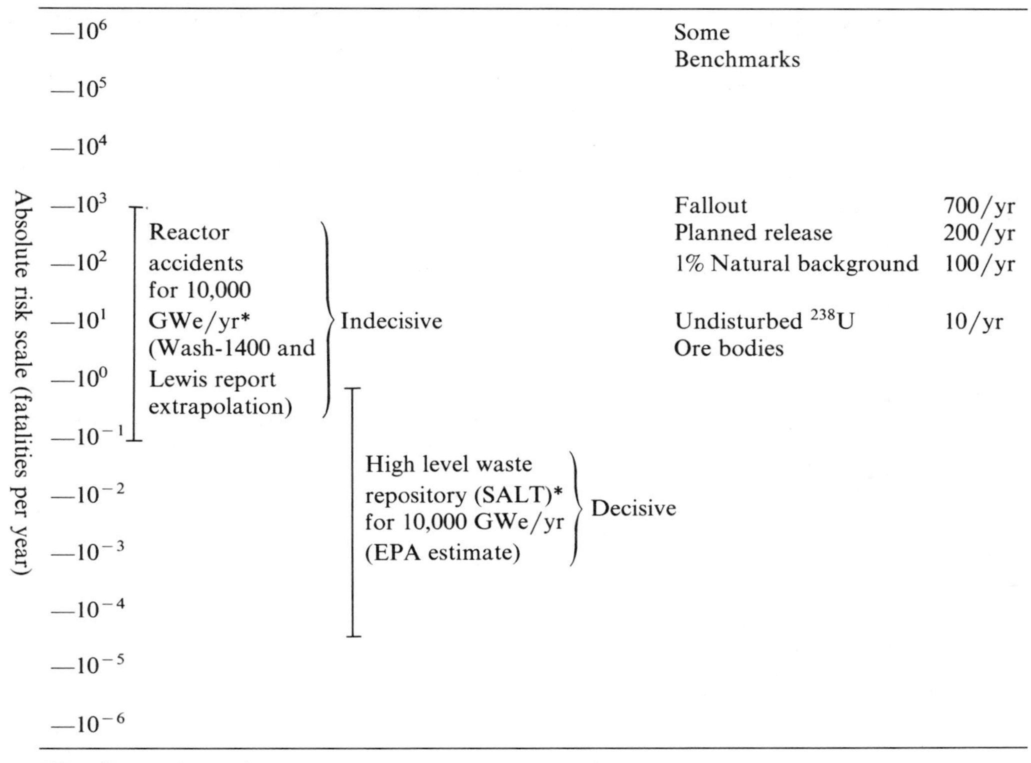

Figure 2 Absolute risks, uncertainty, and benchmarks. Fallout represents worldwide risk from existing weapons fallout: Ce^{137}, C^{-14}, Pu. Planned release is 10,000 GWe/yr based on 40 CFR 190.

nium ore bodies. These benchmarks are only to provide perspective; they do not, by themselves, imply acceptability.

The range of risk estimates for a high-level waste repository for all high-level wastes (10,000 GWe-years) lies well below the benchmarks. Thus a decision on high-level waste is resolvable by probabilistic methods. The range of risk estimates for all nuclear reactor accidents (10,000 GWe-years of operation) is shown based on the United States Nuclear Regulatory Commission Reactor Safety Study [5] and extrapolation from there. The exact range may be argued, but it probably envelopes all of the benchmarks, making any decision based on probabilistic analysis alone indecisive. In this case, although it may be possible to refine the estimates to some extent, it may be impossible to reduce the residual uncertainties to a level for which meaningful decisions can ever be made by this approach. Thus some decisions may be truly indecisive by this approach, while others can be decided. *Indecisive* refers only to a probabilistic solution, since many other approaches, including social–political analyses, may still be effective.

A reasonable management strategy for investigating the probability of rare events is to proceed in two steps. The first would be a preliminary analysis to determine the range of uncertainty bands in probability estimations, which then would be compared to benchmarks to determine whether overlap exists and to what extent. Then the value of information provided by a second, more detailed analysis can be determined. If warranted, the more detailed analysis can then be undertaken.

In the absence of valid risk estimates on

an absolute basis, relative risk estimates can sometimes be effective. In this case, one of several alternatives is chosen as a baseline, and all others are compared to it. In most cases, the same absolute risk uncertainties occur in each one of the alternatives. On a relative basis, these uncertainties cancel out, and the remaining uncertainties among alternatives are much smaller. For example, one might ask questions regarding the comparative risk of earthquakes among alternatives. In this case, the uncertainty of whether it will occur may be ignored; the relative risk of alternatives can be established meaningfully as to whether a risk is greater or lower than the baseline.

NONRANDOM OCCURRENCES

Up to this point, the occurrence of rare events has been considered to be random in nature. In many cases this assumption may not be used. For example, the probability of flooding in the year following a large flood may be higher than at other times because of saturation of the ground and a resultant high water table. In other cases, the building of dams for flood control may decrease the likelihood of flooding, but when a flood does occur the consequences may be of much greater magnitude. In these cases, new structures and homes may have been built as a result of a flood control project opening new land, which depends on the dam for flooding.

Human error contributes to rare occurrences. In the maritime safety area, human error has been identified in over 80% of all accidents.

Techniques to identify margins of safety and levels of redundancy in time and space may be a useful approach for nonrandom events. These techniques use models to assess the levels of redundancy and safety margins to identify those conditions where these levels are lacking. The approach is one of comparative risk assessment in that system weak points are identified, not whether events will actually occur.

HUMAN EVALUATION OF RARE EVENTS

Up to this point, only problems of estimation of the probabilities and consequences of rare events have been addressed, particularly the wide ranges of uncertainty in such measures. However, even with precise information about probabilities and consequences, people evaluate rare events in different ways. In several experiments [3] reported elsewhere, the author provided information on choices between certain events and rare events of large magnitude to a large number of respondents. Essentially, the choices were based on von Neumann–Morgenstern utility* functions and equivalent gambles under a number of different conditions. The probabilities and magnitude of events were specified exactly. Three classes of responses were noted:

1. **Risk Neutral:** Expected value is equated with certain occurrence.
2. **Risk Averse:** The expected value of large consequence events is valued below certain occurrence (i.e., one wants to reduce the risk of a large magnitude event in respect to many small events).
3. **Risk Prone:** The expected value of large consequence events is valued above certain occurrence (i.e., one is willing to take a chance that nothing will happen rather than have many small events occur). If something occurs, however, it is large in respect to the small events.

In quantitative terms, the results were trimodal with variation of two orders of magnitude on either side of risk neutrality. The conclusion is a rather obvious one—even with precise information about probabilities and consequences, people's values about rare events have a wide variation.

The key question is whether the estimation and evaluation uncertainties propagate or cancel in a behavioral sense. Since humankind has dealt with rare events in a reasonably effective manner since the dawn of history, it may be that the uncertainties tend to cancel. If this is the case, a possible means to address rare events on a rational basis may emerge. The real problem is to better understand how the process of merged estimation and evaluation works. Little has been done in this area at present; however, it may hold promise.

References

[1] Page, T. (1979). In *Energy Risk Management*, G. T. Goodman and W. D. Rowe, eds. Academic Press, London, pp. 177–186.

[2] Page, T. (1979). In *Energy Risk Management*, G. T. Goodman and W. D. Rowe, eds. Academic Press, London, pp. 177–186. See also H. W. Lewis, R. J. Budnitz, H. J. C. Kouts, W. B. Loewenstein, W. B. Rowe, F. von Hippel, and F. Zachariasen. Risk Assessment Review Group Report to the U.S. Nuclear Regulatory Commission. *NUREG/CR-0400*, U.S. Nuclear Regulatory Commission, Washington, D.C., September 1978.

[3] Rowe, W. D. (1982). *Corporate Risk Assessment*. Marcel Dekker, New York.

[4] U.S. Environmental Protection Agency (1978). Supporting Documentation for Proposed Standards for High Level Radioactive Wastes. *40 CFR 191*, U.S. Environmental Protection Agency, Washington, D.C., November.

[5] U.S. Nuclear Regulatory Commission (1975). Reactor Safety Study: An Assessment of Accident Risks in U.S. Commercial Nuclear Power Plants. *WASH-1400 (NUREG-75/014)*, U.S. Nuclear Regulatory Commission, Washington, D.C., October.

Bibliography

See the following works, as well as the references just given, for more information on rare-event risk analysis.

Fischhoff, B., Slovic, P., Lichtenstein, S., Read, S., and Combs, B. (1978). In *Policy Sciences*, Vol. 9. Elsevier, Amsterdam, pp. 127–152.

Lewis, H. W., Budnitz, R. J., Kouts, H. J. C., Loewenstein, W. B., Rowe, W. D., von Hippel, F., and Zachariasen, F. (1978). Risk Assessment Review Group Report to the U.S. Nuclear Regulatory Commission. *NUREG/CR-0400*, U.S. Nuclear Regulatory Commission, Washington, D.C., September.

Lowrence, W. (1976). *Of Acceptable Risk*. Kaufman, Los Angeles.

Okrent, D. and Whipple, H. (1977). An Approach to Societal Risk Acceptable Criteria and Risk Management. *UCLA-ENG-7746*, School of Engineering and Applied Science, UCLA, Los Angeles, June.

Otway, H. J. (1975). Risk Assessment and Societal Choices. *IIASA Research Memorandum, RM-75-2*, Schloss Laxenburg, Laxenburg, Austria, February.

Otway, H. J. and Fishbein, M. (1977). Public Attitudes and Decisionmaking. *IIASA Res. Memor., RM-77-54*, Schloss Laxenburg, Laxenburg, Austria, November.

Otway, H. J., Pahner, P. D., and Linnerooth, J. (1975). Social Values in Risk Acceptance. *IIASA Res. Memo. RM-75-54*, Schloss Laxenburg, Laxenburg, Austria, November.

Rowe, W. D. (1977). *An Anatomy of Risk*. Wiley, New York.

Rowe, W. D. (1977). Assessing Risk to Society. Presented to the Symposium on Risk Assessment and Hazard Control, American Chemical Society, New Orleans, March.

Rowe, W. D. (1977). *George Washington Law Rev.*, **45**, 944–968.

Thedeen, T. (1979). In *Energy Risk Management*, G. T. Goodman and W. D. Rowe, eds. Academic Press, London, pp. 169–176.

Thomas, K., Maurer, D., Fishbein, M., Otway, H. J., Hinkle, R., and Wimpson, D. (1978). A Comparative Study of Public Beliefs about Five Energy Systems. *IIASA Res. Memo., RM-78-XX*, Schloss Laxenburg, Laxenburg, Austria.

Wilson, R. (1977). The FDA Criteria for Assessing Carcinogens. Written for the American Industrial Health Council, Washington, D.C.

(COMMUNICATION THEORY
ENVIRONMENTAL STATISTICS
INVERSE SAMPLING
NUCLEAR MATERIALS SAFEGUARDS
POISSON DISTRIBUTION
RISK THEORY)

WILLIAM D. ROWE

RAREFACTION CURVES

BACKGROUND

When one samples information about a population of people, machines, butterflies, stamps, trilobites, and nearly anything else, it is a natural tendency that as more objects are obtained, the number of distinct kinds of objects increases. Rarefaction is a sampling technique used to compensate for the effect of sample size on the number of groups observed in a sample and can be important in comparisons of the diversity* of populations. Starting from a sample of units classified into groups, the rarefaction technique provides the expected number of groups still present when a specified proportion of the units are randomly discarded. In this way a large sample can be "rarefied," or made smaller, to facilitate comparison with a smaller sample.

For example, suppose that I spend two months collecting specimens and find 102 distinct species represented among the 748 individuals collected. If you then spend a week collecting specimens at another location and find only 49 species among 113 specimens, can we conclude by comparing your 49 to my 102 species that your population was less diverse than mine? Of course not. We need to correct for sample size to do a proper comparison because if you had collected for a longer time, you would probably have obtained a larger number of samples and of species. Applying the techniques of rarefaction to the detailed data (the method needs to know how many individuals are in each species), the rarefied number of species in my sample might turn out to be 50.14 species for subsamples of size 113. When compared to your count of 49 species, we could then conclude that the population diversities are not very different.

A brief history of rarefaction begins with work by Sanders [12], who developed a technique to compare deep-sea diversity to shallow-water habitats. Problems of overestimation were noted by Hurlbert [8], Fager [3], and Simberloff [14], with an improved formulation given by Hurlbert and by Simberloff. A formula for the variance of the number of groups present in a rarefied sample was provided by Heck et al. [7], who also considered the determination of sufficient sample size for data collection*.

Sampling properties of the rarefaction measure were explored by Smith and Grassle [15]. Rarefaction has been applied extensively in the study of diversity throughout the fossil record by Raup [10, 11] and by others. Some criticisms and suggestions relating to the application of rarefaction methods were made by Tipper [18]. Upper and lower bounds on rarefaction curves were developed by Siegel and German [13]. Because it is based on sampling from the data, rarefaction is related to the bootstrap* method of Efron [1]; this connection is explored by Smith and van Belle [16].

Rarefaction is related to the ideas of diversity and evenness in populations. A recent overview of diversity measurement may be found in Patil and Taillie [9] with discussion by Good [4] and Sugihara [17]. Rarefaction may be considered as an interpolation process, compared to the more difficult problem of extrapolation as considered by Good and Toulmin [5] and by Efron and Thisted [2], in which the goal is to estimate the number of additional groups that would be observed if a larger sample could be obtained.

DEFINITION AND PROPERTIES OF RAREFACTION

Suppose that we have a situation in which N items are classified into K groups in such a way that each item is in exactly one group and each group contains at least one item. For example, the items might be individual specimens that have been collected and the groups might represent the various species present; for analysis at a higher taxonomic level, the items might be species grouped according to genus.

To describe the situation completely, let the number of items in group i be denoted N_i. The data may be described as follows:

N = total number of items

K = total number of groups

N_i = number of items in group i $(i = 1, \ldots, K)$.

To facilitate computation, we will define M_j

to be the number of groups containing exactly j units ($j \geqslant 1$):

$$M_j = \text{number of } N_i \text{ equal to } j.$$

From these definitions, it follows that

$$\sum_{i=1}^{K} N_i = N, \quad \sum_{j=1}^{\infty} M_j = K, \quad \sum_{j=1}^{\infty} jM_j = N.$$

Now consider a rarefied sample, constructed by choosing a random subsample of n from N items without replacement. Some of the groups may be absent from this subsample. Let X_n denote the (random) number of groups that still contain at least one item from the rarefied sample:

X_n = number of groups still present in a subsample of n items.

It must be true that $X_n \leqslant K$ with strict inequality whenever at least one group is missing from the rarefied sample.

The *rarefaction curve*, $f(n)$, is defined as the expected number of groups in a rarefied sample of size n, and can be computed in several ways:

$$f(n) = E[X_n] = K - \binom{N}{n}^{-1} \sum_{i=1}^{K} \binom{N - N_i}{n}$$

$$= K - \sum_{j=1}^{\infty} M_j \binom{N-j}{n} \binom{N}{n}^{-1}.$$

It is always true that $0 \leqslant f(n) \leqslant K$, $f(0) = 0$, $f(1) = 1$, and $f(N) = K$. Moreover, f is monotone increasing and concave downward.

Because these binomial coefficients can become large and overflow when computers are used, it is preferable to compute directly with the ratio of the two coefficients, which is always between 0 and 1. This ratio may easily be updated using a multiply and a divide to obtain successive terms:

$$\left[\binom{N-(j+1)}{n}\binom{N}{n}^{-1}\right] = \frac{N-n-j}{N-j}\left[\binom{N-j}{n}\binom{N}{n}^{-1}\right].$$

The rarefaction values $f(n)$ are often displayed as a continuous curve even though they are actually discrete values. Consider, for example, the rarefaction curve for $N = 748$ units (species) within $K = 102$ groups (families) of bivalves from Siegel and German [13], Fig. 1. (Data were collected by Gould and are described in ref. 6.)

SAMPLING PROPERTIES

In many situations it is more realistic to suppose that the observed values of items and groups are not fixed but instead represent a sample from a multinomial* population. The expected number of groups represented in a sample of n items from this

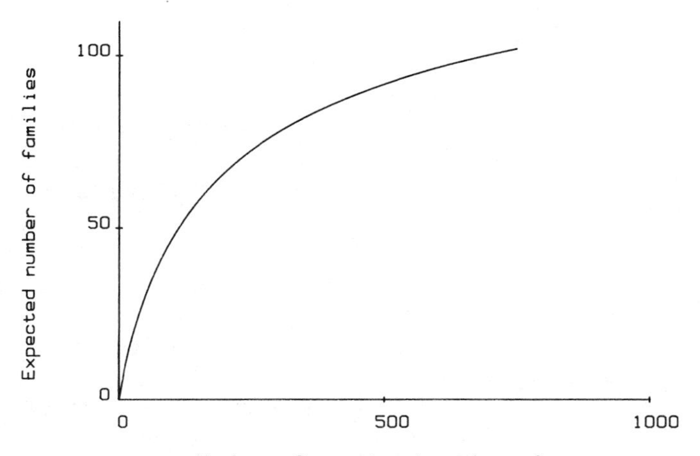

Figure 1

population can be used as a measure of the population diversity. Based on the observed data, the rarefaction curve value $f(n)$ can be used as an estimate of this population diversity measure. Within this context, Smith and Grassle [15] have proven that the rarefaction value is a minimum variance unbiased estimate* (MVUE). They also provide an unbiased estimate of the variance of the estimate which takes into account the sampling variability of the process that generated the data.

References

[1] Efron, B. (1982). *The Jackknife, the Bootstrap, and Other Resampling Plans*. SIAM, Philadelphia.

[2] Efron, B. and Thisted, R. (1976). *Biometrika*, **63**, 435–448.

[3] Fager, E. W. (1972). *Amer. Naturalist*, **106**, 293–310.

[4] Good, I. J. (1982). *J. Amer. Statist. Ass.*, **77**, 561–563.

[5] Good, I. J. and Toulmin, G. H. (1956). *Biometrika*, **43**, 45–63.

[6] Gould, S. J., Raup, D. M., Sepkoski, J. J., Schopf, T. J. M., and Simberloff, D. S. (1977). *Paleobiology*, **3**, 23–40.

[7] Heck, K. L., Jr., van Belle, G., and Simberloff, D. S. (1975). *Ecology*, **56**, 1459–1461.

[8] Hurlbert, S. H. (1971). *Ecology*, **52**, 577–586.

[9] Patil, G. P. and Taillie, C. (1982). *J. Amer. Statist. Ass.*, **77**, 548–561, 565–567.

[10] Raup, D. M. (1975). *Paleobiology*, **1**, 333–342.

[11] Raup, D. M. (1979). *Science*, **206**, 217–218.

[12] Sanders, H. L. (1968). *Amer. Naturalist*, **102**, 243–282.

[13] Siegel, A. F. and German, R. Z. (1982). *Biometrics*, **38**, 235–241.

[14] Simberloff, D. S. (1972). *Amer. Naturalist*, **106**, 414–418.

[15] Smith, W. and Grassle, J. F. (1977). *Biometrics*, **33**, 283–292.

[16] Smith, E. P. and van Belle, G. (1984). *Biometrics*, **40**, 119–129.

[17] Sugihara, G. (1982). *J. Amer. Statist. Ass.*, **77**, 564–565.

[18] Tipper, J. C. (1979). *Paleobiology*, **5**, 423–434.

(BOOTSTRAPPING
DIVERSITY INDICES
ECOLOGICAL STATISTICS
LOGARITHMIC DISTRIBUTION)

ANDREW F. SIEGEL

RASCH, GEORG

Born: September 21, 1901, in Odense, Denmark.

Died: October 19, 1980, in Byrum, Laesø, Denmark.

Contributed to: mathematics, theoretical statistics, psychometrics, philosophy.

Georg Rasch was professor of statistics at the University of Copenhagen from 1962 to 1972. Rasch received his degree in mathematics from the University of Copenhagen in 1925 and worked as a mathematician at the university until 1930, at age 29, when he became a doctor of science on a dissertation concerned with matrix calculations and its applications in differential and difference equation theory [1]. At the time he was considered to be one of the most talented of the new generation of Danish mathematicians. But as no satisfactory position was created for him as a mathematician, he chose to work as a consultant in applied mathematics, primarily data analysis and statistics. In the 1930s he worked with problems in medicine and biology, but he later added psychology and sociology as fields of interest.

In 1935–1936, he visited University College in London, primarily to work with R. A. Fisher*. He was much impressed by Fisher's ideas on the foundation of mathematical statistics and introduced them in Denmark after his return. In the following years he worked primarily at the Danish National Serum Laboratory, where he founded the Bio-Statistical Department and was its director from 1940 to 1956. In this capacity he made many contributions to new scientific developments in biology and medicine. But he had a much more lasting influence on the development of statistics, in both theoretical and applied senses, through the fact that most, if not all, of the next generation of Danish statisticians worked as his assistants at the Serum Laboratory.

In the 1940s and 1950s he had various part-time teaching assignments at the university, but it was not until 1961, when he was almost 60 years old, that he was appointed to a chair in statistics at the University of Copenhagen.

It may seem surprising, but it is nevertheless a fact, that he did not work with applications in education and psychology until the mid-1950s, when he was well into his own fifties. But it was these disciplines that occupied most of his thinking in the 1960s and 1970s, and it was here that he made his most original contributions. As a consultant to the Ministry of Social Affairs, to the Office of Military Psychology, and to the Danish Educational Research Institute, he was faced with the task of extracting information on individuals from intelligence and ability tests. He rejected the traditional statistical methods, primarily based on various factor analytic techniques, and developed new and more exact methods based on latent* trait models as we know them today. The most simple and elegant of these models was fully developed by Rasch in 1960 and now bears his name: the Rasch model*. The model was not invented as a new theoretical development, but was established through careful study of the empirical data with which he worked. He also realized that the model required a new statistical methodology based on the use of conditional probabilities. During the year 1960, in his famous book [2] and in an important paper read at the Berkeley Symposium on Probability and Statistics [3], he presented both a new revolutionary model and an associated fascinating new statistical methodology. The model was further developed in the following years and he showed many important applications of it, but to a remarkable degree the theory was developed within a span of three to four years.

In the 1960s and 1970s there followed a few papers [4, 5], in which he tried to extend his discoveries from 1960 to a more general theory of measurement primarily directed toward the social sciences. It was these ideas that occupied his thinking for the rest of his life.

In his scientific works Rasch combined mathematical skill and a skill for reading empirical evidence in a very unique way. He used mathematics to make ideas precise and to formulate the theoretical framework for the analysis of data in an exact way. But data as they are found in real life were the main source for all his theoretical developments and model formulations. Rasch was thus an early and always eager advocate of controlling the model by statistical methods.

Georg Rasch was a knight of the Danish order of Dannebrog and an honorary member of the Danish Statistical Society.

References

[1] Rasch, G. (1930). *Om Matrixregning og dens Anvendelse på Differens-og Differentialligninger*. Levin og Munksgård, Copenhagen.

[2] Rasch, G. (1960). *Probabilistic Models for Some Intelligence and Attainment Tests*. Danmarks Paedagogiske Institut, Copenhagen.

[3] Rasch, G. (1961). *Proc. 4th Berkeley Symp. Math. Statist. Prob.*, Vol. 5. University of California Press, Berkeley, Calif., pp. 321–333.

[4] Rasch, G. (1972). *Nationaløkon. Tidsskr.*, **110**, 161–196.

[5] Rasch, G. (1974). *Dan. Philos. Yearbook*, **14**, 58–94.

ERLING B. ANDERSEN

RASCH MODEL, THE

BACKGROUND

Originally proposed in the context of oral reading tests by the Danish statistician Georg Rasch [18], the *Rasch model* was an attempt to model the probability of pupil ν

making $a_{\nu i}$ mistakes in test i, consisting of N_i words. Given N_i, Rasch took $a_{\nu i}$ as binomial with parameters $\theta_{\nu i} = \delta_i/\xi_\nu$ and N_i. Then he used the Poisson approximation to the binomial (for large N_i) to get the mean number of mistakes for pupil ν in test i as

$$\lambda_{\nu i} = \frac{N_i \delta_i}{\xi_\nu} = \frac{t_i}{\xi_\nu}, \tag{1}$$

where he labeled t_i as the "impediment of the test" and ξ_ν as the "ability of the pupil." As Cressie and Holland [8] note, the ability of the pupil is not directly observable, and all that can be estimated are *manifest probabilities*, corresponding to the proportion of pupils in a given population who obtain particular patterns of mistakes over a set of tests.

By considering multiple pupils and multiple tests, and by assuming independence of the binomial variates, $\{a_{\nu i}\}$, Rasch showed that the *conditional distribution* of $\{a_{\nu i}\}$, given their totals for each pupil (i.e., $\{a_{\nu +}\}$ where a "+" indicates summation over the corresponding subscript), does not depend on the ability parameters, $\{\xi_\nu\}$. Rasch then suggested using large samples of pupils and this conditional distribution to estimate the "test impediments" $\{t_i\}$ (or equivalently the difficulties $\{\delta_i\}$). With reasonably accurate estimates of test parameters in hand, one can then reverse the role of pupils and tests, and produce estimates of the "ability" parameters *for a new set of pupils*. A critical feature of Rasch's work was his detailed assessment of the adequacy of both the multiplicative form of expression (1), and the various explicit and implicit assumptions inherent in the binomial sampling model.

One of the inadequacies that Rasch observed in the context of several different types of tests (other than reading tests) was the lack of homogeneity of material in different parts. This led him to apply his multiplicative model locally, within tests, allowing different test items to have different difficulties. Instead of observing integer-valued counts $\{a_{\nu i}\}$ as in the original problem, Rasch now had to focus on collections of zeros and ones, where the independence assumption from item to item is far more tenuous.

Rasch then formulated the following model for ability tests, which we give in its log-linear rather than its multiplicative form. In presenting the model, we are not endorsing the popular interpretation of "ability" estimates associated with individuals. Rather, we are trying to set the model in the context in which it was originally proposed.

THE BASIC MODEL

For a test with k problems or items administered to n individuals, we let

$$Y_{ij} = \begin{cases} 1 & \text{if individual } i \text{ answers item } j \text{ correctly.} \\ 0 & \text{otherwise.} \end{cases} \tag{2}$$

Thus we have a two-way table of random variables $\{Y_{ij}\}$ with realizations $\{y_{ij}\}$. An alternative representation of the data is in the form of a $n \times 2^k$ table $\{W_{ij_1j_2\ldots j_k}\}$ where the subscript i still indexes individuals and now $j_1, j_2, \ldots, j_k$ refer to the correctness of the responses on items $1, 2, \ldots, k$, respectively, that is,

$$\{W_{ij_1j_2\ldots j_k}\} = \begin{cases} 1 & \text{if } i \text{ responds } (j_1, j_2, \ldots, j_k), \\ 0 & \text{otherwise.} \end{cases} \tag{3}$$

The simple Rasch [18] model (see also Andersen [2, 3]) for the $\{Y_{ij}\}$ is

$$\log[P(Y_{ij} = 1)/P(Y_{ij} = 0)] = \psi + \mu_i + \nu_j, \tag{4}$$

where

$$\sum \mu_i = \sum \nu_j = 0. \tag{5}$$

Differences of the form $\mu_i - \mu_r$ are typically described as measuring the "relative abilities" of individuals i and r, while those of the form $\nu_j - \nu_s$ are described as measuring the "relative difficulties" of items j and s, although this interpretation is one with which many statisticians take issue. Expression (4) is a *logit* model in the usual contingency table sense for a three-dimensional array, with first layer $\{y_{ij}\}$ and second layer $\{1 - y_{ij}\}$. *See* the discussion of logit models in CONTINGENCY TABLES.

By appealing to the general results for log-linear models applied to contingency table problems, we see that the sufficient statistics* for the Rasch model are $\{y_{i+}\}$ and $\{y_{+j}\}$, corresponding to individual and item parameters, respectively. The maximum likelihood equations are then found by setting these totals equal to the estimates of their expectations. *See* CONTINGENCY TABLES for a discussion of these general results.

As Andersen [2, 3] notes, the logit model of expressions (4) and (5) was developed independently by Birnbaum [4] (and presented by him in an appendix to Lord and Novick [15]) as a special case of the latent trait model

$$\log\left[P(Y_{ij} = 1)/P(Y_{ij} = 0)\right] \\ = (\psi + \mu_i + \nu_j)a_j. \qquad (6)$$

Here the added parameter, a_i, is typically interpreted as a measure of the "discriminating power" of the item. The Rasch model takes all items to have the same "discriminating power" (i.e., $a_i = 1$). *See also* LATENT STRUCTURE ANALYSIS. For a more general class of item response models, see Holland [13].

LIKELIHOOD ESTIMATION

Estimation of the parameters in the basic Rasch model of expression (4) has been the focus of attention for several authors, including Rasch himself. Many of the approaches proposed have been either heuristic, ad hoc, or both. In the end three techniques have emerged in recent years, and all three are based on the method of maximum likelihood: unconditional, conditional, and marginal likelihood approaches.

The unconditional approach solves the likelihood equations described above simultaneously for estimates of both the individual and the item parameters (we refer to these here as UML estimates). Although UML estimates can be calculated using various iterative techniques, they have problematic asymptotic properties. For example, Haberman [12] shows that the UML estimates are inconsistent as $n \to \infty$ and k remains moderate, but that they are consistent when both n and $k \to \infty$. Fischer [11] derives a set of necessary and sufficient conditions for UML estimates by representing the Rasch model in the form of an incomplete version of the Bradley–Terry model* for binary paired comparisons*, which is itself a logit model.

As we noted above, the conditional approach to likelihood estimation (CML) was advocated initially by Rasch, who noted that the conditional distribution of $\{Y_{ij}\}$ given the individual marginal totals $\{y_{i+}\}$, which are the sufficient statistics for the individual parameters, depends only on the item parameters $\{\nu_j\}$. Then each of the row sums $\{y_{i+}\}$ can take only $k+1$ distinct values corresponding to the number of correct responses. Next, we recall the alternate representation of the data in the form of an $n \times 2^k$ array, $\{W_{ij_1j_2\ldots j_k}\}$, as given by expression (3). Adding across individuals we create a 2^k contingency table, X, with entries

$$X_{j_1j_2\ldots j_k} = W_{+j_1j_2\ldots j_k} \qquad (7)$$

and realizations $x = \{x_{j_1j_2\ldots j_k}\}$.

Duncan [9] and Tjur [21] independently noted that we can estimate the item parameters for the Rasch model of expression (3) using the 2^k array, x, and the log-linear model

$$\log m_{j_1j_2\ldots j_k} = \omega + \sum_{s=1}^{k} \delta_{j_s}\nu_s + \gamma_{j_+}, \qquad (8)$$

where the subscript $j_+ = \sum_{s=1}^{k} j_s$, $\delta_{j_s} = 1$ if $j_s = 1$ and is 0 otherwise, and

$$\sum_{p=0}^{k} \gamma_p = 0. \qquad (9)$$

Using a Poisson sampling scheme and the log-linear model of expression (8), Tjur [21] showed that maximum likelihood estimation* of the 2^k contingency table of expected values, $m = \{m_{j_1j_2\ldots j_k}\}$, produces the conditional maximum likelihood estimates of $\{\nu_j\}$

for the original Rasch model. Tjur proved this equivalence by (a) assuming that the individual parameters are independent identically distributed random variables from some completely unknown distribution, π, (b) integrating the conditional distribution of Y given $\{Y_{i+} = y_{i+}\}$ over the mixing distribution π, (c) embedding this "random effects" model in an "extended random model," and (d) noting that the likelihood for the extended model is equivalent to that for expression (8) applied to x.

Although the model of expression (8) appears somewhat complex, Fienberg [10] notes that it is simply a concise representation of the k-dimensional model of quasi-symmetry* preserving one-dimensional marginal totals as first described in Bishop et al. [5, p. 305]. Thus the estimation procedures used for the quasi-symmetry model can also be used for conditional Rasch model estimates. This approach is especially useful for small or moderate values of k.

Andersen [1] showed that the CML estimates of the item parameters are consistent as $n \to \infty$ and have an asymptotic normal distribution. (By the duality of items and individuals, it follows immediately that CML estimates for the individual ability parameters are consistent as $k \to \infty$.) Thus the conditional estimation approach appears to be well suited for problems in which only one set of parameters (e.g., the item parameters), are of interest. But in Rasch's original work he suggested first estimating the item parameters, "eliminating" the individual effects by conditioning, and then estimating the individual parameters, "eliminating" the item effects. Thus his approach, while based on conditional estimation, in effect runs into the same problem as the unconditional estimation approach, with the implicit requirement that both n and k need to tend to ∞ for consistency of the estimates.

One way out of this dilemma is to treat, say, the individual parameters as coming from some underlying distribution of known form but with a small number of unknown parameters (e.g., see Andersen [3], Bock and Aitkin [6], and Sanathanan and Blumenthal [20]). This approach leads to what Bock and Aitkin [6] refer to as the *marginal maximum likelihood* (MML) method, and technically it comes close to being the same as the device used by Tjur [21] and described above. In the MML approach, however, the model is not embedded in a larger one and as a consequence there are some additional restrictions on the parameters of the model. Cressie and Holland [8] describe these as moment inequalities. When these inequalities are satisfied by the CML estimates, the CML and MML estimates coincide. In the context of the marginal likelihood function, the CMLEs are consistent but somewhat inefficient [8].

GOODNESS OF FIT*

In his original monograph, Rasch [18, Chap. 6] suggested a variety of ways of checking the fit of his model, a process he referred to as "controlling the model." In particular for the item analysis problem, he advocated controlling the model by eliminating both sets of parameters through conditioning on the row and column totals of the array y. This conditional distribution under the model is of multivariate hypergeometric form, but how it should be used to get a χ^2 test of fit is unclear.

In his examples, Rasch attempted to examine goodness of fit by grouping individuals with roughly the same scores (i.e., row totals y_{i+}), and then comparing estimated item effects in the different groups using a series of graphical displays. (He used the same approach for individual effects.) Andersen [1] subsequently developed a formal likelihood-ratio version of this approach. In particular, he derived separate CML estimates for the item parameters from each group of individuals with the same overall score, and then compared the maximum value of the likelihood for this more general model to the maximum for the conditional likelihood under the Rasch model. Then minus twice the log-likelihood ratio has an asymptotic χ^2 distribution with $(k-1)(k-2)$

degrees of freedom, if the Rasch model fits, as $n \to \infty$.

When viewed from the contingency table perspective of the preceding section, Andersen's chi-square test can be interpreted as a conditional test of the quasi-symmetry model preserving one-way marginal totals. Kelderman [14] pursues this interpretation and provides alternatives to Andersen's test.

Plackett [17], in a very brief section of the second edition of his monograph on categorical data analysis, notes that the Q-statistic of Cochran [7] can be viewed as a means of testing that the item parameters in the Rasch model are all equal and thus zero (i.e., $\nu_j = 0$ for all j). This observation is intimately related to the 2^k contingency table representation of the conditional likelihood problem, and our original data representation in the form of an $n \times k$ (individual by item) array y is exactly the same representation used by Cochran. If we carry out a conditional test for the equality of marginal proportions in the 2^k table given model (8) (i.e., conditional on the model of quasi-symmetry preserving one-dimensional marginals), we get a test that is essentially equivalent to Cochran's test. But this is also the test for $\{\nu_j = 0\}$ within model (8).

APPLICATIONS

As we noted at the outset, Rasch originally applied his model to the study of mistakes in oral reading tests, but then went on to apply it to various ability and intelligence tests using it as a model for item analysis. This item analysis model has seen widespread application in psychological measurement problems over the past 20 years. In his "Afterword" to the 1980 reprinting of Rasch's monograph, Wright [22] summarizes applications of the Rasch model that others have pursued, including: item banking, test design, and self-tailored testing.

In recent years, the Rasch model has been applied in a variety of other settings. For example, Perline and Wainer [16] develop a Rasch model for the parole of criminal offenders and use it to predict recidivism by treating it as a latent trait. Duncan [9] gives several examples of the application of the Rasch model to survey research problems, and he presents a variety of extensions of the basic model, indicating how they can be represented in a multidimensional contingency table form.

References

[1] Andersen, E. B. (1973). *Psychometrika*, **38**, 123–140.

[2] Andersen, E. B. (1980). *Discrete Statistical Models with Social Science Applications*. North-Holland, Amsterdam.

[3] Andersen, E. B. (1983). *J. Econometrics*, **22**, 215–227.

[4] Birnbaum, A. (1957). On the Estimation of Mental Ability. *Rep. No. 15*, Randolph Air Force Base, U.S. Air Force School of Aviation Medicine, Texas.

[5] Bishop, Y. M. M., Fienberg, S. E., and Holland, P. W. (1975). *Discrete Multivariate Analysis: Theory and Practice*. MIT Press, Cambridge, Mass.

[6] Bock, R. D. and Aitkin, M. (1981). *Psychometrika*, **46**, 443–459.

[7] Cochran, W. G. (1950). *Biometrika*, **37**, 256–266.

[8] Cressie, N. E. and Holland, P. W. (1983). *Psychometrika*, **48**, 129–141.

[9] Duncan, O. D. (1984). In *Survey Measurement of Subjective Phenomena, Vol.* 2, C. F. Turner and E. Martin, eds. Russell Sage Foundation, New York, pp. 367–403.

[10] Fienberg, S. E.(1981). *Bull. Int. Statist. Inst.*, **49**(Book 2), 763–791.

[11] Fischer, G. H. (1981). *Psychometrika*, **46**, 59–78.

[12] Haberman, S. J. (1977). *Ann. Statist.*, **5**, 815–841.

[13] Holland, P. W. (1981). *Psychometrika*, **46**, 79–92.

[14] Kelderman, H. (1983). *Loglinear Rasch Model Tests*. Twente University of Technology, The Netherlands.

[15] Lord, F. M. and Novick, M. R. (1968). *Statistical Theories of Mental Test Scores*. Addison-Wesley, Reading, Mass.

[16] Perline, R. and Wainer, H. (1980). In *Indicators of Crime and Criminal Justice: Quantitative Studies*, S. E. Fienberg and A. J. Reiss, Jr., eds. U.S. Department of Justice, Washington, D.C., pp. 59–62.

[17] Plackett, R. L. (1981). *The Analysis of Categorical Data*, 2nd ed. Charles Griffin, London.

[18] Rasch, G. (1960). *Probabilistic Models for Some Intelligence and Attainment Tests*. Danmarks Paedagogiske Institut, Copenhagen.

[19] Rasch, G. (1980). *Probabilistic Models for Some Intelligence and Attainment Tests* (expanded edition). University of Chicago Press, Chicago.

[20] Sanathanan, L. and Blumenthal, S. (1978). *J. Amer. Statist. Ass.*, **73**, 794–799.

[21] Tjur, T. (1982). *Scand. Statist.*, **9**, 23–30.

[22] Wright, B. (1980). Afterword to G. Rasch [ref. 19, pp. 185–199].

Acknowledgment

The preparation of this entry was supported in part by Contract N00014-80-C-0637 from the Office of Naval Research to Carnegie-Mellon University. Dean Follmann and Paul Holland provided helpful comments on an earlier draft.

(CONTINGENCY TABLES
LATENT STRUCTURE ANALYSIS)

STEPHEN E. FIENBERG

RATES, STANDARDIZED

PURPOSE OF STANDARDIZATION*

Consider J populations, $P_1, P_2, \ldots, P_J$, and an event A (1 or 0 response) that can occur in these populations. For example, let A be ischemic heart disease (IHD), and suppose that we are considering two populations P_1, white males (WM), and P_2, black males (BM). Two questions might be of public interest: (a) What are the overall incidence rates (and prevalence proportions) of IHD in these populations? (b) Are the values of these indices in WM different from those in BM?

Suppose that event A depends (strongly) on a certain (continuous or discrete) characteristic X, called a *risk* or *predicting factor* for event A. For example, blood pressure (BP) is a risk factor for IHD. If the distributions of X in P_1 and P_2 are different, then X is a *confounding factor*. It is, then, appropriate to modify questions (a) and (b) by adding to them the phrase: "after adjusting (controlling) for blood pressure."

Question (a) is concerned with *estimation* of the population effects on incidence (or prevalence) of IHD, while question (b) deals with comparisons of these effects (*testing hypotheses*).

If the overall, unadjusted, and X-adjusted rates (proportions) are different, or perhaps more appropriately, their ratios differ significantly from 1, then X can be considered as a confounding risk factor; lack of this relation indicates that X is either not a risk factor or is not a confounding risk factor.

Methods of adjusting for extraneous factor(s) are referred to as *standardization*.* They can also be used in situations when the response is not necessarily binary but can be measured on a continuous scale. Various standardization techniques are discussed in an excellent article by Kalton [7].

CRUDE AND STANDARDIZED RATES

Consider a fairly common situation in which *age* is the confounding factor (X), and the data—collected over a determined period τ—are grouped in fixed age intervals (strata). For the jth study population and the ith stratum, let N_{ij} denote midperiod population exposed to risk (so that τN_{ij} is the amount of person-years exposed to risk, A_{ij}, say); D_{ij} be the number of deaths; and $\lambda_{ij} = D_{ij}/A_{ij}$ be the estimated age specific death rate (per person, per year) over period τ. For a *model* or *standard* population (S), we denote the corresponding quantities by N_{iS}, D_{iS}, λ_{iS}, and so on. We use the customary notation, $N_{i.} = \sum_{j=1}^{J} N_{ij}$, $D_{i.} = \sum_{j=1}^{J} D_{ij}$, and so on.

Let $w_{ij} = N_{ij}/N_{.j}$, with $\sum_{i=1}^{I} w_{ij} = 1$, represent the proportionate age distribution in the jth population, and similarly, $w_{iS} = N_{iS}/N_{.S}$, for the standard population. Further, let $E_{ij} = A_{ij}\lambda_{iS}$, for $i = 1, 2, \ldots, I$, be the expected number of deaths in the ith stratum and the jth population under the null hypothesis that the standard population (S) is the correct model, and $E_{.j} = \sum_{i=1}^{I} E_{ij}$ be the total number of expected deaths under this model.

For the jth *study population* the following

indices are defined:

1. The overall or *crude* rate,
$$\lambda_j^{(C)} = \sum_{i=1}^{I} w_{ij}\lambda_{iS} = D_{.j}/A_{.j}\,; \qquad (1)$$
2. The age-adjusted or *directly* standardized rate,
$$\lambda_j^{(DS)} = \sum_{i=1}^{I} w_{iS}\lambda_{ij}\,; \qquad (2)$$
3. The *indirectly* standardized rate,
$$\lambda_j^{(IS)} = \frac{D_{.j}}{E_{.j}}\lambda_S^{(C)}, \qquad (3)$$
where $\lambda_S^{(C)} = \sum_{i=1}^{I} w_{iS}\lambda_{iS}$ is the crude rate of the standard population.
4. The ratio
$$D_{.j}/E_{.j} = (\mathrm{SMR})_j, \qquad (4)$$
(usually expressed as a percentage) is called the *standardized mortality ratio* (SMR). In fact, the SMR is used more often than the $\lambda_j^{(IS)}$ given by (3).

If the standard population is arbitrarily selected, we speak about *external* standardization; if the standard population is the mixture of study populations, we speak about *internal* standardization.

VALIDITY OF STANDARDIZATION: ESTIMATION AND TESTING HYPOTHESES

The use of standardized rates in comparative studies of mortality has been criticized on the ground that their values depend on the choice of the standard population, and some misleading results might be obtained if the standard population is chosen incorrectly. Freeman and Holford [5] discuss conditions under which standardized rates have valid meaning relative to certain *models* for age specific rates.

Estimation

It is easy to show [4, 5] that if the models for age-specific rates are *multiplicative*, of the form
$$\lambda_{ij} = \epsilon_i\delta_j\,, \qquad i = 1, \ldots, I; \quad j = 1, \ldots, J, \qquad (5)$$
and
$$\lambda_{iS} = \epsilon_i\delta_S\,, \qquad (5a)$$
where ϵ_i is the contribution of the ith stratum and δ_j is the contribution of the jth population to λ_{ij}, then
$$(\mathrm{SMR})_j = \delta_j/\delta_S = \gamma_j\,, \qquad j = 1, \ldots, J, \qquad (6)$$
that is, the standardized mortality ratio estimates the relative (with respect to the standard population) effect of the jth study population. Clearly, if the standard population is a mixture of the study populations for which (5) holds, the condition (5a), and so also the relation (6), do not hold.

Testing Hypotheses

When the A_{ij}'s are fairly large and the λ_{ij}'s are small, it is not unreasonable to assume that conditionally on the A_{ij}'s, the D_{ij}'s are independent Poisson variables with expected values $E(D_{ij}) \doteq A_{ij}\lambda_{ij}$. In particular, *multiplicative Poisson models*, with
$$E(D_{ij}) \doteq A_{ij}\epsilon_i\delta_j\,, \qquad (7)$$
have attracted the attention of several authors as reasonable and convenient models for fitting different kinds of incidence data [1, 2, 6].

It should be noted, however, that if the standardized rates are used in testing hypothesis $H_0^{(j)}: \lambda_{ij} = \lambda_{iS}$ for all i, while estimation of the population effect δ_j is not required, the conditions (5) and (5a) can be replaced by the milder conditions
$$\lambda_{ij} \leqslant \lambda_{iS} \qquad \text{for all } i, \qquad (8)$$
and internal standardization can also be used [7].

If model (5) and (5a) is valid, the direct and indirect standardized rates are identical. Although some authors have favored direct standardization [3, 5], it seems rather "odd" to interpret this technique in terms of model

Table 1. U.S. White Males: Ischemic Heart Disease Mortality[a]

		WM, 1968				WM, 1970 (Census)			WM, 1972			
(i)	Age Group	$N_{i1} \times 10^{-3}$	D_{i1}	$\lambda_{i1} \times 10^5$	$\gamma_{i1} \times 10^2$	N_{iS}	D_{iS}	$\lambda_{iS} \times 10^5$	$N_{i2} \times 10^{-3}$	D_{i2}	$\lambda_{i2} \times 10^5$	$\gamma_{i2} \times 10^2$
1	30–35	4,789	655	13.7	102.7	4,925,069	656	13.3	5,295	634	12.0	89.9
2	35–40	4,910	2,386	48.6	105.5	4,784,375	2,203	46.0	4,777	1,980	41.4	90.0
3	40–45	5,333	6,572	123.2	106.6	5,194,497	6,006	115.6	5,071	5,652	111.5	96.4
4	45–50	5,235	13,259	253.3	105.4	5,257,619	12,629	240.2	5,205	12,192	234.2	97.5
5	50–55	4,747	21,495	452.8	104.6	4,832,555	20,924	433.0	5,061	20,944	413.8	95.6
6	55–60	4,224	31,913	755.5	104.6	4,310,921	31,134	722.1	4,393	30,162	686.6	95.1
7	60–65	3,494	41,691	1193.2	106.5	3,647,243	40,874	1120.7	3,788	42,052	1110.1	99.1
8	65–70	2,761	48,244	1747.3	102.9	2,807,974	47,694	1098.6	2,971	48,494	1632.2	96.1
9	70–75	2,047	54,860	2680.0	108.6	2,107,552	52,029	2468.7	2,117	52,006	2456.6	99.5
Total		37,540	221,075			37,867,805	214,149		38,678	214,116		

[a] In this table $\tau = 1$.

Table 2. Crude and Adjusted Rates

Rates	U.S. WM, 1968	U.S. WM, 1972
Crude rate, $\lambda_j^{(C)} \times 10^5$	588.91	553.59
Directly standardized rate, $\lambda_j^{(DS)} \times 10^5$	597.43	550.26
Indirectly standardized rate, $\lambda_j^{(IS)} \times 10^5$	597.37	550.17
Standardized mortality ratio, $(\text{SMR})_j \times 10^2$	105.63	97.29

fitting. Freeman and Holford [5] also considered an *additive* model for the rate,

$$\lambda_{ij} = \alpha_i + \beta_j . \quad (9)$$

It seems, however, difficult to find a meaningful interpretation for (9), since λ_{ij} expresses also relative frequency (like probability) of response. Additive models are usually not valid for analysis of rates and proportions [7].

It should also be noted that standardized rates, used as summary indices, can be meaningfully interpreted where applied to large population data, while standardization *techniques* in fitting models and testing for goodness of fit are appropriate for use with sample data.

MULTIPLE STRATIFICATION

There is usually more than one risk factor for which a researcher would like to control. There are, however, some difficulties with multiple stratification. First, the number of strata increases in a multiplicative fashion; if the event is fairly rare, some strata may have zero frequencies. Second, the risk factors are often associated or may interact. Care should be taken in interpretation when adjusting for more than one factor; standardization with respect to each factor separately may exhibit remarkable differences between crude and adjusted rates, while when used jointly, the effects of the factors may cancel each other so that the results are similar to those obtained with no stratification; that is, after double adjustment, there may be no or only small differences between crude and adjusted rates.

Example. Mortality from ischemic heart disease (IHD) in U.S. white male population aged 30–76 in 1968 and in 1972 was compared using standardized rates; the corresponding 1970 (census) population was used as the standard population. The data given in Table 1 are extracted from a larger set (1968–1977) in which a (declining) trend in IHD mortality was investigated; Table 2 gives both the crude and adjusted rates. The ratios $\gamma_{ij} = \lambda_{ij}/\lambda_{iS}$, although not quite constant, exhibit fairly small variation from stratum to stratum; it seems, then, not unreasonable to assume that the SMRs express the (declining) secular effect on the mortality from IHD in white males.

References

[1] Anderson, E. B. (1977). *Scand. J. Statist.*, **4**, 153–158.

[2] Breslow, N. and Day, N. E. (1975). *J. Chronic Dis.*, **28**, 289–303.

[3] Doll, R. and Peto, R. (1981). *J. Natl. Cancer Inst.*, **66**, 1193–1308.

[4] Elandt-Johnson, R. C. (1982). *Inst. Statist. Mimeo Series, No. 1414*, Dept. of Biostatistics, University of North Carolina, Chapel Hill, N.C.

[5] Freeman, D. H. and Holford, T. R. (1980). *Biometrics*, **36**, 195–205.

[6] Gail, M. (1978). *J. R. Statist. Soc. A*, **141**, 224–234.

[7] Kalton, G. (1968). *Appl. Statist.*, **17**, 118–136.

(DEMOGRAPHY
LIFE TABLES
MORBIDITY
STANDARDIZATION
VITAL STATISTICS)

Regina C. Elandt-Johnson

RATIO CONTROL METHOD *See* EDITING STATISTICAL DATA

RATIO CORRELATION

A ratio refers to a composite of two numbers where one number is divided by another and takes the form a/b. Ratios are often expressed in decimal or fractional form and variables computed as ratios are popular in most all disciplines. Dividing one number by another is often done to control or proportion for an extraneous factor (e.g., population size, land mass, body size, total biomass) to obtain per capita, per acre, length to size, and so on, information. For example, demographers will divide number of births by total population to control for population size. A popular measure of economic development is GNP per capita; the gross national product is divided by population size. In criminology, the imprisonment rate is calculated by dividing the number of admissions by the number of crimes. In biology, body circumference or animal tail length may be controlled for by dividing by body weight. In medicine or pharmacology, drug effects may be calculated as percent of control achieved, or the drug results may be divided by nondrug results.

When used for descriptive purposes ratio variables are easy to understand and contain no real fallacies. However, when one or more ratio variables are used in bivariate or multivariate statistical procedures, a controversy is generated that began as far back as the early 1900s with work by Pearson [24, 25] and Yule [34] (see also Maynard [22] and Brown et al. [6]) and continues today to cross the boundaries of a variety of disciplines (see Albrecht [1], Atchley and Anderson [3], Bollen and Ward [5], Chayes [8, 9], Dodson [10], Long [18, 19], Kasarda and Nolan [13], Pendleton et al. [27, 28], Uslaner [32], and Vanderbok [33]). The controversy surrounds the potential for conceptual and statistical "spuriousness" arising from the use of ratio variables that may have similar components (e.g., same or highly correlated numerators, same or highly correlated denominators). "Spuriousness" refers to the result resting not on the relationships between the variables, but on their sharing of equal or highly correlated components. The most common situations appear to be those in which ratio variables with common or highly correlated denominators are used. See also ref. 12 and SPURIOUS CORRELATION.

STATISTICAL CHARACTER OF RATIO CORRELATION

The argument for spuriousness existing when ratio variables are used in parametric statistical analyses rests on equations developed by Pearson [24] to approximate the correlations between ratios that have highly correlated, or identical, denominators. For the situation with highly correlated denominators, let $a =$ the ratio b/c, $d =$ the ratio e/f, and $r_{cf} \neq 1.00$ but is high. What constitutes a "high" correlation has not been empirically investigated, but 0.60 has been implied by Atchley and Anderson [3] and in the literature on multicollinearity*.

The correlation between ratios a and d, where denominators are highly correlated but not identical, is

$$r_{ad} = \frac{w_{be} - w_{ce} - w_{bf} + w_{cf}}{(v_{bc}v_{ef})^{1/2}}, \tag{1}$$

where

$$w_{be} = r_{be}V_bV_e,$$

$$v_{bc} = V_b^2 + V_c^2 - 2r_{bc}V_bV_c,$$

where r_{be}, r_{ce}, r_{bf}, r_{cf}, r_{bc}, and r_{ef} are zero-order correlation* coefficients (i.e., Pearsonian correlation coefficients) measuring the degree of linear association between two variables and takes the well-known form

$$r_{ij} = \frac{\sum(i-\bar{i})(j-\bar{j})}{\left[\sum(i-\bar{i})^2\right]^{1/2}\left[\sum(j-\bar{j})^2\right]^{1/2}}, \tag{2}$$

and V_b, V_e, V_c, V_f, and V_b are coefficients of variation (the ratio of the standard deviation to the mean) for each component variable. If the intercorrelations are set to zero as $r_{be} = r_{ce} = r_{bf} = r_{bc} = r_{ef} = 0$, the "spurious" correlation can then be approximated with

$$r_{ad} = \frac{r_{cf} V_c V_f}{\left(V_b^2 + V_c^2\right)^{1/2} \left(V_e^2 + V_f^2\right)^{1/2}} \qquad (3)$$

for the situation when denominators are highly correlated but not equal.

When the denominators are equal, the correlation between the two ratio variables is defined in the following way. Let $a = b/c$, $g = e/c$, and $r_{cc} = 1.00$, so that

$$r_{ag} = \frac{w_{be} - w_{bc} - w_{ec} + V_c^2}{\left(v_{bc} v_{ec}\right)^{1/2}}. \qquad (4)$$

The degree to which r_{ag} is defined by the two ratios' mutual dependency on the common denominator c can then be defined by setting $r_{be} = r_{ec} = r_{bc} = 0$ and calculating

$$r_{ag} = \frac{V_c^2}{\left(V_b^2 + V_c^2\right)^{1/2} \left(V_e^2 + V_c^2\right)^{1/2}}. \qquad (5)$$

For the special situation where a ratio is correlated with its own denominator, the correlation is approximated by [3, 7]

$$r_{ac} \doteqdot \frac{r_{bc} V_b - V_c}{\left(V_b^2 + V_c^2 - r_{bc} V_b V_c\right)^{1/2}}, \qquad (6)$$

where $a = b/c$. When r_{bc} is set equal to 0, spuriousness is approximated by

$$r_{ac} \doteqdot \frac{V_c}{\left(V_a^2 + V_c^2\right)^{1/2}}. \qquad (7)$$

Of necessary interest are the distributions of variables and ratios used in (1) and (3) to (6). The effect of these distributions on the equations, although important, is a topic for which there is a dearth of research. The few notable exceptions include papers by Atchley et al. [4], Kuh and Meyer [14], and Albrecht [1]. Parametric statistics assume normally distributed data. When ratios are formed from metric data, they generally display distributions that are skewed and leptokurtic* [4]. When the denominator's coefficient of variation increases in size, the distribution of the ratio variable becomes even more skewed and leptokurtic.

It must also be assumed that the numerator is a linear homogeneous function of the denominator [14]. Ratios introduce a degree of nonlinearity, questioning the accuracy of parametric statistics used. To what degree the interaction between nonlinear, nonnormal data, and the size of the coefficients of variation affect statistical spuriousness remains to be explored. Implications are that parametric statistics in general, and the approximation equations (1) and (3) to (6) in particular, are accurate for most situations (see especially Atchley et al. [4] and Atchley and Anderson [3]).

Such is the logic of the statistical argument for determining the degree of spuriousness when two (or more) ratio variables are correlated. Recognizing that multivariate techniques (e.g., regression analyses, factor analyses, discriminant analyses) rely, of course, on the structure of the correlation matrix, the concern of many that such spuriousness can lead to serious data misinterpretations appears justified.

CONCEPTUAL CHARACTER OF RATIO CORRELATION

Beginning as far back, again, as Yule [34] is the argument that using ratio variables with the same or highly correlated components will not, in some situations, result in spuriousness. The argument is one that is best defined by the content of the theoretical scheme the researcher is working with and the sophistication of available measurement techniques, and has been discussed by Kasarda and Nolan [13], Long [18, 19], Macmillan and Daft [20], Logan [17], and Schuessler [30]. It deals essentially with the question of whether the theory itself requires measurement to be of ratio form. If the theory suggests a relationship between two concepts a/c and b/c, and that is the form in which the variables are measured and analyzed. then no spuriousness exists; what is found in the analysis is, in fact, what the theory requires. If, however, the theory suggests a relationship between a and b and the researcher correlates a/c and b/c and generalizes to the theoretical relationship, there may be spuriousness because of the mutual dependency of a and b on c, and c is not part of the theoretical framework. In other words, if the epistemic correlation between the empirical measure and the theoretical concept approaches 1.00 (e.g., concept a is measured by a, and concept b is measured by b) spuriousness probably is not a problem. The lower the epistemic correlation becomes (e.g., concept a is measured by a/c, and concept b is measured by b/c), the greater the likelihood for spuriousness.

It would follow that correlated ratios produce only minor problems when one is testing clearly defined hypotheses (model testing) and comparing R^2's for models to see if they differ significantly. However, when one is model building and using, for example, regression weights to make predictions, the correlated ratios are of paramount importance and correlation factors should be pursued [26, 28].

CORRECTION FACTORS

Little has been done to define successfully effective correction techniques for ratio correlation spuriousness. Among the techniques studied are part and partial correlation by Kuh and Meyer [14], Logan [15–17], Przeworski and Cortes [29], Madansky [21], and Schuessler [30]. Logarithmic transformations have been presented by O'Connor [23], Schuessler [30, 31], Vanderbok [33], and Anderson and Lydic [2], and logit, arcsine, probit, and residual data have been investigated by Pendleton [26, 28] and Logan [16]. Chayes [9] has published a computer program for generating a set of random variables with means and standard deviations identical to the raw data variables but with zero magnitude correlations. When the random variables are converted to ratio form and intercorrelated, they can be used as a measure against the real ratio variable data and a degree of spuriousness can be obtained (see also Bollen and Ward [5]). Anderson and Lydic [2] have written a computer program for analyses of variance and covariance when ratio variables exist.

The correction techniques that appear to hold the greatest promise are the variety of residualizing techniques discussed and formulated by Freeman and Kronenfeld [11], Atchley et al. [4], Logan [17], Pendelton et al. [26–28], Schuessler [31], and Vanderbok [33], and by principal components analysis* [4].

References

[1] Albrecht, G. H. (1978). *Syst. Zool.*, **27**, 67–71. (Comments on Atchley et al. [4].)

[2] Anderson, D. and Lydic, R. (1977). *Biobehav. Rev.*, **1**, 55–57. (Note the availability of a computer program that applies ANOVA and ANCOVA tests when ratio variables exist.)

[3] Atchley, W. R. and Anderson, D. (1978). *Syst. Zool.*, **27**, 71–78. (Response to comments by Dodson [10], Albrecht [1], and others.)

[4] Atchley, W. R., Gaskins, C. T., and Anderson, D. (1976). *Syst. Zool.*, **25**, 137–148. (An excellent examination of the statistical properties of ratio variable correlations.)

[5] Bollen, K. and Ward, S. (1979). *Sociol. Meth. Res.*, **7**, 431–450. (Good general discussion.)

[6] Brown, J. W., Greenwood, M., and Wood, F. (1914). *J. R. Statist. Soc.*, **77**, 317–346. (Historical interest.)

[7] Chayes, F. (1949). *J. Geol.*, **57**, 239–254.

[8] Chayes, F. (1971). *Ratio Correlation*. University of Chicago Press, London. (The only book dealing with ratio variable correlation; an excellent and comprehensive treatise.)

[9] Chayes, F. (1975). In *Concepts in Geostatistics*, R. B. McCammer, ed. Springer-Verlag, New York, pp. 106–136. (Describes a computer program to use in investigating the nature of ratio variable correlations.)

[10] Dodson, P. (1978). *Syst. Zool.*, **27**, 62–67. (Comments on Atchley et al. [4].)

[11] Freeman, J. and Kronenfeld, J. E. (1973). *Social Forces*, **52**, 108–121.

[12] Fuguitt, G. and Lieberson, S. (1974). In *Sociological Methodology, 1973–1974*, H. Costner, ed. Jossey–Bass, San Francisco, pp. 128–144. (Excellent review of the technical components and varieties of ratio variables.)

[13] Kasarda, J. D. and Nolan, P. D. P. (1979). *Social Forces*, **58**, 212–227. (Excellent discussion detailing arguments and examples for the conceptual character of ratio variables.)

[14] Kuh, E. and Meyer, J. R. (1955). *Econometrica*, **23**, 416–440. (Technical reading.)

[15] Logan, C. H. (1971). *Social Prob.*, **19**, 280–284. (The three papers by Logan deal with organizational examples of ratio variable correlations.)

[16] Logan, C. H. (1972). *Social Forces*, **51**, 64–73.

[17] Logan, C. H. (1982). *Social Forces*, **60**, 791–810.

[18] Long, S. B. (1979). *Deterrence Findings: Examining the Impact of Errors in Measurement*. Bureau of Social Science Research, Inc., Washington, D.C. (The two papers by Long discuss criminology examples of the conceptual character of ratio variable correlations.)

[19] Long, S. B. (1979). In *Sociological Methodology, 1980*, K. Schuessler, ed. Jossey–Bass, San Francisco, pp. 37–74.

[20] Macmillan, A. and Daft, R. L. (1979). *Social Forces*, **58**, 228–248.

[21] Madansky, A. (1964). *Econometrica*, **32**, 652–655.

[22] Maynard, G. D. (1910). *Biometrika*, **7**, 276–304. (Historical interest.)

[23] O'Connor, J. F. (1977). *Sociol. Meth. Res.*, **6**, 91–102.

[24] Pearson, K. (1897). *Proc. R. Soc. Lond.*, **60**, 489–498. (The first paper to delineate clearly the statistics of ratio variable correlation; of historical interest.)

[25] Pearson, K. (1910). *J. R. Statist. Soc. A*, **73**, 534–539. (Historical interest.)

[26] Pendleton, B. F. (1984). *Social Sci. Res.*, **13**, 268–286. (A paper detailing a substantive example from demography and supporting a residual approach for correcting statistical spuriousness.)

[27] Pendleton, B. F., Warren, R. D., and Chang, H. C. (1979). *Sociol. Meth. Res.*, **7**, 451–475. (A paper discussing correlated denominators in ratio variables in cross-sectional and longitudinal analyses.)

[28] Pendleton, B. F., Newman, I., and Marshall, R. S. (1983). *J. Statist. Comp. Simul.*, **18**, 93–124. (This paper tests probit, logit, arcsine, and residual approaches to correcting for statistical spuriousness; very extensive references.)

[29] Przeworski, A. and Cortes, F. (1977). *Polit. Methodol.*, **4**, 63–75.

[30] Schuessler, K. (1973). In *Structural Equation Models in the Social Sciences*, A. S. Goldberger and O. D. Duncan, eds. Seminar Press, New York, pp. 201–228.

[31] Schuessler, K. (1974). *Amer. J. Sociol.*, **80**, 379–396.

[32] Uslaner, E. (1977). *Amer. J. Polit. Sci.*, **21**, 183–201.

[33] Vanderbok, W. G. (1977). *Polit. Methodol.*, **4**, 171–184.

[34] Yule, G. U. (1910). *J. R. Statist. Soc.*, **73**, 644–647. (The first paper to describe the conceptual character of ratio variables and situations where spuriousness does not exist; of historical interest.)

(CORRELATION
SPURIOUS CORRELATION)

B. F. PENDLETON

RATIO ESTIMATORS

Ratio estimation dates back to Laplace*. According to Cochran [5], the population of France as of September 22, 1802, was estimated by Laplace by means of a ratio estimator. He used the known total number of registered births during the preceding year in the entire country, X, and in a sample of communes, x, for which the total population count, y, was also obtained to arrive at the ratio estimator $\hat{Y}_r = (y/x)X$ of the population of France, Y.

The ratio estimator of a population total, Y, is extensively used in practice, especially with large-scale surveys containing many items, because of its computational simplicity, applicability to general sampling designs, and increased efficiency through the utilization of concomitant information. Frequently, the parameter of interest is a ratio of totals, $R = Y/X$, rather than the total Y, for example, a population mean or subpopulation (domain) mean or proportion if the total number of units in the population, N, or in the domain, N_1, is unknown. Standard textbooks on sampling provide a detailed account of ratio estimators and their applications; in particular Cochran's [4] book covers the developments up to 1976.

COMBINED RATIO ESTIMATOR

The (combined) ratio estimator of $Y(= y_1 + \cdots + y_N)$ is given by

$$\hat{Y}_r = \frac{\hat{Y}}{\hat{X}}X,$$

where $\hat{Y}$ and $\hat{X}$ are unbiased estimators of

totals Y and X with respect to the sampling design, and $X(= x_1 + \cdots + x_N)$ is known. In the case of simple random sampling (SRS) or any self-weighting design, $\hat{Y} \propto y$ and $\hat{X} \propto x$ and hence $\hat{Y}_r$ reduces to Laplace's estimator, where y and x are the sample totals. The corresponding estimator of R is given by

$$\hat{R} = \hat{Y}/\hat{X}.$$

By taking $x_i = 1$ for all $i = 1, \ldots, N$, $\hat{R}$ reduces to the estimator of the population mean $\overline{Y} = Y/N$ when N is unknown. Similarly, the estimator of a domain mean $\overline{Y}_1 = Y_1/N_1$ is obtained from $\hat{R}$ by using y_i' and x_i' in place of y_i and x_i, respectively, where Y_1 $(= y_1' + \cdots + y_N')$ is the domain total, $N_1 = x_1' + \cdots + x_N'$ and $y_i' = y_i$, $x_i' = 1$ if the ith unit is in the domain; $y_i' = 0$, $x_i' = 0$, otherwise.

An exact upper bound on the absolute relative bias (|bias|/standard error) of $\hat{Y}_r$ or $\hat{R}$ is given by

$$\frac{|B(\hat{Y}_r)|}{\sigma(\hat{Y}_r)} = \frac{|B(\hat{R})}{\sigma(\hat{R})} \leqslant C(\hat{X}),$$

where $C(\hat{X})$ is the coefficient of variation of $\hat{X}$ [7]. Hence the ratio bias is negligible if $C(\hat{X})$ is small. In the case of SRS, $C(\hat{X})$ is of the order $n^{-1/2}$, where n is the sample size. Hence the ratio bias is negligible for large n in this case. Empirical evidence indicates that the ratio bias is likely to be negligible in large-scale surveys, except possibly in the case of a small domain [10].

The approximate mean square error* (MSE) of $\hat{Y}_r$ for large n (which is also the approximate variance) is given by $V(\hat{Y} - R\hat{X}) = V(\hat{U})$, where $\hat{U}$ is the unbiased estimator of the total U of the residuals $u_i = y_i - Rx_i$, $i = 1, \ldots, N$, and V denotes the variance operator. In the case of SRS, empirical evidence [24] indicates that $V(\hat{U})$ can seriously underestimate the exact mean square error of $\hat{Y}_r$ for small sample sizes $(n \leqslant 12)$, but recent empirical work [39] with larger n $(n = 32)$ has shown no such systematic pattern of underestimation.

For large n, the ratio estimator $\hat{Y}_r$ has a smaller MSE than the unbiased estimator $\hat{Y}$ if $\rho(\hat{Y}, \hat{X}) > \frac{1}{2}[C(\hat{X})/C(\hat{Y})]$, where $\rho(\hat{Y}, \hat{X})$ is the correlation* coefficient between $\hat{Y}$ and $\hat{X}$. However, $\hat{Y}_r$ is less efficient, in larger samples, than the (combined) regression estimator

$$\hat{Y}_{\text{reg}} = \hat{Y} + \hat{B}(X - \hat{X}),$$

where $\hat{B}$ is an estimator of the regression* coefficient, B, between $\hat{Y}$ and $\hat{X}$: $B = \operatorname{cov}(\hat{Y}, \hat{X})/V(\hat{X})$. The regression estimator $\hat{Y}_{\text{reg}}$ is computationally cumbersome for general sampling designs due to difficulty in estimating B. Moreover, limited empirical studies [23] for SRS indicate that $\hat{Y}_r$ compares favorably in efficiency to $\hat{Y}_{\text{reg}}$ for small n $(\leqslant 12)$.

Several alternatives to $\hat{Y}_r$ have been proposed in the literature. In particular, when $\rho(\hat{Y}, \hat{X})$ is suspected to be not large and/or $K = C(\hat{X})/C(\hat{Y}) > 1$, the estimator

$$\hat{Y}_{r1} = (1 - W)\hat{Y} + W\hat{Y}_r, \qquad 0 \leqslant W \leqslant 1,$$

may be used in place of $\hat{Y}_r$ or $\hat{Y}$. For SRS, Chakrabarty [3] found that $W = \frac{1}{4}$ is a good overall choice for low correlation $[0.2 < \rho(\hat{Y}, \hat{X}) < 0.4]$ and/or $K > 1$, and $W = \frac{1}{2}$ is a good choice for moderate-to-high correlation $(\geqslant 0.4)$ and $K > 1$, whereas $\hat{Y}_r$ is preferable when $\rho(\hat{Y}, \hat{X}) > 0.8$ and $K \leqslant 1$.

A general class of estimators of Y is given by

$$\hat{Y}_t = t(\hat{Y}, \hat{X})$$

such that $t(Y, X) = Y$ for all Y, where $t(\hat{Y}, \hat{X})$ is a continuous function of $\hat{Y}$ and $\hat{X}$ [36]. This class includes $\hat{Y}_r$ and $\hat{Y}_{r1}$ and several other ratio estimators proposed in the literature.

Turning to the estimation of MSE($\hat{Y}_r$), a class of consistent estimators is given by

$$v_g = v_g(\hat{Y}_r) = \left(\frac{X}{\hat{X}}\right)^g v(\tilde{U}),$$

where $v(\tilde{U})$ is obtained from the formula for the estimator of variance of $\hat{Y}$ by replacing y_i by the residual $\tilde{u}_i = y_i - \hat{R}x_i$, and $g \geqslant 0$ generates the class (Wu [38]). The choices $g = 0$ and 2 lead to classical estimators

$v_0(\hat{Y}_r)$ and $v_2(\hat{Y}_r)$. Using a model-based approach, Royall and Eberhardt [34] arrived at an estimator that is approximately equal to $v_2(\hat{Y}_r)$ under SRS, for large n and $N \gg n$. The choice $g = 1$ leads to a ratio-type estimator $(X/\hat{X})v(\tilde{U})$ of MSE$(\hat{Y}_r)$. The estimator of MSE$(\hat{R})$, when X is unknown, can be obtained from $v_2(\hat{Y}_r)$:

$$v(\hat{R}) = \frac{1}{X^2} v_2(\hat{Y}_r) = \frac{1}{\hat{X}^2} v(\tilde{U}).$$

Suppose that the population is divided into L ($\geqslant 1$) strata and $\hat{Y} = \hat{Y}_1 + \cdots + \hat{Y}_L$, where $\hat{Y}_h$ is an unbiased estimator of the hth stratum total Y_h ($h = 1, \ldots, L$). Also, suppose that $\hat{Y}_h$ can be expressed as the mean of m_h independent and identically distributed random variables $\tilde{y}_{hi}$ with $E(\tilde{y}_{hi}) = Y_h$, as in the case of stratified multistage sampling* in which the first-stage units are sampled with replacement. Then a jackknife* estimator of the variance of $\hat{Y}_r$ is given by

$$v_J = v_J(\hat{Y}_r)$$
$$= X^2 \sum_{h=1}^{L} \frac{m_h - 1}{m_h} \sum_{i=1}^{m_h} (\hat{R}^{hi} - \hat{R})^2,$$

where $\hat{R}^{hi} = \hat{Y}^{hi}/\hat{X}^{hi}$ and $\hat{Y}^{hi}$ is the estimator of Y computed from the sample after omitting $\tilde{y}_{hi}$, that is,

$$\hat{Y}^{hi} = \hat{Y} - (\tilde{y}_{hi} - \hat{Y}_h)/(m_h - 1);$$

$\hat{X}^{hi}$ is defined similarly. Alternative versions to $v_J(\hat{Y}_r)$ have also been proposed (see Krewski and Rao [12]).

Another sample reuse* method for the case $m_h = 2$ for all h and $L \geqslant 2$, called *balanced repeated replication* (BRR), leads to a variance estimator based on a number of half-samples* formed by deleting one of $(\tilde{y}_{h1}, \tilde{x}_{h1})$ and $(\tilde{y}_{h2}, \tilde{x}_{h2})$ from the sample in each stratum. The set of S half-samples satisfies the property $\sum_{j=1}^{S} \delta_{jh}\delta_{jh'} = 0$ $(h \neq h')$, where $\delta_{jh} = \pm 1$ according as whether $(\tilde{y}_{h1}, \tilde{x}_{h1})$ or $(\tilde{y}_{h2}, \tilde{x}_{h2})$ in the hth stratum is in the jth half-sample [14]. Let $(\hat{Y}^{(j)}, \hat{X}^{(j)})$ denote the estimator of (Y, X) based on the jth half-sample and let $\hat{R}^{(j)} = \hat{Y}^{(j)}/\hat{X}^{(j)}$; then a BRR variance estimator is given by

$$v_B = v_B(\hat{Y}_r) = X^2 \left[\frac{1}{S} \sum_{j=1}^{S} (\hat{R}^{(j)} - \hat{R})^2 \right].$$

Again alternative versions to $v_B(\hat{Y}_r)$ are available (see Krewski and Rao [12]).

The confidence intervals for Y are obtained by treating $T = (\hat{Y}_r - Y)/\sqrt{v(\hat{Y}_r)}$ as approximately $N(0, 1)$ or as a t-statistic with $\sum m_h - L$ degrees of freedom, where $v(\hat{Y}_r)$ can be chosen as v_g, v_J, or v_B. Scott and Wu [35] and Krewski and Rao [12] established the asymptotic normality* of T for v_2 under SRS and v_2, v_J, and v_B under stratified sampling (as $L \to \infty$), respectively. Alternative confidence limits for Y can also be obtained by treating

$$Z = \frac{\hat{Y} - R\hat{X}}{\left[v(\hat{Y} - R\hat{X})\right]^{1/2}}$$

as approximately $N(0, 1)$ and solving the resulting quadratic equation for R. This method, well known as *Fieller's method*, takes some account of the skewness of the distribution of $\hat{Y}_r$, unlike the intervals based on T. *See also* FIELLER'S THEOREM.

Empirical study by Kish and Frankel [9] with $L = 6$, 12, and 30 strata indicated that in terms of attained coverage probability for a specified confidence coefficient $1 - \alpha$, v_B performs better than v_J, which in turn performs better than v_2, although the differences are small. Their performance, however, is in the reverse order in terms of the MSE of the variance estimators. Rao and Rao [30] and Wu and Deng [39], among others, studied the preformance of v_0, v_2, and v_J under simple random sampling, employing both analytical and Monte Carlo methods. Their results indicate that v_J is better than v_2, which in turn is better than v_0 in providing reliable t-intervals. On the other hand, v_J is the least stable. Turning to the bias of the variance estimators, both v_0 and v_2 tend to underestimate MSE$(\hat{Y}_r)$, whereas v_J is almost always upward biased. By conditioning on the ancillary statistic $\bar{x} = x/n$, and employing natural populations, Royall and Cumberland [33] and Wu and Deng [39] show that v_J and v_2 perform better than v_0 in tracking the conditional mean square errors $\text{MSE}(\hat{Y}_r \mid \bar{x})$. The performance of v_1 is in between those of v_2 and v_0. P. S. R. S. Rao [29] used v_0 and v_2 to estimate the conditional variance of $\hat{Y}_r$ under a model, and then compared their average biases and average mean square errors.

APPROXIMATELY UNBIASED OR EXACTLY UNBIASED RATIO ESTIMATORS

If the strata totals, X_h, of the x-variables are known, a separate ratio estimator

$$\hat{Y}_{rs} = \sum_{h=1}^{L} \frac{\hat{Y}_h}{\hat{X}_h} X_h = \sum_{h=1}^{L} \hat{Y}_{rh}$$

is sometimes used, since it is more efficient than $\hat{Y}_r$ when the sample sizes in all strata are large and the strata ratios $R_h = Y_h/X_h$ vary considerably. However, the absolute relative bias of $\hat{Y}_{rs}$ is of the order $\sqrt{L}\,\bar{C}(\hat{X}_h)$, where $\bar{C}(\hat{X}_h)$ is the average coefficient of variation of the unbiased estimators $\hat{X}_h$ of X_h, so that the bias in $\hat{Y}_{rs}$ may not be negligible relative to its standard error when the strata sample sizes, m_h, are small and L is large. In such a situation, separate ratio-type estimators that are unbiased or approximately unbiased may be useful when the R_h's differ considerably.

For SRS, Hartley and Ross [7] proposed the following unbiased ratio estimator of Y based on the individual ratios $r_i = y_i/x_i$:

$$\hat{Y}_{\text{HR}} = \bar{r}X + \frac{n(N-1)}{(n-1)}(\bar{y} - \bar{r}\bar{x}),$$

where $\bar{y} = y/n$, $\bar{r} = r/n$, and r is the sample total of the r_i's. Robson [32] obtained the exact variance of $\hat{Y}_{\text{HR}}$ and its unbiased estimator, for finite N, by employing multivariate polykays*. Another unbiased ratio estimator [15], based on the ratios

$$r_i' = (n\bar{y} - y_i)/(n\bar{x} - x_i),\ n > 2,$$

was found to be more efficient than $\hat{Y}_{\text{HR}}$ [24, 26]. This estimator is given by

$$\hat{Y}_{\text{M}} = X\bar{r}' + n(N-n+1)(\bar{y} - \bar{r}'\bar{x}),$$

where $\bar{r}' = \sum r_i'/n$. Both $\hat{Y}_{\text{HR}}$ and $\hat{Y}_{\text{M}}$ readily extend to stratified multistage sampling in which the first-stage units are sampled with replacement [22], but not to general sampling designs, unlike $\hat{Y}_r$. *See also* MICKEY'S UNBIASED RATIO AND REGRESSION ESTIMATORS.

The ratio estimator $(y/x)X$ can also be made unbiased for the total Y by selecting the sample with probability proportional to aggregate size, x [6, 13, 16]. Rao and Vijayan [28] have given the following unbiased estimators of the variance of $\hat{Y}_r = (y/x)X$:

$$v_a(1) = -\sum_{i<j\in s}\sum \frac{a_{ij}}{\pi_{ij}} x_i x_j \left(\frac{y_i}{x_i} - \frac{y_j}{x_j}\right)^2$$

and

$$v_a(2) = \frac{X}{x}\left(\frac{N-1}{n-1} - \frac{X}{x}\right) \times \sum_{i<j\in s}\sum x_i x_j \left(\frac{y_i}{x_i} - \frac{y_j}{x_j}\right)^2,$$

where s denotes the sample of n units,

$$a_{ij} = \frac{X}{\binom{N-1}{n-1}} \sum_{s\supset i,j} \frac{1}{x_s} - 1, \qquad x_s = \sum_{i\in s} x_i$$

and

$$\pi_{ij} = \frac{(n-1)(N-n)}{(N-1)(N-2)} \frac{x_i + x_j}{X} + \frac{(n-1)(n-2)}{(N-1)(N-2)}.$$

The estimator $v_a(2)$ is computationally simpler than $v_a(1)$, although the latter is slightly more efficient.

The jackknife* method was first proposed by Quenouille [20] as a method of reducing bias of estimators of order n^{-1} to order n^{-2}. In the case of ratio estimators this method leads to the following approximately unbiased estimator of Y in SRS:

$$\hat{Y}_{\text{Q}} = n\{1 - (n-1)/N\}\hat{Y}_r - (n-1)(1 - 1/N)\bar{r}'X.$$

The asymptotic bias of $\hat{Y}_{\text{Q}}$ does not involve terms of order n^{-1} and order N^{-1}. Another approximately unbiased estimator, due to

Beale [1], is given by

$$\hat{Y}_{\rm B} = \hat{Y}_r \frac{1 + \left(\frac{1}{n} - \frac{1}{N}\right)\frac{s_{xy}}{\bar{x}\bar{y}}}{1 + \left(\frac{1}{n} - \frac{1}{N}\right)\frac{s_x^2}{\bar{x}^2}},$$

where $s_{xy} = \sum(x_i - \bar{x})(y_i - \bar{y})/(n-1)$ and $s_x^2 = \sum(x_i - \bar{x})^2/(n-1)$. Beale's estimator $\hat{Y}_{\rm B}$ is always nonnegative when all sample pairs (y_i, x_i) are positive, unlike Tin's [37] estimator*,

$$\hat{Y}_{\rm T} = \hat{Y}_r \left[1 + \left(\frac{1}{n} - \frac{1}{N}\right)\left(\frac{s_{xy}}{\bar{x}\bar{y}} - \frac{s_x^2}{\bar{x}^2}\right)\right].$$

Empirical studies and analytical results under regression models indicate that $\hat{Y}_{\rm B}$ and $\hat{Y}_{\rm T}$ have approximately the same efficiency, and are slightly more efficient than $\hat{Y}_{\rm Q}$. The estimators $\hat{Y}_{\rm Q}$, $\hat{Y}_{\rm B}$, and $\hat{Y}_{\rm T}$ readily extend to stratified multistage sampling in which the first-stage units are sampled with replacement, but not to general sampling designs.

Murthy and Nanjamma [17] considered the case of a sample drawn in the form of m independent interpenetrating subsamples each of the same size and selected by the same sample design. Their approximately unbiased ratio estimator of Y is given

$$\hat{Y}_{\rm MN} = \frac{m}{m-1}\frac{\hat{Y}}{\hat{X}}X - \frac{1}{m-1}\left(\frac{1}{m}\sum\frac{\hat{Y}_i}{\hat{X}_i}\right)X,$$

where $\hat{Y}_i$ and $\hat{X}_i$ are the unbiased estimators of Y and X, respectively, from the ith subsample ($i = 1, \ldots, m$) and $\hat{Y} = \sum \hat{Y}_i/m$ and $\hat{X} = \sum \hat{X}_i/m$. Similarly, an exactly unbiased ratio estimator of the Hartley–Ross type is given by

$$\hat{Y}'_{\rm HR} = \frac{1}{m}\sum\frac{\hat{Y}_i}{\hat{X}_i}X + \frac{m}{m-1}\left(\hat{Y} - \frac{1}{m}\sum\frac{\hat{Y}_i}{\hat{X}_i}\hat{X}\right)$$

T. J. Rao [31] provides further discussion of $\hat{Y}_{\rm MN}$ and $\hat{Y}'_{\rm HR}$.

DOUBLE RATIO ESTIMATION

In some applications the parameter of interest is a ratio of ratios, $R = R_1/R_0$, where $R_1 = Y_1/X_1$ and $R_0 = Y_0/X_0$ denote the ratios on the current and previous occasions, respectively. A double ratio estimator of R is given by

$$\hat{R}_d = \hat{R}_1/\hat{R}_0,$$

where $\hat{R}_1 = \hat{Y}_1/\hat{X}_1$ and $\hat{R}_0 = \hat{Y}_0/\hat{X}_0$ are the estimated ratios. Kish [8] used $\hat{R}_d$ to construct indices as measures of social and economic indicators. If the population ratio R_0 is known, a double ratio estimator of R_1 is obtained as

$$\hat{R}_{1d} = \frac{\hat{R}_1}{\hat{R}_0}R_0.$$

Rao [21] used $\hat{R}_{1d}$ in forest surveys to estimate the ratio of number of trees in two diameter classes, given previous complete enumeration data. If the population total X_1 is also known, a double ratio estimator of the total Y_1 is given by

$$\hat{Y}_{1d} = \frac{\hat{R}_1}{\hat{R}_0}(R_0X_1) = \left(\hat{Y}_1\frac{X_1}{\hat{X}_1}\right)\frac{R_0}{\hat{R}_0}.$$

Keyfitz (see Yates [40]) first used $\hat{Y}_{1d}$ to estimate the total labor force, Y_1, from a sample given a current census of production, X_1, and past census of production, X_0, and labor force, Y_0. Rao and Pereira [27] have shown that the absolute relative bias of $\hat{R}_{1d}$ is likely to be larger than that of the single ratio $\hat{R}_1$.

The approximate variance of $\hat{Y}_{1d}$ is given by

$$V(\hat{Y}_{1d}) \doteq V\left(\hat{Y}_1 + \frac{Y_1}{X_0}\hat{X}_0 - R_1\hat{X}_1 - \frac{Y_1}{Y_0}\hat{Y}_0\right)$$
$$= V(\hat{U}_1),$$

where $\hat{U}_1$ is the unbiased estimator of the total U_1 of the residuals $u_{1i} = y_{1i} + (Y_1/X_0)x_{1i} - R_1x_{1i} - (Y_1/Y_0)y_{0i}$. It follows from this formula that $\hat{Y}_{1d}$ will be more efficient in large samples than $(\hat{Y}_1/\hat{X}_1)X_1$ if the approximate correlation between $\hat{R}_1$ and $\hat{R}_0$ is greater than $\frac{1}{2}[C(\hat{R}_0)/C(\hat{R}_1)]$, where $C(\hat{R}_i)$ is the approximate coefficient of variation of $\hat{R}_i$.

Variance estimators similar to $v_g(\hat{Y}_r)$, $v_J(\hat{Y}_r)$, and $v_B(\hat{Y}_r)$ can be constructed. Rao and Pereira [27] obtained confidence limits for Y_1 by the Fieller method, based on $\hat{Y}_{1d}$. They have also given approximately unbiased double ratio estimators and the Hartley–Ross and Mickey-type unbiased double ratio estimators.

RAKING RATIO ESTIMATORS

"Raking" is an iterative method of adjusting estimated cell counts in a two-way table to agree with known marginal population frequencies. The resulting estimators of cell counts and the cell totals of a sample characteristic have the form of ratio estimators and hence are called *raking ratio estimators*. These estimators are usually more efficient than those based only on the sample cell counts or those utilizing only the row marginals or the column marginals, especially when the sample cell frequencies are small. Moreover, the raking ratio estimator of the population total, Y, of a sample characteristic provides sample-population consistency when the sample characteristic is cross-classified against the characteristic with known marginal population counts.

Suppose that in a $I \times J$ table with known marginal population counts $N_{i.}$ and $N_{.j}$ ($i = 1, \ldots, I$: $j = 1, \ldots, J$), the sample cell counts from a simple random sample of size n are denoted by $n_{ij}(\sum\sum n_{ij} = n)$. The parameters of interest include the population cell counts N_{ij} (if unknown), the cell totals Y_{ij} of a sample characteristic and their marginals $Y_{i.}$ and $Y_{.j}$, and the grand total Y. The tth iteration weight, $W_{ij}^{(t)}$, in the raking procedure is defined as

$$W_{ij}^{(t)} = \begin{cases} W_{ij}^{(t-1)} \dfrac{N_{i.}}{\sum_b n_{ib} W_{ib}^{(t-1)}} & \text{if } t \text{ odd} \\ W_{ij}^{(t-1)} \dfrac{N_{.j}}{\sum_a n_{aj} W_{aj}^{(t-1)}} & \text{if } t \text{ even} \\ W^{(0)}, \text{ a constant} & \text{if } t = 0 \end{cases}$$

($a = 1, \ldots, I$; $b = 1, \ldots, J$), where $W^{(0)} = N/n$ in the case of SRS. The procedure converges if all $n_{ij} > 0$, and in practice $t = 4$ to 6 iterations might ensure convergence of $N_{i.}^{(t)}$ and $N_{.j}^{(t)}$ to the marginals $N_{i.}$ and $N_{.j}$, respectively, where $N_{i.}^{(t)} = \sum_b N_{ib}^{(t)}$, $N_{.j}^{(t)} = \sum_a N_{ia}^{(t)}$, and $N_{ij}^{(t)} = W_{ij}^{(t)} n_{ij}$ is the raking ratio estimator of N_{ij}. The corresponding raking ratio estimators of Y_{ij} and Y are given by

$$Y_{ij}^{(t)} = W_{ij}^{(t)} y_{ij}, \qquad Y^{(t)} = \sum\sum W_{ij}^{(t)} Y_{ij},$$

where y_{ij} denotes the sample total in the (i, j)th cell. For $t = 1$, $Y^{(t)}$ reduces to the post-stratified estimator $Y^{(1)} = \sum(N_{i.}/n_{i.})y_{i.}$, where $y_{i.} = \sum_j y_{ij}$.

Rao [24] and Konijn [11] have given the large-sample variance formula for $Y_{ij}^{(t)}$, $t = 1, 2$, including conditional variance formulae for given $n_{i.} = \sum_b n_{ij}$ or given $n_{.j} = \sum_a n_{aj}$. Konijn also derived the corresponding formulae for $Y_{i.}^{(t)} = \sum_b Y_{it}^{(t)}$ and $Y_{.j}^{(t)} = \sum_a Y_{aj}^{(t)}$, $t = 1, 2$. Brackstone and Rao [2] have obtained the following general recurrence formulae for the approximate variance of $Y^{(t)}$:

$$V(Y^{(t)}) \doteq V\left[\sum\sum W_{ij}^{(t-1)}(y_{ij} - R_{.j} n_{ij})\right], \qquad t \text{ even},$$

where $R_{.j} = Y_{.j}/N_{.j}$. Thus $V(Y^{(t)})$ for even t can be obtained from $V(Y^{(t-1)})$ by substituting $y_{ij} - R_{.j} n_{ij}$ for y_{ij} in the latter. For t odd, a similar formula is given by

$$V(Y^{(t)}) \doteq V\left[\sum\sum W_{ij}^{(t-1)}(y_{ij} - R_{i.} n_{ij})\right], \qquad t \text{ odd},$$

where $R_{i.} = Y_{i.}/N_{i.}$. Therefore, given the starting expression $V(Y^{(0)})$, the formula for $V(Y^{(t)})$ can be obtained iteratively for any t, where $Y^{(0)} = (N/n)y$ in the case of SRS and $y = \sum\sum y_{ij}$ is the sample total. Brackstone and Rao [2] extended these results to single-stage cluster sampling*.

A bibliography on raking ratio estimators to 1978 is provided by Oh and Scheuren [18]. They have also described a multivariate extension of raking ratio estimation.

Multivariate Ratio Estimators

In some surveys two or more concomitant variables with known population totals might be available. Then a multivariate ratio estimator of Y, given by

$$\hat{Y}_{\mathrm{mr}} = \sum_{j=1}^{p} w_j(\hat{Y}/\hat{X}_j)X_j = \sum_{i=1}^{p} w_j \hat{Y}_{rj},$$

can be used [19]. Here $\hat{X}_j$ is the unbiased estimator of the population total, X_j, of the jth concomitant variable ($j = 1, \ldots, p \geqslant 2$) and the weights w_j are determined to minimize the approximate variance of $\hat{Y}_{mr}$ subject to $\sum w_j = 1$. The optimal vector of weights $\mathbf{w}^*$ is obtained as

$$\mathbf{w}^* = \mathbf{\Sigma}^{-1}\mathbf{1}/(\mathbf{1}'\mathbf{\Sigma}^{-1}\mathbf{1}),$$

where $\mathbf{1}$ is the p-vector of 1's and $\mathbf{\Sigma}$ is the approximate covariance matrix of the individual ratio estimators $\hat{Y}_{ri}$ with (i, j)th element

$$\sigma_{ij} = \mathrm{cov}(\hat{Y} - R_i\hat{X}_i, \hat{Y} - R_j\hat{X}_j)$$

and $R_i = Y/X_i$. In practice, the optimal weights $\mathbf{w}^*$ are obtained from the estimated covariance matrix, $\hat{\mathbf{\Sigma}}$. As in the case of the regression estimator, $\hat{Y}_{mr}$ is computationally cumbersome for general sampling designs due to difficulty in estimating $\mathbf{\Sigma}$. The estimator $\hat{Y}_{mr}$ is more efficient, in large samples, than $\hat{Y}_{ri}$ for any i or the corresponding multivariate ratio estimator based on q $(< p)$ concomitant variables*.

Olkin [19] generalized the Hartley–Ross unbiased ratio estimator for simple random sampling as follows:

$$\hat{Y}_{\mathrm{HR}}(m) = \sum_{j=1}^{p} w_j \hat{Y}_{\mathrm{HR},j},$$

where $\hat{Y}_{\mathrm{HR},j}$ is the Hartley–Ross unbiased estimator based on the individual ratios $r_{tj} = y_t/x_{tj}$ $(t = 1, \ldots, n)$ corresponding to jth concomitant variable. Similarly, the other unbiased ratio estimators or approximately unbiased ratio estimators can be generalized to the multivariate case.

References

[1] Beale, E. M. L. (1962). *Ind. Org.*, **31**, 51–52.

[2] Brackstone, G. J. and Rao, J. N. K. (1979). *Sankhyā C*, **41**, 97–114.

[3] Chakrabarty, R. P. (1979). *J. Indian Soc. Agric. Statist.*, **31**, 49–62.

[4] Cochran, W. G. (1977). *Sampling Techniques*, 3rd ed. Wiley, New York, Chap. 6.

[5] Cochran, W. G. (1978). In *Contributions to Survey Sampling and Applied Statistics*, H. A. David, ed. Academic Press, New York, pp. 3–10.

[6] Hájek, J. (1949). *Statist. Obzor.*, **29**, 384–394 (in Czech).

[7] Hartley, H. O. and Ross, A. (1954). *Nature (Lond.)*, **174**, 270–271.

[8] Kish, L. (1962). *Amer. Statist. Ass., Proc. Soc. Statist. Sec.*, pp. 190–199.

[9] Kish, L. and Frankel, M. R. (1974). *J. R. Statist. Soc. B*, **36**, 1–37.

[10] Kish, L., Namboodiri, N. K., and Pillai, R. K. (1962). *J. Amer. Statist. Ass.*, **57**, 863–876.

[11] Konijn, H. S. (1981). *Metrika*, **28**, 109–121.

[12] Krewski, D. and Rao, J. N. K. (1981). *Ann. Statist.*, **9**, 1010–1019.

[13] Lahiri, D. B. (1951). *Bull. Int. Statist. Inst.*, No. 2, **33**, 133–140.

[14] McCarthy, P. J. (1969). *Rev. Int. Statist. Inst.*, **37**, 239–264.

[15] Mickey, M. R. (1959). *J. Amer. Statist. Ass.*, **54**, 594–612.

[16] Midzuno, H. (1952). *Ann. Inst. Statist. Math.*, **3**, 99–107.

[17] Murthy, M. N. and Nanjamma, N. S. (1959). *Sankhyā*, **21**, 381–392.

[18] Oh, H. L. and Scheuren, F. J. (1978). *Amer. Statist. Ass., Proc. Sect. Surv. Res. Meth.*, pp. 723–728.

[19] Olkin, I. (1958). *Biometrika*, **45**, 154–165.

[20] Quenouille, M. H. (1956). *Biometrika*, **43**, 353–360.

[21] Rao, J. N. K. (1957). *J. Indian Soc. Agric. Statist.*, **9**, 191–204.

[22] Rao, J. N. K. (1964). *J. Indian Soc. Agric. Statist.*, **16**, 175–188.

[23] Rao, J. N. K. (1968). *J. Indian Statist. Assoc.*, **6**, 160–168.

[24] Rao, J. N. K. (1969). In *New Developments in Survey Sampling*, N. L. Johnson and H. Smith, Jr., eds. Wiley, New York, pp. 213–234.

[25] Rao, J. N. K. (1974). *Tech. Rep.*, Statistics Canada.

[26] Rao, J. N. K. and Kuzik, R. A. (1974). *Sankhyā C*, **36**, 43–58.

[27] Rao, J. N. K. and Pereira, N. P. (1968). *Sankhyā A*, **30**, 83–90.

[28] Rao, J. N. K. and Vijayan, K. (1977). *J. Amer. Statist. Ass.*, **72**, 579–584.

[29] Rao, P. S. R. S. (1981). In *Current Topics in Survey Sampling*, D. Krewski, R. Platek and J. N. K. Rao, eds. Academic Press, New York, pp. 305–316.

[30] Rao, P. S. R. S. and Rao, J. N. K. (1971). *Biometrika*, **58**, 625–630.

[31] Rao, T. J. (1966). *Ann. Inst. Statist. Math.*, **18**, 117–121.

[32] Robson, D. S. (1957). *J. Amer. Statist. Ass.*, **52**, 511–522.

[33] Royall, R. M. and Cumberland, W. G. (1981). *J. Amer. Statist. Ass.*, **76**, 66–77.

[34] Royall, R. M. and Eberhardt, K. R. (1975). *Sankhyā C*, **37**, 43–52.

[35] Scott, A. J. and Wu, C. F. (1981). *J. Amer. Statist. Ass.*, **76**, 98–102.

[36] Srivastava, S. K. (1980). *Canad. J. Statist.*, **8**, 253–254.

[37] Tin, M. (1965). *J. Amer. Statist. Ass.*, **60**, 294–307.

[38] Wu, C. F. (1982). *Biometrika*, **69**, 183–189.

[39] Wu, C. F. and Deng, L. Y. (1983). In *Scientific Inference, Data Analysis and Robustness*, G. E. P. Box et al., eds. Academic Press, New York, pp. 245–277.

[40] Yates, F. (1960). *Sampling Methods for Censuses and Surveys*, 4th ed. Macmillan, New York, p. 271.

(JACKKNIFE METHODS
MICKEY'S UNBIASED RATIO AND REGRESSION ESTIMATORS
PASCUAL ESTIMATOR
QUENOUILLE ESTIMATOR
TIN ESTIMATOR)

J. N. K. RAO

RATIONAL SUBGROUP *See* ACCEPTANCE SAMPLING; CONTROL CHARTS; QUALITY CONTROL, STATISTICAL

RATIO OF BIVARIATE NORMAL VARIABLES

The distribution of the ratio $Y = X_1/X_2$ when the joint distribution of X_1 and X_2 is bivariate normal* arises in statistical theory in a number of ways, among which the following may be mentioned:

1. Estimation of intercept of a regression* line on the axis of the independent (predicting) variable (or more generally, of the value of the predicting variable giving a specified value for the regression)
2. Computation of relative closeness* of estimators (see Johnson [2])

Given that the joint distribution of X_1, X_2 has probability density function (PDF)

$$f_{X_1,X_2}(x_1, x_2) = \frac{1}{2\pi\sigma_1\sigma_2\sqrt{1-\rho^2}} \times \exp\left[-\frac{1}{2(1-\rho^2)}(u_1^2 - 2\rho u_1 u_2 + u_2^2)\right],$$

$$u_i = (x_i - \xi_i)/\sigma_i, \quad i = 1, 2$$

[so that X_j has a normal $N(\xi_j, \sigma_j^2)$ distribution ($j = 1, 2$) and the correlation* between X_1 and X_2 is ρ], we have

$$Y = \frac{X_1}{X_2} = \frac{\xi_1 + \sigma_1 U_1}{\xi_2 + \sigma_1 U_2} = \frac{\sigma_1}{\sigma_2}\frac{U_1 + \xi_1/\sigma_1}{U_2 + \xi_2/\sigma_2}$$
$$= \frac{\sigma_1}{\sigma_2}\frac{U_1 + \delta_1}{U_2 + \delta_2}, \qquad (1)$$

where U_1, U_2 has a *standardized* bivariate normal distribution with correlation ρ, and $\delta_j^{-1} = (\xi_j/\sigma_j)^{-1}$ is the coefficient of variation* of X_j ($j = 1, 2$).

Apart from the multiplying factor σ_1/σ_2, the distribution of Y depends only on δ_1 and δ_2, and the correlation coefficient, ρ.

If δ_2 is large so that $\Pr[U_2 + \delta_2 < 0]$ is negligible, the formula

$$\Pr\left[\frac{U_1 + \delta}{U_2 + \delta} < y\right] \doteq \Phi\left[(y\delta_2 - \delta_1)(y^2 - 2\rho y + 1)^{-1/2}\right] \qquad (2)$$

[where $\Phi(z) = (\sqrt{2\pi})^{-1}\int_{-\infty}^{z} e^{-(1/2)u^2}\,du$] gives good approximate values (see Fieller [1]).

Further details regarding the distribution can be found in the works listed in the Bibliography and in ref. 3.

References

[1] Fieller, E. C. (1932). *Biometrika*, **24**, 428–440.

[2] Johnson, N. L. (1950). *Biometrika*, **37**, 42–49.

[3] Shanmugalingam, S. (1982). *The Statistician*, **31**, 251–258.

Bibliography

See the following works, as well as the references just given, for more information on the topic of ratio of bivariate normal variables.

Hinkley, D. V. (1969). *Biometrika*, **55**, 635–639. [Investigates the accuracy of approximation (2). Correction (1970); *Biometrika*, **57**, 683.]

Marsaglia, G. (1965). *J. Amer. Statist. Ass.*, **60**, 193–204. (Graphs of the density function of the ratio.)

Nicholson, C. (1943). *Biometrika*, **33**, 59–72. (Convenient formulas for computation.)

RAYLEIGH DISTRIBUTION

The Rayleigh distribution was originally derived by Lord Rayleigh (J. W. Strutt) [6] in a problem in acoustics. Let $Y_1, Y_2, \dots, Y_n$ be a sample of size n from a normal distribution $N(0, \sigma^2)$. The probability density function (PDF) of $X = \sqrt{\sum_{i=1}^{n} Y_i^2}$—that is, the distance from the origin to a point $(Y_1, \dots, Y_n)$ in n-dimensional Euclidean space—is

$$f(x; n, \sigma) = \frac{2x^{n-1}\exp\left[-x^2/(2\sigma^2)\right]}{(2\sigma^2)^{n/2}\Gamma(n/2)},$$

$$x > 0, \quad \sigma > 0.$$

This is the Rayleigh distribution [5]. The distribution with PDF $f(x; n, 1)$ is the chi distribution* with n degrees of freedom [3]. Some important properties of the Rayleigh distribution are as follows:

1. The cumulative distribution function is $P(n/2, x^2/(2\sigma^2))$ where
$$P(a, x) = \frac{1}{\Gamma(a)} \int_0^x t^{a-1} e^{-t}\, dt.$$

2. The νth moment about zero is
$$2^{\nu/2}\sigma^{\nu}\Gamma\left(\frac{\nu+n}{2}\right)\Big/\Gamma(n/2).$$

3. The mode is $\sqrt{n-1}\,\sigma$.

4. For fixed n, the distribution belongs to the one-parameter exponential family of distributions*. Suppose that $X_1, \dots, X_N$ is a sample of size N from the distribution. A sufficient statistic* for σ is $\sum_{i=1}^{N} X_i^2$, and its mean and variance are
$$E\left[\sum_{i=1}^{N} X_i^2\right] = nN\sigma^2,$$
$$\text{var}\left[\sum_{i=1}^{N} X_i^2\right] = 2nN\sigma^4.$$
The maximum likelihood* estimator of σ is $\sqrt{\sum_{i=1}^{N} X_i^2/nN}$.

5. The distribution is a special case of type II Bessel function distributions. For further information, see refs. 1, 4, and 7.

The definition of a Rayleigh random vector, based on Miller [5], is as follows. Let $\mathbf{X}_n^{(k)} = (x_{1k}, x_{2k}, \dots, x_{nk})$, $1 \leqslant k \leqslant p$, be n-dimensional normal random vectors. Let $E[\mathbf{X}_n^{(k)}] = \mathbf{A}_n^{(k)}$, $1 \leqslant k \leqslant p$, be the mean vectors. Let $\mathbf{V}_p^{(j)}$, $1 \leqslant j \leqslant n$, be the p-dimensional vector composed of the jth components of $\mathbf{X}_n^{(k)}$. Let $\mathbf{V}_p^{(j)}$, $1 \leqslant j \leqslant n$, be independent with positive-definite covariance matrix $\mathbf{M}_p$ independent of j. Then we shall say that $\mathbf{Z}_{np} = \{\mathbf{X}_n^{(1)}, \mathbf{X}_n^{(2)}, \dots, \mathbf{X}_n^{(p)}\}$ is of *class* $\mathscr{N}_n[\mathbf{M}_p : \mathbf{A}_{np}]$, where $\mathbf{A}_{np}$ is the $n \times p$ matrix whose columns are $\mathbf{A}_n^{(k)}$, $1 \leqslant k \leqslant p$, vectors. If $\mathbf{Z}_{np}$ is of class $\mathscr{N}_n[\mathbf{M}_p : \mathbf{A}_{np}]$, then we say $\mathbf{R}_p = \{R_1, R_2, \dots, R_p\}$, where $R_k = |\mathbf{X}_n^{(k)}| = \sqrt{\sum_{i=1}^{n} x_{ik}^2}$, $1 \leqslant k \leqslant p$, is a *Rayleigh random vector*.

Miller [5] discusses the following: (a) the PDF of $R = |\mathbf{X}_n|$ when $\{\mathbf{X}_n\}$ belongs to class $\mathscr{N}_n[\psi_0 \mathbf{I}_n : \mathbf{A}_n]$, where ψ_0 is a positive constant and $\mathbf{I}_n$ is the $n \times n$ identity matrix, and the νth moment of R; (b) the joint PDF of $R = |\mathbf{X}_n|$, $S = |\mathbf{Y}_n|$ when $\{\mathbf{X}_n, \mathbf{Y}_n\}$ is of class $\mathscr{N}_n[\mathbf{M}_2 : \|\mathbf{A}_n : \mathbf{A}_n\|]$, and the second moment $E[R^{\mu}S^{\nu}]$ when $\mathbf{A}_n$ is zero vector; (c) the PDF of $\mathbf{R}_p$ when $\mathbf{Z}_{np} = \{\mathbf{X}_n^{(1)}, \dots, \mathbf{X}_n^{(p)}\}$ is of class $\mathscr{N}_n[\mathbf{M}_p : \mathbf{0}_{np}]$, where $\mathbf{0}_{np}$ is a $n \times p$ zero matrix; and (d) the PDFs of the products RS and the quotient S/R, where R

$= |\mathbf{X}_n|$, $S = |\mathbf{Y}_m|$, and $\mathbf{X}_n$ and $\mathbf{Y}_m$ are mutually independent and of class $\mathcal{N}_n[\psi_0\mathbf{I}_n : \mathbf{A}_n]$ and $\mathcal{N}_m[\psi_0'\mathbf{I}_m : \mathbf{B}_m]$, respectively, and the νth moments of the product RS and the quotient S/R.

The distributions with $f(x;1,\sigma)$, $f(x;2,\sigma)$ and $f(x;3,\sigma)$ are sometimes called the folded* Gaussian (the half normal distribution), the Rayleigh distribution, and the Maxwell (–Boltzmann) distribution, respectively.

Some important properties of the Rayleigh distribution with PDF $f(x;2,\theta)$ are as follows:

Probability Density Function: $f(x;2,\theta) = (x/\theta^2)\exp[-x^2/(2\theta^2)]$, $x > 0$, $\theta > 0$

Cumulative Distribution Function: $1 - \exp[-x^2/(2\theta^2)]$

νth Moment about Zero: $(\sqrt{2}\,\theta)^\nu(\nu/2)\Gamma(\nu/2)$

Mean: $\sqrt{\pi/2}\,\theta \doteq 1.253331\theta$

Variance: $[(4-\pi)/2]\theta^2 \doteq 0.429204\theta^2$

Skewness: $2(\pi-3)\sqrt{\pi}/(4-\pi)^{3/2} \doteq 0.631110$

Kurtosis: $(32-3\pi^2)/(4-\pi)^2 \doteq 3.24509$

Mode: θ

Median: $\sqrt{\ln 4}\,\theta \doteq 1.177410\theta$

pth Percentile of the Distribution:

$$x(p) = \left[-2\theta^2\ln(1-p)\right]^{1/2}, \quad 1 > p > 0$$

Hazard Function: x/θ^2 (The Rayleigh distribution has been frequently used as a model with increasing hazard function of the form ax, $a > 0$.)

The distribution with PDF $f(x;2,\theta)$ is the special case of the Weibull distribution*.

There are several generalizations of this distribution [5, 9]. For estimation of the parameters of a generalized Rayleigh distribution, see ref. 9. When a random variable X is distributed as the Rayleigh distribution with PDF $f(x;2,\theta)$, the variate X^{-1} is distributed as the inverse Rayleigh distribution with

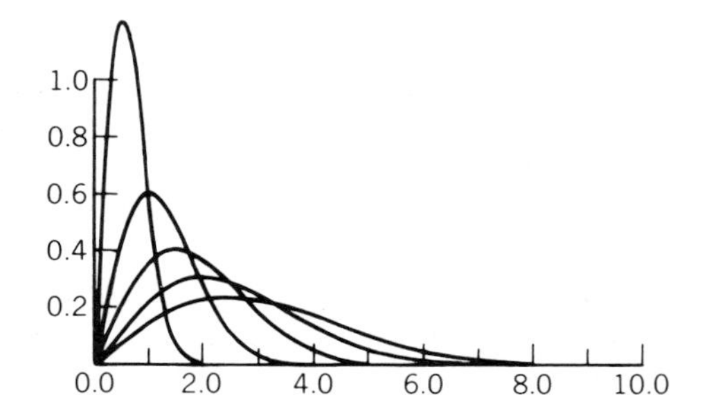

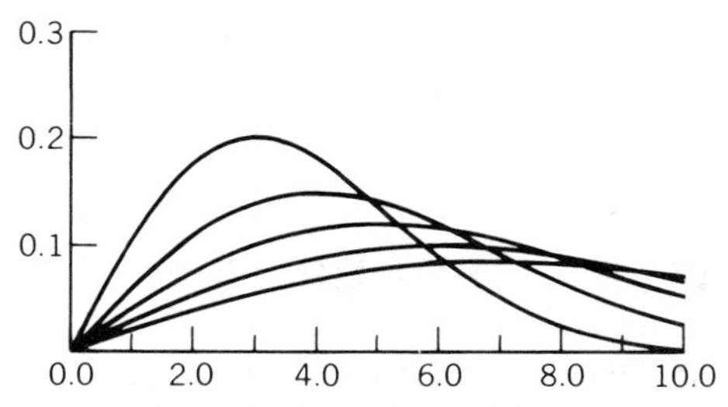

Figure 1 Above: $\theta = 0.5, 1, 1.5, 2, 2.5$; below: $\theta = 3, 4, 5, 6, 7$.

PDF

$$g(x;\theta) = \left[1/(\theta^2x^3)\right]\exp\left[-1/(2\theta^2x^2)\right], \quad x > 0, \quad \theta > 0.$$

For estimation and tests of hypotheses of the parameter in the inverse Rayleigh distribution, see refs. 2 and 8. Figure 1 shows the graphs of the PDF $f(x;2,\theta)$. In drawing the figure we used a program in ref. 1.

References

[1] Hirano, K., Kuboki, H., Aki, S., and Kuribayashi, A. (1983). *Comput. Sci. Monogr., No. 19*, Inst. Statist. Math., Tokyo, Appendix.

[2] Iliescu, D. V. and Voda, V. Gh. (1973). *Studii si Cerc. Mat. Bucuresti*, **25**, 1507–1521 (in Rumanian).

[3] Johnson, N. L. and Kotz, S. (1976). *Distributions in Statistics, Vol. 1: Continuous Univariate Distributions*. Wiley, New York, Chap. 17.

[4] Kotz, S. and Srinivasan, R. (1969). *Ann. Inst. Statist. Math., Tokyo*, **21**, 201–210.

[5] Miller, K. S. (1964). *Multidimensional Gaussian Distributions*. Wiley, New York, Chaps. 2 and 3.

[6] Rayleigh, J. W. S. (1919). *Philos. Mag., 6th Ser.*, **37**, 321–347.

[7] Springer, M. D. (1979). *The Algebra of Random Variables*. Wiley, New York, Chap. 9.

[8] Voda, V. Gh. (1972). *Rep. Statist. Appl. Res. JUSE*, **19**(4), 13–21.

[9] Voda, V. Gh. (1976). *Apl. Mat.*, **21**, 395–419.

(CHI DISTRIBUTION
CHI-SQUARE DISTRIBUTION)

KATUOMI HIRANO

***R*-CHART** *See* CONTROL CHARTS

RECIPROCAL AVERAGING *See* CORRESPONDENCE ANALYSIS

RECIPROCAL DIFFERENCES

These are formed by a process that can be regarded as a sort of inversion of the construction of divided differences.

The reciprocal difference of $f(x)$ for $x = x_0, x_1$ is

$$\rho(x_0, x_1) = \frac{x_0 - x_1}{f(x_0) - f(x_1)};$$

for x_0, x_1, x_2 (a *second-order* reciprocal difference) it is

$$\rho_2(x_0, x_1, x_2) = \frac{x_0 - x_2}{\rho(x_0, x_1) - \rho(x_1, x_2)} + f(x_1);$$

and generally, the nth-order reciprocal difference of $x_0, x_1, \ldots, x_n$ is

$$\rho_n(x_0, x_1, \ldots, x_n) = \frac{x_0 - x_n}{\rho_{n-1}(x_0, \ldots, x_{n-1}) - \rho_{n-1}(x_1, \ldots, x_n)} + \rho_{n-2}(x_1, \ldots, x_{n-1}).$$

Thiele's interpolation* formula expresses $f(x)$ as a continued fraction

$$f(x) \simeq f(x_1) + \cfrac{x - x_1}{\rho(x_1, x_2) + \cfrac{x - x_2}{\rho_2(x_1, x_2, x_3) - f(x_1) + \cfrac{x - x_3}{\rho_3(x_1, x_2, x_3, x_4) - \rho(x_1, x_2) + \cdots}}},$$

the sequence continuing until the numerator $(x - x_{n-1})$ is reached. This gives correct values for $f(x)$ at $x = x_1, \ldots, x_n$ and is an approximation in the form of a ratio of polynomials, as compared with the more conventional polynomial interpolation formulas (Bessel*, Lagrange, Newton*). Further details are available in ref. 1 (Chap. 5).

Reference

[1] Milne-Thompson, L. M. (1933). *Calculus of Finite Differences*. Macmillan, London.

(FINITE DIFFERENCES, CALCULUS OF
INTERPOLATION)

RECIPROCAL DISTRIBUTION

This is a Pareto distribution* with index 1. Its probability density function (PDF) is of the form

$$f_x(x) = x^{-1}/(\log b - \log a) \qquad (0 < a \leqslant x \leqslant b).$$

RECORD LINKAGE AND MATCHING SYSTEMS

Problems of record linkage and matching arise when it is desired to pair the records in two different files or data sets with each other. For example, if the two files contain different types of records for a sample of units from some specified population, it may be desired to match each record in the first file with that of the same or a similar unit in the second file in order to obtain a fuller or more complete record for each unit. In many of the applications of record linkage and matching, the units in each file are individual persons, and in this article it will often be convenient to refer to these units as individuals or persons.

Some of the types of records for a particular group of individuals that have been used in matching studies are federal income tax returns; employment histories maintained by the Social Security Administration; data from a sample survey* of some specific population of individuals, households, or businesses; decennial census data; medical or educational records; and birth or death certificates. For example, data on the age, ethnic background, and sex of a large group of individuals might be matched with income tax data in order to learn about the relationship between taxes and those variables. Also, income tax data for a given year might be matched with the corresponding data for a later year in order to study the changes that have occurred. Of course, record linkage and matching can be carried out for three or more files. However, throughout this article we shall restrict ourselves to problems of matching just two files.

Roughly speaking, there are two different broad categories of matching problems: (1) problems of exact matching in which it is desired to identify pairs of records in the two files that pertain to the same individual, and (2) problems of statistical matching in which it is desired to identify pairs of records in the two files, or groups of records in the two files, that pertain to similar individuals.

EXACT MATCHING

The central purpose of carrying out the exact matching of two different files is often to combine the two different records of the individuals that are present in both files into a single enlarged record. The combining of different records for an individual into one large record in this way raises important issues regarding the individual's rights to privacy and confidentiality. For this reason, matching studies based on government files must often be carried out under rigid restrictions regarding the sharing of data across different agencies.

A second common purpose for the exact matching of two files is the construction of a single master file with duplications eliminated. A related but different purpose in some problems is simply to determine the presence or absence of individuals in the two files. For example, one method of estimating the magnitude of the population undercount in the decennial U.S. census* is to determine whether persons listed in birth records, listings of drivers' licenses, or alien registrations are included in the census files.

When the records in each file contain accurate identifying information such as each individual's name and social security number, the process of exact matching can be carried out essentially without any errors. In general, however, some of the identifying information in a record may be missing, erroneous, or not unique. Matching procedures must then be developed which control both the probability of pairing two records that actually pertain to different individuals and the probability of not pairing two records that actually do pertain to the same individual. Optimal methods must take into account the relative seriousness of each of these two types of errors.

Statistical models and methods for exact matching procedures have been presented by Nathan [13], Tepping [21], DuBois [8], and Fellegi and Sunter [9]. For example, DuBois [8] regards the identifying information in each record as an 11-dimensional vector of which the components are the following variables: the individual's last name, first name, middle initial, place of birth, month of birth, day of birth, and year of birth; the first initial of his or her spouse; the first three digits of his or her social security number, the next two digits of that number, and the final four digits of the number. He develops a model for deciding whether or not two records are linked (i.e., matched), based on which of these components are present in both vectors and which are identical in both vectors. He also presents an example in which approximately 73,000 questionnaires for males in a 1957 American Legion lung cancer study are matched with more than 77,000 death certificates for males in the 1960 death index of the California State Department of Public Health. Fellegi and Sunter [9] consider similar models that make use of somewhat more information. If the last names in the two records are identical, it might be noted whether it is a common or an unusual name. If these names are not identical, it might be noted whether they sound alike or whether the first several letters in the names agree.

Once it has been decided that two records are linked, in the sense that they pertain to the same individual, there often remains the problem of reconciling the discrepancies that were present in the two records. Thus, if the individual's date of birth or reported income is not identical in the two records, it must often be decided which value, if either, it is appropriate to retain in the merged record.

STATISTICAL MATCHING

Problems of statistical matching* arise when there are expected to be relatively few or no individuals whose records are present in both files. For example, the two files might have been obtained from two distinct or independent samples drawn from some large fixed population. The two files might contain a common set of variables X that were observed for all individuals in either of the files, as well as a set of variables Y that were observed only for individuals in the first file and a set of variables Z that were observed only for individuals in the second file. One central purpose of a statistical matching study in this context might be to study relationships between Y and Z in the overall population from which the samples were drawn.

Many of the commonly used procedures have the following general form. One of the two files is regarded as the base file or primary file. For each individual A in the base file, a measure $m_A(B)$ of similarity to A is specified for each individual B in the other file. This measure is based on the similarity between the observed values of the common variables X for A and B, and possibly also on other information contained in the observed values of Y for A or of Z for B. An individual B in the second file for which $m_A(B)$ is a maximum might then be matched with A or, among all individuals B in the second file for which $m_A(B)$ exceeds some fixed level of similarity, one individual might be selected at random to be matched with A, where the probability of selection of any individual B is an increasing function of $m_A(B)$. In some problems, a synthetic "average" individual might be created from those individuals B in the second file for which $m_A(B)$ is sufficiently large, and this synthetic individual matched with A. The matching procedure could either be constrained, in the sense that no individual in the second file can be matched with more than one individual in the base file, or it can be unconstrained, in which case the records in the second file can be used repeatedly in the construction of matches.

A basic assumption that underlies almost all matching procedures either explicitly or implicitly, including procedures having the general form just described, is that Y and Z are conditionally independent given X. One way of thinking about statistical matching is that it is a problem of trying to make inferences about the joint distribution of X, Y, and Z based on information about just the joint marginal distribution of X and Y and the joint marginal distribution of X and Z. Such inferences must be based on some assumption linking these marginal distributions, such as the assumption of *conditional independence* just mentioned. Unfortunately, the two data files typically contain no information regarding the appropriateness of this assumption, and prior information from other sources must be considered.

The methodology of statistical matching can be used to reduce the *response burden* on the individuals participating in a *survey**. Suppose that the total time that would be required for a person to respond carefully to all the items in the survey is so great that many individuals in the population of interest would be expected to refuse to participate. One way to proceed in order to increase the response rate is to have different samples from the population respond to different sets of items, with a common core of items for all the samples. The different samples could then be statistically matched based on the common core of items in order to study relationships among the other items.

One purpose in carrying out a statistical matching procedure for the individuals in a base file is to fill in gaps and missing information in the records of those individuals. From this point of view, statistical matching is a problem of *missing data* and many of the same techniques and *imputation** methods that have been developed for filling in missing data in surveys are relevant here.

EXAMPLES AND DISCUSSIONS OF EXACT AND STATISTICAL MATCHING

Ruggles and Ruggles [18] discuss the shortcomings of trying to use elementary regression methods for imputation purposes, and the relative advantages of using a statistical matching procedure of the general form just described. They present in detail the matching of the Bureau of the Census* 1970 Public Use Sample (the PUS) with the Social Security Administration Longitudinal Employer-Employee Data File (the LEED). Each of these data files contained approximately 2 million individuals. These files had 15 X-variables in common *after* adjustments, realignments, and the creation and imputation of some values of X that could be derived from other observed values. To determine an appropriate similarity measure between values of X in the two files, weights had to be assigned to the 15 X-variables that appropriately reflected their relative importance in the matching process. These weights were constructed by studying the relationships between the X-variables and the Y-variables in the PUS and between the X-variables and the Z-variables in the LEED, and considering the goals of the study in terms of which relationships between Y and Z it was most important to explore.

The subcommittee on Statistical Uses of Administrative Records, Federal Committee on Statistical Methodology [20] describes in detail four specific studies of the matching of administrative records by government agencies. Other examples of statistical matching studies are given by Okner [14] and by Alter [1], who describes the creation of a synthetic data set by matching family units from two Canadian surveys that were carried out for the year 1970: the Survey of Consumer Finances (the SCF) and the Family Expenditure Survey (the FEX). The SCF, containing about 10,000 records, was regarded as the primary file; the FEX, containing about 14,000 records, was the secondary file; and an unconstrained matching procedure was carried out.

Kadane [11] discusses both exact and statistical matching methods when data are assumed to have a *multivariate normal distribution*, and proposes a *Bayesian* approach*. Radner [16] also proposes a formal framework for statistical matching. The Subcommittee on Matching Techniques, Federal Committee on Statistical Methodology [19], presents an excellent overview of exact and statistical matching for research and statistical purposes (rather than for administrative purposes). They describe several examples and give a useful bibliography. Okner [15] also presents a brief but insightful summary of problems of record linkage. Rodgers [17] gives a useful review of statistical matching that includes an evaluation of matching procedures based on simulation* studies. Cassel [2] also carries out an evaluation of statistical matching by means of a simulation study based on a file containing actual data from the *Survey on Living Conditions* carried out by Statistics Sweden*. Klevmarken [12] proposes a method to estimate a linear relation between Y and Z based on statistical matching.

BROKEN SAMPLES

We shall now describe a class of problems of exact matching for which there is an explicit statistical model and a well-defined optimal matching procedure that in some circumstances can be explicitly described. Suppose that U is an r-dimensional random vector, V is an s-dimensional random vector, and U and V have some specified $(r + s)$-dimensional joint distribution. Suppose also that a random sample $(U_1, V_1), \ldots, (U_n, V_n)$ has been drawn from this distribution but that before the sample values $(u_1, v_1), \ldots, (u_n, v_n)$ can be observed, each vector (u_i, v_i) gets broken into the two separate components u_i and v_i. As a result, all that can be observed are the r-dimensional vectors $y_1, \ldots, y_n$ and the s-dimensional vectors $z_1, \ldots, z_n$, where $y_1, \ldots, y_n$ are a random permutation of the vectors $u_1, \ldots, u_n$, and $z_1, \ldots, z_n$ are a random permutation of the vectors $v_1, \ldots, v_n$. The observations $y_1, \ldots, y_n$ and $z_1, \ldots, z_n$ are said to form a *broken random sample* from the joint distribution of U and V.

In this context it is known that $y_1, \ldots, y_n$ and $z_1, \ldots, z_n$ are exactly matched with each other in some unknown rearrangement. The basic problem, in general terms, is to determine a matching procedure that will reproduce as many pairs in the original unbroken sample as possible.

This problem of repairing a broken sample was introduced by DeGroot et al. [7], who discuss the following three different criteria: (1) maximizing the probability of obtaining a completely correct set of n matches, (2) maximizing the probability of correctly matching one particular y_i, and (3) maximizing the expected number of correct matches in the n pairings. Some of the results that they obtain are as follows:

Suppose that $r = s = 1$ (i.e., both U and V are one-dimensional random variables), and that the joint distribution of U and V is a *bivariate normal distribution* with *correlation* ρ. If $\rho > 0$, the procedure that maximizes the probability of obtaining a completely correct set of n matches is always to order the observations $y_1, \ldots, y_n$ and $z_1, \ldots, z_n$ so that $y_1 < \cdots < y_n$ and $z_1 < \cdots < z_n$, and then to pair y_i with z_i for $i = 1, \ldots, n$. If $\rho < 0$, the optimal procedure is always to pair y_1 with z_n, y_2 with z_{n-1}, and in general, y_i with z_{n+1-i} for $i = 1, \ldots, n$. This procedure is called the *maximum likelihood* procedure*.

When $\rho > 0$, in order to maximize the probability of correctly matching y_1, it should always be paired with z_1; and in order to maximize the probability of matching y_n, it should always be paired with z_n. However, for $n \geqslant 3$, there are some broken samples for which the probability of correctly matching y_2 is not maximized by pairing it with z_2. Similarly, when $\rho > 0$, in order to maximize the expected number of correct matches, it is always optimal to pair y_1 with z_1 and y_n with z_n. However, for $n \geqslant 4$, there are some broken samples for which it is *not* optimal to pair y_i with z_i for $i = 2, \ldots, n - 1$. Nevertheless, Monte Carlo* studies show that the maximum likelihood procedure does maximize the expected number of correct matches for the vast majority of broken samples [5]. Furthermore, Zolotuchina and Latishev [23] have derived an asymptotic value for the expected number of correct matches when the maximum likelihood procedure is used, where the expectation is taken over all possible broken samples. They show that this expectation converges to $1/(1 - |\rho|)$ as $n \to \infty$.

Now suppose that the parameters of the bivariate normal distribution are unknown. The problem of making inferences about these parameters based on a broken random sample is considered in DeGroot and Goel [6]. In particular, it is emphasized that a broken sample does contain information about the value of ρ, and procedures for deciding whether ρ is positive or negative are developed.

Extensions of some of these ideas to arbitrary joint distributions of U and V with monotone likelihood ratio* have been given by Chew [3] and Goel [10], and to problems of approximate matching by Yahav [22]. A survey of this topic is presented in DeGroot [4].

Acknowledgment

This research was supported in part by the National Science Foundation under Grant SES-8207295. I am indebted to Miron L. Straf for helpful comments.

References

[1] Alter, H. E. (1974). *Ann. Econ. Social Meas.*, **3**, 373–394.

[2] Cassel, C. M. (1983). *Statistical Review 1983:5 Essays in Honour of Tore E. Dalenius*, L. Lyberg, ed. Statistics Sweden, Stockholm, pp. 55–66.

[3] Chew, M. C., Jr. (1973). *Ann. Statist.*, **1**, 433–445.

[4] DeGroot, M. H. (1980). *Symposia Mathematica, Vol. XXV* (Istituto Nazionale di Alta Matematica, Francesco Severi, Roma). Academic Press, New York, pp. 123–135.

[5] DeGroot, M. H. and Goel, P. K. (1976). *Sankhyā B*, **38**, 14–29.

[6] DeGroot, M. H. and Goel, P. K. (1980). *Ann. Statist.*, **8**, 264–278.

[7] DeGroot, M. H., Feder, P. I., and Goel, P. K. (1971). *Ann. Math. Statist.*, **42**, 578–593.

[8] DuBois, N. S. D'A., Jr. (1969). *J. Amer. Statist. Ass.*, **64**, 163–174.

[9] Fellegi, I. T. and Sunter, A. B. (1969). *J. Amer. Statist. Ass.*, **64**, 1183–1210.

[10] Goel, P. K. (1976). *Ann. Statist.*, **3**, 1364–1369.

[11] Kadane, J. B. (1978). *Compendium of Tax Research.* Office of Tax Analysis, Dept. of the Treasury, Washington, D.C., pp. 159–171.

[12] Klevmarken, N. A. (1983). *Statistical Review 1983:5 Essays in Honour of Tore E. Dalenius*, L. Lyberg, ed. Statistics Sweden, Stockholm, pp. 67–79.

[13] Nathan, G. (1967). *J. Amer. Statist. Ass.*, **62**, 454–469.

[14] Okner, B. (1972). *Ann. Econ. Social Meas.*, **1**, 325–342.

[15] Okner, B. (1974). *Ann. Econ. Social Meas.*, **3**, 347–352.

[16] Radner, D. B. (1979). *Amer. Statist. Ass., 1978 Proc. Social Statist. Sec.*, pp. 503–508.

[17] Rodgers, W. L. (1984). *J. Bus. Econ. Statist.*, **2**, 91–102.

[18] Ruggles, N. and Ruggles, R. (1974). *Ann. Econ. Social Meas.*, **3**, 353–371.

[19] Subcommittee on Matching Techniques, Federal Committee on Statistical Methodology (1980). *Statistical Policy Working Paper 5*, Office of Federal Statistical Policy and Standards, U.S. Dept. of Commerce, Washington, D.C.

[20] Subcommittee on Statistical Uses of Administrative Records, Federal Committee on Statistical Methodology (1980). *Statistical Policy Working Paper 6*, Office of Federal Statistical Policy and Standards, U.S. Dept. of Commerce, Washington, D.C.

[21] Tepping, B. J. (1968). *J. Amer. Statist. Ass.*, **63**, 1321–1332.

[22] Yahav, J. A. (1982). *Statistical Decision Theory and Related Topics III*, Vol. 2, S. S. Gupta and J. O. Berger, eds. Academic Press, New York, pp. 497–504.

[23] Zolotuchina, L. A. and Latishev, K. P. (1978). *Leningrad Otd. Mat. Inst., Akad. Nauk USSR*, **29**, 4–10 (in Russian).

(STATISTICAL MATCHING)

MORRIS H. DEGROOT

RECOVERY OF INTERBLOCK INFORMATION

Recovery of interblock information refers to an estimation technique, first suggested by Yates [13, 14], for extracting treatment information from block totals in incomplete block designs*. In randomized *complete* block designs (RCBD), treatments and blocks are orthogonal and all information about treatment effects is contained in the treatment totals—which comprise the set of sufficient statistics* for the treatment effects. Incomplete block designs (IBDs), characterized by a subset of the treatment levels occurring in each block, result in partial confounding* between the treatment and block effects. Thus the block totals contain information about treatments called interblock information*, which is usually ignored. The treatment information usually obtained from an IBD is the intrablock information*, derived from treatment comparisons within blocks, and referred to as treatment effects *adjusted for blocks*.

By recovering the interblock, in addition to the intrablock information, two independent sets of treatment effect estimates are available. Combined estimates of the treatment effects can then be constructed by weighting the inter- and intrablock estimates by the inverse of their estimated variances. In most cases these combined estimates are preferred over the interblock estimates as described below.

BALANCED INCOMPLETE BLOCK DESIGNS

The simplest setting for recovery of interblock information is with balanced incomplete block designs (BIBDs). Let t be the number of treatments, b the number of blocks, and $k < t$ the block size. The design is specified by the numbers n_{ij}, which determine the number of times treatment i occurs in block j. This $t \times b$ matrix $N = (n_{ij})$ is called the *incidence matrix* of the design. For IBDs, the n_{ij} are 0 or 1, which characterizes a binary design. An equireplicate design is characterized by the additional restriction that treatments are replicated an equal number of times in the entire design (i.e., $r_i = r$). The number of times treatment pairs occur together in blocks (say λ_{il}) is equal for BIBDs, where

$$\lambda_{il} = \sum_{j=1}^{b} n_{ij} n_{lj} .$$

However, BIBDs exist only for values of the parameters (b, k, t, r) satisfying $b \geqslant t$, $bk = rt$ and $\lambda = r(k-1)/(t-1)$ an integer.

INTRABLOCK AND INTERBLOCK ESTIMATES

The model for IBDs is given by

$$y_{ij} = \mu + \tau_i + b_j + e_{ij} ,$$

where $i = 1, \ldots, t$, $j = 1, \ldots, b$, $\mu + \tau_i$ are the treatment means, $b_j \sim N(0, \sigma_b^2)$ and $e_{ij} \sim N(0, \sigma_e^2)$ and b_j and e_{ij} are all independent. The $E(y_{ij}) = \mu + \tau_i$ and $\operatorname{var}(y_{ij}) = \sigma_b^2 + \sigma_e^2$ (see John [6, p. 233]).

The usual treatment estimates for BIBDs are found by solving the normal equations obtained by differentiating

$$\sum_{i=1}^{t} \sum_{j=1}^{b} n_{ij} (y_{ij} - \mu - \tau_i - b_j)^2$$

with respect to μ, τ_i, and b_j, setting to zero, and simplifying using the equalities

$$\sum_{i=1}^{t} n_{ij} = k, \qquad \sum_{j=1}^{b} n_{ij} = r, \qquad \sum_{j=1}^{b} n_{ij} n_{lj} = \lambda,$$

which characterize BIBDs. This yields the intrablock estimates of the treatment effects

$$\hat{\tau}_i = \frac{kQ_i}{\lambda t}, \qquad \text{where } Q_i = y_{i\cdot} - \frac{1}{k} \sum_{j=1}^{b} n_{ij} y_{\cdot j}, \quad i = 1, \ldots, t,$$

and $y_{i\cdot} = \sum_{j=1}^{b} n_{ij} y_{ij}$ and $y_{\cdot j} = \sum_{i=1}^{t} n_{ij} y_{ij}$.

The variances of the intrablock estimates are given by

$$V(\hat{\tau}_i) = \frac{k(t-1)}{\lambda t^2} \sigma_e^2 .$$

The interblock estimates are derived from the model equation for the block totals:

$$y_{\cdot j} = k\mu + \sum_{i=1}^{t} n_{ij} \tau_i + f_j ,$$

where $f_j = kb_j + \sum_{i=1}^{t} n_{ij} e_{ij}$ and $f_j \sim N(0, k^2\sigma_b^2 + k\sigma_e^2)$. Since f_j is an error term with constant variance, ordinary least squares* produces estimates of the parameters which are called the interblock estimates of the treatment effects,

$$\tilde{\tau}_i = \frac{\sum_{j=1}^{b} n_{ij} y_{\cdot j} - kr\bar{y}_{\cdot\cdot}}{r - \lambda} \qquad \text{for } i = 1, \ldots, t.$$

These estimates are independent of the intrablock estimates and have variances

$$V(\tilde{\tau}_i) = \frac{k(t-1)}{t(r-\lambda)} (\sigma_e^2 + k\sigma_b^2).$$

The combined treatment effect estimates are obtained as a weighted combination

$$\tau_i^* = (\alpha_1 \hat{\tau}_i + \alpha_2 \tilde{\tau}_i)/(\alpha_1 + \alpha_2),$$

where α_1 and α_2 are the inverses of the variances or estimated variances of the intra- and interblock estimates, respectively.

HISTORY

Yates [13] first described the recovery of interblock information for three-dimensional lattice designs* and the next year [14] for balanced incomplete block designs (BIBDs). The potential gain in efficiency from using small incomplete blocks is compared with the loss due to the partial confounding of treatment effects with blocks. However, with the recovery of interblock information, there is no appreciable loss in efficiency, and the choice of smaller blocks can result in a gain in efficiency. Yates [15] extended the recovery principle to recovery of interrow and intercolumn information in lattice square designs with both row and column blocking factors. Graybill and Weeks [4] showed that the combined estimator presented by Yates is unbiased and based on a set of minimal sufficient statistics.

Since the combined estimator τ_i^* depends on the variance ratio σ_b^2/σ_e^2 which usually must be estimated, τ_i^* can have higher variance than $\hat{\tau}_i$, the intrablock estimator. Graybill and Weeks (1959) found conditions (BIBD with $b - t > 10$) for which the variance of τ_i^* is uniformly less than that of $\hat{\tau}_i$. Shah [11] provides another criterion for any IBD which ensures that τ_i^* is uniformly better than $\hat{\tau}_i$, and an example not meeting the criterion with τ_i^* having higher variance.

Nair [8], Rao [9, 10], Shah [11], and Stein [12] extended the recovery idea to obtain improved combined estimators under various conditions. Brown and Cohen [3], Khatri and Shah [7], and more recently, Bhattacharya [2] have discussed a family of improved combined estimators for any incomplete block design with more than three blocks.

References

[1] Bement, T. R. and Milliken, G. A. (1977). *J. Amer. Statist. Ass.*, **72**, 157–159. (Describes recovery of interblock information where the blocks can be divided into independent groups, such that within groups the blocks are correlated.)

[2] Bhattacharya, C. G. (1980). *Ann. Statist.*, **8**, 205–211.

[3] Brown, L. D. and Cohen, A. (1974). *Ann. Statist.*, **2**, 963–976.

[4] Graybill, F. A. and Weeks, D. L. (1959). *Ann. Math. Statist.*, **30**, 799–805.

[5] Harville, D. A. (1975). *J. Amer. Statist. Ass.*, **70**, 200–206. (Clear presentation of the combined intra- and interblock analysis in the context of the general mixed linear model. Computational formulas are given for all incomplete block designs.)

[6] John, P. W. M. (1971). *Statistical Design and Analysis of Experiments*. Macmillan, New York, pp. 233–245. (General text for advanced treatment of the analysis and construction of incomplete block designs.)

[7] Khatri, C. G. and Shah, K. R. (1975). *J. Amer. Statist. Ass.*, **70**, 402–406. (Calculates exact variances of combined inter- and intrablock estimates for several procedures. Results are given for several incomplete block designs for a range of ratios σ_b^2/σ_e^2.)

[8] Nair, K. R. (1944). *Sankhyā*, **6**, 383–390.

[9] Rao, C. R. (1947). *J. Amer. Statist. Ass.*, **42**, 541–561.

[10] Rao, C. R. (1956). *Sankhyā*, **17**, 105–114.

[11] Shah, K. R. (1964). *Ann. Math. Statist.*, **35**, 1064–1078.

[12] Stein, C. (1966). In *Research Papers in Statistics*, H. A. David, ed. Wiley, New York, pp. 351–366.

[13] Yates, F. (1939). *Ann. Eugen. (Lond.)*, **9**, 136–156.

[14] Yates, F. (1940). *Ann. Eugen. (Lond.)*, **10**, 317–325.

[15] Yates, F. (1940). *J. Agric. Sci.*, **30**, 672–687.

(BLOCKS, BALANCED INCOMPLETE
BLOCKS, RANDOMIZED COMPLETE
INCOMPLETE BLOCK DESIGNS
POOLING)

JAMES L. ROSENBERGER

RECTANGULAR DISTRIBUTION *See* UNIFORM DISTRIBUTION

RECTANGULAR FREQUENCY POLYGON *See* HISTOGRAMS

RECTIFIED INDEX NUMBER *See* TEST APPROACH TO INDEX NUMBERS

RECTILINEAR TREND *See* TREND

RECURRENCE CRITERION

The wide applicability of the stochastic models known as Markov chains and processes rests to a large extent on the property that under suitable irreducibility assumptions, any particular model can be classified as recurrent or transient and that recurrent models have a variety of useful general properties. For example, if a process in discrete time is classified as an irreducible positive recurrent Markov chain, on a countable state space, then it admits a unique stationary distribution π (*see* MARKOV PROCESSES) which describes the long-term behavior of the process.

The role of a recurrence criterion is to

enable a particular model to be assessed as transient or recurrent, and as null or positive recurrent. Such criteria are of particular importance in the application of the general theory. The fundamental paper by Kendall [3] which first showed this applicability of the theory of Markov chains to queueing* models also carried, in the discussion, the first idea of a recurrence criterion, put forward by F. G. Foster. The idea was amplified in ref. 2, and the criteria discussed below can generally be related to Foster's result.

By a recurrence criterion, we mean a method of checking the recurrence or otherwise of a process from the simple parameters in its definition, which usually means the one-step transition probabilities. For example, simple queueing models as introduced by Kendall are usually defined in terms of a service-time distribution, with mean μ (say), and an interarrival-time distribution with mean λ (say). A queue is recurrent if it empties with probability 1 and transient otherwise. The criterion in this case is that $\lambda \geqslant \mu$ implies recurrence while $\lambda < \mu$ implies transience. This criterion is itself derived from the general criterion below, as originally shown by Foster [2].

Foster's exact result has been rediscovered more than once. The best known example is in Pakes [6], and the recurrence criterion is consequently often called Pakes' theorem in the operations research* literature. Other examples of application of recurrence criteria are in the assessment of computer network models, where recurrence essentially means that programs are run in finite time, and storage theory, where recurrence essentially means that the reservoir modeled returns to finite levels with probability 1. In most cases there is a criticality property: when some parameter is below a critical value, recurrence occurs, and above the critical value transience occurs. In the queueing example above, μ/λ is such a critical value.

Specific examples of criteria are most easily described for a discrete-time countable space irreducible Markov chain $\{X_n\}$. We let $P_i(A) = \Pr(A \mid X_0 = i)$, $\tau_B = \inf(n > 0 : X_n \in B)$. The chain is recurrent if for some one i $P_i(\tau_i < \infty) = 1$ [and then $P_i(\tau_j < \infty) \equiv 1$ for all i, j]; the chain is positive recurrent if $E_i(\tau_i) < \infty$ for some i, when $E_i(\tau_j) < \infty$ for all i, j. Positive recurrence is also equivalent to various other properties: for example, the existence of a stationary probability measure π, and the convergence in some way of the transition probability distributions $P_i(X_n = j)$ to a nonzero set of values, which must then be the probability distribution π (*see* MARKOV PROCESSES). These equivalent definitions are often difficult to verify, since they depend on long-term behavior, although the model itself is defined in terms of the one-step probabilities $P(i, j) = P_i(X_1 = j)$ for each i, j.

The current most general form of a criterion for positive recurrence is given in ref. 10. Suppose that there exists a nonnegative "test function" $g(j)$ on $\{0, 1, \dots, \}$, a number $\epsilon > 0$ and nonnegative integer N such that

$$\begin{aligned} E_i(g(X_1)) &\leqslant g(i) - \epsilon, \qquad & i > N,\\ E_i(g(X_1)) &< \infty, & i \leqslant N. \end{aligned}$$

Then $E_i(\tau_{\{0,\dots,N\}}) \leqslant g(i)/\epsilon$, so $\{X_n\}$ is positive recurrent. The existence of such a function g is also necessary for positive recurrence.

This criterion tells us that provided that we rescale the state space using the function g, then positive recurrence is equivalent to the chain having mean "drift" toward the set $\{0, \dots, N\}$ for some N.

Criteria for recurrence but not necessarily positive recurrence are of less practical value but typically give a "boundary" to the critical class of processes. The current best criterion for recurrence, for which sufficiency was first shown essentially by Kendall in 1951 and the general necessary and sufficient result in ref. 5, is that a nonnegative function g and a nonnegative N exist such that

$$E_i(g(X_1)) \leqslant g(i), \qquad i \geqslant N,$$

with $g(i) \to \infty$ as $i \to \infty$.

In Markov chains which are essentially variations of random walks*, the most common choice of g in the criteria above is $g(i) \equiv i$. For example, in most queueing models, the recurrence criteria for this choice of g lead easily to the criticality result, giving positive recurrence if and only if $\lambda > \mu$. The systematic use of criteria in operations research models is described in ref. 4.

Other recurrence properties follow from similar criteria. The Markov chain $\{X_n\}$ is called *geometrically ergodic* if the transition probabilities converge at a geometric rate to π: if the test function g satisfies the slightly stronger condition

$$E_i(g(X_1)) \leqslant (1-\epsilon)g(i), \qquad i \geqslant N,$$

then [11] $\{X_n\}$ is also geometrically ergodic; other rate of recurrence criteria are also given in ref. 11.

The most widely applied criteria are for chains that "drift" toward either $\{0, \dots, N\}$ or infinity. For chains that may have a mixture of drifts, toward zero for some states and toward infinity for others, criteria for positive recurrence are given in refs. 8 and 9.

Recent work on Markov chain theory has shown that the classical positive recurrence and recurrence definitions extend to very general state spaces [10], and of particular practical importance is the extension to real-valued or Euclidean-space-valued chains. The recurrence criteria described above all extend to much more general contexts, with the set $\{0, \dots, N\}$ typically replaced by a compact set.

Again the identity function is the most common test function when the state space is $(0, \infty)$. Multidimensional chains are intrinsically more difficult to handle. One extension of Foster's results is in ref. 7, illustrating the use of a multidimensional criterion. The more sophisticated use of a one-dimensional criterion in ref. 1 for vector autoregressive time series* processes shows that the use of a quadratic form as a test function recovers known conditions for second-order stationarity from a Markov chain approach.

References

[1] Feigin, P. D. and Tweedie, R. L. (1985). *J. Time Series Anal.*, **6**, 1–14.

[2] Foster, F. G. (1953). *Ann. Math. Statist.*, **24**, 355–360. (Most criteria are extensions of ideas in this paper, which relates to queueing models in particular.)

[3] Kendall, D. G. (1951). *J. R. Statist. Soc. B*, **13**, 151–185. (The earliest reference to recurrence criteria is in the discussion to this paper.)

[4] Laslett, G. M., Pollard, D. B., and Tweedie, R. L. (1978). *Naval Res. Logist. Quart.*, **25**, 455–472.

[5] Mertens, J. F., Samuel-Cahn, E., and Zamir, S. (1978) *J. Appl. Prob.*, **15**, 848–851.

[6] Pakes, A. G. (1969). *Operat. Res.*, **17**, 1058–1061. (Rediscovers the result in ref. 2, widely quoted in the operations research literature.)

[7] Rosberg, Z. (1980). *J. Appl. Prob.*, **17**, 790–801.

[8] Rosberg, Z. (1981). *J. Appl. Prob.*, **18**, 112–121. (This paper and ref. 9 independently derive results in "variable-drift" chains.)

[9] Tweedie, R. L. (1975). *Aust. J. Statist.*, **17**, 96–102.

[10] Tweedie, R. L. (1976). *Adv. Appl. Prob.*, **8**, 737–771. (Partly a review paper, including constructive proofs of results extending those in ref. 2; in a general state-space setting.)

[11] Tweedie, R. L. (1982). In *Papers in Probability, Statistics and Analysis*, J. F. C. Kingman and G. E. H. Reuter, eds. Cambridge University Press, London.

(MARKOV PROCESSES
QUEUEING THEORY
RANDOM WALK
RETURN STATE)

R. L. TWEEDIE

RECURRENT EVENTS *See* RENEWAL THEORY

RECURRENT STATE *See* MARKOV PROCESSES

REDUCED MODEL

The standard basis for statistical inference involves a statistical model and data; *see*, for example, INFERENCE, STATISTICAL: I, II. The model might be: $y_1, \dots, y_n$ independent, identically distributed (i.i.d.) normal (μ, σ^2) with (μ, σ^2) in $R \times R^+$; and the data might be $y_1^0, \dots, y_n^0$. A reduced model obtained from this is: $\bar{y}$ is normal $(\mu, \sigma^2/n)$ and independently $\sum(y - \bar{y})^2$ is $\sigma^2\chi^2$ where χ^2 is chi-square $(n - 1)$; the relevant corresponding data are $(\bar{y}^0, \sum(y^0 - \bar{y}^0)^2)$. This reduced model applies on $R \times R^+$ and is a major simplification of the original model on the n-dimensional sample space R^n.

The *reduction* to the reduced model in the preceding example can be based on the *sufficiency principle*, or on the *invariance principle** (*see also* ROTATION GROUP), or on the *conditionality principle* (*see also* ANCILLARY STATISTICS), or on the *weak likelihood principle*. Principles of statistical inference as just indicated quite commonly lead to a reduced or simplified statistical model.

Reduced models can, however, arise deductively without recourse to principles of inference. As a first example suppose that θ has occurred as a realized value from a prior density $p(\theta)$ and that y comes from the statistical model $f(y \mid \theta)$. The initial model is then $p(\theta)f(y \mid \theta)$ and the observed datum is, say, y^0. The use of probability as part of the modeling process then predicates the *reduced model* $cp(\theta)f(y \mid \theta)$, the conditional density for θ given y^0. For a related discussion, *see* BAYESIAN INFERENCE.

As a second example, consider the *error* or *structural model** $y_1 = \theta + e_1, \ldots, y_n = \theta + e_n$, where $e_1, \ldots, e_n$ is a sample from the normal $(0, \sigma_0^2)$ [or from some given density $f(e)$] and let the related data be, say, $y_1^0, \ldots, y_n^0$. The data allow the calculation of $(e_1 - \bar{e}, \ldots, e_n - \bar{e}) = (y_1^0 - y^0, \ldots, y_n^0 - \bar{y}^0)$; thus all but one degree of freedom for the e's is known. As in the preceding example, this predicates the conditional model: $\bar{y} = \theta + \bar{e}$, where $\bar{e}$ has the conditional distribution $cf(\bar{e} - \theta + y_1^0 - \bar{y}^0)f(\bar{e} - \theta + y_n^0 - \bar{y}^0)$ and the related datum is $\bar{y}^0$. This is a reduction from a model on R^n to a reduced model on R^1. *See* STRUCTURAL INFERENCE for various generalizations.

Bibliography

Box, G. E. P. and Tiao, G. C. (1973). *Bayesian Inference in Statistical Analysis*. Addison-Wesley, Reading, Mass.

Fraser, D. A. S. (1979). *Inference and Linear Models*. DAI, University of Toronto Textbook Store, Toronto.

(ANCILLARY STATISTICS
STRUCTURAL INFERENCE
SUFFICIENT STATISTICS)

D. A. S. FRASER

REDUCIBLE CHAIN *See* MARKOV PROCESSES

REDUCTION OF DATA

Reducing observed data to summary figures is a central part of statistics. Fisher [4, p. 1] referred to the study of methods of the *reduction of data* as being one of the three main aspects of statistics. (The other two are the study of *populations* and the study of *variation*.)

One use of the term is in reducing the dimensions of multivariate data, as in factor analysis* or correlational analyses more generally (e.g., Simon [5]). But a more recent use stems from the fact that it is not unusual for statistical workers to apply analysis techniques to their data without ever having "looked at the data." For example,

> In the analysis of variance*, they may report F-ratios and significance levels, but not the mean values.
>
> In factor analysis*, they may report the factor loadings and amounts of variance accounted for, but not the observed correlations.

As a reaction, there has been renewed emphasis on data analysis. In Tukey's exploratory data analysis* (EDA) the focus is on *exploration* (i.e., finding patterns and exceptions in data that are new to the analyst). *Data reduction* is a more general term used for a boiling down of *any* data, including repetitive kinds such as occur in information systems. A particular aim is to facilitate the comparison of different data sets, so as to lead to the empirical generalizations and lawlike relationships* of ordinary science.

The term "data reduction" has also become associated with a narrow range of rules or procedures designed to help the analyst to see and to communicate the structure of data (e.g., Ehrenberg [1, 2]). They concern (1) the use of averages, (2) rounding, (3) ordering the rows or columns of a table by some measure of size, (4) using columns rather than rows for figures that are to be compared, together with (5) using the layout of the table to guide the eye (e.g., not too many grid lines, and putting figures in single spacing with occasional gaps between rows), and (6) using a verbal summary to guide the reader into the table. Here we illustrate the first four rules further.

Rule 1: Averages. The mean is the most commonly used average. It is easy to calculate routinely (e.g., without having to order the readings), and it is easy to combine the means of different sets of readings. (Arguments are nowadays also put forward for the median*, as being robust to outliers*.) Such a summary measure is useful in comparing different sets of data (even skew data, if the skewness* is of the same form). But many tables of data are reported without averages.

A special emphasis in data reduction is on the average as a visual and mental focus. In Table 1 it is difficult to see the pattern at a glance or to communicate it to anyone else, partly because one is not sure which figure to compare with which.

Table 1 Sales Data in Four Quarters and Four Regions

	Quarter			
Region	I	II	III	IV
North	97.63	92.24	100.90	90.39
South	48.24	42.31	49.98	39.98
East	75.23	75.16	100.11	74.23
West	49.69	57.21	80.19	51.09

Table 2 gives row and column averages and we can first look at these. We see (1) that the regions differ markedly (on average 95 in the north, 45 in the south, etc.), and (2) that there is not much difference between the quarters except that quarter III was high. Now we can look at the individual figures in the body of the table and compare each with its appropriate marginal averages. This shows that the figures in each row are much the same (i.e., close to the row averages) except that quarter III was high in the east and in the west. Similarly, we can see that each column follows much the same high–low–high–low pattern as the column of averages, except again for quarter III.

Table 2 Row and Column Averages

	Quarter				
Region	I	II	III	IV	Average
North	97.63	92.24	100.90	90.39	95.29
South	48.24	42.31	49.98	39.98	45.13
East	75.23	75.16	100.11	74.23	81.18
West	49.69	57.21	80.19	51.09	59.55
Average	67.71	66.73	82.79	63.92	70.29

Rule 2: Rounding. The rule is to round to two *effective digits*. These are defined as digits that vary in the given set of data, that is, ones that help to distinguish one figure from another (carrying an extra digit if the numbers are close to 100, say). In Table 3 the figures are easier to perceive and remember (*see also* NUMERACY).

Table 3 Rounding to Two Effective Digits

	Quarter				
Region	I	II	III	IV	Average
North	98	92	101	90	95
South	48	42	50	39	45
East	75	75	100	74	81
West	50	57	80	51	60
Average	68	67	83	64	70

It has been argued that other than for certain specific cases such as compound interest, rounding to two effective digits does not affect any conclusions or decisions that would be reached from the data. The criticism of possible overrounding can be avoided by giving fuller data in an appendix or data bank.

The reason for such rounding is that it is difficult, for example, to subtract 17.9% from 35.2% in one's head, and virtually impossible to divide one number into the other. Yet rounded to two digits, 18 and 35, we can see that one number is about twice the other. Such mental arithmetic is essential in scanning a table of numbers visually.

Rule 3: Ordering by Size. Ordering the rows and/or columns by some measure of size as in Table 4 helps to bring out the patterns and exceptions. One can use either the marginal averages as the criterion or an external measure (e.g., population size). Where different tables use the same breakdowns (as in much official statistics), it is essential to keep to the same order. In general, some order is better than none.

Table 4 Rows Ordered by Size

	Quarter				
Region	I	II	III	IV	Average
North	98	92	101	90	95
East	75	75	100	74	81
West	50	57	80	51	60
South	48	42	50	39	45
Average	68	67	83	64	70

Rule 4: Use Columns for Comparison. It is visually easier to compare figures which are presented underneath each other (especially in single spacing) rather than across in a row. In Table 4 the quarters are generally similar to each other. But this is easier to see in Table 5 when in comparing the quarters, the relevant digits (the "tens") are close together, whereas in Table 4 the eye had to jump (i.e., in the first row, 9 (8) blank 9 (2) blank 10 (1) blank, etc.). With longer numbers and larger tables the effect is more striking. (The title of Table 5, "Approximately Constant Columns," illustrates Rule 6: giving the reader a verbal summary.)

Table 5 Approximately Constant Columns

	Regions				
Quarter	North	East	South	West	Average
I	98	75	50	48	68
II	92	75	57	42	67
III	101	(100)	(80)	50	83
IV	90	74	51	39	64
Average	95	81	60	45	70

Final Comment. The rules described have been receiving increasing attention but limited application so far. Faced with a typical correlation matrix as in Table 6, analysts tend to turn to techniques such as factor analysis to try and discover patterns. But the application of Rules 2 (with deliberate over-rounding), 3, and 5, as in Table 7, serves to make the pattern self-evident.

Table 6 Correlation Matrix to Five Digits (Correlations Between the Liking Score of 8 U.K. TV Programs)

	PrB	ThW	Tod	WoS	GrS	MoD	Pan	24H
PrB	1.0000	0.1064	0.0653	0.5054	0.4741	0.4732	0.1681	0.1242
ThW	0.1064	1.0000	0.2701	0.1424	0.1321	0.0815	0.3520	0.3946
Tod	0.0653	0.2701	1.0000	0.0926	0.0704	0.0392	0.2004	0.2432
WoS	0.5054	0.1424	0.0926	1.0000	0.6217	0.5806	0.1867	0.1403
GrS	0.4741	0.1321	0.0704	0.6217	1.0000	0.5932	0.1813	0.1420
MoD	0.4732	0.0815	0.0392	0.5806	0.5932	1.0000	0.1314	0.1221
Pan	0.1681	0.3520	0.2004	0.1867	0.1813	0.1314	1.0000	0.5237
24H	0.1242	0.3946	0.2432	0.1403	0.1420	0.1221	0.5237	1.0000

Table 7 The Correlation Rounded and Reordered

Program	WoS	MoD	GrS	PrB	24H	Pan	ThW	Tod
World of Sport		0.6	0.6	0.5	0.1	0.2	0.1	0.1
Match of the Day	0.6		0.6	0.5	0.1	0.1	0.1	0.0
Grandstand	0.6	0.6		0.5	0.1	0.2	0.1	0.1
Professional Boxing	0.5	0.5	0.5		0.1	0.2	0.1	0.1
24 Hours	0.1	0.1	0.1	0.1		0.5	0.4	0.2
Panorama	0.2	0.1	0.2	0.2	0.5		0.4	0.2
This Week	0.1	0.1	0.1	0.1	0.4	0.4		0.3
Today	0.1	0.0	0.1	0.1	0.2	0.2	0.3	

References

[1] Ehrenberg, A. S. C. (1978). *Data Reduction: Analysing and Interpreting Statistical Data* (rev. reprint). Wiley, New York.

[2] Ehrenberg, A. S. C. (1981). *Amer. Statist.*, **35**, 67–71.

[3] Ehrenberg, A. S. C. (1982). *A Primer in Data Reduction*. Wiley, New York.

[4] Fisher, R. A. (1950). *Statistical Methods for Research Workers*, 11th ed. Oliver & Boyd, Edinburgh.

[5] Simon, G. (1977). *J. Amer. Statist. Ass.*, **72**, 367–376.

(GRAPHICAL REPRESENTATION OF DATA
LAWLIKE RELATIONSHIPS
MULTIVARIATE GRAPHICS
NUMERACY
PATTERN RECOGNITION
ROUND-OFF ERROR)

A. S. C. EHRENBERG

REDUNDANCY *See* COHERENT STRUCTURE THEORY

REDUNDANCY ANALYSIS

The interrelationships between two sets of measurements made on the same subjects can be studied by canonical* correlation. Originally developed by Hotelling [9], the canonical correlation is the maximum correlation between linear functions or canonical factors of two sets of variables. An alternative pair of statistics to investigate the interrelationships between two sets of variables are the redundancy measures, developed by Stewart and Love [21]. A redundancy coefficient is an index of the average proportion of variance in the variables in one set that is reproducible from the variables in the other set. Unlike canonical correlation, redundancy measures are nonsymmetric in that a measure can be calculated for each set of variables (predictor and criterion) and need not be equal to each other. Van Den Wollenberg [26] has developed a method of extracting factors that maximize redundancy, as opposed to canonical correlation. DeSarbo [6], Johansson [11], and Israels [10] have developed extensions of this methodology.

CANONICAL CORRELATION

Assume two sets of variables, $\mathbf{x} = [X_1, X_2, \ldots, X_p]'$ (predictor set) and $\mathbf{y} = [Y_1, Y_2, \ldots, Y_q]'$ (criterion set), all measured on the same N observations with sample data matrices $\mathbf{X}(N \times p)$ and $\mathbf{Y}(N \times q)$. The sample correlation matrix can be calculated and partitioned as

$$\mathbf{R} = \begin{bmatrix} \mathbf{R}_{xx} & \mathbf{R}_{xy} \\ \mathbf{R}_{yx} & \mathbf{R}_{yy} \end{bmatrix}, \tag{1}$$

where $\mathbf{R}_{xy} = \mathbf{R}'_{yx}$. In canonical correlation analysis developed by Hotelling [9], canonical factors or linear components $\mathbf{w}'\mathbf{x}$ and $\mathbf{v}'\mathbf{y}$ are derived from these two sets of variables simultaneously in such a way as to maximize the correlation between these linear components. Here the constrained maximand or Lagrangian is

$$\phi = \mathbf{w}'\mathbf{R}_{xy}\mathbf{v} - \tfrac{1}{2}\lambda_1(\mathbf{w}'\mathbf{R}_{xx}\mathbf{w} - 1) - \tfrac{1}{2}\lambda_2(\mathbf{v}'\mathbf{R}_{yy}\mathbf{v} - 1), \quad (2)$$

where λ_1 and λ_2 are Lagrange multipliers*. Anderson [1] derives the solution to (2) in terms of the following eigenstructure equations:

$$(\mathbf{R}_{xx}^{-1}\mathbf{R}_{xy}\mathbf{R}_{yy}^{-1}\mathbf{R}_{yx} - \lambda_1^2\mathbf{I})\mathbf{w} = \mathbf{0} \quad (3)$$

and

$$(\mathbf{R}_{yy}^{-1}\mathbf{R}_{yx}\mathbf{R}_{xx}^{-1}\mathbf{R}_{xy} - \lambda_2^2\mathbf{I})\mathbf{v} = \mathbf{0}, \quad (4)$$

where $\mathbf{I}$ is the identity matrix. The corresponding eigenvalues* λ_1^2 and λ_2^2 are equal because of the dual nature of (3) and (4), and are also equivalent to the squared canonical correlation coefficient, or variance accounted for between the two linear components. After extraction of the first pair of canonical factors, a second pair having maximum correlation can be determined simultaneously, with the restriction that the derived canonical factors are uncorrelated with all canonical factors except with their counterparts in the other set, and so on (*see* CANONICAL ANALYSIS). This continues until $r = \min(p, q)$ pairs of canonical factors are determined.

STEWART AND LOVE [21] REDUNDANCY INDICES

Thus, $\lambda_1^2 = \lambda_2^2 = \lambda^2$ in expressions (3) and (4) indicates that the canonical correlation is a symmetric measure of association between these derived linear components or canonical factors. However, occasions arise where one is interested in exploring how much variance in one set of original variables (e.g., the criterion set $\mathbf{y}$) is accounted for by variation in the other set of original variables (the predictor set $\mathbf{x}$). As several authors have demonstrated (see Stewart and Love [21], Van Den Wollenberg [26], DeSarbo [6], Green [8], Levine [13], and Thompson [22]), canonical correlation analysis gives no information about the explained variance of the variables in one set given the other, since no attention is paid to factor loadings (correlations between the original variables in a set and its associated canonical factors). For example, two minor canonical factors (one from each corresponding battery or set of variables) might correlate very highly, while the explained variance of the variables is quite low, because of the near-zero loadings (or structure correlations) of the variables with these canonical factors. Thus a high canonical correlation does not necessarily imply a high level of communality of the two sets of variables, a fact that can lead to problems concerning interpretation. This is, in fact, very much related to the classical issue of correlation versus regression.

To counteract this problem, Stewart and Love [21] have proposed a measure of explained variance called the *redundancy index*, which is the mean variance of the variables of one set that is explained by a canonical factor of the other set. Their index expresses the proportion of variance accounted for in a battery or set of variables by the multiplication of:

1. The proportion of variance in the original battery or set of variables that is accounted for by that set's canonical factor, times
2. The proportion of variance that this canonical factor shares with the corresponding canonical factor of the other set of variables (i.e., λ_i^2)

Note that canonical correlation only maximizes the second part of this calculation. Also, unlike canonical correlation, redundancy is, in general, nonsymmetric. That is, given a squared canonical correlation value, the associated redundancy of the $\mathbf{y}$ variables will not be equal to that for the $\mathbf{x}$ variables. To illustrate this, let us consider the first pair of canonical factors. We denote the redundancy measure of the predictor set ($\mathbf{x}$) as $\mathrm{RD}(X|Y)_1$, and that of the criterion set as

$\mathrm{RD}(Y|X)_1$, where:

$$\mathrm{RD}(Y|X)_1 = \frac{\mathbf{g}_1'\mathbf{g}_1}{q}\lambda_1^2 \quad (5)$$

$$\mathrm{RD}(X|Y)_1 = \frac{\mathbf{h}_1'\mathbf{h}_1}{p}\lambda_1^2 \quad (6)$$

with $\mathbf{g}_1 = \mathbf{R}_{yy}\mathbf{v}_1$ = structure correlations or factor loadings for the first criterion canonical factor

$\mathbf{h}_1 = \mathbf{R}_{xx}\mathbf{w}_1$ = structure correlations or factor loadings for the first predictor canonical factor

p = number of predictor variables

q = number of criterion variables

λ_1^2 = first squared canonical correlation.

By construction of the canonical factors, Anderson [1] shows that

$$\mathbf{R}_{yx}\mathbf{w}_1 = \lambda_1\mathbf{R}_{yy}\mathbf{v}_1 \quad (7)$$

$$\mathbf{R}_{xy}\mathbf{v}_1 = \lambda_1\mathbf{R}_{xx}\mathbf{w}_1\,. \quad (8)$$

By substitution into expressions (5) and (6), respectively, one obtains

$$\mathrm{RD}(Y|X)_1 = \frac{1}{q}\mathbf{r}_1'\,\mathbf{r}_1 \quad (9)$$

$$\mathrm{RD}(X|Y)_1 = \frac{1}{p}\mathbf{s}_1'\mathbf{s}_1, \quad (10)$$

where $\mathbf{r}_1' = \mathbf{w}_1'\mathbf{R}_{xy}$ = vector of loadings or structure correlations between the original criterion variables ($\mathbf{y}$) and the first canonical factor of the predictor set

$\mathbf{s}_1' = \mathbf{v}_1'\mathbf{R}_{yx}$ = vector of loadings or structure correlations between the original set of predictor variables ($\mathbf{x}$) and the first canonical factor of the criterion set.

This illustrates the fact that $\mathrm{RD}(Y|X)_i \neq \mathrm{RD}(X|Y)_i$, in general, for any ith canonical factor—thus the nonsymmetry of the two indices. The expressions above also indicate that it is thus possible to have a high λ_i and consequently a high shared variance between the two sets of canonical factors, yet find that $\mathbf{v}_i'\mathbf{y}$ accounts for very little of the variance in the criterion set. If so, $\mathrm{RD}(Y|X)_i$ might be small. High redundancy requires both high λ_i and high variance accounted for by that battery's canonical factor.

One can therefore equate redundancy to the mean squared loadings of the original variables of one battery or set on the canonical factor under consideration of the other set. Overall redundancy measures (across all canonical factors) can be formed by merely summing the individual components:

$$\overline{\mathrm{RD}}(Y|X) = \sum_{i=1}^{r}\mathrm{RD}(Y|X)_i \quad (11)$$

$$\overline{\mathrm{RD}}(X|Y) = \sum_{i=1}^{r}\mathrm{RD}(X|Y)_i. \quad (12)$$

Miller [14] and Miller and Farr [17] develop an algorithm for redundancy calculation for general linear components not necessarily restricted to canonical factors, although Tyler [23] presents a counterexample questioning the interpretation of such measures. Cramer and Nicewander [2] question the usefulness of the redundancy measures in expressions (11) and (12), and examine a number of competing "symmetric" measures of multivariate association. Tziner [25], and Dawson-Saunders and Doolen [4] present applications of these redundancy measures.

PROPERTIES OF THE REDUNDANCY MEASURE

As Miller [15], Nicewander and Wood [19, 20], and Gleason [7] demonstrate, $\overline{\mathrm{RD}}(Y|X)$ represents the proprtion of total variance in the criterion set of original variables which is accounted for by the linear prediction of $\mathbf{y}$ by $\mathbf{x}$. More specifically, as Gleason [7] and Tyler [23] show:

$$\overline{\mathrm{RD}}(Y|X) = \frac{\mathrm{tr}\left(\mathbf{R}_{yx}\mathbf{R}_{xx}^{-1}\mathbf{R}_{xy}\right)}{\mathrm{tr}(\mathbf{R}_{yy})}, \quad (13)$$

or equivalently, if one were to compute the squared multiple correlation* R_k^2 for each variable in the criterion set, as regressed on the full set of predictor variables one would find that

$$\overline{\text{RD}}(Y|X) = \sum_{k=1}^{q} \frac{R_k^2}{q}. \qquad (14)$$

Thus the total redundancy of the criterion set, given the predictor set, is nothing more than the average squared multiple correlation of each variable in the criterion set of variables with the full set of predictor variables. Similar conclusions pertain to the relationship of $\overline{\text{RD}}(X|Y)$ to the averaged squared multiple correlation of each predictor variable on the full set of criterion variables. Gleason [7] also shows how the measures generalize easily to other types of cross-product matrices.

From inspection of expression (13), one can easily see that $\overline{\text{RD}}(Y|X)$ is invariant under orthogonal transformation of the criterion set and under nonsingular transformation of the predictor set. As Tyler [24] demonstrates,

$$\overline{\text{RD}}(Y|X) = \overline{\text{RD}}(\mathbf{B}'\mathbf{y}|\mathbf{A}'\mathbf{x}), \qquad (15)$$

where $\mathbf{B}$ is a $(q \times q)$ orthogonal matrix and $\mathbf{A}$ is a $(p \times p)$ nonsingular matrix. Dawson-Saunders and Tatsuoka [5] show that affine transformations of the predictor set of variables results in no alteration of $\overline{\text{RD}}(Y|X)$, but that such affine transformations of the criterion set generally change the value of $\overline{\text{RD}}(Y|X)$.

Miller [16] approximated the sampling distribution of $\overline{\text{RD}}(Y|X)$ via Monte Carlo methods showing that

$$F = \frac{\overline{\text{RD}}(Y|X)}{1 - \overline{\text{RD}}(Y|X)} \frac{(N-p-1)q}{pq} \qquad (16)$$

can be approximated with an F-distribution* with pq and $(N-p-1)q$ degrees of freedom. Dawson-Saunders [3], using Monte Carlo methods*, found that these aggregate redundancy measures exhibit sampling bias mostly affected by sample sizes, although interset correlations, p, and q have some limited effect on the bias. She recommends the use of the Wherry or Olkin–Pratt formulae [12] for correcting the bias.

REDUNDANCY FACTORING ANALYSIS OF VAN DEN WOLLENBERG [26]

It is clear from expressions (5) and (6) that canonical correlation maximizes only one part of the redundancy formula (the λ_i's). Van Den Wollenberg [26] develops a methodology for maximizing redundancy, instead of the canonical correlation. Given the two sets of variables $\mathbf{x}$ and $\mathbf{y}$, he seeks a factor $\mathbf{a} = \mathbf{w}'\mathbf{x}$ with unit variance such that the sum of squared correlations of the $\mathbf{y}$ variables with $\mathbf{a}$ is maximal, and a factor $\mathbf{b} = \mathbf{v}'\mathbf{y}$ for which the same holds in the opposite direction. The sample correlation of the $\mathbf{y}$ variables (criterion) with the factor $\mathbf{a}$ is given by the column vector $\mathbf{Y}'\mathbf{X}\mathbf{w}/N$. Then the sum of squared correlations is equal to the minor product moment. Therefore, he maximizes Z_1 and Z_2 defined as

$$Z_1 = \frac{1}{N^2}\mathbf{w}'\mathbf{X}'\mathbf{Y}\mathbf{Y}'\mathbf{X}\mathbf{w} - u_1\left(\frac{1}{N}\mathbf{w}'\mathbf{X}'\mathbf{X}\mathbf{w} - 1\right) \qquad (17)$$

$$Z_2 = \frac{1}{N^2}\mathbf{v}'\mathbf{Y}'\mathbf{X}\mathbf{X}'\mathbf{Y}\mathbf{v} - u_2\left(\frac{1}{N}\mathbf{v}'\mathbf{Y}'\mathbf{Y}\mathbf{v} - 1\right) \qquad (18)$$

or

$$Z_1 = \mathbf{w}'\mathbf{R}_{xy}\mathbf{R}_{yx}\mathbf{w} - u_1(\mathbf{w}'\mathbf{R}_{xx}\mathbf{w} - 1) \qquad (19)$$

$$Z_2 = \mathbf{v}'\mathbf{R}_{yx}\mathbf{R}_{xy}\mathbf{v} - u_2(\mathbf{v}'\mathbf{R}_{yy}\mathbf{v} - 1). \qquad (20)$$

Setting the partial derivatives with respect to $\mathbf{w}$ and $\mathbf{v}$ equal to zero, and simplifying, one obtains

$$(\mathbf{R}_{xy}\mathbf{R}_{yx} - u_1\mathbf{R}_{xx})\mathbf{w} = \mathbf{0} \qquad (21)$$

$$(\mathbf{R}_{yx}\mathbf{R}_{xy} - u_2\mathbf{R}_{yy})\mathbf{v} = \mathbf{0}. \qquad (22)$$

These characteristic equations can be solved through simple eigenstructure analysis, similar to that of canonical correlation, since both matrix products $\mathbf{R}_{xy}\mathbf{R}_{yx}$ and $\mathbf{R}_{yx}\mathbf{R}_{xy}$, and the matrices $\mathbf{R}_{xx}$ and $\mathbf{R}_{yy}$, are real, symmetric matrices. However, as Van Den Wollenberg [26] points out, the eigenvalues u_1 and u_2 need not be equal, as is the case of canonical correlation, so that one has to compute both eigenstructures. One can interpret u_1 as q times the mean variance of the $\mathbf{y}$ variables that is explained by the first linear factor of the $\mathbf{x}$ variables. A similar interpretation holds for u_2. Subsequent vectors, $\mathbf{w}_j$ and $\mathbf{v}_j$, are obtained from the jth eigenvectors of the characteristic equations in expressions (21) and (22). Note that (unlike canonical correlation), while linear composite factors extracted *within* the same set of original variables are orthogonal to each other, one does not, in general, obtain biorthogonal components from this redundancy factoring analysis (i.e., the factors extracted from one set of variables are not necessarily orthogonal to the components in the other set, since **a** and **b** are determined separately). Van Den Wollenberg [26] suggests following a redundancy factoring analysis with a canonical correlation analysis of the derived redundancy variates to obtain this biorthogonality property. Johansson [11] presents a least squares* extension of the Van Den Wollenberg [26] procedure to obtain biorthogonal factors. DeSarbo [6] presents another extension called canonical/redundancy factoring which maximizes user specified convex combinations of canonical correlation and the two redundancy measures presented in expressions (11) and (12). Israels [10] generalizes redundancy factoring to qualitative variables utilizing optimal scaling techniques. Muller [18] shows the relationship between Van Den Wollenberg's [26] redundancy factoring and multivariate multiple linear regression on rotated component scores.

References

[1] Anderson, T. W. (1958). *Introduction to Multivariate Statistical Analysis*. Wiley, New York.

[2] Cramer, E. M. and Nicewander, W. A. (1979). *Psychometrika*, **44**, 43–54.

[3] Dawson-Saunders, B. (1982). *Educ. Psychol. Meas.*, **42**, 131–143.

[4] Dawson-Saunders, B. and Doolen, D. R. (1981). *J. Med. Educ.*, **56**, 295–300.

[5] Dawson-Saunders, B. and Tatsuoka, M. M. (1983). *Psychometrika*, **48**, 299–302.

[6] DeSarbo, W. S. (1981). *Psychometrika*, **46**, 307–329.

[7] Gleason, T. C. (1976). *Psychol. Bull.*, **83**, 1004–1006.

[8] Green, P. E. (1978). *Analyzing Multivariate Data*. Holt, Rinehart and Winston, New York.

[9] Hotelling, H. (1936). *Biometrika*, **28**, 321–377.

[10] Israels, A. Z. (1984). *Psychometrika*, **49**, 331–346.

[11] Johansson, J. K. (1981). *Psychometrika*, **46**, 93–103.

[12] Kendall, M. G. and Stuart, A. (1967). *The Advanced Theory of Statistics*, Vol. 2. Harper & Row, New York.

[13] Levine, M. S. (1977). *Canonical Analysis and Factor Comparison*. Sage, Beverly Hills, Calif.

[14] Miller, J. (1969). *Doctoral dissertation*, State University of New York at Buffalo, N.Y.

[15] Miller, J. K. (1975). *Psychol. Bull.*, **82**, 207–209.

[16] Miller, J. K. (1975). *Multivariate Behav. Res.*, **10**, 233–244.

[17] Miller, J. K. and Farr, D. S. (1971). *Multivariate Behav. Res.*, **6**, 313–324.

[18] Muller, K. E. (1981). *Psychometrika*, **46**, 139–142.

[19] Nicewander, W. A. and Wood, D. A. (1974). *Psychol. Bull.*, **81**, 92–94.

[20] Nicewander, W. A. and Wood, D. A. (1975). *Psychol. Bull.*, **82**, 210–212.

[21] Stewart, D. and Love, W. (1968). *Psychol. Bull.*, **70**, 160–163.

[22] Thompson, B. (1984). *Canonical Correlation Analysis*. Sage, Beverly Hills, Calif.

[23] Tyler, D. E. (1982). *Multivariate Behav. Res.*, **17**, 131–135.

[24] Tyler, D. E. (1982). *Psychometrika*, **47**, 77–86.

[25] Tziner, A. (1983). *J. Occup. Psychol.*, **56**, 49–56.

[26] Van Den Wollenberg, A. L. (1977). *Psychometrika*, **42**, 207–219.

(CANONICAL ANALYSIS
COMPONENTS ANALYSIS
FACTOR ANALYSIS
MULTIPLE CORRELATION
MULTIVARIATE ANALYSIS)

WAYNE S. DESARBO
KAMEL JEDIDI

REED–FROST CHAIN BINOMIAL MODEL

The study of the flow of a disease through a population can be approached by expressing assumptions about the factors producing the spread of the disease in terms of a mathematical model. Such models have existed since the early nineteenth century. A number of simplifying assumptions are made in setting up the models, partly for mathematical simplicity, partly as a result of limitations on available data.

The spread of infectious diseases has been studied in terms of both deterministic and stochastic models, expressed in discrete and continuous time. Much of the theory and analysis of the models is given in a book by Bailey [2]. Among discrete-time epidemic models, perhaps the simplest and most attractive is the *chain binomial* type of model. There are two sets of chain binomial models, known as the *Greenwood* and *Reed–Frost models*, respectively. In unpublished work, used in class lectures at Johns Hopkins University in the 1930s, Reed and Frost developed their model [1]. In an article, Greenwood demonstrated that the chain binomial model now bearing his name gave a good fit to a set of English measles data [6].

DEFINITIONS AND MODEL PROPERTIES

Chain binomial models can briefly be described as follows. In a closed group of freely intermingling individuals, there are at each stage of the process certain numbers of infectives and susceptibles. It is assumed that the latter will yield a fresh crop of cases at the next stage, distributed in a binomial series. At the moment of infectiousness of any given infective, the chance of adequate contact—sufficient to transmit infection—with any specified susceptible is indicated by a parameter P, and a single attack of the disease confers immunity. At each stage or generation, the actual number of new cases will thus have a binomial distribution* depending on the parameter P. The precise form of this distribution varies according to our choice of certain additional biological assumptions. If, therefore, an epidemic is started by a single case, or by several simultaneously infectious cases, the whole process will continue in a series of stages or generations, governed by a *chain of binomial distributions*. The process will terminate as soon as any stage produces no fresh cases.

To examine the models in a little more detail, we need some notation. Let N be the initial size of the group or household. The process starts by S_0 individuals becoming infected—the primary cases—while $N - S_0$ thus remain susceptible. The latent time, or incubation period, is taken as the discrete-time unit of the process. Let S_t denote the number of infected just prior to time t, and R_t the remaining number of uninfected. In this notation $R_t = S_{t+1} + R_{t+1}$.

The conditional probability of having S_{t+1} individuals infected prior to time $t + 1$, given outcomes s_t and r_t, is then, under the Greenwood model,

$$\Pr[S_{t+1} = s_{t+1} \mid r_t] = \binom{r_t}{s_{t+1}} P^{s_{t+1}} Q^{r_{t+1}}, \quad s_t \geqslant 1, \quad (1)$$

where $P + Q = 1$ and P is the "probability of adequate contact." In the Reed–Frost model, the probability of infection during $(t, t+1)$ is assumed to depend on the number of infectives s_t present in the group, and the corresponding conditional probability is given by

$$\Pr[S_{t+1} = s_{t+1} \mid s_t, r_t] = \binom{r_t}{s_{t+1}} (1 - Q^{s_t})^{s_{t+1}} Q^{s_t r_{t+1}}. \quad (2)$$

Note that $1 - Q^{s_t}$, occurring in (2), is the probability of adequate contact with at least one of the s_t susceptibles. The assumption of a constant probability of infection—independent of the number of infected—is thus the point where the Greenwood model (1) departs from the Reed–Frost model (2).

For small group sizes N, complete enumeration of all possible chains and associated probabilities is feasible. As an illustration, consider two primary cases in a group of size 4 (i.e., $S_0 = 2$ and $N = 4$). Let us write $\{2, 1\}$ for the case in which one individual is infected at the first stage and the process then ceases.

Under the Reed–Frost model (2), we have

the probability

$$\Pr[\{2,1\}] = \binom{2}{1}(1-Q^2)Q^2\binom{1}{0}(1-Q)^0 Q = 2Q^3(1-Q^2),$$

while under the Greenwood model, formula (1) gives

$$\Pr[\{2,1\}] = 2PQ^2.$$

If we are interested in the total number of cases rather than particular chains, we add probabilities for the relevant chains. To illustrate, in a family of four with a single introduction, a total of three cases might have arisen from either $\{1,2\}$ or $\{1,1,1\}$, and the associated probabilities are

$$3P^2Q^3 + 6P^2Q^4$$

and

$$3P^2Q^2 + 6P^2Q^4$$

under the Reed–Frost and Greenwood models, respectively. For household sizes up to $N = 5$, Bailey [2, pp. 243–246] has listed all possible paths for $S_0 = 1,2,3,4$, together with the corresponding model probabilities.

Even for moderate group sizes, the foregoing method of enumeration becomes awkward. Recurrence relations given by Bailey (p. 248) might help, but the main analytical and theoretical tool for the study of chain binomial models was developed by Gani and Jerwood [5], who used the Markov chain technique (see also Bailey, Chap. 8). If we rewrite (1) in the form

$$\Pr[R_{t+1} = r_{t+1} \mid r_t] = \binom{r_t}{r_{t+1}} P^{r_t - r_{t+1}} Q^{r_{t+1}},$$

it is clear that we have a univariate Markov chain for the sequence of random variables R_{t+1}, $t = 0,1,2,\ldots$ Similarly, the Reed–Frost model gives rise to a bivariate Markov chain for the pair of random variables (R_{t+1}, S_{t+1}). For details, the reader is referred to Chapter 8 in Bailey's book or, of course, the original Gani and Jerwood article. The Markov chain format, based solidly on the theory of stochastic processes*, provides convenient computational algorithms for key random variables such as process *duration time* and *total epidemic size*. Recently, Longini [8] has modified the Reed–Frost model to allow infected individuals to become susceptible again. Using Markov chain methods, Longini determines a critical population size needed for a disease to remain *endemic* (continuously present).

Further theoretical analysis of the models presented here is given by von Bahr and Martin-Löf [11] and Ball [3]. By viewing the development of the epidemic disease as a random graph* process, Ball discusses asymptotic behavior and, in particular, *threshold analysis*. The introduction of infectious cases into a community of susceptibles will not give rise to an epidemic if the density of susceptibles is below a certain level, or threshold.

APPLICATIONS

Although reasonable for certain infectious diseases (e.g., measles, mumps, or chickenpox), the assumptions underlying chain binomial models are restrictive. Thus 'constant latent periods' is indeed a restrictive assumption, and extended infectious periods seem to be the rule rather than the exception. Despite their limitations, however, chain binomial models have a didactic value in providing a convenient systematization of epidemiological* ideas and in bringing forth the relative importance of factors involved in an outbreak of an infectious disease, such as the number of introductory cases, the contact intensity, and the effect of isolation of susceptibles.

In her article, Abbey [1] gives an excellent review and reexamines early empirical work on chain binomial models. In ref. 2 (Chap. 14), Bailey discusses in detail the problem of fitting chain binomial models to data from an investigation into measles epidemics in Providence, Rhode Island (1929–1934). A reasonably good fit is obtained when the total number of cases is considered. However, the models give a poor fit when indi-

Table 1 Bailey's [2, pp. 258–259] Modified Greenwood Chains and Providence Measles Data

Type of chain	$\{1\}$	$\{1^2\}$	$\{1^3\}$	$\{1,2\}$	$\{1^4\}$	$\{1^2 2\}$	$\{1,2,1\}$	$\{1,3\}$	Σ
Observed numbers	4	3	1	8	4	3	10	67	100
Fitted values	4.9	2.6	2.0	5.2	2.1	3.1	13.8	66.3	100.0

vidual chains are taken into account. To some extent, the situation can be saved by postulating heterogeneity between families with respect to the contact parameter P. An example of this is shown in Table 1.

The probability of the chain $\{1,2\}$ is $3P^2(1-P)^2$ under the Greenwood model. The parameter P in turn is assumed to follow a beta distribution* with parameters x and y. The expected values of chain probabilities, expressed in P, x, and y, are then computed and Bailey uses maximum likelihood scoring to estimate model parameters.

In a recent application, Saunders [10] attempts to fit a chain binomial model to data from an epidemic outbreak of myxomatosis in Canberra, Australia. The data consist of counts of the number of rabbits first seen with symptoms of the disease on each day of the epidemic. A further recent application of chain binomial models is presented in an article by Poku [9], who studies the risk of streptococcal infection in rheumatic and nonrheumatic families with the aid of the Greenwood model.

The parameters of an epidemic model may be used conceptually for investigations in a purely theoretical framework where it would be impractical to carry out actual experiments. In recent years, research has been done on *control models* for epidemics to assist in the development of effective public health programs, involving, for example, immunization and/or removal of susceptibles. Such control programs for the Greenwood and Reed–Frost models have been suggested by Dayananda and Hogarth [4] and their work has been completed and extended by Lefèvre [7].

References

[1] Abbey, H. (1952). *Hum. Biol.*, **24**, 201–233.

[2] Bailey, N. T. J. (1975). *The Mathematical Theory of Infectious Diseases*, 2nd ed. Charles Griffin, London.

[3] Ball, F. (1983). *J. Appl. Prob.*, **20**, 153–157.

[4] Dayananda, P. W. A. and Hogarth, W. L. (1978). *Math. Biosci.*, **41**, 241–251.

[5] Gani, J. and Jerwood, D. (1971). *Biometrics*, **27**, 591–603.

[6] Greenwood, M. (1931). *J. Hyg. Camb.*, **31**, 336–351.

[7] Lefèvre, C. (1981). *Biom. J.*, **23**, 55–67.

[8] Longini, I. M. (1980). *Math. Biosci.*, **50**, 85–93.

[9] Poku, K. (1979). *Amer. J. Epidemiol.*, **109**, 226–235.

[10] Saunders, I. W. (1979). *Math. Biosci.*, **45**, 1–15.

[11] von Bahr, B. and Martin-Löf, A. (1980). *Adv. Appl. Prob.*, **12**, 319–349.

Bibliography

Recent work on statistical aspects of chain binomial models and extensions thereof is further exemplified by:

Becker, N. (1981). *Biometrics*, **37**, 251–258.

Becker, N. (1981). *Biometrika*, **68**, 133–141.

Saunders, I. W. (1980). *Aust. J. Statist.*, **22**, 307–316.

(EPIDEMIOLOGICAL STATISTICS
MARKOV PROCESSES)

SVEN BERG

REFERENCE SET

An alternative term for sample space*. It tends to be used when some care is needed in (or there is some controversy about) the choice of sample space.

REFLECTION PRINCIPLE

The *reflection principle* or the *method of images* is widely used in probability theory and in mathematical statistics. It has its origin in physics, namely, in geometrical optics, in heat conduction, in diffusion, in the theory of electrical images, and in hydrodynamics.

In probability theory and in mathematical statistics, the reflection principle is used mostly in solving various problems of random walks*, games of chance*, ballots, order statistics*, sequential analysis*, and Brownian motion*.

The reflection principle can be explained in the simplest way by concentrating on random walk processes. Let us suppose that a particle performs an *unrestricted symmetric random walk** on a straight line (see Fig. 1). The particle starts at $x = 0$ and moves along the straight line in a series of steps. In each step, independently of the past journey, the particle moves either a unit distance to the right with probability $\frac{1}{2}$ or a unit distance to the left with probability $\frac{1}{2}$. We can represent the movement of the particle by a path in the space–time diagram (see Fig. 2). The probability that at the end of the nth step the particle reaches the point x is given by

$$P(n,x) = \begin{cases} \binom{n}{k}\dfrac{1}{2^n}, & x = 2k - n; \\ & k = 0, 1, \ldots, n, \\ 0, & \text{otherwise.} \end{cases} \tag{1}$$

For if the particle moves k steps to the right and $n - k$ steps to the left, where $k = 0, 1, \ldots, n$, then $x = 2k - n$ is its final position and the number of such paths is $\binom{n}{k}$, each path having probability $1/2^n$.

Now let us suppose that there is a *reflecting barrier* at the point $x = a$, where a is a positive integer. Every time the particle reaches the point $x = a$, the next step is always taken to the point $x = a - 1$. Then the probability that at the end of the nth step the particle reaches the position x is given by

$$Q_1(n,x) = \begin{cases} P(n,x) + P(n,2a - x) & \text{if } x < a, \\ P(n,a) & \text{if } x = a, \\ 0 & \text{if } x > a, \end{cases} \tag{2}$$

where $P(n,x)$ is defined by (1). To prove formula (2), let us assume that two particles perform random walks on two different lines. The first particle performs an unrestricted random walk on the line; however, the line is bent as is shown in Fig. 3; that is, point $2a - x$ is just below point x if $x < a$. The second particle performs a random walk on the half-line $(-\infty, a]$ in such a way that its position is x if and only if the position of the first particle is either x or $2a - x$. In other words, the second particle moves like the perpendicular projection of the position of the first particle. Since the second particle moves as if it were a reflecting barrier at the point $x = a$, the foregoing model proves (2). Since the point $2a - x$ is the mirror image of

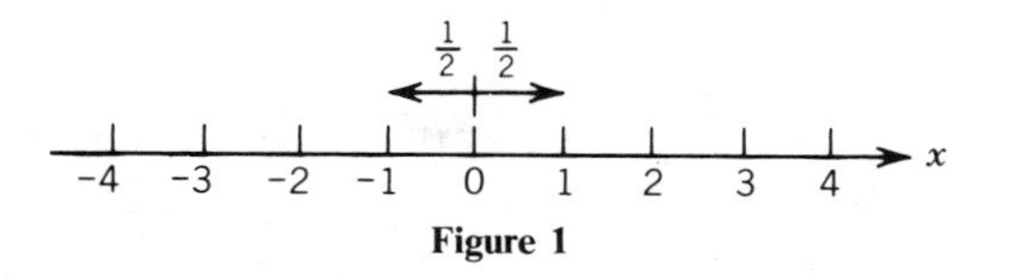

Figure 1

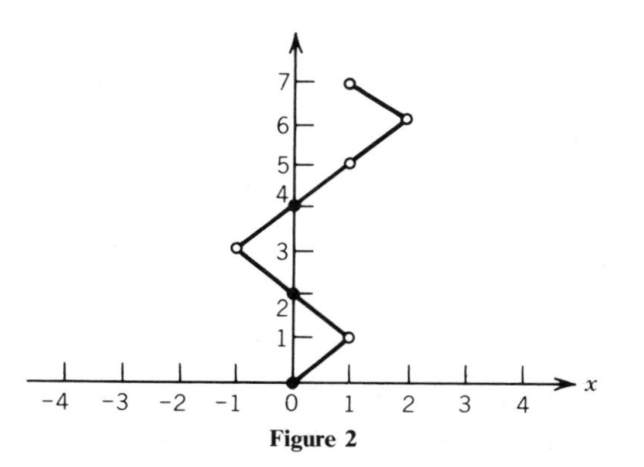

Figure 2

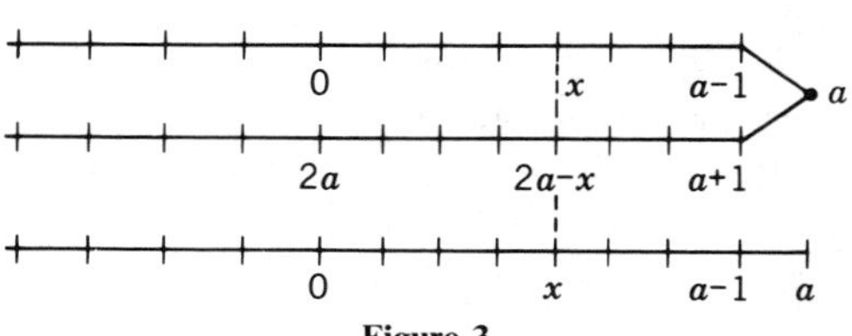

Figure 3

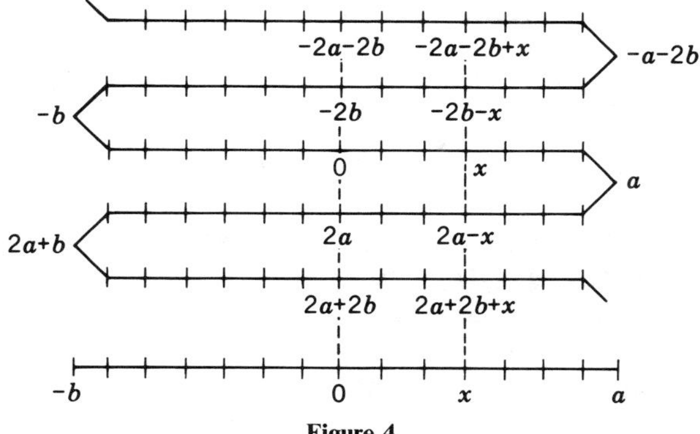

Figure 4

the point x with respect to the point a of the straight line, we say that the probability $Q_1(n, x)$ is obtained by the *reflection principle* or by the *method of images*.

Next let us suppose that there are *two reflecting barriers*, one at the point $x = a$ and another at the point $x = -b$, where a and b are positive integers. Every time the particle reaches the point $x = a$, the next step is taken to the point $x = a - 1$, and every time the particle reaches the point $x = -b$, the next step is taken to the point $x = -b + 1$. Now the probability that at the end of the nth step the particle reaches the position x is given by

$$Q_2(n, x) = \sum_j \{ P(n, 2(a + b)j + x) + P(n, 2(a + b)j + 2a - x) \} \quad (3)$$

for $-b < x < a$, where $P(n, x)$ is defined by (1). We have also

$$Q_2(n, a) = \sum_j P(n, 2(a + b)j + a) \quad (4)$$

and

$$Q_2(n, -b) = \sum_j P(n, 2(a + b)j - b). \quad (5)$$

The formulas above can be proved in the same way as (2). We consider again two particles performing random walks on two different lines. The first line is folded as is shown in Fig. 4; that is, points $2(a + b)j + x$ and $2(a + b)j + 2a - x$, $j = 0, \pm 1, \dots$, are along the same perpendicular line. The second particle performs a random walk on the interval $[-b, a]$ in such a way that its position is determined by the perpendicular projection of the position of the first particle. Since the second particle moves as if it were a reflecting barrier at the point $x = a$ and a reflecting barrier at the point $x = -b$, the foregoing model proves (3) to (5). Since the points $2(a + b)j + x$ and $2(a + b)j + 2a - x$, $j = 0, \pm 1, \dots$, can be obtained from x by repeated reflections in the points a and $-b$ of the straight line, we say again that the probability $Q_2(n, x)$ is obtained by the reflection principle or by the method of images.

It is also customary to consider random walk models in which there is a reflecting barrier at the point $x = a + \frac{1}{2}$ or there are two reflecting barriers at the points $x = a + \frac{1}{2}$ and $x = -b - \frac{1}{2}$, where a and b are positive integers. For these random walks $Q_1(n, x)$ and $Q_2(n, x)$ can be determined in the same way as above, except that in Fig. 3 the first line should be bent as shown in Fig. 5, and in Fig. 4 the folding of the first line should also be changed appropriately.

The reflection principle or the method of images can also be extended to random

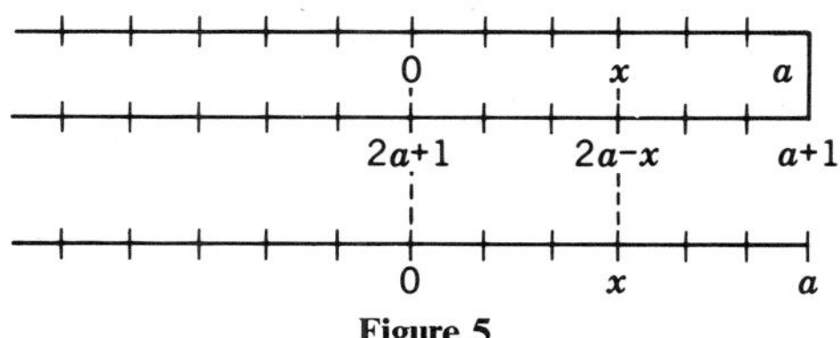

Figure 5

walks in any dimensions. For example, if in a two-dimensional unrestricted symmetric random walk $P(n, x, y)$ denotes the probability that a particle, starting at the origin, in n steps reaches the point (x, y), and if we put reflecting barriers in the lines $x = a$, $x = -b$, $y = c$, and $y = -d$, where a, b, c, and d are positive integers, then $Q(n, x, y)$, the probability that the nth step takes the particle to the point (x, y), where $-b < x < a$ and $-d < y < c$, can be obtained by summing $P(n, x_i, y_i)$ for all pairs (x_i, y_i) $(i = 0, 1, \ldots)$ which are the repeated images of the point (x, y) in the aforementioned four barrier lines.

The solutions of the above-mentioned random walk problems are in perfect analogy with the method of images in geometrical optics. Reflection of light was first studied systematically by Christiaan Huygens* (1629–1695) in 1678 and Isasc Newton* (1642–1727) in 1704. See Huygens [12] and Newton [16]. If a single point-like light source S is placed in front of a polished plane mirror and we want to determine the light intensity in a point P in front of the mirror, we can proceed so that we place a light source S', with equal light output in the mirror image of the original light source, remove the mirror, and calculate the light intensity in P by the superposition of the effects of the two light sources (see Fig. 6). The situation is analogous if a single point-like light source is placed between two parallel polished plane mirrors. The method of images is also widely used in solving heat conduction and diffusion problems (see Fürth [9]). The reflection principle or the method of images can be equally well applied to random walks with absorbing barriers.

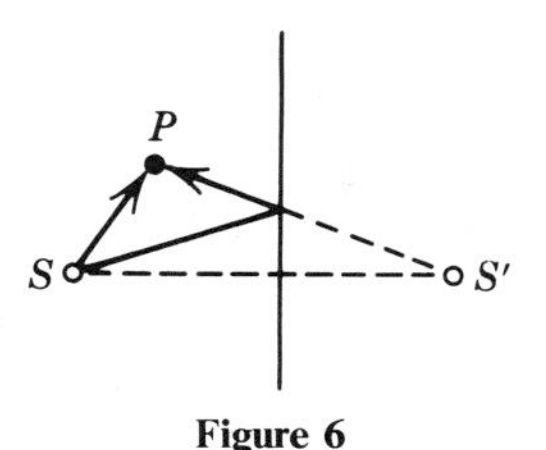

Figure 6

Let us consider again an unrestricted symmetric random walk on a straight line and place an *absorbing barrier* at the point $x = a$, where a is a positive integer. If the particle reaches the point $x = a$, it remains forever at this point. The probability that at the end of the nth step the particle reaches the position x is given by

$$R_1(n, x) = P(n, x) - P(n, 2a - x) \quad (6)$$

for $x < a$, where $P(n, x)$ is defined by (1). If there are *two absorbing barriers*, one at the point $x = a$ and one at the point $x = -b$, where a and b are positive integers, then the probability that at the end of the nth step the particle reaches the point x is given by

$$R_2(n, x) = \sum_j \{P(n, 2(a + b)j + x) - P(n, 2(a + b)j + 2a - x)\} \quad (7)$$

for $-b < x < a$, where $P(n, x)$ is defined by (1). For a proof of formulas (6) and (7) see, for example, Takács [18–20].

The formulas above can also be extended to higher-dimensional random walks having either absorbing barriers or both absorbing and reflecting barriers. We can also extend formulas (6) and (7) for nonsymmetric random walks (see Takács [20, 21]).

The solutions of the foregoing problems for random walks with absorbing barriers are in analogy with the method of electric images, used for the first time by Lord Kelvin, formerly William Thomson (1824–1907) in 1845, and James Clerk Maxwell* (1831–1879) in 1873. See Gray [11, pp. 31, 39–40] and Maxwell [15, pp. 224–249]. The method of electric images is also widely used in hydrodynamics. See Birkhoff [3].

Random walks with reflecting barriers and absorbing barriers were first studied by Marian v. Smoluchowski (1872–1917). See Smoluchowski [17], Chandrasekhar [4], Kac [13, 14], and Feller [8].

Besides random walks, the reflection principle is widely used in other areas of probability theory and mathematical statistics: in the theory of Brownian motion* [2, 5–7], in order statistics* [10], in games of chance* [20], and in ballot* theory [1].

References

[1] André, D. (1887). *C. R. Acad. Sci. Paris*, **105**, 436–437.

[2] Bachelier, L. (1900). *Ann. Sci. École Norm. Sup.*, **17**, 21–86. (English translation: *The Random Character of Stock Market Process*, P. H. Cootner, ed. MIT Press, Cambridge, Mass., 1964, pp. 17–78.)

[3] Birkhoff, G. (1950). *Hydrodynamics*. Princeton University Press, 1950. (Reprinted by Dover, New York, 1955.)

[4] Chandrasekhar, S. (1943). *Rev. Mod. Phys.*, **15**, 1–89. (Reprinted in N. Wax [22, pp. 3–91].)

[5] Dinges, H. (1962). *Z. Wahrscheinl. verw. Geb.*, **1**, 177–196.

[6] Doob, J. L. (1942). *Ann. Math.*, **43**, 351–369. (Reprinted in N. Wax [22, pp. 319–337].)

[7] Doob, J. L. (1949). *Ann. Math. Statist.*, **20**, 393–403.

[8] Feller, W. (1968). *An Introduction to Probability Theory and Its Applications*, Vol. 1, 3rd ed. Wiley, New York.

[9] Fürth, R. (1934). In *Differentialgleichungen der Physik*, Vol. II, 2nd ed., Ph. Frank and R. von Mises, eds. Vieweg, Braunschweig, pp. 526–626.

[10] Gnedenko, B. V. and Rvačeva, E. L. (1952). *Dokl. Akad. Nauk SSSR*, **82**, 513–516. (English translation: *Select. Transl. Math. Statist. Prob.*, **1**, 69–72, 1961.)

[11] Gray, A. (1908). *Lord Kelvin. An Account of His Life and Work*. London. (Reprinted by Chelsea, New York, 1973.)

[12] Huygens, Chr. (1690). *Treatise on Light*. The Hague (in Latin). (English translations: Macmillan, 1912, and Dover, 1962.)

[13] Kac, M. (1945). *Ann. Math. Statist.*, **16**, 62–67.

[14] Kac, M. (1947). *Amer. Math. Monthly*, **54**, 369–391. (Reprinted in N. Wax [22, pp. 295–317] and in *The Chauvenet Papers*, Vol. I, J. C. Abbott, ed. Math. Assoc. America, Washington, D.C., 1978, pp. 253–275.)

[15] Maxwell, J. C. (1873). *A Treatise of Electricity and Magnetism*. (3rd ed., 1891; reprinted by Dover, New York, 1954.)

[16] Newton, I. (1704). *Opticks*. London. (The 4th edition of 1730 was reprinted by Dover, 1952 and 1979.)

[17] Smoluchowski, M. v. (1923). *Abhandlungen über die Brownsche Bewegung und verwandte Erscheinungen*, R. Fürth, ed. Akademische Verlagsgesellschaft, Leipzig. (Ostwald's *Klassiker der exacten Wissenschaften*, **207**.)

[18] Takács, L. (1962). *Z. Wahrscheinl. verw. Geb.*, **1**, 154–158.

[19] Takács, L. (1967). *Combinatorial Methods in the Theory of Stochastic Processes*. Wiley, New York.

[20] Takács, L. (1969). *J. Amer. Statist. Ass.*, **64**, 889–906.

[21] Takács, L. (1979). *SIAM Rev.*, **21**, 222–228.

[22] Wax, N., ed. (1954). *Selected Papers on Noise and Stochastic Processes*. Dover, New York.

(GAMBLING, STATISTICS IN
PASSAGE TIMES
RANDOM WALK)

LAJOS TAKÁCS

REFUSAL RATE

This is not really a rate, but a proportion—the proportion refusing to respond to a request for items of information. There have been many studies of refusal rates observed in particular inquiries. See refs. 1 to 3 for a few examples.

References

[1] De Maio, T. J. (1980). *Public Opinion Quart.*, **44**, 223–233.

[2] O'Neill, M. J., Groves, R. M., and Cannell, C. F. (1979). *Amer. Statist. Ass., Proc. Surv. Res. Sect.*, pp. 280–287.

[3] Wiseman, F. and Macdonald, P. (1978). *Amer. Statist. Ass., Proc. Surv. Res. Sec.*, pp. 252–255.

(CALLBACKS
FOLLOW-UP
NONRESPONSE (IN SAMPLE SURVEYS)
SURVEY SAMPLING)

REGENERATIVE PROCESSES

A stochastic process* X is said to be *regenerative* if there exist stopping times T at which the future $(X_{T+t})_{t\geqslant 0}$ becomes totally independent of the past before T and is a probabilistic replica of X. Thus X regenerates itself, or is reborn, at certain times, and such times T are called *regeneration times*.

The following example shows up often in applications. We shall describe it in the context of reliability*. Consider a machine. Initially, it is new. It works for some random amount of time and fails eventually. When it fails, it is repaired, which takes a random amount of time. When the repair is completed, the machine is as good as new, begins working again, and the whole cycle starts over. It is assumed that the lengths of working and repair periods are all independent, the former have some distribution Φ and the latter another distribution Ψ. Let X_t be set equal to w if the machine is working at time t and to r if it is under repair. Then $X = (X_t)_{t \geqslant 0}$ is a regenerative process; the times $S_1, S_2, \ldots$ of successive repair completions are regeneration times.

For another example, take a recurrent Markov process* X with a discrete state space. Let its initial state be a stable state i, and consider the successive instants at which X jumps into i. By the strong Markov property, at each such time, the future of X is independent of its past and is a probabilistic replica of X. So X is regenerative.

Finally, let X be a standard Brownian motion* and let T be a stopping time for which $X_T = 0$. Then, by the strong Markov property of X, the future $(X_{T+t})_{t \geqslant 0}$ is independent of the past before T and is again a standard Brownian motion. Thus X is regenerative, and every such time T is a regeneration time.

Regenerative processes are abundant in the fields of queueing*, storage, reliability and replacement, and in the theories of Markov and semi-Markov processes*. During the classical period, before 1965, research on them concentrated on cases like the first two examples above, and the main results have been limit theorems of the ergodic* kind. Recent research is concerned more with cases like Brownian motion, and the interest is on the structure of the random set of all regeneration times and on the excursions of the process between regenerations.

Accounts of the classical theory may be found in most textbooks. We illustrate its line of approach and the nature of its results by discussing the reliability problem above.

The main idea is to focus on the regeneration times S_n. Together with $S_0 = 0$, which is a regeneration time trivially, they form a renewal* process $(S_n)_{n=0,1,\ldots}$. Each renewal cycle $[S_n, S_{n+1})$ consists of a working period and a repair period. Thus the distribution F of the length of a cycle is the convolution of Φ and Ψ. The distribution F_n of S_n is the n-fold convolution of F with itself. The expected number of renewals during $[0, t]$ is

$$R(t) = E \sum_{n=0}^{\infty} I_{\{S_n \leqslant t\}} = \sum_{n=0}^{\infty} F_n(t).$$

The function R is called the *renewal function* corresponding to F; it plays the role of a potential operator in renewal theory.

Consider the probability $p(t)$ that the machine is working at time t. It is the sum of $p_n(t)$ over all $n \geqslant 0$, where $p_n(t)$ is the probability that the machine is working at time t and the time t belongs to the nth renewal cycle. Since the regeneration at S_n erases all influence of the past, $p_n(t)$ is the probability that $S_n \leqslant t$ and the nth working period lasts longer than $t - S_n$. Thus

$$\begin{aligned} p(t) &= \sum_{n=0}^{\infty} p_n(t) \\ &= \sum_{n=0}^{\infty} \int_0^t F_n(ds)[1 - \Phi(t-s)] \\ &= \int_0^t R(ds)[1 - \Phi(t-s)]. \end{aligned}$$

In reliability theory, $p(t)$ is called the *availability* of the machine at time t, and its limit as t tends to $+\infty$ is of interest. This is obtained by applying the key renewal theorem, which states that, under reasonable conditions,

$$\lim_{t \to \infty} \int_0^t R(ds) g(t-s) = \frac{1}{m} \int_0^{\infty} g(s) ds,$$

where m is the mean cycle length. The conditions are satisfied if g is a decreasing Riemann integrable function and the distribution F is nonlattice (i.e., time is not discretized). Assuming the mean working time a and the mean repair time b are both finite and the time is not essentially discrete, we see that the key renewal theorem applies to

$p(t)$ and we have

$$\lim_{t\to\infty} p(t) = \frac{1}{m}\int_0^\infty [1 - \Phi(s)]\,ds = \frac{a}{a+b},$$

which is the proportion of time spent working during a cycle.

The principle embedded in the last result holds for arbitrary regenerative processes X. Suppose that X has a finite regeneration time T, let m be the mean of T, and let $m(A)$ be the expected amount of time spent in A during $[0, T)$ for a measurable subset A of the state space of X. Then, assuming that the values T takes do not fall on a lattice and that $m(A) < \infty$, we have

$$\lim_{t\to\infty} P[X_t \in A] = m(A)/m.$$

For instance, for the Markov process X of the second example above, taking $A = \{i\}$ we obtain the limit $m(A)/m$, with $m(A) = 1/q(i)$, where $q(i)$ is the parameter of the exponential sojourn at i and with m the mean return time to i (including the sojourn at i).

The remainder of this article is devoted to describing the recent concerns in regenerative theory during the last 20 years. First, we formulate the concept of regeneration with more precision. Let $(\Omega, \mathbf{F}, P)$ be a complete probability space and let $(\mathbf{F}_t)$ be a right-continuous and complete filtration on it. Let X be a stochastic process taking values in some topological space E and progressively measurable relative to $(\mathbf{F}_t)$. By a functional of X we mean a real-valued random variable that is measurable relative to the σ-algebra generated by X. Thus F is a functional if and only if $F = f(X_{t_1}, X_{t_2}, \ldots)$ for some deterministic measurable function f and some sequence of times t_n, and then we write $F \circ \theta_T$ for $f(X_{T+t_1}, X_{T+t_2}, \ldots)$.

For each outcome ω in Ω, let $M(\omega)$ be a subset of $[0, \infty)$ and suppose that $M(\omega)$ is right-closed and minimal: for every sequence of numbers t_n in $M(\omega)$ decreasing to t, we have $t \in M(\omega)$, and the complement of $M(\omega)$ is a countable union of intervals of the form $[\cdot)$. We suppose that the process 1_M is progressively measurable relative to $(\mathbf{F}_t)$.

The pair (M, X) is said to be a *regenerative system* if

$$E[F \circ \theta_T \mid \mathbf{F}_T] = E[F]$$

for every bounded functional F of X and every finite stopping time T of $(\mathbf{F}_t)$. Also $T(\omega) \in M(\omega)$ for almost all ω are called *regenerative times*. The classical case discussed above restricts M to be a discrete set: $M(\omega)$ consists of $0 = S_0(\omega), S_1(\omega), S_2(\omega), \ldots$ for a renewal process (S_n). By contrast, the example with Brownian motion X allows $M(\omega)$ to be the set of all t such that $X_t(\omega) = 0$ except that the countably many t values for which $X_t(\omega) = 0$ and $X_{t+u}(\omega) \neq 0$ for all $u \leqslant \epsilon$ for some $\epsilon > 0$ are excluded [in order to make $M(\omega)$ minimal].

The following fundamental result, due to Maisonneuve [9], clarifies the structure of regeneration sets M. We restrict the statement to the case of unbounded M, the recurrent case.

Let M be an unbounded regeneration set. Then either $M(\omega)$ consists of $0 = S_0(\omega)$, $S_1(\omega), \ldots$ for almost every ω, or else, again for almost every ω, $M(\omega)$ has no isolated points. In the former case, (S_n) is a renewal process; this is the classical case. In the latter case, there exists a (continuous time parameter) strictly increasing right-continuous process (S_u) with stationary and independent increments such that $M(\omega)$ is the range of the function $u \to S_u(\omega)$ for all ω; we call this the *continuous regeneration case*.

The geometry of the regeneration set can now be deduced from the well-known results on processes with stationary and independent increments. The classical case is easy. In the continuous regeneration case there are three possibilities.

1. S is a compound Poisson process* with a strictly positive drift. Then M has no isolated points; it is a countable union of intervals of form $[\cdot)$ whose lengths are independent and identically distributed exponential random variables. The lengths of the contiguous intervals (which make up the complement of M)

are also independent and identically distributed, with an arbitrary distribution, and these lengths are independent of the lengths of the component intervals of M. If X is a Chung process with initial state i, and if i is a stable recurrent state, then $M = \{t : X_t = i\}$ is such a set.

2. S has an infinite Lévy measure and a strictly positive drift. Then M has no isolated points, it contains no open intervals, its interior is empty, but its Lebesgue measure is infinite; so M is like a generalized Cantor set. If X is a Chung process with initial state i, and if i is an instantaneous recurrent state, then $M = \{t : X_t = i\}$ is such a set.
3. S has an infinite Lévy measure and no drift. Then M has no isolated points, it contains no open intervals, its interior is empty, its Lebesgue measure is zero, but M has the power of the continuum. If X is the standard Brownian motion, then $M = \{t : X_t = 0\}$ is such a set.

In the first two cases, the function $p(t) = P[t \in M]$ is strictly positive. It is called a *p-function* by Kingman [7], who studied its analytical properties in detail. In the last case, $p(t) = 0$ for all t, as shown by Kesten [6].

For the classical case of renewal processes*, we refer the reader to Feller [3], Smith [14], and Çinlar [1]. For continuous regeneration, in addition to Maisonneuve [8], the paper by Fristedt [4] is recommended. The concept of regeneration has been weakened to allow the future after a regeneration time to depend on the current state. Such processes are called semiregenerative or regenerative systems. The simpler case, very close to renewal theory, goes under the names of Markov renewal theory and semi-Markov processes*; an account may be found in Çinlar [1]. For the continuous regeneration case, the fundamental work is Maisonneuve [10]. Such regenerative systems are used extensively in studying Markov processes and their excursions. For more details, see Maisonneuve [11, 12], Maisonneuve and Meyer [13], Jacod [5], and Çinlar and Kaspi [2].

References

[1] Çinlar, E. (1975). *Introduction to Stochastic Processes*. Prentice-Hall, Englewood Cliffs, N.J.

[2] Çinlar, E. and Kaspi, H. (1983). Regenerative systems and Markov additive processes. In *Seminar on Stochastic Processes 1982*. Birkhäuser Boston, Cambridge, Mass., pp. 123–147.

[3] Feller, W. (1966). *An Introduction to Probability Theory and Its Applications*, Vol. 2. Wiley, New York.

[4] Fristedt, B. (1964). Sample functions of stochastic processes with stationary and independent increments. *Advances in Probability and Related Topics, Vol. 3*, P. Ney and S. Port, eds. Marcel Dekker, New York, pp. 241–396.

[5] Jacod, J. (1974). Systèmes régénératifs et processus semi-markoviens. *Z. Wahrscheinl. verw. Geb.*, **31**, 1–23.

[6] Kesten, H. (1969). Hitting probabilities for single points for processes with stationary independent increments. *Mem. Amer. Math. Soc.*, **93**.

[7] Kingman, J. F. C. (1972). *Regenerative Phenomena*. Wiley, New York.

[8] Maisonneuve, B. (1968).

[9] Maisonneuve, B. (1971). Ensembles régénératifs, temps locaux et subordinateurs. *Sémin. Prob. V (Univ. Strasbourg)*, pp. 147–169. *Lect. Notes Math.*, **191**. Springer-Verlag, Berlin.

[10] Maisonneuve, B. (1974). Systèmes régénératifs. *Astérisque*, No. 15, Soc. Math. Fr., Paris.

[11] Maisonneuve, B. (1975). Entrance–exit results for semiregenerative processes. *Z. Wahrscheinl. verw. Geb.*, **32**, 81–94.

[12] Maisonneuve, B. (1979). On the structure of certain excursions of a Markov process. *Z. Wahrscheinl. verw. Geb.*, **47**, 61–67.

[13] Maisonneuve, B. and Meyer, P. A. (1974). Ensembles aléatoires markoviens homogènes, I–V. *Sémin. Prob. (Univ. Strasbourg)*, VIII pp. 172–261. *Lect. Notes Math.*, **381**. Springer-Verlag, Berlin.

[14] Smith, W. L. (1957). Renewal theory and its ramifications. *J. R. Statist. Soc. B*, **20**, 243–302.

(BROWNIAN MOTION
MARKOV PROCESSES
RENEWAL THEORY
SEMI-MARKOV PROCESSES
STOCHASTIC PROCESSES)

E. ÇINLAR

REGRESSION, BAYESIAN

A Bayesian approach to regression analysis is based on the belief that probability distributions may be assigned to all unknown parameters. Bayes' theorem* is then employed to update or revise these distributions as new (sample) information becomes available.

Bayesian regression analysis is an extremely broad topic, and it is not possible to cover all issues of interest in this one entry. This focus will be on the linear multiple regression* model and how a Bayesian statistician makes inferences about parameters and predicts future values of the process. Bayesian approaches to regression model selection are discussed, and the problem of eliciting the distributions of unknown parameters is also considered. Related topics which may be of interest, such as Bayesian approaches to analysis of variance, experimental design, polynomial regression, time-series analysis, and multivariate models will not be considered here, but an annotated bibliography will indicate additional sources of information. *See also* BAYESIAN INFERENCE.

The Bayesian approach to regression analysis was first widely publicized in ref. 7, where the first detailed examination of conjugate* prior distributions also appears. During the 1960s and early 1970s, Bayesian inference in general was based primarily on diffuse (improper) prior distributions, and Bayesian inference in regression was no exception (see, e.g., refs. 1 and 11). Later, after possible marginalization paradoxes arising from improper prior distributions became known, informative prior distributions* became more widely used. The hierarchical linear model developed as a result of very special assumptions about the nature of prior information (see refs. 6 and 8). More recently, Bayesian approaches have been considered in conjunction with more complex problems, such as fitting autoregressive and/or moving-average time-series* models, and selecting a "best" model from a choice of several (e.g., ref. 9). Elicitation of subjective prior distributions has been examined beginning in the late 1970s, thereby addressing the issue of how to make the relatively complex Bayesian regression methodology practical for the typical user.

LINEAR MULTIPLE REGRESSION MODEL

The linear multiple regression model may be written as

$$y_i = \sum_{j=1}^{p} x_{ij}\beta_j + \epsilon_i, \qquad i = 1, \ldots, n,$$

where y is the response or dependent variable and the x_j's are the predictor or independent variables. The unobservable random error terms, ϵ_i, are typically assumed to be uncorrelated with mean zero and common variance σ^2. In the normal linear regression model, the ϵ_i are assumed to be independently normal distributed with mean zero and variance σ^2. The quantities β_j, $j = 1, \ldots, p$ are the unknown regression coefficients, and together with σ^2 constitute the (unknown) parameters of the regression model. The model may be more conveniently stated using matrix notation:

$$\mathbf{y} = \mathbf{X}\boldsymbol{\beta} + \boldsymbol{\epsilon},$$

where

$\mathbf{y} = n \times 1$ vector of observations on the response variable
$\mathbf{X} = n \times p$ matrix of observations on the p predictor variables
$\boldsymbol{\beta} = p \times 1$ vector of regression coefficients
$\boldsymbol{\epsilon} = n \times 1$ vector of error terms.

Note that the term "linear" refers to the fact that the model is linear in $\boldsymbol{\beta}$ and $\boldsymbol{\epsilon}$.

The Bayesian approach to regression analysis views $\boldsymbol{\beta}$ and σ^2 as uncertain quantities having probability distributions associated with them. It allows for the combination of this prior information about the parameters with information obtained from the sample, via Bayes' theorem*, to obtain a posterior distribution* for $\boldsymbol{\beta}$ and σ^2. This distribution constitutes an inference regarding the parameters; Bayesian estimation* or hypothesis-testing* procedures may be used in conjunction with the posterior distribution if desired.

Inferences about Parameters

Let $f(\cdot)$ generically represent a density function. From Bayes' theorem,

$$f(\boldsymbol{\beta}, \sigma^2 \mid \mathbf{y}, \mathbf{X}) = \frac{f(\mathbf{y}, \mathbf{X} \mid \boldsymbol{\beta}, \sigma^2) f(\boldsymbol{\beta}, \sigma^2)}{f(\mathbf{y}, \mathbf{X})}.$$

The normal linear regression model implies a conditional distribution for $\mathbf{y}$ such that $f(\mathbf{y} \mid \mathbf{X}\boldsymbol{\beta}, \sigma^2)$ is multivariate normal with mean $\mathbf{X}\boldsymbol{\beta}$ and covariance matrix $\sigma^2\mathbf{I}$. If $\mathbf{X}$ is fixed, or if $\mathbf{X}$ is random but distributed independently of $\mathbf{B}$ and σ^2, then

$$f(\boldsymbol{\beta}, \sigma^2 \mid \mathbf{y}, \mathbf{X}) = \frac{f(\mathbf{y} \mid \mathbf{X}, \boldsymbol{\beta}, \sigma^2) f(\boldsymbol{\beta}, \sigma^2)}{f(\mathbf{y} \mid \mathbf{X})}.$$

Notice that any randomness in $\mathbf{X}$ is irrelevant for inferences about $\boldsymbol{\beta}$ and σ^2 as long as $\mathbf{X}$ is distributed independently of $\boldsymbol{\beta}$ and σ^2.

To obtain the posterior distribution $f(\boldsymbol{\beta}, \sigma^2 \mid \mathbf{y}, \mathbf{X})$, a choice of prior distribution for $(\boldsymbol{\beta}, \sigma^2)$ must be made. Many choices can lead to mathematical intractabilities, requiring numerical integration. Two tractable choices are considered below.

Case 1: Conjugate Prior Distribution. The conjugate prior distribution for $(\boldsymbol{\beta}, h)$, where $h = \sigma^{-2}$ is called the precision, is the normal-gamma* distribution whose density is given by

$$f(\boldsymbol{\beta}, h \mid \mathbf{b}, \mathbf{N}, v, \delta) \propto e^{-h(\boldsymbol{\beta}-\mathbf{b})'\mathbf{N}(\boldsymbol{\beta}-\mathbf{b})/2} \times h^{(1/2)p} e^{-hv\delta/2} h^{(1/2)\delta - 1}.$$

This distribution designates that $\boldsymbol{\beta}$, given h, is distributed normally with mean $\mathbf{b}$ and precision matrix $h\mathbf{N}$ (or covariance matrix $\sigma^2\mathbf{N}^{-1}$), and that h has a gamma distribution* with shape parameter v and scale parameter δ. The quantities $\mathbf{b}$, $\mathbf{N}$, v, and δ are commonly referred to as hyperparameters. Note that $\mathbf{N}$ must be of rank p for the normal-gamma distribution to be proper. Employing the normal-gamma prior distribution with hyperparameters $\mathbf{b}^*$, $\mathbf{N}^*$, v^*, and δ^* gives a posterior distribution for $(\boldsymbol{\beta}, h)$ which is normal-gamma with hyperparameters

$$\begin{aligned}
\mathbf{b}^{**} &= (\mathbf{N}^* + \mathbf{X}'\mathbf{X})^{-1}(\mathbf{N}^*\mathbf{b}^* + \mathbf{X}'\mathbf{X}\mathbf{b}) \\
&= (\mathbf{N}^* + \mathbf{X}'\mathbf{X})^{-1}(\mathbf{N}^*\mathbf{b}^* + \mathbf{X}'\mathbf{y}), \\
\mathbf{N}^{**} &= \mathbf{N}^* + \mathbf{X}'\mathbf{X}, \\
v^{**} &= \frac{1}{\delta^{**}}(\delta^* v^* + \mathbf{b}^{*\prime}\mathbf{N}^*\mathbf{b}^* + \mathbf{y}'\mathbf{y} \\
&\qquad - \mathbf{b}^{**\prime}\mathbf{N}^{**}\mathbf{b}^{**}), \\
\delta^{**} &= \delta^* + p^* + n - p^{**},
\end{aligned}$$

where $\mathbf{b}$ is the usual least-squares estimate $(\mathbf{X}'\mathbf{X})^{-1}\mathbf{X}'\mathbf{y}$, $p^* = \text{rank}\ (\mathbf{N}^*)$ and $p^{**} = \text{rank}\ (\mathbf{N}^{**})$. These results hold regardless of the rank of $\mathbf{N}^*$ and $\mathbf{X}$; if $\mathbf{N}^*$ and $\mathbf{X}$ are of full rank, then $\delta^{**} = \delta^* + n$.

To make inferences about h, the marginal posterior distribution of h would be used. For a normal-gamma prior distribution, the marginal posterior distribution of h is gamma with parameters v^{**} and δ^{**}. This implies an inverted-gamma distribution* for σ^2. Using squared-error loss for estimation, the Bayes estimate of σ^2 is $\delta^{**}v^{**}/(\delta^{**} - 2)$ as long as $\delta^{**} > 2$.

Inferences about the regression coefficients* can be made based on the marginal posterior distribution of $\boldsymbol{\beta}$. Based on the normal-gamma prior, the marginal posterior distribution of $\boldsymbol{\beta}$ is multivariate Student t with hyperparameters $\mathbf{b}^{**}$, $\mathbf{N}^{**}/v^{**}$, and δ^{**} (i.e., the mean of $\boldsymbol{\beta}$ is $\mathbf{b}^{**}$ and the covariance matrix of $\boldsymbol{\beta}$ is $(\mathbf{N}^{**})^{-1}v^{**}[\delta^{**}/(\delta^{**} - 2)]$). Under squared-error loss, therefore, the Bayes estimate of $\boldsymbol{\beta}$ is $\mathbf{b}^{**}$—a weighted average of the sample location $\mathbf{b}$ and the prior location $\mathbf{b}^*$.

Case 2: Diffuse Prior Distribution. To express prior ignorance or relative lack of prior information, a diffuse prior distribution for $(\boldsymbol{\beta}, \sigma^2)$ may be used. The most common choice is the improper distribution with density

$$f(\boldsymbol{\beta}, \sigma^2) \propto 1/\sigma^2$$

for $-\infty < \beta_i < \infty$, $i = 1, \ldots, p$ and $0 < \sigma^2 < \infty$. This distribution implies that $\log \sigma^2$ is uniform over $(-\infty, \infty)$, as suggested by Jeffreys (see ref. 19). It may be obtained from the normal-gamma conjugate prior by setting $\mathbf{N}^* = \mathbf{0}$ and $\delta^* = 0$. Given this prior specification, the posterior distribution of $(\boldsymbol{\beta}, \sigma^2)$ is normal-gamma with hyperparameters

$$\mathbf{b}^{**} = \mathbf{b},$$
$$\mathbf{N}^{**} = \mathbf{X}'\mathbf{X},$$
$$v^{**} = \frac{1}{\delta^{**}}\left[\mathbf{y}'\mathbf{y} - \mathbf{b}'(\mathbf{X}'\mathbf{X})\mathbf{b}\right],$$
$$\delta^{**} = n - p,$$

where $p = \text{rank}(\mathbf{X})$. This posterior distribution is proper as long as $n > p$. Marginally, $\boldsymbol{\beta}$ has a multivariate Student t-distribution with hyperparameters $\mathbf{b}$, $\mathbf{X}'\mathbf{X}/s^2$, and $n - p$, where s^2 is the usual mean squared error $[\mathbf{y}'\mathbf{y} - \mathbf{b}'(\mathbf{X}'\mathbf{X})\mathbf{b}]/(n - p)$. Under squared-error loss, the Bayes estimate of $\boldsymbol{\beta}$ is $\mathbf{b}$, the usual least-squares estimate.

Illustrative Example. (Adapted from ref. 7.)

Assume that the model under consideration is $\mathbf{y} = \beta_0 + \beta_1\mathbf{X}_1 + \beta_2\mathbf{X}_2 + \boldsymbol{\epsilon}$, where the observations are as in Table 1 ($n = 20$ observations):

Table 1

y	X_1	X_2
10.74126	0.693	0.693
10.98296	1.733	0.693
10.52923	0.693	1.386
11.58911	1.733	1.386
11.79831	0.693	1.792
11.75854	2.340	0.693
11.94535	1.733	1.792
12.33049	2.340	1.386
11.89575	2.340	1.792
10.35612	0.693	0.693
11.16725	0.693	1.386
11.16556	1.733	0.693
11.68194	1.733	1.386
11.41347	0.693	1.792
11.20968	2.340	0.693
11.98168	1.733	1.792
12.32426	2.340	1.386
12.02976	2.340	1.792
11.39302	1.733	1.386
10.89403	0.693	0.693

A least-squares analysis of these data would provide $\mathbf{b}' = (9.770, 0.524, 0.693)$, i.e., $\hat{\mathbf{y}} = 9.770 + 0.524\mathbf{X}_1 + 0.693\mathbf{X}_2$. Also, $\hat{\sigma}^2 = \text{MSE} = 0.07953$, and there are $20 - 3 = 17$ degrees of freedom.

Assume that prior information about β_1 and β_2 is such that

$$E(\beta_1) = E(\beta_2) = 0.5,$$
$$P(0.9 < \beta_1 + \beta_2 < 1.1 \mid \sigma = 0.3) = 0.9,$$
$$P(0.2 < \beta_1 < 0.8 \mid \sigma = 0.3)$$
$$= P(0.2 < \beta_2 < 0.8 \mid \sigma = 0.3) = 0.9.$$

This information implies that

$$\text{var}(\beta_1 \mid \sigma = 0.3) = \text{var}(\beta_2 \mid \sigma = 0.3) = 0.03325912,$$

$$\text{cov}(\beta_1, \beta_2 \mid \sigma = 0.3) = -0.03141139.$$

In addition, assume that β_0 is a priori independent of β_1 and β_2 with $E(\beta_0) = 5$ and

$$P(-10 < \beta_0 < 20 \mid \sigma = 0.3) = 0.9,$$

implying that $\text{var}(\beta_0 \mid \sigma = 0.3) = 83.1478$. Hence the prior information about $\boldsymbol{\beta}$, conditional on σ^2, may be summarized as

$$E(\boldsymbol{\beta}) = \mathbf{b}^* = \begin{pmatrix} 5.0 \\ 0.5 \\ 0.5 \end{pmatrix},$$

$$\text{cov}(\boldsymbol{\beta} \mid \sigma^2) = \sigma^2\mathbf{N}^{*-1}$$
$$= \sigma^2\begin{pmatrix} 924.0 & 0 & 0 \\ 0 & 0.3695 & -0.349 \\ 0 & -0.349 & 0.3695 \end{pmatrix}.$$

To obtain a prior density for σ, assume that a priori $P(\sigma < 0.3) = 0.5$, $P(\sigma > 0.65) = 0.05$. These two equations can be used to solve for δ^* and v^*, yielding $\delta^* = 4$, $v^* = 0.0754$ (i.e., the marginal prior density of h is gamma with $v^* = 0.0754$ and $\delta^* = 4$).

After combining prior and sample information, the posterior hyperparameters are $\mathbf{b}^{**\prime} = (10.028, 0.476, 0.548)$,

$$\mathbf{N}^{**-1} = \begin{pmatrix} 0.12264 & -0.03549 & -0.01391 \\ -0.03549 & 0.07106 & -0.05907 \\ -0.01391 & -0.05907 & 0.08340 \end{pmatrix},$$

$v^{**} = 0.0755$, and $\delta^{**} = 24$. In the posterior distribution, β_0 is no longer independent of β_1 and β_2. The marginal distribution of $\boldsymbol{\beta}$ is multivariate t^* with hyperparameters $\mathbf{b}^{**}$,

$\mathbf{N}^{**}/v^{**}$, and δ^{**} as given above. Any single element of $\boldsymbol{\beta}$ has a univariate t distribution; for example, β_1 is univariate t with mean $b_1^{**} = 0.476$, variance

$$(0.07106)(0.0755)\left[24/(24-2)\right] = 0.00585,$$

and 24 degrees of freedom. The Bayes point estimate of β_1 using squared-error loss is $b_1^{**} = 0.476$. A 90% *credible interval* for β_1 is given by (0.35, 0.60) [i.e., there is a 0.9 probability that β_1 lies in the interval (0.35, 0.60)].

If the diffuse prior had been used for $(\boldsymbol{\beta}, h)$, the posterior mean of β_1 would have been $b_1 = 0.524$ (the least-squares estimate). The posterior variance of β_1 would have been 0.00996, which is the sampling distribution variance of b_1 in the least-squares analysis.

X Not of Full Rank

The matrix $\mathbf{X}'\mathbf{X}$ is singular when the $n \times p$ matrix $\mathbf{X}$ is of rank r with $0 \leqslant r < p$. This may occur in cases of multicollinearity* (i.e., the columns of $\mathbf{X}$ may not be linearly independent). Also, if $n < p$, then $\mathbf{X}$ cannot be of rank p. When $\mathbf{X}'\mathbf{X}$ is singular, not all the regression coefficients will be estimable (in the classical, sampling-theory sense) unless prior information can be added to the sample data. A Bayesian analysis can accommodate singular $\mathbf{X}'\mathbf{X}$ in a straightforward way. Assuming a normal-gamma prior distribution with hyperparameters $\mathbf{b}^*$, $\mathbf{N}^*$, v^*, and δ^* to represent the prior information about $(\boldsymbol{\beta}, h)$, where $\mathbf{N}^*$ is of full rank, the posterior distribution of $(\boldsymbol{\beta}, h)$ will be proper normal-gamma, with hyperparameters as given previously, even though $\mathbf{X}'\mathbf{X}$ is singular.

Prediction

It is often the case in regression analysis that it is desirable to predict values for the as yet unobserved "observations." In the Bayesian approach it is possible to obtain the distribution of these future values, given the existing sample information. Such a distribution is called a *predictive distribution*; it does not depend on any unknown parameters and hence is a marginal distribution with respect to them.

Let $\bar{\mathbf{y}} = (y_{n+1}, y_{n+2}, \ldots, y_{n+k})$ represent the vector of k future observations for which the predictive distribution will be obtained. These observations are assumed to be generated by the model

$$\bar{\mathbf{y}} = \bar{\mathbf{X}}\boldsymbol{\beta} + \bar{\boldsymbol{\epsilon}},$$

where $\bar{\mathbf{X}}$ is a $k \times p$ matrix of given values for the independent variables in the k future periods, and $\bar{\boldsymbol{\epsilon}}$ is a $k \times 1$ vector of k future error terms. The vector $\bar{\boldsymbol{\epsilon}}$ is assumed to be normally distributed with mean $\mathbf{0}$ and covariance matrix $\sigma^2\mathbf{I}$. If data $(\mathbf{y}, \mathbf{X})$ have already been obtained, the predictive density $f(\bar{\mathbf{y}} \mid \bar{\mathbf{X}}, \mathbf{y}, \mathbf{X})$ is desired. If no data are currently available, the predictive density $f(\bar{\mathbf{y}} \mid \bar{\mathbf{X}})$ may be obtained as long as the prior distribution of $(\boldsymbol{\beta}, \sigma^2)$ is proper.

The predictive density may be derived as

$$\begin{aligned} f(\bar{\mathbf{y}} \mid \bar{\mathbf{X}}, \mathbf{y}, \mathbf{X}) &= \int\int f(\bar{\mathbf{y}}, \boldsymbol{\beta}, \sigma^2 \mid \bar{\mathbf{X}}, \mathbf{y}, \mathbf{X})\, d\sigma^2\, d\boldsymbol{\beta} \\ &= \int\int f(\bar{\mathbf{y}} \mid \bar{\mathbf{X}}, \boldsymbol{\beta}, \sigma^2) \\ &\qquad \times f(\boldsymbol{\beta}, \sigma^2 \mid \mathbf{y}, \mathbf{X})\, d\sigma^2\, d\boldsymbol{\beta}, \end{aligned}$$

where $0 < \sigma^2 < \infty$ and $-\infty < \beta_i < \infty$ for $i = 1, \ldots, p$. If $f(\boldsymbol{\beta}, \sigma^2 \mid \mathbf{y}, \mathbf{X})$ is a normal-gamma density, with hyperparameters $\mathbf{b}^*$, $\mathbf{N}^*$, v^*, and δ^*, then the posterior predictive density of $\bar{\mathbf{y}}$ is multivariate Student t with hyperparameters $\mathbf{b}^{**}$, $\mathbf{N}_y/v^{**}$, and δ^{**}, where $\mathbf{N}_y = \mathbf{I} - \bar{\mathbf{X}}(\mathbf{N}^{**})^{-1}\bar{\mathbf{X}}'$. The mean of this distribution is $\bar{\mathbf{X}}\mathbf{b}^{**}$ (for $\delta^{**} > 1$) and its covariance matrix is $\mathbf{N}_y^{-1}v^{**}[\delta^{**}/(\delta^{**} - 2)]$ for $\delta^{**} > 2$. These results apply even if $\bar{\mathbf{X}}$ is of less than full rank. Under squared-error prediction loss, the Bayes prediction of $\bar{\mathbf{y}}$ is $\bar{\mathbf{X}}\mathbf{b}^{**}$. If the prior distribution of $(\boldsymbol{\beta}, \sigma^2)$ had been the diffuse prior $f(\boldsymbol{\beta}, \sigma^2) \propto 1/\sigma^2$, then $f(\bar{\mathbf{y}} \mid \bar{\mathbf{X}}, \mathbf{y}, \mathbf{X})$ would be multivariate Student t with mean $\bar{\mathbf{X}}\mathbf{b}$, the least-squares fitted value. The covariance matrix in this case is

$$\frac{\delta s^2}{\delta - 2}\left[\mathbf{I} + \bar{\mathbf{X}}(\mathbf{X}'\mathbf{X})^{-1}\bar{\mathbf{X}}'\right],$$

where $\delta = n - p$.

The marginal distribution of any element of

$\bar{\mathbf{y}}$ will be univariate Student t. Also, a linear combination of the elements of $\bar{\mathbf{y}}$ will be univariate Student t.

HIERARCHICAL LINEAR REGRESSION MODEL

By considering prior information about the regression coefficients β_i, $i = 1, \ldots, p$, to be of a type that admits a prior distribution which is exchangeable, a regression model involving linearity at more than one stage in a hierarchy may be obtained. The exchangeable prior distribution $f(\boldsymbol{\beta})$ means that the distribution would be unaltered by any permutation of the suffixes. In particular, it implies that $E(\beta_i) = \mu$ for $i = 1, \ldots, p$, and hence the parameters exhibit a linear structure analogous to that assumed for the dependent variable. It may, of course, be necessary to rescale the predictor variables so that the assumption of exchangeability* is reasonable. Levels or stages are added to the hierarchy by introducing distributions over unknown hyperparameters; the number of stages may reflect the type of model under consideration (e.g., fixed effects, random effects).

For example, suppose that in the linear multiple regression model

$$\mathbf{y} = \mathbf{X}\boldsymbol{\beta} + \boldsymbol{\epsilon}$$

it is assumed that $\boldsymbol{\epsilon}$ has a normal distribution with mean vector $\mathbf{0}$ and covariance matrix $\sigma^2\mathbf{I}$, with σ^2 known. Assume also that each β_i, $i = 1, \ldots, p$, is distributed normally with mean μ and known variance σ_β^2. If the distribution of μ is assumed to be diffuse over the entire real line, then the Bayes estimate of $\boldsymbol{\beta}$ is $\{\mathbf{I} + k(\mathbf{X}'\mathbf{X})^{-1}(\mathbf{I} - p^{-1}\mathbf{J})\}^{-1}\mathbf{b}$ under squared-error loss, where $\mathbf{J}$ is a matrix of 1's, $\mathbf{b}$ is the usual least-squares estimate, and $k = \sigma^2/\sigma_\beta^2$. If instead it had been assumed that μ was not random but took the known value 0, then the Bayes estimate of $\boldsymbol{\beta}$ is $\{\mathbf{I} + k(\mathbf{X}'\mathbf{X})^{-1}\}^{-1}\mathbf{b}$ under squared-error loss. These estimates are similar to the ridge regression* estimate, which shrinks $\mathbf{b}$ toward $\mathbf{0}$. Complete details and generalizations are given in refs. 6 and 8.

MODEL SELECTION

It is often important to be able to compare alternative regression models in order to select one that is suitable or most likely to be "true." Various Bayesian approaches may be taken to this problem, which essentially involves a decision regarding which predictor variables to include in the regression equation.

If alternative regression models are stated as competing statistical hypotheses, a Bayesian approach to hypothesis testing may be employed. This approach involves examination of the Bayes factor or posterior-to-prior odds ratio in favor of one hypothesis relative to another. For example, assume that H_i, $i = 0, 1$, are hypotheses specifying two nested normal linear regression models (i.e., the model specified by H_1, say, contains the model specified by H_0). Each model may be written as

$$\mathbf{y} = \mathbf{X}_i\boldsymbol{\beta}_i + \boldsymbol{\epsilon},$$

where $\boldsymbol{\epsilon}$ has a multivariate normal distribution with mean $\mathbf{0}$ and covariance structure $\sigma^2\mathbf{I}$, $\mathbf{X}_i$ is known and of full rank, $\boldsymbol{\beta}_1' = [\boldsymbol{\beta}_0' \; \boldsymbol{\beta}']$, and $\mathbf{X}_1 = [\mathbf{X}_0 \; \mathbf{X}]$. Without loss of generality, the columns of $\mathbf{X}$ may be assumed to be orthogonal to the columns of $\mathbf{X}_0$. The Bayes factor for H_0 against H_1 is

$$B_{01} = f(\mathbf{y} \mid H_0)/f(\mathbf{y} \mid H_1),$$

where

$$f(\mathbf{y} \mid H_i) = \int\int f(\mathbf{y} \mid \mathbf{X}_i, \boldsymbol{\beta}_i, \sigma^2) \times f(\boldsymbol{\beta}_i, \sigma^2 \mid \mathbf{X}_i)\, d\sigma^2\, d\boldsymbol{\beta}_i.$$

In other words, the Bayes factor is a ratio of the predictive densities under the two models. It may be sufficient to examine the Bayes factor to see which hypothesis is more likely, or given a loss structure for the incorrect choice of hypothesis, an optimal decision may be made by minimizing expected loss.

Under a loss structure which considers that taking observations may be costly, a formal decision-theoretic analysis may be used to decide which variables to observe, as long as $\mathbf{X}$ is considered to be stochastic. For example, under squared-error loss for prediction, the subset (indexed by I) of predictor variables to observe is chosen to minimize $E\{y - f(\mathbf{x}_I)\}^2 + C_I$, where y is the true value of the dependent variable, $f(\mathbf{x}_I)$ is its predictor, and C_I is the cost of observing those variables whose subscripts are in the set I. This minimization may be solved by choosing $f(\mathbf{x}_I)$ conditionally on knowing $\mathbf{x}_I$, and then averaging over x_I for fixed I. The optimal predictor for fixed $\mathbf{x}_I$ is $f(\mathbf{x}_I) = E(y \mid \mathbf{x}_I)$. After averaging over $\mathbf{x}_I$, it is seen that the set I must be chosen to minimize $E(\boldsymbol{\beta})' V(\mathbf{x}_J) E(\boldsymbol{\beta}) + C_I$, where $V(\mathbf{x}_J)$ is the covariance matrix of the *unobserved* predictor variables. Full details and an illustration are given in ref. 6. Other Bayesian model selection criteria are discussed in refs. 2 and 9.

ELICITATION

An important aspect of the Bayesian regression problem involves the elicitation or assessment of the prior density for $(\boldsymbol{\beta}, h)$ or $(\boldsymbol{\beta}, \sigma^2)$. Since $f(\boldsymbol{\beta}, h)$ represents a multivariate density, direct interrogation about all the parameters may not be fruitful. An alternative technique involves elicitation of the predictive distribution of y, conditional on sets of values for the predictor variables in the model. The predictive distribution for a single y value is univariate, and it involves observable variables only, not unknown parameters. In addition, such a predictive distribution may be considered without direct reference to either the sampling model or the form of the prior distribution. Once a sampling model and family of prior distributions is imposed, the distribution of $(\boldsymbol{\beta}, \sigma^2)$ may be inferred or "fitted" from the elicited predictive distributions.

To assess the predictive distributions, moments of the distributions may be elicited. Alternatively, fractiles may be easier to elicit; moments could then be calculated from the elicited predictive distributions.

Existing methods for inferring or fitting the distribution $f(\boldsymbol{\beta}, \sigma^2)$ are, to date, quite crude (see refs. 4 and 10 for two possibilities). Additional research is needed so that a more sophisticated inferential procedure can be determined. In particular, a model of elicitation error is needed, and current elicitation research is being aimed in that direction.

RESEARCH DIRECTIONS

Current research in the Bayesian linear regression model extends in many directions. As mentioned previously, improved elicitation techniques is one important goal. Also, Bayesian analysis of more complex models continues (e.g., time-series* models, growth curves*, log-linear models).

Bayesian multivariate analysis* has an obvious and important link with Bayesian regression analysis. In this regard, the development of new families of multivariate distributions suitable for use as prior distributions is important. In addition, an understanding of how Bayesians make inference in the presence of missing observations on some variables will lead to extensions of the Bayesian analysis of the linear model.

References

[1] Box, G. E. P. and Tiao, G. C. (1973). *Bayesian Inference in Statistical Analysis*. Addison-Wesley, Reading, Mass. (An early Bayesian examination of the linear model, particularly in analysis of variance. Diffuse priors assumed throughout.)

[2] Goldstein, M. (1976). *Biometrika*, **63**, 51–58. (Employs a direct specification of a prior distribution and considers estimation, prediction, and model selection.)

[3] Judge, G. G., Hill, R. C., Griffiths, W. E., Lütkepohl, H., and Lee, T.-C. (1982). *Introduction to the Theory and Practice of Econometrics*. Wiley, New York. (Provides an elementary introduction to Bayesian regression for the econometrics researcher or student.)

[4] Kadane, J. B., Dickey, J. M., Winkler, R. L., Smith, W. S., and Peters, S. C. (1980). *J. Amer. Statist. Ass.*, **75**, 845–854. (Describes procedure for eliciting hyperparameters in a conjugate prior distribution for the normal regression model.)

[5] Lindley, D. V. (1968). *J. R. Statist. Soc. B*, **30**, 31–66. (Bayesian linear model selection under

predictive squared-error loss with explicit costs for observation.)

[6] Lindley, D. V. and Smith, A. F. M. (1972). *J. R. Statist. Soc. B*, **34**, 1–41. (Very important paper discussing the Bayesian linear model with the assumption of exchangeability of the regression coefficients.)

[7] Raiffa, H. and Schlaifer, R. (1961). *Applied Statistical Decision Theory*. MIT Press, Cambridge, Mass. (Pioneering work on natural conjugate prior distributions is presented; the Bayesian regression model with conjugate priors is extensively considered.)

[8] Smith, A. F. M. (1973). *J. R. Statist. Soc., B*, **35**, 67–75. (Follow-up to ref. 6.)

[9] Trader, R. L. (1983). *Manag. Sci.*, **29**, 622–632. (The role of predictive distribtuions in regression model selection is considered.)

[10] Winkler, R. L., Smith, W. S., and Kulkarni, R. B. (1978). *Manag. Sci.*, **24**, 977–986. (First discussion of eliciting predictive distributions to infer prior distributions for regression parameters.)

[11] Zellner, A. (1971). *An Introduction to Bayesian Inference in Econometrics*. Wiley, New York. (Bayesian inference in a wide variety of econometric models, including multivariate regression and simultaneous equation models. Diffuse prior distributions employed throughout.)

Bibliography

See the following works, as well as the references just given, for more information on the topic of Bayesian regression.

DeGroot, M. H. (1970). *Optimal Statistical Decisions*. McGraw-Hill, New York. (Brief mention of the Bayesian linear model, using conjugate priors, in Chapter 11.)

Drèze, J. H. (1977). *J. Econometrics*, **6**, 329–354. (Discusses how poly-*t* densities arise in Bayesian regression analysis.)

Halpern, E. F. (1973). *J. Amer. Statist. Ass.*, **68**, 137–143. (Bayesian selection of the degree and coefficients of a polynomial regression model under the assumption of normal errors.)

Judge, G. G., Griffiths, W. E., Hill, R. C. and Lee, T.-C. (1980). *The Theory and Practice of Econometrics*. Wiley, New York.

Leamer, E. E. (1978). *Specification Searches*. Wiley, New York. (Bayesian approach to analysis of nonexperimental data. Title refers to the notion that in nonexperimental situations, a statistical model is not known but must be sought as part of the inferential process.)

Press, S. J. (1982). *Applied Multivariate Analysis*, 2nd ed. Krieger, Malabar, Fla. (Bayesian multivariate analysis, including multivariate regression analysis.)

Smith, A. F. M. and Spiegelhalter, D. J. (1980). *J. R. Statist. Soc. B*, **42**, 213–220.

Smith, A. F. M. and Spiegelhalter, D. J. (1982). *J. R. Statist. Soc. B*, **44**, 377–387. (Examinations of Bayes factors. Reference 15 examines global and local Bayes factors and their roles in model selection. In ref. 16, (improper) diffuse prior distributions are considered in computation of Bayes factors for both linear and log-linear models.)

(BAYESIAN INFERENCE
FIDUCIAL INFERENCE)

RAMONA L. TRADER

REGRESSION COEFFICIENTS

The term "regression" denotes a broad collection of statistical methods, which involve studying more than one characteristic associated with experimental units. The fundamental idea is that if the characteristics are related, knowledge of values of observations on some of the characteristics sheds light on the value or expected value of another characteristic.

Examples abound, and regression methods are used in all types of investigative studies. We concentrate on the role and interpretation of particular parameters appearing in regression equations.

Consider first the case where one explanatory variable X is studied to shed light on one response variable Y. There are two popular statistical models employed to describe pairs (X_i, Y_i) which have a linear relationship. The fixed model assumes that the random values Y_i are matched with X_i, which can either be chosen in advance of gathering the data or can be controlled in the process of data gathering. The conditional expected value of Y given that the explanatory variable has the value X is assumed to be $\beta_0 + \beta_1 X$. The quantities β_0 and β_1 are called regression coefficients. β_0 is the model intercept and β_1 is the model slope.

When the X values cannot be controlled, it may be more appropriate to assume that the pairs (X_i, Y_i) are jointly distributed as a bivariate normal* with parameters, μ_X, μ_Y,

ρ, σ_X, and σ_Y. For the bivariate normal the conditional expected value of Y given X is $\beta_0 + \beta_1 X$, where $\beta_0 = \mu_Y - \rho\sigma_Y/\sigma_X$ and $\beta_1 = \rho\sigma_Y/\sigma_X$. Since β_1 and ρ are thusly related, correlation analysis relative to ρ is essentially the same as regression analysis relative to β_1.

To estimate the regression coefficients, an estimate criterion must be adopted. The most commonly used criterion for the fixed model is least squares (*see* LEAST SQUARES). The objective is to fit a line, denoted by $\hat{Y} = b_0 + b_1 X$, to n pairs (X_i, Y_i) in such a way that the residual sum of squares is minimized [i.e., the expression $\sum_{i=1}^n (Y_i - \hat{Y}_i)^2$ is minimized]. In the simplest version of this problem (referred to as simple linear regression), it is assumed that there is a common variance σ^2 for each of the Y_i. The functions of X_i and Y_i that minimize $\sum_{i=1}^n (Y_i - \hat{Y}_i)^2$ are

$$b_1 = \frac{\sum_i (X_i - \overline{X})(Y_i - \overline{Y})}{\sum_i (X_i - \overline{X})^2} \quad \text{and}$$

$$b_0 = \overline{Y} - b_1 \overline{X}.$$

b_1 is an unbiased estimator of β_1, b_0 is an unbiased estimator of β_0, and in addition an unbiased estimator of σ^2 is the

$$\text{mean squared error} = \text{MSE} = \sum_{i=1}^{n} \frac{(Y_i - \hat{Y})^2}{n-2}.$$

The number $n - 2$ is called the degrees of freedom appropriate for simple linear regression, it being 2 less than the sample size because two parameters, namely β_0 and β_1, are being estimated.

An important regression concept is that of the standard error of the regression coefficient. The standard error of b_i is denoted by S_{b_i} and is the square root of the estimated variance of b_i. S_{b_i} is a measure of how b_i varies from sample to sample. In simple linear regression, S_{b_1} works out to be

$$\sqrt{\frac{\text{MSE}}{\sum_{i=1}^n (X_i - \overline{X})^2}}.$$

If both X_i and Y_i are random variables and are jointly distributed as a bivariate normal, the maximum likelihood criterion is often invoked (*see* MAXIMUM LIKELIHOOD). The estimators of β_0 and β_1 are, upon employing the maximum likelihood* criterion, found to be given by the same expression as obtained through least squares.

A test of the hypothesis that X and Y are linearly unrelated is equivalent to testing the hypothesis that $\beta_1 = 0$. If a bivariate normal model is assumed, the appropriate test statistic is $t = b_1/S_{b_1}$. t is distributed as Student's t with $n - 2$ degrees of freedom. If the fixed model is assumed, then the additional assumption of conditional normality for the Y's given the X's leads to the same test statistic.

An applied setting is helpful to put ideas in perspective. Out of a limitless number of possible settings we choose an economic one. Suppose that $n = 7$ cities of roughly the same size are selected. No two cities are geographically close to one another. Let X_i denote the planned and controlled amount of advertising spent to promote a new product in city i, and let Y_i be the sales total for the product in the ith city. In convenient units of money the data pairs were as follows:

X_i	Y_i
0	1.8
0	2.2
2	2.0
3	4.0
5	3.9
5	5.2
6	4.0

The data yields the following statistics:

$$\overline{Y} = 3.3, \quad b_0 = 1.95, \quad b_1 = 0.45,$$

$$\text{MSE} = 0.562, \quad S_{b_1} = 0.125.$$

The least-squares line is $\hat{Y} = 1.95 + 0.45X$.

Before testing the hypothesis that X and Y are linearly unrelated, the assumptions are spelled out in detail.

1. The X_i values are controlled and exact.
2. The Y_i values are realizations of random variables. The values are rounded to the nearest tenth of a unit for illustrative purposes.
3. The Y_i random variables are mutually independent.
4. For each city i, the conditional random variable Y_i given X_i: (a) has expected value $\beta_0 + \beta_1 X_i$, (b) has common variance σ^2, and (c) is normal.

If $\beta_1 = 0$, $t = b_1/S_{b_1}$ is distributed as a Student t variable with 5 degrees of freedom. The nature of the setting suggests a one-sided critical region. Working with a type I error probability of 0.05, the tabulated critical value is found to be 2.571. The computed value of b_1/S_{b_1} is 3.6014; hence one rejects the hypothesis that advertising and sales are unrelated.

One should be careful with regard to reporting and interpreting findings relative to regression coefficients. As a measure of the linear relationship between X and Y, the coefficient b_1 can be meaningless when the relationship is something other than linear. Even when the relationship between X and Y is linear, the magnitude of b_1 alone does not suffice as an indicator of the strength of the linear relationship. This is because b_1 depends on the units of measurement for both X and Y. b_1X can be written $b_1X = (b_1/c)cX = b_1^* X^*$, where $b_1^* = b_1/c$ and $X^* = cX$. The regression slope is small or large in numerical magnitude depending on the value of c, or said in another way, the magnitude of the slope depends on the unit in which the explanatory variable is measured.

If $Y = KY^*$ with $Y^* = b_0/K + (b_1/K)X$, then we see that the units for the measurement of Y also effect the magnitudes of the regression coefficients. These observations concerning the nature of the magnitude of regression coefficients enlighten one with regard to why b_1 is compared with its standard error in order to judge whether or not b_1 should be declared significantly different from zero.

After one had decided that a regression coefficient is significantly different from zero, one must be careful with regard to reporting the finding. Mary Sue Younger, in *A Handbook for Linear Regression* [1], lists situations in which variables X and Y will be related. X and Y will be related if:

1. X causes Y
2. Y causes X
3. X and Y interact with each other
4. Both are caused by a third variable
5. They act alike just by chance

Furthermore, they can be related spuriously because of a nonrepresentative (biased) sample. If a regression slope is correctly declared significantly different from zero, we do not necessarily know which of the above is the true situation.

In longitudinal studies* it may be appropriate to report that a statistically significant positive regression coefficient implies that when X increases, Y will increase and b_1 indicates the magnitude of the increase. In cross-sectional studies, this sort of statement can be misinterpreted. It could be that a third variable, say Z, is important but not included in the study. It might be that if one increases X without regard to Z, the value of Y will decrease. For example, consider an illustrative situation where Y denotes store sales as a function of the number of checkout employees X. Adding more employees, that is, increasing X, may not in fact increase the sales until one also adds more checkout counters, Z.

In regression situations where there are multiple predictor variables, the role of regression coefficients is more complicated. Consider a least-squares multiple regression* equation

$$\hat{Y} = f(X_1, \ldots, X_k) = b_0 + b_1X_1 + \cdots + b_kX_k.$$

The predictors X_i may be functions of other variables. These functions are of many types. For illustration suppose that the data consist of observations of triplets $(Y_i, W_{1i},$

W_{2i}), suppose that $X_1 = W_1 + W_2$ and $X_2 = W_1 W_2$, and suppose that we fit to the data by least squares* the equation

$$\hat{Y} = f(X_1, X_2) = b_0 + b_1 X_1 + b_2 X_2.$$

We discuss here the use and interpretation of the regression coefficients b_i. They describe to some extent the role played by X_i in explaining the variability of the response variable Y. In most situations part of the variability in Y can be associated with any of several predictor variables. To see this, consider the extreme case where $X_2 = cX_1$; then

$$b_1 X_1 + b_2 X_2 = (b_1 - ac)X_1 + (b_2 + a)X_2$$

for all values of a. In this extreme case the magnitude of $b_1 + b_2 c$ could be important but the value of b_1 alone would be meaningless.

Despite the fact that individually the regression coefficients can be meaningless, collectively they can play an important role in investigative studies. To study their joint properties, it is helpful to express matters in matrix notation. Some notation follows. Let $\mathbf{A}'$ denote the transpose of a matrix $\mathbf{A}$. Let $\mathbf{A}^{-1}$ denote the inverse of a nonsingular matrix $\mathbf{A}$. Let $\mathbf{Y}'$ denote the row vector of response observations $Y_1, Y_2, \ldots, Y_n$. Let

$$\mathbf{X} = \begin{bmatrix} 1 & X_{11} & \cdots & X_{1j} & \cdots & X_{1k} \\ 1 & X_{21} & \cdots & X_{2j} & \cdots & X_{2k} \\ \vdots & \vdots & & \vdots & & \vdots \\ 1 & X_{i1} & \cdots & X_{ij} & \cdots & X_{ik} \\ \vdots & \vdots & & \vdots & & \vdots \\ 1 & X_{n1} & \cdots & X_{nj} & \cdots & X_{nk} \end{bmatrix}$$

and

$$\boldsymbol{\beta} = \begin{bmatrix} \beta_0 \\ \beta_1 \\ \vdots \\ \beta_j \\ \vdots \\ \beta_k \end{bmatrix}.$$

The model assumptions are:

1. Y_i $(i = 1, \ldots, n)$ are normal.
2. The expected value of Y_i, given $X_{i1}, X_{i2}, \ldots, X_{ik}$, is $\beta_0 + \sum_{j=1}^{k} \beta_j X_{ij}$.
3. Y_i are independent with $\text{var}\{Y_i\} = \sigma^2$.

Translated into matrix notation the assumptions can be expressed as:

1. $\mathbf{Y}$ is an n-component multivariate normal vector.
2. The expected value of the vector $\mathbf{Y}$ is $\mathbf{X}\boldsymbol{\beta}$.
3. The covariance matrix of $\mathbf{Y}$ is $\text{cov}(\mathbf{Y}) = \sigma^2 \mathbf{I}$.

Based on these assumptions, the following results can be obtained, provided only that $\mathbf{X}'\mathbf{X}$ is nonsingular.

1. $\mathbf{b} = (\mathbf{X}'\mathbf{X})^{-1}\mathbf{X}'\mathbf{Y}$ is the maximum likelihood estimate of $\boldsymbol{\beta}$.
2. $\mathbf{b}$ is an unbiased estimator of $\boldsymbol{\beta}$.
3. $\hat{\boldsymbol{\beta}} = \mathbf{b}$ is distributed as a $k + 1$ variate normal with covariance matrix $\sigma^2(\mathbf{X}'\mathbf{X})^{-1}$.

Furthermore, even without the normality assumption, the following result holds. Consider an arbitrary linear combination $\mathbf{h}'\boldsymbol{\beta}$ of the regression coefficients. The linear unbiased estimate of $\mathbf{h}'\beta$ with minimum variance is $\mathbf{h}'\hat{\boldsymbol{\beta}} = \mathbf{h}'(\mathbf{X}'\mathbf{X})^{-1}\mathbf{X}'\mathbf{Y}$. For further results, *see* MULTIPLE REGRESSION.

The literature on regression coefficients is enormous and there are many new developments with regard to computational procedures for, and estimation and robustness of, regression coefficients. A few of the possible sources are listed in the references.

Reference

[1] Younger, M. S. (1979). *A Handbook for Linear Regression*. Duxbury, North Scituate, Mass.

Bibliography

See the following works, as well as the reference just given, for more information on the topic of regression coefficients.

Allen, D. M. and Cady, F. B. (1982). *Analyzing Experimental Data by Regression.* Wadsworth, Belmont, Calif.

Cramer, E. M. (1972). *Amer. Statist.*, **26**(4), 26–30.

Chatterjee, S. and Price, B. (1977). *Regression Analysis by Example.* Wiley, New York.

Cowden, D. J. (1958). *J. Amer. Statist. Ass.*, **53**, 144–150; correction: 811 (1959).

Draper, N. and Smith, H. (1981). *Applied Regression Analysis*, 2nd ed. Wiley, New York.

Dutka, A. F. and Ewen, F. J. (1971). *J. Quality Tech.*, **3**, 149–155.

Gallant, A. R. (1975). *Amer. Statist.*, **29**, 73–81.

Graybill, F. A. (1976). *Theory and Application of the Linear Model.* Duxbury, Belmont, Calif.

Gujarati, D. (1970). *Amer. Statist.*, **24**(1), 50–52.

Gujarati, D. (1970). *Amer. Statist.*, **24**(5), 18–22.

Gunst, R. F. and Mason, R. L. (1980). *Regression Analysis and Its Application.* Marcel Dekker, New York.

Hahn, G. J. (1977). *J. Quality Tech.*, **9**, 56–61.

Hahn, G. J. (1977). *J. Quality Tech.*, **9**, 159–165.

Han, C. P. and Bancroft, T. A. (1978). *Commun. Statist. A*, **7**, 47–56.

Hoerl, A. E. and Kennard, R. W. (1970). *Technometrics*, **12**, 55–67.

Hoerl, A. E. and Kennard, R. W. (1970). *Technometrics*, **12**, 69–82.

Jaech, J. L. (1966). *Ind. Quality Control*, **23**(6), 260–264.

Li, C. C. (1964). *Amer. Statist.*, **18**(4), 27–28.

McCable, G. P. (1978). *Technometrics*, **20**, 131–140.

Mosteller, F. and Tukey, J. W. (1977). *Data Analysis and Regression.* Addison-Wesley, Reading, Mass.

Neter, H. and Wasserman, W. (1974). *Applied Linear Statistical Models.* Irwin, Georgetown, Ontario.

(LINEAR REGRESSION
MULTIPLE REGRESSION
RIDGE REGRESSION
SPURIOUS CORRELATION)

ROBERT HULTQUIST

REGRESSION: CONFLUENCE ANALYSIS

Confluence analysis (C.A.) is the name given by Ragnar Frisch [3] to his method for analyzing interrelations between a set of statistical variables observed over time or a cross section. In the early developments of econometrics*, C.A. belongs under the "errors-in-variables" approach. As distinct from "errors in equations", C.A. assumes that the error-free variables are connected by an exact (deterministic, error-free) equation. In the space of the observables the equation defines a hyperplane, and if the errors in the variables are mutually uncorrelated, the plane can be assessed by an oblique least-squares* regression in a direction determined by the error variances. Standardizing the variables to unit variance, the *bunch map* of C.A. is a graphic display of all possible regressions among the variables [8].

The rationale of bunch-map analysis is summarized by Stone et al. [7, p. 300]:

> If an arbitrary set of finite positive weights is assigned to a series of observations, then the direction in which the sum of squares should be minimized will depend on these weights. It has been shown by Frisch [3], in the two-variable case and by Koopmans [5], in the general case that, provided the variance matrix of the errors in the variables is diagonal, the appropriate regression coefficient* (for the equation under analysis) in these circumstances will lie between the largest and smallest values obtained by assuming in turn that each variable . . . has a relative weight of zero. Only one variable in any calculation of the regression estimates may have such a weight, since otherwise the problem is rendered indeterminate.
>
> The method of bunch-map analysis proposed by Frisch consists precisely in calculating the complete set of regression estimates under all such extreme assumptions about the relative weights of the variables. Accordingly, granted the assumption of the preceding paragraph, an error of weighting cannot carry the estimate which would be obtained if the correct weights were known outside the limits set by the largest and smallest ordinary least squares* regressions, although, if the disturbance in the equation is large compared with the errors in the variables, then a large part of the range will be irrelevant and attention may be concentrated on the first elementary regression estimates. In any case, if the limits are close together, then the error of weighting made cannot be large. Thus the presentation of bunch maps, as is done here, serves as an indication of the possible im-

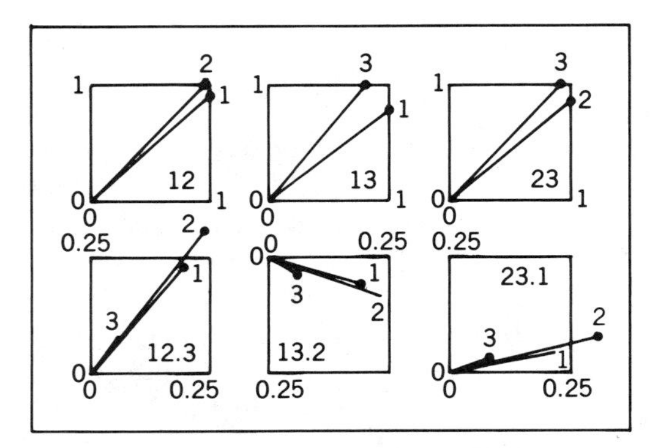

Figure 1 Budgets, 1937–1939.

portance in each analysis of the error of weighting, though it does nothing, of course, to correct this error.

Stone's bunch maps for consumer demand for food [7, p. 387] are shown in Figs. 1 and 2, with notation 1 for food expenditure, 2 for consumer income, and 3 for food price. We see that the regression limits set by the bunch map for demand as dependent on income and price are rather close for the budget data, but not for the time-series data.

Frisch's statistical methods [2, 3] were pioneering in the systematic use of vector and matrix algebra. He emphasizes the dangers of collinearity*, and shows that while multicollinearity renders some regression coefficients indeterminate, others may be determinate.

Characteristic features of C.A. are the symmetric treatment of the variables under analysis and the assumption of a deterministic relationship between the error-free variables. The mainstream of later developments breaks away from these stringent features. Although C.A. is now mainly of historical interest, the spirit of C.A. still hovers over some of the central avenues of multivariate analysis*.

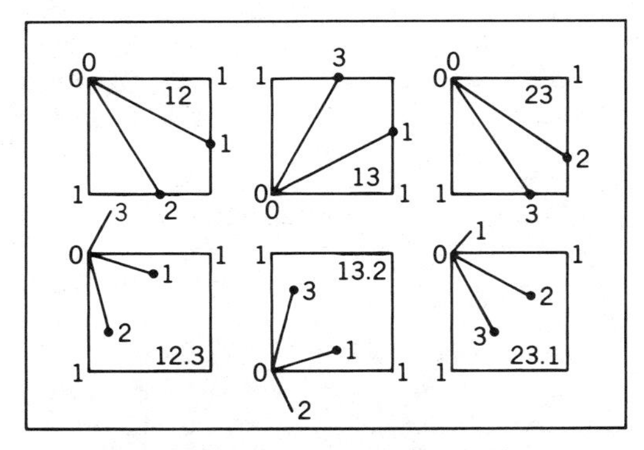

Figure 2 Time series, 1920–1938.

In the statistical analysis of complex phenomena, developments range from data-oriented approaches to theory-oriented modeling. The C.A. idea of symmetric relations is abandoned in explanatory-predictive approaches. On the data-oriented side an approach somewhat in the spirit of C.A. is Sonquist and Morgan's program [6] for the selection of explanatory variables in OLS regression analysis. Part of the story is the encouraging theorem that if the regression $y = a + \sum_1^N b_n x_n + e$ with $s(e) > 0$ is extended by a variable x_{N+1} that is uncorrelated with all x_n, the coefficients b_n will not change; at the other extreme is the discouraging theorem that if we prescribe arbitrarily the coefficient for one of the variables, say $b_1 = 10$, a variable x_{N+1} can be constructed such that when it is included in the regression, b_1 changes to 100. The *ridge regression** of Hoerl and Kennard [4] meets the problem of multicollinearity* among the explanatory variables by an ad hoc modification of $[C_{ik}^{-1}]$, the inverse of the correlation matrix*. To the realm of C.A. belongs also the *covariance selection* criterion of Dempster [1] for an observed small entry C_{ik}^{-1} not to differ significantly from zero, thus providing a criterion for deleting x_k in the regression of x_i on $x_1, \ldots, x_{i-1}, x_{i+1}, \ldots, x_N$.

The transition to "errors in equations" does not broaden the narrow and unrealistic framework of "errors in variables" as long as the equations are treated as deterministic and reversible. The transition to stochastic relations brings a fundamental broadening of concepts and scope if the relations are specified as "predictors," namely in the simplest case $y = \alpha + \beta x + \epsilon$ with $E[y \mid x] = \alpha + \beta x$; that is, the linear conditional expectation is exact, not the relation between the variables. Predictors are irreversible; they can be used both for controlled experiments and nonexperimental data, and under mild supplementary conditions OLS regression gives consistent parameter estimates (Wold [9]). Predictor specification* is the basis of the fix-point method* and soft modeling. Being designed primarily for research contexts that are simultaneously data-rich and theory-poor, soft modeling belongs under the wide realm of C.A.

References

[1] Dempster, A. P. (1972). *Biometrics*, **28**, 157–175.

[2] Frisch, R. (1928). *Nord. Statist. Tidskr.*, **8**, 36–102.

[3] Frisch, R. (1934). Confluence Analysis by Means of Complete Regression Systems. *Publ. No. 5*, Universitetets Økonomiske Institutt. Oslo.

[4] Hoerl, A. E. and Kennard, R. W. (1970). *Technometrics*, **12**, 55–67.

[5] Koopmans, T. C. (1936). Linear Regression Analysis of Economic Time Series. Thesis, University of Leiden. (Published ed. at Bohn, Haarlem, The Netherlands.)

[6] Sonquist, J. A. and Morgan, J. N. (1964). The Detection of Interaction Effects: A Report on a Computer Program for the Selection of Optimal Combinations of Explanatory Variables. *Monogr. No. 35*, Survey Research Center, University of Michigan, Ann Arbor, Mich.

[7] Stone, R., assisted by D. A. Rowe, W. J. Corlett, Renée Hurstfield, and Muriel Potter (1954). *The Measurement of Consumers' Expenditure and Behaviour in the United Kingdom 1920–1938, I*. Cambridge University Press, London.

[8] Wold, H. in association with L. Juréen (1953). *Demand Analysis. A Study in Econometrics*. Wiley, New York.

[9] Wold, H. (1963). *Sankyhā A*, **25**, 211–215.

(ECONOMETRICS
FIX-POINT METHOD
LEAST SQUARES
LINEAR REGRESSION)

HERMAN WOLD

REGRESSION DIAGNOSTICS

Regression diagnostics is the name given to a collection of techniques for detecting disagreement between a regression model and the data to which it is fitted. The major techniques are derived from study of the effect of deletion of one, or sometimes more, observations on various aspects of the fit. For ease of interpretation the results are usually presented graphically, as they are in the example at the end of this entry. The methods are computationally straightforward and have been incorporated in the regression routines of several computer packages. An introduction can be found in Weisberg [17, Chaps. 5, 6]. A fuller treatment, with an emphasis on applications in economics, is given by Belsley et al. [5]. The most advanced treatment is that of Cook and Weisberg [10]. Atkinson [4] provides an intermediate treatment.

A regression model and the data to which it is fitted may disagree for several reasons. There may be "outliers*" due to gross errors in either response or explanatory variables, which may arise, for example, in keypunching or data entry. The linear model may be inadequate to describe the systematic structure of the data, or the response might be better analyzed after a transformation. Another possibility is that the error distribution for the response may be appreciably longer tailed than the normal distribution.

If the error distribution is nonnormal, the methods of robust statistics provide an alternative to least squares*. Developments in robust regression* are explored by Huber [14] and by the discussants to that paper. Some comments on the relationship between robust and diagnostic statistical methods are given at the end of this article. The other departures from the fitted model, which are often hidden by the process of fitting, may be revealed by the effect, on the fitted model and its residuals, of the deletion of one or more observations. Details of deletion of groups of observations, including examples, are given by Cook and Weisberg [9; 10, p. 145]. Here the discussion is confined to the deletion of a single observation.

We start with the linear regression* model $E[\mathbf{Y}] = \mathbf{X}\boldsymbol{\beta}$, where $\mathbf{X}$ is an $n \times p$ matrix of known carriers which are functions of the explanatory variables and $\text{Var}(\mathbf{Y}) = \sigma^2\mathbf{I}$. The effect of deletion can readily be calculated from the fit to the full data for which the least-squares estimate of $\boldsymbol{\beta}$ is

$$\hat{\beta} = (\mathbf{X}^T\mathbf{X})^{-1}\mathbf{X}^T\mathbf{y}.$$

The predicted values are given by

$$\hat{\mathbf{y}} = \mathbf{X}\hat{\boldsymbol{\beta}} = \mathbf{X}(\mathbf{X}^T\mathbf{X})^{-1}\mathbf{X}^T\mathbf{y} = \mathbf{H}\mathbf{y}, \qquad (1)$$

where $\mathbf{H}$, with diagonal elements h_i, is often called the "hat" matrix*. The ordinary residuals*

$$\mathbf{r} = \mathbf{y} - \hat{\mathbf{y}} = (\mathbf{I} - \mathbf{H})\mathbf{y} \qquad (2)$$

have variance $\mathrm{Var}(\mathbf{r}) = \sigma^2(\mathbf{I} - \mathbf{H})$, so that the standardized residuals

$$r_i' = r_i \Big/ \sqrt{s^2(1 - h_i)} \qquad (3)$$

all have the same distribution, where s^2 is the usual mean square residual estimate of σ^2.

The agreement of the ith observation with the fit from the remaining $n - 1$ observations can be checked by comparing the prediction

$$\hat{y}_{(i)} = \mathbf{x}_i^T\hat{\boldsymbol{\beta}}_{(i)}, \qquad (4)$$

where the subscripted i in parentheses is to be read as "with observation i deleted," with the observed value y_i. The t test criterion for this comparison reduces to the "deletion" residual

$$r_i^* = r_i \Big/ \sqrt{s_{(i)}^2(1 - h_i)}, \qquad (5)$$

which can be shown to be a monotone function of the standardized residual r_i'. Unfortunately, nomenclature in the books mentioned in the first paragraph is not standardized. In particular, a studentized residual may be either r_i^* or r_i'.

The quantity h_i, often called a "leverage*" measure, indicates how remote, in the space of the carriers, the ith observation is from the other $n - 1$ observations. As an example, for linear regression with a single explanatory variable

$$h_i = (1/n) + (x_i - \bar{x})^2 \Big/ \sum (x_i - \bar{x})^2.$$

For a balanced experimental design, such as a D-optimum design, all $h_i = p/n$. For a point with high leverage, $h_i \to 1$ and the prediction at x_i will be almost solely determined by y_i, the rest of the data being irrelevant. The ordinary residual r_i will therefore have very small variance. Points with high leverage are often created by errors in entering the values of the explanatory variables. Investigation of the values of h_i and of the deletion residuals r_i^* is therefore one way of checking for such departures. A fuller discussion and examples are given by Hoaglin and Welsch [13] and in LEVERAGE.

Perhaps more important than the identification of outliers is the use of diagnostic methods to identify influential observations*, that is, observations that significantly affect the inferences drawn from the data. Methods for assessing influence are based on the change in the vector of parameter estimates when observations are deleted, given by

$$\Delta\hat{\boldsymbol{\beta}} = \hat{\boldsymbol{\beta}} - \hat{\boldsymbol{\beta}}_{(i)} = (\mathbf{X}^T\mathbf{X})^{-1}\mathbf{x}_i r_i/(1 - h_i). \qquad (6)$$

The individual components of this p vector can be used to determine the effect of observation i on the estimation of β_j. The change can also be scaled by the estimated standard error of $\hat{\beta}_j$.

If, instead of particular components of (6), the vector of estimates is of interest, Cook's [8] distance measure is found by considering the position of $\hat{\boldsymbol{\beta}}_{(i)}$ relative to the confidence region for $\boldsymbol{\beta}$ derived from all the data. Then

$$D_i = (\hat{\boldsymbol{\beta}}_{(i)} - \hat{\boldsymbol{\beta}})^T\mathbf{X}^T\mathbf{X}(\hat{\boldsymbol{\beta}}_{(i)} - \hat{\boldsymbol{\beta}})/ps^2$$
$$= (1/p)r_i'^2\left[h_i/(1 - h_i)\right]. \qquad (7)$$

Modifications of the square root of (7) lead to quantities which are multiples of residuals. One such quantity is the *modified Cook statistic*

$$C_i = \left(\frac{n - p}{p}\,\frac{h_i}{1 - h_i}\right)^{1/2} |r_i^*|. \qquad (8)$$

The relationship of these and other measures of influence is described in INFLUENTIAL DATA.

The leverage measures, residuals, and versions of Cook's statistic can be plotted against observation number to yield index plots. The various kinds of residuals and C_i can also be plotted in any of the ways customary for ordinary residuals, including normal and half-normal plots*. For normal plotting the deletion residuals r_i^*, which have a t distribution, are to be preferred to the standardized residuals r_i' since $r_i'^2$ has a scaled beta distribution. Interpretation of the half-normal plots is often aided by the presence of a simulation envelope. Examples are given by Atkinson [2] and by Cook and Weisberg [10, p. 133].

Use of these plots leads to the detection of individual observations which are in some way different from the rest of the data. For the detection of systematic departures, other plots are needed. Suppose that the model with carriers $\mathbf{X}$ has been fitted and it is desired to determine whether a new carrier $\mathbf{w}$ should be added to the model. The augmented model is

$$E[\mathbf{Y}] = \mathbf{X}\boldsymbol{\beta} + \mathbf{w}\gamma. \tag{9}$$

Although a plot of residuals against $\mathbf{w}$ may indicate the need to include the new carrier, this plot will not, in general, have slope $\hat{\gamma}$. Plots with this desirable property can be derived from least-squares estimation in the partitioned model (9), which yields

$$\hat{\gamma} = \frac{\mathbf{w}^T\left[\mathbf{I} - \mathbf{X}(\mathbf{X}^T\mathbf{X})^{-1}\mathbf{X}^T\right]\mathbf{y}}{\mathbf{w}^T\left[\mathbf{I} - \mathbf{X}(\mathbf{X}^T\mathbf{X})^{-1}\mathbf{X}^T\right]\mathbf{w}}$$

$$= \frac{\mathbf{w}^T(\mathbf{I} - \mathbf{H})\mathbf{y}}{\mathbf{w}^T(\mathbf{I} - \mathbf{H})\mathbf{w}} = \frac{\mathbf{w}^T\mathbf{A}\mathbf{y}}{\mathbf{w}^T\mathbf{A}\mathbf{w}}, \tag{10}$$

a result familiar from the analysis of covariance*. Since $\mathbf{A}$ is idempotent, $\hat{\gamma}$ is the coefficient of regression of the residuals r on the residual variable

$$\overset{*}{\mathbf{w}} = \mathbf{A}\mathbf{w}. \tag{11}$$

A plot of $\mathbf{r}$ against $\overset{*}{\mathbf{w}}$ is called an *added variable plot*. It provides a means of assessing the effect of individual observations on the estimated coefficient $\hat{\gamma}$. Added variable plots are of particular use in the study of transformations. Examples of the use of these plots in the analysis of transformations are given by Atkinson [3] and in TRANSFORMATIONS.

The methods of diagnostic regression can readily be extended from the multiple regression model to nonlinear least squares and to general inference based on the likelihood function so that, for example, generalized linear models* can be included in this framework. If interest is in inference about the vector parameter $\boldsymbol{\theta}$, influence measures can be derived from the distance $\hat{\boldsymbol{\theta}} - \hat{\boldsymbol{\theta}}_{(i)}$, which will in general require the possibly iterative calculation of $n + 1$ sets of maximum likelihood estimates. Approximations to this distance can be found from the quadratic approximation to the log likelihood at $\hat{\boldsymbol{\theta}}$, which yields the "one-step" estimates $\hat{\boldsymbol{\theta}}_{(i)}^1$. One special case of this theory is nonlinear least squares [10, Sec. 5.3] when the one-step estimates of the parameters are found from the model linearized at $\hat{\boldsymbol{\theta}}$.

The choice of one step or fully iterated estimates arises also in the extension of linear regression diagnostics to cover the range of generalized linear models through the use of diagnostics obtained by downweighting observations. Application of these results to the iterative weighted least-squares fitting method used in GLIM* provides easily calculated diagnostics for generalized linear models. The details are given by Pregibon [16] and by McCullagh and Nelder [15, Chap. 11].

As an example of these ideas applied to multiple regression, some plots are given which illuminate the analysis of the stack loss data introduced by Brownlee [6, p. 454]. There are 21 observations on the operation of a plant for the oxidation of ammonia. The response y is 10 times the percentage of NH_3 escaping unconverted into the atmosphere. The three explanatory variables have to do with the conditions of operation of the plant.

If a first-order regression model is fitted to the data, the resulting half-normal plot of the modified Cook statistic C_i (Fig. 1) shows that observation 21 lies outside the simulation envelope. One interpretation is that observation 21 is an outlier and should be rejected. Another is that the observation is informative about model inadequacy and there is, in fact, some evidence that the response should be transformed. If $\log y$ is

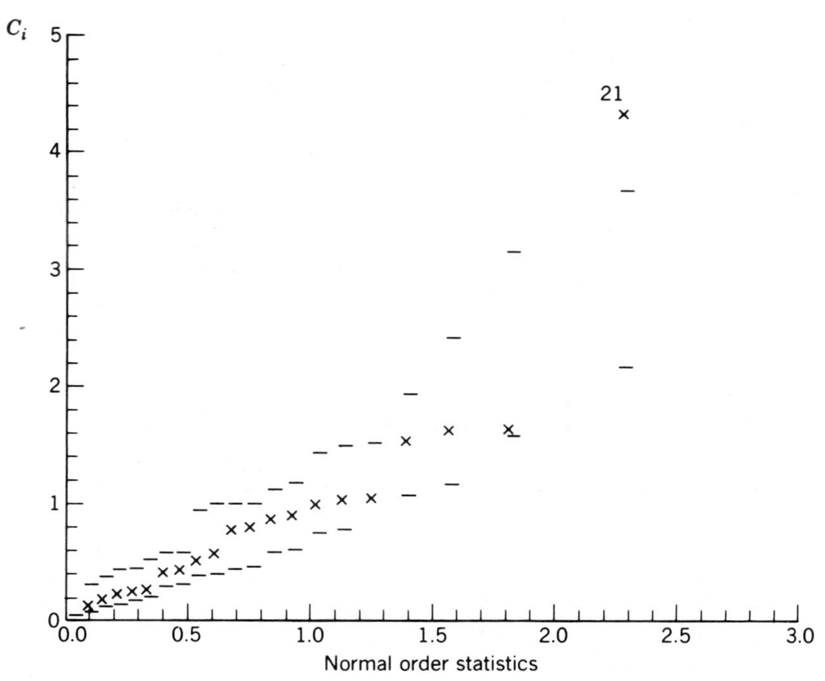

Figure 1 Stack loss data. Half-normal plot of modified Cook statistic C_i [equation (8)]: $\times$, observed values; $-$, envelope from 19 simulations.

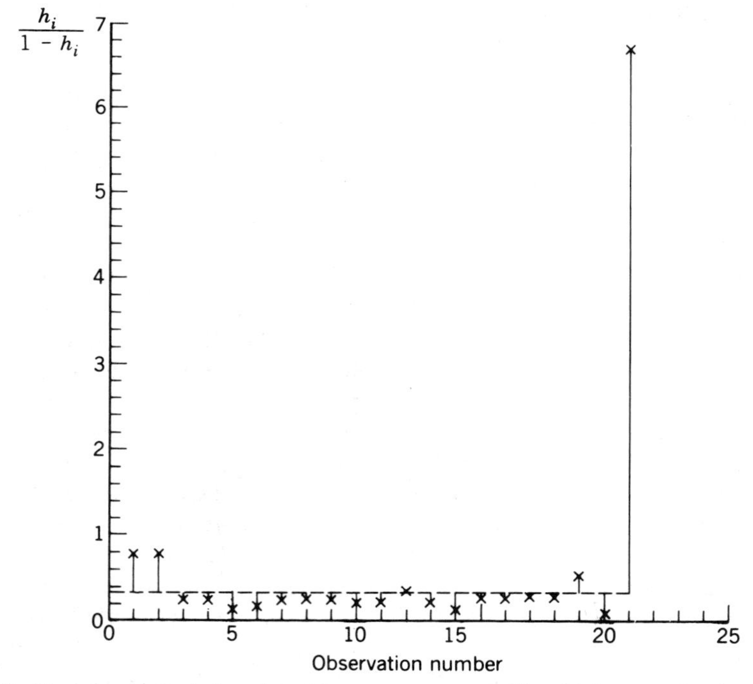

Figure 2 Stack loss data. Index plot of leverage measure $h_i/(1 - h_i)$ for second-order model.

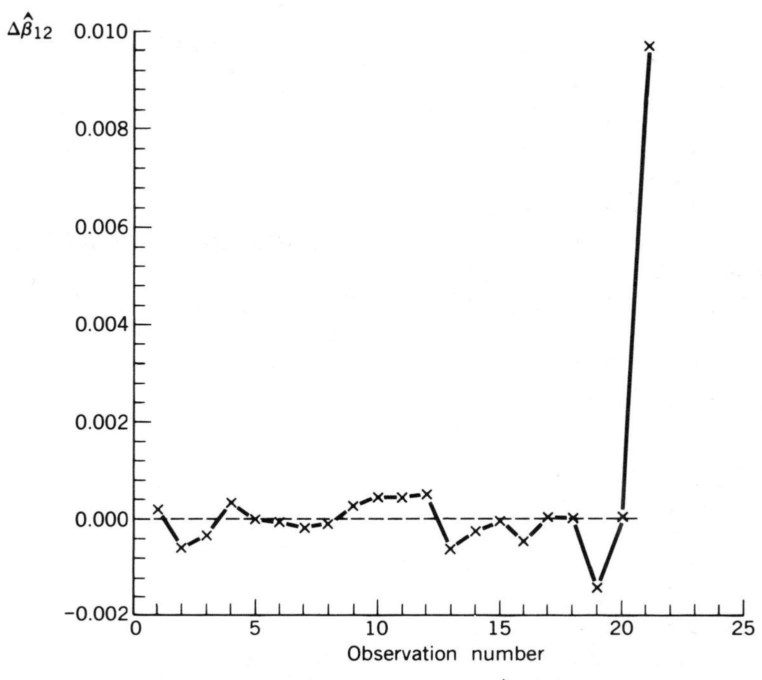

Figure 3 Stack loss data. Index plot of $\Delta\hat{\beta}_{12}$ [equation (6)].

taken as the response, standard techniques lead to a model in which the carriers are x_1, x_2, x_1x_2, and x_1^2. Plots of residuals and of C_i do not reveal any departures from this model and observation 21 is reconciled with the body of the data.

A diagnostic analysis of this fitted model, however, suggests that observation 21 has a special role. For the second-order model $h_{21} = 0.87$, so that the leverage measure $h_{21}/(1 - h_{21}) = 6.65$. The average value of the h_i is p/n, that is, 5/21 or 0.24. The large value for observation 21 is clearly shown by the index plot of $h_i/(1 - h_i)$ (Fig. 2). This figure alerts us to the special nature of observation 21, which is a point of high leverage. Is it influential? This is clearly shown by the index plot of $\Delta\hat{\beta}_{12} = \hat{\beta}_{12} - \hat{\beta}_{12_{(i)}}$ in Fig. 3. Deletion of observation 21 has a large effect on the estimate of β_{12}. The plot for $\Delta\hat{\beta}_{22}$ is similar, although the effect is less pronounced.

One conclusion of this diagnostic analysis, which can be seen from a plot of x_2 against x_1, is that observation 21 lies in a remote part of the space of the explanatory variables. The diagnostic analysis calls attention to the special importance of this one point and to its effect on the fitted second-order model. This rather sharp conclusion can be contrasted with that from the robust analysis of Andrews [1], which suggests that observations 1, 3, 4, and 21 are outliers and should be deleted. The more extensive robust analysis of Chambers and Heathcote [7] demonstrates how the number of observations downweighted in robust fitting depends on the model to be fitted. Both sets of analyses exemplify the difference, stressed by Cook and Weisberg [11], between the identification of important observations which results from diagnostic techniques and the accommodation achieved by robust methods. By this they are denoting the fitting of models in the belief that a small but unidentified fraction of the observations come from some other process.

The use of diagnostic tests and the associated plots coupled with interactive computer graphics means that complicated statistical analyses can now be routinely performed by scientists and engineers with little statistical training or expertise. The combination of the computer and the diagnostic approach may then serve as a substitute for the insight and guile of an experienced statistician. An example of such qualities applied to the analysis of the stack loss data is given by Daniel and Wood [12, Chap. 5]. The next stage in the development of the methods would seem to be an expert or intelligent knowledge-based system in which the results of diagnostic tests are used to guide the statistical analysis with the minimum of human intervention.

References

[1] Andrews, D. F. (1974). *Technometrics*, **16**, 523–531.

[2] Atkinson, A. C. (1981). *Biometrika*, **68**, 13–20.

[3] Atkinson, A. C. (1982). *J. R. Statist. Soc. B*, **44**, 1–36.

[4] Atkinson, A. C. (1985). *Plots, Transformations and Regression*. Oxford University Press, Oxford, England.

[5] Belsley, D. A., Kuh, E., and Welsch, R. E. (1980). *Regression Diagnostics*. Wiley, New York.

[6] Brownlee, K. A. (1965). *Statistical Theory and Methodology*. Wiley, New York.

[7] Chambers, R. L. and Heathcote, C. R. (1981). *Biometrika*, **68**, 21–33.

[8] Cook, R. D. (1977). *Technometrics*, **19**, 15–18.

[9] Cook, R. D. and Weisberg, S. (1980). *Technometrics*, **22**, 495–508.

[10] Cook, R. D. and Weisberg, S. (1982). *Residuals and Influence in Regression*. Chapman & Hall, London.

[11] Cook, R. D. and Weisberg, S. (1983). *J. Amer. Statist. Ass.*, **78**, 74–75.

[12] Daniel, C. and Wood, F. S. (1980). *Fitting Equations to Data*. Wiley, New York.

[13] Hoaglin, D. C. and Welsch, R. (1978). *Amer. Statist.*, **32**, 17–22.

[14] Huber, P. (1983). *J. Amer. Statist. Ass.*, **78**, 66–72.

[15] McCullagh, P. and Nelder, J. A. (1983). *Generalized Linear Models*. Chapman & Hall, London.

[16] Pregibon, D. (1981). *Ann. Statist.*, **9**, 705–724.

[17] Weisberg, S. (1985). *Applied Linear Regression*, 2nd ed. Wiley, New York.

(HAT MATRIX
INFLUENCE FUNCTIONS
INFLUENTIAL DATA
OPTIMUM DESIGN OF EXPERIMENTS
ROBUST REGRESSION)

A. C. ATKINSON

REGRESSION, FRACTIONAL

This is essentially another name for quantile regression. The α-fractional regression of Y on X is

$$q_\alpha(x) = \sup\{y : \Pr[Y \leqslant y \mid X = x] \leqslant \alpha\}.$$

Median regression corresponds to $\alpha = \frac{1}{2}$.

Fractional regression is used in the construction of kernel estimators* of the (usual) regression function

$$E[Y \mid X = x].$$

Bibliography

Schlee, W. (1980). *Z. Angew. Math. Mech.*, **60**, 369–371.

(KERNEL ESTIMATORS)

REGRESSION, INVERSE

Classical regression includes the case where an explanatory variable (X) takes preassigned values or values that are controlled by the experimenter (*see* REGRESSION COEFFICIENTS for a broader definition). In the simple case, a single random response variable Y_i, corresponding to $X = x_i$, has expectation and variance given by

$$E(Y_i) = \alpha + \beta x_i, \qquad v(Y_i) = \sigma^2$$

and the technique of least squares* (*again see* REGRESSION COEFFICIENTS) is usually applied to estimate the parameters α and β. Note that the variables X and Y are not symmetrically defined and the expectation of X for given values of Y is not in consideration. Nevertheless, the term *linear inverse regression* (sometimes called the "wrong regression") is used to refer to the line relating variable X to Y using formal least squares formulae in contrast to the classical regression of Y on X. Thus, with a bivariate sample of n pairs of observations (x_i, y_i; $i = 1, n$), the inverse regression equation

$$X = a + bY$$

has $b = s_{xy}/s_{yy}$ and $a = \bar{x} - b\bar{y}$, where $\bar{x} = \sum_1^n x_i/n$, $\bar{y} = \sum_1^n y_i/n$, and

$$s_{xy} = \sum_1^n (x_i - \bar{x})(y_i - \bar{y}), \quad s_{yy} = \sum_1^n (y_i - \bar{y})^2.$$

Inverse regression arises in the context of calibration*, when it is the prediction of new X values from observed Y values (e.g., at $Y = y_0$) that is of interest. From the above we obtain the inverse regression estimate,

$$\hat{X}_I = a + by_0.$$

Following Eisenhart (quoted in [8]) the use of inverse regression was not advocated in most statistical texts. Interest in the topic was renewed by Krutchkoff [8], who claimed superiority, in mean square error of prediction, for the inverse relative to the classical estimate ($\hat{X}_C$), which is given by

$$\hat{X}_C = (y_0 - \hat{\alpha})/\hat{\beta},$$

where $\hat{\beta} = s_{xy}/s_{xx}$ and $\hat{\alpha} = \bar{y} - \hat{\beta}\bar{x}$, with $\bar{x}, \bar{y}$, and s_{xy} retaining their defined meaning and $s_{xx} = \sum_1^n (x_i - \bar{x})^2$.

To illustrate these estimators, consider the following data from Hunter and Lamboy [6];

X	1	2	3	4	5
Y	1.8	3.1	3.6	4.9	6.0
	1.6	2.6	3.4	4.2	5.9

X	6	7	8	9	10
Y	6.8	8.2	8.8	9.5	10.6
	6.9	7.3	8.5	9.5	10.6

with $n = 20$ observations (i.e., two Y's for each distinct value of X) and

$$\bar{x} = 5.50, \quad \bar{y} = 6.19,$$

$$s_{xx} = 8.6842, \quad s_{yy} = 8.5325, \quad s_{xy} = 8.5737,$$

giving

$$b = 8.5737/8.5325 = 1.0048,$$
$$a = 5.50 - 1.0048(6.19) = -0.720,$$

so that

$$\hat{X}_I = -0.720 + 1.0048y_0.$$

Similarly,

$$\hat{\beta} = 8.5737/8.6842 = 0.9873,$$
$$\hat{\alpha} = 6.19 - 0.9873(5.50) = 0.760,$$

so that

$$\hat{X}_C = (y_0 - 0.760)/0.9873$$
$$= -0.770 + 1.0129y_0.$$

Suppose that a new value of Y arises, $y_0 = 2.0$; then $\hat{X}_I = 1.29$ while $\hat{X}_C = 1.26$. For a value of y_0 nearer the center of the data, say $y_0 = 6.0$, we have $\hat{X}_I = 5.31$ and $\hat{X}_C = 5.31$.

Clearly, the classical estimator is ill-defined if $\hat{\beta}$ is close to zero. Krutchkoff's [8] comparison of $\hat{X}_I$ with $\hat{X}_C$ was made using simulation experiments in which a truncated version of $\hat{X}_C$ was used ($\hat{\beta}$ being replaced by 0.001 whenever $|\hat{\beta}| < 0.001$). This modification has far-reaching theoretical consequences; $\hat{X}_C$ itself has infinite mean square error [14].

A rapid exchange in views, in volumes 10–12 of *Technometrics*, followed Krutchkoff's claim. They were mainly critical of the broad range of the conclusions reached and resulted in a careful qualification of the "superiority" of the method. From Krutchkoff [9]: "I believe the inverse method is far superior to the classical method for calibration within the range of observations when multiple measurements are not used. I agree the classical method is superior for large n when extrapolating out of the range of observation."

Shukla [13] gives analytical results supporting the simulation study but concludes: "For general purposes it is advisable to prefer an estimator with desirable properties (consistency*) for large sample sizes which suggests the use of classical estimators in the absence of prior information about unknown X."

Berkson [2] concluded that since most calibration experiments are based on large numbers of observations and that the ratio of slope to error standard deviation is not too small:

1. There is not much difference when estimating the calibration range.
2. The classical method is better for extrapolation.
3. Interval estimates are available for the classical approach.

The controversy over whether to use $\hat{X}_C$ or $\hat{X}_I$ has, to a large extent, been replaced by a number of proposals that give more attention to a closer specification of the problem. A Bayesian* approach was given by Hoadley [5] and a suggested interval estimate of X, based on $\hat{X}_I$, was studied by Frazier [4] using simulation. Further Bayesian considerations were elaborated by Hunter and Lamboy [6], followed by several discussion papers by various authors indicating that there is still considerable debate in this subject area. Alternative (non-Bayesian) methods are considered by Brown [3] and Ali and Singh [1]; Lwin and Maritz [10] propose a nonlinear estimator, while sequential design aspects are dealt with by Perng and Tong [12]. A likelihood analysis of the problem is dealt with by Minder and Whitney [11] and an approach based on "structural analysis" is considered by Kalotay [7].

References

[1] Ali, M. A. and Singh, N. (1981). *J. Statist. Comput. Simul.*, **14**, 1–15.

[2] Berkson, J. (1969). *Technometrics*, **11**, 649–660.

[3] Brown, G. H. (1979). *Technometrics*, **21**, 575–579.

[4] Frazier, L. T. (1974). *J. Statist. Comput. Simul.*, **3**, 99–103.

[5] Hoadley, B. (1970). *J. Amer. Statist. Ass.*, **65**, 356–369.

[6] Hunter, W. G. and Lamboy, W. F. (1981). *Technometrics*, **23**, 323–328; discussion, 329–350.

[7] Kalotay, A. J. (1971). *Technometrics*, **13**, 761–769.

[8] Krutchkoff, R. G. (1967). *Technometrics*, **9**, 425–439.

[9] Krutchkoff, R. G. (1970). *Technometrics*, **12**, 433.

[10] Lwin, T. and Maritz, J. S. (1980). *Appl. Statist.*, **29**, 135–141.

[11] Minder, Ch. E. and Whitney, J. B. (1975). *Technometrics*, **17**, 463–471.

[12] Perng, S. K. and Tong, Y. L. (1977). *Ann. Statist.*, **5**, 191–196.

[13] Shukla, G. K. (1972). *Technometrics*, **14**, 547–553.

[14] Williams, E. J. (1969). *Technometrics*, **11**, 189–192.

(CALIBRATION
CONSISTENCY
MEASUREMENT ERROR
REGRESSION COEFFICIENTS)

GEORGE W. BROWN

REGRESSION, ISOTONIC *See* ISOTONIC REGRESSION

REGRESSION, LATENT ROOT *See* LATENT ROOT REGRESSION

REGRESSION, LINEAR *See* LINEAR REGRESSION

REGRESSION, MEDIAN

The *median regression* function of a response variable Y on a predicting (set of) variable(s) $\mathbf{X}$ is the median of the conditional distribution of Y, given $\mathbf{X} = \mathbf{x}$. It is, of course, a function of $\mathbf{x}$.

The median regression is sometimes easier to calculate than the usual (expected value) regression. This is so, for example, if a monotonic function $g(x)$ of Y has a conditional distribution which has an explicit formula for the median $m(\mathbf{x})$, say. For continuous variables the conditional median of Y is $g(m(\mathbf{x}))$. The expected values have no such simple relation, in general.

REGRESSION MODELS, TYPES OF

The mathematical representation of the relationship among the variables in a system is called a "mathematical model" of the system. Models range from deterministic (e.g., physical laws) to highly indeterminate predictive models with large predictive errors (e.g., a model representing attitudes toward the work ethic as a function of age, sex, economic level, etc.). The term "regression model" is used to depict any kind of a model whose parameters are estimated from a set of data. These models have a wide variety of forms and degrees of complexity. The purpose here is to outline a classification scheme for regression models indicating mathematical complexity, statistical problems, and some practical guidelines for their use.

In the discussion, the following basic definitions and conditions are needed.

1. There are available n observations, Y_i, on a response Y and n corresponding sets of values on X_{ij} variables ($i = 1, 2, \ldots, n$; $j = 1, 2, \ldots, p$).
2. The X_j's are fixed (not subject to random error). The number of available X's, the definition of each X, the number of X's required for a problem, the levels of any particular X, and so on, are a function of the type of mathematical model involved. Since the X_j's are fixed, any transformation of an X is considered just another X for this formulation (e.g., X^2 is considered as just another X, say X_{11}).
3. Two forms of mathematical models may be fitted to the data:

   Linear (in parameters):

$$Y_i = \beta_0 X_{0i} + \sum_{j=1}^{p} \beta_j X_{ij} + \epsilon_i \qquad (1)$$

   For example,

$$Y_i = \beta_0 + \beta_1 X_{1i} + \epsilon_i,$$

   a simple linear model in one X, of order 1.

$$Y_i = \beta_0 + \beta_1 X_{1i} + \beta_2 X_{2i} + \epsilon_i,$$

   a linear model in 2 X's of order 1

$$Y_i = \beta_0 + \beta_1 X_{1i} + \beta_{11} X_{1i}^2 + \epsilon_i,$$

   a linear regression model in one X of order 2, sometimes called a quadratic model.

$$Y_i = \beta_0 + \beta_1 X_{1i} + \beta_2 \sqrt{X_{2i}} + \beta_3 \ln X_{3i} + \epsilon_i,$$

   a model built by transforming the X's; one to be used with extreme caution.

   Or:

   Nonlinear (in parameters): any mathematical relationship between Y and the X_j's not expressible in form (1). For example,

$$Y_i = \beta_0 + \beta_1 e^{\beta_2 X_i} + \epsilon_i,$$

$$Y_i = \frac{e^{\beta_1 X_{1i}}}{1 + e^{\beta_1 X_{1i}}} + \epsilon_i.$$

4. The $E(\epsilon_i) = 0$, $V(\epsilon_i) = \sigma^2$, $\operatorname{cov}(\epsilon_i, \epsilon_j) = 0$ ($i \neq j$).
5. The assumptions about the distribution of the errors, ϵ_i, depends on a particular problem and its method of solution.

For a discussion of the various parameter estimation methods, *see* LEAST SQUARES, MAXIMUM LIKELIHOOD ESTIMATION, *and* MEAN SQUARED ERROR.

Models can be classified into three main types; functional models, control models, and predictive models. A summary of the features of these three types is shown in Table 1.

FUNCTIONAL MODELS

These are mathematical models representing true cause and effect. They are usually nonlinear in form with extremely small errors. Estimates for parameter values are obtained by iterative numerical analysis* methods. The number, levels, and definition of the X's (causative variables) are completely determined by theory. Regression techniques have little to offer.

Table 1

Definition	Some Areas of Application	Kind of Models	Number and Definition of X Variables	Solution Methods	References for Further Reading
Functional					
Models of true cause and effects	Basic science theory	Nonlinear; rarely linear	Exactly known by definition	Numerical analysis; statistical methods rarely used	Dahlquist and Björck [2]
Control					
Models in which X variables are directly controllable by the experimenter or manager	Applied science laboratory and pilot-plant designed experiments	Both linear and nonlinear	Known but with some choices available under pragmatic constraints	Maximum likelihood; least squares; minimizing the sum of absolute deviations	Box et al. [1] Davies [4] Graybill [6]
Prediction					
Models in which the X variables are unrestricted	Any field where designed experiments are difficult to do and the data are highly intercorrelated (i.e., "messy")	Usually linear; rarely nonlinear	Usually unknown with many X's available for investigation	Least squares; mean square error methods; maximum likelihood rarely used; graphical techniques; residual analysis	Daniel and Wood [3] Draper and Smith [5] Neter and Wasserman [7]

CONTROL MODEL

The difference between a functional model and a control model is the difference between theory and practicality. An advertising executive can formulate a theoretical model relating the impact of TV commercials on sales conditioned by competitors' activities. The choice of a TV commercial, its length, its viewing time, and so on, are under the advertiser's control and represents controllable X's in this framework. However, competitor activity (a fundamental variable in the functional model) is not controllable, and how to handle it is complex. Any experimental situation in which the number, level, and definition of X variables are completely controllable but not necessarily identical to the variables in the functional model, is considered describable by a "control" model. These models can be either linear or nonlinear. The data are usually obtained from a well-designed experiment where the choice of the X's and the levels of the X's ensure that independent estimates of the betas in the control model can be obtained. Then these parameter estimates can be used to control the response.

While the regression procedures for choosing a "best" control model are well known, the rules for stopping at some point can be very complex (e.g., what degree of a polynomial model is adequate for control?). *See* MULTIPLE REGRESSION. Further, the adequacy of the model in controlling the response is very dependent on the size and behavior of the random component that remains unexplained by the model. It is in this area that the assumption of the distribution of the error (condition 5) becomes critical. Usually, a normality assumption with common variance is adequate, *but* it must always be carefully checked.

PREDICTION MODEL

When neither a functional model exists nor can a response be controlled as indicated above, a predictive model can often be obtained which reproduces the main features of the behavior of the response under study.

These problems are usually referred to as "problems with messy data*"—that is, data in which much intercorrelation among the X variables exists. While these models are linear in parameters, the number, levels, and definitions of the available X variables are large. The problem of choosing the X's and determining the "best" predictive model is difficult. It is in this area of modeling that a wide variety of multiple regression* techniques have been proposed, as well as extensive work on judging the adequacy of the model ultimately chosen for predictive purposes. In the former category are such techniques as all possible regressions, best subsets regression, Mallow's C_p statistic, stepwise regression*, stagewise regression, principal components* regression, ridge regression*, and so on. In the category of judging the adequacy of the predictive model, new techniques are residual analysis, Cook's statistic, Mahalanobis* distance statistic, PRESS residuals, jackknifing*, validation methods, and so on.

Special techniques in residual analysis are used to determine the adequacy of the assumptions about error distributions, for example, normal and half-normal* plotting of residuals against predicted values or against the time that each data point was collected. The predictive model must always be checked for any potential deviations from any assumptions before it can be deemed useful.

For a discussion of all these techniques and for other specific items, see the references cited in the bibliography in Draper and Smith, *Applied Regression Analysis* [5].

References

[1] Box, G. E. P., Hunter, W. G., and Hunter, J. S. (1978). *Statistics for Experimenters*. Wiley, New York. (An excellent down-to-earth experimental design book.)

[2] Dahlquist, G. and Björck, A. (1974). *Numerical Methods*. Prentice-Hall, Englewood Cliffs, N.J.

[3] Daniel, C. and Wood, F. S. (1971). *Fitting Equations to Data*. Wiley, New York.

[4] Davies, O. L. (1960). *The Design and Analysis of Industrial Experiments*. Hafner, New York.

[5] Draper, N. R. and Smith, H. (1981). *Applied Regression Analysis*, 2nd ed. Wiley, New York. (An easily read text on regression which includes material on residual analysis and methods of selecting X-variables.)

[6] Graybill, F. A. (1976). *Theory and Application of the Linear Model*. Wadsworth, Belmont, Calif. (An excellent theoretical development of the general linear model.)

[7] Neter, J. and Wasserman, W. (1974). *Applied Linear Statistical Models*. Richard D. Irwin, Homewood, Ill.

(COMPONENT ANALYSIS
C_p STATISTICS
JACKKNIFE METHODS
LINEAR REGRESSION
MULTIPLE LINEAR REGRESSION
RESIDUALS
RIDGE REGRESSION
STEPWISE REGRESSION)

HARRY SMITH, JR.

REGRESSION, MULTIPLE LINEAR *See* MULTIPLE LINEAR REGRESSION

REGRESSION, NONLINEAR *See* NONLINEAR REGRESSION

REGRESSION, PARTIAL *See* PARTIAL REGRESSION

REGRESSION, POLYNOMIAL

The general multiple linear regression* model is given by $Y = \beta_0 + \sum_{i=1}^{p} \beta_i X_i + \epsilon$, where Y is a random (dependent) variable, $X_1, X_2, \ldots, X_p$ are independent (predictor) variables, $\beta_0, \beta_1, \ldots, \beta_p$ are parameters, and ϵ is an unobservable random error. The sample model, based on n observations on $Y, X_1, X_2, \ldots, X_p$, is $\mathbf{Y} = \mathbf{X}\boldsymbol{\beta} + \boldsymbol{\epsilon}$, where $\mathbf{Y}$ is $n \times 1$, $\mathbf{X}$ is $n \times (p+1)$, $\boldsymbol{\beta}$ is $(p+1) \times 1$, and $\boldsymbol{\epsilon}$ is $n \times 1$. A common assumption on $\boldsymbol{\epsilon}$ is that $\boldsymbol{\epsilon} \sim \text{MVN}(\mathbf{0}, \sigma^2\mathbf{I})$; that is, the errors are independent and normally distributed with mean 0 and variance σ^2.

When there is only one predictor X, and the model $Y = \beta_0 + \beta_1 X + \epsilon$ is not adequate in explaining the variation in Y, then $p - 1$ other variables can be defined to yield the *polynomial* or *curvilinear* regression model $Y = \beta_0 + \sum_{i=1}^{p} \beta_i X^i + \epsilon$ or $\mathbf{Y} = \mathbf{X}_p \boldsymbol{\beta} + \boldsymbol{\epsilon}$, where $\mathbf{X}_p = (x_{ij}) = (x_i^{j-1})$. A basic goal in polynomial regression is to determine the smallest-degree model to adequately explain the variation in Y. To accomplish this end it is convenient and advantageous to utilize *orthogonal* polynomials and replace the polynomial model by the *equivalent* orthogonal polynomial model $Y = \sum_{i=0}^{p} \gamma_i \xi_i(X) + \epsilon$, where the $\xi_i(X)$ are orthogonal polynomials of degrees $i = 0, 1, 2, \ldots, p (\xi_0(X) \equiv 1)$. That is, $\xi_i(X)$ and $\xi_j(X)$, $i \neq j$, are such that if X takes on the values $x_1, x_2, \ldots, x_n$, then $\sum_{k=1}^{n} \xi_i(x_k)\xi_j(x_k) = 0$. The sample orthogonal model is $\mathbf{Y} = \mathbf{Q}\boldsymbol{\gamma} + \boldsymbol{\epsilon}$, where $\mathbf{Q} = (q_{ij}) = (\xi_j(x_i))$, $i = 1, \ldots, n$; $j = 0, 1, \ldots, p$. The *normal equations** for the orthogonal model are $\mathbf{Q}^T\mathbf{Q}\hat{\boldsymbol{\gamma}} = \mathbf{Q}^T\mathbf{Y}$, where

$$\mathbf{Q}^T\mathbf{Q} = \text{diag}\left(\sum_{k=1}^{n} [\xi_j(x_k)]^2, j = 0, 1, \ldots, p \right).$$

The maximum likelihood estimates are $\hat{\gamma}_0 = \overline{Y}$ and

$$\hat{\gamma}_j = \left(\sum_{i=1}^{n} \xi_j(x_i) Y_i \right) \Big/ \sum_{k=1}^{n} [\xi_j(x_k)]^2,$$

$j = 1, 2, \ldots, p$. If $\boldsymbol{\epsilon} \sim \text{MVN}(0, \sigma^2\mathbf{I})$, then $H_0 : \gamma_j = 0$ is rejected if

$$|t_j| = |\hat{\gamma}_j| \left\{ \sum_{k=1}^{n} [\xi_j(x_k)]^2 / \hat{\sigma}^2 \right\}^{1/2} \geqslant t(\alpha, n - j - 1),$$

where $t(\alpha, n - j - 1)$ is the $(1 - \alpha/2)$th percentile from a t-distribution with $(n - j - 1)$ degrees of freedom and $\hat{\sigma}^2$ is the residual variance based on a model of degree j.

There are several advantages to using the orthogonal polynomial model, two of these being greater computational accuracy and less computational time. Furthermore, tests on the parameters are independent, which means that only the new parameters have to be estimated and tested when increasing the degree of polynomial. Both polynomial models yield the same estimate $\hat{Y}$. In the nonorthogonal polynomial model the estimates in a model of degree greater than 4 may be inaccurate due to round-off* errors [3, 8].

When using the orthogonal model, however, orthogonal polynomials have to be determined. When the values of X are equally spaced and they occur with equal frequencies, Cooper [4] has a computer subroutine that generates orthogonal polynomials. Tables of orthogonal polynomials such as those in Beyer [2] are also available. Narula [18] has a procedure and computer program to determine orthogonal polynomials when the values of X occur with unequal spacing and unequal frequencies.

The problem of choice of degree has been treated by several investigators. Guttman [13] presented a method for estimating degree which seems to use the lack of fit at a single point. Hager and Antle [14] studied four methods of estimating degree, including Guttman's method and the usual lack-of-fit method. Their study concluded that Guttman's method had no "practical value" and the lack-of-fit method is the best, suggesting that any new method should be compared with it.

Kussmaul [17] discusses the influence of the uncertainty of the degree p on the problem of allocating values of X to minimize the predictive variance, var$(\hat{Y})$. His recommendation is that if the assumed degree is p_0, but one suspects $p = p_1 = p_0 + 1$ or $p_0 + 2$, the observations should be allocated as if $p = p_1$. The recommended allocation at $p + 1$ points for several values of p is presented.

Anderson [1] considers the choice of degree as a sequence of hypothesis tests, where it is assumed that the degree is at least m but at most q. He derives a best (uniformly most powerful) α_i-level test of the hypothesis $\gamma_i = 0$, assuming that $\gamma_{i+1} = \cdots = \gamma_q = 0$. The best test is to reject $\gamma_i = 0$ if

$$\hat{\gamma}_i^2 \sum_{j=1}^{n} \left[\xi_i(x_j)\right]^2 / \hat{\sigma}^2 > t^2(\alpha_i, n - i - 1),$$

where $\hat{\sigma}^2$ is the residual variance based on a polynomial of degree i. The procedure is to test $\gamma_q = 0$, $\gamma_{q-1} = 0, \ldots$ in sequence until one either rejects, say $\gamma_i = 0$, and decides on $\gamma_q = \cdots = \gamma_{i-1} = 0$, $\gamma_i \neq 0$, or one accepts $\gamma_q = \cdots = \gamma_{m+1} = 0$. The same sequence of tests is suggested by Plackett [19, p. 92], where q is selected from an inspection of the data, and by Williams [26, p. 41] after an initial test of all coefficients. The reverse sequence of tests appears to be commonly used, Snedecor [24, Sec. 15.6], starting with the lowest degree and "testing up" until some hypothesis $\gamma_i = 0$ is rejected. A pitfall here is that if some γ_j is large, one could choose a degree that is too low. As a precautionary measure perhaps $\gamma_{i+1} = 0$ and $\gamma_{i+2} = 0$ should be accepted before degree i is selected.

In polynomial regression, linear transformations, $Z = a + bX_i$, on the predictor variables can affect t-tests on the coefficients of lower-order terms. Griepentrog et al. [12] demonstrate this phenomenon. Their results "lend support to those who advocate retention of lower-order terms in polynomial models regardless of their t-ratios."

An observation that could be useful in fitting polynomial models is that of Crouse [5], who shows that in estimating Y in a polynomial model of degree p, "for every Y_t there exists in general at least $(k - 1)$ x-values within the range of the x_t's such that the estimates of Y corresponding to these values are independent of Y_t."

When the columns of the incidence matrix $\mathbf{X}$ in $\mathbf{Y} = \mathbf{X}\boldsymbol{\beta} + \boldsymbol{\epsilon}$, are linearly dependent, $\det(\mathbf{X}^T\mathbf{X}) = 0$. This means that an excessive number of parameters are being used (model overspecification) or the data are not adequate to estimate the postulated model. In such cases one can either modify the model or check the possibility of obtaining data that will allow estimation. When these dependencies occur only approximately, multicollinearity* in the predictor variables is present. Multicollinearity can cause undesirable consequences, the most serious of which is computational error. When multicollinearity is present, one can consider the two options as in the case of linear dependence above or one may resort to *ridge regression**. Useful comments on *ill-conditioning* or multicollinearity can be found in Willan and Watts [25] and MULTICOLLINEARITY.

Polynomial regression models can be defined in terms of k predictors $X_1, \ldots, X_k$. The degree of such a model is the largest sum of exponents of the X_i present in any term. Driscoll and Anderson [7] discuss such *multivariable* models

$$Y = \beta_0 + \sum_{i=1}^{t} \beta_i X_i + \epsilon,$$

$$X_i = \prod_{j=1}^{b} (u_j - c_j)^{n_j},$$

where the c_j are constants and the u_j are basic variables. They discuss centering of the basic variables so as to alleviate multicollinearity among the X_i. They state that "multicollinearity among the predictor variables is a problem with polynomial models since, for example, quadratic terms tend to be highly correlated with linear terms." Griepentrog et al. [12] discuss the effect of linear transformations on tests of the parameters of cross-product (or variable interaction) terms in multivariable polynomial models.

The polynomial model can be examined from a Bayesian* point of view by placing a prior distribution* on the polynomial coefficients. Deaton [6] studies the orthogonal model $\mathbf{Y} = \mathbf{Q}\boldsymbol{\gamma} + \boldsymbol{\epsilon}$ and puts a prior distribution* on $\boldsymbol{\gamma}$, which takes the γ_i to be independent with $\gamma_i \sim N(0, \sigma_i^2)$. He proposes an empirical Bayes method for determining the correct degree, which compares well with the lack-of-fit method. Young [27] puts a multivariate normal prior on $\boldsymbol{\gamma}$; however, his method is concerned only with optimal prediction and yields a polynomial of high degree. Halpern [15] assumes a multivariate prior on $\boldsymbol{\gamma}$ and presents a procedure that can be useful in determining the degree of a polynomial regression function.

The polynomial model can be formulated in a multivariate sense by treating the dependent variable as a $k \times 1$ vector $\mathbf{Y}^T = (Y_1, \ldots, Y_k)$, say. The vectors $\mathbf{Y}_i^T = (Y_{1i}, \ldots, Y_{ki})$, $i = 1, \ldots, n$, are assumed to be independent, each one distributed as a MVN$(\boldsymbol{\mu}, \boldsymbol{\Sigma})$, where both $\boldsymbol{\mu}$ and $\boldsymbol{\Sigma}$ are unknown. The mean of each component Y_i of $\mathbf{Y}^T$ is assumed to have the form $E(Y_i) = \sum_{j=0}^{p} \beta_j x_i^j = \mu_i$, $i = 1, \ldots, k$. Such models are known as *multivariate growth models*, where it is assumed that k correlated measurements are made on each of n individuals at time points $t_1, t_2, \ldots, t_k$ (i.e., $x_i = t_i$). Gafarian [11] considers such a model and develops two methods for constructing confidence bands for a polynomial curve of known degree. Other work related to the problem of confidence bands in multivariate polynomial regression is that of Hoel [16], Rao [21, 22], Elston [9], Elston and Grizzle [10], and Potthoff and Roy [20].

Example. Rubin and Stroud [23] investigated the relationship of high school final average, X, and first-year university average, Y, in a Canadian university. Their study involved nine schools and six matriculation years, or 54 cells. The total sample size for all cells was $n = 973$. Their basic model was $E(Y \mid X) = \alpha_{st} + f_t(X)$, $s = 1, \ldots, 9$; $t = 1, \ldots, 6$, where s denotes school, t denotes year, and $f_t(X)$ denotes a polynomial of at most degree 3. They considered this model after first finding for each of the 973 (Y, X, X^2, X^3)-observations its deviation from its cell mean $(\overline{Y}_{st}, \overline{X}_{st}, \overline{X}_{st}^2, \overline{X}_{st}^3)$. They then compared within-cell regressions in each year and found the cubic term to contribute significantly only in year 4. However, the cubic term was not significant for all years combined, so it was dropped. Each school–year cell had its own intercept and a quadratic regression of Y and X that was the same for all schools but changed from year to year. The quadratic regression of Y on (X, X^2) in year t was $\hat{Y} = \hat{\beta}_{0t} + \hat{\beta}_{1t}X + \hat{\beta}_{2t}X^2$. The coefficient $\hat{\beta}_{2t}$ increased over time. For year 1, $\hat{\beta}_{01} = 1.29762$, $\hat{\beta}_{11} = 0.96160$, and $\hat{\beta}_{21} = -0.001430$, and for year 6, $\hat{\beta}_{06} = 1.96264$, $\hat{\beta}_{16} = 0.61493$, and $\hat{\beta}_{26} = 0.014620$.

References

[1] Anderson, T. W. (1962). *Ann. Math. Statist.*, **33**, 255–265.

[2] Beyer, W. H., ed. (1971). *CRC Basic Statistical Tables*. Chemical Rubber Co., Cleveland, Ohio, pp. 222–235.

[3] Bright, J. W. and Dawkins, C. S. (1965). *Ind. Eng. Chem. Fundam.*, **4**, 93–97.
[4] Cooper, B. E. (1971). *Appl. Statist.*, **20**, 209–213.
[5] Crouse, C. F. (1964). *Biometrika*, **51**, 501–503.
[6] Deaton, L. W. (1980). *Biometrika*, **67**, 111–117.
[7] Driscoll, M. F. and Anderson, D. J. (1980). *Commun. Statist. A*, **9**, 821–836.
[8] Dutka, A. F. and Ewens, F. J. (1971). *J. Quality Tech.*, **3**, 149–155.
[9] Elston, R. C. (1964). *Biometrics*, **20**, 643–647.
[10] Elston, R. C. and Grizzle, J. E. (1962). *Biometrics*, **18**, 148–159.
[11] Gafarian, A. V. (1978). *Technometrics*, **20**, 141–149.
[12] Griepentrog, G. L., Ryan, J. M., and Smith, L. D. (1982). *Amer. Statist.*, **36**, 171–174.
[13] Guttman, I. (1967). *J. R. Statist. Soc. B*, **29**, 83–100.
[14] Hager, H. and Antle, C. (1968). *J. R. Statist. Soc. B*, **30**, 469–471.
[15] Halpern, E. F. (1973). *J. Amer. Statist. Ass.*, **68**, 137–143.
[16] Hoel, P. G. (1954). *Ann. Math. Statist.*, **25**, 534–542.
[17] Kussmaul, K. (1969). *Technometrics*, **11**, 677–682.
[18] Narula, S. C. (1978). *J. Quality Tech.*, **10**, 170–179.
[19] Plackett, R. L. (1960). *Principles of Regression Analysis*. Oxford University Press, Oxford.
[20] Potthoff, R. F. and Roy, S. N. (1964). *Biometrika*, **51**, 313–326.
[21] Rao, C. R. (1959). *Biometrika*, **46**, 49–58.
[22] Rao, C. R. (1965). *Biometrika*, **52**, 447–458.
[23] Rubin, D. B. nad Stroud, T. W. F. (1977). *J. Educ. Statist.*, **2**, 139–155.
[24] Snedecor, G. W. (1956). *Statistical Methods*, 5th ed. Iowa State University Press, Ames, Iowa.
[25] Willan, A. W. and Watts, D. G. (1978). *Technometrics*, **20**, 407–412.
[26] Williams, E. J. (1959). *Regression Analysis*. Wiley, New York.
[27] Young, A. S. (1977). *Biometrika*, **64**, 309–317.

(COMPONENT ANALYSIS
GENERAL LINEAR MODEL
METHOD OF LEAST ABSOLUTE VALUES
MULTICOLLINEARITY
MULTIPLE LINEAR REGRESSION
NONLINEAR REGRESSION
REGRESSION COEFFICIENTS
REGRESSION: CONFLUENCE ANALYSIS
REGRESSION DIAGNOSTICS
REGRESSION TO THE MEAN
RIDGE REGRESSION
STEPWISE REGRESSION)

BENJAMIN S. DURAN

REGRESSION, PRINCIPAL COMPONENT *See* COMPONENT ANALYSIS

REGRESSION, REPEATED MEDIAN LINE METHOD

This is a robust alternative to ordinary least squares* for fitting a regression line, proposed by Siegel [1]. Given n pairs of data points (x_i, y_i) $(i = 1, \ldots, n)$ from which it is desired to fit a regression line

$$y = a + bx,$$

the procedure consists of the folllowing steps:

Step 1: Take *all* possible pairs of points $(x_i, y_i), (x_j, y_j)$ $(j \neq i)$ and calculate the "pairwise slopes"

$$b_{ij} = \frac{y_j - y_i}{x_j - x_i} \qquad (j \neq i).$$

Step 2: For each i, obtain $\text{median}_{j \neq i}(b_{ij}) = b_i$ $(j \neq i)$.

Step 3: The required estimate of slope is $\text{median}\,(b_i) = b_{\text{RM}}$.

Note that any one b_{ij} contributes to the calculation of both b_i and b_j. The intercept (a) is estimated as

$$\underset{i}{\text{median}}\,(y_i - b_{\text{RM}} x_i).$$

Reference

[1] Siegel, A. F. (1982). *Biometrika*, **69**, 242–244.

(LEAST SQUARES
REGRESSION (VARIOUS ENTRIES)
ROBUSTIFICATION AND ROBUST SUBSTITUTES)

REGRESSION, RIDGE *See* RIDGE REGRESSION

REGRESSIONS, SWITCHING

The general switching model may be written as

$$y_i = \beta_1' x_{1i} + u_{1i} \qquad \text{if } z_i > 0 \tag{1}$$

$$y_i = \beta_2' x_{2i} + u_{2i} \qquad \text{if } z_i \leqslant 0 \tag{2}$$

$$z_i = \beta_3' x_{3i} + u_{3i}, \tag{3}$$

where x_{1i}, x_{2i}, x_{3i} are exogenous variable vectors, and $u_i = (u_{1i}, u_{2i}, u_{3i})$ is a vector of unobserved disturbances with $E(u_i) = 0$, $E(u_i u_i') = \Omega$, $E(u_i u_j') = 0$ for $i \neq j$. The variable z_i may or may not be observed; if it is not, we also do not observe whether the observed value of y was generated from (1) or from (2). If $\omega_{13} = \omega_{23} = 0$, the switching model is said to exhibit *exogenous switching*; otherwise, it exhibits *endogenous switching*.

Model (1) to (3) yields numerous special models that are commonly used: (a) $\omega_{11} = \omega_{22} = 0$, $\beta_1' x_{1i} \equiv 1$, $\beta_2' x_{2i} \equiv 0$ yields the *probit model*; (b) $\beta_1' x_{1i} + u_{1i} \equiv \beta_3' x_{3i} + u_{3i}$, $\beta_2' x_{2i} + u_{2i} \equiv 0$ yields the *tobit model*; (c) $z_i = \beta_2' x_{2i} - \beta_1' x_{1i} + u_{2i} - u_{1i}$ gives the *disequilibrium model*; (d) x_{3i} = a vector constant with respect to i gives the λ-switching regression model; and (e) $\omega_{33} = 0$ yields the deterministic switching regression model.

The motivations for the various switching models are several and vary substantially from case to case. Case (a) is relevant whenever the investigator observes only a discrete yes/no decision. An example is provided by attempting to explain college-going behavior. The investigator observes a vector of exogenous variables x_i (family income, parents' educational level, etc.) and a decision variable y_i ($i = 1$ for yes, 0 for no). If u_i is a random error, the full model is

$$y_i = \begin{cases} 1 & \text{if } \beta' x_i + u_1 > 0 \\ 0 & \text{otherwise.} \end{cases}$$

Case (b) is often relevant in the theory of the consumer. Purchases by consumer i, q_i, may be thought to depend linearly on price p_i income, y_i, as in $q_i = \beta_0 + \beta_1 p_i + \beta_2 y_i + u_i$. However, the functional form posited could produce negative q_i's, which make no sense in economic terms. Hence all $q_i \leqslant 0$ are censored at zero, whereas positive q_i's are observed.

Case (c) is one in which the quantities of a commodity demanded and supplied are given by demand and supply equations but in which the price is rigid and thus the market does not clear. Thus the demand and supply functions are

$$D_t = \beta_1' x_{1t} + u_{1t},$$

$$S_t = \beta_2' x_{2t} + u_{2t},$$

but neither D_t and S_t are observed; what is observed is

$$Q_t = \min(D_t, S_t).$$

Case (d) can be written as

$$y_i = \begin{cases} \beta_1' x_{1i} + u_{1i} & \text{with probability } \lambda \\ \beta_2' x_{2i} + u_{2i} & \text{with probability } 1 - \lambda. \end{cases} \tag{4}$$

Parameters may be estimated by the method of moments* [2, 3], although even in the simple case when $\beta_1' x_{1i} = \mu_1$, $\beta_2' x_{2i} = \mu_2$, this requires the solution of a ninth-degree polynomial. Estimation is also possible with normal errors by maximizing the likelihood function

$$L = \prod_{i=1}^{n} \left\{ \frac{\lambda}{\sqrt{2\pi}\,\sigma_1} \exp\left[-\frac{1}{2\sigma_1^2} (y_i - \beta_1' x_{1i})^2 \right] + \frac{1-\lambda}{\sqrt{2\pi}\,\sigma_2} \exp\left[-\frac{1}{2\sigma_2^2} (y_i - \beta_2' x_{2i})^2 \right] \right\}.$$

The parameters of finite mixtures of normals are identifiable [18]. Also [10] if the likelihood function possesses an interior maximum, it corresponds under general conditions to a consistent root of the likelihood equations. However, the likelihood function is unbounded in parameter space. This can easily be shown by selecting values of, say β_1 so that for some i, say $\bar{i}$, $y_{\bar{i}} - \beta_1' x_{1\bar{i}} = 0$ and then considering the behavior of L over a sequence of points characterized by $\sigma_1^2 \to 0$.

Numerical optimization typically breaks down if an iterative algorithm gets near a point of unboundedness. Quandt and Ramsey [13] have suggested that an estimating method not prone to this difficulty is obtained by minimizing the squared difference between the theoretical and sample moment generating functions*. An important generalization of their approach is discussed by Schmidt [14], who points out that it is more efficient to obtain the generalized least-squares estimates.

Several of these special cases share interesting characteristics, such as the fact that they contain latent variables, that their likelihood functions may be unbounded in parameter space, and that estimates can often be computed via the E-M algorithm.

Cases (d) and (e) discussed below are the "classical" switching regressions. A special form of (3) is given by

$$y_i = \begin{cases} \beta_1' x_{1i} + u_{1i} & \text{if } i < i^* \\ \beta_2' x_{2i} + u_{2i} & \text{otherwise,} \end{cases} \tag{4}$$

where i^* is unknown. This is the case of a discrete change in the structure of the economy at an unknown point. If u_{1i}, u_{2i} are normally distributed, the parameters β_1 and β_2 and i^* may be estimated by first maximizing the likelihood conditional on i^*,

$$L(y \mid i^*) = (2\pi)^{-n/2} \sigma_1^{-i^*} \sigma_2^{-(n-i^*)} \times \exp\left[-\frac{1}{2\sigma_1^2} \sum_{i=1}^{i^*} (y_i - \beta_1' x_{1i})^2 - \frac{1}{2\sigma_2^2} \sum_{i=i^*+1}^{n} (y_i - \beta_2' x_{2i})^2 \right]$$

and then choosing as the estimate for i^* the value that maximizes $L(y \mid i^*)$. For an application, see ref. 17.

One may also wish to use a likelihood ratio test* for testing the null hypothesis that no switch took place. The likelihood ratio is $\lambda = \hat{\sigma}_1^{i^*} \hat{\sigma}_2^{(n-i^*)} / \hat{\sigma}^n$, where $\hat{\sigma}$ is the estimated standard deviation of the residuals from the regression over the entire sample. Unfortunately, the distribution of $-2\log\lambda$ is very complicated. Feder [5] notes that the distribution is that of the maximum of a large number of correlated χ^2 variables. More recently, Freeman [6] has found that the distribution is well approximated by a Pearson type III curve. This model may also be estimated by Bayesian methods; the posterior probability density function (PDF) for i^* is discussed in ref. 11 and 15.

If the two equations of (4) hold under the slightly more general conditions $\beta_3' x_{3i} > 0$ and $\beta_3' x_{3i} \leqslant 0$, respectively, it is possible to define a composite regression

$$y_i = (1 - D_i)\beta_1' x_{1i} + D_i \beta_2' x_{2i} + (1 - D_i) u_{1i} + D_i u_{2i},$$

where $D_i = 0$ if $\beta_3' x_{3i} > 0$ and $D_i = 1$ otherwise. Since obtaining least-squares estimates for this equation is not tractable as stated, workable procedures use approximations that consist of replacing D_i with a smooth function with the appropriate qualitative behavior [8, 15].

Testing the null hypothesis that the regression coefficients* in (4) are stable ($\beta_1 = \beta_2$) in a context in which x_{1i} and x_{2i} have the same number of components can also be accomplished by a test based on recursive residuals [1]. Define $\hat{\beta}_i$ to be the least-squares estimate based on the first i observations and let X_i be a matrix the rows of which are the regressor variables for the first i observations. Finally, let

$$w_i = \frac{y_i - \beta_{i-1}' x_i}{\left[1 + x_i'(X_{i-1}' X_{i-1})^{-1} x_i\right]^{1/2}}, \qquad i = k+1, \dots, n,$$

where x_i has k components. It can be shown that under H_0 the vector $w' = (w_{k+1}, \dots, w_n)$ is distributed as $N(0, \sigma^2 I)$, where σ^2 is the variance of u_i. The tests are based on the departure from zero of either the CUSUM $C_i = \sum_{j=k+1}^{i} w_j / s$, $i = k+1, \dots, n$, or the CUSUM of squares

$$C_i^* = \sum_{j=k+1}^{i} w_j^2 \Big/ \sum_{j=k+1}^{n} w_j^2, \qquad i = k+1, \dots, n,$$

where $s^2 = \sum_{j=k+1}^{n} w_j^2/(n-k)$. Using C_i, H_0 is rejected if the sequence of C_i's crosses either the line between $(k, 0.948(n-k)^{1/2})$ and $(k, 2.844(n-k)^{1/2})$ or the line between $(k, -0.948(n-k)^{1/2})$ and $(k, -2.844(n-k)^{1/2})$. The confidence interval for C_i^* is $(i-k)/(n-k) \pm C_0^*$, where C_0^* is taken from Durbin [4, Table 1]. Garbade [7] has shown that the test based on C_i^* is more powerful than that based on C_i and that neither does very well if the true generating mechanism is given by a variable parameter regression model in which $y_i = \beta_i' x_i + u_i$, $u_i \sim N(0, \sigma^2)$, $\beta_i = \beta_{i-1} + p_i$, $p_i \sim N(0, \sigma^2 P)$. The Brown–Durbin–Evans test has also been used with some success by Hwang [9] in a slightly different context to test the hypothesis of equilibrium against the alternative of disequilibrium.

References

[1] Brown, R. L., Durbin, J., and Evans, J. M. (1975). Technique for testing the constancy of regression relations over time. *J. R. Statist. Soc. B*, **37**, 149–192.

[2] Cohen, A. C. (1967). Estimation in mixtures of two normal distributions. *Technometrics*, **9**, 15–28.

[3] Day, N. E. (1969). Estimating the components of a mixture of normal distributions. *Biometrika*, **56**, 463–474.

[4] Durbin, J. (1969). Tests for serial correlation in regression analysis based on the periodogram of least squares residuals. *Biometrika*, **56**, 1–15.

[5] Feder, P. I. (1975). The log likelihood ratio in segmented regressions. *Ann. Statist.*, **3**, 84–97.

[6] Freeman, J. M. (1983). Sampling experiments and the Quandt statistic. *Commun. Statist. Theor. Meth.*, **12**, 1879–1888.

[7] Garbade, K. (1977). Two methods for examining the stability of regression coefficients. *J. Amer. Statist. Ass.*, **72**, 56–63.

[8] Goldfeld, S. M. and Quandt, R. E. (1976). *Studies in Nonlinear Estimation*. Ballinger, Cambridge, Mass., pp. 3–36.

[9] Hwang, H.-S. (1980). A test of a disequilibrium model. *J. Econometrics*, **12**, 319–334.

[10] Kiefer, N. M. (1978). Discrete parameter variations: efficient estimation of a switching regression model. *Econometrica*, **46**, 427–434.

[11] Otani, K. (1982). Bayesian estimation of the switching regression model with autocorrelated errors. *J. Econometrics*, **18**, 239–250.

[12] Quandt, R. E. (1982). Econometric disequilibrium models. *Econometric Rev.*, **1**, 1–63.

[13] Quandt, R. E. and Ramsey, J. B. (1978). Estimating mixtures of normal distributions and switching regressions. *J. Amer. Statist. Ass.*, **73**, 730–752.

[14] Schmidt, P. (1982). An improved version of the Quandt-Ramsey MGF estimator for mixtures of normal distributions and switching regressions. *Econometrica*, **50**, 501–516.

[15] Tishler, A. and Zang, I. (1979). A switching regression method using inequality conditions. *J. Econometrics*, **11**, 247–258.

[16] Tsurumi, H. (1982). A Bayesian and maximum likelihood analysis of a gradual switching regression in a simultaneous equation framework. *J. Econometrics*, **19**, 165–182.

[17] White, L. J. (1976). Searching for the critical industrial concentration ratio: an application of the "switching of regimes" technique. In *Studies in Nonlinear Estimation*, S. M. Goldfeld and R. E. Quandt, eds. Ballinger, Cambridge, Mass., pp. 61–76.

[18] Yakowitz, S. J. (1970). Unsupervised learning and the identification of finite mixtures. *IEEE Trans. Inf. Theory*, **IT-16**, 330–338.

(ECONOMETRICS
ESTIMABILITY
IDENTIFIABILITY)

RICHARD E. QUANDT

REGRESSION, STRUCTURAL *See* STRUCTURAL INFERENCE

REGRESSION TO THE MEAN

Regression to the mean was first described by Sir Francis Galton* [7] in 1886. He reported his findings:

> It is some years since I made an extensive series of experiments on the produce of seeds of different size but of the same species . . . . It appeared from these experiments that the offspring did not tend to resemble their parent seeds in size, but to be always more mediocre than they—to be smaller than the parents, if the parents were large; to be larger than the parents, if the parents were very small . . . .
>
> The experiments showed further that the filial regression towards mediocrity was directly proportional to the parental deviation from it.

Galton called this phenomenon "regression toward mediocrity," later replacing "mediocrity" with "mean."

Galton's observation on peas can be

quantified in what is now a familiar way. If X = size of the parent pea and Y = size of the offspring pea, and μ_x, μ_y, σ_x^2, σ_y^2, and ρ represent the means, variances, and correlation*, respectively, then

$$E(Y \mid X = x) = \mu_y = \rho\sigma_y(x - \mu_x)/\sigma_x .$$

Thus the expected size of the offspring is "directly proportional to the parental deviation (from the mean)." This is, of course, the well-known simple linear regression equation.

In current statistical literature, regression to the mean is used to identify the phenomenon that a variable that is extreme on its first measurement will tend to be closer to the center of the distribution for a later measurement. For example, in a screening program for hypertension, only persons with high blood pressure are asked to return for a second measure. On the average, the second measure will be less than the first [1]. If $\mu_x = \mu_y = \mu$, $\sigma_x^2 = \sigma_y^2 = \sigma^2$, and X and Y are bivariate normally distributed, then

$$E(X - Y \mid Y > k) = c_1\sigma(1 - \rho) > 0,$$

where

$$c_1 = \frac{\phi\{(k - \mu_x)/\sigma_x\}}{1 - \Phi\{(k - \mu_x)/\sigma_x\}},$$

where $\phi(\cdot)$ is the probability density function (PDF) of the standard normal distribution and $\Phi(\cdot)$ is the corresponding cumulative distribution function (CDF). For a simple derivation of this formula, see Cutter [1]. Suppose, for example, that in some population blood pressure is normally distributed with mean 80 mm, variance 100 mm^2, and that the correlation between two blood pressure measures on a patient is 0.8. If a patient is selected because his or her blood pressure is 90 mm, the regression to the mean will be

$$c_1\sigma(1 - \rho) = \phi(1)\sigma(1 - \rho)/[1 - \Phi(1)]$$
$$= 3.5 \text{ mm}.$$

Of course, if selection is made for small values of X, the mean of Y will be larger.

Other examples of regression to the mean are easy to identify. Suppose that a highway survey is conducted and intersections with a large number of traffic accidents are chosen for the installation of traffic signals. On the average, there will be a reduction in the number of traffic accidents at these intersections, even if the traffic light has no effect on the rate of accidents. It will thus be difficult to estimate the effect of installation of the traffic signal. For other examples of regression to the mean, see Ederer [4] and Healy and Goldstein [9].

The examples of regression to the mean given above are in two categories: the estimation of treatment effect in uncontrolled studies with selection of extreme experimental units and the design of screening programs to identify extreme values. If the selection of subjects to be measured a second time is made at random, the first observation is an unbiased predictor of the subsequent observation (i.e., they have the same expected value). However, if the selection is based on extreme values, the first observation is not an unbiased predictor of the second. Hence regression to the mean is a form of selection bias*, and methods of predicting future observations must take into account the selection process.

The best method to address the regression to the mean problem in a study designed to measure treatment effect is to conduct a designed experiment with random allocation of experimental units to the treatment of interest and a control treatment. The difference between the two treatment means will then be an unbiased estimate of the treatment effect regardless of how much regression to the mean occurs (i.e., the control group can be used to estimate the effect of regression to the mean and any difference observed between the two groups can be attributed to the treatment). If it is impossible to conduct a controlled experiment, James [10] gives formulae for estimating treatment effects using the method of moments*. Alternatively, maximum likelihood* estimation for the truncated bivariate normal distribution can be used (see, e.g., Johnson and Kotz [11]).

The effect of regression to the mean can be reduced by using the mean of several measurements. Let $Y_i = U + e_i$ be the ith measure on an individual, $i = 1, 2, \ldots,$ where U is the individuals "true" value and e_i is the error in measuring U. Assume that U is normally distributed with mean μ and variance σ^2, e_i is normally distributed with mean 0 and variance γ^2, and that e_i is independent of both e_j $(i \neq j)$ and U. Under this simple model the variance of $\overline{Y}_n$, the mean of n observations, is $\sigma^2 + \gamma^2/n = \delta^2$ (say). It follows that

$$E(\overline{Y}_n - Y_{n+1} \mid \overline{Y}_n > k) = c_2\gamma^2/(n\delta) \to 0 \quad \text{as } n \to \infty,$$

where

$$c_2 = \phi\{(k - \mu)/\sigma\} \Big/ [1 - \Phi\{(k - \mu)/\sigma\}]$$

and Y_{n+1} is the $(n + 1)$st observation. Thus if the average of two or more measures is used, the regression to the mean will be reduced. Gardner and Heady [8] and Davis [3] prove this result and give examples of its use.

The conditional distribution of U given Y_i is normal with mean $\mu + \rho(Y_i - \mu)$ and variance $\rho\gamma^2$, where $\rho = \sigma^2/(\sigma^2 + \gamma^2)$. Thus for a squared-error loss function, the Bayes estimate* of U is $U_i' = \mu + \rho(Y_i - \mu)$, an estimate that shrinks the observed value Y_i toward μ. It follows that

$$E(U_1' - Y_2 \mid Y_1 > k) = 0.$$

Thus the use of the Bayes estimate provides a predictor of future observations for which there is no regression to the mean. Returning to the blood pressure example, where $\mu = 80$, $\sigma^2 = 100$ and $\rho = 0.8$, the Bayes estimator of U for a patient with observed blood pressure 90 is $80 + 0.8(90 - 80) = 88$.

Efron and Morris [5] propose the use of the James–Stein estimator* for U'. This will also provide a prediction of future observation which is free of regression to the mean. A very easily read version of their work appeared in *Scientific American* [6].

Until recently, all the work on regression to the mean has used the assumption of a normal distribution. Das and Mulder [2] have derived formulae for regression to the mean when this assumption is relaxed. As before, let $Y_i = U + e_i$ be the ith measure on an individual. Retain the assumption that e_i is normally distributed, but let U have an arbitrary distribution. If the probability density function of Y is $g(y)$, Das and Mulder show that

$$E(Y_1 - Y_2 \mid Y_1 = y) = (1 - \rho)(\sigma^2 + \gamma^2)\frac{d}{dy}\ln[g(y)],$$

where σ^2, γ^2, and ρ are defined above. If the distribution of U and hence Y_i is normal, this reduces to $(1 - \rho)(y - \mu)$. It is interesting to note that the regression effect is positive for measurement values y, where $g(y)$ is decreasing; negative for measurement values y, where $g(y)$ is increasing; and zero when $dg(y)/dy = 0$. Thus the regression is not in general to the mean. If g is unimodal, the regression will be to the mode rather than the mean. It seems, then, that the phrase regression to the mean may ultimately be replaced with "regression to the mode."

References

[1] Cutter, G. R. (1976). *Amer. Statist.*, **30**, 194–197.

[2] Das, P. and Mulder, P. C. H. (1983). *Statist. Neerlandica*, **37**, 15–20.

[3] Davis, C. E. (1976). *Amer. J. Epidemiol.*, **104**, 493–498.

[4] Ederer, F. (1972). *J. Chronic Dis.*, **25**, 277–289.

[5] Efron, B. and Morris, C. (1973). *J. Amer. Statist. Ass.*, **68**, 117–130.

[6] Efron, B. and Morris, C. (1974). *Sci. Amer.*, **236**, 119–127.

[7] Galton, F. (1886). *J. Anthrop. Inst.*, **15**, 246–263.

[8] Gardner, M. J. and Heady, J. A. (1973). *J. Chronic Dis.*, **26**, 781–795.

[9] Healy, M. J. R. and Goldstein, H. (1978). *Ann. Hum. Biol.*, **5**, 277–280.

[10] James, K. E. (1973). *Biometrics*, **29**, 121–130.

[11] Johnson, N. L. and Kotz, S. (1972). *Distributions in Statistics: Continuous Multivariate Distributions*. Wiley, New York, p. 116.

(BIVARIATE NORMAL DISTRIBUTION
CORRELATION
GALTON, FRANCIS
TRUNCATION)

C. E. DAVIS

REGRESSION VARIABLES, SELECTION OF

This problem arises when we want to explain our data adequately by a subset of possible regression variables. The objective can be specified:

1. For descriptive purposes, simply to identify factors of importance in some process or phenomena
2. For prediction or control, to include only effective variables for reducing the error of prediction; or simply to reduce the number of variables for securing the stability of the regression equation

Traditionally, testing procedures have been used for the former objective [8]; for the latter, "criterion procedures" have been developed since the mid-1960s [3, 16]. For either objective, if any prior knowledge of the underlying phenomena is available, we should first reduce the number of variables as much as possible by making use of such prior knowledge. Such a reduction will result in selecting a stable model as well as in saving computation time.

One basic statistic is the residual sum of squares. Let us consider a simple multiple regression equation,

$$y = X\beta + \epsilon,$$

with the disturbance $\epsilon \sim N(0, \sigma^2 I)$. Here $\beta' = (\beta_0, \beta_1, \ldots, \beta_{K-1})$ is the vector of regression parameters, and $y' = (y_1, \ldots, y_n)$ is the vector of n samples. We assume that the first column vector of the design matrix X is $x_0' = (1, \ldots, 1)$.

We call the regression model above the "full model," which is denoted by a set, $\bar{p} = \{1, \ldots, K-1\}$, of all nonzero indexes of the variables. Since the variable x_0 is always included in a model, a set $p = \{j_1, \ldots, j_{k-1}\}$ specifies a submodel which includes k variables x_0 and $x_{j_1}, \ldots, x_{j_{k-1}}$. The simplest submodel is the constant-term model $\underline{p} = \{\cdot\}$, in which only the variable x_0 is included. If no prior knowledge of the underlying phenomena is available, we have to select a model from all possible models between $\underline{p}$ and $\bar{p}$. Such selection is called "subset selection."

The residual sum of squares for a model p is

$$\text{RSS}(p) = \|y - X\hat{\beta}(p)\|^2,$$

where $\hat{\beta}(p)$ is the ordinary least-squares* estimate of β under the model p, with undefined entries being 0. Since the value of RSS(p) decreases as more variables are included in a model p, a simple minimization of RSS(p) fails to give a parsimonious model.

The most intuitive procedure is to select p for which the multiple correlation coefficient*

$$R^2(p) = 1 - \frac{\text{RSS}(p)}{\text{RSS}(\underline{p})},$$

Table 1 Multiple Correlation Coefficients for Hald's Data

Model	{1}	{2}	{3}	{4}	{1,2}
$R^2(p)$	0.534	0.666	0.286	0.675	0.979
$\bar{R}^2(p)$	0.492	0.636	0.221	0.645	0.974
Model	{1,3}	{1,4}	{2,3}	{2,4}	{3,4}
$R^2(p)$	0.548	0.972	0.847	0.680	0.935
$\bar{R}^2(p)$	0.458	0.967	0.816	0.616	0.922
Model	{1,2,3}	{1,2,4}	{1,3,4}	{2,3,4}	{1,2,3,4}
$R^2(p)$	0.982	0.982	0.981	0.973	0.982
$\bar{R}^2(p)$	0.976	0.976	0.975	0.964	0.974

Source. Draper and Smith [8].

or the adjusted R^2,

$$\bar{R}^2(p) = 1 - \frac{(n-1)\mathrm{RSS}(p)}{(n-k)\mathrm{RSS}(\underline{p})},$$

is close enough to 1. However, it is not a good idea to simply select a model that maximizes $R^2(p)$ or $\bar{R}^2(p)$. Since the multiple correlation coefficient $R^2(p)$ monotonically increases as the number of variables is increased, and even the adjusted multiple correlation coefficient* $\bar{R}^2(p)$ changes very little after a sufficient number of variables are included in a model p, it is hard to find a unique model. To illustrate this, we refer to Table 1. The values are calculated from the well-known Hald data with sample size $n = 13$. The maximum of $R^2(p)$ is attained with three models, $\{1,2,3\}$, $\{1,2,4\}$, and $\{1,2,3,4\}$, and that of $\bar{R}^2(p)$ with two models, $\{1,2,3\}$ and $\{1,2,4\}$.

TESTING PROCEDURE

Let $\{j_1^*, \dots, j_{K-k+1}^*\}$ be the complement of $p = \{j_1, \dots, j_{k-1}\}$. The model p is then specified by hypothesis,

$$H_p : \beta_{j_1^*} = \beta_{j_2^*} = \cdots = \beta_{j_{K-k+1}^*} = 0.$$

We can test H_p for an alternative H_{p+} by a statistic

$$F(p, p+) = \frac{\mathrm{RSS}(p) - \mathrm{RSS}(p+)}{(\mathrm{RSS}(p+))/(n-k-1)},$$

which is distributed as F with degrees of freedom $(1, n-k-1)$ under the null hypothesis H_p. Here $p+ = p \cup \{j\}$ is a model when a variable x_j is adjoined to the model p. We can then select a model by combining such testing procedures for all possible p's. Well-known procedures for doing this are forward inclusion or backward elimination*, or a mixed stepwise inclusion and elimination procedure [8, 29, 30]. For example, in backward* elimination, starting from the full model $\bar{p}$, variables are eliminated one at a time. At any step where the current model is p, if

$$\min_j F(p - \{j\}, p)$$

is insignificant, then the most insignificant variable j is eliminated from p. If it is significant, the elimination process is terminated.

It is well known that such procedures often yield different results, even if the significance levels are the same. In fact, for Hald's data, backward elimination yields a selection $\{1,2\}$, but forward inclusion yields a selection $\{1,2,4\}$ for the same level 0.10. Mantel [18] recommended the use of backward elimination rather than forward inclusion, because of the economy of effort. On the other hand, forward inclusion is recommended from the viewpoint of the simplicity of computation and stopping rules. Another controversial point is how to choose many significance levels required. The most widely used level is 10% or 5%, or the same critical level 2 or 2.5% for any tests in each step. But overall power as well as the type I error rate are unknown unless the order of entry of the variables into the model is specified explicitly before applying a procedure. It is easily seen that the order of entry differs with observations. To avoid such difficulties, Aitkin [1] or McKay [19] proposed an application of a simultaneous testing* procedure. But it requires much calculation to obtain a set of significance levels.

CRITERION PROCEDURE

For prediction or control, many types of criteria have been proposed. The first group of criteria is based on RSS(p). Each criterion can be represented by the following final prediction error (FPE_α) criterion:

$$\mathrm{FPE}_\alpha(p) = \mathrm{RSS}(p) + \alpha k\,\mathrm{RSS}(\bar{p})/(n-K), \tag{1}$$

which is an extension of a criterion proposed by Akaike [2] or almost equivalently, the general information criterion,

$$C(\alpha, p) = \log \mathrm{RSS}(p) + \alpha k, \tag{2}$$

proposed by Atkinson [6] as an extension of Akaike's* information criterion (AIC) [3]. The most common procedure is to select a

Table 2 Asymptotic Distribution of the Model Selected

1	2	3	4	5	6	7	8	9	10
0.0	0.0	0.724	0.115	0.059	0.036	0.024	0.017	0.013	0.012

model, or possibly more than one model, so as to minimize one of the criteria above.

If our main concern is the prediction error, a natural choice of α might be 2, since $\text{FPE}_2(p)$ is an unbiased estimate of the prediction error,

$$E(\|y^* - X\hat{\beta}(p)\|^2) - n\sigma^2,$$

where y^* is a vector of future observations for the same sampling points.

The C_p criterion, proposed by Mallows [17] (*see* C_p STATISTICS),

$$C_p = \text{RSS}(p)/\hat{\sigma}^2 + 2k - n,$$

is equivalent to the FPE_2 when $\text{RSS}(\bar{p})/(n - K)$ is used as $\hat{\sigma}^2$, an estimate of σ^2. He suggests, not only the minimum C_p procedure, but also a procedure called the C_p plot. A set of models is selected so that the value of C_p is close enough to k.

Some theoretical analysis of the behavior of the procedure minimizing (1) or (2) can be found in Shibata [23, 24]. He showed that the procedure with a fixed α has a tendency toward overestimation and is not consistent as an estimate of the true model. We use the term "true model" here only for convenience. Table 2 is an example of the asymptotic distribution of the model selected from the models $1 = \{1\}$, $2 = \{1,2\}, \ldots, 10 = \{1,2,\ldots,10\}$. Here $\alpha = 2$ and the true model is assumed to be $3 = \{1,2,3\}$.

He also showed [24] that in terms of the prediction error, only FPE_2 or its equivalents are asymptotically efficient if the true number of variables is assumed very large (mathematically infinite). On the other hand, Hannan and Quinn [11] showed that if α is a function of n, a necessary and sufficient condition for strong consistency of procedure is that $\alpha > 2c \log\log n$ for some $c > 1$. The criterion with $\alpha = 2c \log\log n$, which is called the HQ criterion, is the most conservative among all consistent criteria with the form of (1) or (2), having a tendency to overestimate for small samples. From the Bayesian viewpoint, Schwarz [21] proposed the choice $\alpha = \log n$, which is called a Bayesian information criterion (BIC). The BIC satisfies the foregoing consistency condition. A summary of criteria procedures is given in Table 3, and the values of each criterion for Hald's data are given in Table 4.

One of the interesting theoretical results is that of C. Stone [28]. He showed that the selection procedures based on (1) or (2) are all asymptotically locally admissible. His result implies that no superior procedure exists among such criterion procedures if the true number of variables is assumed to be fixed by an increase of the sample size. However, we should note that the results given above are all obtained in a framework of asymp-

Table 3 Criterion Procedures

Name	Criterion	Consistency	Asymptotic Efficiency[a]
FPE	$\simeq \text{FPE}_2(p)$	X	O
AIC	$C(2, p)$	X	O
C_p	$\text{FPE}_2(p)$	X	O
BIC	$C(\log n, p)$	O	X
HQ	$C(2c \log\log n, p)$	O	X

[a] In the sense of Shibata [24].

Table 4 Values of Each Criterion for Hald's Data

Model	{1}	{2}	{3}	{4}	{1, 2}
FPE_2	1289.63	930.37	1963.40	907.83	93.70[a]
AIC	11.14	10.81	11.57	10.78	10.05[a]
BIC	12.27	11.94	12.70	11.91[b]	11.75[a]
Model	{1, 3}	{1, 4}	{2, 3}	{2, 4}	{3, 4}
FPE_2	1262.96	110.65	451.30	904.76	211.59
AIC	13.11	10.31[b]	12.03	12.77	11.17
BIC	14.81	12.01	13.72	14.46	12.86
Model	{1, 2, 3}	{1, 2, 4}	{1, 3, 4}	{2, 3, 4}	{1, 2, 3, 4}
FPE_2	95.96	95.83[b]	98.70	121.70	107.70
AIC	11.19	11.87	11.93	12.30	13.87
BIC	14.13	14.13	14.19	14.56	16.69

[a]The minimum.
[b]The second minimum.

totic theory, so that for small samples the foregoing theorems do not work as well.

The same model $\{1,2\}$ is selected by the three procedures above, but the second minimum behaves differently. Although the AIC and the FPE_2 are asymptotically equivalent, the second minimum is attained by different models $\{1,4\}$ and $\{1,2,4\}$, respectively. These models are selected by backward elimination and by forward inclusion, respectively. In this example, the BIC seems likely to underestimate the model.

The mean squared error of prediction (MSEP) criterion proposed by Allen [4] is similar to the FPE_2 but based on the prediction error at a specific point x. Another group of criteria might include cross-validation. Allen [5] proposed the use of the prediction sum-of-squares criterion,

$$\text{PRESS}(p) = \sum_{1}^{n} (y_i - \hat{y}_i(-i))^2,$$

where $\hat{y}_i(-i)$ is the prediction of y_i under the model p, based on all observations except the ith. This criterion can be rewritten as

$$\sum_{1}^{n} \left[(y_i - \hat{y}_i)/(1 - a_i) \right]^2,$$

where $\hat{y} = X\hat{\beta}(p)$ is an ordinary least-squares predictor and $a_i = x_i'(X'X)^{-1}x_i$. This is a special case of cross-validation which is extensively investigated by Stone [26]. This criterion has a wide applicability, but not much is known about its behavior. An interesting analysis is by Stone [27]. He showed that the cross-validation criterion is asymptotically equivalent to the AIC criterion or the general information criterion with $\alpha = 2$, that is, $C(2, p)$, for independent identically distributed samples.

COMPUTATIONAL TECHNIQUE

Since there are $2^K - 1$ possible submodels, it is indispensable to use an efficient computational technique for comparing a number of models. If all possible models are searched and if a direct calculation is applied, the number of operations is of the order K^3. However, making use of a sweeping technique as in Garside [10], we can keep it of order K^2. The algorithms developed by Schatzoff et al. [20] require less than half as much computation as that described by Garside. Furnival and Wilson [9] propose a more efficient procedure in which a Gaussian elimination technique is employed in a "regression tree." The root of the tree is the full model $\bar{p}$ and each terminal node represents one of $2^K - 1$ submodels. In each node, two branches are generated by pivoting on the first regression variable or deleting the first row and column vectors of $X'X$. If we want to avoid possible ill-conditioning of $X'X$, we can apply the Householder transform directly to the design matrix X itself. If our aim is to select a model so as to minimize criterion (1) or (2), it is enough to calculate RSS(p). We can further reduce computation time by skipping some submodels which never attain the minimum of the criterion. Hocking and Leslie [13] propose such a computation technique for the minimum C_p procedure. LaMotte and Hocking [14] further refined this technique. The reader can consult Seber [22, Chap. 12] for more detail.

LITERATURE

Large-scale simulation studies are presented in Dempster et al. [7]. Discussions from the viewpoint of the Bayesian are given, for example, in Atkinson [6], Lindley [15], and Smith and Spiegelhalter [25]. An extensive survey can be found in Hocking [12]. Wilkinson and Dallal [31] examine the effect of the selection on the distribution of the multiple correlation coefficient $R^2(p)$ when selection by forward inclusion is applied.

References

[1] Aitkin, M. A. (1974). *Technometrics*, **16**, 221–227. (Simultaneous test.)

[2] Akaike, H. (1969). *Ann. Inst. Statist. Math.*, **21**, 243–247. [Final prediction error (FPE) criterion.]

[3] Akaike, H. (1973). In *Second International Symposium on Information Theory*, B. N. Petrov and F. Csáki, eds. Akadémia Kiadó, Budapest, pp. 267–281. [Akaike's information criterion (AIC).]

[4] Allen, D. M. (1971). *Technometrics*, **13**, 469–481. [Mean squared error of prediction (MSEP) criterion.]

[5] Allen, D. M. (1974). *Technometrics*, **16**, 125–127. [Ridge type estimate and selection of variables, prediction sum of squares (PRESS) criterion.]

[6] Atkinson, A. C. (1978). *Biometrika*, **65**, 39–48. (Simple Bayesian formula is misleading.)

[7] Dempster, A. P., Schatzoff, M., and Wermuth, N. (1977). *J. Amer. Statist. Ass.*, **72**, 77–106. (Simulation study.)

[8] Draper, N. R. and Smith, H. (1966). *Applied Regression*. Wiley, New York. (Good introduction.)

[9] Furnival, G. M. and Wilson, R. W., Jr. (1974). *Technometrics*, **16**, 499–511. [Computational techniques (leaps and bounds).]

[10] Garside, M. J. (1965). *Appl. Statist.*, **14**, 196–200. (Computational techniques.)

[11] Hannan, E. J. and Quinn, B. G. (1979). *J. R. Statist. Soc. B*, **41**, 190–195. (Consistency of criterion procedure.)

[12] Hocking, R. R. (1976). *Biometrics*, **32**, 1–49. (Excellent overview with an extensive bibliography.)

[13] Hocking, R. R. and Leslie, R. N. (1967). *Technometrics*, **9**, 531–540. (Computational techniques.)

[14] LaMotte, L. R. and Hocking, R. R. (1970). *Technometrics*, **12**, 83–93. (Computational efficiency.)

[15] Lindley, D. V. (1968). *J. R. Statist. Soc. B*, **30**, 31–53. (Discussion from Bayesian point of view.)

[16] Mallows, C. L. (1964). Joint Statist. Meet., Los Angeles.

[17] Mallows, C. L. (1973). *Technometrics*, **15**, 661–675. (C_p criterion.)

[18] Mantel, N. (1970). *Technometrics*, **12**, 621–625. (Use of stepdown procedures in variable selection.)

[19] McKay, R. J. (1977). *J. R. Statist. Soc. B*, **39**, 371–380. (Simultaneous test.)

[20] Schatzoff, M., Tsao, R., and Fienberg, S. (1968). *Technometrics*, **10**, 769–779. (Computational technique.)

[21] Schwarz, G. (1978). *Ann. Statist.*, **6**, 461–464. [Bayesian information criterion (BIC).]

[22] Seber, G. A. F. (1977). *Linear Regression Analysis*. Wiley, New York. (Nice introductory book.)

[23] Shibata, R. (1976). *Biometrika*, **63**, 117–126. (Asymptotic behavior of AIC.)

[24] Shibata, R. (1981). *Biometrika*, **68**, 45–54. (Optimality of a criterion procedure.)

[25] Smith, A. F. M. and Spiegelhalter, D. J. (1980). *J. R. Statist. Soc. B*, **42**, 213–220. (Criterion procedures and Bayesian method.)

[26] Stone, M. (1974). *J. R. Statist. Soc. B*, **36**, 111–133. (Cross-validatory choice.)

[27] Stone, M. (1977). *J. R. Statist. Soc. B*, **39**, 44–47. (Asymptotic equivalence of cross validation and the AIC.)

[28] Stone, C. J. (1981). *Ann. Statist.*, **9**, 475–485. (Admissibility.)

[29] Thompson, M. L. (1978). *Int. Statist. Rev.*, **46**, 1–19. [Extensive survey (Part I; Testing).]

[30] Thompson, M. L. (1978). *Int. Statist. Rev.*, **46**, 129–146. [Extensive survey (Part II; Criterion procedure).]

[31] Wilkinson, L. and Dallal, G. E. (1981). *Technometrics*, **23**, 377–380. [Effect of the selection on the distribution of $R^2(p)$.]

(AKAIKE'S CRITERION
BACKWARD ELIMINATION
C_p STATISTICS
MULTIPLE LINEAR REGRESSION
STEPWISE REGRESSION)

RITEI SHIBATA

REGRESSION, WRONG *See* REGRESSION, INVERSE